Features and Assessments of Pain, Anesthesia, and Analgesia

Features and Assessments of Pain, Anesthesia, and Analgesia
The Neuroscience of Pain, Anesthetics, and Analgesics

Edited by

Rajkumar Rajendram
Department of Medicine, King Abdulaziz Medical City, King Abdulaziz International Medical Research Center, Ministry of National Guard - Health Affairs, Riyadh, Saudi Arabia

College of Medicine, King Saud bin Abdulaziz University of Health Sciences, Riyadh, Saudi Arabia

Vinood B. Patel
Centre for Nutraceuticals, School of Life Sciences, University of Westminster, London, United Kingdom

Victor R. Preedy
King's College London, London, United Kingdom

Colin R. Martin
Institute for Health and Wellbeing, University of Suffolk, United Kingdom

Academic Press is an imprint of Elsevier
125 London Wall, London EC2Y 5AS, United Kingdom
525 B Street, Suite 1650, San Diego, CA 92101, United States
50 Hampshire Street, 5th Floor, Cambridge, MA 02139, United States
The Boulevard, Langford Lane, Kidlington, Oxford OX5 1GB, United Kingdom

Copyright © 2022 Elsevier Inc. All rights reserved.

No part of this publication may be reproduced or transmitted in any form or by any means, electronic or mechanical, including photocopying, recording, or any information storage and retrieval system, without permission in writing from the publisher. Details on how to seek permission, further information about the Publisher's permissions policies and our arrangements with organizations such as the Copyright Clearance Center and the Copyright Licensing Agency, can be found at our
website: www.elsevier.com/permissions.

This book and the individual contributions contained in it are protected under copyright by the Publisher (other than as may be noted herein).

Notices
Knowledge and best practice in this field are constantly changing. As new research and experience broaden our understanding, changes in research methods, professional practices, or medical treatment may become necessary.

Practitioners and researchers must always rely on their own experience and knowledge in evaluating and using any information, methods, compounds, or experiments described herein. In using such information or methods they should be mindful of their own safety and the safety of others, including parties for whom they have a professional responsibility.

To the fullest extent of the law, neither the Publisher nor the authors, contributors, or editors, assume any liability for any injury and/or damage to persons or property as a matter of products liability, negligence or otherwise, or from any use or operation of any methods, products, instructions, or ideas contained in the material herein.

Library of Congress Cataloging-in-Publication Data
A catalog record for this book is available from the Library of Congress

British Library Cataloguing-in-Publication Data
A catalogue record for this book is available from the British Library

ISBN 978-0-12-818988-7

SET ISBN 978-0-12-821066-6

For information on all Academic Press publications
visit our website at https://www.elsevier.com/books-and-journals

Printed in Great Britain
Last digit is the print number: 10 9 8 7 6 5 4

Publisher: Nikki Levy
Acquisitions Editor: Natalie Farra
Editorial Project Manager: Timothy Bennett
Production Project Manager: Paul Prasad Chandramohan
Cover Designer: Miles Hitchen

Typeset by STRAIVE, India

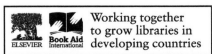

Contents

Contributors   xix
Preface   xxv

Part I
Setting the scene: General aspects of anesthesia, analgesics and pain

1. The concept of multimorphic cancer pain: A new approach from diagnosis to treatment

Antoine Lemaire

Introduction	3
Assessing and reassessing cancer pain and patients	4
Assessing cancer pain	4
Key facts of cancer pain etiologies	4
Global assessment of patients	5
Other agents of interest: Benefits of supportive care	6
Applications to other areas: From supportive care in cancer to supportive medicine	6
The concept of multimorphic cancer pain	6
Key facts of criteria defining the multimorphic nature of cancer pain	7
Treatment strategies in multimorphic cancer pain	7
Key facts of WHO cancer pain management guiding principles	8
Drug-based management of nociceptive cancer-related pain	8
Summary points: Strong opioids in cancer pain management	10
Drug-based treatments for cancer-related neuropathic pain	10
Management of bone pain	10

Interventional and surgical analgesic techniques	11
Non-drug-based approaches and noninvasive techniques	11
Palliative and end-of-life situations	11
Conclusion	11
Acknowledgments	11
References	11

2. Recent advances in the linkage of attachment and pain: A new review

Alessandro Failo

Introduction	15
Meanings and measurement of attachment	15
Interlinking attachment and pain	18
Attachment and pain: Change in relationship into different conditions considered	18
Applications to other areas	21
Other agents of interest	21
Mini-dictionary of terms	22
Key facts	22
Summary points	22
References	23

3. The management of pain in older people

Felicity Veal and Kelsey Ng

Introduction	27
Pain perception in the elderly	28
Assessing pain in older people	28
Pharmacokinetic and pharmacodynamic changes affecting older people	29
Overview of management of pain in older people	30
Paracetamol	31
Opioids	32
Adjuvant analgesics	32
Summary of the pharmacological management of pain in older people	33

v

vi Contents

Application to other areas	33
Other agents of interest	33
Mini-dictionary of terms	33
Summary points	33
References	34

4. Anesthesia and body mass: Epidural depth and beyond

Mehmet Canturk

Introduction	37
Preoperative assessment	37
Assessment of respiratory system	37
Assessment of cardiovascular system	38
Assessment of endocrine and musculoskeletal systems	38
Intraoperative management	39
Induction of anesthesia, anesthetic agents, and muscle relaxants	39
Thiopental sodium	39
Propofol	40
Opioids	40
Neuromuscular-blocking agents and antagonists	40
Volatile anesthetics	40
Regional anesthesia	40
Cessation of anesthesia and postoperative care	42
Applications to other areas	42
Mini-dictionary of terms	42
Key facts of anesthesia and body mass: Epidural depth and beyond	42
Summary points	42
References	43

5. Anesthetics and analgesic activities of herbal medicine: Review of the possible mechanism of action

U.G. Chandrika and Ureshani Karunarathna

Introduction	47
Local anesthetics herbs	47
Mechanisms of local anesthetics	47
General anesthetics herbs	50
Passiflora incarnata	50
Valerian officinalis	51
Genus *Panax*	51
Hypericum perforatum	52
Potential hazards of anesthetics herbal medicine	53
Application of anesthetics herbal medicine	53
Conclusion	53
Applications to other areas	53

Mini-dictionary of terms	54
Key facts	54
Summary	54
References	55

6. Analgesia-first sedation and nonopioid multimodal analgesia in the intensive care unit

John W. Devlin and Paul M. Szumita

Introduction	57
Pain and its assessment in critically ill adults	58
Assessment-driven protocols	58
Opioid choice, route of administration, and dosing	59
Analgosedation	62
Multimodal nonopioid analgesia use	63
Applications to other areas	65
Mini-dictionary of terms	65
Summary points	66
References	66

7. The multidisciplinary Acute Pain Service: Features and experiences

Turi Stefano, Deni Francesco, Marmiere Marilena, Meani Renato, and Beretta Luigi

Introduction	69
The history of Acute Pain Services	70
Basic characteristics of an Acute Pain Service	71
Acute Pain Service and clinical outcome	72
The diffusion of Acute Pain Services	74
The main clinical experiences regarding the establishment of an Acute Pain Service	74
Conclusion	76
Application to other areas	76
Other agents of interest	76
Key facts—Mini-dictionary	76
Summary points	77
References	77

8. Monitoring anesthesia: Electroencephalography and beyond

Mengmeng Chen and Wangning ShangGuan

Introduction	79
Raw electroencephalography	80
Influence on raw EEG	81
Bispectral index	82
Limitations and influence on BIS	82
Evoked potential	83

Somatosensory evoked potential and visual
evoked potential 83
Auditory evoked potential 83
Entropy 84
Narcotrend 84
Cerebral state index 86
Applications to other areas 86
Mini-dictionary of terms 87
Key facts 87
Key facts of raw EEG 87
Summary points 87
References 87

Part II
The syndromes of pain

9. Cluster headache and pain: Features and treatments

R.B. Brandt, J. Haan, G.M. Terwindt, and R. Fronczek

Introduction 93
Epidemiology 93
Rhythmicity 94
Pathophysiology 95
Trigeminal nerve 95
Autonomic system 95
Hypothalamus 95
Genetics 95
Trigeminal autonomic cephalalgias 96
Treatment 98
Pharmacological treatments 98
Neuromodulatory treatments 100
Applications to other areas 100
Other agents of interest 101
Conclusion 101
Mini-dictionary of terms 101
Key facts/summary points 101
References 102

10. Migraine and pain: Features and treatments

Javier Díaz de Terán and Alfonso Gil-Martínez

Features 105
Chronic migraine 106
Treatments 106
Acute migraine treatment 106
Nonspecific acute treatment 107
Specific acute treatment 107
Future 108

Prophylactic migraine treatment 108
β-Blockers 109
Antiepileptics (also known as
neuromodulators) 109
Antidepressants 109
Calcium channel blockers 110
Angiotensin converting enzyme
inhibitors, and angiotensin II
receptor blockers 110
OnabotulinumtoxinA (Botox) 110
CGRP monoclonal antibodies 110
Nonpharmacological approach 110
Exercise based on cognitive behavioral
therapy 110
Conclusion 113
Applications to other areas 113
Mini-dictionary of terms 113
Key facts 113
Summary points 114
References 114

11. Complex regional pain syndrome

C. Ryan Phillips, Derek M. Miletich, and Lynita Mullins

Introduction 117
Diagnosis 117
Pathophysiology 118
Management 118
Pharmacologic treatment 119
Anticonvulsants 119
Antidepressants 119
Antihypertensives and α-adrenergic
antagonists 119
Antiinflammatory drugs 120
Bisphosphonates 120
Calcitonin 121
Cannabis 121
IVIG 121
Naltrexone 121
NMDA receptor antagonists 121
Opioids 121
Interventional treatment 121
Sympathetic nerve blocks 121
Intravenous regional anesthesia 122
Spinal cord stimulation 122
Applications to other areas 122
Other agents of interest 122
Mini-dictionary of terms 123
Key facts 123
Summary points 123
References 124

viii Contents

12. Phantom limb pain

Derek M. Miletich, Lynita Mullins, and C. Ryan Phillips

Introduction	127
Diagnosis	127
Pathophysiology	128
Peripheral	128
Spinal	130
Supraspinal (brain)	130
Management	132
Prevention	132
Treatment	133
Mini-dictionary of terms	136
Key facts	136
Summary points	136
Disclaimer	136
References	136

13. Painful diabetic neuropathy: The roles of microglia

Che Aishah Nazariah Ismail and Idris Long

Introduction	139
Painful diabetic neuropathy	140
Microglia	141
Microglial activation during neuropathic pain	143
Central glia-neuronal interaction	143
Conclusion	144
Applications to other areas	145
Other agents of interest	145
Mini-dictionary of terms	146
Key facts	146
Key facts of Pregabalin	146
Key facts of duloxetine	146
Summary points	146
References	147

14. Maternal deprivation and nociception

Liciane Fernandes Medeiros, Dirson João Stein, Bettega Costa Lopes, and Iraci L.S. Torres

Introduction	149
Development of nociceptive pathways	150
Maternal deprivation	151
Maternal deprivation and nociception: Preclinical perspectives	151
Maternal deprivation and nociception: Signaling pathways changes	154
Early life stress in humans including maternal deprivation	156
Environmental enrichment as a potential intervention against the effects of MD	156
Applications to other areas	158
Other agents of interest	158
Mini-dictionary of terms	159
Key facts	159
Key facts of maternal deprivation	159
Key facts of early-life stress	159
Summary points	159
References	159

15. Giving birth and pain

Pelin Corman Dincer

Introduction	163
Modes of delivery	163
Stages of delivery	164
Pain treatment in giving birth	165
Pharmacologic treatment	165
Neuraxial analgesia and anesthesia	167
Postpartum pain	168
Nonpharmacological treatment	169
Application to other areas	170
Mini-dictionary of terms	170
Key facts	170
Key facts of giving birth and pain	170
Summary points	170
References	171

16. Abdominal pain in gastroparesis

Olubunmi Oladunjoye, Asad Jehangir, Adeolu Oladunjoye, Anam Qureshi, Zubair Malik, and Henry P. Parkman

Introduction	173
Types of gastroparesis	174
Symptoms of gastroparesis	174
Abdominal pain in gastroparesis	174
Symptom assessment in gastroparesis	176
Physical examination of gastroparesis patients	177
Diagnosis of gastroparesis	177
Treatment of gastroparesis	177
Dietary modifications	177
Pharmacologic treatments	178
Endoscopic and surgical treatments	179
Treatment of abdominal pain in gastroparesis	180
Neuromodulators	180
Antispasmodics	182

Cannabinoids	182
Nonsteroidal antiinflammatory drugs	183
Opiate analgesics	183
Celiac plexus block	184
Treatment of chronic abdominal wall pain	184
Conclusion	184
Mini-dictionary of terms	184
Key facts	184
Summary points	185
References	185

17. Appendicitis and related abdominal pain

Marcos Prada-Arias

Introduction	189
Etiology	189
Pathophysiology	190
Natural history models	190
Typical presentation	192
Atypical presentation	192
Inflammatory markers	192
Clinical prediction rules and imaging tests	193
Management	193
Surgical approach and complications	194
Applications to other areas	194
Other agents of interest	194
Mini-dictionary of terms	197
Key facts	197
Summary points	197
References	198

18. Ovarian hormones, site-specific nociception, and hypertension

Bruna Maitan Santos,
Glauce Crivelaro Nascimento, and
Luiz Guilherme S. Branco

Introduction	201
Nociception mechanisms and modulation in hypertension: An overview	202
Nociception and blood pressure	203
Orofacial nociception	204
Sex differences and ovarian hormones	204
Normotensive patients and rodent models	204
Hypertensive patients and spontaneously hypertensive rats (SHR) rodent model	205
Applications to other areas	207
Other agents of interest	207
Mini-dictionary of terms	207
Key facts	208
Key facts of sex differences in pain	208

Summary points	208
References	208

19. Linking the heart and pain: Physiological and psychophysiological mechanisms

Dmitry M. Davydov

Introduction	211
The history of the subject	212
Aspects related to (i, ii) a hypertension-related hypoalgesia phenomenon in chronic and acute pain conditions and (iii) a pain-killing etiology of the essential (primary) hypertension	212
Aspects related to (iv) the effectiveness of pain management prediction and monitoring by cardiovascular measures and to (v) effective pain control by cardiovascular mechanisms with pain-killing effects	213
An aspect related to (vi) the pain-o-meter technology	214
An aspect related to (vii) the baroreflex control of blood volume and pressure fluctuations in response to posture or body position changes in patients with pain	214
Summarizing the current knowledge of heart and pain relationships	215
Applications to other areas	217
Other areas of interest	219
Mini-dictionary of terms	219
Key facts of the issue associated with heart and pain interactions and related challenges	220
Summary points	220
References	221

20. Chronic pain in military veterans

Ariel Baria, Nancy Liu, Quinn Wonders, and Sanjog Pangarkar

Introduction	225
Theoretical models of chronic pain	226
Military basic training and rates of injury	226
Combat-related injuries in military personnel	227
Types of chronic pain in veterans	227
Comorbid conditions in veterans with chronic pain	227

x Contents

Psychological comorbidities and social
factors in veterans with chronic pain 228
Depression, anxiety, PTSD, substance use
disorder 228
Socioeconomic challenges 228
Pain assessment and treatment 229
Nonpharmacologic pain modalities 229
Conclusion 231
Applications to other areas 231
Other agents of interest 231
Mini-dictionary of terms 231
Key facts of chronic pain in military veterans 232
Summary points 232
References 232

21. Nociception during surgery

Munetaka Hirose

Introduction 235
Nociceptive pathway during surgery 236
Monitoring of nociception under general
anesthesia 237
Intraoperative nociception and
postoperative complications 239
Effects of intraoperative nociception on
postoperative pain 240
Applications to other areas 240
Monitoring of nociception in the awake state 240
Somatosensory processing and monitoring
of nociception: A hypothesis 241
Other agents of interest 242
Mini-dictionary of terms 242
Key facts 242
Key facts of the Somato-sympathetic reflex 242
Summary points 242
References 242

22. Breast cancer and nociceptione

*Amanda Spring de Almeida and
Gabriela Trevisan*

Introduction 247
Treatments for breast cancer-induced pain 247
Limitations of cancer pain treatments 248
Bone pain in breast cancer 249
Treatments of bone pain in breast cancer 249
Side effects observed for the treatments of
bone cancer pain 249
Animal models of breast cancer pain: Breast
inoculation models 250
Animal models of breast cancer pain: A brief
history of bone inoculation models 251
Animal models of breast cancer pain:
Distinct bone inoculation models 251

Possible pharmacological targets for breast
cancer pain relief 251
Cancer pain treatment by
nonpharmacological treatments 252
Conclusions 252
Applications to other areas 252
Other agents of interest 252
Mini-dictionary of terms 253
Key facts 253
Key facts of breast cancer-induced pain 253
Key facts of chronic pain management 253
Summary points 253
References 254

23. Postoperative pain after rhinoplasty and rhinologic surgery

*Andre Shomorony, Arron M. Cole,
Matthew Kim, and Anthony P. Sclafani*

Introduction 257
Rhinoplasty 257
Septoplasty 257
Rhinologic surgery 258
Intraoperative analgesia and anesthesia 258
Rhinoplasty 258
Septoplasty 259
Rhinologic surgery 259
Postoperative pain management 259
Sources of postoperative pain 259
Assessment of pain after surgery 260
Modalities of management of
postoperative pain 260
Conclusion 262
Applications to other areas 262
Other agents of interest 262
Mini-dictionary of terms 262
Key facts 262
Summary points 263
References 263

24. Pain response, neonates, and venipuncture

Hardeep Kaur and Gaurav Mahajan

Introduction 265
Pathophysiology of pain and CNS effects 265
Pain assessment tools 266
Problems in assessing neonatal pain 268
Pain relief measures 268
Conclusion 271
Application to other areas 272
Other agents of interest 272
Mini-dictionary of terms 272
Key facts 272

Summary points	272
Conflict of interest	273
References	273
Further reading	273

25. Carpal tunnel syndrome and pain

Rodrigo Núñez-Cortés,
Carlos Cruz-Montecinos, Claudio Tapia,
Paula Pino Pommer, and Sofía Pérez-Alenda

Introduction	275
Median nerve compression	275
Applications to other areas	276
Considerations for evaluation	276
Conservative treatment	278
Other agents of interest	278
Surgical treatment	278
Psychosocial aspects	280
Mini-dictionary of terms	280
Key facts	281
Summary points	281
Acknowledgments	281
References	281

26. Pain and HIV

Sara Pullen

Introduction	285
Etiology of chronic pain in HIV	285
HIV-associated distal symmetric polyneuropathy	286
Current pain management	286
HIV, pain, and opioid use	286
Nonpharmacologic approaches to HIV-related pain	287
Future steps	288
Applications to other areas	288
Mini-dictionary of terms	289
Key facts	289
Key facts of HIV and pain	289
Summary points	289
References	289

27. Pain mechanisms in computer and smartphone users

Alberto Marcos Heredia-Rizo,
Pascal Madeleine, and Grace P.Y. Szeto

Introduction	291
Origins and mechanisms of neck-shoulder pain	291

Low-back pain mechanisms in computer and smartphone users	292
Individual risks factors	293
Physical risks factors for the development of pain	293
Psychosocial risks factors	294
Physical activity to prevent and manage musculoskeletal pain in computer and smartphone users	295
Ergonomics to prevent occupational musculoskeletal pain in computer and smartphone users	297
Applications to other areas	297
Other agents of interest	297
Mini-dictionary of terms	298
Key facts	298
Key facts of risk factors	298
Key facts of interventions in computer users	298
Summary points	298
References	298

Part III
Interlinking anesthesia, analgesics and pain

28. Health professionals' and lay people's positions regarding the use of analgesics in cancer cases

Etienne Mullet and Paul Clay Sorum

Introduction	305
Studies of clinicians and the public	305
Mapping general practitioner's positions regarding the use of opioids	306
Mapping lay people's and health professionals' positions on the use of analgesics in postoperative pain management	307
Mapping lay people's and health professionals' positions regarding the use of morphine to relieve intense pain	309
Mapping lay people's and health professionals' positions regarding temporary or terminal sedation	310
General discussion	312
Applications to other areas	313
Mini-dictionary of terms	313
Key facts	313
Summary points	314
References	314

29. Manual compression at myofascial trigger points ameliorates musculoskeletal pain

Kouichi Takamoto, Susumu Urakawa, Shigekazu Sakai, Taketoshi Ono, and Hisao Nishijo

Introduction	317
Link between MTrPs and musculoskeletal pain	318
Low-back pain	318
Neck pain	318
Shoulder pain	318
Knee pain	320
Other body regions	320
Diagnosis of MTrPs	320
Effects of compression at MTrPs for musculoskeletal pain	320
Low-back pain	320
Neck pain	322
Shoulder pain	323
Knee pain	323
Pain in other body regions	323
Possible mechanisms of MTrP compression effects	323
Applications to other areas	325
Mini-dictionary of terms	326
Key facts of MTrP compression for musculoskeletal pain	326
Summary points	326
References	326

30. Multimodal analgesia and postsurgical pain

Martina Rekatsina, Antonella Paladini, Giorgia Saltelli, and Giustino Varrassi

Introduction	329
Definition and classification of PSP	329
Incidence of PSP and CPSP	329
Pathophysiology of postoperative pain	330
Peripheral sensitization, central sensitization, and neuroplastic changes in the brain after incision	331
Importance of PSP control	331
Risk factors for CPSP	332
Evaluation of pain intensity	332
PSP management: Enhanced recovery after surgery programs (ERAS)-preemptive, preventive, and multimodal analgesia (MA)	333
Framework of multimodal analgesia (MA)	333
Nonpharmacological interventions	334

Systemic nonopioid analgesics	334
Paracetamol	334
Nonsteroidal antiinflammatory drugs (NSAIDs)	334
Other agents of interest	335
Ketamine	335
Systemic magnesium	335
Systemic lidocaine	335
α_2-Agonists	335
Glucocorticoids	336
Systemic opioid analgesics	336
Regional anesthesia, local anesthesia, and site infiltration with local anesthetics	336
Neuraxial analgesia (epidural and spinal)	336
Transversus abdominis plane blocks	337
Paravertebral blocks	337
Local infiltration	337
Populations that need special consideration in treating PSP	337
Children	337
Obese patients and patients with chronic obstructive pulmonary disease (COPD)	337
Patients receiving long-term opioids	338
Application of MA to other areas	338
Conclusions	338
Application to other areas	338
Other agents of interest	338
Key facts	338
Mini-dictionary of terms	339
Summary points	339
References	339

31. Pain, ultrasound-guided Pecs II block, and general anesthesia

A.A. Gde Putra Semara Jaya, Marilaeta Cindryani, and Tjokorda Gde Agung Senapathi

Introduction	343
Acute and chronic postoperative pain in breast surgery	343
General anesthesia: A brief description	344
Technical description of ultrasound-guided Pecs II block	345
Clinical application ultrasound-guided Pecs II block	347
Pectoral nerves, intercostal nerve, and long thoracic nerve	348
Applications to other areas	348
Other agents of interest	348
Mini-dictionary of terms	349
Key facts of ultrasound-guided Pecs II block	350

32. Pain control during prostate biopsy and evolution of local anesthesia techniques

Mustafa Suat Bolat, Önder Cinar, Ali Batur (Furkan), Ramazan Aşcı, and Recep Büyükalpelli

Summary points	351
References	351
Introduction	353
Why is prostate cancer significant?	353
Neuroanatomy of the prostate	354
History of prostate biopsy	355
Different numbers and sites of infiltration	357
Applications of local anesthetics to other areas	359
Mini dictionary	359
Key facts	360
Summary points	360
References	360

33. Pain reduction in cosmetic injections: Fillers and beyond

Hamid Reza Fallahi, Roya Sabzian, Seied Omid Keyhan, and Dana Zandian

Introduction	363
Needle fear	363
Psychological intervention	364
Vibration	364
Vibration safety	366
Disadvantages	366
Local anesthesia	366
Disadvantages	367
Injecting anesthetics techniques	368
Vapocoolant anesthesia	368
Applications to other areas	369
Other agents of interest	370
Mini-dictionary of terms	370
Key facts	370
Summary points	370
References	370

34. Anesthesia and combat-related extremity injury

Robert (Trey) H. Burch, III

Introduction	373
Anesthesia and war	374
Acute pain	374
Regional anesthesia in extremity injury	375
Osseointegration and targeted muscle reinnervation	376
Damage control resuscitation	376
Conclusion	377
Applications to other areas	377
Other agents of interest	377
Mini-dictionary of terms	377
Key facts	377
Key facts of ketamine	377
Key facts of point of injury care	378
Summary points	378
References	378

35. Spinal anesthesia: Applications to cesarean section and pain

Reyhan Arslantas

Introduction	381
History of spinal anesthesia in the field of obstetrics	381
Spinal technique	381
Anatomy	382
Preparation	382
Choice of spinal medication	383
Bupivacaine	383
Ropivacaine	384
Levobupivacaine	384
Chloroprocaine	384
Lidocaine	384
Intrathecal opioids	384
Lipophilic opioids	384
Hydrophilic opioids	384
Other adjuvants	385
Continuous spinal anesthesia	385
Intraoperative management	385
Recovery from spinal anesthesia	386
Postoperative complications	386
Conclusion	387
Application to other areas	387
Other agents of interest	387
Mini-dictionary of terms	387
Key facts	387
Summary points	387
References	388

36. Postoperative pain management: Truncal blocks in thoracic surgery

Gulbin Tore Altun

Introduction	391
Pain management for thoracic surgery patients within ERAS (enhanced recovery after surgery) protocols	392
Multimodal analgesia	392
Truncal blocks	392
Pectoral block type I and II	392

xiv Contents

Introduction	392
Indications	392
Technique	393
Complications	394
Serratus anterior plane block (SAPB)	394
Introduction	394
Indications	394
Technique	394
Complications	395
Paravertebral block (PVB)	395
Introduction	395
Indications	395
Technique	395
Complications	396
Intercostal block	396
Introduction	396
Indications	397
Technique	397
Complications	397
Retrolaminar block (RLB)	398
Introduction	398
Indications	398
Technique	398
Complications	399
Erector spinae plane block (ESPB)	399
Introduction	399
Indications	399
Technique	399
Complications	400
Truncal blocks contraindications	400
Truncal blocks postprocedure checks	400
Equipment	400
Conclusion	401
Application to other areas	401
Other agents of interest	401
Mini-dictionary of terms	401
Key facts	402
Summary points	402
References	402

37. Postoperative pain management: Truncal blocks in general surgery

Gulbin Tore Altun

Introduction	405
Truncal blocks	406
Transversus abdominis plane block	406
Introduction	406
Indications	406
Technique	407
Complications	408
Rectus sheath block	408
Introduction	408

Indications	409
Technique	409
Complications	409
Ilioinguinal nerve and iliohypogastric nerve blocks	409
Introduction	409
Indications	409
Technique	410
Complications	410
Paravertebral block	410
Introduction	410
Indications	410
Technique	411
Complications	411
Intercostal block	411
Introduction	411
Indications	412
Technique	412
Complications	412
Quadratus lumborum block	412
Introduction	412
Indications	412
Technique	413
Complications	414
Erector spinae plane block	414
Introduction	414
Indications	414
Technique	414
Complications	415
Transversalis fascia plane block	415
Introduction	415
Indications	415
Technique	415
Complications	415
Contraindications in truncal blocks	415
Preparation for truncal blocks	416
Equipment	416
Postprocedure	416
Conclusion	416
Application to other areas	416
Other agents of interest	416
Mini-dictionary of terms	416
Key facts	417
Summary points	417
References	417

38. Linking analgesia, epidural oxycodone, pain, and laparoscopy

Merja Kokki and Hannu Kokki

Introduction	421
Laparoscopy	422
Pain after laparoscopy	422

Pain treatment after laparoscopy	423
Interventions during surgery to prevent postoperative pain	423
Epidural analgesia	425
Oxycodone	425
Oxycodone central nervous penetration	427
Epidural oxycodone	428
Applications to other areas	428
Other agents of interest	429
Mini-dictionary of terms	429
Key facts of epidural oxycodone	430
Summary points	430
References	430

39. Levobupivacaine features and linking in infiltrating analgesia

D. Bagatin, T. Bagatin, J. Nemrava, K. Šakić, L. Šakić, J. Deutsch, E. Isomura, M. Malić, M. Šarec Ivelj, and Z. Kljajić

Pharmacological basis for choice of local anesthetic for central and peripheral blocks	433
Local anesthetic properties	434
The mechanism of LA action	434
Pharmacokinetics	435
Absorption	435
Distribution, biotransformation, and excretion	436
Local anesthetic toxicity	436
Pharmacology and use of levobupivacaine	437
Our local infiltration analgesia (LIA) body protocol	438
Discussion	438
Conclusions	439
Applications to other areas	440
Other agents of interest	440
Mini-dictionary of terms	440
Key facts of levobupivacaine	441
Summary points	441
References	442

Part IV
Assessments, screening, and resources

40. The pain catastrophizing scale: Features and applications

Turgay Tuna

Introduction	445
Features	446

Applications	446
Prevention	446
Research	447
Treatment	447
Summary-conclusion	448
Applications to other areas	448
Mini-dictionary of terms	448
Key facts	448
Conflict of interest	448
References	448

41. Pain-related behavioral scales among a low back pain population: A narrative review

Dalyah Alamam

Introduction	451
Methodology	452
Results	452
Discussion	453
Limitations	457
Conclusions	457
Clinical implication/Applications to other areas	457
Mini-dictionary of terms	457
Key facts	458
Summary points	458
Appendix	458
References	459

42. The analgesia nociception index: Features and application

Sonia Bansal and Kamath Sriganesh

Introduction	463
Basic principle of the ANI	464
Features of the ANI monitor	465
Interpretation of ANI values	466
Clinical applications	467
Detection of intraoperative nociception	467
Postoperative pain assessment	468
Pain assessment in pediatric patients	469
Pain assessment in ICU setting	469
Effect on opioid consumption	469
Comparison with other objective analgesia/ nociception monitors	470
Confounders for ANI	470
Conclusions	470
Mini-dictionary of terms	471
Key facts of ANI	471
Summary points	471
References	471

xvi Contents

43. Pain, anesthetics and analgesics/back pain evaluation questionnaires

Jun Komatsu

Introduction	475
Back pain evaluation questionnaires for disease-specific measures	476
Oswestry disability index (ODI)	476
Roland-Morris disability questionnaire (RMQ)	476
Japanese Orthopaedic Association (JOA) score for diseases with LBP	476
Japanese Orthopaedic Association back pain evaluation questionnaire (JOABPEQ)	476
Conclusion	478
Suggestions for practical procedures	481
Applications to other areas	481
Other agents of interest	481
Comprehensive measure of general health status	481
Psychiatric problems measure in orthopedic patients	482
Mini-dictionary of terms	482
Key facts	482
Summary points	484
References	484

44. The back pain functional scale: Features and applications

Meltem Koç and Kılıçhan Bayar

Introduction	487
Applications to other areas	489
Other agents of interest	489
Mini-dictionary of terms	490
Key facts of back pain functional scale	490
Summary points	490
References	490

45. Cognitive impairment, pain, and analgesia

Vanesa Cantón-Habas, José Manuel Martínez-Martos, Manuel Rich-Ruiz, María Jesús Ramirez-Éxposito, and María del Pilar Carrera-González

Introduction	493
Applications to other areas	493
Mini-dictionary of terms	494
Key facts	494

Physiological and pathological aging: Cognitive impairment	494
Biology of aging	494
Biology of pain	495
Pain in the elderly	496
Neurodegenerative disorders: Alzheimer's disease	496
Pain and Alzheimer's disease	497
Pain in the elderly with advanced dementia	498
Future	502
References	504

46. Biomarkers in endometriosis-associated pain

Deborah Margatho and Luis Bahamondes

Background	507
Prevalence	509
Diagnosis	509
Classification	509
Treatment	510
Biomarkers	510
Blood biomarkers	510
Biomarkers of angiogenesis, growth factors, and growth factor receptors	510
Apoptosis markers	511
Cell adhesion molecules and other matrix-related proteins	511
Cytoskeleton molecules	511
Molecules involved in DNA repair/telomere maintenance	511
High-throughput molecular markers	511
Hormonal markers	511
Immune system and inflammatory biomarkers	511
Oxidative stress markers	518
Posttranscriptional regulators of gene expression	518
Tumor markers	518
Urine biomarkers	518
Endometrial biomarkers	518
Angiogenesis and growth factors/cell adhesion molecules/DNA repair molecules/endometrial and mitochondrial proteome/posttranscriptional regulators of gene expression	519
Hormonal markers	519
Inflammatory and myogenic markers	520
Neural markers	520
Tumor markers	520
Discussion	522
Mini-dictionary of terms	523

Key facts of endometriosis	523
Summary points	523
Applications to other areas	523
References	523

47. Biomarkers in bladder pain syndrome: A new narrative

Thais F. de Magalhaes and Jorge Haddad

Introduction	527
Challenges in IC/BPS knowledge	528
Biomarkers	528
Bladder wall biopsy specimens	531
Urine	532
Serum	532
Stool	532
IC/BPS biomarkers and their potential relationship to disease pathophysiology	532
NGF	532
Inflammatory and angiogenic markers	533
APF and HB-EGF	533
Etio-S	533
Urothelial dysfunction	533
GP51	534
Microbiome	534
Applications to other areas	534
Other agents of interest	534
Mini-dictionary of terms	535
Key facts	535
Key facts of biomarkers in bladder pain syndrome	535
Summary points	535
References	535

48. Biomarkers of statin-induced musculoskeletal pain: Vitamin D and beyond

Michele Malaguarnera

Introduction	539
Statin associated muscle symptoms	539
Cholesterol and vitamin D	541

SAMS biomarkers	541
Creatine	542
Lactic acid	542
Creatinine	542
Troponin	542
Myoglobin	543
Fatty acid-binding proteins	543
Coenzyme Q_{10} (CoQ_{10})	543
Enzymes	543
Urinary biomarkers	545
Micro RNA	545
Conclusion	545
Other agents of interest	545
Applications to other areas	546
Mini-dictionary of terms	546
Key facts	546
Summary points	547
References	547

49. Performance-based and self-reported physical fitness in musculoskeletal pain

Cristina Maestre-Cascales, Javier Courel-Ibáñez, and Fernando Estévez-López

Physical fitness	551
Musculoskeletal disorders	551
Widespread musculoskeletal pain	552
Assessment of physical fitness in widespread musculoskeletal pain	552
Applications to other areas	552
Other agents of interest	557
Mini-dictionary of terms	557
Key facts	559
Summary points	559
References	559

Index	563

Contributors

Numbers in parenthesis indicate the pages on which the authors' contributions begin.

Dalyah Alamam (451), Physiotherapy, Health Rehabilitation Sciences Department, Collage of Applied Medical Sciences, King Saud University, Riyadh, Saudi Arabia; Arthritis and Musculoskeletal Research Group, Sydney School of Health Sciences, Faculty of Medicine and Health, The University of Sydney, NSW, Australia

Amanda Spring de Almeida (247), Graduate Program in Pharmacology, Federal University of Santa Maria (UFSM), Santa Maria, RS, Brazil

Gulbin Tore Altun (391, 405), Memorial Health Group, Department of Anesthesiology and Reanimation, Istanbul, Turkey

Reyhan Arslantas (381), Department of Anesthesiology and Reanimation, Taksim Training and Research Hospital, Istanbul, Turkey

Ramazan Aşcı (353), Department of Urology, Ondokuzmayıs University, Samsun, Turkey

D. Bagatin (433), Polyclinic Bagatin, Department of Surgery and Anestesiology with Ranimatology, Zagreb, Split; Faculty of Dental Medicine and Health Osijek, Department of Surgery and Anesthesiology With Reanimatology, Depatment of Dental Medicine 1, Univeristy Josip Juraj Strossmayer, Osijek, Croatia

T. Bagatin (433), Polyclinic Bagatin, Department of Surgery and Anestesiology with Ranimatology, Zagreb, Split; Faculty of Dental Medicine and Health Osijek, Department of Surgery and Anesthesiology With Reanimatology, Depatment of Dental Medicine 1, Univeristy Josip Juraj Strossmayer, Osijek, Croatia

Luis Bahamondes (507), Department of Obstetrics and Gynaecology, University of Campinas Medical School, Campinas, SP, Brazil

Sonia Bansal (463), Department of Neuroanaesthesia and Neurocritical Care, National Institute of Mental Health and Neurosciences, Bengaluru, India

Ariel Baria (225), Los Angeles, CA, United States

Ali Batur (Furkan) (353), Department of Urology, Selçuk University, Konya, Turkey

Kılıçhan Bayar (487), Department of Physiotherapy and Rehabilitation, Faculty of Health Sciences, Muğla Sıtkı Koçman University, Muğla, Turkey

Mustafa Suat Bolat (353), Department of Urology, Gazi State Hospital, Samsun, Turkey

Luiz Guilherme S. Branco (201), Department of Basic and Oral Biology, School of Dentistry of Ribeirao Preto, University of Sao Paulo, Sao Paulo, Brazil

R.B. Brandt (93), Department of Neurology, Leiden University Medical Center, Leiden, The Netherlands

Robert (Trey) H. Burch III (373), Department of Anesthesiology, Walter Reed National Military Medical Center, Bethesda, MD, United States

Recep Büyükalpelli (353), Department of Urology, Ondokuzmayıs University, Samsun, Turkey

Vanesa Cantón-Habas (493), Department of Nursing, Pharmacology and Physiotherapy, Faculty of Medicine and Nursing, University of Córdoba; Maimónides Institute for Biomedical Research (IMIBIC), University of Córdoba, Reina Sofia University Hospital, Córdoba, Spain

Mehmet Canturk (37), Department of Anesthesiology and Reanimation, Karadeniz Eregli Government Hospital, Ereğli, Turkey

María del Pilar Carrera-González (493), Department of Nursing, Pharmacology and Physiotherapy, Faculty of Medicine and Nursing, University of Córdoba; Maimónides Institute for Biomedical Research (IMIBIC), University of Córdoba, Reina Sofia University Hospital, Córdoba; Experimental and Clinical Physiopathology Research Group, Department of Health Sciences, Faculty of Experimental and Health Sciences, University of Jaén, Jaén, Spain

U.G. Chandrika (47), Department of Biochemistry, Faculty of Medical Sciences, University of Sri Jayewardenepura, Nugeggoda, Sri Lanka

Mengmeng Chen (79), Department of Anesthesiology and Perioperative Medicine, The Second Affiliated Hospital and Yuying Children's Hospital of Wenzhou Medical University, Wenzhou, China

Önder Cinar (353), Department of Urology, Bülent Ecevit University, Zonguldak, Turkey

Marilaeta Cindryani (343), Department of Anesthesiology and Intensive Care, Sanglah Hospital, Faculty of Medicine, Udayana University, Denpasar, Bali, Indonesia

Arron M. Cole (257), Department of Otolaryngology—Head and Neck Surgery, NewYork-Presbyterian Hospital, New York, NY, United States

Javier Courel-Ibáñez (551), Department of Physical Activity and Sport, University of Murcia, San Javier, Spain

Carlos Cruz-Montecinos (275), Department of Physical Therapy, Faculty of Medicine, University of Chile, Santiago, Chile; Department of Physiotherapy, Physiotherapy in Motion Multispeciality Research Group (PTinMOTION), University of Valencia, Valencia, Spain

Dmitry M. Davydov (211), University of Jaén Hospital, Jaén, Spain; Wyższa Szkoła Społeczno-Przyrodnicza im. Wincentego Pola w Lublinie / Vincent Pol University in Lublin, Poland; Laboratory of Neuroimmunopathology, Institute of General Pathology and Pathophysiology, Russian Academy of Sciences, Moscow, Russia

J. Deutsch (433), Polyclinic Bagatin, Department of Surgery and Anestesiology with Ranimatology, Zagreb, Split, Croatia

John W. Devlin (57), Department of Pharmacy and Health Systems Sciences, Bouve College of Health Sciences, Northeastern University; Division of Pulmonary and Critical Care Medicine, Brigham and Women's Hospital, Boston, MA, United States

Javier Díaz de Terán (105), Department of Neurology; Unit or Division of Physiotherapy, Hospital La Paz Institute for Health Research, Madrid, Spain

Pelin Corman Dincer (163), Department of Anesthesiology and Reanimation, School of Medicine, Marmara University, Istanbul, Turkey

Fernando Estévez-López (551), Department of Pediatrics, Wilhelmina Children's Hospital, University Medical Center Utrecht, Utrecht, The Netherlands

Alessandro Failo (15), Department of Psychology and Cognitive Sciences, University of Trento, Rovereto, Italy

Hamid Reza Fallahi (363), Dental Research Center, Research Institute of Dental Sciences, Shahid Beheshti University of Medical Sciences, Tehran; Founder and Director of Maxillofacial Surgery and Implantology and Biomaterial Research Foundation (www.maxillogram.com), Isfahan, Iran

Deni Francesco (69), Department of Anesthesia and Intensive Care, IRCCS Ospedale San Raffaele, Milano, Italy

R. Fronczek (93), Department of Neurology, Leiden University Medical Center, Leiden, The Netherlands

Alfonso Gil-Martínez (105), Department of Physiotherapy, Health Sciences Faculty, University of La Salle Madrid; Unit or Division of Physiotherapy, Hospital La Paz Institute for Health Research, Madrid, Spain

J. Haan (93), Department of Neurology, Leiden University Medical Center, Leiden; Department of Neurology, Alrijne Hospital, Leiderdorp, The Netherlands

Jorge Haddad (527), Urogynecology Division, Discipline of Gynecology, Clinics Hospital at University of Sao Paulo, Sao Paulo, SP, Brazil

Alberto Marcos Heredia-Rizo (291), Department of Physiotherapy, Faculty of Nursing, Physiotherapy and Podiatry, University of Seville, Seville, Spain

Munetaka Hirose (235), Department of Anesthesiology and Pain Medicine, Hyogo College of Medicine, Nishinomiya, Hyogo, Japan

Che Aishah Nazariah Ismail (139), Department of Physiology, School of Medical Sciences, Health Campus, Universiti Sains Malaysia, Kota Bharu, Kelantan, Malaysia

E. Isomura (433), Polyclinic Bagatin, Department of Surgery and Anestesiology with Ranimatology, Zagreb, Split, Croatia

A.A. Gde Putra Semara Jaya (343), Department of Anesthesiology and Intensive Care, Mangusada Hospital, Faculty of Medicine, Udayana University, Badung, Bali, Indonesia

Asad Jehangir (173), Gastroenterology Section, Department of Medicine, Temple University School of Medicine, Philadelphia, PA; Gastroenterology Section, Medical College of Georgia at Augusta University, Augusta, GA, United States

Ureshani Karunarathna (47), Department of Pharmacy and Pharmaceutical Sciences, Faculty of Allied Health Sciences, University of Sri Jayewardenepura, Nugeggoda, Sri Lanka

Hardeep Kaur (265), Department of Pediatrics, Division of Pediatric Pulmonology and Intensive Care, All India Institute of Medical Sciences, New Delhi, India

Seied Omid Keyhan (363), Founder and Director of Maxillofacial Surgery and Implantology and Biomaterial Research Foundation (www.maxillogram.com), Isfahan, Iran

Matthew Kim (257), Department of Otolaryngology—Head and Neck Surgery, Westchester Medical Center, Valhalla, NY, United States

Z. Kljajić (433), Polyclinic Bagatin, Department of Surgery and Anestesiology with Ranimatology, Zagreb, Split, Croatia

Meltem Koç (487), Department of Physiotherapy and Rehabilitation, Faculty of Health Sciences, Muğla Sıtkı Koçman University, Muğla, Turkey

Hannu Kokki (421), Department of Anaesthesiology and Intensive Care, Kuopio University Hospital; Faculty of Health Sciences, Clinical Medicine, University of Eastern Finland, Kuopio, Finland

Merja Kokki (421), Department of Anaesthesiology and Intensive Care, Kuopio University Hospital; Faculty of Health Sciences, Clinical Medicine, University of Eastern Finland, Kuopio, Finland

Jun Komatsu (475), Department of Medicine for Motor Organs, Juntendo University Graduate School of Medicine; Department of Orthopaedic Surgery, Juntendo Tokyo Koto Geriatric Medical Center, Tokyo, Japan

Antoine Lemaire (3), Pain and Palliative Medicine, Head of Oncology and Medical Specialties Department, Valenciennes General Hospital, Valenciennes, France

Nancy Liu (225), Los Angeles, CA, United States

Idris Long (139), Biomedicine programme, School of Health Sciences, Health Campus, Universiti Sains Malaysia, Kubang Kerian, Kelantan, Malaysia

Bettega Costa Lopes (149), Postgraduate Program in Physiology, Federal University of Rio Grande do Sul, Porto Alegre, RS, Brazil

Beretta Luigi (69), Department of Anesthesia and Intensive Care, IRCCS Ospedale San Raffaele, Milano, Italy

Pascal Madeleine (291), Sport Sciences—Performance and Technology, Department of Health Science and Technology, Aalborg University, Aalborg, Denmark

Cristina Maestre-Cascales (551), LFE Research Group, Department of Health and Human Performance, Polytechnic University of Madrid, Madrid, Spain

Thais F. de Magalhaes (527), Urogynecology Division, Discipline of Gynecology, Clinics Hospital at University of Sao Paulo, Sao Paulo, SP, Brazil

Gaurav Mahajan (265), Department of Gastroenterology, Postgraduate Institute of Medical Sciences and Research, Chandigarh, India

Michele Malaguarnera (539), Research Centre "The Great Senescence", University of Catania, Catania, Italy

M. Malić (433), Polyclinic Bagatin, Department of Surgery and Anestesiology with Ranimatology, Zagreb, Split, Croatia

Zubair Malik (173), Gastroenterology Section, Department of Medicine, Temple University School of Medicine, Philadelphia, PA, United States

Deborah Margatho (507), Department of Obstetrics and Gynaecology, University of Campinas Medical School, Campinas, SP, Brazil

Marmiere Marilena (69), Department of Anesthesia and Intensive Care, IRCCS Ospedale San Raffaele, Milano, Italy

José Manuel Martínez-Martos (493), Experimental and Clinical Physiopathology Research Group, Department of Health Sciences, Faculty of Experimental and Health Sciences, University of Jaén, Jaén, Spain

Liciane Fernandes Medeiros (149), Postgraduate Program in Health and Human Development, University La Salle, Canoas, RS, Brazil

Derek M. Miletich (117, 127), CDR, MC, Department of Anesthesiology and Pain Medicine, Naval Medical Center San Diego, San Diego, CA, United States

Etienne Mullet (305), Department of Ethics, Institute of Advanced Studies (EPHE), Plaisance, France

Lynita Mullins (117, 127), CDR, MC, Department of Pain Medicine, Naval Medical Center Camp Lejeune, NC, United States

Glauce Crivelaro Nascimento (201), Department of Basic and Oral Biology, School of Dentistry of Ribeirao Preto, University of Sao Paulo, Sao Paulo, Brazil

J. Nemrava (433), Polyclinic Bagatin, Department of Surgery and Anestesiology with Ranimatology, Zagreb, Split; Faculty of Dental Medicine and Health Osijek, Department of Surgery and Anesthesiology With Reanimatology, Depatment of Dental Medicine 1, Univeristy Josip Juraj Strossmayer, Osijek, Croatia

Kelsey Ng (27), Hong Kong

Hisao Nishijo (317), System Emotional Science, Faculty of Medicine, University of Toyama, Toyama, Japan

Rodrigo Núñez-Cortés (275), Department of Physical Therapy, Faculty of Medicine, University of Chile, Santiago, Chile; Department of Physiotherapy, Physiotherapy in Motion Multispeciality Research Group (PTinMOTION), University of Valencia, Valencia, Spain

Adeolu Oladunjoye (173), Division of Medical Critical Care, Boston Children's Hospital, Boston, MA, United States

Olubunmi Oladunjoye (173), Department of Medicine, Reading Hospital-Tower Health System, Reading, PA, United States

Taketoshi Ono (317), System Emotional Science, Faculty of Medicine, University of Toyama, Toyama, Japan

Antonella Paladini (329), Department MESVA, University of L'Aquila, L'Aquila, Italy

Sanjog Pangarkar (225), Los Angeles, CA, United States

Henry P. Parkman (173), Gastroenterology Section, Department of Medicine, Temple University School of Medicine, Philadelphia, PA, United States

Sofía Pérez-Alenda (275), Department of Physiotherapy, Physiotherapy in Motion Multispeciality Research Group (PTinMOTION), University of Valencia, Valencia, Spain

C. Ryan Phillips (117, 127), CDR, MC, Department of Anesthesiology and Pain Medicine, Naval Medical Center San Diego, San Diego, CA, United States

Paula Pino Pommer (275), Traumatology Department, Hospital Clínico La Florida, Santiago, Chile

Marcos Prada-Arias (189), Department of Pediatric Surgery, Vigo University Hospital Álvaro Cunqueiro, Vigo, Spain Rare Diseases and Pediatric Medicine Research Group, Galician Sur Health Research Institute, Carretera Clara Campoamor, Vigo, Spain

Sara Pullen (285), Department of Rehabilitation Medicine, Emory University School of Medicine, Atlanta, GA, United States

Anam Qureshi (173), Department of Medicine, Reading Hospital-Tower Health System, Reading, PA; Department of Medicine, Medical College of Georgia at Augusta University, Augusta, GA, United States

María Jesús Ramirez-Éxposito (493), Experimental and Clinical Physiopathology Research Group, Department of Health Sciences, Faculty of Experimental and Health Sciences, University of Jaén, Jaén, Spain

Martina Rekatsina (329), Chronic Pain Clinical Fellow, Whipps Cross University Hospital, Barts Health NHS Trust, London, United Kingdom

Meani Renato (69), Department of Anesthesia and Intensive Care, IRCCS Ospedale San Raffaele, Milano, Italy

Manuel Rich-Ruiz (493), Department of Nursing, Pharmacology and Physiotherapy, Faculty of Medicine and Nursing, University of Córdoba; Maimónides Institute for Biomedical Research (IMIBIC), University of Córdoba, Reina Sofia University Hospital, Córdoba; Experimental and Clinical Physiopathology Research Group, Department of Health Sciences, Faculty of

Experimental and Health Sciences, University of Jaén, Jaén, Spain

Roya Sabzian (363), Dental Students Research Center, School of Dentistry, Isfahan University of Medical Sciences, Isfahan, Iran

Shigekazu Sakai (317), System Emotional Science, Faculty of Medicine, University of Toyama, Toyama, Japan

K. Šakić (433), Polyclinic Bagatin, Department of Surgery and Anestesiology with Ranimatology, Zagreb, Split; Faculty of Dental Medicine and Health Osijek, Department of Surgery and Anesthesiology With Reanimatology, Depatment of Dental Medicine 1, Univeristy Josip Juraj Strossmayer, Osijek, Croatia

L. Šakić (433), Faculty of Dental Medicine and Health Osijek, Department of Surgery and Anesthesiology With Reanimatology, Depatment of Dental Medicine 1, Univeristy Josip Juraj Strossmayer, Osijek, Croatia

Giorgia Saltelli (329), Sant'Andrea Hospital, "La Sapienza" University of Roma, Roma, Italy

Bruna Maitan Santos (201), Department of Physiology, School of Medicine of Ribeirão Preto, University of Sao Paulo, Sao Paulo, Brazil

M. Šarec Ivelj (433), Polyclinic Bagatin, Department of Surgery and Anesthesiology with Ranimatology, Zagreb, Split; Faculty of Dental Medicine and Health Osijek, Department of Surgery and Anesthesiology With Reanimatology, Depatment of Dental Medicine 1, Univeristy Josip Juraj Strossmayer, Osijek, Croatia

Anthony P. Sclafani (257), Department of Otolaryngology—Head and Neck Surgery, Weill Cornell Medical College, New York, NY, United States

Tjokorda Gde Agung Senapathi (343), Department of Anesthesiology and Intensive Care, Sanglah Hospital, Faculty of Medicine, Udayana University, Denpasar, Bali, Indonesia

Wangning ShangGuan (79), Department of Anesthesiology and Perioperative Medicine, The Second Affiliated Hospital and Yuying Children's Hospital of Wenzhou Medical University, Wenzhou, China

Andre Shomorony (257), Department of Otolaryngology—Head and Neck Surgery, NewYork-Presbyterian Hospital, New York, NY, United States

Paul Clay Sorum (305), Department of Internal Medicine and Pediatrics, Albany Medical College, Albany, NY, United States

Kamath Sriganesh (463), Department of Neuroanaesthesia and Neurocritical Care, National Institute of Mental Health and Neurosciences, Bengaluru, India

Turi Stefano (69), Department of Anesthesia and Intensive Care, IRCCS Ospedale San Raffaele, Milano, Italy

Dirson João Stein (149), Postgraduate Program in Medical Sciences, Federal University of Rio Grande do Sul, Porto Alegre, RS, Brazil

Grace P.Y. Szeto (291), School of Medical and Health Sciences, Tung Wah College, Hong Kong, China

Paul M. Szumita (57), Department of Pharmacy, Brigham and Women's Hospital, Boston, MA, United States

Kouichi Takamoto (317), System Emotional Science, Faculty of Medicine, University of Toyama, Toyama; Department of Sports and Health Sciences, Faculty of human sciences, University of East Asia, Yamaguchi, Japan

Claudio Tapia (275), Department of Physical Therapy, Faculty of Medicine, University of Chile, Santiago, Chile

G.M. Terwindt (93), Department of Neurology, Leiden University Medical Center, Leiden, The Netherlands

Iraci L.S. Torres (149), Laboratory of Pain Pharmacology and Neuromodulation: Preclinical Research, Porto Alegre, RS, Brazil

Gabriela Trevisan (247), Graduate Program in Pharmacology, Federal University of Santa Maria (UFSM), Santa Maria, RS, Brazil

Turgay Tuna (445), University Hospital Erasme, Free University of Brussels, Anderlecht, Belgium

Susumu Urakawa (317), Department of Musculoskeletal Functional Research and Regeneration, Graduate School of Biomedical and Health Sciences, Hiroshima University, Hiroshima, Japan

Giustino Varrassi (329), Paolo Procacci Foundation, Rome, Italy

Felicity Veal (27), Unit for Medication Outcomes and Research and Education, School of Pharmacy and Pharmacology, University of Tasmania, Hobart, TAS, Australia

Quinn Wonders (225), Los Angeles, CA, United States

Dana Zandian (363), Dental Research Center, Research Institute of Dental Sciences, Shahid Beheshti University of Medical Sciences, Tehran; Director of Maxillofacial Surgery and Implantology and Biomaterial Research Foundation(www.maxillogram.com), Isfahan, Iran

Preface

The etiology of pain is complex and multifactorial. This complexity is magnified by the use of analgesics and local or general anesthetics. While analgesics can reduce pain, general anesthetics reduce consciousness and local anesthetics reduce localized pain. Analgesics may be used after surgery performed under anesthesia. However, anesthesia is not without risk. Depending on patient-related factors such as comorbid disease, anesthesia may be associated with significant morbidity or even mortality. Adverse events during or after anesthesia may necessitate the use of other pharmacological agents such as vasopressors, inotropes, sedatives, antiarrhythmics, or antiemetics.

The perception of pain itself results from a multifaceted interaction between illness beliefs, age, gender, time of onset, stress, socioeconomic status, and other factors. To a certain extent, one could argue that pain itself is helpful in treating disease. It can be considered the "sixth vital sign" as it indicates the need for assessment by a healthcare professional and the need for clinical investigations, diagnosis, and appropriate medication. The pain associated with myocardial infarction is a good example of this. One needs to consider though that some acute and chronic pain can lead to psychological distress and reduced quality of life. In the long term, chronic persistent pain can impact significantly on the family unit and many diseases present with pain. There are a plethora of pharmacological agents currently available for pain management. Furthermore, studies showing the beneficial effects of plant or natural extracts provide the foundation for further rigorous studies in clinical trials.

The neuroscience of pain in one condition may be relevant to understanding the pain observed in other conditions. The onset of pain, the cause of the pain, and the administration of analgesia or anesthesia is a continuing scientific spectrum. At each point, there is a firm scientific basis with established literature and also an ongoing drive to discover new facts and data. Hitherto, such material on the neuroscience of pain, anesthesia, and analgesia has been sporadic and/or written for the experts who specialize in narrow and focused areas. For example, the expert in the use of general anesthetics in the surgical setting may not necessarily be an expert in the molecular biology of neurons activated in the pain process. The cellular biologist may be aware of neither the science underpinning the provision of anesthesia nor the adverse outcomes that the anesthesiologist may encounter in the clinical setting. To address the aforementioned issues, the editors have compiled *The Neuroscience of Pain, Anesthetics, and Analgesics*.

The Neuroscience of Pain, Anesthetics, and Analgesics is divided into three books:

Book 1: *Features and Assessments of Pain, Anesthesia, and Analgesia*
Book 2: *Treatments, Mechanisms, and Adverse Reactions of Anesthetics and Analgesics*
Book 3: *The Neurobiology, Physiology, and Psychology of Pain*

This book, *Features and Assessments of Pain, Anesthesia, and Analgesia*, is divided into four parts. *Part I, Setting the Scene: General Aspects of Anesthesia, Analgesics, and Pain*, covers topics on anesthesia and body mass, management of pain in older people, concept of multimorphic cancer pain, and multidisciplinary acute pain service. *Part II, The Syndromes of Pain*, covers topics on cluster headache and pain, migraine and pain, phantom limb pain, birth and pain, breast cancer and nociception, and Carpal tunnel syndrome. *Part III, Interlinking Anesthesia, Analgesics, and Pain*, covers topics on multimodal analgesia and postsurgical pain, spinal anesthesia and cesarean section, postoperative pain management and truncal blocks in surgery, and anesthesia and combat-related extremity injury. *Part IV, Assessments, Screening, and Resources*, covers topics on the Pain Behavior Scale, the Pain Catastrophizing Scale, The Analgesia Nociception Index, Back Pain Functional Scale, and serum biomarkers for headache.

Each chapter has:

- *An Abstract (published online)*
- *Key Facts*
- *Mini-Dictionary of Terms*
- *Applications to Other Areas*

- *Other Agents of Interest*
- *Summary Points*

The section *Key Facts* focuses on areas of knowledge written for the novice. The *Mini-Dictionary of Terms* explains terms that are frequently used in the chapter. The *Applications to Other Areas* describes other fields of neuroscience, pain science, analgesia, and anesthesia that the chapter may be relevant to. The *Summary Points* encapsulates each chapter in a succinct way.

The Neuroscience of Pain, Anesthetics, and Analgesics is designed for research and teaching purposes. It is suitable for neurologists and anesthesiologists as well as those interested in pain and its interlink with pain relief. It is valuable as a personal reference book and also for academic libraries that cover the neuroscience of pain, anesthetics, and analgesics. Contributions are from leading national and international experts including those from world-renowned institutions.

Rajkumar Rajendram, Victor R. Preedy, Vinood B. Patel, and Colin R. Martin
(Editors)

Part I

Setting the scene: General aspects of anesthesia, analgesics and pain

Chapter 1

The concept of multimorphic cancer pain: A new approach from diagnosis to treatment

Antoine Lemaire

Pain and Palliative Medicine, Head of Oncology and Medical Specialties Department, Valenciennes General Hospital, Valenciennes, France

Abbreviations

DN4	neuropathic pain 4 questionnaire
ECOG-PS	eastern cooperative oncology group performance status
NPSI	neuropathic pain symptom inventory
NRS	numerical rating scale
NSAIDs	nonsteroidal anti-inflammatory drugs
OIC	opioid induced constipation
PCA	patient-controlled analgesia
VAS	visual analogue scale
VRS	verbal rating scale
WHO	world health organization

Introduction

Cancer pain remains a major public health issue at the international level, with wide disparities from one country to another. It is estimated that 55% of patients undergoing cancer treatment and 66% of patients with advanced, metastatic, or terminal cancer are in pain. Pain in cancer survivors, which occurs at distance from the illness and after the curative treatment, concerns 39% of patients (van den Beuken-van Everdingen, Hochstenbach, Joosten, Tjan-Heijnen, & Janssen, 2016). These data have evolved little in more than 40 years (Van Den Beuken-Van Everdingen et al., 2007), which underlines the complexity of managing cancer pain. Cancer is still one of the main causes of morbidity and mortality in the world, with a constantly rising incidence (Frankish, 2003; World Health Organization, 2018).

Despite the available tools for optimal management of cancer pain, combining the best understanding of physiopathological mechanisms with drug-based, non-drug-based, and interventional treatments, via pain assessment, pain still remains underdiagnosed, insufficiently assessed, and undertreated, including in the most severe cases (Fallon et al., 2018). Access notably to opioids, the reference analgesic treatment, is unfortunately still insufficient in many countries (Frankish, 2003), and most patients are not referred to a multidisciplinary team specializing in pain (Institut national du Cancer (INCA), 2012).

Improved survival of cancer patients—certain forms of cancer are even becoming chronic diseases—and both the complexity and multiplicity of the physiopathological and etiological mechanisms involved are key factors in understanding cancer pain. Here, we shall describe the concept of multimorphic cancer pain from an exhaustive and innovative point of view that breaks with the standard models (Lemaire et al., 2019). Pain management remains one of the fundamental pillars of early supportive care in cancer, carrying the organizational and structural archetypes and having the main objective: the optimization of the patients' quality of life through a targeted, multimodal, and personalized approach (Lemaire et al., 2019).

Features and Assessments of Pain, Anesthesia, and Analgesia. https://doi.org/10.1016/B978-0-12-818988-7.00010-8
Copyright © 2022 Elsevier Inc. All rights reserved.

4 **PART | I** Setting the scene: General aspects of anesthesia, analgesics and pain

Assessing and reassessing cancer pain and patients

Assessing the cancer pain and the concerned patients corresponds to a dynamic approach that must constantly be renewed over time (Bennett et al., 2018; Frankish, 2003; Lemaire et al., 2019; World Health Organization, 2018) given the multi-morphism of the pain (or rather, the various types of cancer pain) and its ability to change as time progresses, as we shall see later. At every stage of the disease, and as soon as the diagnosis is made (NICE, 2004), each medical or paramedical consultation, including noncancer follow-up, is thus an occasion to ensure the stability of the pain and adopt an optimal analgesic treatment approach, as well as an environment, in the broadest sense of the term, that is best suited to caring for the patient (healthcare teams, home treatment teams, family and friends, etc.).

Assessing cancer pain

The first stage is to assess the cancer pain, as the principles of targeted, multimodal, and personalized management will depend on it. The different types of cancer pain effectively have several specificities that need to be sought out. As in other chronic pain situations, assessment must be multidimensional: sensory-discriminatory, cognitive, emotional, and behavioral. Communication with patients and their entourage is a major element in the assessment of cancer pain, which needs to be qualitative, something that requires time (Bennett et al., 2018; Fallon et al., 2018; Street et al., 2014). It is essential that the patient be associated as a full participant in his or her own management by means of dynamic therapeutic education, both for assessment and subsequent follow-up of the pain on analgesia.

Standardized pain assessment tools can be recommended, such as the *visual analogue scale (VAS), the verbal rating scale (VRS), or the numerical rating scale (NRS)* (Fallon et al., 2018; Minello et al., 2019; World Health Organization, 2018). In certain specific cases such as cognitive disorders or a patient's inability to communicate (coma, multiple disabilities, etc.), it is recommended to use heteroassessment pain scales based on observation, which will allow the pain follow-up over time and its assessment on analgesia (Fallon et al., 2018). Neuropathic pain requires specific assessment scales such as the *DN4* or the *NPSI* (Attal, Bouhassira, & Baron, 2018).

The causes of the pain must be sought very carefully *(Key Facts of cancer pain etiologies)* (Bennett et al., 2018; Fallon et al., 2018), bearing in mind that there may be many in addition to the cancer itself and/or its metastases. These causes can also be associated with each other or over time, providing one of the elements for understanding the multimorphic nature of pain, that is, its ability to change over time. In terms of semantics, it is possible to talk of *cancer-related pain*, when the pain is linked to the tumoral disease itself (primary tumor and/or metastases), or *cancer pain* in a more global manner when referring to all the possible pain, linked to the tumor, or its treatments, including at distance from the cure (so-called persistent pain).

Key facts of cancer pain etiologies (Fallon et al., 2018; Lemaire et al., 2019)

- The **tumor** itself or/and **metastasis** (including painful paraneoplastic syndromes).
- **Acute procedural pain**: diagnostic intervention, biopsies, endoscopies, blood sampling, central line position, injections, punctures, wound care, etc.
- **Iatrogenic pain causes**: surgery, chemotherapies, hormonal therapies, targeted therapies, radiotherapy, chemoembolizations, thermal ablations or cryotherapy, adverse events from other treatments including supportive medications.
- **Comorbidity-related pain**: headache/migraines, fibromyalgia, postherpetic neuralgia, diabetic neuropathy, osteoarthritis, arthrosis, etc.
- **Pain in cancer survivors**: postchirurgical or radiotherapy pain, anticancer drug-related pain, follow-up procedures, postherpetic neuralgia.

The physiopathological mechanism of pain needs to be specified as it will allow to better guide the analgesic therapies, in the broadest sense of the term (Table 1). It may be the question of somatic nociceptive pain (for example: bone damage) or visceral pain (for example: pancreatic damage), pain associated with tissue lesions (other than the nervous system), or stimulation of the nociceptors on a functional nervous system (World Health Organization, 2018). Neuropathic expression can be frequent, through dysfunction or damage to the central or peripheral nervous system: direct compression by a tumor, neurotoxicity, central hypersensitization, dysfunction in the sympathetic system, etc. (World Health Organization, 2018). Once again, these mechanisms can coexist and vary over time, thus resulting in mixed, overlapped, combined, or associated pain (Minello et al., 2019), illustrating the multimorphic nature of pain.

TABLE 1 Neural mechanisms of cancer pain types (World Health Organization, 2018).

Type			Neural mechanism
Nociceptive	Visceral		Stimulation of pain receptors on normal sensory nerve endings
	Somatic		
Neuropathic	Nerve compression		Stimulation of nervi nervorum
	Nerve injury	Peripheral	Lowered firing threshold of sensory nerves (deafferentation pain)
		Central	Injury to central nervous system
		Mixed	Peripheral and central injury
	Sympathetically maintained		Dysfunction of sympathetic system

There are often multiple localizations of pain in cancer, and the painful areas can present different physiopathological mechanisms, particularly in the advanced forms of the disease when in most cases there are at least two different types of pain (Portenoy & Koh, 2010).

It is essential to assess the characteristics of the pain, particularly its intensity, duration, the factors that trigger or relieve it, any associated symptoms, and the efficacy of treatments. Even if the definition has not always reached a consensus, breakthrough cancer pain is currently defined as an episode of severe pain that occurs in patients receiving a stable opioid regimen for persistent pain sufficient to provide at least mild sustained analgesia (Fallon et al., 2018). It is thus pain that is of rapid onset, attaining a peak in just a few minutes, and generally calming down in less than an hour. Its physiopathology can be complex, as well as nociceptive, neuropathic, or mixed (Krakowski et al., 2003). This type of pain is characteristic of cancer pain and is thus dissociated from background pain. The latter is defined as a pain present for ≥12 h/day during the previous week (or would be present if not taking analgesia) (Davies et al., 2009). If this background pain is adequately controlled, that is, pain rated as "none" or "mild," but not "moderate" or "severe" for ≥12 h/day during previous week (Davies et al., 2009), and the patient presents transitory exacerbation of the pain, it is then possible to speak of breakthrough cancer pain. If this is not the case, careful assessment of the pain will allow to rebalance the background analgesic treatment. Cancer pain may present in acute, subacute, or chronic forms (Fallon et al., 2018; Lemaire et al., 2019).

This pain assessment should always be carried out in parallel to the assessment not only of the efficacy of the analgesic treatments, but also of their tolerance and the patient's compliance with the treatment, as we shall discuss later.

Global assessment of patients

Assessing patients in the context of follow-up of their pain symptomatology should nevertheless not be limited to the pain itself: It is essential to adopt a global approach to patients in order to target certain factors in particular, whether they are intrinsic or extrinsic, so as to propose optimal management. The assessment is demanding (Bennett et al., 2018; Fallon et al., 2018; Minello et al., 2019), but carrying it out in an exhaustive manner is necessary for a high-quality management and will avoid limiting it to a simple, reductive equation: pain = morphine. Using a multidisciplinary cancer supportive care team is recommended in the most complex situations. Assessing patients and their entourage is effectively not reserved exclusively for doctors: The additional input of other healthcare providers from the multidisciplinary team is essential, and certain models have even shown the advantages of delegating certain tasks usually reserved for doctors to carers (Ferrell et al., 2015; Higginson & Booth, 2011; Hui, Hannon, Zimmermann, & Bruera, 2018; Tuggey & Lewin, 2014).

The aim of this multidisciplinary global assessment is to understand the multidimensional needs of patients and their entourage, be they physical (such as the symptoms including pain), emotional (such as anxiety or depression), socioeconomic (for example, family caregivers, relationships, living situation, and financial aspects), and spiritual or informational (for example, advance care plans, treatments and disease comprehension, etc.) (Hui et al., 2018).

The management of cancer pain is one of the fundamental pillars of supportive care in cancer and is thus part of a complementary approach to the care specific to cancer (Hui, 2014; Hui et al., 2013). Since the foundation studies on early palliative care, which were the first to demonstrate the impact of early intervention by a multidisciplinary palliative care team on the quality of life of patients with metastatic cancer, the concepts of supportive care and palliative care have often been the subject of debate, but the definitions are clear (Hui, 2014; Hui et al., 2013; Hui & Bruera, 2015; MASCC, 2020). Supportive care in cancer is "the prevention and management of the adverse effects of cancer and its treatment. This includes management of both physical and psychological symptoms and adverse events across the continuum of the cancer experience from diagnosis, through anti-cancer treatment, to post-treatment care. Enhancing rehabilitation, secondary cancer prevention, survivorship, and end-of-life care are integral to supportive care" (MASCC, 2020). For our part, we retain the fact that the management of cancer pain is an integral part of supportive care as soon as the cancer diagnosis is made. Palliative care is thus one aspect of supportive care in the cancer realm, provided as soon as cancer patients can no longer be cured (Hui et al., 2013; MASCC, 2020). It is interesting to measure the positive impacts of the different models for early supportive care at different levels, whether in terms of optimized management of the symptoms, improving quality of life and satisfaction, increasing survival, assessing the level of home care, or reducing healthcare costs (Hui et al., 2018; Hui & Bruera, 2015). The most recent models recommend earlier intervention in supportive care in a targeted and timely manner, from patient screening (Hui et al., 2018), which is perfectly suited to the assessment and management of cancer pain.

Other agents of interest: Benefits of supportive care (MASCC, 2020)

- Alleviation of symptoms and complications of cancer
- Prevention or reduction of treatment toxicities
- Improved communication between patients and caregivers
- Increased tolerance, and thus benefits, of active therapy
- Easing of the emotional burden for patients and caregivers
- Psychosocial support for cancer survivors

A full clinical examination, and if necessary paraclinical explorations (biology or imaging), may be proposed to understand the mechanisms of pain or to diagnose any comorbidities. As with the pain, specific evaluation scales can be used to assess symptoms, quality of life, or functional status. It is important not to consider—in a stigmatizing way—that pain must systematically be linked to cancer, as well as to retain an analytical approach so as to eliminate any possible differential diagnoses that may be curable, such as painful diseases that have no link to the cancer or its treatments (for example, arthrosis, inflammatory rheumatisms, migraines, diabetic neuropathy, etc.). These diagnostic traps are common in situations in which pain is the reason behind the consultation at the emergency room: Diagnoses of curable medical emergencies might then be evoked following analysis of the situation, such as a pulmonary embolism or appendicular peritonitis (Burnod et al., 2019). Special attention must be paid to all the treatments a patient is taking, as well as any comorbidity, during the global assessment.

Applications to other areas: From supportive care in cancer to supportive medicine

Following the philosophy of supportive care in cancer, the concept of supportive medicine shares the same philosophy but extends to other chronic conditions: "*We believe that in parallel to the specific management of cancer or chronic conditions, an exhaustive model focusing on specialized, advance management of the symptoms with a sometimes-major impact in the care pathway (pain, fatigue, depression, malnutrition, etc.), and other support activities (dietary, psychological and physical well-being, socio-environmental aspects etc.) could be set up without major upheaval to healthcare structures. With this comprehensive, transdisciplinary and exhaustive approach, as soon as a chronic condition is diagnosed in a specialized centre, the intervention of an interdisciplinary supportive care team as part of the continued relations between those taking part should make it possible to optimize the healthcare pathway and the prevention of avoidable complications, all whilst also avoiding extended hospitalizations and encouraging a significant level of in-home care. This approach also makes it possible to establish a functional and synergic link between specialists of the disease in question, organ specialists and the teams intervening in-home, preventing management from being divided between specialists. On discharge from the establishment, this follow-up can persist through the supportive care team by means of regular contacts with the private practice care networks, with benefits for the patient in terms of quality of care, and in the use of resources. (…) This type of supportive medicine model is a relevant response to the constant development in the levels of over-specialization in the management of chronic conditions such as cancer, and makes it possible to provide the right expertise, to the right patient, by the right teams and at the right time.*" (Lemaire et al., 2019).

The concept of multimorphic cancer pain

Presented for the first time in a series of articles devoted to cancer pain in 2019 (Allano et al., 2019; Burnod et al., 2019; George et al., 2019; Lemaire, 2019a, 2019b; Lemaire et al., 2019; Maindet et al., 2019; Minello et al., 2019), this concept aims to propose an approach that breaks with standard models to make it easier to understand and manage cancer pain: "*From an etymological point of view, the term multimorphic refers to the possibility of adopting several forms at the same time, and of changing form. This term seems to us to be adapted to the dynamic definition that we have sought to give to cancer pain: this type of pain can effectively evolve in how it presents, in relation to the different factors, whether or not they are linked to the cancer and its management*" (Lemaire et al., 2019). This definition reveals the potential complexity of cancer pain as a specific nosological entity, particularly in situations of chronic cancer or progress to a palliative stage.

As already mentioned, cancer pain must thus be analyzed on the basis of a continuum throughout the patient's care pathway as it may change over time in relation to the various disruptive factors described in the following section and which may potentially influence the pain and analgesic balance at any time (*Key Facts of Criteria defining the multimorphic nature of cancer pain*).

Key facts of criteria defining the multimorphic nature of cancer pain (Lemaire et al., 2019)

- Factors influencing the complexity of cancer pain:
  - Components of the pain (nociceptive, neuropathic, and nociplastic for associated pain)
  - Etiopathogenic mechanisms (cancer, its treatments, and other causes)
  - Presentation of the pain (intensity, duration, background, exacerbations including breakthrough cancer pain, and pain emergencies)
- Intrinsic factors of variability over time concerning the cancer:
  - Type of cancer and its stage on diagnosis
  - Progression of the "chronic illness" cancer (cure, sequelae, relapse, metastases, and palliative progression)
  - Evolution in the treatments and complications (cancer treatments, supportive treatments, and complications)
- Extrinsic factors of variability over time concerning the state of health:
  - Environmental factors (ethnodemographic, cultural, spiritual, and socioeconomic factors; earliness, level of access to care, and abandonment factors; and communication)
  - Interindividual factors: genetics, variability factors of pain thresholds, immunity, and metabolism
  - Intraindividual factors: motivation, risk factors, comorbidities and multimorbidities, intercurrent treatments, treatment compliance, and treatment education

Assessing both multimorphic cancer pain and patients is thus an exhaustive and rigorous process, one that continues throughout the care pathway, from diagnosis to the postcancer period, via chronic metastatic phases or palliative situations that can be difficult to manage. From this multidisciplinary assessment will result a treatment approach that is tirelessly targeted, multimodal, and personalized so as to be able to adapt it to the patients, providing them with optimal quality of life (Bennett et al., 2018; Lemaire et al., 2019).

Treatment strategies in multimorphic cancer pain

The analgesic strategy deployed by the World Health Organization in the context of cancer-related pain has allowed, since 1986 (revised in 1996), to recommend analgesics—particularly opioids—in the context of nociceptive cancer pain, and in accordance with a classification in analgesic power levels proportional to the intensity of the pain (World Health Organization, 1986, 1996). Several studies esteem that if this strategy, called the WHO analgesic ladder, were applied, 75% to 90% of cancer pain could be relieved (Grond, Zech, Schug, Lynch, & Lehmann, 1991; Zech, Grond, Lynch, Hertel, & Lehmann, 1995). This strategy has even considerably exceeded the context of cancer pain to become a reference in the management of nociceptive pain in general, notably at the pedagogical level. Today, this approach is nevertheless considered by the WHO itself as being poorly adapted to the management of cancer pain, given that the methods for assessing pain and the treatments available have progressed considerably, with greater scientific knowledge (4). Furthermore, as we have just seen, multimorphic cancer pain is more complex given the progress made in anticancer treatments and the increase in life expectancy for cancer patients, coupled with the aging of the population (Lemaire, 2019a, 2019b; Lemaire et al., 2019; World Health Organization, 2018). It is essential that the supportive care—and thus the management of cancer pain—be integrated into real healthcare policies with a minimum of knowledge and treatments available in all countries (World Health Organization, 2018).

In this section, we will present the analgesic treatment strategies that are recommended and pertinent, whether they are drug-based, non-drug-based, or interventional. Anticancer treatments such as surgery, chemotherapy, or radiotherapy are

8 **PART | I** Setting the scene: General aspects of anesthesia, analgesics and pain

obviously the first line of analgesic treatment, given that their aim is to treat the etiology of the pain—that is, the cancer or its metastases. These treatments will not be discussed here in detail, but it is necessary to remember that they exemplify the paradox of being the first analgesics, while being potentially the source of persistent pain, notably neuropathic (Fallon et al., 2018; Lemaire et al., 2019).

Key facts of WHO cancer pain management guiding principles (World Health Organization, 2018)

- The goal of optimum management of pain is to reduce pain to levels which allow an acceptable quality of life.
- Global assessment of the person should guide treatment, recognizing that individuals experience and express pain differently.
- Safety of patients, carers, healthcare providers, communities, and society must be assured.
- A pain management plan includes pharmacological treatments and may include psychosocial and spiritual care.
- Analgesics, including opioids, must be accessible: both available and affordable.
- Administration of analgesic medicine should be given "by mouth," "by the clock," "for the individual," and with "attention to detail."
- Cancer pain management should be integrated as a part of cancer care.

Drug-based management of nociceptive cancer-related pain

Analgesic drugs for nociceptive cancer-related pain

The preferred analgesic drugs for nociceptive cancer-related pain remain classified in three groups:

- Mild analgesics: nonsteroidal antiinflammatories (NSAIDs, including ibuprofen) and paracetamol (the former WHO step 1), available for oral, rectal, or injectable delivery depending on the specialties
- Weak opioids (the former WHO step 2): available for oral, rectal, or injectable delivery depending on the specialties. These drugs include tramadol, codeine, opium powder, and dihydrocodeine
- Strong opioids (the former WHO step 3): available for oral, transmucosal, transdermal, and injectable delivery. These drugs notably include morphine, methadone, oxycodone, hydromorphone, fentanyl, buprenorphine, and oxymorphone

These different treatments were previously recommended (World Health Organization, 1986, 1996) in relation to the intensity of the pain and in successive steps (step 1, then 2, then 3, with synergic associations possible between steps 1 and 2, or steps 1 and 3). This progressive strategy has thus been abandoned, notably because it could delay access to a satisfactory analgesic with prescription from the outset of strong opioids when the situation justified it, with analgesia limited in time with regard to step 2 (Minello et al., 2019; Ventafridda, Tamburini, Caraceni, De Conno, & Naldi, 1987; World Health Organization, 2018).

Initiation of cancer-related pain relief

Initiation of cancer-related pain relief in adults, including the elderly, can be achieved by prescribing mild analgesics or opioids—strong or weak—alone or in combination, depending on the initial clinical assessment and the intensity of the pain. When the pain is moderate to severe, a combination is recommended to obtain relief as quickly as possible, and to save on opioids. Coformulations of combined opioid and nonopioid analgesics are nevertheless not recommended when the pain has not been stabilized as they do not allow fine titration of the drugs, but beyond that, there is also a risk of toxicity in the long-term linked to mild analgesics, in particular liver failure for paracetamol and gastrotoxicity and nephrotoxicity for NSAIDs (World Health Organization, 2018).

Maintenance of cancer-related pain relief

Maintenance of cancer-related pain relief must respond to several qualitative imperatives and go through weak or strong opioids. The choice of compound depends not only on the clinical assessment and intensity of the pain, with the aim of providing the patient with rapid relief, but also on the comorbidity profile and any possible drug intolerances. When the pain is mild to moderate, weak opioids can be used if necessary, in association with mild analgesics, but strong opioids at small doses are also effective and well tolerated (Fallon et al., 2018).

Which strong opioid should be chosen, and according to what modalities?

Strong opioids are now recommended as the first line from the outset for moderate to severe pain, and it has been shown that their efficacy/tolerance profile is perfectly pertinent, in particular for oral morphine (Wiffen, Wee, & Moore, 2016), but also for transdermal fentanyl (Wiffen, Wee, Derry, Bell, & Moore, 2017). On the other hand, transdermal formulations are not recommended in situations in which fine titration is needed, and they are reserved for pain that has already been stabilized (Fallon et al., 2018). The oral formulation remains the reference, but it is not rare to use other parenteral routes when the oral route is no longer possible (for example, dysphagia, severe mucositis, vomiting, neurological disorders, etc.) or when fine, rapid titration is needed to relieve the patient as quickly as possible (Fallon et al., 2018; World Health Organization, 2018). The subcutaneous route is recommended as the first line as an alternative to the oral route over the venous route, which can be reserved for cases in which the subcutaneous route is contraindicated (Fallon et al., 2018; World Health Organization, 2018).

Whether it is the oral or the parenteral route, the principle of continuous analgesia by opioids is the rule: a maintenance dose to control the background pain (slow release opioids by oral or transdermal formulations; continuous parenteral perfusion, including PCA) and rapid-action back-up doses in case of acute exacerbation (immediate release opioids or rapid onset fentanyl; parenteral bolus injection), with subsequent adjustment of the maintenance dose depending on the rescue doses taken by the patient. Regardless of the administration route, it is necessary to perform an individual titration of the opioid selected not only to find the minimum effective dose for each patient, but also to relieve them as quickly as possible without any major adverse events (Fallon et al., 2018; George et al., 2019).

The choice of a strong opioid is guided by the compounds and formulations available in each country, as well as by the following criteria (George et al., 2019):

- Age, comorbidities (particularly renal and hepatic), and past history (including addiction and misuse)
- Concomitant treatments and risk of potential drug interactions
- Characteristics of the multimorphic pain, associated with cancer, identified during the initial interdisciplinary assessment
- Pharmacological properties of envisaged drug(s)
- Patient's representations and preferences.

We shall retain that fentanyl and buprenorphine are recommended in cases of severe kidney failure (Fallon et al., 2018). Opioids should be used at lower doses for elderly (>75 years) or fragile patients (Eastern Cooperative Oncology Group Performance Status (ECOG-PS) >2) (Pergolizzi et al., 2008). In case of breakthrough cancer pain, and when the background pain has been stabilized, it is possible to use strong standard immediate release oral opioids, particularly when the breakthrough cancer pain is predictable or of slow onset (Fallon et al., 2018; George et al., 2019; World Health Organization, 2018). When the breakthrough cancer pain is unpredictable or of rapid onset, immediate release opioids—including rapid action fentanyl transmucosal formulations—are relevant because their pharmacokinetic properties adjust to the kinetics of the breakthrough cancer pain (Fallon et al., 2018; George et al., 2019).

Adverse events associated with opioids

Finally, adverse events are common to all opioids, whether weak or strong. These effects can modify quality of life and be responsible for poor treatment compliance. Specific information on the side effects, as well as their prevention and management, must thus be provided. It is essential that opioids not be stigmatized as being responsible for all the symptoms presented by the patient, and that an analytic approach be retained to eliminate other differential diagnoses, such as neuropsychiatric or metabolic disorders, kidney failure, or tumor extension (Delorme et al., 2004; Fallon et al., 2018; National Cancer Institute, 2015). Digestive adverse events are among the most common. Nausea and vomiting occur at the start of the treatment and generally stop after the first week. They can also be treated with antiemetics (Fallon et al., 2018; George et al., 2019). Opioid-induced bowel dysfunction is a catch-all term associated with opioid receptor binding in the gastrointestinal tract. Within this term, we can cite opioid-induced constipation (OIC), which is a frequent and non–dose-dependent adverse event of all opioids. Standard laxative or targeted treatments can and must be proposed as both prophylaxis and treatment. The other adverse events can include drowsiness, and more rarely pruritus, psychological dependency, or opioid-induced hyperalgesia (George et al., 2019).

Opioid switching

In cases of insufficient efficacy in a well-conducted opioid strategy, or persistent adverse events with a negative impact on the patient's quality of life, a change of opioid can be proposed (Cherny et al., 2001; Fallon et al., 2018; George et al., 2019; Mercadante, 1999). This means switching one opioid for another, on the basis of theoretical equianalgesia between compounds, while focusing on safety to avoid the risk of overdose and taking into account the pharmacokinetic particularities of each treatment (Cherny et al., 2001; Fallon et al., 2018; George et al., 2019; Mercadante, 1999). There is no rule that allows to prefer one strong opioid to another, but methadone has an advantage in cases of intractable pain thanks to its mechanism of action which is original and different from that of other opioids. Finally, it may be necessary to envisage a change of route of administration for the opioid in certain situations, while always respecting the rules of equianalgesia (Fallon et al., 2018; George et al., 2019).

Cessation of opioid use

When the cause of the pain has been treated by anticancer treatments, it is important to be able to decrease, or even stop, the use of opioids and not continue to administer them in the long term without justification, to avoid any additional iatrogeny. This withdrawal can be difficult to achieve and may require the intervention of a supportive care team, particularly when dependency phenomena have been observed (World Health Organization, 2018).

Summary points: Strong opioids in cancer pain management

- The historical WHO analgesic ladder is no longer recommended.
- Oral morphine is the first choice of compound for moderate to severe pain.
- In cases where the oral route is impossible, the subcutaneous route is recommended.
- The venous route is indicated when the subcutaneous route is contraindicated, and is useful for doing a rapid titration to relieve the patient quickly.
- The choice of strong opioid is based on the clinical analysis of the situation, the intensity of the pain, and the profile of the patient.
- Buprenorphine and fentanyl are recommended in cases of severe kidney failure.
- Transdermal forms are reserved for cases of stabilized pain.
- Methadone is useful in cases of complex and intractable pain but, as it is difficult to handle, its use is reserved for specialized teams.
- The rules of titration must be applied to attain the minimum effective dose without wasting time.
- The rules of conversion and equianalgesia must be applied when switching between strong opioids.

Drug-based treatments for cancer-related neuropathic pain

From the outset, it is important to distinguish between cancer-related neuropathic pain which occurs "as a direct consequence of a cancer-induced injury to the somatosensory system" and other forms of neuropathic pain, notably those that are residual following cancer treatments (neurotoxic chemotherapy, surgery, or radiotherapy) (Mulvey et al., 2014). Both types can nevertheless occur throughout the care pathway, even coexisting simultaneously, perfectly illustrating the multimorphic nature of pain (Lemaire et al., 2019). For cancer-related neuropathic pain, a judicious combination of opioids and adjuvant treatments can be recommended when opioids alone are not enough (Fallon et al., 2018; World Health Organization, 2018). Here, we can notably cite tricyclic antidepressants, serotonin and noradrenaline reuptake inhibitors, or certain antiepileptic drugs that can be used at minimum effective doses. Steroids are also some of the most useful adjuvant therapies in cases of nerve compression by the tumor, but they are not antineuropathic drugs in themselves (World Health Organization, 2018).

As for chronic neuropathic pain that remains following cancer treatments, when this pain persists after the disease has been cured, it must be dealt with according to the standard strategies recommended for all forms of chronic neuropathic pain. The use of ketamine or medical cannabis is not currently recommended in daily practice for this indication, even if their use by specialist teams can be useful, as can the management of opioid-induced hyperalgesia with ketamine in palliative situations (Fallon et al., 2018).

Management of bone pain

Beyond the analgesic strategy using opioids, pain linked to bone metastases for example can benefit from targeted external radiotherapy, radioisotope therapy, bisphosphonates, or targeted therapies like monoclonal antibodies (Fallon et al., 2018; World Health Organization, 2018). It is important to be able to establish early on, and in a multidisciplinary approach, the indications for these particular treatments so as to not introduce them too late in the patient's clinical history, notably in cases of multiple bone metastases.

Interventional and surgical analgesic techniques

When pain persists despite well-conducted multimodal analgesia, or in certain specific indications, interventional and surgical pain relief techniques are indicated to improve the patient's quality of life. This can notably include intrathecal drug delivery, peripheral nerve block, neurolytic blockade, spinal neurolytic blocks, spinal cord stimulation for neuropathic pain, or even cordotomy (Allano et al., 2019; Fallon et al., 2018). The indications for these techniques will also go through a multidisciplinary decision-making process following careful assessment of the clinical situation by specialized teams.

Non-drug-based approaches and noninvasive techniques (Maindet et al., 2019)

Various complementary non-drug-based approaches are widely practiced around the world in the context of cancer pain, acting on different aspects of the pain. These approaches can include (nonexhaustive list) physical exercise, physical therapy, yoga, acupuncture, supportive-expressive group therapy, hypnosis, virtual reality, music therapy, mindfulness meditation, cognitive-behavioral therapy, electrical nerve stimulation, and photobiomodulation (or low-level laser therapy). Proof of the efficacy of these techniques on pain is nevertheless frequently insufficient and very heterogeneous, making it difficult to recommend them using the same methodologies as drugs or invasive approaches.

Palliative and end-of-life situations

In cases of evolved cancer that has resisted curative therapies, or in complex end-of-life situations, if the drug-based, non-drug-based, and interventional analgesic strategies have been well conducted but have failed to provide the patient with relief, we talk of refractory pain. This type of pain can justify deep and continuous sedation until death, following a rigorous multidisciplinary decision-making process that calls on the skills of specialist teams in an appropriate environment, and without forgetting the ethical aspects of the situation (Fallon et al., 2018).

Conclusion

Managing cancer pain is one of the pillars of supportive care in cancer, the ultimate aim of which is to improve the patient's quality of life as a complement to the cancer treatments. This should be done in a timely and targeted approach from the time of diagnosis of the disease. This management is demanding and requires multidisciplinary expertise to offer to the patient a targeted, multimodal, and personalized management. It involves understanding multimorphic cancer pain as a nosological entity in its own right, leading to treatment strategies being proposed that are the best suited to each patient and each situation.

Acknowledgments

The author would like to thank Dr. Viorica Braniste, M.D. & Ph.D., for her continuous support.

References

Allano, G., George, B., Minello, C., Burnod, A., Maindet, C., & Lemaire, A. (2019). Strategies for interventional therapies in cancer-related pain—A crossroad in cancer pain management. *Supportive Care in Cancer, 27*(8), 3133–3145. https://doi.org/10.1007/s00520-019-04827-9.

Attal, N., Bouhassira, D., & Baron, R. (2018). Diagnosis and assessment of neuropathic pain through questionnaires. *The Lancet Neurology, 17*(5), 456–466. https://doi.org/10.1016/S1474-4422(18)30071-1.

Bennett, M. I., Eisenberg, E., Ahmedzai, S. H., Bhaskar, A., O'Brien, T., Mercadante, S., ... Morlion, B. (2018). Standards for the management of cancer-related pain across Europe. A position paper from the EFIC Task Force on Cancer Pain. *European Journal of Pain, 23*(4), 660–668. https://doi.org/10.1002/ejp.1346.

Burnod, A., Maindet, C., George, B., Minello, C., Allano, G., & Lemaire, A. (2019). A clinical approach to the management of cancer-related pain in emergency situations. *Supportive Care in Cancer, 27*(8), 3147–3157. https://doi.org/10.1007/s00520-019-04830-0.

Cherny, B. N., Ripamonti, C., Pereira, J., Davis, C., Fallon, M., Mcquay, H., & Mercadante, S. (2001). Morphine: An evidence-based report. *Journal of Clinical Oncology, 19*(9), 2542–2554.

Davies, A. N., Dickman, A., Reid, C., Stevens, A.-M., Zeppetella, G., & Science Committee of the Association for Palliative Medicine of Great Britain and Ireland. (2009). The management of cancer-related breakthrough pain: Recommendations of a task group of the science Committee of the Association for palliative medicine of Great Britain and Ireland. *European Journal of Pain, 13*(4), 331–338. https://doi.org/10.1016/j.ejpain.2008.06.014.

Delorme, T., Wood, C., Bartaillard, A., Pichard, E., Dauchy, S., Orbach, D., ... Collin, E. (2004). 2003 clinical practice guideline: Standards, Options and Recommendations for pain assessment in adult and children with cancer (summary report). *Bulletin du Cancer, 91*(5), 419–430.

Fallon, M., Giusti, R., Aielli, F., Hoskin, P., Rolke, R., Sharma, M., & Ripamonti, C. I. (2018). Management of cancer pain in adult patients: ESMO clinical practice guidelines†. *Annals of Oncology, 29*(Suppl 4), iv166–iv191. https://doi.org/10.1093/annonc/mdy152.

Ferrell, B., Sun, V., Hurria, A., Cristea, M., Raz, D. J., Kim, J. Y., … Koczywas, M. (2015). Interdisciplinary palliative care for patients with lung cancer. *Journal of Pain and Symptom Management, 50*(6), 758–767. https://doi.org/10.1016/j.jpainsymman.2015.07.005.

Frankish, H. (2003). 15 million new cancer cases per year by 2020, says WHO. *Lancet, 361*, 1278.

George, B., Minello, C., Allano, G., Maindet, C., Burnod, A., & Lemaire, A. (2019). Opioids in cancer-related pain: Current situation and outlook. *Supportive Care in Cancer, 27*(8), 3105–3118. https://doi.org/10.1007/s00520-019-04828-8.

Grond, S., Zech, D., Schug, S. A., Lynch, J., & Lehmann, K. A. (1991). Validation of World Health Organization guidelines for cancer pain relief during the last days and hours of life. *Journal of Pain and Symptom Management, 6*(7), 411–422. https://doi.org/10.1016/0885-3924(91)90039-7.

Higginson, I. J., & Booth, S. (2011). The randomized fast-track trial in palliative care: Role, utility and ethics in the evaluation of interventions in palliative care? *Palliative Medicine, 25*(8), 741–747. https://doi.org/10.1177/0269216311421835.

Hui, D. (2014). Definition of supportive care. *Current Opinion in Oncology, 26*(4), 372–379. https://doi.org/10.1097/CCO.0000000000000086.

Hui, D., & Bruera, E. (2015). Models of integration of oncology and palliative care. *Annals of Palliative Medicine, 4*(3), 89–98. https://doi.org/10.3978/j.issn.2224-5820.2015.04.01.

Hui, D., De La Cruz, M., Mori, M., Parsons, H. A., Kwon, J. H., Torres-Vigil, I., … Bruera, E. (2013). Concepts and definitions for "supportive care," "best supportive care," "palliative care," and "hospice care" in the published literature, dictionaries, and textbooks. *Supportive Care in Cancer, 21*(3), 659–685. https://doi.org/10.1007/s00520-012-1564-y.

Hui, D., Hannon, B., Zimmermann, C., & Bruera, E. (2018). Improving patient and caregiver outcomes in oncology: Teambased, timely, and targeted palliative care. *CA: A Cancer Journal for Clinicians, 68*(5), 356–376. https://doi.org/10.1038/s41395-018-0061-4.

Institut national du Cancer (INCA). (2012). *Synthèse de l'enquête nationale 2010 sur la prise en charge de la douleur chez des patients adultes atteints de cancer.* www.e-cancer.fr/content/download/63502/571325/file/ENQDOUL12.pdf.

Krakowski, I., Theobald, S., Balp, L., Bonnefoi, M. P., Chvetzoff, G., Collard, O., … FNCLCC. (2003). Summary version of the standards, options and recommendations for the use of analgesia for the treatment of nociceptive pain in adults with cancer (update 2002). *British Journal of Cancer, 89* (Suppl 1), S67–S72. https://doi.org/10.1038/sj.bjc.6601086.

Lemaire, A. (2019a). Special articles: Managing cancer pain—Key issues and lessons learned to optimize patient care. *Supportive Care in Cancer, 27*(8), 3089. https://doi.org/10.1007/s00520-019-04853-7.

Lemaire, A. (2019b). Modeling cancer pain: "The times they are a-changin'". *Supportive Care in Cancer, 27*(8), 3091–3093. https://doi.org/10.1007/s00520-019-04832-y.

Lemaire, A., George, B., Maindet, C., Burnod, A., Allano, G., & Minello, C. (2019). Opening up disruptive ways of management in cancer pain: The concept of multimorphic pain. *Supportive Care in Cancer, 27*(8), 3159–3170. https://doi.org/10.1007/s00520-019-04831-z.

Maindet, C., Burnod, A., Minello, C., George, B., Allano, G., & Lemaire, A. (2019). Strategies of complementary and integrative therapies in cancer-related pain—Attaining exhaustive cancer pain management. *Supportive Care in Cancer, 27*(8), 3119–3132. https://doi.org/10.1007/s00520-019-04829-7.

MASCC. (2020). *Consensus on the core ideology of MASCC. Definition of supportive care.* https://www.mascc.org/index.php?option=com_content&view%C2%BCarticle&id=493:mascc-strategic-plan&catid=30:navigation.

Mercadante, S. (1999). Opioid rotation for cancer pain: Rationale and clinical aspects. *Cancer, 86*(9), 1856–1866. https://doi.org/10.1002/(SICI)1097-0142(19991101)86:9<1856::AID-CNCR30>3.0.CO;2-G.

Minello, C., George, B., Allano, G., Maindet, C., Burnod, A., & Lemaire, A. (2019). Assessing cancer pain—The first step toward improving patients' quality of life. *Supportive Care in Cancer, 27*(8), 3095–3104. https://doi.org/10.1007/s00520-019-04825-x.

Mulvey, M. R., Rolke, R., Klepstad, P., Caraceni, A., Fallon, M., Colvin, L., … Bennett, M. I. (2014). Confirming neuropathic pain in cancer patients: Applying the NeuPSIG grading system in clinical practice and clinical research. *Pain, 155*(5), 859–863. https://doi.org/10.1016/j.pain.2013.11.010.

National Cancer Institute. (2015). *Cancer Pain (PDQ®): Health Professional Version.* Retrieved from https://www.cancer.gov/about-cancer/treatment/side-effects/pain/pain-hp-pdq#_3.

NICE. (2004). *Improving supportive and palliative care for adults with cancer: Guidance and guidelines.* NICE. Retrieved from https://www.nice.org.uk/guidance/csg4.

Pergolizzi, J., Böger, R. H., Budd, K., Dahan, A., Erdine, S., Hans, G., … Sacerdote, P. (2008). Opioids and the management of chronic severe pain in the elderly: Consensus statement of an international expert panel with focus on the six clinically most often used World Health Organization step III opioids (buprenorphine, fentanyl, hydromorphone, methadone, morphine, oxycodone). *Pain Practice, 8*(4), 287–313. https://doi.org/10.1111/j.1533-2500.2008.00204.x.

Portenoy, R., & Koh, M. (2010). Cancer pain syndromes. In C. C. U. Press, E. Bruera, & R. K. Portenoy (Eds.), *Vol. 4. Cancer pain. Assessment and management* (pp. 53–88).

Street, R. L., Tancredi, D. J., Slee, C., Kalauokalani, D. K., Dean, D. E., Franks, P., & Kravitz, R. L. (2014). A pathway linking patient participation in cancer consultations to pain control. *Psycho-Oncology, 23*(10), 1111–1117. https://doi.org/10.1002/pon.3518.

Tuggey, E. M., & Lewin, W. H. (2014). A multidisciplinary approach in providing transitional care for patients with advanced cancer. *Annals of Palliative Medicine, 3*(3), 139–143. https://doi.org/10.3978/j.issn.2224-5820.2014.07.02.

van den Beuken-van Everdingen, M. H. J., De Rijke, J. M., Kessels, A. G., Schouten, H. C., Van Kleef, M., & Patijn, & J. (2007). Prevalence of pain in patients with cancer: A systematic review of the past 40 years. *Annals of Oncology, 18*, 1437–1449. https://doi.org/10.1093/annonc/mdm056.

van den Beuken-van Everdingen, M. H. J., Hochstenbach, L., Joosten, E. A. J., Tjan-Heijnen, V. C. G., & Janssen, D. J. A. (2016). Update on prevalence of pain in patients with cancer: Systematic review and meta-analysis. *Journal of Pain and Symptom Management, 51*, 1070–1090. https://doi.org/10.1016/j.jpainsymman.2015.12.340.

Ventafridda, V., Tamburini, M., Caraceni, A., De Conno, F., & Naldi, F. (1987). A validation study of the WHO method for cancer pain relief. *Cancer, 59*(4), 850–856.

Wiffen, P. J., Wee, B., Derry, S., Bell, R. F., & Moore, R. A. (2017). Opioids for cancer pain—An overview of cochrane reviews. *Cochrane Database of Systematic Reviews, 2017*(7). https://doi.org/10.1002/14651858.CD012592.pub2.

Wiffen, P. J., Wee, B., & Moore, R. A. (2016). Oral morphine for cancer pain. *Cochrane Database of Systematic Reviews, 2016*(4). https://doi.org/10.1002/14651858.CD003868.pub4.

World Health Organization. (1986). *World Health Organization cancer pain relief*. Retrieved from http://apps.who.int/iris/bitstream/handle/10665/43944/9241561009_eng.pdf.

World Health Organization. (1996). *Cancer pain relief* (2nd ed.). Retrieved from http://apps.who.int/iris/bitstream/handle/10665/37896/9241544821.pdf?sequence=1&isAllowed=y.

World Health Organization. (2018). *WHO Guidelines for the pharmacological and radiotherapeutic management of cancer pain in adults and adolescents*.

Zech, D. F. J., Grond, S., Lynch, J., Hertel, D., & Lehmann, K. A. (1995). Validation of World Health Organization guidelines for cancer pain relief: A 10-year prospective study. *Pain, 63*(1), 65–76. https://doi.org/10.1016/0304-3959(95)00017-M.

Chapter 2

Recent advances in the linkage of attachment and pain: A new review

Alessandro Failo

Department of Psychology and Cognitive Sciences, University of Trento, Rovereto, Italy

List of abbreviations

IWM Internal Working Model
SSD Somatic Symptom Disorders

Introduction

The need to form attachments with others is one of the first fundamental skills that has a profound effect on the social and emotional development of each individual.

Attachment is a central construct that offers a conceptual and methodological framework for the study of interpersonal relationships across lifespans. This is relevant in particular in threat situations such as pain, which can thus activate specific attachment behaviors (for recent reviews see Failo, Giannotti, & Venuti, 2019; Romeo, Tesio, Castelnuovo, & Castelli, 2017).

It is crucial early in life and can be view as a natural explanation of the quality of support-related interactions with the main caregiver (usually mother, of course) that have enduring influences, giving that, these early regulatory processes can have long-term shaping effects in relationships (Hofer, 1994).

Building on Bowlby's postulations, attachment theory offers a real adaptive and maladaptive functioning clarification point of view. Over the years, it has had a few variants.

Since the early theorization, (Bowlby, 1982, 1988), Bowlby claimed that we are born with an innate psychobiological system, called the attachment behavioral system, that pushes us to seek proximity to significant others (caregivers or persons) in certain situations, particularly when the primary attachment figure is unavailable or a threat is present (e.g., pain). Several years later, Shaver and Mikulincer (2009), deeply analyzing the original Bowlby model, suggested that it is necessary to factor in other key components of this theory, such as interindividual differences on how this attachment system works and how these differences are internalized into relatively stable mental representations, as a schemas, already called "internal working models" (IWM) of Self (as the unique and individual's perception of the self, defines the extent to which individuals consider themselves worthy of support and proximity) and Others (the perception of people and the world, influencing the individual's confidence of receiving support from them and the environment) both, not only from a physical perspective but also symbolic's. Further are relevant the individual differences in these IWMs and the importance how to measure their, given that define our "attachment style", as it guide different responses in the relationship compared to that of another person and vice versa (Collins & Feeney, 2004; Collins & Ford, 2010).

Meanings and measurement of attachment

For easier "navigation" in this topic, we begin by briefly explaining how important are measures wof "attachment style." Due to the fact that attachment theory is a lifespan developmental approach, numerous methods for the measurement of individual differences in quality of attachment have been developed over the last four decades. In this scenario, observational procedures, are considered the most suitable to assess patterns of attachment in infancy, toddlerhood, and preschool (see the very early study of Ainsworth & Wittig, 1969 and a recent overview by Farnfield & Holmes, 2014).

Features and Assessments of Pain, Anesthesia, and Analgesia. https://doi.org/10.1016/B978-0-12-818988-7.00027-3
Copyright © 2022 Elsevier Inc. All rights reserved.

Furthermore, to attempt to explain anomalous behavior in children, the pioneering work of Ainsworth on three patterns of attachment (one called "secure" and two called "insecure") was extended and different methods of classification have been proposed. The most known is the addition of the "disorganized pattern" by Main and Solomon (1990) to describe a lack of coherent strategy in a child's attachment behaviors.

From middle childhood to adolescence, narratives and interviews are used to assess attachment representations that are no longer connected exclusively to the sensorimotor system (Dykas, Woodhouse, Cassidy, & Waters, 2006; Psouni & Apetroaia, 2014).

To understand adult representations of childhood attachment relations, Main and colleagues (George, Kaplan, & Main, 1985; Main & Goldwyn, 1988) developed a specific interview, but the first attempt to describe a classification system of adult attachment styles was from Bartholomew and Horowitz (1991). They found one "secure" pattern (characterized by a sense of worthiness combined with an expectation that other people are generally accepting and responsive) and three "insecure" patterns (preoccupied, fearful-avoidant, and dismissive-avoidant).

Starting from another developmental perspective, Crittenden (2000a, 2000b, 2006) extended the traditional classification, developing a new approach that split from the mainstream of attachment theory. In her model, she describes a wide array of substrategies on the basis of a dimensional information processing approach, pointing out the role of adaption from danger. In this perspective, self-protective strategies are in dynamic interaction with brain maturation and ongoing experience, accounting for a potential discontinuity of attachment classification over time (Shah & Strathearn, 2014).

Despite the well-established differences between the two classification systems, a continuity of the secure pattern was found with reference to the intergenerational transmission of attachment (Shah, Fonagy, & Strathearn, 2010), suggesting the predictive and protective value of security across generations.

It is difficult to summarize a streamlined vision of the complex attachment theory. As said, it can be useful to use the secondary attachment strategies term (Cassidy & Kobak, 1988). This provides an understanding within a dimensional concept, starting from secure attachment (a balance between them) (Fig. 1), where attachment insecurity stands out between the "weight" of attachment anxiety and the "weight" of attachment avoidance. This two-dimensional space with different weights reflects both the person's sense of attachment security and the ways in which he or she deals with threats and distress. In avoidance dimension is relevant deactivation strategies (primarily underestimation of threat), as avoiding help and controlling or lacking emotion to avoid further negative experiences, given they tend to have negative representation of others (classically called "dismissing") (Fig. 2). Who tend more toward anxious dimension (classically called "preoccupied") use hyperactivating strategies (catastrophizing and hypervigilance) in an effort to cope with difficulties, like are exaggeration of the threats and intensification of the demands for care, because they have negative representation of the self and fear of abandonment (Bartholomew & Horowitz, 1991; Brennan, Clark, & Shaver, 1998; Griffin & Bartholomew, 1994; Mikulincer & Shaver, 2007; Shaver & Mikulincer, 2002) (Fig. 3).

Those who have "weight," which can vary but has a tendency to be high on both anxious and avoidance attachment dimensions, appear to have a fearful attachment style (Fig. 4), which means they may engage hyperactivating strategies for threat cues (like those with anxious/preoccupied attachment) and simultaneously deactivating strategies of emotion regulation (like those with avoidance/dismissing attachment) (Mikulincer & Shaver, 2012). Therefore, they have a negative model of self and others.

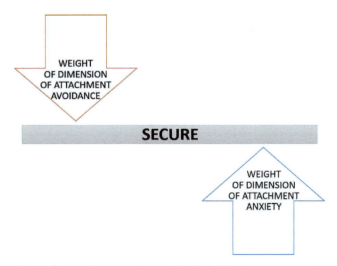

FIG. 1 Secure attachment pattern: the "weight" of attachment anxiety and the "weight" of attachment avoidance are in balance.

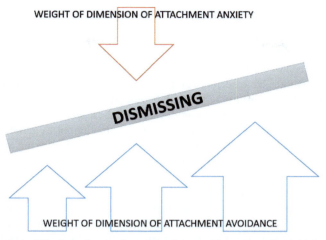

FIG. 2 Insecure attachment patterns: in the "dismissing" style, the weight of the avoidance dimension is higher.

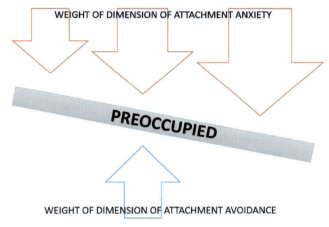

FIG. 3 Insecure attachment patterns: in the "proccupied"style, the weight of the anxiety dimension is higher.

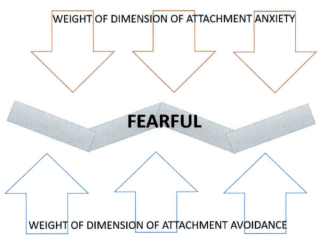

FIG. 4 Insecure attachment patterns: in the "fearful" style, the weights of both dimensions can vary, with a tendency to stay high.

Interlinking attachment and pain

It is well known that there are several psychological predisposing factors and mechanisms that play a role in developing and maintaining chronic pain. Mostly over the last 15 years, several studies have suggested that attachment is a specific and sensitive construct that we should keep in mind to better explain individual difference between recurrent pain and chronic experiences. Recently, several new studies have been developed to explore specific conditions such unexplained chest pain (Bolat et al., 2019), vulvodynia (Charbonneau-Lefebvre, Vaillancourt-Morel, Brassard, Steben, & Bergeron, 2019), scleroderma (Hicks & Kearney, 2019), and others. In the next section, some of these studies will be scrutinized to do this update of the literature. But first, because now we know something more about attachment in general, we do another step.

Over the years, several models have been developed for conceptualizing and studying attachment patterns in different pain conditions; this is useful in better understanding patient experiences. New models have appeared about every 10 years, since the first one in early 1980.

The first attempt to relate attachment theory to pain was proposed by Kolb (1982), who pointed out how patients with persistent pain often had insecure attachment and dependent personalities, possibly due to the lack of a secure base in childhood. He affirmed that starting from this consideration, therapists acting as a substitute attachment figure can be a facilitator for patients with chronic pain in reducing separation anxiety while fostering hope to face illness and difficulties related to pain.

Another interesting model, utilizing an interpersonal perspective, was theorized by Mikail, Henderson, and Tasca (1994). Taking a further step and relying on both clinical experiences and literature evidence, they suggested that an insecure attachment presents a vulnerability to the development of chronic pain, with poor coping strategies and low self-efficacy to face stress due to maladaptive mental representations. They pointed out that the four different attachment styles influence how individuals face pain and also a patient's behaviors toward his or her physician. Also within this conceptualization, we can probably say that a person with a secure attachment copes better with pain and is less likely to develop chronic pain.

Going ahead with our brief models applied overview, it is interesting Meredith's heuristic model (Meredith, Ownsworth, & Strong, 2008) that, by reviewing several studied in adult population with chronic pain, has put forward that attachment may impact directly on pain-related outcomes. It seems that individuals high in anxious attachment have a particular tendency to catastrophize their pain and emphasize their negative feelings to elicit more support from others. Therefore, there are several variables that mediate these relations, such as perception of self (pain self-efficacy) and social support (others can help me) as well as appraisal of pain (pain threat).

Another significant contribution is from Kozlowska (2009) who, applying Crittenden's approach (2000a, 2000b, 2006), proposed a model based on several observational studies with children, claiming that different attachment relationships shape pain-signaling behavior because they also depend upon changes in the specific contexts in which they occur. Therefore, these associations may not be generalized to all contexts. Furthermore, Kozlowska and Khan (2011) subsequently tested a multimodal intervention with this model for idiopathic chronic pain that has shown positive outcomes in the family and in school settings.

More recently, Romeo et al. (2017), through a specific review of the literature, suggested that several recent studies started to move from the individual attachment pattern toward integrating a relationship's context. Therefore, they proposed to move toward a more elaborate model that considers that insecure attachment style is associated with chronic pain, but on the other hand, it is also mediated by different psychological aspects as well as interplay between partner attachment styles.

Fig. 5 shows a possible overall model that considers the recent theorizations.

Attachment and pain: Change in relationship into different conditions considered

To gain a better understanding of the literature, we suggest that readers start to explore pain conditions between those with secure attachment versus insecure attachment (i.e., subjects who are preoccupied, fearful, and dismissing) (see Table 1). Subsequently, when the reader becomes more confident, it is possible to examine the specific associations, keeping in mind the importance of considering the two dimensions toward anxious or avoidance.

Evidence in child and adolescent age

The association between attachment and pain in the developmental age shows the relevance of examining this association at different ages.

In fact, during infancy and toddlerhood, the different dyadic patterns of interactions between the adult and infant, given their preeminent and central role in regulation, influence both a child's expression to pain and management of pain

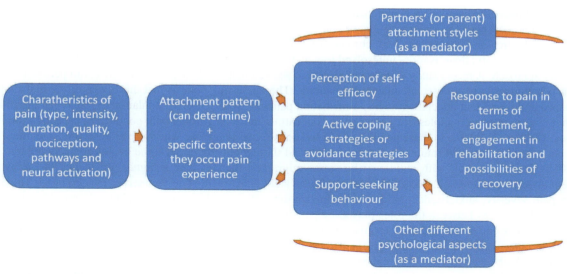

FIG. 5 Simplification of an overall model permits a better understanding of attachment in pain experience. Adaptation and integration of concepts presented in models of Kozlowska, 2009; Meredith et al., 2008; Romeo et al., 2017.

TABLE 1 Typical individual features of people (in a lifespan perspective) with pain based on their attachment style. This is a simplification, in effect it should be borne in mind the importance of underling hyperativation o deattivation strategies, rather than attachment pattern tout-court.

Secure	Insecure
These infants have more interaction with their parents to regulate distress by close physical contact with them	During venipuncure, these infants may have a greater distress level
Parents of these children are usually more engaged in pain-reducing behaviors	In children, this factor is relevant in contributing to the maintenance of the chronic pain condition, also to intensifying the pain experience or hindering effective rehabilitation
In adolescents, there are lower levels of pain severity, depression, pain catastrophizing, and anxiety	In adolescents who have poor physical and mental health, this attachment has effects on cognitive appraisals and coping strategies. Furthermore, this attachment is associated with pain severity among adolescents
These adults cope better with pain and are less likely to develop chronic pain, may have more pain self-efficacy, and actively seek support and stronger alliances with healthcare professionals with implications for pain outcome	These adults have greater pain intensity and disability-related, that contribute to the development of emotion regulation disturbances, especially in patients with somatic symptom disorder. They tend to have greater healthcare utilization in a dysfunctional way. Seems that partner presence to facing pain not have beneficial effect.

Synthesis from: Donnelly & Jaaniste, 2016; Hicks & Kearney, 2019; Horton, Riddell, Flora, Moran, & Pederson, 2015; Horton, Riddell, Moran, & Lisi, 2016; Krahé et al., 2015; Meredith et al., 2008; Pritchett, Minnis, Puckering, Rajendran, & Wilson, 2013; Sinclair, Meredith, & Strong, 2020; Tremblay & Sullivan, 2010; Wolff et al., 2011.

(Riddell & Racine, 2009). This is especially evident during immunizations, where it has been found that avoidance infants were significantly less stressed before a procedure than secure infants. This is probably because they may have minimized distress signals preneedle in keeping with an avoidance strategy while secure infants were engaged in more interaction with their parents (usually the mother) to help them regulate distress by close physical contact (Horton et al., 2015, 2016; O'Neill et al., 2021). Further evidence is that fear predicts an increase in distress only in infants with a disorganized attachment classification (Wolff et al., 2011). Furthermore, parents of securely attached children are usually more engaged in pain-reducing behaviors versus parents of insecurely attached children (Pritchett et al., 2013).

When a child ages, the focus moves to other conditions. For example, pain behaviors following an everyday pain incident, security, and avoidance are not completely related to pain behavior displayed in a child (Walsh, McGrath, & Symons, 2008).

20 **PART | I** Setting the scene: General aspects of anesthesia, analgesics and pain

Children with conversion disorders (e.g., medically unexplained pain) face these problems with extreme psychological inhibition or psychological coercion-preoccupation, or both strategies (Kozlowska & Khan, 2011). It also seems that patterns of attachment in children and adolescents with chronic functional pain are characterized with at-risk patterns of attachment. Further, in their life story, there is more unresolved loss and trauma (Ratnamohan & Kozlowska, 2017), as Crittenden theorized in her approach. A recent study (Sinclair et al., 2020) showed that children and adolescents with persistent pain have particular behaviors related to insecure attachment patterns. The interesting point is that this is one of the first studies that hypothesized and found that attachment is like a mediating pathway from sensory modulation to functional disability.

If we move more toward adolescent age, studies concentrate more on chronic pain conditions, especially on primary headaches (Failo et al., 2019). In these populations, it has been found that those who have an insecure attachment tend to amplify both the perception of traumatic events experienced throughout life but also the evolution of migraines (Faedda et al., 2018). In more detail, with children who suffer migraines, there is also a higher prevalence of the avoidance attachment style (Esposito et al., 2013), but also the ambivalent attachment style that serves as a common vulnerability factor on pain severity, anxiety, depression, and somatization symptoms in young migraine sufferers (Tarantino et al., 2017). Some differences between the roles played by perceived maternal and paternal attachment security and presence of anxiety disorders, can play a role in this pain conditions (Barone et al., 2016; McWilliams, 2017; Tarantino et al., 2017).

Instead, in other condition such as unexplained chest pain, no one association between different attachments has been found, but rather high levels of emotional and behavioral problems (Bolat et al., 2019). In adolescents with different subtypes of pain (severity and frequency), anxiety, helplessness, and cognitive rumination about pain catastrophizing are factors that mediate the relation between secure, preoccupied, and fearful attachment styles and depression (Tremblay & Sullivan, 2010).

Relevant discoveries and topics in adult age

In adult age, the associations mentioned above are different and can be analyzed at different levels. Some of these are well known and deeply studied.

If we start from the belief that an understanding of basic research is essential for the foundation of applied science, it becomes more clear as to why perception and social support in experimental pain can be useful in real life and concrete situations. The empirical studies on experimentally induced pain of a correlational nature offer mixed indications on the role of attachment, from a link to avoidance dimensions (more pain tolerance, less facial expression of pain) or rather anxious dimensions (tendency to catastrophize and feel more pain), but still both are related to insecure attachment, as stated in the last review available (Meredith, 2013). Also, in the few subsequent studies, these considerations are still valid (Andersen, Ravn, Manniche, & O'Neill, 2018; Hurter, Paloyelis, Williams, & Fotopoulou, 2014). Further, reviews in the literature are beginning to study the effects of partner presence on pain ratings and pain-related neural processing; a person with higher attachment avoidance may not have a beneficial effect from a partner's presence (Krahé et al., 2015).

It is well known that attachment concept is dynamic (as is pain), and therefore it may be modified a little bit in some way over time. However, the literature has always been focused on observational correlational studies, forgetting the importance of assessing repeated observations of the same variable over time by longitudinal study designs.

Andersen, Sterling, Maujean, and Meredith (2019) developed the first longitudinal study to assess the impact of attachment insecurity on disability after a whiplash injury. They found that attachment orientations are stable traits over 1 year. Furthermore, according with Meredith's model, attachment dimensions (avoidance and anxiety) were moderators in the association between pain intensity and physical disability. In particular, the avoidance dimension was a predictor of past medically unexplained chronic pain. Once again, adult attachment insecurity is positively associated with medically unexplained chronic pain (McWilliams, 2017). In the same direction, another recent work contributed to the development of a longitudinal perspective. Charbonneau-Lefebvre et al. (2019) examined, for the first time, the effects of attachment and pain self-efficacy on vulvodynia (an idiopathic chronic pain condition). They found that self-efficacy is a mediator of the attachment and pain association. In particular, both attachment dimensions (avoidance and anxious) predicted pain intensity, but only the anxious dimension predicted pain self-efficacy in women with provoked vestibulodynia. This supported Meredith's model, which suggests that people with anxious attachment styles tend to live with the conviction that they are not able to manage their chronic pain.

Finally, we must not forget pain in different parts of the body due to somatic symptom disorders (SSD), often accompanied by excessive thoughts, feelings, and specific behaviors. A recent review (Okur Güney, Sattel, Witthöft, & Henningsen, 2019) confirmed previous findings that patients suffering from different SSDs, such as bodily pain, fatigue, and functional symptoms, have similar emotional regulation difficulties such as reduced emotional awareness, reflection capacity, and rigid emotional attention. Thus, it is important to recall that insecure attachment style and emotional dysregulation are positively correlated. In particular, anxious attachment patterns and these two variables (pain and emotional regulation) also appear to contribute to maladaptive behavior and affective problems (Liu & Ma, 2019).

Applications to other areas

In this chapter, we have reviewed how attachment is related to different pain conditions. We have described in detail the fact that insecurely attached patients have different needs and characteristics. In the following section, we describe how some attachment-based interventions can be integrated in interdisciplinary multimodal pain treatment programs.

Although research underlining the importance of attachment dimensions in individuals with chronic pain given that reflected their maladaptive habitual approaches to managing activity (Andrews, Meredith, Strong, & Donohue, 2014) few study applied specific attachment-based interventions.

It seems clear that active support and effective multidimensional treatment for patients with chronic pain disorders are complex.

Probably one of the first approaches in this direction was by Peilot et al. (2014). They adjusted a cognitive therapy approach based on attachment and mindfulness in a small group of patients, mainly women, with chronic pain and psychiatric comorbidity. They found that the main changes in patient well-being and in the better management of pain were due to a meaningful process of interaction with the therapist, such as feeling as if they were being listened to, their experiences were legitimate, and being told that there was time to gain trust. Therefore, it seems that a trust relationship with health professionals, oriented on the attachment theory principles, can serve as a base to engage in other important rehabilitation treatments.

Another important step carried out by Pfeifer et al. (2016, 2018, 2019) was the first attachment-oriented approach for people with different pain conditions. It is fundamental pointed out, that the program were not provided alone, but in an outpatient interdisciplinary multimodal pain therapy setting, provided to the patients during intensive four weeks (including physiotherapy, ergotherapy, music/dance therapy, individual and group psychotherapy). It is worth pointing out how attachment-based approaches have been really employed by all healthcare professionals with diverse backgrounds, and not only from psychotherapists. The findings support that this working alliance can provide a secure base for patients, but also help in better understanding and dealing with patient motivations and needs. Thus, the utility of incorporating attachment-informed interventions with multimodal pain therapies can be a further step toward efficacy change. Lastly, health care professionals, through this deeper knowledge of possible interpersonal problems, can be more prepared to plan target interventions on these important factors.

Other agents of interest

In the above parts, we have extensively discussed how attachment dimensions can determine the adoption of emotion regulation strategies, coping strategies, and social interactions. We also talked about how attachment patterns can connect with pain behaviors and, in part, with pain perception. At this point, it seems useful to connect these considerations with neuroscience, given that some studies have been oriented toward the relationship between brain functions (from an anatomical and connection point of view), attachment, and sensory modulation. These considerations elicit another question: is there a neurobiological base that can connect attachment and pain in a broader sense?

We can face this question from two main perspectives: pain that evolves from acute to chronic or pain that carries other meanings (such emotional pain, loss, etc.).

About the transition, as Ratnamohan and Kozlowska recently wrote (2017), there is a reorganization of the neural areas involved in the sensory processing of pain toward brain regions encoding emotional and motivational subjective states (Apkarian, 2008; Vachon-Presseau et al., 2016). This reorganization overlaps with the mechanism involved in the activation of the stress system (Chrousos, 2009) and the attachment system (Mikulincer, Gillath, & Shaver, 2002). Furthermore, both attachment deactivating and hyperactivating strategies contribute to the disregulation of the stress system and to increased pain sensitivity (Donnelly & Jaaniste, 2016). It also was discovered that brain circuitry interfacing with different kinds of acute and chronic pain conditions involves areas commonly thought to be essential in emotional learning and memory, reward, and specific behaviors (Apkarian, 2008).

The other point of view is that studies typically investigate how we encode and retrieve personal information as well as the accumulation of social memories (Laurita, Hazan, & Spreng, 2019).

Eisenberger and her research group are fruitful researchers in this field. She hypothesized that experiences of physical pain and social pain (social rejection, exclusion, or loss) may have a common set of neural substrates (Eisenberger, 2012). Her group found that an attachment figure who serves as a safety signal during an experience of physical pain leads to greater activity in a region implicated in safety signaling and fear extinction, and reducing reports of pain intensity and pain-related neural activity (Eisenberger et al., 2011).

Therefore, it is not surprising that a growing number of studies point out that attachment patterns vary in how a person receives and modulates sensory information (called sensory processing styles) across the lifespan (Branjerdporn, Meredith, Strong, & Green, 2019; Jerome & Liss, 2005; Kerley, Meredith, & Harnett, 2021; Lee & Park, 2020; Meredith, Bailey, Strong, & Rappel, 2016; Sinclair et al., 2020).

These considerations are important because some experiences that are not commonly considered related to "pain" (or not called in this way) can later increase sensitivity to physical pain. A good example to understand this could be patients with somatoform pains or syndromic presentations who experience pain without a clear medical explanation. In fact, in these cases, patients often report early experiences of social/emotional pain such as social exclusion or rejection, emotional abuse, loss, interpersonal conflict, and experiences of ostracism. Therefore, it is likely that there is a substantial overlap between physical and social pain (Eisenberger, 2012; Sturgeon & Zautra, 2016).

As the concept of a neuromatrix was applied to chronic pain, future research could increase our understanding of the flows of information between brain substrates and what disfunctional attachment patterns contribute in different kinds of pain experiences.

Mini-dictionary of terms

Attachment. Specific pshycological construct, fundamental for emotional and social development, refers to personal interpetations about self and interpersonal relationships across the lifespan.

Internal working Model. Mental representations of the inner world about self, other, and reciprocal roles between the person and the world.

Attachment pattern or strategies. Depends on theorization. Classically, there are four patterns, but a new approach considers several more divisions. Overall, patterns or strategies reflect the individual approach to solve difficulties and face problems in life.

Secondary attachment strategies. They could be viewed as a two-dimensional space that reflects both the person's sense of attachment security and ways to deal with threats and distress.

Bowlby's attachment theory. It is the first theorization that explains this innate psychobiological system called "attachment style," which is mainly divided between "secure" and "insecure."

Crittenden's developmental prospective. It is an extension of the traditional classification of attachment by developing a new approach to describe a wide array of substrategies.

Key facts

- Attachment behavior is a well-known fundamental and innate psychobiological system that plays a role when we face different threats.
- Important features include the perception of self-efficacy, the use of different coping strategies, and support-seeking behavior.
- Insecure attachment is a well-known disfunctional "behavior pattern" characterized with heightened threat perceptions and distress.
- Several studies have investigated the associations between attachment during the lifespan and pain.
- There are attachment-based factors that can play a role as a vulnerability feature to the onset of chronic pain.
- Much like there is no "pain generator" or "pain center," there is no "attachment system" localized in the brain.

Summary points

- This chapter focuses on recent advances on the linkage of attachment and pain, starting from an overall explanation.
- It is getting ever clearer that attachment is a specific and sensitive construct to better explain individual pain differences, given also the lasty studies developed to explore specific conditions.
- Over the years, several models have been developed for conceptualizing and studying attachment style in different pain conditions; the latest of these was theorized in 2017.
- In the last few years, the first attachment-oriented treatment approach for people with different pain conditions was integrated with multimodal pain therapies.
- The experiences of physical pain and social pain (social rejection, exclusion, or loss) have a common set of neural substrates; attachment figures can serve as a "safety signal" during both type of experiences.

References

Ainsworth, M. D. S., & Wittig, B. A. (1969). Attachment and exploratory behavior of one-year-olds in a strange situation. B. M. Foss (Ed.), *Determinants of infant behavior* (Vol. 4, pp. 111–136). London: Methuen.

Andersen, T. E., Ravn, S. L., Manniche, C., & O'Neill, S. (2018). The impact of attachment insecurity on pain and pain behaviors in experimental pain. *Journal of Psychosomatic Research*, *111*, 127–132.

Andersen, T. E., Sterling, M., Maujean, A., & Meredith, P. (2019). Attachment insecurity as a vulnerability factor in the development of chronic whiplash associated disorder—A prospective cohort study. *Journal of Psychosomatic Research*, *118*, 56–62.

Andrews, N. E., Meredith, P. J., Strong, J., & Donohue, G. F. (2014). Adult attachment and approaches to activity engagement in chronic pain. *Pain Research & Management*, *19*, 317–327.

Apkarian, V. (2008). Pain perception in relation to emotional learning. *Current Opinion in Neurobiology*, *18*, 464–468.

Barone, L., Lionetti, F., Dellagiulia, A., Galli, F., Molteni, S., & Balottin, U. (2016). Behavioural problems in children with headache and maternal stress: Is children's attachment security a protective factor? *Infant and Child Development*, *25*, 502–515.

Bartholomew, K., & Horowitz, L. M. (1991). Attachment styles among young adults: A test of a four-category model. *Journal of Personality and Social Psychology*, *61*, 226–244.

Bolat, N., Eliacik, K., Yavuz, M., Kanik, A., Mertek, H., Guven, B., et al. (2019). Adolescent mental health, attachment characteristics, and unexplained chest pain: A case-control study. *Psychiatry & Clinical Psychopharmacology*, *29*, 487–491.

Bowlby, J. (1982). *Attachment and loss: Vol. I attachment* (2nd ed.). New York: Basic Books (original work published 1969).

Bowlby, J. (1988). *A secure base: Parent-child attachment and healthy human development*. New York: Basic Books.

Branjerdporn, G., Meredith, P., Strong, J., & Green, M. (2019). Sensory sensitivity and its relationship with adult attachment and parenting styles. *PLoS One*, *14*, e0209555.

Brennan, K. A., Clark, C. L., & Shaver, P. R. (1998). Self-report measurement of adult attachment: An integrative overview. In J. A. Simpson, & W. S. Rholes (Eds.), *Attachment theory and close relationships* (pp. 46–76). New York: Guilford Press.

Cassidy, J., & Kobak, R. R. (1988). Avoidance and its relation to other defensive processes. In J. Belsky, & T. Nezworski (Eds.), *Clinical implications of attachment* (pp. 300–323). Hillsdale, NJ: Erlbaum.

Charbonneau-Lefebvre, V., Vaillancourt-Morel, M.-P., Brassard, A., Steben, M., & Bergeron, S. (2019). Self-efficacy mediates the attachment-pain association in couples with provoked Vestibulodynia: A prospective study. *Journal of Sexual Medicine*, *16*, 1803–1813.

Chrousos, G. P. (2009). Stress and disorders of the stress system. *Nature Reviews Endocrinology*, *5*, 374–381.

Collins, N. L., & Feeney, B. C. (2004). Working models of attachment shape perceptions of social support: Evidence from experimental and observational studies. *Journal of Personality and Social Psychology*, *287*, 363–383.

Collins, N. L., & Ford, M. B. (2010). Responding to the needs of others: The caregiving behavioral system in intimate relationships. *Journal of Personality and Social Psychology*, *27*, 235–244.

Crittenden, P. M. (2000a). A dynamic-maturational exploration of the meaning of security and adaptation. Empirical, cultural and theoretical considerations. In P. M. Crittenden, & A. H. Claussen (Eds.), *The organization of attachment relationships. Maturation, culture and context* (pp. 358–383). Cambridge: Cambridge University Press.

Crittenden, P. M. (2000b). A dynamic-maturational approach to continuity and change in pattern of attachment. In P. M. Crittenden, & A. H. Claussen (Eds.), *The organization of attachment relationships. Maturation, culture and context* (pp. 343–357). Cambridge: Cambridge University Press.

Crittenden, P. M. (2006). A dynamic-maturational model of attachment. *Australia and New Zealand Journal of Family Therapy*, *27*, 105–115.

Donnelly, T. J., & Jaaniste, T. (2016). Attachment and chronic pain in children and adolescents. *Children*, *3*, 1–14.

Dykas, M. J., Woodhouse, S. S., Cassidy, J., & Waters, H. S. (2006). Narrative assessment of attachment representations: Links between secure base scripts and adolescent attachment. *Attachment & Human Development*, *8*, 221–240.

Eisenberger, N. I. (2012). The neural bases of social pain: Evidence for shared representations with physical pain. *Psychosomatic Medicine*, *74*, 126–135.

Eisenberger, N. I., Master, S. L., Inagaki, T. K., Taylor, S. E., Shirinyan, D., Lieberman, M. D., et al. (2011). Attachment figures activate a safety signal-related neural region and reduce pain experience. *Proceedings of the National Academy of Sciences of the United States of America*, *108*, 11721–11726.

Esposito, M., Parisi, L., Gallai, B., Marotta, R., Di Dona, A., Lavano, S. M., et al. (2013). Attachment styles in children affected by migraine without aura. *Neuropsychiatric Disease and Treatment*, *9*, 1513–1519.

Faedda, N., Natalucci, G., Piscitelli, S., Fegatelli, A. D., Verdecchia, P., & Guidetti, V. (2018). Migraine and attachment type in children and adolescents: What is the role of trauma exposure? *Neurological Sciences*, *39*, 109–110.

Failo, A., Giannotti, M., & Venuti, P. (2019). Associations between attachment and pain: From infant to adolescent. *SAGE Open Medicine*, *7*, 1–12.

Farnfield, S., & Holmes, P. (2014). Introduction. In P. Holmes, & S. Farnfield (Eds.), *The Routledge handbook of attachment: Implications and interventions* (pp. 1–31). London: Routledge/Taylor & Francis Group.

George, C., Kaplan, N., & Main, M. (1985). *Adult attachment interview*. Unpublished manuscriptBerkeley, CA: University of California.

Griffin, D. W., & Bartholomew, K. (1994). Models of the self and other: Fundamental dimensions underlying measures of adult attachment. *Journal of Personality and Social Psychology*, *67*, 430–445.

Hicks, R. E., & Kearney, K. (2019). Pain in relation to emotion regulatory resources and self-compassion: A non-randomized correlational study involving recollected early childhood experiences and insecure attachment. *Health Psychology Report*, *1*, 19–31.

Hofer, M. A. (1994). Early relationships as regulators of infant physiology and behavior. *Acta Paediatrica*, *397*(Suppl), 9–18.

Horton, R. E., Riddell, R. P., Flora, D., Moran, G., & Pederson, D. (2015). Distress regulation in infancy: Attachment and temperament in the context of acute pain. *Journal of Developmental and Behavioral Pediatrics, 36*, 35–44.

Horton, R., Riddell, R. P., Moran, G., & Lisi, D. (2016). Do infant behaviors following immunization predict attachment? An exploratory study. *Attachment & Human Development, 18*, 90–99.

Hurter, S., Paloyelis, Y., Williams, A. C., & Fotopoulou, A. (2014). Partners´ empathy increases pain ratings: Effects of perceived empathy and attachment style on pain report and display. *Journal of Pain, 15*, 934–944.

Jerome, E. M., & Liss, M. (2005). Relationships between sensory processing style, adult attachment, and coping. *Personality and Individual Differences, 38*, 1341–1352.

Kerley, L., Meredith, P. J., Harnett, P., et al. (2021). Families of children in pain: Are attachment and sensory processing patterns related to parent functioning? *Journal of Child and Family Studies, 30*, 1554–1566. https://doi.org/10.1007/s10826-021-01966-.

Kolb, L. C. (1982). Attachment behavior and pain complaints. *Psychosomatics: Journal of Consultation and Liaison Psychiatry, 23*, 413–425.

Kozlowska, K. (2009). Attachment relationships shape pain-signaling behavior. *The Journal of Pain, 10*, 1020–1028.

Kozlowska, K., & Khan, R. (2011). A developmental, body-oriented intervention for children and adolescents with medically unexplained chronic pain. *Clinical Child Psychology and Psychiatry, 16*, 575–598.

Krahé, C., Paloyelis, Y., Condon, H., Jenkinson, P. M., Williams, S. C. R., & Fotopoulou, A. (2015). Attachment style moderates partner presence effects on pain: A laser-evoked potentials study. *Social Cognitive and Affective Neuroscience, 10*, 1030–1037.

Laurita, A. C., Hazan, C., & Spreng, R. N. (2019). An attachment theoretical perspective for the neural representation of close others. *Social Cognitive and Affective Neuroscience, 14*, 237–251.

Lee, O., & Park, G.-A. (2020). The mediating effects of attachment styles on the relationship between sensory processing styles and interpersonal problems in healthy university students. *Occupational Therapy International, 2020*, 1–6.

Liu, C., & Ma, J. L. (2019). Adult attachment style, emotion regulation, and social networking sites addiction. *Frontiers in Psychology, 10*, 2352.

Main, M., & Goldwyn, R. (1988). *Adult attachment classification system.* Version 3.2. Unpublished manuscriptBerkeley, CA: University of California.

Main, M., & Solomon, J. (1990). Procedures for identifying infants as disorganized/disoriented during the Ainsworth strange situation. In M. T. Greenberg, D. Cicchetti, & E. M. Cummings (Eds.), *The John D. and Catherine T. MacArthur Foundation series on mental health and development. Attachment in the preschool years: Theory, research, and intervention* (pp. 121–160). Chicago: University of Chicago Press.

McWilliams, L. (2017). Adult attachment insecurity is positively associated with medically unexplained chronic pain. *European Journal of Pain, 21*, 1378–1383.

Meredith, P. J. (2013). A review of the evidence regarding associations between attachment theory and experimentally induced pain. *Current Pain and Headache Reports, 17*, 326–334.

Meredith, P. J., Bailey, K. J., Strong, J., & Rappel, G. (2016). Adult attachment, sensory processing, and distress in healthy adults. *American Journal of Occupational Therapy, 70*, 7001250010p1-8.

Meredith, P. J., Ownsworth, T., & Strong, J. (2008). A review of the evidence linking adult attachment theory and chronic pain: Presenting a conceptual model. *Clinical Psychology Review, 28*, 407–429.

Mikail, S. F., Henderson, P. R., & Tasca, G. A. (1994). An interpersonally based model of chronic pain: An application of attachment theory. *Clinical Psychology Review, 14*, 1–16.

Mikulincer, M., Gillath, O., & Shaver, P. R. (2002). Activation of the attachment system in adulthood: Threatrelated primes increase the accessibility of mental representations of attachment figures. *Journal of Personality and Social Psychology, 83*, 881–895.

Mikulincer, M., & Shaver, P. R. (2007). *Attachment in adulthood: Structure, dynamics, and change.* New York: Guilford.

Mikulincer, M., & Shaver, P. R. (2012). An attachment perspective on psychopathology. *World Psychiatry: Official Journal of the World Psychiatric Association, 11*, 11–15.

Okur Güney, Z. E., Sattel, H., Witthöft, M., & Henningsen, P. (2019). Emotion regulation in patients with somatic symptom and related disorders: A systematic review. *PLoS One, 14*, e0217277.

O'Neill, M. C., Pillai Riddell, R., Bureau, J. F., Deneault, A. A., Garfield, H., & Greenberg, S. (2021). Longitudinal and concurrent relationships between caregiver-child behaviours in the vaccination context and preschool attachment. *Pain, 162*(3), 823–834. https://doi.org/10.1097/j.pain.0000000000002091.

Peilot, B., Andréll, P., Samuelsson, A., Mannheimer, C., Frodi, A., & Sundler, A. J. (2014). Time to gain trust and change—Experiences of attachment and mindfulness-based cognitive therapy among patients with chronic pain and psychiatric co-morbidity. *International Journal of Qualitative Studies on Health and Well-Being, 9*, 24420.

Pfeifer, A., Amelung, D., Gerigk, C., Schroeter, C., Ehrenthal, J., Neubauer, E., et al. (2016). Study protocol—Efficacy of an attachment-based working alliance in the multimodal pain treatment. *BMC Psychology, 4*, 10.

Pfeifer, A. C., Gómez-Penedo, J. M., Ehrenthal, J. C., Neubauer, E., Amelung, D., Schroeter, C., et al. (2018). Impact of attachment behavior on the treatment process of chronic pain patients. *Journal of Pain Research, 11*, 2653–2662.

Pfeifer, A. C., Meredith, P., Schröder-Pfeifer, P., Gomez Penedo, J. M., Ehrenthal, J. C., Schroeter, C., et al. (2019). Effectiveness of an attachment-informed working alliance in interdisciplinary pain therapy. *Journal of Clinical Medicine, 8*, 364.

Pritchett, R., Minnis, H., Puckering, C., Rajendran, G., & Wilson, P. (2013). Can behaviour during immunisation be used to identify attachment patterns? A feasibility study. *International Journal of Nursing Studies, 50*, 386–391.

Psouni, E., & Apetroaia, A. (2014). Measuring scripted attachment-related knowledge in middle childhood: The secure base script test. *Attachment & Human Development, 16*, 22–41.

Ratnamohan, L., & Kozlowska, K. (2017). When things get complicated: At-risk attachment in children and adolescents with chronic pain. *Clinical Child Psychology & Psychiatry, 22*, 588–602.

Riddell, P. R., & Racine, N. (2009). Assessing pain in infancy: The caregiver context. *Pain Research and Management, 14*, 27–32.

Romeo, A., Tesio, V., Castelnuovo, G., & Castelli, L. (2017). Attachment style and chronic pain: Toward an interpersonal model of pain. *Frontiers in Psychology, 8*, 284.

Shah, P. E., Fonagy, P., & Strathearn, L. (2010). Is attachment transmitted across generations? The plot thickens. *Clinical Child Psychology and Psychiatry, 15*(3), 329–345.

Shah, P. E., & Strathearn, L. (2014). Similarities and differences of the ABCD model and the DMM classification systems for attachment: A practitioner's guide. In P. Holmes, & S. Farnfields (Eds.), *The guidebook to attachment theory and interventions* (pp. 73–88). London: Routledge.

Shaver, P. R., & Mikulincer, M. (2002). Attachment-related psychodynamics. *Attachment & Human Development, 4*, 133–161.

Shaver, P. R., & Mikulincer, M. (2009). An overview of adult attachment theory. In J. H. Obegi, & E. Berant (Eds.), *Attachment theory and research in clinical work with adults* (pp. 17–45). New York: The Guilford Press.

Sinclair, C., Meredith, P., & Strong, J. (2020). Pediatric persistent pain: Associations among sensory modulation, attachment, functional disability, and quality of life. *American Journal of Occupational Therapy, 74*(2), 1.

Sturgeon, J. A., & Zautra, A. J. (2016). Social pain and physical pain: Shared paths to resilience. *Pain management, 6*, 63–74.

Tarantino, S., De Ranieri, C., Dionisi, C., Gagliardi, V., Paniccia, M. F., Capuano, A., et al. (2017). Role of the attachment style in determining the association between headache features and psychological symptoms in migraine children and adolescents. An analytical observational case–control study. *Headache, 57*, 266–275.

Tremblay, I., & Sullivan, M. J. L. (2010). Attachment and pain outcomes in adolescents: The mediating role of pain catastrophizing and anxiety. *The Journal of Pain, 11*, 160–171.

Vachon-Presseau, E., Centeno, M. V., Ren, W., Berger, S. E., Tétreault, P., Ghantous, M., et al. (2016). The emotional brain as a predictor and amplifier of chronic pain. *Journal of Dental Research, 95*, 605–612.

Walsh, T. M., McGrath, P. J., & Symons, D. K. (2008). Attachment dimensions and young children's response to pain. *Pain Research and Management, 13*, 33–40.

Wolff, N. J., Darlington, A. S., Hunfeld, J. A., Tharner, A., van Ijzendoorn, M. H., Bakermans-Kranenburg, M. J., et al. (2011). The influence of attachment and temperament on venipuncture distress in 14-month-old infants: The generation R study. *Infant Behavior and Development, 34*, 293–302.

Chapter 3

The management of pain in older people

Felicity Veal[a] and Kelsey Ng[b]

[a]*Unit for Medication Outcomes and Research and Education, School of Pharmacy and Pharmacology, University of Tasmania, Hobart, TAS, Australia,*
[b]*Hong Kong*

Abbreviations

COX	cyclooxygenase
G	gram
KG	kilogram
Mg	milligram
NSAID	nonsteroidal antiinflammatory drug
SNRI	serotonin and noradrenaline reuptake inhibitor
TCA	tricyclic antidepressant
TENS	transcutaneous electrical nerve stimulation

Introduction

Pain is common throughout life, with virtually everyone experiencing at least one episode of acute pain. However, the prevalence of chronic or persistent pain is less common, with about 20% of the world's population living with chronic pain (Dahlhamer et al., 2016). The prevalence of chronic pain increases up to the age of 85 years, after which it begins to decline (Fayaz, Croft, Langford, Donaldson, & Jones, 2016; Meana, Cho, & DesMeules, 2004; Schofield, 2018). Up to 40% of older people (aged 65 years or older) experience chronic pain, and up to 80% of aged care facility residents experience persistent pain (Gibson, 2007; Schofield et al., 2019).

This high prevalence is due to an increased likelihood of being diagnosed with one or more pain-causing condition as we age. The most common cause of pain in older people is a musculoskeletal disorder, particularly affecting the back, hip, and knee (Abdulla et al., 2013; Schofield, 2018; Schofield et al., 2019). Common pathologies that cause pain in older people are as follows: osteoarthritis, osteoporosis, neuropathic conditions such as diabetic peripheral neuropathy, and as a consequence of cancer treatment (Schofield et al., 2019).

Pharmacological and nonpharmacological strategies can be used to manage chronic pain in older people; however as these patients are almost universally excluded from clinical trials, it is difficult to extrapolate the safety and efficacy data to this population. Additionally, comorbidities and polypharmacy are common in this population (Maher, Hanlon, & Hajjar, 2014; Schofield et al., 2019; Sharifi, Hasanloei, & Mahmoudi, 2014). This, in combination with the pharmacokinetic and pharmacodynamic changes that occur with aging, increases the likelihood of drug-drug and drug-disease interactions (Abdulla et al., 2013; Maher et al., 2014; Schofield et al., 2019).

Additionally, pain management is further complicated in older people by factors associated with identification and assessment of pain. Older people tend to be less willing to report pain, due to fear that it could indicate worsening of an existing disease state or occurrence of a new disease, not wanting to be a nuisance, belief that it is a natural part of aging, or concern regarding medication such as side effects and addiction (Schofield et al., 2019). These perceptions regarding pain often result in suboptimal management, meaning older people are more likely to suffer and experience greater levels of disability associated with their pain than younger people (Schofield et al., 2019). Cognitive impairment can also be a barrier to identification of pain, with studies suggesting that those with significant cognitive impairment suffer higher pain intensity and are less likely to receive pharmacological treatment than those without (Schofield, 2018). There are also other communication-related factors, such as patients being hard of hearing or not speaking the same language as the health care professionals or carers looking after them, that may also affect their ability to comprehend and appropriately respond to questions regarding pain management.

Features and Assessments of Pain, Anesthesia, and Analgesia. https://doi.org/10.1016/B978-0-12-818988-7.00008-X
Copyright © 2022 Elsevier Inc. All rights reserved.

This chapter will provide a general guidance on the use of analgesic in older people. However, it is important to remember that older people are very heterogeneous, with no two patients alike. Age in itself is not the sole factor that should determine how a patient is treated. Some older people are very robust, active, and healthy, whereas others may be frail and have multiple comorbidities; and these other factors should be considered when determining the appropriate pain management for an individual patient.

Pain perception in the elderly

Pain is a subjective experience, and some patients cope with higher levels of pain than others, which is true regardless of age. Pain should also be considered within the context of the biopsychosocial model of health, with changes in physiology being only one component affecting pain perception and experience. As we age, there are a number of changes throughout the body that have the potential to affect the way a person perceives pain (Gibson & Farrell, 2004), with studies indicating variable sensitivity to experimental pain (Gibson & Helme, 2001; Lautenbacher, Kunz, Strate, Nielsen, & Arendt-Nielsen, 2005; Lautenbacher, Peters, Heesen, Scheel, & Kunz, 2017). However, research suggests reduced reporting of pain by older patients in some situations, including postoperatively, visceral infection, malignancy, and atypical presentation of myocardial infarction with pain not always being the primary symptom (Gibson & Helme, 2001; Rittger et al., 2011), and it is unclear if the sensitivity to this pain reduces with age or just the reporting of it.

Some studies indicate that the pain threshold becomes higher with aging; however, this appears to only be seen with lower intensity pain (Gibson & Farrell, 2004; Lautenbacher et al., 2017), with pain tolerance thresholds (i.e. the amount of pain a person can tolerate) not significantly differing with age (Lautenbacher et al., 2017). Some studies have also identified that there is a reduction in production of endogenous analgesic substance which may make older people more sensitive to pain (Riley 3rd, King, Wong, Fillingim, & Mauderli, 2010; Washington, Gibson, & Helme, 2000). Additionally, we do not understand the experience of pain in dementia and whether dementia affects the efficacy of analgesics. Consequently, patients with dementia need to be monitored particularly closely for changes in behavior to ascertain if management strategies have been successful.

Overall, it is likely that age in itself is not a significant contributor to changes in pain perception and sensitivity and that, irrespective of age, patients should be questioned about their pain and the associated level of disability to ensure that pain in older people is not ignored or underestimated.

Assessing pain in older people

The assessment of pain in older people is complicated by both system and patient-specific factors, including cognition, communication, knowledge of the patient, and a patient's reluctance to report or complain about pain (Schofield, 2018; Schofield et al., 2019). The goal of pain management should be two-fold, reduction of pain intensity and improved functionality. This is particularly important in older people as the disability associated with pain increases with age (Schofield, 2018), and consequently, pain assessment should be multidimensional in its approach.

For patients with either no cognitive impairment or mild to moderate impairment, the best and most reliable method of assessing pain intensity is self-report (Schofield, 2018). This can be done using a number of different scales or tools: Some are disease or joint specific, and others are more generic. The most common scales used to measure pain intensity are descriptors (Fig. 1) or the numerical rating scale (NRS) (Fig. 2), and for people who are hard of hearing, who do not speak the same language as the health care practitioner, who have some degree of cognitive decline, or who may not be able to communicate as well, the FACES pain scale can be used (Wong & Baker, 1983) (Fig. 3). However, due to some patient

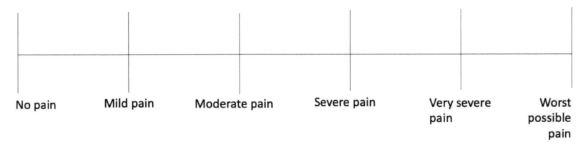

FIG. 1 Simple descriptive pain intensity scale.

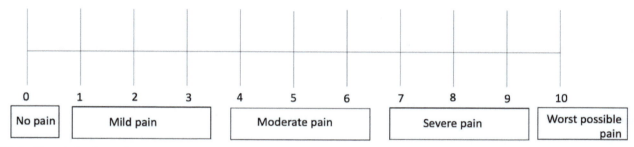

FIG. 2 Numerical rating scale.

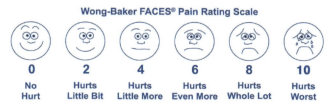

FIG. 3 Wong-Baker FACES Pain Rating Scale (Wong & Baker, 1983).

beliefs regarding pain, the phrases used to assess pain intensity sometimes have to be altered to elicit an accurate response (Schofield, 2018), for example, "pain" may need to be changed to "discomfort." Multidimensional assessment of pain and function can be undertaken by a number of tools including the Multidimensional Pain Inventory (MPI), McGill Pain Questionnaire, Brief Pain Inventory (BPI), Short Form (SF-36), and Western Ontario and McMaster Universities Osteoarthritis Index (WOMAC) for hip and knee pain (Schofield, 2018). These tools however have limited capacity to be used in people with more severe cognitive impairment due to the complexity of the questions.

For people with moderate to severe cognitive impairment, once they are no longer able to comprehend or respond to questions regarding pain, then observational tools should be utilized. These tools may however have a variable efficacy at identifying pain (Lichtner et al., 2014). In addition, they are time consuming to utilize. To assess changes in behavior, the carer must know the patient and their normal behaviors, and there is a need for assessment at a number of time points. For example, a patient may not have pain at rest, but when being moved or during activities of daily living, they may suffer with significant pain. Observational pain scores frequently used and validated in this population include the Abbey Pain Scale, PAINAD, and Doloplus2 (Schofield, 2018). With the aid of newer technology, assessment of pain in nonverbal patients through digital applications is now being undertaken, for example PainChek, which uses photos as well as other domains related to pain to determine a severity score (Atee, Hoti, Parsons, & Hughes, 2017; PainChek, 2019).

Pharmacokinetic and pharmacodynamic changes affecting older people

Pharmacokinetics focuses on what the body does to the drug, whereas pharmacodynamics is the study of how a drug affects the body, including both efficacy and adverse events profile (Holford & Sheiner, 1982). Pharmacokinetics is the study of how drugs move through the body; includes three key aspects, bioavailability, distribution, and clearance; and encompasses the metabolism and elimination of the drug (Holford & Sheiner, 1982). As noted previously, both pharmacokinetics and pharmacodynamics change as a person ages (Abdulla et al., 2013; Schofield et al., 2019; The Australian Pain Society, 2018) and with their level of frailty (Hilmer & Gnjidic, 2017). Pharmacokinetic changes may influence the dose and timing of medications, whereas pharmacodynamic issue may not only affect the dose but also make a medicine less suitable for a certain group of patients, such as NSAIDs in those with renal impairment. Table 1 shows the difference in a number of pharmacokinetic parameters between an older person and a younger person. It is important to note that pharmacokinetic changes are specific to each drug, with some having increased bioavailability and others decreased bioavailability. However, as a general rule, older patients will require lower dose or less frequent dosing than a younger person for initial management, before titrating to effect.

Table 1 shows that there is not a significant change in the bioavailability of oral paracetamol tablets based on age; however, there is a reduction in volume of distribution, likely attributed to changes in body composition, with older people

30 PART | I Setting the scene: General aspects of anesthesia, analgesics and pain

TABLE 1 Pharmacokinetic changes of oral paracetamol with age.

Pharmacokinetic factor	Younger people	Robust older person
Bioavailability	79–98	72–95
Volume of distribution	0.77–1.40 L/kg	0.74–1.08 L/kg
Clearance	0.28–0.7 L/h/kg	0.2–0.38 L/h/kg Note: The clearance of frail older people is 46.8% lower than younger people and 32.4% lower than robust older people.

Based on data from Mian, P., Allegaert, K., Spriet, I., Tibboel, D., & Petrovic, M. (2018). Paracetamol in older people: Towards evidence-based dosing? *Drugs & Aging 35*(7), 603–624.

generally having a higher fat content. The greatest difference can be seen with clearance, when comparing younger people and older people, with older robust people, and more significantly frailer older people, having reduced clearance.

Overview of management of pain in older people

There is a dearth of published literature in older people, especially frail older people, to support the long-term safety and efficacy of pharmacological agents for the management of pain (O'Brien & Wand, 2020; Schofield et al., 2019). Similarly, there is limited evidence to support the best way to use of nonpharmacological management strategies including massage, heat packs, TENS, psychosocial interventions, and exercise programs in older people (Park & Hughes, 2012). However, there is evidence to suggest that psychosocial interventions may particularly improve pain reporting and pain interference (Schofield et al., 2019). While nonpharmacological management is recommended by a number of guidelines, either alone or in conjunction with pharmacological therapy (Schofield et al., 2019; The Australian Pain Society, 2018), further evidence is required as to how a number of these modalities are best utilized. For pharmacological management, there are a number of principles that are recommended when treating older people (Schofield et al., 2019, The Australian Pain Society, 2018).

- Start with a low dose and titrate slowly.
- Change one drug at a time, where possible.
- Multimodal analgesia is recommended (using more than one class of analgesic together to provide a synergistic effect).
- Treat according to the pain experience.

Fig. 4 summarizes the order of treatment for chronic pain in older people. By undertaking these principles, there is a lower risk of adverse effects occurring. Similarly, treating according to the pain experience is likely to result in improved pain outcomes and in some cases lower daily doses. For example, a patient who has short duration acute pain associated with activities of daily living would not be a good candidate for a patch or sustained release preparation as this may expose the

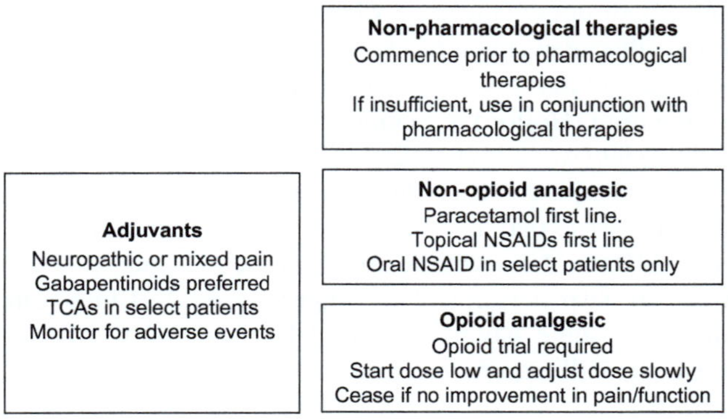

Non-pharmacological therapies
Commence prior to pharmacological therapies
If insufficient, use in conjunction with pharmacological therapies

Adjuvants
Neuropathic or mixed pain
Gabapentinoids preferred
TCAs in select patients
Monitor for adverse events

Non-opioid analgesic
Paracetamol first line.
Topical NSAIDs first line
Oral NSAID in select patients only

Opioid analgesic
Opioid trial required
Start dose low and adjust dose slowly
Cease if no improvement in pain/function

FIG. 4 The sequence of management of pain in older people.

person to higher levels of medication than providing an "as required" preparation given prior to painful activities occurring. Whereas, for residents who may be unable to verbalize pain, regular dosing of medication may be more suitable than as required dosing as there is no need for a patient to request it. The next section shall discuss the main analgesics used in practice.

Paracetamol

Paracetamol (also known as acetaminophen) is one of the most commonly prescribed analgesics worldwide. The mode of action of paracetamol is not fully understood; however, some proposed mechanisms include inhibition of prostaglandin H2 synthesis via cyclooxygenase-2 (COX-2), as well as the activation of descending serotonergic pathways (Sharma & Mehta, 2013).

The evidence to support the use of paracetamol in chronic pain is variable, with some studies indicating that paracetamol is no more effective than placebo for a number of conditions including back pain and osteoarthritis (Leopoldino et al., 2019; Machado et al., 2015). However, it should be noted that the placebo effect is not insignificant in pain management (Colloca, 2019), and thus, paracetamol is likely to provide some analgesic benefit to some people. Consequently, paracetamol should be trialed, with function and pain assessed after an adequate time period to identify if the patient finds it beneficial. Additionally, particularly in the acute setting, paracetamol has been shown to be opioid sparing (Martinez et al., 2017), and potentially, this could also be the case in chronic pain, although further evidence is required. Thus, it may be worthwhile continuing paracetamol in patients who find there is modest benefit if additional analgesics are also added in a multimodal fashion as recommended by guidelines.

Despite being regarded as one of the safest analgesics, adverse events associated with paracetamol administration have been observed, particularly in supratherapeutic doses (daily doses exceeding 4 g). These include hepatic and renal failure, cardiac events (including tachycardia and cardiac injury), and gastrointestinal bleeds (Ralapanawa, Jayawickreme, Ekanayake, & Dissanayake, 2016; Roberts et al., 2016). Paracetamol hepatotoxicity accounts for over 50% of overdose-related acute liver failure in the United States of America (Roberts et al., 2016). Currently, dose reduction (to a maximum dose of 2 g/day) is recommended for people who are frail, have a body weight less than 50 kg, are malnourished, taking specific concomitant enzyme inducers that induce Cytochrome P450 2E1 isoenzymes (Kalsi, Wood, Waring, & Dargan, 2011), or have significant liver dysfunction. However, it is unclear if the risk of adverse events is greater in older people generally, and thus, questions have been raised as to whether the maximum dose should be reduced for all older people (No authors listed, 2018).

Nonsteroidal antiinflammatory drugs (NSAIDs)

NSAIDs such as ibuprofen and celecoxib work by inhibiting the COX enzymes in order to block the production of prostaglandins, which cause inflammation. NSAIDs are commonly prescribed for mild to moderate pain and are generally found to be more efficacious than paracetamol in both acute and chronic pain; however, their use in older people is significantly limited due to their adverse effects profile. NSAIDs, even at therapeutic doses, are associated with increased risk of gastrointestinal, cardiovascular events, and nephrotoxicity (Davis & Robson, 2016; Wongrakpanich, Wongrakpanich, Melhado, & Rangaswami, 2018). NSAIDs are also contraindicated in many conditions including heart failure, moderate to severe renal dysfunction, and active gastric bleeds (Solomon, 2020). One study (Page & Henry, 2000) found that NSAIDs' consumption elevated the risk of first hospital admission due to congestive heart failure by 110%. They also have many drug interactions including warfarin, angiotensin-converting enzyme (ACE) inhibitors, and lithium. NSAIDs can be an effective analgesic in carefully selected patients; however, they should be used at the lowest effective dose for the shortest period of time to minimize the risk of adverse events (Abdulla et al., 2013; Schofield et al., 2019; The Australian Pain Society, 2018).

NSAIDs are often used topically in older patients to avoid the risk of adverse events associated with them when taken orally, as there is minimal systematic absorption. They are recommended as first line for the short-term management of osteoarthritis in older people (Schofield et al., 2019). A Cochrane Review found that topical NSAIDs provide pain relief for more people than placebo with minimal side effects (Derry, Conaghan, Da Silva, Wiffen, & Moore, 2016). However, this review concluded that the evidence to support safety was of very low quality. Consequently, care still needs to be taken in older people as they may experience increased systematic absorption due to thinner skin and increased sensitivity to medications.

Opioids

Opioid analgesics (e.g., morphine, oxycodone) mimic the effect of endogenous opioids at opioid receptors throughout the peripheral and central nervous system to produce analgesia. Some opioids, such as tapentadol and tramadol, also act on additional pathways to inhibit of neuronal reuptake of noradrenaline and/or serotonin. There is a significant body of literature to support use of opioids in acute and end of life pain. However, opioids have become increasingly utilized for the management of chronic pain, despite limited evidence regarding efficacy and concerns around their safety in the longer term. Due to the lack of evidence, it is recommended that for the management of chronic pain, opioids should be used for the shortest possible period of time or used intermittently at the lowest possible dose (Schofield et al., 2019). Furthermore, if ongoing opioids are considered appropriate, an opioid trial should be undertaken to see if any benefit is provided, and if no benefit to function or pain has been identified, the opioid should be ceased or switched (Bannister et al., 2020).

As with all drugs, the pharmacokinetics and pharmacodynamics of opioids change in older people, and this makes older people more likely to experience adverse events. Due to the increased sensitivity of older people to side effects, it is also recommended to start opioids at a lower dose than that for younger patients. Side effects of opioids include constipation, which is more common in older people due to comorbidities and coprescribed therapies, nausea, vomiting, and respiratory depression. There is also an increased risk of falls and fractures (Rolita, Spegman, Tang, & Cronstein, 2013), which is particularly of note in this population, due to the morbidity and mortality associated with these events in older people.

Adjuvant analgesics

Adjuvant analgesics are a group of medications with a primary indication that these are not for pain management, but nonetheless have analgesic properties that may be particularly useful in pain with a neuropathic component. The most commonly used adjuvants are anti-epileptic drugs, particularly the gabapentinoids (pregabalin and gabapentin), tricyclic antidepressants (TCAs), and serotonin and noradrenalin reuptake inhibitors (SNRIs), particularly duloxetine.

Gabapentinoids are considered to be a first-line adjuvant option for the management of pain in older people (Schofield et al., 2019), as they have better safety profile and are associated with fewer drug interactions than TCAs. There is evidence to support the use of gabapentin for diabetic peripheral neuropathy and postherpetic neuralgia, and pregabalin in diabetic peripheral neuropathy, postherpetic neuralgia, and mixed neuropathic pain. However, a substantial proportion of patients who trial a gabapentinoid will not achieve clinically relevant improvements in their pain (Derry et al., 2019; Wiffen et al., 2017). Older people are more prone to the adverse effects of gabapentinoids, even at usual therapeutic doses which may limit their use. Common adverse effects of gabapentinoids include dizziness, drowsiness, altered mental state, and somnolence. There is also a growing body of evidence of the risk of addiction and misuse associated with these medicines (Evoy, Morrison, & Saklad, 2017), which may limit their use.

TCAs are often regarded as first-line drugs for neuropathic pain despite lacking high-quality trial evidence. Although TCAs provide their analgesic effect at relatively low doses, care still needs to be taken in older people as they are particularly prone to the anticholinergic side effects (Schofield et al., 2019). The elderly, especially those with renal impairment, are more prone to these adverse effects, and this is compounded when patients take multiple anticholinergic medicines. If a TCA is indicated, it is recommended to start with a low dose and titrate slowly with close monitoring (Schofield et al., 2019).

Duloxetine, an SNRI, has been found to be effective in the management of a number of neuropathic pain conditions including diabetic neuropathy and fibromyalgia (Lunn, Hughes, & Wiffen, 2014) as well as osteoarthritis (Abou-Raya, Abou-Raya, & Helmii, 2012). However, side effects are common at the doses required for the management of pain, and this may limit the use of these medicines. However, in patients who have comorbid depression, duloxetine may be a suitable first-line agent.

Topical lidocaine and capsaicin are recommended for second-line use in the management of neuropathic pain (Schofield et al., 2019). The evidence to support capsaicin is of moderate to low quality; however, high-dose (8%) capsaicin has been found to reduce neuropathic pain compared to the control, and in patients who found it effective, it also improved other parameters including sleep (Derry, Rice, Cole, Tan, & Moore, 2017). Some reviews have also indicated that it is effective in reducing osteoarthritis-related pain as well as postherpetic neuralgia; however, again the quality of the evidence was rated low (Guedes, Castro, & Brito, 2018; Yong et al., 2017). The evidence to support the use of lidocaine patches is of low quality (Derry et al., 2017). Both of these medications should be reserved for patients who have been unable to tolerate or had no relief with other neuropathic pain agents.

Summary of the pharmacological management of pain in older people

The aforementioned analgesic drugs/drug classes (paracetamol, NSAIDs, opioids, and adjuvant analgesics) have their respective therapeutic benefits and toxicity profiles. One commonality between them is that older patients are more prone to adverse effects, even at normal therapeutic doses. Consequently, the "start low and go slow" principle to management is a key to reducing the likelihood of adverse events. Furthermore, the evidence to support these medications as well as their respective safety profiles, when used long term for chronic pain, is lacking. It is important for this reason to set clear therapeutic goals regarding trialing these medicines and if they do not work, considering both function and pain intensity, then trialing alternatives. The pain experience is not homogenous and neither is this population. Each patient and their needs should be individually considered, and pain management is tailored specifically to them.

Application to other areas

Although the information presented in this chapter is focused on the management of pain in older people, the vast majority of the information apply also to younger adult populations. Further care must be taken in older people, but pharmacological management remains mostly the same, with nonopioid analgesics being used preferentially to opioid analgesics, opioid trials being recommended, and ensuring that management is undertaken in a holistic way.

Other agents of interest

In this chapter, we have discussed the most common pharmacological agents used in the management of chronic pain. One group of agents that was not discussed was the cannabinoids. There has been significant interest in the use of cannabinoids for the management of chronic pain in recent years. At this point in time, there is modest evidence to support the use of cannabinoids in noncancer pain (Beedham, Sbai, Allison, Coary, & Shipway, 2020; Johal et al., 2020), and a systematic review of patients with advanced cancer found no evidence to support their use for cancer-related pain (Boland, Bennett, Allgar, & Boland, 2020). However, older people are insufficiently represented in current clinical trial data, additionally current data are of low quality, using variable routes of administration, formulations, and doses, and thus the true efficacy and adverse events profile is unlikely to be representative of this population (Beedham et al., 2020). One systematic review, evaluating safety specifically in older patients, identified a higher rate of adverse events associated with cannabinoids than the control (van den Elsen et al., 2014). Consequently, there are currently insufficient safety and efficacy data to recommend the use of cannabinoids in this population.

Mini-dictionary of terms

Frailty: A state of increased vulnerability to disease, morbidity and mortality due to aging related decline in the functioning of all body systems (Hilmer & Gnjidic, 2017; Xue, 2011).
Gabapentinoids: A class of adjuvants that includes gabapentin and pregabalin.
Older person: People aged 65 years and older.
Polypharmacy: Taking five or more chronic medications (Masnoon, Shakib, Kalisch-Ellett, & Caughey, 2017).

Summary points

- Older people are a heterogeneous population, and there is no single way to manage their pain.
- Chronic pain is often undermanaged in this population, which increases the likelihood of disability and pain interference.
- Patients' perception of pain being a normal part of aging or wanting to be stoic reduces the likelihood of pain being reported.
- Although there are changes in the physiology related to pain perception with age, these generally do not result in clinically significant difference in pain experience or disability.
- Pain management in older people is complicated by comorbidities, polypharmacy, changes in pharmacodynamics and pharmacokinetics, as well as a lack of trial data to support management.
- Both nonpharmacological and pharmacological management strategies should be utilized when managing pain in older people.

34 PART | I Setting the scene: General aspects of anesthesia, analgesics and pain

- Side effects to medications are more likely than in younger people, so medications should be started low and dose titration undertaken slowly.
- Self-report is the best method to determine pain severity and should be the primary method unless a patient has moderate to severe cognitive impairment.

References

Abdulla, A., Adams, N., Bone, M., Elliot, A. M., Gaffin, J., Jones, D., et al. (2013). Guidance on the management of pain in older people. *Age and Aging*, *42*(i), i1–i57.

Abou-Raya, S., Abou-Raya, A., & Helmii, M. (2012). Duloxetine for the management of pain in older adults with knee osteoarthritis: Randomised placebo-controlled trial. *Age and Ageing*, *41*(5), 646–652.

Atee, M., Hoti, K., Parsons, R., & Hughes, J. D. (2017). Pain assessment in dementia: Evaluation of a point-of-care technological solution. *Journal of Alzheimer's Disease*, *60*(1), 137–150.

Bannister, K., Buchser, E., Casale, R., Chenot, J. F., Chumbley, G., Drewes, A. M., et al. (2020). *Updated position paper on the appropriate use of opioid for chronic non-cancer pain of the European Pain Federation EFIC*. Retrieved from https://europeanpainfederation.eu/wp-content/uploads/2020/06/Opioids-Position-Paper-2020-full.docx. (Accessed 26 June 2020).

Beedham, W., Sbai, M., Allison, I., Coary, R., & Shipway, D. (2020). Cannabinoids in the older person: A literature review. *Geriatrics (Basel)*, *5*(1), 2.

Boland, E. G., Bennett, M. I., Allgar, V., & Boland, J. W. (2020). Cannabinoids for adult cancer-related pain: Systematic review and meta-analysis. *BMJ Supportive & Palliative Care*, *10*(1), 14.

Colloca, L. (2019). The placebo effect in pain therapies. *Annual Review of Pharmacology and Toxicology*, *59*, 191–211.

Dahlhamer, J., Lucas, J., Zelaya, C., Nahin, R., Mackey, S., DeBar, L., et al. (2016). Prevalence of chronic pain and high-impact chronic pain among adults—United States. *MMWR. Morbidity and Mortality Weekly Report*, *67*(36), 1001–1006. https://doi.org/10.15585/mmwr.mm6736a2external.

Davis, A., & Robson, J. (2016). The dangers of NSAIDs: Look both ways. *The British Journal of General Practice*, *66*(645), 172–173.

Derry, S., Bell, R. F., Straube, S., Wiffen, P. J., Aldington, D., & Moore, R. A. (2019). Pregabalin for neuropathic pain in adults. *The Cochrane Database of Systematic Reviews*, *1*(1), CD007076.

Derry, S., Conaghan, P., Da Silva, J. P., Wiffen, P. J., & Moore, R. A. (2016). Topical NSAIDs for chronic musculoskeletal pain in adults. *Cochrane Database of Systematic Reviews*, *4*. https://doi.org/10.1002/14651858.CD007400.pub3, CD007400.

Derry, S., Rice, A. S., Cole, P., Tan, T., & Moore, R. A. (2017). Topical capsaicin (high concentration) for chronic neuropathic pain in adults. *The Cochrane Database of Systematic Reviews*, *1*(1), Cd007393.

Derry, S., Wiffen, P. J., Kalso, E. A., Bell, R. F., Aldington, D., Phillips, T., et al. (2017). Topical analgesics for acute and chronic pain in adults—An overview of cochrane reviews. *The Cochrane Database of Systematic Reviews*, *5*(5), Cd008609.

Evoy, K. E., Morrison, M. D., & Saklad, S. R. (2017). Abuse and misuse of pregabalin and gabapentin. *Drugs*, *77*(4), 403–426.

Fayaz, A., Croft, P., Langford, R. M., Donaldson, L. J., & Jones, G. T. (2016). Prevalence of chronic pain in the UK: A systematic review and meta-analysis of population studies. *BMJ Open*, *6*(6), e010364.

Gibson, S. (2007). IASP global year against pain in older persons: Highlighting the current status and future perspectives in geriatric pain. *Expert Review of Neurotherapeutics*, *7*(6), 627–635.

Gibson, S., & Farrell, M. (2004). A review of age differences in the neurophysiology of nociception and the perceptual experience of pain. *The Clinical Journal of Pain*, *20*(4), 227–239.

Gibson, S. J., & Helme, R. D. (2001). Age-related differences in pain perception and report. *Clinics in Geriatric Medicine*, *17*(3), 433–456 (v–vi).

Guedes, V., Castro, J. P., & Brito, I. (2018). Topical capsaicin for pain in osteoarthritis: A literature review. *Reumatologia Clinica*, *14*(1), 40–45.

Hilmer, S. N., & Gnjidic, D. (2017). Prescribing for frail older people. *Australian Prescriber*, *40*, 174–178.

Holford, N. H., & Sheiner, L. B. (1982). Kinetics of pharmacologic response. *Pharmacology & Therapeutics*, *16*, 143–166.

Johal, H., Devji, T., Chang, Y., Simone, J., Vannabouathong, C., & Bhandari, M. (2020). Cannabinoids in chronic non-cancer pain: A systematic review and meta-analysis. *Clinical Medicine Insights*, *13*, 1179544120906461.

Kalsi, S. S., Wood, D. M., Waring, W. S., & Dargan, P. I. (2011). Does cytochrome P450 liver isoenzyme induction increase the risk of liver toxicity after paracetamol overdose? *Open Access Emergency Medicine*, *3*, 69–76.

Lautenbacher, S., Kunz, M., Strate, P., Nielsen, J., & Arendt-Nielsen, L. (2005). Age effects on pain thresholds, temporal summation and spatial summation of heat and pressure pain. *Pain*, *115*(3), 410–418.

Lautenbacher, S., Peters, J. H., Heesen, M., Scheel, J., & Kunz, M. (2017). Age changes in pain perception: A systematic-review and meta-analysis of age effects on pain and tolerance thresholds. *Neuroscience & Biobehavioral Reviews*, *75*, 104–113.

Leopoldino, A. O., Machado, G. C., Ferreria, P. H., Puinheiro, M. B., Day, R. O., McLachlan, A. J., et al. (2019). Paracetamol versus placebo for knee and hip osteoarthritis. *The Cochrane Database of Systematic Reviews*, *2*(2). https://doi.org/10.1002/14651858.CD013273, CD013273.

Lichtner, V., Dowding, D., Esterhuizen, P., Closs, S. J., Long, A. F., Corbett, A., et al. (2014). Pain assessment for people with dementia: A systematic review of systematic reviews of pain assessment tools. *BMC Geriatrics*, *14*, 138.

Lunn, M. P., Hughes, R. A., & Wiffen, P. J. (2014). Duloxetine for treating painful neuropathy, chronic pain or fibromyalgia. *The Cochrane Database of Systematic Reviews*, *1*, CD007115.

Machado, G. C., Maher, C. G., Ferreira, P. H., Pinheiro, M. B., Lin, C. W., Day, R. O., et al. (2015). Efficacy and safety of paracetamol for spinal pain and osteoarthritis: Systematic review and meta-analysis of randomised placebo controlled trials. *BMJ*, *350*, h1225.

Maher, R. L., Hanlon, J., & Hajjar, E. R. (2014). Clinical consequences of polypharmacy in elderly. *Expert Opinion on Drug Safety, 13*(1), 57–65.

Martinez, V., Beloeil, H., Marret, E., Fletcher, D., Ravaud, P., & Trinquart, L. (2017). Non-opioid analgesics in adults after major surgery: Systematic review with network meta-analysis of randomized trials. *British Journal of Anaesthesia, 118*(1), 22–31.

Masnoon, N., Shakib, S., Kalisch-Ellett, L., & Caughey, G. E. (2017). What is polypharmacy? A systematic review of definitions. *BMC Geriatrics, 17*(1), 230.

Meana, M., Cho, R., & DesMeules, M. (2004). Chronic pain: The extra burden on Canadian women. *BMC Women's Health, 4*(1), S17.

No authors listed. (2018). What dose of paracetamol for older people? *Drug and Therapeutics Bulletin, 56*(6), 69–72.

O'Brien, M. D. C., & Wand, A. P. F. (2020). A systematic review of the evidence for the efficacy of opioids for chronic non-cancer pain in community-dwelling older adults. *Age and Ageing, 49*(2), 175–183.

Page, J., & Henry, D. (2000). Consumption of NSAIDs and the development of congestive heart failure in elderly patients: An underrecognized public health problem. *Archives of Internal Medicine, 160*(6), 777–784.

PainChek. (2019). *Pain asssessment protocol.* Retrieved from https://www.painchek.com/wp-content/uploads/2019/08/3668-Pain-Assessment-Protocol-Poster-Update-v5.pdf. (Accessed 4 June 2020).

Park, J., & Hughes, A. K. (2012). Non-pharmacological approaches to the management of chronic pain in community-dwelling older adults: A review of empirical evidence. *Journal of the American Geriatrics Society, 60*(3), 555–568.

Ralapanawa, U., Jayawickreme, K. P., Ekanayake, E. M. M., & Dissanayake, A. M. S. D. M. (2016). A study on paracetamol cardiotoxicity. *BMC Pharmacology and Toxicology, 17*(1), 30.

Riley, J. L., 3rd, King, C. D., Wong, F., Fillingim, R. B., & Mauderli, A. P. (2010). Lack of endogenous modulation and reduced decay of prolonged heat pain in older adults. *Pain, 150*(1), 153–160.

Rittger, H., Rieber, J., Breithardt, O. A., Dücker, M., Schmidt, M., Abbara, S., et al. (2011). Influence of age on pain perception in acute myocardial ischemia: A possible cause for delayed treatment in elderly patients. *International Journal of Cardiology, 149*(1), 63–67.

Roberts, E., Delgado Nunes, V., Buckner, S., Latchem, S., Constanti, M., Miller, P., et al. (2016). Paracetamol: Not as safe as we thought? A systematic literature review of observational studies. *Annals of the Rheumatic Diseases, 75*(3), 552–559.

Rolita, L., Spegman, A., Tang, X., & Cronstein, B. N. (2013). Greater number of narcotic analgesic prescriptions for osteoarthritis is associated with falls and fractures in elderly adults. *Journal of the American Geriatrics Society, 61*(3), 335–340.

Schofield, P. (2018). The assessment of pain in older people: UK National Guidelines. *Age and Ageing, 47*(Suppl 1), i1–i22.

Schofield, P., Dunham, M., Martin, D., Bellamy, G., Francis, S., Sookhoo, D., et al. (2019). *National Guidelines for the management of pain in older adults consultation document England.*

Sharifi, H., Hasanloei, M. A. C., & Mahmoudi, J. (2014). Polypharmacy-induced drug-drug interactions: Threats to patient safety. *Drug Research, 64*(12), 633–637.

Sharma, C. V., & Mehta, V. (2013). Paracetamol: Mechanisms and updates. *Continuing Education in Anaesthesia, Critical Care & Pain, 14*(4), 153–158.

Solomon, D. H. (2020). In P. W. UpToDate (Ed.), *Patient education: Nonsteroid antiinflammatory drugs (NSAIDs) (beyond the basics).* Walktham, MA: UpToDate.

The Australian Pain Society (Ed.). (2018). *Pain in residential aged care facilities management strategies* (2nd ed.). Sydney: The Australian Pain Society.

van den Elsen, G. A., Ahmed, A. I., Lammers, M., Kramers, C., Verkes, R. J., van der Marck, M. A., et al. (2014). Efficacy and safety of medical cannabinoids in older subjects: A systematic review. *Ageing Research Reviews, 14*, 56–64.

Washington, L. L., Gibson, S. J., & Helme, R. D. (2000). Age-related differences in the endogenous analgesic response to repeated cold water immersion in human volunteers. *Pain, 89*(1), 89–96.

Wiffen, P. J., Derry, S., Bell, R. F., Rice, A. S., Tölle, T. R., Phillips, T., et al. (2017). Gabapentin for chronic neuropathic pain in adults. *The Cochrane Database of Systematic Reviews, 6*(6), Cd007938.

Wong, D., & Baker, C. (1983). *Wong-Baker FACES pain rating scale.*

Wongrakpanich, S., Wongrakpanich, A., Melhado, K., & Rangaswami, J. (2018). A comprehensive review of non-steroidal anti-inflammatory drug use in the elderly. *Aging and Disease, 9*(1), 143–150.

Xue, Q. L. (2011). The frailty syndrome: Definition and natural history. *Clinics in Geriatric Medicine, 27*(1), 1–15.

Yong, Y. L., Tan, L. T.-H., Ming, L. C., Chan, K.-G., Lee, L.-H., Goh, B.-H., et al. (2017). The effectiveness and safety of topical capsaicin in postherpetic neuralgia: A systematic review and meta-analysis. *Frontiers in Pharmacology, 7*, 538.

Chapter 4

Anesthesia and body mass: Epidural depth and beyond

Mehmet Canturk

Department of Anesthesiology and Reanimation, Karadeniz Eregli Government Hospital, Ereğli, Turkey

Abbreviations

BMI body mass index (kg/m^2)
OSA obstructive sleep apnea
OHS obesity hypoventilation syndrome
TOF train of four (to determine the depth of neuromuscular block)

Introduction

The number of people with increased body mass (overweight and/or obese) is increasing every year as a pandemic, which is a public health problem that anesthesiologists have to deal with either for obesity surgery or other types of surgeries. The prevalence of obesity has tripled since 1975, and 40 million children under the age of five were overweight in 2018. This fact compels the anesthesiologists to take care of more and more patients with increased body mass and manage the comorbidities brought with the increased body mass.

The body mass index (BMI) is a ratio defined as the division of patient weight in kilograms divided by the square of the height of the patient in meters (kg/m^2). Although the BMI is accepted for the classification of nutritional status of a patient (Table 1) worldwide, it cannot determine the cause of increased body mass as exceptional body mass increases due to ascites or huge tumoral bulk.

As anesthesiologists, we have to deal with patients with increased body mass with an increasing frequency every next year. Besides the increased body mass, the anesthesiologists have to manage the respiratory, cardiovascular, endocrine, and musculoskeletal comorbidities related to increased body mass.

Preoperative assessment

Preoperative assessment of a patient with increased body mass is very important for a stable perioperative management. A detailed anamnesis including the patients' previous anesthesia experience, respiratory and cardiovascular system examination, detailed history of metabolic diseases (diabetes, hypothyroidism, hyperlipidemia, etc.), and the medications (both prescribed and under the counter) that the patient is taking must be questioned to achieve a stable perioperative period.

Assessment of respiratory system

The common respiratory system comorbidities include asthma, restrictive lung disease, obstructive sleep apnea (OSA), and obesity hypoventilation syndrome (OHS). OSA is thought to be more frequent in patients with increased body mass scheduled for surgery, but its prevalence is underestimated since not all patients are taking overnight polysomnography test (Sun et al., 2015). Screening tests such as STOP-Bang (Chung & Elsaid, 2009; Yegneswaran & Chung, 2009) and overnight pulse oximetry (Chung et al., 2012) have limited diagnostic success for OSA. Postoperative opioid use for the relief of pain may result in fatal complications in OSA patients.(Boushra, 1996; Lee et al., 2015; Orlov, Ankichetty, Chung, & Brull, 2013). Obese patients frequently have decreased blood leptin levels, which decreases the sensitivity to carbon dioxide retention, thereby causing obesity hypoventilation syndrome (OHS) (O'Donnell et al., 1999). OHS patients usually have daytime hypoventilation and hypoxemia (PaO2 < 65 mmHg at room air).

Features and Assessments of Pain, Anesthesia, and Analgesia. https://doi.org/10.1016/B978-0-12-818988-7.00035-2
Copyright © 2022 Elsevier Inc. All rights reserved.

TABLE 1 Body mass index classification defined by the WHO.	
BMI (kg/m^2)	Classification
<18.5	Underweight
18.5–24.9	Normal weight
25.0–29.9	Overweight
30.0–34.9	Obesity, Class I
35.0–39.9	Obesity, Class II
>40	Obesity, Class III

WHO, World Health Organization; *BMI*, body mass index.

Airway management during induction of anesthesia is more complicated in obese patients as compared to normal population (Karkouti, Rose, Wigglesworth, & Cohen, 2000). The increased bulk mass of peripharyngeal fat tissue, small oral cavity, and limited mouth opening make mask ventilation difficult (Siyam & Benhamou, 2002). The increased neck diameter (>42 cm at cricoid cartilage level) (Domi & Laho, 2012) and the reduced joint movements due to diabetes limit the neck extension during laryngoscopy (Nguyen et al., 2002; Riley et al., 1997; Sahin, Bilgen, Tasbakan, Midilli, & Basoglu, 2014). Because of the increased trunk mass and decreased pulmonary and chest compliances, a restrictive respiratory pattern is present (Lazarus, Sparrow, & Weiss, 1997). Moreover, there is a decreased functional residual capacity and expiratory reserve volume (Jones & Nzekwu, 2006; Pelosi et al., 1997). Therefore, patients with increased body mass become vulnerable to hypoxia during the induction phase of general anesthesia and to the increased risk of postoperative atelectasis.

Assessment of cardiovascular system

Peripheral venous access is generally so difficult and sometimes impossible in obese patients that a central venous line is inserted to achieve an intravenous route due to the increased subcutaneous fat and connective tissue accompanying to the increased body mass. And also noninvasive blood pressure measurements are problematic since appropriate cuff must be provided for a reliable blood pressure measurement. Even though an appropriate cuff size is provided, it may be difficult to keep the cuff in the right position on the patients' arm. For some of the selected cases, invasive arterial blood pressure measurement is preferred to provide a route for arterial blood gas sampling (DeMaria, Portenier, & Wolfe, 2007). Ultrasound can be used to assist central venous and arterial cannulation in selected cases to increase the success rate in first attempt (Blanco, 2019).

Venous thromboembolism is a frightening complication that has to be considered and managed properly. Either mechanical (external pressure stockings) or pharmacological (low-molecular-weight heparin or warfarin) or both antithrombotic prophylaxis modalities should be provided for the prevention of thromboembolic complications in patients with increased body mass (Venclauskas, Maleckas, & Arcelus, 2018; Yong, Thurston, Singh, & Allaire, 2019).

Hypertension, cardiomegaly, arrhythmia, left- and/or right-sided heart failure, and coronary artery disease can be observed in obese patients. These complications may develop in time as a consequence of increased blood volume (Miller & Borlaug, 2020). The heart increases the cardiac output with increasing body mass, which results in hypertrophy of the heart. Moreover, the persistence of increased muscle work leads to cardiomyopathy, conduction defects, and heart failure. The right-sided heart failure should be expected in patients with OHS since chronic daytime hypoxia results in pulmonary hypertension that may progress to right heart hypertrophy and failure at the end (Terla, Rajbhandari, Kurian, & Pesola, 2019).

Assessment of endocrine and musculoskeletal systems

Type II diabetes frequently coincides with the increase in body mass. The control of plasma glucose level < 200 mg/dL is extremely important in patients undergoing surgery; thus, hyperglycemia impairs the normal function of immune cells (leukocyte chemotaxis and granulocyte phagocytic activity are disturbed) and delays wound healing (Portou et al., 2020; Sunahara, Sannomiya, & Martins, 2012). Moreover, it increases the incidence of wound infection (Iannantuoni et al.,

2019). Fasting blood glucose sampling may guide to regulate the plasma glucose level of the patient during the preparation for the surgery.

Calcium levels may be lower than normal levels either due to inadequate intake, due to medications used to control hypertension, or both. The preoperative blood sample may guide to determine the preoperative level of plasma calcium level. Either of the reasons can be managed by regulating the appropriate intake of calcium with diet (Teegarden, 2003).

Preoperative assessment of musculoskeletal system is important for anesthesiologist not to harm the patient while he/she is under anesthesia. Range of motion may be limited in the joints due to increased weight carried by the joint especially on the hip and knee joints (Park, Ramachandran, Weisman, & Jung, 2010). Patient may have had joint replacement surgery before the scheduled surgery, and therefore, the joint movements can be limited. Hyperabduction of a prosthetic hip joint may cause dislocation of the femoral head and may also result in sciatic nerve injury. Forcing the knee for hyperflexion may dislocate the prosthetic knee joint or may cause fracture on the bonny shaft. Atlanto-occipital joint movement may be hindered due to increased adipose tissue around the neck and also as a result of pathologic glycosylation of proteins that result in stiff tissue in diabetic patients. Increased body mass also increases the risk of ulnar nerve damage, brachial plexus injuries, and nerve root damage due to malpositioning (Hewson, Bedforth, & Hardman, 2018; Sawyer, Richmond, Hickey, & Jarrratt, 2000; Warner, Warner, & Martin, 1994).

Intraoperative management

The anesthesia work station must be double checked before accepting patient in to the operation room. Appropriate airway devices must be prepared and ready for use in case of difficult ventilation or intubation. Although difficult mask ventilation is expected in obese patients, difficult intubation does not correlate with increasing body mass (Saasouh et al., 2018).

Monitorization of the patient should be tailored according to the scheduled surgical procedure. At least one intravenous line must be secured before the induction of anesthesia. Ultrasound guidance may facilitate the insertion of a central venous catheter if the cannulation of peripheral veins is problematic due to the aforementioned reasons in patients with increased body mass. A suitable sized cuff must be prepared to achieve a reliable noninvasive blood pressure measurement or an arterial line must be secured preferably with ultrasound guidance, and continuous arterial blood pressure measurements are obtained.

Obese patients are generally uncomfortable in supine position since the compliance of the chest wall and lungs and also the functional residual capacity decrease in supine position, leading to an increased ventilation-perfusion mismatch (Mandal & Hart, 2012). Moreover, compression of abdominal organs on the diaphragm limits the movement of diaphragm during inspiration, increasing the work of respiration that adds on to the hypoxemia. Therefore, patient must be kept in reverse Trendelenburg position, and the upper body must be in ramped position (Semler et al., 2017). Supplemental high flow nasal oxygen is suggested to increase the safe apnea time (time from the induction of anesthesia to the 95% oxygen saturation measured with pulse oximetry or a 6-min apnea time) in morbidly obese patients (Wong et al., 2019).

Induction of anesthesia, anesthetic agents, and muscle relaxants

All anesthetic agents used for the induction of normal weight patients can be used in patients with an increased body mass. However, the dosing of the anesthetic agents and neuromuscular blocking agents is more challenging in obese patients. The volume of distribution for the lipophilic agents increases with the increased adipose tissue. Meanwhile, the cardiac output also increases in patients with increased body mass, altering the pharmacodynamics and pharmacokinetics of an anesthetic agent. Herein, the most commonly used anesthetic agents are concerned.

Thiopental sodium

Thiopental sodium is a highly lipophilic agent. Following intravenous bolus dose of thiopental sodium, it rapidly distributes to the highly perfused organs such as brain, liver, lung, and kidneys. Its loss of effect is by its redistribution to peripheral organs. The increased adipose tissue in obese patients increases the volume of distribution of thiopental sodium (Jung, Mayersohn, Perrier, Calkins, & Saunders, 1982). The increased volume of distribution means lower drug plasma concentrations but a longer elimination time. Thus, the induction dose of thiopental is adjusted according to ideal body weight.

Propofol

Propofol is a highly lipophilic anesthetic agent with a short duration of action and a fast onset of action. There still is a controversy for the intravenous bolus dose of propofol for induction of general anesthesia. Some authors suggest that the induction bolus dose should be calculated according to ideal body weight, and some others suggest that lean body weight should be used to calculate the dose of propofol needed for induction of anesthesia (Domi & Laho, 2012; Ingrande, Brodsky, & Lemmens, 2011; Ingrande & Lemmens, 2010; Tsui, Murtha, & Lemmens, 2017).

Opioids

Fentanyl is one of the most commonly used drugs to suppress the adrenergic response during the laryngoscopy to prevent inadvertent increases in blood pressure of the patient, to decrease the response to surgical stimuli, and to control postoperative pain. The fentanyl dose is calculated according to total body weight (Barras & Legg, 2017; Ingrande & Lemmens, 2010), although its clearance correlates well with lean body weight. Remifentanil is a short acting opioid, and its dose is regulated according to ideal body weight (Egan et al., 1998).

Neuromuscular-blocking agents and antagonists

Except for succinylcholine and atracurium, all neuromuscular agents' doses are adjusted according to ideal body weight, but succinylcholine and atracurium are adjusted according to total body weight; thus, plasma pseudocholinesterase level increases (the enzyme that inactivates succinylcholine) with the increasing body mass (Domi & Laho, 2012; Lemmens & Brodsky, 2006; Mandal & Hart, 2012).

There is no alteration in the use or dosing of neostigmine. Total dose should not exceed 5 mg, and the suggested dose is 0.04 to 0.08 mg/kg; however, the time between the neostigmine bolus and the time to reach 0.9 TOF (train of four) ratio increase with increasing body mass (Lemmens & Ingrande, 2013; Suzuki, Masaki, & Ogawa, 2006).

Sugammadex is a cyclodextrin derivative agent that encapsulates the steroidal muscle relaxants with a higher affinity to rocuronium than vecuronium. One limitation for the use of sugammadex for muscle relaxation reversal is the renal insufficiency since the compound formed by the neuromuscular agent and sugammadex is excreted through kidneys. Dose adjustment must be done according to the depth of neuromuscular block. For the reversal of a deep block, ideal body weight should be used for dose adjustment (Loupec et al., 2016; Nightingale et al., 2015).

Volatile anesthetics

Although choosing the volatile agent for the maintenance of general anesthesia depends mostly on the availability of the anesthetic agent at the operation room, clinical literature suggests desflurane in obese patients. Isoflurane, sevoflurane, and desflurane are the available anesthetics agents in use in daily practice. Lipid solubility is highest for isoflurane and least for desflurane. In a recent metaanalysis, desflurane provided faster verbal contact and earlier extubation (Singh, Borle, McGavin, Trikha, & Sinha, 2017) in bariatric surgical patients. The effect of sevoflurane and desflurane was reported to be comparable on respiratory mechanics during the intraoperative period (Ozturk, Demiroluk, Abitagaoglu, & Ari, 2019). Increased adipose tissue in parallel to increased body mass in obese patients results in prolonged release of volatile anesthetic agents during awaking from anesthesia. Since desflurane has the smallest distribution in adipose tissue, it provides the fastest recovery when compared to sevoflurane or isoflurane.

Regional anesthesia

Regional anesthesia seems to be more advantageous to keep the normal physiology of the patient during surgical procedures than general anesthesia. The risks of apnea during induction, difficulty in mask ventilation, difficult intubation and airway management, and postoperative sore throat risks are dismissed by administering regional anesthesia instead of general anesthesia. Regional anesthesia and analgesia are also used as a part of multimodal analgesia that enables the use of decreased opioid dose for postoperative analgesia. Decreasing the need for postoperative opioid consumption for the control of postoperative pain decreases the prevalence of inadvertent postoperative respiratory events (Thorell et al., 2016), Besides these advantages, it is more difficult to achieve a successful regional anesthesia in obese patients when compared to non obese (Cotter et al., 2004). Anatomical landmarks are obscured with the additional adipose tissue. It is difficult to localize the midline and so forth by traditional palpation methods, but anesthesiologists can overcome these problems with ultrasound

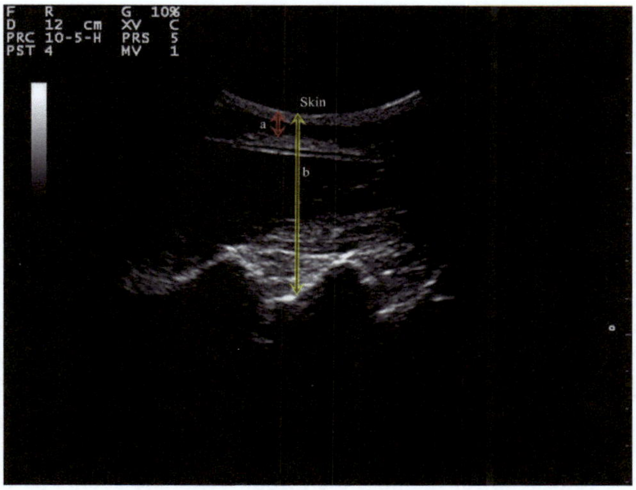

FIG. 1 Paramedian sagittal view of vertebral canal, lean patient. (a) subcutaneous and adipose tissue depth, (b) epidural depth.

guidance (Carvalho, Seligman, & Weiniger, 2019). Sometimes, the depth of epidural space from the skin is so deep that epidural space cannot be localized with a standard needle in obese patients. The depth of epidural space has a strong correlation with the BMI of the patient, both in normal population and parturients (Canturk, Karbancioglu Canturk, Kocaoglu, & Hakki, 2019; Canturk, Kocaoglu, & Hakki, 2019; Galbraith, Wallace, & Devitt, 2019; Kim et al., 2011; Ravi, Kaul, Kathuria, Gupta, & Khurana, 2011). The depth of epidural space in parasagittal oblique view with ultrasound is presented in Figs. 1 and 2. In both figures, the distance of subcutaneous and adipose tissue is represented with "a," and the epidural space is represented with arrow "b." With the increasing body mass, the adipose tissue and subcutaneous tissue under the skin are getting thicker as presented in Figs. 1 and 2. Another concern is the spread of local anesthetic agent in spinal anesthesia. When compared to nonobese patients, spinal anesthesia done with the same amount of local anesthetic agent results in a higher level of spinal anesthesia in obese patients (Ingrande, Brodsky, & Lemmens, 2009).

Hypotension is a common complication of neuraxial anesthesia. Obese patients are more vulnerable to hemodynamic changes; thus, cardiac function may be restricted due to aforementioned reasons, cardiac output may be reduced due to vena cava compression by intraabdominal organs in supine position, and intravenous volume may be reduced due to preoperative fasting, under the effect of antihypertensive drugs, and due to diabetes mellitus. When these findings are gathered, a safe anesthesia approach should be tailored individually to every patient appropriate for the scheduled surgery.

Another limitation of regional anesthesia in obese patients undergoing surgery is the timing of antithrombotic prophylaxis and the type of agent used for this purpose. Guidelines by the American Society of Regional Anesthesia and European

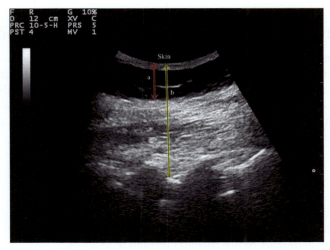

FIG. 2 Paramedian sagittal view of vertebral canal, obese patient. (a) subcutaneous and adipose tissue depth, (b) epidural depth.

42 PART | I Setting the scene: General aspects of anesthesia, analgesics and pain

Society of regional Anesthesia have specified the timing of regional anesthesia interventions with regard to the antithrombotic agent use. The surgeon, anesthesiologist, and the ward nurse responsible for the care of the patient must be in close coordination to prevent complications.

Performing peripheral blocks is more difficult in obese patients, but ultrasound guidance may facilitate the performance of a safe and effective procedure (Schroeder et al., 2012). Increased body mass and the increased subcutaneous adipose tissue limit the visibility of target nerves in ultrasound; therefore, performing deep peripheral nerve blocks in more difficult in obese patients (Marhofer, Pilz-Lubsczyk, Lonnqvist, & Fleischmann, 2014).

Cessation of anesthesia and postoperative care

Extubation and postoperative care of the patients with increased body mass are very important since this patient population is vulnerable to cardiopulmonary complications (Chen, Deng, & Li, 2019). Sudden cardiac death is more common in overweight and obese patients. Patients should be monitored in the postoperative care unit and be followed up until the patient reaches the preoperative values for vital signs. Supplemental oxygen and ramped position must be provided.

Patients should also be in close follow-up for the possible thromboembolic events despite the preoperative and perioperative measures.

An effective multimodal analgesia should be planned and tailored for the type of surgery and the patient. Preferably, a multimodal analgesia combined with regional analgesia techniques will provide effective analgesia free of opioids. Use of opioids should be minimized or if possible prevented to control the postoperative pain in obese patients (Veiga de Sa, Cavaleiro, & Campos, 2020).

Applications to other areas

Anesthetic management of a patient with increased body mass initiates with the preoperatie visit till the patients is discharged. Increased body mass is generally coincided with several comorbidities that must be strictly diagnosed and treated before starting with the management of anesthetic care. Keeping abreast of principal anesthetic management for a patient with increased body mass index will increase the patient safety during the hospitalization period. Meanwhile a careful preasthetic evaluation may revalate the undiagnosed comorbidities.

Mini-dictionary of terms

BMI: Ratio of patients' weight in kilograms divided by the square of patients' height in meters (kg/m^2).
Obesity: A body mass index ≥ 30 is accepted as obesity.
Neuraxial block: Epidural, spinal, and combined spinal epidural anesthesia.
Pseudocholinesterase: An enzyme degrading succinylcholine in plasma.
Epidural depth: The distance from skin to ligamentum flavum.

Key facts of anesthesia and body mass: Epidural depth and beyond

- Obesity is a growing pandemic in the world.
- Number of obese patients whom anesthesiologist must provide perioperative care to is increasing.
- Detailed preoperative assessment of the patient is obligatory for an excellent perioperative care.
- Anesthesia method should be decided on patient and surgery basis and tailored individually.
- Regional anesthesia techniques should be considered whenever possible for obese patients to avoid the serious complications of general anesthesia.

Summary points

- Preoperative visit is extremely important for a safe perioperative period.
- Patient comorbidities and medications must be well documented.
- Patient positioning is an important issue in obese patients.
- Regional anesthesia may be a preferable choice to avoid complications.
- Appropriate dosing of the anesthetic agents is vital.

References

Barras, M., & Legg, A. (2017). Drug dosing in obese adults (Review). *Australian Prescriber*, *40*(5), 189–193. https://doi.org/10.18773/austprescr.2017.053.

Blanco, P. (2019). Ultrasound-guided peripheral venous cannulation in critically ill patients: A practical guideline (Review). *The Ultrasound Journal*, *11*(1), 27. https://doi.org/10.1186/s13089-019-0144-5.

Boushra, N. N. (1996). Anaesthetic management of patients with sleep apnoea syndrome (Review). *Canadian Journal of Anaesthesia*, *43*(6), 599–616. https://doi.org/10.1007/BF03011774.

Canturk, M., Karbancioglu Canturk, F., Kocaoglu, N., & Hakki, M. (2019). Abdominal girth has a strong correlation with ultrasound-estimated epidural depth in parturients: A prospective observational study (Observational study). *Journal of Anesthesia*, *33*(2), 273–278. https://doi.org/10.1007/s00540-019-02621-9.

Canturk, M., Kocaoglu, N., & Hakki, M. (2019). Abdominal girth has a strong correlation with actual and ultrasound estimated epidural depth. *Turkish Journal of Medical Sciences*, *49*(6), 1715–1720. https://doi.org/10.3906/sag-1902-115.

Carvalho, B., Seligman, K. M., & Weiniger, C. F. (2019). The comparative accuracy of a handheld and console ultrasound device for neuraxial depth and landmark assessment (Comparative study). *International Journal of Obstetric Anesthesia*, *39*, 68–73. https://doi.org/10.1016/j.ijoa.2019.01.004.

Chen, H., Deng, Y., & Li, S. (2019). Relation of body mass index categories with risk of sudden cardiac death (Meta-analysis systematic review). *International Heart Journal*, *60*(3), 624–630. https://doi.org/10.1536/ihj.18-155.

Chung, F., & Elsaid, H. (2009). Screening for obstructive sleep apnea before surgery: Why is it important? (Review). *Current Opinion in Anaesthesiology*, *22*(3), 405–411. https://doi.org/10.1097/ACO.0b013e32832a96e2.

Chung, F., Liao, P., Elsaid, H., Islam, S., Shapiro, C. M., & Sun, Y. (2012). Oxygen desaturation index from nocturnal oximetry: A sensitive and specific tool to detect sleep-disordered breathing in surgical patients (Research support, Non-U.S. Gov't). *Anesthesia and Analgesia*, *114*(5), 993–1000. https://doi.org/10.1213/ANE.0b013e318248f4f5.

Cotter, J. T., Nielsen, K. C., Guller, U., Steele, S. M., Klein, S. M., Greengrass, R. A., & Pietrobon, R. (2004). Increased body mass index and ASA physical status IV are risk factors for block failure in ambulatory surgery—An analysis of 9,342 blocks (Evaluation study research support, Non-U.S. Gov't). *Canadian Journal of Anaesthesia*, *51*(8), 810–816. https://doi.org/10.1007/BF03018454.

DeMaria, E. J., Portenier, D., & Wolfe, L. (2007). Obesity surgery mortality risk score: Proposal for a clinically useful score to predict mortality risk in patients undergoing gastric bypass (Multicenter study). *Surgery for Obesity and Related Diseases*, *3*(2), 134–140. https://doi.org/10.1016/j.soard.2007.01.005.

Domi, R., & Laho, H. (2012). Anesthetic challenges in the obese patient (Review). *Journal of Anesthesia*, *26*(5), 758–765. https://doi.org/10.1007/s00540-012-1408-4.

Egan, T. D., Huizinga, B., Gupta, S. K., Jaarsma, R. L., Sperry, R. J., Yee, J. B., & Muir, K. T. (1998). Remifentanil pharmacokinetics in obese versus lean patients (Comparative study research support, Non-U.S. Gov't). *Anesthesiology*, *89*(3), 562–573. https://doi.org/10.1097/00000542-199809000-00004.

Galbraith, A. S., Wallace, E., & Devitt, A. (2019). Examining the association of body mass index and the depth of epidural space, radiation dose exposure and fluoroscopic screening time during transforaminal nerve block injection: A retrospective cohort study. *Irish Journal of Medical Science*, *188*(1), 295–302. https://doi.org/10.1007/s11845-018-1845-7.

Hewson, D. W., Bedforth, N. M., & Hardman, J. G. (2018). Peripheral nerve injury arising in anaesthesia practice (Review). *Anaesthesia*, *73*(Suppl 1), 51–60. https://doi.org/10.1111/anae.14140.

Iannantuoni, F., Diaz-Morales, N., Escribano-Lopez, I., Sola, E., Roldan-Torres, I., Apostolova, N., ... Victor, V. M. (2019). Does glycemic control modulate the impairment of NLRP3 inflammasome activation in type 2 diabetes? (Research support, Non-U.S. Gov't). *Antioxidants & Redox Signaling*, *30*(2), 232–240. https://doi.org/10.1089/ars.2018.7582.

Ingrande, J., Brodsky, J. B., & Lemmens, H. J. (2009). Regional anesthesia and obesity (Review). *Current Opinion in Anaesthesiology*, *22*(5), 683–686. https://doi.org/10.1097/ACO.0b013e32832eb7bd.

Ingrande, J., Brodsky, J. B., & Lemmens, H. J. (2011). Lean body weight scalar for the anesthetic induction dose of propofol in morbidly obese subjects (Comparative study randomized controlled trial research support, Non-U.S. Gov't). *Anesthesia and Analgesia*, *113*(1), 57–62. https://doi.org/10.1213/ANE.0b013e3181f6d9c0.

Ingrande, J., & Lemmens, H. J. (2010). Dose adjustment of anaesthetics in the morbidly obese (Review). *British Journal of Anaesthesia*, *105*(Suppl 1), i16–i23. https://doi.org/10.1093/bja/aeq312.

Jones, R. L., & Nzekwu, M. M. (2006). The effects of body mass index on lung volumes. *Chest*, *130*(3), 827–833. https://doi.org/10.1378/chest.130.3.827.

Jung, D., Mayersohn, M., Perrier, D., Calkins, J., & Saunders, R. (1982). Thiopental disposition in lean and obese patients undergoing surgery (Comparative study research support, Non-U.S. Gov't). *Anesthesiology*, *56*(4), 269–274. https://doi.org/10.1097/00000542-198204000-00007.

Karkouti, K., Rose, D. K., Wigglesworth, D., & Cohen, M. M. (2000). Predicting difficult intubation: A multivariable analysis (Research support, Non-U.S. Gov't). *Canadian Journal of Anaesthesia*, *47*(8), 730–739. https://doi.org/10.1007/BF03019474.

Kim, L. K., Kim, J. R., Shin, S. S., Kim, I. J., Kim, B. N., & Hwang, G. T. (2011). Analysis of influencing factors to depth of epidural space for lumbar transforaminal epidural block in korean. *Korean Journal of Pain*, *24*(4), 216–220. https://doi.org/10.3344/kjp.2011.24.4.216.

Lazarus, R., Sparrow, D., & Weiss, S. T. (1997). Effects of obesity and fat distribution on ventilatory function: The normative aging study. *Chest*, *111*, 891–898 (Research support, Non-U.S. Gov't P.H.S.).

Lee, L. A., Caplan, R. A., Stephens, L. S., Posner, K. L., Terman, G. W., Voepel-Lewis, T., & Domino, K. B. (2015). Postoperative opioid-induced respiratory depression: A closed claims analysis (Research support, Non-U.S. Gov't). *Anesthesiology*, *122*(3), 659–665. https://doi.org/10.1097/ALN.0000000000000564.

Lemmens, H. J., & Brodsky, J. B. (2006). The dose of succinylcholine in morbid obesity (Randomized controlled trial). *Anesthesia and Analgesia, 102*(2), 438–442. https://doi.org/10.1213/01.ane.0000194876.00551.0e.

Lemmens, H. J., & Ingrande, J. (2013). Pharmacology and obesity (Review). *International Anesthesiology Clinics, 51*(3), 52–66. https://doi.org/10.1097/AIA.0b013e31829a4d56.

Loupec, T., Frasca, D., Rousseau, N., Faure, J. P., Mimoz, O., & Debaene, B. (2016). Appropriate dosing of sugammadex to reverse deep rocuronium-induced neuromuscular blockade in morbidly obese patients (Randomized controlled trial research support, Non-U.S. Gov't). *Anaesthesia, 71*(3), 265–272. https://doi.org/10.1111/anae.13344.

Mandal, S., & Hart, N. (2012). Respiratory complications of obesity (Research support, Non-U.S. Gov't review). *Clinical Medicine (London, England), 12*(1), 75–78. https://doi.org/10.7861/clinmedicine.12-1-75.

Marhofer, P., Pilz-Lubsczyk, B., Lonnqvist, P. A., & Fleischmann, E. (2014). Ultrasound-guided peripheral regional anaesthesia: A feasibility study in obese versus normal-weight women (Comparative study evaluation study). *International Journal of Obesity, 38*(3), 451–455. https://doi.org/10.1038/ijo.2013.119.

Miller, W. L., & Borlaug, B. A. (2020). Impact of obesity on volume status in patients with ambulatory chronic heart failure. *Journal of Cardiac Failure, 26*(2), 112–117. https://doi.org/10.1016/j.cardfail.2019.09.010.

Nguyen, N. T., Ho, H. S., Fleming, N. W., Moore, P., Lee, S. J., Goldman, C. D., … Wolfe, B. M. (2002). Cardiac function during laparoscopic vs open gastric bypass. *Surgical Endoscopy, 16*(1), 78–83 (Clinical Trial Comparative Study Randomized Controlled Trial).

Nightingale, C. E., Margarson, M. P., Shearer, E., Redman, J. W., Lucas, D. N., Cousins, J. M., … Griffiths, R. (2015). Peri-operative management of the obese surgical patient 2015: Association of Anaesthetists of Great Britain and Ireland Society for Obesity and Bariatric Anaesthesia (Letter practice guideline). *Anaesthesia, 70*(7), 859–876. https://doi.org/10.1111/anae.13101.

O'Donnell, C. P., Schaub, C. D., Haines, A. S., Berkowitz, D. E., Tankersley, C. G., Schwartz, A. R., & Smith, P. L. (1999). Leptin prevents respiratory depression in obesity (Research support, U.S. Gov't, P.H.S.). *American Journal of Respiratory and Critical Care Medicine, 159*(5 Pt 1), 1477–1484. https://doi.org/10.1164/ajrccm.159.5.9809025.

Orlov, D., Ankichetty, S., Chung, F., & Brull, R. (2013). Cardiorespiratory complications of neuraxial opioids in patients with obstructive sleep apnea: A systematic review (Research support, Non-U.S. Gov't review systematic review). *Journal of Clinical Anesthesia, 25*(7), 591–599. https://doi.org/10.1016/j.jclinane.2013.02.015.

Ozturk, M. C., Demiroluk, O., Abitagaoglu, S., & Ari, D. E. (2019). The effect of sevoflurane, desflurane and propofol on respiratory mechanics and integrated pulmonary index scores in laparoscopic sleeve gastrectomy. A randomized trial. *Saudi Medical Journal, 40*(12), 1235–1241. https://doi.org/10.15537/smj.2019.12.24693.

Park, W., Ramachandran, J., Weisman, P., & Jung, E. S. (2010). Obesity effect on male active joint range of motion. *Ergonomics, 53*(1), 102–108. https://doi.org/10.1080/00140130903311617.

Pelosi, P., Croci, M., Ravagnan, I., Cerisara, M., Vicardi, P., Lissoni, A., & Gattinoni, L. (1997). Respiratory system mechanics in sedated, paralyzed, morbidly obese patients. *Journal of Applied Physiology, 82*(3), 811–818. https://doi.org/10.1152/jappl.1997.82.3.811.

Portou, M. J., Yu, R., Baker, D., Xu, S., Abraham, D., & Tsui, J. (2020). Hyperglycaemia and ischaemia impair wound healing via toll-like receptor 4 pathway activation in vitro and in an experimental murine model. *European Journal of Vascular and Endovascular Surgery, 59*(1), 117–127. https://doi.org/10.1016/j.ejvs.2019.06.018.

Ravi, K. K., Kaul, T. K., Kathuria, S., Gupta, S., & Khurana, S. (2011). Distance from skin to epidural space: Correlation with body mass index (BMI). *Journal of Anaesthesiology Clinical Pharmacology, 27*(1), 39–42.

Riley, R. W., Powell, N. B., Guilleminault, C., Pelayo, R., Troell, R. J., & Li, K. K. (1997). Obstructive sleep apnea surgery: Risk management and complications. *Otolaryngology and Head and Neck Surgery, 117*(6), 648–652. https://doi.org/10.1016/s0194-5998(97)70047-0.

Saasouh, W., Laffey, K., Turan, A., Avitsian, R., Zura, A., You, J., … Ruetzler, K. (2018). Degree of obesity is not associated with more than one intubation attempt: A large centre experience. *British Journal of Anaesthesia, 120*(5), 1110–1116. https://doi.org/10.1016/j.bja.2018.01.019.

Sahin, M., Bilgen, C., Tasbakan, M. S., Midilli, R., & Basoglu, O. K. (2014). A clinical prediction formula for apnea-hypopnea index. *International Journal of Otolaryngology, 2014*, 438376. https://doi.org/10.1155/2014/438376.

Sawyer, R. J., Richmond, M. N., Hickey, J. D., & Jarrratt, J. A. (2000). Peripheral nerve injuries associated with anaesthesia (Review). *Anaesthesia, 55*(10), 980–991. https://doi.org/10.1046/j.1365-2044.2000.01614.x.

Schroeder, K., Andrei, A. C., Furlong, M. J., Donnelly, M. J., Han, S., & Becker, A. M. (2012). The perioperative effect of increased body mass index on peripheral nerve blockade: An analysis of 528 ultrasound guided interscalene blocks. *Revista Brasileira de Anestesiologia, 62*(1), 28–38. https://doi.org/10.1016/S0034-7094(12)70100-9.

Semler, M. W., Janz, D. R., Russell, D. W., Casey, J. D., Lentz, R. J., Zouk, A. N., … Rice, T. W. (2017). A multicenter, randomized trial of ramped position vs sniffing position during endotracheal intubation of critically ill adults (Multicenter study randomized controlled trial research support, N.I.H., Extramural). *Chest, 152*(4), 712–722. https://doi.org/10.1016/j.chest.2017.03.061.

Singh, P. M., Borle, A., McGavin, J., Trikha, A., & Sinha, A. (2017). Comparison of the recovery profile between Desflurane and sevoflurane in patients undergoing bariatric surgery—A meta-analysis of randomized controlled trials (Comparative study meta-analysis review). *Obesity Surgery, 27*(11), 3031–3039. https://doi.org/10.1007/s11695-017-2929-6.

Siyam, M. A., & Benhamou, D. (2002). Difficult endotracheal intubation in patients with sleep apnea syndrome. *Anesthesia and Analgesia, 95*(4), 1098–1102. https://doi.org/10.1097/00000539-200210000-00058.

Sun, Z., Sessler, D. I., Dalton, J. E., Devereaux, P. J., Shahinyan, A., Naylor, A. J., … Kurz, A. (2015). Postoperative hypoxemia is common and persistent: A prospective blinded observational study (Observational study research support, Non-U.S. Gov't). *Anesthesia and Analgesia, 121*(3), 709–715. https://doi.org/10.1213/ANE.0000000000000836.

Sunahara, K. K., Sannomiya, P., & Martins, J. O. (2012). Briefs on insulin and innate immune response (Research support, Non-U.S. Gov't). *Cellular Physiology and Biochemistry, 29*(1–2), 1–8. https://doi.org/10.1159/000337579.

Suzuki, T., Masaki, G., & Ogawa, S. (2006). Neostigmine-induced reversal of vecuronium in normal weight, overweight and obese female patients (Research support, Non-U.S. Gov't). *British Journal of Anaesthesia, 97*(2), 160–163. https://doi.org/10.1093/bja/ael142.

Teegarden, D. (2003). Calcium intake and reduction in weight or fat mass (Review). *The Journal of Nutrition, 133*(1), 249S–251S. https://doi.org/10.1093/jn/133.1.249S.

Terla, V., Rajbhandari, G. L., Kurian, D., & Pesola, G. R. (2019). A case of right ventricular dysfunction with right ventricular failure secondary to obesity hypoventilation syndrome (Case reports). *American Journal of Case Reports, 20,* 1487–1491. https://doi.org/10.12659/AJCR.918395.

Thorell, A., MacCormick, A. D., Awad, S., Reynolds, N., Roulin, D., Demartines, N., ... Lobo, D. N. (2016). Guidelines for perioperative Care in Bariatric Surgery: Enhanced recovery after surgery (ERAS) society recommendations (Review). *World Journal of Surgery, 40*(9), 2065–2083. https://doi.org/10.1007/s00268-016-3492-3.

Tsui, B. C., Murtha, L., & Lemmens, H. J. (2017). Practical dosing of propofol in morbidly obese patients (Editorial comment). *Canadian Journal of Anaesthesia, 64*(5), 449–455. https://doi.org/10.1007/s12630-017-0853-9.

Veiga de Sa, A., Cavaleiro, C., & Campos, M. (2020). Haemodynamic and analgesic control in a perioperative opioid-free approach to bariatric surgery— A case report (Case reports). *Indian Journal of Anaesthesia, 64*(2), 141–144. https://doi.org/10.4103/ija.IJA_620_19.

Venclauskas, L., Maleckas, A., & Arcelus, J. I. (2018). European guidelines on perioperative venous thromboembolism prophylaxis: Surgery in the obese patient (Practice guideline systematic review). *European Journal of Anaesthesiology, 35*(2), 147–153. https://doi.org/10.1097/EJA.0000000000000703.

Warner, M. A., Warner, M. E., & Martin, J. T. (1994). Ulnar neuropathy. Incidence, outcome, and risk factors in sedated or anesthetized patients (Research support, Non-U.S. Gov't). *Anesthesiology, 81*(6), 1332–1340.

Wong, D. T., Dallaire, A., Singh, K. P., Madhusudan, P., Jackson, T., Singh, M., ... Chung, F. (2019). High-flow nasal oxygen improves safe apnea time in morbidly obese patients undergoing general anesthesia: A randomized controlled trial (Randomized controlled trial research support, Non-U.S. Gov't). *Anesthesia and Analgesia, 129*(4), 1130–1136. https://doi.org/10.1213/ANE.0000000000003966.

Yegneswaran, B., & Chung, F. (2009). The importance of screening for obstructive sleep apnea before surgery (Case reports letter). *Sleep Medicine, 10*(2), 270–271. https://doi.org/10.1016/j.sleep.2008.02.009.

Yong, P. J., Thurston, J., Singh, S. S., & Allaire, C. (2019). Directive clinique N(o) 386 - Chirurgie gynecologique chez les patientes obeses. *Journal of Obstetrics and Gynaecology Canada, 41*(9), 1371–1388. e1377 https://doi.org/10.1016/j.jogc.2019.04.006.

Chapter 5

Anesthetics and analgesic activities of herbal medicine: Review of the possible mechanism of action

U.G. Chandrika[a] and Ureshani Karunarathna[b]

[a]*Department of Biochemistry, Faculty of Medical Sciences, University of Sri Jayewardenepura, Nugeggoda, Sri Lanka,* [b]*Department of Pharmacy and Pharmaceutical Sciences, Faculty of Allied Health Sciences, University of Sri Jayewardenepura, Nugeggoda, Sri Lanka*

Abbreviations

GA	general anesthetics
GABA	γ-aminobutyric acid receptors
LA	local anesthetics
NMDA	*N*-methyl-d-aspartic acid

Introduction

Pain is a sensory experience that is frequently but not always associated with nerve or tissue damage (Muir III, 2009). Thus, analgesic drugs act on the peripheral and central nervous system in various ways to minimize the pain (Carlini, 2003; Tsuchiya, 2015). The word analgesic is derived from Greek words: an (without) and algia (pain) (Vaishnavi, Rao, Venkatesh, Hepcykalarani, & Prema, 2020). It merely means the absence of pain without losing consciousness. Thus, the analgesia system is mediated by the periaqueductal gray matter (in the midbrain), the nucleus raphe magnus (in the medulla), and the pain-inhibitory neurons within the dorsal horns of the spinal cord (Kumar, Shete, & Akbar, 2010).

Besides modern medicine, a wide variety of plants have been found to be used as analgesic or anesthetic drugs. Use of herbal remedies as anesthetics and analgesic drugs is dated back to 40–90 CE. However, in the last few decades, global interest in herbal or alternative medicine has increased (Cheng, Hung, & Chiu, 2002). Almeida, Navarro, and Barbosa-Filho (2001) have listed 166 plants belonging to 79 families with an analgesic activity which act on the central nervous system. Depending on the mechanism of action, these anesthetics and analgesic herbs can be categorized as general anesthesia and local anesthesia (Vaishnavi et al., 2020).

This chapter is focused on a few widely used herbs which show anesthetics and analgesic activities under two categories of local and general anesthetics herbs.

Local anesthetics herbs

Mechanisms of local anesthetics

Although local anesthetics (LA) are generally a diverse group, they usually contain an aromatic ring and an ionizable amino group. The charge on the LA molecule may vary depending on the pH and/or the local environment near to it that may be responsible for the effect of the anesthetic of these LA (Kopeć, Telenius, & Khandelia, 2013).

LA will block the transmission of nerve impulses from a specific part of the body (Vaishnavi et al., 2020). LA interact with voltage-gated Na^+ channel to inhibit sensory and motor function. Anesthetic molecules bind to intracellular or cell-interior-binding sites on Na+ channels embedded in membranes after they penetrated the lipid barriers of nerve sheaths and diffused across the lipid bilayers of cell membranes. Also, LA can act on membrane-constituting lipids and modify the

Features and Assessments of Pain, Anesthesia, and Analgesia. https://doi.org/10.1016/B978-0-12-818988-7.00003-0
Copyright © 2022 Elsevier Inc. All rights reserved.

48 PART | I Setting the scene: General aspects of anesthesia, analgesics and pain

physicochemical properties of neuronal and cardiomyocyte membranes. It directly depresses the functions of neuronal membranes, while it indirectly inhibits the activity of Na^+ channels by adjusting the physicochemical properties such as fluidity, order, microviscosity, or elasticity of lipid membranes and surrounding channel proteins. Just like synthetic LA, phytochemicals are also expected to affect the activity of voltage-gated Na^+ channels through the common mechanisms (Catterall & Mackie, 2011; Tsuchiya, 2015). The potency to produce anesthesia of anesthetics is closely linked with its ability to penetrate membrane lipids. Thus, there is a correlation between anesthetic activity and the lipid/water partition coefficient (Tsuchiya, 2001).

This chapter is focused on providing an overview of selected medicinal herbs used in local anesthetics namely *Erythroxylum coca*, *Syzygium aromaticum*, *Cinchona officinalis*, and *Spilanthes acmella* and on exploring their uses, active metabolites, and evidence for the mechanism of activity.

Erythroxylum coca

Family *Erythroxylaceae*

Common name—Cocaine The earliest use and cultivation of coca date back to several thousand years BC. In 1860, Albert Niemann isolated an active substance from *Erythroxylum coca* and named it as "cocaine" (Tsuchiya, 2017). Cocaine is abundant in the leaves of the coca shrub (Catterall & Mackie, 2011). It is the first local anesthetics that originates from a specific plant alkaloid (Vaishnavi et al., 2020).

In 1898, Richard Willstätter identified the structure of cocaine as methyl (1R,2R,3S,5S)-3-(benzoyloxy)-8-methyl-8-azabicyclo [3.2.1] octane-2-carboxylate (Fig. 1) (Tsuchiya, 2017).

Cocaine is a highly lipophilic compound that has local anesthetics as well as sympathomimetic properties (Albuquerque & Kurth, 1993). Cocaine reversibly blocks the action potential conduction of neuron, or it prevents the generation and conduction of the nerve impulse. The primary site of action of cocaine is the cell membrane (Vaishnavi et al., 2020). Thus, cocaine blocks the sodium channel of the axonal membrane, and it prevents the increase in the permeability to sodium ions during the initial depolarization of the axon membrane (Brain & Coward, 1989). Albuquerque and Kurth (1993) have studied the effect of cocaine on cerebral arterioles in newborn pigs. The data obtained by them implicate the Na^+ or K^+ blocking properties of cocaine as a mechanism for the contractile response of the cerebral arterioles.

Preliminary, in addition to the blockage of the action potential conduction off the neuron, cocaine induces local vasoconstriction. Secondary, it inhibits the reuptake of the local neurotransmitter norepinephrine in the brain (Catterall & Mackie, 2011).

Syzygium aromaticum

Family Myrtaceae

Common name—Clove Cloves are the aromatic flower buds of a tree *Syzygium aromaticum*. Clove has been traditionally used as a carminative, antiemetic, toothache remedy, and counterirritant (Ebadi, 2006). Clove oil is the best-known herbal product that has been used to obtain transient relief from toothache (Catterall & Mackie, 2011). It was known to the Chinese as early as 266 BCE (Spinella, 2001).

Clove buds contain eugenol, eugenyl acetate, tannins, oleanolic acid, vanillin, chromene-eugenin, β-caryophyllene, methyl-n-amyl ketone, and esters (Vaishnavi et al., 2020). The main constituent of clove oil is eugenol [2-methoxy-4-(2-propenyl) phenol] (Fig. 2) (Guenette, Beaudry, Marier, & Vachon, 2006).

The anesthetic action of clove is attributable to eugenol and its natural analogs. Eugenol is a phenolic compound. Eugenol has not explicitly been shown to block Na^+ channels, but its phenolic structure suggests that it does (Spinella, 2001).

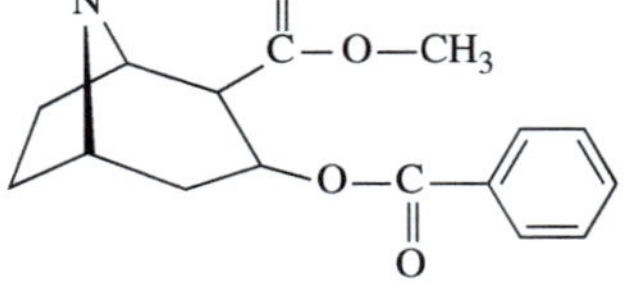

FIG. 1 Structure of cocaine.

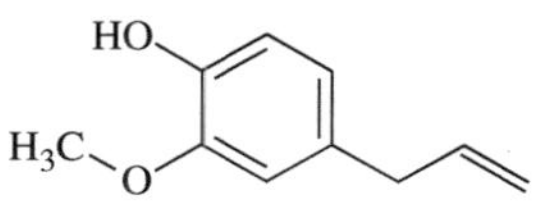

FIG. 2 Structure of eugenol.

It is believed that the mechanism of inhibition of nerve and muscle activity by eugenol is associated to its membrane-stabilizing potential. Brodin and Røed (1984) observed that eugenol is worked as a local anesthetic drug at low concentrations through reversible compound-action potential inhibition, increasing the threshold of the nerve without affecting the resting membrane potential of the muscle. Moreover, eugenol shows both presynaptic and postsynaptic effects on end-plate potentials and miniature end-plate potentials at the neuromuscular junction. Further, the analgesic effect of eugenol may also be caused by d-tubocurarine-like effects that block the neuromuscular transmitter substance, acetylcholine. This acetylcholine-involved pain mechanism is suggested by Avery and Rapp (1958).

The clove oil's effect is inhibition of prostaglandin synthesis through both cyclooxygenase and lipoxygenase pathways (Spinella, 2001). Ohkubo and Kitamura (1997) have found that low millimolar concentrations of eugenol activate a Ca^{2+} channel through capsaicin receptor or independent mechanism.

According to Yang et al. (2003), analgesic properties of eugenol would be associated with its antagonistic activity on the vanilloid receptor and transient receptor potential vanilloid. In addition to that, Wie et al. (1997) and Aoshima and Hamamoto (1999) have reported that the anesthetics activities of eugenol may be mediated by GABAergic and glutamatergic (NMDA) mechanisms.

Cinchona officinalis
Family Rubiaceae

Cinchona species are native to the tropical forest of western South America, while a few species are reported in Central America and Jamaica.

The dried bark of *Cinchona* species contains compounds such as quinine, quinidine, cinchonine, cinchonidine, quinic acid, keno-tannic acid, qinovin, and kinova-tannic acid (Vaishnavi et al., 2020).

The major chemical constituents of *Cinchona* species are quinine and quinidine, and cinchonine and cinchonidine. The aqueous extract of *C. officinalis* has significant local anesthetics and antipyretic activities in guinea pigs (Li & Tian, 2016). Quinine is a cinchona alkaloid that belongs to the aryl amino alcohol group of drugs, as shown in Fig. 3 (Achan et al., 2011).

Quinine blocks neuromuscular transmitter substance through inhibiting the phosphodiesterase activity in the skeletal muscle cytosol. Quinidine that is a close relative to quinine has also been reported to interact with both depolarizing and nondepolarizing muscle relaxants. It produces prolonged neuromuscular blockade (SaaChai & Lin, 2013). Quinidine blocks myocardial Na^+ channels (Debnath et al., 2018).

Hiatt (1949) has demonstrated the depressor effect of quinine on the circulation that is caused primarily by peripheral vasodilatation.

Spilanthes acmella
Family Asteraceae

These herbs originated in Tropical Africa and South America and are now widely distributed in the tropics and subtropics, including Tropical America, North Australia, Africa, India, and Sri Lanka (Prachayasittikul, Prachayasittikul, Ruchirawat, & Prachayasittikul, 2013).

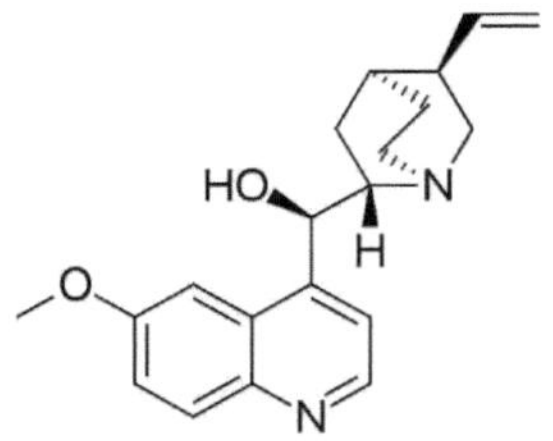

FIG. 3 Structure of quinine.

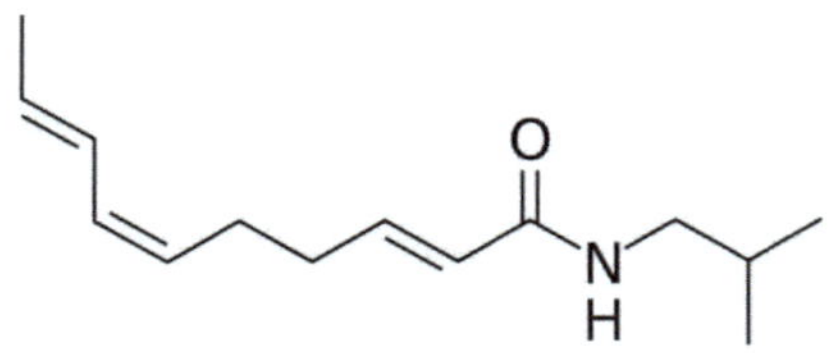

FIG. 4 Structure of spilanthol.

Spilanthes species contains spilanthol, isobutyl amide, choline, tannins, and resin. Spilanthol or affinin ((2E,6Z,8E)-N-isobutyl-2,6,8decatrienamide (molecular formula $C_{14}H_{23}NO$) (Fig. 4) is the most abundant isolate of the plant species (Barbosa, de Carvalho, Smith, & Sabaa-Srur, 2016). It is an alkamide that contains a different number of unsaturated hydrocarbons. It was first discovered in 1945 from *S. acmella* (Prachayasittikul et al., 2013).

The flower heads of *Spilanthes* produced an unusual tingling sensation in the mouth, which increases the saliva flow with an anesthetic effect. These effects are due to spilanthol (Vaishnavi et al., 2020).

Plants which contain spilanthol are also called as toothache plants because of their analgesic activity. The spilanthol of *S. acmella* extracts was reported to display local anesthetic activity through the blockage of Na^+ channels (Dubey, Maity, Singh, Saraf, & Saha, 2013; Prachayasittikul et al., 2013). *S. acmella* aerial aqueous extract exhibited significant local anesthetic activity in animal models through intracutaneous wheal in guinea pigs and plexus anesthesia in frogs. This activity could be due to the presence of alkamides (Chakraborty, Devi, Sanjebam, Khumbong, & Thokchom, 2010).

General anesthetics herbs

General anesthesia (GA) suppresses central nervous system activity, resulting in unconsciousness (Kopeć et al., 2013). The target sites of general anesthetics are believed to be nerve membrane proteins (Tsuchiya, 2001). The mechanism of action of GAs involves membranes and protein receptors. Active sites of GAs bind to the inhibitory γ-aminobutyric acid type A receptors (GABA$_A$) and *N*-methyl-D-aspartic acid (NMDA) receptors (Chau, 2010; Tsuchiya, 2017).

GABA is the primary inhibitory neurotransmitter (20%–50%) in the central nervous system, and it is released from GABAergic neurons and operates via ionotropic GABA$_A$ and GABA$_C$ receptors and metabotropic GABA$_B$ receptors (Tsang & Xue, 2004).

Thus, currently used GA and GA-related drugs (sedatives, anxiolytics, or anesthetic adjuncts) are considered to target GABA$_A$ and NMDA receptors (Chau, 2010). The ligand gate formed by binding inhibitory neurotransmitter of GABA to GABA$_A$ receptors allowed the influx of Cl^- into postsynaptic neurons. It inhibits the neuronal excitability through synaptic phase currents and extrasynaptic tonic currents (Olsen & Li, 2011).

NMDA receptors that consist of NR1 subunit combined with NR2 subunits are activated by glutamate and glycine. The activated NMDA receptor opens the ion channels for nonselective to positively charged cations, resulting in neuronal excitation. Activation of NMDA subtypes of glutamate receptors also caused cellular damages by neuronal ischemia (Tsuchiya, 2017).

In addition to receptor proteins, GAs act on membrane-constituting lipids and modify the membranous organization, dynamics, and physicochemical properties. It will change the conformation of membrane-embedded proteins which resulted in a direct disturbance of the function of neuronal membranes and indirect modulation of the receptor activity (Patel, Patel, & Roth, 2011; Tsuchiya, 2015, 2017).

Here, we focus on *Passiflora incarnata, Valerian officinalis,* and Genus *Panax* (Gingseng) as general anesthetic herbs. Thus, here we briefly explain their uses, active metabolites, and evidence for the mechanism of activity.

Passiflora incarnata

Family—Passifloraceae

Common name—Passion flower

Passiflora incarnata has long been used for its sedative action, and it has been used worldwide for anxiety, neuralgia, and insomnia (Speroni & Minghetti, 1988).

Flavonoids, carboline alkaloids, cyanogenic glycosides, and terpene derivatives have been identified from *P. incarnata* species (Speroni & Minghetti, 1988). Moreover, *P. incarnata* contains C-glycosyl flavones such as vitexin, isovitexin, schaftoside, isoschaftoside and isovitexin-2-O-glucoside phenols, glycosylflavonoids, and cyanogenic compounds

(Wohlmuth, Penman, Pearson, & Lehmann, 2010). Although the active compounds responsible for the therapeutic properties of the *P. incarnata* have not been identified (Appel et al., 2011; Grundmann, Wang, McGregor, & Butterweck, 2008; Speroni & Minghetti, 1988; Wohlmuth et al., 2010), alkaloid components are assumed to be responsible for its beneficial health effects through $GABA_A$ receptor modulation. These alkaloids are very likely to belong to harmala alkaloids that were found in *Peganum harmala* (Nitrariaceae) (Tsuchiya, 2017). Dhawan, Kumar, and Sharma (2002) have reported that the bioactivity of *P. incarnata* might be responsible for its phytoconstituent, which has benzoflavone nucleus.

Many pharmacological effects of *P. incarnata* are mediated through modulation of the GABA system that includes affinity to $GABA_A$ and $GABA_B$ receptors, and effects on GABA uptake. The contribution of GABAergic system to the analgesic effects of *P. incarnata* is comparable to that of diazepam (Appel et al., 2011). In addition to that, Lolli et al. (2007) reported the involvement of $GABA_A$ system in the anxiolytic activity of *Passiflora actinia* using the elevated plus-maze test in mice.

Valerian officinalis

Family Valerianaceae

V. officinalis been used medicinally for 2000 years. It was first used in a treatment for brain disorder in the late 16th century, and it is found in North America, Europe, and Asia (Nandhini, Narayanan, & Ilango, 2018). The sedative activity of Valerian roots has been known from the time of ancient Greece and Rome (Caron & Riedlinger, 1999).

V. officinalis contains alkaloids, such as valerine, chatinine, and valeriana alkaloid XI, and volatile oils such as bornyl acetate, β-caryophyllene, valeranone, valerenal, bornyl isovalerate, and valerenic acid. But the exact chemical compounds responsible for the activities have not been identified. Among them, volatile oils are thought to be the main active ingredient (Jadhav, Thorat, Kadam, & Sathe, 2009; Murti, Kaushik, Sangwan, & Kaushik, 2011). The sesquiterpene valerenic acid (Fig. 5) was recently identified as a $GABA_A$ receptor modulator (Khom et al., 2007; Neuhaus et al., 2008).

Valerian inhibits the reuptake of GABA in the presynaptic vesicles and enhances the release of GABA. The constituents in the essential oil of valerian attributed to its effect in the central nervous system; thus, it is thought to be contributing to sedative effects also (Cowlishaw, 2007; Santos, Ferreira, Cunha, Carvalho, & Macedo, 1994). However, the transport system that crosses the blood-brain barrier, valerenic acid, and its derivatives (acetoxyvalerenic acid and hydroxyvalerenic) are still unknown (Neuhaus et al., 2008). Moreover, isolated valerenic acid failed to show the documented sedative effects, though once valeric acid is considered to be responsible for the sedative effects of *V. officinalis* (Murti et al., 2011).

Genus *Panax*

Family Araliaceae

Common name—Ginseng

Panax species have been considered as a valuable herbal remedy for a long period of human history. *Panax ginseng*, *Panax quinquefolius* (Xiyangshen, American ginseng), *Panax notoginseng* (Sanqi), and *Panax japonicus* (Japanese ginseng) belong to the Genus *Panax* (Leung & Wong, 2010). Ginseng is consumed as a raw herb, a powder, or made into ginseng tea. It is believed to be "heat raising" for the hematological and circulatory systems. It can stimulate the body, increasing the ability to cope with fatigue and physical stress (Cheng et al., 2002).

At the end of 2012, phytochemical investigations on 11 different *Panax* species have resulted in the isolation of 289 pure saponins. These saponins are known as steroid receptors to modulate the central nervous system (Yang, Hu, Wu, Ye, & Guo, 2014). The saponin constituents of *Panax* that are known as ginsenosides are the main bioactive ingredients (Fig. 6). Ginsenosides are the oligosaccharide glycosides of either dammarane or oleanane type triterpenoids (Leung & Wong, 2010).

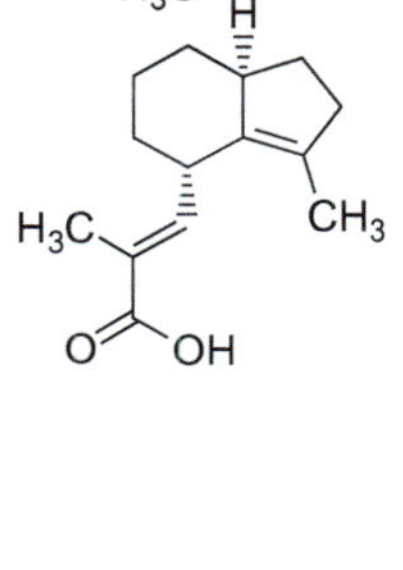

FIG. 5 Structure of valerenic acid.

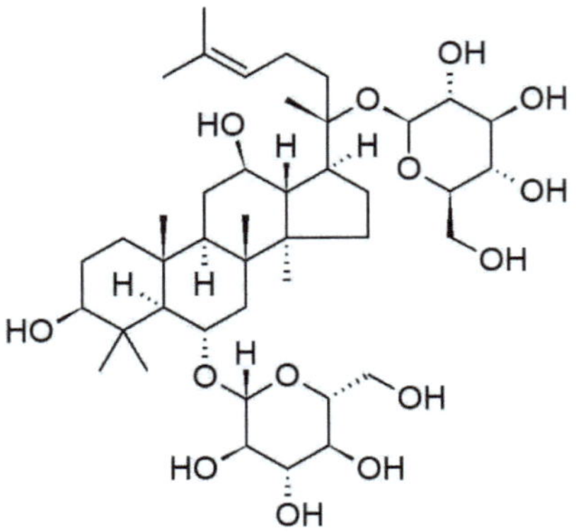

52 PART | I Setting the scene: General aspects of anesthesia, analgesics and pain

FIG. 6 Structure of ginsenosides.

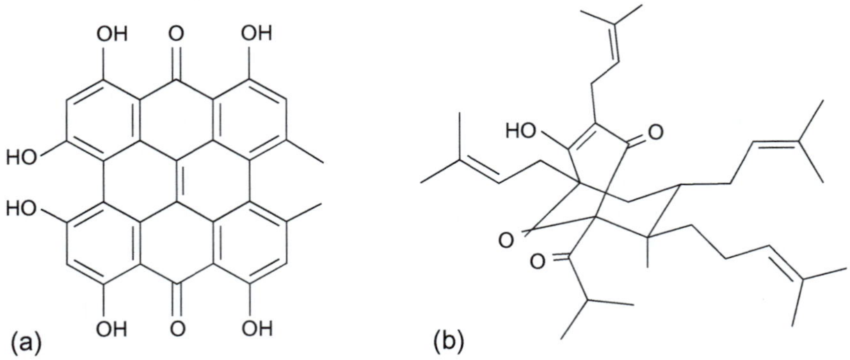

FIG. 7 Structure of (A) hyperforin and (B) hypericin.

Ginsenosides are the active constituents found in ginseng. The ginsenoside Rb-1 has been shown to block the inactive state of sodium channels and prevent the abnormal influx of sodium during ischemic episodes. It also found to be useful in regulating GABAergic transmission in the brainstem (Ebadi, 2006).

P. ginseng was found as $GABA_A$ receptor potentiating tincture (Hoffmann et al., 2016). Moreover, the study done on human recombinant $GABA_A$ receptor channel activity expressed in Xenopus oocytes using a two-electrode voltage lamp technique suggests that ginsenosides regulated $GABA_A$ receptor expressed in Xenopus oocyte, which might be the regulated pharmacological actions of *P. ginseng* (Choi et al., 2003).

Hypericum perforatum

Family Hypericaceae

Common name—St. John's wort (hypericum, millepertuis)

H. perforatum is native to Europe and Asia (Barnes, Anderson, & Phillipson, 2001), and it is a popular antidepressant drug in many European countries, while the United States and Australia regulated it as a dietary supplement (Nathan, 2001).

St. John's wort has been used for centuries as a medicinal herb for the activities such as antiinflammatory, sedative, analgesic, diuretic, antimalarial, antidepressant, and vulnerary action (Nathan, 1999).

Naphthodianthrones (hypericin and pseudohypericin), acylphloroglucinols (hyperforin and adhyperforin), flavonol glycosides (quercetin, quercitrin, isoquercitrin, rutin, hyperoside, kampferol, leteolin, and myricetin), biflavones, proanthocyanidins (procycanidin B2), and phenyl propanes (chlorogenic acid and caffeic acid) are identified as medicinally important compounds in St. John wort (Nathan, 1999). Further, he has reported that the activity related to *Hypericum* might be due to the multiple compounds present in it. However, Barnes et al. (2001) have reported hyperforin (Fig. 7A) and hypericin (Fig. 7B) as the major active constituents of *H. perforatum*.

St. John's wort can produce sedation and confusion through inhibition of synaptic reuptake of serotonin, noradrenaline, and dopamine (Cheng et al., 2002). It has also been shown to affect several synaptic mechanisms of GABAergic and glutamatergic neurotransmission, with moderate affinities for $GABA_A$, $GABA_B$, and NMDA receptors (Nathan, 1999, 2001).

Potential hazards of anesthetics herbal medicine

Although the use of herbal medicines has increased dramatically over the past few years, still, some herbs are known to cause problems, especially when large doses are taken (Abebe, 2002; Pradhan & Pradhan, 2011).

Thus, toxicity due to overdose; contamination by other medicinal plants; mistaken plants; physiological changes on bodily systems; and adverse drug interactions are some of the harmful effects that can be caused by herbal medicines. Moreover, coagulation disorders, cardiovascular side effects, water and electrolyte disturbances, endocrine effects, hepatotoxicity, and prolongation of the effects of anesthetic agents are associated with the use of anesthetic herbs (Cheng et al., 2002). For example, cannabis may cause cardiac arrhythmias, cardiac and respiratory depression, or even pulmonary edema. Thus, if it is used in conjunction with an anesthetics, it can cause severe complications (Dickerson, 1980).

Although the positive information with the easy access to herbal medicines has boosted the growth of alternative medicine, the herbal remedies are not always "gentler" and "safer" than pharmaceutical drugs (Cheng et al., 2002).

Application of anesthetics herbal medicine

About 25% of drugs listed in the pharmacopoeias of developed countries were isolated from plant origin, and another 25% are modifications of molecules first found in plants (Dippenaar, 2015). The majority of currently used anesthetic drugs are derived from or associated with plants (Tsuchiya, 2017). Debnath et al. (2018) has shown the alkaloids in discovering new drugs.

Moreover, since the phytochemicals (terpenoids, alkaloids, and flavonoids) show the mechanistic requirements to interact with receptors, channels, and membranes with the characteristic molecular structures, they become potential agents in the development of new local anesthetic, general anesthetic, antinociceptive, analgesic, or sedative drugs (Tsuchiya, 2017). Guimarães, Quintans, and Quintans-Júnior (2013) have shown the potential of development of new analgesic drugs from monoterpenes isolated from plants.

However, clinical efficacy, adverse action, pharmacokinetics, and pharmacodynamics of these phytochemicals need to be studied collectively with structural modification to improve the activity or reduce the toxicity.

Conclusion

Besides the modern system of medicine, herbal medicines are also popular as analgesic or anesthetics drugs. These herbs can either block the nerve conduction (local anesthesia) or suppress the central nervous system (general anesthesia). *Erythroxylum coca*, *Syzygium aromaticum*, *Cinchona officinalis*, and *Spilanthes acmella* are used as local anesthetic herbs, while *Passiflora incarnata*, *Valerian officinalis*, and Gingseng are used as general anesthetic herbs. Active compounds and mechanism of action of these plants are well studied.

Applications to other areas

In addition to the anaesthetics activities of *E. coca*, *Syzygium aromaticum*, *C. officinalis*, *Spilanthes acmella*, *P. incarnata*, *V. officinalis*, and Ginseng, which we have reviewed in this chapter, are also reported to exhibit different pharmacological activities. *E. coca* enhances work capacity, including the reduction of fatigue and mitigate thirst and hunger (Bhardwaj & Anuja, 2018) while clove oil from *Syzygium aromaticum* shows antiinflammatory, antibacterial activities, anticancer, antiischemic, antihistaminic, and antianaphylactic effects. Further, clove oil from inhibits thromboxane biosynthesis and to display other pharmacologic properties, such as neuroprotective (Guenette et al., 2006). The alkaloids found in *C. officinalis* were known as a remedy for malaria, and they showed antibacterial, antiarrhythmic, antipyretic, human platelet aggregation inhibition and decreasing the excitability of motor endplate. It also used to treat ophthalmia and internal hemorrhoids (Insanu et al., 2019) *Spilanthes acmella* has been used as acute or long-term treatment for microbial infections: oral pathogenic microorganisms, dental caries, periodontosis, gum disease, gum bleeding, and/or plaque reduction. It also possesses diuretic, antinociceptive, antihyperalgesic activity, vasorelaxant and antioxidant activities. In addition, spilanthol is used as a flavor for carbonated beverages and also used in the antiwrinkle products (Barbosa et al., 2016; Dias et al., 2012). *P. incarnata* has potential effects for the treatment of some diseases such as anxiety, opiates withdrawal, insomnia,

54 PART | I Setting the scene: General aspects of anesthesia, analgesics and pain

attention-deficit hyperactivity disorder and cancer (Appel et al., 2011). In addition, it is used for asthma, muscle spasms, and many other diseases related to respiratory, and cardiovascular systems in the homeopathic medicine (Dhawan et al., 2002). *V. officinalis* used for the treatment of antispasmodic, anthelmintic, diuretic, diaphoretic, and emmenagogue, rheumatism, low-grade fevers, aphrodisiac, and hysteria in addition to the varied nervous disorders (Nandhini et al., 2018). Ginseng, the root of *Panax ginseng* is a traditional medicine and tonic (Choi et al., 2003), that modulates blood pressure, metabolism and immune functions (Leung & Wong, 2010).

Mini-dictionary of terms

Herbal Medicine: Type of medicine that uses roots, stems, leaves, flowers, or seeds of plants to improve health, prevent disease, and treat illness.

Local anesthetics: The reversible loss of sensation in a defined area of the body. This loss of sensation is achieved by the topical application or injection of agents that block the sodium channels that facilitate nerve impulses in tissue.

General anesthetics: An anesthetic that affects the whole body and usually causes a loss of consciousness.

GABAergic mechanism: Pathways that originate in the brain can inhibit nociceptive neurons in the dorsal horn, including spinothalamic tract (STT) cells. Several different inhibitory neurotransmitters are used by the endogenous analgesia system. One of these is gamma-aminobutyric acid (GABA). GABA can be released either directly by the axons of brainstem neurons that descend to the spinal cord or indirectly by the excitation of GABAergic inhibitory interneurons in the spinal cord through the release of excitatory transmitters from descending axons. There are several mechanisms for GABAergic inhibitory actions. These include pre- and postsynaptic inhibition following actions of GABA on $GABA_A$ or $GABA_B$ receptors.

Antagonistic activity: Antagonism refers to the action of any organism that suppresses or interferes with the normal growth and activity of a plant pathogen, such as the main parts of bacteria or fungi. These organisms can be used for pest control and are referred to as biological control agents.

Sympathomimetic: A sympathomimetic is any drug that causes an effect similar to that produced by stimulation of the sympathetic nervous system.

Cyclooxygenase Pathway: The cyclooxygenase (COX) enzyme system is the major pathway catalyzing the conversion of arachidonic acid into prostaglandins (PGs). PGs are lipid mediators implicated in a variety of physiological and pathophysiological processes in the kidney, including renal hemodynamics, body water and sodium balance, and the inflammatory injury characteristic in multiple renal diseases.

Prostaglandins generated from arachidonate by the action of cyclooxygenase (COX) isoenzymes and their biosynthesis are blocked by nonsteroidal antiinflammatory drugs (NSAIDs), including those selective for inhibition of COX-2.

Lipoxygenase pathway: The chemical reactions and pathways by which an unsaturated fatty acid (such as arachidonic acid or linolenic acid) is converted to other compounds, and in which the first step is hydroperoxide formation catalyzed by lipoxygenase.

Key facts

- A wide variety of plants have been found to be used as analgesic or anesthetics drugs either as local or as general anesthetic drug.
- Both in vitro and in vivo experiments have been carried out to study the analgesic or anesthetic activity of herbs.
- The active compounds of many anesthetic herbs are also identified.
- There is a potential for the development of new anesthetic drugs from these well-studied anesthetics herbs.
- The activity of anesthetics phytochemicals can be improved through collective studies on their clinical efficacy, adverse action, pharmacokinetics, and pharmacodynamics.

Summary

- Anesthetics can be used as analgesics to relieve pain.
- Local anesthesia blocks the nerve conduction, while general anesthesia drugs suppress the central nervous system.
- Herbal anesthetics herbs can be categorized as local anesthetic herbs and general anesthetic herbs.
- Local anesthetic herbs interact with voltage-gated Na^+ channel, whereas general anesthesia interacts with membranes and protein receptors to inhibit sensory and motor function.
- *Erythroxylum coca*, *Syzygium aromaticum*, *Cinchona officinalis*, and *Spilanthes acmella* act as local anesthetics herbs.

- *Passiflora incarnata, Valerian officinalis,* and Gingseng act as general anesthetics herbs.
- The phytochemicals (terpenoids, alkaloids, flavonoids, and monoterpenes) isolated from plants can be used in the development of new anesthetic drugs.

References

Abebe, W. (2002). Herbal medication: Potential for adverse interactions with analgesic drugs. *Journal of Clinical Pharmacy and Therapeutics, 27*(6), 391–401.

Achan, J., Talisuna, A. O., Erhart, A., Yeka, A., Tibenderana, J. K., Baliraine, F. N., … D'Alessandro, U. (2011). Quinine, an old anti-malarial drug in a modern world: Role in the treatment of malaria. *Malaria Journal, 10*(1), 144.

Albuquerque, M. L. C., & Kurth, C. D. (1993). Cocaine constricts immature cerebral arterioles by a local anesthetic mechanism. *European Journal of Pharmacology, 249*(2), 215–220.

Almeida, R. N., Navarro, D. S., & Barbosa-Filho, J. M. (2001). Plants with central analgesic activity. *Phytomedicine, 8*(4), 310–322.

Aoshima, H., & Hamamoto, K. (1999). Potentiation of GABAA receptors expressed in Xenopus oocytes by perfume and phytoncid. *Bioscience, Biotechnology, and Biochemistry, 63*(4), 743–748.

Appel, K., Rose, T., Fiebich, B., Kammler, T., Hoffmann, C., & Weiss, G. (2011). Modulation of the γ-aminobutyric acid (GABA) system by *Passiflora incarnata* L. *Phytotherapy Research, 25*(6), 838–843.

Avery, J. K., & Rapp, R. (1958). Demonstration of cholinesterase in teeth. *Stain Technology, 33*(1), 31–37.

Barbosa, A. F., de Carvalho, M. G., Smith, R. E., & Sabaa-Srur, A. U. (2016). Spilanthol: Occurrence, extraction, chemistry and biological activities. *Revista Brasileira de Farmacognosia, 26*(1), 128–133.

Barnes, J., Anderson, L. A., & Phillipson, J. D. (2001). St John's wort (*Hypericum perforatum* L.): A review of its chemistry, pharmacology and clinical properties. *The Journal of Pharmacy and Pharmacology, 53,* 583–600.

Bhardwaj, A., & Anuja, K. (2018). *Homeopathic remedies.* In K. Misra, P. Sharma, & A. Bhardwaj (Eds.), *Management of high altitude pathophysiology* (pp. 217–229). Elsevier Inc.

Brain, P. F., & Coward, G. A. (1989). A review of the history, actions, and legitimate uses of cocaine. *Journal of Substance Abuse, 1,* 431–451.

Brodin, P., & Røed, A. (1984). Effects of eugenol on rat phrenic nerve and phrenic nerve-diaphragm preparations. *Archives of Oral Biology, 29*(8), 611–615.

Carlini, E. A. (2003). Plants and the central nervous system. *Pharmacology Biochemistry and Behavior, 75*(3), 501–512.

Caron, M. F., & Riedlinger, J. E. (1999). Valerian: A practical review for clinicians. *Nutrition in Clinical Care, 2*(4), 250–257.

Catterall, W. A., & Mackie, K. (2011). Local anesthetics. In *Goodman & Gilman's the pharmacological basis of therapeutics* (pp. 565–582). New York, NY: McGraw-Hill.

Chakraborty, A., Devi, B. R. K., Sanjebam, R., Khumbong, S., & Thokchom, I. S. (2010). Preliminary studies on local anesthetic and antipyretic activities of *Spilanthes acmella* Murr. in experimental animal models. *Indian Journal of Pharmacology, 42*(5), 277–279.

Chau, P. L. (2010). New insights into the molecular mechanisms of general anaesthetics. *British Journal of Pharmacology, 161,* 288–307.

Cheng, B., Hung, C. T., & Chiu, W. (2002). Herbal medicine and anaesthesia. *Hong Kong Medical Journal, 8*(2), 123–130.

Choi, S., Choi, S., Lee, J., Whiting, P. J., Lee, S., & Nah, S. (2003). Effects of ginsenosides on GABAA receptor channels expressed in Xenopus oocytes. *Archives of Pharmacal Research, 26*(1), 28–33.

Cowlishaw, P. (2007). A survey and review of herbal medicine and anaesthesia. *Anaesthesia Points West, 40*(1), 39–42.

Debnath, B., Singh, W. S., Das, M., Goswami, S., Singh, M. K., Maiti, D., & Manna, K. (2018). Role of plant alkaloids on human health: A review of biological activities. *Materials Today Chemistry, 9,* 56–72.

Dhawan, K., Kumar, S., & Sharma, A. (2002). Comparative anxiolytic activity profile of various preparations of *Passiflora incarnata* linneaus: A comment on medicinal plants' standardization. *The Journal of Alternative & Complementary Medicine, 8*(3), 283–291.

Dias, A. M. A., Santos, P., Seabra, I. J., Júnior, R. N. C., Braga, M. E. M., & De Sousa, H. C. (2012). Spilanthol from *Spilanthes acmella* flowers, leaves and stems obtained by selective supercritical carbon dioxide extraction. *The Journal of Supercritical Fluids, 61,* 62–70. https://doi.org/10.1016/j.supflu.2011.09.020.

Dickerson, S. J. (1980). Cannabis and its effect on anesthesia. *AANA Journal, 48*(526), e528.

Dippenaar, J. M. (2015). Herbal and alternative medicine: The impact on anesthesia. *Southern African Journal of Anaesthesia and Analgesia, 21*(1), 15–20.

Dubey, S., Maity, S., Singh, M., Saraf, S. A., & Saha, S. (2013). Phytochemistry, pharmacology and toxicology of Spilanthes acmella: A review. *Advances in Pharmacological Sciences, 2013.* https://doi.org/10.1155/2013/423750, 423750.

Ebadi, M. S. (2006). In M. Ebadi (Ed.), *Pharmacodynamic basis of herbal medicine* (p. 90). CRC Press, Taylor & Francis Group: CRC Press.

Grundmann, O., Wang, J., McGregor, G. P., & Butterweck, V. (2008). Anxiolytic activity of a phytochemically characterized *Passiflora incarnata* extract is mediated via the GABAergic system. *Planta Medica, 74,* 1769–1773.

Guenette, S. A., Beaudry, F., Marier, J. F., & Vachon, P. (2006). Pharmacokinetics and anesthetic activity of eugenol in male Sprague–Dawley rats. *Journal of Veterinary Pharmacology and Therapeutics, 29*(4), 265–270.

Guimarães, A. G., Quintans, J. S., & Quintans-Júnior, L. J. (2013). Monoterpenes with analgesic activity—A systematic review. *Phytotherapy Research, 27*(1), 1–15.

Hiatt, E. P. (1949). Sympatholytic effects of quinine and quinidine. *American Journal of Physiology, 160*(1), 212–216.

Hoffmann, K. M., Herbrechter, R., Ziemba, P. M., Lepke, P., Beltrán, L., Hatt, H., ... Gisselmann, G. (2016). Kampo medicine: Evaluation of the pharmacological activity of 121 herbal drugs on GABAA and 5-HT3A receptors. *Frontiers in Pharmacology, 7*, 219.

Insanu, M., Aziz, S., Fidrianny, I., Hartati, R., Elfahmi, S., & Wirasutisna, K. R. (2019). Natural anthraquinone from the bark of *Cinchona officinalis* L. *Rasayan Journal of Chemistry, 12*(2), 519–522. https://doi.org/10.31788/RJC.2019.1221831.

Jadhav, V. M., Thorat, R. M., Kadam, V. J., & Sathe, N. S. (2009). Herbal anesthetics. *Journal of Pharmacy Research, 2*(8), 1242–1244.

Khom, S., Baburin, I., Timin, E., Hohaus, A., Trauner, G., Kopp, B., & Hering, S. (2007). Valerenic acid potentiates and inhibits GABAA receptors: Molecular mechanism and subunit specificity. *Neuropharmacology, 53*(1), 178–187.

Kopeć, W., Telenius, J., & Khandelia, H. (2013). Molecular dynamics simulations of the interactions of medicinal plant extracts and drugs with lipid bilayer membranes. *The FEBS Journal, 280*(12), 2785–2805.

Kumar, M., Shete, A., & Akbar, Z. (2010). A review on analgesic: From natural sources. A review on analgesic: From natural sources. *International Journal of Pharmaceutical & Biological Archives, 1*(2), 95–100.

Leung, K. W., & Wong, A. S. T. (2010). Pharmacology of ginsenosides: A literature review. *Chinese Medicine, 5*(1), 20.

Li, Y., & Tian, J. (2016). Evaluation of local anesthetic and antipyretic activities of cinchona alkaloids in some animal models. *Tropical Journal of Pharmaceutical Research, 15*(8), 1663–1666.

Lolli, L. F., Sato, C. M., Romanini, C. V., Villas-Boas, L. D. B., Santos, C. A. M., & de Oliveira, R. M. (2007). Possible involvement of GABAA-benzodiazepine receptor in the anxiolytic-like effect induced by Passiflora actinia extracts in mice. *Journal of Ethnopharmacology, 111*(2), 308–314.

Muir, W. W., III. (2009). Physiology and pathoPhysiology of pain. In J. S. Gaynor, & W. W. Muir (Eds.), *Handbook of veterinary pain management-E-book* (p. 13). Elsevier Health Sciences.

Murti, K., Kaushik, M., Sangwan, Y., & Kaushik, A. (2011). Pharmacological properties of *Valeriana officinalis*—A review. *Pharmacology, 3*, 641–646.

Nandhini, S., Narayanan, K. B., & Ilango, K. (2018). Valeriana officinalis: A review of its traditional uses, phytochemistry and pharmacology. *Asian Journal of Pharmaceutical and Clinical Research, 11*(1), 36–41.

Nathan, P. J. (1999). The experimental and clinical pharmacology of St John's Wort (*Hypericum perforatum* L.). *Molecular Psychiatry, 4*(4), 333–338.

Nathan, P. J. (2001). Hypericum perforatum (St John's Wort): A non-selective reuptake inhibitor? A review of the recent advances in its pharmacology. *Journal of Psychopharmacology, 15*(1), 47–54.

Neuhaus, W., Trauner, G., Gruber, D., Oelzant, S., Klepal, W., Kopp, B., & Noe, C. R. (2008). Transport of a GABAA receptor modulator and its derivatives from *Valeriana officinalis* L. s. l. across an in vitro cell culture model of the blood-brain barrier. *Planta Medica, 74*(11), 1338–1344.

Ohkubo, T., & Kitamura, K. (1997). Eugenol activates Ca^{2+}-permeable currents in rat dorsal root ganglion cells. *Journal of Dental Research, 76*(11), 1737–1744.

Olsen, R. W., & Li, G. D. (2011). GABA A receptors as molecular targets of general anesthetics: Identification of binding sites provides clues to allosteric modulation. *Canadian Journal of Anesthesia, 58*(2), 206–215.

Patel, P. M., Patel, H. H., & Roth, D. M. (2011). General anesthetics and therapeutic gases. In L. L. Brunton, A. B. Chabner, & B. C. Knollmann (Eds.), *Goodman & Gilman's pharmacological basis of therapeutics* (12th ed., pp. 527–564). New York, NY, USA: McGraw-Hill.

Prachayasittikul, V., Prachayasittikul, S., Ruchirawat, S., & Prachayasittikul, V. (2013). High therapeutic potential of *Spilanthes acmella*: A review. *EXCLI Journal, 12*, 291.

Pradhan, S. L., & Pradhan, P. S. (2011). Ayurvedic medicine and anaesthesia. *Indian Journal of Anaesthesia, 55*(4), 334.

SaaChai, T., & Lin, J. (2013). Anesthetic aspect of malaria disease: A brief review. *Middle East Journal of Anesthesiology, 21*(4), 457–462.

Santos, M. S., Ferreira, F., Cunha, A. P., Carvalho, A. P., & Macedo, T. (1994). An aqueous extract of valerian influences the transport of GABA in synaptosomes. *Planta Medica, 60*(03), 278–279.

Speroni, E., & Minghetti, A. (1988). Neuropharmacological activity of extracts from Passiflora incarnata. *Planta Medica, 54*(06), 488–491.

Spinella, M. (2001). *The psychopharmacology of herbal medicine: Plant drugs that alter mind, brain, and behavior* (pp. 325–328). MIT Press.

Tsang, S. Y., & Xue, H. (2004). Development of effective therapeutics targeting the GABAA receptor: Naturally occurring alternatives. *Current Pharmaceutical Design, 10*(9), 1035–1044.

Tsuchiya, H. (2001). Structure-specific membrane-fluidizing effect of propofol. *Clinical and Experimental Pharmacology and Physiology, 28*(4), 292–299.

Tsuchiya, H. (2015). Membrane interactions of phytochemicals as their molecular mechanism applicable to the discovery of drug leads from plants. *Molecules, 20*(10), 18923–18966.

Tsuchiya, H. (2017). Anesthetic agents of plant origin: A review of phytochemicals with anesthetic activity. *Molecules, 22*(8), 1369.

Vaishnavi, G., Rao, M. D. S., Venkatesh, P., Hepcykalarani, D., & Prema, R. (2020). A review on anesthetic herbs. *Research Journal of Pharmacognosy and Phytochemistry, 12*(1), 52–56.

Wie, M. B., Won, M. H., Lee, K. H., Shin, J. H., Lee, J. C., Suh, H. W., ... Kim, Y. H. (1997). Eugenol protects neuronal cells from excitotoxic and oxidative injury in primary cortical cultures. *Neuroscience Letters, 225*(2), 93–96.

Wohlmuth, H., Penman, K. G., Pearson, T., & Lehmann, R. P. (2010). Pharmacognosy and chemotypes of passionflower (*Passiflora incarnata* L.). *Biological and Pharmaceutical Bulletin, 33*(6), 1015–1018.

Yang, B. H., Piao, Z. G., Kim, Y. B., Lee, C. H., Lee, J. K., Park, ... Oh, S. B. (2003). Activation of vanilloid receptor 1 (VR1) by eugenol. *Journal of Dental Research, 82*(10), 781–785.

Yang, W. Z., Hu, Y., Wu, W. Y., Ye, M., & Guo, D. A. (2014). Saponins in the genus Panax L.(Araliaceae): A systematic review of their chemical diversity. *Phytochemistry, 106*, 7–24.

Chapter 6

Analgesia-first sedation and nonopioid multimodal analgesia in the intensive care unit

John W. Devlin[a,c] and Paul M. Szumita[b]

[a]*Department of Pharmacy and Health Systems Sciences, Bouve College of Health Sciences, Northeastern University, Boston, MA, United States,*
[b]*Department of Pharmacy, Brigham and Women's Hospital, Boston, MA, United States,* [c]*Division of Pulmonary and Critical Care Medicine, Brigham and Women's Hospital, Boston, MA, United States*

Abbreviations

ABCDEF	A for Assessment, Prevention, and Manage pain B for Both Spontaneous Awakening Trials and Spontaneous Breathing Trials C for Choice of Analgesia and Sedation D for Delirium Assess, Prevent, and Manage E for Early Mobility and Exercise and F for Family Engagement and Empowerment
BPS	behavioral pain scale
CI	confidence interval
CPOT	critical care pain observation tool
ECMO	extracorporeal membrane oxygenation
ERAS	enhanced recovery after surgery
ICU	intensive care unit
IV	intravenous
NI	nonintubated
NRS	numeric rating scale
NSAID	nonsteroidal antiinflammatory drug
ORADE	opioid-related adverse event
PADIS	pain, agitation/sedation, immobility and sleep
RCT	randomized-controlled trial
SCCM	society of critical care medicine
SMD	standard mean difference
U.S.	United States

Introduction

Maintaining critically ill adults in a comfortable and calm state is an important goal of intensive care unit (ICU) care. Pain management is complex in the intensive care unit (ICU); pain patterns are highly individual (e.g., acute, chronic, and acute on chronic), they arise from different sources (e.g., somatic, visceral, and neuropathic), and patients with critical illness (compared to other populations) more often have subjective perceptions and variable tolerability and response to analgesic therapy (Barr, Puntillo, Ely, Gelinas, & Dasta, 2013; Devlin, Skrobik, & Gélinas, 2018). A consistent approach to pain assessment and management is therefore paramount given the unique features of critically ill adults that include impaired communication, altered mental status, mechanical ventilation, procedures and use of invasive devices, sleep disruption, and immobility/mobility status (Devlin et al., 2018).

Analgesics, both opioids and nonopioids, are usually effective in reaching treating (and preventing) pain. However, opioids, if their use is not carefully titrated to pain score or other goals of therapy (e.g., sedation scores, respiratory rate, or degree of ventilatory compliance), are also associated with important safety concerns, including delirium, constipation,

Features and Assessments of Pain, Anesthesia, and Analgesia. https://doi.org/10.1016/B978-0-12-818988-7.00018-2
Copyright © 2022 Elsevier Inc. All rights reserved.

58 PART | I Setting the scene: General aspects of anesthesia, analgesics and pain

slower liberation from mechanical ventilation, withdrawal, and the development of chronic pain states (Devlin et al., 2018). This chapter focuses on current evidence and recommendations from the 2018 Society of Critical Care Medicine (SCCM) PADIS (Pain, Agitation/Sedation, Delirium, Immobility, and Sleep Disruption) guidelines to provide clinicians with strategies to optimize pain assessment and treatment in the ICU (Devlin et al., 2018).

After reviewing pain assessment in critically ill adults, we will describe the importance of using analgesics in a protocolized fashion, summarize the data of how opioids can provide analgosedation (as opposed to just pain treatment), and provide strategies on how nonopioid analgesics, when administered using a multimodal approach, can improve pain control and reduce opioid consumption.

Pain and its assessment in critically ill adults

Pain is frequent in critically ill adults, occurring in up to half of patients at rest, regardless of surgical intervention, and in up to 75% of patients during common bedside procedures (Devlin et al., 2018). Pain and its control remain the highest priority concern among survivors of critical illness. A patient's self-report of pain is the reference standard for pain assessment in the ICU for patients who can communicate reliably. Patient self-reported pain remains the gold standard for pain assessment in the ICU. The 0–10 Numeric Rating Scale (NRS) administered either verbally or visually is both valid and feasible in this population (Devlin et al., 2018). Among critically ill adults unable to self-report pain, and in whom behaviors are observable, the Behavioral Pain Scale in intubated (BPS) and nonintubated (BPS-NI) patients and the Critical-Care Pain Observation Tool (CPOT) demonstrate the greatest validity and reliability for monitoring pain (Barr et al., 2013; Devlin et al., 2018). When appropriate, and when the patient is unable to self-report, family can be involved in their loved one's pain assessment process (Devlin et al., 2018). Vital signs (e.g., heart rate, respiratory rate, oxygen saturation, or end-tidal CO_2) are not valid indicators for pain in critically ill adults and should only be used as cues to initiate further assessment using appropriate and validated assessment methods (Devlin et al., 2018).

Multiple barriers to optimize pain control exist in the ICU. The ability for clinicians to recognize, measure, and treat pain in a reliable and consistent fashion remains problematic in adult critical care settings (Devlin et al., 2018; Sjostrom, Jakobsson, & Haljamae, 2000). In clinical practice, nurses judge themselves as more accurate assessors of patient pain than the patients themselves (Rose, Smith, Gelinas, et al., 2012; Watt-Watson, Stevens, Garfinkel, Streiner, & Gallop, 2001) despite the overwhelmingly evidence demonstrating the consistent use of a standardized approach to pain assessment and treatment improves care quality, reduces complications, and reduces healthcare costs (Chanques et al., 2007). In addition, nurses may not titrate analgesic based on their pain assessments (Rose et al., 2012; Watt-Watson et al., 2001). Nursing beliefs surrounding opioid use and baseline patient characteristics (e.g., gender, age, race) seem to influence pain assessment and analgesic provision more than evidence-based data and clinical data (Glynn & Ahern, 2000). These factors are important to recognize when analgesic protocols are being formulated, implemented, and evaluated in the ICU.

All critically ill adults should be routinely evaluated for pain, and analgesia should be considered when pain is identified. The 2018 SCCM PADIS guidelines specifically note: "management of pain for adult intensive care unit (ICU) patients should be guided by routine pain assessment and pain should be treated before a sedative agent is considered" (Devlin et al., 2018). Management of pain in the critical care setting is the first component of the ABCDEF bundle (i.e., A = assess, prevent, and manage pain), a quality improvement bundle whose use improves both ICU and post-ICU outcomes (Pun, Balas, Barnes-Daly, et al., 2019).

Assessment-driven protocols

The PADIS guidelines also conditionally recommend the use of an assessment-driven, protocol-based, stepwise approach for pain management in critically ill adults given this practice will improve pain scores, reduce sedative use, improve time spent at the target sedation goal, allow earlier liberation from mechanical ventilation, reduce ICU and hospital length of stay, and decrease mortality rates (Brook, Ahrens, Schaiff, et al., 1999; Chanques, Jaber, Barbotte, et al., 2006; Devlin et al., 2018; Payen, Bosson, Chanques, et al., 2009). For patients requiring an opioid, this approach favors the use of intermittent intravenous (IV) boluses rather than continuous IV infusions. An example of an ICU assessment-driven pain protocol is presented in Fig. 1. Importantly, the use of assessment-driven pain protocols has not yet been shown to reduce nosocomial infections, constipation, hypotension, bradycardia, or opioid consumption (Brook et al., 1999; Chanques et al., 2006; Devlin et al., 2018; Payen et al., 2009).

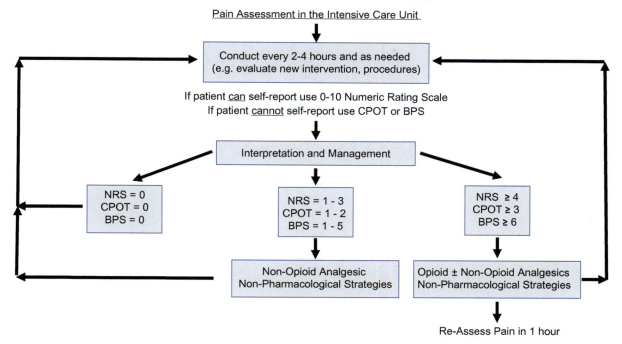

FIG. 1 Example of an assessment-driven pain protocol. This protocol describes how critically ill adults should be assessed for pain, how assessments should be interpreted, and how plans for analgesia should be formulated, evaluated, and revised.

Pain associated with patient activity or procedures known to cause pain should be assessed and treated through targeted, preemptive analgesic administration (Barr et al., 2013; Devlin et al., 2018). Little data exist about the effect of these protocols on the prevalence of opioid withdrawal (Sneyers, Duceppe, Frenette, et al., 2020).

The 2013 SCCM PAD guidelines (Barr et al., 2013), which predated the 2019 PADIS guidelines, suggest IV opioids be considered as the first-line class of drugs to treat nonneuropathic pain (Devlin et al., 2018). However, over the past decade, the idea that opioids should always be the first-line analgesic in all ICU patients has come under some fire given the sparse evidence to support this practice. Despite these concerns about data quality and the limited ICU patient populations evaluated, the guideline statement about IV opioid equivalence can be interpreted that all opioids are equally effective in providing analgesia (Barr et al., 2013; Devlin et al., 2018).

Opioid choice, route of administration, and dosing

Recommendations for opioid choice, route of administration, and dosing have not changed from the 2013 SCCM PAD guidelines (Barr et al., 2013). Opioid regimens should be individualized to patients; a number of different factors need to be considered (Table 1) (Barr et al., 2013; Devlin et al., 2018; Devlin & Roberts, 2011). The pharmacokinetic and pharmacodynamic properties differ among those opioids most commonly administered in the ICU (i.e., fentanyl, hydromorphone, morphine, and remifentanil) (Barr et al., 2013; Devlin & Roberts, 2011). Fentanyl is highly distributed to fat and can result in unpredictable serum concentrations among patients of different body habitus. Its clearance is reduced in patients with severe liver disease or congestive heart failure (Choi, Ferrell, Vasilevskis, et al., 2016). In patients with renal dysfunction, morphine's active morphine 6-glucuronide metabolite will accumulate, increasing the risk for seizures and myoclonus (Obeng, Hamadeh, & Smith, 2017). Although remifentanil has a short half-life and thus is more easily titrated, the tachyphylaxis and hyperalgesia associated with prolonged infusions limit its use to durations of less than 2 days (Lim, Nquyen, Qian, et al., 2018). Hydromorphone is less lipophilic than fentanyl and is a viable option, particularly for obese patients and patients with severe liver disease (Guignard, Bossard, Coste, et al., 2000).

Clinicians might consider rotating opioids for ICU patients who require prolonged infusions, in an effort to reduce accumulation and toxicity and optimize response (Landolf, Rivosecchi, & Goméz, 2020). The rationale for conversion from continuous fentanyl to continuous hydromorphone in one critically ill cohort entailed improved ventilatory compliance, reduced sedative exposure, and favorable PK/PD properties, although a prospective analysis is needed to evaluate clinical

TABLE 1 Summary of randomized-controlled trials evaluating analgosedation in critically ill adults.

Reference	Population	Intervention	Control	Primary Outcome	Results			Comments
					Inter'n	Control	P value	
Chiochotti, Kessler, Kirkham, et al. (2002)	MV, mixed medical-surgical[a]	Remifentanil 9–60 µg/kg/h + prn midazolam ($n = 74$)	Morphine 0.045–0.3 mg/kg/h + prn midazolam ($n = 78$)	% study hours with optimal sedation	83%	84%	>.05	Double-blind; midazolam required in <25% of patients; daily midazolam use 2× greater in morphine group
Dahaba, Grabner, Rehak, List, and Metzler (2004)	MV, orthopedic/general surgery[a]	Remifentanil 9–12 µg/kg/h + prn midazolam ($n = 20$)	Morphine 0.04–0.6 mg/kg/h + prn midazolam ($n = 78$)	% study hours with optimal sedation	78%	68%	.04	Double-blind; Remifentanil associated with shorted direction of mechanical ventilation ($P = .04$) and ICU length of stay ($P = .01$); Midazolam use greater in morphine group.
Muellejans et al. (2004)	MV, mixed medical-surgical[a]	Remifentanil 9–12 µg/kg/h + prn propofol ($n = 77$)	Fentanyl 1–2 µg/kg/h + prn propofol ($n = 75$)	% study hours with optimal sedation	88%	89%	>.05	Double-blind; % of patient requiring propofol similar between remifentanil and fentanyl groups (35% vs 50%, $P > .05$)
Rauf, Vohra, Fernandez-Jimenez, O'Keeffe, and Forrest (2005)	MV, off-pump cardiac surgery[a]	Remifentanil 6 µg/kg/h + propofol + morphine ($n = 10$)	Normal saline (placebo) + propofol + morphine ($n = 10$)	Morphine use in first hour after propofol D/C	8.15 mg	3.29 mg	<.01	Double-blind; pain and sedation scores similar between the two groups.
Carrer, Brocchi, Candini, et al. (2007)	MV, post-surgical[a]	Morphine 0.01 mg/kg/h + remifentanil 0.06–6.6 µg/kg/k + prn diazepam ($n = 50$)	Morphine 0.01–0.08 mg/kg/h + prn diazepam ($n = 50$)	% of patients not requiring diazepam	60%	25%	<.001	Nonblinded; Diazepam required in fewer remifentanil patients (28% vs 60%, $P < .01$); Time of ventilator, duration of ICU stay and ICU mortality similar between the 2 groups.
Karabinis, Mandragos, Stergiopoulos, et al. (2004)	MV, severe neurologic injury	Remifentanil 9–60 µg/kg/h + prn propofol (days 1–3); prn midazolam (≥ day 4) ($n = 84$)	Propofol (days 1–3) or midazolam (≥ day 4) and fentanyl 0.1–7.9 µg/kg/h ($n = 37$) or morphine 0–6.8 mg/kg/h ($n = 40$)	Time to be able to conduct neurologic assessment	Shorter time to conduct neurologic assessment with remifentanil compared to either fentanyl ($P < .01$) or morphine ($P = .02$)			Nonblinded; Time to extubation (after wakeful) shorter with remifentanil vs morphine (1 vs. 1.9 h; $P = .001$) but not fentanyl vs morphine (1 vs 0.7 h; $P = .47$); Both groups had optimal sedation on >95% of ICU days.

Study	Population	Intervention	Control	Primary outcome	Intervention result	Control result	P value	Comments
Breen, Karabinis, Malbrain, et al. (2005)	MV, mixed medical-surgical[a]	Remifentanil 9–60 µg/kg/h + prn midazolam IVP (n = 57)	Midazolam IVP + morphine IVP (n = 48)	Time to extubation	Remifentanil associated with a significantly reduced time on ventilator (vs midazolam) of 53.5 h (P = .003)			Nonblinded; time spent optimally sedated similar between groups.
Muellejans, Matthey, Scholpp, et al. (2006)	MV, cardiac surgery[a]	Remifentanil 6–60 u/kg/h + prn propofol (n = 39)	Midazolam 0.02–0.4 mg/kg/h and fentanyl IVP/1–7 µg/kg/h (n = 33)	ICU length of stay	46 h	62 h	P < .05	Nonblinded; Propofol added to 54% of remifentanil-managed patients. Time spent with adequate sedation similar between two groups. Overall ICU costs similar between two groups.
Rozendaal, Spronk, Snellen, et al. (2009)	MV, medical-surgical[a]	Remifentanil up to 12 µg/kg/h + prn propofol (n = 96)	Propofol, midazolam, or lorazepam + morphine or fentanyl (n = 109)	Duration of mechanical ventilation	3.9 days	5.1 days	P > .05	Nonblinded; Time spent at adequate level of sedation similar between groups; 69% of remifentanil group received propofol.
Strom, Martinussen, Toft, et al. (2010)	MV, mixed medical-surgical[a]	Morphine IVP 2.5–5.0 mg prn (n = 55)	Propofol 0.7 mg/kg/h + morphine 2.5–5.0 mg IVP (n = 58)	Days free of mechanical ventilation in the ICU	13.8 days	9.6 days	P = .019	Nonblinded; no difference in self-extubations or ventilator-associated pneumonia; all patients managed with 1:1 nursing ratios; sitter use common.
Tanios, Nguyen, Park, et al. (2019)	MV, mixed medical-surgical[a]	Fentanyl IVP and infusion to reach CPOT ≤2 and RASS 0 to −2 + prn propofol (n = 29)	Midazolam IVP and infusion to reach RASS 0 to −2 + prn fentanyl to maintain CPOT ≤2 (PS, n = 30); (PS + DSI n = 31)	Days free of mechanical ventilation in 28 days	24 days	24 days (PS); PDS + DSI (24 days)	P = .62	Nonblinded; time spent in first 48 h without pain or agitation similar between the three groups. Nursing perceived workload higher with fentanyl infusion intervention group. Only 5 fentanyl patients required short propofol course.
Olsen, Nedergaard, Strom, et al. (2020)	MV, mixed medical-surgical	No sedation (n = 349)	Light sedation (n = 351)	All-cause mortality at 90 days after randomization.	42%	37%	P = .65	Multicenter, nonblinded RCT; no difference in ICU-free days or ventilator-free days. The no-sedation group had 1 additional day free from coma or delirium. Patient in the no-sedation group had a higher APACHE II score at baseline.

This table provides a summary of the patient populations, a description of study interventions and controls, the results for the primary outcome, and key comments for the 12 published randomized controlled trials evaluating analgosedation in critically ill adults. CPOT, critical care pain observation tool; DSI, daily sedation interruption; ICU, intensive care unit; Inter'n, intervention; IVP, intravenous push; h, hour; kg, kilogram; MV, mechanical ventilation; PS, protocolized sedation; RASS, Richmond Agitation Sedation Scale; µg, microgram.

[a]Patients with acute neurological condition/injury excluded.

62 **PART | I** Setting the scene: General aspects of anesthesia, analgesics and pain

outcomes (Calderon, Pernia, Ysasi, et al., 2002; Fine & Portenoy, 2009). Prolonged exposure to opioids in the ICU setting may lead to tolerance, dependence, and/or withdrawal (Sneyers et al., 2020). Fentanyl patches, patient-controlled analgesia, and enteral methadone have each been shown to effectively wean critically ill adults off opioid infusions by maintaining pain control and minimizing withdrawal effects (Al-Qadheeb, Roberts, & Griffin, 2012; Chiu, Contreras, & Mehta, 2017; Chlan, Weinert, Skaar, et al., 2010; Kovacevic, Szumita, Dube, et al., 2018; Samala, Bloise, & Davis, 2014).

The approach to pain management in ICU patients with opioid use disorder, including patients managed with methadone or buprenorphine, is complex given that these patients may be more susceptible to pain but also higher tolerant to opioid therapy (Al-Qadheeb, Roberts, Griffin, et al., 2012). For opioid-abstinent patients with opioid use disorder who do not want to receive opioid therapy during their ICU stay, the increasing literature focused on opioid-sparing or opioid-free pain management after surgery should be considered (Karamchandani, Klick, Linskey-Dougherty, et al., 2019; Raheemullah & Lembke, 2019).

Analgosedation

Analgosedation is defined as either analgesia-first sedation (i.e., an analgesic, usually an opioid, is used before a sedative to reach the sedation goal) or analgesia-based sedation (i.e., an analgesic, usually an opioid, is used instead of a sedative to reach the sedative goal) (Devlin et al., 2018; Gabriel, Swisher, Sztain, et al., 2019). Analgosedation takes advantage of the analgesic, sedating, and respiratory depressant properties of opioids to optimize pain management and facilitate mechanical ventilation without the potential safety concerns associated with sedative use (Barr et al., 2013; Devlin et al., 2018).

Twelve published randomized-controlled trials have evaluated the use of an analgesia-based sedation in critically ill adults (Table 2) (Breen et al., 2005; Carrer et al., 2007; Chiochotti et al., 2002; Dahaba et al., 2004; Devabhakthuni, Armahizer, Dasta, et al., 2012; Karabinis et al., 2004; Muellejans et al., 2004, 2006; Rauf et al., 2005; Rozendaal

TABLE 2 Comparison of opioids commonly used in the ICU.

Medication	Time to onset (minute)	Half-life	Influence of context-sensitive half-life on duration of effect	Impact of organ failure on clinical duration of effect	Practical considerations
Fentanyl	1	2–4 h	Yes: significant	Hepatic	• Administration of CYP450 3A4/5 inhibitors prolongs clinical effect • Highly fat soluble; accumulation of effect with obesity • Rarely: (1) Serotonin syndrome; (2) Chest wall rigidity
Hydromorphone	5–10	2–3 h	Not applicable	Hepatic	• Useful substitute for fentanyl or morphine in patients with end-stage hepatic or renal failure
Morphine	5–10	3–4 h	Not applicable	Renal & Hepatic	• Histamine release may lead to bronchospasm and/or hypotension • Metabolite accumulation in renal dysfunction may lead to central nervous system toxicity • Cholecystitis
Remifentanil	1–3	3–10 min	Yes: minor	Renal (low)	• Opioid-associated tachyphylaxis and/or hyperalgesia common • May increase ammonia levels • Use ideal body (vs actual) body weight to dose

This table compares time of onset, half-life, duration of effect, impact of organ dysfunction on duration of effect and practical considerations for the use of fentanyl, hydromorphone, morphine, and remifentanil in critically ill adults. *ICU,* intensive care unit.

et al., 2009; Strom et al., 2010; Tanios et al., 2019). Only four of the trials were blinded (Chiochotti et al., 2002; Dahaba et al., 2004; Devabhakthuni et al., 2012; Muellejans et al., 2004). The ICU populations studied varied considerably; seven trials evaluated mixed-medical surgical patients (Dahaba et al., 2004; Devabhakthuni et al., 2012; Karabinis et al., 2004; Muellejans et al., 2006; Olsen et al., 2020; Rozendaal et al., 2009; Strom et al., 2010; Tanios et al., 2019), four postsurgical (Breen et al., 2005; Chiochotti et al., 2002; Muellejans et al., 2004; Rauf et al., 2005), and one with severe neurologic injury (Carrer et al., 2007). Remifentanil, an ultra-short-acting opioid, was studied alone in eight trials (Breen et al., 2005; Carrer et al., 2007; Chiochotti et al., 2002; Dahaba et al., 2004; Devabhakthuni et al., 2012; Karabinis et al., 2004; Muellejans et al., 2004, 2006) and in combination with morphine in one (Rauf et al., 2005). Two studies evaluated morphine alone (Rozendaal et al., 2009; Tanios et al., 2019), and the sole US trial evaluated fentanyl alone (Strom et al., 2010). Control group patients in seven trials were managed with continuously infused morphine or fentanyl (Breen et al., 2005; Carrer et al., 2007; Chiochotti et al., 2002; Dahaba et al., 2004; Devabhakthuni et al., 2012; Muellejans et al., 2006; Rauf et al., 2005) and in seven trials with continuously infused midazolam or propofol (Breen et al., 2005; Carrer et al., 2007; Muellejans et al., 2004, 2006; Rozendaal et al., 2009; Strom et al., 2010; Tanios et al., 2019). Four of the trials used time spent on mechanical ventilation-related primary endpoint (Karabinis et al., 2004; Muellejans et al., 2006; Rozendaal et al., 2009; Strom et al., 2010), three evaluated ICU time spent with optimal sedation (Chiochotti et al., 2002; Dahaba et al., 2004; Devabhakthuni et al., 2012), one used ICU length of stay (Breen et al., 2005), one used 90-day mortality (Tanios et al., 2019), and the trial enrolling neurologically injured patients used time to be able to conduct a neurologic assessment after opioid infusion discontinuation (Carrer et al., 2007).

Few studies reported Opioid-Related Adverse Drug Events (ORADES) or potential safety concerns. Importantly, 11 RCTs were single center (Breen et al., 2005; Carrer et al., 2007; Chiochotti et al., 2002; Dahaba et al., 2004; Devabhakthuni et al., 2012; Karabinis et al., 2004; Muellejans et al., 2004, 2006; Rauf et al., 2005; Rozendaal et al., 2009; Strom et al., 2010), the analgesic-first approach had already been in use before trial enrollment in many studies, the use of nonopioid analgesics was not protocolized, and a few patients were managed with either protocolized or daily sedative interruption. The most recent and largest RCT published to date ($n = 710$ patients) enrolled patients at multiple centers and did not observe a difference in 90-day mortality or ventilator-free days (Tanios et al., 2019). However, the trial did report a strong signal of reduced coma- and delirium-free days and fewer major thromboembolic events in the morphine-alone (no-sedation) treatment group.

Overall, the results of these trials cannot be formally synthesized given the analog-sedative intervention, control group, ICU patient population, and outcomes evaluated differ substantially between the studies (Devlin et al., 2018; Gabriel et al., 2019). Overall, although results across the studies vary widely, use of analgosedation was associated with fewer days of mechanical ventilation and an improved ability to maintain wakefulness. Despite the limitations and heterogeneity of these trials, the 2018 SCCM PADIS guideline provided only a conditional (vs. a strong) recommendation for the use of analgosedation in critically ill adults (Devlin et al., 2018).

Only one trial was conducted in the United States, where ICU practices are often different from Europe, and remifentanil (an ultra-short-acting opioid) is rarely used (Strom et al., 2010). In this trial, the use of analgosedation was associated with great daily nurse workload but no difference in the ability to reach pain and/or sedation goals, liberate patients from mechanical ventilation, or days spent with delirium. Although not well evaluated in RCTs, concerns exist that analgesia-based sedation may accentuate ORADEs (Devlin et al., 2018; Gabriel et al., 2019). Analgesia-based sedation may be better suited for surgical patients than medical patients; it should be used with caution in patients requiring continuous gabaminergic sedation (e.g., those with acute alcohol withdrawal, neuromuscular blockade, status epilepticus, or refractory elevated intracranial pressure) (Devlin et al., 2018; Gabriel et al., 2019).

Multimodal nonopioid analgesia use

While opioids remain the mainstay analgesic in critically ill adults, safety concerns associated with their use, particularly sedation, respiratory depression, and ileus, are of important concern for ICU clinicians (Devlin et al., 2018). The goal of multimodal analgesia, defined as the combined use of analgesics working by different mechanisms and at different nervous system sites, is to produce a degree of analgesia that is greater than that which can be achieved with a single analgesic (usually an opioid). Given that many ORADES are dose related, administration of lower opioid doses may improve patient safety. Multimodal analgesic approaches are increasingly being used in perioperative settings (e.g., as part of Enhanced Recovery after Surgery [ERAS] protocols) (Kehlet & Dahl, 1993; Ljungqvist, Scott, & Fearon, 2017; Olsen et al., 2020) and were rigorously evaluated in the PADIS guidelines (Devlin et al., 2018). In an era of ever-increasing concern about the overuse of opioids during acute hospitalization and their potential role in escalating posthospital opioid use disorder, multimodal analgesic efforts in the ICU have been focused on reducing opioid exposure.

TABLE 3 Nonopioid analgesics recommended for use in the ICU by the 2018 PADIS guidelines.

Medication/ Medication class	2018 PADIS recommendation	Primary efficacy results from ICU trials	Clinically relevant safety concerns in the ICU	Practical considerations
Acetaminophen	Suggest using as an adjunct to an opioid to decrease pain intensity and opioid consumption for pain management in critically ill adults (conditional recommendation, very low quality of evidence).	• Decrease in pain intensity • Decrease in opioid consumption	• Liver toxicity • Nausea and vomiting • Hypotension (intravenous route)	May be given by enteral, rectal, or IV route.
Ketamine	Suggest using low-dose ketamine (0.5 mg/kg IV push × 1 followed by 1- to 2-ug/kg/min infusion) as an adjunct to opioid therapy when seeking to reduce opioid consumption in postsurgical adults admitted to the ICU (conditional recommendation, very low quality of evidence).	• Decrease in opioid consumption	• Nausea • Delirium • Sedation • Tachycardia • Hypertension	• Clinical effects are dose-dependent. • Monitor for emergence delirium. • Monitor for excess secretions. • Catecholamine depletion associated with hypotension secondary to reduced cardiac output. • May increase intracranial pressure.
Nefopam	Suggest using nefopam (if available) either as an adjunct or replacement for an opioid to reduce opioid use and their safety concerns for pain management in critically ill adults (conditional recommendation, very low quality of evidence).	• Similar patient-self-reported pain intensity • Less nausea	• Tachycardia • Seizures • Delirium	• Nonopioid analgesic that exerts its effect by inhibiting dopamine, noradrenaline, and serotonin recapture in both the spinal and supraspinal spaces. • A 20-mg dose has an analgesic effect comparable to 6 mg of IV morphine • No effects on hemostasis, vigilance, ventilatory drive or intestinal motility.
Neuropathic agents (i.e., gabapentin, carbamazepine, pregabalin)	• Recommend using a neuropathic pain medication (e.g., gabapentin, carbamazepine, pregabalin) with opioids for neuropathic pain management in critically ill adults (strong recommendation, moderate quality of evidence). • Suggest using a neuropathic pain medication (e.g., gabapentin, carbamazepine, pregabalin) with opioids for pain management in ICU adults after cardiovascular surgery (conditional recommendation, low quality of evidence).	• Reduction in pain intensity for patients with Guillain-Barre syndrome • Reduction in opioid exposure in cardiovascular surgery	• Somnolence • Dizziness • Atrial fibrillation • Edema • Dose adjustment for renal dysfunction (gabapentin, pregabalin)	• Carbamazepine: many drug interactions. • Use with caution in patients with hepatic disease. • Gabapentin and pregabalin: use with caution in elderly patients ± renal dysfunction

This table summarizes the recommendations from the 2018 PADIS guidelines for the use of acetaminophen, ketamine, nefopam, and neuropathic agents as adjuvants to opioids in critically ill adults. Primary efficacy results, safety concerns and practical consideration for the use of each agent are also summarized. *ICU*, intensive care unit; *IV*, intravenous; *kg*, kilogram; *mg*, milligram; *µg*, microgram.

Among the six agents or classes of nonopioid analgesics evaluated in the PADIS guidelines as adjuvants for opioids when treating pain, four were recommended (i.e., acetaminophen, ketamine, neuropathic agents, and nefopam) (Devlin et al., 2018) (Table 3). While nefopam is widely used in Europe, it is not currently available in the United States (Young & Buvanendran, 2012). The PADIS guideline made conditional recommendations against the use of IV lidocaine or COX-1-specific NSAIDS as adjuncts to opioids in critically ill adults given the very low quality of evidence supporting the use of either agent and potential safety concerns with the use of either agent in the ICU (Devlin et al., 2018). In critically ill adults, the risk for lidocaine-associated neurologic toxicities and NSAID-related bleeding and acute kidney injury is increased. It should be noted that the guideline panel struggled with providing recommendations for how best to incorporate nonopioid analgesics in routine ICU care, defining the ICU settings and patient populations most appropriate for specific nonopioid strategies, and the optimal way to evaluate potential benefit, and most importantly, define risks (Devlin et al., 2018).

Since the PADIS guidelines have been published, one recent systematic review and metaanalysis addressed the efficacy and safety of a larger number of nonopioid adjunctive analgesics for adults admitted to the ICU (Kim, Kim, Choi, et al., 2014). The authors identified 33 eligible trials comparing a nonopioid analgesic as an opioid adjuvant compared to opioid therapy alone [acetaminophen ($n = 7$), dexmedetomidine ($n = 11$), ketamine ($n = 4$), nefopam ($n = 3$), neuropathic pain medications (i.e., carbamazepine, gabapentin, or pregabalin) ($n = 8$), an NSAID ($n = 6$), or tramadol ($n = 2$)]. The use of any analgesic adjuvant and an opioid (vs. an opioid alone) led to reductions in patient-reported pain scores at 24 h (standard mean difference [SMD] -0.94, 95% confidence interval [CI] -1.37 to -0.50, low certainty) and decreased opioid consumption (in oral morphine equivalents over 24 h, mean difference [MD] 27.25 mg less, 95% CI 19.80 mg to 34.69 mg less, low certainty).

In terms of individual medications, reductions in opioid use were demonstrated with dexmedetomidine (MD 10.21 mg less, 95% CI 1.06 mg to 19.37 mg less, low certainty), nefopam (MD 70.89 mg less, 95% CI 64.46 mg to 77.32 mg less, low certainty), NSAIDs (MD 11.07 mg less, 95% CI 2.7 mg to 19.44 mg less, low certainty), paracetamol (MD 45.51 mg less, 95% CI 24.31 mg to 66.71 mg less, very low certainty), carbamazepine (MD 54.69 less, 95% CI 40.39 mg to 68.99 mg less, moderate certainty), ketamine (MD 36.81 mg less, 95% CI 27.32 mg to 46.30 mg less, moderate certainty), and tramadol (MD 22.14 mg less, 95% CI 6.67 mg to 37.61 mg less, moderate certainty). The authors concluded clinicians should consider using adjunct agents to limit opioid exposure and improve pain scores in critically ill patients. The impact of these multimodal adjuncts on other clinically relevant outcomes, such as length of mechanical ventilation, length of ICU stay, and ORADEs, requires further research (Devlin et al., 2018).

Despite this newer evidence, it is important to highlight that the use of nonopioid analgesics (as an adjuvant to opioid use) has been poorly evaluated in medical ICU patients. For example, preliminary data suggest that ketamine might not have the opioid-sparing effects in medical patients that it has been reported to have in the surgical critically ill (Garber, Droege, Carter, Harger, & Mueller, 2019; Wheeler, Grilli, Centofanti, et al., 2020). In one cohort study of 104 medical critically ill adults, the addition of a ketamine infusion (2–8 μg/kg/min) did not affect fentanyl doing requirements (Wheeler et al., 2020). When low-dose ketamine was added to a sedation protocol for venovenous extracorporeal membrane oxygenation (ECMO) in patients with acute lung injury, opioid requirements remained unchanged (Garber et al., 2019).

Regional analgesia, such as neuraxial analgesia, nerve blocks, epidural blocks, and peripheral nerve blocks, is routinely used in the non-ICU perioperative setting. These techniques, although not formally evaluated in PADIS, likely have a place in ICU practice when multimodal analgesic approaches are being constructed despite the current limited evidence supporting their use in the ICU.

Applications to other areas

Pain is prevalent after surgery and trauma in patients who do not require admission to the intensive care unit. While patients in non-ICU settings would not require analgesedation, a multimodal approach to pain management with nonopioid analgesics is common in both inpatient and outpatient patient settings. Robust data demonstrates multimodal analgesic approaches will reduce opioid use and improve pain control.

Mini-dictionary of terms

ABCDEF bundle: The multimodal ABCDEF (A for Assessment, Prevention, and Manage pain; B for Both Spontaneous Awakening Trials and Spontaneous Breathing Trials; C for Choice of Analgesia and Sedation; D for Delirium Assess, Prevent, and Manage; E for Early Mobility and Exercise; and F for Family Engagement and Empowerment) Bundle is a

66 **PART | I** Setting the scene: General aspects of anesthesia, analgesics and pain

proven ICU care approach that reduces delirium, shortens mechanical ventilation duration, prevents post-ICU syndrome, and reduces healthcare costs.

Analgosedation: It refers to either analgesia-first sedation (i.e., an analgesic, usually an opioid, is before a sedative to reach the sedation goal) or analgesia-based sedation (i.e., an analgesic, usually an opioid, is used instead of a sedative to reach the sedative goal).

Behavioral Pain Scale (BPS): A pain assessment tool validated for use in critically ill adults who cannot self-report pain.

Critical-Care Pain Observation Tool (CPOT): A pain assessment tool validated for use in critically ill adults who cannot self-report pain.

Enhanced Recovery after Surgery (ERAS) protocol: These multimodal perioperative protocols, widely implemented in hospitals around the globe, have been developed to optimize pain control, reduce opioid use, and enhance cognitive and physical function after major surgical procedures.

Opioid-Related Adverse Drug Events (ORADES): Adverse effects (e.g., respiratory depression, nausea/vomiting, constipation) associated with opioid administration.

Pain, Agitation/Sedation, Delirium, Immobility and Disrupted Sleep Practice Guidelines for Critically Ill Adults (PADIS): These rigorous practice guidelines were developed using GRADE criteria by 32-member interprofessional panel of international experts and critical care survivors and make 36 recommendations surrounding the assessment, prevention and treatment of pain, agitation, delirium, immobility, and disrupted sleep in critically ill adults.

Summary points

- Pain, and its control, remains the highest priority concern among survivors of critical illness.
- A patient's self-report of pain should be the reference standard for pain assessment in the ICU over behavioral assessment tools like the BPS or CPOT.
- A pain-first, assessment-driven, protocolized management approach should be used for all adults admitted to the ICU.
- The choice of analgesic to treat pain in critically ill adults requires careful agent selection based on known medication properties and patient-specific characteristics.
- Analgosedation, by taking advantage of the analgesic, sedating, and respiratory depressant properties of opioids to optimize pain management and mechanical ventilation, improves pain control and reduces excessive sedation but has not been consistently shown to reduce the duration of mechanical ventilation or ICU length of stay.
- A nonopioid, multimodal analgesia strategy has been shown to improve pain control and reduce opioid use, particularly in surgical critically ill adults.

References

Al-Qadheeb, N. S., Roberts, R. J., Griffin, R., et al. (2012). Impact of enteral methadone on the ability to wean off continuously infused opioids in critically ill, mechanically ventilated adults: A case-control study. *The Annals of Pharmacotherapy, 46*, 1160–1166.

Barr, J., Puntillo, K., Ely, E. W., Gelinas, C., Dasta, J. F., et al. (2013). Clinical practice guidelines for the management of pain, agitation, and delirium in adult patients in the intensive care unit. *Critical Care Medicine, 41*, 263–306.

Breen, D., Karabinis, A., Malbrain, M., et al. (2005). Decreased duration of mechanical ventilation when comparing analgesia-based sedation using remifentanil with standard hypnotic-based sedation for up to 10 days in intensive care unit patients: A randomised trial. *Critical Care, 9*(3), R200–R210.

Brook, A. D., Ahrens, T. S., Schaiff, R., et al. (1999). Effect of a nursing-implemented sedation protocol on the duration of mechanical ventilation. *Critical Care Medicine, 27*, 2609–2615.

Calderon, E., Pernia, A., Ysasi, A., et al. (2002). Acute selective tolerance to remifentanil after prolonged infusion. *Revista Española de Anestesiología y Reanimación, 49*, 421–423.

Carrer, S., Brocchi, A., Candini, M., et al. (2007). Short term analgesia-based sedation in the intensive care unit: Morphine vs. rernifentanil and morphine. *Minerva Anestesiologica, 73*, 327–732.

Chanques, G., Jaber, S., Barbotte, E., et al. (2006). Impact of systematic evaluation of pain and agitation in an intensive care unit. *Critical Care Medicine, 34*, 1691–1699.

Chanques, G., Sebbane, M., Barbotte, E., Viel, E., Eledjam, J. J., & Jaber, S. (2007). A prospective study of pain at rest: Incidence and characteristics of an unrecognized symptom in surgical and trauma versus medical intensive care unit patients. *Anesthesiology, 107*, 858–860.

Chiochotti, T., Kessler, P., Kirkham, A., et al. (2002). Remifentanil vs. morphine for patients in intensive care unit who need short-term mechanicalventilation. *Journal of the Medical Association of Thailand, 85*, S848–S857.

Chiu, A. W., Contreras, S., Mehta, S., et al. (2017). Iatrogenic opioid withdrawal in critically ill patients: A review of assessment tools and management. *The Annals of Pharmacotherapy, 51*, 1099–1111.

Chlan, L. L., Weinert, C. R., Skaar, D. J., et al. (2010). Patient-controlled sedation: A novel approach to sedation management for mechanically ventilated patients. *Chest, 138*, 1045–1053.

Choi, L., Ferrell, B. A., Vasilevskis, E. E., et al. (2016). Population pharmacokinetics of fentanyl in the critically ill. *Critical Care Medicine, 44*, 64–72.

Dahaba, A. A., Grabner, T., Rehak, P. H., List, W. F., & Metzler, H. (2004). Remifentanil versus morphine analgesia and sedationformechanically ventilated critically ill patients: A randomized double blind study. *Anesthesiology, 101*, 640–646.

Devabhakthuni, S., Armahizer, M. J., Dasta, J. F., et al. (2012). Analgosedation: A paradigm shift in intensive care unit sedation practice. *The Annals of Pharmacotherapy, 46*(4), 530–540.

Devlin, J. W., & Roberts, R. J. (2011). Pharmacology of commonly used analgesics and sedatives in the ICU: Benzodiazepines, propofol and opioids. *Anesthesiology Clinics, 29*(4), 567–580.

Devlin, J. W., Skrobik, Y., Gélinas, C., et al. (2018). Clinical practice guidelines for the prevention and management of pain, agitation/sedation, delirium, immobility, and sleep disruption in adult patients in the ICU. *Critical Care Medicine, 46*, e825–e873.

Fine, P. G., & Portenoy, R. K. (2009). Establishing "best practices" for opioid rotation: Conclusions of an expert panel. *Journal of Pain and Symptom Management, 38*, 418–425.

Gabriel, R. A., Swisher, M. W., Sztain, J. F., et al. (2019). State of the art opioid-sparing strategies for post-operative pain in adult surgical patients. *Expert Opinion on Pharmacotherapy, 20*, 949–961.

Garber, P. M., Droege, C. A., Carter, K. E., Harger, N. J., & Mueller, E. W. (2019). Continuous infusion ketamine for adjunctive analgosedation in mechanically ventilated, critically ill patients. *Pharmacotherapy, 39*, 288–296.

Glynn, G., & Ahern, M. (2000). Determinants of critical care nurses' pain management behaviour. *Australian Critical Care, 13*, 144–151.

Guignard, B., Bossard, A. E., Coste, C., et al. (2000). Acute opioid tolerance: Intraoperative remifentanil increases postoperative pain and morphine requirement. *Anesthesiology, 93*, 409–417.

Karabinis, A., Mandragos, K., Stergiopoulos, S., et al. (2004). Safety and efficacy of analgesia-based sedation with remifentanil versus standard hypnotic-based regimens in intensive care unit patients with brain injuries: A randomised, controlled trial. *Critical Care, 8*(4), R268–R280.

Karamchandani, K., Klick, J. C., Linskey-Dougherty, M., et al. (2019). Pain management in trauma patients affected by the opioid epidemic: A narrative review. *Journal of Trauma and Acute Care Surgery, 87*, 430–439.

Kehlet, H., & Dahl, J. B. (1993). The value of "multimodal" or "balanced analgesia" in postoperative pain treatment. *Anesthesia and Analgesia, 77*, 1048–1056.

Kim, K., Kim, W. J., Choi, D. K., et al. (2014). The analgesic efficacy and safety of nefopam in patient-controlled analgesia after cardiac surgery: A randomized, double-blind, prospective study. *The Journal of International Medical Research, 42*, 684–692.

Kovacevic, M. P., Szumita, P. M., Dube, K. M., et al. (2018). Transition from continuous infusion fentanyl to hydromorphone in critically ill patients. *Journal of Pharmacy Practice, 33*(2), 129–135.

Landolf, K. M., Rivosecchi, R. M., & Gomez, H. (2020). Comparison of hydromorphone versus fentanyl-based sedation in extracorporeal membrane oxygenation: A propensity-matched analysis. *Pharmacotherapy, 40*(5), 389–397.

Lim, K. H., Nquyen, N. N., Qian, Y., et al. (2018). Frequency, outcomes, and associated factors for opioid-induced neurotoxicity in patients with advanced cancer receiving opioids in inpatient palliative care. *Journal of Palliative Medicine, 21*(12), 1698–1704.

Ljungqvist, O., Scott, M., & Fearon, K. C. (2017). Enhanced recovery after surgery: A review. *JAMA Surgery, 152*, 292–298.

Muellejans, B., Lopez, A., Cross, M. H., Bonome, C., Morrison, L., & Kirkham, A. J. (2004). Remifentanil versus fentanyl for analgesia based sedation to provide patient comfort in the intensive care unit: A randomized, double-blind controlled trial. *Critical Care, 8*, R1–R11.

Muellejans, B., Matthey, T., Scholpp, J., et al. (2006). Sedation in the intensive care unit with remifentanil/propofol versus midazolam/fentanyl: A randomised, open-label, pharmacoeconomic trial. *Critical Care, 10*(3), R91.

Obeng, A. O., Hamadeh, I., & Smith, M. (2017). Review of opioid pharmacogenetics and considerations for pain management. *Pharmacotherapy, 37*, 1105–1121.

Olsen, H. T., Nedergaard, H. K., Strom, T., et al. (2020). Nonsedation or light sedation in critically ill, mechanically ventilated patients. *The New England Journal of Medicine, 382*, 1103–1111.

Payen, J. F., Bosson, J. L., Chanques, G., et al. (2009). Pain assessment is associated with decreased duration of mechanical ventilation in the intensive care unit: A post hoc analysis of the DOLOREA study. *Anesthesiology, 111*, 1308–1316.

Pun, B. T., Balas, M. C., Barnes-Daly, M. A., et al. (2019). Caring for critically ill patients with the ABCDEF bundle: Results of the ICU liberation collaborative in over 15,000 adults. *Critical Care Medicine, 47*(1), 3–14.

Raheemullah, A., & Lembke, A. (2019). Initiating opioid agonist treatment for opioid use disorder in the inpatient setting: A teachable moment. *JAMA Internal Medicine, 179*, 427–428.

Rauf, K., Vohra, A., Fernandez-Jimenez, P., O'Keeffe, N., & Forrest, M. (2005). Remifentanil infusion association with fentanyl-propofol anaesthesia in patients undergoing cardiac surgery: Effects on morphine requirement and postoperative analgesia. *British Journal of Anaesthesia, 95*, 611–615.

Rose, L., Smith, O., Gelinas, C., et al. (2012). Critical care nurses' pain assessment and management practices: A survey in Canada. *American Journal of Critical Care, 21*, 251–269.

Rozendaal, F. W., Spronk, P. E., Snellen, F. F., et al. (2009). Remifentanil-propofol analgo-sedation shortens duration of ventilation and length of ICU stay compared to a conventional regimen: A centre randomised, cross-over, open-label study in the Netherlands. *Intensive Care Medicine, 35*(2), 291–298.

Samala, R. V., Bloise, R., & Davis, M. P. (2014). Efficacy and safety of a six-hour continuous overlap method for converting intravenous to transdermal fentanyl in cancer pain. *Journal of Pain and Symptom Management, 48*, 126–132.

Sjostrom, B., Jakobsson, E., & Haljamae, H. (2000). Clinical competence in pain assessment. *Intensive & Critical Care Nursing, 16*, 273–282.

Sneyers, B., Duceppe, M. A., Frenette, A. J., et al. (2020). Strategies for the prevention and treatment of iatrogenic withdrawal from opioids and benzodiazepines in critically ill neonates, children and adults: A systematic review of clinical studies. *Drugs, 80*(12), 1211–1233.

Strom, T., Martinussen, T., Toft, P., et al. (2010). A protocol of no sedation for critically ill patients receiving mechanical ventilation: A randomised trial. *Lancet, 375*(9713), 475–480.

Tanios, M., Nguyen, H. M., Park, H., et al. (2019). Analgesia-first sedation in critically ill adults: A U.S. pilot, randomized controlled trial. *Journal of Critical Care, 53*, 107–113.

Watt-Watson, J., Stevens, B., Garfinkel, P., Streiner, D., & Gallop, R. (2001). Relationship between nurses' pain knowledge and pain management outcomes for their postoperative cardiac patients. *Journal of Advanced Nursing, 36*, 535–545.

Wheeler, J., Grilli, R., Centofanti, J., et al. (2020). Adjuvant analgesic use in the critically ill: A systematic review and meta-analysis. *Critical Care Explorations, 2*(7), e0157.

Young, A., & Buvanendran, A. (2012). Recent advances in multimodal analgesia. *Anesthesiology Clinics, 30*, 91–100.

Chapter 7

The multidisciplinary Acute Pain Service: Features and experiences

Turi Stefano, Deni Francesco, Marmiere Marilena, Meani Renato, and Beretta Luigi
Department of Anesthesia and Intensive Care, IRCCS Ospedale San Raffaele, Milano, Italy

Abbreviations

APS   Acute Pain Service
ERAS   Enhanced Recovery After Surgery
IV   intravenous
PCA   patient controlled analgesia

Introduction

The International Association for the Study of Pain (IASP) defines pain as "An unpleasant sensory and emotional experience." Moreover, pain is nowadays regarded as a vital parameter, as well as heart rate, oxygen saturation, blood pressure, and diuresis.

The importance of achieving an adequate and complete pain control in the postoperative period is widely accepted (Wu, Rowlingson, Partin, et al., 2005). As a matter of fact, clinicians should obtain a satisfying postoperative analgesia not only to preserve patients from experiencing a psychological trauma, but also to prevent possible serious clinical damages. An insufficient pain control can be associated with increase in blood pressure and heart rate and, more generally, with a hyperactivation of the sympathetic nervous system (Basbaum, Bautista, Scherrer, et al., 2009; Kehlet, 1989). Moreover, the release of cytokines and proinflammatory agents associated with this condition can increase the risk of developing thrombotic complications such as myocardial infarction, deep venous thrombosis, or pulmonary embolism (Engquist, Brandt, Fernandes, et al., 1977). In addition, a patient with poorly controlled pain is also incapable of moving from the bed, breathing properly, and coughing, inevitably delaying the return to preoperative physical conditions. This is particularly true for patients undergoing thoracic or upper-gastrointestinal surgery (Colucci, Fiore, Paisani, et al., 2015).

In recent years, the Enhanced Recovery After Surgery (ERAS) approach obtained important clinical results in managing patients in the perioperative period. The main purpose of ERAS is to preserve the surgical patient's functional capacity, in order to allow for a faster return to preoperative physical conditions (Bardram, Funch-Jensen, Jensen, et al., 1995). The adoption of ERAS protocols has been associated with a reduction in clinical postsurgical, cardiovascular, and pulmonary complications (Greco, Capretti, Beretta, et al., 2014), leading to its progressive diffusion in many different surgical specialties (Ljungqvist, Scott, & Fearon, 2017).

The adequate control of pain throughout the perioperative period is one of the cornerstones of ERAS philosophy (Table 1) (Joshi & Kehlet, 2019), and the ability to obtain optimal postoperative pain control with oral analgesics is considered fundamental for hospital discharge (Table 2) (Braga, Scatizzi, Borghi, et al., 2018). Nevertheless, an extensive observational multicenter European survey showed that postoperative pain management is usually suboptimal, and improvements are needed in surgical wards (Benhamou, Berti, Brodner, et al., 2008).

During the last decades, the medical and scientific communities made many efforts trying to optimize existing drugs and improve analgesic techniques to ensure adequate pain control. However, these elements alone are not sufficient and must be included into a wider, systematic organization.

Features and Assessments of Pain, Anesthesia, and Analgesia. https://doi.org/10.1016/B978-0-12-818988-7.00041-8
Copyright © 2022 Elsevier Inc. All rights reserved.

TABLE 1 Perioperative Enhanced Recovery After Surgery (ERAS) interventions.
ERAS interventions
Preoperative
Preadmission patient education
Medical preoperative optimization
Carbohydrate load
Selective premedication
Intraoperative
Antibiotic prophylaxis
Opioid-sparing multimodal analgesia
Avoid fluid overload
PONV prophylaxis
Avoid hypothermia
Postoperative
Thromboembolic prophylaxis
Postoperative analgesia
Early mobilization
Early termination of IV fluid infusion
Early removal of urinary catheter
PONV, postoperative nausea vomiting; *IV*, intravenous.

TABLE 2 Clinical conditions to define the patient as fit for discharge.
Criteria for patient's discharge
Autonomous mobilization
Oral fluid and solid intake
Adequate pain control with oral analgesics
Recovery of bowel function
Absence of complications
The achievement of pain control with oral analgesics is necessary to obtain the hospital discharge.

The history of Acute Pain Services

The need to create multidisciplinary systems to manage acute postoperative pain dates to the early 1960s (Avery-Jones, 1966), and in the following decade, many scientific editorials confirmed the importance of designing standardized organizational protocols. The first Acute Pain Services (APS) were created in 1985 in Germany and in the United States (Maier, Kibbel, Mercker, et al., 1994; Petrakis, 1989).

Two were the main models of APS: the "American" model described by Brian Ready in 1986 and the "Swedish" model proposed by Rawal in 1994 (Rawal & Berggren, 1994; Ready, Oden, Chadwick, et al., 1988). As it occurs in other clinical organizational systems, they both presented strengths and potential disadvantages.

The American model is mainly conducted by the anesthesiologist. It is based on a team of hyperspecialized doctors, trained to tailor the analgesic protocol according to patients' individual characteristics. On the other hand, this system can be very expensive (personnel costs) and can create a separation between the medical and the nursing staff, possibly leading to a defective integration between patients' pain management and clinical observation.

On the other hand, the Swedish model was based on the creation of the Acute Pain Nurse, a nurse qualified in postoperative pain care. This new professional figure allows for the management of a high number of patients, with lower staff costs compared to the American model. Nevertheless, it requires the formulation of an adequate educational program to properly train the Acute Pain Nurse, focusing on the accurate definition of pain management protocols and all the conditions in which it is necessary to activate the medical staff.

These two different models, developed at the beginning of the APS history, were modified through the years and optimized and adapted to the resources available in each single healthcare facility.

Basic characteristics of an Acute Pain Service

Recent national and international guidelines defined the minimum criteria to fulfil in order to activate an APS (Table 3) (Savoia, Alampi, Amantea, et al., 2010).

First of all, it is fundamental to establish a clear organization to the service, nominating the head and his collaborators. They must be reachable throughout the day, including night shifts, weekends, and nonworking days.

The members of APS must monitor the efficacy of postoperative analgesia using defined scales for pain intensity (e.g., the Numerical Rating Scale, NRS) and note it down on the medical record at fixed intervals. If pain results inadequately controlled, further treatment must always be provided and recorded, especially if moderate or severe. The incidence of possible side effects related to the analgesic treatment has to be observed, detected, and promptly treated, when necessary (Fig. 1).

Acute Pain Services must be carried out by multidisciplinary teams to work at their best: All the different professionals (anesthesiologists, surgeons, nurses, and physiotherapists) involved in patient management throughout the perioperative period should take part in all clinical decisions. Moreover, members of the APS team should promote the creation of shared guidelines and defined programs for addressing postoperative pain. Continuous programs of education about the importance of achieving an adequate postoperative analgesia should also be performed. The adherence to protocols, the obtained results, and the need of changing current clinical practice according to an evidence-based evaluation should be periodically analyzed with multidisciplinary audits (Feldheiser, Aziz, Baldini, et al., 2016). However, there is no organizational model superior to others, as long as the APS respects the abovementioned characteristics, using available resources.

More in detail, the APS team must design shared clinical protocols and improve educational pathways to reach complete awareness of pain, evaluate its intensity at patients' bedside, and monitor the occurrence of adverse drug reactions, eventually treating them as needed. APS performance should be critically analyzed by means of periodical meetings.

The activity of the APS team can also begin in the preoperative period. Indeed, postoperative pain, as well as its intensity and main characteristics, could be predicted in advance, allowing to determine specific pain management protocols according to the type of surgery. Sharing information with the patient about the type of perioperative analgesic treatment

TABLE 3 Basic characteristics of an Acute Pain Service.

Basic characteristics of an acute pain service
Presence of dedicated personnel
Identification of the leading professional figure
Periodical evaluation of postoperative pain
Detection of side effects possibly related to the analgesic therapy
Nonstop activity (included night shifts, weekends, and nonworking days)
Definition of multidisciplinary programs and guidelines
Continuous personnel education and training
Periodical multidisciplinary audits for an evidence-based evaluation of obtained results

In this table, we report the main elements that an efficient Acute Pain Service (APS) must develop. Despite the diffusion of APS, only few hospitals present complete organizational models for postoperative pain treatment.

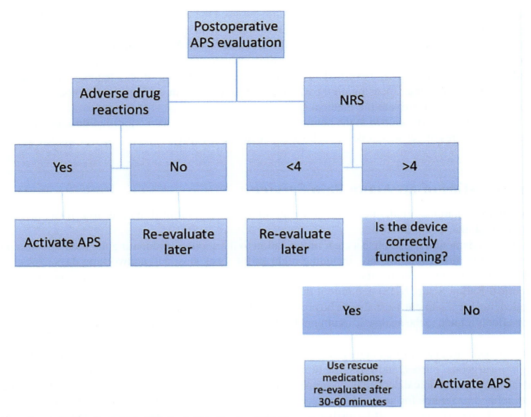

FIG. 1 Postoperative evaluation flow chart. *APS*, Acute Pain Service; *NRS*, Numerical Rating Scale.

and the technique the anesthesiologist chooses to perform can reduce anxiety before surgery, improve doctor-patient relationship, and increase patients' compliance to the analgesic treatment.

Preoperative pain optimization and immediate postoperative pain treatment are both essential to prevent chronic postsurgical pain (Fletcher, Stamer, Pogatzki-Zahn, et al., 2015). The analgesic protocol must be adapted as much as possible to the specific characteristics of each single patient, possibly improving pain control in the preoperative setting (as in a sort of "analgesic prehabilitation").

Acute Pain Service and clinical outcome

Is the creation of an APS eventually associated with an important clinical advantage?

It is not so easy to give an answer to this type of questions, given the difficulties in performing randomized-controlled trials comparing APS to non-APS postoperative pain management regimes. So far, studies were designed to compare hospital activity before and after the introduction of APS. These analyses are certainly biased, since the possible clinical advantages detected in the second period of time might not be due to the creation of an APS, but to other clinical factors, such as, for example, the improvement in anesthesiologic and surgical techniques through the years (Tsui, Law, Fok, et al., 1997).

In this regard, Werner et al. analyzed 154 papers (audit trials, 40%; surveys, 11%; reviews, 12%; expert opinions, 38%) investigating the clinical advantages deriving from the activity of APS published between 1986 and 2000 (Werner, Soholm, Rotboll-Nielsen, et al., 2002). It appears that the introduction of an APS has generally been associated with a reduction in the intensity of acute postoperative pain and in the incidence of analgesic therapy–related side effects, such as postoperative nausea and vomiting (PONV) and urinary retention (Miaskwoski, Crews, Ready, et al., 1999; Stacey, Rudy, & Nelhaus, 1997). However, the level of evidence is still not strong due to the low power of the included studies, since the improvements in clinical practice could be related to medical progress. Moreover, the pain intensity reduction observed in hospitals with an APS can be the result of a possible placebo effect.

As stated at the beginning of the chapter, pain is also an emotional experience and thus its perceived intensity can be increased in conditions of fear or anxiety. Consequently, daily visits from members of the APS team might reassure the patient, making them feel safe in case of poor pain control, and help with the postoperative recovery (Fig. 2).

Another reason for the reduction of pain intensity in hospitals with an APS could be the increasing awareness, among all professionals involved in patient care, of the importance of an adequate pain management (Harmer & Davies, 1998). The presence of APS, indeed, comes with the creation of multidisciplinary guidelines and educational programs specifically focused on the understanding of all the aspects of pain, leading to a progressive cultural change through the years.

Many authors underline the ability of implementing the use of very effective but potentially dangerous analgesic techniques also in the context of general wards thanks to the strict monitoring provided by APS. The positioning of an epidural catheter, for example, is one of the cornerstones in the treatment of severe acute postoperative pain. However, this technique can be associated with the development of rare but serious neurological complications, such as epidural hematoma, thus making clinical monitoring and immediate reaction mandatory (Carli & Klubien, 1999; Mcleod, Davies, & Colvin, 1995). At the same time, epidural infusions usually include local anesthetics and strong opioids: The appearance of side effects due to these drugs should be monitored as well.

The intravenous administration of strong opioids provides a valid alternative to central neuraxial blocks in case of severe pain. The safest technique for administering such drugs in general wards is represented by patient-controlled analgesia (PCA) systems (Peng, Ren, Qin, et al., 2016). With this term, we usually refer to simple machines that can be handled directly by patient according to the perceived intensity of pain. Generally, the system can be programmed by a trained professional in terms of drug bolus to be administered, maximum dosage allowed per unit of time (generally 1 h), and lockout interval. These characteristics, added to the possibility of drug self-administration by the patient as needed, make this system extremely safe also in a general ward. PCA methods include both intravenous administration of drugs, such as morphine or oxycodone, as well as new sublingual sufentanil tablet systems (Turi, Deni, Lombardi, et al., 2019). In both cases, patients can autotitrate the pain medication dose according to individual perception and intensity variation throughout the postoperative days. An effective and well-organized APS is necessary to manage these devices and check for the correct programming, function, and use. Many studies and guidelines consider PCA absolutely superior to a continuous elastomeric infusion (Chou, Gordon, de Leon-Casasola, et al., 2016).

In conclusion, it can be very difficult to determine if the reduction in measured postoperative pain intensity in hospitals with an APS can be attributable to the diffusion of both neuraxial blocks and PCA in the general ward as well. However, the role of APS in preventing serious adverse events such as epidural hematoma, meningitis, epidural abscess, or severe respiratory depression has not been completely established, probably due to the extremely low incidence of these types of complications.

FIG. 2 Timing of postoperative pain evaluation.

74 **PART | I** Setting the scene: General aspects of anesthesia, analgesics and pain

Another possible question is if the application of an APS can be associated with a reduction in costs (Zenz & Tryba, 1996).

It is quite difficult to perform this type of analysis, both because it is necessary to consider several elements related to the management of patient and because, as explained at the beginning of this chapter, there are several organizational models of APS with different associated costs. We also stated that an adequate postoperative pain control is fundamental to allow a faster return to preoperative physical conditions before being discharged home, and an organized system like APS can help in this context.

Several studies demonstrated how a single item of the ERAS approach is not enough to improve patients' recovery and hospital discharge. Conversely, the higher is the adherence to an elevated number of ERAS items, the higher will be the obtained clinical advantage (Braga et al., 2018).

In conclusion, we can say that the presence of an APS, specifically of a well-organized structure for managing postoperative pain, is associated with a clinical advantage (faster discharge and possible cost-savings) only if associated to a detailed and complete program of enhanced recovery.

The diffusion of Acute Pain Services

Since the first editorials theorizing the creation of organized systems for the management of acute postoperative pain were published during the 1960s and 1970s, the number of APS increased all over the world, as described by countless surveys and analyses. Despite the increase in the number of hospitals offering an APS, the percentage of structures able to fulfil all the necessary criteria is still limited.

A recent German survey revealed that 80% of hospitals arranged an APS, but less than half were able to accomplish the minimum objectives required for an effective organizational model of pain management. In particular, the allocation of dedicated personnel, the organization of patient care during nights and weekends, the presence of written protocols for postoperative pain management, and regular assessments documenting pain intensity are not performed in every center. More in detail, only 25% had a dedicated pain nurse, and 9% did not receive any input from a pain specialist (Erlenwein, Koschwitz, Pauli-magnus, et al., 2016).

Rocket et al. analyzed the organizational model of inpatient pain management in UK hospitals: only 15% (20 hospitals) provided APS during night shifts and 29% (39 hospitals) during the weekends (Rockett, Vanstone, Chand, et al., 2017). Many centers also integrated the APS with chronic pain care, particularly for those patients who presented persistent post-surgical pain. The authors concluded that it is very difficult to define the national level of APS in the United Kingdom due to the extensive variability between different hospitals.

Two different surveys (POPSI and POPSI-2) analyzed the diffusion of APS in Italy in 2006 and 2012, respectively (Coluzzi, Mattia, Savoia, et al., 2015). Despite the presence of APS in 51% of the consulted hospitals, the number of centers satisfying the minimum organizational criteria was much more limited. The absence of an adequate training of involved professionals and the lack of time and of resources were identified as the main problems preventing the diffusion of APS.

Nasir et al. launched a survey to 301 US hospitals (200 nonteaching and 101 teaching centers): 74% of those who responded offered an organized APS to surgical patients (Nasir, Howard, Joshi, et al., 2011). The service was significantly more formally organized in academic hospitals with respect to nonteaching ones. At the same time, postoperative pain management protocols were more carefully designed in academic hospitals, when compared to nonteaching centers. Intravenous patient-controlled analgesia (IV-PCA) was more commonly supervised by surgeons (75%), while epidural analgesia and peripheral nerve block infusions were exclusively handled by anesthesiologists.

We can conclude that the number of APS increased in many countries during the last decades. Despite this, a high heterogeneity is still detectable between different medical centers, since many of them do not fulfil the minimum basic criteria required for establishing an adequate pain management organization.

The main clinical experiences regarding the establishment of an Acute Pain Service

In this section, we present the main published experiences describing the activity of APS (Table 4), and Table 5 lists commonly used or experimental agents for pain control.

At IRCCS San Raffaele Hospital, Milan, Italy, the APS was established more than 20 years ago. This service is mainly managed by anesthesiologists and residents in anesthesia. Moreover, nurses are adequately trained to monitor pain intensity and possible side effects of analgesic techniques. Multidisciplinary protocols and shared flow charts have been defined for the correct management of postoperative pain. We published our clinical experience in 2011 (Deni, Greco, Turi, et al., 2019). A total of 17,913 adult patients were followed by APS over a 10-year period. The majority of them (41%) underwent

TABLE 4 The main published experiences describing the activity of Acute Pain Services.

Authors	Number of patients	Journal	Year
Paul JE	35,384	*Anesthesiology*	2014
Popping DM	18,925	*British Journal of Anaesthesiology*	2008
Deni F	17,913	*Journal of Pain Practice*	2019
Flisberg P	2696	*Acta Anesthesiologica Scandinavica*	2003
Syngelin FJ	1338	*Journal of Clinical Anesthesia*	1999

TABLE 5 List of commonly used or experimental agents for pain control and/or anesthesia.

Commonly used agents in our clinical experience, according to the predicted postoperative pain intensity

Mild	NSAIDs Acetaminophen
Moderate	NSAIDs/acetaminophen Association with weak opioids
Severe	Strong opioids PCA (morphine, oxycodone, sublingual sufentanil) Consider locoregional techniques (peridural/paravertebral block, abdominal wall block, intrathecal morphine)

Switch from intravenous to oral administration, when possible
Consider local infiltration of surgical wounds with anesthetics

NSAIDs, nonsteroidal antiinflammatory drugs; *PCA*, patient controlled analgesia.

abdominal surgery. Epidural analgesia was used in 7653 cases (43%), while 9239 (52%) patients used intravenous PCA. Combined epidural analgesia and intravenous PCA was used in 87 patients (0.5%). A total of 456 (2.5%) perineural catheters were placed. Numerical Rating Scale for pain was daily assessed by residents in anesthesia. Mean pain intensity was already low on postoperative day 1 (pain at rest 1 ± 1.3, pain upon movement 2.1 ± 1.7), and further decreased on postoperative day 2 (pain at rest 0.8 ± 1.1, pain upon movement 1.9 ± 1.6) and day 3 (pain at rest 0.6 ± 0.9, pain upon movement 1.7 ± 1.5). A total of 2117 patients (12%) experienced some drug reactions: The majority of them presented just one adverse event, and 94 patients presented more than one (just two of them reported three adverse reactions). Overall, side effects were significantly ($P < .001$) more common after epidural analgesia (1276, 16%) rather than during PCA system use (793, 9%), perineural catheters (35, 8%), or other IV drugs administration (13, 3%). We reported 163 (2%) probable or confirmed dural punctures during epidural catheter placement. For perineural catheters, the most common event was paresthesia (25, 5%). No epidural hematoma, epidural abscess, or meningitis was reported in our experience with epidural catheters, and no permanent modification in sensory or motor functions was observed.

Paul et al. compared the incidence of adverse events on an acute pain service in three hospitals, before and after the introduction of a formal root cause analysis process (Paul, Buckley, McLean, et al., 2014). On a cohort of 35,384 patients in a 7-year interval, a significant reduction in the overall adverse event rate, in terms of respiratory depression, severe hypotension, and PCA pump programming errors, was observed. Specifically, 165 episodes of respiratory depression, 4 epidural abscesses, 2 spinal hematomas, and 2 deaths were reported.

Popping et al. designed an observational study on a cohort of 18,925 patients between 1998 and 2006, followed in a university hospital by an acute pain service (Popping, Zahn, van Aken, et al., 2008). A total of 14,223 patients received patient-controlled epidural analgesia (PCEA), 1591 intravenous patient-controlled analgesia (IV-PCA), 1737 continuous brachial plexus block, and 1374 continuous femoral or sciatic nerve block. Epidural analgesia and peripheral nerve blocks provided a superior pain relief compared to IV-PCA. The authors reported three cases of epidural hematoma, two cases of epidural abscess, one case of bacterial meningitis, and one case of respiratory depression.

Flisberg et al. published their clinical experience on 2696 patients undergoing major surgery and receiving either epidural or intravenous analgesia for postoperative pain relief in the surgical ward. Patients with epidural analgesia experienced less pain compared to those with IV-PCA. Twenty episodes of respiratory depression and one epidural hematoma were reported (Flisberg, Rudin, Linner, et al., 2003).

Syngelin et al. published their clinical experience with 1338 patients scheduled for elective unilateral total hip arthroplasty (Syngelyn & Gouverneur, 1999). Pain intensity, rescue analgesia, satisfaction score, technical problems, and side effects were collected and reported by the acute pain service. Peridural analgesia, IV-PCA, and "3 in 1" continuous block provided comparable pain relief. No serious adverse events were registered.

Conclusion

More than 100 million people in the United States and Europe and 312 million people worldwide undergo surgery every year. The number of surgical procedures increased by 34% between 2004 and 2012, and it is expected to grow further in the next years (Richebè, Capdevilla, & Rivat, 2018). Consequently, the number of patients potentially experiencing postoperative pain could be extremely high.

The correct management of analgesia in the postoperative period is fundamental to allow patients' recovery and to reduce postsurgical morbidity and mortality. In order to meet this important objective, the creation and diffusion of efficient and organized multidisciplinary acute pain services are mandatory.

Application to other areas

Perioperative medicine includes all health-related interventions from the moment of diagnosis and surgical planning to patients' full recovery. A successful perioperative strategy relies on the collaboration of different medical professionals in a multidisciplinary team, and postoperative pain control represents one of its cornerstones.

Effective analgesia is associated with lower rates of postoperative complications, reduced length of hospital stay, and faster return to presurgical functionality. A well-organized APS, integrated with all the other clinical professionals involved in surgical patients' care, is essential to achieve optimal control of postoperative pain.

The ultimate goal is to obtain a "pain-free hospital," in which pain management is considered a priority. Thus, it is essential to encourage the adoption of recent effective pharmacological strategies to prevent pain, the application of new analgesic techniques (e.g., abdominal wall blocks, peripheral nerve blocks, and other types of regional anesthesia), and the organization of courses for medical and nursing staff to provide proper education and training. The adoption of shared clinical protocols tailored according to surgical procedures and patients' characteristics, as well as the careful observation of possible adverse events, should become the standard of care.

In conclusion, all the elements related to the organization of an APS should be included in the process of perioperative care.

Other agents of interest

Being pain a multidimensional experience, an Acute Pain Service team should be able to tackle both its physical and emotional aspects.

The entire perioperative period should be exploited to correctly manage patients' pain, somehow performing an "analgesic prehabilitation." Indeed, poorly controlled preoperative pain is associated with less satisfactory postsurgical analgesia and with a higher risk of developing chronic postoperative pain. Patients should be assessed at least 4 weeks before elective surgery and eventually treated according to their individual characteristics. In addition, the adoption of such strategy could improve doctor-patient relationship and, consequently, reduce patients' anxiety. Follow-up visits should be scheduled after hospital discharge to detect the presence of persistent postsurgical pain and to intervene as soon as possible.

The creation of an international APS association could promote data gathering regarding the relation between different types of surgery and postoperative pain intensity as well as the adopted analgesic techniques, their efficacy, and observed side effects. In this way, optimal standards of postoperative pain management could be defined.

Key facts—Mini-dictionary

Adequate pain control is fundamental for a fast and successful hospital discharge in surgical patients.

The need to establish an adequate organizational system for the management of postoperative pain dates to the late 1950s.

The first Acute Pain Service teams were conceived in the United States and Germany in the 1980s.

Originally, the Acute Pain Service structure hinged on anesthesiologists (the "American" model) or acute pain nurses (the "Swedish" model).

The minimum criteria an Acute Pain Service has to fulfil are defined by international guidelines.

The main duties of Acute Pain Services are the evaluation of the intensity of pain, its treatment, and the monitoring of potential drug reactions.

The Acute Pain Service team should be multidisciplinary and needs continuous education and training.

Summary points

1. Even if pain is a frequent and predictable complication occurring during the postoperative period, it still remains undertreated. According to US National Institute of Health, less than 50% of surgical patients receive adequate pain relief.
2. Pain may amplify patients' autonomic reflexes and modify endocrine response, leading to an increase in cortisol and catecholamine levels.
3. Abdominal or thoracic pain may impair patients' ability to cough, contributing to the development of pulmonary complications.
4. Pain may reduce mobilization, resulting in increased length of hospital stay and costs.
5. The APS approach has been validated by clinical trials and quality improvement commissions, and creating a multidisciplinary team is a winning strategy.
6. The teams should include nurses, physiotherapists, surgeons, and anesthesiologists, the latter acquiring a leading role thanks to their specific skills in pain management.
7. The APS team:
   - defines protocols to treat pain;
   - performs a periodical evaluation of pain intensity, as well as therapeutic efficacy and adverse drug reactions;
   - promotes multidisciplinary audits.

Conclusion 1. APS is fundamental to modern-era high-quality hospitals and is beneficial to patients, providing adequate postoperative pain relief without major complications.

Conclusion 2. Modern-era medicine needs a role expansion for the APS into the preoperative and postdischarge phases to allow for a continuum in perioperative pain management.

References

Avery-Jones, F. (1966). Post-operative pain. In *Bailey H and Pie's surgical handicraft* (pp. 197–206). Bristol, UK: Wright.

Bardram, L., Funch-Jensen, P., Jensen, P., et al. (1995). Recovery after laparoscopic colonic surgery with epidural analgesia and early oral nutrition and mobilization. *Lancet, 345*(8952), 763–764.

Basbaum, A. I., Bautista, D. M., Scherrer, G., et al. (2009). Cellular and molecular mechanisms of pain. *Cell, 139*, 267–284.

Benhamou, D., Berti, M., Brodner, G., et al. (2008). Postoperative analgesic observational survey (PATHOS): A practice pattern study in 7 central/southern European countries. *Pain, 136*(1–2), 134–141.

Braga, M., Scatizzi, M., Borghi, F., et al. (2018). Identification of core items in the enhanced recovery pathway. *Clinical Nutrition ESPEN, 25*, 139–144.

Carli, F., & Klubien, K. (1999). Thoracic epidurals: Is analgesia all we want? *Canadian Journal of Anesthesia, 46*, 409–414.

Chou, R., Gordon, D. B., de Leon-Casasola, O. A., et al. (2016). Management of postoperative pain: A clinical practice guideline from the American Pain Society, the American Society of Regional Anesthesia and Pain Medicine, and the American Society of Anesthesiologists' Committee on Regional Anesthesia, Executive Committee, and Administrative Council. *Journal of Pain, 17*, 131–157.

Colucci, D. B., Fiore, J. F., Jr., Paisani, D. M., et al. (2015). Cough impairment and risk of postoperative pulmonary complications after open upper abdominal surgery. *Respiratory Care, 60*, 673–678.

Coluzzi, F., Mattia, C., Savoia, G., et al. (2015). Postoperative pain surveys in Italy from 2006 and 2012 (POPSI-1 and POPSI-2). *European Review for Medical and Pharmacological Sciences, 19*(22), 4261–4269.

Deni, F., Greco, M., Turi, S., et al. (2019). Acute Pain Service: A ten-year experience. *Journal of Pain Research, 19*(6), 586–593.

Engquist, A., Brandt, M. R., Fernandes, A., et al. (1977). The blocking effect of epidural analgesia on the adrenocortical and hyperglycemic responses to surgery. *Acta Anaesthesiologica Scandinavica, 21*(4), 330–335.

Erlenwein, J., Koschwitz, R., Pauli-magnus, D., et al. (2016). A follow-up on Acute Pain Services in Germany compared to international survey data. *European Journal of Pain, 20*(6), 874–883.

Feldheiser, A., Aziz, O., Baldini, G., et al. (2016). Enhanced recovery after surgery (ERAS) for gastrointestinal surgery, part 2: Consensus for anesthesia practice. *Acta Anaesthesiologica Scandinavica, 60*, 289–334.

Fletcher, D., Stamer, U. M., Pogatzki-Zahn, E., et al. (2015). Chronic postsurgical pain in Europe. An observational study. *European Journal of Anaesthesiology, 32*, 725–734.

Flisberg, P., Rudin, A., Linner, R., et al. (2003). Pain relief and safety after major surgery. A prospective study of epidural and intravenous analgesia in 2696 patients. *Acta Anaesthesiologica Scandinavica, 47*(4), 457–465.

Greco, M., Capretti, G., Beretta, L., et al. (2014). Enhanced recovery programs in colo-rectal surgery: A meta-analysis of randomized studies. *World Journal of Surgery, 38*(6), 1531–1541.

Harmer, M., & Davies, K. A. (1998). The effect of education, assessment and a stanardised prescription on postoperative pain management: The value of clinical audit in the establishment of acute pain services. *Anesthesia, 53*, 424–430.

Joshi, G. P., & Kehlet, H. (2019). Postoperative pain management in the era of ERAS: An overview. *Best Practice & Research. Clinical Anaesthesiology, 33*(3), 259–267.

Kehlet, H. (1989). The stress response to surgery: Release mechanisms and the modifying effect of pain relief. *Acta Chirurgica Scandinavica. Supplementum, 550*, 22–28.

Ljungqvist, O., Scott, M., & Fearon, K. C. (2017). Enhance recovery after surgery: A review. *JAMA, 152*(3), 192–198.

Maier, C., Kibbel, K., Mercker, S., et al. (1994). Postoperative pain at general nursing stations: An analysis of eight years' experience at an anesthesiological acute pain service. *Anaesthesist, 43*, 385–397.

Mcleod, G. A., Davies, H. T., & Colvin, J. R. (1995). Shaping attitudes to postoperative pain release: The role of acute pain team. *Journal of Pain and Symptom Management, 10*, 30–34.

Miaskwoski, C., Crews, J., Ready, L. B., et al. (1999). Anesthesia-based pain services improve the quality of postoperative pain management. *Pain, 80*, 23–29.

Nasir, D., Howard, J. E., Joshi, G. P., et al. (2011). A survey of acute pain service structure and function in the United States hospitals. *Pain Research and Treatment, 2011*, 934932.

Paul, J. E., Buckley, N., McLean, F., et al. (2014). Hamilton acute pain service safety study: Using root cause analysis to reduce the incidence of adverse events. *Anesthesiology, 120*(1), 97–109.

Peng, L., Ren, L., Qin, P., et al. (2016). The impact of patient-controlled analgesia on prognosis of patients receiving major abdominal surgery. *Minerva Anestesiologica, 82*(8), 827–838.

Petrakis, J. K. (1989). Acute pain services in a community hospital. *Clinical Journal of Pain, 5*(Suppl 1), S34–S41.

Popping, D. M., Zahn, P. K., van Aken, H. K., et al. (2008). Effectiveness and safety of postoperative pain management: A survey of 18 925 consecutive patients between 1998 and 2006 (2nd revision): A database analysis of prospectively raised data. *British Journal of Anaesthesia, 101*(6), 832–840.

Rawal, N., & Berggren, L. (1994). Organization of acute pain services: A low-cost model. *Pain, 57*(1), 117–123.

Ready, L. B., Oden, R., Chadwick, H. S., et al. (1988). Development of an anesthesiology-based postoperative pain management service. *Anesthesiology, 68*(1), 100–106.

Richebè, P., Capdevilla, X., & Rivat, C. (2018). Persistent postsurgical pain: Pathophysiology and preventative pharmacologic considerations. *Anesthesiology, 129*, 590–607.

Rockett, M., Vanstone, R., Chand, J., et al. (2017). A survey of acute pain services in UK. *Anaesthesia, 72*(10), 1237–1242.

Savoia, G., Alampi, D., Amantea, B., et al. (2010). Postoperative pain treatment SIAARTI recommendations 2010. Short version. *Minerva Anestesiologica, 76*(8), 657–667.

Stacey, B. R., Rudy, T. E., & Nelhaus, D. (1997). Management of patient-controlled analgesia: A comparison of primary surgeons and a dedicated pain service. *Anesthesia and Analgesia, 85*, 130–134.

Syngelyn, F. J., & Gouverneur, J. M. (1999). Postoperative analgesia after total hip arthroplasty: i.v. PCA with morphine, patient-controlled epidural analgesia, or continuous "3-in-1" block? A prospective evaluation by our acute pain service in more than 1,300 patients. *Journal of Clinical Anesthesia, 11*(7), 550–554.

Tsui, S. L., Law, S., Fok, M., et al. (1997). Postoperative analgesia reduces morbidity and mortality after esophagectomy. *American Journal of Surgery, 173*, 472–478.

Turi, S., Deni, F., Lombardi, G., et al. (2019). Sufentanil sublingual tablet system (SSTS) for the management of postoperative pain after major abdominal and gynecological surgery within an ERAS protocol: An observational study. *Journal of Pain Research, 12*, 2313–2319.

Werner, M. U., Soholm, L., Rotboll-Nielsen, P., et al. (2002). Does an acute pain service improve postoperative outcome? *Anesthesia and Analgesia, 95*(5), 1361–1372.

Wu, C. L., Rowlingson, A. J., Partin, A. W., et al. (2005). Correlation of postoperative pain to quality of recovery in the immediate postoperative period. *Regional Anesthesia and Pain Medicine, 30*, 516–522.

Zenz, M. W., & Tryba, M. (1996). Economic aspects of pain therapy. *Current Opinion in Anaesthesiology, 9*, 430–435.

Chapter 8

Monitoring anesthesia: Electroencephalography and beyond

Mengmeng Chen and Wangning ShangGuan

Department of Anesthesiology and Perioperative Medicine, The Second Affiliated Hospital and Yuying Children's Hospital of Wenzhou Medical University, Wenzhou, China

Abbreviations

EEG	electroencephalography
BIS	bispectral index
REM	rapid eye movement
USB	universal serial bus
AAGA	accidental awareness
SSEP	somatosensory evoked potentials
AEP	auditory evoked potential
VEP	visual evoked potential
SLR	short or early latency response
MLR	middle latency response
LLR	long latency response
MLAEP	mid-latency auditory evoked potential
BAEP	brainstem auditory evoked potential
SE	state entropy
RE	reaction entropy
NI	narcotrend index
TIVA	total intravenous anesthesia
CSI	cerebral state index

Introduction

With the development of anesthesia technology in surgery, more and more attention has been paid to the monitoring of anesthesia depth. Anesthesia monitoring methods are also increasing. The most primitive anesthesia depth monitoring is the raw electroencephalography (EEG), and then a variety of monitoring methods have been extended on the basis of the EEG. In this chapter, we will introduce different types of EEG monitoring methods.

EEG was first described by Hans Berger in 1929 shortly before EEG monitoring under anesthesia (Berger, 1929). Raw EEG has limited use in the measurement of depth of anesthesia at the beginning. Only until 1937, the investigations of EEG started under anesthesia (Gibbs). EEG is referred as a noninvasive neurophysiologic measurement of the activity of the brain. The raw EEG monitoring and its analysis can be applied to determine the level of sedation and analgesia during general anesthesia. Although raw EEG data are very reliable, anesthesiologists usually pay less attention during anesthesia monitoring comparing to bispectral index (BIS) or other EEG monitors (Hagihira, 2015). Usually, anesthesiologists assess the depth of anesthesia based on clinical interpretations in the underlying changes of physiological symptoms, such as heart rate, respiration rates, blood pressure, eye movement, and physical responses to stimulation from the surgical incision (Whyte & Booker, 2003). However, the raw EEG should be put in a standard position, because some EEG monitors are not credible in special populations through preset algorithms and software. For anesthesiologists, it is necessary to understand the raw EEG changes under anesthesia. The brain monitoring in anesthesia has brought upon many clinical advantages in modern practice.

Features and Assessments of Pain, Anesthesia, and Analgesia. https://doi.org/10.1016/B978-0-12-818988-7.00025-X
Copyright © 2022 Elsevier Inc. All rights reserved.

Raw electroencephalography

The classification of the raw EEG activity and its waveforms is based relatively on the degree of localization. Generally, the normal routine EEG monitoring uses 21 electrodes including 19 scalp electrodes and the remaining 2 electrodes which are placed at the ear lobes at specific position, whereas in EEG monitoring under general anesthesia, 4-set frontal electrodes are used which are respectively illustrated in Figs. 1 and 2.

There are five main types for EEG waves: delta, theta, alpha, beta, and gamma waves, with four of them being shown in Fig. 3. Each raw EEG wave has its own frequency waveforms and amplitudes that are manifested during awaken period, the different stages of sleep, and under the state of general anesthesia. A frequency range of 0.5–4 Hz (Hertz) and amplitude of 20–200 µV are distinct features of the delta wave which is known to generate from the thalamus. The frequency range of theta wave is 4–7 Hz with amplitude of 20–100 µV and is mainly produced from the neocortex and the hippocampus. Alpha wave has a frequency range of 8–12 Hz with amplitude of 20–60 µV and is generated by the thalamus. The cortex generates beta waves in the frequency of 13–30 Hz and amplitude of 2–20 µV. Gamma wave has a frequency range of 20–70 Hz with very low amplitude of 3–5 µV and is brought by the thalamic nucleus. Delta, theta, and alpha wave frequencies are mainly seen during different stages of general anesthesia and are usually manifested under different anesthetic agent. Beta and gamma waves are known as fast oscillations EEG waves that are mainly associated with paradoxical rapid eye movement (REM) sleep and wakefulness.

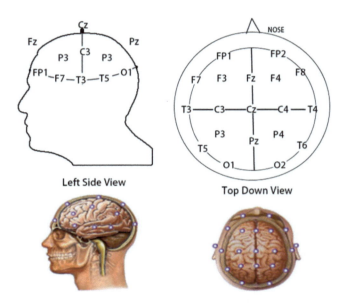

FIG. 1 Electrodes positioning during routine EEG monitoring.

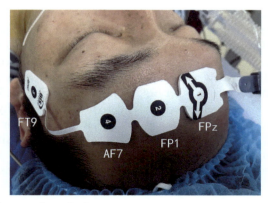

FIG. 2 Frontal electrodes positioning during general anesthesia.

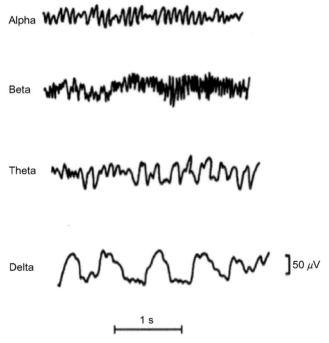

FIG. 3 Waveforms of the EEG of the human brain.

Actually, the raw EEG waves during general anesthesia are complicated and inconsistent. Usually, high frequency with low amplitude waves is seen in moderate level of anesthesia. Moreover, beta activation occurs at low anesthetic doses where the EEG desynchronizes and increases to higher frequencies of 13–30 Hz.

Influence on raw EEG

It has been shown that different anesthetics affect EEG waveforms (Purdon, Sampson, Pavone, & Brown, 2015). An increase in the concentration of anesthetic such as inhaled anesthetics or propofol would decrease the frequency and increase the amplitude of the dominant EEG. With increasing dosage of anesthetic drugs, the spindle wave (alpha wave range) becomes dominant under surgical anesthesia. At deeper level, this activity decreased and the theta and delta waves become dominant. At even deeper level, the EEG waveform manifests into "burst and suppression" mode and eventually becomes flat. Under harmful stimuli (surgical incisions), the EEG waveform changes, but this damage is not always reflected in the BIS or other processed EEG index. Spindle waves are sufficiently sensitive to harmful stimuli: Under surgical anesthesia, they disappear when harmful stimuli are applied, and they reappear when sufficient analgesic is given. Therefore, observing the spindle wave during anesthesia could provide a reference for the effect of anesthesia. During deep anesthesia, the "burst and suppression" mode becomes apparent, manifested as extreme activity, as a high-frequency large-amplitude wave (burst), with flat traces (suppression). This pattern (excluding cerebral ischemia or other factors) indicates that the level of anesthesia is too deep.

The raw EEG reflects the metabolic activity of cells in the brain in terms of energy. Therefore, these changes in energy state of the brain cells can strongly influence the raw EEG waves such as low-oxygen concentration in the brain, decreased brain perfusion (low cerebral blood flow), choice of anesthetic agent, hypothermia, and hyperthermia. Mild hypothermia causes slight changes in the EEG, with minor swings in the raw EEG frequencies to theta and beta activity, whereas hyperthermia induces high oscillation in the frequency waves (Kochs, 1995). Some studies have demonstrated that how the raw EEG influenced by age, increased anesthetic concentration, noxious stimulus, and hemodynamics disturbances (MacKay, Sleigh, & Voss, 2010).

In a recent research (Beekoo et al., 2019), the raw EEG was investigated in different ages from 1 month to 80 years old. The electroencephalogram was recorded continuously by means of a universal serial bus (USB) over a period of 10 min using the Aspect A-2000 BIS monitor (version XP). Raw EEG data, BIS values, and end-target expiratory concentration of sevoflurane were recorded. The research indicated that the raw EEG provides real-time information about the actual depth of anesthesia when standard monitors provide misleading values. Observing the raw EEG waveforms provides accurate reliable data which can be used to assess the depth of anesthesia in different age and may reduce brain injury during general anesthesia.

Bispectral index

Most anesthesiologists are not familiar with the original raw EEG data; hence, they cannot interpret the information to determine the anesthetic-related changes. Raw EEG is of limited use in the measurement of anesthesia depth in actual monitor. In the actual anesthesia monitor, anesthesiologists prefer to use BIS to monitor the depth of anesthesia. On the one hand, BIS monitoring is more convenient; on the other hand, the digital display of BIS monitoring is more intuitive. Nowadays, anesthesia under BIS monitoring is very common (Dutta et al., 2019; Laitio et al., 2008a). Meanwhile, BIS-guided closed-loop control anesthesia has been extensively applied and proven to provide good anesthesia (Absalom, Sutcliffe, & Kenny, 2002; Puri et al., 2016).

BIS monitor takes and analyzes a complex EEG signals and processes the signals into numbers 1–100 to indicate the state and sedation of the cerebral cortex. A BIS value below 60 is indicative of no awareness during the surgical procedure. The calculation of BIS value is a multivariate statistical model (Glass et al., 1997; Rampil, 1998). The correspondence between BIS and EEG is displayed in Fig. 4. Controlling BIS value in the range of 40–60 during anesthesia could improve postoperative recovery. BIS monitoring is generally considered to be able to prevent the incidence of accidental awareness (AAGA) during the anesthesia (Myles, Leslie, McNeil, Forbes, & Chan, 2004). However, most authors also agree that BIS is poor at detecting wakefulness. The applications of BIS in different groups have some differences.

Limitations and influence on BIS

Owing to the BIS calculation rule which is more suited to adults, it may not be suitable for children. Tokuwaka et al. stated that under sevoflurane anesthesia, BIS is unreliable in children aged less than 2 years old when assessing the depth of anesthesia (Tokuwaka et al., 2015). The recent research reported that under general anesthesia with 1.0 MAC sevoflurane and the use of

FIG. 4 The correspondence between BIS and EEG.

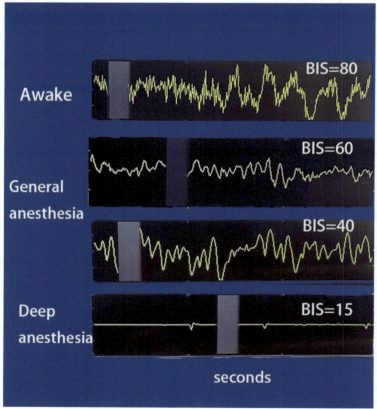

opioids and muscle relaxant according to each patient's weight, some pediatric patients presented with a higher BIS value, and theta wave (4–8 Hz) was observed in all subjects (Beekoo et al., 2019). Alpha and beta waves were absent, and BIS value was statistically correlated to the raw EEG wave frequency. Wodey et al. (2005a) explained how at the steady state under 1 MAC of sevoflurane, in a whole population of 100 children, the BIS values ranged from 20 to 74 (median 40) and how the EEG bispectrum as well as the BIS (Aspect XP) measured at 1 MAC sevoflurane appears to be strongly related to both the age and weight of children. The study showed a similar finding in pediatric cases with BIS value in the range of 24–74 and with 14 children less than 2 years old having a BIS value greater than 60 even though the raw EEG wave showed theta coherence (Beekoo et al., 2019). In another study, Tirel et al. (2006) suggested that BIS values are related to age of children regardless of the volatile anesthetic agent used and that, in children, the difference in BIS values for different agents at the same MAC can be explained by the specific effect on the EEG bispectrum induced by each anesthetic agent. The other groups presented a much lower BIS value in the range of 40–60 under 1.0 MAC of sevoflurane which is suggestive of its accuracy in such cases but not reliable in pediatric patients. The inaccuracy of the brain monitors in pediatric patients can be explained by the changes in the brain during development such as apoptosis, neurogenesis, axonal-dendritic growth, and synaptogenesis which begins after birth, with synaptic density peaking at around 2 years old (Wodey et al., 2005b). These developments are not reflected by the monitors as the BIS algorithm is derived only for adults. Hence, assessing the raw EEG wave in children is much more precise than relying on the monitors (Brown, Purdon, & Van Dort, 2011).

BIS monitoring has many benefits, but it is also affected by many factors such as muscle relaxants. Onset of muscle paralysis may result in a BIS fall (Schuller, Newell, Strickland, & Barry, 2015). The reversal of nondepolarizing muscle relaxant like sugammadex or neostigmine results in rises in BIS value (Dahaba et al., 2012). Others, such as specific anesthetics or some serious clinical conditions of the patient, such as epilepsy and so on, will have varying degrees of impact on the BIS value (Ohshima et al., 2007; Ozcengiz, Unlügenç, Güneş, & Karacaer, 2012).

Evoked potential

Somatosensory evoked potential and visual evoked potential

Evoked potentials include somatosensory evoked potentials (SSEP), auditory evoked potential (AEP), visual evoked potential (VEP), and others. Evoked potentials were a localized potential change produced in nervous system when affected by direct or external stimuli (electric, light, sound, etc.).

During surgery, SSEP is considered essential to assess and monitor peripheral nerve and plexus regions. Thus, the SSEP has been applied in detecting sciatic nerve injury and related position nerve during hip surgery. In recent years, novel automated SSEP device has been applied in cardiac surgery to monitor peripheral nerve injury, and it has been shown that the use of intraoperative SSEP monitoring significantly reduces the incidence of nerve injury. In terms of the effects of narcotic drugs on SSEP monitoring, early research indicated that some general anesthetics, especially most inhalational anesthetics (such as nitrous oxide, isoflurane, etc.), would inhibit SSEP. Intravenous anesthetics and opioids such as propofol and remifentanil are not as effective at inhibiting SSEP as inhaled anesthetics. During spinal surgery, it has been proved that anesthetic drugs such as lidocaine or dexmedetomidine have no effect on intraoperative SSEP monitoring. However, Meng, Wang, Zhao, & Guo (2015) has reported that small doses of etomidate (less than 0.3 mg/kg) had dose-related amplification effects on SSEP.

VEP monitoring has limited application in the operating. VEP monitoring is more sensitive to intravenous and inhalation anesthetics than other evoked potentials. There is currently no unified standard for VEP monitoring.

Auditory evoked potential

AEP is an evoked potential generated and recorded by the central nerves in different planes of auditory receptors after receiving external acoustic stimulation. It objectively checks the auditory pathway from the cochlea to the cortex. According to the latency of the reaction, it can be divided into short or early latency response (SLR), middle latency response (MLR), and long latency response (LLR). Mid-latency auditory evoked potentials (MLAEP) and the coherent frequency of the AEP are the most promising for the monitoring depth of anesthesia. One of the most common AEP is the brainstem auditory evoked potential (BAEP). BAEP is a more sensitive and objective indicator of brainstem damage. Any injury involving the auditory channel will affect BAEP. BAEP changes when the brainstem is even slightly damaged without clinical signs.

The monitoring of BAEPs is particularly important during the operations involving auditory nerve pathways. It is also indispensable in neurosurgery that may damage the brainstem. BAEPs are widely applied in neurosurgery such as acoustic neuroma resection, basilar aneurysm resection, and exploration for skull base lesion (Gentili, Lougheed, Yamashiro, &

Corrado, 1985; Harper et al., 1992). Although the monitoring BAEPs are simple and sensitive, it cannot be applied in the deafness (conductive and sensorineural hearing disorder) on the operative side.

The research about AEP and anesthesia mostly focuses on MLAEP, which can be used to indicate the depth of anesthesia. MLAEP appears 10–50 ms after stimulation, composed of a series of waves of No, Po, Na, Pa, and Nb, related with the thalamus and the original hearing perceptual cortical electrical activity. Many intravenous and inhaled anesthetic (except ketamine and nitrous oxide) can prolong the incubation period of Pa and Nb, and decrease the amplitude of vibration; while encountering the surgical stimulation or tracheal intubation, the latency of Pa and Nb becomes shorter and the amplitude increases (Bonhomme, Llabres, Dewandre, Brichant, & Hans, 2006). Schneider et al. found a combination of EEG and AEP detectors, compared to BIS, has high predictive capacity of distinguishing conscious and unconscious. Studies on AEP and implicit memory have only been reported in the early literature (Schneider, Hollweck, Ningler, Stockmanns, & Kochs, 2005). Schwender et al. studied the occurrence of implicit memory during the administration of several different anesthetic agents (Schwender, Kaiser, Klasing, Peter, & Pöppel, 1994). They observed that implicit memory occurred only in patients in whom the latency increase in Pa was less than 12 ms. Mantzaridis and Kenny suggested that the auditory evoked potential index (AEP idx) can be a reliable indicator of potential awareness instead of latencies and amplitudes (Mantzaridis & Kenny, 1997). Maybe applying AEP is a potential method to prevent awareness during anesthesia. The studies on AEP in recently years are rare. Even though AEP has a certain role in anesthesia monitoring, its use is still limited due to its expensive equipment and difficult interpretation.

Entropy

Entropy is a physical concept and was first entropy as a physical concept. What it expresses is the irregularity of information. The more irregular the signal is, the higher is the entropy value. When the signal is completely regular, the entropy value is 0. Spectral entropy is widely applied in anesthesia depth monitoring and is divided into state entropy (SE) and reaction entropy (RE). SE is calculated according to EEG, reflecting the degree of inhibition of the cerebral cortex. RE comes from the integrated calculation of EEG and electromyogram (EMG), reflecting the joint effect of EEG and frontal muscles. The SE value varies from 0 to 91, and the RE varies from 0 to 100. The range of both entropies is 40–60 under anesthesia (Vakkuri et al., 2005).

Compared with traditional clinical monitoring, reasonable use of entropy monitoring can be a better guide of the dosage of anesthetic (Hor, Van Der Linden, Hert, Mélot, & Bidgoli, 2013; Tewari, Bhadoria, Wadhawan, Prasad, & Kohli, 2016). Entropy-guided monitoring to assess the depth of anesthesia may allow the hypnotic drugs to be administered more accurately, thereby reducing the consumption of anesthetics. In terms of awareness monitoring, there is a good correlation between entropy and BIS, which can effectively monitor the depth of anesthesia. Laitio et al. found that during induction of xenon anesthesia, BIS and entropy showed a delay to detect the loss of response, but both monitors performed well in distinguishing conscious and unconscious states during steady-state anesthesia (Laitio et al., 2008b). However, more evidence showed that compared with BIS, entropy made more falsely indication in transition between consciousness and unconsciousness (Pilge et al., 2015). Weil et al. put forward that RE predicted a motor response to noxious stimulations (Weil, Passot, Servin, & Billard, 2008). High values (RE > 55) should be avoided before the stimulation s in order to decrease the risk of motor response. In terms of hemodynamics, whether entropy monitoring has an advantage over traditional clinical standard monitoring remains controversial. Although some authors believe that entropy monitoring can better guide hemodynamic stability, others believe that there is no difference between the two (Gruenewald et al., 2007; Riad, Schreiber, & Saeed, 2007).

However, entropy will be affected by factors such as age or anesthetics. Older age under sevoflurane anesthesia is associated with a shift to faster, more irregular, and less predictable (Kreuzer et al., 2020). What kind of specific changes of entropy index would be in different ages under varied anesthesia is still less studied. Entropy-guided anesthesia should be used more cautiously in the elderly and children. As for anesthetics, entropy is not sensitive to nitrous oxide, and SE and RE values would not change under nitrous oxide (Anderson & Jakobsson, 2004). The addition of ketamine will also affect its correct measurement of anesthesia depth. Nevertheless, the research on the entropy index is still not thorough enough, and its advantages and limitations still have a lot of research space.

Narcotrend

Narcotrend is a new anesthesia sedation monitoring system. It automatically interprets EEG during the anesthesia. The classification algorithms are based on visually classified EEG. In 2000, the classification of the depth of anesthesia was started using 6 stages and 15 subgrading systems, called Narcotrend classification. From the wake state to very deep anesthesia, A, B0-2, C0-2, D0-2, E0-2, and F0-1 are distinguished. The appropriate depth of anesthesia is expressed as D0,

D1, D2, E0, and E1. F0 and F1 represent burst suppression and isoelectric EEG, respectively. Later, these stages have been transformed into a digital scale, from 0 to 100, called Narcotrend Index (NI). The correspondence between the Narcotrend stage and NI is shown in Table 1. Narcotrend monitoring has an objective and finer sedation depth classification and cheaper consumables. Due to these reasons, Narcotrend monitoring has been widely used in European and American countries.

Narcotrend monitoring collects and monitors of bioelectricity (original EEG) at any position in the brain in real time, with automatic analysis and classification of EEG, and it uses NI for anesthesia awareness and depth. Much evidence has showed that modern electroencephalographic parameters, such as BIS or Narcotrend, are reliable indicators for the anesthetic states than classic one (Schmidt et al., 2003). They could provide a reliable production between awake versus steady-state anesthesia and reaction/extubation. Using Narcotrend-guided anesthesia effectively reduced the consumption of propofol during anesthesia and the recovery time (Kreuer, Biedler, Larsen, Altmann, & Wilhelm, 2003).

TABLE 1 The correspondence between the Narcotrend stage and the Narcotrend index (NI).

Nacotrend stage	Narcotrend Index (NI)
A	100-95
B0	94-90
B1	89-85
B2	84-80
C0	79-75
C1	74-70
C2	69-65
D0	64-57
D1	56-47
D2	46-37
E0	36-27
E1	26-20
E2	19-13
F0	12-5
F1	4-0

A: awake state; B: light sedation state; C: deep sedation state; D: general anesthesia state; E: deep anesthesia state; F: burst suppression state.

Regarding the application of Narcotrend in children, some studies indicated that in most infants and young children from the age of 4 months, NI had a significant correlation and acceptable prediction probability for minimal alveolar sevoflurane concentration (Dennhardt et al., 2018). However, another study (Dennhardt et al., 2017) found that children below 2 years of age displayed higher NI values and more pronounced interindividual variation during the TIVA of propofol (Dennhardt et al., 2017). This was consistent with the earlier research indicating that the NI was age-related and younger children had higher NI values (Münte et al., 2009). There is still no uniform standard for NI within children, but almost all authors believe that the use of raw EEG in children's EEG monitoring produces more accurate result.

Cerebral state index

The cerebral state index (CSI) was put forward by E.W.J. and P.M. and was based on the four subparameters of the EEG signal. Three of them were from EEG spectral analysis, and the fourth was the burst suppression rate calculated by formula.

A prospective study has indicated that the application of CSI in propofol general anesthesia could be a good assessment of anesthesia grade compared to the Observer's Assessment of Alertness and Sedation score. In this study, the authors applied BIS and the A-line ARX Index monitoring. Both the propofol burst suppression and the normal propofol sedation samples were tested at the same time (Jensen et al., 2006). Though both indicators were showed as numbers from 0 to 100, CSI is often compared with BIS, and the calculation algorithms were different. Because CSI has a lower cost than BIS, some hospitals prefer using CSI as an alternative than BIS. Many studies have compared the CSI and the BIS in different states of different anesthetics. Overall, the performance of CSI monitoring and BIS monitoring showed similar advantages during propofol induction. The sensitivity of the CSI response to sevoflurane anesthesia is still controversial, and different studies have given opposite results. Some studies have pointed out that the CSI was more sensitive than the BIS monitoring during sevoflurane anesthesia, while others have stated that the rate of change of CSI on sevoflurane is slower. This may be due to some conditions in the operation room which also affected the CSI. Though some studies have indicated the different dynamic characteristics of these monitors, others suggest that BIS may be a more useful indicator for assessing moderate anesthetic levels, while CSI may be better to assess deeper anesthetic levels. Whether the CSI has advantages among different EEG monitoring ways still needs to be further explored. Similar to BIS monitoring, the CSI monitoring in sedation may also not be applicable in pediatric patients.

As for the influence of anesthetic agent to CSI, Pilge et al. have analyzed the raw EEG data from deep to light anesthesia under two different anesthetic agents and found that the CSI performance during propofol anesthesia was superior to sevoflurane anesthesia (Pilge et al., 2012). Therefore, the author inferred that the algorithm of the CSI may be not independent of anesthetics agents. Other studies have shown that the use of the muscle relaxant under propofol anesthesia has no effect on CSI.

Applications to other areas

The raw EEG can be used in patients of special ages, such as very old patients or long-term neonatal surgery. Also, raw EEG monitoring can be applied to identify brain states under different anesthetics. BIS monitoring is mainly used to prevent intraoperative awareness. The application of BIS monitoring can be used in anesthesia outside the operating room, such as long time endoscopy or endoscopic treatment.

About evoked potential, few studies on evoked potentials published in recent years. BAEP are usually continuously monitored during neurosurgery to prevent brainstem damage and monitor neuronal function to improve the prognosis of neurosurgery. SSEP may be applied to monitor nerve function in nerve transplantation operations such as brachial plexus surgery. In some delicate operations, such as nerve suture under the microscope to monitor nerve function. AEP can monitor implicit memory. In recent years, most studies believe that anesthetics in a short period of time have basically no effect on children's neurodevelopment. But it is about whether anesthetics will affect children's implicit memory and how it will affect them are still unknown. AEP can be applied in the study of perioperative neuroprotection in children.

The entropy and BIS have a good correlation, which is suitable for various occasions where BIS is applicable. The entropy index can also predict the timing of extubation and accurately predict the state of consciousness and body movements during skin incision.

Compared with BIS, the price of Narcotrend and CSI is cheaper, and it is suitable for monitoring the depth of anesthesia in some primary hospitals.

Mini-dictionary of terms

Rapid eye movement (REM) sleep: Human sleep can be divided into two parts: rapid eye movement (REM) sleep and nonrapid eye movement (NREM) sleep. During the REM period, the frequency of brain waves becomes faster and the amplitude becomes lower. At the same time, it also shows increased heart rate, increased blood pressure, muscle relaxation, erection of the penis, and the eyeballs keep swinging.

Burst and suppression: The electroencephalogram is characterized by high-amplitude fulminant activity and low voltage or electrical suppression alternately. Mainly high-amplitude slow waves, sometimes compound spike waves, sharp waves and fast waves, lasting 0.5–1 s. Between the two burst periods, there is a low voltage or electric suppression period lasting 5–20 s, and the amplitude is less than 10 µV. It is a manifestation of extensive damage or inhibition of the cerebral cortex and cortex.

Bispectral index (BIS): BIS refers to the determination of the linear components (frequency and power) of the EEG, and the selection of various EEG signals that can represent different levels of sedation. After standardization and digitization, it is finally transformed into a simple quantitative index.

Accidental awareness (AAGA) during the anesthesia: Patients recover of consciousness during general anesthesia. The patient has a certain degree of perception and memory of the surrounding environment or sound. After general anesthesia, the patient can recall what happened during the operation and can tell whether there is pain or not. Anesthetic awareness is a serious complication of general anesthesia, which can cause serious psychological and mental disorders to patients.

Evoked potential: Evoked potential refers to the specific stimulus given to the nervous system (from receptors to the cerebral cortex), or the brain to process the stimulus (positive or negative) information, which can be detected in the system and the corresponding parts of the brain. Stimulus has a relatively fixed time interval (time-locked relationship) and a specific phase of the bioelectric response.

Key facts
Key facts of raw EEG

- In terms of monitoring the depth of anesthesia, the raw EEG is the most objective but also the most complicated.
- Unlike other monitoring methods, the raw EEG is suitable for the elderly and children in monitoring the depth of anesthesia.
- Raw EEG provides real time information about the actual depth of anesthesia when standard monitors provide misleading values.
- Observing the raw EEG waveforms provides accurate reliable data which can be used to assess the depth of anesthesia in different age and may reduce brain injury during general anesthesia.
- Raw EEG monitoring can help to avoid adverse events in clinical practice, such as profound anesthesia, light anesthesia, brain damage or brain death and hypotension.

Summary points

- The raw EEG is the most basic monitoring tool and has a wide range of application, but it is relatively complex.
- The other monitoring methods mentioned in this chapter are calculated based on the formula.
- The EEG monitoring will be affected by different anesthetics.
- Most calculation formulas are based on the EEG of adults, so values such as BIS may not be applicable in children and the elderly.
- Different monitoring methods have their own applicable scope, and there should be different choices in different operations and different patients.

References

Absalom, A. R., Sutcliffe, N., & Kenny, G. N. (2002). Closed-loop control of anesthesia using bispectral index: Performance assessment in patients undergoing major orthopedic surgery under combined general and regional anesthesia. *Anesthesiology, 96*, 67–73.

Anderson, R. E., & Jakobsson, J. G. (2004). Entropy of EEG during anaesthetic induction: A comparative study with propofol or nitrous oxide as sole agent. *British Journal of Anaesthesia, 92*, 167–170.

Beekoo, D., Yuan, K., Dai, S., Chen, L., Di, M., Wang, S., et al. (2019). Analyzing electroencephalography (EEG) waves provides a reliable tool to assess the depth of sevoflurane anesthesia in pediatric patients. *Medical Science Monitor, 25*, 4035–4040.

Berger, H. (1929). Ueber das Elektrenkephalogramm des Menschen. *Archivfür Psychiatrie und Nervenkrankheiten, 87*, 527.

Bonhomme, V., Llabres, V., Dewandre, P. Y., Brichant, J. F., & Hans, P. (2006). Combined use of Bispectral index and A-Lin e autoregressive index to assess anti-nociceptive component of balanced anaesthesia during lumbar arthrodesis. *British Journal of Anaesthesia, 96*, 353–360.

Brown, E. N., Purdon, P. L., & Van Dort, C. J. (2011). General anesthesia and altered states of arousal: A systems neuroscience analysis. *Annual Review of Neuroscience, 34*, 601–628.

Dahaba, A. A., Bornemann, H., Hopfgartner, E., Ohran, M., Kocher, K., Liebmann, M., et al. (2012). Effect of sugammadex or neostigmine neuromuscular block reversal on bispectral index monitoring of propofol/remifentanilanaesthesia. *British Journal of Anaesthesia, 108*, 602–606.

Dennhardt, N., Arndt, S., Beck, C., Boethig, D., Heiderich, S., Schultz, B., et al. (2018). Effect of age on narcotrend index monitoring during sevoflurane anesthesia in children below 2 years of age. *Paediatric Anaesthesia, 28*, 112–119.

Dennhardt, N., Boethig, D., Beck, C., Heiderich, S., Boehne, M., Leffler, A., et al. (2017). Optimization of initial propofol bolus dose for EEG narcotrend index-guided transition from sevoflurane induction to intravenous anesthesia in children. *Paediatric Anaesthesia, 27*, 425–432.

Dutta, A., Sethi, N., Sood, J., Panday, B. C., Gupta, M., Choudhary, P., et al. (2019). The effect of dexmedetomidine on propofol requirements during anesthesia administered by bispectral index-guided closed-loop anesthesia delivery system: A randomized controlled study. *Anesthesia and Analgesia, 129*, 84–91.

Gentili, F., Lougheed, W. M., Yamashiro, K., & Corrado, C. (1985). Monitoring of sensory evoked potentials during surgery of skull base tumours. *Canadian Journal of Neurological Science, 55*, 336–340.

Glass, P. S., Bloom, M., Kearse, L., Rosow, C., Sebel, P., & Manberg, P. (1997). Bispectral analysis measures sedation and memory effects of propofol, midazolam, isoflurane, and alfentanil in healthy volunteers. *Anesthesiology, 86*, 836–847.

Gruenewald, M., Zhou, J., Schloemerkemper, N., Meybohm, P., Weiler, N., Tonner, P. H., et al. (2007). M-entropy guidance vs standard practice during propofol-remifentanil anaesthesia: A randomised controlled trial. *Anaesthesia, 62*, 1224–1229.

Hagihira, S. (2015). Changes in the electroencephalogram during anaesthesia and their physiological basis. *British Journal of Anaesthesia.* https://doi.org/10.1093/bja/aev212.

Harper, C. M., Harner, S. G., Slavit, D. H., Litchy, W. J., Daube, J. R., Beatty, C. W., et al. (1992). Effect of BAER monitoring on hearing preservation during acoustic neuroma resection. *Neurology, 55*, 1551–1553.

Hor, E. T., Van Der Linden, P., Hert, D. S., Mélot, C., & Bidgoli, J. (2013). Impact of entropy monitoring on volatile anesthetic uptake. *Anesthesiology, 118*, 868–873.

Jensen, E. W., Litvan, H., Revuelta, M., Rodriguez, B. Z., Caminal, P., Martinez, P., et al. (2006). Cerebral state index during propofol anesthesia: A comparison with the bispectral index and the A-line ARX index. *Anesthesiology, 105*, 28–36.

Kochs, E. (1995). Electrophysiological monitoring and mild hypothermia. *Journal of Neurosurgical Anesthesiology, 7*, 222–228.

Kreuer, S., Biedler, A., Larsen, R., Altmann, S., & Wilhelm, W. (2003). Narcotrend monitoring allows faster emergence and a reduction of drug consumption in propofol-remifentanil anesthesia. *Anesthesiology, 99*, 34–41.

Kreuzer, M., Stern, M. A., Hight, D., Berger, S., Schneider, G., Sleigh, J. W., et al. (2020). Spectral and entropic features are altered by age in the electroencephalogram in patients under sevoflurane anesthesia. *Anesthesiology, 132*, 1003–1016.

Laitio, R. M., Kaskinoro, K., Särkelä, M. O., Kaisti, K. K., Salmi, E., Maksimow, A., et al. (2008b). Bispectral index, entropy, and quantitative electroencephalogram during single-agent xenon anesthesia. *Anesthesiology, 108*, 63–70.

Laitio, R. M., Kaskinoro, K., Särkelä, M. O., Kaisti, K. K., Salmi, E., Maksimow, A., et al. (2008a). Bispectral index, entropy, and quantitative electroencephalogram during single-agent xenon anesthesia. *Anesthesiology, 108*, 63–70.

MacKay, E. C., Sleigh, J. W., & Voss, L. J. (2010). Episodic waveforms in the electroencephalogram during general anaesthesia: A study of patterns of response to noxious stimuli. *Anaesthesia and Intensive Care, 38*, 102–112.

Mantzaridis, H., & Kenny, G. N. C. (1997). Auditory evoked potential index: A quantitative measure of changes in auditory evoked potentials during general anaesthesia. *Anaesthesia, 55*, 1030–1036.

Meng, X. L., Wang, L. W., Zhao, W., & Guo, X. Y. (2015). Effects of different etomidate doses on intraoperative somatosensory-evoked potential monitoring. *Irish Journal of Medical Science, 184*, 799–803.

Münte, S., Klockars, J., Gils, M. V., Hiller, A., Winterhalter, M., Quandt, C., et al. (2009). The narcotrend index indicates age-related changes during propofol induction in children. *Anesthesia and Analgesia, 109*, 53–59.

Myles, P. S., Leslie, K., McNeil, J., Forbes, A., & Chan, M. T. (2004). Bispectral index monitoring to prevent awareness during anaesthesia: The B-aware randomised controlled trial. *Lancet, 29*, 1757–1763.

Ohshima, N., Chinzei, M., Mizuno, K., Hayashida, M., Kitamura, T., Shibuya, H., et al. (2007). Transient decreases in bispectral index without associated changes in the level of consciousness during photic stimulation in an epileptic patient. *British Journal of Anaesthesia, 98*, 100–104.

Ozcengiz, D., Unlügenç, H., Güneş, Y., & Karacaer, F. (2012). The effect of dexmedetomidine on bispectral index monitoring in children. *Middle East Journal of Anaesthesiology, 214*, 613–618.

Pilge, S., Kreuzer, M., Karatchiviev, V., Kochs, E. F., Malcharek, M., & Schneider, G. (2015). Differences between state entropy and bispectral index during analysis of identical electroencephalogram signals: A comparison with two randomised anaesthetic techniques. *European Journal of Anaesthesiology, 32*, 354–365.

Pilge, S., Kreuzer, M., Kochs, E. F., Zanner, R., Paprotny, S., & Schneider, G. (2012). Monitors of the hypnotic component of anesthesia—Correlation between bispectral index and cerebral state index. *Minerva Anestesiologica, 78*, 636–645.

Purdon, P. L., Sampson, A., Pavone, K. J., & Brown, E. N. (2015). Clinical electroencephalography for anesthesiologists. Part I. Background and basic signatures. *Anesthesiology, 123*, 937–960.

Puri, G. D., Mathew, P. J., Biswas, I., Dutta, A., Sood, J., Gombar, S., et al. (2016). A multicenter evaluation of a closed-loop anesthesia delivery system: A randomized controlled trial. *Anesthesia and Analgesia, 122*, 106–114.

Rampil, I. J. (1998). A primer for EEG signal processing in anesthesia. *Anesthesiology, 89*, 980–1002.

Riad, W., Schreiber, M., & Saeed, A. B. (2007). Monitoring with EEG entropy decreases propofol requirement and maintains cardiovascular stability during induction of anaesthesia in elderly patients. *European Journal of Anaesthesiology, 24*, 684–688.

Schmidt, G. N., Bischoff, P., Standl, T., Jensen, K., Voigt, M., & Esch, J. S. A. (2003). Narcotrend and bispectral index monitor are superior to classic electroencephalographic parameters for the assessment of anesthetic states during propofol-remifentanil anesthesia. *Anesthesiology, 99*, 1072–1077.

Schneider, G., Hollweck, R., Ningler, M., Stockmanns, G., & Kochs, E. F. (2005). Detection of consciousness by electroencephalogram and auditory evoked potentials. *Anesthesiology, 103*, 934–943.

Schuller, P. J., Newell, S., Strickland, P. A., & Barry, J. J. (2015). Response of bispectral index to neuromuscular block in awake volunteers. *British Journal of Anesthesia, 115*, 95–103.

Schwender, D., Kaiser, A., Klasing, S., Peter, K., & Pöppel, E. (1994). Midlatency auditory evoked potentials and explicit and implicit memory in patients undergoing cardiac surgery. *Anesthesiology, 80*, 493–501.

Tewari, S., Bhadoria, P., Wadhawan, S., Prasad, S., & Kohli, A. (2016). Entropy vs standard clinical monitoring using total intravenous anesthesia during transvaginal oocyte retrieval in patients for in vitro fertilization. *Journal of Clinical Anesthesia, 34*, 105–112.

Tirel, O., Wodey, E., Harris, R., Bansard, J. Y., Ecoffey, C., & Senhadji, L. (2006). The impact of age on bispectral index values and EEG bispectrum during anaesthesia with desflurane and halothane in children. *British Journal of Anaesthesia, 96*, 480–485.

Tokuwaka, J., Satsumae, T., Mizutani, T., Yamada, K., Inomata, S., & Tanaka, M. (2015). The relationship between age and minimum alveolar concentration of sevoflurane for maintaining bispectral index below 50 in children. *Anaesthesia, 70*, 318–322.

Vakkuri, A., Yli-Hankala, A., Sandin, R., Mustola, S., Høymork, S., Nyblom, S., et al. (2005). Spectral entropy monitoring is associated with reduced propofol use and faster emergence in propofol-nitrous oxide-alfentanil anesthesia. *Anesthesiology, 103*, 274–279.

Weil, G., Passot, S., Servin, F., & Billard, V. (2008). Does spectral entropy reflect the response to intubation or incision during propofol-remifentanil anesthesia? *Anesthesia and Analgesia, 106*, 152–159.

Whyte, S. D., & Booker, P. D. (2003). Monitoring depth of anaesthesia by EEG. *BJA CEPD Reviews, 3*, 106–110.

Wodey, E., Tirel, O., Bansard, J. Y., Terrier, A., Chanavaz, C., Harris, R., et al. (2005a). Impact of age on both BIS values and EEG bispectrum during anaesthesia with sevoflurane in children. *British Journal of Anaesthesia, 94*, 810–820.

Wodey, E., Tirel, O., Bansard, J. Y., Terrier, A., Chanavaz, C., Harris, R., et al. (2005b). Impact of age on both BIS values and EEG bispectrum during anaesthesia with sevoflurane in children. *British Journal of Anaesthesia, 94*, 810–820.

Part II

The syndromes of pain

Chapter 9

Cluster headache and pain: Features and treatments

R.B. Brandt[a], J. Haan[a,b], G.M. Terwindt[a], and R. Fronczek[a]

[a]*Department of Neurology, Leiden University Medical Center, Leiden, The Netherlands,* [b]*Department of Neurology, Alrijne Hospital, Leiderdorp, The Netherlands*

List of abbreviations

CGRP	calcitonin gene-related protein
DBS	deep brain stimulation
GON	greater occipital nerve
NVNS	noninvasive vagus nerve stimulation
SPG	sphenopalatine ganglion
SSN	superior salivatory nucleus
SUNA	short-lasting unilateral neuralgiform headache attacks with cranial autonomic symptoms
SUNCT	short-lasting unilateral neuralgiform headache attacks with conjunctival injection and tearing
SUNHA	short-lasting unilateral neuralgiform headache attacks
TAC	trigeminal autonomic cephalalgia
TCC	trigeminal cervical complex
TCN	trigeminal caudal nucleus

Introduction

Cluster headache is a primary headache disorder consisting of unilateral headache attacks that last 15–180 min and may occur up to 8 times a day (Table 1) (Headache Classification Committee, 2018). The pain experienced during these headache attacks is considered as one of the most intense forms of pain known, and therefore it is often referred to as "suicide headache" with suicidal ideations occurring in 55% of patients (Rozen & Fishman, 2012). The pain is localized mostly periorbital and/or temporal and often described as an intense stabbing sensation.

Besides the pain, hallmarks of cluster headache are prominent autonomic features ipsilateral to the pain and/or restlessness (Torelli & Manzoni, 2003). Autonomic features may include red eye, tearing, rhinorrhea, nasal congestion, miosis, ptosis, eyelid edema, forehead, and facial sweating.

Cluster headache can be distinguished in a chronic (20%) and an episodic (80%) form. In episodic cluster headache, the attacks occur in "bouts" lasting weeks to months, alternating with attack-free periods lasting months. In the chronic form, patients do not experience sustained attack-free periods (Manzoni et al., 1983).

Epidemiology

Cluster headache prevalence is 1 per 1000 persons with an age of onset between 20 and 40 years. Cluster headache was regarded as a "male" disease with a male to female ratio of 3:1. This ratio appears to be declining since the 1980s to as low as 2:1 (Frederiksen, Lund, Barloese, Petersen, & Jensen, 2019), likely due to an improved recognition in woman.

Although timely and correct diagnosis is vital to receive adequate treatment, diagnostic delay in cluster headache was high with a median time to diagnosis of 5 years, and with only 20% of patients receiving the correct diagnosis at first visit (Rozen & Fishman, 2012). Luckily, time to diagnosis appears to be improving to 0.9 years in the 2010 decade (Frederiksen et al., 2019).

Features and Assessments of Pain, Anesthesia, and Analgesia. https://doi.org/10.1016/B978-0-12-818988-7.00022-4
Copyright © 2022 Elsevier Inc. All rights reserved.

TABLE 1 Diagnostic criteria for cluster headache according to the international classification of headache disorders third edition (ICHD-3) (Headache Classification Committee, 2018).

Diagnostic criteria for cluster headache

A. At least five attacks fulfilling criteria B-D

B. Severe or very severe unilateral orbital, supraorbital and/or temporal pain lasting 15–180 min (when untreated)

C. Either or both of the following:

   1. at least one of the following symptoms or signs, ipsilateral to the headache:
- conjunctival injection and/or lacrimation
- nasal congestion and/or rhinorrhea
- eyelid edema
- forehead and facial sweating
- miosis and/or ptosis

   2. a sense of restlessness or agitation

D. Occurring with a frequency between one every other day and 8 per day

E. Not better accounted for by another ICHD-3 diagnosis

Diagnostic criteria for episodic cluster headache

A. Attacks fulfilling criteria for *Cluster headache* and occurring in bouts (cluster periods)

B. At least two cluster periods lasting from 7 days to 1 year (when untreated) and separated by pain-free remission periods of ≥3 months

Diagnostic criteria for chronic cluster headache

A. Attacks fulfilling criteria for *Cluster headache*, and criterion B below

B. Occurring without a remission period, or with remissions lasting <3 months, for at least 1 year

Cluster headache is often associated with psychiatric comorbidity which may complicate treatment and increase burden. Cluster headache patients show a three times higher odds to develop depression than controls (Louter et al., 2016). Other frequent comorbidities are anxiety, aggressive behavior, and sleeping problems. A striking association exists between cluster headache and smoking, with a majority (60%–95%) patients who smoke. Due to this evident association, a possible pathophysiological role is attributed to first and second hand smoking (Rozen, 2018).

Rhythmicity

A fascinating aspect of cluster headache is its rhythmicity, highlighted by the episodic nature in which cluster periods occur where they appear to follow an annual rhythm in more than 50% of patients (Barloese et al., 2015) with most cluster periods beginning in the spring and fall with the changing of the seasons (Naber et al., 2019). This preponderance for spring and fall is likely related to daylight changes. This hypothesis is emphasized by the fact that this rhythmicity is lower in patients living closer to the equator, thus experiencing less change in seasonal daylight (Naber et al., 2019). Rhythmicity also occurs in individual cluster attacks, as they may occur in a predictable 24-h rhythm with most attacks occurring between 00:00 and 04:00 h (de Coo et al., 2019).

Patients may have a fear of going to bed, actively postponing sleep or using voluntary sleep deprivation to try to avoid attacks. Since attacks often occur around 1.5 h after sleep onset, a link with REM-sleep has been suggested (Barloese, Jennum, Lund, & Jensen, 2015). The exact nature of the link between cluster headache and sleep has never been fully elucidated, with conflicting study results. One of the first studies investigating this phenomenon showed the occurrence of attacks in both slow–wave-sleep and REM-sleep. Later studies did not show any relationship between sleep stages and attacks but showed the presence of a continuing or slowly recovering disturbance of sleep outside the bout rather than a disturbance secondary to attacks (Lund et al., 2019). This intrinsic sleep disturbance in patients was highlighted in another study where no differences in sleep parameters between patients inside and outside a cluster period were observed, but where all patients showed differences in sleep parameters compared to controls (Lund, Snoer, Jennum, Jensen, & Barloese, 2018).

Pathophysiology

Cluster headache is most often regarded as a neurovascular disease (Goadsby, 2002) although its exact cause remains unknown. Several structures have been identified which are thought to play key roles in attacks. First, the pain is perceived in the region of the ophthalmic division of the trigeminal nerve. Second, autonomic symptoms occur with a parasympathetic activation and a sympathetic deficit. Third, the hypothalamus has been identified as a possible "attack generator" (Hoffmann & May, 2018).

Trigeminal nerve

The trigeminal nerve consists of peripheral axons projecting to the dura mater and cerebral vessels and central axons projecting to the trigeminal caudal nucleus (TCN) in the trigeminocervical complex (TCC). The TCC acts as a relay station, projecting peripheral signals to cerebral structures involved in pain processing such as the thalamus and the cerebral cortex, causing the experience of pain upon activation. Since the pain experienced during a cluster headache attack is strictly unilateral, activation of the trigeminal nerve is supposedly unilateral.

Activation of the trigeminal nerve causes Calcitonin Gene-Related Protein (CGRP) release from the central and perivascular afferents. CGRP exhibits strong vasoactive effects causing vasodilation and appears to lower activation threshold in the TCN. Elevated serum CGRP levels in spontaneous and triggered cluster headache attacks were found (Fanciullacci, Alessandri, Figini, Geppetti, & Michelacci, 1995; Goadsby & Edvinsson, 1994). However, trigeminal nerve activation alone is not sufficient as demonstrated by continuation of attacks after trigeminal nerve root section (Matharu & Goadsby, 2002). CGRP infusion may cause a cluster headache attacks, but interestingly this attack provocation only occurred in chronic patients or episodic patients who were in a bout indicating a fluctuating susceptibility to CGRP (Vollesen et al., 2018).

Autonomic system

Unilateral autonomic features are one of the hallmarks of cluster headache attacks. These features are due to a sympathetic and parasympathetic imbalance. The trigeminal nerve is connected with the superior salivatory nucleus (SSN) from which parasympathetic nerve fibers project to the periphery via the sphenopalatine ganglion (SPG). Activation of these neurons causes release of neuropeptides (VIP and PACAP), called the trigeminal autonomic reflex. Notably, this reflex can only be induced via stimulation of the ophthalmic division of the trigeminal nerve (May, Büchel, Turner, & Goadsby, 2001). Although the autonomic features appear to be a consequence of attacks, recent successes in SPG stimulation and blockade suggest a larger role in cluster headache (Barloese et al., 2018; Mojica, Mo, & Ng, 2017).

Hypothalamus

The hypothalamus is regarded as "attack generator" in cluster headache. PET-CT and fMRI studies showed activation of the hypothalamus during both triggered and spontaneous attacks (May, Bahra, Buchel, Frackowiak, & Goadsby, 1998; Sprenger et al., 2004). Interestingly, no hypothalamic activation was shown during direct painful nociceptive stimulation in the region of the trigeminal nerve showing that it likely to be causal and not a consequence of trigeminal pain itself.

The apparent relation between sleep and attacks, and the characteristically circadian and annual rhythm in which attacks occur suggests involvement of the biological clock in cluster headache. The biological clock resides in the suprachiasmatic nucleus in the anterior part of the hypothalamus (Saper, Scammell, & Lu, 2005), which is shown to be enlarged in patients (Arkink et al., 2017). Regulation of the biological clock is partly due to melatonin. Production and excretion of this hormone takes places in the pineal gland and is suppressed by light, causing melatonin peak levels in the night. Melatonin peaks were shown to be diminished or absent in cluster headache patients further implying hypothalamic dysfunction (Waldenlind, Gustafsson, Ekbom, & Wetterberg, 1987).

Genetics

A genetic component in the pathophysiology of cluster headache has long been suggested. This suggestion was strengthened by epidemiological studies indicating a 5–18 times increased risk to develop cluster headache in first degree relatives of cluster headache patients, whereas second degree relatives exhibit a 1- to 3-fold increased risk (Gibson, Santos, Lund, Jensen, & Stylianou, 2019). Small genetic association studies focusing on the hypocretin receptor-2 gene

96 PART | II The syndromes of pain

(Weller et al., 2015), PER3 clock gene (Ofte, Tronvik, & Alstadhaug, 2016), CACNA1A gene (Haan et al., 2001), and PACAP receptor gene variant (Ran et al., 2017) were inconclusive. A gene expression study showed moderate gene expression differences and indicated a possible role for intracellular signaling and inflammatory genes in cluster headache (Eising et al., 2017). Large Genome Wide Association Studies (GWAS) and next-generation sequencing will shed more light on the role of complex genetics in cluster headache.

Trigeminal autonomic cephalalgias

Cluster headache is part of the "trigeminal autonomic cephalalgias" (TACs), an umbrella term for five distinctive primary headaches with overlapping pathophysiological and clinical features. These headaches are characterized by a unilateral, mostly periorbital and/or temporal headache with ipsilateral cranial autonomic symptoms (Table 2). Among cluster headache, the others include "Short-lasting unilateral neuralgiform headache attacks with conjunctival injection and tearing (SUNCT) or autonomic symptoms" (SUNA), paroxysmal hemicranias, and hemicrania continua.

TABLE 2 Diagnostic criteria for paroxysmal hemicrania according to the international classification of headache disorders third edition (ICHD-3) (Headache Classification Committee, 2018).

Diagnostic criteria for paroxysmal hemicrania:

A. At least 20 attacks fulfilling criteria B-E

B. Severe unilateral orbital, supraorbital and/or temporal pain lasting 2–30 min

C. Either or both of the following:

    1. at least one of the following symptoms or signs, ipsilateral to the headache:
- conjunctival injection and/or lacrimation
- nasal congestion and/or rhinorrhea
- eyelid edema
- forehead and facial sweating
- miosis and/or ptosis

    2. a sense of restlessness or agitation

D. Occurring with a frequency of >5 per day

E. Prevented absolutely by therapeutic doses of indomethacin

F. Not better accounted for by another ICHD-3 diagnosis.

Diagnostic criteria for short-lasting unilateral neuralgiform headache attacks (SUNHA):

A. At least 20 attacks fulfilling criteria B–D

B. Moderate or severe unilateral head pain, with orbital, supraorbital, temporal and/or other trigeminal distribution, lasting for 1–600 s and occurring as single stabs, series of stabs or in a saw-tooth pattern

C. At least one of the following five cranial autonomic symptoms or signs, ipsilateral to the pain:
- conjunctival injection and/or lacrimation
- nasal congestion and/or rhinorrhea
- eyelid edema
- forehead and facial sweating
- forehead and facial flushing
- sensation of fullness in the ear
- miosis and/or ptosis

D. Occurring with a frequency of at least one a day

E. Not better accounted for by another ICHD-3 diagnosis

Diagnostic criteria for hemicrania continua:

A. Unilateral headache fulfilling criteria B-D

B. Present for >3 months, with exacerbations of moderate or greater intensity

TABLE 2 Diagnostic criteria for paroxysmal hemicrania according to the international classification of headache disorders third edition (ICHD-3)—cont'd

C. Either or both of the following:
   1. at least one of the following symptoms or signs, ipsilateral to the headache:
      - conjunctival injection and/or lacrimation
      - nasal congestion and/or rhinorrhea
      - eyelid edema
      - forehead and facial sweating
      - miosis and/or ptosis
   2. a sense of restlessness or agitation, or aggravation of the pain by movement

D. Responds absolutely to therapeutic doses of indomethacin

E. Not better accounted for by another ICHD-3 diagnosis

In SUNCT and SUNA ultra short-lasting attacks (1–600 s) occur in single or multiple stabs or in a sawtooth pattern with a frequency of ~30 per day. During attacks, the pain is most often experienced in the ophthalmic region of the trigeminal nerve. Estimated prevalence range is 0.001%–0.1% (Giorgio Lambru, Rantell, Levy, & Matharu, 2019). Like trigeminal neuralgia, SUNCT and SUNA can be triggered by cutaneous or intraoral stimuli, although a characteristic refractory period and a mandibular/maxillary distribution of the pain are more specific features of trigeminal neuralgia.

Paroxysmal hemicrania has an unknown prevalence and is characterized by short-lasting (2–30 min) attacks with an average of ~10 attacks per day (Cittadini, Matharu, & Goadsby, 2008). Because of the severe pain and cranial autonomic symptoms, paroxysmal hemicrania may be easily confused with cluster headache, also due to the overlap in attack duration. However, as opposed to cluster headache, paroxysmal hemicrania exhibits a response to indomethacin treatment with patients showing complete remission or a significant reduction in attack frequency (Baraldi, Pellesi, Guerzoni, Cainazzo, & Pini, 2017).

Hemicrania continua are characterized by a continuous mild to moderate, strictly unilateral, mostly periorbital, headache with exacerbations of varying duration (minutes to days) and intensity. During the exacerbations, ipsilateral cranial autonomic symptoms and, in 60% of cases, migrainous features occur, complicating the differentiation between migraine and hemicrania continua. However, as in paroxysmal hemicrania, hemicrania continua are indomethacin responsive. The headache of hemicrania continua is side-locked and does not show remission periods between attacks (Mehta, Chilakamarri, Zubair, & Kuruvilla, 2018; Prakash & Patel, 2017).

Differentiation between the TACs can be difficult. Especially paroxysmal hemicrania and cluster headache can show remarkable overlap. Since no laboratory test or imaging technique can differentiate between the TACs, the clinical history is important. In general, TACs can be distinguished by differences in attack duration and attack frequency (Fig. 1). Furthermore, attack pattern (e.g., series of stabs, nightly attacks, etc.) and response to treatment can help distinguish between them (Table 3). Finally, especially, the near-absolute response to indomethacin can confirm the diagnosis of a hemicrania variant.

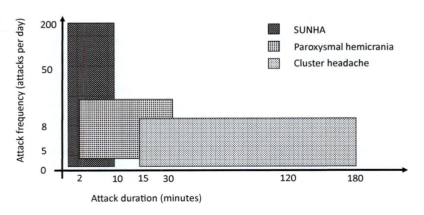

FIG. 1 Attack frequency and duration of the different TACs.

98　**PART** | **II** The syndromes of pain

TABLE 3 Differentiation between the different TACs.

	Cluster headache	Paroxysmal hemicrania	SUNHA	Hemicrania continua
Attack frequency	1–8 attacks daily	>5 attacks daily	1–200 daily	Continuous pain, fluctuating excacerbations
Attack duration	15–180 min	2–30 min	1–600 s	Varying excacerbation durations (minutes–days)
Restlessness	>90%	80%	30%	50%
Circadian pattern	Evident	Very rare	Rare	Very rare
Treatment response				
100% oxygen	>75%	No effect	No effect	No effect
Sumatriptan s.c.	>75%	20%	<10%	No effect
Indometacine	No effect	100%	No effect	100%

Treatment

Treatment of cluster headache consists of two principles. First, the extreme pain of a cluster headache attack requires a fast-acting acute treatment. Second, cluster headache attacks always need prophylactic therapy. Although treatment should be aimed to obtain attack freedom, this cannot always be achieved since (chronic) cluster headache can be very therapy-resistant (Brandt, Doesborg, Haan, Ferrari, & Fronczek, 2020) (Table 4).

Pharmacological treatments

Acute treatment

Sumatriptan appears to be the most effective acute treatment with 75% of patients achieving pain-freedom within 15 min when sumatriptan is used subcutaneously (Ekbom et al., 1993). At present, the recommended dosage is 6 mg although lower dosages (2–3 mg) may be effective in a subgroup of patients as well (Gregor et al., 2005). Triptan-related long-term side effects are negligible, even for patients using more than two injections per day for over longer time periods (Leone & Proietti Cecchini, 2016). If subcutaneous injections are not tolerated by the patient, sumatriptan nasal spray may be advised although its onset of effect lasts longer due to slower drug resorption.

High-flow oxygen can also be used as attack treatment. High-flow oxygen therapy most likely inhibits dural inflammation and neuronal activation in the trigeminocervical complex (Simon Akerman, Holland, Lasalandra, & Goadsby, 2009). Since no adverse events have been reported in treatment with high-flow oxygen, it is recommended for all patients as a first-choice treatment. Recommended flow rates are between 12 and 15 L/min although a recent study suggested the use of an even higher flow rate using "demand valve oxygen," supporting flow rates up to 160 L/min (Petersen, Barloese, Lund, & Jensen, 2017).

Prophylactic treatment

Only a very limited number of RCTs has been conducted regarding prophylactic therapy in cluster headache. Most prophylactic drugs are prescribed off-label and their use is based on a low level of evidence. Verapamil up to 720 to 960 mg daily is commonly regarded as the first-choice prophylactic therapy in cluster headache.

The mechanism of action of verapamil in cluster headache is not fully elucidated. Verapamil is a calcium antagonist, preventing slow calcium influx into the cell. Presynaptic calcium channels are present on trigeminovascular neurons. Blocking these channels prevents CGRP-release, thereby possibly modifying trigeminal nociceptive transmission and consequently inhibiting the hyperresponsive state (Akerman, Williamson, & Goadsby, 2003; Petersen, Barloese, Snoer, Soerensen, & Jensen, 2019). A placebo-controlled RCT with verapamil (Leone et al., 2000) showed decrease in daily attack frequency. Unfortunately, patients may experience, mainly cardiac related, side effects even at low dosages, such as arrhythmias, bradycardia, and palpitations (Lanteri-Minet, Silhol, Piano, & Donnet, 2011). Therefore, EKG monitoring is advised pretreatment and before or after dose increases (Koppen et al., 2016).

TABLE 4 Treatment recommendations.

Acute treatment	
Pharmacological	
Sumatriptan	6 mg subcutaneous up to 8 times a day
High flow oxygen	7–12 L/min or demand valve oxygen
Neuromodulatory	
NVNS	Stimulation during attacks
Prophylactic treatment	
Pharmacological	
Verapamil	Up to 960 mg daily
Lithium	Serum concentration between 0.7 and 1.2 mmol/L
Topiramate	Up to 200 mg daily
Neuromodulatory	
Occipital nerve stimulation	Continuous stimulation
SPG modulation	Stimulation during and outside attacks
Deep brain stimulation	Last treatment option
Transitional treatment	
Pharmacological	
Frovatriptan	2.5–5 mg daily
Prednisone	Different dosage available
Neuromodulatory	
GON-injection	Single injection

Other prophylactic treatment options include lithium and topiramate. Two RCTs have been conducted studying lithium with only one trial showing positive results, although the endpoint was probably set too soon in the negative trial (Bussone et al., 1990; Steiner, Hering, Couturier, Davies, & Whitmarsh, 1997). Since adequate lithium serum levels (between 0.8 and 1.2 mmol/L) have to be titrated carefully, the effect of lithium therapy often takes rather long to achieve. Side effects of lithium are common and can be severe, often causing discontinuation of therapy. They include tremor, dizziness, and nausea. Moreover, hypothyroidism and kidney dysfunction can occur during long-term therapy. Evidence of the efficacy of topiramate is scarce with only a few small retrospective and prospective open-label trials. In clinical practice, topiramate may be added to (the maximum tolerated) verapamil and may be helpful.

Recently, galcanezumab, an anti-CGRP monoclonal antibody, has shown effectiveness as prophylactic therapy for episodic cluster headache (Goadsby et al., 2019). Longer trials are needed to establish long-term efficacy and adverse events, also for chronic cluster headache.

Transitional treatment

Unfortunately, prophylactic treatment often does not exert its effect within the first days to weeks since it needs to be slowly titrated to the correct dosage. During this titration period, transitional therapy is often needed.

Although evidence from RCTs is lacking, high doses of corticosteroids are shown to be highly effective but the reason for this efficacy is unknown. No consensus has been reached regarding the mode of administration, dosage, and treatment duration. Since long-term use of corticosteroids is not recommended, prednisolone is mostly administered during a short duration, starting at a high dose of 60 mg per day and tapered down fast (D'Arrigo, Di Fiore, Galli, & Frediani, 2018). An alternative therapy is frovatriptan where some effect was shown in a case series (Siow, Pozo-Rosich, & Silberstein, 2004).

Neuromodulatory treatments

Noninvasive vagus nerve stimulation

Noninvasive vagus nerve stimulation (NVNS) is a relatively new attack treatment in cluster headache. Two RCTs have been conducted and a pooled analysis showed some efficacy in episodic cluster headache. Almost half of patients exhibited a positive response in >50% of attacks. Although probably not as effective as traditional attack treatment, NVNS is easy to use, minimally invasive, and can also be used in patients with cardiovascular disease (De Coo et al., 2019).

SPG stimulation/blockade

The sphenopalatine ganglion is the final common parasympathetic pathway. Interestingly, both blockade and stimulation of the SPG can result in attack cessation and prevention. This implicates that the SPG may be a key pathophysiological component in cluster headache. Patients may experience long-lasting effects of SPG modulation, making this a promising treatment for chronic patients (Narouze, Kapural, Casanova, & Mekhail, 2009; Schoenen et al., 2013).

GON stimulation/blockade

Local steroid injection in the region of the Greater Occipital Nerve (GON-injection) is a therapeutic intervention existing since the 1960s for the treatment of chronic local neuropathic pain. A case series and two small RCTs suggested its efficacy in both episodic and chronic cluster headache (Ambrosini et al., 2005; Leroux et al., 2011). It has been suggested that the beneficial effect is due to a systemic effect of the injected corticosteroids (Ambrosini et al., 2005), but this seems highly unlikely. The most plausible mechanism of action is supposedly through modification of trigeminal nociceptive transmission. GON-injection is mainly used as transitional treatment during the titration of oral preventive medication. However, a single GON-injection might be a good first-choice therapy for episodic cluster headache, removing the need for high dosages of verapamil. Repeated GON-injection might be a viable treatment option for chronic cluster headache as suggested by two open-label trials (Gantenbein, Lutz, Riederer, & Sandor, 2012; G. Lambru et al., 2014), especially for patients that do not respond to, or suffer from debilitating side effects from, regular preventive drugs.

Occipital Nerve Stimulation (ONS) has been studied in several open-label studies showing good results in medically intractable chronic cluster headache patients. Its efficacy appears to be persistent in long-term follow-up (Leone, Proietti Cecchini, Messina, & Franzini, 2017). A recently completed double-blind RCT studying the efficacy of occipital nerve stimulation in medically intractable chronic cluster headache showed promising results (Wilbrink et al., 2013).

Deep brain stimulation

The first report of deep brain stimulation in cluster headache was published in the new century. A 39-year-old male with intractable chronic cluster headache received DBS in the posterior hypothalamus resulting in attack freedom (Leone, Franzini, & Bussone, 2001). Despite the invasive and risky procedure, multiple case series have been published reporting a long-lasting reduction of 48%–100% in attack frequency (Vyas et al., 2019). Three stimulation targets have been identified: the posterior hypothalamus, the ventral tegmental area, and the floor of the third ventricle. The exact mechanism of action is not yet know, but DBS probably exerts its effect through functional neuromodulation, as suggested by the mean of 42 days before onset of pain relief (Vyas et al., 2019). Because of the invasive nature of DBS and its possible severe complications, it is commonly regarded as last resort treatment option. In order to be eligible for this treatment, patients must be diagnosed with refractory chronic cluster headache, failing all other treatment options.

Applications to other areas

Cluster headache is part of the group of TACs that show pathophysiological and clinical overlap. Other TACs might benefit from advances made in the understanding of cluster headache pathophysiology and treatment. For example, drugs used in the treatment of cluster headache frequently have shown effect in other TACs as well. The same is true for the shift from pharmacological treatment to neuromodulation. Research in cluster headache therapy, for instance ONS and GON-injection, may lead to understanding of the functional basis of these therapies, and of pain neuroplasticity in general. Neuroimaging studies in cluster headache have shown changes in pain processing areas, which also occur in other chronic pain patients.

Other agents of interest

Methysergide and ergotamine are two of the oldest effective treatments of cluster headache. Unfortunately, methysergide is no longer available and ergotamine can only be used for a limited duration due to possibly severe vascular side effects. Because of the apparent hypothalamic involvement in cluster headache, hypothalamic-controlled hormones such as melatonin and testosterone have been tried as treatment. A possible effect of melatonin was found in a small RCT (Leone, D'Amico, Moschiano, Fraschini, & Bussone, 1996). Although with low evidence, low testosterone levels have been found in cluster headache patients, and testosterone replacement therapy was tried with clomiphene, an androgen production stimulating agent, in two case reports (Rozen, 2015). A plethora of other drugs, such as pizotiphene, gabapentin, and sodium valproate, are used with only very limited evidence available.

Cluster headache patients report significantly higher drug use when compared to the general population (de Coo et al., 2018). Although some cluster headache patients use illicit drugs to treat their headache, no scientific evidence for effectiveness exists. Only one case series suggested promising effects of psilocybin and LSD in the treatment of cluster headache (Sewell, Halpern, & Pope Jr., 2006).

Conclusion

Cluster headache is part of the trigeminal autonomic cephalalgias (TACs), a group of primary headaches that share distinctive clinical and pathophysiological characteristics with unilateral attacks of excruciating pain accompanied by autonomic symptoms. Cluster headache can appear in the episodic variant, in which attacks mostly occur in bouts, and the chronic variant, where attacks occur without a remission period, or with remissions lasting <3 months, for at least 1 year. Three key structures have been identified to play a role in its pathophysiology: the hypothalamus, as attack generator; the trigeminal vascular system, leading to the pain; and the (para)sympathetic autonomic system, leading to typical ipsilateral autonomic features. First priority in cluster headache treatment is effective attack treatment with 100% oxygen and/or subcutaneous sumatriptan. Prophylactic treatment is urgently needed in long periods of episodic and in chronic cluster headache. First choice is verapamil, with lithium or topiramate as alternative options. If these options fail, other, lower-level evidence pharmacological options can be tried, but neuromodulatory options can be considered as well. Since attack freedom cannot always be achieved, cluster headache therapy needs to be highly individualized and effect/side effect ratio needs to be carefully evaluated.

Mini-dictionary of terms

ECH: Episodic cluster headache is the most frequent form of cluster headache (80%) with attacks occurring in cluster of several weeks to months separated by a period of attack freedom lasting months to years.

CCH: Chronic cluster headache occurs in approximately 20% of cluster headache patients and is characterized by recurring attacks without an attack-free period of 3 months.

TACs: Trigeminal autonomic cephalalgia is an umbrella term for a group of headaches that share distinctive clinical and pathophysiological characteristics consisting of unilateral headache attacks by ipsilateral autonomic symptoms.

Attack treatment: Acute treatment aimed to rapidly relief the pain. Due to the short attacks and the aim to achieve attack relieve as soon as possible, attack treatment mostly consists of *subcutaneous* administered sumatriptan and 100% oxygen.

Transitional treatment: Transitional treatment is aimed to "bridge" the period between start of prophylactic treatment and onset of effect. Transitional treatment, therefore, is only effective for days to weeks.

Prophylactic treatment: The primary goal of prophylactic treatment is to achieve attack freedom. However, since this is not always possible, we strive to achieve the best effect/side effect ratio.

Key facts/summary points

- Cluster headache is characterized by attacks of excruciating unilateral periorbital pain accompanied by ipsilateral autonomic symptoms, such as lacrimation, nasal congestion, miosis, and conjunctival injection.
- Cluster headache is categorized in an episodic and a chronic form, with episodic cluster headache being the most frequent (80%).
- Although prevalence of cluster headache is around 1 in 1000 persons, it is a relatively unknown disorder, leading to diagnostic delay and inadequate treatment.

- Pathophysiology of cluster headache has not yet been elucidated. However, several key structures (hypothalamus, trigeminovascular system, and autonomic system) are identified with the hypothalamus often being regarded as the "attack generator."
- Cluster headache attacks often occur during sleep and on predictable times and patients can experience up to 8 attacks per day.
- Cluster headache is one of five trigeminal autonomic cephalalgias, a group of headaches that is characterized by unilateral attacks of excruciating pain accompanied by ipsilateral autonomic symptoms.
- Cluster headache treatment is divided in acute treatment (oxygen and sumatriptan) and prophylactic treatment (verapamil, lithium).
- Neuromodulatory treatment options are increasingly effective in acute and prophylactic treatment of cluster headache.

References

Akerman, S., Holland, P. R., Lasalandra, M. P., & Goadsby, P. J. (2009). Oxygen inhibits neuronal activation in the trigeminocervical complex after stimulation of trigeminal autonomic reflex, but not during direct dural activation of trigeminal afferents. *Headache*, *49*(8), 1131–1143.

Akerman, S., Williamson, D. J., & Goadsby, P. J. (2003). Voltage-dependent calcium channels are involved in neurogenic dural vasodilatation via a presynaptic transmitter release mechanism. *British Journal of Pharmacology*, *140*(3), 558–566.

Ambrosini, A., Vandenheede, M., Rossi, P., Aloj, F., Sauli, E., Pierelli, F., et al. (2005). Suboccipital injection with a mixture of rapid- and long-acting steroids in cluster headache: A double-blind placebo-controlled study. *Pain*, *118*(1–2), 92–96.

Arkink, E. B., Schmitz, N., Schoonman, G. G., van Vliet, J. A., Haan, J., van Buchem, M. A., et al. (2017). The anterior hypothalamus in cluster headache. *Cephalalgia*, *37*(11), 1039–1050.

Baraldi, C., Pellesi, L., Guerzoni, S., Cainazzo, M. M., & Pini, L. A. (2017). Therapeutical approaches to paroxysmal hemicrania, hemicrania continua and short lasting unilateral neuralgiform headache attacks: A critical appraisal. *The Journal of Headache and Pain*, *18*(1).

Barloese, M. C., Jennum, P. J., Lund, N. T., & Jensen, R. H. (2015). Sleep in cluster headache - beyond a temporal rapid eye movement relationship? *European Journal of Neurology*, *22*(4). 656-e640.

Barloese, M. C., Lund, N. L. T., Petersen, A. S., Rasmussen, M., Jennum, P., & Jensen, R. (2015). Sleep and chronobiology in cluster headache. *Cephalalgia*, *35*(11), 969–978.

Barloese, M. C., Petersen, A. S., Stude, P., Jürgens, T., Jensen, R. M., & May, A. (2018). Sphenopalatine ganglion stimulation for cluster headache, results from a large, open-label European registry. *The Journal of Headache and Pain*, *19*(1).

Brandt, R. B., Doesborg, P. G. G., Haan, J., Ferrari, M. D., & Fronczek, R. (2020). Pharmacotherapy for cluster headache. *CNS Drugs*.

Bussone, G., Leone, M., Peccarisi, C., Micieli, G., Granella, F., Magri, M., et al. (1990). Double blind comparison of lithium and verapamil in cluster headache prophylaxis. *Headache*, *30*(7), 411–417.

Cittadini, E., Matharu, M. S., & Goadsby, P. J. (2008). Paroxysmal hemicrania: A prospective clinical study of 31 cases. *Brain*, *131*(4), 1142–1155.

D'Arrigo, G., Di Fiore, P., Galli, A., & Frediani, F. (2018). High dosage of methylprednisolone in cluster headache. *Neurological Sciences*, *39*(Suppl 1), 157–158.

De Coo, I. F., Marin, J. C., Silberstein, S. D., Friedman, D. I., Gaul, C., McClure, C. K., et al. (2019). Differential efficacy of non-invasive vagus nerve stimulation for the acute treatment of episodic and chronic cluster headache: A meta-analysis. *Cephalalgia*, *39*(8), 967–977.

de Coo, I. F., Naber, W. C., Wilbrink, L. A., Haan, J., Ferrari, M. D., & Fronczek, R. (2018). Increased use of illicit drugs in a Dutch cluster headache population. *Cephalalgia*. 333102418804160.

de Coo, I. F., van Oosterhout, W. P. J., Wilbrink, L. A., van Zwet, E. W., Ferrari, M. D., & Fronczek, R. (2019). Chronobiology and sleep in cluster headache. *Headache*, *59*(7), 1032–1041.

Eising, E., Pelzer, N., Vijfhuizen, L. S., Vries, B. D., Ferrari, M. D., t'Hoen, P. A. C., et al. (2017). Identifying a gene expression signature of cluster headache in blood. *Scientific Reports*, *7*(1), 40218.

Ekbom, K., Monstad, I., Prusinski, A., Cole, J. A., Pilgrim, A. J., & Noronha, D. (1993). Subcutaneous sumatriptan in the acute treatment of cluster headache: A dose comparison study. The Sumatriptan cluster headache study group. *Acta Neurologica Scandinavica*, *88*(1), 63–69.

Fanciullacci, M., Alessandri, M., Figini, M., Geppetti, P., & Michelacci, S. (1995). Increase in plasma calcitonin gene-related peptide from the extracerebral circulation during nitroglycerin-induced cluster headache attack. *Pain*, *60*(2), 119–123.

Frederiksen, H. H., Lund, N. L., Barloese, M. C., Petersen, A. S., & Jensen, R. H. (2019). Diagnostic delay of cluster headache: A cohort study from the Danish cluster headache survey. *Cephalalgia*. 333102419863030.

Gantenbein, A. R., Lutz, N. J., Riederer, F., & Sandor, P. S. (2012). Efficacy and safety of 121 injections of the greater occipital nerve in episodic and chronic cluster headache. *Cephalalgia*, *32*(8), 630–634.

Gibson, K. F., Santos, A. D., Lund, N., Jensen, R., & Stylianou, I. M. (2019). Genetics of cluster headache. *Cephalalgia*, *39*(10), 1298–1312.

Goadsby, P. J. (2002). Pathophysiology of cluster headache: A trigeminal autonomic cephalgia. *Lancet Neurology*, *1*(4), 251–257.

Goadsby, P. J., Dodick, D. W., Leone, M., Bardos, J. N., Oakes, T. M., Millen, B. A., et al. (2019). Trial of Galcanezumab in prevention of episodic cluster headache. *The New England Journal of Medicine*, *381*(2), 132–141.

Goadsby, P. J., & Edvinsson, L. (1994). Human in vivo evidence for trigeminovascular activation in cluster headache. Neuropeptide changes and effects of acute attacks therapies. *Brain*, *117*(Pt 3), 427–434.

Gregor, N., Schlesiger, C., Akova-Ozturk, E., Kraemer, C., Husstedt, I. W., & Evers, S. (2005). Treatment of cluster headache attacks with less than 6 mg subcutaneous sumatriptan. *Headache, 45*(8), 1069–1072.

Haan, J., Van Vliet, J., Kors, E., Terwindt, G., Vermeulen, F., Van Den Maagdenberg, A., et al. (2001). No involvement of the calcium channel gene (CACNA1A) in a family with cluster headache. *Cephalalgia, 21*(10), 959–962.

Headache Classification Committee. (2018). The international classification of headache disorders, 3rd edition. *Cephalalgia, 38*(1), 1–211.

Hoffmann, J., & May, A. (2018). Diagnosis, pathophysiology, and management of cluster headache. *Lancet Neurology, 17*(1), 75–83.

Koppen, H., Stolwijk, J., Wilms, E. B., van Driel, V., Ferrari, M. D., & Haan, J. (2016). Cardiac monitoring of high-dose verapamil in cluster headache: An international Delphi study. *Cephalalgia, 36*(14), 1385–1388.

Lambru, G., Abu Bakar, N., Stahlhut, L., McCulloch, S., Miller, S., Shanahan, P., et al. (2014). Greater occipital nerve blocks in chronic cluster headache: A prospective open-label study. *European Journal of Neurology, 21*(2), 338–343.

Lambru, G., Rantell, K., Levy, A., & Matharu, M. S. (2019). A prospective comparative study and analysis of predictors of SUNA and SUNCT. *Neurology, 93*(12), e1127–e1137.

Lanteri-Minet, M., Silhol, F., Piano, V., & Donnet, A. (2011). Cardiac safety in cluster headache patients using the very high dose of verapamil (>/=720 mg/day). *The Journal of Headache and Pain, 12*(2), 173–176.

Leone, M., D'Amico, D., Frediani, F., Moschiano, F., Grazzi, L., Attanasio, A., et al. (2000). Verapamil in the prophylaxis of episodic cluster headache: A double-blind study versus placebo. *Neurology, 54*(6), 1382–1385.

Leone, M., D'Amico, D., Moschiano, F., Fraschini, F., & Bussone, G. (1996). Melatonin versus placebo in the prophylaxis of cluster headache: A double-blind pilot study with parallel groups. *Cephalalgia, 16*(7), 494–496.

Leone, M., Franzini, A., & Bussone, G. (2001). Stereotactic stimulation of posterior hypothalamic gray matter in a patient with intractable cluster headache. *New England Journal of Medicine, 345*(19), 1428–1429.

Leone, M., & Proietti Cecchini, A. (2016). Long-term use of daily sumatriptan injections in severe drug-resistant chronic cluster headache. *Neurology, 86*(2), 194–195.

Leone, M., Proietti Cecchini, A., Messina, G., & Franzini, A. (2017). Long-term occipital nerve stimulation for drug-resistant chronic cluster headache. *Cephalalgia, 37*(8), 756–763.

Leroux, E., Valade, D., Taifas, I., Vicaut, E., Chagnon, M., Roos, C., et al. (2011). Suboccipital steroid injections for transitional treatment of patients with more than two cluster headache attacks per day: A randomised, double-blind, placebo-controlled trial. *Lancet Neurology, 10*(10), 891–897.

Louter, M. A., Wilbrink, L. A., Haan, J., van Zwet, E. W., van Oosterhout, W. P., Zitman, F. G., et al. (2016). Cluster headache and depression. *Neurology, 87*(18), 1899–1906.

Lund, N. L. T., Snoer, A. H., Jennum, P. J., Jensen, R. H., & Barloese, M. C. J. (2018). Sleep in cluster headache revisited: Results from a controlled actigraphic study. *Cephalalgia.* 333102418815506.

Lund, N. L. T., Snoer, A. H., Petersen, A. S., Beske, R. P., Jennum, P. J., Jensen, R. H., et al. (2019). Disturbed sleep in cluster headache is not the result of transient processes associated with the cluster period. *European Journal of Neurology, 26*(2), 290–298.

Manzoni, G. C., Terzano, M. G., Bono, G., Micieli, G., Martucci, N., & Nappi, G. (1983). Cluster headache- -clinical findings in 180 patients. *Cephalalgia, 3*(1), 21–30.

Matharu, M. S., & Goadsby, P. J. (2002). Persistence of attacks of cluster headache after trigeminal nerve root section. *Brain, 125*(Pt 5), 976–984.

May, A., Bahra, A., Buchel, C., Frackowiak, R. S., & Goadsby, P. J. (1998). Hypothalamic activation in cluster headache attacks. *Lancet, 352*(9124), 275–278.

May, A., Büchel, C., Turner, R., & Goadsby, P. J. (2001). Magnetic resonance angiography in facial and other pain: Neurovascular mechanisms of trigeminal sensation. *Journal of Cerebral Blood Flow and Metabolism, 21*, 1171–1176.

Mehta, A., Chilakamarri, P., Zubair, A., & Kuruvilla, D. E. (2018). Hemicrania continua: A clinical perspective on diagnosis and management. *Current Neurology and Neuroscience Reports, 18*(12), 95.

Mojica, J., Mo, B., & Ng, A. (2017). Sphenopalatine ganglion block in the management of chronic headaches. *Current Pain and Headache Reports, 21*(6).

Naber, W. C., Fronczek, R., Haan, J., Doesborg, P., Colwell, C. S., Ferrari, M. D., et al. (2019). The biological clock in cluster headache: A review and hypothesis. *Cephalalgia.* 333102419851815.

Narouze, S., Kapural, L., Casanova, J., & Mekhail, N. (2009). Sphenopalatine ganglion radiofrequency ablation for the management of chronic cluster headache. *Headache: The Journal of Head and Face Pain, 49*(4), 571–577.

Ofte, H. K., Tronvik, E., & Alstadhaug, K. B. (2016). Lack of association between cluster headache and PER3 clock gene polymorphism. *The Journal of Headache and Pain, 17*(1).

Petersen, A. S., Barloese, M. C., Lund, N. L., & Jensen, R. H. (2017). Oxygen therapy for cluster headache. A mask comparison trial. A single-blinded, placebo-controlled, crossover study. *Cephalalgia, 37*(3), 214–224.

Petersen, A. S., Barloese, M. C. J., Snoer, A., Soerensen, A. M. S., & Jensen, R. H. (2019). Verapamil and cluster headache: Still a mystery. A narrative review of efficacy, mechanisms and perspectives. *Headache: The Journal of Head and Face Pain, 59*(8), 1198–1211.

Prakash, S., & Patel, P. (2017). Hemicrania continua: clinical review, diagnosis and management. *Journal of Pain Research, 10*, 1493–1509.

Ran, C., Fourier, C., Michalska, J. M., Steinberg, A., Sjostrand, C., Waldenlind, E., et al. (2017). Screening of genetic variants in ADCYAP1R1, MME and 14q21 in a Swedish cluster headache cohort. *The Journal of Headache and Pain, 18*(1), 88.

Rozen, T. D. (2015). Clomiphene citrate as a preventive treatment for intractable chronic cluster headache: A second reported case with long-term follow-up. *Headache, 55*(4), 571–574.

Rozen, T. D. (2018). Linking cigarette smoking/tobacco exposure and cluster headache: A pathogenesis theory. *Headache, 58*(7), 1096–1112.

Rozen, T. D., & Fishman, R. S. (2012). Cluster headache in the United States of America: Demographics, clinical characteristics, triggers, suicidality, and personal burden. *Headache, 52*(1), 99–113.

Saper, C. B., Scammell, T. E., & Lu, J. (2005). Hypothalamic regulation of sleep and circadian rhythms. *Nature, 437*(7063), 1257–1263.

Schoenen, J., Jensen, R. H., Lantéri-Minet, M., Láinez, M. J., Gaul, C., Goodman, A. M., et al. (2013). Stimulation of the sphenopalatine ganglion (SPG) for cluster headache treatment. Pathway CH-1: A randomized, sham-controlled study. *Cephalalgia, 33*(10), 816–830.

Sewell, R. A., Halpern, J. H., & Pope, H. G., Jr. (2006). Response of cluster headache to psilocybin and LSD. *Neurology, 66*(12), 1920–1922.

Siow, H. C., Pozo-Rosich, P., & Silberstein, S. D. (2004). Frovatriptan for the treatment of cluster headaches. *Cephalalgia, 24*(12), 1045–1048.

Sprenger, T., Boecker, H., Tolle, T. R., Bussone, G., May, A., & Leone, M. (2004). Specific hypothalamic activation during a spontaneous cluster headache attack. *Neurology, 62*(3), 516–517.

Steiner, T. J., Hering, R., Couturier, E. G., Davies, P. T., & Whitmarsh, T. E. (1997). Double-blind placebo-controlled trial of lithium in episodic cluster headache. *Cephalalgia: An International Journal of Headache, 17*(6), 673–675.

Torelli, P., & Manzoni, G. C. (2003). Pain and behaviour in cluster headache. A prospective study and review of the literature. *Functional Neurology, 18*(4), 205–210.

Vollesen, A. L. H., Snoer, A., Beske, R. P., Guo, S., Hoffmann, J., Jensen, R. H., et al. (2018). Effect of infusion of calcitonin gene-related peptide on cluster headache attacks: A randomized clinical trial. *JAMA Neurology, 75*(10), 1187–1197.

Vyas, D. B., Ho, A. L., Dadey, D. Y., Pendharkar, A. V., Sussman, E. S., Cowan, R., et al. (2019). Deep brain stimulation for chronic cluster headache: A review. *Neuromodulation: Technology at the Neural Interface, 22*(4), 388–397.

Waldenlind, E., Gustafsson, S. A., Ekbom, K., & Wetterberg, L. (1987). Circadian secretion of cortisol and melatonin in cluster headache during active cluster periods and remission. *Journal of Neurology, Neurosurgery, and Psychiatry, 50*(2), 207–213.

Weller, C. M., Wilbrink, L. A., Houwing-Duistermaat, J. J., Koelewijn, S. C., Vijfhuizen, L. S., Haan, J., et al. (2015). Cluster headache and the hypocretin receptor 2 reconsidered: A genetic association study and meta-analysis. *Cephalalgia, 35*(9), 741–747.

Wilbrink, L. A., Teernstra, O. P., Haan, J., van Zwet, E. W., Evers, S. M., Spincemaille, G. H., et al. (2013). Occipital nerve stimulation in medically intractable, chronic cluster headache. The ICON study: Rationale and protocol of a randomised trial. *Cephalalgia, 33*(15), 1238–1247.

Chapter 10

Migraine and pain: Features and treatments

Javier Díaz de Terán[a,c] and Alfonso Gil-Martínez[b,c]

[a]*Department of Neurology, Hospital La Paz Institute for Health Research, Madrid, Spain,* [b]*Department of Physiotherapy, Health Sciences Faculty, University of La Salle Madrid, Madrid, Spain,* [c]*Unit or Division of Physiotherapy, Hospital La Paz Institute for Health Research, Madrid, Spain*

Abbreviations

AEs	adverse events
CGRP	calcitonin gene–related peptide
CSD	cortical spreading depression
mAbs	monoclonal antibodies
mBS	modified Borg scale
NSAIDs	nonsteroidal antiinflammatory drugs

Features

Based on recent studies, migraine is the most disabling primary headache and the second most disabling condition worldwide (Stovner et al., 2018). It is a familiar, episodic, and complex sensory processing disorder (Goadsby & Holland, 2019) that is associated with a constellation of symptoms, with headache being the hallmark.

Migraine is a common disorder as it affects 18% of women and 6% of men, while chronic migraine affects 2% of the global population, and it is extremely burdensome condition for the patients, families, and the society (Burch, Buse, & Lipton, 2019).

Migraine has two major types (Vincent & Wang, 2018):

(1) *Migraine without aura:* It is a recurrent headache disorder occurring in attacks lasting from 4 to 72 h. The headache is usually unilaterally located, throbbing, moderate or severe in intensity, and is aggravated by routine physical activity in association with nausea and/or photophobia and phonophobia.

(2) *Migraine with aura*: It is a recurring attack, lasting 5–60 min, of fully reversible visual, sensory, or other central nervous system symptoms that often progresses gradually and is usually accompanied by headaches and the consequent migraine symptoms described earlier.

The migraine attack consists of four overlapping phases (Dodick, 2018a):

(1) *Premonitory phase*: Presence of nonpainful symptoms appearing hours or days before the onset of the headache (Karsan, Bose, & Goadsby, 2018). These symptoms can include yawning, mood changes, difficulty concentrating, neck stiffness, fatigue, thirst, and elevated frequency of micturition (Maniyar, Sprenger, Monteith, Schankin, & Goadsby, 2015).

(2) *Aura:* About one third of the patients with migraine, especially women, suffer this transient focal neurological symptoms before or during some of their headaches which is called aura (Manzoni & Stovner, 2010). Visual aura is the most common type (90%) followed by sensory (30%–54%) and language aura (31%) (Kissoon & Cutrer, 2017). Motor, brainstem, and retinal aura are atypical and therefore far less often. In 1944, Leao described the cortical spreading depression (CSD) by showing that stimulation of the cortical surface produced hyperexcitation followed by suppression that migrated over the cortical surface at a slow rate of 3–4 mm per minute (Leao, 1944). According to currently available evidence, the symptoms of aura are most probably related to CSD that can induce activation of the vascular trigeminal leading to headache stage of the migraine episode (Goadsby et al., 2017).

Features and Assessments of Pain, Anesthesia, and Analgesia. https://doi.org/10.1016/B978-0-12-818988-7.00013-3
Copyright © 2022 Elsevier Inc. All rights reserved.

(3) *Headache*: This phase is caused by the activation of the trigeminal sensory pathways which generates the throbbing pain of migraine (Dodick, 2018a). The intensity of the headache increases progressively or is explosive at the onset and disrupts daily activities.

The headache is usually associated with nausea and vomiting with an aversion to touch (allodynia), light (photophobia), sound (phonophobia), and smell (osmophobia) (Dodick, 2018b).

(4) *Postdrome:* The most frequent symptoms in this phase are tiredness, drowsiness, difficulty in concentrating, and hypersensitivity to noise (Bose, Karsan, & Goadsby, 2018). The greater the intensity of the pain, the more intense and prolonged these symptoms will be. This phase is colloquially known among patients as the "migraine hangover."

Chronic migraine

This "migraine chronification" (formerly called transformation) has been conceptualized in terms of a transitional model (Fig. 1) that is often used to talk about migraine as a dynamic disease that can progress in both directions (Bigal & Lipton, 2008). In this regard, it is important to identify risk factors and protective factors.

Patients with episodic migraine may become chronic in their pain, and this fact can be explained as a threshold problem: A genetic predisposition in combination with ambient factors (stressful life events, obesity depression, etc.), and frequent headache pain, lower the threshold of migraine attacks, thus enhancing the risk of chronic migraine (May & Schulte, 2016).

Chronic migraine is defined as a headache occurring on 15 or more days/month for more than 3 months, which, on at least 8 days/month, has the features of migraine headache (International Headache Society, 2018).

Treatments

Inadequate treatment of migraine attack has a huge socioeconomic impact and also increases the risk of transformation of migraine into its chronic forms (Lipton et al., 2015).

It should be explained to the patient that migraine is a recurrent and episodic disease that currently has no cure and that in general allows an adequate quality of life when it is known.

Acute migraine treatment

A tailored approach must be provided to each patient. One of the most important aspects is to teach the patient to identify their migraine attacks because early treatment is essential to get an adequate response to end the attack. If the attack is not stopped in the first 20 min, the first set of neurons in the trigeminal ganglion suffers changes that make them hypersensitive to the changing pressure (throbbing). If the pain continues from 60 to 120 min, the second group of neurons in the network located in the spinal trigeminal nucleus undergoes changes that convert them into an independent pain generator of headache without inputs (allodynia) (Burstein, Jakubowski, & Rauch, 2011).

It is well known that response to treatment is higher in those patients who have not developed allodynia and therefore central sensitization, so in the migraine attack, time is also brain.

A stratified treatment must be carried out from the beginning, choosing the drug according to the severity of the symptoms, route of administration characteristics, and comorbidity of the patient.

Migraine therapy can be divided into specific, nonspecific, and adjuvant treatments.

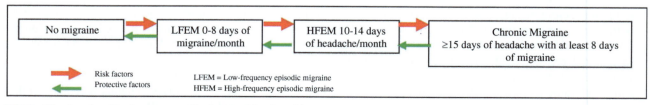

FIG. 1 Migraine chronification conceptualized as transitional model.

TABLE 1 Nonspecific migraine acute treatment.	
Treatment	Recommended dose
Acetaminophen	500–100 mg
Acetylsalicylic acid	500–1000 mg
Ibuprofen	600–1200 mg
Naproxen sodium	550–1100 mg
Dexketoprofen	25–50 mg oral or 50 mg parenteral

Nonspecific acute treatment

There is good quality evidence supporting the use of acetaminophen and nonsteroidal antiinflammatory drugs (NSAIDs) such as acetylsalicylic acid, ibuprofen, diclofenac, and dexketoprofen (Kirthi, Derry, & Moore, 2013; Law, Derry, & Moore, 2013; Rabbie, Derry, & Moore, 2013; Yang, Xu, Chen, Chen, & Xie, 2019). These therapies can control mild migraine attacks and auras on their own. These treatments are not expensive and well tolerated compared to triptans. In the specific case, paracetamol (acetaminophen) is less potent and may be a useful first-choice drug for acute migraine in those with restriction to, or who cannot tolerate, NSAIDs or aspirin (Derry & Moore, 2013). It is generally only recommended in gestational migraine, during adolescence-childhood and in attacks without a severe level of disability (Table 1).

Adjuvant medications are primarily antiemetic/neuroleptics and dopamine D2 receptor antagonists (domperidone, metoclopramide, and chlorpromazine) that are necessary in patients with nausea or vomiting, which also supports the absorption of the rest of the treatment (Marmura, 2012; Marmura, Silberstein, & Schwedt, 2015). When using these treatments, you should monitor the potential extrapyramidal side effects and concern over potentially permanent tardive dyskinesia, sedation, and orthostatic hypotension.

It is strongly recommended to avoid morphs and combinations of analgesics with barbiturates, codeine, tramadol, and/or caffeine, because it is associated with headache chronification and development of medication overuse headache (Ezpeleta, 2015).

Specific acute treatment

Triptans are agonists at the 5-HT1B/D receptor with a postulated mechanism of action—selective intracranial vessel vasoconstriction (5-HT1B), peripheral neuronal inhibition (5-HT1D), and presynaptic dorsal horn stimulation (5-HT1D)—making a second-order brainstem neuronal inhibition (Loder, 2010).

Triptans are specific drugs with proven efficacy and safety in several clinical trials; however due to its vasoconstriction effect, they are contraindicated in patients with uncontrolled hypertension, coronary, cerebrovascular, and peripheral vascular disease (Dodick et al., 2004; Johnston & Rapoport, 2010). The most frequent side effects are palpitations, neck or chest tightness, dysgeusia, and laryngeal discomfort, and patients should always be warned about these side effects before prescribing these drugs. Despite these effects, it should be pointed out that they are extraordinarily safe at the vascular level (Dodick et al., 2004).

There are seven triptans currently available, and the choice of one or the other must be individualized based on time of migraine onset (night or day), severity onset (rapidly or progressive), presence and timing of nausea or vomiting, levels of disability, and frequency and pattern of attacks (Gladstone & Dodick, 2004) (Table 2).

It is also possible to combine triptans with an NSAID or acetaminophen, and using alternative modes of administration such as injectables or nasal spray may be associated with better outcomes than standard-dose triptan tablets (Cameron et al., 2015).

Triptans have a much better adverse effects and safety profile than ergotics which are currently disused drugs and should not be prescribed to newly diagnosed migraines.

TABLE 2 Triptans currently available.

Treatment	Formulation and dosage	For who?
Eletriptan	40 mg tablet	Migraine associated with moderate-to-severe intensity and recurrence
Rizatriptan	10 mg tablet Orally disintegrating tablet	Migraine associated with moderate-to-severe intensity attacks of high intensity and short duration. Pain or gastrointestinal symptoms which evolve rapidly
Sumatriptan	50 mg tablet 10/20 mg nasal spray 6 mg injection	Severe pain attacks resistant to the oral and nasal route. Oral resistant attacks. Pain or gastrointestinal symptoms which evolve rapidly. Potential risk of pregnancy. Children and teenagers
Zolmitriptan	5/2.5 mg tablet 5/2.5 mg orally disintegrating tablet 5/2.5 mg nasal spray	Standard attacks. Attacks which occur during sleep. Pain or gastrointestinal symptoms which evolve rapidly
Almotriptan	12.5 mg tablet	Standard attack. Intolerance to other triptans
Frovatriptan	2.5 mg tablet	Mild migraines of long duration (>24 h) and/or frequent headache recurrence. Menstrual migraine as prophylaxis. Intolerance to other triptans
Naratriptan	2.5 mg tablet	Mild migraines of long duration (>24 h) and/or frequent headache recurrence

Future

5-HT1F receptor agonists (DITANS)

The agonism of the 5-HT1F (serotonin) receptor has been shown to be effective for acute treatment of migraine, has anti-inflammatory effects, and does not produce vasoconstriction, making it a possible future treatment option in patients with cardiovascular risk factors (Vila-Pueyo, 2018).

In this regard, Lasmiditan has been recently approved by the Food and Drugs Administration, and phase 3 studies are still ongoing in Europe. Clinical trials showed that Lasmiditan is an effective acute treatment for migraine, and the majority of adverse events (AEs) were mild to moderate, with the most common being dizziness (16.3% in the 200 mg group, 12.5% in the 100 mg group), fatigue, nausea, paresthesia, and somnolence (Kuca et al., 2018).

CGRP receptor antagonists (GEPANTS)

Ubrogepant is an oral, small-molecule calcitonin gene–related peptide (CGRP) receptor antagonist for acute migraine treatment. In the studies, it showed significantly higher than placebo pain freedom rate and absence of most bothersome symptom at 2 h, and frequency of AEs of <5% (Dodick et al., 2019; Lipton, Dodick, et al., 2019).

In a similar way, Rimegepant has been shown to be effective and well-tolerated in the acute treatment of migraine, and it is also under investigation for migraine prevention (Gao et al., 2019; Lipton, Croop, et al., 2019).

Prophylactic migraine treatment

Preventive treatment is used to reduce the frequency, duration, or severity of migraine attacks, making them easier to control with acute treatment. Ultimately, the goal is to improve quality of life and reduce the impact of migraine on patient functionality.

Preventive treatment of migraine should be considered in the following cases (Carville, Padhi, Reason, & Underwood, 2012; Ezpeleta, 2015; González Oria, 2019; Silberstein, 2015):

1. Frequent headaches (four or more attacks per month or eight or more headache days per month)
2. Failure, contraindications, side effects, or abuse of acute medications
3. Patient preference
4. Presence of prolonged auras, such as hemiplegic migraine, and brainstem aura because they don't usually respond to acute treatment

5. Impact on the patient's quality of life and interference in their daily life despite correct treatment with lifestyle modification strategies and acute treatment of migraine
6. Menstrual migraine

Migraine treatment should be individualized based on patients' comorbidities, preference, lifestyle, age, and gender. A preventive migraine drug is considered successful if it reduces migraine attack frequency or days by at least 50% within 3 months (Silberstein, 2015).

Basic principles of preventive treatment of migraine (Carville et al., 2012; Ezpeleta, 2015; Silberstein, 2015) are as follows:

(a) Start the treatment at a low dose and increase it slowly until therapeutic effects develop, the ceiling dose is reached, or AEs become intolerable.
(b) The chosen treatment should be maintained for a minimum of 3 months, and generally after 6–12 months, try to withdraw the drug slowly.
(c) It is important to warn patients that treatments often take up to a month and a half to start working.
(d) In women of childbearing age, avoid teratogenic drugs.
(e) Set realistic expectations for improvement and adverse effects from treatment.

β-Blockers

According to previously randomized-clinical trials published, beta-blockers, especially metoprolol, propranolol, and timolol, are effective for episodic migraine prevention, while atenolol and nadolol are probably effective (Silberstein et al., 2012). Beta-blockers are one of the most commonly used class of drugs in the preventive treatment of episodic migraine and are about 50% effective in producing a greater than 50% reduction in attack frequency (Silberstein, 2015). Of these, Propranolol is the most evidence-based and widely used drug, although it has the disadvantage of not being taken once daily like the other beta-blockers. This class of drugs should be prescribed to patients who have high blood pressure, anxiety, essential tremor, or hyperthyroidism as a comorbidity associated with migraine. They are contraindicated in cases of heart failure, symptomatic hypotension, severe diabetes, second- or third-degree atrioventricular block, chronic obstructive lung disease, Raynaud disease, and peripheral vascular disease.

Antiepileptics (also known as neuromodulators)

Although several antiepileptic drugs are commonly used in clinical practice, sodium valproate and topiramate are the ones that have been shown to be effective in migraine prophylaxis.

Numerous studies have showed that topiramate is effective as a preventive treatment for chronic and episodic migraine (Diener et al., 2007; Linde, Mulleners, Chronicle, & McCrory, 2013b). It is a very useful drug but is often poorly tolerated by patients. Its most frequent adverse effects are anorexia, weight loss, diarrhea, cognitive impairment, depression, and paresthesia. Side effects are usually associated with higher doses, and it should be considered that doses >100 mg/day interact with oral contraceptives.

Numerous studies have shown that valproate (divalproex sodium, sodium valproate, and valproic acid) is an effective treatment in reducing the intensity and attacks of episodic migraine (Linde, Mulleners, Chronicle, & Mccrory, 2013a). It is contraindicated in pregnancy as well as in the presence of liver, pancreatic, or hematological disease. Available evidence suggests that other antiepileptic drugs such as pregabalin, gabapentin, and lamotrigine are not effective in the preventive treatment of migraine (Linde, Mulleners, Chronicle, & Mccrory, 2013c; Linde, Mulleners, Chronicle, & Mccrory, 2013d). Although lamotrigine could be considered as an aura-preventive treatment.

Antidepressants

It should be considered a second line when there is no response to those listed above or as adjuvants to them. It is of choice for patients who associate depression, insomnia, or a tension headache component with their migraine. In this class of drugs, amitriptyline is the only tricyclic antidepressant with proven evidence in the treatment of episodic migraine (Xu et al., 2017).

Although selective serotonin or noradrenaline reuptake inhibitors are not drugs of choice in migraine, venlafaxine has been shown in one study to be effective (Ozyalcin et al., 2005) and may be an option in patients with associated major psychiatric disorder.

Calcium channel blockers

According to the available evidence from this pharmacological group, Flunarizine, a nonspecific calcium channel blocker is the only effective treatment in the prophylaxis of episodic migraine (Stubberud, Flaaen, McCrory, Pedersen, & Linde, 2019).

Angiotensin converting enzyme inhibitors, and angiotensin II receptor blockers

Although the available evidence is limited, lisinopril and overall candesartan are effective for episodic migraine prevention and may be an option in those migraineurs who associate with high blood pressure but have contraindicated the use of beta-blockers (Dorosch, Ganzer, Lin, & Seifan, 2019).

OnabotulinumtoxinA (Botox)

The PREEMPT trials concluded that OnabotulinumtoxinA reduced headache, migraine, and pain-free days, and was a safe and well-tolerated treatment for chronic migraine (Rendas-Baum et al., 2014) Subsequently, clinical practice studies have supported its use, demonstrating improvement in quality of life and other parameters in addition to being a cost-effective treatment by reducing direct and indirect costs (Gago-Veigaa et al., 2017). Botulinum toxin is not recommended for preventive treatment in episodic migraine. The recommended dose is 155 IUI quarterly, and the increase to 195 IUI may be assessed if there has been a suboptimal response in duration or frequency. Adverse effects are generally rare. Knowing the functional anatomy of the PREEMPT muscles is fundamental to achieving efficacy and safety in clinical trials. It is important to evaluate the facies before injection in order to anticipate the risk of undesirable outcome. Botulinum toxin must be administered in hospital by a clinician with experience in headaches, which can make access difficult. In addition, because of its cost, in many countries it is reserved for patients who do not tolerate or have failed other lines of treatment (Fig. 2, Table 3).

CGRP monoclonal antibodies

The CGRP was discovered more than 30 years ago (Edvinsson, 2015), and as a result, we are currently facing a new era in the treatment of migraine with the arrival of the first specific preventive treatments for migraine. CGRP is important in the pathophysiology of migraine because it mediates the transmission of trigeminovascular pain and plays a role as a vasodilator in neurogenic inflammation.

The treatment against CGRP is an adequate strategy to deal with migraine based on the following previous investigation:

1. The CGRP is released during a migraine attack in the jugular venous system (Goadsby, Edvinsson, & Ekman, 1990).
2. CGRP levels decrease after taking triptans (Durham, 2006).
3. CGRP infusion triggers a migraine attack (Hansen, Hauge, Olesen, & Ashina, 2010).
4. Olcegepant provided evidence that targeting CGRP was an effective migraine treatment strategy (Olesen et al., 2004).
5. Patients with chronic migraine have high levels of CGRP (Cernuda-Morollón et al., 2013).

There are currently three monoclonal antibodies (mAbs) against CGRP that have been approved as preventive treatment for episodic and chronic migraine: Fremanezumab and galcanezumab bind to the CGRP ligand itself, while erenumab is the only one which binds the receptor. All of them are prepared for subcutaneous monthly administration (fremanezumab could also be quarterly). A fourth mAb called eptinezumab is waiting for approval, also binds to CGRP ligand, and is only intravenous. During the clinical trials, a very low incidence of AEs was reported, and most common reported AEs were injection site erythema, injection site induration, diarrhea or constipation, nasopharyngitis, and upper respiratory tract infection (Ceriani, Wilhour, & Silberstein, 2019).

Nonpharmacological approach

Exercise based on cognitive behavioral therapy

Nowadays, most of the treatments for headache patients are based on preventive and/or abortive drugs (acute treatment). Moreover, there are important differences between some first-world countries; for example, in the United States, 23% of patients with chronic headache use medication daily (Kristoffersen, Grande, Aaseth, Lundqvist, & Russell, 2012), while in

Migraine and pain: Features and treatments **Chapter | 10** 111

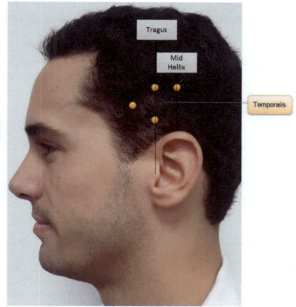

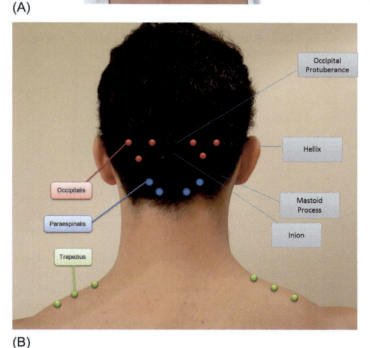

FIG. 2 Fixed-site, fixed-dose PREEMPT injection site locations. (A) Anterior region Botox injection sites. (B) Posterior region Botox injection sites.

northern Europe, only 9% do it (Haag, 2011). In addition to this variability and without losing sight of the efficacy of the drugs, the dangers posed by the high degree of self-medication by patients, the high proportion of drug overuse, and the use of treatments with poor or no scientific evidence should be taken into account.

Precisely, this overuse of drugs could be related to the lack of prescription of exercise in headache patients. This is so much so that despite the supporting evidence, in many public health systems and up to date, its application in patients with headaches within the service portfolio is not contemplated.

In recent years, therapeutic exercise (TE) has been postulated as an emerging therapy in the field of headaches; however, the prescription of TE in headache patients requires knowledge and some experience in its management. In addition, treatment by exercise requires that there should be a commitment on the part of the patient, since its effects are mostly

TABLE 3 Mostly used prophylactic treatments for migraine.

Drug	Daily dose	Comments
Propranolol	40–240 mg	Episodic migraine Risk of bradycardia, fatigue, and erectile dysfunction Of choice if anxiety or high blood pressure
Topiramate	50–200 mg	Episodic and chronic migraine Start 25 mg at bedtime and increase 25 mg per week attempt to reach 50–100 mg. Once or twice/day Of choice if overweight, epilepsy, or anxiety Avoid if depression, elderly, glaucoma, or renal lithiasis
Valproate	500–2000 mg	Episodic migraine Start 300–500 mg. The usual dosage is 600–1000 mg Teratogenic Gastrointestinal effects, alopecia, tremor, and weight gain are common
Amitriptyline	10–75 mg	Episodic migraine Start 10 mg at bedtime and increase 12.5–25 every week Of choice if insomnia, depression, or tensional type headache associated Antimuscarinic adverse events and sedating QT interval prolongation, weight gain, and drowsiness Caution in elderly
Flunarizine	5–10 mg	Episodic migraine Start with 2.5 mg at bedtime, usually effective with 5 mg. Good option in vestibular migraine Weight gain, fatigue, extrapyramidal effects, and depression are common
Candesartan	8–16 mg	Episodic migraine. Start in the morning. Of choice if high blood pressure and nonresponder to beta-blockers
OnabotulinumtoxinA	155–195 UI	Chronic migraine Of choice if previous history of failure to other preventive treatments, poor adherence, contraindication, or poor tolerance Subcutaneous every 3 months at hospital

observed in the medium to long term. The biopsychosocial complexity of headache patients means that many of them do not adhere adequately to the treatments and end up abandoning them. Therefore, as with other approaches, in the case of exercise, it is a challenge for patients to adhere to treatment, and it will be necessary to work to strengthen an adequate therapeutic alliance between patient and physical therapist.

It is also important to consider that TE is not always synonymous with physical exercise or physical activity, since there are other types of nonphysical but mental exercise, such as exercises by observation of actions or motor imagery. Actually, exercise is any action that involves the activation of the different central and/or peripheral mechanisms that carry and involve movement, even if it does not occur. It should also be borne in mind that many pathologies that affect the somatosensory system could directly have influence (through areas of association) on the motor center and therefore alter motor control in different executive (cortex), regulatory (basal ganglia and cerebellum), intermediate structures (diencephalon and brainstem), and/or inferiors structures like the spinal cord.

There are multiple characteristics of patients with headaches that invite us to think about different indications for the use of TE. In the case of chronic headaches, it has been shown that, as occurs in many types of persistent pain, there is an involvement of different medullary and supramedullary areas such as the motor cortex and other areas related to movement such as prefrontal cortex or the supplementary motor cortex among others. More specifically, according to the available evidence from the latest systematic reviews on the effectiveness of exercise in headache patients, the types of exercises used have been aerobic exercise, strength exercise, motor control and stabilization exercise, and neural mobilization exercise (Gil-Martínez et al., 2013).

There is currently limited evidence of the use of exercise in patients with migraines. Perhaps aerobic type exercise, not explained as a modulator but as a positive adaptation of avoidance behavior, therefore can be recommended as a safe and economical strategy within the treatment of migraines (Krøll, Sjödahl Hammarlund, Gard, Jensen, & Bendtsen, 2018). In the TE prescription for migraine, factors such as intensity, duration, frequency, type, and warm-up period should be controlled with the intention of increasing the beneficial effects and also preventing injuries and side effects that may include headache by effort (Koseoglu, Yetkin, Ugur, & Bilgen, 2015). On the other hand, currently, the relationship between neck pain and disability and the fact of suffering migraines are well known. In this case, the use of isometric-type exercises aimed at the cervical muscles could be useful.

The positive action of exercise on migraine is generally related to neurochemical factors, psychological states, and increased cardiovascular and cerebrovascular capacity. In this way, the population of patients with low blood beta endorphin levels, good physical condition, and high motivation could obtain significant benefits from exercise treatment (Koseoglu et al., 2015).

Some previous works in migraine patients without cardiovascular disease, such as those of Varkey et al., have proposed sessions between 30 and 45 min of intense walking or light running controlling the rate of respiration perceived by the patient corresponding to the intensity of the exercise (modified Borg scale (mBS)). The exercise was divided into 10 min of warm-up (slightly breathless, but conversation is possible; corresponds to 11–13 mBS), 30 min of resistance training (breathless, but conversation in short sentences; corresponding to 14–16 mBS), and 5 min of cooling down (slightly breathless, but conversation possible; corresponds to 11–13 mBS).

Conclusion

Migraine is a complex brain dysfunction, beyond a simple headache. Its treatment is based on lifestyle, acute treatment of the attack, and, in some patients, prophylactic drugs (antidepressants, antihypertensive, antiepileptic, and Botox). The CGRP is a neurotransmitter highly involved in its physiopathology, and new therapies have it as a target. From the rehabilitation point of view, general and specific exercises (active approach) are essential to maintain the pain (frequency and intensity) under control.

Applications to other areas

Some of these treatments can also be used in patients with neuropathic pain such as neuromodulatory drugs and antidepressants. Also in other comorbidities frequently present in patients with migraine such as other central sensitization syndromes like fibromyalgia or temporomandibular joint disorders.

Mini-dictionary of terms

1. Migraine: It is a familiar, episodic, and complex sensory processing disorder that is associated with a constellation of symptoms, with headache being the hallmark.
2. Triptans: These are specific drugs with proven efficacy and safety in the migraine attack. They are agonists at the 5-HT1B/D receptor.
3. Preventive treatment: It is used to reduce the frequency, duration, or severity of migraine attacks, making them easier to control with acute treatment. Ultimately, the goal is to improve quality of life and reduce the impact of migraine on patient functionality.
4. CGRP (calcitonin gene–related peptide): It is important in the pathophysiology of migraine because it mediates the transmission of trigeminovascular pain and plays a role as a vasodilator in neurogenic inflammation. It is the main therapeutic target of new antimigraine therapies.
5. Aerobic exercise: It is a physical activity, typically prolonged and of moderate intensity (e.g., walking, jogging, cycling), that involves the use of oxygen in the muscles to provide the needed energy.
6. Motor control: It is a broad term that describes the general ability of a person to initiate and direct muscle function and voluntary movements.

Key facts

- Migraine is the most disabling primary headache and the second most disabling condition worldwide. It has two major types (with aura/without aura). The migraine attack consists of four overlapping phases (premonitory phase, aura, headache, and postdrome).
- The treatment of the migraine must be individualized. The style of life and the treatment of the migraine attack must be insisted on. In some patients with important impact, preventive treatments should be used.

114 PART | II The syndromes of pain

- Among the prophylactic treatments, there are oral drugs (antiepileptics, beta-blockers, antidepressants, calcium blockers, etc.), botulinum toxin injections, and therapies against CGRP.
- There are currently four monoclonal antibodies (mAbs) against CGRP as preventive treatment for episodic and chronic migraine: Fremanezumab, eptinezumab, and galcanezumab bind to the CGRP ligand itself, while erenumab is the only one which binds the receptor.
- The positive effects of exercise on migraine are generally related to neurochemical factors, psychological states, and increased cardiovascular and cerebrovascular capacity.
- The upper neck region is an important area to focus on during physical treatments.

Summary points

- Migraine is a global brain dysfunction of an episodic nature that can become chronic and generate a high level of disability.
- The treatment of the migraine attack is based on nonspecific (NSAIDs and antidopaminergic) and specific (triptans) drugs. The triptans are safe and effective, and there are different types and dosages indicated for each patient.
- The CGRP is a neurotransmitter very involved in the physiopathology of migraine. Research over the last 20 years has shown that it is released during a migraine attack, patients with chronic migraine have high levels, and it decreases after triptan injection. This has led to the development of new therapies against CGRP as a preventive treatment for migraine.
- Aerobic type exercise can be recommended as a safe and economical strategy for the treatment of patients with migraines. However, this approach should be driven by specialized physiotherapist.
- In addition, specific neck exercise, such as motor control exercises, and manual therapy on upper neck region should help to modulate migraine intensity and frequency.

References

Bigal, M. E., & Lipton, R. B. (2008). Clinical course in migraine: Conceptualizing migraine transformation. *Neurology*, *71*(11), 848–855. https://doi.org/10.1212/01.wnl.0000325565.63526.d2.

Bose, P., Karsan, N., & Goadsby, P. J. (2018). The migraine postdrome. *CONTINUUM Lifelong Learning in Neurology*, *24*(4-Headache), 1023–1031. https://doi.org/10.1212/CON.0000000000000626.

Burch, R. C., Buse, D. C., & Lipton, R. B. (2019). Migraine: Epidemiology, burden, and comorbidity. *Neurologic Clinics*, *37*(4), 631–649. https://doi.org/10.1016/j.ncl.2019.06.001.

Burstein, R., Jakubowski, M., & Rauch, S. D. (2011). The science of migraine. *Journal of Vestibular Research*, *21*(6), 305–314. https://doi.org/10.3233/VES-2012-0433.

Cameron, C., Kelly, S., Hsieh, S. C., Murphy, M., Chen, L., Kotb, A., ... Wells, G. (2015). Triptans in the acute treatment of migraine: A systematic review and network meta-analysis. *Headache*, *55*(S4), 221–235. https://doi.org/10.1111/head.12601.

Carville, S., Padhi, S., Reason, T., & Underwood, M. (2012). Diagnosis and management of headaches in young people and adults: Summary of NICE guidance. *BMJ*, *345*(7876), 1–5. https://doi.org/10.1136/bmj.e5765.

Ceriani, C. E. J., Wilhour, D. A., & Silberstein, S. D. (2019). Novel medications for the treatment of migraine. *Headache*, *59*(9), 1597–1608. https://doi.org/10.1111/head.13661.

Cernuda-Morollón, E., Larrosa, D., Ramón, C., Vega, J., Martínez-Camblor, P., & Pascual, J. (2013). Interictal increase of CGRP levels in peripheral blood as a biomarker for chronic migraine. *Neurology*, *81*(14), 1191–1196. https://doi.org/10.1212/WNL.0b013e3182a6cb72.

Derry, S., & Moore, R. A. (2013). Paracetamol (acetaminophen) with or without an antiemetic for acute migraine headaches in adults. *Cochrane Database of Systematic Reviews*, *2013*(30). https://doi.org/10.1002/14651858.CD009455.pub2.

Diener, H. C., Bussone, G., Van Oene, J. C., Lahaye, M., Schwalen, S., & Goadsby, P. J. (2007). Topiramate reduces headache days in chronic migraine: A randomized, double-blind, placebo-controlled study. *Cephalalgia*, *27*(7), 814–823. https://doi.org/10.1111/j.1468-2982.2007.01326.x.

Dodick, D. W. (2018a). A phase-by-phase review of migraine pathophysiology. *Headache*, *58*, 4–16. https://doi.org/10.1111/head.13300.

Dodick, D. W. (2018b). Migraine. *The Lancet*, *391*(10127), 1315–1330. https://doi.org/10.1016/S0140-6736(18)30478-1.

Dodick, D. W., Lipton, R. B., Ailani, J., Lu, K., Finnegan, M., Trugman, J. M., & Szegedi, A. (2019). Ubrogepant for the treatment of migraine. *New England Journal of Medicine*, *381*(23), 2230–2241. https://doi.org/10.1056/NEJMoa1813049.

Dodick, D., Lipton, R. B., Martin, V., Papademetriou, V., Rosamond, W., VanDenBrink, A. M., ... Saiers, J. (2004). Consensus statement: Cardiovascular safety profile of triptans (5-HT 1B/1D agonists) in the acute treatment of migraine. *Headache*, *44*(5), 414–425. https://doi.org/10.1111/j.1526-4610.2004.04078.x.

Dorosch, T., Ganzer, C. A., Lin, M., & Seifan, A. (2019). Efficacy of angiotensin-converting enzyme inhibitors and angiotensin receptor blockers in the preventative treatment of episodic migraine in adults. *Current Pain and Headache Reports*, *23*(11). https://doi.org/10.1007/s11916-019-0823-8.

Durham, P. L. (2006). Calcitonin gene-related peptide (CGRP) and migraine. *Headache*, *46*(Suppl 1), 1–9. https://doi.org/10.1111/j.1526-4610.2006.00483.x.

Edvinsson, L. (2015). The journey to establish CGRP as a migraine target: A retrospective view. *Headache, 55*(9), 1249–1255. https://doi.org/10.1111/head.12656.

Ezpeleta, D. P.-R. (2015). *Guías diagnósticas y terapeúticas de la Sociedad Española de Neurología 2015. Guía práctica clínica en cefaleas.* Sociedad Española de Neurología.

Gago-Veigaa, A., Santos-Lasaosab, S., Cuadradoc, M., Guerreroe, A., Irimia, F. P., Láinez, J., ... Pozo-Rosich, P. (2017). Evidencia y experiencia de bótox en migraña crónica: Recomendaciones para la práctica clínica diaria A.B. *Neurología, 1*(1), 1–6.

Gao, B., Yang, Y., Wang, Z., Sun, Y., Chen, Z., Zhu, Y., & Wang, Z. (2019). Efficacy and safety of rimegepant for the acute treatment of migraine: Evidence from randomized controlled trials. *Frontiers in Pharmacology, 10*(January), 1577.

Gil-Martínez, A., Kindelan-Calvo, P., Agudo-Carmona, D., Muñoz-Plata, R., López-de-Uralde-Villanueva, I., & Touche, R. L. (2013). Ejercicio terapéutico como tratamiento de las migrañas y cefaleas tensionales: Revisión sistemática de ensayos clínicos aleatorizados. *Revista de Neurologia, 57*(10), 433–443.

Gladstone, J. P., & Dodick, D. W. (2004). Acute migraine: Which triptan? *Practical Neurology, 4,* 6–19.

Goadsby, P. J., Edvinsson, L., & Ekman, R. (1990). Vasoactive peptide release in the extracerebral circulation of humans during migraine headache. *Annals of Neurology, 28*(2), 183–187. https://doi.org/10.1002/ana.410280213.

Goadsby, P. J., & Holland, P. R. (2019). An update: Pathophysiology of migraine. *Neurologic Clinics, 37*(4), 651–671. https://doi.org/10.1016/j.ncl.2019.07.008.

Goadsby, P. J., Holland, P. R., Martins-Oliveira, M., Hoffmann, J., Schankin, C., & Akerman, S. (2017). Pathophysiology of migraine: A disorder of sensory processing. *Physiological Reviews, 97*(2), 553–622. https://doi.org/10.1152/physrev.00034.2015.

González Oria, C., & [®] 2019: Grupo de Estudio de Cefaleas de la Sociedad Andaluza de Neurología (SANCE). (2019). Tratamientos en migraña. In C. González Oria, C. Jurado Cobo, & J. Viguera Romero (Eds.), *Guía Oficial De Cefaleas 2019 Grupo De Estudio De Cefaleas De La Sociedad Andaluza De Neurología (Sance)* Medea, Medical Education Agency S.L. ISBN: 978-84-09-09716-6. In press.

Haag, G. (2011). Letter to the editor concerning Bendtsen L, Evers S, Linde M, Mitsikostas DD, Sandrini G, Schoenen J. EFNS guideline on the treatment of tension-type headache—Report of an EFNS task force. Eur J Neurol 2010; 17: 1318–1325. *European Journal of Neurology, 18*(7), e80–e82 (author reply e85) https://doi.org/10.1111/j.1468-1331.2011.03383.x.

Hansen, J. M., Hauge, A. W., Olesen, J., & Ashina, M. (2010). Calcitonin gene-related peptide triggers migraine-like attacks in patients with migraine with aura. *Cephalalgia, 30*(10), 1179–1186. https://doi.org/10.1177/0333102410368444.

International Headache Society. (2018). Headache Classification Committee of the International Headache Society (IHS) The International Classification of Headache Disorders, 3rd edition. *Cephalalgia, 38*(1), 1–211. https://doi.org/10.1177/0333102417738202.

Johnston, M. M., & Rapoport, A. M. (2010). Triptans for the management of migraine. *Drugs, 70*(12), 1505–1518. https://doi.org/10.2165/11537990-000000000-00000.

Karsan, N., Bose, P., & Goadsby, P. J. (2018). The migraine premonitory phase. *CONTINUUM Lifelong Learning in Neurology, 24*(4-Headache), 996–1008. https://doi.org/10.1212/CON.0000000000000624.

Kirthi, V., Derry, S., & Moore, R. A. (2013). Aspirin with or without an antiemetic for acute migraine headaches in adults (Review). *Cochrane Database of Systematic Reviews, 2013*(4). https://doi.org/10.1002/14651858.CD009455.pub2.

Kissoon, N. R., & Cutrer, F. M. (2017). Aura and other neurologic dysfunction in or with migraine. *Headache, 57*(7), 1179–1194. https://doi.org/10.1111/head.13101.

Koseoglu, E., Yetkin, M. F., Ugur, F., & Bilgen, M. (2015). The role of exercise in migraine treatment. *Journal of Sports Medicine and Physical Fitness, 55* (9), 1029–1036.

Kristoffersen, E. S., Grande, R. B., Aaseth, K., Lundqvist, C., & Russell, M. B. (2012). Management of primary chronic headache in the general population: The Akershus study of chronic headache. *Journal of Headache and Pain, 13*(2), 113–120. https://doi.org/10.1007/s10194-011-0391-8.

Krøll, L. S., Sjödahl Hammarlund, C., Gard, G., Jensen, R. H., & Bendtsen, L. (2018). Has aerobic exercise effect on pain perception in persons with migraine and coexisting tension-type headache and neck pain? A randomized, controlled, clinical trial. *European Journal of Pain, 22*(8), 1399–1408. https://doi.org/10.1002/ejp.1228.

Kuca, B., Silberstein, S. D., Wietecha, L., Berg, P. H., Dozier, G., & Lipton, R. B. (2018). Lasmiditan is an effective acute treatment for migraine: A phase 3 randomized study. *Neurology, 91*(24), E2222–E2232. https://doi.org/10.1212/WNL.0000000000006641.

Law, S., Derry, S., & Moore, R. A. (2013). Naproxen with or without an antiemetic for acute migraine headaches in adults. *Cochrane Database of Systematic Reviews, 2013*(10). https://doi.org/10.1002/14651858.CD009455.pub2.

Leao, A. A. P. (1944). Spreading depression of activity in the cerebral cortex. *Journal of Neurophysiology, 7*(6), 359–390. https://doi.org/10.1152/jn.1944.7.6.359.

Linde, M., Mulleners, W., Chronicle, E., & Mccrory, D. (2013a). Valproate (valproic acid or sodium valproate or a combination of the two) for the prophylaxis of episodic migraine in adults (Review). *Cochrane Database of Systematic Reviews,* (6). https://doi.org/10.1002/14651858.CD010611.www.cochranelibrary.com.

Linde, M., Mulleners, W., Chronicle, E., & McCrory, D. (2013b). Topiramate for the prophylaxis of episodic migraine in adults. *Cochrane Database of Systematic Reviews, 90*(1), 24.

Linde, M., Mulleners, W. M., Chronicle, E. P., & Mccrory, D. C. (2013c). Antiepileptics other than gabapentin, pregabalin, topiramate, and valproate for the prophylaxis of episodic migraine in adults. *Cochrane Database of Systematic Reviews, 2013*(6). https://doi.org/10.1002/14651858.CD010608.

Linde, M., Mulleners, W. M., Chronicle, E. P., & Mccrory, D. C. (2013d). Gabapentin or pregabalin for the prophylaxis of episodic migraine in adults. *Cochrane Database of Systematic Reviews, 2013*(6). https://doi.org/10.1002/14651858.CD010609.

Lipton, R. B., Croop, R., Stock, E. G., Stock, D. A., Morris, B. A., Frost, M., ... Goadsby, P. J. (2019). Rimegepant, an oral calcitonin gene-related peptide receptor antagonist, for migraine. *New England Journal of Medicine, 381*(2), 142–149. https://doi.org/10.1056/NEJMoa1811090.

Lipton, R. B., Dodick, D. W., Ailani, J., Lu, K., Finnegan, M., Szegedi, A., & Trugman, J. M. (2019). Effect of ubrogepant vs placebo on pain and the most bothersome associated symptom in the acute treatment of migraine: The achieve ii randomized clinical trial. *Journal of the American Medical Association, 322*(19), 1887–1898. https://doi.org/10.1001/jama.2019.16711.

Lipton, R. B., Fanning, K. M., Serrano, D., Reed, M. L., Cady, R., & Buse, D. C. (2015). Ineffective acute treatment of episodic migraine is associated with new-onset chronic migraine. *Neurology, 84*(7), 688–695. https://doi.org/10.1212/WNL.0000000000001256.

Loder, E. W. (2010). Triptan therapy in migraine. *New England Journal of Medicine, 363*(14), 1377–1378. https://doi.org/10.1056/NEJMc1008580.

Maniyar, F. H., Sprenger, T., Monteith, T., Schankin, C. J., & Goadsby, P. J. (2015). The premonitory phase of migraine – What can we learn from it? *Headache, 55*(5), 609–620. https://doi.org/10.1111/head.12572.

Manzoni, G. C., & Stovner, L. J. (2010). Epidemiology of headache. In *Vol. 97. Handbook of clinical neurology* (1st ed.). Elsevier B.V. https://doi.org/10.1016/S0072-9752(10)97001-2.

Marmura, M. J. (2012). Use of dopamine antagonists in treatment of migraine. *Current Treatment Options in Neurology, 14*(1), 27–35. https://doi.org/10.1007/s11940-011-0150-9.

Marmura, M. J., Silberstein, S. D., & Schwedt, T. J. (2015). The acute treatment of migraine in adults: The American Headache Society evidence assessment of migraine pharmacotherapies. *Headache, 55*(1), 3–20. https://doi.org/10.1111/head.12499.

May, A., & Schulte, L. H. (2016). Chronic migraine: Risk factors, mechanisms and treatment. *Nature Reviews Neurology, 12*(8), 455–464. https://doi.org/10.1038/nrneurol.2016.93.

Olesen, J., Diener, H. C., Husstedt, I. W., Goadsby, P. J., Hall, D., Meier, U., ... Lesko, L. M. (2004). Calcitonin gene-related peptide receptor antagonist BIBN 4096 BS for the acute treatment of migraine. *New England Journal of Medicine, 350*(11), 1104–1110. https://doi.org/10.1056/NEJMoa030505.

Ozyalcin, S. N., Talu, G. K., Kiziltan, E., Yucel, B., Ertas, M., & Disci, R. (2005). The efficacy and safety of venlafaxine in the prophylaxis of migraine. *Headache, 45*(2), 144–152. https://doi.org/10.1111/j.1526-4610.2005.05029.x.

Rabbie, R., Derry, S., & Moore, R. A. (2013). Ibuprofen with or without an antiemetic for acute migraine headaches in adults. *Cochrane Database of Systematic Reviews, 2017*(9). https://doi.org/10.1002/14651858.CD008039.pub3.

Rendas-Baum, R., Yang, M., Varon, S. F., Bloudek, L. M., DeGryse, R. E., & Kosinski, M. (2014). Validation of the Headache Impact Test (HIT-6) in patients with chronic migraine. *Health and Quality of Life Outcomes, 12*, 117. https://doi.org/10.1186/s12955-014-0117-0.

Silberstein, S. D. (2015). Preventive migraine treatment. *CONTINUUM Lifelong Learning in Neurology, 21*(4), 973–989. https://doi.org/10.1212/CON.0000000000000199.

Silberstein, S. D., Holland, S., Freitag, F., Dodick, D. W., Argoff, C., & Ashman, E. (2012). Evidence-based guideline update: Pharmacologic treatment for episodic migraine prevention in adults report of the quality standards subcommittee of the American Academy of Neurology and the American Headache Society. *Neurology, 78*(17), 1337–1345. https://doi.org/10.1212/WNL.0b013e3182535d20.

Stovner, L. J., Nichols, E., Steiner, T. J., Abd-Allah, F., Abdelalim, A., Al-Raddadi, R. M., ... Murray, C. J. L. (2018). Global, regional, and national burden of migraine and tension-type headache, 1990–2016: A systematic analysis for the Global Burden of Disease Study 2016. *Lancet Neurology, 17*(11), 954–976. https://doi.org/10.1016/S1474-4422(18)30322-3.

Stubberud, A., Flaaen, N. M., McCrory, D. C., Pedersen, S. A., & Linde, M. (2019). Flunarizine as prophylaxis for episodic migraine: A systematic review with meta-analysis. *Pain, 160*(4), 762–772. https://doi.org/10.1097/j.pain.0000000000001456.

Vila-Pueyo, M. (2018). Targeted 5-HT 1F therapies for migraine. *Neurotherapeutics*, 1–13. https://doi.org/10.1007/s13311-018-0615-6.

Vincent, M., & Wang, S. (2018). Headache Classification Committee of the International Headache Society (IHS) The International Classification of Headache Disorders, 3rd edition. *Cephalalgia, 38*(1), 1–211. https://doi.org/10.1177/0333102417738202.

Xu, X. M., Yang, C., Liu, Y., Dong, M. X., Zou, D. Z., & Wei, Y. D. (2017). Efficacy and feasibility of antidepressants for the prevention of migraine in adults: A meta-analysis. *European Journal of Neurology, 24*(8), 1022–1031. https://doi.org/10.1111/ene.13320.

Yang, B., Xu, Z., Chen, L., Chen, X., & Xie, Y. (2019). The efficacy of dexketoprofen for migraine attack: A meta-analysis of randomized controlled studies. *Medicine (Baltimore), 46*(September), 2–7.

Chapter 11

Complex regional pain syndrome

C. Ryan Phillips[a,*,†,‡], Derek M. Miletich[a,*,†,‡], and Lynita Mullins[b,*,†,‡]

[a]CDR, MC, Department of Anesthesiology and Pain Medicine, Naval Medical Center San Diego, San Diego, CA, United States, [b]CDR, MC, Department of Pain Medicine, Naval Medical Center Camp Lejeune, NC, United States

List of abbreviations

CRPS	complex regional pain syndrome
NMDA	N-methyl-D-aspartic acid
NSAIDs	nonsteroidal antiinflammatory drugs
SNRI	serotonin norepinephrine reuptake inhibitor
TCA	tricyclic antidepressant

Introduction

Any text proposing treatment strategies for CRPS must start with a brief history of the syndrome and how it came to be recognized as a distinct entity. Many historical accounts of this disease process begin with Silas Weir Mitchell's description of severe burning pain near gunshot wounds in US Civil War soldiers which he called "causalgia" in writings in 1872. James Evans then described a similar condition in patients with sympathetic nervous system abnormalities, and he created the term "reflex sympathetic dystrophy" in 1942 (Shim, Rose, Halle, & Shekane, 2019). In 1994, the International Association for the Study of Pain (IASP) prompted the renaming of reflex sympathetic dystrophy and causalgia to "complex regional pain syndrome" Type I and Type II, respectively, to improve clinical recognition of the disorder. However, this led to over diagnosis and the modification of the 1994 IASP criteria in 2003 into what we now know as the Budapest criteria (Harden et al., 2013).

CRPS is a disorder characterized by chronic disabling, regional pain disproportionate in time or degree to its inciting event (usually after trauma) and is a diagnosis of exclusion. All other potential etiologies for the patient's signs and symptoms should be evaluated and reasonably ruled out first, as significant harm can result if more serious diagnoses (including those of vascular, neurological, and neoplastic etiologies) are not pursued and left untreated (Chang, McDonnell, & Gershwin, 2019). Early diagnosis and intervention in CRPS are associated with improved outcome and function. Seminal studies on CRPS prevalence show that resolution was highly variable with rates between 74% of patients in the first year (Sandroni, Benrud-Larson, McClelland, & Low, 2003) and 36% at 6 years (De Mos et al., 2007). However, recurrence of CRPS is also not uncommon with estimates of recurrence ranging 10%–30%, with the higher rates occurring in younger patients, including children (Zyluk, 2004).

Diagnosis

CRPS is diagnosed through the IASP's Budapest criteria, and this chapter will focus on its clinical criteria (Table 1) with a sensitivity of 0.85 and specificity of 0.69 (Harden et al., 2013). Type I CRPS is that in which no specific nerve injury can be identified, while there is a distinct nerve lesion in Type II. As noted in Table 1, pain is associated with sensory, vasomotor disturbances, sudomotor disturbances, edema, and/or trophic changes with the potential for various complications. Almost

---

* The views expressed herein are those of the author(s) and do not necessarily reflect the official policy or position of the Department of the Navy, Department of Defense, or the U.S. Government.

† Neither the Department of the Navy nor any other component of the Department of Defense has approved, endorsed, or authorized this product.

‡ I am a military service member. This work was prepared as part of my official duties. Title 17, U.S.C. §105 provides that copyright protection under this title is not available for any work of the U.S. Government. Title 17, U.S.C., §101 defines a U.S. Government work as a work prepared by a military service member or employee of the U.S. Government as part of that person's official duties.

Features and Assessments of Pain, Anesthesia, and Analgesia. https://doi.org/10.1016/B978-0-12-818988-7.00001-7
Copyright © 2022 Elsevier Inc. All rights reserved.

TABLE 1 Budapest criteria for CRPS.

Pain out of proportion to inciting event

No other diagnosis that better explains the signs and symptoms

1 Symptom in 3+ and 1 Sign in 2+ categories

   Sensory: Hyperalgesia, allodynia

   Vasomotor: Temperature asymmetry (>1°C on exam), skin color changes, skin color asymmetry

   Sudomotor/edema: Edema, sweating changes, sweating asymmetry

   Motor/trophic: Decreased range of motion, motor dysfunction (weakness, tremor, dystonia), trophic changes (hair, nail, skin)

all organ systems are involved during the course of CRPS and most treating clinicians are unaware of the disease, its potential to spread, and do not fully understand the mechanisms of CRPS and its systemic effects (Schwartzman, 2012).

While the Budapest criteria require the absence of a diagnosis that better explains the symptoms, a 2014 study showed that only 27% of patients diagnosed with CRPS by family physicians and community specialists had the diagnosis confirmed by an expert specialist. Even more, 28% of those diagnosed had a diagnosable neuropathic pain condition and an additional 28% had a diagnosable musculoskeletal pain condition. Finally, 12% were diagnosed with a psychiatric disorder (e.g., Conversion or Factitious) (Mailis-Gagnon, Lakha, Allen, Deshpande, & Harden, 2014). Therefore, reviewing a complete list of differential diagnoses is crucial—as mentioned above these may be of neurogenic, vasogenic, or of various other causes. Workup may include the following: history, physical exam, neurological exam, vascular exam, MRI, triple phase bone scan, nerve conduction velocity and electromyogram, duplex ultrasound, quantitative sensory testing, autonomic function testing, temperature measurement, and sympathetic blockade (Hanling, Fowler, & Phillips, 2019).

Pathophysiology

While the pathophysiology of CRPS is not completely understood, there are three pathophysiologic pathways which are present in varying degrees for any individual patient: aberrant inflammatory mechanisms, vasomotor dysfunction, and maladaptive neuroplasticity. Some of these changes can be induced in a healthy volunteer with 4 weeks of immobilization, and these patients will then demonstrate mild signs of CRPS without pain (Terkelsen, Bach, & Jensen, 2008). The degree to which individual contributing mechanism(s) may differ between patients and even within one patient's course varies over time.

Neurogenic inflammation occurs when cutaneous nociceptors are activated and cause retrograde depolarization of small-diameter primary afferents through the axon reflex with a release of neuropeptides substance P and calcitonin–gene-related peptide (CGRP) from skin sensory terminals (Holzer, 1998). Accordingly, patients with CRPS demonstrate higher levels of CGRP (Birklein, Schmelz, Schifter, & Weber, 2001) and substance P (Schinkel et al., 2006) in their serum.

Vasomotor dysfunction attempts to explain the variation in the warm and cold types of CRPS. The warm type is typically seen after the initial injury and for several months afterward, while the cold type is not usually seen until several years later; however, there are many patients who have a cold type CRPS within weeks of onset. These effects may be related to sympathetic vasoconstrictor neurons and density of α-adrenergic receptors, which have been shown to differ in the affected limb vs. unaffected limb (Marinus et al., 2011).

Maladaptive neuroplasticity is a term to describe the significant changes in the central nervous system (CNS) that are present in CRPS. Central sensitization is another term for this abnormal activity in the CNS. When the spinal nociceptive neurons become sensitized, they fire in response to less stimulation or even in the absence of stimulation. The neuronal NMDA receptor can be blocked to decrease this excess sensitization (Marinus et al., 2011). Apart from pain sensitization, there is also a motor sensitization which presents as dystonia and cannot be reversed with NMDA antagonists (e.g., ketamine), but this can be improved with the intrathecal GABA type B agonist baclofen (van Rijn et al., 2009).

Management

Given the variable signs and symptoms of CRPS and the multiple potential causes described above, there is a variety of potential treatments. Anecdotal evidence suggests early treatment, particularly rehabilitation, is helpful in limiting the disorder.

However, this has not been proven in clinical studies and many treatments have poor results when evaluated for therapy in large cohorts. This lack of evidence should not be taken as a failure of interventions and medications for CRPS, but rather, as further evidence for the multiple subtypes and courses a patient may experience despite no gold standard for diagnosis.

Initial management of CRPS must include a multidisciplinary treatment focus of functional restoration—a term that includes the treatment of the biomedical and psychosocial aspects. It is critical for the clinician to evaluate patients for the root causes for each of these aspects, and then design a treatment plan that comprehensively addresses them, including physical therapy (PT), occupational therapy (OT), recreational therapy, vocational rehabilitation, and behavioral health treatments when indicated (Stanton-Hicks et al., 2002). This program emphasizes physical activity (reanimation), desensitization, and normalization of sympathetic tone in the affected limb and involves a steady progression from the gentlest, least invasive interventions to the goal of holistic rehabilitation and improved quality of life (Harden et al., 2013). While important in treatment, these techniques are well described in other texts and are outside the scope of this chapter.

Pharmacologic treatment

Pharmacologic treatment plans for CRPS often begin with oral medications based on clinician experience and trials for other neuropathic pain conditions. When possible, utilization of the fewest number of drugs is recommended to decrease side effects, cost, and poor compliance; however, it is frequently necessary to add medications to adequately treat the variety of symptoms. When polypharmacy is utilized, it should be done in a thoughtful manner to address different mechanisms of action and for prophylaxis of symptoms vice rescue medications (Harden et al., 2013) (see Table 2).

Anticonvulsants

This class of medications can be divided into two groups: gabapentin and pregabalin (poor anticonvulsants) being the first, and more typical anticonvulsants being the second. Gabapentin has evidence of efficacy in CRPS and other neuropathic conditions, but it requires dosing three times per day at doses that frequently cause significant sedation (Tan, Duman, Taşkaynatan, Hazneci, & Kalyon, 2007; Van de Vusse, Stomp-van den Berg, Kessels, & Weber, 2004). Pregabalin is a similar compound and has the advantages of requiring only twice per day dosing and causing less sedation, but it has only anecdotal evidence of efficacy in CRPS. The other typical anticonvulsants have either negative studies or have not been evaluated for the treatment of CRPS. Carbamazepine, used to treat trigeminal neuralgia, was not found to be beneficial for CRPS compared to placebo (Harke, Gretenkort, Ladleif, Rahman, & Harke, 2001). Oxcarbazepine, lamotrigine, levetiracetam, and topiramate have not been evaluated in clinical trials; however, our clinic routinely uses topiramate when other medications have failed or are not tolerated.

Antidepressants

Tricyclic antidepressants (TCAs) and serotonin-norepinephrine reuptake inhibitors (SNRIs) have not been studied for CRPS, but they have excellent evidence in the treatment of other chronic pain conditions. The TCAs not only increase the concentration of pain modulating neurotransmitters in the brain, but they also provide a peripheral sodium channel blockade which may inhibit excessive nerve firing in CRPS (Woolf & Max, 2001). Data for the use of the TCAs in CRPS have been extrapolated from a review study evaluating efficacy of various drugs in polyneuropathy (Sindrup & Jensen, 2000). Although there are no studies evaluating any antidepressants specifically for CRPS, in our clinical experience, amitriptyline, nortriptyline, and duloxetine are useful for the management of pain and sleep symptoms related to CRPS and other chronic pain conditions.

Antihypertensives and α-adrenergic antagonists

Nifedipine, a calcium-channel blocker, and phenoxybenzamine were shown in one study to be effective for management of early CRPS, but patients with chronic CPRS (longer than 3 months in this study) showed improvement rates around 35% (Muizelaar, Kleyer, Hertogs, & DeLange, 1997). An older study showed phenoxybenzamine did result in resolution of CRPS-related severe burning pain within 18 days of administration (Ghostine, Comair, Turner, Kassell, & Azar, 1984). Clonidine, another routinely used agent in this class, does not appear to be efficacious for treatment of CRPS.

TABLE 2 Medications for CRPS.

Category	Drug (oral unless otherwise noted)	Dose	Duration (if indicated)	Number needed to treat (NNT) (Birklein et al., 2001)
Anticonvulsants	Gabapentin	300 mg once daily titrated slowly to a max of 900 mg three times daily		4.1
	Pregabalin	75 mg once daily titrated slowly to 150 mg twice daily		
	Topiramate	50 mg once daily titrated slowly to 200 mg once daily		2.5
Antidepressants	Amitriptyline	10–100 mg once daily at night		2.6
	Nortriptyline	10–100 mg once daily at night		2.6
	Duloxetine	20–60 mg once daily		
Antihypertensive	Nifedipine	20–60 mg once daily	Until symptom resolution	
α-Adrenergic antagonist	Phenoxybenzamine	10 mg three times daily titrated slowly up to max of 40 mg three times daily	6–12 weeks	
Antiinflammatory	Prednisolone	40 mg once daily	4 weeks	
	Prednisone	10 mg three times per day	up to 12 weeks	
	Vitamin C	500–1000 mg once daily	50 days	
Bisphosphonates	Alendronate	40 mg once daily	8 weeks	
	Pamidronate intravenous	60 mg once		
	Clodronate intravenous	300 mg once daily	10 days	
	Neridronate intravenous	100 mg once daily every 3 days	4 infusions	

Antiinflammatory drugs

Nonsteroidal antiinflammatory drugs (NSAIDs), corticosteroids, cyclooxygenase (COX)-2 inhibitors, and free-radical scavengers have been used to decrease the inflammatory component of CRPS, and they can be used for both prophylaxis and rescue. NSAIDs have very little positive data in the literature, but corticosteroids have some evidence for benefit. Prednisolone and prednisone both have been studied and shown to have significant positive effects that last beyond the duration of treatment (Christensen, Jensen, & Noer, 1982; Kalita, Misra, Kumar, & Bhoi, 2016; Kalita, Vajpayee, & Misra, 2006). Vitamin C is a low-risk medication that has some evidence supporting supplementation for the prevention of CRPS following distal radius fractures and extremity surgery, and it is a reasonable option to implement in acute established CRPS as well (Chen, Roffey, Dion, Arab, & Wai, 2016; Meena, Sharma, Gangary, & Chowdhury, 2015; Zollinger, Tuinebreijer, Breederveld, & Kreis, 2007).

Bisphosphonates

The bisphosphonates are antiosteoclastic agents developed to prevent or reverse bone loss, but they were found to reduce pain in patients in addition to improving their bone density. Clinical trials have evaluated oral alendronate and intravenous pamidronate, clodronate, and neridronate with pain improvement gains out to 3 months posttreatment (Chevreau, Romand, Gaudin, Juvin, & Baillet, 2017).

Calcitonin

In multiple studies, calcitonin has been shown to be generally ineffective specifically for CRPS whether it is administered as an intranasal spray or subcutaneously (Tran, Duong, Bertini, & Finlayson, 2010).

Cannabis

There are multiple studies that show significant benefit to cannabis through a variety of administration routes, at different cannabinoid ratios, and for multiple neuropathic conditions, including one that included CRPS patients with allodynia (Lee, Grovey, Furnish, & Wallace, 2018; Nurmikko et al., 2007).

IVIG

Intravenous immunoglobulin (IVIG) infusions at low doses were shown to have benefit in two small studies, but a larger, more recent study showed no difference from placebo after two infusions (Goebel et al., 2017).

Naltrexone

There is some clinical experience, including our own, in using low-dose naltrexone (LDN) for the treatment of CRPS. This medication is a long-acting opioid receptor antagonist at doses of 50 to 150 mg; however, LDN is administered at much smaller amounts (1.5–4.5 mg/day). When administered in low doses (below 5 mg), naltrexone inhibits microglial activation in the central nervous system via Toll-like receptor 4 (TLR4) antagonism and appears to enhance production of endogenous opioids and opioid receptors through its transient opioid antagonism (Trofimovitch & Baumrucker, 2019).

NMDA receptor antagonists

Ketamine is the most commonly used of the N-methyl-D-aspartic acid (NMDA) receptor antagonists, and it has robust data for efficacy that seems to persist for up to 3 months following infusion and has been reported to lead to remission of symptoms after 5-day infusions (Schwartzman et al., 2009; Zhao, Wang, & Wang, 2018). There are no clinical trials evaluating oral ketamine, but oral use of another NMDA antagonist, dextromethorphan, has been shown to provide pain relief in CRPS and other neuropathic pain conditions (Cohen, Chang, Larkin, & Mao, 2004; Shaibani, Pope, Thisted, & Hepner, 2012).

Opioids

Opioids are unlikely to be an appropriate long-term treatment strategy for CRPS due to concerns for tolerance, opioid-induced hyperalgesia, adverse effects with increasing doses, and minimal evidence for the use of opioids in chronic neuropathic pain; however, they do have a role in the treatment of acute pain flares. There are several opioids that have additional effects besides pure mu-agonism, although they do not have clinical trials to support use in CRPS. Of these, buprenorphine is a mixed mu agonist antagonist opioid that also has NMDA receptor antagonism and orphan-related ligand-1 receptor agonism, and there is anecdotal evidence for better pain control in patients with refractory CRPS (Onofrio et al., 2016). Methadone is a unique opioid that also has NMDA antagonist effect and can help with hyperalgesia. Tramadol is a partial opioid agonist and has additional properties of serotonin and norepinephrine reuptake inhibition with evidence for benefit in neuropathic pain conditions (Sindrup et al., 1999).

Interventional treatment
Sympathetic nerve blocks

As CRPS was historically seen as a sympathetically maintained pain condition, the block of sympathetic ganglia was commonly used for diagnosis and treatment purposes. Although it is difficult to determine the degree of effect from sympathetic nerve block (SNB), it may typically be repeated until there is no apparent further clinical improvement. The common SNB sites are the stellate ganglion for upper extremity CRPS and the lumbar sympathetic chain for lower extremity CRPS. SNB with local anesthetic alone or with adjuncts can provide excellent pain relief to enable patients to engage with prescribed PT/OT, and these injections may provide relief significantly longer (out to 4 weeks or more) than the duration of the local anesthetic (typically only a few hours). The vast majority of patients (85%) who respond to SNB will have at least a 50% reduction in pain for at least 1 week. Local anesthetics typically used for these procedures are 1% lidocaine, 0.25% bupivacaine, or 0.2% ropivacaine. The dose is from 5 mL to 20 mL depending on the site. Corticosteroids, triamcinolone, and dexamethasone have also been routinely added by some practitioners as well. Of note, limb temperature before the block does not predict efficacy of a SNB (Cheng et al., 2019).

Ablation of the sympathetic ganglia has also been evaluated for longer term management of symptoms. Because it produces a temporary ablation, botulinum toxin has been explored as an agent for administration in sympathetic block with promising results for future use, with 5000 units of botulinum toxin type B providing up to 10 weeks of significant pain reduction (Lee, Lee, et al., 2018). Radiofrequency and chemical sympathetic ablations are typically reserved for the terminally ill due to fears of postsympathectomy neuralgia which can present up to 2 years later; however, there are very limited data as to the rate of this condition. A small study comparing thermal radiofrequency to phenol for ablation of the lumbar sympathetic ganglia for CRPS showed significant improvement in pain over 4 months with only one patient (out of 20 total) developing postsympathectomy neuralgia (Manjunath, Jayalakshmi, Dureja, & Prevost, 2008). In spite of this, the risk of increased neuropathic pain and minimal data showing long-term efficacy relegates sympathetic ablation to a treatment of last resort at this time.

Intravenous regional anesthesia

There is strong evidence in the form of a quality metaanalysis showing that intravenous regional anesthesia and other medications administered in that manner have a lack of proven effect (Perez et al., 2010).

Spinal cord stimulation

As with many chronic neuropathic pain conditions, spinal cord stimulation (SCS) remains an option for many patients despite the significant cost, invasiveness of the implant, and controversial data. It appears that implant within 1 year of onset of symptoms offers a better chance at effective pain relief. Additionally, the need for medications to manage pain following SCS is decreased while return to work is possible for some patients (Gopal, Fitzgerald, & McCrory, 2016). While SCS provides the best pain relief with the fewest side effects for many patients in our experience, it remains difficult to use because of the restrictions placed by insurance providers despite the long-term cost effectiveness compared to orthopedic surgeries (which have been shown to be only as effective as sham surgeries for treatment of CRPS).

Applications to other areas

Treatment of CRPS remains very difficult, as it is a condition which has variable symptoms over the course of the disease and may appear to have different etiologies (inflammatory, neuropathic, or spastic) at different points in time, and to varying degrees for those affected. The treatments for CRPS can be used for other neuropathic conditions like phantom limb pain and other neuropathies (e.g., diabetic polyneuropathy). The knowledge gained from a better understanding of the processes involved in CRPS may provide opportunities for the development of treatments which can be utilized for these conditions as well.

Other agents of interest

Integration of virtual reality into PT/OT modalities, especially graded motor imagery, may have future applicability for CRPS. Many other agents are being investigated too, as patients with CRPS are frequently willing to try anything given their severe, persistent pain. More research is needed on novel therapies like hyperbaric oxygen therapy (HBOT), ozone treatment, immunomodulators (e.g., lenalidomide and thalidomide), plasma exchange therapy, and polydeoxyribocucleotide. Less invasive manners of delivering electricity to the central nervous system like transcranial magnetic stimulation (TMS), transcranial direct current stimulation (tDCS), and cranial electrotherapy stimulation (CES) are being investigated for their potential impact on pain, mood, and function in CRPS. Another method of the use of electricity is the resurgence of peripheral nerve stimulation (PNS) with new devices that require more research for their role in CRPS management. The most invasive therapy for treating CRPS is surgical sympathectomy, which still requires more research at this time.

Mini-dictionary of terms

Central sensitization—nervous system condition involving wind-up which causes the nervous system to have a persistent state of high reactivity.

Complex regional pain syndrome—form of chronic pain (lasting greater than 6 months) usually affecting a limb either without a confirmed nerve injury (Type I) or with (Type II) resulting in prolonged and excessive pain with changes in the connective tissues (e.g., skin, muscle, bone) and nervous system.

Functional restoration—multidisciplinary approach with PT/OT working in concert with behavioral health and pain physician to promote desensitization and return to activities.

Postsympathectomy neuralgia—sudden, intense neuropathic pain following ablation of sympathetic ganglia; onset may be delayed several months or years postprocedure.

Sympathectomy—blockade of the sympathetic nervous system transmission through selected sympathetic ganglia; temporary with local anesthetic and botulinum toxin; persistent with radiofrequency, chemical ablation, and surgery.

Key facts

CRPS can occur with (Type II) or without (Type I) a known nerve injury, even a very mild injury such as a sprained wrist or ankle. Diagnosis is made through the application of the Budapest criteria, and it is a diagnosis of exclusion based on signs and symptoms. Treatment begins with a multidisciplinary team and early, aggressive PT/OT. Medications and invasive procedures are used to enable better engagement in PT/OT activities.

Multiple therapies are available, and the process of finding the correct combination of medications or intervention for any one patient can be lengthy. Clinical focus identifies all root causes and therapies addressing the mechanisms of each cause. It is important to understand and explain to patients that complete relief of pain is unlikely in CRPS and that restoration of function is the primary goal of treatment. Recognizing and addressing subsequent mental health conditions that arise from the altered lifestyle of developing CRPS is imperative for the treating provider.

Summary points

- CRPS is a condition encompassing the historic terms reflex sympathetic dystrophy (Type I) and causalgia (Type II) and involves three major pathophysiologic pathways—neurogenic inflammation, vasomotor dysfunction, and maladaptive neuroplasticity.
- TCAs and SNRIs are effective medications not only for pain associated with CRPS but also for the mental health comorbidities that tend to arise in this diagnosis.
- Corticosteroids have evidence of improved pain control beyond the duration of treatment, and vitamin C is a potential option for the prevention of CRPS following distal radius fractures and extremity surgery (and given its safety, may be considered following other sprains and fractures).
- Of the bone-modifying drugs, the bisphosphonates have decent evidence for pain reduction and symptom management.
- Cannabis is useful for pain control but has a potential for significant side effects in addition to the legal issues in many locations.
- Low-dose naltrexone needs more research as most evidence at this time is anecdotal.
- Ketamine can produce excellent results in some patients, but it may require long-term infusion therapy and has a dose-limiting side effect profile.
- Opioids are indicated for acute pain, especially following surgery or injury, even in CRPS, but long-term opioid therapy is not recommended for CRPS due to concerns for opioid-induced hyperalgesia and tolerance; however, the opioids which are not pure mu-opioid agonists (e.g., buprenorphine, methadone, and tramadol) are better options in CRPS.
- Many interventional procedures have been utilized, and SNB with local anesthetic has the best evidence, although ongoing research for botulinum toxin and sympathetic ablation is underway.
- The best indications for SCS are CRPS and failed back surgery syndrome, and it can result in improved pain control with fewer medications and return to function and work in selected patients.

References

Birklein, F., Schmelz, M., Schifter, S. A., & Weber, M. (2001). The important role of neuropeptides in complex regional pain syndrome. *Neurology, 57*(12), 2179–2184.

Chang, C., McDonnell, P., & Gershwin, M. E. (2019). Complex regional pain syndrome–false hopes and miscommunications. *Autoimmunity Reviews, 18*(3), 270–278.

Chen, S., Roffey, D. M., Dion, C. A., Arab, A., & Wai, E. K. (2016). Effect of perioperative vitamin C supplementation on postoperative pain and the incidence of chronic regional pain syndrome. *The Clinical Journal of Pain, 32*(2), 179–185.

Cheng, J., Salmasi, V., You, J., Grille, M., Yang, D., Mascha, E. J., et al. (2019). Outcomes of sympathetic blocks in the management of complex regional pain syndrome: A retrospective cohort study. *Anesthesiology, 131*(4), 883–893.

Chevreau, M., Romand, X., Gaudin, P., Juvin, R., & Baillet, A. (2017). Bisphosphonates for treatment of complex regional pain syndrome type 1: A systematic literature review and meta-analysis of randomized controlled trials versus placebo. *Joint, Bone, Spine, 84*(4), 393–399.

Christensen, K. J. E. L. D., Jensen, E. M., & Noer, I. V. A. N. (1982). The reflex dystrophy syndrome response to treatment with systemic corticosteroids. *Acta Chirurgica Scandinavica, 148*(8), 653–655.

Cohen, S. P., Chang, A. S., Larkin, T., & Mao, J. (2004). The intravenous ketamine test: A predictive response tool for oral dextromethorphan treatment in neuropathic pain. *Anesthesia & Analgesia, 99*(6), 1753–1759.

De Mos, M., De Bruijn, A. G. J., Huygen, F. J. P. M., Dieleman, J. P., Stricker, B. C., & Sturkenboom, M. C. J. M. (2007). The incidence of complex regional pain syndrome: A population-based study. *Pain, 129*(1–2), 12–20.

Ghostine, S. Y., Comair, Y. G., Turner, D. M., Kassell, N. F., & Azar, C. G. (1984). Phenoxybenzamine in the treatment of causalgia: Report of 40 cases. *Journal of Neurosurgery, 60*(6), 1263–1268.

Goebel, A., Bisla, J., Carganillo, R., Frank, B., Gupta, R., Kelly, J., et al. (2017). Low-dose intravenous immunoglobulin treatment for long-standing complex regional pain syndrome: A randomized trial. *Annals of Internal Medicine, 167*(7), 476–483.

Gopal, H., Fitzgerald, J., & McCrory, C. (2016). Spinal cord stimulation for FBSS and CRPS: A review of 80 cases with on-table trial of stimulation. *Journal of Back and Musculoskeletal Rehabilitation, 29*(1), 7–13.

Hanling, S. R., Fowler, I. M., & Phillips, C. R. (2019). Complex regional pain syndrome. In C. C. Buckenmaier, M. L. Kent, J. C. Brookman, P. J. Tighe, E. R. Mariano, & D. A. Edwards (Eds.), *Acute pain medicine* (pp. 262–270). New York, NY: Oxford University Press.

Harden, R. N., Oaklander, A. L., Burton, A. W., Perez, R. S., Richardson, K., Swan, M., et al. (2013). Complex regional pain syndrome: Practical diagnostic and treatment guidelines. *Pain Medicine, 14*(2), 180–229.

Harke, H., Gretenkort, P., Ladleif, H. U., Rahman, S., & Harke, O. (2001). The response of neuropathic pain and pain in complex regional pain syndrome I to carbamazepine and sustained-release morphine in patients pretreated with spinal cord stimulation: A double-blinded randomized study. *Anesthesia & Analgesia, 92*(2), 488–495.

Holzer, P. (1998). Neurogenic vasodilatation and plasma leakage in the skin. *General Pharmacology: The Vascular System, 30*(1), 5–11.

Kalita, J., Misra, U., Kumar, A., & Bhoi, S. K. (2016). Long-term prednisolone in post-stroke complex regional pain syndrome. *Pain Physician, 19*(8), 565–574.

Kalita, J., Vajpayee, A., & Misra, U. K. (2006). Comparison of prednisolone with piroxicam in complex regional pain syndrome following stroke: A randomized controlled trial. *QJM, 99*(2), 89–95.

Lee, G., Grovey, B., Furnish, T., & Wallace, M. (2018). Medical cannabis for neuropathic pain. *Current Pain and Headache Reports, 22*(1), 8.

Lee, Y., Lee, C., Choi, E., Lee, P., Lee, H. J., & Nahm, F. (2018). Lumbar sympathetic block with botulinum toxin type A and type B for the complex regional pain syndrome. *Toxins, 10*(4), 164.

Mailis-Gagnon, A., Lakha, S. F., Allen, M. D., Deshpande, A., & Harden, R. N. (2014). Characteristics of complex regional pain syndrome in patients referred to a tertiary pain clinic by community physicians, assessed by the Budapest clinical diagnostic criteria. *Pain Medicine, 15*(11), 1965–1974.

Manjunath, P. S., Jayalakshmi, T. S., Dureja, G. P., & Prevost, A. T. (2008). Management of lower limb complex regional pain syndrome type 1: An evaluation of percutaneous radiofrequency thermal lumbar sympathectomy versus phenol lumbar sympathetic neurolysis—A pilot study. *Anesthesia & Analgesia, 106*(2), 647–649.

Marinus, J., Moseley, G. L., Birklein, F., Baron, R., Maihöfner, C., Kingery, W. S., et al. (2011). Clinical features and pathophysiology of complex regional pain syndrome. *The Lancet Neurology, 10*(7), 637–648.

Meena, S., Sharma, P., Gangary, S. K., & Chowdhury, B. (2015). Role of vitamin C in prevention of complex regional pain syndrome after distal radius fractures: A meta-analysis. *European Journal of Orthopaedic Surgery & Traumatology, 25*(4), 637–641.

Muizelaar, J. P., Kleyer, M., Hertogs, I. A., & DeLange, D. C. (1997). Complex regional pain syndrome (reflex sympathetic dystrophy and causalgia): Management with the calcium channel blocker nifedipine and/or the α-sympathetic blocker phenoxybenzamine in 59 patients. *Clinical Neurology and Neurosurgery, 99*(1), 26–30.

Nurmikko, T. J., Serpell, M. G., Hoggart, B., Toomey, P. J., Morlion, B. J., & Haines, D. (2007). Sativex successfully treats neuropathic pain characterised by allodynia: A randomised, double-blind, placebo-controlled clinical trial. *Pain, 133*(1–3), 210–220.

Onofrio, S., Vartan, C. M., Nazario, M., DiScala, S., Cuevas-Trisan, R., & Melendez-Benabe, J. (2016). The use of transdermal buprenorphine in complex regional pain syndrome: A report of two cases. *Journal of Pain & Palliative Care Pharmacotherapy, 30*(2), 124–127.

Perez, R. S., Zollinger, P. E., Dijkstra, P. U., Thomassen-Hilgersom, I. L., Zuurmond, W. W., Rosenbrand, K. C., et al. (2010). Evidence based guidelines for complex regional pain syndrome type 1. *BMC Neurology, 10*(1), 20.

Sandroni, P., Benrud-Larson, L. M., McClelland, R. L., & Low, P. A. (2003). Complex regional pain syndrome type I: Incidence and prevalence in Olmsted county, a population-based study. *Pain, 103*(1–2), 199–207.

Schinkel, C., Gaertner, A., Zaspel, J., Zedler, S., Faist, E., & Schuermann, M. (2006). Inflammatory mediators are altered in the acute phase of posttraumatic complex regional pain syndrome. *The Clinical Journal of Pain, 22*(3), 235–239.

Schwartzman, R. J. (2012). Systemic complications of complex regional pain syndrome. *Neuroscience & Medicine, 3*(03), 225.

Schwartzman, R. J., Alexander, G. M., Grothusen, J. R., Paylor, T., Reichenberger, E., & Perreault, M. (2009). Outpatient intravenous ketamine for the treatment of complex regional pain syndrome: A double-blind placebo controlled study. *Pain, 147*(1–3), 107–115.

Shaibani, A. I., Pope, L. E., Thisted, R., & Hepner, A. (2012). Efficacy and safety of dextromethorphan/quinidine at two dosage levels for diabetic neuropathic pain: A double-blind, placebo-controlled, multicenter study. *Pain Medicine, 13*(2), 243–254.

Shim, H., Rose, J., Halle, S., & Shekane, P. (2019). Complex regional pain syndrome: A narrative review for the practicing clinician. *British Journal of Anaesthesia, 123*(2), e424–e433.

Sindrup, S. H., Andersen, G., Madsen, C., Smith, T., Brøsen, K., & Jensen, T. S. (1999). Tramadol relieves pain and allodynia in polyneuropathy: A randomised, double-blind, controlled trial. *Pain, 83*(1), 85–90.

Sindrup, S. H., & Jensen, T. S. (2000). Pharmacologic treatment of pain in polyneuropathy. *Neurology, 55*(7), 915–920.

Stanton-Hicks, M. D., Burton, A. W., Bruehl, S. P., Carr, D. B., Harden, R. N., Hassenbusch, S. J., et al. (2002). An updated interdisciplinary clinical pathway for CRPS: report of an expert panel. *Pain Practice, 2*(1), 1–16.

Tan, A. K., Duman, I., Taşkaynatan, M. A., Hazneci, B., & Kalyon, T. A. (2007). The effect of gabapentin in earlier stage of reflex sympathetic dystrophy. *Clinical Rheumatology, 26*(4), 561–565.

Terkelsen, A., Bach, F., & Jensen, T. (2008). Experimental forearm immobilization in humans induces cold and mechanical hyperalgesia. *Anesthesiology, 109*(2), 297.

Tran, D. Q., Duong, S., Bertini, P., & Finlayson, R. J. (2010). Treatment of complex regional pain syndrome: A review of the evidence. *Canadian Journal of Anesthesia/Journal canadien d'anesthésie, 57*(2), 149–166.

Trofimovitch, D., & Baumrucker, S. J. (2019). Pharmacology update: Low-dose naltrexone as a possible nonopioid modality for some chronic, nonmalignant pain syndromes. *American Journal of Hospice & Palliative Medicine, 36*(10), 907–912.

Van de Vusse, A. C., Stomp-van den Berg, S. G., Kessels, A. H., & Weber, W. E. (2004). Randomised controlled trial of gabapentin in complex regional pain syndrome type 1 [ISRCTN84121379]. *BMC Neurology, 4*(1), 13.

van Rijn, M. A., Munts, A. G., Marinus, J., Voormolen, J. H., de Boer, K. S., Teepe-Twiss, I. M., et al. (2009). Intrathecal baclofen for dystonia of complex regional pain syndrome. *Pain, 143*(1–2), 41–47.

Woolf, C. J., & Max, M. B. (2001). Mechanism-based pain diagnosis issues for analgesic drug development. *Anesthesiology: The Journal of the American Society of Anesthesiologists, 95*(1), 241–249.

Zhao, J., Wang, Y., & Wang, D. (2018). The effect of ketamine infusion in the treatment of complex regional pain syndrome: A systemic review and meta-analysis. *Current Pain and Headache Reports, 22*(2), 12.

Zollinger, P. E., Tuinebreijer, W. E., Breederveld, R. S., & Kreis, R. W. (2007). Can vitamin C prevent complex regional pain syndrome in patients with wrist fractures?: A randomized, controlled, multicenter dose-response study. *JBJS, 89*(7), 1424–1431.

Zyluk, A. (2004). Complex regional pain syndrome type I. Risk factors, prevention and risk of recurrence. *The Journal of Hand Surgery, 29*(4), 334–337.

Chapter 12

Phantom limb pain

Derek M. Miletich[a,*,†,‡], Lynita Mullins[b,*,†,‡], and C. Ryan Phillips[a,*,†,‡]

[a]*CDR, MC, Department of Anesthesiology and Pain Medicine, Naval Medical Center San Diego, San Diego, CA, United States,* [b]*CDR, MC, Department of Pain Medicine, Naval Medical Center Camp Lejeune, NC, United States*

Abbreviations

C-FINE	composite flat interface nerve electrode
DRG	dorsal root ganglion
PLP	phantom limb pain
PNS	peripheral nerve stimulation
RLP	residual limb pain
SCS	spinal cord stimulation
TMR	targeted muscle reinnervation
TMS	transcranial magnetic stimulation

Introduction

Limb amputation has been a part of medicine dating back 45,000 years and is not an uncommon event. Postamputation pain is highly prevalent, with 60%–80% experiencing PLP within the first week after amputation, and is more common if pain was preexisting in the amputated limb prior to its amputation (Kehlet, Jensen, & Woolf, 2006). PLP usually diminishes with time, and if it persists beyond 6 months, prognosis for spontaneous recovery is poor (Jackson & Simpson, 2004).

In 2008, it was estimated 1 in 190 US citizens were living with limb loss, and this number may double by 2050. In the United States, 54% of amputations are due to vascular disease (two-thirds of those cases have comorbid diabetes), 45% due to trauma, and 2% due to malignancy. Approximately 65% of amputations are in the lower extremities, with more than half considered "major" amputations (i.e., not a digit), while only 8% of upper extremity amputations are considered major (Urits et al., 2019).

The term "Phantom Limb Pain" was first credited to Dr. Silas Weir Mitchell, a US Civil War physician; however, the concept was present in writings across the world as early as the 16th century. The incidence of PLP was estimated as high as 90% of all amputations in the Civil War. By contrast, the PLP incidence in World War II dropped to only 5%. It is hypothesized that this era attributed symptoms of PLP to psychiatric causes and/or malingering (Finger & Hustwit, 2003). A history of reporting inconsistencies and relative lack of research has contributed to a limited understanding of this disease process and its underlying mechanisms.

Diagnosis

PLP is defined by two components—painful sensations which are (1) in the limb affected by amputation and (2) referred to the missing limb as if it were still present. There are two additional conditions which may be confused or misdiagnosed in PLP. They are residual limb pain (RLP), which is only the first component of the definition, and phantom sensation, which is the second component without pain. In clinical practice, the picture may be quite complicated, as any one or all three of

---

*. The views expressed herein are those of the author(s) and do not necessarily reflect the official policy or position of the Department of the Navy, Department of Defense, or the U.S. Government.

†. Neither the Department of the Navy nor any other component of the Department of Defense has approved, endorsed, or authorized this product.

‡. We are military service members. This work was prepared as part of my official duties. Title 17, U.S.C. §105 provides that copyright protection under this title is not available for any work of the U.S. Government. Title 17, U.S.C., §101 defines a U.S. Government work as a work prepared by a military service member or employee of the U.S. Government as part of that person's official duties.

Features and Assessments of Pain, Anesthesia, and Analgesia. https://doi.org/10.1016/B978-0-12-818988-7.00034-0
Copyright © 2022 Elsevier Inc. All rights reserved.

these conditions can be present in the same patient. Therefore, comprehensive symptom evaluation is needed to develop an optimal treatment plan that addresses all factors. To assist in this effort, clinicians may wish to develop a series of questions, such as those from the Phantom Limb Questionnaire (Michael & Christine, 2018; Urits et al., 2019). See Fig. 1 for a Post-Amputation Symptom Clarification Algorithm which that be utilized to clarify these definitions.

Pathophysiology

The exact mechanism behind PLP is not fully understood; however, this condition appears likely to be from a combination of both peripheral factors and central sensitization of the nervous system (i.e., a peripheral trauma that generates a neurochemical cascade that sweeps centrally until cortical brain structures are recruited). This involvement of cortical structures may be responsible for the complex and "vivid" nature of PLP and the associated phenomenon of telescoping, wherein the phantom limb is perceived to shorten in length over time.

In analyzing the pathophysiology of PLP, three main levels will be the focus:

Peripheral
Spinal
Supraspinal (brain)

Clinicians should consider this three-tiered approach for management of PLP (both prevention and treatment, as discussed later), based on which levels specifically apply to the patient's clinical case (see Table 1).

Peripheral

Peripheral pain receptors and signaling systems are essential for human survival. They serve as early warning systems to prevent injury or to help mitigate injury after it has already occurred. Hyperactivity of these systems can be problematic because it results in remodeling of the nervous system and predisposes an individual to be in a chronic pain state. At least four peripheral mechanisms can be identified which may contribute to PLP:

Afferent C-fibers
Neuromas
Dorsal root ganglion
Postganglionic/sympathetic sprouting onto neuromas/dorsal root ganglion (DRG)

Afferent C-fibers

Afferent C-fibers are specialized nerve fibers that specifically send pain signals to the spinal cord from a peripheral site after a noxious stimulus (such as trauma/amputation). These C-fibers have increased activity post amputation and can continue

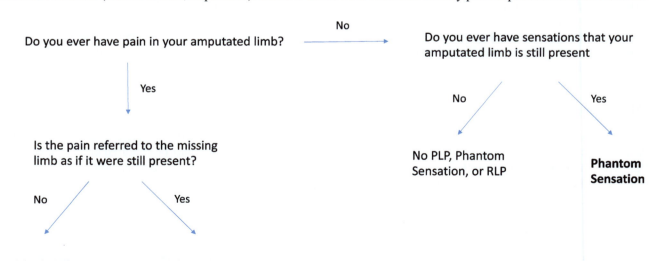

FIG. 1 Postamputation symptom clarification algorithm.

TABLE 1 Management approaches to PLP.

Modalities	Peripheral	Spinal	Supraspinal
Prevention			
Systemic opioids		X	X
Epidural analgesia		X	
Peripheral regional anesthesia	X		
TMR	X		X
Treatment			
Systemic pharmacotherapy			
Ketamine		X	X
Opioids		X	X
Therapeutic modalities			
Mirror therapy/GMI			X
EMDR			X
TENS		X	
Acupuncture	X	X	X
Invasive interventions			
PNS	X	X	X
SCS	X	X	X
DRG stimulation	X	X	X
TMR	X		X
Noninvasive interventions			
TMS			X
tDCS/microcurrent electrotherapy	X	X	X
Other treatments			
C-FINE prosthetics	X	X	X
Cryoneurolysis	X		

Prevention and treatment based on anatomic target levels.

to send pain signals to the spinal cord even though there is no longer any primary noxious stimulus present. C-fibers are "slow" pain fibers, meaning their signal conduction is relatively slow compared to other types of nerve fibers (Aβ, etc.). The sensation typically characterized by their activity is a burning sensation (Forstenpointner et al., 2019).

Neuromas

Neuromas are a collection of proliferated neuronal growth cones after trauma, and are a source of ectopic neuronal depolarization (i.e., they cause proximal nerve firing even without distal stimulation). When a nerve is severed (such as through amputation), the nerve axon will attempt to regenerate and grow distally toward its site of original innervation. Neuromas occur when the axon grows chaotically due to scar tissue, and in the case of amputation, the absence of distal tissue in which to regenerate. In the context of PLP, when these neuromas form in the area proximal to the amputation, due to the factors discussed earlier, they can affect both the perception of sensation and pain coming from the missing limb, even though no stimulation is actually present (Dumanian et al., 2019).

Dorsal root ganglion

The DRGs are groups of cell bodies responsible for the transmission of sensory messages from multiple specialized receptors to the CNS for a response. In PLP, the DRG may respond with a state of increased sensitivity causing allodynia and hyperalgesia in the residual limb and may contribute toward a perception of increased signaling coming from the affected amputated limb, resulting in continual awareness by the individual even when there are no noxious stimuli (Collins et al., 2018).

Postganglionic/sympathetic sprouting onto neuromas/DRG

This process can occur after traumatic nerve injury (e.g., amputation). This then sets up the neuroma/DRG so that it is sensitive to adrenergic tone (i.e., emotional or physical stress). PLP is known to be aggravated by states of emotional or physical stress, and so this aforementioned mechanism may be an important factor (Dumanian et al., 2019).

Spinal

In chronic pain states, such as those propagated through abnormal signaling from peripheral sources, the spinal cord can go through central sensitization, in which functional and structural changes occur. Three of these mechanisms will be discussed:

Altered interneuron control
Bulbospinal pathways
Increased activity of N-methyl-D-aspartic acid (NMDA) receptor systems

Altered interneuron control

Altered interneuron control of the dorsal horn of the spinal cord is theorized to occur via a change in GABA-A and glycine receptor functioning (Yaksh, 2019). Interneurons are inhibitory neurons in the spinal cord and serve to decrease afferent signaling, which normally results in appropriately modulating pain signals transmitted to the brain. If this interneuron network is altered, however, so that it no longer is as efficacious in inhibition of the dorsal horn, pain transmission will pathologically be increased, and the individual will experience chronic pain.

Bulbospinal pathways

Bulbospinal pathways allow the brainstem to exert control over the dorsal horn, specifically through the medullary raphe and the locus coeruleus (via the neurotransmitters serotonin and noradrenaline, respectively). Raphe spinal terminals excite dorsal horn wide dynamic range (WDR) neurons and facilitate wind up (progressively increased excitability of spinal cord neurons in response to repeated stimuli above a critical rate). Conversely, coeruleus spinal terminals inhibit dorsal horn WDR neurons through $\alpha2$ adrenergic receptors and decrease pain transmission (Ossipov, Morimura, & Porreca, 2014). Imbalances in these networks or their neurotransmitters can result in a chronic pain state. PLP likely is modulated according to the balance of these systems, and they should be considered when developing management strategies.

Increased activity of NMDA receptor systems

Increased activity of the NMDA receptor systems is also thought to contribute to wind up and result in a chronic pain state. NMDA is a receptor for the excitatory neurotransmitter glutamate, which is released when there is a noxious peripheral stimulus. Ongoing noxious stimulus (e.g., amputation) acutely enhances NMDA receptor function, resulting in hyperalgesia, neuropathic pain, and reduced functionality of opioid receptors. These effects can directly impact the clinical course and treatment response of PLP, and therefore are an important area of focus when assessing treatment options (Hachisuka et al., 2018; Lodge, Watkins, Bortolotto, Jane, & Volianskis, 2019).

Supraspinal (brain)

Pain is both a sensory and emotional experience, and as such, its ultimate perception and meaningfulness/impact on the functioning of an individual are determined at the supraspinal level; two areas will be the focus for this chapter:

Cortical
Subcortical

Cortical

Cortical reorganization occurs after the loss of afferent input to the cerebral cortex, such as by limb amputation. The sensory and motor strips of the cerebral cortex have a homunculus in which the components of the body are mapped out. Because the brain is incredibly plastic and seeks to optimize efficiency, if it stops receiving input from a particular region of the body, it will reallocate that region to its neighbors on the homunculus (see Fig. 2). As a result, PLP may correlate with the degree of this reorganization in the somatosensory cortex secondary to loss of afferent input (Molina-Rueda et al., 2019).

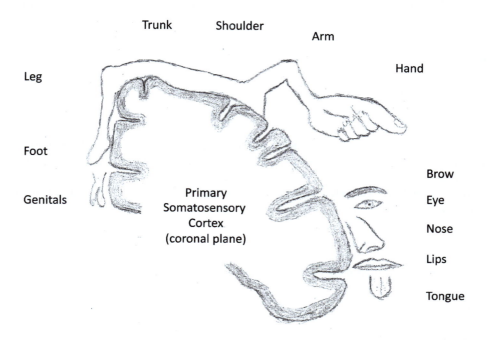

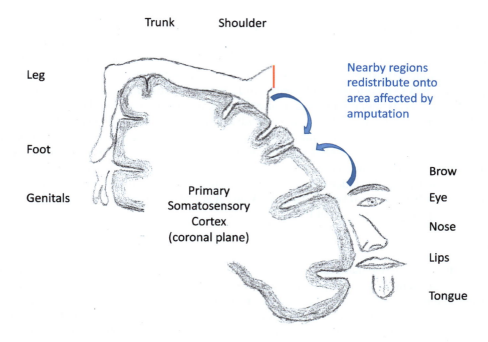

FIG. 2 Homunculus reorganization following amputation. Normal vs postamputation somatotopic distribution of homunculus.

Subcortical

Subcortical reorganization occurs through changes in pathways in how the brain integrates pain both somatically and emotionally. These regions and pathways include the thalamus, primary and secondary somatosensory cortices, insula, anterior cingulate, and dorsal prefrontal cortex. After a significant physical and/or emotional trauma (of which an amputation may be both), these pathways may be altered and cause abnormal processing, resulting in dysfunction in resolving both the physical and emotional components of pain. Since the loss of a limb will intrinsically involve both of these components, these brain regions should also be a focus when assessing and developing a treatment plan for PLP (Molina-Rueda et al., 2019).

Management

Management of PLP can be broken down into two components:

Prevention
Treatment

Logically, prevention of PLP is ideal whenever possible, and proper preamputation care combined with high-quality peri-amputation care potentially can accomplish this. In PLP, prevention may not always be an attainable goal due to the unpredictable nature of amputation itself (in which preparation is not always possible) and variability in quality of aftercare throughout medical systems. In these cases, effective treatment is needed to mitigate established PLP signs and symptoms, with the goal of resolution to the maximum degree possible.

Prevention

The following modalities have shown promise and/or efficacy for helping to prevent PLP from occurring:

Systemic pharmacotherapy
Epidural analgesia
Peripheral regional anesthesia
Targeted muscle reinnervation

Systemic pharmacotherapy

This modality is often desirable from a clinical standpoint because of its easy accessibility, and because special provider skillsets and interventional tools are often unnecessary. Unfortunately, in regard to prevention of PLP, the current data do not show many options for systemic pharmacotherapy. Among the limited agents and limited research available for the prevention of PLP, opioids appear to show some promise. In a small study with patient-controlled analgesia opiates administered 48 h preoperatively and 48 h postoperatively, only 3 of 13 patients were found to have PLP at 6 months compared to 9 of 12 in the control group receiving intramuscular/oral opiates (Karanikolas et al., 2011).

Other typical chronic pain medications do not show similar promise for prevention currently. Ketamine, commonly used in the management of general chronic pain, has conflicting data as to the benefit of use in prevention of PLP (Ahuja, Thapa, & Ghai, 2018). Another NMDA receptor antagonist, memantine, which can also be used for treatment of chronic pain, also failed to show significant clinical benefit in prevention of PLP (Maier et al., 2003). Gabapentinoids, a mainstay in treating chronic pain, also failed to show evidence in the prevention of PLP. Additionally, NSAIDs and acetaminophen do not show any benefit for prevention of PLP (Ahuja et al., 2018). Our practice is to utilize ketamine infusions in the perioperative setting for prevention of PLP.

Epidural analgesia

There are only a few studies available that evaluated the effect of epidural analgesia (bupivacaine, ketamine, etc.) on the occurrence of PLP following amputation. Despite this paucity of data, epidural catheters have become a mainstay of perioperative care. Data from the available studies tend to agree that epidural analgesia provides benefit in the short term, and possibly out to 6 months, but the occurrence of PLP is similar between the intervention and control groups at 12+ months (Ahuja et al., 2018; Lambert et al., 2001; Wilson, Nimmo, Fleetwood-Walker, & Colvin, 2008). Yousef and Aborahma (2017) recently published data showing significant difference in PLP rates at 1 year for patients who received calcitonin as part of the epidural catheter infusion.

Peripheral regional anesthesia

Research on the use of regional anesthesia for amputation is primarily available for the lower extremity; however, upper extremity postamputation pain can be managed with regional anesthesia as well. The limited research shows similar rates of PLP development despite its use in the upper extremity (Enneking, Scarborough, & Radson, 1997). Regional anesthesia in the lower extremity appears to be more efficacious than epidural analgesia as it decreases pain in the acute postoperative period and reduces risk of PLP at 1 year (Ahuja et al., 2018). For lower extremity amputations, it is our common practice to place preoperative femoral and sciatic nerve catheters for preemptive analgesia and continue the catheters for at least 72 h.

Targeted muscle reinnervation

TMR can be performed as a preventive measure at the time of initial amputation or at a later date for a patient with RLP or PLP from a neuroma. It involves surgically rerouting and coapting the distal aspect of a transected nerve (such as after amputation) to a motor nerve that innervates an intact adjacent muscle. The mechanism by which this can be used to reduce PLP is thought to be through neuroma prevention and cortical reorganization (Aternali & Katz, 2019). A recent randomized control trial suggests superior results compared to neuroma excision, with >50% reduction in pain score with TMR vs slight increase in pain score with neuroma excision, at 1 year (Dumanian et al., 2019).

Treatment

The following modalities have shown promise and/or efficacy in treating established PLP:

Pharmacotherapy
Therapeutic modalities/psychological/other treatments
Invasive interventions
Noninvasive interventions

Pharmacotherapy

Ketamine

Ketamine has been shown to be effective at reducing PLP to <10% of the average baseline pain; however, data on long-term efficacy are lacking (Richardson & Kulkarni, 2017). Our clinical experience has shown that daily 1-h infusions of 0.5 mg/kg in an inpatient setting immediately following traumatic amputation can lead to significant pain reduction or resolution.

Opioids

Opioids are shown to be effective in reducing PLP pain by approximately 50% (Wu et al., 2002, 2008). However, as PLP is a chronic condition, and opioid therapy is typically intended to be a short-term solution for acute pain, it is not an optimal treatment strategy if other effective options can be utilized.

Gabapentinoids/tricyclic antidepressants

Gabapentin, an anticonvulsant and analgesic medication that acts as a ligand for the a2d subunit of neuronal voltage–dependent calcium channels, is a mainstay treatment for many neuropathic chronic pain conditions; however, in treatment of PLP, it shows mixed results for pain improvement compared to placebo (Collins et al., 2018. Similarly, tricyclic antidepressants, which are also mainstay treatments in managing chronic pain, show scant evidence for efficacy in managing PLP (Alviar, Hale, & Lim-Dungca, 2016; Robinson et al., 2004).

Therapeutic modalities/psychological/other treatments

Nonmedical treatment modalities are mentioned in higher precedence in this chapter because these treatments can be among the most effective strategies for managing PLP, carry little or no additional risks, and can be very cost effective; therefore, they should be given strong consideration by the clinician.

Mirror therapy and graded motor imagery

Mirror therapy may be among the least expensive and most effective modalities for treatment of PLP, with >50% reduction in PLP seen in some studies; however, high-quality data are still lacking (Aternali & Katz, 2019). This treatment entails the use of a mirror box placed facing the unaffected limb (in single amputees), which conditions the brain into believing that the amputated limb has returned. It is theorized that this causes the somatosensory cortex to return closer to a preamputation baseline (Collins et al., 2018). Virtual reality treatment modalities may provide a more immersive experience for patients and additionally offer treatment options for dual amputees for whom mirror therapy would be impossible. It should be noted that mirror therapy, while effective for some patients, has been known to potentially paradoxically aggravate PLP and/or worsen telescoping phenomenon (Urits et al., 2019), so the clinician should monitor patient progress closely and modify or discontinue if these events occur.

Of note, graded motor imagery (GMI) is a larger intervention strategy that incorporates mirror therapy and utilizes a graded sequence of approaches including left/right judgments (implicit motor imagery) and imagined movements (explicit motor imagery) (Limakatso, Madden, Manie, & Parker, 2020). In addition to desensitization techniques employed by therapists, engaging patients in mental imagery phantom limb exercises has been proposed to help treat PLP and influence cortical reorganization.

Eye movement desensitization and reprocessing therapy

Eye movement desensitization and reprocessing therapy (EMDR) is a specialized type of psychotherapy in which patients process traumatic or otherwise problematic symptoms through mentally visualizing a disturbing image associated with the traumatic experience and then counterbalancing with an alternative positive image while systematically alternating eye movement left and right by following external cues (e.g., alternating lights). This bilateral eye movement is theorized to assist the brain into more effectively integrating and resolving the traumatic experience because of engagement of both cerebral hemispheres. Patients with PLP have shown to have mean pain severity decreased from 7.15 to 2.25 over 24 months, compared to control showing an increase in pain severity from 6.8 to 8.06 (Rostaminejad, Behnammoghadam, Rostaminejad, Behnammoghadam, & Bashti, 2017). This treatment impacts the supraspinal level with specific targeting theorized on the thalamus, cerebral cortex, and other brain structures associated with pain (Landin-Romero, Moreno-Alcazar, Pagani, & Amann, 2018). EMDR is primarily based on the patient's own internal processing instead of dialogue with a therapist, which may offer a substantial benefit for patients, as stigma, shame, or other avoidance can often be a major deterrent from seeking mental healthcare.

Transcutaneous electrical nerve stimulation

Transcutaneous electrical nerve stimulation therapy is mainstay treatment for chronic musculoskeletal pain, and involves using an electrical current applied at or near the painful area in order to stimulate Aβ afferent neurons and impede pain signal transmission to the brain, which is one of the tenets of the Gate Theory model of pain. Data supporting efficacy in treatment specifically for PLP are mixed (Petersen, Nanivadekar, Chandrasekaran, & Fisher, 2019), but as a low-cost and low-risk modality, it may be worthwhile to consider when formulating a holistic treatment strategy for PLP.

Acupuncture

Acupuncture has been shown to be of potential benefit for treatment of PLP, although there is limited high-quality evidence at this time as most publications are case studies (Trevelyan, Turner, Summerfield-Mann, & Robinson, 2016). Its therapeutic effect in PLP may be related to somatosensory cortex and sympathetic/parasympathetic actions (Tseng, Chen, & Lee, 2014), and there is evidence suggesting beneficial effects on the CNS including neurochemical, neurogenesis, memory, and cerebral blood flow (Chavez et al., 2017).

Invasive interventions

Invasive interventional modalities can be very effective in treating pain that is unable to be relieved through other mechanisms, but they may not be appropriate for all patients because they involve surgery, implanting a device (often permanently) into the body, and risk infection, and/or carry significant financial cost. Careful screening should be performed to ensure that risks and benefits align with the patient's needs and situation.

Peripheral nerve stimulation

Direct inhibition of primary nociceptive afferents through electrical stimulation by permanently implanted leads near a peripheral nerve can reduce central sensitization. There is evidence of effectiveness specifically in PLP, but high-quality data are limited (Cornish & Wall, 2015).

Removable peripheral nerve stimulation (PNS) systems are a new modality in which the lead is removed after a finite time period (e.g., after 60 days). Although the exact mechanism of action is not fully understood at this time, it is theorized to be similar to traditional PNS, but due to smaller leads being used, there is improved stimulation of Aβ neurons and improved cortical reorganization. Recent research has shown significantly sustained improvement (10 months post removal) in PLP (Gilmore et al., 2019), which is comparable to traditional PNS. Additionally, the small lead size and external, nonpermanent implantation may be more ideal for some amputees as traditional systems may not be implantable due to inherently limited tissue space.

Spinal cord stimulation

Spinal cord stimulation (SCS) is an implantable neuromodulation therapy that can be very effective for managing intractable pain and offers potential application in PLP patients as well. The mechanism of action is not completely understood, but it is theorized to work through dual mechanisms of antidromic and orthodromic neuronal activation which releases neurotransmitters and regulates WDR neurons. Thalamic, cortical, and sympathetic nervous system effects are also described with the use of SCS (Deogaonkar, Zibly, & Slavin, 2014; Reddy, 2019).

Dorsal root ganglion stimulation

DRG stimulation is a newer type of neuromodulation that has a similar mechanism to SCS; however, it allows for increased precision in treatment at the affected DRG, which may also allow for fewer adverse outcomes such as paresthesias. Because of these factors, it is theorized that DRG may be more effective than SCS in treating chronic pain conditions such as PLP; however, evidence is still currently limited (Deer et al., 2017).

Noninvasive interventions

These modalities may be appealing for some patients, due to the inherent noninvasive quality; however, there still may be certain risks (e.g., seizure) particularly when the CNS is targeted, and additionally may be financially expensive and/or have limited insurance coverage, and so careful patient screening should still be performed.

Transcranial magnetic stimulation

TMS involves the use of an external magnet to induce electrical current/depolarization in the brain (Morales-Quezada, 2017), which when executed in a repetitive fashion may result in cortical modulation. In treating PLP, targeting specific regions of the brain (e.g., somatosensory cortex) appears important in efficacy. This modality is useful because it may be able to specifically target the supraspinal level of pain not adequately accessible through other means.

Transcranial direct current stimulation and microcurrent electrotherapy stimulation

There is limited research at present for transcranial direct current stimulation (tDCS) and microcurrent electrotherapy stimulation, which involve the use of weak electrical currents applied transcutaneously to the scalp and/or body to modulate cortical imbalances (such as those seen with PLP) (Morales-Quezada, 2017; Taylor, Anderson, Riedel, Lewis, & Bourguignon, 2013). Advantages include relative low cost, minimal discomfort, and patient self-administration allowing flexibility and access on demand. Further, these have shown promise in mental health conditions including depression and anxiety and may be of particular benefit in helping to treat these comorbidities often associated with chronic pain conditions such as PLP.

Application to other areas

PLP is a complicated neuropathic pain condition and many of the treatments, both medications and procedures, which have shown benefit for PLP likely have benefit in other large neuron disease processes. Furthering research on neuronal disruption in PLP will apply to patients who have suffered injuries with nerve transection without amputation as well.

Other treatments of interest

Peripheral nerve cuff integration into prosthetics

Advanced prosthetic research utilizing a type of peripheral nerve cuff called a C-FINE is implanted onto affected afferent nerves in an amputated limb to restore natural sensation (Charkhkar et al., 2018) and shows great promise for restoring functionality for amputees despite its experimental phase. This technology has the potential to impact the peripheral, spinal, and supraspinal factors associated with PLP and may be of particular interest in management strategies.

Cryoneurolysis

Cryoneurolysis is a prolonged nerve block lasting a few months through the deliberate application of cold temperatures to nerves for therapeutic purposes. This intervention appears feasible and safe, and small pilot studies have demonstrated promising pain relief results for PLP (Moesker, Karl, & Trescot, 2014; Prologo et al., 2017).

Mini-dictionary of terms

Acupuncture: medical treatment strategy in which needles are placed in specific points on the body to achieve a desired physiological outcome.

Central sensitization: pain hypersensitivity, allodynia, and hyperalgesia caused by excitability in central nociceptive pathways.

Homunculus: a mapping of the body's parts onto specific regions of the brain.

Phantom limb pain: pain perceived in a region of the body no longer present.

Key facts

PLP is a phenomenon of pain perceived by a region of the body that is no longer present (i.e., after amputation). It is a distinct phenomenon from two similar but clinically different conditions, phantom sensation (a nonpainful perception of a region of the body which is no longer there) and RLP (pain felt in the part of the limb that remains after amputation), and care must be taken to correctly identify which condition(s) the patient is suffering from so that an appropriate treatment plan can be developed.

Central sensitization appears to be a significant component in the phenomenon of PLP. As such, therapeutic modalities addressing central sensitization will likely assist in a comprehensive treatment plan.

Summary points

PLP remains a difficult condition to manage, but research shows that pain free/minimal pain states can be attained with appropriate prevention and treatment plans.

Pretreatment to reduce pain before amputation and in the periamputation aftercare period appears effective in reducing later incidence of PLP.

Even if pretreatment before amputation is not possible, when maximal pain control can be achieved in the immediate days post amputation, evidence shows that incidence and severity of PLP may be reduced.

PLP has at least three loci, the peripheral, spinal, and supraspinal levels of the nervous system, and for optimal treatment, the patient must be carefully evaluated for each of these potential factors, and provided a tailored treatment plan addressing them.

Supraspinal factors can be a significant component for an individual's PLP, and there are several treatment options available which may address this level, including psychotherapy, TMS, acupuncture, and tDCS/microcurrent electrotherapy (see Table 1 for full list).

Disclaimer

The views expressed herein are those of the author(s) and do not necessarily reflect the official policy or position of the Department of the Navy, Department of Defense, or the US Government.

References

Ahuja, V., Thapa, D., & Ghai, B. (2018). Strategies for prevention of lower limb post-amputation pain: A clinical narrative review. *Journal of Anaesthesiology Clinical Pharmacology*, *34*(4), 439.

Alviar, M. J. M., Hale, T., & Lim-Dungca, M. (2016). Pharmacologic interventions for treating phantom limb pain. *Cochrane Database of Systematic Reviews*, (10), 1–53.

Aternali, A., & Katz, J. (2019). Recent advances in understanding and managing phantom limb pain. *F1000Research*, *8*(F1000), 1–11.

Charkhkar, H., Shell, C. E., Marasco, P. D., Pinault, G. J., Tyler, D. J., & Triolo, R. J. (2018). High-density peripheral nerve cuffs restore natural sensation to individuals with lower-limb amputations. *Journal of Neural Engineering*, *15*(5), 056002.

Chavez, L. M., Huang, S. S., MacDonald, I., Lin, J. G., Lee, Y. C., & Chen, Y. H. (2017). Mechanisms of acupuncture therapy in ischemic stroke rehabilitation: A literature review of basic studies. *International Journal of Molecular Sciences*, *18*(11), 2270.

Collins, K. L., Russell, H. G., Schumacher, P. J., Robinson-Freeman, K. E., O'Conor, E. C., Gibney, K. D., ... Tsao, J. W. (2018). A review of current theories and treatments for phantom limb pain. *The Journal of clinical investigation*, *128*(6), 2168–2176.

Cornish, P., & Wall, C. (2015). Successful peripheral neuromodulation for phantom limb pain. *Pain Medicine*, *16*(4), 761–764.

Deer, T. R., Levy, R. M., Kramer, J., Poree, L., Amirdelfan, K., Grigsby, E., ... Scowcroft, J. (2017). Dorsal root ganglion stimulation yielded higher treatment success rate for CRPS and causalgia at 3 and 12 months: Randomized comparative trial. *Pain*, *158*(4), 669–681.

Deogaonkar, M., Zibly, Z., & Slavin, K. V. (2014). Spinal cord stimulation for the treatment of vascular pathology. *Neurosurgery Clinics of North America*, *25*(1), 25–31.

Dumanian, G. A., Potter, B. K., Mioton, L. M., Ko, J. H., Cheesborough, J. E., Souza, J. M., ... Kuiken, T. A. (2019). Targeted muscle reinnervation treats neuroma and phantom pain in major limb amputees: A randomized clinical trial. *Annals of Surgery*, *270*(2), 238–246.

Enneking, F. K., Scarborough, M. T., & Radson, E. A. (1997). Local anesthetic infusion through nerve sheath catheters for analgesia following upper extremity amputation: Clinical report. *Regional Anesthesia and Pain Medicine*, *22*(4), 351–356.

Finger, S., & Hustwit, M. P. (2003). Five early accounts of phantom limb in context: Pare, Descartes, Lemos, Bell, and Mitchell. *Neurosurgery*, *52*(3), 675–686.

Forstenpointner, J., Naleschinski, D., Wasner, G., Hüllemann, P., Binder, A., & Baron, R. (2019). Sensitized vasoactive C-nociceptors: Key fibers in peripheral neuropathic pain. *Pain Reports*, *4*(1), 1–8.

Gilmore, C., Ilfeld, B., Rosenow, J., Li, S., Desai, M., Hunter, C., ... Cohen, S. (2019). Percutaneous peripheral nerve stimulation for the treatment of chronic neuropathic postamputation pain: A multicenter, randomized, placebo-controlled trial. *Regional Anesthesia & Pain Medicine*, *44*(6), 637–645.

Hachisuka, J., Omori, Y., Chiang, M. C., Gold, M. S., Koerber, H. R., & Ross, S. E. (2018). Wind-up in lamina I spinoparabrachial neurons: A role for reverberatory circuits. *Pain*, *159*(8), 1484.

Jackson, M. A., & Simpson, K. H. (2004). Pain after amputation. *Continuing Education in Anaesthesia, Critical Care and Pain*, *4*(1), 20–23.

Karanikolas, M., Aretha, D., Tsolakis, I., Monantera, G., Kiekkas, P., Papadoulas, S., ... Filos, K. S. (2011). Optimized perioperative analgesia reduces chronic phantom limb pain intensity, prevalence, and frequency: A prospective, randomized, clinical trial. *The Journal of the American Society of Anesthesiologists*, *114*(5), 1144–1154.

Kehlet, H., Jensen, T. S., & Woolf, C. J. (2006). Persistent postsurgical pain: Risk factors and prevention. *The Lancet*, *367*(9522), 1618–1625.

Lambert, A. W., Dashfield, A. K., Cosgrove, C., Wilkins, D. C., Walker, A. J., & Ashley, S. (2001). Randomized prospective study comparing preoperative epidural and intraoperative perineural analgesia for the prevention of postoperative stump and phantom limb pain following major amputation. *Regional Anesthesia and Pain Medicine*, *26*(4), 316–321.

Landin-Romero, R., Moreno-Alcazar, A., Pagani, M., & Amann, B. L. (2018). How does eye movement desensitization and reprocessing therapy work? A systematic review on suggested mechanisms of action. *Frontiers in Psychology*, *9*, 1395.

Limakatso, K., Madden, V. J., Manie, S., & Parker, R. (2020). The effectiveness of graded motor imagery for reducing phantom limb pain in amputees: A randomised controlled trial. *Physiotherapy*, *109*, 65–74.

Lodge, D., Watkins, J. C., Bortolotto, Z. A., Jane, D. E., & Volianskis, A. (2019). The 1980s: D-AP5, LTP and a decade of NMDA receptor discoveries. *Neurochemical Research*, *44*(3), 516–530.

Maier, C., Dertwinkel, R., Mansourian, N., Hosbach, I., Schwenkreis, P., Senne, I., ... Tegenthoff, M. (2003). Efficacy of the NMDA-receptor antagonist memantine in patients with chronic phantom limb pain–results of a randomized double-blinded, placebo-controlled trial. *Pain*, *103*(3), 277–283.

Michael, C. O. J., & Christine, S. (2018). Phantom limb pain—A review of mechanisms, therapy, and prevention. *Annals of Anesthesia and Pain Medicine*, *1*, 1002.

Moesker, A. A., Karl, H. W., & Trescot, A. M. (2014). Treatment of phantom limb pain by cryoneurolysis of the amputated nerve. *Pain Practice*, *14*(1), 52–56.

Molina-Rueda, F., Navarro-Fernández, C., Cuesta-Gómez, A., Alguacil-Diego, I. M., Molero-Sánchez, A., & Carratalá-Tejada, M. (2019). Neuroplasticity modifications following a lower-limb amputation: A systematic review. *PM & R: The Journal of Injury, Function, and Rehabilitation*, *11*(12), 1326–1334.

Morales-Quezada, L. (2017). Noninvasive brain stimulation, maladaptive plasticity, and bayesian analysis in phantom limb pain. *Medical Acupuncture*, *29*(4), 220–228.

Ossipov, M. H., Morimura, K., & Porreca, F. (2014). Descending pain modulation and chronification of pain. *Current Opinion in Supportive and Palliative Care*, *8*(2), 143.

Petersen, B. A., Nanivadekar, A. C., Chandrasekaran, S., & Fisher, L. E. (2019). Phantom limb pain: Peripheral neuromodulatory and neuroprosthetic approaches to treatment. *Muscle & Nerve*, *59*(2), 154–167.

Prologo, J. D., Gilliland, C. A., Miller, M., Harkey, P., Knight, J., Kies, D., ... Dariushnia, S. (2017). Percutaneous image-guided cryoablation for the treatment of phantom limb pain in amputees: A pilot study. *Journal of Vascular and Interventional Radiology*, *28*(1), 24–34.

Reddy, R. (2019). *Science of neuromodulation [class handout]*. San Diego, CA: University of California San Diego NEU 265.

Richardson, C., & Kulkarni, J. (2017). A review of the management of phantom limb pain: Challenges and solutions. *Journal of Pain Research*, *10*, 1861.

Robinson, L. R., Czerniecki, J. M., Ehde, D. M., Edwards, W. T., Judish, D. A., Goldberg, M. L., ... Jensen, M. P. (2004). Trial of amitriptyline for relief of pain in amputees: Results of a randomized controlled study. *Archives of Physical Medicine and Rehabilitation*, *85*(1), 1–6.

Rostaminejad, A., Behnammoghadam, M., Rostaminejad, M., Behnammoghadam, Z., & Bashti, S. (2017). Efficacy of eye movement desensitization and reprocessing on the phantom limb pain of patients with amputations within a 24-month follow-up. *International Journal of Rehabilitation Research*, *40*(3), 209–214.

Taylor, A. G., Anderson, J. G., Riedel, S. L., Lewis, J. E., & Bourguignon, C. (2013). A randomized, controlled, double-blind pilot study of the effects of cranial electrical stimulation on activity in brain pain processing regions in individuals with fibromyalgia. *Explorer*, *9*(1), 32–40.

Trevelyan, E. G., Turner, W. A., Summerfield-Mann, L., & Robinson, N. (2016). Acupuncture for the treatment of phantom limb syndrome in lower limb amputees: A randomised controlled feasibility study. *Trials*, *17*(1), 519.

Tseng, C. C., Chen, P. Y., & Lee, Y. C. (2014). Successful treatment of phantom limb pain and phantom limb sensation in the traumatic amputee using scalp acupuncture. *Acupuncture in Medicine*, *32*(4), 356–358.

Urits, I., Seifert, D., Seats, A., Giacomazzi, S., Kipp, M., Orhurhu, V., Viswanath, O. (2019). Treatment strategies and effective management of phantom limb–ssociated pain. Current Pain and Headache Reports, 23(9), 1–7.

Wilson, J. A., Nimmo, A. F., Fleetwood-Walker, S. M., & Colvin, L. A. (2008). A randomised double blind trial of the effect of pre-emptive epidural ketamine on persistent pain after lower limb amputation. *Pain*, *135*(1–2), 108–118.

Wu, C. L., Agarwal, S., Tella, P. K., Klick, B., Clark, M. R., Haythornthwaite, J. A., ... Raja, S. N. (2008). Morphine versus mexiletine for treatment of postamputation pain: A randomized, placebo-controlled, crossover trial. *The Journal of the American Society of Anesthesiologists*, *109*(2), 289–296.

Wu, C. L., Tella, P., Staats, P. S., Vaslav, R., Kazim, D. A., Wesselmann, U., & Raja, S. N. (2002). Analgesic effects of intravenous lidocaine and morphine on postamputation pain: A randomized double-blind, active placebo-controlled, crossover trial. *Anesthesiology*, *96*(4), 841–848.

Yaksh, T. (2019). *Biology of pain processing: Post nerve injury pain state [class handout]*. San Diego, CA: University of California San Diego NEU 265.

Yousef, A. A., & Aborahma, A. M. (2017). The preventive value of epidural calcitonin in patients with lower limb amputation. *Pain Medicine*, *18*(9), 1745–1751.

Chapter 13

Painful diabetic neuropathy: The roles of microglia

Che Aishah Nazariah Ismail[a] and Idris Long[b]

[a]Department of Physiology, School of Medical Sciences, Health Campus, Universiti Sains Malaysia, Kota Bharu, Kelantan, Malaysia, [b]Biomedicine programme, School of Health Sciences, Health Campus, Universiti Sains Malaysia, Kubang Kerian, Kelantan, Malaysia

Abbreviations

ATP	adenosine triphosphate
BDNF	brain-derived neurotrophic factor
CCR2	cysteine-cysteine chemokine receptor-2
cGMP	cyclic guanosine monophosphate
CGRP	calcitonin gene–related peptide
Cl⁻	chloride ion
CNS	central nervous system
CX₃ CR1	CX_3 chemokine receptor 1
DM	diabetes mellitus
DN	diabetic neuropathy
DRG	dorsal root ganglion
GABA	γ-aminobutyric acid
GABA$_A$	γ-aminobutyric acid type A
GABA$_B$	γ-aminobutyric acid type B
IL-1β	interleukin-1β
KCC2	potassium-chloride contransporter-2
KCl	potassium chloride
MCP-1	monocyte chemoattractant protein-1
NF-κB	nuclear factor-κB
NGF	nerve growth factor
NMDA	N-methyl-D-aspartate
PDN	painful diabetic neuropathy
ROS	reactive oxygen species
TLRs	Toll-like receptors
TrkB	tropomyosin-related kinase B
Tyr	tyrosine kinase

Introduction

Neuropathic pain is one of the critical problems in clinical medicine as it is not easy to cure. It is pathological and defined as a chronic pain state resulting from injury or disease of neurons in the peripheral or central nervous system (CNS) (Deng et al., 2020). Neuropathic pain may result either from acute events (e.g., amputation and spinal cord injury) or from systemic disease (e.g., diabetes, viral infection, and cancer) (Ji, Chamessian, & Zhang, 2017; Zhuo, Wu, & Wu, 2011).

One of the devastating diseases classified under neuropathic pain is diabetic peripheral neuropathy (DPN). Diabetic neuropathy is a late complication of diabetes mellitus (DM) of either Type I or II. A study from the Mayo Clinic revealed that diabetic neuropathies are common in diabetic patients affecting approximately 66% with insulin-dependent DM (Type I DM) and 59% in patients with non-insulin-dependent DM (Type II DM) (Sadosky, McDermott, Brandenburg,

Features and Assessments of Pain, Anesthesia, and Analgesia. https://doi.org/10.1016/B978-0-12-818988-7.00029-7
Copyright © 2022 Elsevier Inc. All rights reserved.

140 PART | II The syndromes of pain

TABLE 1 Adverse symptoms of diabetes Mellitus.

Type	Symptoms
Sensorimotor	Paresthesia (numbness and tingling in stocking and, less commonly, glove distribution; progresses to hypoesthesia) Painful neuropathy: dysesthesia with lancinating or burning pain; may be accompanied by anorexia and depression
Focal motor and compression	Weakness or loss of sensation in the distribution of nerve
Autonomic gastrointestinal	Gastroparesis (early satiety, nausea, vomiting) Diabetic diarrhea (nocturnal, with incontinence)
Genitourinary	Impotence, retrograde ejaculation
Cardiac	Overflow incontinence Dizziness

Long-term complications of diabetes mellitus.
Adapted from Nathan (1993).

& Strauss, 2008). Moreover, painful diabetic neuropathy (PDN) is reported to affect approximately 18% of adult diabetic patients compared with a minimum of 30% of patients with overall DPN (Spallone & Greco, 2013). Table 1 illustrated the long-term complication of diabetes mellitus.

In the year 1991, the glial cell has become the center of researchers' attention since an animal model of neuropathic pain was reported to stimulate spinal astrocyte activation (Garrison, Dougherty, Kajander, & Carlton, 1991). During that period, the drugs experimented also demonstrated attenuation toward numbers of astrocytic activation in the rat model of sciatic nerve constriction injury (Garrison, Dougherty, & Carlton, 1994). Beginning from that, glial cell activation was believed to be strongly connected with the development of neuropathic pain. To be specific, microglial activation in neuropathic pain has been widely investigated, and research discovered evidence that glial cell is the key players in the development and maintenance of several types of neurodegenerative diseases. The discovery of spinal cord glial cells, which strongly implicate pain processing, enhances the understanding of pain, including understanding the glia-neuron interactions. As microglia could modulate pain, it is vital to understand the mechanisms of microglial activation on the pathogenesis of neuropathic pain, especially on PDN. Inhibiting the microglial activation could potentially be one of the possible ways to combat the development of PDN. Therefore, understanding this mechanism leading to the pathogenesis of PDN should be well studied.

Painful diabetic neuropathy

PDN results from the prolonged effect of hyperglycemia in diabetic patients. PDN or also known as (painful) peripheral diabetic neuropathy is commonly complained by the diabetic patients that limit many aspects of life and daily routine (Sun et al., 2012). It results from the damage of the peripheral nerve induced by DM. Approximately 30% of diabetic patients would develop diabetic neuropathy that results from the destruction of peripheral nerve–induced DM (Wang, Couture, & Hong, 2014). PDN is one of the most devastating complications of DM and the leading cause, together with microangiopathy, which leads to foot amputation in diabetic patients (Obrosova, 2009). It is a progressive disorder that causes a continual decrease in peripheral sensation, and in severe condition, it leads to the complete loss of sensation. The diabetic patient with PDN usually complains of bilateral tingling or burning sensations (paraesthesia) that spontaneously increase or decrease in intensity over time. They may also have heightened sensitivity to stimuli such as touch (allodynia), temperature changes, or apply pressure that results in excruciating pain with the slightest changes (Kline, Caroll, & Malnar, 2003).

Several suggested pathological mechanisms contribute to PDN in diabetic patients such as chronic nerve damage (i.e., peripheral neuropathy) due to secondary vascular diseases, microglial inflammation in central nerve system (CNS), dysregulation of potassium-chloride cotransporter 2 (KCC2) activity, modified growth factor, and sodium channel expression in the dorsal root ganglion (DRG) neurons (Tan et al., 2012). Other than that, the pathogenesis of PDN also includes interactions of chronic hyperglycemia, generation of AGEs, impaired insulin signaling, increased aldose reductase, enhancement of Na^+ currents in peripheral nerve, oxidative stress, inflammation, hypertension, and thalamic dysfunction (Obrosova, 2009; Wang et al., 2014). The pathological mechanisms of PDN due to hyperglycemia are illustrated in Fig. 1.

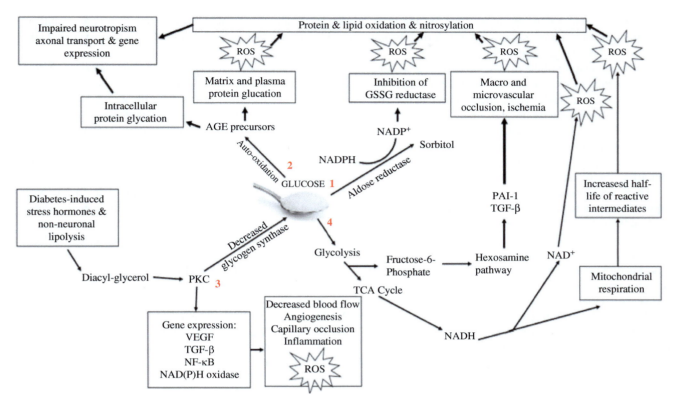

FIG. 1 Effects of hyperglycemia. Hyperglycemia stimulates several signaling mechanisms in cells. As illustrated above, four main pathways that cause cell damage downstream of hyperglycemia are (1) additional glucose shifts to the polyol pathway that reduces cytosolic NADPH, followed by GSH. (2) Excess glucose also undergoes auto-oxidation to yield AGEs that damage protein function and also stimulate RAGEs that use ROS as the second messengers. (3) PKC activation may further enhance hyperglycemia and worsen the tissue hypoxia condition. (4) Excessive and slowing of the electron transfer chain cause escape of reactive intermediates to produce O^{2-} and activation of NADH oxidase that may also yield O^{2-}. The products of mechanisms of injury in each case are ROS, which may damage the protein and gene function. *NADPH*, nicotinamide adenine dinucleotide phosphate; *GSH*, glutathione; *AGEs*, advanced glycation end products; *RAGEs*, receptor for AGEs; *ROS*, reactive oxygen species; *PKC*, protein kinase C; *TCA*, trichloroacetic acid; *PAI-1*, plasminogen activator inhibitor-1; O^{2-}, superoxide anion; *TGF-β*, transforming growth factor-β; *VEGF*, vascular endothelial growth factor; *NF-κB*, nuclear factor-κB. *(Adapted from Feldman, 2003.)*

It is challenging to treat PDN due to the intolerable adverse medication effects and the development of tolerance to the medication used. The numerous therapies utilized to treat PDN are tricyclic antidepressants amitriptyline, selective serotonin and noradrenalin reuptake inhibitor duloxetine, anticonvulsant such as pregabalin and gabapentin, capsaicin, opioid group of tramadol, morphine, oxycodone and most recently, tapentadol (Greigh, Tesfaye, Selvarajah, & Wilkinson, 2014). However, some of these drugs have intolerable adverse effects or were unsuccessful in being useful in a prolonged duration of treatment.

Diabetic neuropathic pain causes inflammation in the periphery, occurrence of hyperglycemia, and production of reactive oxygen species (ROS) that mainly implicates the local microenvironment in the spinal cord. These modifications hugely trigger the resting and sessile microglia to the activated form of microglia. The activated microglia are then synthesized and release proinflammatory cytokines (e.g., IL-1β, TNF-α, IL-6) and neuroactive molecules that are capable of initiating the hyperactivity of the spinal cord pain–related neurons (Wang et al., 2014).

Microglia

Microglia stem from a monocytic cellular lineage of mesodermal (myeloid) origin and passes through the CNS during fetal development (Chan, Kohsaka, & Rezaie, 2007). They provide trophic support for primary sensory neurons, express transporter that controls neurotransmitter levels in the extracellular space, and share some astroglial markers such as glial fibrillary acidic protein (GFAP) (Hanani, 2005). Microglia are in a resting and sessile state under normal conditions. However, when there are any signs of a brain lesion or nervous system dysfunction, together with the initiation of biochemical changes in the peripheral tissues, the microglia are immediately activated. Moreover, it is highly postulated that microglia play a more significant role during the initiation phase of pain before the other glial cells like astrocytes take place in the later stage of pain (Li et al., 2019).

A marked feature of the microglial cells is their capability of generating and secreting numerous substances, including growth factors, cytokines, complement factors, lipid mediators, extracellular matrix components, enzymes, free radicals, neurotoxins, nitric oxide (NO), and prostaglandins. Although the low concentration of these mediators may be advantageous in tissue regeneration and repair, their high levels are reported to lead to neuronal degeneration (Fu, Light, & Maixner, 2001; Minghetti & Levi, 1998). A previous study demonstrated that the use of a microglial activator, such as fluorocitrate, abolished the IL-1β and TNF-α levels (Talbot, Chahmi, Dias, & Couture, 2010). Microglia have been shown to mediate hyperalgesia as they releases the aforementioned neuroactive substances as well as cyclooxygenase and NO that can affect the development and maintenance of hyperalgesia after formalin injection (Fu et al., 2001).

There are several contradictions to the signals that mediate microglial activation in neuropathic pain. It has been reported that the resting microglia possibly do not have fast electrical or chemotactic responses to calcitonin gene–related peptide (CGRP), substance P, glutamate, gamma-aminobutyric acid (GABA), glycine, serotonin, noradrenaline, and many more (Chen, Koga, Li, & Zhuo, 2010; Wu & Zhuo, 2008). It is believed that microglia possibly detect neuronal signals from other means such as adenosine triphosphate (ATP) and its receptors (P2X and P2Y receptors), fractalkine and CX_3 chemokine receptor 1 (CX_3CR1), monocyte chemoattractant protein-1 (MCP-1), TLR2, 3, and 4, and chemokine receptor type 2 (CCR2). Among all, the spinal microglia demonstrate rapid chemotaxis in response to ATP through the purinergic signaling pathway (Chen et al., 2010). Moreover, microglia express both ionotropic receptors (e.g., P2X4 and P2X7) and metabotropic receptors (e.g., P2Y6 and P2Y12) (James & Butt, 2002). The activation of P2X4 in microglia induces brain-derived neurotrophic factor (BDNF) release (Trang, Beggs, Wan, & Salter, 2009), while the activation of P2X7 in microglia facilitates IL-1β release (Clark et al., 2010). These explain the relation of microglia with the release of IL-1β and signaling pathway with BDNF in neuropathic pain as illustrated in Fig. 2.

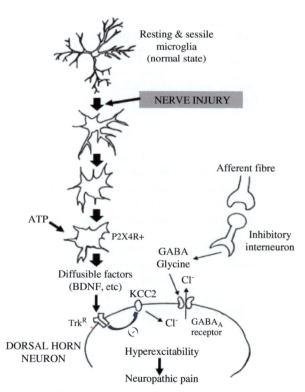

FIG. 2 Shifts of resting microglia to activated microglia. P2X4 receptors of the microglia are activated by ATP and, in turn, produce bioactive diffusible factors such as BDNF. BDNF then downregulates KCC2 via TrkB and leads to an elevation of intracellular chloride ion (Cl^-). This effect causes the collapse of the transmembrane anion gradient in the spinal dorsal horn and, in turn, stimulates the depolarization of dorsal horn neurons after the activation of GABA and glycine. The development of the dorsal horn hyperexcitability induced by the activated microglia may be accountable for the occurrence of neuropathic pain. *ATP*, extracellular adenosine triphosphate; *BDNF*, brain-derived neurotrophic factor; *KCC2*, potassium-chloride transporter; *TrkB*, tropomyosin-related kinase B; *GABA*, γ-aminobutyric acid; (−), inhibition. *(Adapted from Tsuda, 2016.)*

Microglial activation during neuropathic pain

Several mechanisms are contributing to the activation of microglia during PDN. The hyperglycemic condition in diabetic neuropathy may induce the release of ROS, interleukin-8 (IL-8), and PKC phosphorylation, hence activating the TLR/NF-Kb and purinergic signaling pathway. The production of ROS induces the release of IL-1β, TNF-α, inducible nitric oxide synthase (iNOS), and many more from many sources. The release of these products modifies the structural and functional properties of microglia that subsequently lead to its activation during diabetic neuropathy (Wang et al., 2014). Moreover, the activation of JAK2/STAT3 signaling transduction during the pathogenesis of PDN also induces microglia activation (Li et al., 2019).

Moreover, the enhanced glutamate activity in diabetic neuropathy activates NMDA receptors and also leads to the activation of p38 of mitogen-activated protein kinase (MAPK), extracellular-signal-regulated kinase (ERK), and Jun-nuclear kinase (JNK) in the microglia (Daulhac et al., 2006). The resulting pain hypersensitivity contributes to the activation of microglia in PDN. Besides, the activation of p38β, which is merely expressed in microglia, may result in microglial activation during diabetic neuropathy. The class of MAPK mediates cellular responses to external inputs (Deng et al., 2020). Signaling molecule p38 MAPK functions in allowing cytokine synthesis and promoting apoptosis (Deng et al., 2020) and is produced by microglia (Clark et al., 2010). The phosphorylation of p38 (only in microglia) has been shown in PDN but not in the non-PDN condition (Daulhac et al., 2006) and contributes to the occurrence of pain hypersensitivity.

Furthermore, the reduced KCC2 expression in the spinal cord leads to the increased basal and formalin-induced level of GABA in the diabetic rat (Malmberg, O'Connor, Glennon, Cesena, & Calcutt, 2006; Morgado, Pereira-Terra, Cruz, & Tavares, 2011). KCC2 plays a role in maintaining a low level of intracellular chloride ion (Cl^-) concentration. But the binding of $GABA_A$ receptors by GABA induces the influx of Cl^-, thus leading to the inhibition of neuronal activity. On the other hand, the low concentration of KCC2 may disturb anion homeostasis, therefore causing an outflow rather than inflow of Cl^- when GABA binds to the receptors in the spinal cord neurons. These mechanisms result in the depolarization of membrane potential and disinhibition (Coull et al., 2005). It is justified that the low KCC2 expression may activate microglia as the administration of minocycline (microglial inhibitor) diminishes the mechanical hyperalgesia in the STZ-diabetic rats (Morgado et al., 2011). Microglia also produce proinflammatory mediators such as IL-1β, IL-6, and TNF-α not only during the normal state (Hanisch, 2002) but also during neuropathic pain (Tsuda, Inoue, & Salter, 2005) and specifically during the pathogenesis of PDN (Pabreja, Dua, Sharma, Padi, & Kulkarni, 2011; Wang et al., 2014).

Central glia-neuronal interaction

In many years, research has focused on the roles of neurons in the pathogenesis of neuropathic pain without relating the associations with other cell types. Synaptic plasticity, one of the neuronal mechanisms involved in the development of neuropathic pain, is strongly modulated by the classical neuron-derived neurotransmitters and immune markers released from glial cells residing in CNS. Recent researches indicated that potent neuroimmune communications in CNS are an essential phenomenon that causes PDN. Microglia are not silent elements of the nervous system but play a significant role in the modulation of neurotransmission.

During neuropathic pain, the activated microglia are capable of transmitting the abnormal pain stimuli to the spinal cord neurons by secreting neuromodulators and proinflammatory cytokines IL-1β, IL-6, and TNF-α that implicate neuronal firing and intracellular signaling. In previous studies, IL-1β has been demonstrated to be involved in the signal transduction cascade with NMDA receptors in the ascending nociceptive projection (Guo et al., 2007; Viviani et al., 2003). In an in vitro study, the administration of IL-1β to slices of brain stem augmented the phosphorylation of NMDA receptors in the regions that are involved in trigeminal nociceptive processing (Guo et al., 2007). Guo et al. (2007) also revealed that IL-1β selectively blocked NR1-phosphorylated Ser896 (P-ser896-NR1) on the NMDA receptor in which TNF-α did not affect that receptor at the tested dose. It is proposed that the effect of IL-1β on the NMDA receptor is specifically downstream to the glial activation. The IL-1β signaling also contributes to the augmented activity of neuronal NMDA receptors and subsequently contributes to pain hypersensitivity (Ren & Dubner, 2008). The neuron-microglia interactions in the spinal dorsal horn during diabetic neuropathic pain are illustrated in Fig. 3.

Also, the vigorous production of IL-6 occurs in dorsal root ganglia (DRG) during the occurrence of peripheral nerve injury. Some of the cytokines are possibly carried to central terminals of primary afferent nerves. IL-6 binds to its receptor on the microglial cell surface and leads to the phosphorylation of Janus tyrosine kinase-2 (JAK2) to initiate its activation. Subsequently, the activated JAK2 promotes the dimerization of activated signal transducer and activator of transcription (STAT3) that further translocates into the nucleus and binds to the DNA at a specific region to regulate the transcription of

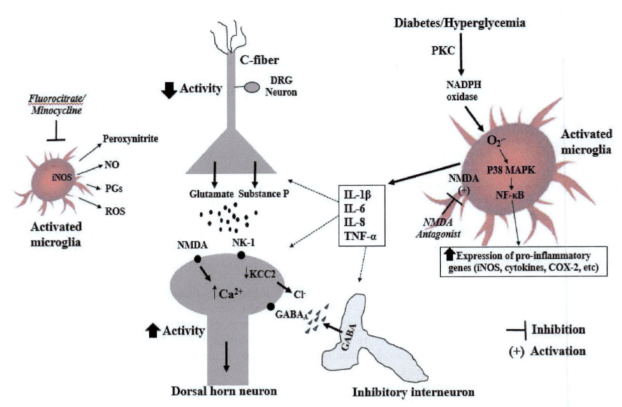

FIG. 3 Neuron-microglia interactions in the spinal dorsal horn during DPN. The robust primary afferent drive leads to the release of glutamate from the presynaptic membrane of the injured primary nociceptor. Glutamate then binds to the NMDA receptor to cause its activation. The relief of Mg^{2+} block from the NMDA receptor leads to the increased influx of Ca^{2+} into the neuron, which then causes conformational changes in DREAM protein at the transcription gene to cause its removal. The activation of the NMDA receptor may also initiate several inflammatory cascades leading to the activation of immune cells such as microglia. The activated microglia, through its P2X7 activation, leads to the synthesis and release of free radicals (ROS) and proinflammatory cytokines. Meanwhile, the activation of P2X4 of microglia initiates the synthesis and release of BDNF. These signaling neuromodulators then act on the NMDA receptor to cause its persistent activation. These repeated interactions between neuron and microglia (nonneuron) may eventually lead to the development of neuropathic pain. *GABA*, γ-aminobutyric acid; *IL*, interleukin; *iNOS*, inducible nitric oxide synthase; *KCC2*, potassium chloride cotransporter 2; *MAPK*, mitogen-activated protein kinase; *NADPH*, nicotinamide adenine dinucleotide phosphate; *NK-1*, neurokinin/substance P receptor-1; *NF-κB*, transcriptional nuclear factor kappa B; *NMDA*, N-methyl-D-aspartate; *NO*, nitric oxide; *PGs*, prostaglandins; *PKC*, protein kinase C; *ROS*, reactive oxygen species; *TNF-α*, tumor necrosis factor-α. *(Adapted from Wang, D., Couture, R., & Hong, Y. (2014). Activated microglia in the spinal cord underlies diabetic neuropathic pain. European Journal of Pharmacology, 728, 59–66.)*

inflammatory factor genes. The release of activating substances including ATP, other proinflammatory mediators (IL-1β and TNF-α), free radicals, inducible nitric oxide (iNOS), and prostaglandins from microglia then interacts with neurons regulating the Ca^{2+} influx of NMDA receptor to cause central sensitization and facilitate the development of neuropathic pain symptoms during PDN (Li et al., 2019; Nicolas et al., 2013; Propiolek-Barczyk & Mika, 2016) (Fig. 4). The previous researches revealed the upregulation of JAK/STAT3 signaling after the occurrence of PDN in the animal model (Li et al., 2019), and the inhibition of this signaling pathway attenuated the mechanical allodynia and thermal hyperalgesia in other models of neuropathic pain (Dominiguez, Rivat, Pommier, Mauborgne, & Pohl, 2008).

Conclusion

As a summary, in normal conditions, the noxious stimuli cause microglia to acquire a reactive phenotype and change their morphology and functioning as a neuroprotective by degrading, internalizing, and removing these harmful stimuli. However, due to prolonged hyperglycemia, it is suggested that the activation goes beyond physiological control. In this stage, microglia foster neuroinflammatory responses and are responsible for releasing many cytokines and proinflammatory mediators through many signaling pathways, including TLR-4/NF-κβ, JAK-STAT, and purinergic signaling pathways, and initiating the pathogenesis of PDN.

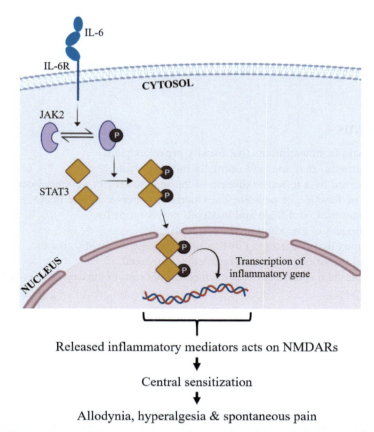

FIG. 4 JAK/STAT3 pathway in microglia during DPN. Following the robust release of proinflammatory cytokines from peripheral nerve during the occurrence of DPN, some of the cytokines are carried to the primary afferent nerve terminals to bind to its receptor, which initiates the activation of JAK2. Activated JAK2 stimulates the dimerization of STAT3, which further translocates into the nucleus to bind on the DNA on a specific region to regulate the transcription of inflammatory factor genes. Consequently, the release of inflammatory mediators, ROS, and prostaglandins such as IL-18, TNF-α, nitric oxide, and others from the activated microglia further act on neurons to regulate the influx of Ca^{2+} into the neurons via NMDARs to cause central sensitization that contributes to the development of PDN symptoms such as allodynia, hyperalgesia, and allodynia. *JAK2*, Janus tyrosine kinase-2; *STAT3*, signal transducer and activator of transcription; *DNA*, deoxyribonucleic acid; *ROS*, reactive oxygen species; *IL-18*, interleukin-18; *TNF-α*, tumor necrosis factor-α; Ca^{2+}, calcium ion; *NMDARs*, N-methyl-D-aspartate receptors. *(Adapted from Nicolas, C. S., Amici, M., Bortolotto, Z. A., Doherty, A., Csaba, Z., Fafouri, A., et al. (2013). The role of JAK-STAT signaling within the CNS. JAK-STAT, 2(1), e22925.)*

Applications to other areas

The understanding of the roles of microglial on the mechanism of PDN provides a rationale to explore the other pathways that can be targeted to combat the development and to alleviate the symptoms of PDN. Current drug treatments have been shown to ease the symptoms of PDN but failed to treat the underlying causes of PDN. Therefore, a better understanding of the molecular mechanisms underlying the development and progression of PDN is needed for early diagnosis and intervention as well as for understanding the failure of existing treatments. It is assumed that 3.6 million Malaysians have diabetes, the highest rate of incidence in Asia and one of the highest in the world. One of the complications of DM is PDN that will be diminishing quality of life, disrupt sleep, and can lead to depression and decreasing productivity and increasing the burden to the economics of our country. The positive impact of this study may improve the quality of life and healthcare of the patient, increasing productivity, and can also reduce the economic burden.

Other agents of interest

Minocycline (microglial activation inhibitor) has demonstrated that it can exert as an antimicrobial agent, antiapoptotic, antiinflammatory, attenuation of proteolysis, angiogenesis and tumor spreading, and inhibition of glial cells activation (microglia and astrocytes). Minocycline has been suggested as a treatment for chronic pain, and the neuroprotective effect of minocycline on several neurodegenerative disease models, spinal cord injury models, and many more pathological disease models has been discovered. It is challenging to treat PDN due to the intolerable adverse medication effects and the development of tolerance to the medication used. The numerous therapies utilized to treat PDN over the years, such as tricyclic antidepressants, selective serotonin, noradrenaline reuptake inhibitors, and antiepileptic drugs such as capsaicin and gabapentin have some intolerable adverse effects or are unsuccessful in being useful in a prolonged duration of treatment.

Mini-dictionary of terms

Hyperalgesia: Increased pain from a stimulus that usually provokes pain.
Allodynia: Pain due to a stimulus that does not normally provoke pain.
Neuropathic pain: Pain caused by a lesion or disease of the somatosensory nervous system.
Neuropathy: Disturbance of function or pathological change in a nerve: in one nerve, mononeuropathy; in several nerves, mononeuropathy multiplex; if diffuse and bilateral, polyneuropathy.
Nociception: The neural process of encoding noxious stimuli.
Noxious stimulus: A stimulus that is damaging or threatens damage to healthy tissues.
Paraesthesia: An abnormal sensation, whether spontaneous or evoked.
Central sensitization: Increased responsiveness of nociceptive neurons in the central nervous system to their standard or subthreshold afferent input.

Key facts

Key facts of Pregabalin

- Pregabalin (3-isobutyl gamma-aminobutyric acid gamma-analog) is a GABAergic drug. Pregabalin can be used as antiepileptic, inhibits partial seizures, is anxiolytic for anxiety disorder, and is also prescribed for painful diabetic neuropathy.
- Therapy potential of Pregabalin can be mediated via high affinity to bind with the alpha2-delta subunit (a2δ) of voltage-gated calcium channels at the presynaptic terminals and modulate the release of excitatory neurotransmitter such as glutamate.

Key facts of duloxetine

- Duloxetine is a serotonin-noradrenaline reuptake inhibitor that is usually used to treat depression but also can alleviate allodynia in several neuropathic pain models. This drug has been approved by the US Food and Drug Administration (FDA) to be used to treat painful diabetic neuropathy.
- The mechanism of action of duloxetine is due to the inhibition of transporters for serotonin and noradrenaline reuptake and makes these neurotransmitter concentrations increase and promote the persistence of their activity. Serotonin binds to its receptor, 5-HT2A receptor, and modulates the mechanism of neuropathic and inflammatory pain.

Summary points

- Neuropathic pain may result either from acute events (e.g., amputation and spinal cord injury) or from systemic disease (e.g., diabetes, viral infection, and cancer), and one of them is diabetic peripheral neuropathy (DPN).
- The numerous therapies utilized to treat DPN such as amitriptyline, duloxetine, pregabalin and gabapentin, capsaicin, opioid group of tramadol, morphine, oxycodone, and tapentadol have intolerable adverse effects or are unsuccessful in being useful in a prolonged duration of treatment.
- There is much evidence that glial cells, especially microglia, are the key players in the creation and maintenance of several types of neurodegenerative diseases and neuropathic pain.
- Microglia detect neuronal signals from its surface receptors such as purinergic receptors (P2X and P2Y receptors), fractalkine and CX3 chemokine receptor 1 (CX3CR1), monocyte chemoattractant protein-1 (MCP-1), TLR2, 3, and 4, and chemokine receptor type 2 (CCR2) and initiate the mechanism of PDN through TLR-4/NF-κβ and purinergic signaling pathway.
- Inhibiting the microglial activation could potentially be one of the possible ways to combat the development of PDN.

References

Chan, W., Kohsaka, S., & Rezaie, P. (2007). The origin and cell lineage of microglia—New concepts. *Brain Research Reviews, 53*(2), 344–354.

Chen, T., Koga, K., Li, X.-Y., & Zhuo, M. (2010). Spinal microglial motility is independent of neuronal activity and plasticity in adult mice. *Molecular Pain, 6*(1), 19.

Clark, A. K., Stainland, A. A., Marchand, F., Kaan, T. K., McMahon, S. B., & Malcangio, M. (2010). P2X7-dependent release of interleukin-1β and nociception in the spinal cord following lipopolysaccharide. *Journal of Neuroscience, 30*(2), 573–582.

Coull, J. A., Beggs, S., Boudreau, D., Boivin, D., Tsuda, M., Inoue, K., et al. (2005). BDNF from microglia causes the shift in neuronal anion gradient underlying neuropathic pain. *Nature, 438*(7070), 1017–1021.

Daulhac, L., Mallet, C., Courteix, C., Etinne, M., Duroux, E., Privat, A.-M., et al. (2006). Diabetes-induced mechanical hyperalgesia involves spinal mitogen-activated protein kinase activation in neurons and microglia via N-methyl-D-aspartate-dependent mechanisms. *Molecular Pharmacology, 70*(4), 1246–1254.

Deng, Y., Yang, L., Xie, Q., Yang, F., Li, G., Zhang, G., et al. (2020). Protein kinase A is involved in neuropathic pain by activating the p38MAPK pathway to mediate spinal cord cell apoptosis. *Mediators of Inflammation, 2020*(6420425), 1–17.

Dominiguez, E., Rivat, C., Pommier, B., Mauborgne, A., & Pohl, M. (2008). JAK/STAT3 pathway is activated in spinal cord microglia after peripheral nerve injury and contributes to neuropathic pain development in the rat. *Journal of Neurochemistry, 107*, 50–60.

Feldman, E. L. (2003). Oxidative stress and diabetic neuropathy: a new understanding of an old problem. *Journal of Clinical Investigation, 111*(14), 431–433.

Fu, K.-Y., Light, A. R., & Maixner, W. (2001). Long-lasting inflammation and long-term hyperalgesia after subcutaneous formalin injection into the rat hind paw. *Journal of Pain, 2*(1), 2–11.

Garrison, C. J., Dougherty, P. M., & Carlton, S. M. (1994). GFAP expression in the lumbar spinal cord of naive and neuropathic rats treated with MK-801. *Experimental Neurology, 129*(2), 237–243.

Garrison, C., Dougherty, P., Kajander, K., & Carlton, S. (1991). Staining of glial fibrillary acidic protein (GFAP) in the lumbar spinal cord increases following a sciatic nerve constriction injury. *Brain Research, 565*(1), 1–7.

Greigh, M., Tesfaye, S., Selvarajah, D., & Wilkinson, I. A. (2014). In D. W. Zochodne, & R. A. Malik (Eds.), *Diabetes, and the nervous system: Vol. 126. Handbook of clinical neurology* Elsevier B.V. (3rd series).

Guo, W., Wang, H., Watanabe, M., Shimizu, K., Zou, S., Lagraize, S. C., et al. (2007). Glial–cytokine–neuronal interactions underlying the mechanisms of persistent pain. *Journal of Neuroscience, 27*(22), 6006–6018.

Hanani, M. (2005). Satellite glial cells in sensory ganglia: From form to function. *Brain Research Reviews, 48*(3), 457–476.

Hanisch, U. K. (2002). Microglia as a source and target of cytokines. *Glia, 40*(2), 140–155.

James, G., & Butt, A. M. (2002). P2Y and P2X purinoceptor mediated Ca^{2+} signaling in glial cell pathology in the central nervous system. *European Journal of Pharmacology, 447*(2), 247–260.

Ji, R.-R., Chamessian, A., & Zhang, Y.-Q. (2017). Pain regulation by non-neuronal cells and inflammation. *Science, 354*(6312), 572–577.

Kline, K. M. M., Caroll, D. G., & Malnar, K. F. (2003). Painful diabetic peripheral neuropathy relieved with the use of oral topiramate. *Southern Medical Journal, 96*(6), 602–605.

Li, C.-D., Zhao, J. Y., Chen, J.-L., Lu, J.-H., Zhang, M. B., Huang, Q., et al. (2019). Mechanism of the JAK2/STAT3-CAV-1-NR2B signaling pathway in painful diabetic neuropathy. *Endocrine, 64*(1), 55–66.

Malmberg, A. B., O'Connor, W. T., Glennon, J. C., Cesena, R., & Calcutt, N. A. (2006). Impaired formalin-evoked changes in spinal amino acid levels in diabetic rats. *Brain Research, 1115*(1), 48–53.

Minghetti, L., & Levi, G. (1998). Microglia as effector cells in brain damage and repair: Focus on prostanoids and nitric oxide. *Progress in Neurobiology, 54*(1), 99–125.

Morgado, C., Pereira-Terra, P., Cruz, C., & Tavares, I. (2011). Minocycline completely reverses mechanical hyperalgesia in diabetic rats through microglia-induced changes in the expression of the potassium chloride cotransporter 2 (KCC2) at the spinal cord. *Diabetes, Obesity and Metabolism, 13*(2), 150–159.

Nathan, D. M. (1993). Long-term complications of diabetes mellitus. *New England Journal of Medicine, 328*(23), 1676–1685. https://doi.org/10.1056/NEJM199306103282306.

Nicolas, C. S., Amici, M., Bortolotto, Z. A., Doherty, A., Csaba, Z., Fafouri, A., et al. (2013). The role of JAK-STAT signaling within the CNS. *JAK-STAT, 2*(1), e22925.

Obrosova, I. G. (2009). Diabetes and the peripheral nerve. *Biochimica et Biophysica Acta, 1792*(10), 931–940.

Pabreja, K., Dua, K., Sharma, S., Padi, S. S., & Kulkarni, S. K. (2011). Minocycline attenuates the development of diabetic neuropathic pain: Possible anti-inflammatory and antioxidant mechanisms. *European Journal of Pharmacology, 661*(1), 15–21.

Propiolek-Barczyk, K., & Mika, J. (2016). Targeting the microglial signaling pathways: Modulation of neuropathic pain. *Current Medicinal Chemistry, 23*, 2908–2928.

Ren, K., & Dubner, R. (2008). Neuron-glia crosstalk gets serious: Role in pain hypersensitivity. *Current Opinion in Anaesthesiology, 21*(5), 570.

Sadosky, A., McDermott, A. M., Brandenburg, N. A., & Strauss, M. (2008). A review of the epidemiology of painful diabetic peripheral neuropathy, postherpetic neuralgia, and less commonly studied neuropathic pain conditions. *Pain Practice, 8*(1), 45–56.

Spallone, V., & Greco, S. (2013). Painful and painless diabetic neuropathy: One disease or two? *Current Diabetes Reports, 13*(4), 533–549.

Sun, W., Miao, B., Wang, X.-C., Duan, J.-H., Wang, W.-T., Kuang, F., et al. (2012). Reduced conduction failure of the main axon of polymodal nociceptive C-fibres contributes to painful diabetic neuropathy in rats. *Brain, 135*(2), 359–375.

Talbot, S., Chahmi, E., Dias, J. P., & Couture, R. (2010). Key role for spinal dorsal horn microglial kinin B 1 receptor in early diabetic pain neuropathy. *Journal of Neuroinflammation*, *7*(1), 36.

Tan, A., Samad, O. A., Fischer, T. Z., Zhao, P., Persson, A. K., & Waxman, S. G. (2012). Maladaptive dendritic spine remodeling contributes to diabetic neuropathic pain. *Journal of Neuroscience*, *32*(20), 6795–6807.

Trang, T., Beggs, S., Wan, X., & Salter, M. W. (2009). P2X4 receptor-mediated synthesis and release of brain-derived neurotrophic factor in microglia is dependent on calcium and p38mitogen-activated protein kinase activation. *Journal of Neuroscience*, *29*(11), 3518–3528.

Tsuda, M., Inoue, K., & Salter, M. W. (2005). Neuropathic pain and spinal microglia: A big problem from molecules in 'small' glia. *Trends in Neurosciences*, *28*(2), 101–107.

Tsuda, M. (2016). Microglia in the spinal cord and neuropathic pain. *Journal of Diabetes Investigation*, *7*(1), 17–26. https://doi.org/10.1111/jdi.12379.

Viviani, B., Bartesaghi, S., Gardoni, F., Vezzani, A., Behrens, M., Bartfai, T., et al. (2003). Interleukin-1β enhances NMDA receptor-mediated intracellular calcium increase through activation of the Src family of kinases. *Journal of Neuroscience*, *23*(25), 8692–8700.

Wang, D., Couture, R., & Hong, Y. (2014). Activated microglia in the spinal cord underlies diabetic neuropathic pain. *European Journal of Pharmacology*, *728*, 59–66.

Wu, L.-J., & Zhuo, M. (2008). Resting microglial motility is independent of synaptic plasticity in the mammalian brain. *Journal of Neurophysiology*, *99*(4), 2026–2032.

Zhuo, M., Wu, G., & Wu, L.-J. (2011). Neuronal and microglial mechanisms of neuropathic pain. *Molecular Brain*, *4*(31), 31.

Chapter 14

Maternal deprivation and nociception

Liciane Fernandes Medeiros[a], Dirson João Stein[b], Bettega Costa Lopes[c], and Iraci L.S. Torres[d]

[a]Postgraduate Program in Health and Human Development, University La Salle, Canoas, RS, Brazil, [b]Postgraduate Program in Medical Sciences, Federal University of Rio Grande do Sul, Porto Alegre, RS, Brazil, [c]Postgraduate Program in Physiology, Federal University of Rio Grande do Sul, Porto Alegre, RS, Brazil, [d]Laboratory of Pain Pharmacology and Neuromodulation: Preclinical Research, Porto Alegre, RS, Brazil

Abbreviations

AMPA	α-amino-3-hydroxy-5-methyl-4-isoxazolepropionic acid receptor
BDNF	brain-derived neurotrophic factor
CNS	central nervous system
EE	environmental enrichment
ELS	early-life stress
GABA-A	γ-aminobutyric acid type A receptor
GR	glucocorticoid receptor
HPA	hypothalamic–pituitary–adrenal axis
KMC	kangaroo mother care method
MD	maternal deprivation
NGF	neuronal growth factor
NICU	neonatal intensive care unit
NMDA	N-methyl-D-aspartate receptor
PND	postnatal day
RVM	rostral ventromedial medulla
SHRP	stress-hyporesponsive period
VH	visceral hypersensitivity

Introduction

There is growing evidence that neonatal adverse experiences, such as injury and suffering, have long-lasting consequences, increasing vulnerability, and the risk of developing several disorders throughout life, including chronic pain conditions (Petersen, Joseph, & Feit, 2014). It is also well established that some sensory experiences in the first years of life, a prominent CNS development period, can influence the CNS structure and function, and consequently, the sensations and behaviors in adulthood (Williams & Lascelles, 2020).

Neonatal chronic stress exposure such as maternal deprivation (MD), a condition in which neonates are separated from their mother, generates lifelong effects, leading to HPA axis dysfunction, making them more susceptible to stress-related dysfunctions, such as anxiety and poor cognition, as well as altered nociceptive responses (Chen & Jackson, 2016). For example, newborn infants hospitalized in NICU are exposed to significant stressful situations, including pain and reduced maternal care. In addition, neglected children present a higher risk of developing effective disorders and changes in social behavior (Petersen et al., 2014).

Considering the difficulty of conducting clinical trials in this population, preclinical studies that simulate human situations are very important. Therefore, animal models of MD are used to mimic human infant early-life stress (ELS), leading to neural, hormonal, and behavioral changes, similar to those reported in humans (Marco et al., 2015). In addition, they can be used to study the molecular mechanisms of its long-term consequences on social behavior.

It is important to note that during the first years of life, there is a close interaction between the mother and her offspring. The newborn is especially vulnerable to MD, this being for him a traumatic situation. In this period, the CNS is in intense development and becomes more vulnerable to harmful stimuli, during which long-term stressors may cause negative consequences on the functional activity of the nociceptive system (Schwaller & Fitzgerald, 2014). In this way, the importance of appropriate mother–neonate interactions becomes evident for newborn's appropriate brain development.

Features and Assessments of Pain, Anesthesia, and Analgesia. https://doi.org/10.1016/B978-0-12-818988-7.00046-7
Copyright © 2022 Elsevier Inc. All rights reserved.

In this chapter, we first describe the development of nociceptive pathways. Second, we contextualize MD as a powerful early-life stressor. Third, we summarize results from preclinical studies and possible signaling pathway changes. Fourth, we describe clinical settings regarding ELS consequences. Fifth, we report nonpharmacological interventions described in preclinical and clinical studies for improving MD-induced nociceptive changes.

Development of nociceptive pathways

In neonates, the nociceptive system undergoes constant maturation during the first postnatal days (PND) or weeks that overlaps with the maturation period of the HPA axis. Thus, tactile stimuli by manipulation or painful interventions during the maturation period may promote long-lasting effects that can be directly observed in the nociceptive pathways or indirect pathways like HPA axis changes.

The nociceptive pathways are still in development after birth, with an intense reorganization of connections, until reaching maturity (Fig. 1). In the dorsal horn of the spinal cord, reorganizations are observed during the first two or three postnatal weeks in rodents; for example, the cutaneous receptive fields of dorsal horn decrease in size during the first two postnatal weeks (Torsney & Fitzgerald, 2002). Also, there is a decreased input of myelinated A-fibers in the spinal cord, while C fibers are strengthened with age (Fitzgerald, 2005); and the descending inhibition is not entirely functional at birth, reaching adult levels only at PND22–24 (Marsh, Hatch, & Fitzgerald, 1997). Interestingly, the maturity of nociceptive pathways also involves other areas of the CNS, like the brainstem rostral ventromedial medulla (RVM), that exclusively facilitates spinal pain transmission at PND21; this descending facilitation of spinal nociception is mediated by μ-opioid receptors (Hathway, Vega-Avelaira, & Fitzgerald, 2012). At PND28, the RVM elicits both facilitation and inhibition responses. The imbalance between excitatory and inhibitory signaling may be involved with the pain felt by newborns and neonates at the development stage, including feeling more pain than adults.

Changes at postnatal periods are accompanied by modifications in the number or functionality of receptors; for example, NMDA and AMPA receptors were found in the neonatal dorsal horn when compared with adults (Jakowec, Fox, Martin, & Kalb, 1995). It is interesting to note that the activation of the GABAA receptor is exclusively inhibitory at PND6–7, and before that, this receptor is functionally excitatory (Fitzgerald, 2005). In addition, shifts in the opioid as μ and κ receptors are predominant at PND6, with a delay in the δ-receptor development (Marsh et al., 1997).

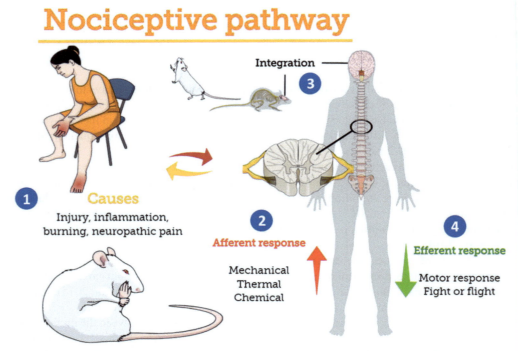

FIG. 1 Nociceptive pathway. (1) After an injury, inflammation, burning, or neuropathic pain, free ending terminals in periphery detect and carry these stimuli through dorsal nerves to the spinal cord. (2) Then, this information goes up to thalamic regions, and (3) finally to cortical brain regions where the nociception is coded, integrated, and an efferent response is driven (4).

Different studies have characterized the translational aspects regarding CNS development between humans and rodents. The rat brain at PND1–10 is equivalent to the third gestation trimester in humans, or PND7 is similar to humans at birth (Andrews & Fitzgerald, 1997). Moreover, the maturation of visual cortical synaptic proteins at the PND11 rat pup is equivalent to a term human infant (Pinto, Jones, Williams, & Murphy, 2015), while the rat cortex at PND14 is similar to the 29-week gestation human infant (Clancy, Darlington, & Finlay, 2001).

Maternal deprivation

Maternal deprivation is the separation of mother and infant during the so-called stress-hyporesponsive period (SHRP), which needs to be prolonged or repeated for persistent effects on the neuroendocrine regulation of the HPA axis (Fig. 2). The HPA axis plays a characteristic response elicited by stress input during the lifetime; however, this axis is still in development after birth, and this period is called the SHRP. In rodents, this period covers PND3–14 in rats and PND1–12 in mice (Schmidt et al., 2003). During the SHRP, a reduced adrenocorticotropic hormone and glucocorticoid release can be observed in response to lasting stressors. Alterations of the environmental conditions and the disruption of neonatal mother–infant interactions during this period can produce changes in the neurobiology, physiology, and emotional behavior of rats (de Oliveira et al., 2017).

Animal models have the advantage of allowing controlling the age of onset, frequency, and duration of the episodes. These animal models involve deprivation of maternal care, nutrition, and warmth that may yield anxiety-like behavior and aggressive behavior in rodents. Some factors, such as treatment during handling, number of separation episodes, and separation time, are critical to variability in neonate pain sensitivity. The terms MD and Maternal Separation have interchangeably been used in the literature to describe experimental manipulations that involve removal of pups from the dam for different intervals (single or repeated, 1–24 h) during the preweaning period. In this chapter, we will use the acronym MD to refer to both terms.

Maternal deprivation and nociception: Preclinical perspectives

Unpredictable MD protocols are the most used paradigms in investigations of the interactions between environmental and neurobiological mechanisms of ELS, providing a robust set of tools that enable unraveling the interactions through a wide range of cellular and molecular techniques (Table 1).

Preclinical studies have shown that early-life MD alters nociception in adulthood, both at the molecular and behavioral levels (Salberg, Noel, Burke, Vinall, & Mychasiuk, 2020). However, the results are quite heterogeneous. For example, Burke et al. (2013) showed that MD female, but not male, rats exhibited thermal hypoalgesia and mechanical allodynia

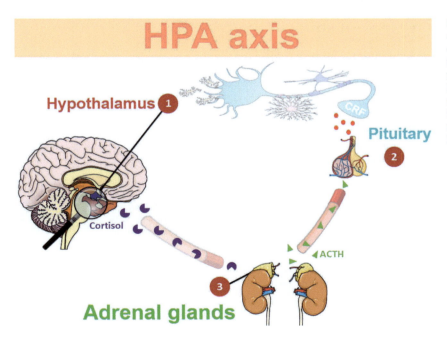

FIG. 2 Hypothalamic–pituitary–adrenal (HPA) axis. (1) Stressful conditions lead to corticotropin-releasing factor (CRF) release from the hypothalamus. (2) Then, CRF acts upon anterior pituitary stimulating the adrenocorticotropic hormone (ACTH) release. (3) After that, ACTH acts upon suprarenal glands, triggering cortisol/corticosterone release. (4) Finally, cortisol/corticosterone sets a negative feedback loop upon the hypothalamus and anterior pituitary release.

152 PART | II The syndromes of pain

TABLE 1 Nociceptive outcome.

Author	Strain	MD protocol	Outcomes	Treatment	Effect	Possible mechanism
Uhelski & Fuchs (2010)	Male rats (strain NS)	Time NS (PND2–15), W. NS	MD increased nociceptive response in the tonic phase of formalin test, and displayed higher anxiety-like behavior	No	–	–
Burke et al. (2013)	Male and female Sprague-Dawley	24 h (PND9–P10 W. PND22)	Enhanced nociceptive response compared to males subjected to a neuropathic pain model	No	–	–
Nishinaka, Nakamoto, and Tokuyama (2015)	Male and female ddY	6 h/d (PND15–P21 W. PND22)	When MD was followed by social isolation and PSL mice displayed bilateral hyperalgesia	No	–	–
Juif et al. (2016)	Male and female Sprague-Dawley	3 h/d (2–12) W. NS	C-type sensory neurons in MD rats were hypoexcitable at P14 revealing an inhibitory barrage in the spinal cord dorsal horn	No	–	–
Yasuda et al. (2016)	Male Sprague-Dawley	3 h/d (2–14) W. PND22	Orofacial hyperalgesia, higher corticosterone levels (serum); increased P2X3R in the TG	Mifepristone or P2X3R antagonist administration	Reduced mechanical allodynia	Possible P2X3 receptors involvement
Vilela, Vieira, Giusti-Paiva, and Silva (2017)	Male and female Wistar rats	3 h/d (2–15) W. PND22	Exacerbated mechanical and thermal hyperalgesia and increased sensitivity to formalin or CFA in adulthood	No	–	–
Genty, Tetsi Nomigni, Anton, and Hanesch (2017)	Male Sprague-Dawley	3 h/d (2–12) W. PND21	Mechanical and thermal hyperalgesia were less pronounced in CFA and CCI pain models	No	–	–
Melchior et al. (2018)	Male and female Wistar rats	3 h/d (2–12) W. PND21	Hypersensitive to noxious mechanical and thermal stimuli and inflammatory pain	OT, SAHA or allopregnanolone (i.p. PND2–12)	Reduced mechanical and thermal hypersensitivity	Hormonal and epigenetic
Ströher et al. (2019)	Male and female Wistar rats	3 h/d (1–10) W. PND21	There was no hypersensitivity at PND21; however, at PND43 only males were hyperalgesic	No	–	–
Mohtashami Borzadaran, Joushi, Taheri Zadeh, Sheibani, and Esmaeilpour (2020)	Male Wistar rats	3 h/d (1–21) W. PND22	MD rats are more sensitive to pain in the hotplate test and formalin test	EE (PND22–60)	EE restored pain sensitivity to normal level	–

compared with nondeprived counterparts. Furthermore, MD female, but not male, rats exhibited enhanced nociceptive response following a peripheral nerve injury that has been related to distinct neuroinflammatory profiles. Surprisingly, two studies suggested a putative resilience role of MD against inflammatory and neuropathic pain models. Briefly, MD animals were prone to a faster recovery of mechanical and thermal hypersensitivity (Genty et al., 2017). Meanwhile, Juif et al. (2016) showed evidence for a deleterious impact of perinatal stress exposure on the maturation of the sensory-spinal nociceptive system that may contribute to the nociceptive hypersensitivity in early adulthood. C fiber–mediated excitation of spinal cord neurons could be observed at PND14 in controls but not in MD rats.

Early-life MD may affect the offspring in a sex-dependent manner. Variation in maternal care is a critical influence in development, and differences in licking/grooming behavior provided to pups after MD episodes may affect pain sensitivity in adulthood. It is known that rat mothers lick and groom male more than female pups, making males more susceptible to maternal care changes (Moore & Morelli, 1979). Moreover, MD anticipates the maturation of the inhibitory nociceptive pathway, induces a delayed reflex response, and alters anxiety-like and nociceptive behaviors in a sex-specific manner in rats. MD male and female presented higher thermal thresholds at PND21, although at PND43 the male rats became hyperalgesic (Ströher et al., 2019). Experiencing MD also increased pain sensitivity in rats subjected to chemical and mechanical stimuli after Complete Freund's Adjuvant injection. Further, females appeared more sensitive than males to thermal stimuli (Vilela et al., 2017). On the other hand, a meta-analysis indicated that early MD induces an increase in the nociceptive thermal threshold of rodent pups, reflecting reduced pain sensitivity, indicating that MD has a powerful impact on neonate responsiveness to nociceptive stimuli (Chen & Jackson, 2016).

The discrepancies between studies may be easily explained through methodological differences, such as MD protocol, pain model, tissue assessed, sex, and strain. Altogether, the results of these studies show that MD in early life is a powerful environmental condition that influences epigenetic, neuroendocrine, and behavioral outcomes related to nociception in the affected offspring (Fig. 3).

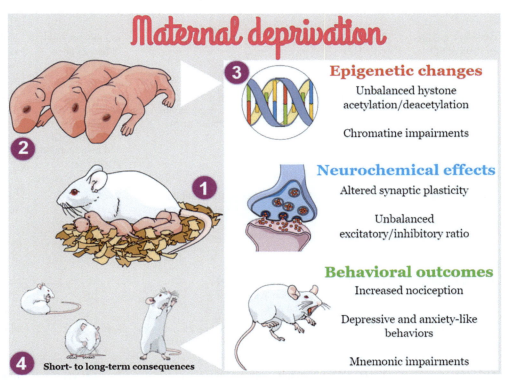

FIG. 3 Maternal deprivation: preclinical evidences. (1) Pups are not deprived from maternal care. (2) Then, maternal deprivation (MD) is provided during the stress-hyporesponsive period. (3) After that, as a consequence of MD, widespread epigenetic, neurochemical effects, and behavioral responses are triggered. (4) Finally, short- to long-term outcomes are displayed during adult life.

Maternal deprivation and nociception: Signaling pathways changes

It is well known that changes elicited by MD protocols include alterations in several signaling pathways. In general, the central sensitization of the nociceptive pathways displayed by MD occurs as a result of: 1. sustained afferent nociceptors excitability; 2. modulation of different receptors; and 3. impairments in descending inhibitory modulatory control. Signaling mechanisms involved in the behavioral long-term nociceptive effects triggered by MD are not well elucidated; thus, many different pathways may be modulated (Table 1). On the other hand, MD-induced visceral hypersensitivity (VH) has been extensively studied regarding the intrinsic mechanisms modulated by MD (Table 2).

Neonatal MD induces long-lasting and differential changes in the opioid system, an important pathway for nociceptive modulation in the neonatal period. Ploj and Nylander (2003) using short MD protocol found increased δ-receptor density in the amygdala, while kappa and opioid receptor–like 1 were not changed, suggesting a putative role of this receptor in the altered emotional behavior. Other studies have also shown that MD changes the levels of growth factors in CNS areas. Ströher et al. (2019) found reduced hypothalamic BDNF and NGF levels in deprived rats, while de Oliveira et al. (2017) showed an increase in cortical NGF levels.

TABLE 2 Visceral outcome.

Author	Strain	MD protocol	Outcomes	Treatment	Effect	Possible mechanism
Zhang et al. (2009)	Male Sprague-Dawley	Period NS (2–14) W. NS	VH, increased glutamate, p-ERK/CREB	Paeoniflorin, MK-801 or ERK phosphorylation inhibitor	Reduced glutamate concentration	A1R adenosine receptor
Gosselin et al. (2010)	Male Sprague-Dawley	3 h/d (2–12) W. NS	VH	Riluzole	Reduced gastrointestinal sensitivity	Increases excitatory amino acid transporter type 1 (EAAT-1) in the DH
Moloney et al. (2012)	Male WT/BALB/c or BALC/c KO GABAB(1B)	3 h/d (1–14) W. PND21	VH	No	KO did not yield different effects from WT	probably GABAB(1B) receptor is not involved
Tsang, Zhao, Wu, Sung, and Bian (2012)	Male Sprague-Dawley	3 h/d (2–14) W. NS	VH, increased NGF and trkA receptor, CGRP, substance P in lumbosacral and DRG	K252a (NGF antagonist) from PND2 to 14	Pain and neuronal activation were restored by attenuated NGF and neuropeptide production	Blockage of trkA signaling
Hu et al. (2013)	Male Sprague-Dawley	3 h/d (2–15) W. PND21–22	VH, increased Nav1.8 expression and TTX-R currents in colon DRG neurons	CBS inhibitor AOAA	Decreased Nav1.8 expression and TTX-R sodium currents	CBS-H2S signaling
Felice et al. (2014)	Male Sprague-Dawley	3 h/d (2–12) W. PND21	VH and neural activation in the caudal ACC	No	–	–
Chen et al. (2015)	Male Sprague-Dawley	3 h/d (3–21) W. PND21	VH hippocampal LTP p-PKCζ	Saline/ZIP into CA1 hippocampus bilaterally	Reduced VH	PKCζ inhibition
Chen et al. (2017)	Male Sprague-Dawley	3 h/d (3–21) W. PND21	VH, increased GluR2 and hippocampal LTP	Intrahippocampal injections of CNQX	Reduced VH	Blockage glutamate receptor

TABLE 2 Visceral outcome—cont'd

Author	Strain	MD protocol	Outcomes	Treatment	Effect	Possible mechanism
Moloney, Stilling, Dinan, and Cryan (2015)	Male Sprague-Dawley	3 h/d (2–12) W. PND21	VH	SAHA (i.p. 5d [P60–64])	Reduced VH	Reduced histone deacetylase acetylation
Miquel et al., 2016	Male C57Bl/AJ	3 h/d (2–14) W. PND21	VH	*Faecalibacterium prausnitzii* commensal bacteria	Reduced VH	Reinforcement of intestinal epithelial barrier
Zhou et al. (2016)	Male Sprague-Dawley	3 h/d (2–16) W. PND21	VH, increased corticosterone (serum), and nefastin-1/NUCB2 expression in the amygdala	Anti-nesfatin-1/NUCB2 into the amygdala	Reduced VH	Nesfatin-1/NUCB2 pathway in the amygdala
Pierce, Eller-Smith, and Christianson (2018)	Female C57Bl/6	3 h/d (1–21) W. PND22	Bladder sensitivity, and mast cell degranulation	Wheel running exercise	Exercise normalized bladder sensitivity and cell degranulation, increased BDNF mRNA hippocampus	HPA axis modulation
O'Mahony et al. (2020)	Male Sprague-Dawley	3 h/d (2–12) W. PND21	VH and increased corticosterone levels (serum)	MFGM + Blend	Reduced VH, and corticosterone levels (serum)	Ameliorated microbiota
Theofanous et al. (2020)	Male C57BL6J; mice lacking ephrinB2 in Na v1.8 nociceptive neurons; and ephrin-B2 knock-in mice	3 h/d (2 – 21) W. NS	VH	No	–	Ephrin-B2/EphB1 spinal cord signaling pathway
Fan et al. (2020)	Male Sprague-Dawley	3 h/d (3–21) W. NS	VH	BDNF inhibitor i.t. (ANA-12)	Reduced visceral sensitivity	BDNF reduction and TrkB signaling inhibition

A1R: adenosine type 1 receptor; *ACC:* anterior cingulate cortex; *ANA-12:* N-[2-[[(hexahydro-2-oxo-1H-azepin-3-yl)amino]carbonyl]phenyl]-benzo[b]thiophene-2-carboxamide; *AOAA:* O-(carboxymethyl) hydroxylamine hemihydrochloride; *BDNF:* brain-derived neurotrophic factor; *CA1:* cornu ammonis 1; *CBS:* cystathionine β-synthase; *CCI:* chronic constriction injury; *CFA:* complete Freund Adjuvant; *CGRP:* calcitonin gene-related peptide; *CNQX:* 6-cyano-7-nitroquinoxaline-2,3-dione; *CREB:* cAMP response element-binding protein; *DH:* dorsal horn; *DRG:* dorsal root ganglia; *EE:* environmental enrichment; *ERK:* extracellular signal regulated kinase; *GluR2:* glutamate ionotropic receptor AMPA type subunit 2; *h/d:* hours per day; *H2S:* hydrogen sulfide; *HPA:* hypothalamus-pituitary-axis; *K252a:* (9S-(9α,10β,12α))-2,3,9,10,11,12-hexahydro-10-hydroxy-10-(methoxycarbonyl)-9-methyl-9,12-epoxy-1H-diindolo [1,2,3-fg:3′,2′,1′-kl]pyrrolo[3,4-i][1,6]benzodiazocin-1-one; *KO:* knockout; *LTP:* long-term potentiation; *MFGM:* milk fat globule membrane; *Nav1.8:* tetrodotoxin (TTX)-resistant voltage-gated channels; *NGF:* nerve growth factor; *NS:* not specified; *NUCB2:* nucleobindin 2; *OT:* oxytocin; *P2 × 3R:* purinergic subunit 2 × 3 receptor; *PND:* postnatal day; *p-PKCζ:* protein kinase C zeta - phosphorylated; *SAHA:* suberanilohydroxamic acid; *TG:* trigeminal ganglia; *TrkA:* tropomyosin receptor kinase A; *TrkB:* tropomyosin receptor kinase B; *TTX-R:* tetrodotoxin-sensitive voltage-gated sodium channels; *VH:* visceral hyperalgesia; *W:* weaning; *WT:* wild-type; *ZIP:* PKCζ pseudosubstrate-derived ζ-inhibitory peptide.

MD induces chronic visceral pain due to an increased neuronal excitability associated with potentiation of tetrodotoxin-resistant sodium channel currents and upregulation of NaV1.8 expression (Hu et al., 2013), and c-fos activation in limbic regions (Felice et al., 2014). Pierce et al. (2018) showed that female mice subjected to MD displayed an increase of bladder sensitivity and lower mRNA levels to corticotropin-releasing factor receptor 1 and glucocorticoid receptor (GR).

Zhou et al. (2016) found an association between VH and MD. Such effects were attributed to increased nefastin-1/NUCB2 expression in the amygdala, which was reversed by GR inhibition. In the hippocampus, Chen et al. (2017) highlighted the role of GluR2 receptors in the MD-induced hyperalgesia and showed that intrahippocampal AMPAR antagonism abolished the MD-induced VH. Also, colorectal distention–evoked pain can be mediated by glutamate release, activation of NMDA receptor, and ERK/CREB signaling in the spinal cord and anterior cingulate cortex (Zhang et al., 2009).

Another important pathway involves Ephrin-B2/EphB receptors, widely expressed in dorsal horn neurons. It was shown that conditional knockout of ephrin-B2 in Nav 1.8 neurons reduces VH in deprived mice (Theofanous et al., 2020). It was also proposed in the putative involvement of Toll-like receptor 4 in the VH induced by MD (Tang et al., 2017). MD-induced VH is related to an increase in the activity of BDNF/TrkB/PKCζ downstream signaling, an effect reversed by intrathecal TrkB antagonism (Fan et al., 2020), or by hippocampal ζ-pseudosubstrate inhibitory peptide (Chen et al., 2015).

Hormonal responses may be involved in MD alterations. For instance, oxytocin, allopregnanolone, and hydroxamic acid–based vorinostat [SAHA, a nonselective histone deacetylase inhibitor] administration reverted the nociceptive effects induced by MD (Melchior et al., 2018). The effect provided by SAHA administration in deprived male rats with VH also decreased the pain scores (Moloney et al., 2015).

Early life stress in humans including maternal deprivation

Neonates experience a wide range of stressful situations during their first days of life, such as needle pricks, and extensive manipulations that impose ELS. For example, the NICU is a stressful environment due to bright light, noise, handling by professionals, MD, and others. Thus, despite high technology and the qualification of health teams, the factors mentioned earlier collaborate for changes in the sleep cycle, the emergence of stress, discomfort, and pain throughout life.

Previous studies have described that ELS is related to alterations in pain sensitivity during adulthood (Walker et al., 2018; Waller et al., 2020). Furthermore, atypical amygdala–medial prefrontal cortex connectivity was found in children who experienced MD (Gee et al., 2013). In addition, ELS may trigger psychiatric disorders and negative effects on health like smoking, overeating, and substance abuse (Danese et al., 2009) besides cardiovascular and heart disease in adulthood (Murphy, Cohn, & Loria, 2017).

Extremely preterm showed reduced thermal threshold sensitivity at age 18–20 years compared to term-born children (Walker et al., 2018). Also, sex-dependent effects were found between extremely preterm males and females undergoing surgery, like reduced pain threshold and increased sensitivity to prolonged noxious cold, respectively (Walker et al., 2018). A recent study highlights a link between ELS and long-term pain sensitivity, where less pressure pain sensitivity at the age of 22 years was associated with more problematic behavior at an early age (Waller et al., 2020).

Long-term effects after ELS may produce a widespread response, including negative results, as it is shown that non-nociceptive handling in preterm infants does not trigger pain responses (Rodrigues & Guinsburg, 2013). However, it is important to highlight that noxious or innocuous stimuli events occurring during the development of CNS may represent long-term effects on cognition, pain transmission/modulation, behavior, and others (Williams & Lascelles, 2020) (Fig. 4).

Environmental enrichment as a potential intervention against the effects of MD

The lack of brain stimulation by a rich environment may impair cognitive development in both humans and nonhuman animals. Environmental enrichment (EE) relates to an individual's stimulating physical and social surroundings, a method used in preclinical and clinical settings aiming to improve the deleterious effects of various harmful events (including MD) and a powerful intervention for recovery from many kinds of stress. The positive effects are more prominent when applied early in life, but EE is also effective in adolescence, adulthood, and aging, although to a lesser degree. Pain sensitivity can be differentially modulated by distinct EE protocols. Few studies however investigated EE as a potential intervention to prevent or treat nociceptive alterations induced by MD. Kimura, Mattaraia, and Picolo (2019) have shown that, although a simple EE is able to decrease anxiety-like behavior, only improved EE is able to abolish acute and chronic pain behavior in rats. When MD rats experience EE, their pain sensitivity is restored at the normal level (Mohtashami Borzadaran et al., 2020).

Similarly to that applied in laboratory animals, EE has been successfully employed in human infants deprived from their mothers. Music, the use of hammocks, and the employment of the Kangaroo Mother Care (KMC) method in NICU have shown beneficial effects to the preterm brains and their behavior (Fig. 5). For instance, Lordier et al. (2019) showed that exposure to music in the NICU has significantly increased coupling between brain networks previously shown to be decreased in premature infants, turning their brain architecture closely similar to those of full-term newborns. Additionally, the use of hammock in NICU favored the sleep of preterm infants and can be used to reduce the stress levels in very

Maternal deprivation and nociception **Chapter | 14** 157

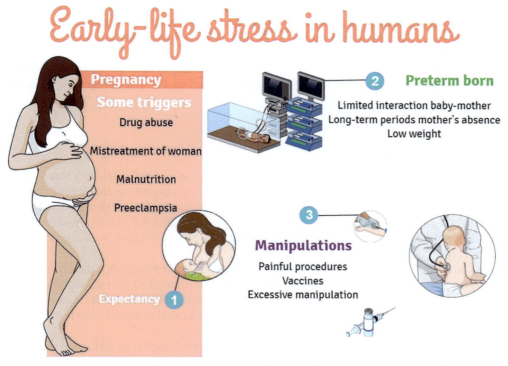

FIG. 4 Early-life stress in humans. (1) During pregnancy, some unwanted issues such as drug abuse, mistreatment of women, malnutrition, and preeclampsia are conveyed to the fetus. (2). Then, it may cause unwanted effects, as preterm born (leading to a spending time into the incubator, less time spent with the mother, and low weight). (3). Finally, painful procedures including pinpricks, vaccines, and excessive manipulation may be harmful to newborn development.

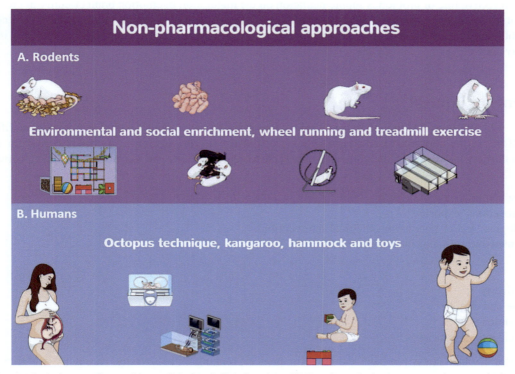

FIG. 5 Nonpharmacological approaches used in preclinical and clinical settings. (A) In rodents, for instance, presenting several stimuli such as environmental enrichment, increased social interaction, wheel running, or treadmill exercise might attenuate these deleterious effects. (B) In humans, the lack of contact with mother can be replaced by using techniques such as octopus, hammock, and kangaroo, providing an enrichment environment to the child.

low-birth-weight preterm newborns (Jesus, Oliveira, & Azevedo, 2018). Compared with conventional neonatal care, the KMC method was found to reduce mortality and severe infections, among others, and increased weight and head circumference gain, proving to be an effective and safe alternative for low-birth-weight infants, mainly in resource-limited countries (Conde-Agudelo & Díaz-Rossello, 2016).

While there are few studies investigating the effects of EE on pain, it is still an open question which protocol is the most effective for treating nociceptive changes induced by MD, particularly in humans, and further investigations are required, especially to examine long-term therapeutic effects.

Applications to other areas

In this chapter, we have reviewed the nociceptive effects displayed by MD in early life. Research that addresses the effects of neonatal maternal deprivation on nociceptive thresholds and behavior in adulthood can contribute to many areas of study and application. Areas such as neuroscience, pharmacology, and pediatrics can be favored with the knowledge generated in these preclinical and clinical studies. These studies can be focused in the search for new pharmacological and nonpharmacological therapies, which can act in the prevention or treatment of the effects resulting from exposure to stressful events in the neonatal, such as maternal deprivation.

Preclinical studies have shown that the disruption of mother–pup interactions in the first days of life might prime irreversible effects in the offspring development period and throughout life. Additionally for developing pain and trouble in coping with pain, these unwanted alterations can include the increase of the probability for anxiogenic and depressive-like symptoms, memory and metabolic impairments, epigenetic mechanisms, as well as disorders in the central and peripheral nervous system (Chen & Jackson, 2016).

It is important to note that early pain experiences can alter the perceptual, cognitive, and social deployment due to the early modifications in somatosensory development (Williams & Lascelles, 2020).

In addition, other studies showed that the challenges experienced in early life can alter phenotypic shifts of immune cells leading to an exacerbated reaction later in adult life and contributing to the increase of the vulnerability to neurological diseases (Viviani et al., 2014).

Growing evidence also shed light on the involvement of intestinal microbiota in the effects of MD. In general, it also seems to be a protective role of probiotics supplementation on the maturation of emotion brain circuitry in preventing premature maturation in infants stressed (O'Mahony et al., 2020). In this way, it is important to increase the knowledge of neonatal neurobiology and consequence of neonatal stress exposure, since its events can involve different implications in biological systems.

Other agents of interest

In this chapter, we have described the influence of MD in the offspring development period and throughout life, inducing irreversible effects in the nociceptive response. Other neonatal stress type exposures, a critical period for neurodevelopment, since the nervous system is immature and vulnerable to neuroplasticity can also alter nociception, between others, include:

Hypoxia-ischemia during childbirth
Medical procedures exposure
Pharmacological and/or nonpharmacological interventions
Neonatal intensive care unit internment
Surgery
Abuse
Extreme poverty
Child abuse and neglect
Domestic violence
Serious accidents or injuries

Studies show that neonatal stress can have serious consequences, relating to the severity, chronicity, and timing of abuse and neglect experiences, to varying degrees, with neural, biological, and behavioral sequelae of abuse and neglect.

Mini-dictionary of terms

Early-life stress: adverse events that occur in a period of CNS development or maturation, and it can be noxious and innocuous stimuli.
Environmental Enrichment: nonpharmacological intervention using enriched environment to minimize adverse events triggered by MD or ELS.
Maternal deprivation: separation of mother and infant during the stress-hyporesponsive period, which needs to be lasting or repeated for immediate and persistent effects on the neuroendocrine regulation of the HPA axis.
Nociceptive signaling: transmission of the nociceptive stimulus through a complex system, including the afferent pathway, brain centers of integration/modulation, and the efferent system.
Stress-hyporesponsive period: a period of reduced adrenal corticosterone and pituitary adrenocorticotropic hormone release in response to stress lasting in the rat at PND4–14.

Key facts

Key facts of maternal deprivation

- MD displays long-term behavioral nociceptive changes in humans and rodents.
- MD applied during the stress-hyporesponsive period alters HPA axis response.
- Nociceptive signaling maturation is changed after MD.
- Hypersensitivity behavior is triggered by MD in a sex-dependent manner.
- Peripheral and central mechanisms are modulated by MD.

Key facts of early-life stress

- ELS includes needle pricks, excessive manipulations, bright light, noise, maternal deprivation, surgery, labor experience, and others.
- Extremely preterm, preterm, and term newborns display different pain hypersensitivity in adulthood.
- ELS alters the pain hypersensitivity in a sex-dependent manner.
- ELS triggers long-term effects on cognition, pain transmission/modulation, behavior, and others.

Summary points

- Maternal deprivation induces changes in the nociceptive response in adulthood.
- Maternal deprivation generates neurochemical and behavioral lifelong effects.
- Maternal deprivation leads to Hypothalamic–Pituitary–Adrenal (HPA) dysfunction.
- Maternal deprivation makes infants more susceptible to stress-related dysfunctions, such as memory impairment, poor cognition, and anxiety.
- Maternal deprivation induces metabolic impairments and epigenetic alterations.
- Maternal deprivation is related to disorders in the central and peripheral nervous system neurochemistry alterations.
- Nonpharmacological approaches are used to mitigate the harmful effects triggered by MD, both in humans and nonhumans.

References

Andrews, K., & Fitzgerald, M. (1997). Biological barriers to paediatric pain management. *Clinical Journal of Pain, 13*, 138–143.

Burke, N. N., Llorente, R., Marco, E. M., Tong, K., Finn, D. P., Viveros, M. P., et al. (2013). Maternal deprivation is associated with sex-dependent alterations in nociceptive behavior and neuroinflammatory mediators in the rat following peripheral nerve injury. *The Journal of Pain, 14*, 1173–1184.

Chen, A., Bao, C., Tang, Y., Luo, X., Guo, L., Liu, B., et al. (2015). Involvement of protein kinase ζ in the maintenance of hippocampal long-term potentiation in rats with chronic visceral hypersensitivity. *Journal of Neurophysiology, 113*, 3047–3055.

Chen, A., Chen, Y., Tang, Y., Bao, C., Cui, Z., Xiao, M., et al. (2017). Hippocampal AMPARs involve the central sensitization of rats with irritable bowel syndrome. *Brain and Behavior: A Cognitive Neuroscience Perspective, 7*, e00650.

Chen, L., & Jackson, T. (2016). Early maternal separation and responsiveness to thermal nociception in rodent offspring: A meta-analytic review. *Behavioural Brain Research, 299*, 42–50.

Clancy, B., Darlington, R. B., & Finlay, B. L. (2001). Translating developmental time across mammalian species. *Neuroscience, 105*, 7–17.

Conde-Agudelo, A., & Díaz-Rossello, J. L. (2016). Kangaroo mother care to reduce morbidity and mortality in low birthweight infants. *The Cochrane Database of Systematic Reviews, 8.* https://doi.org/10.1002/14651858.CD002771.pub4, CD002771.

Danese, A., Moffitt, T. E., Harrington, H., Milne, B. J., Polanczyk, G., Pariante, C. M., et al. (2009). Adverse childhood experiences and adult risk factors for age-related disease: Depression, inflammation, and clustering of metabolic risk markers. *Archives of Pediatrics & Adolescent Medicine, 163,* 1135–1143.

de Oliveira, C., Scarabelot, V. L., Vercelino, R., Silveira, N. P., Adachi, L., Regner, G. G., et al. (2017). Morphine exposure and maternal deprivation during the early postnatal period alter neuromotor development and nerve growth factor levels. *International Journal of Developmental Neuroscience, 63,* 8–15.

Fan, F., Tang, Y., Dai, H., Cao, Y., Sun, P., Chen, Y., et al. (2020). Blockade of BDNF signalling attenuates chronic visceral hypersensitivity in an IBS-like rat model. *European Journal of Pain, 24,* 839–850.

Felice, V. D., Gibney, S. M., Gosselin, R. D., Dinan, T. G., O'Mahony, S. M., & Cryan, J. F. (2014). Differential activation of the prefrontal cortex and amygdala following psychological stress and colorectal distension in the maternally separated rat. *Neuroscience, 267,* 252–262.

Fitzgerald, M. (2005). The development of nociceptive circuits. *Nature Reviews Neuroscience, 6,* 507–520.

Gee, D. G., Gabard-Durnam, L. J., Flannery, J., Goff, B., Humphreys, K. L., Telzer, E. H., et al. (2013). Early developmental emergence of human amygdala-prefrontal connectivity after maternal deprivation. *Proceedings of the National Academy of Sciences of the United States of America, 110,* 15638–15643.

Genty, J., Tetsi Nomigni, M., Anton, F., & Hanesch, U. (2017). Maternal separation stress leads to resilience against neuropathic pain in adulthood. *Neurobiology of Stress, 8,* 21–32.

Gosselin, R. D., O'Connor, R. M., Tramullas, M., Julio-Pieper, M., Dinan, T. G., & Cryan, J. F. (2010). Riluzole normalizes early-life stress-induced visceral hypersensitivity in rats: Role of spinal glutamate reuptake mechanisms. *Gastroenterology, 138*(7), 2418–2425.

Hathway, G. J., Vega-Avelaira, D., & Fitzgerald, M. (2012). A critical period in the supraspinal control of pain: Opioid-dependent changes in brainstem rostroventral medulla function in preadolescence. *Pain, 153,* 775–783.

Hu, S., Xu, W., Miao, X., Gao, Y., Zhu, L., Zhou, Y., et al. (2013). Sensitization of sodium channels by cystathionine β-synthetase activation in colon sensory neurons in adult rats with neonatal maternal deprivation. *Experimental Neurology, 248,* 275–285.

Jakowec, M. W., Fox, A. J., Martin, L. J., & Kalb, R. G. (1995). Quantitative and qualitative changes in AMPA receptor expression during spinal cord development. *Neuroscience, 67,* 893–907.

Jesus, V. R., Oliveira, P., & Azevedo, V. (2018). Effects of hammock positioning in behavioral status, vital signs, and pain in preterms: A case series study. *Brazilian Journal of Physical Therapy, 22,* 304–309.

Juif, P. E., Salio, C., Zell, V., Melchior, M., Lacaud, A., Petit-Demouliere, N., et al. (2016). Peripheral and central alterations affecting spinal nociceptive processing and pain at adulthood in rats exposed to neonatal maternal deprivation. *European Journal of Neuroscience, 44,* 1952–1962.

Kimura, L. F., Mattaraia, V., & Picolo, G. (2019). Distinct environmental enrichment protocols reduce anxiety but differentially modulate pain sensitivity in rats. *Behavioural Brain Research, 364,* 442–446.

Lordier, L., Meskaldji, D. E., Grouiller, F., Pittet, M. P., Vollenweider, A., Vasung, L., et al. (2019). Music in premature infants enhances high-level cognitive brain networks. *Proceedings of the National Academy of Sciences of the United States of America, 116,* 12103–12108.

Marco, E. M., Llorente, R., López-Gallardo, M., Mela, V., Llorente-Berzal, Á., Prada, C., et al. (2015). The maternal deprivation animal model revisited. *Neuroscience and Biobehavioral Reviews, 51,* 151–163.

Marsh, D. F., Hatch, D. J., & Fitzgerald, M. (1997). Opioid systems and the newborn. *British Journal of Anaesthesia, 79,* 787–795.

Melchior, M., Juif, P. E., Gazzo, G., Petit-Demoulière, N., Chavant, V., Lacaud, A., et al. (2018). Pharmacological rescue of nociceptive hypersensitivity and oxytocin analgesia impairment in a rat model of neonatal maternal separation. *Pain, 159,* 2630–2640.

Miquel, S., Martín, R., Lashermes, A., Gillet, M., Meleine, M., Gelot, A., et al. (2016). Anti-nociceptive effect of Faecalibacterium prausnitzii in non-inflammatory IBS-like models. *Scientific Reports, 6,* 19399.

Mohtashami Borzadaran, F., Joushi, S., Taheri Zadeh, Z., Sheibani, V., & Esmaeilpour, K. (2020). Environmental enrichment and pain sensitivity; a study in maternally separated rats. *International Journal of Developmental Neuroscience.* https://doi.org/10.1002/jdn.10031.

Moloney, R. D., O'Leary, O. F., Felice, D., Bettler, B., Dinan, T. G., & Cryan, J. F. (2012). Early-life stress induces visceral hypersensitivity in mice. *Neuroscience Letters, 512*(2), 99–102. https://doi.org/10.1016/j.neulet.2012.01.066.

Moloney, R. D., Stilling, R. M., Dinan, T. G., & Cryan, J. F. (2015). Early-life stress-induced visceral hypersensitivity and anxiety behavior is reversed by histone deacetylase inhibition. *Neurogastroenterology and Motility, 27,* 1831–1836.

Moore, C. L., & Morelli, G. A. (1979). Mother rats interact differently with male and female offspring. *Journal of Comparative and Physiological Psychology, 93,* 677–684.

Murphy, M. O., Cohn, D. M., & Loria, A. S. (2017). Developmental origins of cardiovascular disease: Impact of early life stress in humans and rodents. *Neuroscience and Biobehavioral Reviews, 74,* 453–465.

Nishinaka, T., Nakamoto, K., & Tokuyama, S. (2015). Enhancement of nerve-injury-induced thermal and mechanical hypersensitivity in adult male and female mice following early life stress. *Life Sciences, 121,* 28–34.

O'Mahony, S. M., McVey Neufeld, K. A., Waworuntu, R. V., Pusceddu, M. M., Manurung, S., Murphy, K., ... Cryan, J. F. (2020). The enduring effects of early-life stress on the microbiota-gut-brain axis are buffered by dietary supplementation with milk fat globule membrane and a prebiotic blend. *European Journal of Neuroscience, 51*(4), 1042–1058. https://doi.org/10.1111/ejn.14514.

Petersen, A. C., Joseph, J., & Feit, M. (2014). *Committee on child maltreatment research, policy, and practice for the next decade: Phase II.* Board on Children, Youth, and Families, Committee on Law and Justice, Institute of Medicine, & National Research Council.

Pierce, A. N., Eller-Smith, O. C., & Christianson, J. A. (2018). Voluntary wheel running attenuates urinary bladder hypersensitivity and dysfunction following neonatal maternal separation in female mice. *Neurourology and Urodynamics, 37*, 1623–1632.

Pinto, J. G., Jones, D. G., Williams, C. K., & Murphy, K. M. (2015). Characterizing synaptic protein development in human visual cortex enables alignment of synaptic age with rat visual cortex. *Frontiers in Neural Circuits, 9*, 3.

Ploj, K., & Nylander, I. (2003). Long-term effects on brain opioid and opioid receptor like-1 receptors after short periods of maternal separation in rats. *Neuroscience Letters, 345*, 195–197.

Rodrigues, A. C., & Guinsburg, R. (2013). Pain evaluation after a non-nociceptive stimulus in preterm infants during the first 28 days of life. *Early Human Development, 89*, 75–79.

Salberg, S., Noel, M., Burke, N. N., Vinall, J., & Mychasiuk, R. (2020). Utilization of a rodent model to examine the neurological effects of early life adversity on adolescent pain sensitivity. *Developmental Psychobiology, 62*, 386–399.

Schmidt, M. V., Enthoven, L., van der Mark, M., Levine, S., de Kloet, E. R., & Oitzl, M. S. (2003). The postnatal development of the hypothalamic-pituitary-adrenal axis in the mouse. *International Journal of Developmental Neuroscience, 21*, 125–132.

Schwaller, F., & Fitzgerald, M. (2014). The consequences of pain in early life: Injury-induced plasticity in developing pain pathways. *European Journal of Neuroscience, 39*, 344–352.

Ströher, R., de Oliveira, C., Costa Lopes, B., da Silva, L. S., Regner, G. G., Richardt Medeiros, H., et al. (2019). Maternal deprivation alters nociceptive response in a gender-dependent manner in rats. *International Journal of Developmental Neuroscience, 76*, 25–33.

Tang, H. L., Zhang, G., Ji, N. N., Du, L., Chen, B. B., Hua, R., et al. (2017). Toll-like receptor 4 in paraventricular nucleus mediates visceral hypersensitivity induced by maternal separation. *Frontiers in Pharmacology, 8*, 309.

Theofanous, S. A., Florens, M. V., Appeltans, I., Denadai Souza, A., Wood, J. N., Wouters, M. M., et al. (2020). Ephrin-B2 signaling in the spinal cord as a player in post-inflammatory and stress-induced visceral hypersensitivity. *Neurogastroenterology and Motility, 32*(4), e13782.

Torsney, C., & Fitzgerald, M. (2002). Age-dependent effects of peripheral inflammation on the electrophysiological properties of neonatal rat dorsal horn neurons. *Journal of Neurophysiology, 87*, 1311–1317.

Tsang, S. W., Zhao, M., Wu, J., Sung, J. J., & Bian, Z. X. (2012). Nerve growth factor-mediated neuronal plasticity in spinal cord contributes to neonatal maternal separation-induced visceral hypersensitivity in rats. *European Journal of Pain, 16*(4). https://doi.org/10.1016/j.ejpain.2011.07.005. 463–72.

Uhelski, M. L., & Fuchs, P. N. (2010). Maternal separation stress leads to enhanced emotional responses to noxious stimuli in adult rats. *Behavioral Brain Research, 212*(2), 208–212.

Vilela, F. C., Vieira, J. S., Giusti-Paiva, A., & Silva, M. (2017). Experiencing early life maternal separation increases pain sensitivity in adult offspring. *International Journal of Developmental Neuroscience, 62*, 8–14.

Viviani, B., Boraso, M., Valero, M., Gardoni, F., Marco, E. M., Llorente, R., et al. (2014). Early maternal deprivation immunologically primes hippocampal synapses by redistributing interleukin-1 receptor type I in a sex dependent manner. *Brain, Behavior, and Immunity, 35*, 135–143.

Walker, S. M., Melbourne, A., O'Reilly, H., Beckmann, J., Eaton-Rosen, Z., Ourselin, S., et al. (2018). Somatosensory function and pain in extremely preterm young adults from the UK EPICure cohort: Sex-dependent differences and impact of neonatal surgery. *British Journal of Anaesthesia, 121*, 623–635.

Waller, R., Smith, A. J., O'Sullivan, P. B., Slater, H., Sterling, M., & Straker, L. M. (2020). The association of early life stressors with pain sensitivity and pain experience at 22 years. *Pain, 161*, 220–229.

Williams, M. D., & Lascelles, B. (2020). Early neonatal pain—A review of clinical and experimental implications on painful conditions later in life. *Frontiers in Pediatrics, 8*, 30.

Yasuda, M., Shinoda, M., Honda, K., Fujita, M., Kawata, A., Nagashima, H., et al. (2016). Maternal separation induces orofacial mechanical allodynia in adulthood. *Journal of Dental Research, 95*, 1191–1197.

Zhang, X. J., Chen, H. L., Li, Z., Zhang, H. Q., Xu, H. X., Sung, J. J., et al. (2009). Analgesic effect of paeoniflorin in rats with neonatal maternal separation-induced visceral hyperalgesia is mediated through adenosine A(1) receptor by inhibiting the extracellular signal-regulated protein kinase (ERK) pathway. *Pharmacology, Biochemistry, and Behavior, 94*, 88–97.

Zhou, X. P., Sha, J., Huang, L., Li, T. N., Zhang, R. R., Tang, M. D., et al. (2016). Nesfatin-1/NUCB2 in the amygdala influences visceral sensitivity via glucocorticoid and mineralocorticoid receptors in male maternal separation rats. *Neurogastroenterology and Motility, 28*, 1545–1553.

Chapter 15

Giving birth and pain

Pelin Corman Dincer

Department of Anesthesiology and Reanimation, School of Medicine, Marmara University, Istanbul, Turkey

Abbreviations

CEI	continuous epidural infusion
CSE	combined spinal epidural block
etCO$_2$	end tidal carbon dioxide
GA	general anesthesia
IM	intramuscular
IV	intravenous
NSAIDs	nonsteroidal antiinflammatory drugs
PCA	patient-controlled analgesia
PCEA	patient-controlled epidural analgesia
PIEB	programmed intermittent epidural boluses
SpO$_2$	peripheral oxygen saturation
TENS	transcutaneous electrical nerve stimulation
US	ultrasound

Introduction

Birth isn't something we suffer, but something we actively do and exult in.

—Sheila Kitzinger

Female bodies are designed to give birth, and for all pregnant people, labor and delivery are natural processes, but the mode of delivery varies like cesarean section, medicated, or unmedicated vaginal delivery.

As births can be unpredictable and because vaginal birth is feared by some women, a growing number of births are done with intervention and require pain treatment.

The definition of pain is revised by the International Association for the Study in 2020 as "An unpleasant sensory and emotional experience associated with, or resembling that associated with, actual or potential tissue damage" (Raja et al., 2020). In the notes section, it is stated that personal experience is influenced by biological, physiological, and social factors; the concept of pain is learnt through life experiences. Therefore, every individual's experience of pain could be different, and differences in cultural circumstances, motivations, emotions, age, and previous deliveries of the parturient affect the mode of delivery.

Satisfaction with pain relief during labor was found to be associated with the feeling of personal control; therefore during pregnancy, women should be educated about the aspects of childbirth including pain relief methods and modes of delivery (McCrea & Wright, 1999).

This chapter provides information about pain associated with labor and delivery, and treatment options.

Modes of delivery

1. Vaginal delivery
   a. *Spontaneous vaginal delivery* occurs when no medications, techniques, or devices are used to induce the labor and delivery of the baby.
   b. *Assisted/instrumental vaginal delivery* uses instruments like forceps or vacuum extractors in delivering the baby.
   c. *Induced vaginal delivery* artificially stimulates uterine contractions to promote the spontaneous onset of labor and delivery. Pharmacologic (oxytocin, prostaglandins, misoprostol, mifepristone, and epidural analgesia),

---

Features and Assessments of Pain, Anesthesia, and Analgesia. https://doi.org/10.1016/B978-0-12-818988-7.00005-4
Copyright © 2022 Elsevier Inc. All rights reserved.

nonpharmacologic surgical and mechanical methods (i.e., membrane sweep, amniotomy, cervical dilators, and acupressure), and combined (pharmacologic and nonpharmacologic) techniques are used to induce labor. Medical reasons, past due date, or maternal request mandate *Scheduled Induction* (Mozurkewich et al., 2011).

    d. *Vaginal birth after C-section (VBAC)* may be a delivery option if the parturient had one prior low transverse uterine incision (C-section) with no other uterine incisions.

    Shorter hospital stays, lower infection rates, and quicker recovery are the benefits of vaginal delivery.

2. Cesarean Section

It is the operational delivery of the baby.

a. Scheduled cesarean
b. Unplanned cesarean

Stages of delivery

First stage: It begins with uterine contractions, and during this stage, the cervix dilates to 10 cm (approximately 4 in.). It is the longest stage of the delivery, and the duration is shorter in multiparous women. Distension of the cervix and low uterine segments, isometric contraction of the uterus, and ischemia of uterine and cervical tissues cause the pain in this stage. The impulses are conducted to the spinal cord by C fibers. The pain is more visceral in the beginning, but with cervix dilation, it becomes more somatic.

Second stage: The baby passes through the birth canal. It usually lasts 10–40 min. An episiotomy is made if necessary. The parturient actively takes part in this stage, and she pushes and relaxes as told by the medical staff. The uterine contractions, cervical stretching, distention of vaginal, and perineal tissues cause both visceral and somatic pain. Ischemia of the tissues causes activation of the nociceptors. The impulses are conducted to the spinal cord by Aδ and C fibers.

In the early stages of labor, T11–T12 dermatomes are affected, but as the pain intensity and tissues affected increase, T10–L1 dermatomes and eventually S2–S4 dermatomes also involve in pain sensation (Jurna, 1993) (Fig. 1).

Uterine contractions and uteroplacental perfusion are affected by the increased catecholamines and glucocorticoids. The pain caused by the contractions, the pressure on the cervix, and the stretching of the birth canal and vagina are lessened to some degree by the endorphins released by the pituitary and placenta.

Third stage: It is the delivery of the placenta. It is relatively short (app. 20 min) and painless.

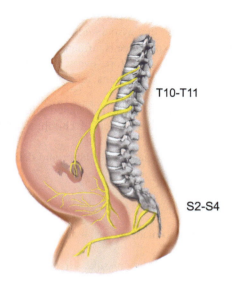

FIG. 1 Pain pathways in pregnant women.

Pain treatment in giving birth

Pharmacologic treatment (Table 1)

Systemic drug administration is performed when women are not willing to have a locoregional analgesic technique and/or locoregional techniques are contraindicated. It is the easiest way to achieve pain control, but it is not as effective as locoregional techniques and may cause sedation and respiratory depression. Also, drugs used by this method have the possibility of passing the placenta. Systemic drug administration costs are lower than the other treatment regimes. Parenteral opioids provide some pain relief during labor, but drowsiness, nausea, vomiting, and slowing of the gastric emptying can be encountered; therefore, adding antiemetic drugs should be considered.

The Cochrane review included studies done in uncomplicated pregnancies who are giving birth at 37–42 weeks, and due to the study designs, the evidence on analgesic effects of opioids and satisfaction was graded as low or very low quality, and they concluded that further research is needed to assess the effect on the newborn and maternal satisfaction with the opioid use (Smith, Burns, & Cuthbert, 2018). Therefore, we must assess every patient's medical history and physical examination with utmost care, get information about the fetus wellbeing, and if any complications are expected during delivery, then form an individualized analgesia plan and discuss it with the parturient.

Satisfaction with analgesia, pain scores, cesarean section, or assisted vaginal birth numbers were not different between TENS (transcutaneous electrical nerve stimulation) using and systemic opioid-administered groups, with higher nausea and vomiting in the latter one (Smith, Burns, & Cuthbert, 2018).

They can be used subcutaneously, intravenously, or intramuscularly. Intravenous patient-controlled analgesia (IV PCA) pumps provide better satisfaction in pain relief than single shots (Weibel et al., 2017). Intravenous administration is less painful at injection and has a faster onset time and less variability in absorption as compared to the other injection routes. After 1–2 h of single-dose opioid administration, additional doses can be administered. The dosage must be tailored according the parturient, and close monitoring of the patient and the fetus is essential (Table 2).

Parenteral **opioids** commonly used in labor include morphine, diamorphine, buprenorphine, fentanyl, alfentanil, sufentanil, remifentanil, pethidine (meperidine), tramadol, and meptazinol. Worldwide, pethidine is the most commonly used opioid during labor (Smith, Burns, & Cuthbert, 2018). Pain management was better with intramuscular pethidine as compared to intramuscular tramadol (Smith, Burns, & Cuthbert, 2018). Normeperidine, active metabolite of meperidine, has a long half-life and can cause neonatal respiratory depression.

Fentanyl, pethidine, alfentanil, nalbuphine, remifentanil, and pentazocine are used in IV PCAs (Douma, Verwey, Kam-Endtz, van der Linden, & Stienstra, 2010; Erskine, Dick, Morrell, Vital, & van den Heever, 1985; Morley-Forster, Reid, & Vandeberghe, 2000; Podlas & Breland, 1987). During second stage of labor, the efficacy of morphine and three different doses of fentanyl administered via PCA were investigated, and 50 µg fentanyl bolus with 6-min lockout time was found to be effective (Castro et al., 2003; Erskine et al., 1985). Short-acting opioids should be preferred in IV PCAs with close

TABLE 1 Pharmacologic treatments.

Systemic drug administration	Parenteral	Opioids
		Mixed opioid agonist/antagonist
		Phencyclidine derivative
		Other
	Inhalation	Nitrous oxide
		Fluorine derivatives
Locoregional techniques	Neuraxial blocks	Spinal
		Epidural
		Combined epidural–spinal
		Dural puncture epidural
	Nerve blocks	Pudendal
		Paracervical

166 PART | II The syndromes of pain

TABLE 2 Parenteral drug administration routes and dosages.

Drug	Route	Dose	Onset of action	Duration of action
Remifentanil	PCA	10–30 µg 2-min lockout	30–60 s	3–4 min
Fentanyl	iv	25–50 µg	2–4 min	30–60 min
	PCA	50 µg loading dose 10–25 µg bolus 5–12-min lockout		
Meperidine	iv	25–50 mg	5 min	2–3 h
	im	50–100 mg	10–15 min	
Nalbuphine	iv	10–20 mg	2–3 min	2–4 h
	im		10–15 min	
Butorphanol	iv	1–2 mg	5–10 min	4–6 h
	im		30–60 min	
Ketamine	iv	10–20 mg	<1 min	5 min

im, intramuscular; *iv*, intravenous; *PCA*, patient-controlled analgesia.

monitoring of the parturient including SpO_2, $etCO_2$ (if available), sedation scores, and respiratory rate. Also, to prevent additive effect, additional analgesics should not be used. The European RemiPCA SAFE Network hospitals use IV remifentanil PCA with bolus doses of 10–30 µg, 2-min lockout interval, and no basal infusion; the infusion is recommended to be stopped 5–10 min prior to cord clamping and providing supplemental oxygen if $SpO_2 < 94\%$ (Melber et al., 2019).

Parenteral **mixed opioid agonist/antagonists** commonly used in labor include nalbuphine, butorphanol, and pentazocine. They have relatively longer duration of action. Nalbuphine causes less nausea and vomiting than meperidine. Duration of action of Nalbuphine administered intravenously/subcutaneously/intramuscularly is 2–4 h (American College of & Gynecologists' Committee on Practice, 2019). Butorphanol when administered intramuscularly (1–2 mg) has a late onset of action, i.e., 30–60 min, but has a long duration (4–6 h) (American College of & Gynecologists' Committee on Practice, 2019).

Mixed opioid agonist/antagonist and partial opioid agonists/opioids should not be administered to the same patient as the analgesic effect will be diminished.

Phencyclidine derivative is used in sedoanalgesia in small doses without compromising the respiratory status. Laryngospasm can be seen due to increased airway secretions. Ketamine shortens the first stage of labor and provides effective analgesia and satisfaction (Joel et al., 2014). Intermittent boluses or a slowly injected loading dose followed by an IV infusion can be used. Joel et al. suggested the loading dose to be injected over 30 min.

Nonopioid drugs are used with other drugs and pain management procedures to enhance pain relief. When used for labor pain, they may reduce the opioid drug usage and negate the undesirable effects of the opioids. They have a ceiling effect in analgesia, and tolerance or physical dependence is not seen. Acetaminophen and aspirin are widely used analgesic antipyretics. *Nonsteroidal antiinflammatory drugs (NSAIDs)* inhibit prostaglandin production and formation of proinflammatory substances, which have a role in decreasing in pain sensation. *Sedatives* such as barbiturates, the phenothiazine derivatives, hydroxyzine, and benzodiazepines were used for pain relief (American College of & Gynecologists' Committee on Practice, 2019). The phenothiazine derivatives also have antiemetic effect. *Antihistamines* by blocking H1 receptors cause sedation and can be used as an adjunct to opioids (Othman, Jones, & Neilson, 2012).

Subanesthetic concentrations of **inhalation anesthetics** can be used in labor. The parturient self-administers it by using a hand-held face mask over her nose and mouth and stays awake. The onset and duration of action is short. Fluorine derivatives provided better pain relief and lower pain intensity scores but more drowsiness as compared to *nitrous oxide* in the first stage of labor (Klomp et al., 2012).

Locoregional techniques

Pudendal nerve block: The pudendal nerve is a sensory and motor nerve. It arises from the sacral plexus and forms S2–S4 from spinal nerve roots, and in pudendal canal, it divides into inferior rectal and perineal nerves. In the second stage of labor, bilateral blockage of the pudendal nerve may ease the pain caused by the distention of the perineum and vagina (Jones et al., 2012).

Before the widespread use of epidural blocks, infiltration of the pudendal nerve was used especially in instrumental deliveries. In cases when sacral region is not efficiently blocked by epidural intervention, pudendal nerve block, with lidocaine 1% or bupivacaine 0.25%, could be added for pain relief (Ghanavatian & Derian, 2020; Jones et al., 2012).

For obstetric indications, most commonly transvaginal approach is used. Allergy to the local anesthetic, infection at the injection site, bleeding disorders, and altered anatomy are absolute and relative contraindications. Discomfort at the injection site, pudendal nerve damage, pudendal artery puncture, systemic local anesthetic toxicity, and rectum or bladder injections can be encountered (Ghanavatian & Derian, 2020).

Paracervical block: The use of this block has lost its popularity as it may lead to fetal bradycardia or even fetal death (Ranta, Jouppila, Spalding, Kangas-Saarela, & Jouppila, 1995; Rosen, 2002). Local anesthetic drugs are injected to 2–6 sites at a depth of 3–7 mm alongside the vaginal portion of the cervix. In this infiltration technique, several injection points can be used (3, 6, 9, 12 o'clock, etc.). Ranta et al. injected 0.25% bupivacaine (5 mL each site) superficially (3 mm depth) to 3–9 o'clock positions and concluded that a low-dose superficial technique has an effective analgesic effect with minimal fetal and neonatal side effects, but close fetal monitoring is essential (Ranta et al., 1995). Repeated doses can be required to achieve pain relief. This block does not cause parasympathetic or sympathetic response, weakness in the lower extremities, any changes in blood pressure, or cardiac return. The incidence of instrumental delivery is also not effected (Ranta et al., 1995).

Neuraxial analgesia and anesthesia

These methods are used for labor analgesia or operative anesthesia. They require a qualified anesthesia specialist and close monitoring of the parturient. Fetal heart monitoring should be done according the institutional protocol. The risks and benefits should be considered according to the case, and the best option should be offered to the parturient. COVID-19 is not a contraindication, and on the contrary, it should be advised if operative anesthesia is planned. Neuraxial analgesia may be initiated at any stage during labor, but the dermatomes aimed to be blocked differ according the stage of the labor. In the early phase of the first stage, T10-L1 dermatomes, in the late phase of the first stage and in the second stage S2-4 dermatomes must be blocked.

An increase in maternal temperature, which is not associated with infection, is seen with epidural analgesia, and it is most commonly encountered in nulliparous women and in whom the duration of epidural is long (American College of & Gynecologists' Committee on Practice, 2019). Although the mechanism is not known, poor neonatal outcome in the form of neonatal encephalopathy and cerebral palsy is suggested to be associated with the use of labor epidural analgesia or with intrapartum pyrexia (Labor & Maguire, 2008; Qiu et al., 2020). Canadian Anesthesiologists' Society published a statement emphasizing that Qiu et al. (2020) study had several methodological limitations, and current Canadian anesthetic care should not be changed by this study (McKeen, Zaphiratos, & Canadian Anesthesiologists, 2020).

The duration of the first stage of labor is shortened with intrathecal drug (opioid alone or with local anesthetic) as compared to systemic drugs administration, and the second stage of labor is prolonged with epidural techniques (American College of & Gynecologists' Committee on Practice, 2019).

In patients with prior history of failed or difficult neuraxial block, spinal deformities, vertebral surgery, morbid obesity, or inexperience with neuraxial placement (i.e., trainees), the use of neuraxial ultrasound (US) is gaining popularity. Assessing the depth of epidural space, location of the midline, and accurate identification of desired interspaces can be achieved by visualization of the lumbar region. Traditional palpation and landmark technique may not identify the actual spinal interspace correctly; the usage of US reduces the incidence of erroneous interspace location and the risk of traumatic procedures (Perlas, Chaparro, & Chin, 2016). A catheter in a lower interspace will provide better analgesia as it may cover sacral nerve roots better than a catheter placed in a higher interspace.

Spinal anesthesia is used for cesarean delivery or when the delivery is to happen in an hour (American College of & Gynecologists' Committee on Practice, 2019). Local anesthetic drug (ropivacaine or bupivacaine) with or without opioid (fentanyl or sufentanil) is injected to the subarachnoid space. It has a fast onset time and produces motor and sensory block (Table 3).

Continuous spinal analgesia can be performed after accidental dural puncture and in patients with difficult epidural catheter placement estimation. Postdural puncture headache, meningitis, abscess, hematoma, arachnoiditis, or cauda equina syndrome may be encountered.

Epidural block is considered the gold-standard method for labor pain relief (Nanji & Carvalho, 2020). The catheter could be placed at any stage of the labor; early or late initiation of the infusion does not affect the maternal or fetal outcomes

168 PART | II The syndromes of pain

TABLE 3 Drugs used in spinal anesthesia.

	Alone	With opioid	With local anesthetics
Bupivacaine	2.5–5 mg	2–3 mg	
Ropivacaine	3–4 mg	2–4 mg	
Fentanyl	15–25 µg		10–15 µg
Sufentanil	5–8 µg		2.5–5 µg
Morphine (preservative free)			0.1 mg

(Sng et al., 2014). It is important to check the function of the catheter during labor. In inadequate or incomplete pain relief after checking the catheter, additional top-up medications or systemic drugs could be administered. One of the advantages of this method is that they can be used in cesarean anesthesia as well.

Combined spinal epidural block (CSE) has faster onset of pain relief and higher maternal satisfaction, fewer break-through pain, fewer unilateral blocks, and catheter failure than epidural block alone (Nanji & Carvalho, 2020). As a smaller gauge spinal needle is used, postspinal headache occurrence is low. The spinal portion of this block may cause maternal hypotension, opioid-induced pruritus, and fetal heart rate changes, especially bradycardia. Fetal bradycardia is treated with correction of the maternal hypotension and application of uterus relaxants like nitroglycerine or terbutaline (Nanji & Carvalho, 2020).

Dural puncture epidural block is a relatively new described method. In this block, dura is punctured with a small gauge spinal needle as in CSE, but no medication is given through that needle and then the epidural catheter is placed. The drugs applied via the epidural catheter slowly translocate to the intrathecal space through that hole. Relief from sacral spread, onset, and bilateral pain is improved by this method (Cappiello, O'Rourke, Segal, & Tsen, 2008).

After catheter insertion, intermittent boluses, continuous epidural infusions (CEI), patient-controlled epidural analgesia boluses (PCEA), or programmed intermittent epidural boluses (PIEB) of local anesthetic drugs with or without lipophilic opioids can be administered. Intermittent boluses are given at preset times or when pain occurs which is not as efficient as the other methods. PCEA has advantages of causing less lower extremity motor block, less consumption of local anesthetic medication, less top-up drug administration, and higher satisfaction of pain relief as compared to CEI, but it has been shown that a background infusion added to PCEA has better results than PCEA alone (Halpern & Carvalho, 2009). PIEB delivers automated fix boluses of local anesthetic with or without opioid at preset times. Combining PIEB and PCEA for better pain relief, less top-ups, and unilateral blockade is suggested (Nanji & Carvalho, 2020; Roofthooft et al., 2020). However, a background infusion may prolong the second stage of labor and increase the incidence of instrumental delivery (Koyyalamudi et al., 2016). Computer-integrated PCEAs titrate the background infusions according the demand dose frequency of the previous hour (Koyyalamudi et al., 2016).

Large bolus doses of diluted ropivacaine and bupivacaine (up to 0.125% bupivacaine or 0.2% ropivacaine) with background infusion rates between 2 and 10 mL/h and more than 5 mL bolus doses can be used effectively for PCEA in labor (Halpern & Carvalho, 2009). When using a diluted local anesthetic, a lipophilic opioid must be added to epidural solution to maintain the desired analgesia and pain relief. Higher doses of opioids may cause pruritus so the recommended doses of fentanyl and sufentanil are 2–3 µg/mL and 0.2–0.4 µg/mL, respectively (Nanji & Carvalho, 2020).

General anesthesia (GA) is no longer the first choice in cesarean delivery. In emergency cases when neuraxial methods cannot be performed, in cases when neuraxial methods are contraindicated, or regional analgesia is ineffective for a cesarean delivery, GA is done.

Parturient with suspected or confirmed COVID-19 should be advised to have pharmacologic treatment rather than general anesthesia as the aerosolization of airway secretions happens during induction of anesthesia and extubation of the trachea, and the physiologic changes in pregnancy may exacerbate COVID-19 symptoms. In these patients, deep sedation and respiratory depression should also be avoided.

Postpartum pain

Acute pain management includes continuation of the PCEA if epidural catheter exists. Truncal blocks (i.e., ilioinguinal, iliohypogastric nerve blocks, transversus abdominis plane block) and wound infiltration can also be used for acute pain control following Caesarean section. Intravenous, oral, rectal, or parenteral opioids, NSAIDs, and acetaminophen can also be administered intermittently or via IV PCA. Adverse effects can be seen in the mother and breastfeeding newborn due to opioids, but pain may lead to unwanted effects like shallow breathing and atelectasis, a decrease in motility, and reduction in the breastfeeding quality. Postoperative pain intensity does not change between spinal and general anesthesia in emergency and elective settings (Arslantas & Umuroglu, 2019).

Chronic pain after delivery It is suggested that pregnancy or delivery or endogenous secretion of oxytocin has a protective effect on the response to physical injury, and the incidence of chronic pain following delivery is 1.8% at 6 months and 0.3% at 12 months (Eisenach et al., 2013; Landau, Bollag, & Ortner, 2013). The chronic pain after cesarean delivery is neuropathic, but the nature of pain after vaginal delivery is yet to be studied.

Postpartum depression is seen in 11.2% of women after delivery, and acute postpartum pain but not the mode of delivery is associated with the risk of depression (Eisenach et al., 2008). Efficient pain control is important in preventing postpartum depression and chronic pain formation.

Nonpharmacological treatment

These treatment options have less negative effect on fetus than pharmacological methods, and their usage is gaining popularity as adjunct therapies to pharmacological methods. The efficacy of these methods still needs to be studied.

Acupuncture or acupressure decreases pain for the first 30 min by 11%, but the need of an analgesic drug or method was lower when compared to other pain relief methods. They can be used as an adjunct treatment (Cho, Lee, & Ernst, 2010).

Hypnotherapy: Self-hypnosis treatment did not change the intrapartum epidural analgesic usage but decreased the use of analgesia (Downe et al., 2015; Madden, Middleton, Cyna, Matthewson, & Jones, 2016). High-quality randomized clinical trials are needed to recommend hypnosis in delivery units. Parturient with a history of severe psychological disturbances is not eligible for this method.

Yoga: Although appropriately designed studies are needed to determine the effects of yoga during pregnancy, some reviews revealed that yoga done in the peripartum period has beneficial effects on pain sensation and maternal comfort without changing the analgesic usage (Babbar, Parks-Savage, & Chauhan, 2012).

Birth ball exercise is studied in the antenatal and labor periods. Pain scores were significantly lower in the first stage of labor (Makvandi, Latifnejad Roudsari, Sadeghi, & Karimi, 2015).

Hydrotherapy water immersion decreases neuraxial analgesia usage and shortens the labor when used in the first stage of labor, but there are limited data regarding its usage during the second stage of labor (Cluett, Burns, & Cuthbert, 2018).

Intracutaneous or subcutaneous sterile water injection (water block) is used for low back pain and significantly decreases pain scores. In this method, 0.5 cc sterile water is injected at four sites: on each posterior superior iliac spine, 3 cm lower, and 1 cm inner than that point (Almassinokiani et al., 2020).

Warm showers also cause relaxation and a decrease in pain scores (Lee, Liu, Lu, & Gau, 2013).

Manual healing methods are simple, safe, and reduce pain and length of labor, but research is needed to show its efficacy and efficiency in labor pain (Smith et al., 2018).

Transcutaneous electronic nerve stimulation (TENS) electrodes can be placed on the lower back or on the acupuncture points. Pain scales are not different between TENS and control groups (Dowswell, Bedwell, Lavender, & Neilson, 2009).

Biofeedback can be used in labor pain, but there is insufficient evidence on its effectiveness (Barragan Loayza, Sola, & Juando Prats, 2011).

Aromatherapy: There is a lack of regulation of aromatherapy products, so only trained professionals must deliver aromatherapy as essential oils are potentially dangerous. It can reduce patient anxiety and labor pain but must be used as an adjunct method (Tabatabaeichehr & Mortazavi, 2020).

Audio analgesia uses music, white noise, or environmental sounds that help the women to relax. Review revealed no difference in pain intensity, satisfaction with pain relief, and operational delivery rates with this method (Smith, Levett, Collins, & Crowther, 2011).

Breathing techniques are widely used. Rhythmic breathing without hyperventilating enables relaxation and a reduction in anxiety and duration of the first stage of labor (Cicek & Basar, 2017).

Hot-cold packs can be applied alone or intermittently. Pain intensity and length of labor decrease with thermal manual methods (Smith et al., 2018).

Labor support: Partner, family, friend, or doula may provide support to the parturient. With doula support, nonmedical labor pain management and labor induction methods were increased (Kozhimannil, Johnson, Attanasio, Gjerdingen, & McGovern, 2013).

Application to other areas

Pain management is important as some of the postoperative complications are associated with poor postoperative pain control. The drugs and blocs mentioned in this section are not specific to labor pain. The dosages and concentrations of the drugs can be modified according the surgery and the patients' health status. All abdominal surgeries especially lower abdominal ones may benefit regional and truncal blocks mentioned in this chapter. There are various obstetric and gynecologic procedures other than labor which can be done in hospital or in day-clinics. In developing countries and in places with limited resources physicians seek effective and economic pain control methods with minimal side effects. Methods that do not require specific equipment like ultrasound machine, block needle, etc., and lead to less hospitalization are sought. Nonpharmacological methods have wide indication areas and can be utilized effectively.

Mini-dictionary of terms

Dermatome: It is an area of skin, and the sensation of this area is achieved by a single spinal nerve root.

Patient-controlled analgesia: The patient administers his pain relief medication by pressing a button of a specially designed infusion pump used to deliver medications. Intravenous and epidural administration routes are widely used for labor analgesia.

Neuraxial anesthesia: The administration of medication into the subarachnoid or epidural space to produce anesthesia and analgesia.

Nonpharmacological pain treatment for labor: It refers to the complementary and alternative therapies administered to parturient.

Pharmacological pain treatment for labor: Administering drugs to the parturient according the planned route (i.e., parenteral, neuraxial, nerve blocks, inhalation).

Key facts

Key facts of giving birth and pain

- The pain of labor in the first stage is mediated by T10 to L1 spinal segments, and in the second stage, it is mediated by T12 to L1 and S2 to S4 spinal segments.
- Visceral pain during labor occurs during the early first stage and the second stage of childbirth, and it is a dull pain that is felt in the lower abdomen, sacrum, and back.
- Somatic pain during labor occurs during the late first stage and the second stage; it is a sharp pain localized to the vagina, rectum, and perineum.
- Pharmacological treatments reduce pain perception and pain transmission.
- Nonpharmacological treatments are easy to administer, have minimal side effects, and are associated with high maternal satisfaction.

Summary points

- The pain of labor is severe, and available treatment options should be discussed with the parturient.
- Every individual's experience of pain could be different, and differences in cultural circumstances, motivations, emotions, age, and previous deliveries affect the mode of delivery and analgesia requirements.
- Neuraxial analgesia provides better analgesia than nonpharmacological treatments.
- Pharmacological treatments may cause nausea, vomiting, dizziness, and drowsiness; therefore, vital parameters and sedation level of the parturient must be monitored.
- Systemically applied drugs may pass the placenta and cause side effects on the fetus, and newborns might need close monitoring for respiratory depression and sedation.
- Nonpharmacological treatments have minimal side effects and are easy to administer; they may be used as an adjunct to pharmacological treatments.

References

Almassinokiani, F., Ahani, N., Akbari, P., Rahimzadeh, P., Akbari, H., & Sharifzadeh, F. (2020). Comparative analgesic effects of intradermal and subdermal injection of sterile water on active labor pain. *Anesthesiology and Pain Medicine, 10*(2). https://doi.org/10.5812/aapm.99867, e99867.

American College of O., & Gynecologists' Committee on Practice, B.-O. (2019). ACOG practice bulletin no. 209: Obstetric analgesia and anesthesia. *Obstetrics and Gynecology, 133*(3), e208–e225. https://doi.org/10.1097/AOG.0000000000003132.

Arslantas, R., & Umuroglu, T. (2019). Comparing the effects of general and spinal anesthesia on the postoperative pain intensity in patients undergoing emergent or elective cesarean section. *Marmara Medical Journal,* 62–67. https://doi.org/10.5472/marumj.570905.

Babbar, S., Parks-Savage, A. C., & Chauhan, S. P. (2012). Yoga during pregnancy: A review. *American Journal of Perinatology, 29*(6), 459–464. https://doi.org/10.1055/s-0032-1304828.

Barragan Loayza, I. M., Sola, I., & Juando Prats, C. (2011). Biofeedback for pain management during labour. *Cochrane Database of Systematic Reviews, 6.* https://doi.org/10.1002/14651858.CD006168.pub2, CD006168.

Cappiello, E., O'Rourke, N., Segal, S., & Tsen, L. C. (2008). A randomized trial of dural puncture epidural technique compared with the standard epidural technique for labor analgesia. *Anesthesia and Analgesia, 107*(5), 1646–1651. https://doi.org/10.1213/ane.0b013e318184ec14.

Castro, C., Tharmaratnam, U., Brockhurst, N., Tureanu, L., Tam, K., & Windrim, R. (2003). Patient-controlled analgesia with fentanyl provides effective analgesia for second trimester labour: A randomized controlled study. *Canadian Journal of Anaesthesia, 50*(10), 1039–1046. https://doi.org/10.1007/BF03018370.

Cho, S. H., Lee, H., & Ernst, E. (2010). Acupuncture for pain relief in labour: A systematic review and meta-analysis. *BJOG, 117*(8), 907–920. https://doi.org/10.1111/j.1471-0528.2010.02570.x.

Cicek, S., & Basar, F. (2017). The effects of breathing techniques training on the duration of labor and anxiety levels of pregnant women. *Complementary Therapies in Clinical Practice, 29*, 213–219. https://doi.org/10.1016/j.ctcp.2017.10.006.

Cluett, E. R., Burns, E., & Cuthbert, A. (2018). Immersion in water during labour and birth. *Cochrane Database of Systematic Reviews, 5.* https://doi.org/10.1002/14651858.CD000111.pub4, CD000111.

Douma, M. R., Verwey, R. A., Kam-Endtz, C. E., van der Linden, P. D., & Stienstra, R. (2010). Obstetric analgesia: A comparison of patient-controlled meperidine, remifentanil, and fentanyl in labour. *British Journal of Anaesthesia, 104*(2), 209–215. https://doi.org/10.1093/bja/aep359.

Downe, S., Finlayson, K., Melvin, C., Spiby, H., Ali, S., Diggle, P., et al. (2015). Self-hypnosis for intrapartum pain management in pregnant nulliparous women: A randomised controlled trial of clinical effectiveness. *BJOG, 122*(9), 1226–1234. https://doi.org/10.1111/1471-0528.13433.

Dowswell, T., Bedwell, C., Lavender, T., & Neilson, J. P. (2009). Transcutaneous electrical nerve stimulation (TENS) for pain relief in labour. *Cochrane Database of Systematic Reviews, 2.* https://doi.org/10.1002/14651858.CD007214.pub2, CD007214.

Eisenach, J. C., Pan, P., Smiley, R. M., Lavand'homme, P., Landau, R., & Houle, T. T. (2013). Resolution of pain after childbirth. *Obstetric Anesthesia Digest, 33*(4), 183. https://doi.org/10.1097/01.aoa.0000436294.46263.9f.

Eisenach, J. C., Pan, P. H., Smiley, R., Lavand'homme, P., Landau, R., & Houle, T. T. (2008). Severity of acute pain after childbirth, but not type of delivery, predicts persistent pain and postpartum depression. *Pain, 140*(1), 87–94. https://doi.org/10.1016/j.pain.2008.07.011.

Erskine, W. A., Dick, A., Morrell, D. F., Vital, M., & van den Heever, J. (1985). Self-administered intravenous analgesia during labour. A comparison between pentazocine and pethidine. *South African Medical Journal, 67*(19), 764–767. https://www.ncbi.nlm.nih.gov/pubmed/3992403.

Ghanavatian, S., & Derian, A. (2020). *Pudendal nerve block.* StatPearls. https://www.ncbi.nlm.nih.gov/pubmed/31855362.

Halpern, S. H., & Carvalho, B. (2009). Patient-controlled epidural analgesia for labor. *Anesthesia and Analgesia, 108*(3), 921–928. https://doi.org/10.1213/ane.0b013e3181951a7f.

Joel, S., Joselyn, A., Cherian, V. T., Nandhakumar, A., Raju, N., & Kaliaperumal, I. (2014). Low-dose ketamine infusion for labor analgesia: A double-blind, randomized, placebo controlled clinical trial. *Saudi Journal of Anaesthesia, 8*(1), 6–10. https://doi.org/10.4103/1658-354X.125897.

Jones, L., Othman, M., Dowswell, T., Alfirevic, Z., Gates, S., Newburn, M., et al. (2012). Pain management for women in labour: An overview of systematic reviews. *Cochrane Database of Systematic Reviews,* (3). https://doi.org/10.1002/14651858.CD009234.pub2, CD009234.

Jurna, I. (1993). Labor pain-causes, pathways and issues. *Schmerz, 7*(2), 79–84. https://doi.org/10.1007/BF02527864 (Geburtsschmerz-Entstehung, Erregungsleitung und Folgen.).

Klomp, T., van Poppel, M., Jones, L., Lazet, J., Di Nisio, M., & Lagro-Janssen, A. L. (2012). Inhaled analgesia for pain management in labour. *Cochrane Database of Systematic Reviews, 9.* https://doi.org/10.1002/14651858.CD009351.pub2, CD009351.

Koyyalamudi, V., Sidhu, G., Cornett, E. M., Nguyen, V., Labrie-Brown, C., Fox, C. J., et al. (2016). New labor pain treatment options. *Current Pain and Headache Reports, 20*(2), 11. https://doi.org/10.1007/s11916-016-0543-2.

Kozhimannil, K. B., Johnson, P. J., Attanasio, L. B., Gjerdingen, D. K., & McGovern, P. M. (2013). Use of nonmedical methods of labor induction and pain management among U.S. women. *Birth, 40*(4), 227–236. https://doi.org/10.1111/birt.12064.

Labor, S., & Maguire, S. (2008). The pain of labour. *Reviews in Pain, 2*(2), 15–19. https://doi.org/10.1177/204946370800200205.

Landau, R., Bollag, L., & Ortner, C. (2013). Chronic pain after childbirth. *International Journal of Obstetric Anesthesia, 22*(2), 133–145. https://doi.org/10.1016/j.ijoa.2013.01.008.

Lee, S. L., Liu, C. Y., Lu, Y. Y., & Gau, M. L. (2013). Efficacy of warm showers on labor pain and birth experiences during the first labor stage. *Journal of Obstetric, Gynecologic, and Neonatal Nursing, 42*(1), 19–28. https://doi.org/10.1111/j.1552-6909.2012.01424.x.

Madden, K., Middleton, P., Cyna, A. M., Matthewson, M., & Jones, L. (2016). Hypnosis for pain management during labour and childbirth. *Cochrane Database of Systematic Reviews, 5.* https://doi.org/10.1002/14651858.CD009356.pub3, CD009356.

Makvandi, S., Latifnejad Roudsari, R., Sadeghi, R., & Karimi, L. (2015). Effect of birth ball on labor pain relief: A systematic review and meta-analysis. *The Journal of Obstetrics and Gynaecology Research, 41*(11), 1679–1686. https://doi.org/10.1111/jog.12802.

McCrea, B. H., & Wright, M. E. (1999). Satisfaction in childbirth and perceptions of personal control in pain relief during labour. *Journal of Advanced Nursing, 29*(4), 877–884. https://doi.org/10.1046/j.1365-2648.1999.00961.x.

McKeen, D. M., Zaphiratos, V., & Canadian Anesthesiologists, S. (2020). Lack of evidence that epidural pain relief during labour causes autism spectrum disorder: A position statement of the Canadian Anesthesiologists' Society. *Canadian Journal of Anaesthesia.* https://doi.org/10.1007/s12630-020-01840-z (absence de preuve voulant que le recours a la peridurale pour soulager la douleur Durant le travail cause le trouble du spectre de l'autisme.).

Melber, A. A., Jelting, Y., Huber, M., Keller, D., Dullenkopf, A., Girard, T., et al. (2019). Remifentanil patient-controlled analgesia in labour: Six-year audit of outcome data of the RemiPCA SAFE network (2010-2015). *International Journal of Obstetric Anesthesia, 39*, 12–21. https://doi.org/10.1016/j.ijoa.2018.12.004.

Morley-Forster, P. K., Reid, D. W., & Vandeberghe, H. (2000). A comparison of patient-controlled analgesia fentanyl and alfentanil for labour analgesia. *Canadian Journal of Anaesthesia, 47*(2), 113–119. https://doi.org/10.1007/BF03018845.

Mozurkewich, E. L., Chilimigras, J. L., Berman, D. R., Perni, U. C., Romero, V. C., King, V. J., et al. (2011). Methods of induction of labour: A systematic review. *BMC Pregnancy and Childbirth, 11*, 84. https://doi.org/10.1186/1471-2393-11-84.

Nanji, J. A., & Carvalho, B. (2020). Pain management during labor and vaginal birth. *Best Practice & Research. Clinical Obstetrics & Gynaecology, 67*, 100–112. https://doi.org/10.1016/j.bpobgyn.2020.03.002.

Othman, M., Jones, L., & Neilson, J. P. (2012). Non-opioid drugs for pain management in labour. *Cochrane Database of Systematic Reviews, 7*. https://doi.org/10.1002/14651858.CD009223.pub2, CD009223.

Perlas, A., Chaparro, L. E., & Chin, K. J. (2016). Lumbar Neuraxial ultrasound for spinal and epidural anesthesia: A systematic review and meta-analysis. *Regional Anesthesia and Pain Medicine, 41*(2), 251–260. https://doi.org/10.1097/AAP.0000000000000184.

Podlas, J., & Breland, B. D. (1987). Patient-controlled analgesia with nalbuphine during labor. *Obstetrics and Gynecology, 70*(2), 202–204. https://www.ncbi.nlm.nih.gov/pubmed/3601283.

Qiu, C., Lin, J. C., Shi, J. M., Chow, T., Desai, V. N., Nguyen, V. T., et al. (2020). Association between epidural analgesia during labor and risk of autism spectrum disorders in offspring. *JAMA Pediatrics, 174*(12), 1168–1175. https://doi.org/10.1001/jamapediatrics.2020.3231.

Raja, S. N., Carr, D. B., Cohen, M., Finnerup, N. B., Flor, H., Gibson, S., et al. (2020). The revised International Association for the Study of Pain definition of pain: Concepts, challenges, and compromises. *Pain.* https://doi.org/10.1097/j.pain.0000000000001939.

Ranta, P., Jouppila, P., Spalding, M., Kangas-Saarela, T., & Jouppila, R. (1995). Paracervical block—A viable alternative for labor pain relief? *Acta Obstetricia et Gynecologica Scandinavica, 74*(2), 122–126. https://doi.org/10.3109/00016349509008919.

Roofthooft, E., Barbe, A., Schildermans, J., Cromheecke, S., Devroe, S., Fieuws, S., et al. (2020). Programmed intermittent epidural bolus vs. patient-controlled epidural analgesia for maintenance of labour analgesia: A two-centre, double-blind, randomised studydagger. *Anaesthesia, 75*(12), 1635–1642. https://doi.org/10.1111/anae.15149.

Rosen, M. A. (2002). Paracervical block for labor analgesia: A brief historic review. *American Journal of Obstetrics and Gynecology, 186*(5 Suppl nature), S127–S130. https://doi.org/10.1067/mob.2002.121812.

Smith, L. A., Burns, E., & Cuthbert, A. (2018). Parenteral opioids for maternal pain management in labour. *Cochrane Database of Systematic Reviews, 6*. https://doi.org/10.1002/14651858.CD007396.pub3, CD007396.

Smith, C. A., Levett, K. M., Collins, C. T., Armour, M., Dahlen, H. G., & Suganuma, M. (2018). Relaxation techniques for pain management in labour. *Cochrane Database of Systematic Reviews, 3*. https://doi.org/10.1002/14651858.CD009514.pub2, CD009514.

Smith, C. A., Levett, K. M., Collins, C. T., & Crowther, C. A. (2011). Relaxation techniques for pain management in labour. *Cochrane Database of Systematic Reviews, 12*. https://doi.org/10.1002/14651858.CD009514, CD009514.

Smith, C. A., Levett, K. M., Collins, C. T., Dahlen, H. G., Ee, C. C., & Suganuma, M. (2018). Massage, reflexology and other manual methods for pain management in labour. *Cochrane Database of Systematic Reviews, 3*. https://doi.org/10.1002/14651858.CD009290.pub3, CD009290.

Sng, B. L., Leong, W. L., Zeng, Y., Siddiqui, F. J., Assam, P. N., Lim, Y., et al. (2014). Early versus late initiation of epidural analgesia for labour. *Cochrane Database of Systematic Reviews, 10*. https://doi.org/10.1002/14651858.CD007238.pub2, CD007238.

Tabatabaeichehr, M., & Mortazavi, H. (2020). The effectiveness of aromatherapy in the management of labor pain and anxiety: A systematic review. *Ethiopian Journal of Health Sciences, 30*(3), 449–458. https://doi.org/10.4314/ejhs.v30i3.16.

Weibel, S., Jelting, Y., Afshari, A., Pace, N. L., Eberhart, L. H., Jokinen, J., et al. (2017). Patient-controlled analgesia with remifentanil versus alternative parenteral methods for pain management in labour. *Cochrane Database of Systematic Reviews, 4*. https://doi.org/10.1002/14651858.CD011989.pub2, CD011989.

Chapter 16

Abdominal pain in gastroparesis

Olubunmi Oladunjoye[a], Asad Jehangir[b,d], Adeolu Oladunjoye[c], Anam Qureshi[a,e], Zubair Malik[b], and Henry P. Parkman[b]

[a]*Department of Medicine, Reading Hospital-Tower Health System, Reading, PA, United States,* [b]*Gastroenterology Section, Department of Medicine, Temple University School of Medicine, Philadelphia, PA, United States,* [c]*Division of Medical Critical Care, Boston Children's Hospital, Boston, MA, United States,* [d]*Gastroenterology Section, Medical College of Georgia at Augusta University, Augusta, GA, United States,* [e]*Department of Medicine, Medical College of Georgia at Augusta University, Augusta, GA, United States*

List of abbreviations

COX-2	cyclooxygenase-2
CNS	central nervous system
ED	emergency department
EPS	extrapyramidal side effects
FDA	Food and Drug Administration
GCSI	gastroparesis cardinal symptom index
GCSI-DD	gastroparesis cardinal symptom index-daily diary
GES	gastric electric stimulators
GI	gastrointestinal
IBS	irritable bowel syndrome
NIDDK	National Institute of Diabetes and Digestive and Kidney Diseases
NPQ	Neuropathic Pain Questionnaire
PAGI-SYM	patient assessment of gastrointestinal symptoms
U.S.	United States
5-HT	5-hydroxytryptamine (serotonin)

Introduction

Over a third (41%) of the general population in the United States experience gastrointestinal (GI) symptoms, the most common of which is abdominal pain (22%) (Sandler, Stewart, Liberman, Ricci, & Zorich, 2000). Abdominal pain is the leading primary GI diagnosis for emergency department (ED) visits in the United States and poses a significant healthcare economic burden (Myer et al., 2013). Though often secondary to a benign etiology, abdominal pain can also be a symptom of an acute pathology. Detailed history, physical examination, and often additional investigations are needed to determine the underlying cause(s) of abdominal pain. On history/examination, the chronicity (acute or chronic) and localization (epigastric, right upper quadrant, left upper quadrant, lower, or diffuse) can help narrow the differential of abdominal pain. Chronic upper abdominal pain may be secondary to gastrointestinal etiologies (peptic ulcer disease, gastritis, irritable bowel syndrome, functional dyspepsia, or gastroparesis), pancreaticobiliary etiologies (pancreatitis or biliary colic), or cardiac etiologies (cardiac ischemia), among other causes.

Abdominal pain can occur in patients with gastroparesis; often it can be unrelated to the gastroparesis condition. However, abdominal pain can be an underrecognized symptom of gastroparesis (Hoogerwerf, Pasricha, Kalloo, & Schuster, 1999). Gastroparesis is a chronic disorder of the stomach characterized by delayed gastric emptying without evidence of any mechanical obstruction causing upper GI symptoms. Delayed gastric emptying in gastroparesis is secondary to abnormal neuromuscular function of the GI tract (Reddivari & Mehta, 2019). This neuromuscular dysfunction may be due to abnormalities in several elements including autonomic nervous system, enteric neurons, interstitial cells of Cajal, and smooth muscle cells (Camilleri et al., 2018). Gastroparesis has a female preponderance and may affect nearly 2% of the general population (Rey et al., 2012). There has been an increasing healthcare burden due to gastroparesis, secondary to a rise in the number of ED visits and hospitalizations (Jehangir & Parkman, 2018a; Wang, Fisher, & Parkman, 2008).

Features and Assessments of Pain, Anesthesia, and Analgesia. https://doi.org/10.1016/B978-0-12-818988-7.00007-8
Copyright © 2022 Elsevier Inc. All rights reserved.

Types of gastroparesis

Gastroparesis patients are often classified into four categories: idiopathic, diabetic, postsurgical, and atypical (Jehangir & Parkman, 2018b). Diabetes is the most prevalent known etiology of gastroparesis (Jehangir and Parkman, 2018b). Gastroparesis is more commonly seen in type 1 diabetes (up to 40%) than type 2 diabetes (up to 20%) (Jones et al., 2001; Reddivari & Mehta, 2019). Over half of the patients with gastroparesis have idiopathic gastroparesis (i.e. gastroparesis with no obvious etiology) (Jehangir et al., 2018b). Some patients may develop gastroparesis postsurgically due to vagal nerve injury (Jehangir et al., 2018b), or in atypical cases secondary to other disorders (Jehangir & Parkman, 2018a; Parkman, Hasler, & Fisher, 2004a). Some medications, particularly opiate analgesics, have also been implicated as iatrogenic causes of gastroparesis (Table 1).

Symptoms of gastroparesis

Gastroparesis symptoms can present at any age; the average age of presentation is in the 40s (Jehangir et al., 2018a; Parkman et al., 2011a). Symptoms of gastroparesis include nausea, vomiting, early satiety, postprandial fullness, and in some patients, upper abdominal discomfort or pain (Table 2). Nausea and vomiting are the most common predominant symptoms in gastroparesis patients. Patients with diabetic gastroparesis have more severe nausea and vomiting than patients with idiopathic gastroparesis (Camilleri et al., 2018). Abdominal pain appears more prominently in idiopathic and postsurgical gastroparesis. (Jehangir et al., 2018a, 2018b; Parkman et al., 2011a). The symptoms of gastroparesis, including abdominal pain, worsen in severity postprandially and also progressively worsen during the course of the day (Shahsavari et al., 2019). The symptoms of gastroparesis frequently impair the quality of lives of the patients and also have considerable burden on their caregivers (Jehangir, Collier, Shakhatreh, Malik, & Parkman, 2019; Yu et al., 2017).

Abdominal pain in gastroparesis

When questioned, the majority of gastroparesis patients report abdominal pain (Parkman et al., 2019). About two thirds (66%) of gastroparesis patients have at least moderate abdominal pain, and a third (34%) report severe abdominal pain (Hasler et al., 2013; Parkman, Wilson, et al., 2019). Abdominal pain is the predominant symptom in about a fifth (21%) of gastroparesis patients (Hasler et al., 2013). Nearly half (43%) of gastroparesis patients report experiencing abdominal pain daily, and over a third (38%) have constant abdominal pain (Cherian et al., 2010; Krause & Backonja, 2003). Female gender, nonwhite race, and idiopathic etiology without infectious prodrome are associated with more severe abdominal pain in gastroparesis (Friedenberg, Kowalczyk, & Parkman, 2013; Parkman et al., 2019). Abdominal pain in gastroparesis is typically located in the epigastric/periumbilical region and generally described as "cramping or sickening" pain (Cherian et al., 2010; Hoogerwerf et al., 1999; Yu et al., 2017).

In general, pain can be categorized into neuropathic and nociceptive pain. Neuropathic pain is secondary to direct nerve injury or sensitization from injury or inflammation of tissue. Nociceptive pain results from a noxious stimulus activating receptive pathways by damaging or threatening tissue damage. Nociceptive pain can be visceral pain or somatic/nonvisceral pain (e.g., musculoskeletal pain). The etiology of abdominal pain in gastroparesis may be multifactorial with neuropathic and/or nociceptive components (Jehangir, Abdallah, & Parkman, 2019) (Fig. 1). A study performed to characterize the abdominal pain in gastroparesis revealed that of 32 patients recruited, nearly two thirds of patients (62.5%), had somatic abdominal pain, over a third (35.5%) of the patients had neuropathic abdominal pain, and a fourth (25.8%) of the patients had visceral abdominal pain (Jehangir, Abdallah, & Parkman, 2019). This implies that abdominal pain in gastroparesis patients may be due to diverse causes including somatic etiologies (e.g., abdominal wall pain, myofascial pain, postsurgical trapped nerve), neuropathic etiologies (e.g., autonomic neuropathy), and visceral etiologies (possibly due to due delayed gastric emptying, gastric distension).

Abdominal pain in gastroparesis correlates poorly with the severity of delay in gastric emptying, suggesting that the abdominal pain is usually from other causes. In some of these patients, abdominal pain may be from gastric hypersensitivity, possibly from gastric distension (Karamanolis, Caenepeel, Arts, & Tack, 2007; Tack, 2002). Central mechanisms may be important in some patients with abdominal pain as studies using positron emission tomography have shown altered central processing to gastric distension in patients with dyspeptic symptoms (Van Oudenhove et al., 2010). Some gastroparesis patients may have other visceral causes to explain their symptom of abdominal pain. Among patients with chronic pancreatitis, over a third (44%) may have delayed gastric emptying, suggesting that the visceral abdominal pain in some

TABLE 1 Different types of gastroparesis.

Idiopathic gastroparesis[a]

Diabetic gastroparesis

Postsurgical gastroparesis[b]
- Fundoplication
- Roux-en-Y surgery
- Billroth II
- Gastrectomy
- Pancreatic surgery
- Heart-lung transplant

Atypical gastroparesis
Endocrine disorders
- Hypothyroidism

Autoimmune disorders
- Myasthenia gravis
- Sjogren's syndrome
- Systemic lupus erythematosus

Neurological disorders
- Parkinson's disease
- Multiple sclerosis

Connective tissue disorders
- Systemic sclerosis
- Ehlers–Danlos syndrome
- Polymyositis/dermatomyositis

Degenerative disorders
- Amyloidosis

Eating disorders
- Anorexia nervosa
- Bulimia nervosa

Iatrogenic causes
- Opioids
- Anticholinergics
- Glucagon-like peptide 1 (GLP-1) analogues (e.g., exenatide, liraglutide)
- Amylin analogues (e.g., pramlintide)
- Calcium channel blockers
- Alpha 2 agonists (e.g., clonidine)
- Dopamine agonists
- Phenothiazine
- Cyclosporine
- Octreotide
- Lithium
- Progesterone

[a]*Some patients with idiopathic gastroparesis may have onset of gastroparesis symptoms post viral infection including cytomegalovirus, Epstein–Barr virus, varicella zoster virus, herpes zoster, Norwalk virus, Hawaiian virus, and rotavirus infections.*
[b]*In the past, postsurgical gastroparesis was generally seen after ulcer surgery, now more commonly after fundoplication for gastroesophageal reflux disease treatment and bariatric surgeries for obesity treatment.*

gastroparesis patients may be contributed by chronic pancreatitis (Chowdhury, Forsmark, Davis, Toskes, & Verne, 2003). Determining the underlying cause of abdominal pain in gastroparesis patients may help devise appropriate treatment strategies (Jehangir, Abdallah, & Parkman, 2019).

Abdominal pain often affects the quality of lives of gastroparesis patients (Yu et al., 2017). Over two thirds (74%) of gastroparesis patients have nocturnal abdominal pain frequently impairing their sleep (Cherian et al., 2010; Krause & Backonja, 2003). Severe abdominal pain in gastroparesis patients associates with higher anxiety, depression, and somatization scores (Parkman, Wilson, et al., 2019; Hasler et al., 2013). Half of all these gastroparesis patients who report abdominal pain as an important symptom of gastroparesis are hypervigilant to their pain (Jehangir, Abdallah, & Parkman, 2019).

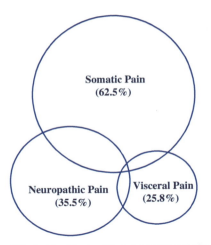

FIG. 1 Abdominal pain in gastroparesis may be multifactorial with somatic, neuropathic and visceral components Jehangir, Abdallah, and Parkman (2019).

TABLE 2 The frequency of common upper gastrointestinal symptoms in patients with gastroparesis.

Symptom	Frequency
Stomach fullness	97.3%
Nausea	92.9%–96%
Postprandial fullness	96%
Loss of appetite	91.1%
Bloating	93.8%
Abdominal pain	89.3%–90%
Abdominal distension	88.4%
Early satiety	85.7%
Retching	70.7%
Vomiting	67.9%–68.9%

Frequency of symptoms as reported in Cherian, Sachdeva, Fisher, and Parkman (2010); Hoogerwerf et al. (1999); Jehangir et al. (2018b); Parkman, Wilson, et al. (2019).

Symptom assessment in gastroparesis

There are validated questionnaires to assess symptom severity in gastroparesis. One of the most commonly used questionnaires includes Patient Assessment of Gastrointestinal Symptoms (PAGI-SYM), in which patients rate the severity of 20 common upper GI symptoms, including upper and lower abdominal pain, over the past 2 weeks on a 0–5 scale (0 = none, 1 = very mild, 2 = mild, 3 = moderate, 4 = severe, and 5 = very severe). PAGI-SYM symptoms are categorized into six-symptom subscales: nausea/vomiting subscale, postprandial fullness/early satiety subscale and bloating subscale, upper abdominal pain subscale, lower abdominal pain subscale, and heartburn/regurgitation subscale. PAGI-SYM has good reproducibility and reliability in assessing symptoms of gastroparesis, dyspepsia, and gastroesophageal reflux disease (Rentz et al., 2004). The Gastroparesis Cardinal Symptom Index (GCSI) consists of three subscales of PAGI-SYM (nausea/vomiting subscale, postprandial fullness/early satiety subscale and bloating subscale); the GCSI is contained in the larger PAGI-SYM. The PAGI-SYM and GCSI use a 2-week recall of symptoms. To better capture ongoing symptoms of gastroparesis, which could be either persistent or with periodic worsening and/or improvement, the Gastroparesis Cardinal Symptom Index-Daily Diary (GCSI-DD) has been developed (Hasler, 2011; Revicki et al., 2009). This questionnaire grades the severity of nausea, early satiety, postprandial fullness, and upper abdominal pain on a 0–4 scale (0 = none, 1 = mild, 2 = moderate, 3 = severe, and 4 = very severe), and also records the number of episodes of vomiting to assess these symptoms over the past 24 h (Revicki et al., 2009). GCSI-DD, but not GCSI, assesses the symptom of abdominal pain. The neuropathic pain questionnaire (NPQ) can help assess the type of abdominal pain in gastroparesis, with sensitivity and specificity of 66.6% and 74.4%, respectively, to distinguish neuropathic and nociceptive pain (Cherian et al., 2010; Krause & Backonja, 2003). The pain vigilance and awareness questionnaire can help measure vigilance to abdominal pain in gastroparesis patients with abdominal pain (Jehangir, Abdallah, & Parkman, 2019; McCracken, 1997).

Physical examination of gastroparesis patients

On physical examination, succussion splash (a sloshing sound heard on gastric auscultation during side-to-side movement of the patient or rapid palpation of the stomach) may indicate fluid in the stomach from either delayed gastric emptying or gastric outlet obstruction (Parkman, Hasler, & Fisher, 2004b). On examination of gastroparesis patients with abdominal pain, Carnett's sign may help characterize the abdominal pain. Nearly two thirds of gastroparesis patients have a positive Carnett's sign suggesting somatic pain (Jehangir, Abdallah, & Parkman, 2019). The test was first described by Carnett (1926). This is performed by identifying the point of maximum tenderness on abdominal exam. The patient is then asked either by raising and extending both legs or the upper body unaided by the patient's arms to tense the abdominal wall. A positive test (worsening pain on raising legs and/or upper body) suggests that the source of the pain is from the abdominal wall (muscle contraction increases the pain), and a negative test suggests that the source of the pain is intraabdominal.

Diagnosis of gastroparesis

It may be difficult to differentiate between gastroparesis and other upper GI disorders based on symptomatology alone. To exclude other causes of dyspetic symptoms, an esophagogastroduodenoscopy (EGD) is usually performed to look for especially ulcer disease. If clinically indicated, an abdominal obstruction series, abdominal ultrasound, or abdominal/pelvic computerized tomography can be done (Parkman et al., 2004a). Metabolic changes should be investigated including hypothyroidism and diabetes, and pregnancy in females (Camilleri, Parkman, Shafi, Abell, & Gerson, 2013). Other blood tests that should be checked include complete blood count and comprehensive metabolic panel (Jehangir et al., 2018a; H. Parkman et al., 2004a). Among patients with gastroparesis, neurologic or autoimmune disorders that can delay gastric emptying should be considered (Camilleri et al., 2013).

Gastric emptying can be assessed with gastric emptying scintigraphy, gastric emptying breath test, and wireless motility capsule (Camilleri et al., 2012; Szarka et al., 2008). Some patients with symptoms of gastroparesis may have normal overall gastric emptying but impaired gastric accommodation (Park et al., 2017). Assessment of gastric accommodation can be assessed using a single-photon emission computed tomography or magnetic resonance imaging (Bouras et al., 2002; Fidler et al., 2009), or indirectly with the use of gastric emptying scintigraphy or liquid nutrient drink tests.

There is overlap of gastroparesis with functional dyspepsia (Parkman et al., 2011b; Stanghellini & Tack, 2014). Gastric emptying assessment may help distinguish gastroparesis and functional dyspepsia, although of patients meeting criteria for functional dyspepsia, about 25% can have delayed gastric emptying (Vanheel et al., 2017).

Treatment of gastroparesis

Common treatment strategies for gastroparesis patients include dietary modification and pharmacologic treatment such as prokinetic and antiemetic medications. Endoscopic treatments such as pyloric Botulinum toxin injections and pyloric balloon dilations usually do not cause persistent symptom improvement. In cases refractory to medical management, surgical treatments such as pyloric surgeries and gastric stimulations can be considered. Some patients may need enteral nutrition through feeding tubes to maintain their caloric intake (Table 3). Antiemetics, prokinetics, and surgical treatment options for gastroparesis may not directly improve abdominal pain, though it may indirectly help alleviate abdominal pain in some patients by decreasing nausea/vomiting and subsequent excessive contractions of abdominal wall muscles. In this section, we discuss the common treatment strategies used for gastroparesis patients and their effects on abdominal pain.

Dietary modifications

Dietary modification is regarded as the first line of management of gastroparesis. Gastroparesis patients are instructed to eat small frequent meals to maintain adequate caloric intake and avoid high-fiber and indigestible foods as these can delay

178 PART | II The syndromes of pain

TABLE 3 Management of patients with gastroparesis symptoms, including symptom assessment, diagnostic work-up, and treatment options.

Symptom assessment	Diagnostic work-up	Treatment options
Questionnaires – PAGI-SYM – GCSI – GSCI-DD Physical examination – Succussion splash	Serological – CBC – CMP – TSH – HbA1c Endoscopic – EGD Imaging – GES – GEBT – WMC	Dietary modification – Small frequent meals – Low fiber, low fat diet Pharmacological options – Metoclopramide[a] – Domperidone[a] – Erythromycin Endoscopic options – Botulinum toxin injection – Pyloric balloon dilation Surgical options – Pyloromyotomy[a] – Gastric Stimulation[a] – Feeding tubes (G-, J-, G-J tubes)

[a]*May help improve abdominal pain in some gastroparesis patients.*
Abbreviations: CBC, *complete blood count;* CMP, *complete metabolic panel;* EGD, *esophagogastroduodenoscopy;* G, *gastric;* GCSI, *gastroparesis cardinal symptom index;* GCSI-DD, *gastroparesis cardinal symptom index- daily diary;* GEBT, *gastric emptying breath test;* GES, *gastric emptying scintigraphy;* GJ, *gastrojejunal;* HbA1C, *hemoglobin a1c;* J, *jejunal;* PAGI-SYM, *patient assessment of gastrointestinal symptoms;* TSH, *thyroid stimulating hormone;* WMC, *wireless motility capsule.*

gastric emptying (Camilleri et al., 2013; Consortium, 2011). Patients unable to tolerate solid foods are encouraged to take liquid diet since gastric emptying to liquids may be preserved in some patients with gastroparesis (Camilleri et al., 2013).

Pharmacologic treatments

In addition to dietary modification, most gastroparesis patients also need pharmacological treatments to manage their symptoms. The available pharmacological options are classified broadly into two groups: prokinetics and antiemetic. Prokinetic agents include metoclopramide, erythromycin, and domperidone (McCallum & George, 2001). Metoclopramide and domperidone also have antiemetic properties.

Prokinetic medications

Metoclopramide: Metoclopramide is a dopamine type 2 receptor antagonist with both peripheral (mainly the stomach) and central nervous system (CNS) actions. In a randomized, double-blind, placebo-controlled study by Perkel et al., metoclopramide improved symptoms of gastroparesis including epigastric pain (Perkel, Moore, Hersh, & Davidson, 1979). In another randomized placebo-controlled trial of gastroparesis patients, metoclopramide nasal spray improved gastrointestinal symptoms including epigastric pain in females, but not males (Parkman, Carlson, & Gonyer, 2015). Metoclopramide can have several side effects including extrapyramidal side effects (EPS), anxiety, restlessness, hyperprolactinemia, and QT prolongation (Reddivari & Mehta, 2019). In the United States, metoclopramide is suggested to be only taken for a maximum of 3 months due to the risk of tardive dyskinesia (Camilleri et al., 2018). However, some patients may need it for longer periods for their chronic condition and due to paucity of other pharmacological treatment options for gastroparesis.

Domperidone: Domperidone is a peripheral dopamine receptor antagonist. A recent study of 115 gastroparesis patient treated with domperidone found improvement in gastroparesis symptoms in over half (60%) of patients, including severity of abdominal pain (Schey et al., 2016). One study suggested that domperidone may help improve abdominal pain severity by decreasing visceral hypersensitivity (Bradette, Pare, Douville, & Morin, 1991). Domperidone is less likely to cause EPS than metoclopramide because it does not readily pass the blood–brain barrier. Unfortunately, domperidone has been associated with prolongation of QT interval with concerns of cardiac arrhythmias. Other side effects of domperidone include hyperprolactinemia (causing breast discharge and change in menstrual function) and headaches (Reddivari & Mehta, 2019). Its use is restricted in the United States and is prescribed to patients with symptoms refractory to other pharmacological options. The U.S. Food and Drug Administration (FDA) requires an investigational new drug application, which is a request to use this drug for special situations. An Institutional Review Board approval is also required before the medication can be commenced. The patient needs to pay for the medication, as insurance does not cover this. Before and during treatment, potassium and magnesium levels and an electrocardiogram need to be monitored because of concerns of adverse side effects (Jehangir et al., 2018a).

Erythromycin: Erythromycin is a macrolide antibiotic that also acts on gastroduodenal motilin receptors. This action is responsible for increasing antral contractility and initiation of the migrating motor complex in the upper GI tract. Erythromycin increases the amplitude of the stomach fundal area. Although this medication is not approved for treatment of gastroparesis, it is used as an off-label medication for gastroparesis, usually at low doses. Erythromycin was found to improve symptoms of gastroparesis in 43% of patients (Maganti, Onyemere, & Jones, 2003). A study by Arts et al. showed that erythromycin may improve postprandial bloating in gastroparesis patients, but not other symptoms including abdominal pain (Arts, Caenepeel, Verbeke, & Tack, 2005). Erythromycin can cause side effects such as ototoxicity and GI side effects (nausea, vomiting, diarrhea) (Reddivari & Mehta, 2019). Erythromycin is associated with tachyphylaxis if used over a long period of time, where its efficacy can diminish. Stopping the medication for a time period, and then restarting, may help if this develops. A week drug holiday period every month is often used to prevent this.

Endoscopic and surgical treatments

Botulinum toxin: Botulinum toxin, a potent inhibitor of acetylcholine, may decrease pyloric sphincter spasm in some patients. Botulinum toxin A is more commonly used in gastroparesis patients. A study using Botulinum toxin injection in 179 gastroparesis patients found a decrease in gastroparesis symptoms in 52% of the patients, with higher efficiency in patients receiving 200 units of Botulinum toxin compared to 100 units (Coleski, Anderson, & Hasler, 2009). However, randomized placebo-controlled trials have not shown benefit of botulinum toxin injections for gastroparesis patients (Friedenberg, Palit, Parkman, Hanlon, & Nelson, 2008), and botulinum injections are not routinely recommended for management of gastroparesis patients by the American College of Gastroenterology (Camilleri et al., 2013). In the authors' experience, symptomatic improvement with Botulinum toxin injection can occur in some patients. Treatment response generally lasts up to 6 months (Reichenbach et al., 2019). A recent article suggests that those with impaired pyloric compliance (<10 mm^2/mm Hg as assessed on Endoscopic Functional Luminal Imaging Probe) have better improvement of gastroparesis symptoms and quality of life with pyloric Botulinum toxin injection, than patients with normal pyloric compliance (Desprez et al., 2019). Response to botulinum toxin injection may identify gastroparesis patients who could benefit from surgical options aimed at pylorus such as pyloroplasty and pyloromyotomy.

Pyloroplasty/pyloromyotomy: Pyloroplasty and pyloromyotomy (laparoscopic or peroral pyloroplasty or gastric peroral endoscopic myotomy) may be utilized in gastroparesis patients with symptoms refractory to medical management (Camilleri et al., 2018). In a study of 107 gastroparesis patients who had laparoscopic pyloroplasty, the majority (90%) of patients had improvement in gastric emptying (Shada et al., 2016). These patients had improvement of several important symptoms of gastroparesis on 3-month follow-up, including abdominal pain (Shada et al., 2016). In a recent single-center study, 25 patients undergoing pyloric surgeries for refractory symptoms of gastroparesis had improvement of symptoms including abdominal pain (Zoll et al., 2019). Endoscopic performance of the pyloromyotomy (G-POEM) might improve morbidity of the procedure compared to laparoscopic approach.

Gastric electrical stimulation: Gastric electrical stimulators are implantable neurostimulators using stimulating electrodes placed surgically into the gastric muscularis propria (Reddivari & Mehta, 2019). The Enterra GES (Medtronic, Inc.) delivers high-frequency, low-energy signals with short pulses (Jehangir et al., 2018a). GES is made up of a pair of leads which are connected to a pulse generator (Reddivari & Mehta, 2019). Enterra GES was approved by the FDA in 2000 as a humanitarian use device to treat chronic intractable (drug refractory) nausea and vomiting due to diabetic or idiopathic gastroparesis in patients 18–70 years of age (Camilleri et al., 2018). GES can be effective for symptoms of nausea and vomiting in gastroparesis patients (Camilleri et al., 2013; Zoll et al., 2018). In a recent study of refractory gastroparesis cases, GES improved symptoms in up to 75% of patients, and 43% of refractory cases of gastroparesis had at least moderate improvement in their symptoms (Heckert, Sankineni, Hughes, Harbison, & Parkman, 2016). Numerous studies show that GES is more effective for diabetic gastroparesis patients than nondiabetics (Jehangir et al., 2018a). A study by Maranki et al. showed that GES improves primarily from nausea and vomiting and may not be effective for gastroparesis patients with a main symptom of abdominal pain (Maranki et al., 2008). GES was also more effective in diabetic gastroparesis rather than idiopathic gastroparesis. However, some studies have suggested that GES may improve severity of abdominal pain in some gastroparesis patients (Heckert et al., 2016; Zoll et al., 2019).

Jejunostomy and venting gastrostomy tubes: If gastroparesis patients are unable to maintain intake orally, enteral feeding via jejunostomy tube (or J-tube) or gastrojejunal tube (G-J tube) should be considered. Indications for enteral nutrition include: (1) unintentional loss of 10% or more of the usual body weight during a period of 3–6 months, and/ or (2) repeated hospitalizations for refractory symptoms (Camilleri et al., 2013). After placement of J-tubes, gastroparesis patients with severe refractory symptoms report improvement of nausea/vomiting, nutritional status, overall health, and hospitalization rates (Fontana & Barnett, 1996). Complications from J-tube can include intestinal obstruction, dislodgement, wound infection, and cellulitis. A venting gastrostomy tube (G-tube) to decompress the distended stomach may help alleviate the symptoms (Kim & Nelson, 1998).

Gastrectomy: In end-stage gastroparesis patients, gastrectomy can be considered to remove the gastroparetic stomach. Typically, this is only considered in patients who have failed several other pharmacological and surgical treatments given the risk of postoperative adverse events. In subtotal gastrectomy, 70% of the stomach including the antrum and pylorus are resected, with closure of the duodenum and restoration of continuity with a Roux-en-Y jejunal loop. In a study by Zehetner et al., subtotal gastrectomy more commonly improved symptoms compared to GES (87% vs 63%, respectively); however, the postoperative 30-day morbidity was higher (23% vs 8%, respectively). Patients who failed to respond to GES had favorable outcomes with subsequent subtotal gastrectomy (Zehetner et al., 2013). In the authors' practice, we try not to resort to gastrectomy, as patients can have postsurgical problems or also have small bowel motility issues that then contribute to persistent symptoms after surgery.

Treatment of abdominal pain in gastroparesis

Treatment of abdominal pain in gastroparesis is challenging as the cause of abdominal pain is often not clear. In addition, the commonly used pharmacological and surgical treatments for gastroparesis may not directly improve abdominal pain. Hence, gastroparesis patients with moderate to severe abdominal pain often do not report symptom improvement on long-term follow-up (Pasricha et al., 2015). Treatment of abdominal pain in gastroparesis often needs a multidisciplinary approach to address diverse etiologies of abdominal pain as well as other disorders these patients may have including anxiety, depression, and hypervigilance. Here, we discuss various pharmacological options that may be utilized in gastroparesis patients with abdominal pain. We also emphasize that if patients with gastroparesis have abdominal pain, one should entertain other etiologies of abdominal pain and not assume it is from their gastroparesis (see Table 4). One important question is if the abdominal pain is chronic, associated with their other gastroparesis symptoms, or is it a new symptom, not associated with their gastroparesis symptoms, suggesting an unrelated cause.

Neuromodulators

Antidepressants

Tricyclic antidepressants (TCAs) besides being used to treat depression are given for a variety of pain conditions, including migraine headaches and fibromyalgia. TCAs may help reduce severity of abdominal pain in patients with dyspeptic symptoms, possibly through their central effects and unrelated to their antidepressant effects (Table 5). TCAs are of two types: tertiary or secondary amines. Secondary amines have fewer anticholinergic effects and less gastrointestinal side effects. In a study by Mertz et al., amitriptyline, a tertiary amine, decreased abdominal pain in patients with functional dyspepsia (Mertz et al., 1998). In a randomized, comparator-controlled, multicenter study by Drossman et al. of 431 adults with functional bowel disorders, including functional abdominal pain, patients taking desipramine, a secondary amine, had significant improvement of symptoms compared to placebo per protocol analysis, but not by intention-to-treat analysis (Drossman et al., 2003). The GpCRC nortriptyline study for patients with Idiopathic Gastroparesis (NORIG Trial) did not show significant improvement of gastroparesis symptoms with nortriptyline, another secondary amine, after 15 weeks (Parkman et al., 2013). Although there was initial improvement in abdominal pain after 3 weeks with low dose (10 mg), this was not sustained over time (Parkman et al., 2013).

Mirtazapine, an atypical antidepressant with central adrenergic and serotonergic activity, and direct antiemetic activity through 5-HT3 antagonist activity, may be useful in some gastroparesis patients with abdominal pain. Recent studies have shown beneficial effects of mirtazapine in patients with gastroparesis and functional dyspepsia (Malamood, Roberts, Kataria, Parkman, & Schey, 2017; Tack et al., 2016). Other studies suggest that it might be helpful for nausea and vomiting symptoms (Bhattacharjee, Doleman, Lund, & Williams, 2019).

TABLE 4 Assessment of gastroparesis patients with abdominal pain, including investigations to consider.

Symptom assessment	Neuropathic Pain Questionnaire
	  — Can help identify neuropathic abdominal pain
	Pain Vigilance and Awareness Questionnaire
	  — Can help identify patients who are hypervigilant to pain

Physical examination	Carnett's sign — To assess for possible somatic (e.g., musculoskeletal) etiologies
Serological work-up	Trypsin, amylase, lipase — To assess for pancreatic cause of pain (i.e. chronic pancreatitis)

Imaging	Abdominal obstruction series
	  — To rule out partial intestinal obstruction
	Ultrasonography
	  — To rule out hepatobiliary etiologies
	Computerized tomography
	  — To assess for pancreatic and hepatobiliary etiologies

Others	*Helicobacter pylori* urea breath or stool antigen testing — Some functional dyspepsia patients may have symptom improvement with *H. pylori* treatment

TABLE 5 Treatments used in patients with abdominal pain as an important symptom of gastroparesis.

Neuromodulators
Antidepressants
  Tricyclic antidepressants
    — Nortriptyline
    — Desipramine
    — Amitriptyline
    — Imipramine
    — Doxepin
  Atypical antidepressants
    — Mirtazapine
  5-Hydroxytryptamine 1 A receptor agonist
    — Buspirone

Antiepileptic agents
  — Gabapentin
  — Pregabalin

Antipsychotics
  — Haloperidol
  — Olanzapine

Antispasmodics
  — Dicyclomine
  — Hyoscyamine

Nonsteroidal antiinflammatory drugs
  — Indomethacin
  — Ketorolac

Cannabinoids
  — Dronabinol
  — Natural cannabinoids

Celiac plexus block

Specific treatments for abdominal wall pain
  — Transdermal lidocaine patches
  — Heating pads
  — Cutaneous nerve blocks

Buspirone, a serotonin 5-HT$_{1A}$ receptor agonist approved for anxiety, also relaxes proximal stomach in healthy individuals (Van Oudenhove, Kindt, Vos, Coulie, & Tack, 2008). In a randomized, double-blind, placebo-controlled, crossover study of 17 patients, buspirone improved symptoms in patients with functional dyspepsia by increasing gastric accommodation (Tack, Janssen, Masaoka, Farré, & Van Oudenhove, 2012). The National Institutes of Health (NIH) GpCRC is performing a larger double-blinded, placebo-controlled trial to study if buspirone can improve symptoms in gastroparesis patients, particularly with early satiety.

Antiepileptics

Antiepileptics, like gabapentin and pregabalin, are often used for peripheral neuropathic pain in diabetics (Eisenberg, River, Shifrin, & Krivoy, 2007). These drugs bind to the α-2 δ subunit of voltage-gated calcium channels blocking release of nociceptive neurotransmitters like substance P, norepinephrine, and glutamate, and subsequently producing antinociceptive effects. A third of gastroparesis patients with abdominal pain have neuropathic pain (Jehangir, Abdallah, & Parkman, 2019). The efficacy of gabapentin and pregabalin for abdominal pain in gastroparesis patients has not been assessed in clinical trials. These medications may be beneficial in treating abdominal pain in gastroparesis patients given their less significant effects on gastrointestinal motility. Pregabalin has been shown to be helpful in patients with irritable bowel syndrome (IBS). In a study by Saito et al., 85 patients were randomized to receive 225 mg of pregabalin or placebo twice daily for 12 weeks (Saito et al., 2019). The patients receiving pregabalin had improvement of abdominal pain, bloating, and diarrhea compared with the patients receiving placebo (Saito et al., 2019). Future studies assessing the efficacy of gabapentin and pregabalin in gastroparesis patients with abdominal pain of neuropathic origin will be helpful.

Antipsychotics

Antipsychotics like haloperidol may be used to manage gastroparesis patients with severe symptoms including abdominal pain, particularly in the acute ED setting. In a study by Ramirez et al., 52 diabetic gastroparesis patients presenting to the ED with nausea, vomiting, or abdominal pain who received 5 mg of intramuscular haloperidol required lower doses of opioid analgesics (median morphine equivalent doses of analgesia 6.8 mg vs 10.8 mg, $P < 0.01$) and less frequent hospitalizations (9.6% vs 26.9%, $P = 0.02$) compared to their most recent previous presentation to the ED when they did not receive haloperidol (Ramirez, Stalcup, Croft, & Darracq, 2017). In a randomized, double-blind, placebo-controlled trial of gastroparesis patients presenting to the ED, 15 patients receiving 5 mg of haloperidol intravenously had significant reductions in the severity of abdominal pain and nausea 1 h after the administration of haloperidol (both $P \leq 0.01$), while 18 patients receiving placebo did not have significant reduction in the severity of these symptoms (Roldan et al., 2017). In a recent retrospective study at our tertiary care center, over half (56.6%) of 280 patients presenting with nausea, vomiting, and/or abdominal pain to the ED who received haloperidol (average dose of 2.5 mg, mostly intravenously) were successfully discharged home (Shahsavari et al., 2021 [Abstract]). These patients had more severe symptoms at presentation, and more frequently reported opioid and cannabinoid use compared to patients presenting with similar symptoms to the ED who did not receive haloperidol.

Antispasmodics

Antispasmodics, like dicyclomine and hyoscyamine, may be used to address abdominal pain in some patients with gastroparesis who also have symptoms suggestive of IBS. Some patients with IBS experience epigastric pain corresponding to the location of the stomach and transverse colon. These antispasmodics, with their anticholinergic properties, help in the relaxation of gastrointestinal smooth muscles (Roblin et al., 2009) that may lessen the severity of abdominal pain. Hyoscyamine can be given in an orally dissolving tablet that might be beneficial in patients with gastroparesis. However, these anticholinergic medications may further impair gastric emptying.

Cannabinoids

A third of gastroparesis patients report cannabinoid use for symptoms of gastroparesis with majority perceiving benefit (Jehangir & Parkman, 2019). Younger gastroparesis patients with higher severity of gastroparesis symptoms including abdominal pain are more likely to report cannabinoid use (Parkman et al., 2019). A recent single-center study showed significant improvement of abdominal pain in 24 gastroparesis patients with refractory symptoms who were prescribed synthetic or natural cannabinoids (Barbash, Mehta, Siddiqui, Chawla, & Dworkin, 2019). However, cannabinoid use may also cause cannabinoid hyperemesis syndrome and has been associated with other acute side effects (anxiety, paranoia, impaired motor coordination, etc.) and long-term side effects (e.g., cognitive impairment, psychotic disorders, and potential for abuse and addiction) (Volkow, Baler, Compton, & Weiss, 2014). The regulations for medical use of marijuana vary across different states. In Pennsylvania, medical marijuana can often be given for neuropathic conditions and for severe chronic or intractable pain, but not for gastroparesis per se (Pennsylvania Department of Health, 2021). Further studies on the safety and efficacy of cannabinoid use in gastroparesis will be helpful.

Nonsteroidal antiinflammatory drugs

Nonsteroidal antiinflammatory drugs (NSAIDs) may be used for brief periods in selective gastroparesis patients for mild to moderate abdominal pain. Oral indomethacin has been shown to reduce hyperglycemia-associated gastric tachyarrhythmias (Hasler, Soudah, Dulai, & Owyang, 1995). Ketorolac can be given intramuscularly or intravenously in hospitalized patients. Chronic NSAID use may be associated with gastrointestinal (e.g., peptic ulcer disease), and extragastrointestinal (e.g., renal dysfunction) side effects, because of which these medications should be used cautiously for short intervals to address abdominal pain in selective gastroparesis patients. Cyclooxygenase-2 (COX-2) inhibitors like celecoxib have better dyspepsia-related tolerability than NSAIDs (Goldstein et al., 2002). COX-2 inhibitors were intended to reduce the side effects, but they can also have gastrointestinal (gastritis or heartburn) and extragastrointestinal side effects (hypertension, heart failure, etc.).

Opiate analgesics

In today's era of opioid problems, one avoids use of opioid analgesics in patients with gastroparesis. Physicians should be cautious about opioid use in gastroparesis as opioids can cause nausea and vomiting through their central effects on chemoreceptor trigger zone. Opioids can also delay gastric emptying and have been associated with higher severity of symptoms in gastroparesis patients (Jehangir & Parkman, 2017). Hence, opioids should be avoided in gastroparesis patients when possible. Unfortunately, many gastroparesis patients on chronic opioids use them for extragastrointestinal causes, and in our experience, these patients may not be able to come off opioids (Jehangir and Parkman, 2017).

A considerable minority of patients with gastroparesis who have abdominal pain report chronic opioid use, particularly patients with abdominal pain as their predominant symptom (Jehangir et al., 2017). In a study from Temple University Hospital on 223 patients referred for evaluation of gastroparesis, 43 (19.3%) patients reported chronic (>1 month) scheduled (at least once daily) opioid use (Jehangir et al., 2017). These gastroparesis patients with chronic opioid use had a higher severity of several important symptoms of gastroparesis including upper abdominal pain ($P < 0.01$) compared to gastroparesis patients without opioid use (Jehangir et al., 2017). These opioid-using gastroparestics also had less work productivity and more frequent hospitalizations in the prior 1 year (Jehangir et al., 2017). In another study by the NIDDK Gastroparesis Consortium interviewing 583 patients with gastroparesis, 41% of the patients reported opioid use (Hasler et al., 2019). These opioid-using gastroparesis patients had more severe symptoms (including abdominal pain), greater levels of gastric retention, more frequent hospitalizations, and worse quality of life compared to nonopioid users (Hasler et al., 2019). The deleterious effects of opioids were more pronounced with potent opioids (e.g., morphine, oxycodone, etc.) than weak opioids (e.g., tramadol, codeine, etc.) (Hasler et al., 2019).

Some patients taking opioids chronically may develop narcotic bowel syndrome (NBS). NBS involves worsening abdominal pain associated with chronic or increasing use of opioids, likely mediated at the CNS level (Szigethy, Schwartz, & Drossman, 2014). Treatment of NBS includes gradual tapering and eventual detoxification from opioids, utilizing nonopioid pharmacotherapy during detoxification phase such as antidepressants, benzodiazepines, and clonidine (Szigethy et al., 2014).

A considerable proportion (40%–90%) of patients on chronic opioids experience opioid-induced constipation (OIC) (Chey et al., 2014). OIC may not respond to the routinely used laxatives. Peripherally acting μ-opioid receptor antagonists (PAMORAs), e.g., methylnaltrexone and naloxegol, may be used for OIC in some patients with chronic pain who are unable to come off opioids (Pergolizzi Jr et al., 2017). In a 4-week, double-blind, randomized, placebo-controlled study assessing the effect of subcutaneous methylnaltrexone (12 mg daily or every other day) in 460 patients, methylnaltrexone (both daily and every other day dosing) resulted in greater improvement in constipation-related quality of life compared to placebo (Michna et al., 2011). In two identical phase 3, double-blind studies on patients on opioids for chronic noncancer pain experiencing OIC (total participants 1352), naloxegol (12.5 mg or 25 mg) resulted in higher response rate than placebo

Celiac plexus block

Postganglionic sympathetic innervation to the stomach and small intestine is provided by the celiac and superior mesenteric ganglia. Celiac plexus block, performed by instilling ethanol and bupivacaine into the plexus via fluoroscopic, computed tomography, or endoscopic ultrasound guidance, has been utilized for chronic abdominal pain from chronic pancreatitis and intraabdominal malignancy (Singh & Toskes, 2003). Anecdotal use of celiac plexus block for abdominal pain in gastroparesis patients has been reported.

Treatment of chronic abdominal wall pain

Many patients with gastroparesis have a positive Carnett's sign, suggesting abdominal wall pain (Jehangir, Abdallah, & Parkman, 2019). Some of these patients may benefit from transdermal lidocaine patches, heating pads, and cutaneous nerve blocks to address their somatic abdominal pain.

Conclusion

Abdominal pain is frequently present in gastroparesis and may have multifactorial etiologies, including causes unrelated to their gastroparesis. A detailed assessment of these patients may help elucidate the underlying cause(s) of abdominal pain in these patients (somatic, neuropathic, and/or visceral). This may help direct appropriate treatment strategies that could improve outcomes of these patients. Unfortunately, in many patients, the cause of the abdominal pain is not known. The commonly used pharmacological treatments for gastroparesis, and surgical treatments reserved for more severe cases, may not directly improve abdominal pain in these patients. Various treatment modalities that may be utilized in attempt to reduce abdominal pain in gastroparesis depend on the underlying etiology; these include antidepressants, antipsychotics, antiepileptics, and treatments directed toward chronic abdominal wall pain.

Mini-dictionary of terms

1. Gastroparesis: Delayed gastric emptying in the absence of mechanical obstruction.
2. PAGI-SYM: Patient Assessment of Gastrointestinal Symptoms questionnaire in which patients rate the severity of 20 common upper GI symptoms, including upper and lower abdominal pain, over the past 2 weeks on a 0–5 scale (0 = none, 1 = very mild, 2 = mild, 3 = moderate, 4 = severe, and 5 = very severe).
3. Neuropathic pain: Pain that is secondary to direct nerve injury or sensitization from injury or inflammation of tissue.
4. Nociceptive pain: Pain that results from a noxious stimulus activating receptive pathways by damaging or threatening tissue damage.
5. Neuromodulators: antidepressants, antipsychotics, antiepileptics, and other nervous system-targeted medications.

Key facts

- The common symptoms of gastroparesis include nausea, vomiting, early satiety, and postprandial fullness. Sometimes, patients can have abdominal pain.
- Abdominal pain is an underrecognized symptom of gastroparesis. It should be kept in mind that other disorders can cause abdominal pain in patients with gastroparesis.
- Patients with idiopathic and postsurgical gastroparesis present with more severe abdominal pain than diabetic and atypical forms of gastroparesis.
- Diagnosis of gastroparesis is confirmed by delayed gastric emptying in the absence of mechanical obstruction.
- Common treatment strategies for gastroparesis patients include dietary modification and pharmacologic treatment such as prokinetic and antiemetic medications. Surgical treatment options are considered for refractory cases.
- Abdominal pain in gastroparesis is often difficult to manage. Neuromodulators are often used, avoiding the use of narcotic analgesics.

Summary points

- This chapter focuses on abdominal pain in gastroparesis, a chronic disorder of the stomach characterized by delayed gastric emptying without evidence of any mechanical obstruction, causing upper gastrointestinal symptoms.
- Many gastroparesis patients report abdominal pain which may be multifactorial with neuropathic and/or nociceptive components.
- Carnett's sign on physical examination can help characterize abdominal pain in gastroparesis.
- Treatment of abdominal pain in gastroparesis is challenging as the cause of abdominal pain is often not clear.
- Determining the underlying cause of abdominal pain in gastroparesis patients may help direct appropriate treatment strategies.
- The commonly used pharmacological treatments for gastroparesis, and surgical treatments for more severe cases, may not directly improve the abdominal pain.
- Various treatment modalities that may be utilized to address abdominal pain in gastroparesis depending on the underlying etiology include antidepressants, antipsychotics, antiepileptics, and treatments directed toward chronic abdominal wall pain.

References

Pennsylvania Department of Health. (2021). *Medical marijuana patient and caregiver resources*. Retrieved from https://www.health.pa.gov/topics/programs/Medical%20Marijuana/Pages/Patients.aspx.

Arts, J., Caenepeel, P., Verbeke, K., & Tack, J. (2005). Influence of erythromycin on gastric emptying and meal related symptoms in functional dyspepsia with delayed gastric emptying. *Gut, 54*(4), 455–460.

Barbash, B., Mehta, D., Siddiqui, M. T., Chawla, L., & Dworkin, B. (2019). Impact of cannabinoids on symptoms of refractory gastroparesis: A single-center experience. *Cureus, 11*(12).

Bhattacharjee, D., Doleman, B., Lund, J., & Williams, J. (2019). Mirtazapine for postoperative nausea and vomiting: Systematic review, meta-analysis, and trial sequential analysis. *Journal of Perianesthesia Nursing, 34*(4), 680–690.

Bouras, E., Delgado-Aros, S., Camilleri, M., Castillo, E., Burton, D., Thomforde, G., et al. (2002). SPECT imaging of the stomach: Comparison with barostat, and effects of sex, age, body mass index, and fundoplication. *Gut, 51*(6), 781–786.

Bradette, M., Pare, P., Douville, P., & Morin, A. (1991). Visceral perception in health and functional dyspepsia. *Digestive Diseases and Sciences, 36*(1), 52–58.

Camilleri, M., Chedid, V., Ford, A. C., Haruma, K., Horowitz, M., Jones, K. L., et al. (2018). Gastroparesis. *Nature Reviews Disease Primers, 4*(1), 1–19.

Camilleri, M., Iturrino, J., Bharucha, A. E., Burton, D., Shin, A., JEONG, I. D., et al. (2012). Performance characteristics of scintigraphic measurement of gastric emptying of solids in healthy participants. *Neurogastroenterology & Motility, 24*(12), 1076–e1562.

Camilleri, M., Parkman, H. P., Shafi, M. A., Abell, T. L., & Gerson, L. (2013). Clinical guideline: Management of gastroparesis. *The American Journal of Gastroenterology, 108*(1), 18.

Carnett, J. (1926). Intercostal neuralgia as a cause of abdominal pain and tenderness. *Surgery, Gynecology & Obstetrics, 24*, 625–632.

Cherian, D., Sachdeva, P., Fisher, R. S., & Parkman, H. P. (2010). Abdominal pain is a frequent symptom of gastroparesis. *Clinical Gastroenterology and Hepatology, 8*(8), 676–681.

Chey, W. D., Webster, L., Sostek, M., Lappalainen, J., Barker, P. N., & Tack, J. (2014). Naloxegol for opioid-induced constipation in patients with non-cancer pain. *New England Journal of Medicine, 370*(25), 2387–2396.

Chowdhury, R. S., Forsmark, C. E., Davis, R. H., Toskes, P. P., & Verne, G. N. (2003). Prevalence of gastroparesis in patients with small duct chronic pancreatitis. *Pancreas, 26*(3), 235–238.

Coleski, R., Anderson, M. A., & Hasler, W. L. (2009). Factors associated with symptom response to pyloric injection of botulinum toxin in a large series of gastroparesis patients. *Digestive Diseases and Sciences, 54*(12), 2634–2642.

Consortium, N. G. C. R. (2011). Bloating in gastroparesis: Severity, impact, and associated factors. *The American Journal of Gastroenterology, 106*(8), 1492.

Desprez, C., Melchior, C., Wuestenberghs, F., Zalar, A., Jacques, J., Leroi, A.-M., et al. (2019). Pyloric distensibility measurement predicts symptomatic response to intrapyloric botulinum toxin injection. *Gastrointestinal Endoscopy, 90*(5), 754–760. e751.

Drossman, D. A., Toner, B. B., Whitehead, W. E., Diamant, N. E., Dalton, C. B., Duncan, S., et al. (2003). Cognitive-behavioral therapy versus education and desipramine versus placebo for moderate to severe functional bowel disorders. *Gastroenterology, 125*(1), 19–31.

Eisenberg, E., River, Y., Shifrin, A., & Krivoy, N. (2007). Antiepileptic drugs in the treatment of neuropathic pain. *Drugs, 67*(9), 1265–1289.

Fidler, J., Bharucha, A. E., Camilleri, M., Camp, J., Burton, D., Grimm, R., et al. (2009). Application of magnetic resonance imaging to measure fasting and postprandial volumes in humans. *Neurogastroenterology & Motility, 21*(1), 42–51.

Fontana, R., & Barnett, J. (1996). Jejunostomy tube placement in refractory diabetic gastroparesis: A retrospective review. *American Journal of Gastroenterology, 91*(10).

Friedenberg, F. K., Kowalczyk, M., & Parkman, H. P. (2013). The influence of race on symptom severity and quality of life in gastroparesis. *Journal of Clinical Gastroenterology, 47*(9), 757–761.

Friedenberg, F. K., Palit, A., Parkman, H. P., Hanlon, A., & Nelson, D. B. (2008). Botulinum toxin a for the treatment of delayed gastric emptying. *American Journal of Gastroenterology, 103*(2), 416–423.

Goldstein, J. L., Eisen, G., Burke, T., Pena, B., Lefkowith, J., & Geis, G. (2002). Dyspepsia tolerability from the patients' perspective: A comparison of celecoxib with diclofenac. *Alimentary Pharmacology & Therapeutics, 16*(4), 819–827.

Hasler, W. L. (2011). Gastroparesis: Pathogenesis, diagnosis and management. *Nature Reviews Gastroenterology & Hepatology, 8*(8), 438.

Hasler, W. L., Soudah, H. C., Dulai, G., & Owyang, C. (1995). Mediation of hyperglycemia-evoked gastric slow-wave dysrhythmias by endogenous prostaglandins. *Gastroenterology, 108*(3), 727–736.

Hasler, W. L., Wilson, L. A., Nguyen, L. A., Snape, W. J., Abell, T. L., Koch, K. L., et al. (2019). Opioid use and potency are associated with clinical features, quality of life, and use of resources in patients with gastroparesis. *Clinical Gastroenterology and Hepatology, 17*(7), 1285–1294. e1281.

Hasler, W., Wilson, L., Parkman, H., Koch, K., Abell, T., Nguyen, L., et al. (2013). Factors related to abdominal pain in gastroparesis: Contrast to patients with predominant nausea and vomiting. *Neurogastroenterology & Motility, 25*(5), 427–e301.

Heckert, J., Sankineni, A., Hughes, W. B., Harbison, S., & Parkman, H. (2016). Gastric electric stimulation for refractory gastroparesis: A prospective analysis of 151 patients at a single center. *Digestive Diseases and Sciences, 61*(1), 168–175.

Hoogerwerf, W., Pasricha, P. J., Kalloo, A. N., & Schuster, M. (1999). Pain: The overlooked symptom in gastroparesis. *The American Journal of Gastroenterology, 94*(4), 1029–1033.

Jehangir, A., Abdallah, R. T., & Parkman, H. P. (2019). Characterizing abdominal pain in patients with gastroparesis into neuropathic and nociceptive components. *Journal of Clinical Gastroenterology, 53*(6), 427–433.

Jehangir, A., Collier, A., Shakhatreh, M., Malik, Z., & Parkman, H. P. (2019). Caregiver burden in gastroparesis and GERD: Correlation with disease severity, healthcare utilization and work productivity. *Digestive Diseases and Sciences, 64*(12), 3451–3462.

Jehangir, A., & Parkman, H. P. (2017). Chronic opioids in gastroparesis: Relationship with gastrointestinal symptoms, healthcare utilization and employment. *World Journal of Gastroenterology, 23*(40), 7310.

Jehangir, A., & Parkman, H. (2018a). Gastroparesis. In *Encyclopedia of gastroenterology* (2nd ed., pp. 720–730). Cambridge, MA: Academic Press.

Jehangir, A., & Parkman, H. P. (2018b). Rome IV diagnostic questionnaire complements patient assessment of gastrointestinal symptoms for patients with gastroparesis symptoms. *Digestive Diseases and Sciences, 63*(9), 2231–2243.

Jehangir, A., & Parkman, H. P. (2019). Cannabinoid use in patients with gastroparesis and related disorders: Prevalence and benefit. *American Journal of Gastroenterology, 114*(6), 945–953.

Jones, K. L., Russo, A., Stevens, J. E., Wishart, J. M., Berry, M. K., & Horowitz, M. (2001). Predictors of delayed gastric emptying in diabetes. *Diabetes Care, 24*(7), 1264–1269.

Karamanolis, G., Caenepeel, P., Arts, J., & Tack, J. (2007). Determinants of symptom pattern in idiopathic severely delayed gastric emptying: Gastric emptying rate or proximal stomach dysfunction? *Gut, 56*(1), 29–36.

Kim, C. H., & Nelson, D. K. (1998). Venting percutaneous gastrostomy in the treatment of refractory idiopathic gastroparesis. *Gastrointestinal Endoscopy, 47*(1), 67–70.

Krause, S. J., & Backonja, M.-M. (2003). Development of a neuropathic pain questionnaire. *The Clinical Journal of Pain, 19*(5), 306–314.

Maganti, K., Onyemere, K., & Jones, M. P. (2003). Oral erythromycin and symptomatic relief of gastroparesis: A systematic review. *The American Journal of Gastroenterology, 98*(2), 259–263.

Malamood, M., Roberts, A., Kataria, R., Parkman, H. P., & Schey, R. (2017). Mirtazapine for symptom control in refractory gastroparesis. *Drug Design, Development and Therapy, 11*, 1035.

Maranki, J. L., Lytes, V., Meilahn, J. E., Harbison, S., Friedenberg, F. K., Fisher, R. S., et al. (2008). Predictive factors for clinical improvement with Enterra gastric electric stimulation treatment for refractory gastroparesis. *Digestive Diseases and Sciences, 53*(8), 2072–2078.

McCallum, R. W., & George, S. J. (2001). Gastric dysmotility and gastroparesis. *Current Treatment Options in Gastroenterology, 4*(2), 179–191.

McCracken, L. M. (1997). "Attention" to pain in persons with chronic pain: A behavioral approach. *Behavior Therapy, 28*(2), 271–284.

Mertz, H., Fass, R., Kodner, A., Yan-Go, F., Fullerton, S., & Mayer, E. (1998). Effect of amitryptiline on symptoms, sleep, and visceral perception in patients with functional dyspepsia. *The American Journal of Gastroenterology, 93*(2), 160–165.

Michna, E., Blonsky, E. R., Schulman, S., Tzanis, E., Manley, A., Zhang, H., et al. (2011). Subcutaneous methylnaltrexone for treatment of opioid-induced constipation in patients with chronic, nonmalignant pain: A randomized controlled study. *The Journal of Pain, 12*(5), 554–562.

Myer, P. A., Mannalithara, A., Singh, G., Singh, G., Pasricha, P. J., & Ladabaum, U. (2013). Clinical and economic burden of emergency department visits due to gastrointestinal diseases in the United States. *American Journal of Gastroenterology, 108*(9), 1496–1507.

Park, S.-Y., Acosta, A., Camilleri, M., Burton, D., Harmsen, S. W., Fox, J., et al. (2017). Gastric motor dysfunction in patients with functional gastroduodenal symptoms. *American Journal of Gastroenterology, 112*(11), 1689–1699.

Parkman, H. P., Carlson, M. R., & Gonyer, D. (2015). Metoclopramide nasal spray reduces symptoms of gastroparesis in women, but not men, with diabetes: Results of a phase 2B randomized study. *Clinical Gastroenterology and Hepatology, 13*(7), 1256–1263. e1251.

Parkman, H., Hasler, W., & Fisher, R. (2004a). American Gastroenterological Association medical position statement: Diagnosis and treatment of gastroparesis. *Gastroenterology, 127*(5), 1589–1591.

Parkman, H. P., Hasler, W. L., & Fisher, R. S. (2004b). American Gastroenterological Association technical review on the diagnosis and treatment of gastroparesis. *Gastroenterology, 127*(5), 1592–1622.

Parkman, H. P., Sharkey, E. P., Nguyen, L. A., Yates, K. P., Abell, T. L., Hasler, W. L., et al. (2019). Marijuana use in patients with symptoms of gastroparesis: Prevalence, patient characteristics, and perceived benefit. *Digestive Diseases and Sciences*, 1–10.

Parkman, H. P., Van Natta, M. L., Abell, T. L., McCallum, R. W., Sarosiek, I., Nguyen, L., et al. (2013). Effect of nortriptyline on symptoms of idiopathic gastroparesis: The NORIG randomized clinical trial. *JAMA, 310*(24), 2640–2649.

Parkman, H. P., Wilson, L. A., Hasler, W. L., McCallum, R. W., Sarosiek, I., Koch, K. L., et al. (2019). Abdominal pain in patients with gastroparesis: Associations with gastroparesis symptoms, etiology of gastroparesis, gastric emptying, somatization, and quality of life. *Digestive Diseases and Sciences, 64*(8), 2242–2255.

Parkman, H. P., Yamada, G., Van Natta, M. L., Yates, K., Hasler, W. L., Sarosiek, I., et al. (2019). Ethnic, racial, and sex differences in etiology, symptoms, treatment, and symptom outcomes of patients with gastroparesis. *Clinical Gastroenterology and Hepatology, 17*(8), 1489–1499. e1488.

Parkman, H., Yates, K., Hasler, W., Nguyen, L., Pasricha, P., Snape, W., et al. (2011a). National Institute of Diabetes and Digestive and Kidney Diseases Gastroparesis Clinical Research Consortium. Clinical features of idiopathic gastroparesis vary with sex, body mass, symptom onset, delay in gastric emptying, and gastroparesis severity. *Gastroenterology, 140*(1), 101–115.

Parkman, H. P., Yates, K., Hasler, W. L., Nguyen, L., Pasricha, P. J., Snape, W. J., et al. (2011b). Clinical features of idiopathic gastroparesis vary with sex, body mass, symptom onset, delay in gastric emptying, and gastroparesis severity. *Gastroenterology, 140*(1), 101–115. e110.

Pasricha, P. J., Yates, K. P., Nguyen, L., Clarke, J., Abell, T. L., Farrugia, G., et al. (2015). Outcomes and factors associated with reduced symptoms in patients with gastroparesis. *Gastroenterology, 149*(7), 1762–1774. e1764.

Pergolizzi, J. V., Jr., Raffa, R. B., Pappagallo, M., Fleischer, C., Pergolizzi, J., III, Zampogna, G., et al. (2017). Peripherally acting μ-opioid receptor antagonists as treatment options for constipation in noncancer pain patients on chronic opioid therapy. *Patient Preference and Adherence, 11*, 107.

Perkel, M. S., Moore, C., Hersh, T., & Davidson, E. D. (1979). Metoclopramide therapy in patients with delayed gastric emptying. *Digestive Diseases and Sciences, 24*(9), 662–666.

Ramirez, R., Stalcup, P., Croft, B., & Darracq, M. A. (2017). Haloperidol undermining gastroparesis symptoms (HUGS) in the emergency department. *The American Journal of Emergency Medicine, 35*(8), 1118–1120.

Reddivari, A. K. R., & Mehta, P. (2019). Gastroparesis. In *StatPearls [Internet]* StatPearls Publishing.

Reichenbach, Z. W., Stanek, S., Patel, S., Ward, S. J., Malik, Z., Parkman, H. P., et al. (2019). Botulinum toxin A improves symptoms of gastroparesis. *Digestive Diseases and Sciences*, 1–9.

Rentz, A., Kahrilas, P., Stanghellini, V., Tack, J., Talley, N., Trudeau, E., et al. (2004). Development and psychometric evaluation of the patient assessment of upper gastrointestinal symptom severity index (PAGI-SYM) in patients with upper gastrointestinal disorders. *Quality of Life Research, 13*(10), 1737–1749.

Revicki, D., Camilleri, M., Kuo, B., Norton, N., Murray, L., Palsgrove, A., et al. (2009). Development and content validity of a gastroparesis cardinal symptom index daily diary. *Alimentary Pharmacology & Therapeutics, 30*(6), 670–680.

Rey, E., Choung, R. S., Schleck, C. D., Zinsmeister, A. R., Talley, N. J., & Locke, G. R., III. (2012). Prevalence of hidden gastroparesis in the community: The gastroparesis" iceberg". *Journal of Neurogastroenterology and Motility, 18*(1), 34.

Roblin, X., Biroulet, L. P., Phelip, J. M., Nancey, S., Flourie, B., Heidelbaugh, J. J., et al. (2009). Corrigendum: An evidence-based position statement on the management of irritable Bowel Syndrome. *The American Journal of Gastroenterology, 104*(1), S1–S35.

Roldan, C. J., Chambers, K. A., Paniagua, L., Patel, S., Cardenas-Turanzas, M., & Chathampally, Y. (2017). Randomized controlled double-blind trial comparing haloperidol combined with conventional therapy to conventional therapy alone in patients with symptomatic gastroparesis. *Academic Emergency Medicine, 24*(11), 1307–1314.

Saito, Y. A., Almazar, A. E., Tilkes, K. E., Choung, R. S., Van Norstrand, M. D., Schleck, C. D., et al. (2019). Randomised clinical trial: Pregabalin vs placebo for irritable bowel syndrome. *Alimentary Pharmacology & Therapeutics, 49*(4), 389–397.

Sandler, R. S., Stewart, W. F., Liberman, J. N., Ricci, J. A., & Zorich, N. L. (2000). Abdominal pain, bloating, and diarrhea in the United States. *Digestive Diseases and Sciences, 45*(6), 1166–1171.

Schey, R., Saadi, M., Midani, D., Roberts, A. C., Parupalli, R., & Parkman, H. P. (2016). Domperidone to treat symptoms of gastroparesis: Benefits and side effects from a large single-center cohort. *Digestive Diseases and Sciences, 61*(12), 3545–3551.

Shada, A. L., Dunst, C. M., Pescarus, R., Speer, E. A., Cassera, M., Reavis, K. M., et al. (2016). Laparoscopic pyloroplasty is a safe and effective first-line surgical therapy for refractory gastroparesis. *Surgical Endoscopy, 30*(4), 1326–1332.

Shahsavari, D., Reznick-Lipina, K., Malik, Z., Weiner, M., Jehangir, A., Repanshek, Z. D., et al. (2021). Haloperidol use in the emergency department for gastrointestinal symptoms: Nausea, vomiting, and abdominal pain. *Clinical and Translational Gastroenterology, 12*(6), e00362.

Shahsavari, D., Yu, D., Jehangir, A., Lu, X., Zoll, B., & Parkman, H. P. (2019). Symptom variability throughout the day in patients with gastroparesis. *Neurogastroenterology & Motility*, e13740.

Singh, V. V., & Toskes, P. P. (2003). Medical therapy for chronic pancreatitis pain. *Current Gastroenterology Reports, 5*(2), 110–116.

Stanghellini, V., & Tack, J. (2014). Gastroparesis: Separate entity or just a part of dyspepsia? *Gut, 63*(12), 1972–1978.

Szarka, L. A., Camilleri, M., Vella, A., Burton, D., Baxter, K., Simonson, J., et al. (2008). A stable isotope breath test with a standard meal for abnormal gastric emptying of solids in the clinic and in research. *Clinical Gastroenterology and Hepatology, 6*(6), 635–643. e631.

Szigethy, E., Schwartz, M., & Drossman, D. (2014). Narcotic bowel syndrome and opioid-induced constipation. *Current Gastroenterology Reports, 16* (10), 410.

Tack, J. (2002). Drink tests in functional dyspepsia. *Gastroenterology, 122*(7), 2093–2094.

Tack, J., Janssen, P., Masaoka, T., Farré, R., & Van Oudenhove, L. (2012). Efficacy of buspirone, a fundus-relaxing drug, in patients with functional dyspepsia. *Clinical Gastroenterology and Hepatology, 10*(11), 1239–1245.

Tack, J., Ly, H. G., Carbone, F., Vanheel, H., Vanuytsel, T., Holvoet, L., et al. (2016). Efficacy of mirtazapine in patients with functional dyspepsia and weight loss. *Clinical Gastroenterology and Hepatology, 14*(3), 385–392. e384.

Van Oudenhove, L., Kindt, S., Vos, R., Coulie, B., & Tack, J. (2008). Influence of buspirone on gastric sensorimotor function in man. *Alimentary Pharmacology & Therapeutics*, *28*(11 – 12), 1326–1333.

Van Oudenhove, L., Vandenberghe, J., Dupont, P., Geeraerts, B., Vos, R., Dirix, S., et al. (2010). Regional brain activity in functional dyspepsia: A H215O-PET study on the role of gastric sensitivity and abuse history. *Gastroenterology*, *139*(1), 36–47.

Vanheel, H., Carbone, F., Valvekens, L., Simren, M., Tornblom, H., Vanuytsel, T., et al. (2017). Pathophysiological abnormalities in functional dyspepsia subgroups according to the Rome III criteria. *American Journal of Gastroenterology*, *112*(1), 132–140.

Volkow, N. D., Baler, R. D., Compton, W. M., & Weiss, S. R. (2014). Adverse health effects of marijuana use. *New England Journal of Medicine*, *370*(23), 2219–2227.

Wang, Y. R., Fisher, R. S., & Parkman, H. P. (2008). Gastroparesis-related hospitalizations in the United States: Trends, characteristics, and outcomes, 1995–2004. *American Journal of Gastroenterology*, *103*(2), 313–322.

Yu, D., Ramsey, F. V., Norton, W. F., Norton, N., Schneck, S., Gaetano, T., et al. (2017). The burdens, concerns, and quality of life of patients with gastroparesis. *Digestive Diseases and Sciences*, *62*(4), 879–893.

Zehetner, J., Ravari, F., Ayazi, S., Skibba, A., Darehzereshki, A., Pelipad, D., et al. (2013). Minimally invasive surgical approach for the treatment of gastroparesis. *Surgical Endoscopy*, *27*(1), 61–66.

Zoll, B., Jehangir, A., Edwards, M. A., Petrov, R., Hughes, W., Malik, Z., et al. (2019). Surgical treatment for refractory gastroparesis: Stimulator, pyloric surgery, or both? *Journal of Gastrointestinal Surgery*, 1–8.

Zoll, B., Zhao, H., Edwards, M. A., Petrov, R., Schey, R., & Parkman, H. P. (2018). Outcomes of surgical intervention for refractory gastroparesis: A systematic review. *Journal of Surgical Research*, *231*, 263–269.

Chapter 17

Appendicitis and related abdominal pain

Marcos Prada-Arias

Department of Pediatric Surgery, Vigo University Hospital Álvaro Cunqueiro, Vigo, Spain Rare Diseases and Pediatric Medicine Research Group, Galician Sur Health Research Institute, Carretera Clara Campoamor, Vigo, Spain

Abbreviations

ANC   absolute neutrophil count
CRP   C-reactive protein
CT   computed tomography
MRI   magnetic resonance imaging
RTC   randomized clinical trial
US   ultrasound
WBC   white blood cell count

Introduction

Appendicitis is one of the most common causes of the acute abdomen and one of the major indications for emergency surgery worldwide (Stewart et al., 2014).

Recent evidences show that the appendix is a component of the immune system that would be especially designed, preserving mutualistic microbial biofilms, to fight threats, such as frequent gastrointestinal infections, that no longer apply to humans in a postindustrial society. Since appendicitis is a disease associated with industrialized countries and urban areas, some authors have postulated that it could be the result of appendix overreactivity due to lack of stimulation, similar to other processes related with the immune system (Laurin, Everett, & Parker, 2011; Smith et al., 2009). Other evidence of the immune role of the appendix comes from studies showing the increasingly likely negative correlation between appendicitis and ulcerative colitis (Frisch, Pedersen, & Andersson, 2009).

Appendicitis can affect people of any age but is most common between the ages of 10 and 20 years, being uncommon in children under 5 years and adults over 50 years. Approximately 8 to 11 per 10,000 people will experience appendicitis annually, with an overall lifetime risk of 7% to 8%, although with significant geographic differences. Most studies show a slight male predominance with an overall rate ratio of 1.4:1 (Addiss, Shaffer, Fowler, & Tauxe, 1990; Bhangu, Søreide, Di Saverio, Assarsson, & Drake, 2015). In children, the incidence increases from an annual rate of 1 to 3 per 10,000 children between birth and 4 years old to 12 to 23 per 10,000 children between 5 and 14 years old (Andersen, Paerregaard, & Larsen, 2009; Buckius et al., 2012).

Etiology

Appendicitis is a complex disease with a probably multifactorial etiology, but its cause remains hidden in most cases. Luminal obstruction is considered the main factor implicated but is not always identified (Arnbjörnsson & Bengmark, 1984). The mechanism of obstruction varies depending upon the patient's age. In the young, lymphoid follicular hyperplasia due to specific or unspecific infection is thought to be the main cause. These follicles reach their maximal size during adolescence, when the peak incidence of appendicitis occurs (Bundy et al., 2007). In older patients, luminal obstruction is more likely to be caused by appendicolith, but it is also found in histologically normal appendix and absent in many appendicitis cases (Nitecki, Karmeli, & Sarr, 1990). Other identified but infrequent causes of obstruction are fibrosis, parasites, Crohn's disease, neoplasia (carcinoid, adenocarcinoma, lymphoma), mucocele, and twist of the organ itself (Rabah, 2007). Dietary and genetic factors also seem to be involved. Vegetable intake appears to be a protective factor, possibly through an

---

Features and Assessments of Pain, Anesthesia, and Analgesia. https://doi.org/10.1016/B978-0-12-818988-7.00037-6
Copyright © 2022 Elsevier Inc. All rights reserved.

189

effect on the gut flora (Andersen et al., 2009), and genetic factors would be associated with gender and racial differences, and the increased risk observed in twins and in people with a positive family history (Bhangu et al., 2015).

Pathophysiology

Following primary or secondary appendiceal obstruction, the lumen fills with mucus and distends, increasing luminal, and intramural pressure, resulting in occlusion of the small vessels and stasis of lymphatic flow. As the appendix becomes engorged, the visceral unmyelinated (C) afferent fibers entering the spinal cord at T8–T10 are stimulated (Fig. 1), leading to vague and cramping periumbilical pain that occurs initially in the course of the disease. The influx of neutrophils causes a fibronopurulent reaction on the serosal surface (uncomplicated, simple, or supurative appendicitis), irritating the surrounding parietal peritoneum, that results in stimulation of myelinated (A-delta) somatic fibers, causing a well-localized sharp pain, that occurs later in the course of disease (Birnbaum & Wilson, 2000). This migration of pain from periumbilical region to right iliac fossa is a cardinal symptom of appendicitis (Andersson, 2004). Vascular compromise and bacterial invasion of the wall, due to intraluminal bacterial overgrowth with rupture of the mucosal barrier, cause the progression of inflammation and ischemia, and finally necrosis and perforation (complicated or gangrenous and perforated appendicitis) (Bundy et al., 2007). The bacteria include the usual aerobic and anaerobic gut flora, being *Escherichia coli*, *Peptostreptococcus*, *Bacteroides fragilis*, and *Pseudomonas* species, the common organisms involved (Bennion et al., 1990). Recently, common oral bacteria such as *Fusobacterium*, *Gemella*, or *Parvimonas* have been found in molecular studies of appendices, and their abundance seems to correspond to disease severity (Bhangu et al., 2015; Zhong, Brower-Sinning, Firek, & Morowitz, 2014). Neuroproliferation in the appendix in association with an increase in neuropeptides has been observed in patients with clinical diagnosis of appendicitis in the absence of acute inflammation. This disorder, known as neurogenic appendicopathy, may be quite common in children and may play a role in some cases of functional abdominal pain (Sesia, Mayr, Bruder, & Haecker, 2013).

Natural history models

The perforation rate is correlated with the duration of the disease, rarely occurs in the first 12 to 24 h of symptoms (10%–20%) and becomes common after 48 h (75%) and 72 h (90%) (Nitecki et al., 1990). The reported rates of perforation are slightly higher in men (18%) than women (13%), and vary significantly by age, from 50% to 100% in children under 5 years to 10% to 20% in adolescents and adults, and 40% to 70% in the elderlies (Colvin, Bachur, & Kharbanda, 2007; Segev et al., 2015). Higher perforation rates in younger children and elderlies are associated with delayed presentation and diagnosis due

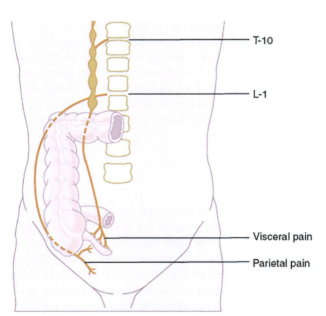

FIG. 1 Transmission of pain from the appendix. Visceral and parietal transmission of pain signals from the appendix. *From Guyton and Hall Textbook of Medical Physiology, 12th ed. (2011) with permission.*

to nonspecific or subtle clinical signs (Humes, Speake, & Simpson, 2007). Similarly, the negative appendectomy rate varies with age, but is more affected by gender. The highest frequency is reported in children under 5 years (up to 17%) and in postmenarchal women (up to 5%), while rates of 1% to 2% are described in older children and adults males (Bachur, Hennelly, Callahan, Chen, & Monuteaux, 2012). The inverse relation between perforation and negative appendectomy is classically referred as a measure of the management quality, but negative appendectomy rates have decreased in the past decade without increasing of perforation rates, which is likely attributable to the increased utilization of imaging tests (Drake et al., 2012).

The traditional model of the natural history of appendicitis affirms that patients with appendicitis will eventually progress to perforation if left untreated. The increasing perforation rates with the duration of symptoms are considered a proof of the progressive character of disease. In this model, the proportion of perforations is considered a measure of the management quality of patients. An alternative model has been proposed in which simple and perforated appendicitis are two separate entities with different behavior, so most perforations would occur early and the increase in the proportion of perforations over time would be mainly explained by selection of patients with perforated appendicitis due to spontaneous resolution of simple appendicitis (Fig. 2). In this alternative model, perforation rates are therefore a questionable measure of patient management quality (Andersson, 2007; Livingston, Woodward, Sarosi, & Haley, 2007).

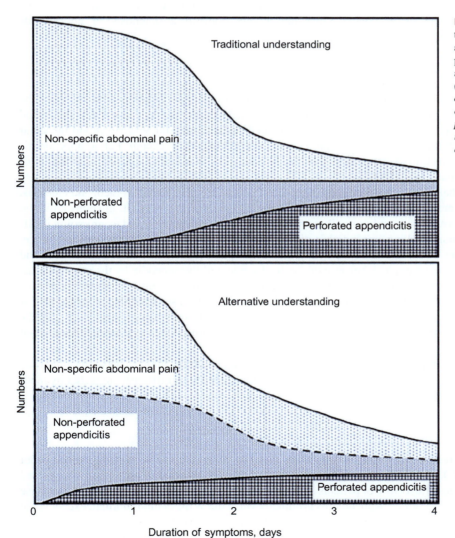

FIG. 2 The traditional and alternative models of the natural history of appendicitis. The traditional and the proposed alternative understanding of the progression of appendicitis and nonspecific abdominal pain over time. *From Andersson, R.E. (2007). The natural history and traditional management of appendicitis revisited: Spontaneous resolution and predominance of prehospital perforations imply that a correct diagnosis is more important than an early diagnosis. World Journal of Surgery, 31(1), 86–92, with permission.*

Typical presentation

The diagnosis of appendicitis is predominantly clinical. Abdominal pain is the most common symptom and is reported in almost all confirmed cases of appendicitis. In the typical presentation, the patient describes the onset of abdominal pain as the first symptom, as periumbilical cramping pain (visceral pain), which intensifies during the first hours, with subsequent migration to the right lower quadrant as the inflammation progresses, becoming constant and sharp (parietal pain). Migratory pain within 24 h of symptom onset only occurs in 50% to 60% of patients, but it is the strongest symptom associated with diagnosis of appendicitis. Other frequent signs and symptoms are anorexia, nausea, vomiting, and fever, which generally occur after the onset of pain (Andersson, 2004; Bundy et al., 2007). On examination, right lower quadrant or McBurney's point tenderness is a reliable clinical sign of acute appendicitis (Andersson et al., 1999). Classical signs of peritoneal irritation may be present such as: involuntary muscle guarding with abdominal palpation; rebound tenderness (increased pain with sudden release of pressure applied to the lower right quadrant for 10–15 s); iliopsoas sign (pain on extension of the right hip, when a retrocecal inflamed appendix irritates iliopsoas muscle); Rovsing sign (pain in the right lower quadrant with palpation of the left side); obturator sign (pain on flexion and internal rotation of the right hip, when a pelvic inflamed appendix causes irritation of the obturator internus muscle); increased pain with movement, such as coughing, walking, or shifting position in bed. The diagnostic accuracy of iliopsoas, Rovsing, and obturator signs has not been well defined and, in general, they are not used by experienced clinicians (Bundy et al., 2007).

Many pediatric surgeons feel that testing for rebound tenderness is often unnecessary, since it is very painful in children who have appendicitis, may be falsely positive, and abdominal pain can also be elicited by asking the child to cough or to jump, or with abdominal percussion (Samuel, 2002). Signs of peritoneal irritation are the most reliable clinical findings, but there are no physical signs, taken alone or in concert, that definitively confirm a diagnosis of appendicitis (Andersson, 2004).

Atypical presentation

The clinical presentation can be influenced by anatomical position of the appendix. Right lower quadrant tenderness may be less obvious when the appendix is in the retrocecal or pelvic position. Retrocecal appendicitis may cause a dull abdominal ache since the appendix does not come into contact with the anterior parietal peritoneum, and tenderness and muscular guarding to abdominal palpation are often absent because of protection from the overlying cecum. Pelvic appendicitis can cause suprapubic pain associated with urinary frequency, dysuria, tenesmus, or diarrhea as a result of irritation of bladder or rectum (Andersson, 2004; Bundy et al., 2007).

The evaluation can be particularly challenging in children younger than 5 years, postmenarchal women, and elderlies, because of typical symptoms, and signs are often not present and specific findings are difficult to elicit. In young children, the clinical findings frequently overlap with other conditions (Bundy et al., 2007), communication and examination is difficult, and diarrhea is relatively common, making appendicitis difficult to differentiate from gastroenteritis (Nance, Adamson, & Hedrick, 2000; Prada-Arias et al., 2020). In women, gynecologic pathologies can mimic appendicitis and in cases of pregnancy, the presentation is often atypical, because of the displacement of the appendix by the gravid uterus. Postmenarchal women should be queried regarding the possibility of pregnancy, and a pelvic examination should be always performed (Humes et al., 2007). Elderlies tend to have a diminished inflammatory response, resulting in subtle or less remarkable findings on history and physical exam, so patients and clinicians may minimize them (Segev et al., 2015). Importantly, a high index of suspicion for the diagnosis of appendicitis should be maintained when evaluating these special groups of patients (Colvin et al., 2007).

In the past, analgesia was discouraged in patients with suspected appendicitis, including children, because of the mistaken belief that pain control would mask symptoms, but diagnosis of appendicitis is not significantly impacted by medications for pain control (Anderson & Collins, 2008).

Inflammatory markers

Although limited in their ability to differentiate appendicitis from other causes of abdominal pain, white blood cell count (WBC), absolute neutrophil count (ANC), and C-reactive protein (CRP) are typically obtained in appendicitis-suspected cases. WBC and ANC are elevated in up to 80% to 96% of patients with appendicitis, but this finding is nonspecific because it is present in many other abdominal pain conditions. Appendicitis is unlikely when the WBC count is normal, except in the very early course of the illness, and WBC counts are higher in complicated cases. Sensitivity and specificity of CRP elevation range widely (Andersson, 2004; Bundy et al., 2007; Kharbanda, Cosme, Liu, Spitalnik, & Dayan, 2011), being more sensitive than WBC for patients after the first 24 h of onset of symptoms. CRP levels greater than 30–50 mg/L are helpful to predict perforation (Kharbanda et al., 2011), and other biomarkers like procalcitonin, fibrinogen, and bilirubin also can predict perforation with moderate sensitivity and specificity (Gavela, Cabeza, Serrano, & Casado-Flores, 2012; Prada-Arias et al., 2017). CRP and WBC combination achieve better accuracy to discriminate appendicitis (Kwan & Nager, 2010), and CRP increase at the repeat exam, independent of the observed level, is a strong predictor of appendicitis, and vice versa for a decrease (Andersson et al., 2000). Many other inflammatory markers have been studied to diagnose appendicitis, but none of them has a diagnostic accuracy higher than WBC, ANC, and CRP. Currently, no inflammatory marker alone or in combination can identify appendicitis with high specificity and sensitivity, so they should not be used in isolation to make or exclude the diagnosis (Bhangu et al., 2015). The clinician should also obtain a pregnancy test (urine beta-human chorionic gonadotropin) in postmenarchal women to aid in the diagnosis of ectopic pregnancy and to guide imaging decisions. A urinalysis is usually performed to identify alternative conditions such as a urinary infection or nephrolithiasis. However, many patients with appendicitis will have pyuria, and less commonly hematuria, due to irritation of the ureter or bladder (Humes et al., 2007).

Clinical prediction rules and imaging tests

Several clinical prediction rules or scores have been proposed to increase the predictive ability of symptoms, signs, and laboratory tests and standardize the assessment for appendicitis; the modified Alvarado Score and the Appendicitis Inflammatory Response Score, in adults (Andersson, Kolodziej, & Andersson, 2017; Bhangu et al., 2015) and the Pediatric Appendicitis Score and the refined Low-Risk Appendicitis Score, in children (Kharbanda et al., 2018; Samuel, 2002), are the most prospectively studied and currently used in clinical practice. These scores are unable to identify patients who must be operated, and their convenience lies in the ability to categorize patients into groups of low, moderate, and high risk of appendicitis, identifying patients who may benefit from diagnostic imaging and/or surgical consultation. Clinical pathways that utilize these scores have the potential to achieve acceptable diagnostic accuracy and more efficient use of imaging tests (Saucier, Huang, Emeremni, & Pershad, 2014).

Abdominal ultrasound (US) demonstrates a high diagnostic accuracy of appendicitis, but lower than computed tomography (CT), and it is the preferred imaging exam in children, pregnant women, and when the CT is not readily available. Advantages of US include the lack of ionizing radiation and intravenous contrast, but its performance is highly variable and depends on patient-specific features (body habitus, discomfort, appendix location) and operator experience. The normal appendix is frequently not visualized on US, which increases the rate of indeterminate exams. A negative US result indicates that a normal appendix has been visualized and appendicitis is highly unlikely, but a nondiagnostic result does not rule out appendicitis (Kessler et al., 2004). Abdominal CT is recommended as the preferred imaging test in adults by its high diagnostic accuracy and low rate of nondiagnostic exams, because it is able to visualize the normal appendix in almost all cases. Some advantages of CT include a greater availability of expert radiologist, a better tolerance to exam by the majority of patients, and a very good evaluation of other pathologies. The most important disadvantages of CT are exposure to ionizing radiation and intravenous contrast. Nonvisualization of the appendix decreases but does not eliminate the likelihood of appendicitis (Doria et al., 2006). Magnetic resonance imaging (MRI) should be used in pregnant women and older children who can cooperate with the exam, to minimize ionizing radiation exposure. Intravenous contrast can be administered to improve accuracy if images without contrast prove nondiagnostic. Diagnostic accuracy is comparable to CT, but the exam is worse tolerated than US or CT (Barger & Nandalur, 2010).

Management

The dominant management strategy for appendicitis remains operative, with either open or laparoscopic appendectomy (Di Saverio et al., 2020). A short in-hospital surgical delay up to 24 h is not associated with an increased risk of perforation (Sallinen et al., 2016). Prophylactic antibiotics are important in preventing wound infection, and intraabdominal abscess following appendectomy and postoperative antibiotics are unnecessary in cases of uncomplicated appendicitis (Andersen, Kallehave, & Andersen, 2005). Patients who present at night and will not undergo appendectomy until the next morning should be given intravenous antibiotics as soon as possible, rather than waiting until just before surgery (Bhangu et al., 2015).

The management of perforated appendicitis depends on the nature of the perforation. Unstable patients with free perforation and generalized peritonitis require preoperative resuscitation and emergency appendectomy. Stable patients with perforation contained as an appendiceal mass (phlegmon or abscess) in imaging studies can be treated with immediate surgery or initial nonoperative management, associating percutaneous image-guided drainage of abscesses, since both approaches are safe (Cheng et al., 2017). However, immediate surgery in patients with appendiceal mass and long duration of symptoms (more than 4–5 days) increases morbidity due to the extensive dissection required by dense adhesions and inflammation that may lead to injury of adjacent structures, making initial nonoperative management safer and more effective than immediate surgery (Mentula, Sammalkorpi, & Leppäniemi, 2015).

In complicated appendicitis, a minimum of 3–5 days of broad-spectrum intravenous empirical antibiotics with activity against gram-negative and anaerobic organisms is recommended, although the duration of treatment should be based on clinical criteria, such as fever, pain, return of bowel function, and WBC count (Di Saverio et al., 2020; Yu et al., 2017).

Surgical approach and complications

Open and laparoscopic appendectomies have been compared in several randomized clinical trials (RCT), systematic reviews, and meta-analyses. The laparoscopic approach is associated with a lower rate of wound infections, less postoperative pain, shorter duration of hospital stay, quicker return to full function, and fewer adhesive bowel obstructions. The open approach is associated with a lower rate of intraabdominal abscesses and a shorter operative time. The choice of operative approach is determined by surgeon preference with consideration of personal experience, patient condition, and local resources. Evidence suggests that laparoscopy is preferable in cases of uncertain diagnosis, since it allows inspection of other abdominal organs. Two groups of patients in which laparoscopy is especially indicated are postmenarchal women, since laparoscopy can reveal other causes of pelvic pathology, and obese patients, since open appendectomy may require larger, morbidity-prone incisions (Di Saverio et al., 2020; Lin, Lai, & Lai, 2014).

Interval appendectomy after successful nonoperative treatment of appendiceal mass is usually performed laparoscopically in 8–12 weeks. In children, this could not be indicated because of the low risk of recurrence (lower than 20%), low risk of missing carcinoid tumor (lower than 1%), and significant number of surgical complications (3%–4%) (Hall, Jones, Eaton, Stanton, & Burge, 2011). Since the incidence of appendiceal neoplasms is high (7%–13%) in adult over 40 years of age with complicated appendicitis, interval appendectomy and colonic screening (colonoscopy and/or CT) are recommended in these patients when they are managed without surgery (Mällinen et al., 2019).

The most common complication following appendectomy is surgical infection; either wound infection (3%–10%) or abdominal abscess (9%), and mainly associated with perforated appendicitis. Mortality rate of appendicitis range from less than 1% in general population to 4% in the elderlies (Bhangu et al., 2015; Storm-Dickerson & Horattas, 2003).

Applications to other areas

The evaluation of abdominal pain requires knowledge of its responsible mechanisms, its wide differential diagnosis, and its usual clinical presentations. From a pathophysiological point of view, abdominal pain may have a visceral, parietal, or referred origin, which allows us to understand its cause, quality, and location. Visceral afferent unmyelinated (C) fibers are stimulated by ischemia, chemical damage by inflammatory mediators, spasm of the smooth muscle, overdistention of the viscus, and stretching of the connective tissue of the viscus or solid organs capsule, and produces a dull, poorly localized pain in the periumbilical or suprapubic region in relation to the visceral organ affected. Conversely, peritoneal parietal stimuli such as ischemia, chemical damage by inflammatory mediators, and stretching of the connective tissue are transmitted directly by myelinated fibers (A-Delta) and produce sharper and localized pain directly over the painful area. Another type of abdominal pain is perceived at a site far from the causal tissue, at the cutaneous dermatomes sharing the same spinal cord level as the visceral inputs. This referred pain mechanism is also responsible of skin hyperalgesia and increased muscle tone of the abdominal wall, sometimes observed in patients with an acute abdomen (Hall, 2011).

The main conditions that can simulate appendicitis, producing lower or suprapubic abdominal pain, referring to its responsible mechanism and clinical key features, are presented in tables, differentiating adults (Table 1), women (Table 2), and children (Table 3). Other common and rare causes of abdominal pain are presented in Table 4 (Natesan, Lee, Volkamer, & Thoureen, 2016; Penner & Majumdar, 2010; Yang, Chen, & Wu, 2013).

Other agents of interest

Although appendectomy is very safe and effective, it carries some risks and complications associated with general anesthesia and surgery, so interest in nonsurgical treatment has increased in recent years. Several RCT performed in adults have suggested that it is feasible to treat uncomplicated appendicitis nonsurgically with antibiotics (Nikolaidis et al., 2006), since most patients respond clinically and compared with operated patients, they have lower or similar pain scores, require less analgesic doses, have a faster return to work, and they do not have a higher perforation rate. Approximately, 90% of these

TABLE 1 Causes of lower abdominal pain in adults.

Process	Mechanism	Clinical key feature
Appendicitis	Appendix inflammation	Periumbilical pain initially that migrates to the right lower quadrant
Diverticulitis	Diverticulum inflammation	Left lower quadrant pain usually constant and present for several days prior to presentation
Nephrolithiasis	Ureter distension and spams	Generally flank pain, but may have back or abdominal pain
Pyelonephritis	Renal inflammation	Flank pain and costovertebral angle tenderness
Acute urinary retention	Bladder distension and spams	Suprapubic pain and inability to pass urine
Cystitis	Bladder inflammation	Suprapubic pain, dysuria, urgency, frequency and/or hematuria
Infectious colitis	Bowel inflammation, distension and spams	Diarrhea

Main causes of lower abdominal pain, mechanism involved, and clinical key features, in adult patients
Data from Penner, R. M., & Majumdar, S. R. (2010). Acute Abdominal Pain. In *Practical Gastroenterology and Hepatology: Small and Large Intestine and Pancreas* (pp. 129–134). Oxford, UK: Wiley-Blackwell; Natesan, S., Lee, J., Volkamer, H., & Thoureen, T. (2016). Evidence-Based Medicine Approach to Abdominal Pain. *Emergency Medicine Clinics of North America, 34*(2), 165–190.

TABLE 2 Causes of pelvic abdominal pain in women.

Process	Mechanism	Clinical key features
Ectopic pregnancy	Fallopian tube distension or peritoneal inflammation	First trimester vaginal bleeding
Pelvic inflammatory disease	Genital tract inflammation	Pelvic organ tenderness and cervical discharge
Ovarian torsion	Ovarian ischemia	Sudden onset of unilateral pain and adnexal mass
Ruptured ovarian cyst	Ovarian distension and capsule rupture	Sudden onset of unilateral pain
Ovulatory pain (Mittelsmerz)	Ovarian distension and capsule rupture	Occurs mid-cycle
Endometriosis	Fallopian tube or ovarian inflammation	Dysmenorrhea, dyspareunia and excessive menstrual bleeding
Endometritis	Uterus inflammation	Proceeded by pelvic inflammatory disease

Main causes of pelvic abdominal pain, mechanism involved, and clinical key features, in postmenarchal women patients.
Data from Penner, R. M., & Majumdar, S. R. (2010). Acute Abdominal Pain. In *Practical Gastroenterology and Hepatology: Small and Large Intestine and Pancreas* (pp. 129–134). Oxford, UK: Wiley-Blackwell; Natesan, S., Lee, J., Volkamer, H., & Thoureen, T. (2016). Evidence-Based Medicine Approach to Abdominal Pain. *Emergency Medicine Clinics of North America, 34*(2), 165–190.

patients are able to avoid surgery during the initial admission, and about 70% during the first year. The other 30% eventually require appendectomy for recurrence of appendicitis or abdominal pain (Salminen et al., 2015; Sartelli et al., 2018; Vons et al., 2011). However, it is not clear whether the success in avoiding immediate surgery is justified in relation to potential recurrence or the risk of missed appendiceal neoplasm, especially in older adults (Harnoss et al., 2017), although it could be an alternative treatment when surgery is contraindicated (Wilms, de Hoog, de Visser, & Janzing, 2011). Even though there are few studies in children and the level of evidence is still low, the results of nonsurgical treatment seem similar to those observed in adults. However, larger RCT are needed to determine the safety and efficacy of nonsurgical treatment of uncomplicated appendicitis in children, and it is currently only recommended to be indicated as part of an ongoing RCT (Maita, Andersson, Svensson, & Wester, 2020; Patkova et al., 2020).

A recent RCT compared supportive care alone versus antibiotics in adult patients with CT-verified uncomplicated appendicitis, and showed that patients treated with supportive care alone did just as well as patients treated with antibiotics, with a similar percentage of failure and recurrence. This study indicates that perhaps supportive care, not antibiotics, is

TABLE 3 Causes of lower abdominal pain in children.

Process	Mechanism	Clinical features
Appendicitis	Appendix inflammation	Periumbilical pain that migrates to the right lower quadrant
Gastroenteritis	Bowel inflammation	Diarrhea, vomiting
Constipation	Bowel distension and spasms	Fecal retention
Mesenteric adenitis	Mesenteric inflammation	Abdominal pain, fever
Intussusception	Bowel distension, spasms and ischemia	Intermittent colicky pain
Infection tract urinary	Bladder or ureter inflammation	Suprapubic pain, dysuria
Volvulus/Malrotation	Intestinal ischemia	Biliary vomiting
Nonspecific abdominal pain	Neuroimmune?	Right lower quadrant abdominal pain
Abdominal trauma	Bowel, mesenteric or solid organ injury	Trauma history
Pharingotonsillitis	Mesenteric inflammation (adenitis)	Sore throat, fever
Pneumonia	Diaphragmatic irritation	Cough, tachypnea, fever
Schönlein–Henoch purpura	Bowel inflammation (vasculitis)	Symmetric purpuric rash
Tiflitis	Ileocecal inflammation	Neutropenia

Main causes of lower abdominal pain, mechanism involved, and clinical key features, in children patients.
Data from Yang, W.-C., Chen, C.-Y., & Wu, H.-P. (2013). Etiology of non-traumatic acute abdomen in pediatric emergency departments. *World Journal of Clinical Cases, 1*(9), 276–284; Natesan, S., Lee, J., Volkamer, H., & Thoureen, T. (2016). Evidence-Based Medicine Approach to Abdominal Pain. *Emergency Medicine Clinics of North America, 34*(2), 165–190.

TABLE 4 Other causes of abdominal pain.

Upper abdominal pain	Diffuse pain	Less common causes
Right upper quadrant pain	Obstruction	Abdominal aortic aneurysm
Biliary colic	Perforation	Abdominal compartment syndrome
Cholecystitis	Mesenteric ischemia	Abdominal migraine
Cholangitis	Gastroenteritis	Acute intermittent porphyria
Hepatitis	Crohn's disease	Angioedema
Perihepatitis	Ulcerative colitis	Celiac axis syndrome
Liver abscess	Spontaneous peritonitis	Epiploic appendagitis
Budd–Chiari syndrome	Malignancy	Familial Mediterranean fever
Portal vein thrombosis	Celiac disease	Herpes zoster
Epigastric pain	Ketoacidosis	Hypercalcemia
Myocardial infarction	Adrenal insufficiency	Hypothyroidism
Pancreatitis	Foodborne disease	Lead poisoning
Peptic ulcer disease	Irritable bowel syndrome	Meckel's diverticulum
Gastroesophageal reflux	Constipation	Narcotic bowel syndrome
Gastritis/Gastropathy	Lactose intolerance	Paroxismal nocturnal hemoglobinuria
Left upper quadrant pain		Pulmonary etiologies
Splenomegaly		Renal infarction
Splenic infarction		Rib pain
Splenic abscess		Wandering spleen
Splenic rupture		

Main causes of upper and diffuse abdominal pain, and less common causes of abdominal pain.
Data from Penner, R. M., & Majumdar, S. R. (2010). Acute Abdominal Pain. In *Practical Gastroenterology and Hepatology: Small and Large Intestine and Pancreas* (pp. 129–134). Oxford, UK: Wiley-Blackwell; Yang, W.-C., Chen, C.-Y., & Wu, H.-P. (2013). Etiology of non-traumatic acute abdomen in pediatric emergency departments. *World Journal of Clinical Cases, 1*(9), 276–284; Natesan, S., Lee, J., Volkamer, H., & Thoureen, T. (2016). Evidence-Based Medicine Approach to Abdominal Pain. *Emergency Medicine Clinics of North America, 34*(2), 165–190.

responsible for the success of nonoperative management, so patients whose appendicitis is destined to resolve will improve with or without antibiotics (Park, Kim, & Lee, 2017). This is another evidence that appendicitis not always progress to perforation, and a milder form that can resolve spontaneously exists. The goal of future research is to learn how to distinguish between these different forms of appendicitis; the severe form would likely benefit from appendectomy, and the milder form could likely be treated nonsurgically.

Mini-dictionary of terms

Appendiceal phlegmon: appendiceal inflammatory tumor that involves adjacent tissues as bowel, mesentery, omentum, or peritoneum.

Appendicolith: stone made of feces, undigested food or other foreign material into the appendix.

Biofilms: adherent colonies of microorganisms growing within a self-produced matrix of extracellular polymeric substances.

Clinical prediction rules: clinical decision-making tools that use three or more variables from history, physical examination, or simple tests to provide the probability of an outcome or suggest a diagnostic or therapeutic action.

McBurney's point: point over the right side of the abdomen that is one-third of distance from the anterior superior iliac spine to the umbilicus, and corresponds to the most common location of the base of the appendix.

Protein C reactive: acute-phase protein of hepatic origin, released into the bloodstream during inflammation.

Vermiform appendix: thin and blind-ended tube of variable length, attached to and opening into the cecum of humans and some other mammals.

Key facts

An alternative model of natural history of appendicitis has been proposed in which there are two separate forms of appendicitis: a severe form that progresses to early perforation and a milder form that can resolve spontaneously.

The clinical history of vague and cramping periumbilical pain (visceral pain) that migrates to the right iliac fossa, becoming sharp and constant (parietal pain), is a cardinal symptom of appendicitis.

High index of suspicion for the diagnosis should be maintained when evaluating groups of patients with frequent atypical presentation, like young children, postmenarchal women, and the elderlies.

Preoperative prophylactic antibiotics are important in preventing infectious complications, and the duration of antibiotic treatment in complicated cases should be based on clinical criteria.

Laparoscopic appendectomy is associated with a lower rate of wound infections, less postoperative pain, shorter duration of hospital stay, quicker return to full function, and fewer adhesive bowel obstructions, than open appendectomy.

Nonsurgical management of the appendiceal mass (phlegmon or abscess), especially in cases of long duration of symptoms, is safer and more effective than immediate surgery.

Uncomplicated appendicitis cases can be safely and effectively treated nonsurgically in adults and probably in children, but it does not improve surgical treatment.

Summary points

Appendicitis can affect people of any age but is most common between the ages of 10 and 20 years, being uncommon in children under 5 years and adults over 50 years.

Approximately 8 to 11 per 10,000 people will experience appendicitis annually, with an overall lifetime risk of 7% to 8%.

Appendicitis is a complex disease with a probably multifactorial etiology, with genetic and acquired factors involved, such as obstruction or infection.

There are two forms of appendicitis: uncomplicated, simple, or supurative appendicitis, and complicated or gangrenous and perforated appendicitis.

Lower abdominal pain is the main symptom, clinically implying a broad differential diagnosis, and there are no physical signs or inflammatory markers that definitively confirm diagnosis.

Abdominal ultrasound and computed tomography have achieved greater diagnostic accuracy, and magnetic resonance imaging is an effective alternative.

Surgery is the dominant strategy for the treatment of appendicitis, and laparoscopic appendectomy seems to be the most recommended approach.

References

Addiss, D. G., Shaffer, N., Fowler, B. S., & Tauxe, R. V. (1990). The epidemiology of appendicitis and appendectomy in the United States. *American Journal of Epidemiology, 132*(5), 910–925.

Andersen, B. R., Kallehave, F. L., & Andersen, H. K. (2005). Antibiotics versus placebo for prevention of postoperative infection after appendicectomy. *Cochrane Database of Systematic Reviews*, CD001439.

Andersen, S. B., Paerregaard, A., & Larsen, K. (2009). Changes in the epidemiology of acute appendicitis and appendectomy in Danish children 1996–2004. *European Journal of Pediatric Surgery, 19*(05), 286–289.

Anderson, M., & Collins, E. (2008). Analgesia for childrenwith acute abdominal pain and diagnostic accuracy. *Archives of Disease in Childhood, 93*(11), 995–997.

Andersson, R. E. (2004). Meta-analysis of the clinical and laboratory diagnosis of appendicitis. *British Journal of Surgery, 91*(1), 28–37.

Andersson, R. E. (2007). The natural history and traditional management of appendicitis revisited: Spontaneous resolution and predominance of prehospital perforations imply that a correct diagnosis is more important than an early diagnosis. *World Journal of Surgery, 31*(1), 86–92.

Andersson, R. E., Hugander, A. P., Ghazi, S. H., Ravn, H., Offenbartl, S. K., Nyström, P. O., et al. (1999). Diagnostic value of disease history, clinical presentation, and inflammatory parameters of appendicitis. *World Journal of Surgery, 23*(2), 133–140.

Andersson, R. E., Hugander, A. P., Ravn, H., Offenbartl, S. K., Ghazi, S. H., Nyström, P. O., et al. (2000). Repeated clinical and laboratory examinations in patients with an equivocal diagnosis of appendicitis. *World Journal of Surgery, 24*(4), 479–485.

Andersson, M., Kolodziej, B., & Andersson, R. E. (2017). Randomized clinical trial of appendicitis inflammatory response score-based management of patients with suspected appendicitis. *British Journal of Surgery, 104*(11), 1451–1461.

Arnbjörnsson, E., & Bengmark, S. (1984). Role of obstruction in the pathogenesis of acute appendicitis. *The American Journal of Surgery, 147*(3), 390–392.

Bachur, R. G., Hennelly, K., Callahan, M. J., Chen, C., & Monuteaux, M. C. (2012). Diagnostic imaging and negative appendectomy rates in children: Effects of age and gender. *Pediatrics, 129*(5), 877–884.

Barger, R. L., & Nandalur, K. R. (2010). Diagnostic performance of magnetic resonance imaging in the detection of appendicitis in adults. *Academic Radiology, 17*(10), 1211–1216.

Bennion, R., Baron, E., Thompson, J., Downes, J., Summanen, P., Talan, D., et al. (1990). The bacteriology of gangrenous and perforated appendicitis—Revisited. *Annals of Surgery, 211*(2), 165–171.

Bhangu, A., Søreide, K., Di Saverio, S., Assarsson, J. H., & Drake, F. T. (2015). Acute appendicitis: Modern understanding of pathogenesis, diagnosis, and management. *The Lancet, 386*(10000), 1278–1287.

Birnbaum, B. A., & Wilson, S. R. (2000). Appendicitis at the millennium. *Radiology, 215*(2), 337–348.

Buckius, M. T., McGrath, B., Monk, J., Grim, R., Bell, T., & Ahuja, V. (2012). Changing epidemiology of acute appendicitis in the United States: Study period 1993-2008. *Journal of Surgical Research, 175*(2), 185–190.

Bundy, D. G., Byerley, J. S., Liles, E. A., Perrin, E. M., Katznelson, J., & Rice, H. E. (2007). Does this child have appendicitis? *JAMA, 298*(4), 438–451.

Cheng, Y., Xiong, X., Lu, J., Wu, S., Zhou, R., & Cheng, N. (2017). Early versus delayed appendicectomy for appendiceal phlegmon or abscess. *Cochrane Database of Systematic Reviews*, CD011670.

Colvin, J. M., Bachur, R., & Kharbanda, A. (2007). The presentation of appendicitis in preadolescent children. *Pediatric Emergency Care, 23*(12), 849–855.

Di Saverio, S., Podda, M., De Simone, B., Ceresoli, M., Augustin, G., Gori, A., et al. (2020). Diagnosis and treatment of acute appendicitis: 2020 Update of the WSES Jerusalem guidelines. *World Journal of Emergency Surgery, 15*(1), 27.

Doria, A. S., Moineddin, R., Kellenberger, C. J., Epelman, M., Beyene, J., Schuh, S., et al. (2006). US or CT for diagnosis of appendicitis in children and adults? A meta-analysis. *Radiology, 241*(1), 83–94.

Drake, F. T., Florence, M. G., Johnson, M. G., Jurkovich, G. J., Kwon, S., Schmidt, Z., et al. (2012). Progress in the diagnosis of appendicitis. *Annals of Surgery, 256*(4), 586–594.

Frisch, M., Pedersen, B. V., & Andersson, R. E. (2009). Appendicitis, mesenteric lymphadenitis, and subsequent risk of ulcerative colitis: Cohort studies in Sweden and Denmark. *BMJ, 338*(2), b716.

Gavela, T., Cabeza, B., Serrano, A., & Casado-Flores, J. (2012). C-reactive protein and procalcitonin are predictors of the severity of acute appendicitis in children. *Pediatric Emergency Care, 28*(5), 416–419.

Hall, J. (2011). *Guyton and hall textbook of medical physiology* (12th ed.). Philadelphia: Saunders Elsevier (Chapter 48, Unit IX).

Hall, N. J., Jones, C. E., Eaton, S., Stanton, M. P., & Burge, D. M. (2011). Is interval appendicectomy justified after successful nonoperative treatment of an appendix mass in children? A systematic review. *Journal of Pediatric Surgery, 46*(4), 767–771.

Harnoss, J. C., Zelienka, I., Probst, P., Grummich, K., Müller-Lantzsch, C., Harnoss, J. M., et al. (2017). Antibiotics versus surgical therapy for uncomplicated appendicitis. *Annals of Surgery, 265*(5), 889–900.

Humes, D., Speake, W. J., & Simpson, J. (2007). Appendicitis. *BMJ Clinical Evidence, 2007*, 0408.

Kessler, N., Cyteval, C., Gallix, B., Lesnik, A., Blayac, P.-M., Pujol, J., et al. (2004). Appendicitis: Evaluation of sensitivity, specificity, and predictive values of US, Doppler US, and laboratory findings. *Radiology, 230*(2), 472–478.

Kharbanda, A. B., Cosme, Y., Liu, K., Spitalnik, S. L., & Dayan, P. S. (2011). Discriminative accuracy of novel and traditional biomarkers in children with suspected appendicitis adjusted for duration of abdominal pain. *Academic Emergency Medicine, 18*(6), 567–574.

Kharbanda, A. B., Vazquez-Benitez, G., Ballard, D. W., Vinson, D. R., Chettipally, U. K., Kene, M. V., et al. (2018). Development and validation of a novel pediatric appendicitis risk calculator (pARC). *Pediatrics, 141*(4), e20172699.

Kwan, K. Y., & Nager, A. L. (2010). Diagnosing pediatric appendicitis: Usefulness of laboratory markers. *American Journal of Emergency Medicine, 28*(9), 1009–1015.

Laurin, M., Everett, M. L., & Parker, W. (2011). The Cecal appendix: One more immune component with a function disturbed by post-industrial culture. *Anatomical Record, 294*(4), 567–579.

Lin, H.-F., Lai, H.-S., & Lai, I.-R. (2014). Laparoscopic treatment of perforated appendicitis. *World Journal of Gastroenterology, 20*(39), 14338–14347.

Livingston, E. H., Woodward, W. A., Sarosi, G. A., & Haley, R. W. (2007). Disconnect between incidence of nonperforated and perforated appendicitis. *Annals of Surgery, 245*(6), 886–892.

Maita, S., Andersson, B., Svensson, J. F., & Wester, T. (2020). Nonoperative treatment for nonperforated appendicitis in children: A systematic review and meta-analysis. *Pediatric Surgery International, 36*(3), 261–269.

Mällinen, J., Rautio, T., Grönroos, J., Rantanen, T., Nordström, P., Savolainen, H., et al. (2019). Risk of appendiceal neoplasm in periappendicular abscess in patients treated with interval appendectomy vs follow-up with magnetic resonance imaging. *JAMA Surgery, 154*(3), 200–2007.

Mentula, P., Sammalkorpi, H., & Leppäniemi, A. (2015). Laparoscopic surgery or conservative treatment for appendiceal abscess in adults? A randomized controlled trial. *Annals of Surgery, 262*(2), 237–242.

Nance, M. L., Adamson, W. T., & Hedrick, H. L. (2000). Appendicitis in the young child: A continuing diagnostic challenge. *Pediatric Emergency Care, 16*(3), 160–162.

Natesan, S., Lee, J., Volkamer, H., & Thoureen, T. (2016). Evidence-based medicine approach to abdominal pain. *Emergency Medicine Clinics of North America, 34*(2), 165–190.

Nikolaidis, P., Hammond, N., Marko, J., Miller, F. H., Papanicolaou, N., & Yaghmai, V. (2006). Incidence of visualization of the normal appendix on different MRI sequences. *Emergency Radiology, 12*(5), 223–226.

Nitecki, S., Karmeli, R., & Sarr, M. G. (1990). Appendiceal calculi and fecaliths as indications for appendectomy. *Surgery, Gynecology & Obstetrics, 171*(3), 185–188.

Park, H. C., Kim, M. J., & Lee, B. H. (2017). Randomized clinical trial of antibiotic therapy for uncomplicated appendicitis. *British Journal of Surgery, 104*(13), 1785–1790.

Patkova, B., Svenningsson, A., Almström, M., Eaton, S., Wester, T., & Svensson, J. F. (2020). Nonoperative treatment versus appendectomy for acute nonperforated appendicitis in children. *Annals of Surgery, 271*(6), 1030–1035.

Penner, R. M., & Majumdar, S. R. (2010). Acute abdominal pain. In *Practical gastroenterology and hepatology: Small and large intestine and pancreas* (pp. 129–134). Oxford, UK: Wiley-Blackwell.

Prada-Arias, M., Gómez-Veiras, J., Vázquez, J. L., Salgado-Barreira, Á., Montero-Sánchez, M., & Fernández-Lorenzo, J. R. (2020). Appendicitis or non-specific abdominal pain in pre-school children: When to request abdominal ultrasound? *Journal of Paediatrics and Child Health, 56*(3), 367–371.

Prada-Arias, M., Vázquez, J. L., Salgado-Barreira, Á., Gómez-Veiras, J., Montero-Sánchez, M., & Fernández-Lorenzo, J. R. (2017). Diagnostic accuracy of fibrinogen to differentiate appendicitis from nonspecific abdominal pain in children. *The American Journal of Emergency Medicine, 35*(1), 66–70.

Rabah, R. (2007). Pathology of the appendix in children: An institutional experience and review of the literature. *Pediatric Radiology, 37*(1), 15–20.

Sallinen, V., Akl, E. A., You, J. J., Agarwal, A., Shoucair, S., Vandvik, P. O., et al. (2016). Meta-analysis of antibiotics versus appendicectomy for non-perforated acute appendicitis. *British Journal of Surgery, 103*(6), 656–667.

Salminen, P., Paajanen, H., Rautio, T., Nordström, P., Aarnio, M., Rantanen, T., et al. (2015). Antibiotic therapy vs appendectomy for treatment of uncomplicated acute appendicitis. *JAMA, 313*(23), 2340.

Samuel, M. (2002). Pediatric appendicitis score. *Journal of Pediatric Surgery, 37*(6), 877–881.

Sartelli, M., Baiocchi, G. L., Di Saverio, S., Ferrara, F., Labricciosa, F. M., Ansaloni, L., et al. (2018). Prospective observational study on acute appendicitis worldwide (POSAW). *World Journal of Emergency Surgery, 13*(1), 19.

Saucier, A., Huang, E. Y., Emeremni, C. A., & Pershad, J. (2014). Prospective evaluation of a clinical pathway for suspected appendicitis. *Pediatrics, 133*(1), e88–e95.

Segev, L., Keidar, A., Schrier, I., Rayman, S., Wasserberg, N., & Sadot, E. (2015). Acute appendicitis in the elderly in the twenty-first century. *Journal of Gastrointestinal Surgery, 19*(4), 730–735.

Sesia, S., Mayr, J., Bruder, E., & Haecker, F.-M. (2013). Neurogenic appendicopathy: Clinical, macroscopic, and histopathological presentation in pediatric patients. *European Journal of Pediatric Surgery, 23*(03), 238–242.

Smith, H. F., Fisher, R. E., Everett, M. L., Thomas, A. D., Randal Bollinger, R., & Parker, W. (2009). Comparative anatomy and phylogenetic distribution of the mammalian cecal appendix. *Journal of Evolutionary Biology, 22*(10), 1984–1999.

Stewart, B., Khanduri, P., McCord, C., Ohene-Yeboah, M., Uranues, S., Vega-Rivera, F., et al. (2014). Global disease burden of conditions requiring emergency surgery. *British Journal of Surgery, 101*(1), e9–e22.

Storm-Dickerson, T. L., & Horattas, M. C. (2003). What have we learned over the past 20 years about appendicitis in the elderly? *The American Journal of Surgery, 185*(3), 198–201.

Vons, C., Barry, C., Maitre, S., Pautrat, K., Leconte, M., Costaglioli, B., et al. (2011). Amoxicillin plus clavulanic acid versus appendicectomy for treatment of acute uncomplicated appendicitis: An open-label, non-inferiority, randomised controlled trial. *The Lancet, 377*(9777), 1573–1579.

Wilms, I. M., de Hoog, D. E., de Visser, D. C., & Janzing, H. M. (2011). Appendectomy versus antibiotic treatment for acute appendicitis. *Cochrane Database of Systematic Reviews*, CD008359.

Yang, W.-C., Chen, C.-Y., & Wu, H.-P. (2013). Etiology of non-traumatic acute abdomen in pediatric emergency departments. *World Journal of Clinical Cases, 1*(9), 276–284.

Yu, M.-C., Feng, Y.-J., Wang, W., Fan, W., Cheng, H.-T., & Xu, J. (2017). Is laparoscopic appendectomy feasible for complicated appendicitis? A systematic review and meta-analysis. *International Journal of Surgery, 40*, 187–197.

Zhong, D., Brower-Sinning, R., Firek, B., & Morowitz, M. J. (2014). Acute appendicitis in children is associated with an abundance of bacteria from the phylum fusobacteria. *Journal of Pediatric Surgery, 49*(3), 441–446.

Chapter 18

Ovarian hormones, site-specific nociception, and hypertension

Bruna Maitan Santos[a], Glauce Crivelaro Nascimento[b], and Luiz Guilherme S. Branco[b]

[a]*Department of Physiology, School of Medicine of Ribeirão Preto, University of Sao Paulo, Sao Paulo, Brazil,* [b]*Department of Basic and Oral Biology, School of Dentistry of Ribeirao Preto, University of Sao Paulo, Sao Paulo, Brazil*

List of abbreviations

CNS   Central nervous system.
IASP   International Association for the Study of Pain
SHR   Spontaneously hypertensive rat
TMD   Temporomandibular disorder

Introduction

The population of older adults is increasing rapidly ("WHO | The World Health Report 1999 - Making a Difference,", 2013) and thus health concerns such as pain management, hypertension control, and physiological changes after menopause is a ripe and urgent issue (Yach, Hawkes, Gould, & Hofman, 2004). For instance, most community-dwelling older adults or nursing home residents report pain that precludes normal function on a daily basis (AGS Panel on Persistent Pain in Older Persons, 2002; Won et al., 2004). Similarly, prevalence of arterial hypertension is also rising together with aging of the population (Benetos, Petrovic, & Strandberg, 2019). These conditions are interconnected, for instance, the identification of variables that modulate pain in hypertension and the understanding of their mechanisms are essential for comprehensive pain assessment and personalized pain management (Fillingim, 2017).

Hypertension is a multifactorial and complex disorder that triggers severe organ damage and dysfunction (Bankir, Bichet, & Bouby, 2010; Hayashi, Saruta, Goto, & Ishii, 2010). The classical clinical sign for fitting in hypertension criteria is defined as a blood pressure greater than 140/90 mmHg or the use of antihypertensive drugs (Whitworth, 2003). Intriguingly, hypertension has been associated with hypoalgesia, i.e., a decrease in pain perception to a previously or typically considered painful stimulus (Ghione, Rosa, Mezzasalma, & Panattoni, 1988; Guasti et al., 2002) and, by extension, experimental hypertension with hyponociception, i.e., a decrease in neuronal encoding of a noxious stimuli (Santos et al., 2019). Hypertension is also a cardiovascular risk factor for several cardiovascular conditions including acute myocardial infarction and cardiac arrest. One of the features of acute myocardial infarction, for example, is pain, and hypertension increases the probability of silent myocardial infarction in patients (Kannel, Dannenberg, & Abbott, 1985). Hence, pain response must be evaluated carefully in hypertensive patients and all the variables that modulate pain must be remembered and evaluated for correct diagnosis of myocardial infarction.

The concept of pain was defined by IASP as an "unpleasant sensory and emotional experience associated with, or resembling that associated with, actual or potential tissue damage" (Raja et al., 2020). This definition has an important value to discriminate pain from nociception which involves encoding a noxious stimulus by a specific pathway activated inducing a behavioral and autonomic response, inasmuch as pain is the psychological event (Fig. 1). Escape reaction caused by a noxious mechanical stimulus or increasing blood pressure after a noxious stimulus are examples of behavioral and autonomic events. Therefore, pain studies are conducted mostly in humans that can express their emotional experiences and explain this phenomenon. Nociception assessment can be evaluated in clinical and preclinical studies; however, animal's studies have the advantage to study separate or combined variables modulating nociception. Nociception is assessed in these models by applying one of a variety of types of noxious stimulus including mechanical, chemical, electric, heat, and cold stimulus in the local of interest with a certain degree of intensity and duration (Deuis, Dvorakova, & Vetter, 2017). This noxious stimulus can be applied only once (acute) or chronically (Fig. 2). Not only these variables can modulate nociception, but a plenty of other variables modulate nociception, including psychological factors (see next section).

Features and Assessments of Pain, Anesthesia, and Analgesia. https://doi.org/10.1016/B978-0-12-818988-7.00045-5
Copyright © 2022 Elsevier Inc. All rights reserved.

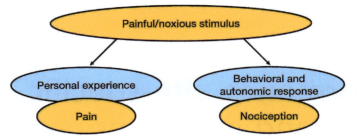

FIG. 1 Differences in pain and nociception nomenclatures. Pain and nociception are complementary concepts (but not interchangeable) about noxious/painful stimulus processing information which compile psychological, behavioral, and autonomic events.

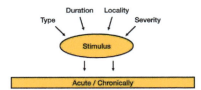

FIG. 2 Variables of a noxious stimulus. A noxious stimulus variates dependent upon some own characteristics such as type, duration, severity, and the site of stimulation and can be applied at the same local acutely or chronically.

Altogether, this chapter brings to light the proposed mechanisms related to hyponociception in hypertension in the literature (Section "Nociception mechanisms and modulation in hypertension: An overview") and discusses about some variables that modulate nociception in hypertension such as the locality of the stimulus, the sex of the individual, and the ovarian hormones (Sections "Nociception and blood pressure" and "Sex differences and ovarian hormones").

Nociception mechanisms and modulation in hypertension: An overview

The link between hyponociception and hypertension has great preclinical and clinical interest, but its mechanism is insufficiently understood. It seems to start with a biological defense mechanism through a baroreceptor reflex which lowers blood pressure by controlling heart rate and peripheral vasoconstriction and by downmodulating nociceptive responses by activation of descending inhibitory pathways, inducing endogenous opioids production, release, and by the direct action at the nociceptor terminal nerves (Chiang, Huang, Lin, Chan, & Chia, 2019; Saccò et al., 2013). In the same line, the overlap between the opioid and noradrenergic transmission systems can play a role in this putative mechanism. However, this hypothesis is not totally recognized (France et al., 2005), and more research is required to clarify the mechanism of hypertension-related hyponociception.

While a reduction in perception of an acute noxious stimulus initially plays a role in hypertension as a body adaptive defense, the pathophysiological mechanism is significantly different in chronic pain in which such adaptive relationship enclosed by blood pressure and nociception is considerably the opposite (Saccò et al., 2013). Most of the preclinical models of hypertension in rodents observed a decrease in acute nociception along with an increase in persistent nociception induced by inflammation in an inherited hypertension (Taylor et al., 1995; Taylor, Roderick, St Lezin, & Basbaum, 2001).

Few studies have addressed the orofacial nociception in murine hypertension models (Ghione, 1996b; Santos et al., 2019). The unique composition of the trigeminocervical system with anatomic, neural, and molecular variances indicates a specific and singular pain mechanism from that innervated by the dorsal ganglion of the spinal cord (Lopes, Denk, & McMahon, 2017). Orofacial pain, controlled by this complex neural arrangement of the trigeminal nerve, is of ongoing scientific interest due to its high prevalence in the worldwide population, culminating in the commitment of the clinicians and researchers to pinpoint its clinical and physiological nuances for a well-managed diagnosis and treatment (Hargreaves, 2011). Many pain states limited to the orofacial region proceed with sex-specific differences in predominance, but also with differences dependent upon the chronicity of the stimuli and its severity (Mogil, 2012). Despite that, Ghione (1996a) observed no differences in overall acute orofacial nociception between hypertensive women and men. Corroborating, paw mechanical nociception in rats is sex-dependent; i.e., male normotensive and hypertensive rats have higher nociceptive threshold than female normotensive and hypertensive rats, respectively (Santos et al., 2019). However, sex differences were not observed in orofacial nociception of hypertensive rats (Santos et al., 2019).

In fact, the hypertension-related hyponociception also includes the neuroendocrine system, and, specially, the hypothalamic-pituitary-adrenal axis, besides elevating activation of the sympathetic system through adrenal glands

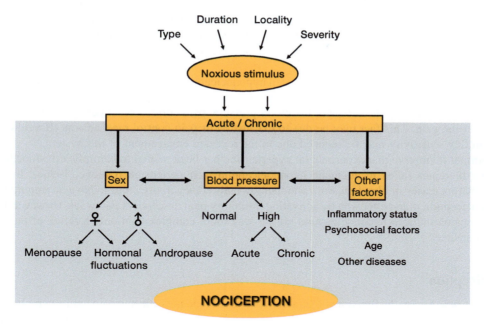

FIG. 3 Diagram of the variables acting on nociception modulation. A simple diagram of the variables acting on nociception focusing on the key role of the characteristics of each noxious stimulus, the sex of each individual and their blood pressure.

(Saccò et al., 2013). Undeniably, sex exerts a considerable influence on the modulation of this sensory stimulus. Essentially, women have an increased nociceptive threshold (Bartley & Fillingim, 2013; Fillingim, 2000; Fillingim, King, Ribeiro-Dasilva, Rahim-Williams, & Riley, 2009; Greenspan et al., 2007) and are more prone to chronic pain development (Fillingim, 2000; Mogil, 2012) compared with men. Some other variables can interfere in nociception including age, the inflammatory status, psychosocial factors, and other diseases. To clarify the difficulties of clinicians to understand nociception in each patient, we designed a simple diagram with some of the variables that can modulate nociception and the interconnection between them (Fig. 3) focusing on the role of the noxious stimulus, the sex of the individual, and blood pressure.

Nociception and blood pressure

It has become progressively clear that pain is a varied experience that differs extensively depending on the affected tissue (muscle, skin, viscera, joint) and the mechanism of damage (inflammatory, neuropathic, thermal, mechanical) (Yam et al., 2018). This is not different when considering blood pressure–involved pain response. There is an essential influence of the type of pain, the stage of the painful symptoms, and sexual differences in this interaction (Santos et al., 2019; Taylor et al., 2001). The neurobiology involved in the impact of these different factors still needs to be investigated, although we have some well-known mechanisms established.

Although not all of the findings on blood pressure–nociception interactions are explained by the baroreflex, it is important to understand the involvement of this key homeostatic reflex. Baroreceptors are stretch-sensitive mechanoreceptors localized mostly in the aortic arch and the carotid sinus region playing a major role in blood pressure regulation and also antinociception, as a means to control blood pressure arousal in the presence of noxious stimulus. Baroreceptor denervation, for example, inhibits antinociception induced by increased blood pressure (Thurston & Randich, 1990). Furthermore, electric stimulation of the nucleus tractus solitaries, the first synaptic station of the afferent fibers arising from the baroreceptors ameliorates hypernociception (Aicher & Randich, 1990). Furthermore, the role of the descending inhibitory pathways activated by the baroreflex-inducing antinociception was verified by blocking the spinal cord transmission and thus inhibiting antinociception (Thurston & Randich, 1990). The role of the noradrenergic descending pathway was confirmed using a noradrenergic antagonist by intrathecal administration (Thurston & Randich, 1990).

Similar data were obtained in many studies using a genetic model of hypertension, the spontaneously hypertensive rat (SHR) (Hoffmann, Plesan, & Wiesenfeld-Hallin, 1998; Santos et al., 2019; Taylor et al., 2001; Zamir, Simantov, & Segal, 1980). Comparable to rats with experimental hypertension, the hyponociception of SHR seems to depend upon undamaged baroreceptor connections to the nucleus of the solitary tract (Maixner et al., 1982) and unrestricted spinal transmission (Randich & Robertson, 1994).

Another descending pathway that causes antinociception in experimental hypertensive models is the opioidergic pathway. Hyponociception in SHR has been found to be reversible by one classical opioid antagonist, naloxone (Zamir et al., 1980). Again, Hoffmann et al. (1998) showed that SHR were more responsive to morphine than Wistar-Kyoto rats, indicating an upregulated opioidergic system.

It is interesting to observe that blood pressure–related hyponociception interconnects with adaptive stress-related hyponociception. While it is possible to find evidence of reduced acute and subchronic nociceptive responses under conditions of high blood pressure, regarding chronic pain, the blood pressure–pain correlation is positive, i.e., the higher the blood pressure, the higher the pain. This fact may be due to exhaustion of pain control mechanisms (Bruehl & Chung, 2004). In instance, baroreflex sensitivity may be reduced due to the repeated engagement of this system.

It is also important to observe that experimental hypertension has been related to baroreflex dysfunction and that antihypertensive drug therapies failed to attenuate hyponociception in hypertension (Ghione, 1996a). In summary, these findings suggest that increases in blood pressure and thereby baroreflex activity seem to have an important role in development of hyponociception in hypertension by activating inhibitory pain descendent pathways. However, the mechanisms underlying hypertension-induced hyponociception are far way more complex than only a reflex of the increased blood pressure in hypertension.

Orofacial nociception

As to different locations of pain, the functional interaction between blood pressure and pain has been replicated in studies on odontogenic pain. Trigeminal-regulated orofacial pain conditions include temporomandibular disorders (TMDs), migraines, cluster headaches, tension-type headaches, burning mouth syndrome, neuralgias, dental-associated conditions, head and neck cancers, and certain idiopathic pain syndromes (Crandall, 2018).

It was found that untreated hypertensive individuals have higher pain threshold all along electrical stimulation of dental pulp than individuals with normal blood pressure levels (Zamir & Shuber, 1980); besides, pain tolerance are higher in hypertensive patients than in normotensive individuals in this condition (Guasti et al., 1995). In addition, the administration of antihypertensive drugs resulted in a significant reduction in nociception threshold and tolerance in response to electrical pulpal stimulation (Guasti et al., 2002). Still on the relationship of orofacial nociception and increased blood pressure, Santos et al. (2019) observed that SHRs have reduced orofacial nociception.

Studies about the cardiovascular system and clinical pain conditions report that chronic pain conditions such as temporomandibular disorders and chronic musculoskeletal complaints are less prevalent in individuals with higher resting blood pressure values as compared to those with low blood pressure values (Hagen et al., 2002; Stovner & Hagen, 2009). It is important to highlight here that the chronicity of pain in the temporomandibular region related to the increase in blood pressure behaves differently from pain in other regions as mentioned above. In fact, hypertensive individuals have a lower probability of developing temporomandibular disorders as compared to patients with low blood pressure at resting conditions (Diatchenko, Nackley, Slade, Fillingim, & Maixner, 2006), while fibromyalgia patients have higher resting systolic and diastolic blood pressure as compared to pain-free controls (Kulshreshtha, Gupta, Yadav, Bijlani, & Deepak, 2012).

In the orofacial region, there is an important sex specificity implicated. Distinct sexual differences seen in orofacial pain conditions are mediated by hormone-based mechanisms. In fact, changes in gonadal hormones such as estrogen, progesterone, and androgens are shown to be associated with changes in pain experience in many orofacial pain conditions (Shinal & Fillingim, 2007). For example, women using hormones therapy refer to more severe orofacial pain compared to women not using hormones (Wise, Riley, & Robinson, 2000). Animal studies are in agreement with the hormone-dependency of this effect since estrogen replacement treatment in male or ovariectomized female rats increase excitability of innervation in the temporomandibular joint (Cairns, Sim, Bereiter, Sessle, & Hu, 2002; Flake, Bonebreak, & Gold, 2005).

In briefly, several aspects are noteworthy when evaluating orofacial nociception such as the differences between sex, the gonadal hormonal status, and the noxious stimulus per se. More attention should be given to the disparities in site-specific nociception since orofacial nociception has its own neuronal processing mechanism encoding a noxious stimulus.

Sex differences and ovarian hormones

Normotensive patients and rodent models

There is plenty of evidence in the practical clinics and literature that overall pain is a sex-dependent complex phenomenon. The current understanding regarding nociception and sex differences suggests that women have a lower threshold to detect a potentially noxious stimulus than men (Mogil, 2012). Moreover, there is no doubt that hyperalgesia (Gregory,

Gibson-Corley, Frey-Law, & Sluka, 2013) and chronic pain (Moloney et al., 2016) are also elevated in females compared to males. Pain state conditions are also more prevalent in women, and their pain responses last longer and with higher intensity than men (Moloney et al., 2016).

Unfortunately, there is a vast difference in the amount of research using male preclinical or animal studies, and most of the pain mechanisms were described in male murine models. Even with this lack of basic knowledge between male and female physiology, most of the studies assume that sex differences observed in pain responses are partially sex hormones-related. A study conducted with kids lower than 12 years old found no differences in nociception between sexes. After this phase of life, when sex hormones levels get high, it becomes clear the sex differences in nociceptive response, and thus the sex hormones' key role (Boerner, Birnie, Caes, Schinkel, & Chambers, 2014). Three sex hormones seem to play the major role in sex differences in nociception and this chapter focused on them: testosterone, progesterone, and estrogen.

Testosterone levels are undoubtedly higher in males than in females. It is well established that this hormone downmodulates nociception and hyperalgesia, and it has a protective effect in TMJ hypernociception in males (Fanton, Macedo, Torres-Chávez, Fischer, & Tambeli, 2017). Moreover, it seems that the antinociceptive and antihyperalgesic role of testosterone in males establishes in their development. Intriguingly, testosterone can preserve only the antihyperalgesic effect in male adulthood (Borzan & Fuchs, 2006).

Progesterone and estrogen are ovarian hormones that fluctuate dependent upon the phase of the menstrual cycle in women and dependent upon the estrous cycle phase in experimental rodent models (Butcher, Collins, & Fugo, 1974). The fluctuations in ovarian hormones production complicate the correct analysis of the pain mechanisms in women at the reproductive stage and, therefore, remain unclear. Even with the complexity to evaluate the mechanism itself, it is established that ovarian hormones play a crucial role in menstrual/estrous cycle–dependent nociceptive responses, and therefore, it changes across the menstrual cycle (Fillingim, 2000). In experimental rodent models, the estrous cycle is divided into four phases, each one with specific hormonal fluctuations. The proestrus and the estrous phases are the most studied ones. The proestrus phase has a hormonal profile of elevated levels of estrogens and a later rise in levels of progesterone. The estrus phase is swiftly after the proestrus phase and it begins when the progesterone peak levels decline accompanied by a low-grade production of estrogens (Butcher et al., 1974). It was found a lower pain threshold and consequently, hypernociception in the proestrus and estrus phases compared to the other phases of the estrous cycle and male rats (Moloney et al., 2016).

The influence of these hormones in nociception is often verified by analyzing the effects after gonadectomy. Gonadectomy is a surgical procedure for removal of either the testes (orchiectomy) or the ovaries (ovariectomy) causing gonadal hormones production loss. Often, gonadectomy increases nociception and pain in preclinical and clinical studies (Kuba & Quinones-Jenab, 2005). It is essential to emphasize that ovariectomy mimics the menopausal stage in women adulthood, and the ovarian hormones replacement can prevent some of the consequences induced by ovarian hormones depletion. For example, estradiol and progesterone replacement in ovariectomized rodents induce hyponociception in most cases (Kuba & Quinones-Jenab, 2005).

Progesterone replacement alone is involved in nociceptive and neuropathic pain downmodulation (González et al., 2019). Estrogen replacement alone induces antinociception in rats (Lawson, Nag, Thompson, & Mokha, 2010; Mannino, South, Quinones-Jenab, & Inturrisi, 2007). On the contrary, the incidence of TMD increases in postmenopausal women undergoing estrogen replacement therapy, and hormone replacement therapy is associated with enhanced pain sensitivity in postmenopausal women (Sarajari & Oblinger, 2010). However, a well-conducted study revealed that the regime of estrogen replacement differently regulates estrogen hypo- or hypernociceptive effect explaining in partially how such discrepancies in estrogen replacement findings occur to some extent (Zhang et al., 2020).

Hypertensive patients and spontaneously hypertensive rats (SHR) rodent model

As aforementioned (Section "Orofacial nociception"), the view that women are more sensitive to pain is a common belief between clinical and research staff. However, despite all the knowledge supporting this assertion, this standpoint is relatively straightforward, given the myriad of variables that modulate pain responses. One of these variables that can undoubtedly modulate pain responses not only in the woman but also in man individuals is the effect of the acute surge in blood pressure and chronic increase in blood pressure, i.e., hypertension. Both acute and chronic blood pressure disturbances decrease nociception in hypertensive patients. However, no sex differences were found in pain evaluated in hypertensive patients (Ghione, 1996a).

Although the absence of sex differences in pain in hypertension is observed in the clinical studies involving hypertensive patients, the difficulties in conducting an adequate standardized clinical study, such as the small experimental groups, the differences in social parameters, and the nonidentified variables, have an impact on the effect of sex in pain responses of hypertensive people (see Table 1). One example is the individual variability in blood pressure inside each group and the strong correlation between blood pressure and pain threshold (Ghione et al., 1988). These factors would weaken the effect of sex differences in hypertensive patients. Moreover, most of the studies performed in humans evaluate orofacial nociception, which has a unique pain mechanism. These issues bring to light the importance of preclinical research to identify and assess sex differences in nociceptive responses using rodent models to induce hypertension.

TABLE 1 Clinical and preclinical studies used in the discussion of the relation hypertension × pain × sex differences in this chapter.

Author, year	Type of study	Type of pain analyzed	Hypertension × Pain × Sex specificity findings
Chiang et al. (2019)	Clinical	Postoperative pain	It was not observed hypertensive hypoanalgesia effect in the male patients but it was observed a "reverse" hypertensive hypoanalgesia effect in the female patients
France et al. (2005)	Clinical	Nociceptive knee flexion reflex	Endogenous opioid blockade was associated with increased pain ratings in women but with increased pain threshold in men
Guasti et al. (2002)	Clinical	Dental pain (pulpar test)	No sex specificity was investigated
Hagen et al. (2002)	Clinical	Headache	High systolic and diastolic pressures were associated with reduced risk of non-migrainous headache for both sexes
Guasti et al. (1995)	Clinical	Dental pain (pulpar test)	No sex specificity was investigated
Rau et al. (1994)	Clinical	Thermal and mechanical pain	No sex specificity was investigated
Ghione et al. (1988)	Clinical	Dental pain (pulpar test)	No significant effects could be detected by analysis of variance for sex, age, and family history of hypertension
Zamir and Shuber (1980)	Clinical	Dental pain (pulpar test)	Hypertensive man had a higher threshold for sensation of pain in the tooth-pulp test than normotensive man controls
Santos et al. (2019)	Preclinical	Mechanical and inflammatory nociception in temporomandibular joint or paw	Female SHR had higher mechanical nociception than male SHR only in the paw, but it had higher formalin-induced orofacial nociception than male SHR. The absence of ovarian hormones caused an increase in mean arterial pressure and a decrease in paw nociception in female SHR
Taylor et al. (2001)	Preclinical	Thermal and inflammatory nociception in paw	No sex specificity was investigated
Taylor et al. (1995)	Preclinical	Inflammatory nociception in paw	No sex specificity was investigated
Dworkin et al. (1994)	Preclinical	Achilles tendon reflexes	No sex specificity was investigated
Randich and Robertson (1994)	Preclinical	Thermal nociception in paw	No sex specificity was investigated
Aicher and Randich (1990)	Preclinical	Thermal nociception in paw	No sex specificity was investigated
Thurston and Randich (1990)	Preclinical	Thermal nociception in paw	No sex specificity was investigated
Randich and Maixner (1981)	Preclinical	Thermal nociception in paw	No sex specificity was investigated
Zamir et al. (1980)	Preclinical	Thermal and mechanical nociception in paw	No sex specificity was investigated

A table containing studies about hypertension, pain, and sex differences used in the chapter. The types of pain (considering stimulus and local) were highlighted.

One of the most employed models to induce hypertension in preclinical studies is the SHR model. It is well established that male SHR have higher mean arterial pressure compared to female SHR of the same age (Riley, Robinson, Wise, Myers, & Fillingim, 1998). Also, high blood pressure has a strong correlation with hyponociception in SHR and hypertensive patients. The magnitude of this effect is probably different between sexes in SHR since no sex differences were observed in nociception in the tail and orofacial area (Reckelhoff, Zhang, Srivastava, & Granger, 1999; Santos et al., 2019). On the contrary, paw nociception had a sex-dependent component in SHR, i.e., male hypertensive rats have decreased nociception compared to female SHR (Santos et al., 2019). It is important to note that sex differences in nociception are dependent not only on the blood pressure levels but also on the site of the harmful stimulus in SHR.

The lack of estrogen induced by ovariectomy in SHR induces an increase in sympathetic activity followed by an increase in blood (Ito, Hirooka, Kimura, Sagara, & Sunagawa, 2006). This increase in blood pressure after ovariectomy correlates with the increase in paw nociception, but not with orofacial nociception in ovariectomized SHR. However, in an acute inflammatory condition induced with a chemical irritant, it was observed that female SHR had greater behavioral nociceptive responses compared to male SHR (Santos et al., 2019), indicating that inflammatory status has a key role in inducing sex differences in orofacial nociception.

Of note, sex hormones have a role in the immune system, which has an enormous impact on pain modulation. Thus, sex hormones can modulate pain responses by acting indirectly in the immune system (Totsch & Sorge, 2017). Hypertension induces a chronic inflammatory state which seems to be sex hormones–dependent (Iwasa et al., 2014; Rizzo et al., 2009). Altogether, it is remarkable the strong interconnection between sex hormones, pain, and immune system and, more importantly, hypertension.

Applications to other areas

In this chapter, we have reviewed hypertension-related hyponociception and the critical relationship between hypertension and sex specificities in nociception. We must also point out that one of the main features of myocardial infarction is pain. Indeed, hypertension-associated hypoalgesia may be harmful especially in silent myocardial ischemia and undetected myocardial infarction, both of which are more predominant in hypertensive people (Ghione, 1996a). About orofacial pain, women are more likely to have chronic orofacial pain than men and are more inclined to having jaw pain as an advice sign of acute myocardial infarction, increasing the risk of make mistakes in diagnosis.

Through a behavioral analysis, we can also highlight psychiatric comorbidity as one of the negative long-term impacts on health outcomes in individuals with cardiovascular diseases (Liu et al., 2016). The presence of depression and/or anxiety facilitates considerably disease burdens on adults with hypertension (Wallace, Zhao, Misra, & Sambamoorthi, 2018). About sex differences in this field, both men and women with hypertension are more likely to have a diagnosis of depression and anxiety disorder, with slightly greater risk for men than women (Graham & Smith, 2016).

Other agents of interest (250–500 words)

Not applicable.

Mini-dictionary of terms

Hyperalgesia: Psychological response to a painful stimulus that is not normally painful, induced by lowering nociceptor threshold level.

Nociception: Mechanism by which peripheral noxious stimuli are transmitted to the CNS inducing behavioral and autonomic response.

Acute Pain: It starts with an injury, and algogenic substances are synthesized and released on the lesioned tissue stimulating nerve endings (nociceptors) of myelinated thin or amyelinated fibers. Its natural evolution is remission. It is related to the permanence or appearance of neurovegetative changes (warning signs).

Chronic Pain: It is more than a symptom; it is the disease that persists, does not disappear after healing the lesion, or is related to chronic pathological processes.

Arterial Hypertension: It is a chronic disease in which the blood pressure in the arteries is constantly elevated. At long term is one of the main risk factors for a number of serious diseases from a long-term perspective. It stems from nonspecific genetic and lifestyle factors or can be comorbidity from a recognized disease.

Ovariectomy: It is the surgical removal of one (unilateral) or both ovaries (bilateral).

Key facts

Key facts of sex differences in pain

Bullet 1: Sex differences have been recognized for decades in a number of fields of research.

Bullet 2: Women report more severe levels of pain, more frequent pain in more areas of the body, and pain of longer duration than men.

Bullet 3: Testosterone levels have been associated with prevalence of cluster headaches and fibromyalgia.

Bullet 4: Estrogen levels have been linked to irritable bowel syndrome, temporal mandibular joint disorder, and rheumatoid arthritis.

Bullet 5: Pain is multifactorial and the differences between men and women could be attributed to anatomical, physiological, biological, or psychosocial factors.

Bullet 6: Lack of sex differences in some trials may be due to issues of power, population demographics, social variables, or other unidentified variables.

Summary points

- Bullet of Summary Point 1: This chapter focuses on the relation of hypertension and pain and its sex specificities.
- Bullet of Summary Point 2: In acute pain, hypertension has been proven to increase pain threshold. In chronic pain, hypertension may be associated with the worsening of pain.
- Bullet of Summary Point 4: Central noradrenergic and opioidergic pathways are important players in the descending pain inhibitory pathways and regulation of the cardiovascular system.
- Bullet of Summary Point 3: Multiple factors are considered responsible for sex differences in pain perception and for the great prevalence of chronic pain conditions in women. Biological factors such as sex hormones are thought to be one of the main mechanisms explaining these differences.
- Bullet of Summary Point 5: Orofacial nociception is positively correlated with arterial blood pressure in hypertensive patients, but no sex differences were found. In experimental hypertension model, the locality and the type of the noxious stimulus combined to the effect of blood pressure in nociception play an important role in sex differences in hypertension-induced antinociception.

References

AGS Panel on Persistent Pain in Older Persons. (2002). The management of persistent pain in older persons. *Journal of the American Geriatrics Society, 50*, 205–224.

Aicher, S. A., & Randich, A. (1990). Antinociception and cardiovascular responses produced by electrical stimulation in the nucleus tractus solitarius, nucleus reticularis ventralis, and the caudal medulla. *Pain, 42*, 103–119.

Bankir, L., Bichet, D. G., & Bouby, N. (2010). Vasopressin V2 receptors, ENaC, and sodium reabsorption: A risk factor for hypertension? *American Journal of Physiology. Renal Physiology, 299*, F917–F928.

Bartley, E. J., & Fillingim, R. B. (2013). Sex differences in pain: A brief review of clinical and experimental findings. *British Journal of Anaesthesia, 111*, 52–58.

Benetos, A., Petrovic, M., & Strandberg, T. (2019). Hypertension management in older and frail older patients. *Circulation Research, 124*, 1045–1060.

Boerner, K. E., Birnie, K. A., Caes, L., Schinkel, M., & Chambers, C. T. (2014). Sex differences in experimental pain among healthy children: A systematic review and meta-analysis. *Pain, 155*, 983–993.

Borzan, J., & Fuchs, P. N. (2006). Organizational and activational effects of testosterone on carrageenan-induced inflammatory pain and morphine analgesia. *Neuroscience, 143*, 885–893.

Bruehl, S., & Chung, O. Y. (2004). Interactions between the cardiovascular and pain regulatory systems: An updated review of mechanisms and possible alterations in chronic pain. *Neuroscience and Biobehavioral Reviews, 28*, 395–414.

Butcher, R. L., Collins, W. E., & Fugo, N. W. (1974). Plasma concentration of LH, FSH, prolactin, progesterone and estradiol-17β throughout the 4-day estrous cycle of the rat. *Endocrinology, 94*, 1704–1708.

Cairns, B. E., Sim, Y., Bereiter, D. A., Sessle, B. J., & Hu, J. W. (2002). Influence of sex on reflex jaw muscle activity evoked from the rat temporomandibular joint. *Brain Research, 957*, 338–344.

Chiang, H.-L., Huang, Y.-C., Lin, H.-S., Chan, M.-H., & Chia, Y.-Y. (2019). Hypertension and postoperative pain: A prospective observational study. *Pain Research and Management, 2019*, 8946195.

Crandall, J. A. (2018). An introduction to orofacial pain. *Dental Clinics of North America, 62*, 511–523. W.B. Saunders, October 1.

Deuis, J. R., Dvorakova, L. S., & Vetter, I. (2017). Methods used to evaluate pain behaviors in rodents. *Frontiers in Molecular Neuroscience, 10*, 284. Frontiers Media S.A., September 6.

Diatchenko, L., Nackley, A. G., Slade, G. D., Fillingim, R. B., & Maixner, W. (2006, August). Idiopathic pain disorders—Pathways of vulnerability. *Pain*, *123*, 226–230. Pain.

Dworkin, B. R., Elbert, T., Rau, H., Birbaumer, N., Pauli, P., Droste, C., & Brunia, C., H. (1994). Central effects of baroreceptor activation in humans: Attenuation of skeletal reflexes and pain perception. *Proceedings of the National Academy of Sciences of the United States of America*, *91*(14), 6329–6333. https://doi.org/10.1073/pnas.91.14.6329.

Fanton, L. E., Macedo, C. G., Torres-Chávez, C. E., Fischer, L., & Tambeli, C. H. (2017). Activational action of testosterone on androgen receptors protects males preventing temporomandibular joint pain. *Pharmacology Biochemistry Behavior*, *152*, 30–35.

Fillingim, R. B. (2000). Sex, gender, and pain: Women and men really are different. *Current Review of Pain*, *4*, 24–30.

Fillingim, R. B. (2017). Individual differences in pain: Understanding the mosaic that makes pain personal. *Pain*, *158*, S11–S18.

Fillingim, R. B., King, C. D., Ribeiro-Dasilva, M. C., Rahim-Williams, B., & Riley, J. L. (2009). Sex, gender, and pain: A review of recent clinical and experimental findings. *The Journal of Pain*, *10*, 447–485.

Flake, N. M., Bonebreak, D. B., & Gold, M. S. (2005). Estrogen and inflammation increase the excitability of rat temporomandibular joint afferent neurons. *Journal of Neurophysiology*, *93*, 1585–1597.

France, C. R., Al'Absi, M., Ring, C., France, J. L., Brose, J., Spaeth, D., ... Wittmers, L. E. (2005). Assessment of opiate modulation of pain and nociceptive responding in young adults with a parental history of hypertension. *Biological Psychology*, *70*, 168–174.

Ghione, S. (1996a). Hypertension-associated hypalgesia. Evidence in experimental animals and humans, pathophysiological mechanisms, and potential clinical consequences. *Hypertension (Dallas, Tex. : 1979)*, *28*, 494–504.

Ghione, S. (1996b). Hypertension-associated hypalgesia. *Hypertension*, *28*, 494–504.

Ghione, S., Rosa, C., Mezzasalma, L., & Panattoni, E. (1988). Arterial hypertension is associated with hypalgesia in humans. *Hypertension (Dallas, Tex. : 1979)*, *12*, 491–497.

González, S. L., Meyer, L., Raggio, M. C., Taleb, O., Coronel, M. F., Patte-Mensah, C., & Mensah-Nyagan, A. G. (2019). Allopregnanolone and progesterone in experimental neuropathic pain: Former and new insights with a translational perspective. *Cellular and Molecular Neurobiology*, *39*, 523–537. Springer New York LLC, May 1.

Graham, N., & Smith, D. J. (2016). Comorbidity of depression and anxiety disorders in patients with hypertension. *Journal of Hypertension*, *34*, 397–398. Lippincott Williams and Wilkins, March 1.

Greenspan, J. D., Craft, R. M., LeResche, L., Arendt-Nielsen, L., Berkley, K. J., Fillingim, R. B., ... Consensus Working Group of the Sex, Gender, and Pain SIG of the IASP. (2007). Studying sex and gender differences in pain and analgesia: A consensus report. *Pain*, *132*, S26–S45.

Gregory, N. S., Gibson-Corley, K., Frey-Law, L., & Sluka, K. A. (2013). Fatigue-enhanced hyperalgesia in response to muscle insult: Induction and development occur in a sex-dependent manner. *Pain*, *154*, 2668–2676.

Guasti, L., Cattaneo, R., Rinaldi, O., Rossi, M. G., Bianchi, L., Gaudio, G., ... Venco, A. (1995). Twenty-four–hour noninvasive blood pressure monitoring and pain perception. *Hypertension*, *25*, 1301–1305.

Guasti, L., Zanotta, D., Mainardi, L. T., Petrozzino, M. R., Grimoldi, P., Garganico, D., ... Cerutti, S. (2002). Hypertension-related hypoalgesia, autonomic function and spontaneous baroreflex sensitivity. *Autonomic Neuroscience*, *99*, 127–133.

Hagen, K., Stovner, L. J., Vatten, L., Holmen, J., Zwart, J. A., & Bovim, G. (2002). Blood pressure and risk of headache: A prospective study of 22 685 adults in Norway. *Journal of Neurology Neurosurgery and Psychiatry*, *72*, 463–466.

Hargreaves, K. M. (2011). Orofacial pain. *Pain*, *152*, S25–S32.

Hayashi, K., Saruta, T., Goto, Y., & Ishii, M. (2010). Impact of renal function on cardiovascular events in elderly hypertensive patients treated with efonidipine. *Hypertension Research*, *33*, 1211–1220.

Hoffmann, O., Plesan, A., & Wiesenfeld-Hallin, Z. (1998). Genetic differences in morphine sensitivity, tolerance and withdrawal in rats. *Brain Research*, *806*, 232–237.

Ito, K., Hirooka, Y., Kimura, Y., Sagara, Y., & Sunagawa, K. (2006). Ovariectomy augments hypertension through rho-kinase activation in the brain stem in female spontaneously hypertensive rats. *Hypertension*, *48*, 651–657.

Iwasa, T., Matsuzaki, T., Tungalagsuvd, A., Munkhzaya, M., Kawami, T., Kato, T., ... Irahara, M. (2014). Effects of ovariectomy on the inflammatory responses of female rats to the central injection of lipopolysaccharide. *Journal of Neuroimmunology*, *277*, 50–56.

Kannel, W. B., Dannenberg, A. L., & Abbott, R. D. (1985). Unrecognized myocardial infarction and hypertension: The Framingham study. *American Heart Journal*, *109*, 581–585.

Kuba, T., & Quinones-Jenab, V. (2005). The role of female gonadal hormones in behavioral sex differences in persistent and chronic pain: Clinical versus preclinical studies. *Brain Research Bulletin*, *66*.

Kulshreshtha, P., Gupta, R., Yadav, R. K., Bijlani, R. L., & Deepak, K. K. (2012). A comprehensive study of autonomic dysfunction in the fibromyalgia patients. *Clinical Autonomic Research*, *22*, 117–122.

Lawson, K. P., Nag, S., Thompson, A. D., & Mokha, S. S. (2010). Sex-specificity and estrogen-dependence of kappa opioid receptor-mediated antinociception and antihyperalgesia. *Pain*, *151*, 806–815.

Liu, N., Pan, X., Yu, C., Lv, J., Guo, Y., Bian, Z., ... Peng, Y. (2016). Association of major depression with risk of ischemic heart disease in a mega-cohort of Chinese adults: The China Kadoorie Biobank Study. *Journal of the American Heart Association*, *5*. https://doi.org/10.1161/JAHA.116.004687.

Lopes, D. M., Denk, F., & McMahon, S. B. (2017). The molecular fingerprint of dorsal root and trigeminal ganglion neurons. *Frontiers in Molecular Neuroscience*, *10*, 304.

Maixner, W., Touw, K. B., Brody, M. J., Gebhart, G. F., John, P., & L. (1982). Factors influencing the altered pain perception in the spontaneously hypertensive rat. *Brain Research*, *237*, 137–145.

Mannino, C. A., South, S. M., Quinones-Jenab, V., & Inturrisi, C. E. (2007). Estradiol replacement in ovariectomized rats is antihyperalgesic in the formalin test. *The Journal of Pain: Official Journal of the American Pain Society, 8*, 334–342.

Mogil, J. S. (2012). Sex differences in pain and pain inhibition: Multiple explanations of a controversial phenomenon. *Nature Reviews Neuroscience, 13*, 859–866.

Moloney, R. D., Sajjad, J., Foley, T., Felice, V. D., Dinan, T. G., Cryan, J. F., & O'Mahony, S. M. (2016). Estrous cycle influences excitatory amino acid transport and visceral pain sensitivity in the rat: Effects of early-life stress. *Biology of Sex Differences, 7*, 1–8.

Raja, S. N., Carr, D. B., Cohen, M., Finnerup, N. B., Flor, H., Gibson, S., ... Vader, K. (2020). The revised International Association for the Study of Pain definition of pain: Concepts, challenges, and compromises. *Pain, 161*(9), 1976–1982.

Randich, A., & Robertson, J. D. (1994). Spinal nociceptive transmission in the spontaneously hypertensive and Wistar-Kyoto normotensive rat. *Pain, 58*, 169–183.

Randich, A., & Maixner, W. (1981). Acquisition of conditioned suppression and responsivity to thermal stimulation in spontaneously hypertensive, renal hypertensive and normotensive rats. *Physiology & Behavior, 27*, 585–590. https://doi.org/10.1016/0031-9384(81)90226-2.

Rau, H., Brody, S., Larbig, W., Pauli, P., Vohringer, M., Harsch, B., ... Birbaumer, P. (1994). Effects of PRES baroreceptor stimulation on thermal and mechanical pain threshold in borderline hypertensives and normotensives. *Psychophysiology, 31*(5), 480–485. https://doi.org/10.1111/j.1469-8986.1994.tb01051.x.

Reckelhoff, J. F., Zhang, H., Srivastava, K., & Granger, J. P. (1999). Gender differences in hypertension in spontaneously hypertensive rats. *Hypertension, 34*, 920–923.

Riley, J. L., Robinson, M. E., Wise, E. A., Myers, C. D., & Fillingim, R. B. (1998). Sex differences in the perception of noxious experimental stimuli: A meta-analysis. *Pain, 74*, 181–187.

Rizzo, M., Corrado, E., Coppola, G., Muratori, I., Novo, G., & Novo, S. (2009). Markers of inflammation are strong predictors of subclinical and clinical atherosclerosis in women with hypertension. *Coronary Artery Disease, 20*, 15–20.

Saccò, M., Meschi, M., Regolisti, G., Detrenis, S., Bianchi, L., Bertorelli, M., ... Caiazza, A. (2013). The relationship between blood pressure and pain. *The Journal of Clinical Hypertension, 15*, 600–605.

Santos, M. B., Nascimento, G. C., Capel, C. P., Borges, G. S., Rosolen, T., Sabino, J. P. J., ... Branco, L. G. S. (2019). Sex differences and the role of ovarian hormones in site-specific nociception of SHR rats. *American Journal of Physiology—Regulatory, Integrative and Comparative Physiology, 317*, R223–R231.

Sarajari, S., & Oblinger, M. M. (2010). Estrogen effects on pain sensitivity and neuropeptide expression in rat sensory neurons. *Experimental Neurology, 224*, 163–169.

Shinal, R. M., & Fillingim, R. B. (2007, January). Overview of orofacial pain: Epidemiology and gender differences in orofacial pain. *Dental Clinics of North America, 51*, 1–18. Dent Clin North Am.

Stovner, L. J., & Hagen, K. (2009). Hypertension-associated hypalgesia: A clue to the comorbidity of headache and other pain disorders. *Acta Neurologica Scandinavica, 120*, 46–50.

Taylor, B. K., Alex Peterson, M., Basbaum, A. I., Taylor, B. K., Peterson, M. A., & Basbaum, A. I. (1995). Exaggerated cardiovascular and behavioral nociceptive responses to subcutaneous formalin in the spontaneously hypertensive rat. *Neuroscience Letters, 201*, 9–12.

Taylor, B. K., Roderick, R. E., St Lezin, E., & Basbaum, A. I. (2001). Hypoalgesia and hyperalgesia with inherited hypertension in the rat. *American Journal of Physiology. Regulatory, Integrative and Comparative Physiology, 280*, R345–R354.

Thurston, C. L., & Randich, A. (1990). Acute increases in arterial blood pressure produced by occlusion of the abdominal aorta induces antinociception: Peripheral and central substrates. *Brain Research, 519*, 12–22.

Totsch, S. K., & Sorge, R. E. (2017). Immune system involvement in specific pain conditions. *Molecular Pain, 13*. 174480691772455.

Wallace, K., Zhao, X., Misra, R., & Sambamoorthi, U. (2018). The humanistic and economic burden associated with anxiety and depression among adults with comorbid diabetes and hypertension. *Journal of Diabetes Research, 2018*, 4842520.

Whitworth, J. A. (2003). 2003 World Health Organization (WHO)/International Society of Hypertension (ISH) statement on management of hypertension. *Journal of Hypertension, 21*, 1983–1992. J Hypertens, November.

WHO. (2013). *The world health report 1999—Making a difference*. WHO.

Wise, E. A., Riley, J. L., & Robinson, M. E. (2000). Clinical pain perception and hormone replacement therapy in postmenopausal women experiencing orofacial pain. *The Clinical Journal of Pain, 16*(2), 121–126. https://doi.org/10.1097/00002508-200006000-00005.

Won, A. B., Lapane, K. L., Vallow, S., Schein, J., Morris, J. N., & Lipsitz, L. A. (2004). Persistent nonmalignant pain and analgesic prescribing patterns in elderly nursing home residents. *Journal of the American Geriatrics Society, 52*, 867–874.

Yach, D., Hawkes, C., Gould, C. L., & Hofman, K. J. (2004). The global burden of chronic diseases: Overcoming impediments to prevention and control. *Journal of the American Medical Association, 291*, 2616–2622.

Yam, M. F., Loh, Y. C., Tan, C. S., Adam, S. K., Manan, N. A., & Basir, R. (2018). General pathways of pain sensation and the major neurotransmitters involved in pain regulation. *International Journal of Molecular Sciences, 19*. MDPI AG, August 1.

Zamir, N., & Shuber, E. (1980). Altered pain perception in hypertensive humans. *Brain Research, 201*, 471–474.

Zamir, N., Simantov, R., & Segal, M. (1980). Pain sensitivity and opioid activity in genetically and experimentally hypertensive rats. *Brain Research, 184*, 299–310.

Zhang, W., Wu, H., Xu, Q., Chen, S., Sun, L., Jiao, C., ... Chen, X. (2020). Estrogen modulation of pain perception with a novel 17β-estradiol pretreatment regime in ovariectomized rats. *Biology of Sex Differences, 11*. https://doi.org/10.1186/s13293-019-0271-5.

Chapter 19

Linking the heart and pain: Physiological and psychophysiological mechanisms

Dmitry M. Davydov[a,b,c]

[a]*University of Jaén Hospital, Jaén, Spain,* [b]*Wyższa Szkoła Społeczno-Przyrodnicza im. Wincentego Pola w Lublinie / Vincent Pol University in Lublin, Poland,* [c]*Laboratory of Neuroimmunopathology, Institute of General Pathology and Pathophysiology, Russian Academy of Sciences, Moscow, Russia*

Introduction

An acute pain is elicited by noxious signals from the peripheral nociceptors with an immediate behavioral or psychophysiological response to escape from a source of nociception (nociceptive flexion or withdrawal reflex), later followed by additional physiological (e.g., cardiovascular, endocrine, metabolic, immune, hemostatic) and cognitive-emotional (pain sensation) stress responses to promote somatic healing and behavioral recovering. Repeats of the noxious signals may differently condition these two kinds of early (escape) and later (stress) responses (Harvie, Moseley, Hillier, & Meulders, 2017; Madden et al., 2016). In some individuals, an outcome is habituation (i.e., decreases in responding) of the nociceptive flexion with sensitization (i.e., increases in responding) of the stress responses. However, in other individuals the result is the sensitization of the nociceptive withdrawal with the habituation of the stress responses (Slepian et al., 2016). The regulation of the two (early and later) stages in the responses to acute pain and the two corresponding ways (i.e., habituation and sensitization) of conditioning in response to persistent pain may be associated with individual predisposition to either resilience or adaptation processes in the interaction between cardiovascular and pain control mechanisms and their failure in some individuals (Davydov, 2020). The predisposition to a higher pain resilience mechanism increasing the pain threshold was considered to be related to an individual predisposition to increased cardiac baroreflex as a key cardiovascular homeostatic reflex indicated by increased baroreflex-related bradycardia associated with empowered vagus activity (i.e., a higher central inhibitory activity) (Davydov, Naliboff, Shahabi, & Shapiro, 2018; Duschek, Heiss, Buechner, & Schandry, 2009; Duschek, Mück, & Reyes Del Paso, 2007; Reyes del Paso, Montoro, Muñóz Ladrón de Guevara, Duschek, & Jennings, 2014). The predisposition to a higher pain adaptation mechanism increasing pain tolerance was considered to be related to an individual predisposition to attenuated cardiac baroreflex with a central resetting of baroreceptor activation to a higher operating blood pressure point (i.e., to a higher sensory tolerance) (Davydov et al., 2018; Lautenschläger et al., 2015). These predispositions are related to central mechanisms regulating gain and gating processes in the nucleus tractus solitarii in response to neural impulses originating from baroreceptors (Davydov et al., 2018; Pickering, Boscan, & Paton, 2003). Moreover, the latter adaptation mechanism is considered to be the main one in pain chronification through promoting permanent physiological alterations (e.g., baroreflex resetting for hypertension and reduced heart rate variability associated with increased sympathetic and reduced vagus activities), behavioral adjustments (e.g., functional disability or restrictions in behavioral functioning), and cognitive-emotional modifications (e.g., central sensitization) conditioning most pain-related stimuli as fearful or undesired (Davydov, 2020; Davydov et al., 2018). However, an overadaptation to pain in some predisposed people with exaggerated cardiovascular responses can lead to a higher endorphin release with mood elevation, transferring the fearful pain-induced risk-avoidance behavior to fearless risk-taking and antisocial activities, vigorous exercise, verbal aggression, fighting, and nonsuicidal self-injury as a result of unconditioned or Pavlovian conditioned reflex to pain-related situations as desired (Betensky & Contrada, 2010; Harmon-Jones, Summerell, & Bastian, 2018).

Features and Assessments of Pain, Anesthesia, and Analgesia. https://doi.org/10.1016/B978-0-12-818988-7.00011-X
Copyright © 2022 Elsevier Inc. All rights reserved.

The history of the subject

Historically, the view that the cardiovascular system and systems regulating pain perception are closely intertwined was predicted by a variety of findings in the 1970s. These findings showed that a direct activation of the baroreceptors determining bradycardia resulted not only in a blood pressure decrease, but also in generalized inhibitory effects (e.g., decreased muscle tone, diminished cortical activity, a sleep-like state) (Randich & Maixner, 1984). This bradycardia is an intrinsic reflex related to the response of mechano- or high pressure baro-receptors, mainly located in the walls of the carotid sinuses, aortic arch, carotid arteries, and the bifurcation of the brachiocephalic and subclavian arteries to mechanical distention associated with arterial blood pressure increases. This baroreceptor activation was found to increase the afferent activity in the carotid sinus and aortic depressor nerves (within the glossopharyngeal and the vagus nerves, respectively) transmitting its inhibitory, gating, and braking effects through nucleus tractus solitarius to multiple structures of the central nervous system. However, other preliminary studies have favored the view that increased arterial blood pressure itself, for example, by sympathomimetics or other mechanisms related to experimental and genetic forms of hypertension, can induce hypoalgesia by acting directly on the central nervous system and its endogenous opioid system, rather than acting through afferent neural inputs to the central nervous system (i.e., excluding the baroreflex mechanism inducing bradycardia) (Randich & Maixner, 1984).

Later, studies from Dworkin et al. (Dworkin, Filewich, Miller, Craigmyle, & Pickering, 1979), Randich and Hartunian (Randich & Hartunian, 1983) and Randich and Maixner (Randich & Maixner, 1984) directly confirmed the first mechanism, concluding that, "… *systems involved in the regulation of blood pressure are physiologically linked to systems involved in the regulation of pain*" (Randich & Hartunian, 1983). In three separate experiments in rats, Dworkin et al. (Dworkin et al., 1979) and Randich and Hartunian (Randich & Hartunian, 1983) showed that the degree of analgesia (assessed by a tail-flick latency or a run distance on a treadwheel as two different models of acute pain escape-avoidance behavior) during infusions of phenylephrine to induce a rapid increase in the blood pressure level was significantly correlated with the magnitude of reflex bradycardia, but not with arterial and central venous blood pressure increases or the dose of phenylephrine. Moreover, a later study from Randich and Maixner (Randich & Maixner, 1984) showed that the degree of bradycardia as an indicator of the magnitude of the baroreflex response is also not critical for predicting the antinociception gain. They considered that the stimulation of the afferent component of baroreflex was more important for regulating the antinociception mechanism than the gain of the efferent cardiac response to it. Several aspects of these early experiments of interactions between pain and cardiovascular control systems were found to deserve further consideration and have been elaborated upon in follow-up studies by other researchers.

Aspects related to (i, ii) a hypertension-related hypoalgesia phenomenon in chronic and acute pain conditions and (iii) a pain-killing etiology of the essential (primary) hypertension

First, because the baroreceptor reflex corrects rapid blood pressure increases through increases in vagal tone and withdrawals of sympathetic tone, the arterial blood pressure increases can also be related to the degree of analgesia, but only when these correctional mechanisms are failing to adequately lower arterial blood pressure, as in essential (primary) hypertension (Nascimento Rebelatto, Alburquerque-Sendín, Guimarães, & Salvini, 2017). This may happen when pain becomes persistent or chronic (i.e., becomes related to more central or affective than to peripheral or nociceptive sources) and this in turn decreases the gain of baroreflex cardiac responses and resets baroreceptors for correcting blood pressure fluctuations at a higher steady state (Davydov, 2020). This consideration was supported several decades later (Davydov, Naliboff, Shahabi, & Shapiro, 2016; Davydov et al., 2018; Davydov & Perlo, 2015).

Second, the findings proposed that initial increases in arterial blood pressure during the development of hypertension could contribute to the production of analgesia or hypoalgesia through the baroreflex mechanisms elucidated in the experiments, that is, that hypertension may be considered a protective factor against the intensity of future acute pain. Indeed, higher preoperative arterial blood pressure was found to provide reduced pain ratings a day and 2 days after surgery (France & Katz, 1999). The sustained increase of blood pressure was proposed to have the pain-killing effect through descending pain inhibitory (antinociceptive) processes associated with enhanced noradrenergic activity (Boscan, Pickering, & Paton, 2002; Davydov et al., 2016, 2018; Lautenschläger et al., 2015). Moreover, in longitudinal studies, hypertension was also found to be a protective factor against the risk of pain chronification (e.g., against new cases of chronic headache, migraine, low back pain) (Hagen et al., 2002; Heuch, Heuch, Hagen, & Zwart, 2014). However, in cases of already presented chronic pain, elevated systolic blood pressure predicted the increase of chronic pain severity at follow-up (e.g., in the low back) (Leino-Arjas, Solovieva, Kirjonen, Reunanen, & Riihimäki, 2006). In this case, the baroreflex resetting in response to the sustained increase of systolic blood pressure may determine the decrease in vagal tone triggering a central pain sensitization mechanism (Davydov, 2017; Davydov et al., 2018). In contrast, elevated diastolic blood pressure as an indicator of increased systemic vascular resistance in response to baroreflex resetting predicted the decrease of chronic pain severity at follow-up (Davydov, 2020; Davydov et al., 2018).

Third, these early experiments also considered that the endogenous pain inhibition by these mechanisms could be important for the etiology or pathophysiology of hypertension (i.e., for the development of so-called essential hypertension) as a disease of adaptation to pain with hypoalgesia as a risk factor or a predictive marker of predisposition to hypertension. In other words, this mechanism may be the background for resetting the blood pressure level to a higher steady state as an instrumentally conditioned or learned process for adapting to painful conditions (i.e., to decrease chronic pain severity) if nociceptive activity becomes sustained or frequent (Davydov, 2020; Davydov et al., 2018). Some longitudinal studies have presented findings supporting this consideration (Campbell et al., 2002; Rau & Elbert, 2001). This mechanism may also explain the lack of compliance to antihypertension treatment in some hypertensive subjects because removing the blood pressure-related pain inhibition by antihypertensive treatment would not be well tolerated by patients.

Aspects related to (iv) the effectiveness of pain management prediction and monitoring by cardiovascular measures and to (v) effective pain control by cardiovascular mechanisms with pain-killing effects

Fourth, the individual effectiveness of pain management in its sensorial and affective domains can be predicted and monitored by baroreflex-related cardiovascular measures such as the increase of baroreceptor sensitivity to inhibitions in the case of interventions with predominantly physical engagement (e.g., walking), but its decrease in the case of interventions with predominantly sensorial engagement (e.g., yoga) (Davydov, Shahabi, & Naliboff, 2019).

Fifth, the early findings proposed that baroreflex-related mechanisms could be used in pain control and management. For example, a recent study showed that the effectiveness of treatment of patients with predominant somatic or affective symptoms of a chronic pain syndrome may be improved by personalizing nonpharmacological interventions using pretreatment cardiovascular phenotyping of patients with respect to predominant mechanisms of baroreflex regulation of blood pressure and knowledge of specific effects of the interventions on these regulation mechanisms (Davydov et al., 2019). Other studies showed that the stimulation of carotid baroreceptors by external suction in the neck chamber or by blood pressure control using alpha1-sympathomimetic agents such as midodrine significantly decreases acute pain intensity in people if it produces strong baroreflex-related bradycardia (Duschek et al., 2009; Reyes del Paso et al., 2014). Other findings showed that pain processing fluctuates during normal breathing and that pain feeling can be inhibited by different mechanisms within the central nervous system during either expirations or inspirations modulating the vagus nerve activity (Arsenault, Ladouceur, Lehmann, Rainville, & Piché, 2013; Iwabe, Ozaki, & Hashizume, 2014). Applying these findings in yoga-like breathing (Prânayâma) manipulations (e.g., compared to spontaneous breathing, breath holding, or retention following a deep inhalation as in yoga's *Antara Kumbhaka,* and at some other respiration phases, as in yoga's *Bahya Kumbhaka*) was found to reduce the experience of pain and increase the nociception reflex by (i) affecting blood volume redistribution resulting in a higher stroke volume and blood pressure that stimulates baroreceptors activating a pain inhibitory or an antinociceptive resilience (arousal-dampening) mechanism, or (ii) a sensory distraction process associated with baroreflex-related vasoconstriction and diastolic blood pressure increase inducing another antinociceptive or pain adaptation (arousal-enhancing) mechanism resetting sensory tolerance thresholds to a higher steady state (Bjerre et al., 2011; del Reyes Paso, Muñoz Ladrón de Guevara, & Montoro, 2015; Jafari et al., 2016; Lautenschläger et al., 2015). Water deprivation and hypohydration through decreasing blood volume lead to baroreflex unloading, thus increasing acute pain sensitivity and impacts on the development of chronic pain such as headaches, but more water intake through increasing blood volume determines baroreflex loading and thus reduces pain intensity and duration (Bear, Philipp, Hill, & Mündel, 2016; Cheuvront et al., 2012; Spigt, Weerkamp, Troost, van Schayck, & Knottnerus, 2012). Additionally, a direct stimulation of the afferent branch of the vagus nerve, one of the nerves regulating cardiovascular activity through the baroreflex, has successfully been explored in the treatment of chronic pain syndromes (Busch et al., 2013; Yuan & Silberstein, 2017). It was found to have resolving effects on systemic, central, and peripheral inflammatory processes (by reducing inflammatory signaling in the spleen and hypothermia as in experimental sepsis, inhibiting cytokines in the hypothalamus as in endotoxemic conditions, and locally as in experimental arthritis, respectively) as frequent sources of chronic and acute pain (Amorim et al., 2020; Bassi et al., 2015; Brognara et al., 2018). Moreover, some effective analgesics such as nonsteroidal antiinflammatory drugs are proposed to have a supplementary hypertension-related hypoalgesic mechanism (earlier considered their side effect) that increases their total pain-killing effect (Davydov, 2020).

An aspect related to (vi) the pain-o-meter technology

Sixth, an objective pain-o-meter developed with the cardiovascular measures was recently proposed for clinical application (Perlo & Davydov, 2016). During decades of subsequent studies of the interrelationships between the cardiovascular and pain-regulating systems, their complexity indicated in the earlier studies was often overlooked (Bruehl, Walker, & Smith, 2017; Davydov, 2017; Ledowski, Reimer, Chavez, Kapoor, & Wenk, 2012). Many scientific publications on the relationships have been devoted to mainly a reduced respiratory sinus arrhythmia or a power of the heart rate variability associated with respiration (i.e., in its high frequency band) as an indicator of reduced cardiac vagal reactivity in chronic pain populations (Koenig et al., 2016; Tracy et al., 2016). Moreover, more severe chronic pain was found to be associated with a greater reduction in this heart rate variability (Walker et al., 2017). However, this measure was found to be a nonspecific, clinically significant indicator of the energy capacity that is used for physiological, affective, cognitive, and behavioral coping activities against different external and internal challenges besides pain, that is, a nonspecific indicator of severity in coping with any challenge (Davydov, 2017). Other scientists considered that more severe pain as a stressor should evoke a more generalized sympathetic arousal accumulated in simple hemodynamic and neuroendocrine responses as vital signs according to the common assumption that "pain triggers a sympathetic stress response," but failed to confirm this proposal in studies of postoperative pain severity with the final conclusion that, "...*the monitoring of hemodynamic or autonomic parameters to predict states of pain appears not to be useful and remains highly questionable...*" (Ledowski et al., 2012). Inconsistent results of the correspondence of higher clinical pain severity to higher overall sympathetic drive to the heart and vessels, including the assessment of its control by baroreflex, were also obtained in studies of patients with chronic pain (Bruehl et al., 2018; Zamunér et al., 2015).

These unreliable results demonstrated that the simplified physiological models such as pain intensity relation to simple sympathetic cardiovascular overarousal and simplified statistical models of direct effects between them were not accurate and thus not suitable for the development of the pain-o-meter metric. Nevertheless, these results have had a dramatic implication on research in the area–the cardiovascular studies have been pushed to the periphery of mainstream research of specific pain-o-meter technologies in contrast to the brain-mapping approach. However, compared with simple hemodynamic measures such as baroreceptor sensitivity, heart rate variability, and blood pressure associated with simple statistical models of their relationship to pain measures, only a few studies in pain research have used statistical analyses with complex indirect parallel and serial mediation models applied to complex measures of spontaneous baroreceptor reflex. The latter measures include (a) asymmetrically increased baroreflex bradycardia or decreased baroreflex tachycardia as an indicator of increased pain threshold, (b) decreased baroreflex bradycardia as an indicator of increased pain sensitivity, (c) reduced parasympathetic "braking" of sympathetic activity as an indicator of either (i) increased pain sensitization if it is associated with reduction of heart rate variability or (ii) increased pain tolerance if it is associated with peripheral vasoconstriction (increasing diastolic blood pressure), i.e., various baroreflex components as specific metrics of pain severity or intensity in its different (nociceptive, affective, and cognitive) domains (Davydov et al., 2016, 2018). In all these cases, the regulation of blood pressure becomes asymmetric, favoring the predominant dampening of either blood pressure increases or decreases that shift the arousal homeostasis down or up from the healthy optimum (Fig. 1) (Davydov, 2018; Davydov, Shapiro, Goldstein, & Chicz-DeMet, 2007). Moreover, some other physiological reflexes such as pupil size fluctuations as a specific measure of pain intensity (Charier et al., 2017) were found to correspond to the same baroreflex activities responding to blood pressure fluctuations (Calcagnini, Giovannelli, Censi, Bartolini, & Barbaro, 2001).

An aspect related to (vii) the baroreflex control of blood volume and pressure fluctuations in response to posture or body position changes in patients with pain

Last but not least, pain-related effects on the baroreflex symmetry mechanism, which diminish baroreflex tachycardia in response to blood pressure decreases, may determine, as a byproduct, the symptoms of orthostatic intolerance (a profound blood pressure fall during standing) and the postural tachycardia syndrome in some patients with chronic pain syndromes as one of multiple distinct pathophysiological subtypes within them (Mustafa et al., 2012). Indeed, clinical findings showed that patients with various chronic pain syndromes frequently suffer from orthostatic intolerance and postural tachycardia syndrome (Durham, McDonald, Hutchinson, & Newton, 2015; Furlan et al., 2005). Accordingly, symmetry and asymmetry in cardiovascular (e.g., heart rate and blood pressure) responses to clino- and ortho-static challenges were suggested for measuring and monitoring pain severity in its particular nociceptive, affective, and cognitive domains by wearable devices at inpatient and outpatient clinics (Davydov, 2018; Davydov & Perlo, 2015; Davydov & Sokolova, 2015). These responses are considered to be proxy measures of baroreceptor activation and inhibition mechanisms diurnally associated with every low and high arousal condition such as laying/standing and waking/sleeping states (Davydov, Shapiro, & Goldstein, 2010).

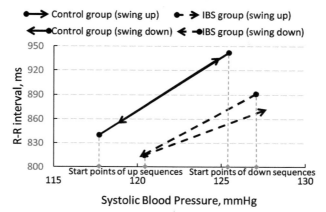

FIG. 1 Schematic representation of spontaneous cardiac baroreflex responses drawn using real data obtained by 20-min continuous recordings of finger systolic blood pressure and electrocardiogram in groups of control subjects and patients with irritable bowel syndrome (IBS) during resting baseline recording. Computer program scans the data for sequences of at least three beats of consecutively increasing or decreasing systolic pressures, which are accompanied by changes in the same direction of the RR intervals of the subsequent beats with several lags. The plot shows average linear regression slopes of baroreflex-related responses of RR interval to systolic blood pressure ramps up and down. Vertical split (gain asymmetry) and diagonal shift (steady-state change) in baroreflex were found between its start and end points in IBS patients. *From Davydov, D. M., Naliboff, B., Shahabi, L., & Shapiro, D. (2018). Asymmetries in reciprocal baroreflex mechanisms and chronic pain severity: Focusing on irritable bowel syndrome. Neurogastroenterology and Motility, 30, e13186. with permission.*

This corresponds with a consideration that in the supine position, signals from baroreceptors, if they are correctly conditioned, should inhibit the cortical arousal of frontal-central regions for a relaxing response. In the standing position, the inhibition of baroreceptors, if they are correctly conditioned, should stimulate cortical arousal for a prolonged activation response. The balanced reciprocity of baroreceptor sensitivity to activation and inhibition during different body positions (standing, sitting, lying) is maintained by asymmetry between bradycardiac and tachycardiac baro-responses as a resilience mechanism counteracting adaptive gravity-induced blood volume movements between the upper and lower body parts (Davydov, 2018; Davydov et al., 2018; Davydov & Czabak-Garbacz, 2017). According to the physiological resilience concept, the resilience mechanisms should regulate the capacity for arousal dampening and arousal-promoting functions (i) through increasing bradycardiac baro-sensitivity or baro-responsivity to blood pressure rises with higher tolerance to its falls (i.e., with tachycardiac hyporesponsivity) and (ii) through decreasing bradycardiac baro-sensitivity or baro-responsivity to blood pressure rises with lower tolerance to its falls (i.e., with tachycardiac hyperresponsivity), respectively. Experimental clino-orthostatic studies in patients with chronic pain syndromes showed that these evolutionary-developed physiological resilience mechanisms (i.e., cardiovascular responses at different stages of the clino-orthostatic challenge) were specifically disturbed by the severity of pain intensity, pain distress, and pain catastrophizing in patients and by comparing the patients with healthy subjects (Davydov, 2018; Davydov & Czabak-Garbacz, 2017; Davydov & Perlo, 2015). In these cases, the longer duration of pain or pain chronicity can transfer mainly the cardiac regulation of the blood pressure level within the normal fluctuation range presented in healthy people to the overriding vascular mechanism of its regulation in patients with chronic pain indicated by asymmetrically enlarged vasoconstriction (extra resistance of the vascular system) in responses to orthostatic stress that do not correspond to the homeostatic level of systolic blood pressure in posture (Figs. 2A–C). Thus, the orthostatic extra response of the vascular system in patients with pain may be used as an indicator of the individual grade of adaptation or coping with it.

Summarizing the current knowledge of heart and pain relationships

In summary, the subjective acute pain intensity sensation is negatively related to blood pressure, but this relationship may be masked by correct compensation effects of baroreflex on the heart rate (i.e., baroreflex bradycardia associated with high baroreceptor sensitivity to rises of blood pressure). The subjective acute pain intensity sensation is positively related to heart rate (i.e., tachycardia syndrome), but this relationship may be masked by centrally inhibited baroreflex bradycardia as a compesantion for inhibited baroreflex tachycardia. And the subjective acute pain intensity sensation is negatively related to baroreflex bradycardia (baroreceptor sensitivity to rises of blood pressure), that is, to the gain of the central inhibition effect as the common core mechanism of pain and cardiovascular regulation. When the pain sensation becomes persistent, it first determines a persistent increase of systolic blood pressure level as a response to sustained or frequent nociceptive activity (i.e., a pain-elicited hypertension) after about a year of chronic pain duration (Figs. 2A and 3) (Davydov et al.,

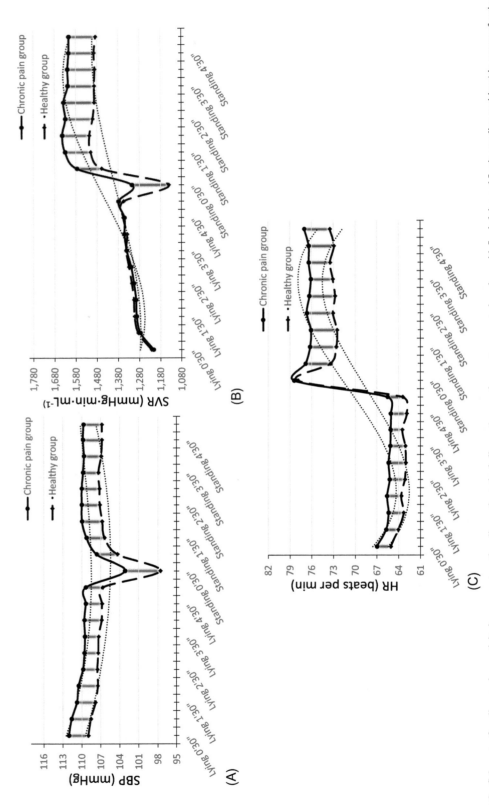

FIG. 2 Schemata of cardiovascular reactivity in response to active clino-orthostatic challenges (averaged across two repeated sessions with 5-min lying and 5-min standing positions) in groups of patients with chronic pain (disease duration from 1 to 28 years) and healthy people, represented for 30 s means of averaged beat-by-beat (A) systolic blood pressure (SBP), (B) systemic vascular resistance (SVR), and (C) heart rate (HR) additionally approximated with cubic polynomial functions (dotted lines). Unpublished data presented at the Second International Congress on Controversies in Fibromyalgia in Berlin by Contreras-Merino, A. M., Davydov, D. M., & del Paso, G. A. R. (2020). Blunted cardiovascular responses to orthostatic and clinostatic posture changes in fibromyalgia: Further evidence of autonomic abnormalities in the disorder. *Clinical and Experimental Rheumatology, 38* (1, Suppl. 123), S117–S117.

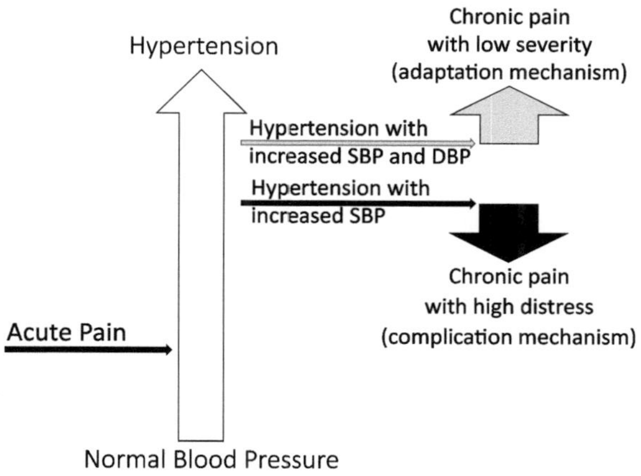

FIG. 3 Response of blood pressure to acute pain and two ways of chronic pain severity regulation in hypertension as its outcome. *From Davydov, D. M. (2020). Impact of antihypertensive treatment on resiliency to clinical pain. Journal of Hypertension, 38, 961–967, with permission.*

2016). This in turn initiates an adaptive central resetting of baroreceptor activation for an additional, sustained increase of diastolic blood pressure associated with high systemic vascular resistance (Fig. 2B) induced by enhanced noradrenergic activity with the pain-killing effect (interpreted as a pain adaptation mechanism). Longer pain duration may further determine an asymmetrically strengthened bradycardiac compared with a tachycardiac baro-response (interpreted as a later pain resilience mechanism related to the ability of baroreceptor activation to suppress central noradrenergic pathways) (Fig. 1) that decreases usual pain complaints (Davydov et al., 2018). These pain adaptation and pain resilience mechanisms may produce temporal pain relief (decrease of pain severity and related complaints) in patients with the most chronic pain syndromes. From an evolutionary viewpoint, these adaptation hypertensive and resilience bradycardiac mechanisms may help in the recovery of physical functioning after impairments.

However, a longer pain duration may also decrease an absolute gain of bradycardiac baro-responses (interpreted as a later pain decompensation mechanism associated with decreased vagus reactivity). This increases the usual pain complaints in patients (i) through the mechanism associated with the decrease of power of the low frequency of heart rate variability coupled with somatic complaints (somatic pain aggravation mechanism) and (ii) through the mechanism associated with the decrease of power of the high frequency of heart rate variability coupled with affective complaints (affective pain aggravation mechanism) (Davydov et al., 2016). Both mechanisms are related to the impaired vagus regulation of energy supply indicated by a decreased muscle-heart reflex and a decreased respiratory cardiac arrhythmia, normally providing adequate oxygen to fulfill the metabolic demands of muscles in active mode and the brain and body in default or relaxing mode, respectively (Al-Ani et al., 1997; Davydov, 2017; Roach & Sheldon, 2018). This decrease in vagal reactivity coupled with negative somatic or affect comorbidity to chronic pain syndromes (i.e., high distress; Fig. 3) is considered to be a mechanism for triggering an avoidance coping behavior. From an evolutionary viewpoint, these pain aggravation mechanisms are part of a general or nonspecific mechanism augmenting reality delivered by specific sensory (visceral, proprioceptive, and nociceptive) harm-related signaling to consciousness if it is insufficient for behavior change for coping with the problem. This mechanism is especially enhanced in individuals with deficits in sensory processing such as those with 'difficulty to identify feelings' subtype of alexithymia construct (Di Tella et al., 2017; Kano, Hamaguchi, Itoh, Yanai, & Fukudo, 2007; Pollatos, Dietel, Gündel, & Duschek, 2015).

Applications to other areas

Some issues discussed in the chapter may be incorporated in more advanced guidelines for the management of high blood pressure. For example, an individual history of chronic pain should be taken into account for less-aggressive blood pressure regulation in general or, at least, for selecting specific antihypertensive drugs in patients with hypertension. For example, antihypertensive therapies that decrease both systolic and diastolic blood pressures would be expected to increase (i) chronic pain severity by affecting the adaptation to pain mechanism and (ii) the probability of new cases of chronic pain by affecting the pain resilience mechanism. Therefore, this could be a disadvantage for pain control by pain-killing drugs and can contribute to the increased risk of seeking more-aggressive painkilling medication (Fig. 4). In contrast, the antihypertensive therapies that decrease only systolic blood pressure with improvement in vagus activity are proposed to decrease chronic pain severity and may help in pain control. Moreover, using information introduced in the chapter, new

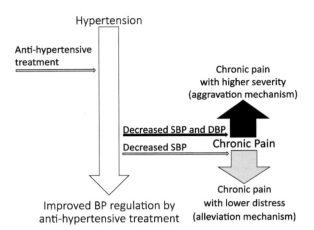

FIG. 4 Blood pressure and two ways of chronic pain severity regulation in response to antihypertensive treatment. *From Davydov, D. M. (2020). Impact of antihypertensive treatment on resiliency to clinical pain. Journal of Hypertension, 38, 961–967, with permission.*

antihypertensive and antihypotensive drug and nondrug treatment approaches can be developed for improving acute and chronic pain management using different pain resilience and pain adaptation mechanisms related to cardiovascular regulation systems. For example, direct and indirect blood pressure manipulations can increase pain resilience or pain adaptation (i.e., pain threshold or pain tolerance, respectively) through the above-mentioned baroreflex mechanisms. For instance, a study in patients with fibromyalgia showed that at least three different cardiovascular mechanisms regulating the blood pressure level are affected in the chronic pain syndrome adapting the patients to pain. Respectively, these cardiovascular mechanisms can be sensitive to the blood pressure lowering effects of the three main groups of antihypertensive drugs thus disrupting their adaptation to pain effect with increase of chronic pain severity and number of new cases of chronic pain. These groups are diuretics with effects on body fluid volume, beta-blockers with effects on heart rate, and angiotensin-converting enzyme inhibitors and angiotensin II receptor blockers with effects on peripheral vasoconstriction (Fig. 2) (Contreras-Merino, Davydov, & del Paso, 2020; Davydov, 2020). Thus, the potency in pain control of new pain-killing drugs and nondrug treatments can be increased through their additional effects on blood pressure regulation mechanisms. Moreover, some effects of blood pressure regulation mechanisms on pain control may be associated with their effects on innate and adaptive immune systems and inflammation processes (Harrison et al., 2011). For example, a recent study showed that an increase in inflammation, as in rheumatoid arthritis, was associated with an increase in systolic blood pressure at low levels of inflammation. However, at higher levels of inflammation, an increase in inflammation was associated with an increase in systolic blood pressure, and when the inflammation was mitigated, the blood pressure increased as the inflammation decreased (Yu et al., 2018).

Thus, this chapter presented an example for the application of a more holistic, health-centric alternative in medicine in contrast to the disease-oriented approach. In general, the health-centric approach considers that the good health outcome of a particular treatment should not only be evaluated with respect to recovery from a specific disorder such as chronic pain syndrome, but according to its harmful or positive effects on health in general, that is, on health recovery and health promotion of the most vital systems regulating body and mind activity such as blood pressure control (Davydov, 2018). The contemporary crisis in chronic pain treatment may be an example of the wrong way associated with the disease-oriented approach (Davydov, 2020). While current guidelines of antihypertension treatment have helped to decrease the mean blood pressure in most Western countries, they ignore that the suggested aggressive approach of blood pressure regulation could increase the need for pain control in chronic pain patients with hypertension, increase incidences of pain chronification, and thus, increase the risk of seeking more aggressive painkilling medication similar to what has been observed in the opioid epidemic in the United States. Moreover, clinical cases of so-called essential (primary) hypertension are not evaluated for the history of chronic pain syndrome or phenotypes with risk for impaired pain tolerance or pain threshold as the risk factors for this kind of hypertension. Thus, the current widely accepted orientation on a specific outcome in disease treatment may contribute to the increased risk of comorbidities, multimorbidities, and chronification of the diseases.

The topic of this chapter is also relevant to the regulation of lifestyle behaviors such as cigarette smoking or tobacco use and alcohol consumption that can increase chronic pain severity in somatic and affective domains through effects on processes associated with baroreceptor activation and heart rate variability, respectively (Davydov et al., 2018). Caffeine consumption also affects baroreflex that can determine the greater severity of chronic pain syndrome in a follow-up. Regular physical activity has different (protective and exacerbating) effects on pain intensity or severity that may be related to different autonomic (metabolic) phenotypes of people predisposed to the regulation of the heart and pain interactions through different afferent, central, or efferent vagal mechanisms (Davydov et al., 2018, 2019; Davydov & Nurbekov, 2016).

Other areas of interest

As was presented in the chapter, the brain integrates cardiovascular and pain systems by intertwining stimulatory nociceptive and inhibitory baroreceptive signals coming through the respective peripheral (bottom-up) sensory afferent pathways. However, a variety of findings have shown that the inhibitory effect of baroreceptor activation with bradycardia response to a blood pressure increase is a more generalized arousal-dampening mechanism. For example, this mechanism can protect the mood and promote mental resiliency against affective disorders in the face of strong aversive events (Davydov, 2018; Davydov, Stewart, Ritchie, & Chaudieu, 2010; Davydov, Stewart, Ritchie, & Chaudieu, 2012). Thus, this mechanism can inhibit or resolve various physical sources and psychological companions of pain (e.g., peripheral, central, and systemic inflammation, increased muscle tone or spasticity, negative emotions and ruminations, insomnia, and agitation) (Bassi et al., 2015; Brognara et al., 2018; Davydov, Shapiro, Cook, & Goldstein, 2007; Davydov, Shapiro, & Goldstein, 2010; Delgado, Vila, & Reyes del Paso, 2014; McCubbin et al., 2018; Randich & Maixner, 1984).

Respiratory sinus arrhythmia or a heart rate variability associated with respiration as an evolutionary developed mechanism controlled by the vagus nerve for more effective oxygen uptake per breath while minimizing the work done by the heart was discussed in the chapter as an indicator of energy resources for coping with cognitive and affective components of pain (Davydov, 2017; Davydov et al., 2016, 2018; Walker et al., 2017). However, this mechanism of energy metabolism regulation was also found to be a factor determining effective coping with or protection against many other physical (e.g., autoimmune diseases), affective (e.g., depression), and cognitive (e.g., attention deficit and hyperactivity disorder) challenges, claiming energy resources for recovery after responses to various sources of stress besides pain (Balzarotti, Biassoni, Colombo, & Ciceri, 2017; Davydov, 2017; Kemp & Quintana, 2013; Koopman et al., 2016).

In contrast to the bottom-up cardiovascular mechanisms of pain inhibition in its nociceptive and central (e.g., affective) domains, there are some central mechanisms related to a full or partial detachment of afferent pain stimuli from their awareness (i.e., from processing of nociceptive stimuli) that can reduce the acute pain intensity and severity of persistent or chronic pain (Kohl, Rief, & Glombiewski, 2013; Schreiber et al., 2014). These mechanisms may include more or less conscious processes competing for attentional and other central processing resources at the moment. For example, such cognitive coping strategies as cognitive or attention distraction can be used by people with a high predisposition to the externally oriented thinking facet of the alexithymia trait for coping with pain and related distress (Davydov, Luminet, & Zech, 2013; Kano et al., 2007; Pollatos et al., 2015). Another central detaching mechanism that was evolutionarily developed for the management of competing needs such as hunger, thirst, defecation/urination, noxious stimuli escape, etc., can prevent unconditional and conditional pain signals from spreading widely within the brain (i.e., from pain processing) and can thus protect the activity related to a more important basic or nonbasic need at the moment (Ponomarenko & Korotkova, 2018). This central mechanism was called "dominant" by Alexey Ukhtomsky who experimentally showed its connection to a temporary excitation in the central nervous system associated with a specific need/stimulation determining the organism to the related specific activity (e.g., hunger with food seeking and intake) (Davydov, 1991; Ukhtomsky, 2002; Uryvaev et al., 1991). This "dominant" inhibits behaviors related to other stimuli (e.g., a noxious stimulation with escape/avoidance and pain relief seeking behaviors), transferring related central excitation in support of the current "dominant" consequently amplifying (reinforcing) it (i.e., the noxious stimulation amplifies food seeking in the example).

Mini-dictionary of terms

Baroreflex or baroreceptor reflex: An intrinsic reflex mediated by the parasympathetic and sympathetic nervous systems with, respectively, faster bradycardia and slower vasodilation responses to the stimulation of mechano- or high pressure baro-receptors mainly located in the walls of the carotid sinuses, aortic arch, carotid arteries, and the bifurcation of the brachiocephalic and subclavian arteries evoked by mechanical distention associated with arterial blood pressure increases. Spontaneous baroreflex responses to regular blood pressure fluctuations evoked by regular respiratory movements modulate the heart rate with bradycardia and tachycardia regular fluctuations in the low frequency band around 1 Hz. Baroreflex-related variables are considered to be indicators of the effectiveness of the regulation of information processes in the organism associated mainly with inhibitory systems.

A muscle-heart reflex: A vagus withdrawal mechanism of a mechanical component of the exercise pressor reflex providing adequate oxygen to fulfill the metabolic demands of spontaneously or voluntary activated muscles without excessive pressure variations. This can be used as an indicator of the effectiveness of energy regulation in the organism (an effective energy supply) in active conditions. This reflex is related to irregular minibursts of only tachycardia responses to muscle contractions presented within the same frequency band of heart rate variability around 1 Hz, and should not be mixed up with the regular heart rate variability as an indicator of baroreflex activity.

Respiratory sinus arrhythmia or high frequency of heart rate variability: A regular variation in intervals between heartbeats associated with respiration as an evolutionary developed physiological mechanism controlled by the vagus nerve for more effective oxygen uptake per breath while minimizing the work done by the heart. The respiratory sinus arrhythmia is considered to be an indicator of the effectiveness of energy regulation in the organism (an effective energy supply) in default, relaxing, and recovery conditions.

Resilience: A series of mechanisms contributing to survival as well as mental and physical health by maintaining a pro-active biological or psychosocial stability in response to potential risks. One of these mechanisms is habituation: a non-associative learning process relating to disengagement or detachment from repeating challenges such as pain with inhibited responses keeping homeostasis with preserved standing points. Another example is the parasympathetic braking mechanism regulating (inhibiting) sympathetic activity for keeping an arousal within the homeostatic range.

Adaptation: A series of mechanisms contributing to survival as well as mental and physical fitness by maintaining reactive biological or psychosocial flexibility in response to imposed conditions with their acceptance (or attachment to) and transferring of homeostasis to new, adjusted standing points. One of these mechanisms is the tolerance to pain or its acceptance. Another example is the hypertension-related hypoalgesia mechanism related to the individual predisposition to attenuated cardiac baroreflex with a central resetting of the baroreceptor activation to a higher operating blood pressure point (i.e., to a higher sensory tolerance to the mechanical distention of arteries when arterial blood pressure increases).

Sensitization: A nonassociative learning process related to profound engagement or attachment to repeating challenges such as pain with amplified stress responses affecting homeostasis with preserved standing points.

Resiliency: An overtly positive outcome, that is, a result of successful coping with stress and adversity without the development of specific disorders that can be reached by both resilience and adaptation mechanisms but with different detrimental or side effects on the total health if the balance between the resilience and adaptation mechanisms is not recovered after coping with the challenges.

Key facts of the issue associated with heart and pain interactions and related challenges

- By 2020, the number of experimental and clinical publications on the topic included in PubMed with the keywords pain AND (blood pressure OR hypertension OR heart rate) as the major topic is about 1800, confirming its scientific and clinical importance.
- Some cardiovascular measures have been successfully exploited in California for about 5 years as metrics of pain severity and pain-related mind and body impairment during mandated medical-legal examinations of injured workers with litigated disability and injury claims (as evident from an interview of Dr. Solomon (Sandy) Perlo for WorkCompCentral).
- An announced opioid epidemic in pain management in the United States may in part be provoked by a more aggressive antihypertensive treatment suggested by respective guidelines some decades ago.
- However, the knowledge of heart and pain interrelations has not yet been implemented for regular clinical application, as is evident from the clinical recommendations for both hypertension and pain treatment.

Summary points

- The cardiovascular system and systems regulating pain perception are closely interrelated.
- Acute and persistent noxious stimuli and pain-related feelings and thoughts moderate mechanisms regulating cardiovascular activity (e.g., increase in essential hypertension incidence).
- Individual responses to noxious stimuli are modified (diminished or augmented) by mechanisms regulating cardiovascular activity (e.g., by increased or decreased gain in bradycardic baro-responses to blood pressure increases).
- Hypertension associated with only a systolic blood pressure increase determines higher pain severity through several pronociceptive mechanisms.
- Hypertension with both systolic and diastolic blood pressure elevation is associated with a decrease of pain intensity and severity through an antinociceptive mechanism.
- Antihypertensive drugs can increase chronic pain severity as well as the probability of new cases of chronic pain, and thus can be disadvantageous for pain control by pain-killing drugs.

- Pain can be controlled by antihypotension drugs and nondrug treatments.
- Pain control effectiveness can be predicted and monitored by specific cardiovascular measures.
- An individual level of suffering from pain in its different nociceptive, affective, and cognitive domains can be assessed by various indicators of affected regulation of blood pressure.
- Pain-o-meter technology can be developed using indicators of common mechanisms regulating cardiovascular and pain systems.

References

Al-Ani, M., Robins, K., Al-Khalidi, A. H., Vaile, J., Townend, J., & Coote, J. H. (1997). Isometric contraction of arm flexor muscles as a method of evaluating cardiac vagal tone in man. *Clinical Science, 92*, 175–180.

Amorim, M. R., de Deus, J. L., Pereira, C. A., da Silva, L. E. V., Borges, G. S., Ferreira, N. S., ... Branco, L. G. S. (2020). Baroreceptor denervation reduces inflammatory status but worsens cardiovascular collapse during systemic inflammation. *Scientific Reports, 10*, 1–13.

Arsenault, M., Ladouceur, A., Lehmann, A., Rainville, P., & Piché, M. (2013). Pain modulation induced by respiration: Phase and frequency effects. *Neuroscience, 252*, 501–511.

Balzarotti, S., Biassoni, F., Colombo, B., & Ciceri, M. R. (2017). Cardiac vagal control as a marker of emotion regulation in healthy adults: A review. *Biological Psychology, 130*, 54–66.

Bassi, G. S., Brognara, F., Castania, J. A., Talbot, J., Cunha, T. M., Cunha, F. Q., ... Salgado, H. C. (2015). Baroreflex activation in conscious rats modulates the joint inflammatory response via sympathetic function. *Brain, Behavior, and Immunity, 49*, 140–147.

Bear, T., Philipp, M., Hill, S., & Mündel, T. (2016). A preliminary study on how hypohydration affects pain perception. *Psychophysiology, 53*, 605–610.

Betensky, J. D., & Contrada, R. J. (2010). Depressive symptoms, trait aggression, and cardiovascular reactivity to a laboratory stressor. *Annals of Behavioral Medicine, 39*, 184–191.

Bjerre, L., Andersen, A. T., Hagelskjær, M. T., Ge, N., Mørch, C. D., & Andersen, O. K. (2011). Dynamic tuning of human withdrawal reflex receptive fields during cognitive attention and distraction tasks. *European Journal of Pain, 15*, 816–821.

Boscan, P., Pickering, A. E., & Paton, J. F. R. (2002). The nucleus of the solitary tract: An integrating station for nociceptive and cardiorespiratory afferents. *Experimental Physiology, 87*, 259–266.

Brognara, F., Castania, J. A., Dias, D. P. M., Lopes, A. H., Fazan, R., Kanashiro, A., ... Salgado, H. C. (2018). Baroreflex stimulation attenuates central but not peripheral inflammation in conscious endotoxemic rats. *Brain Research, 1682*, 54–60.

Bruehl, S., Olsen, R. B., Tronstad, C., Sevre, K., Burns, J. W., Schirmer, H., ... Rosseland, L. A. (2018). Chronic pain-related changes in cardiovascular regulation and impact on comorbid hypertension in a general population: The Tromsø study. *Pain, 159*, 119–127.

Bruehl, S., Walker, L. S., & Smith, C. A. (2017). Reply. *Pain, 158*, 2497–2498.

Busch, V., Zeman, F., Heckel, A., Menne, F., Ellrich, J., & Eichhammer, P. (2013). The effect of transcutaneous vagus nerve stimulation on pain perception—An experimental study. *Brain Stimulation, 6*, 202–209.

Calcagnini, G., Giovannelli, P., Censi, F., Bartolini, P., & Barbaro, V. (2001). Baroreceptor-sensitive fluctuations of heart rate and pupil diameter. In *Vol. 1. Engineering in medicine and biology society: Proceedings of the 23rd annual international conference of the IEEE* (pp. 600–603).

Campbell, T. S., Ditto, B., Séguin, J. R., Assaad, J.-M., Pihl, R. O., Nagin, D., & Tremblay, R. E. (2002). A longitudinal study of pain sensitivity and blood pressure in adolescent boys: Results from a 5-year follow-up. *Health Psychology, 21*, 594–600.

Charier, D. J., Zantour, D., Pichot, V., Chouchou, F., Barthelemy, J.-C. M., Roche, F., & Molliex, S. B. (2017). Assessing pain using the variation coefficient of pupillary diameter. *The Journal of Pain, 18*, 1346–1353.

Cheuvront, S. N., Ely, B. R., KeneWck, R. W., Buller, M. J., Charkoudian, N., & Sawka, M. N. (2012). Hydration assessment using the cardiovascular response to standing. *European Journal of Applied Physiology, 112*, 4081–4089.

Contreras-Merino, A. M., Davydov, D. M., & del Paso, G. A. R. (2020). Blunted cardiovascular responses to orthostatic and clinostatic posture changes in fibromyalgia: Further evidence of autonomic abnormalities in the disorder. *Clinical and Experimental Rheumatology, 38*, S117.

Davydov, D. M. (1991). *Physiological analysis of awareness and unawareness of motivation for verbal responses.* Ph.D. thesis Moscow: I.M. Sechenov First Moscow State Medical University (MSMU).

Davydov, D. M. (2017). Cardiac vagal tone as a reliable index of pain chronicity and severity. *Pain, 158*, 2496–2497.

Davydov, D. M. (2018). Health in medicine: The lost graal. *Journal of Psychosomatic Research, 111*, 22–26.

Davydov, D. M. (2020). Impact of antihypertensive treatment on resiliency to clinical pain. *Journal of Hypertension, 38*, 961–967.

Davydov, D. M., & Czabak-Garbacz, R. (2017). Orthostatic cardiovascular profile of subjective well-being. *Biological Psychology, 123*, 74–82.

Davydov, D. M., Luminet, O., & Zech, E. (2013). An externally oriented style of thinking as a moderator of responses to affective films in women. *International Journal of Psychophysiology, 87*, 152–164.

Davydov, D. M., Naliboff, B., Shahabi, L., & Shapiro, D. (2016). Baroreflex mechanisms in irritable bowel syndrome: Part I. Traditional indices. *Physiology & Behavior, 157*, 102–108.

Davydov, D. M., Naliboff, B., Shahabi, L., & Shapiro, D. (2018). Asymmetries in reciprocal baroreflex mechanisms and chronic pain severity: Focusing on irritable bowel syndrome. *Neurogastroenterology and Motility, 30*, e13186.

Davydov, D. M., & Nurbekov, M. K. (2016). Central and peripheral pathogenetic forms of type 2 diabetes: A proof-of-concept study. *Endocrine Connections, 5*, 55–64.

Davydov, D. M., & Perlo, S. (2015). Cardiovascular activity and chronic pain severity. *Physiology & Behavior, 152*, 203–216.

Davydov, D. M., Shahabi, L., & Naliboff, B. (2019). Cardiovascular phenotyping for personalized lifestyle treatments of chronic abdominal pain in irritable bowel syndrome: A randomized pilot study. *Neurogastroenterology and Motility, 31*, e13710.

Davydov, D. M., Shapiro, D., Cook, I. A., & Goldstein, I. (2007). Baroreflex mechanisms in major depression. *Progress in Neuro-Psychopharmacology & Biological Psychiatry, 31*, 164–177.

Davydov, D. M., Shapiro, D., & Goldstein, I. B. (2010). Relationship of resting baroreflex activity to 24-hour blood pressure and mood in healthy people. *Journal of Psychophysiology, 24*, 149–160.

Davydov, D. M., Shapiro, D., Goldstein, I. B., & Chicz-DeMet, A. (2007). Moods in everyday situations: Effects of combinations of different arousal-related factors. *Journal of Psychosomatic Research, 62*, 321–329.

Davydov, D. M., & Sokolova, M. E. (2015). Mobile health indicators: The future belongs to them? *Intellectual Capital, 1*, 24–35.

Davydov, D. M., Stewart, R., Ritchie, K., & Chaudieu, I. (2010). Resilience and mental health. *Clinical Psychology Review, 30*, 479–495.

Davydov, D. M., Stewart, R., Ritchie, K., & Chaudieu, I. (2012). Depressed mood and blood pressure: The moderating effect of situation-specific arousal levels. *International Journal of Psychophysiology, 85*, 212–223.

del Reyes Paso, G. A., Muñoz Ladrón de Guevara, C., & Montoro, C. I. (2015). Breath-holding during exhalation as a simple manipulation to reduce pain perception. *Pain Medicine (Malden, Mass.), 16*, 1835–1841.

Delgado, L. C., Vila, J., & Reyes del Paso, G. A. (2014). Proneness to worry is negatively associated with blood pressure and baroreflex sensitivity: Further evidence of the blood pressure emotional dampening hypothesis. *Biological Psychology, 96*, 20–27.

Di Tella, M., Ghiggia, A., Tesio, V., Romeo, A., Colonna, F., Fusaro, E., … Castelli, L. (2017). Pain experience in fibromyalgia syndrome: The role of alexithymia and psychological distress. *Journal of Affective Disorders, 208*, 87–93.

Durham, J., McDonald, C., Hutchinson, L., & Newton, J. L. (2015). Painful temporomandibular disorders are common in patients with postural orthostatic tachycardia syndrome and impact significantly upon quality of life. *Journal of Oral & Facial Pain and Headache, 29*, 152–157.

Duschek, S., Heiss, H., Buechner, B., & Schandry, R. (2009). Reduction in pain sensitivity from pharmacological elevation of blood pressure in persons with chronically low blood pressure. *Journal of Psychophysiology, 23*, 104–112.

Duschek, S., Mück, I., & Reyes Del Paso, G. A. (2007). Relationship between baroreceptor cardiac reflex sensitivity and pain experience in normotensive individuals. *International Journal of Psychophysiology, 65*, 193–200.

Dworkin, B. R., Filewich, R. J., Miller, N. E., Craigmyle, N., & Pickering, T. G. (1979). Baroreceptor activation reduces reactivity to noxious stimulation: Implications for hypertension. *Science, 205*, 1299–1301.

France, C. R., & Katz, J. (1999). Postsurgical pain is attenuated in men with elevated presurgical systolic blood pressure. *Pain Research and Management, 4*, 100–103.

Furlan, R., Colombo, S., Perego, F., Atzeni, F., Diana, A., Barbic, F., … Sarzi-Puttini, P. (2005). Abnormalities of cardiovascular neural control and reduced orthostatic tolerance in patients with primary fibromyalgia. *The Journal of Rheumatology, 32*, 1787–1793.

Hagen, K., Stovner, L. J., Vatten, L., Holmen, J., Zwart, J.-A., & Bovim, G. (2002). Blood pressure and risk of headache: A prospective study of 22 685 adults in Norway. *Journal of Neurology, Neurosurgery, and Psychiatry, 72*, 463–466.

Harmon-Jones, C., Summerell, E., & Bastian, B. (2018). Anger increases preference for painful activities. *Motivation Science, 4*, 301–314.

Harrison, D. G., Guzik, T. J., Lob, H. E., Madhur, M. S., Marvar, P. J., Thabet, S. R., … Weyand, C. M. (2011). Inflammation, immunity, and hypertension. *Hypertension, 57*, 132–140.

Harvie, D. S., Moseley, G. L., Hillier, S. L., & Meulders, A. (2017). Classical conditioning differences associated with chronic pain: A systematic review. *Journal of Pain, 18*, 889–898.

Heuch, I., Heuch, I., Hagen, K., & Zwart, J. A. (2014). Does high blood pressure reduce the risk of chronic low back pain? The Nord-Trøndelag health study. *European Journal of Pain, 18*, 590–598.

Iwabe, T., Ozaki, I., & Hashizume, A. (2014). The respiratory cycle modulates brain potentials, sympathetic activity, and subjective pain sensation induced by noxious stimulation. *Neuroscience Research, 84*, 47–59.

Jafari, H., Van de Broek, K., Plaghki, L., Vlaeyen, J. W. S., Van den Bergh, O., & Van Diest, I. (2016). Respiratory hypoalgesia? Breath-holding, but not respiratory phase modulates nociceptive flexion reflex and pain intensity. *International Journal of Psychophysiology, 101*, 50–58.

Kano, M., Hamaguchi, T., Itoh, M., Yanai, K., & Fukudo, S. (2007). Correlation between alexithymia and hypersensitivity to visceral stimulation in human. *Pain, 132*, 252–263.

Kemp, A. H., & Quintana, D. S. (2013). The relationship between mental and physical health: Insights from the study of heart rate variability. *International Journal of Psychophysiology, 89*, 288–296.

Koenig, J., Falvay, D., Clamor, A., Wagner, J., Jarczok, M. N., Ellis, R. J., … Thayer, J. F. (2016). Pneumogastric (Vagus) nerve activity indexed by heart rate variability in chronic pain patients compared to healthy controls: A systematic review and meta-analysis. *Pain Physician, 19*, E55–E78.

Kohl, A., Rief, W., & Glombiewski, J. A. (2013). Acceptance, cognitive restructuring, and distraction as coping strategies for acute pain. *Journal of Pain, 14*, 305–315.

Koopman, F. A., Tang, M. W., Vermeij, J., de Hair, M. J., Choi, I. Y., Vervoordeldonk, M. J., … Tak, P. P. (2016). Autonomic dysfunction precedes development of rheumatoid arthritis: A prospective cohort study. *eBioMedicine, 6*, 231–237.

Lautenschläger, G., Habig, K., Best, C., Kaps, M., Elam, M., Birklein, F., & Krämer, H. H. (2015). The impact of baroreflex function on endogenous pain control—A microneurography study. *The European Journal of Neuroscience, 42*, 2996–3003.

Ledowski, T., Reimer, M., Chavez, V., Kapoor, V., & Wenk, M. (2012). Effects of acute postoperative pain on catecholamine plasma levels, hemodynamic parameters, and cardiac autonomic control. *Pain, 153*, 759–764.

Leino-Arjas, P., Solovieva, S., Kirjonen, J., Reunanen, A., & Riihimäki, H. (2006). Cardiovascular risk factors and low-back pain in a long-term follow-up of industrial employees. *Scandinavian Journal of Work, Environment & Health, 32*, 12–19.

Madden, V. J., Harvie, D. S., Parker, R., Jensen, K. B., Vlaeyen, J. W. S., Moseley, G. L., & Stanton, T. R. (2016). Can pain or hyperalgesia be a classically conditioned response in humans? A systematic review and meta-analysis. *Pain Medicine, 17*, 1094–1111.

McCubbin, J. A., Nathan, A., Hibdon, M. A., Castillo, A. M., Graham, J. G., & Switzer, F. S., 3rd. (2018). Blood pressure, emotional dampening, and risk behavior: Implications for hypertension development. *Psychosomatic Medicine, 1*.

Mustafa, H. I., Raj, S. R., Diedrich, A., Black, B. K., Paranjape, S. Y., Dupont, W. D., … Robertson, D. (2012). Altered systemic hemodynamic and baroreflex response to angiotensin II in postural tachycardia syndrome. *Circulation. Arrhythmia and Electrophysiology, 5*, 173–180.

Nascimento Rebelatto, M., Alburquerque-Sendín, F., Guimarães, J. F., & Salvini, T. F. (2017). Pressure pain threshold is higher in hypertensive compared with normotensive older adults: A case–control study. *Geriatrics and Gerontology International, 17*, 967–972.

Perlo, S., & Davydov, D. M. (2016). "Chronic pain and the brain" impairment: Introducing a translational neuroscience-based metric. *Pain Medicine, 17*, 799–802.

Pickering, A. E., Boscan, P., & Paton, J. F. R. (2003). Nociception attenuates parasympathetic but not sympathetic baroreflex via NK1 receptors in the rat nucleus tractus solitarii. *The Journal of Physiology, 551*, 589–599.

Pollatos, O., Dietel, A., Gündel, H., & Duschek, S. (2015). Alexithymic trait, painful heat stimulation, and everyday pain experience. *Frontiers in Psychiatry, 6*, 139.

Ponomarenko, A., & Korotkova, T. (2018). Hunger is a gatekeeper of pain in the brain. *Nature, 556*, 445–446.

Randich, A., & Hartunian, C. (1983). Activation of the sinoaortic baroreceptor reflex arc induces analgesia: Interactions between cardiovascular and endogenous pain inhibition systems. *Physiological Psychology, 11*, 214–220.

Randich, A., & Maixner, W. (1984). Interactions between cardiovascular and pain regulatory systems. *Neuroscience and Biobehavioral Reviews, 8*, 343–367.

Rau, H., & Elbert, T. (2001). Psychophysiology of arterial baroreceptors and the etiology of hypertension. *Biological Psychology, 57*, 179–201.

Reyes del Paso, G. A., Montoro, C., Muñóz Ladrón de Guevara, C., Duschek, S., & Jennings, J. R. (2014). The effect of baroreceptor stimulation on pain perception depends on the elicitation of the reflex cardiovascular response: Evidence of the interplay between the two branches of the baroreceptor system. *Biological Psychology, 101*, 82–90.

Roach, D., & Sheldon, R. (2018). Origins of the power of the low frequency heart rate variability bandwidth. *Journal of Electrocardiology, 51*, 422–427.

Schreiber, K. L., Campbell, C., Martel, M. O., Greenbaum, S., Wasan, A. D., Borsook, D., … Edwards, R. R. (2014). Distraction analgesia in chronic pain patients the impact of catastrophizing. *Anesthesiology, 121*, 1292–1301.

Slepian, P. M., France, C. R., Rhudy, J. L., Himawan, L. K., Güereca, Y. M., Kuhn, B. L., & Palit, S. (2016). Behavioral inhibition and behavioral activation are related to habituation of nociceptive flexion reflex, but not pain ratings. *The Journal of Pain, 18*, 349–358.

Spigt, M., Weerkamp, N., Troost, J., van Schayck, C. P., & Knottnerus, J. A. (2012). A randomized trial on the effects of regular water intake in patients with recurrent headaches. *Family Practice, 29*, 370–375.

Tracy, L. M., Ioannou, L., Baker, K. S., Gibson, S. J., Georgiou-Karistianis, N., & Giummarra, M. J. (2016). Meta-analytic evidence for decreased heart rate variability in chronic pain implicating parasympathetic nervous system dysregulation. *Pain, 157*, 7–29.

Ukhtomsky, A. A. (2002). Dominant. In *Papers of different years 1887–1939*. St. Petersburg: Piter.

Uryvaev, I. V., Davydov, D. M., & Gavrilenko, A. I. (1991). Study of food dominants by means of a verbal test. *Fiziologiia Cheloveka, 17*, 67–72.

Walker, L. S., Ston, A. L., Smith, C. A., Bruehl, S., Garber, J., Puzanovova, M., & Diedrich, A. (2017). Interacting influences of gender and chronic pain status on parasympathetically mediated heart rate variability in adolescents and young adults. *Pain, 158*, 1509–1516.

Yu, Z., Kim, S. C., Vanni, K., Huang, J., Desai, R., Murphy, S. N., … Liao, K. P. (2018). Association between inflammation and systolic blood pressure in RA compared to patients without RA. *Arthritis Research and Therapy, 20*, 107.

Yuan, H., & Silberstein, S. D. (2017). Vagus nerve stimulation and headache. *Headache: The Journal of Head and Face Pain, 57*, 29–33.

Zamunér, A. R., Barbic, F., Dipaola, F., Bulgheroni, M., Diana, A., Atzeni, F., … Furlan, R. (2015). Relationship between sympathetic activity and pain intensity in fibromyalgia. *Clinical and Experimental Rheumatology, 33*, S53–S57.

Chapter 20

Chronic pain in military veterans

Ariel Baria, Nancy Liu, Quinn Wonders, and Sanjog Pangarkar
Los Angeles, CA, United States

Abbreviations

BPI	brief pain inventory
BPS	biopsychosocial
CBT	cognitive behavioral therapy
CDC	Centers for Disease Control and Prevention
CIH	complementary and integrated health
CPSP	chronic postsurgical pain
DoD	Department of Defense
DVPRS	defense and veterans pain rating scale
MBSR	mindfulness-based stress reduction
NRS	numerical rating scale
OA	osteoarthritis
OEF	operation enduring freedom
OIF	operation Iraqi freedom
PEG	pain enjoyment and general activity
PTSD	posttraumatic stress disorder
SUD	substance use disorder
US	United States
VACO	VA Central Office
VAS	visual analog scale
VHA	Veterans Health Administration
VVC	VA video conference

Introduction

Chronic pain affects 18% to 20% of the world's population, or roughly 1.5 billion people (Sá et al., 2019). The prevalence among military veterans is suspected to be higher, yet published data from many countries is unavailable. Decades of military conflict, often involving multiple tours-of-duty, have led to wide-ranging consequences for many service members, including combat and noncombat-related injuries (Gallagher, 2016). These injuries may persist and develop into chronic pain after a soldier returns to civilian life. In addition, more military personnel are surviving catastrophic wounds from blast injuries, burns, shrapnel, traumatic brain injury, and/or spinal cord injuries because of new technologies in protective gear, early medical care, and advanced warfare (i.e., drones and cruise missiles) (Gallagher, 2016; Gauntlett-Gilbert & Wilson, 2013).

In the United States, the prevalence of chronic pain in the general population is 30%; however, this rate is higher (50%) for veterans enrolled in the Veterans Health Administration (VHA) (Gallagher, 2016). Because of prolonged engagement in multiple theaters of operation (i.e., Operation Enduring Freedom (OEF)/Operation Iraqi Freedom (OIF)), veterans experience high rates of chronic pain from musculoskeletal pain conditions, headaches, and traumatic blast injuries (Gallagher, 2016; Macey, Weimer, Grimaldi, Dobscha, & Morasco, 2013). Compared to the general population, veterans with chronic pain tend to have more complex conditions with higher rates of mental health disorders and social problems that include depression, posttraumatic stress disorder (PTSD), functional disabilities, social isolation, and substance use disorders (SUD) (Gallagher, 2016; Kerns, Otis, Rosenberg, & Reid, 2003).

Features and Assessments of Pain, Anesthesia, and Analgesia. https://doi.org/10.1016/B978-0-12-818988-7.00048-0
Copyright © 2022 Elsevier Inc. All rights reserved.

Given the high prevalence and impact of chronic pain in military veterans, this chapter describes common pain conditions along with the psychological and social factors, which contribute to the pain experience. In addition, comprehensive assessments and treatment approaches are reviewed that are specific to military veterans with chronic pain.

Theoretical models of chronic pain

Various theoretical models exist that describe the chronic pain experience (Gallagher, 2016). The most heuristic of these frameworks is the biopsychosocial (BPS) model of chronic pain adapted for military veterans (Baria, Pangarkar, Abrams, & Miaskowski, 2019). Based on this model, the interaction between biology, psychology, and sociology is described. Fig. 1 displays the BPS model for veterans with chronic pain and outlines the most common conditions associated with this experience. The biological factors are identified as the most prevalent type of injuries seen in military veterans. The psychological factors include depression, anxiety, posttraumatic stress disorder, and substance use disorders. Last, social factors in military veterans that influence pain include financial difficulties, disability, homelessness, and limited access to care/isolation. These factors contribute to the pain experience of many military veterans and serves as a framework for clinical practice and research.

Military basic training and rates of injury

Two million injuries from overuse and acute musculoskeletal trauma are reported across US military services annually (Jones & Hauschild, 2015). These injuries result in more military discharges than any other single medical cause (Nindl, Williams, Deuster, Butler, & Jones, 2013). The cause of these injuries includes rigorous exercise, physical training, and/or sports injuries (Bedno et al., 2014). For example, a survey of US Army trainees found road marching (23% for women, 24% for men), physical training (24% of injuries for women, 26% for men), and obstacle courses (4% for women, 5% for men) the most common causes of injury in military personnel (Knapik, Graham, Rieger, Steelman, & Pendergrass, 2013).

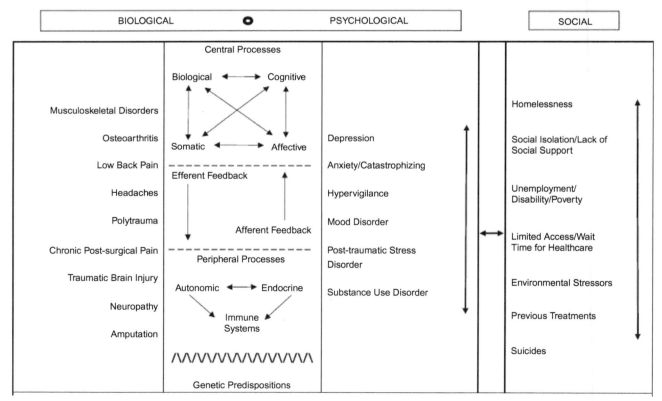

FIG. 1 Biopsychosocial model of chronic noncancer pain in veterans. A conceptual model of the biopsychosocial model of chronic noncancer pain in veterans. *(Requested permission from Oxford Academic (Pain Medicine).)*

Musculoskeletal injuries, including low back pain, overuse or stress syndromes, stress fractures, sprains and strains (i.e., muscle, ankle, and knee), tendonitis, iliotibial band syndrome, and patellofemoral syndrome, are common causes of chronic pain in active duty military personnel (Kaufman, Brodine, & Shaffer, 2000). The risks of these injuries are higher depending on the frequency and intensity of the physical activity (i.e., paratroopers, heavy machinery operators). In addition, more strenuous activity and an individual's level of physical fitness, personal health risk behaviors (e.g., sedentary lifestyles, tobacco use), and sociodemographic characteristics (e.g., age, sex) can influence the severity and risks for these injuries (Jones & Hauschild, 2015).

Combat-related injuries in military personnel

For combat-related injuries, more military personnel (90%) are surviving catastrophic injuries from penetrating wounds and shrapnel or mine blast injuries related to improvised explosive devices than those serving during the Vietnam war (Casey, Demers, Deben, Nelles, & Weiss, 2015; Gallagher, 2016; Phillips et al., 2016). The survivors of these injuries report chronic pain related to musculoskeletal disorders, headache, brain injury, and other neuropathic pain conditions. Collectively known as "polytrauma," these chronic pain conditions are described as two or more life-threatening injuries resulting in physical/functional disability, cognitive, psychological, and social impairment (Clark, Bair, Buckenmaier, Gironda, & Walker, 2007; Gallagher, 2016; Lew et al., 2009; Phillips et al., 2016). Based on the Polytrauma System of Care Registry (Phillips et al., 2016), headaches (59%) and low back pain (33%) are the most common pain complaints. In OEF/OIF veterans (Lew et al., 2009), other common locations for chronic pain include the shoulder (21%), neck (19%), and knee (18%). These pain conditions were associated with physical disabilities, psychological distress, and vocational issues that persisted after the veteran returned to civilian life (Gallagher, 2016; Lew et al., 2009).

Types of chronic pain in veterans

As active duty military personnel retire into civilian life, chronic pain remains highly prevalent. A comparison of pain severity between veterans versus nonveterans found 65.5% of veterans reported chronic pain, of which, 9.1% was rated severe (Nahin, 2017). The most common areas affected were peripheral joints and the spine. Similarly, in a veteran cohort study where over 5 million were surveyed (Goulet et al., 2016), the most common types of chronic pain diagnoses were nontraumatic joint disorder (26.5%), back disorder (25.4%), and osteoarthritis (OA) (20.9%). Arthritis was more prevalent in veterans than nonveterans, especially those between 45 and 65 years of age (40.3% in women and 36.0% in men) and 18 to 44 years of age (17.3% in men and 11.6% in women) (Murphy et al., 2014). In addition, female veterans experienced an adjusted OA incidence rate that was nearly 20% higher than men, and African American veterans had a higher incidence of OA than all other race categories (Cameron, Hsiao, Owens, Burks, & Svoboda, 2011).

Comorbid conditions in veterans with chronic pain

Chronic pain may have considerable consequences to a veteran's overall health and quality of life. Due to physical dysfunction, deconditioning, prolonged splinting, immobilization, and/or obesity, veterans with chronic pain may develop a number of concurrent health conditions that affect the endocrine, cardiovascular, and musculoskeletal systems. Based on the VA and Department of Defense (DoD) Weight Management Guidelines (Department of Veterans Affairs, Department of Defense, 2014), 78% of all veterans and 61% to 83% of DoD personnel are overweight or obese. Obesity is a common complication of immobility and deconditioning which may worsen chronic pain. In the veteran population, physical deconditioning and obesity contribute to numerous comorbidities such as hypertension, diabetes, coronary artery disease, and obstructive sleep apnea (Goulet et al., 2016).

Veterans with chronic pain are more likely to experience higher rates of surgical complication as a result of overall poor health, psychosocial comorbidities, and decreased functional status compared to veterans without chronic pain (Deyo, Hickam, Duckart, & Piedra, 2013; Hadlandsmyth et al., 2018; Kubat, Giori, Hwa, & Eisenberg, 2016; Owens, Williams, & Wolf, 2015; Rozet et al., 2014). These perioperative and postoperative complications were seen in shoulder rotator cuff (Owens et al., 2015), lumbar spine (Deyo et al., 2013), knee arthroscopic (Rozet et al., 2014), and bariatric surgeries (Kubat et al., 2016). Other surgical complications seen in veterans with chronic pain include the development of chronic postsurgical pain (CPSP) (Rozet et al., 2014), prolonged opioid use (Hadlandsmyth et al., 2018), and difficulty tapering off opioids postoperatively (Mudumbai et al., 2016). Preoperative opioid use was an important predictor or risk for these complications to develop.

Psychological comorbidities and social factors in veterans with chronic pain

Nearly half of veterans with a pain diagnoses have at least one mental health comorbidity (Kerns et al., 2003). The most common comorbidities include depression, anxiety, posttraumatic stress disorder (PTSD), and substance use disorder (Baria et al., 2019). In addition, social factors such as unemployment, financial problems, social isolation, and limited access to pain care contribute to the pain experience of veterans.

Depression, anxiety, PTSD, substance use disorder

Comorbid pain and depression have a bidirectional relationship in that chronic pain has been shown to cause depressive symptoms, and depression can worsen pain perception (Muñoz et al., 2005). Among veterans with cooccurring chronic pain and depression, pain intensity has also been correlated with worse depressive symptoms (Muñoz et al., 2005; Tripp, Curtis Nickel, Landis, Wang, & Knauss, 2004). The prevalence of depression among veterans is 14% (Olenick, Flowers, & Diaz, 2015), while the prevalence of depression among veterans with chronic pain was as high as 57% (Baria et al., 2019).

Similarly, the prevalence of anxiety among veterans was estimated at 4.8% (Kerns et al., 2003); however, among veterans with chronic pain, this prevalence ranges between 6.9% and 53% (Baria et al., 2019). Comorbid anxiety can also lead to maladaptive pain behaviors including catastrophizing and hypervigilance, which contribute to feelings of helplessness, vulnerability, and hopelessness (Outcalt et al., 2014; Phillips et al., 2016; Terry, Moeschler, Hoelzer, & Hooten, 2016).

PTSD and chronic pain frequently co-occur and share the familiar processes of fear, avoidance, anxiety, sensitivity, and catastrophizing (Otis, Keane, & Kerns, 2003). The prevalence of PTSD is between 4% and 30% in veterans (i.e., OEF/OIF, Vietnam veterans, and other combat veterans) (Alschuler & Otis, 2012; Gibson, 2012); however, the prevalence of co-occurring PTSD and chronic pain can be has high as 80% (Baria et al., 2019). Affected veterans often report high pain scores and have comorbid conditions that require increased health care resources (Gibson, 2012; Phifer et al., 2011; Phillips et al., 2016; Runnals et al., 2013; Smeeding, Bradshaw, Kumpfer, Trevithick, & Stoddard, 2010). In addition, the concomitant diagnoses of PTSD, depression, and chronic pain in veterans could triple the likelihood of disability outcomes and suicidal thoughts and behaviors (Outcalt et al., 2015).

SUD can be characterized by misuse or abuse of alcohol, opioids, marijuana, and other illicit drugs (Bennett, Elliott, & Golub, 2013; Gibson, 2012; Olenick et al., 2015; Phillips et al., 2016). Among veterans with chronic pain, substance use disorder prevalence can be as high as 52% to 77% (Bennett et al., 2013; Gaither et al., 2016; Gros, Szafranski, Brady, & Back, 2015; Grossbard, Malte, Saxon, & Hawkins, 2014; Lovejoy, Dobscha, Turk, Weimer, & Morasco, 2016; Morasco, Corson, Turk, & Dobscha, 2011; Morasco & Dobscha, 2008; Morasco, Duckart, & Dobscha, 2011; Naliboff et al., 2011; Phillips et al., 2016; Trafton, Oliva, Horst, Minkel, & Humphreys, 2004; Whitehead et al., 2008). Veterans with chronic pain and SUD typically have more severe medical and mental health disorders (i.e., depression and anxiety) that compounds social, behavioral, and cognitive problems (Morasco, Corson, et al., 2011; Trafton et al., 2004).

Socioeconomic challenges

Unique to pain care in veterans are socioeconomic factors (i.e., financial problems, poverty, unemployment/disability, homelessness, social isolation, and limited access to care) that contribute to the challenges of managing this disease (Baria et al., 2019; Dahlhamer et al., 2018). The prevalence of poverty in veterans ranged from 6.9% to 9.4%, with higher rates seen among veterans with chronic pain, between 35 and 54 years of age, and post-9/11 veterans (Baria et al., 2019). Further, contributing to the poverty level among veterans is the unemployment rate (5.1%). Veterans have challenges gaining employment due to: military skills not translating into civilian work-life, employer discrimination, poor socioeconomic demographic of persons who enlist, and various physical dysfunction, chronic pain, and mental health disorders (Smith, 2015). Veterans who are most likely to be unemployed are female veterans, homeless, over 45 years of age, ethnically diverse, and/or disabled. Furthermore, 22% (4.6 million) of veterans have a service-connected disability that is often associated with their physical, psychological, and social problems (Baria et al., 2019).

Chronic pain disproportionately affects homeless veterans who account for 11% of the entire US homeless population (39,000–63,000/day) (Gabrielian, Yuan, Andersen, Rubenstein, & Gelberg, 2014). At the same time, 1.4 million veterans are at risk for homelessness due to poverty, lack of social support, and poor living conditions (Baria et al., 2019; Gabrielian et al., 2014). In addition, up to 50% of homeless veterans experience serious mental illness, 70% have past or active SUD, and 59.3% have chronic pain. As such, these veterans have difficulty receiving medical/pain care because of lack of transportation, fragmentation of heath care services, lack of trust in clinicians, social isolation, poor social support, and other competing primary care needs (O'Toole, Johnson, Aiello, Kane, & Pape, 2016). Moreover, barriers to pain care can be attributed to social isolation and poor social support experienced by veterans because of frequent relocations, deployments required in military service, and difficulties with reintegration to civilian status (Bennett et al., 2013). These factors all contribute to the difficulties in managing veterans with chronic pain.

Pain assessment and treatment

Based on the BPS model of chronic pain in veterans, a comprehensive pain assessment coupled with a multidisciplinary team approach are important facets in the management of this vulnerable population. The diagnosis and treatment of pain starts with a careful history and physical exam. Prior to the evaluation, assessment tools are often employed to gather information and clarify the symptoms experienced by a patient. In the United States, the most common assessment tools for pain are the numerical rating scale (NRS) and the visual analog scale (VAS). These ordinal scales provide a subjective measure of pain intensity and help categorize whether a patient is experiencing mild, moderate, or severe pain (Kim, 2017). Pain scales that were developed to assess active duty military members and veterans include the Defense and Veterans Pain Rating Scale (DVPRS) and the Pain, Enjoyment, and General Activity (PEG) scale. As illustrated in Fig. 2, the DVPRS uses color coding, descriptions, and faces to represent pain levels. In the 2.0 version, clinical assessment quantifies the impact of pain on activity, sleep, mood, and stress (Polomano et al., 2016). In addition, as displayed in Fig. 3 provided by Krebs et al. (2009), the PEG was derived from the Brief Pain Inventory (BPI) and assesses pain intensity along with emotional and physical function (Krebs et al., 2009). Both of these scales are aimed to provide clinicians an easy to perform, multidimensional pain measure that can be performed quickly.

The management of chronic pain in veterans requires multimodal treatment including pharmacological, nonpharmacological, rehabilitation therapies, interventional pain procedures, psychological evaluation, and integrative health approaches provided by an interdisciplinary team. These treatments should include primary and secondary prevention strategies since a significant number of injuries and trauma occur while in active duty training, combat, and noncombat settings. Early diagnosis and treatment by DoD and VA, in concert with improved coordination, education, and collaboration across the continuum of care, may improve long-term outcomes.

Clinical practice guidelines and policies developed by VA Central Office (VACO), DoD, and other government agencies include the VA Stepped Care Model (Gallagher, 2016), the National Pain Strategy (U.S. Department of Health and Human Services Interagency Pain Research Coordinating Committee, 2016), Opioid Safety Initiatives (OSI) (U.S. Department of Veterans Affairs, 2020), VA/DoD Opioid Prescribing Guidelines (U.S. Department of Veterans Affairs, 2020), and the Center for Disease Control and Prevention (CDC) Guideline for Prescribing Opioids for Chronic Pain (Centers for Disease Control and Prevention, 2019). These VACO initiatives expanded multimodal pain treatments (i.e., acupuncture, integrative health & complementary alternative medicine, chiropractic care, interventional pain procedures) to veterans with chronic pain and provided alternatives to long-term opioid use. In addition, telehealth and VA video conference (VVC) have contributed to delivering timely access to pain care for all veterans including those living in rural or geographically isolated areas (Hadlandsmyth et al., 2018; Rozet et al., 2014; Sá et al., 2019).

For many veterans with chronic pain, dedicated rehabilitation, mental health, vocational, and social services are paramount to recovery and promote functional independence and quality of life. These essential services should remain a backbone to pain management because of their ability to increase physical function, reduce long-term disability, address the multitude of social problems, improve mental health, and reduce adverse effects of opioid therapy (Gallagher, 2016).

Nonpharmacologic pain modalities

Within the framework of the BPS model of chronic pain, an important aspect to holistic pain care for veterans is complementary and integrated health (CIH) modalities. A national survey on the prevalence of CIH approaches in VHA care (Taylor, Hoggatt, & Kligler, 2019) found 52% of veterans had tried a CIH modality (i.e., trialing massage (44%), chiropractic (37%), mindfulness (24%), and yoga (25%). The two most common reasons for utilizing CIH therapy were to manage pain and stress. In addition, Taylor et al. (Taylor et al., 2019; Taylor, Hoggatt, & Kligler, 2019) studied the type of CIH approaches being used to manage pain between 2010 and 2013 and found meditation, yoga, acupuncture, chiropractic, guided imagery, biofeedback, tai chi, massage, and hypnosis were used by approximately 27% of veterans for chronic muscular pain. Other commonly used CIH treatments were cognitive behavioral therapy (CBT) and mindfulness-based stress reduction (MBSR).

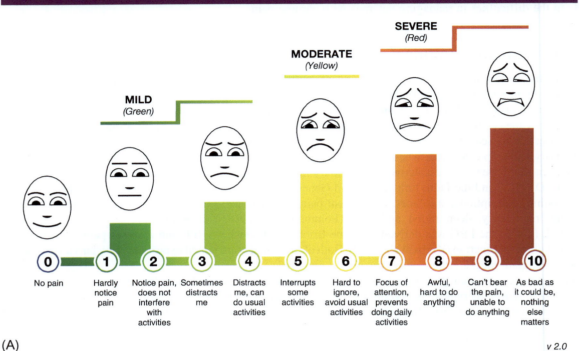

FIG. 2 Defense and veterans pain rating scale. For clinicians to evaluate the biopsychosocial impact of pain. *(Requested permission from Oxford Academic (Pain Medicine).)*

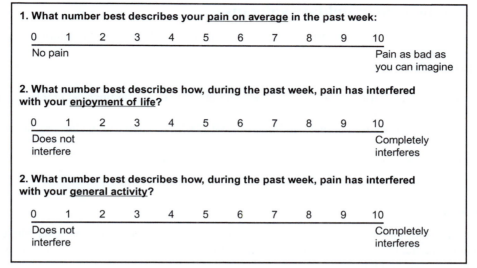

FIG. 3 PEG scale assessing pain intensity and interference (pain, enjoyment, general activity). The score is best used to track an individual's change in pain, mood, and function over time. *(Open access permission from "The development and initial validation of the PEG, a three-item scale assessing pain intensity and interference" Krebs (2009).)*

Conclusion

Chronic pain is a highly prevalent condition for many veterans after completing military service. Injuries may occur from multiple sources and include training, combat and noncombat trauma, as well sequelae from these injuries. In addition to the multitude of physical injuries and chronic pain conditions experienced by military veterans, numerous psychological and social factors may contribute to the challenge of managing this disease. Depression, anxiety, PTSD, SUD, as well as financial problems, disability, homelessness, social isolation, and limited access to care are factors that may compound and/or worsen the overall pain experience. To this end, pain management in military veterans requires a biopsychosocial framework that utilizes a multimodal approach and expertise from an interdisciplinary team.

Applications to other areas

The biopsychosocial model for chronic pain is applicable to the general population and provides a holistic framework for the evaluation and treatment of pain. Understanding the interaction between the physical, psychological, and social aspects of the pain experience will allow for comprehensive management of this complex disease.

Other agents of interest

Not applicable.

Mini-dictionary of terms

- **Biopsychosocial**—the complex interaction among the biological, psychological, and social factors associated with health
- **Complementary and integrative health**—products and practices that are currently not part of mainstream medical practice
- **Interdisciplinary**—relating to more than one branch of knowledge
- **Chronic pain**—pain that persists or recurs beyond the usual course of acute illness or injury or greater than 3 to 6 months and affects the individual's well-being
- **Chronic postsurgical pain**—pain lasting more than 3 to 6 months after surgery

Key facts of chronic pain in military veterans

- The biological injuries related to military service include musculoskeletal disorders, osteoarthritis, low back pain, headaches, TBI, and polytrauma.
- Veterans with chronic pain exhibit high rates of mental health disorders that include depression, anxiety, PTSD, and SUD.
- Veterans with chronic pain experience high rates of social problems that include financial problems, disability, homelessness, social isolation, and limited access to care.
- Nonpharmacological treatments for chronic pain include CBT, MBSR, acupuncture, chiropractic care, and rehabilitation therapies.
- Specific tools used to assess pain in veterans include the DVPRS and PEG.

Summary points

- Worldwide prevalence of chronic pain is 1.5 billion (18%–20% of the population); however, the exact number in worldwide military veterans is unknown. The prevalence of chronic pain in US military veterans is 50% of the US veteran population or 1.44 million.
- Common types of injuries seen in military veterans include physical/basic training injuries, combat and noncombat trauma, chronic military injuries, and sequelae from these injuries.
- Psychological factors found in veterans with chronic pain include depression, anxiety, PTSD, and SUD.
- Social factors found in veterans with chronic pain include financial problems, disability, homelessness, social isolation, and limited access to care.
- Specific tools for pain assessment in veterans include the DVPRS and PEG scales. It is important to treat pain using multimodal approach including nonpharmacological and complementary and integrative health interventions.

References

Alschuler, K. N., & Otis, J. D. (2012). Coping strategies and beliefs about pain in veterans with comorbid chronic pain and significant levels of posttraumatic stress disorder symptoms. *European Journal of Pain, 16*(2), 312–319. https://doi.org/10.1016/j.ejpain.2011.06.010.

Baria, A. M., Pangarkar, S., Abrams, G., & Miaskowski, C. (2019). Adaption of the biopsychosocial model of chronic noncancer pain in veterans. *Pain Medicine, 20*(1), 14–27. https://doi.org/10.1093/pm/pny058.

Bedno, S., Hauret, K., Loringer, K., Kao, T. C., Mallon, T., & Jones, B. (2014). Effects of personal and occupational stress on injuries in a young, physically active population: A survey of military personnel. *Military Medicine, 179*(11), 1311–1318. https://doi.org/10.7205/MILMED-D-14-00080.

Bennett, A. S., Elliott, L., & Golub, A. (2013). Opioid and other substance misuse, overdose risk, and the potential for prevention among a sample of OEF/OIF veterans in New York City. *Substance Use & Misuse, 48*(10), 894–907. https://doi.org/10.3109/10826084.2013.796991.

Cameron, K. L., Hsiao, M. S., Owens, B. D., Burks, R., & Svoboda, S. J. (2011). Incidence of physician-diagnosed osteoarthritis among active duty United States military service members. *Arthritis and Rheumatism, 63*(10), 2974–2982. https://doi.org/10.1002/art.30498.

Casey, K., Demers, P., Deben, S., Nelles, M. E., & Weiss, J. S. (2015). Outcomes after long-term follow-up of combat-related extremity injuries in a multidisciplinary limb salvage clinic. *Annals of Vascular Surgery, 29*(3), 496–501. https://doi.org/10.1016/j.avsg.2014.09.035.

Centers for Disease Control and Prevention. (2019). *CDC guideline for prescribing opioids for chronic Pain.* Centers for Disease Control and Prevention, Opioid Overdose. https://www.cdc.gov/drugoverdose/prescribing/guideline.html.

Clark, M. E., Bair, M. J., Buckenmaier, C. C., Gironda, R. J., & Walker, R. L. (2007). Pain and combat injuries in soldiers returning from Operations Enduring Freedom and Iraqi Freedom: Implications for research and practice. *Journal of Rehabilitation Research and Development, 44*(2), 179–194. https://doi.org/10.1682/jrrd.2006.05.0057.

Dahlhamer, J., Lucas, J., Zelaya, C., Nahin, R., Mackey, S., DeBar, L., … Helmick, C. (2018). Prevalence of chronic pain and high-impact chronic pain among adults—United States, 2016. *MMWR. Morbidity and Mortality Weekly Report, 67*(36), 1001–1006. https://doi.org/10.15585/mmwr.mm6736a2.

Department of Veterans Affairs, Department of Defense. (2014). *VA/DoD clinical practice guideline for screening and management of overweight and obesity. VA/DoD clinical practice guidelines.* April 18 https://www.healthquality.va.gov/guidelines/CD/obesity/CPGManagementOfOverweightAndObesityFINAL041315.pdf.

Deyo, R. A., Hickam, D., Duckart, J. P., & Piedra, M. (2013). Complications after surgery for lumbar stenosis in a veteran population. *Spine, 38*(19), 1695–1702. https://doi.org/10.1097/BRS.0b013e31829f65c1.

Gabrielian, S., Yuan, A. H., Andersen, R. M., Rubenstein, L. V., & Gelberg, L. (2014). VA health service utilization for homeless and low-income Veterans: A spotlight on the VA Supportive Housing (VASH) program in greater Los Angeles. *Medical Care, 52*(5), 454–461. https://doi.org/10.1097/MLR.0000000000000112.

Gaither, J. R., Goulet, J. L., Becker, W. C., Crystal, S., Edelman, E. J., Gordon, K., ... Fiellin, D. A. (2016). The effect of substance use disorders on the association between guideline-concordant long-term opioid therapy and all-cause mortality. *Journal of Addiction Medicine*, *10*(6), 418–428. https://doi.org/10.1097/ADM.0000000000000255.

Gallagher, R. M. (2016). Advancing the pain agenda in the veteran population. *Anesthesiology Clinics*, *34*(2), 357–378. https://doi.org/10.1016/j.anclin.2016.01.003.

Gauntlett-Gilbert, J., & Wilson, S. (2013). Veterans and chronic pain. *British Journal of Pain*, *7*(2), 79–84. https://doi.org/10.1177/2049463713482082.

Gibson, C. A. (2012). Review of posttraumatic stress disorder and chronic pain: The path to integrated care. *Journal of Rehabilitation Research and Development*, *49*(5), 753–776. https://doi.org/10.1682/jrrd.2011.09.0158.

Goulet, J. L., Kerns, R. D., Bair, M., Becker, W. C., Brennan, P., Burgess, D. J., ... Brandt, C. A. (2016). The musculoskeletal diagnosis cohort: Examining pain and pain care among veterans. *Pain*, *157*(8), 1696–1703. https://doi.org/10.1097/j.pain.0000000000000567.

Gros, D. F., Szafranski, D. D., Brady, K. T., & Back, S. E. (2015). Relations between pain, PTSD symptoms, and substance use in veterans. *Psychiatry*, *78*(3), 277–287. https://doi.org/10.1080/00332747.2015.1069659.

Grossbard, J. R., Malte, C. A., Saxon, A. J., & Hawkins, E. J. (2014). Clinical monitoring and high-risk conditions among patients with SUD newly prescribed opioids and benzodiazepines. *Drug and Alcohol Dependence*, *142*, 24–32. https://doi.org/10.1016/j.drugalcdep.2014.03.020.

Hadlandsmyth, K., Vander Weg, M. W., McCoy, K. D., Mosher, H. J., Vaughan-Sarrazin, M. S., & Lund, B. C. (2018). Risk for prolonged opioid use following total knee arthroplasty in veterans. *The Journal of Arthroplasty*, *33*(1), 119–123. https://doi.org/10.1016/j.arth.2017.08.022.

Jones, B. H., & Hauschild, V. D. (2015). Physical training, fitness, and injuries: Lessons learned from military studies. *Journal of Strength and Conditioning Research*, *29*(Suppl. 11), S57–S64. https://doi.org/10.1519/JSC.0000000000001115.

Kaufman, K. R., Brodine, S., & Shaffer, R. (2000). Military training-related injuries: Surveillance, research, and prevention. *American Journal of Preventive Medicine*, *18*(3 Suppl), 54–63. https://doi.org/10.1016/s0749-3797(00)00114-8.

Kerns, R. D., Otis, J., Rosenberg, R., & Reid, M. C. (2003). Veterans' reports of pain and associations with ratings of health, health-risk behaviors, affective distress, and use of the healthcare system. *Journal of Rehabilitation Research and Development*, *40*(5), 371–379. https://doi.org/10.1682/jrrd.2003.09.0371.

Kim, T. K. (2017). Practical statistics in pain research. *The Korean Journal of Pain*, *30*(4), 243–249. https://doi.org/10.3344/kjp.2017.30.4.243.

Knapik, J. J., Graham, B. S., Rieger, J., Steelman, R., & Pendergrass, T. (2013). Activities associated with injuries in initial entry training. *Military Medicine*, *178*(5), 500–506. https://doi.org/10.7205/MILMED-D-12-00507.

Krebs, E. E., Lorenz, K. A., Bair, M. J., Damush, T. M., Wu, J., Sutherland, J. M., ... Kroenke, K. (2009). Development and initial validation of the PEG, a three-item scale assessing pain intensity and interference. *Journal of General Internal Medicine*, *24*(6), 733–738. https://doi.org/10.1007/s11606-009-0981-1.

Kubat, E., Giori, N. J., Hwa, K., & Eisenberg, D. (2016). Osteoarthritis in veterans undergoing bariatric surgery is associated with decreased excess weight loss: 5-year outcomes. *Surgery for Obesity and Related Diseases: Official Journal of the American Society for Bariatric Surgery*, *12*(7), 1426–1430. https://doi.org/10.1016/j.soard.2016.02.012.

Lew, H. L., Otis, J. D., Tun, C., Kerns, R. D., Clark, M. E., & Cifu, D. X. (2009). Prevalence of chronic pain, posttraumatic stress disorder, and persistent postconcussive symptoms in OIF/OEF veterans: Polytrauma clinical triad. *Journal of Rehabilitation Research and Development*, *46*(6), 697–702. https://doi.org/10.1682/jrrd.2009.01.0006.

Lovejoy, T. I., Dobscha, S. K., Turk, D. C., Weimer, M. B., & Morasco, B. J. (2016). Correlates of prescription opioid therapy in Veterans with chronic pain and history of substance use disorder. *Journal of Rehabilitation Research and Development*, *53*(1), 25–36. https://doi.org/10.1682/JRRD.2014.10.0230.

Macey, T. A., Weimer, M. B., Grimaldi, E. M., Dobscha, S. K., & Morasco, B. J. (2013). Patterns of care and side effects for patients prescribed methadone for treatment of chronic pain. *Journal of Opioid Management*, *9*(5), 325–333. https://doi.org/10.5055/jom.2013.0175.

Morasco, B. J., Corson, K., Turk, D. C., & Dobscha, S. K. (2011). Association between substance use disorder status and pain-related function following 12 months of treatment in primary care patients with musculoskeletal pain. *The Journal of Pain: Official Journal of the American Pain Society*, *12*(3), 352–359. https://doi.org/10.1016/j.jpain.2010.07.010.

Morasco, B. J., & Dobscha, S. K. (2008). Prescription medication misuse and substance use disorder in VA primary care patients with chronic pain. *General Hospital Psychiatry*, *30*(2), 93–99. https://doi.org/10.1016/j.genhosppsych.2007.12.004.

Morasco, B. J., Duckart, J. P., & Dobscha, S. K. (2011). Adherence to clinical guidelines for opioid therapy for chronic pain in patients with substance use disorder. *Journal of General Internal Medicine*, *26*(9), 965–971. https://doi.org/10.1007/s11606-011-1734-5.

Mudumbai, S. C., Oliva, E. M., Lewis, E. T., Trafton, J., Posner, D., Mariano, E. R., ... Clark, J. D. (2016). Time-to-cessation of postoperative opioids: A population-level analysis of the veterans affairs health care system. *Pain Medicine*, *17*(9), 1732–1743. https://doi.org/10.1093/pm/pnw015.

Muñoz, R. A., McBride, M. E., Brnabic, A. J., López, C. J., Hetem, L. A., Secin, R., & Dueñas, H. J. (2005). Major depressive disorder in Latin America: The relationship between depression severity, painful somatic symptoms, and quality of life. *Journal of Affective Disorders*, *86*(1), 93–98. https://doi.org/10.1016/j.jad.2004.12.012.

Murphy, L. B., Helmick, C. G., Allen, K. D., Theis, K. A., Baker, N. A., Murray, G. R., ... Barbour, K. E. (2014). Arthritis among veterans—United States, 2011-2013. *MMWR. Morbidity and Mortality Weekly Report*, *63*(44), 999–1003.

Nahin, R. L. (2017). Severe pain in veterans: The effect of age and sex, and comparisons with the general population. *The Journal of Pain: Official Journal of the American Pain Society*, *18*(3), 247–254. https://doi.org/10.1016/j.jpain.2016.10.021.

Naliboff, B. D., Wu, S. M., Schieffer, B., Bolus, R., Pham, Q., Baria, A., ... Shekelle, P. (2011). A randomized trial of 2 prescription strategies for opioid treatment of chronic nonmalignant pain. *The Journal of Pain: Official Journal of the American Pain Society*, *12*(2), 288–296. https://doi.org/10.1016/j.jpain.2010.09.003.

Nindl, B. C., Williams, T. J., Deuster, P. A., Butler, N. L., & Jones, B. H. (2013). Strategies for optimizing military physical readiness and preventing musculoskeletal injuries in the 21st century. *U.S. Army Medical Department Journal*, 5–23.

Olenick, M., Flowers, M., & Diaz, V. J. (2015). US veterans and their unique issues: Enhancing health care professional awareness. *Advances in Medical Education and Practice, 6*, 635–639. https://doi.org/10.2147/AMEP.S89479.

Otis, J. D., Keane, T. M., & Kerns, R. D. (2003). An examination of the relationship between chronic pain and post-traumatic stress disorder. *Journal of Rehabilitation Research and Development, 40*(5), 397–405. https://doi.org/10.1682/jrrd.2003.09.0397.

O'Toole, T. P., Johnson, E. E., Aiello, R., Kane, V., & Pape, L. (2016). Tailoring care to vulnerable populations by incorporating social determinants of health: The veterans health administration's "homeless patient aligned care team" program. *Preventing Chronic Disease, 13*. https://doi.org/10.5888/pcd13.150567, E44.

Outcalt, S. D., Ang, D. C., Wu, J., Sargent, C., Yu, Z., & Bair, M. J. (2014). Pain experience of Iraq and Afghanistan Veterans with comorbid chronic pain and posttraumatic stress. *Journal of Rehabilitation Research and Development, 51*(4), 559–570. https://doi.org/10.1682/JRRD.2013.06.0134.

Outcalt, S. D., Kroenke, K., Krebs, E. E., Chumbler, N. R., Wu, J., Yu, Z., & Bair, M. J. (2015). Chronic pain and comorbid mental health conditions: Independent associations of posttraumatic stress disorder and depression with pain, disability, and quality of life. *Journal of Behavioral Medicine, 38*(3), 535–543. https://doi.org/10.1007/s10865-015-9628-3.

Owens, B. D., Williams, A. E., & Wolf, J. M. (2015). Risk factors for surgical complications in rotator cuff repair in a veteran population. *Journal of Shoulder and Elbow Surgery, 24*(11), 1707–1712. https://doi.org/10.1016/j.jse.2015.04.020.

Phifer, J., Skelton, K., Weiss, T., Schwartz, A. C., Wingo, A., Gillespie, C. F., ... Ressler, K. J. (2011). Pain symptomatology and pain medication use in civilian PTSD. *Pain, 152*(10), 2233–2240. https://doi.org/10.1016/j.pain.2011.04.019.

Phillips, K. M., Clark, M. E., Gironda, R. J., McGarity, S., Kerns, R. W., Elnitsky, C. A., ... Collins, R. C. (2016). Pain and psychiatric comorbidities among two groups of Iraq and Afghanistan era Veterans. *Journal of Rehabilitation Research and Development, 53*(4), 413–432. https://doi.org/10.1682/JRRD.2014.05.0126.

Polomano, R. C., Galloway, K. T., Kent, M. L., Brandon-Edwards, H., Kwon, K. N., Morales, C., & Buckenmaier, C. (2016). Psychometric testing of the defense and veterans pain rating scale (DVPRS): A new pain scale for military population. *Pain Medicine, 17*(8), 1505–1519. https://doi.org/10.1093/pm/pnw105.

Rozet, I., Nishio, I., Robbertze, R., Rotter, D., Chansky, H., & Hernandez, A. V. (2014). Prolonged opioid use after knee arthroscopy in military veterans. *Anesthesia and Analgesia, 119*(2), 454–459. https://doi.org/10.1213/ANE.0000000000000292.

Runnals, J. J., Van Voorhees, E., Robbins, A. T., Brancu, M., Straits-Troster, K., Beckham, J. C., & Calhoun, P. S. (2013). Self-reported pain complaints among Afghanistan/Iraq era men and women veterans with comorbid posttraumatic stress disorder and major depressive disorder. *Pain Medicine, 14*(10), 1529–1533. https://doi.org/10.1111/pme.12208.

Sá, K. N., Moreira, L., Baptista, A. F., Yeng, L. T., Teixeira, M. J., Galhardoni, R., & de Andrade, D. C. (2019). Prevalence of chronic pain in developing countries: Systematic review and meta-analysis. *Pain Reports, 4*(6). https://doi.org/10.1097/PR9.0000000000000779, e779.

Smeeding, S. J., Bradshaw, D. H., Kumpfer, K., Trevithick, S., & Stoddard, G. J. (2010). Outcome evaluation of the Veterans Affairs Salt Lake City Integrative Health Clinic for chronic pain and stress-related depression, anxiety, and post-traumatic stress disorder. *Journal of Alternative and Complementary Medicine, 16*(8), 823–835. https://doi.org/10.1089/acm.2009.0510.

Smith, D. L. (2015). The relationship of disability and employment for veterans from the 2010 Medical Expenditure Panel Survey (MEPS). *Work, 51*(2), 349–363. https://doi.org/10.3233/WOR-141979.

Taylor, S. L., Herman, P. M., Marshall, N. J., Zeng, Q., Yuan, A., Chu, K., ... Lorenz, K. A. (2019). Use of complementary and integrated health: A retrospective analysis of U.S. veterans with chronic musculoskeletal pain nationally. *Journal of Alternative and Complementary Medicine, 25*(1), 32–39. https://doi.org/10.1089/acm.2018.0276.

Taylor, S. L., Hoggatt, K. J., & Kligler, B. (2019). Complementary and integrated health approaches: What do veterans use and want. *Journal of General Internal Medicine, 34*(7), 1192–1199. https://doi.org/10.1007/s11606-019-04862-6.

Terry, M. J., Moeschler, S. M., Hoelzer, B. C., & Hooten, W. M. (2016). Pain catastrophizing and anxiety are associated with heat pain perception in a community sample of adults with chronic pain. *The Clinical Journal of Pain, 32*(10), 875–881. https://doi.org/10.1097/AJP.0000000000000333.

Trafton, J. A., Oliva, E. M., Horst, D. A., Minkel, J. D., & Humphreys, K. (2004). Treatment needs associated with pain in substance use disorder patients: Implications for concurrent treatment. *Drug and Alcohol Dependence, 73*(1), 23–31. https://doi.org/10.1016/j.drugalcdep.2003.08.007.

Tripp, D. A., Curtis Nickel, J., Landis, J. R., Wang, Y. L., & Knauss, J. S. (2004). Predictors of quality of life and pain in chronic prostatitis/chronic pelvic pain syndrome: Findings from the National Institutes of Health Chronic Prostatitis Cohort Study. *BJU International, 94*(9), 1279–1282. https://doi.org/10.1111/j.1464-410X.2004.05157.x.

U.S. Department of Health and Human Services Interagency Pain Research Coordinating Committee. (2016). *National pain strategy a comprehensive population health-level strategy for pain. National pain strategy overview.* https://www.iprcc.nih.gov/sites/default/files/HHSNational_Pain_Strategy_508C.pdf.

U.S. Department of Veterans Affairs. (2020). *VHA pain management.* U.S. Department of Veterans Affairs. https://www.va.gov/painmanagement/.

Whitehead, A. J., Dobscha, S. K., Morasco, B. J., Ruimy, S., Bussell, C., & Hauser, P. (2008). Pain, substance use disorders and opioid analgesic prescription patterns in veterans with hepatitis C. *Journal of Pain and Symptom Management, 36*(1), 39–45. https://doi.org/10.1016/j.jpainsymman.2007.08.013.

Chapter 21

Nociception during surgery

Munetaka Hirose

Department of Anesthesiology and Pain Medicine, Hyogo College of Medicine, Nishinomiya, Hyogo, Japan

Abbreviations

ANI	analgesia nociception index
BIS	bispectral index
CRP	C-reactive protein
CVI	composite variability index
ERAS	enhanced recovery after surgery
GABA	γ-aminobutylic acid
GPCR	G protein-coupled receptor
HR	heart rate
HRV	heart rate variability
Na_v	voltage-gated sodium channel
NFRT	nociception flexion reflex threshold
NoL	nociception level
NR	nociceptive response
NTS	nucleus tractus solitaries
PAG	periaqueductal gray
MAC	minimum alveolar concentration
MLR	mesencephalic locomotor region
PI	perfusion index
RE/SE	response entropy/state entropy
RVLM	rostal ventrolateral medulla
RVMM	rostral ventromedial medulla
SBP	systolic blood pressure
SPI	surgical pleth index
TrkA	tropomyosine receptor kinase A
TRP	transient receptor potential

Introduction

Nociception is defined as the neural processes of encoding and processing noxious stimuli (Loeser & Treede, 2008), with the transfer of signals from nociceptors in the peripheral organs to the brain. During surgery, surgical invasion activates peripheral nociceptors, leading to activation of the autonomic nervous system and hypothalamic-pituitary-adrenal axis, in addition to inflammatory and immune responses. Greater noxious stimulation during surgery activates these responses more, inducing postoperative complications (Dobson, 2015).

Pain is conscious perception of nociceptive information (Brown, Pavone, & Naranjo, 2018). Pain is not clinically obvious in patients under general anesthesia, since general anesthetics induce unconsciousness. Noxious stimuli, however, induce changes in autonomic responses or the electroencephalogram in unconscious patients under general anesthesia. Surgical invasion augments these responses, and anesthesia inversely suppresses them. The degree of these responses depends on the balance between surgical invasiveness, namely nociception, and anesthetic effects, namely antinociception. Several methods to monitor the balance between nociception and antinociception during surgery under general anesthesia, using changes in autonomic responses or the electroencephalogram, have been developed (Table 1) (Gruenewald & Ilies, 2013; Hight et al., 2019; Hirose et al., 2018; Ledowski, 2019; Ooba, Ueki, Kariya, Tatara, & Hirose, 2020).

Features and Assessments of Pain, Anesthesia, and Analgesia. https://doi.org/10.1016/B978-0-12-818988-7.00049-2
Copyright © 2022 Elsevier Inc. All rights reserved.

TABLE 1 Nociception monitors.

Devise	Sources of measurement	Disadvantage
Analgesia nociception index (ANI)	Heart rate variability (HRV)	Nonspecific responses to vasoactive agents
Nociception level (NoL)	Accelerometry Plethysmographic pulse wave Skin galvanic response Temperature	
Nociceptive response (NR)	Heart rate Perfusion index Systolic blood pressure	
Surgical pleth index (SPI)	Heart beat interval Plethysmographic amplitude	
Composite variability index (CVI)	Bispectral index (BIS), Electromyogram	Nonspecific responses to sedatives and muscle relaxants
Nociceptive flexion reflex threshold (NFRT)	RIII-reflex	
qNOX	Electroencephalogram Electromyogram	
Response entropy/state entropy (RE/SE)	Electroencephalogram Electromyogram	

Nociceptive pathway during surgery

Surgical incision increases sympathetic activity, with increases in blood pressure and heart rate and decrease in skin blood flow, in patients under general anesthesia (Mashimo et al., 1997). These autonomic responses are mediated through nociceptors responding to mechanical, thermal, or chemical stimuli (Ringkamp, Raja, Campbell, & Meyer, 2013). Surgical invasion induces mechanical stimulation, which activates Piezo2 mechanoreceptors (Zhang, Wang, Geng, Zhou, & Xiao, 2019) and also induces inflammation, which activates chemical receptors of transient receptor potential (TRP) channels, tropomyosine receptor kinase A (TrkA), voltage-gated sodium channels (Na_v), and several metabotropic G protein-coupled receptors (GPCRs) (Dai, 2016; Hirose, 2018; Ringkamp et al., 2013; Rostock, Schrenk-Siemens, Pohle, & Siemens, 2018).

The nociceptive signal ascends from the nociceptors along myelinated Aδ fibers and unmyelinated C fibers in the neospinothalamic and paleospinothalamic tracts, to the thalamus and the cerebral cortex, respectively. These ascending projections send collaterals to the sympathetic regions at the spinal level, the rostral ventrolateral medulla (RVLM), rostral ventromedial medulla (RVMM), the nucleus tractus solitarius (NTS) in the medulla, and the periaqueductal gray (PAG) matter in the midbrain (Brown et al., 2018; Cividjian et al., 2017) (Fig. 1).

In 1971, electrical stimulation of the somatic nociceptive nerve was found to elicit sympathetic reflex discharges, consisting of two components, which were recorded from the sympathetic trunk in cats under chloralose anesthesia. The early component, which is called the A-reflex, was evoked via the spinal reflex, and had a latency of 9.8 msec. The late component, which is called the C-reflex, was evoked via the supraspinal reflex, and had a latency of 64 msec (Sato & Schmidt, 1971). In these somato-sympathetic reflexes, somatic afferents from the skin and muscle along Aδ and C nerve fibers activate the spinoreticular tract, increasing blood pressure via neurons of the RVLM (Morrison & Reis, 1989). The viscero-sympathetic reflex evoked by noxious stimuli of visceral organs is also mediated by neurons of the RVLM (Koganezawa, Shimomura, & Terui, 2010). Activation of somato-sympathetic reflex is inhibited by baroreceptor inputs in the RVLM (McMullan, Pathmanandavel, Pilowsky, & Goodchild, 2008).

Several anesthetics reportedly suppress the somato-sympathetic reflexes. Propofol was reported to suppress both the A-reflex and C-reflex, decreasing blood pressure and heart rate in dogs (Whitwam, Galletly, Ma, & Chakrabarti, 2000). The C-reflex, which was elicited by electrical stimulation of the peroneal nerve and recorded from the cardiac sympathetic nerve in cats under chloralose and urethane anesthesia, was also suppressed by midazolam at the inframidbrain region, including the RVLM, with reduction of blood pressure and heart rate (Iida, Iwasaki, Kato, Saeki, & Ogawa, 2007).

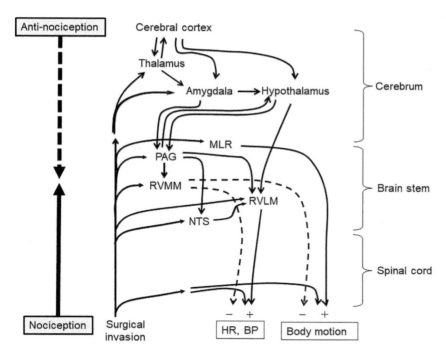

FIG. 1 Nociceptive pathways during surgery. *BP*, blood pressure; *HR*, heart rate; *MLR*, mesencephalic locomotor region; *NTS*, nucleus tractus solitarius; *PAG*, periaqueductal gray matter; *RVLM*, rostral ventrolateral medulla; *RVMM*, rostral ventromedial medulla.

Fentanyl and inhaled anesthetics (e.g., isoflurane, desflurane, halothane, sevoflurane) also suppressed the somatosympathetic reflexes (Ganjoo, Farber, Schwabe, Kampine, & Schmeling, 1996; Ma et al., 1998; Ma, Sapsed-Byrne, Chakrabarti, Ridout, & Whitwam, 1998; Pac-Soo, Ma, Wang, Chakrabarti, & Whitwam, 1999).

Withdrawal reflexes are motor responses to noxious stimuli. Noxious stimuli activate the neurons in the mesencephalic locomotor region (MLR), facilitating locomotor circuits in the spinal cord, with inhibitory modulation from the RVMM (Fig. 1). Half of patients under general anesthesia using minimum alveolar concentrations (MAC) of inhaled anesthetics respond to noxious stimuli with movement. This suggests that the effects of inhaled anesthetics on the activities of these networks are likely associated with the MAC of each volatile anesthetic (Jinks, Bravo, Satter, & Chan, 2010).

Monitoring of nociception under general anesthesia

Surgical invasion under general anesthesia activates peripheral nociceptors, inducing autonomic responses, including increases in blood pressure and heart rate, and decrease in peripheral blood flow (Cividjian et al., 2017; Mashimo et al., 1997). Therefore, nociceptive levels under general anesthesia have been assessed using changes in heart rate and blood pressure during surgery. These hemodynamic parameters, however, do not solely reflect intraoperative nociception (Ben-Israel, Kliger, Zuckerman, Katz, & Edry, 2013). Then, many nociception monitors have been developed (Cividjian et al., 2017; Gruenewald & Ilies, 2013; Ledowski, 2019) (Table 1). Several nociception monitors under general anesthesia evaluate the variables derived from autonomic responses, which are representative of changes in the balance between nociception and antinociception. The analgesia nociception index (ANI), which is calculated using heart rate variability (Logier, Jeanne, Tavernier, & De Jonckheere, 2006), and the surgical pleth index (SPI), which is derived from photoplethysmographic pulse wave amplitude and heart beat interval (Huiku et al., 2007) are two such monitoring methods. The nociceptive response (NR), an index of physiological responses to the balance between nociception and antinociception, is calculated using the three variables of heart rate (HR), systolic blood pressure (SBP), and perfusion index (PI), using the following hemodynamic equation:

$$NR = -1 + \frac{2}{1 + e^{-0.01HR - 0.02SBP + 0.17PI}}$$

(Hirose et al., 2018; Ooba, Ueki, Kariya, Tatara, & Hirose, 2020). Three coefficients of HR, SBP, and PI in the equation, which are −0.01, −0.02, and +0.17, were determined using ordinal logistic analysis in the order of noxious stimuli during skin incision for tympanoplasty (minor noxious stimuli), laparoscopic cholecystectomy (moderate noxious stimuli), and open gastrectomy (severe noxious stimuli) (Hirose et al., 2018). Noxious stimuli under general anesthesia increase SPI and NR values, and decrease the ANI value (Funcke et al., 2017; Hirose et al., 2018; Ledowski, 2019).

A NR value of 0.70 is estimated to represent no noxious stimulus, values between 0.70 and 0.75 represent minor noxious stimuli, such as a very small incision, 0.75 to 0.85 indicates moderate noxious stimuli, such as those elicited by laparoscopic surgery, 0.85 to 0.90 indicates severe noxious stimuli as with open abdominal surgery, and values above 0.90 represent extremely noxious stimuli (Hirose et al., 2018). Fig. 2A and B show the changes in NR values before and after intubation, at skin incision, and at the end of surgery in patients undergoing mastectomy and video-assisted thoracoscopic surgery. During general anesthesia, the NR value typically decreases from 0.9 to 0.6–0.7 after induction of anesthesia, increasing again after tracheal intubation. It also increases at the time of skin incision, thereafter remaining at approximately 0.7 to 0.9 during surgery (Fig. 2A and B). In patients undergoing video-assisted thoracoscopic surgery under general anesthesia, a thoracic paravertebral block, which suppresses nociceptive transmission at the peripheral nerve, reportedly prevents an increase in NR values after skin incision (Miyawaki et al., 2019).

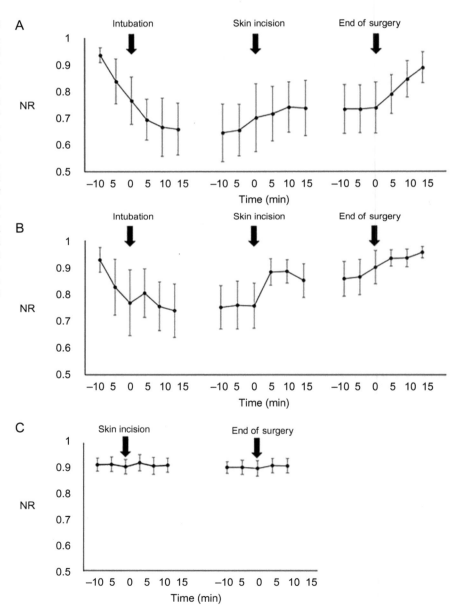

FIG. 2 Nociceptive response (NR) values before and after intubation, at skin incision, and at the end of surgery in patients undergoing mastectomy (A) or video-assisted thoracoscopic surgery (B) under general anesthesia. NR values in patients undergoing skin surgery under regional anesthesia in the awake state (C). In patients undergoing mastectomy ($n = 17$), anesthesia was induced with propofol, fentanyl, and remifentanil, followed by insertion of a supraglottic airway, and maintained using a volatile anesthetic with remifentanil (A). In patients undergoing video-assisted thoracoscopic surgery ($n = 10$), on the other hand, anesthesia was induced with propofol, fentanyl, remifentanil, and rocuronium, followed by endotracheal intubation using a double-lumen tracheal tube, and maintained with target-controlled infusions of propofol and remifentanil (B). Skin surgery was performed under infiltration anesthesia with 1% lidocaine ($n = 10$) (C). Data represent the mean ± SD.

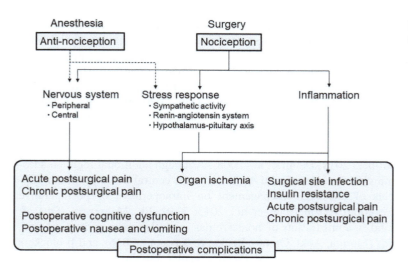

FIG. 3 Relationship between the balance of nociception and antinociception and postoperative complications.

Intraoperative nociception and postoperative complications

Surgical invasion induces stress responses and inflammation. These responses are associated with postoperative complications, including postoperative pain, postoperative cognitive dysfunction, organ ischemia, surgical site infection, and insulin resistance (Fig. 3). Higher degrees of surgical invasiveness are reportedly one of the risk factors for postoperative complications (e.g., older age, obesity, longer duration of surgery, preoperative comorbidities) (Lee et al., 2014). Minimally invasive abdominal surgery leads to fewer postoperative complications compared to open abdominal surgery (Dos Reis, Andrade, Frumovitz, Munsell, & Ramirez, 2018; Li et al., 2010; Xourafas, Pawlik, & Cloyd, 2019). To evaluate associations between intraoperative nociception and postoperative complications, an objective index of intraoperative nociception was developed using the averaged values of NR from the start to end of surgery (Fig. 4). Higher levels of mean NR value, of 0.83 or more throughout surgery, were associated with the severity of postoperative complications

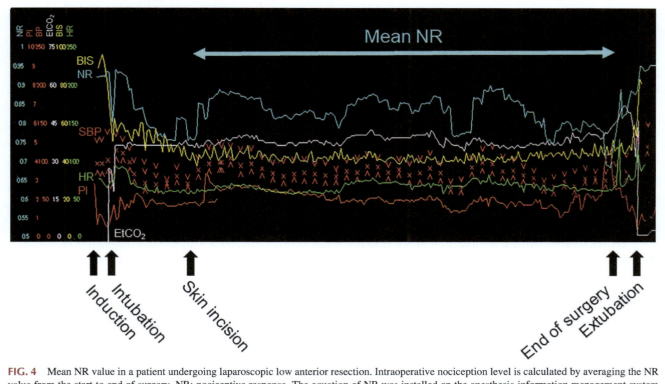

FIG. 4 Mean NR value in a patient undergoing laparoscopic low anterior resection. Intraoperative nociception level is calculated by averaging the NR value from the start to end of surgery. NR; nociceptive response. The equation of NR was installed on the anesthesia information management system (ORSYS, Philips Japan, Tokyo, Japan).

in patients undergoing gastrointestinal surgery (Ogata et al., 2019, 2021). Therefore, intraoperative management that suppresses nociception throughout surgery might reduce the incidence of postoperative complications. Further study is needed to examine the effect on postoperative complications of nociception monitoring–guided anesthesia management.

C-reactive protein (CRP) is an acute phase reactant that increases postoperatively in response to inflammation or tissue damage during surgery. Longer duration of surgery, higher intraoperative nociception, and higher level of preoperative CRP were risk factors for early increases in postoperative CRP levels (Kawasaki et al., 2020; Nakamoto & Hirose, 2019). Higher levels of serum CRP after surgery correlate with higher incidence of postoperative complications (Dos Reis et al., 2018; Kröll et al., 2018; Matsunaga et al., 2017; Straatman et al., 2018; Xourafas et al., 2019). Limited surgery and enhanced recovery after surgery (ERAS) reportedly reduce postoperative CRP levels (Cabellos Olivares, Labalde Martínez, Torralba, Rodríguez Fraile, & Atance Martínez, 2018; Demir et al., 2008). Although additional regional anesthesia during general anesthesia, which blocks the transmission of nociceptive signals to the central nervous system, suppresses stress responses during surgery, the effects of anesthetic management on intraoperative or postoperative inflammatory responses, including CRP levels, are controversial (Fant et al., 2013; Kuchálik, Magnuson, Tina, & Gupta, 2017; Naito et al., 1992). On the other hand, administration of nonsteroidal antiinflammatory drugs (Esme, Kesli, Apiliogullari, Duran, & Yoldas, 2011), steroids (Maruta, Aoki, Omoto, Iizuka, & Kawaura, 2016), and β-blockers (Kim et al., 2015) was reported to reduce postoperative CRP levels. Hence, administration of a combination of these agents during prophylactic anesthetic management for the prevention of postoperative complications could be expected to reduce postoperative CRP levels.

Effects of intraoperative nociception on postoperative pain

Surgical incision induces inflammation, which activates pro-nociceptive systems causing acute postsurgical pain (Richebé, Capdevila, & Rivat, 2018). Higher levels of CRP after surgery are reportedly associated with higher acute postsurgical pain (Esme et al., 2011; Roje, Racic, Kardum, & Selimovic, 2011). Inflammation is also a potential risk factor for the transformation of acute postsurgical pain to chronic postsurgical pain (Voscopoulos & Lema, 2010). Both postoperative CRP concentration and the intensity of acute postsurgical pain correlate with the increase in pain intensity of chronic postsurgical pain in patients undergoing mastectomy (Hashimoto et al., 2018).

To confirm the hypothesis that suppression of intraoperative nociception reduces acute postsurgical pain, a randomized-controlled study was performed to examine the effect of ANI-guided fentanyl administration during sevoflurane anesthesia for lumbar discectomy and laminectomy on acute postsurgical pain. In this study, patients who received ANI-guided fentanyl administration experienced less acute postsurgical pain (Upton, Ludbrook, Wing, & Sleigh, 2017). On the other hand, a meta-analysis to confirm this hypothesis showed no effect of nociception monitoring as a guide to opioid administration during general anesthesia on postoperative pain (Gruenewald & Dempfle, 2017). No definite effects of nociceptive monitoring as a guide for opioid administration on postoperative pain after surgery under general anesthesia have been confirmed.

Applications to other areas

Monitoring of nociception in the awake state

Mischkowski, Palacios-Barrios, Banker, Dildine, and Atlas (2019) studied the effects of psychological processes on autonomic responses after noxious stimuli in awake healthy volunteers and found that the experience of pain affects the autonomic responses in the awake state. It is plausible that there are discrepancies in autonomic responses to noxious stimuli between the unconscious state under general anesthesia and the conscious state. The SPI value is known to be higher in the conscious state without noxious stimulation than that under general anesthesia with noxious stimulation (Ryu et al., 2018). The NR value in the awake state is also higher than that in the unconscious state, being also higher than the level obtained during skin incision under general anesthesia (Fig. 2C). Additionally, there are some discrepancies between conscious and unconscious states in terms of the changes in ANI values caused by noxious stimuli. Although noxious stimuli decrease the ANI value under general anesthesia (Funcke et al., 2017), the ANI value in the conscious state is lower than that under the hypnotic state without noxious stimulation (Boselli et al., 2018). Moreover, noxious stimuli induce only few changes in the ANI value in the conscious state (Issa et al., 2017).

Why are these SPI or NR values in the awake state without noxious stimulation higher than the levels with noxious stimulation under general anesthesia? The question suggests that the changes in values of nociceptive indices that are based on autonomic responses cannot be simply explained by changes in the nociceptive-antinociceptive balance in the awake state.

Somatosensory processing and monitoring of nociception: A hypothesis

Lichtner et al. (2018) examined the transmission of somatosensory stimuli from the spinal cord to the brain with and without propofol anesthesia in humans, showing that although suppression of somatosensory processing induced by moderate noxious stimulation depended on the propofol concentration, intense noxious stimuli induce somatosensory cortical processing even under deep propofol anesthesia. They also found that somatosensory signal transmission depends on the intensity of noxious stimuli under the same propofol concentration. Similar results were reported during remifentanil anesthesia (Lichtner et al., 2018). Therefore, increase in the depth of general anesthesia likely suppresses somatosensory processing in the brain in a dose-dependent manner. Conversely, somatosensory processing under general anesthesia likely occurs in an intensity-dependent manner.

Fig. 5A shows the relationship between depth of general anesthesia and somatosensory processing at a constant level of noxious stimulation, and Fig. 5B shows the relationship between intensity of noxious stimulation and somatosensory processing under a constant depth of general anesthesia. Under general anesthesia, as seen in the shaded area in Fig. 5C, the competence of somatosensory processing in the central nervous system might correspond to changes in SPI and NR values. Somatosensory processing can reach the cerebral cortex before anesthesia induction, when the NR is 0.9, being suppressed after induction of anesthesia, when the NR decreases to 0.7. At the time of tracheal intubation or skin incision, somatosensory processing exceeds the antinociceptive level due to the high intensity of the nociceptive stimuli, resulting in an increase in NR to 0.8 (Fig. 5C).

If nociceptive index values can be explained by the nociceptive–antinociceptive balance, the value in the awake state without noxious stimulation should be equal to that under general anesthesia without noxious stimulation. Further, the value in the awake state should change according to the level of noxious stimulation. Considering the fact that transmission of somatosensory stimuli in the central nervous system depends on the balance between depth of anesthesia and intensity of

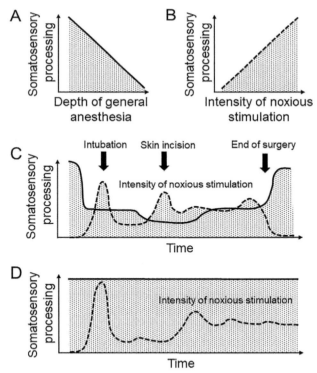

FIG. 5 Proposed associations between somatosensory processing, depth of general anesthesia (A), and intensity of noxious stimulation (B). Noxious stimulation induces dramatic changes in somatosensory processing under general anesthesia (C). On the other hand, noxious stimulation induces few changes in somatosensory processing in the conscious state under regional anesthesia (D). The *bold line* shows the depth of general anesthesia, the *dotted line* shows the intensity of noxious stimulation, and the *shaded area* represents the competence of somatosensory processing in the central nervous system. During surgery under general anesthesia, which suppresses the transmission of somatosensory stimuli in the central nervous system, somatosensory processing changes in accordance with the balance between depth of general anesthesia (antinociception) and intensity of noxious stimulation (nociception) (C). During surgery under regional anesthesia, somatosensory processing remains at a high level, since somatosensory signals from sites other than the anesthetic region can reach the cerebral cortex in the awake state (D).

242 PART | II The syndromes of pain

noxious stimuli, somatosensory processing in the conscious state would always reach the parts of the brain with higher-order functions regardless of the intensity of noxious stimuli (Fig. 5D). Since nociceptive indices using autonomic responses likely represent the competence of somatosensory processing in the central nervous system, the SPI and NR values in the awake state are likely maintained higher than that in the unconscious state.

Other agents of interest

Neurons in the RVLM contain catecholamine, enkephalin, neuropeptide Y, substance P, and glutamate. In the RVMM, neurons contain γ-aminobutylic acid (GABA), glycine, serotonin, substance P, and glutamate (Stornetta, 2009). Microinjection of glutamate antagonist, kynurenic acid, suppressed the somato-sympathetic reflex induced by electric stimulation of the intercostal nerves (Zanzinger, Czachurski, Offner, & Seller, 1994), and also suppressed the viscero-sympathetic reflex induced by gallbladder stimulation with bradykinin in cats (Zhou, Fu, Tjen, Guo, & Longhurst, 2006). Receptors of these agents are activated or suppressed by multimodal general anesthesia, using opioid, ketamine, magnesium, dexmedetomidine, lidocaine, propofol, and inhaled anesthetics (Brown et al., 2018).

Mini-dictionary of terms

- **Intraoperative nociception:** Neural processes involved in encoding noxious stimuli during surgery.
- **Nociceptor:** A high-threshold sensory receptor encoding noxious stimuli.
- **Noxious stimulus**: An actually or potentially damaging-to-tissue stimulus, inducing afferent inputs through nociceptors.
- **Somato-sympathetic reflex:** The neural arc mediating autonomic responses after stimulation of afferent somatic sensory nerves.
- **Surgical stress:** Systemic responses to surgical invasion, including activation of the sympathetic nervous and endocrine systems.

Key facts

Key facts of the Somato-sympathetic reflex

- Noxious stimulation of afferent somatosensory nerves causes sympathetic responses, increasing blood pressure and heart rate, and decreasing peripheral blood flow.
- The somato-sympathetic reflex represents the neural arc via the spinal cord and the brain stem.
- Baroreceptor reflexes affect somato-sympathetic reflexes.
- Anesthetics attenuate somato-sympathetic reflexes.
- Neurons in the rostral ventrolateral medulla (RVLM) mediate both the somato-sympathetic reflex from somatic nociceptors and the viscero-sympathetic reflex from visceral nociceptors.

Summary points

- This chapter focuses on sympathetic responses to noxious stimuli during surgery.
- Monitoring of nociception estimates the balance between nociception and antinociception, using variables evoked in autonomic responses or the electroencephalogram after noxious stimulation.
- The nociceptive response (NR), an index used in nociceptive monitoring, is increased by nociceptive stimuli, being decreased by anesthetics during general anesthesia.
- Higher levels of the averaged values of NR from the start to end of surgery (mean NR) are associated with postoperative complications.
- The utility of monitoring nociception in the awake state is controversial.

References

Ben-Israel, N., Kliger, M., Zuckerman, G., Katz, Y., & Edry, R. (2013). Monitoring the nociception level: A multi-parameter approach. *Journal of Clinical Monitoring and Computing, 27*, 659–668.

Boselli, E., Musellec, H., Martin, L., Bernard, F., Fusco, N., Guillou, N., et al. (2018). Effects of hypnosis on the relative parasympathetic tone assessed by ANI (Analgesia/Nociception Index) in healthy volunteers: A prospective observational study. *Journal of Clinical Monitoring and Computing, 32*, 487–492.

Brown, E. N., Pavone, K. J., & Naranjo, M. (2018). Multimodal general anesthesia: Theory and practice. *Anesthesia and Analgesia, 127*, 1246–1258.

Cabellos Olivares, M., Labalde Martínez, M., Torralba, M., Rodríguez Fraile, J. R., & Atance Martínez, J. C. (2018). C-reactive protein as a marker of the surgical stress reduction within an ERAS protocol (Enhanced Recovery After Surgery) in colorectal surgery: A prospective cohort study. *Journal of Surgical Oncology, 117*, 717–724.

Cividjian, A., Petitjeans, F., Liu, N., Ghignone, M., de Kock, M., & Quintin, L. (2017). Do we feel pain during anesthesia? A critical review on surgery-evoked circulatory changes and pain perception. *Best Practice & Research. Clinical Anaesthesiology, 31*, 445–467.

Dai, Y. (2016). TRPs and pain. *Seminars in Immunopathology, 38*, 277–291.

Demir, A., Bige, O., Saatli, B., Solak, A., Saygili, U., & Onvural, A. (2008). Prospective comparison of tissue trauma after laparoscopic hysterectomy types with retroperitoneal lateral transsection of uterine vessels using ligasure and abdominal hysterectomy. *Archives of Gynecology and Obstetrics, 2008* (277), 325–330.

Dobson, G. P. (2015). Addressing the global burden of trauma in major surgery. *Frontiers in Surgery, 2*, 43.

Dos Reis, R., Andrade, C. E. M. C., Frumovitz, M., Munsell, M., & Ramirez, P. T. (2018). Radical hysterectomy and age: Outcomes comparison based on a minimally invasive vs an open approach. *Journal of Minimally Invasive Gynecology, 25*, 1224–1230.

Esme, H., Kesli, R., Apiliogullari, B., Duran, F. M., & Yoldas, B. (2011). Effects of flurbiprofen on CRP, TNF-α, IL-6, and postoperative pain of thoracotomy. *International Journal of Medical Sciences, 8*, 216–221.

Fant, F., Tina, E., Sandblom, D., Andersson, S. O., Magnuson, A., Hultgren-Hörnkvist, E., et al. (2013). Thoracic epidural analgesia inhibits the neurohormonal but not the acute inflammatory stress response after radical retropubic prostatectomy. *British Journal of Anaesthesia, 110*, 747–757.

Funcke, S., Sauerlaender, S., Pinnschmidt, H. O., Saugel, B., Bremer, K., Reuter, D. A., et al. (2017). Validation of innovative techniques for monitoring nociception during general anesthesia: A clinical study using tetanic and intracutaneous electrical stimulation. *Anesthesiology, 127*, 272–283. .

Ganjoo, P., Farber, N. E., Schwabe, D., Kampine, J. P., & Schmeling, W. T. (1996). Desflurane attenuates the somatosympathetic reflex in rats. *Anesthesia and Analgesia, 83*, 55–61.

Gruenewald, M., & Dempfle, A. (2017). Analgesia/nociception monitoring for opioid guidance: Meta-analysis of randomized clinical trials. *Minerva Anestesiologica, 83*, 200–213.

Gruenewald, M., & Ilies, C. (2013). Monitoring the nociception-anti-nociception balance. *Best Practice & Research. Clinical Anaesthesiology, 27*, 235–247.

Hashimoto, K., Tsuji, A., Takenaka, S., Ohmura, A., Ueki, R., Noma, H., et al. (2018). C-reactive protein level on postoperative day one is associated with chronic postsurgical pain after mastectomy. *Anesthesiology and Pain Medicine, 8*, e79331.

Hight, D. F., Gaskell, A. L., Kreuzer, M., Voss, L. J., García, P. S., & Sleigh, J. W. (2019). Transient electroencephalographic alpha power loss during maintenance of general anaesthesia. *British Journal of Anaesthesia, 122*, 635–642.

Hirose, M. (2018). In S. Choi (Ed.), *NGF. Encyclopedia of signaling molecules* (2nd ed., pp. 3475–3479). Springer.

Hirose, M., Kobayashi, Y., Nakamoto, S., Ueki, R., Kariya, N., & Tatara, T. (2018). Development of a hemodynamic model using routine monitoring parameters for nociceptive responses evaluation during surgery under general anesthesia. *Medical Science Monitor, 24*, 3324–3331.

Huiku, M., Uutela, K., van Gils, M., Korhonen, I., Kymäläinen, M., Meriläinen, P., et al. (2007). Assessment of surgical stress during general anaesthesia. *British Journal of Anaesthesia, 98*, 447–455.

Iida, R., Iwasaki, K., Kato, J., Saeki, S., & Ogawa, S. (2007). Reflex sympathetic activity after intravenous administration of midazolam in anesthetized cats. *Anesthesia and Analgesia, 105*, 832–837.

Issa, R., Julien, M., Décary, E., Verdonck, O., Fortier, L. P., Drolet, P., et al. (2017). Evaluation of the analgesia nociception index (ANI) in healthy awake volunteers. *Canadian Journal of Anaesthesia, 64*, 828–835.

Jinks, S. L., Bravo, M., Satter, O., & Chan, Y. M. (2010). Brainstem regions affecting minimum alveolar concentration and movement pattern during isoflurane anesthesia. *Anesthesiology, 112*, 316–324.

Kawasaki, Y., Park, S., Miyamoto, K., Ueki, R., Kariya, N., Tatara, T., et al. (2020). Modified model for prediction of early C-reactive protein levels after gastrointestinal surgery: A prospective cohort study. *PLoS One, 15*, e0239709.

Kim, Y., Hwang, W., Cho, M. L., Her, Y. M., Ahn, S., & Lee, J. (2015). The effects of intraoperative esmolol administration on perioperative inflammatory responses in patients undergoing laparoscopic gastrectomy: A dose-response study. *Surgical Innovation, 22*, 177–182.

Koganezawa, T., Shimomura, Y., & Terui, N. (2010). The viscerosympathetic response in rabbits is mediated by GABAergic and glutamatergic inputs into the sympathetic premotor neurons of the rostral ventrolateral medulla. *Experimental Physiology, 95*, 1061–1070.

Kröll, D., Nakhostin, D., Stirnimann, G., Erdem, S., Haltmeier, T., Nett, P. C., et al. (2018). C-reactive protein on postoperative day 1: A predictor of early intra-abdominal infections after bariatric surgery. *Obesity Surgery, 27*, 2760–2766.

Kuchálik, J., Magnuson, A., Tina, E., & Gupta, A. (2017). Does local infiltration analgesia reduce peri-operative inflammation following total hip arthroplasty? A randomized, double-blind study. *BMC Anesthesiology, 17*, 63.

Ledowski, T. (2019). Objective monitoring of nociception: A review of current commercial solutions. *British Journal of Anaesthesia, 123*, e312–e321.

Lee, K. G., Lee, H. J., Yang, J. Y., Oh, S. Y., Bard, S., Suh, Y. S., et al. (2014). Risk factors associated with complication following gastrectomy for gastric cancer: Retrospective analysis of prospectively collected data based on the Clavien-Dindo system. *Journal of Gastrointestinal Surgery, 18*, 1269–1277.

Li, X., Zhang, J., Sang, L., Zhang, W., Chu, Z., Li, X., et al. (2010). Laparoscopic versus conventional appendectomy—A meta-analysis of randomized controlled trials. *BMC Gastroenterology, 10*, 129.

Lichtner, G., Auksztulewicz, R., Kirilina, E., Velten, H., Mavrodis, D., Scheel, M., et al. (2018). Effects of propofol anesthesia on the processing of noxious stimuli in the spinal cord and the brain. *NeuroImage, 172*, 642–653.

Lichtner, G., Auksztulewicz, R., Velten, H., Mavrodis, D., Scheel, M., Blankenburg, F., et al. (2018). Nociceptive activation in spinal cord and brain persists during deep general anaesthesia. *British Journal of Anaesthesia, 121*, 291–302.

Loeser, J. D., & Treede, R. D. (2008). The Kyoto protocol of IASP basic pain terminology. *Pain, 137*, 473–477.

Logier, R., Jeanne, M., Tavernier, B., & De Jonckheere, J. (2006). Pain/analgesia evaluation using heart rate variability analysis. *Conference Proceedings: Annual International Conference of the IEEE Engineering in Medicine and Biology Society, 1*, 4303–4306.

Ma, D., Sapsed-Byrne, S. M., Chakrabarti, M. K., Ridout, D., & Whitwam, J. G. (1998). Synergism between sevoflurane and intravenous fentanyl on A delta and C somatosympathetic reflexes in dogs. *Anesthesia and Analgesia, 87*, 211–216.

Ma, D., Wang, C., Pac Soo, C. K., Chakrabarti, M. K., Lockwood, G. G., & Whitwam, J. G. (1998). The effect of sevoflurane on spontaneous sympathetic activity, A delta and C somatosympathetic reflexes, and associated hemodynamic changes in dogs. *Anesthesia and Analgesia, 86*, 1079–1083.

Maruta, K., Aoki, A., Omoto, T., Iizuka, H., & Kawaura, H. (2016). The effect of steroid therapy on postoperative inflammatory response after endovascular abdominal aortic aneurysm repair. *Annals of Vascular Diseases, 9*, 168–172.

Mashimo, T., Zhang, P., Kamibayashi, T., Inagaki, Y., Ohara, A., Yamatodani, A., et al. (1997). Laser Doppler skin blood flow and sympathetic nervous responses to surgical incision during halothane and isoflurane anesthesia. *Anesthesia and Analgesia, 85*, 291–298.

Matsunaga, T., Saito, H., Murakami, Y., Kuroda, H., Fukumoto, Y., & Osaki, T. (2017). Serum level of C-reactive protein on postoperative day 3 is a predictive indicator of postoperative pancreatic fistula after laparoscopic gastrectomy for gastric cancer. *Asian Journal of Endoscopic Surgery, 1*, 382–387.

McMullan, S., Pathmanandavel, K., Pilowsky, P. M., & Goodchild, A. K. (2008). Somatic nerve stimulation evokes qualitatively different somatosympathetic responses in the cervical and splanchnic sympathetic nerves in the rat. *Brain Research, 1217*, 139–147.

Mischkowski, D., Palacios-Barrios, E. E., Banker, L., Dildine, T. C., & Atlas, L. Y. (2019). Pain or nociception? Subjective experience mediates the effects of acute noxious heat on autonomic responses—Corrected and republished. *Pain, 160*, 1469–1481.

Miyawaki, H., Ogata, H., Nakamoto, S., Kaneko, T., Ueki, R., Kariya, N., et al. (2019). Effects of thoracic paravertebral block on nociceptive levels after skin incision during video-assisted thoracoscopic surgery. *Medical Science Monitor, 25*, 3140–3145.

Morrison, S. F., & Reis, D. J. (1989). Reticulospinal vasomotor neurons in the RVL mediate the somatosympathetic reflex. *The American Journal of Physiology, 256*, R1084–R1097.

Naito, Y., Tamai, S., Shingu, K., Shindo, K., Matsui, T., Segawa, H., et al. (1992). Responses of plasma adrenocorticotropic hormone, cortisol, and cytokines during and after upper abdominal surgery. *Anesthesiology, 77*, 426–431.

Nakamoto, S., & Hirose, M. (2019). Prediction of early C-reactive protein levels after non-cardiac surgery under general anesthesia. *PLoS One, 2020*(14), e0226032.

Ogata, H., Matsuki, Y., Okamoto, T., Ishimoto, D., Onoe, K., Ueki, R., et al. (2021). Intra-operative nociceptive responses and postoperative major complications after gastrointestinal surgery under general anaesthesia: A prospective cohort study. *European Journal of Anaesthesiology.* https://doi.org/10.1097/EJA.0000000000001505.

Ogata, H., Nakamoto, S., Miyawaki, H., Ueki, R., Kariya, N., Tatara, T., et al. (2019). Association between intraoperative nociception and postoperative complications in patients undergoing laparoscopic gastrointestinal surgery. *Journal of Clinical Monitoring and Computing, 34*(3), 575–581. https://doi.org/10.1007/s10877-019-00347-3.

Ooba, S., Ueki, R., Kariya, N., Tatara, T., & Hirose, M. (2020). Mathematical evaluation of responses to surgical stimuli under general anesthesia. *Scientific Reports, 10*, 15300.

Pac-Soo, C. K., Ma, D., Wang, C., Chakrabarti, M. K., & Whitwam, J. G. (1999). Specific actions of halothane, isoflurane, and desflurane on sympathetic activity and A delta and C somatosympathetic reflexes recorded in renal nerves in dogs. *Anesthesiology, 91*, 470–478.

Richebé, P., Capdevila, X., & Rivat, C. (2018). Persistent postsurgical pain: Pathophysiology and preventative pharmacologic considerations. *Anesthesiology, 129*, 590–607.

Ringkamp, M., Raja, S. N., Campbell, J. N., & Meyer, R. A. (2013). Peripheral mechanisms of cutaneous nociception. NGF. In S. B. McMahon, M. Koltzenburg, I. Tracey, & D. C. Turk (Eds.), *Wall and Melzack's textbook of pain* (6th ed., pp. 1–30). Elsevier Saunders.

Roje, Z., Racic, G., Kardum, G., & Selimovic, M. (2011). Is the systemic inflammatory reaction to surgery responsible for post-operative pain after tonsillectomy, and is it "technique-related"? *Wiener Klinische Wochenschrift, 123*, 479–484.

Rostock, C., Schrenk-Siemens, K., Pohle, J., & Siemens, J. (2018). Human vs. mouse nociceptors—Similarities and differences. *Neuroscience, 387*, 13–27.

Ryu, K. H., Song, K., Lim, T. Y., Choi, W. J., Kim, Y. H., & Kim, H. S. (2018). Does Equi-minimum alveolar concentration value ensure equivalent analgesic or hypnotic potency?: A comparison between Desflurane and sevoflurane. *Anesthesiology, 128*, 1092–1098.

Sato, A., & Schmidt, R. F. (1971). Spinal and supraspinal components of the reflex discharges into lumbar and thoracic white rami. *The Journal of Physiology, 212*, 839–850.

Stornetta, R. L. (2009). Neurochemistry of bulbospinal presympathetic neurons of the medulla oblongata. *Journal of Chemical Neuroanatomy, 38*, 222–230.

Straatman, J., Cuesta, M. A., Tuynman, J. B., Veenhof, A. A. F. A., Bemelman, W. A., & van der Peet, D. L. (2018). C-reactive protein in predicting major postoperative complications are there differences in open and minimally invasive colorectal surgery? Substudy from a randomized clinical trial. *Surgical Endoscopy, 32*, 2877–2885.

Upton, H. D., Ludbrook, G. L., Wing, A., & Sleigh, J. W. (2017). Intraoperative "analgesia nociception index"-guided fentanyl administration during sevoflurane Anesthesia in lumbar discectomy and laminectomy: A randomized clinical trial. *Anesthesia and Analgesia, 125*, 81–90.

Voscopoulos, C., & Lema, M. (2010). When does acute pain become chronic? *British Journal of Anaesthesia, 105*(Suppl 1), i69–i85.

Whitwam, J. G., Galletly, D. C., Ma, D., & Chakrabarti, M. K. (2000). The effects of propofol on heart rate, arterial pressure and adelta and C somato-sympathetic reflexes in anaesthetized dogs. *European Journal of Anaesthesiology, 17*, 57–63.

Xourafas, D., Pawlik, T. M., & Cloyd, J. M. (2019). Early morbidity and mortality after minimally invasive liver resection for hepatocellular carcinoma: A propensity-score matched comparison with open resection. *Journal of Gastrointestinal Surgery, 23*, 1435–1442.

Zanzinger, J., Czachurski, J., Offner, B., & Seller, H. (1994). Somato-sympathetic reflex transmission in the ventrolateral medulla oblongata: Spatial organization and receptor types. *Brain Research, 656*, 353–358.

Zhang, M., Wang, Y., Geng, J., Zhou, S., & Xiao, B. (2019). Mechanically activated piezo channels mediate touch and suppress acute mechanical pain response in mice. *Cell Reports, 26*, 1419–1431.

Zhou, W., Fu, L. W., Tjen, A. L. S. C., Guo, Z. L., & Longhurst, J. C. (2006). Role of glutamate in a visceral sympathoexcitatory reflex in rostral ventrolateral medulla of cats. *American Journal of Physiology. Heart and Circulatory Physiology, 291*, H1309–H1318.

Chapter 22

Breast cancer and nociceptione

Amanda Spring de Almeida and Gabriela Trevisan
Graduate Program in Pharmacology, Federal University of Santa Maria (UFSM), Santa Maria, RS, Brazil

Abbreviations

AI	aromatase inhibitors
AIMSS	aromatase inhibitors associated musculoskeletal symptoms
BDNF	brain-derived neurotrophic factor
CB1	cannabinoid receptor 1
CB2	cannabinoid receptor 2
CIBP	cancer-induced bone pain
COX	cyclooxygenase
CPSP	chronic postsurgical pain
CRNP	cancer-related neuropathic pain
DRG	dorsal root ganglion
EBR	external beam radiotherapy
IASP	International Association for the Study of Pain
IL-1	interleukin-1
NGF	nerve growth factor
NMDA	N-methyl D-aspartate
NSAID	Nonsteroidal anti-inflammatory drugs
OIC	opioid-induced constipation
OIH	opioid-induced hyperalgesia
RANKL	receptor activator of nuclear factor kappa-B ligand
SSRIs	selective serotonin reuptake inhibitors
SNRIs	serotonin-norepinephrine reuptake inhibitors
TCAs	tricyclic antidepressants
TNFα	tumor necrosis factor-α
TRP	transient receptor potential
TRPA1	transient receptor potential ankyrin 1
TRPV1	transient receptor potential vanilloid 1
WHO	World Health Organization

Introduction

Breast carcinoma is a frequent public health concern globally, even with different and innovative therapies to decrease the prevalence and mortality (Zhu, Ge, Yu, & Wang, 2015). This type of tumor is the most commonly diagnosed malignancies in women; also, it is the leading cause of death by cancer in females (Bray et al., 2018). Nearby 40% to 89% of patients with breast cancer will have chronic pain (Bokhari & Sawatzky, 2009; Liepe, 2018).

Chronic pain management in patients with active breast cancer needs to be individualized and multidisciplinary; also, pain induced by cancer treatment should be evaluated and controlled. Here, we focused on the clinical treatments, animal models, and novel mechanisms studied for pain in active breast cancer (Hui & Bruera, 2014; Portenoy, 2011).

Treatments for breast cancer-induced pain

The control of cancer pain usually follows the World Health Organization (WHO) analgesic ladder, and this strategy often reduces distress and suffering if adequately addressed (Anekar & Cascella, 2020). The management using the

---

Features and Assessments of Pain, Anesthesia, and Analgesia. https://doi.org/10.1016/B978-0-12-818988-7.00032-7
Copyright © 2022 Elsevier Inc. All rights reserved.

WHO analgesic scale involves various medicines or interventional approaches and originally included three steps; currently, it has four steps.

The first step includes nonopioid analgesics, such as nonsteroidal antiinflammatory drugs (NSAID) and acetaminophen that are used for mild pain, also adjuvants can be added to the therapy. In the second step, the control of mild-to-moderate pain is done by utilizing weak opioids (tramadol, hydrocodone, and codeine). For moderate-to-severe pain management in the third step, strong opioids (morphine, oxycodone, tapentadol, methadone, buprenorphine, fentanyl, hydromorphone, and oxymorphone) should be administered. In the second and third steps, nonopioid analgesics can be used, as well as adjuvants. Also, all the drugs should be used orally and administered at regularly scheduled intervals.

The adjuvant therapies englobe a large group of compounds from different pharmacological classes that should help to control various symptoms further than pain. Anticonvulsants (gabapentin and pregabalin), topical lidocaine or capsaicin, tricyclic antidepressants (TCAs, including nortriptyline and amitriptyline), serotonin-norepinephrine reuptake inhibitors (SNRIs, venlafaxine, and duloxetine), selective serotonin reuptake inhibitors (SSRIs; e.g., sertraline, escitalopram, citalopram), bisphosphonates, and corticosteroids can be used as adjuvants.

The fourth step (interventional step) includes different nonpharmacological and interventional practices to the relief of severe pain, and they can be used in combination with potent opioids and other pain medicines. These therapies are often included in cancer pain treatment when the clinician detects a failure to accomplish adequate analgesia even when high doses of opioids were used, or there is intolerable side effects development. This last step includes neuromodulation approaches, neurosurgical strategies, intrathecal or local administrations of analgesics or local anesthetics, epidural analgesia, palliation radiotherapy, nerve block, physiotherapy, occupational therapy, and ablative procedures.

Also, for cancer-related neuropathic pain (CRNP), antidepressants (TCAs or SNRIs) or anticonvulsants (gabapentin and pregabalin) can be used. Moreover, topical agents (lidocaine or capsaicin high-concentration patches) can be helpful as second-line therapies. Tramadol is also considered as a second-line therapy and potent opioids as third-line drugs.

The analgesic ladder should be used in a bidirectional approach, and de-escalation of drugs can be done if chronic pain reduces or resolves (Anekar & Cascella, 2020; Hui & Bruera, 2014; Liu, Zheng, Tan, & Meredith, 2017; Portenoy, 2011).

Limitations of cancer pain treatments

For better control of cancer pain, it is relevant to manage the side effects of treatments or change the analgesic drug administered. NSAID administration leads to gastrointestinal toxicity, then if the patient has advanced age (more than 60 years), history of peptic ulcer disease, and gastrointestinal bleeding, NSAID therapy should be carefully evaluated to avoid the upper digestive tract bleeding and perforation. Renal toxicity is hampered by preventing the use of NSAID in older patients or individuals with renal insufficiency or compromised fluid status. Moreover, if the patient has a history or risk of cardiovascular disease, the treatment should be terminated if hypertension or congestive heart failure happens or exacerbates.

Acetaminophen has limited analgesic efficacy and did not present antiinflammatory activity. Also, it shows hepatic toxicity as a significant adverse effect. Chronic administration of this compound should be done in the maximum of 3 g per day to prevent liver damage; also anaphylaxis is a relevant issue (Anekar & Cascella, 2020; Fallon & Colvin, 2013; Hui & Bruera, 2014; Liu et al., 2017; Portenoy, 2011; Swarm et al., 2019).

Opioid-induced constipation (OIC) may be controlled prophylactically using increased dietary fiber intake, hydration, exercise, administration of a stimulating laxative, or the coadministration of a peripherally acting μ-opioid receptor antagonist. Moreover, opioids induce other less common side effects, including nausea, vomiting, drowsiness, mental clouding, itch, urinary retention, and hypogonadism. Also, opioid misuse is a problem that clinicians should be aware (Anekar & Cascella, 2020; Fallon & Colvin, 2013; Hui & Bruera, 2014; Liu et al., 2017; Portenoy, 2011; Swarm et al., 2019).

For the SSRIs, drug interactions should be observed to prevent serotonin syndrome. The treatment with TCAs is also limited by the induction of side effects, such as drowsiness, urinary retention, dry mouth, and constipation. Moreover, these compounds may lead to cardiotoxicity, orthostatic hypotension, and liver toxicity, limiting their use in elderly patients.

The SNRIs-induced side effects are commonly well tolerated and usually reduced after continuous administration, including drowsiness, nausea, dizziness, constipation, dry mouth, anorexia, and sexual dysfunction. As venlafaxine treatment may induce a risk for high arterial pressure, systematic monitoring is suggested. Also, duloxetine should not be used in patients with hepatic disease, and venlafaxine dosage needs to be decreased in patients with hepatic or renal disorder.

For gabapentin and pregabalin, the most frequent side effects are drowsiness and dizziness, but weight gain, nausea, dry mouth, peripheral edema, and ataxia are also described.

The side effects for topical lidocaine are minimal, such as skin inflammation, but it should not be used in patients treated with oral class I antiarrhythmic compounds (Anekar & Cascella, 2020; Fallon & Colvin, 2013; Hui & Bruera, 2014; Liu et al., 2017; Portenoy, 2011; Swarm et al., 2019).

Bone pain in breast cancer

Bone metastasis is frequently accompanied by skeletal-related events, such as hypercalcemia, bone marrow aplasia, pathologic skeletal fracture, spinal cord compression, nerve damage, and bone pain. All these factors together reduce mobility, cause bone/muscle mass loss, and reduce life expectancy.

Chronic pain caused by metastasis to bone is frequently referred to as cancer-induced bone pain (CIBP), and it has distinctive and intricate pathophysiology. It is maintained by neuropathic and nociceptive mechanisms (Zajączkowska, Kocot-Kępska, Leppert, & Wordliczek, 2019). Osteogenic pain is poorly localized, and it is an indication of bone metastasis and has a deep dull characteristic. At night, it can be worse, also, it is a continuous pain that is not reduced by resting or lying down and often increases during movement. Thus, refractory CIBP often appears in more than half of metastatic breast cancer patients.

Breast cancer bone pain often increases in intensity with time, but if the tumor continues to grow, causing bone alteration, incident or breakthrough pain can occur. This type of pain is described as a brief burst of severe pain secondary to continuous bone pain cancer. It is provoked by movement or shows spontaneously; also, it is difficult to manage. The management of breakthrough pain is often done with potent opioids (Coleman, 2001; Li et al., 2014; Macedo et al., 2017; Mantyh, 2013; Wong & Pavlakis, 2011; Zajączkowska et al., 2019).

Treatments of bone pain in breast cancer

Osteogenic pain treatment should be controlled by a multimodal approach, using pharmacological and nonpharmacological strategies. Causal anticancer treatment includes radiotherapy, surgery, and systemic therapy; these approaches will decrease tumor growth and tissue infiltration–reducing pain. Surgical procedures can be used to reduce bone pain in patients resistant to palliative analgesic therapy, including neurodestruction or neuromodulation. Neurodestruction induces the disruption of the pain pathways to the brain from the spinal cord, and this can be done in individual cases (as a spinal metastatic disease). It is useful if the life expectancy is short (2–3 months) since the analgesic effect persists for a short time. Neuromodulation is also used for intractable pain from a neuropathic origin, and spinal cord electrical stimulation can be useful for this set of patients.

The systemic administration of bone-targeted radiopharmaceuticals (samarium-153 and strontium-89) is used to reduce refractory and diffuse bone pain in particular cases. These compounds have better efficacy in osteoblastic metastasis where the new reactive bone is produced, but for the relief of breast cancer-induced bone pain, the evidence is still scarce.

In localized metastatic bone pain caused by a mass effect or destruction by cancer cells, the treatment may be done using external beam radiotherapy (EBR). Frequently, it used a short treatment timetable, and results are obtained rapidly (1–6 weeks), also if pain reappears, retreatment is feasible. EBR is extremely useful for pain in patients with a risk for pathologic fracture or patients with spinal cord compression causing neurological difficulties.

Also, systemic treatment can be used to control bone pain; osteoclast inhibitors (bisphosphonates and denosumab) slow down osteolysis and reduce bone pain. Thus, patients presenting widespread poorly localized and nonmechanical bone pain can be managed using the administration of bisphosphonates (clodronate, pamidronate, zoledronate, and ibandronate). Denosumab is a human monoclonal antibody that inhibits the receptor activator of nuclear factor-kappa-B ligand (RANKL), decreasing osteoclasts maturation and proliferation, and then osteoclast-mediated bone destruction. Thus, these bone-targeted compounds reduce bone pain and delay skeletal-related events. In this view, it is suggested to begin the treatment when the metastasis is detected even if the patient is still without pain.

Pain relief is also obtained with analgesic treatment, using nonopioid (NSAID and acetaminophen) and opioid drugs based on the WHO analgesic ladder. Adjuvants are also useful to reduce bone cancer pain, including glucocorticoids (dexamethasone and prednisone) that are used in advanced illness. Glucocorticoids reduce inflammation and swelling and are the most frequently used adjuvant analgesic for the management of cancer bone pain.

Side effects observed for the treatments of bone cancer pain

EBR induces pain reduction for focal metastatic bone pain, but the radiation exposure to vulnerable structures such as the gut, lung, kidney, and liver needs to be decreased. The treatment with radiopharmaceuticals (samarium-153 and strontium-89) may cause distinct side effects, including myelosuppression, renal toxicity, and pain flare.

The bisphosphonates drugs are the preferred treatment when bone metastasis is present; this therapy induces osteonecrosis of the jaw as a severe rare side effect, but usually, these compounds are well tolerated. Bisphosphonates should not be used to patients with renal insufficiency, and the most frequent side effects are flu-like symptoms (arthralgia, myalgia, fever, and weakness), nausea, peripheral edema, dyspnea, and nausea.

250 **PART** | **II** The syndromes of pain

Denosumab increases the risk of osteonecrosis of the jaw, and dental measurements should be taken to prevent this side effect. The most common side effects are nausea, weakness, and diarrhea. However, denosumab treatment may be used in patients with renal failure, but this compound increases the infection rate in patients with early breast cancer.

Moreover, for glucocorticoids (dexamethasone and prednisone), treatment side effects after a long period of administration include immunosuppression, hyperglycemia, hypertension, psychosis, and gastric ulcers. Thus, these compounds should be used for a short period if possible (Ahmad et al., 2018; Coleman, 2001; Macedo et al., 2017; Mantyh, 2013; Portenoy, 2011; Swarm et al., 2019; Zajączkowska et al., 2019).

Animal models of breast cancer pain: Breast inoculation models

There are several models that study the mechanisms of pain caused by metastatic breast cancer. Still, most of them use arthrotomy (or ectopic implantation) models where tumor cells are inoculated directly into the bone. But these models ignore the early stages of the metastatic cascade (Horas, Zheng, Zhou, & Seibel, 2015), not showing what happens to breast cancer patients. Thus, a different model was standarized in animals, where breast cancer cells were inoculated directly into the mammary gland, and after a few days, bone metastasis is observed (de Almeida et al., 2019; Lelekakis et al., 1999; Lofgren et al., 2018) (Table 1).

Lofgren et al. (2018) inoculated 4T1 cells on the fourth mammary gland in female Balb/c mice and observed the development of nociception through the lowering of mechanical threshold and the development of thermal hyperalgesia (Lofgren et al., 2018).

TABLE 1 Arthrotomy models for studying cancer-induced bone pain by breast tumor cell inoculation.

Cells	Local of inoculation	Syngeneic	Nociception test	Reference
NCTC 2472	Femur	X	Nonnoxious mechanical stimulation	Schwei et al. (1999)
MRMT-1	Tibia	X	Mechanical allodynia; mechanical hyperalgesia	Medhurst et al. (2002)
4T1	Femur	X	Spontaneous pain (limb use)	Falk and Dickenson (2014)
4T1	Femur	X	Mechanical allodynia; thermal hyperalgesia; spontaneous pain (limb use)	Zhao et al. (2013)
4T1	Femur	X	Mechanical, cold and heat hypersensibility; spontaneous pain (walking)	Abdelaziz et al. (2014)
MATBIII	Tibia	X	Ongoing pain; evoked hypersensitivity	Remeniuk et al. (2015)
MATBIII	Patella		Tactile hypersensitivity; spontaneous pain (limb use)	Havelin et al. (2017)
66.1	Femur	X	Spontaneous pain (guarding/flinching); mechanical hypersensitivity	Forte et al. (2016)
66.1	Femur	X	Spontaneous pain (flinching and guarding)	Zhang et al. (2018)
Walker 256	Tibia	X	Mechanical allodynia; movement-evoked pain (limb use)	Wang et al. (2019)
Walker 256	Tibia	X	Mechanical hyperalgesia	Zhang et al. (2020)
MRMT-1	Tibia	X	Mechanical allodynia	Tomotsuka et al. (2014)
MDA-MB-231	Femur		Mechanical allodynia	Ungard et al. (2014)

Moreover, recently, it was described that 20 days of the injection of 4T1 cells in the mammary gland of female BALB/c mice, the animals have already presented mechanical allodynia, cold allodynia, and changes in the grimace scale (a measure of spontaneous pain in rodents). Besides, the presence of bone metastasis and hypercalcemia was observed after 20 days of cell injection. In this study, some analgesics on the analgesic pain scale have been tested (acetaminophen, naproxen, codeine, and morphine), in addition to a cannabinoid agonist (WIN 55212,2), and all produced antinociception in this model of breast cancer pain (de Almeida et al., 2019).

Animal models of breast cancer pain: A brief history of bone inoculation models

The first study that described a murine model of CIBP was published in 1999 by Schwei et al. (1999); it initiated the animal studies to determine the neurochemical mechanisms that originate this type of pain. In this model, osteolytic sarcoma cells (NCTC 2472) were implanted directly into the femur of C3H/HeJ normal and B6C3-Fe-a/a wild-type adult mice to induce bone destruction and nociception (Schwei et al., 1999).

After, Medhurst et al. (2002) developed the first model of CIBP in rats using female Sprague-Dawley rats and injecting MRMT-1 cells. After the intratibial inoculation of the syngeneic rat mammary gland carcinoma cells, rats developed mechanical allodynia and showed a discrepancy of weight-bearing among hind paws. Also, in this study, they evaluated the effect of morphine, which showed a reduction in nociceptive parameters (Medhurst et al., 2002).

Animal models of breast cancer pain: Distinct bone inoculation models

To date, diverse models of CIBP may be produced by breast cancer cells injection into the bone. The 4T1 mammary cancer cells inoculation in the femur and tibia of Balb/c mice induced mechanical and cold allodynia, as well as other pain-related behaviors (Abdelaziz, Stone, & Komarova, 2014; Falk & Dickenson, 2014; Zhao et al., 2013). The injection of 4T1-luc2 breast cancer cells into the femur of male and female mice induced pain-related behavior related to limb use, and female mice had an earlier onset of nociception. Still, male and female mice develop similar bone degradation (Falk & Dickenson, 2014). In another study, using female Balb/c mice, there was development of the spontaneous pain (reduction in limb use), mechanical allodynia, and heat hyperalgesia after 4T1 cells injection into the femur (Zhao et al., 2013). Another study demonstrated that 4T1 cells tibial inoculation in female Balb/c mice led to the development of mechanical allodynia, cold hypersensitivity, heat hyperalgesia, and spontaneous nociceptive behaviors (Abdelaziz et al., 2014).

Another syngeneic model that is widely used to induce cancer-induced bone nociception is mammary gland carcinoma Walker 256 cells; this cell line has elevated metastasis capacity and is usually injected into the tibial intramedullary space. After the inoculation of Walker 256 tumor cells, Sprague-Dawley rats may develop mechanical allodynia, heat hyperalgesia, movement-evoked pain caused by the limb use, and ongoing pain behaviors (Ding et al., 2017; Slosky, Largent-Milnes, & Vanderah, 2015; Wang, Jiang, Wu, Tang, & Xu, 2019).

Also, the injection of naturally occurring murine breast adenocarcinoma cell line 66.1 in female Balb/c mice femur can be used as a model of CIBP. The implantation of 66.1 into the femur intramedullary space induces nociception, observed as mechanical allodynia and spontaneous nociception (guarding and flinching of the injected limb) (Forte et al., 2016; Slosky et al., 2015).

Moreover, the injection into the femur of human breast adenocarcinoma cell line MDA-MB-231 in female athymic Balb/c (nu/nu homozygous nude mice, immunodeficient mice) caused mechanical allodynia and a reduction in weight-bearing on the injected limb (Ungard, Seidlitz, & Singh, 2014).

In another study, the intratibial injection of breast adenocarcinoma cell line 13,762 MAT BIII in Fischer F344/NhSD female and male rats caused bone loss, mechanical allodynia, and impaired cancer-injected limb use, also tumor-induced ongoing pain behavior and movement-induced breakthrough pain (Havelin et al., 2017; Remeniuk et al., 2018).

Possible pharmacological targets for breast cancer pain relief

To enhance the clinical management of breast cancer pain, different groups of researchers have been studying the underlying mechanisms of this painful condition (van den Beuken-van Everdingen et al., 2007). Here, we described some of the current findings based on animal models of CIBP or orthotopic injection of breast cancer cells.

Neurotrophins, as nerve growth factor (NGF) and brain-derived neurotrophic factor (BDNF), have been studied in the induction of CIBP. Using different models of CIBP by injecting breast cancer cells, it was observed that NGF is involved in nerve sprouting caused by tumor, cancer pain (nociceptor sensitization), and neuroma induction (Mantyh et al., 2010). Besides, BDNF is involved in CIBP by modulating the N-methyl D-Aspartate (NMDA) receptors. Thus, this neurotrophin contributed to central sensitization and nociception in distinct CIBP models induced by breast cancer cell injection (Tomotsuka et al., 2014; Wang, Cvetkov, Chance, & Moiseenkova-bell, 2012).

Moreover, tumor necrosis factor-α (TNFα), a pro-inflammatory cytokine, is involved in CIBP after breast cancer inoculation, and the levels of this cytokine increase in cancer-bearing animals (Lozano-Ondoua et al., 2013). Also, the level of other pro-inflammatory cytokines and chemokines (interleukine-6, macrophage inflammatory protein-1 or CCL2, and macrophage inflammatory protein-1a or CCL3) were increased in the samples of bone marrow extrudate in CIBP animals (Lozano-Ondoua et al., 2013). Besides, the level of interleukin-1beta (IL-1β) was increased in the tibia and spinal cord in a model of CIBP after breast cancer cell injection, and an IL-1 receptor antagonist induced antinociceptive effect (Baamonde et al., 2007).

Moreover, in breast cancer patients with pain and sensitivity, it was detected an increase in transient receptor potential vanilloid 1 (TRPV1)-positive intraepidermal nerve fibers that correlates with NGF immunoreactivity (Gopinath et al., 2005). Similarly, the expression of TRPV1 was an increase in small- and medium-sized dorsal root ganglion (DRG) of rats after intratibial implantation of breast cancer cells, as well as the TRPV1 current density was enhanced DRG neurons (Li et al., 2014). TRPV1 antagonist reduces CIBP after breast cancer cell injection, a similar effect was detected in mice with genetic deletion of TRPV1 (Ghilardi et al., 2005). Also, using in a model of metastatic breast cancer by orthotopic injection of 4T1 cells, it was described the participation of TRP ankyrin 1 (TRPA1) in nociception in mice, this is a cation channel usually coexpressed with TRPV1 in peptidergic sensory fibers (de Almeida et al., 2020).

The cannabinoid receptor 2 (CB2) is also studied for pain relief in cancer pain. Recently, it was described the antinociceptive effect of a CB2 receptor agonist in a CIBP after breast cancer cell injection, the administration of this compound also improved the quality of life of mice and diminished bone remodeling (Lozano-Ondoua et al., 2013). Also, a peripheral type 1 cannabinoid receptor (CB1) agonist showed an antinociceptive effect in a CIBP model, and this compound decreases bone loss (Zhang et al., 2018).

Cancer pain treatment by nonpharmacological treatments

A study demonstrated that the use of wrist-ankle acupuncture in rats, with bone cancer pain, has similar results to those of electroacupuncture, relieving mechanical hyperalgesia and suppressing the expression of serotonin receptors, in addition to increasing the expression of opioid μ receptors and endomorphin-1 in the spinal cord (Zhang et al., 2020). Other articles have already been published showing the effectiveness of acupuncture in cancer pain associated or not with drug use (Lu et al., 2020; Vickers et al., 2018).

Conclusions

Finally, treatments for advanced breast cancer-induced pain should be improved, especially to control CIBP. Besides, these new treatments would be more attractive if they could reduce pain and control tumor growth and decrease skeletal-related events at the same time. However, as the pain relief in breast cancer patients is mainly done using opioids, the side effects of these compounds should be carefully observed. Also, the continuous development of animal models for CIBP and breast cancer-induced pain is still necessary to find other targeted therapies.

Applications to other areas

In this chapter, we have examined the current treatment options for active breast cancer pain. However, not only the cancer proliferation and metastasis may induce pain in this type of cancer, also aromatase inhibitors (AI), including exemestane, letrozole, and anastrozole that are medicines used for breast cancer therapy induce pain, but the mechanisms are not fully understood. These third-generation AI are used as adjuvant endocrine therapy in a subset of breast cancer postmenopausal patients (estrogen-sensitive breast cancer). Aromatase cytochrome P450 induces the conversion of androgens to estrogens. The administration of AIs may cause dose-limiting side effects named AI-associated musculoskeletal symptoms (AIMSS). Usually, patients show morning stiffness accompanied by pain in joints, such as hands, shoulders, and knees. Also, these medicines may cause neuropathic pain (Fusi et al., 2014; Laroche et al., 2014).

Other agents of interest

In this chapter, we have discussed the treatments of breast cancer-induced pain, and the novel targets that are being developed using animal models. Moreover, as the initial diagnosis of breast cancer and effective therapies are evolving, pain is also a frequent concern in breast cancer survivors. In this setting of patients, pain may be observed as neuropathic pain and arthralgias caused by distinct causes (Moore, 2020). Breast cancer surgery may also lead to a neuropathic pain named chronic postsurgical pain (CPSP). This is a frequent type of pain found in breast cancer survivors (Khan, Ladha, Abdallah, & Clarke, 2020). Here, we included other agents of interest for the control of CPSP in breast cancer survivors: selective serotonin-norepinephrine reuptake inhibitors (duloxetine and venlafaxine) and gabapentinoids (gabapentin and pregabalin).

Mini-dictionary of terms

- **Osteogenic pain**: Painful bone-related condition.
- **Ongoing pain**: The painful process that lasts an extended period.
- **Nociception**: Physiological pain component, which consists of the processes of transduction, transmission, and modulation of neural signals generated in response to a harmful external stimulus.
- **Antinociception**: The reduced ability to perceive nociception in animal models.
- **Hyperalgesia**: The exaggerated sensitivity to pain induced by a harmful stimulus.
- **Allodynia**: The hypersensitivity evoked by a previous nonnoxious stimulus.
- **Orthotopic**: A model where tumor cells are implanted in the standard anatomical site.
- **Arthrotomy**: A surgery in which, through an incision, the joint is opened and exposed to cause the injection of tumor cells.

Key facts

Key facts of breast cancer-induced pain

- The International Association for the Study of Pain (IASP) shows that 40% to 89% of patients who have breast cancer develop pain.
- Nearby 70% of patients with advanced breast cancer will suffer significant bone pain during their disease, as the bone is the most frequent location for metastatic disease.

Key facts of chronic pain management

- To achieve pain control, it is essential to adequately assess pain degree, balance the ideal dosage with the adverse effects of the compounds, and use opioid rotation to have a better analgesic effect without side effects.
- Constipation is observed in 60% to 90% of cancer patients treated with opioids; thus, the management of opioid-induced constipation should be done prophylactically.
- Approximately 40% of cancer patients have limited analgesic efficacy because the compounds used have dose-limiting side effects, or patients are refractory to the available therapies.

Summary points

- Breast carcinoma is the most detected cancer in women; also, active breast cancer usually induces cancer pain.
- The management of breast cancer pain is typically done using the World Health Organization analgesic ladder, which is composed of four steps of treatment.
- However, the treatment of pain should also follow the primary cause of this symptom; thus, neuropathic pain and bone pain management should follow adequate recommendations.
- Cancer-induced bone pain is a frequent concern in advanced breast cancer, and it is treated using pharmacological and nonpharmacological strategies.
- Usually, chronic pain caused by cancer is managed using opioid-based pharmacotherapy, but constipation and other opioid-induced side effects should be controlled.
- The research on breast cancer pain is done using arthrotomy implantation models where tumor cells are injected directly into the bone or the injection of tumor cells on the mammary gland of mice or rats.
- Diverse new pharmacological targets have been studied for breast cancer pain management, including neurotrophins, pro-inflammatory cytokines, chemokines, transient receptor potential vanilloid 1, transient receptor potential ankyrin 1, and cannabinoid receptors.

References

Abdelaziz, D. M., Stone, L. S., & Komarova, S. V. (2014). Osteolysis and pain due to experimental bone metastases are improved by treatment with rapamycin. *Breast Cancer Research and Treatment, 143*, 227–237.

Ahmad, I., Ahmed, M. M., Ahsraf, M. F., Naeem, A., Tasleem, A., Ahmed, M., et al. (2018). Pain management in metastatic bone disease: A literature review. *Cureus, 10*, e3286.

Anekar, A. A., & Cascella, M. (2020). WHO analgesic ladder. In *StatPearls* StatPearls Publishing.

Baamonde, A., Curto-Reyes, V., Juárez, L., Meana, Á., Hidalgo, A., & Menéndez, L. (2007). Antihyperalgesic effects induced by the IL-1 receptor antagonist anakinra and increased IL-1β levels in inflamed and osteosarcoma-bearing mice. *Life Sciences, 81*, 673–682.

Bokhari, F., & Sawatzky, J. A. V. (2009). Chronic neuropathic pain in women after breast cancer treatment. *Pain Management Nursing, 10*, 197–205.

Bray, F., Ferlay, J., Soerjomataram, I., Siegel, R. L., Torre, L. A., & Jemal, A. (2018). Global cancer statistics 2018: GLOBOCAN estimates of incidence and mortality worldwide for 36 cancers in 185 countries. *CA: a Cancer Journal for Clinicians, 68*, 394–424.

Coleman, R. E. (2001). The nature of bone metastases. Metastatic bone disease: Clinical features, pathophysiology and treatment strategies. *Cancer Treatment Reviews, 27*, 165–176.

de Almeida, A. S., Rigo, F. K., de Prá, S. D. T., Milioli, A. M., Dalenogare, D. P., Pereira, G. C., … Trevisan, G. (2019). Characterization of cancer-induced nociception in a murine model of breast carcinoma. *Cellular and Molecular Neurobiology, 39*, 605–617.

de Almeida, A. S., Rigo, F. K., de Prá, S. D. T., Milioli, A. M., Pereira, G. C., Lückemeyer, D. D., … Trevisan, G. (2020). Role of transient receptor potential ankyrin 1 (TRPA1) on nociception caused by a murine model of breast carcinoma. *Pharmacological Research, 152*. https://doi.org/10.1016/j.phrs.2019.104576, 104576.

Ding, Z., Xu, W., Zhang, J., Zou, W., Guo, Q., Huang, C., … Song, Z. (2017). Normalizing GDNF expression in the spinal cord alleviates cutaneous hyperalgesia but not ongoing pain in a rat model of bone cancer pain. *International Journal of Cancer, 140*, 411–422.

Falk, S., & Dickenson, A. H. (2014). Pain and nociception: Mechanisms of cancer-induced bone pain. *Journal of Clinical Oncology, 32*, 1647–1654.

Fallon, M. T., & Colvin, L. (2013). Neuropathic pain in cancer. *British Journal of Anaesthesia, 111*, 105–111.

Forte, B. L., Slosky, L. M., Zhang, H., Arnold, M. R., Staatz, W. D., Hay, M., … Vanderah, T. W. (2016). Angiotensin-(1-7)/Mas receptor as an antinociceptive agent in cancer-induced bone pain. *Pain, 157*, 2709–2721.

Fusi, C., Materazzi, S., Benemei, S., Coppi, E., Trevisan, G., Marone, I. M., … Nassini, R. (2014). Steroidal and non-steroidal third-generation aromatase inhibitors induce pain-like symptoms via TRPA1. *Nature Communications, 5*.

Ghilardi, J. R., Röhrich, H., Lindsay, T. H., Sevcik, M. A., Schwei, M. J., Kubota, K., … Mantyh, P. W. (2005). Selective blockade of the capsaicin receptor TRPV1 attenuates bone cancer pain. *Journal of Neuroscience, 25*, 3126–3131.

Gopinath, P., Wan, E., Holdcroft, A., Facer, P., Davis, J. B., Smith, G. D., … Anand, P. (2005). Increased capsaicin receptor TRPV1 in skin nerve fibres and related vanilloid receptors TRPV3 and TRPV4 in keratinocytes in human breast pain. *BMC Women's Health, 5*, 2.

Havelin, J., Imbert, I., Sukhtankar, D., Remeniuk, B., Pelletier, I., Gentry, J., … King, T. E. (2017). Mediation of movement-induced breakthrough cancer pain by IB4-binding nociceptors in rats. *Journal of Neuroscience, 37*, 5111–5122.

Horas, K., Zheng, Y., Zhou, H., & Seibel, M. J. (2015). Animal models for breast cancer metastasis to bone: Opportunities and limitations. *Cancer Investigation, 33*, 459–468.

Hui, D., & Bruera, E. (2014). A personalized approach to assessing and managing pain in patients with cancer. *Journal of Clinical Oncology, 32*, 1640–1646.

Khan, J. S., Ladha, K. S., Abdallah, F., & Clarke, H. (2020). Treating persistent pain after breast cancer surgery. *Drugs, 80*, 23–31.

Laroche, F., Coste, J., Medkour, T., Cottu, P. H., Pierga, J. Y., Lotz, J. P., … Perrot, S. (2014). Classification of and risk factors for estrogen deprivation pain syndromes related to aromatase inhibitor treatments in women with breast cancer: A prospective multicenter cohort study. *Journal of Pain, 15*, 293–303.

Lelekakis, M., Moseley, J. M., Martin, T. J., Hards, D., Williams, E., Ho, P., … Anderson, R. L. (1999). A novel orthotopic model of breast cancer metastasis to bone. *Clinical & Experimental Metastasis, 17*, 163–170.

Li, Y., Cai, J., Han, Y., Xiao, X., Meng, X. L., Su, L., … Wan, Y. (2014). Enhanced function of TRPV1 via up-regulation by insulin-like growth factor-1 in a rat model of bone cancer pain. *European Journal of Pain (United Kingdom), 18*, 774–784.

Liepe, K. (2018). 188 re-HEDP therapy in the therapy of painful bone metastases. *World Journal of Nuclear Medicine, 17*, 133–138.

Liu, W. C., Zheng, Z. X., Tan, K. H., & Meredith, G. J. (2017). Multidimensional treatment of cancer pain. *Current Oncology Reports, 19*, 10.

Lofgren, J., Miller, A. L., Lee, C. C. S., Bradshaw, C., Flecknell, P., & Roughan, J. (2018). Analgesics promote welfare and sustain tumour growth in orthotopic 4T1 and B16 mouse cancer models. *Laboratory Animals, 52*, 351–364.

Lozano-Ondoua, A. N., Hanlon, K. E., Symons-Liguori, A. M., Largent-Milnes, T. M., Havelin, J. J., Ferland, H. L., … Vanderah, T. W. (2013). Disease modification of breast cancer-induced bone remodeling by cannabinoid 2 receptor agonists. *Journal of Bone and Mineral Research, 28*, 92–107.

Lu, W., Giobbie-Hurder, A., Freedman, R. A., Shin, I. H., Lin, N. U., Partridge, A. H., … Ligibel, J. A. (2020). Acupuncture for chemotherapy-induced peripheral neuropathy in breast cancer survivors: A randomized controlled pilot trial. *The Oncologist, 25*, 310–318.

Macedo, F., Ladeira, K., Pinho, F., Saraiva, N., Bonito, N., Pinto, L., & Gonçalves, F. (2017). Bone metastases: An overview. *Oncology Reviews, 11*, 321.

Mantyh, P. (2013). Bone cancer pain: Causes, consequences, and therapeutic opportunities. *Pain, 154*, S54–S62.

Mantyh, W. G., Jimenez-Andrade, J. M., Stake, J. I., Bloom, A. P., Kaczmarska, M. J., Taylor, R. N., … Mantyh, P. W. (2010). Blockade of nerve sprouting and neuroma formation markedly attenuates the development of late stage cancer pain. *Neuroscience, 171*(2), 588–598. https://doi.org/10.1016/j.neuroscience.2010.08.056.

Medhurst, S. J., Walker, K., Bowes, M., Kidd, B. L., Glatt, M., Muller, M., … Urban, L. (2002). A rat model of bone cancer pain. *Pain, 96*, 129–140.

Moore, H. C. F. (2020). Breast cancer survivorship. *Seminars in Oncology, 47*, 222–228.

Portenoy, R. K. (2011). Treatment of cancer pain. *The Lancet, 377*, 2236–2247.

Remeniuk, B., King, T., Sukhtankar, D., Nippert, A., Li, N., Li, F., ... Porreca, F. (2018). Disease modifying actions of interleukin-6 blockade in a rat model of bone cancer pain. *Pain, 159*, 684–698.

Remeniuk, B., Sukhtankar, D., Okun, A., Navratilova, E., Xie, J. Y., King, T., & Porreca, F. (2015). Behavioral and neurochemical analysis of ongoing bone cancer pain in rats. *Pain, 156*(10), 1864–1873. https://doi.org/10.1097/j.pain.0000000000000218.

Schwei, M. J., Honore, P., Rogers, S. D., Salak-Johnson, J. L., Finke, M. P., Ramnaraine, M. L., ... Mantyh, P. W. (1999). Neurochemical and cellular reorganization of the spinal cord in a murine model of bone cancer pain. *Journal of Neuroscience, 19*, 10886–10897.

Slosky, L. M., Largent-Milnes, T. M., & Vanderah, T. W. (2015). Use of animal models in understanding cancer-induced bone pain. *Cancer Growth and Metastasis, 8*, 47–62.

Swarm, R. A., Paice, J. A., Anghelescu, D. L., Are, M., Bruce, J. Y., Buga, S., ... Gurski, L.A. (2019). Adult cancer pain, version 3.2019. *Journal of the National Comprehensive Cancer Network, 17*, 977–1007.

Tomotsuka, N., Kaku, R., Obata, N., Matsuoka, Y., Kanzaki, H., Taniguchi, A., ... Morimatsu, H. (2014). Up-regulation of brain-derived neurotrophic factor in the dorsal root ganglion of the rat bone cancer pain model. *Journal of Pain Research, 7*, 415–423.

Ungard, R. G., Seidlitz, E. P., & Singh, G. (2014). Inhibition of breast cancer-cell glutamate release with sulfasalazine limits cancer-induced bone pain. *Pain, 155*, 28–36.

van den Beuken-van Everdingen, M. H. J., de Rijke, J. M., Kessels, A. G., Schouten, H. C., van Kleef, M., & Patijn, J. (2007). High prevalence of pain in patients with cancer in a large population-based study in the Netherlands. *Pain, 132*, 312–320.

Vickers, A. J., Vertosick, E. A., Lewith, G., MacPherson, H., Foster, N. E., Sherman, K. J., ... Linde, K. (2018). Acupuncture for chronic pain: Update of an individual patient data meta-analysis. *Journal of Pain, 19*, 455–474.

Wang, L., Cvetkov, T. L., Chance, M. R., & Moiseenkova-bell, V. Y. (2012). Identification of in vivo disulfide conformation of TRPA1 Ion channel. *Journal of Biological Chemistry, 287*, 6169–6176.

Wang, W., Jiang, Q., Wu, J., Tang, W., & Xu, M. (2019). Upregulation of bone morphogenetic protein 2 (Bmp2) in dorsal root ganglion in a rat model of bone cancer pain. *Molecular Pain, 15*, 1–10.

Wong, M. H., & Pavlakis, N. (2011). Optimal management of bone metastases in breast cancer patients. *Breast Cancer: Targets and Therapy, 3*, 35–60.

Zajączkowska, R., Kocot-Kępska, M., Leppert, W., & Wordliczek, J. (2019). Bone pain in cancer patients: Mechanisms and current treatment. *International Journal of Molecular Sciences, 20*, 6047.

Zhang, C., Xia, C., Zhang, X., Li, W., Miao, X., & Zhou, Q. (2020). Wrist-ankle acupuncture attenuates cancer-induced bone pain by regulating descending pain-modulating system in a rat model. *Chinese Medicine (United Kingdom), 15*, 13.

Zhang, H., Lund, D. M., Ciccone, H. A., Staatz, W. D., Ibrahim, M. M., Largent-Milnes, T. M., ... Vanderah, T. W. (2018). Peripherally restricted cannabinoid 1 receptor agonist as a novel analgesic in cancer-induced bone pain. *Pain, 159*, 1814–1823.

Zhao, J., Zhang, H., Liu, S. B., Han, P., Hu, S., Li, Q., ... Wang, Y. Q. (2013). Spinal interleukin-33 and its receptor ST2 contribute to bone cancer-induced pain in mice. *Neuroscience, 253*, 172–182.

Zhu, X., Ge, C., Yu, Y., & Wang, P. (2015). Advances in cancer pain from bone metastasis. *Drug Design, Development and Therapy, 9*, 4239–4245.

Chapter 23

Postoperative pain after rhinoplasty and rhinologic surgery

Andre Shomorony[a], Arron M. Cole[a], Matthew Kim[b], and Anthony P. Sclafani[c]

[a]*Department of Otolaryngology—Head and Neck Surgery, NewYork-Presbyterian Hospital, New York, NY, United States,* [b]*Department of Otolaryngology—Head and Neck Surgery, Westchester Medical Center, Valhalla, NY, United States,* [c]*Department of Otolaryngology—Head and Neck Surgery, Weill Cornell Medical College, New York, NY, United States*

Abbreviations

ESS	endoscopic sinus surgery
GA	general anesthesia
MAC	monitored anesthesia care
MME	morphine milligram equivalents
NSAIDs	nonsteroidal antiinflammatory drugs
VAS	visual analog scale

Introduction

Rhinoplasty

Rhinoplasty is one of the most commonly performed facial plastic surgical procedures, with roughly 55,000 performed in the United States in 2020 (American Society of Aesthetic Plastic Surgery, 2019). It can be performed as either a cosmetic or as a functional procedure (to improve breathing). Characteristics such as nasal deviation, bony vault width, and nasal tip shape, rotation, and projection, among others, can be manipulated to improve both form and function. Open and closed approaches are possible, the former involving a small external incision at the base of the nose in addition to internal incisions, such as those used in the latter technique.

Rhinoplasty begins by making internal (and external, if an open approach is used) incisions, and the skin soft tissue envelope is elevated in avascular planes (preperichondrial over the tip and mid dorsum, and subperiosteal over the upper dorsum) to expose the lower and upper lateral cartilages, dorsal septum, and nasal bones. Manipulation of the septum, including removal and preservation of septal cartilage for use as grafting material as needed, can then be performed. Alternatively, cartilage may be harvested from the rib or conchal bowl of the ear to provide additional tissue. After cartilaginous manipulation, osteotomies to alter the bony anatomy may take place. The nasal bones are separated from each other, and the midline dorsal septum and lateral osteotomies are performed in the nasofacial junction through the ascending process of the maxilla to fully mobilize the nasal bones. As opposed to soft tissue and cartilage dissection, osteotomies are performed with a mallet and osteotome (chisel sharpened on both sides). Closure of incisions, including placement of septal quilting stitches, intranasal splint placement, and external nasal cast placement, can then be performed at the discretion of the surgeon.

Septoplasty

Septoplasty may be performed as a standalone procedure, or it may be combined with rhinoplasty (termed septorhinoplasty) or with sinus surgery. Surgery begins with insertion of anesthetic-soaked pledgets for decongestion, followed by anesthetic injection into the subperichondrial and subperiosteal planes of the nasal septum. In the endonasal approach, an anterior or posterior incision is made extending from the dorsalmost to the caudalmost point of the cartilaginous septum, providing

Features and Assessments of Pain, Anesthesia, and Analgesia. https://doi.org/10.1016/B978-0-12-818988-7.00019-4
Copyright © 2022 Elsevier Inc. All rights reserved.

PART | II The syndromes of pain

appropriate exposure to septal deviations, followed by elevation of bilateral mucoperichondrial flaps. In an external approach, the skin soft tissue envelope is first elevated from the nasal tip, providing direct visualization of the anterior and dorsal septum. Once the septal flaps are properly developed, deviated pieces of cartilage may be resected, while maintaining large enough segments of cartilage caudally and dorsally. Cartilage pieces are removed, straightened by morselizing or scoring of their surface, and replaced between the septal flaps. Mucoperichondrial incisions are then closed using absorbable sutures. Some surgeons choose to place packing or splints.

Rhinologic surgery

While septorhinoplasties commonly focus on cosmetic outcomes, rhinologic surgery is performed primarily to address functional disorders of the nose and sinuses. One such condition, chronic rhinosinusitis, affects between 5% and 12% of the general population and is characterized by nasal congestion, mucopurulent nasal drainage, facial pain and pressure, and decreased sense of smell (Dietz de Loos et al., 2019). Patients who fail maximal medical therapy for such conditions may be good candidates for rhinologic surgery, most commonly endoscopic sinus surgery (ESS), which may be coupled with septoplasty, the straightening of a deviated septum, and turbinoplasty, the reduction of submucosal erectile tissue within the inferior turbinates.

Endoscopic sinus surgery begins with decongestion and local anesthesia of the nasal cavities. A portion of the ethmoid bone, the uncinate process is gently removed with a scalpel or forceps, allowing the maxillary sinus natural ostium (medial wall of maxillary sinus) to be identified and enlarged. The surgeon then proceeds with marsupialization of the anterior and posterior ethmoid cells, exercising caution not to disrupt the medial orbital wall and keeping the orbit protected. Depending on preoperative symptoms and extent of sinus inflammation on endoscopy or imaging, the sphenoid sinus may then be entered and its ostium enlarged, and finally the frontal sinus may be similarly accessed and manipulated. The bones of the sinuses are generally quite thin and weak, allowing removal of mucosa and bone with a precise powered debrider or delicate forceps. The frontal sinus ostium and intersinus septum are exceptions and may require a high-speed powered bone drill. Once surgery is completed, hemostasis is achieved and a pledget or spacer may be inserted into the middle meatus and sphenoethmoidal recess.

Intraoperative analgesia and anesthesia

Preoperative administration of a single dose of gabapentinoids (1200 mg gabapentin or 400 mg pregabalin) has been shown in several studies to reduce self-reported pain scores in the immediate postoperative period, while also reducing overall opioid and nonopioid analgesic need. These drugs are well tolerated, and overall side effects such as postoperative nausea, vomiting, dizziness, sedation, and blurred vision are low when compared to the control group (Park, Kim, Ko, Lee, & Hwang, 2016).

Intraoperative endoscopic sphenopalatine ganglion block can be performed at the beginning of nasal surgery in order to improve postoperative pain after septorhinoplasty. Injection of bupivacaine into the pterygopalatine fossa at the tail of the middle turbinate bilaterally can accomplish effective sphenopalatine ganglion nerve block and has been shown to reduce pain scores within 24 h of surgery and extending to up to 7 days postoperatively (Ekici & Alagöz, 2019). Additionally, intraoperative bleeding is also reduced, which has implications for possible reduced operative time and thus decreased anesthetic duration.

In an academic setting, general anesthesia (GA) is common for rhinoplasty, septoplasty, and endoscopic sinus surgery. This is not necessarily the case, however, in general practice. Ambulatory surgery centers often employ monitored anesthesia care (MAC), given faster and more efficient patient turnover. Historically, GA was associated with increased vasodilation—associated with increased bleeding and bruising—but this is largely no longer the case. While GA techniques have improved, decreasing the relative disadvantages of GA compared to MAC, GA is by no means a requirement. An appropriate MAC case requires excellent vasoconstriction, as well as topical and local anesthesia.

Rhinoplasty

After the patient has received an anxiolytic and analgesic (MAC) or GA has been induced, two slim cottonoids soaked in 4% lidocaine (MAC) or 0.05% oxymetazoline are gently placed in each side of the nose. One cottonoid is placed parallel to the nasal floor between the inferior turbinate and the septum and extends to the posterior inferior turbinate to treat the sphenopalatine nerve and artery. The other cottonoid is placed against the superior septum parallel to the nasal dorsum to block branches of the anterior and posterior ethmoidal nerves and arteries. Local anesthetics (typically 1% xylocaine with 1:100,000 epinephrine) can be injected at the infraorbital foramen (for infraorbital nerve block), across the mid dorsum (to block terminal branches of the anterior ethmoidal nerves innervating the nasal tip), and at the nasal base (to block branches of the nasopalatine nerve providing sensation to the columella, nasal sill, and tip). Supplemental local anesthetic is also injected along the nasomaxillary junction to ensure adequate analgesia during lateral osteotomies.

Septoplasty

In addition to intranasal cottonoids, local anesthetic is injected below the septal perichondrium and periosteum to assist surgery and to provide anesthesia by directly blocking branches of the anterior and posterior ethmoidal and sphenopalatine nerves.

Use of analgesic-infused nasal packing has also been shown to have positive effects on postoperative pain control in septoplasty. Merocel, a dehydrated hydroxylated polyvinyl acetate sponge, is often used as a packing product following septoplasty. When compared to infusion with normal saline, lidocaine or tramadol solution-infused packs have been shown to decrease pain within 24 h postoperatively. Given that the nasal packings typically used in septorhinoplasty themselves cause significant postoperative discomfort until removed at follow-up visit, this is a potentially effective means of improving postoperative comfort for patients undergoing these procedures (Simsek, Musaoglu, & Uluat, 2019). This improvement in comfort, however, usually lasts up to 12 h, after which the infused anesthetic drains out and its effect wears off.

Rhinologic surgery

Due to its unique dual local anesthetic and vasoconstrictive properties, one of the preferred methods of local anesthesia in endoscopic sinus surgery is a liquid solution of topical cocaine (usually 4%), placed onto pledgets that are inserted bilaterally into the nares. Other topical agents commonly used in endoscopic sinus surgery also include epinephrine, phenylephrine, and oxymetazoline. Alternatively, surgeons may opt for a sphenopalatine ganglion block using bupivacaine, shown in some studies to be a safe and effective way to reduce immediate postoperative pain (Rezaeian, Hashemi, & Dokhanchi, 2019).

Rhinologic surgery requires effective vasoconstriction and decongestion, and intraoperative pain control and postoperative pain amelioration begin with directed treatment of the primary neurovascular structures of the nose and sinuses. When general anesthesia is employed, intravenous agents such as propofol and remifentanil are commonly employed to achieve hemodynamic stability while allowing for controlled hypotension and avoiding the vasodilatory effects of the inhaled anesthetics, thus optimizing surgical visibility. Remifentanil also helps to eliminate the use of intraoperative fentanyl, allowing for a rapid and smooth emergence and potentially promoting a quicker transition to an opiate-free postoperative pain regimen (Saxena & Nekhendzy, 2020). While most sinus surgery occurs under general anesthesia, a number of studies report similar surgical outcomes and pain scores between ESS performed under local anesthesia and under general anesthesia. A study by Thaler et al. revealed a mean pain rating score of 14.5/100 for patients undergoing ESS under local anesthesia and 20.1/100 for those undergoing ESS under general anesthesia, yet no statistically significant difference between the two groups (Thaler, Gottschalk, Samaranayake, Lanza, & Kennedy, 1997). Nonetheless, data linking specific local or general anesthetic agents used intraoperatively to postoperative pain scores are still limited.

Postoperative pain management
Sources of postoperative pain
Septorhinoplasty

While the operative components of a septorhinoplasty may vary widely based upon surgical goals and patient anatomical factors, certain steps are associated with greater postoperative pain. Patients undergoing osteotomies, or directed fracturing of the nasal bones to attain desirable structural results, note increased edema and bruising postoperatively. In patients for whom cartilage must be grafted from the rib or ear, additional sites of potential postoperative pain are introduced. Patients also complain of postoperative nasal congestion and pressure, and intranasal dryness and crusting.

Rhinologic surgery

Postoperative care following sinus surgery aims to promote mucosal healing and regeneration, reduce the risk of infection and chronic inflammation, and alleviate short-term symptoms such as nasal congestion, bleeding, and discharge. This is

260 PART | II The syndromes of pain

primarily accomplished by a combination of nasal irrigations, antibiotics and/or intranasal corticosteroids, nasal packing, and debridement of the nasal passages and sinus cavities a few weeks after surgery (Eloy, Andrews, & Poirrier, 2017). While these measures seem to address most of the symptomatic burden in sinus surgery patients, patients may experience nasal fullness, pressure, discomfort, or pain, and physicians commonly prescribe a course of pain medications as part of the patient's postoperative care.

The etiology of pain after sinus surgery is incompletely understood, but it is thought to partially correlate with the degree of innervation of the anatomical structures affected by surgery. Manipulation of the septum and lateral nasal wall, which are largely innervated by the nasopalatine nerve, a branch of the maxillary nerve, likely results in postoperative pain. Similarly, the medial orbital wall, orbital periosteum, and the entire anterior skull base innervated by the anterior ethmoidal nerves are particularly sensitive to pain.

Assessment of pain after surgery

The literature involving perioperative pain management after nasal surgery relies on self-reported pain scores using a visual analog scale (VAS) that allows patients to rate perceived pain levels on a 0- to 100-point scale. Number of pain pills required is another commonly reported metric, presented in number of pills, or in morphine milligram equivalents (MME) for opioid pain medications.

Septorhinoplasty

Many studies assessing postoperative pain in septoplasty and rhinoplasty patients show that patients experience mild-to-moderate pain in the postoperative period. In a study by Szychta et al., patients who underwent septoplasty only reported minimal pain at all times after surgery (Szychta & Antoszewski, 2010). Using 30 on the VAS scale as a cutoff for mild-to-moderate pain, Sclafani et al. found that rhinoplasty patients experienced moderate pain on postoperative days 1 and 2, while septoplasty-only patients experience moderate pain on the day of surgery and postoperative day 1, but mild pain thereafter (Sclafani, Kim, Kjaer, Kacker, & Tabaee, 2019).

Peak pain in patients undergoing either or both procedures is typically reported between the day of surgery and postoperative day 1, with a steady decline thereafter. These findings are consistent with larger studies investigating pain levels after various otorhinolaryngological procedures, including endoscopic sinus surgery, thyroidectomy, lymph node biopsy, and Mohs reconstruction (Kim et al., 2020).

Rhinologic surgery

Pain resulting from rhinologic surgery has been demonstrated to be minimal to mild, and to peak early in the postoperative course. In a prospective study of patients undergoing ESS, Friedman and colleagues reported that most patients experienced a mean postoperative change of three or fewer points on a 10-point pain scale, and that pain was highest at 6 h after surgery, when comparing mean pain rating scores at baseline, 2 h, 6 h, and 24 h after surgery (Friedman, Venkatesan, Lang, & Caldarelli, 1996). In an analysis of pain during the first 6 postoperative days after sinonasal surgery, Wise and colleagues also revealed pain to be most significant on postoperative day 1, corresponding to a pain score of 3.61 on a 10-point scale (Wise, Wise, & Delgaudio, 2005). Most recently, Riley and colleagues also showed that pain after rhinologic surgery peaked on postoperative days 1 and 2, and tapered rapidly over 2 weeks. Moreover, this study revealed postoperative pain scores to be higher for septoplasties than for ESS (Riley et al., 2018).

Modalities of management of postoperative pain

Septorhinoplasty

Nonopioid analgesics

Mounting evidence suggests that the use of multimodal pain control with the addition of nonsteroidal antiinflammatory drugs (NSAIDS) and acetaminophen to postoperative septoplasty and rhinoplasty pain regimens can significantly improve acute pain control. In a randomized control trial by Nguyen et al. in 2019, patients who received opioid medication for ibuprofen as primary analgesic treatment for postoperative pain after common otorhinolaryngological procedures, including septoplasty and rhinoplasty, noted no difference in pain scores between groups (Nguyen et al., 2019). Similarly, patients dosed with IV ibuprofen or paracetamol (acetaminophen) during general anesthesia while undergoing septorhinoplasty were noted to have significantly lower pain intensity scores and opioid consumption rates at all-time points following surgery (Çelik, Kara, Koc, & Yayik, 2018). Of note, no increase in rate of bleeding complications was noted in either of these studies. Though surgeon preference may preclude routine use of NSAIDs in some circumstances, given their longstanding association with increased perioperative bleeding risk, they should be considered a viable option in effective pain management and decreased risk of opioid dependence in the postoperative period.

Opioid analgesics

Opioid pain medications are very commonly prescribed after many otorhinolaryngology surgical procedures, including septoplasty and rhinoplasty. In the United States, prescribing patterns vary widely, and a study by Schwartz et al. surveying prescribers following common otorhinolaryngological procedures revealed that an average of 19 to 22 pills were prescribed for postoperative patients following septoplasty and rhinoplasty. Hydrocodone-acetaminophen and oxycodone-acetaminophen were the most commonly prescribed opioid pain medications (Schwartz, Naples, Kuo, & Falcone, 2018). In light of the ongoing opioid epidemic, there have been concerted efforts by providers to reduce opioid medication prescription.

Notably, legislative changes aimed at reducing opioid use have had an impact on prescribing patterns with reductions by up to half following such changes (Aulet, Trieu, Landrigan, & Millay, 2019). Studies suggest that 10 to 15 typical dose (i.e., 5 mg oxycodone) opioid tablets or roughly 100 MME are sufficient to control postoperative pain for septoplasty and rhinoplasty patients (Patel et al., 2018; Rock et al., 2019).

Rhinologic surgery

Postoperative pain after rhinologic surgery is most often preemptively managed through a combination of opioid and non-opioid over-the-counter medications, to be taken by the patient on an as-needed basis. Due to the ongoing concern for opioid misuse, overdose, and diversion, judicious postoperative prescription of opioids has been increasingly emphasized, and a growing body of literature—in both the fields of otolaryngology and surgery at large—has focused on more accurately determining postoperative opioid requirements, elucidating the role of multimodal pain regimens, and demonstrating the safety of nonopioid alternatives (Zheng, Riley, Kim, Sclafani, & Tabaee, 2020).

Nonopioid analgesics

A prospective study by Wu et al. suggested that NSAIDs could be safely used in the postoperative care of sinus surgery patients without a statistically significant increase in the risk of bleeding complications (Wu et al., 2020). A systematic review by Svider et al. demonstrated a bleeding risk of 1.6% associated with acetaminophen use after rhinologic surgery, vs. 0.8% associated with NSAID use. The study further supported the favorable role of gabapentin after ESS, which was only associated with nausea and vomiting or dizziness in less than 10% of studied patients (Svider et al., 2018).

Opioid analgesics

Similarly to postoperative pain levels, postoperative opioid requirements after rhinologic surgery are low, as shown by a number of recent reports. In a 2018 prospective study of patients undergoing ESS with or without septoplasty or turbinate reduction, Riley et al. demonstrated that the median postoperative opioid use corresponded to 11.3 morphine milligram equivalents (MME), and that most patients were completely off opioid analgesics by postoperative day 3 (Riley et al., 2018). According to a retrospective analysis by Raikundalia et al., 78% of patients undergoing ESS with or without septoplasty required 5 tablets of acetaminophen 300 mg/codeine 30 mg (or 22.5 MME) (Raikundalia et al., 2019). Pruitt et al. also showed a low requirement of postoperative opioid tablets in pediatric sinus surgery patients—albeit slightly greater than in the adult population—of on average 5 opioid tablets (Pruitt, Swords, Russell, Rollins, & Skarda, 2019). In addition to the low postoperative opioid requirements, these recent reports convey a persistent pattern of overprescription of opioids. Pruitt et al. revealed an average of 30 prescribed tablets vs. 5 used (Pruitt et al., 2019), Riley et al. revealed a median of 30 prescribed tablets, with a median of 27 left over (Riley et al., 2018), and numerous other studies demonstrate that a significant proportion of the population does not require any opioid medications at all (Locketz et al., 2019; Raikundalia et al., 2019; Sethi et al., 2019; Wise et al., 2005) (15%–73% in these reports). Taken together, these findings highlight the need for sinus surgeons to reevaluate their opioid prescription patterns, in order to minimize unnecessary prescriptions and opioid misuse.

Conclusion

As illustrated by the studies included in this chapter, postoperative pain after both septorhinoplasty and rhinologic surgery tends to be mild. Such pain, in fact, may be more accurately characterized as "discomfort," and is often fully managed via a combination of moisture, humidification, and over-the-counter pain relievers. An important component of the preoperative process, therefore, is a full discussion between the physician and the patient regarding what postoperative care entails and the nature and degree of discomfort that may be anticipated. The patient should be aware that excruciating pain after a rhinoplasty, endoscopic sinus surgery, septoplasty or turbinate reduction is atypical and warrants further evaluation. Coupled with the low degree of postoperative pain after these surgeries is the similarly low requirement for postoperative opioid medications, shown here to typically correspond to no more than 5 opioid tablets taken in the first 2 or 3 postoperative days. Pain management after these surgeries should therefore be primarily based on over-the-counter nonopioids, such as acetaminophen, taken as first-line agents, with opioids used as breakthrough or rescue agents as needed.

Applications to other areas

In this chapter, we have introduced the basic surgical steps involved in rhinoplasty, septoplasty, and rhinologic surgery, described the likely etiologies of pain associated with them, characterized the different commonly used modalities of intraoperative and postoperative anesthesia and analgesia, and analyzed numerous studies investigating the degree of postoperative pain and pain medication requirement after these surgical procedures.

This work emphasizes a number of metrics for assessing and representing pain levels—such as the visual analog scale—and for accurately conveying amounts of administered pain medications, such as morphine milligram equivalents. These metrics have made the investigation of pain in the context of surgery much more reliable and have allowed for a recent expansion of the body of literature on this topic. We therefore strongly support the continued use of such metrics in other nonsurgical fields within medicine.

Our work also has wider applications for the field of otorhinolaryngology and surgery at large. As shown by Kim et al., various otorhinolaryngological surgical procedures result in similarly low levels of pain (Kim et al., 2020), and an effort should be made to manage their associated postoperative pain with a similar degree of caution as we emphasize in this chapter. As part of the preoperative process, providers should have a full discussion with their patients describing the expected level of postoperative pain and emphasizing that discomfort should be brief and self-limited. While our chapter does not focus on surgeries outside the field of otorhinolaryngology, providers in other surgical fields should exercise similar caution. By and large, elective procedures may be managed with over-the-counter pain relievers as first-line agents, to be taken on an as-needed basis, with stronger opioids used as rescue agents.

Other agents of interest

This chapter has primarily framed postoperative pain management after rhinoplasty and rhinologic surgery in terms of over-the-counter agents, such as NSAIDs, and opioids.

Within the group of NSAIDs that were not included in this chapter's discussion, the agents **ketorolac** and **rofecoxib** deserve attention, having also been studied in the context of endoscopic sinus surgery (Church, Stewart IV, O-Lee, & Wallace, 2006; Moeller, Pawlowski, Pappas, Fargo, & Welch, 2012).

Mini-dictionary of terms

- **Osteotomy**: surgical cutting of a piece of bone.
- **Pterygopalatine fossa**: fossa in the skull located posterior to the maxilla and close to the apex of the orbit.
- **Sphenopalatine ganglion**: a nerve bundle mostly branching off the trigeminal ganglion, innervating facial sensations such as pain of the interior surface of the nose.
- **Turbinates**: bony structures covered in mucosa that are present on the lateral nasal walls, bilaterally, and serve to regulate airflow through the nasal passages.

Key facts

- Approximately 55,000 rhinoplasties take place in the United States each year (American Society of Aesthetic Plastic Surgery, 2019)

- Approximately 250,000 sinus surgeries take place in the United States each year (Pynnonen & Davis, 2014)
- Approximately 260,000 septoplasties take place in the United States each year (Pynnonen & Davis, 2014)
- Chronic rhinosinusitis affects between 5% and 12% of the population (Dietz de Loos et al., 2019)
- Nearly 40% of all US adults used prescription opioids in 2015 (Han et al., 2017)

Summary points

- Septorhinoplasty and rhinologic surgery may be performed under MAC or GA.
- Postoperative pain after both septorhinoplasty and rhinologic surgery tends to be mild.
- Most patients require fewer than 5 opioid tablets in total after septorhinoplasty or rhinologic surgery.
- NSAIDs effectively lower the number of opioid tablets needed after rhinoplasty and rhinologic surgery.
- Postoperative pain after septorhinoplasty and rhinologic surgery should be primarily based on over-the-counter non-opioids, such as acetaminophen, taken as first-line agents, with opioids used as breakthrough or rescue agents as needed.

References

American Society of Aesthetic Plastic Surgery. (2019). *Aesthetic plastic surgery—National databank statistics 2019*. https://www.surgery.org/sites/default/files/Aesthetic-Society_Stats2019Book_FINAL.pdf.

Aulet, R. M., Trieu, V., Landrigan, G. P., & Millay, D. J. (2019). Changes in opioid prescribing habits for patients undergoing rhinoplasty and septoplasty. *JAMA Facial Plastic Surgery, 21*(6), 487–490. https://doi.org/10.1001/jamafacial.2019.0937.

Çelik, E. C., Kara, D., Koc, E., & Yayik, A. M. (2018). The comparison of single-dose preemptive intravenous ibuprofen and paracetamol on postoperative pain scores and opioid consumption after open septorhinoplasty: A randomized controlled study. *European Archives of Oto-Rhino-Laryngology, 275* (9), 2259–2263. https://doi.org/10.1007/s00405-018-5065-6.

Church, C. A., Stewart, C., IV, O-Lee, T. J., & Wallace, D. (2006). Rofecoxib versus hydrocodone/acetaminophen for postoperative analgesia in functional endoscopic sinus surgery. *Laryngoscope, 116*(4), 602–606. https://doi.org/10.1097/01.MLG.0000208341.30628.16.

Dietz de Loos, D., Lourijsen, E. S., Wildeman, M. A. M., Freling, N. J. M., Wolvers, M. D. J., Reitsma, S., et al. (2019). Prevalence of chronic rhinosinusitis in the general population based on sinus radiology and symptomatology. *Journal of Allergy and Clinical Immunology, 143*(3), 1207–1214. https://doi.org/10.1016/j.jaci.2018.12.986.

Ekici, N. Y., & Alagöz, S. (2019). The effectiveness of endoscopic sphenopalatine ganglion block in management of postoperative pain after septal surgery. *International Forum of Allergy & Rhinology, 9*(12), 1521–1525. https://doi.org/10.1002/alr.22411.

Eloy, P., Andrews, P., & Poirrier, A. L. (2017). Postoperative care in endoscopic sinus surgery: A critical review. *Current Opinion in Otolaryngology & Head and Neck Surgery, 25*(1), 35–42. https://doi.org/10.1097/MOO.0000000000000332.

Friedman, M., Venkatesan, T. K., Lang, D., & Caldarelli, D. D. (1996). Bupivacaine for postoperative analgesia following endoscopic sinus surgery. *Laryngoscope, 106*(11), 1382–1385. https://doi.org/10.1097/00005537-199,611,000-00014.

Han, B., Compton, W. M., Blanco, C., Crane, E., Lee, J., & Jones, C. M. (2017). Prescription opioid use, misuse, and use disorders in U.S. adults: 2015 national survey on drug use and health. *Annals of Internal Medicine, 167*, 293–301. https://doi.org/10.7326/M17-0865.

Kim, M., Kacker, A., Kutler, D. I., Tabaee, A., Stewart, M. G., Kjaer, K., et al. (2020). Pain and opioid analgesic use after otorhinolaryngologic surgery. *Otolaryngology–Head and Neck Surgery, 163*, 1178–1185. https://doi.org/10.1177/0194599820933223.

Locketz, G. D., Brant, J. D., Adappa, N. D., Palmer, J. N., Goldberg, A. N., Loftus, P. A., et al. (2019). Postoperative opioid use in sinonasal surgery. *Otolaryngology–Head and Neck Surgery, 160*(3), 402–408. https://doi.org/10.1177/0194599818803343.

Moeller, C., Pawlowski, J., Pappas, A. L., Fargo, K., & Welch, K. C. (2012). The safety and efficacy of intravenous ketorolac in patients undergoing primary endoscopic sinus surgery: A randomized, double-blinded clinical trial. *International Forum of Allergy & Rhinology, 2*(4), 342–347. https://doi.org/10.1002/alr.21028.

Nguyen, K. K., Liu, Y. F., Chang, C., Park, J. J., Kim, C. H., Hondorp, B., et al. (2019). A randomized single-blinded trial of ibuprofen- versus opioid-based primary analgesic therapy in outpatient otolaryngology surgery. *Otolaryngology–Head and Neck Surgery, 160*(5), 839–846. https://doi.org/10.1177/0194599819832528.

Park, I. J., Kim, G., Ko, G., Lee, Y. J., & Hwang, S. H. (2016). Does preoperative administration of gabapentin/pregabalin improve postoperative nasal surgery pain? *The Laryngoscope, 126*(10), 2232–2241. https://doi.org/10.1002/lary.25951.

Patel, S., Sturm, A., Bobian, M., Svider, P. F., Zuliani, G., & Kridel, R. (2018). Opioid use by patients after rhinoplasty. *JAMA Facial Plastic Surgery, 20* (1), 24–30. https://doi.org/10.1001/jamafacial.2017.1034.

Pruitt, L. C. C., Swords, D. S., Russell, K. W., Rollins, M. D., & Skarda, D. E. (2019). Prescription vs. consumption: Opioid overprescription to children after common surgical procedures. *Journal of Pediatric Surgery, 54*(11), 2195–2199. https://doi.org/10.1016/j.jpedsurg.2019.04.013.

Pynnonen, M. A., & Davis, M. M. (2014). Extent of sinus surgery, 2000 to 2009: A population-based study. *Laryngoscope, 124*(4), 820–825. https://doi.org/10.1002/lary.24335.

Raikundalia, M. D., Cheng, T. Z., Truong, T., Kuchibhatla, M., Ryu, J., Abi Hachem, R., et al. (2019). Factors associated with opioid use after endoscopic sinus surgery. *Laryngoscope, 129*(8), 1751–1755. https://doi.org/10.1002/lary.27921.

Rezaeian, A., Hashemi, S. M., & Dokhanchi, Z. S. (2019). Effect of sphenopalatine ganglion block with bupivacaine on postoperative pain in patients undergoing endoscopic sinus surgery. *Allergy & Rhinology*, *10*. https://doi.org/10.1177/2152656718821282, 215265671882128.

Riley, C. A., Kim, M., Sclafani, A. P., Kallush, A., Kjaer, K., Kacker, A. S., et al. (2018). Opioid analgesic use and patient-reported pain outcomes after rhinologic surgery. *International Forum of Allergy & Rhinology*, *9*, 339–344. https://doi.org/10.1002/alr.22260.

Rock, A. N., Akakpo, K., Cheresnick, C., Zmistowksi, B. M., Essig, G. F., Chio, E., et al. (2019). Postoperative prescriptions and corresponding opioid consumption after septoplasty or rhinoplasty. *Ear, Nose, & Throat Journal*. https://doi.org/10.1177/0145561319866824, 145561319866824.

Saxena, A., & Nekhendzy, V. (2020). Anesthetic considerations for functional endoscopic sinus surgery. *Journal of Head and Neck Anesthesia*, *4*(2), e25. https://doi.org/10.1097/hn9.0000000000000025.

Schwartz, M. A., Naples, J. G., Kuo, C. L., & Falcone, T. E. (2018). Opioid prescribing patterns among otolaryngologists. *Otolaryngology–Head and Neck Surgery*, *158*(5), 854–859. https://doi.org/10.1177/0194599818757959.

Sclafani, A. P., Kim, M., Kjaer, K., Kacker, A., & Tabaee, A. (2019). Postoperative pain and analgesic requirements after septoplasty and rhinoplasty. *The Laryngoscope*, *129*(9), 2020–2025. https://doi.org/10.1002/lary.27913.

Sethi, R. K. V., Miller, A. L., Bartholomew, R. A., Lehmann, A. E., Bergmark, R. W., Sedaghat, A. R., et al. (2019). Opioid prescription patterns and use among patients undergoing endoscopic sinus surgery. *Laryngoscope*, *129*(5), 1046–1052. https://doi.org/10.1002/lary.27672.

Simsek, T., Musaoglu, I. C., & Uluat, A. (2019). The effect of lidocaine and tramadol in nasal packs on pain after septoplasty. *European Archives of Oto-Rhino-Laryngology*, *276*(6), 1663–1669. https://doi.org/10.1007/s00405-019-05306-x.

Svider, P. F., Nguyen, B., Yuhan, B., Zuliani, G., Eloy, J. A., & Folbe, A. J. (2018). Perioperative analgesia for patients undergoing endoscopic sinus surgery: an evidence-based review. *International Forum of Allergy & Rhinology*, *8*(7), 837–849. https://doi.org/10.1002/alr.22107.

Szychta, P., & Antoszewski, B. (2010). Assessment of early post-operative pain following septorhinoplasty. *Journal of Laryngology and Otology*, *124*(11), 1194–1199. https://doi.org/10.1017/S0022215110001519.

Thaler, E. R., Gottschalk, A., Samaranayake, R., Lanza, D. C., & Kennedy, D. W. (1997). Anesthesia in endoscopic sinus surgery. *American Journal of Rhinology*, *11*(6), 409–414. https://doi.org/10.2500/105065897780914929.

Wise, S. K., Wise, J. C., & Delgaudio, J. M. (2005). Evaluation of postoperative pain after sinonasal surgery. *American Journal of Rhinology*, *19*, 471–477.

Wu, A. W., Walgama, E. S., Genç, E., Ting, J. Y., Illing, E. A., Shipchandler, T. Z., et al. (2020). Multicenter study on the effect of nonsteroidal anti-inflammatory drugs on postoperative pain after endoscopic sinus and nasal surgery. *International Forum of Allergy & Rhinology*, *10*(4), 489–495. https://doi.org/10.1002/alr.22506.

Zheng, Z., Riley, C. A., Kim, M., Sclafani, A., & Tabaee, A. (2020). Opioid prescribing patterns and usage after rhinologic surgery: A systematic review. *American Journal of Otolaryngology and Head and Neck Surgery*, *41*(4), 102,539. https://doi.org/10.1016/j.amjoto.2020.102539.

Chapter 24

Pain response, neonates, and venipuncture

Hardeep Kaur[a] and Gaurav Mahajan[b]

[a]*Department of Pediatrics, Division of Pediatric Pulmonology and Intensive Care, All India Institute of Medical Sciences, New Delhi, India,* [b]*Department of Gastroenterology, Postgraduate Institute of Medical Sciences and Research, Chandigarh, India*

Abbreviations

PIPP	premature infant pain profile
FLACC	Face Legs Activity Cry Consolability scale
TA	topical anesthesia
LBW	low birth weight

"What cannot be cured, must be endured" is an old adage but not applicable in the modern era. Although unable to express effectively, neonates do feel and remember pain. To provide an effective pain relief to the youngest human beings is the need of the hour.

Introduction

Pain is difficult to define as it has both sensory and emotional components which are very subjective. Nevertheless, traditionally pain is defined as an unpleasant sensory and emotional experience which may or may not be associated with any potential injury to any cell/tissue of the body (2001). Neonates and infants are nonverbal and hence cannot express their pain to their caregivers. This has led to underestimation of their pain and have them subjected to plenty of painful experiences in early life which may have a catastrophic effect on growing brain. Scientific society had the belief that immature brain does not feel pain until recent research proved otherwise (Anand, 2000; Stevens & Franck, 1995).

The communication of neonatal pain is based on certain physiological changes in their heart rate and other vital parameters as well as changes in behavior cues which are again difficult to interpret by the caregiver making them extremely subjective and difficult to standardize (Lander, 1990). Despite all these difficulties, few pain scales have been developed and validated such as premature infant pain profile (PIPP) (Figs. 1 and 2), Face Legs Activity Cry Consolability scale (FLACC) (Fig. 3), and CRIES scale (Fig. 4). This chapter highlights the various aspects of neonatal pain including the problems encountered by the clinicians to interpret neonatal pain and subsequently adopt effective pain relief measures.

Pathophysiology of pain and CNS effects

Nociceptors carry the pain sensation via lateral spinothalamic tract to central nervous system (Fig. 5). The brain then stimulates the sympathetic nervous system, resulting in physiological responses such as increased heart rate and respirations, epinephrine release, and other systemic responses. However, the pain threshold of humans is different from each other. Earlier, it was thought that this system is not well developed in neonates especially the premature ones because of immaturity of nervous system. But now, it is already proven that thalamus, which is responsible for perception of pain, is well developed by 20–24 weeks and hence even the preterm neonates feel and express pain (Craig, Whitfield, Grunau, Linton, & Hadjistavropoulos, 1993; Grunau & Craig, 1987; Lawrence et al., 1993). The neonates experience acute pain during invasive procedures such as heel prick. They also experience chronic inflammatory pain such as of osteomyelitis and nerve damage. In response to pain, the physiological changes occur due to sympathetic stimulation. Apart from sympathetic stimulation, certain hormonal changes also occur such as release of cortisol and aldosterone in addition to epinephrine. All these cause increase in heart rate, blood pressure, blood glucose, and vasoconstriction. Elevated cortisol can even suppress the immune system for a long period making them prone to infections.

Features and Assessments of Pain, Anesthesia, and Analgesia. https://doi.org/10.1016/B978-0-12-818988-7.00047-9
Copyright © 2022 Elsevier Inc. All rights reserved.

Process	Indicator	0	1	2	3	Score
Chart	Gestational Age (at that time)	≥ 36 wks	32 ≤ age < 36	28 ≤ wks < 32	< 28 wks	
Observe Infant 15 seconds Heart rate: Oxygen Saturation:	Behavioural state	Active/Awake Eye open Facial movements Crying with eyes open/closed	Quiet/awake Eyes open No facial movements	Active/sleep Eyes closed Facial movements	Quiet/sleep Eyes closed No facial movements	
Observe infant 30 seconds	Heart rate Max:	0 – 4 beats/ min increase	5 - 14 beats/ min increase	15 - 24 beats/ min increase	25 beats/ min or more increase	
	Oxygen saturation Min:	0% - 2.4% decrease	2.5% - 4.9% decrease	5% - 7.4% decrease	7.5% or more decrease	
	Brow bulge	None 0% - 9% of time(<3sec)	Minimum 10% - 39% of time(>=3 to <12 sec)	Moderate 40% - 69% of time(>=12 to <21)	Maximum 70% of time or more(>=21 sec or more)	
	Eye squeeze	None 0% - 9% of time(<3 sec)	Minimum 10% - 39% of time(>=3 to <12 sec)	Moderate 40% - 69% of time(>=12 to <21 sec)	Maximum 70% of time or more(>=21 sec or more)	
	Nasolabial furrow	None 0% - 9% of time(<3sec)	Minimum 10% - 39% of time(>=3 to <12 sec)	Moderate 40% - 69% of time(>=12 to 21 sec)	Maximum 70% of time or more(>=21 sec or more)	

FIG. 1 Premature infant pain profile scale (PIPP).

Pain transmission in neonates is via unmyelinated nerves which cause slow transmission of pain signals. This may further enhance the pain perceived by neonates (Anand, 2001). Neonates are also subjected to various other stimuli in neonatal ICUs (NICUs) such as noise, ambient light, and repeated touch. All these unavoidable stimuli can add to the pain perceived by neonates and they may develop hyperalgesia (Simons et al., 2003). Recurrent and chronic pain may also inhibit the development of pain inhibitory pathways. Recurrent and chronic pain can cause disruption of capillary blood flow in growing brain, resulting in poor synaptic connections and disorganization of sensory nervous system.

Pain assessment tools

The tools most commonly used for assessment of neonatal pain response are: Premature Infant Pain Profile, Neonatal Facial Coding System, and the Neonatal Infant Pain Scale CRIES scale (Craig et al., 1993; Grunau & Craig, 1987; Lawrence et al., 1993). The PIPP is a behavioral and physiological assessment tool, which provides a measure of the both term and preterm infant's response to pain. The PIPP scale is designed to assess pain during and soon after acute invasive procedures. PIPP is the only tool that accounts for the infant's gestational age, thus allowing the distinction between behavioral differences among full-term and preterm infants.

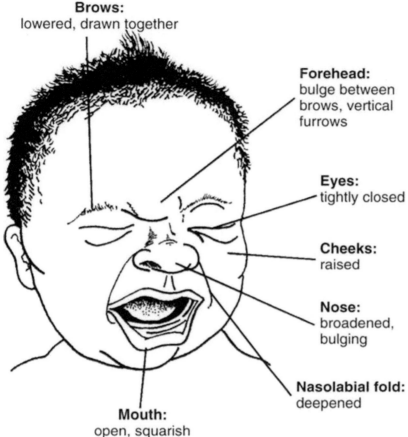

FIG. 2 Facial expression PIPP scale.

FIG. 3 Face leg activity cry consolability scale (FLACC).

Category	Scoring		
	0	1	2
Face	No particular expression or smile	Occasional grimace or frown, withdrawn, disinterested	Frequent to constant quivering chin, clenched jaw
Legs	Normal position or relaxed	Uneasy, restless, tense	Kicking or legs drawn up
Activity	Lying quietly, normal position, moves easily	Squirming, shifting back and forth, tense	Arched, rigid or jerking
Cry	No cry (awake or asleep)	Moans or whimpers: occasional complaint	Crying steadily, screams or sobs, frequent complaints
Consolability	Content, relaxed	Reassured by occasional touching, hugging or being talked to; distractable	Difficult to console

Each of the five categories is scored from 0 – 2, resulting in total range of 0 – 10, FLACC = Face, Leg, Activity, Cry, Consolability

	DATE/TIME						
Crying - Characteristic cry of pain is high pitched. 0 – No cry or cry that is not high-pitched 1 - Cry high pitched but baby is easily consolable 2 - Cry high pitched but baby is inconsolable							
Requires O₂ for SaO₂ < 95% - Babies experiencing pain manifest decreased oxygenation. Consider other causes of hypoxemia, e.g., oversedation, atelectasis, pneumothorax) 0 – No oxygen required 1 – < 30% oxygen required 2 – > 30% oxygen required							
Increased vital signs (BP* and HR*) - Take BP last as this may awaken child making other assessments difficult 0 – Both HR and BP unchanged or less than baseline 1 – HR or BP increased but increase in < 20% of baseline 2 – HR or BP is increased > 20% over baseline.							
Expression - The facial expression most often associated with pain is a grimace. A grimace may be characterized by brow lowering, eyes squeezed shut, deepening naso-labial furrow, or open lips and mouth. 0 – No grimace present 1 – Grimace alone is present 2 – Grimace and non-cry vocalization grunt is present							
Sleepless - Scored based upon the infant's state during the hour preceding this recorded score. 0 – Child has been continuously asleep 1 – Child has awakened at frequent intervals 2 – Child has been awake constantly							
TOTAL SCORE							

FIG. 4 Cries scale.

Problems in assessing neonatal pain

Clinician encounters problems in measuring and relieving the neonatal pain in NICUs due to following reasons:

(A) Lack of self-report by neonates (Merskey & Bogduk, 1994).
(B) Difficulty in interpreting the behavior cues by health care providers (McIntosh, 1997).
(C) Difficulty in assessing the physiological changes displayed by neonates.
(D) Lack of pain measuring scales which are validated and standardized.
(E) Lack of consensus or guidelines to assess and relief neonatal pain.
(F) Lack of safe pharmacological agents to relief neonatal pain (McGrath & Finley, 1996).
(G) Lack of adequate research in this field because parents may not be willing to take part in scientific research considering the fact that most neonates who are admitted to NICUs are usually sick.
(H) Lack of cohort studies which can identify long-term ill effects of chronic pain.
(I) Lack of defined rules in medical ethics to consider neonatal pain relief.
(J) Inadequate training of clinicians, neonatologists, and paramedical staff to identify and relieve neonatal pain.

Pain relief measures

Most of the analgesics used to relieve pain in adults have some serious side effects. Opioids can give rise to severe respiratory depression especially in preterm infants who have immature metabolic system to eliminate these drugs. Similarly, NSAIDs can lead to pulmonary hypertension, risk of bleeding, risk of thrombocytopenia, and necrotizing enterocolitis. Other agents such as IV paracetamol have not been validated and its safe dose is not recommended. Remifentanil is the latest drug under study as it has promising effects with faster excretion and lesser side effects and lesser drug dependence.

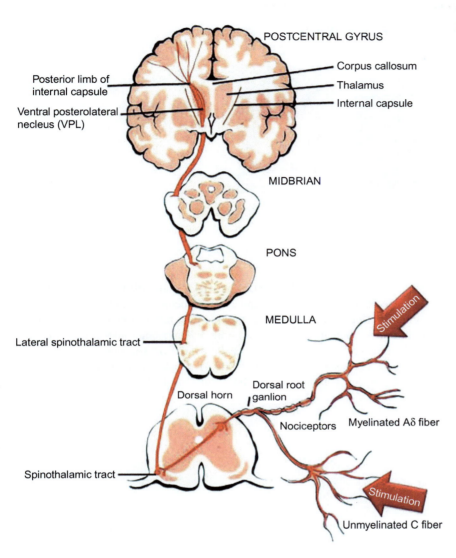

FIG. 5 Spinothalamic tract.

To be safe, many clinicians prefer using nonpharmacological agents to provide some pain relief. Few examples are use of oral sucrose, cuddling, music therapy, rocking the baby, breast feeding, and kangaroo care. Cochrane review has recently established that the use of oral sucrose can be an effective method to relieve neonatal pain during minor procedure such as venipuncture.

Despite the potential side effects of opioids and NSAIDs, these agents are used to relieve postoperative pain of major surgeries and major procedure (Anand, 2001). Recent studies have thrown some light on potential use of topical anesthetic agents without having systemic side effects of NSAIDs and opioids (Gibbins, Stevens, & Asztalos, 2003; Kaur, Gupta, & Kumar, 2003; Long et al., 2003). Most commonly used topical anesthetic agents are lidocaine, eutectic mixture of local anesthetics (EMLA), and prilocaine (Melhuish & Payne, 2006). Topical anesthetic agents are easy to apply before minor procedure. They are safe, effective, and free from systemic side effects as mentioned earlier. However, they may cause local redness or irritation to skin or even blanching or blistering (Lehr et al., 2005; Nestor, 2006; Taddio et al., 2011). EMLA is considered unsafe in preterm neonates because of presence of prilocaine. This agent can cause meth-hemoglobinemia in preterm babies. Four percent tetracaine gel is not very popular as it has long application time before the procedure. Lidocaine has been in use as surface anesthesia in various forms and strengths for a long time now. We have few newer preparations such as 4% liposomal lidocaine which was first available since 2003. Four percent liposomal lidocaine chemically is 2-diethylaminoaceto-2′6′-xylidide and acts by blocking nerve impulse conduction by interfering with voltage-gated sodium channels. Main advantage over EUTECTIC mixture is a shorter onset of action and no risk of methemoglobinemia

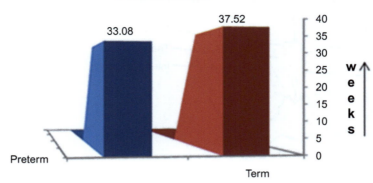

FIG. 6 Period of gestation (weeks).

in preterm babies. Kaur et al. have used this agent in a study conducted at tertiary care level using PIPP scale and authors have found it as an effective agent. PIPP is a validated pain scale to measure and compare neonatal pain in both term and preterm babies (Kaur et al., 2018). This study used 4% liposomal lidocaine to provide pain relief in both groups during venipuncture. Authors measured pain response during venipuncture with and without 4% liposomal lidocaine and compared the results in the end. The average age at the time of first venipuncture was $0.60\,h \pm 0.75$ s in preterm group and $0.54\,h \pm 0.90$ s in term group. The average period of gestation was 33.08 ± 2.41 weeks in preterm group and 37.52 ± 0.73 weeks in term group (Fig. 6). The various components of the results of PIPP scale are shown in Tables 1–6.

TABLE 1 Behavior state score before and after anesthesia.

				P-value	
Behavior state score (mean ± 2SD)	Preterm	Term	Total (n = 100)	Preterm	Term
Before anesthesia	1.86 ± 0.98	2.16 ± 1.07	50	0.150	0.150
After anesthesia	1.82 ± 1.17	1.82 ± 1.17	50	0.800	0.800

TABLE 2 Heart rate and saturation score before and after anesthesia.

				P-value	
	Preterm	Term	Total (n = 100)	Preterm	term
Increase in h score (beats/min)					
Before anesthesia	16.02 ± 9.21	16.60 ± 8.70	50	0.72	0.01
After anesthesia	15.36 ± 8.99	11.84 ± 7.82	50	0.72	0.01
Decrease in O_2 saturation (%)					
Before anesthesia	2.90 ± 1.61	2.52 ± 1.13	50	0.001	0.23
After anesthesia	1.78 ± 2.35	2.30 ± 0.97	50	0.001	0.30

TABLE 3 brow bulge score before and after anesthesia.

Brow bulge score (s)	Preterm	Term	Total	P-value	
				Preterm	term
Before anesthesia	15.70 ± 8.64	19.22 ± 7.49	50	0.23	0.006
After anesthesia	13.60 ± 8.93	14.54 ± 8.98	50	0.23	0.006

TABLE 4 Eye squeeze score before and after anesthesia.

Eye squeeze score (s)	Preterm	Term	Total ($n = 100$)	P-value	
				Preterm	Term
Before anesthesia	16.64 ± 9.09	18.58 ± 7.69	50	0.17	0.05
After anesthesia	14.14 ± 9.29	15.42 ± 8.48	50	0.17	0.05

TABLE 5 nasolabial furrow score before and after anesthesia.

Nasolabial furrow squeeze (s)	Preterm	Term	Total ($n = 100$)	P-value	
				Preterm	Term
Before anesthesia	16.52 ± 9.46	18.56 ± 8.63	50	0.22	0.01
After anesthesia	14.28 ± 8.82	14.14 ± 8.34	0	0.22	0.01

TABLE 6 Total pain score before and after anesthesia.

Total score	Preterm mean ± 2 SD	Term mean ± 2 SD	Total ($n = 100$)	P-value	
				Preterm	Term
Before anesthesia	11.28 ± 3.72	11.54 ± 2.84	50	0.01	0.001
After anesthesia	9.58 ± 3.39	9.04 ± 2.97	50	0.01	0.001

Conclusion

Pain management for neonates is a much-debated topic in the medical field. New research has recently brought about a change in how neonate's and infant's perception of pain is viewed. The capability of the new born infant to feel and remember pain has been underestimated, leading to unnecessary painful experiences in this population. Very few pain measuring tools are validated and no ethical guidelines exist in most parts of the world to deal with issues of neonatal pain. Pain scoring should be a part of routine monitoring in neonatal intensive care units. Caregivers should be trained to assess neonates for pain using suitable multidimensional tools. Besides reducing the exposure to painful procedures as low as possible, wherever possible, some pain-relieving measure (pharmacological or nonpharmacological) should be used. National and international pain societies should lay down guidelines for adopting pain relieving measures and encourage further research on this topic to fill the gaps in our existing knowledge on this subject.

Application to other areas

(1) Topical anesthetic agents can be used to relieve procedural pain before lumbar puncture for CSF study, insertion of intercostal chest tube drainage, pleurocentesis, insertion of PICC line.
(2) There are many randomized controlled trials being conducted on neonates every year. Most of these trials require handling of neonates apart from conducting painful procedures on them. Use of these pain-relieving measures can improve participation of neonate in these trials as it will allay the pain of newborn babies as well as the anxiety of parents.
(3) Neonatal pain relief measures can improve the overall quality of care in terms of parent's satisfaction, less morbidity to patient as there will be less fluctuation of vital parameters. This can become an effective quality control model for NICUs in the world.

Other agents of interest

(1) 5% lidocaine
(2) EMLA (eutectic mixture of local anesthetic)
(3) Prilocaine and tetracaine
(4) Opioids
(5) NSAIDs

The first three of the above mentioned agents are topical anesthetic agents. EMLA is known to cause methemoglobinemia and hence less frequently used especially in preterm babies. Opioids and NSAIDs are given intravenously to relieve pain after major surgeries. These can be given for long periods in the form of infusion.

Mini-dictionary of terms

(1) Preterm—Baby born before 37-week period of gestation.
(2) Very low birth weight—Birth weight less than 1500 g.
(3) Low birth weight (LBW)—Birth weight less than 2500 kg.
(4) FLACC scale—Face leg activity cry consolability scale used to assess the neonatal pain during procedures.
(5) Premature infant pain profile (PIPP)—A validated tool to assess the neonatal pain in both term and preterm babies.
(6) Topical anesthesia (TA)—Agents which are applied superficially over skin to relieve pain.
(7) Venipuncture—Puncturing the veins for intravenous cannulation or blood sampling.

Key facts

(1) Neonatal pain is a much-debated topic in present era of medicine. Their lack of expression of pain leads to underestimation of neonatal pain.
(2) Most neonatal intensive care units do not have a written pain-relieving policy as this is not usually perceived as an important part of intensive care. Reason for this may be inability to measure neonatal pain and lack of self-report by neonate.
(3) Besides giving opioids to relieve postsurgery pain, there are several safer options to relieve minor procedural pain like topical anesthetic agents.
(4) We need to develop evidence-based protocol and carry out further research on this topic to improve the overall quality of neonatal care in the world.

Summary points

(1) Pain is an unpleasant sensory and emotional experience associated with actual or potential tissue damage. Until recently, neonates were not considered to feel pain till research proved that pain perception is well developed even in preterm neonate.
(2) Being nonverbal, neonatal pain is often underestimated.
(3) It is difficult to assess neonatal pain due to subjectivity involved in assessing their behavior and expression.
(4) Most part of the world provides only postsurgical pain relief in NICU. Procedural pain can be relieved by using topical anesthesia before minor procedure.
(5) Topical anesthetic agents are safe, less costly, and very effective in relieving pain of minor procedures.

(6) Further research is needed to assess the safety and effectiveness of various topical anesthetic agents to relieve neonatal pain.

(7) Scientific world needs to be sensitized to adopt a more humane approach and develop future guidelines in relieving neonatal pain.

Conflict of interest

All authors have none to declare.

References

Anand, K. J. (2000). Pain, plasticity, and premature birth: A prescription for permanent suffering? *Nature Medicine, 6*(9), 971–973.

Anand, K. J. (2001). Consensus statement for the prevention and management of pain in the newborn. *Archives of Pediatrics & Adolescent Medicine, 155*, 173–180.

Craig, K. D., Whitfield, M. F., Grunau, R. V., Linton, J., & Hadjistavropoulos, H. D. (1993). Pain in the preterm neonate: Behavioural and physiological indices. *Pain, 52*(3), 287–299.

Gibbins, S., Stevens, B., & Asztalos, E. (2003). Assessment and management of acute pain in high-risk neonates. *Expert Opinion on Pharmacotherapy, 4*, 475–483.

Grunau, R., & Craig, K. (1987). Pain expression in neonates: Facial action and cry. *Pain, 28*, 395–410.

Kaur, G., Gupta, P., & Kumar, A. (2003). A randomized trial of eutectic mixture of local anesthetics during lumbar puncture in newborns. *Archives of Pediatrics & Adolescent Medicine, 157*, 1065–107020.

Kaur, H., et al. (2018). Study of pain response in neonates during venipuncture with a view to analyse utility of topical anaesthetic agent for alleviating pain. *Medical Journal, Armed Forces India.* https://doi.org/10.1016/j.mjafi.2017.12.009.

Lander, J. (1990). Clinical judgments in pain management. *Pain, 42*, 15–22.

Lawrence, J., Alcock, D., McGrath, P., Kay, J., MacMurray, B., & Dulberg, C. (1993). The development of a tool to assess neonatal pain. *Neonatal Network, 12*, 59–66.

Lehr, V. T., Cepeda, E., Frattarelli, D. A., Thomas, R., LaMothe, J., & Aranda, J. V. (2005). Lidocaine 4% cream compared with lidocaine 2.5% and prilocaine 2.5% or dorsal penile block for circumcision. *American Journal of Perinatology, 22*(5), 231–237.

Long, C. P., McCafferty, D. F., Sittlington, N. M., Halliday, H. L., Woolfson, A. D., & Jones, D. S. (2003). Randomized trial of novel tetracaine patch to provide local anaesthesia in neonates undergoing venepuncture. *British Journal of Anaesthesia, 91*, 514–518.

McGrath, P. J., & Finley, G. A. (1996). Attitudes and beliefs about medication and pain management in children. *Journal of Palliative Care, 12*, 46–50.

McIntosh, N. (1997). Pain in the newborn, a possible new starting point. *European Journal of Pediatrics, 156*, 173–177.

Melhuish, S., & Payne, H. (2006). Nurses' attitudes to pain management during routine venepuncture in young children. *Pediatric Nursing, 18*(2), 20–23.

Merskey, H., & Bogduk, N. (1994). Classification of chronic pain. In *IAS 77 task force on taxonomy* (2nd ed., pp. 209–214). Seattle: ISAP Press.

Nestor, M. S. (2006). Safety of occluded 4% lidocaine cream. *Journal of Drugs in Dermatology, 5*(7), 618–620.

Simons, S. H., van Dijk, M., Anand, K. S., Roofthooft, D., van Lingen, R. A., & Tibboel, D. (2003). Do we still hurt newborn babies? A prospective study of procedural pain and analgesia in neonates. *Archives of Pediatrics & Adolescent Medicine, 157*, 1058–1064.

Stevens, B. J., & Franck, L. (1995). Special needs of preterm infants in the management of pain and discomfort. *Journal of Obstetric, Gynecologic, and Neonatal Nursing, 24*, 856–862.

Taddio, A., Shah, V., Stephens, D., Parvez, E., Hogan, M. E., Kikuta, A., et al. (2011). Effect of lidocaine and sucrose alone and in combination for venipuncture pain in newborns. *Pediatrics, 127*(4), e940–e947.

Further reading

AAP & APS (American Academy of Pediatrics & American Pain Society). (2001). The assessment and management of acute pain in infants, children and adolescents. *Pediatrics, 108*, 793–797.

Cignacco, E., Hamers, J., Stoffel, L., Van Lingen, R., Schutz, N., Muller, R., et al. (2008). Routine procedures in NICUs: Factors influencing pain assessment and ranking by pain intensity. *Swiss Medical Weekly, 138*(33), 484–491.

Farrington, E. A., McGuinness, G. A., Johnson, G. F., Erenberg, A., & Leff, R. D. (1993). Continuous intravenous morphine infusion in postoperative newborn infants. *American Journal of Perinatology, 10*, 84–87.

Kaur, H., & Mahajan, G. (2019). A comprehensive analysis of neonatal pain and measures to reduce pain. *Pediatric Critical Care Medicine, 6*(1), 43–48. https://doi.org/10.21304/2019.0601.00483.

Modi, N., & Glover, V. (2000). Fetal pain and stress. In K. J. S. Anand, B. J. Stevens, & P. J. McGrath (Eds.), *Pain in neonates: Pain research and clinical management* (2nd rev. ed., pp. 217–227). Amsterdam: Elsevier.

Sayed, E., Taddio, A., Fallah, S., Silva, N., & Moore, A. (2007). Safety profile of morphine following surgery in neonates. *Journal of Perinatology, 27*, 444–447.

Stevens, B., Jhonston, C., Petryshen, P., & Taddio, A. (1996). Premature infant pain profile: Development & initial validation. *Clinical Journal of Pediatrics, 12*(1), 13–22.

Chapter 25

Carpal tunnel syndrome and pain

Rodrigo Núñez-Cortés[a,b], Carlos Cruz-Montecinos[a,b], Claudio Tapia[a], Paula Pino Pommer[c], and Sofía Pérez-Alenda[b]

[a]Department of Physical Therapy, Faculty of Medicine, University of Chile, Santiago, Chile, [b]Department of Physiotherapy, Physiotherapy in Motion Multispeciality Research Group (PTinMOTION), University of Valencia, Valencia, Spain, [c]Traumatology Department, Hospital Clínico La Florida, Santiago, Chile

Abbreviations

2PD	two-point discrimination
CNS	central nervous system
CTQ-FS	Boston Carpal Tunnel Questionnaire functional scale
CTQ-SSS	Boston Carpal Tunnel Questionnaire-symptom severity scale
CTR	carpal tunnel release
CTS	carpal tunnel syndrome
DASH	Disabilities of the Arm, Shoulder and Hand
PCS	pain catastrophizing scale

Introduction

Carpal tunnel syndrome (CTS) is a compressive peripheral neuropathy characterized by pain, tingling sensation, and paresthesia in the region of the median nerve (Keith et al., 2009). CTS is the most common cause of pain due to peripheral nerve entrapment (Nora, Becker, Ehlers, & Gomes, 2005). Symptoms often lead to functional impairment of the upper limb (Nazari, Shah, MacDermid, & Woodhouse, 2017) and decrease the quality of life of patients (Bickel, 2010). Anatomically, the carpal tunnel is formed by the carpal bones at the base and the transverse carpal ligament as the volar boundary. It contains the flexor tendons of the hand and the median nerve. The latter is the most superficial structure of the carpal tunnel (Fig. 1), being the most vulnerable to compression. Median nerve may be compressed by an increase in the volume of the contents (flexor synovitis, intracanal tumors) or by a decrease in the surrounding volume (thickening of the flexor retinaculum).

The prevalence of CTS in the United States ranges between 3.7% and 12%, being higher in women and people over 50 years of age (Dale et al., 2013; Luckhaupt et al., 2013). The incidence rate of CTS worldwide has recently increased, being higher in the working population, with a rate of 23 per 1000 individuals-year (Dale et al., 2013). However, incidence rates may increase based on the use of different diagnostic criteria and populations.

Median nerve compression

A wide variety of anatomopathological factors are involved in the development of CTS, including high carpal tunnel pressure and ischemic changes within the nerve that increase neural compression (Uchiyama et al., 2010). It has been suggested that high pressure is due to disruption in intraneural blood flow that contributes to intraneural edema and fibrosis (Kuo et al., 2013; Uchiyama et al., 2010). In addition, significant compression of the median nerve can alter normal gliding kinematics (Erel et al., 2003). The literature suggests that median nerve excursion evaluated through ultrasound imaging is reduced in people with CTS compared to healthy controls (Ellis, Blyth, & Arnold, 2016). Prolonged ischemia and mechanical involvement can induce eventual effects such as demyelination and ultimately axonal degeneration (Schmid, Fundaun, & Tampin, 2020).

CTS is associated with loss of function in both unmyelinated and myelinated sensory axons, together with a clear reduction in the density of intraepidermal nerve fibers, which is unrelated to the severity of the electrodiagnostic assessment (Schmid, Bland, Bhat, & Bennett, 2014). This is important, since the American Academy of Orthopedic Surgeon's Clinical

Features and Assessments of Pain, Anesthesia, and Analgesia. https://doi.org/10.1016/B978-0-12-818988-7.00031-5
Copyright © 2022 Elsevier Inc. All rights reserved.

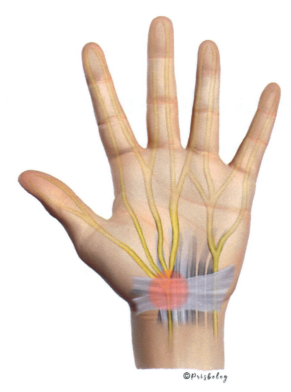

FIG. 1 Median nerve in the carpal tunnel. The illustration shows the transverse carpal ligament and the median nerve as the most superficial structure of the carpal tunnel.

Practice Guidelines indicates that electrodiagnostic assessment may be useful, but not necessary to establish CTS diagnosis (Graham et al., 2016). In addition, it has been proposed that the benefits of surgery might be unrelated to the degree of electrodiagnostic value (Rivlin, Kachooei, Wang, & Ilyas, 2018).

Applications to other areas

Peripheral nerve injury triggers changes in the central nervous system, which include central immune-inflammatory mechanisms, central sensitization and changes in cortical representations, widespread hypersensitivity, and alterations related to conditioned pain modulation (Schmid et al., 2020). Evidence suggests that central and peripheral sensitization mechanisms are involved in the etiology of CTS pain (de-la-Llave-Rincón, Puentedura, & Fernández-de-las-Peñas, 2012) and some cases present with generalized hyperalgesia (Fernandez-de-las-Penas et al., 2009). The extent of sensory symptoms (e.g., hyperalgesia) outside the median nerve region is common in CTS and could be secondary to a sensitization process in the central spinal cord (Zanette, Cacciatori, & Tamburin, 2010). Although the extent of pain in the upper extremity has not been shown to be associated with the severity and clinical, psychological, or psychophysical variables in women with CTS (Fernández-de-Las-Peñas, de-la-Llave-Rincón, et al., 2019; Fernández-de-las-Peñas, Falla, et al., 2019), psychosocial aspects such as catastrophic thinking have shown to be linked with wider areas of pain and paresthesia in the Katz hand diagram (Moradi et al., 2015), so an in-depth study of the mechanisms of pain is needed, as well as understanding its relationship with the clinical evolution and prognosis. Fig. 2 summarizes the multidimensionality of factors linked to CTS.

With regard to pain, due to the various etiopathological factors involved, quantitative sensory tests have proven to be a useful tool helping in the differential diagnosis and prognosis in musculoskeletal pain disorders (Uddin & MacDermid, 2016). Common differential diagnoses include cervical radiculopathy, diabetic or polyneuropathy, thoracic outlet syndrome, and pronator teres syndrome.

Considerations for evaluation

The patient's medical history, the presence of risk factors, and the characteristics of the symptoms are key aspects for establishing a suitable differential diagnosis. Several risk factors are associated with the development of CTS. Intrinsic risk

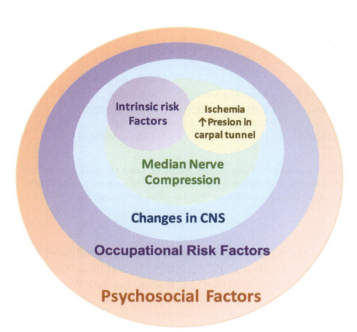

FIG. 2 Multidimensional etiology of CTS. This figure summarizes the key aspects related to the causes of carpal tunnel syndrome, linking the intrinsic (i.e., age, sex, obesity) and pathoanatomic risk factors that lead to compression of the median nerve and the presentation of symptoms in the patient. Changes in the central nervous system (CNS) such as central sensitization are involved in the extent of symptoms, which may be aggravated by occupational risk factors (e.g., vigorous manual force) and modulated by psychosocial factors (e.g., depression, catastrophizing).

factors with a stronger association include obesity, age, and the female sex (Erickson et al., 2019). On the other hand, the most relevant occupational risk factor is vigorous manual exertion, i.e., exposure to manual forces greater than 4 on the Borg CR10 Scale (Harris-Adamson et al., 2015).

Typical symptoms of CTS include hypoesthesia and paresthesia in the distribution of the median nerve. CTS can be acute or chronic. The acute condition is rare, being caused by trauma which generates an increase in pressure inside the canal or direct compression of the nerve, such as in the case of fracture of the distal radius (Gelberman, Rydevik, Pess, Szabo, & Lundborg, 1988). While there is no consensus on the clinical classification of CTS, it is commonly classified based on the severity (i.e., mild, moderate, severe). According to the evidence collected by Erickson et al. (2019), the frequency of symptoms appears to be a factor that distinguishes mild CTS from moderate CTS, i.e., mild referring to more intermittent symptoms and moderate to more constant or repetitive signs. Additionally, severe CTS is characterized by the presence of muscle atrophy in the thenar eminence (Harris-Adamson et al., 2015).

The severity of symptoms can be assessed using the Boston Carpal Tunnel Questionnaire-symptom severity scale (CTQ-SSS), which has demonstrated excellent internal consistency and reliability (Levine et al., 1993). This scale also enables evaluating changes experienced by patients undergoing nonsurgical treatment (Cheung, MacDermid, Walton, & Grewal, 2014) and its baseline value is a good predictor of failure for conservative treatments (Boyd et al., 2005). Erickson et al. (2019) concluded that there is moderate amount of evidence supporting the use of CTQ-SSS for the diagnosis of CTS. Such evaluation will further require a comprehensive physical examination, use of the Katz hand diagram, the Phalen test, the Tinel sign, the carpal compression test, and sensory assessment with the Semmes-Weinstein Monofilament test and two-point discrimination (2PD) (Erickson et al., 2019). The Boston Carpal Tunnel Questionnaire functional scale (CTQ-FS) or the Disabilities of the Arm, Shoulder and Hand (DASH) Assessment Scale is recommended for functional assessment (Erickson et al., 2019).

In terms of grip strength, these are relevant functional variables to consider given their relationship with quality of life (Bickel, 2010), satisfaction (Bae, Kim, Yoon, Kim, & Ho, 2018), and ability to return to work (Ludlow, Merla, Cox, & Hurst, 1997). However, strength should not be used to evaluate short-term changes (less than 3 months) after surgery; this is because surgery can produce scar-related pain during the first 3 months and reduce strength (Erickson et al., 2019). In addition, clinical and functional assessment should be accompanied by the identification of potential psychosocial risk factors (see Section "Psychosocial aspects"). A comprehensive assessment that addresses these aspects will allow clinicians to make more appropriate decisions and deliver greater benefits to patients. Fig. 3 summarizes the key points of the evaluation.

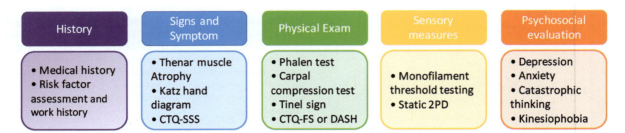

FIG. 3 Evaluation. Diagram illustrating key aspects of carpal tunnel syndrome assessment. These aspects should be aided by other tools for the assessment of quality of life and differential diagnosis optimization. *CTQ-SSS*, Boston Carpal Tunnel questionnaire-symptom severity scale; *CTQ-FS*, Boston Carpal Tunnel questionnaire functional scale; *DASH*, Disabilities of the Arm, Shoulder and Hand; *2PD*, two-point discrimination.

Conservative treatment

Conservative or surgical treatment of CTS may be indicated depending on both the severity and temporality of the condition. A recent meta-analysis that included 10 randomized-controlled trials with 1028 participants established that both surgical and conservative interventions are effective in the management of CTS (Shi, Bobos, Lalone, Warren, & MacDermid, 2020). Conservative treatments involving physiotherapy and occupational therapy for mild and moderate CTS include the use of orthoses, physical agents, manual therapy, therapeutic exercises, education, ergonomic interventions, and multimodal interventions (Erickson et al., 2019). Of all such interventions, the most popular approach for the treatment of carpal tunnel syndrome is education, followed closely by night splints and ergonomic modifications (Parish et al., 2020). Patient education includes aspects related to pathology, risk identification, self-management of symptoms, and symptom-aggravating activities and postures (Erickson et al., 2019). This intervention should be part of a multimodal treatment regime that also addresses psychosocial aspects (Núñez-Cortés et al., 2019).

The use of orthoses for the treatment of symptoms derived from CTS is widely accepted among clinicians. In general, orthotics are recommended in patients with mild-to-moderate symptoms. Their main purpose is to immobilize the wrist, stop the movement of the median nerve and tendons, and decrease the internal pressure of the carpal tunnel. Orthosis design includes the use of different materials, locations (i.e., volar, dorsal, or ulnar), and modifying the angles in neighboring joints (i.e., thumb, metacarpophalangeal joint). Current recommendations for the use of splints in CTS include night-time use with the wrist in a neutral position (i.e., sagittal and frontal) to relieve short-term symptoms and improve functionality. Using a splint during the day, as well as adjusting the wrist position and including the metacarpophalangeal joint, should be considered depending on a successful outcome when using the night splint to control of symptoms. Physical agents recommended to treat symptoms in mild and moderate CTS include surface heat, microwave or shortwave diathermy, interferential current, and phonophoresis. The main effect is the short-term relief of symptoms (Erickson et al., 2019).

Other agents of interest

A Cochrane Review reported that there is limited evidence of the positive effect of therapeutic exercise and manual therapy (Page, O'Connor, Pitt, & Massy-Westropp, 2012). However, recent randomized-controlled trials suggest the effectiveness of long-term manual therapy (Fernández-de-Las-Peñas et al., 2020). While several conservative treatments (e.g., splints, oral drugs, infiltrations, electrotherapy techniques, specific manual techniques, and neural gliding exercises) have been shown to improve symptoms and function in patients with mild and moderate CTS (Jiménez Del Barrio et al., 2018; Shi et al., 2020), the superiority of one treatment over another has not been demonstrated. Therefore, the choice for a conservative treatment, such as physiotherapy and/or occupational therapy, should account for the preference of patients, the expertise of clinicians, and the current recommendations for the treatment of CTS (Erickson et al., 2019; Parish et al., 2020; Schmid et al., 2020).

Surgical treatment

Surgical treatment of carpal tunnel syndrome has shown excellent medium- and long-term results (Huisstede, van den Brink, Randsdorp, Geelen, & Koes, 2018). Surgery is indicated in patients for whom conservative treatment has failed, in acute cases, such as trauma causing increased pressure inside the canal and in severe cases with persistent hypoesthesia of the median nerve region and motor impairment (Urits et al., 2019). It is also indicated in patients suffering from secondary compression of the median nerve, either due to tumors or ganglion cysts within the canal. It is estimated that between 57% and 66% of patients undergoing conservative treatment eventually need surgery after 1–3 years (Burton, Chesterton, Chen, & van der Windt, 2016). Each year, approximately 600,000 carpal tunnel release (CTR) surgeries are performed in the United States, being one of the most frequent operations on the wrist-hand segment (Pourmemari, Heliövaara, Viikari-Juntura, & Shiri, 2018) and representing a important burden for health care systems (Ingram, Mauck, Thompson, & Calandruccio, 2018).

This intervention consists of the surgical release of the median nerve, by sectioning the retinaculum of the flexors. This increases the space available for the structures within the canal, reducing the pressure on the median nerve. With regard to the surgical technique, traditionally, this is performed through an incision, of variable size, in the palmar region of the hand, exposing the flexor retinaculum and subsequently, the interior of the carpal tunnel. This technique has been shown to have excellent long-term results (Louie et al., 2013). However, more recently, interest has been focused on reducing the size of the incision, which has introduced new approaches using mini-open (1 cm) or endoscopic techniques. The main reason is to minimize the discomfort caused by the palmar scar on the hand, which may remain for as long as 3 months. The meta-analysis performed by Sayegh and Strauch shows better short-term results in terms of returning to work and grip strength in endoscopic techniques, but these are not maintained 6 months later (Sayegh & Strauch, 2015). Even though the authors describe less scar pain using the endoscopic technique, a higher risk of median nerve injury is reported. Surgery can be carried out under general, regional, or local anesthesia, the latter two with or without sedation. Studies have shown that patients are equally satisfied with surgery under sedation as with local anesthesia, but present a higher risk of nausea and vomiting, as well as a longer hospital stay when sedation is used (Davison, Cobb, & Lalonde, 2013).

In terms of the results after surgery, about 70%–88% of patients are satisfied and symptom-free (De Kleermaeker, Meulstee, Bartels, & Verhagen, 2019; De Kleermaeker, Meulstee, Claes, Bartels, & Verhagen, 2019; Louie et al., 2013). Approximately 60% of patients experience complete symptom relief after 7–9 months, while 35% report only partial recovery. According to De Kleermaeker et al. good clinical results after 8 months are maintained up to 9 years after surgery (De Kleermaeker, Meulstee, Bartels, et al., 2019), with 81.6% of patients reporting favorably. The rate of complications is between 1% and 25%, most of them being minor and transient. The most feared complication is nerve injury, which occurs in only 0.06% of patients (Karl, Gancarczyk, & Strauch, 2016). On the other hand, reoperation rates due to recurrence are between 1% and 5% (Westenberg et al., 2020), in connection with which male patients, smokers, with rheumatoid arthritis and who have undergone endoscopic surgery are the population presenting the highest risk of needing a new surgical intervention. Fig. 4 summarizes the main interventions recommended for the treatment of CTS, according to the assessment of symptoms in patients with carpal tunnel syndrome.

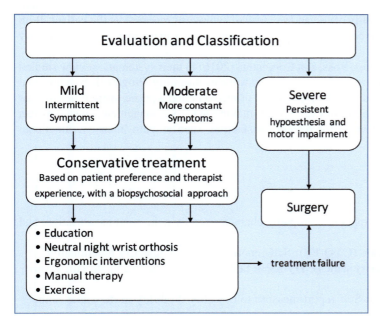

FIG. 4 Treatment. Diagram summarizing the main treatment interventions recommended according to the assessment of symptoms in patients with carpal tunnel syndrome.

Psychosocial aspects

Occupational biomechanical risk factors such as vibration, repetitive, or forced movement of the wrist and hand play an important role in the cause of CTS (Kozak et al., 2015). However, psychosocial factors combined with these biomechanical elements may also have a bearing on the development of CTS, e.g., work-related stress, psychological distress, or high psychological demand for women (Harris-Adamson et al., 2016; Kozak et al., 2015; Rigouin et al., 2014; Roquelaure, Mariel, Dano, Fanello, & Penneau-Fontbonne, 2001).

Some authors argue that psychological variables (e.g., depression, anxiety) are the main risk factors associated with the increase in CTS rates, ranking ahead of biomedical and demographic factors (Nunez, Vranceanu, & Ring, 2010). The need to closely examine these risk factors could have a major impact, given the relevance of the additional costs associated with mental health disorders and the prolonged recovery periods due to complications (e.g., postoperative pain) (Gause, Nunnery, Chhabra, & Werner, 2020; Ingram et al., 2018).

As for pain-related maladaptive cognitive factors, catastrophic thinking has proven to be one of the most important predictors of upper extremity-specific disability (De, Vranceanu, & Ring, 2013) and has been associated with wider areas of pain and paresthesia in the hand diagram (Moradi et al., 2015). In addition, higher scores on the Pain Catastrophizing Scale (PCS) are related to lower gripping and pinching strength (Núñez-Cortés, Cruz-Montecinos, Antúnez-Riveros, & Pérez-Alenda, 2020) and lower patient satisfaction reported after surgery (Mosegaard, Stilling, & Hansen, 2020). Kinesiophobia is another cognitive factor that has been shown to be an important predictor of disability (De et al., 2013). Both kinesiophobia and catastrophizing have proved to hinder the recovery of patients with CTS (Wilkens, Lans, Bargon, Ring, & Chen, 2018).

In terms of emotional and psychological factors, depression and anxiety have a high association with the severity of symptoms and patient satisfaction, before and after surgery (Crijns et al., 2019; Shin, Yoon, Kim, & Kim, 2018). For example, Jerosch-Herold et al. (2017) reported a significant relationship between anxiety and electrodiagnostic severity. In addition, a lower level of anxiety has also been described as a significant predictor of better surgery results (Jerosch-Herold, Shepstone, Houghton, Wilson, & Blake, 2019).

Moreover, the depression levels of CTS patients are strongly associated with pain intensity after surgery (Vranceanu, Jupiter, Mudgal, & Ring, 2010). The level of depression is also a predictor of long-term clinical outcomes (i.e., pain intensity or severity of symptoms) after physical therapy in women with CTS (Fernández-de-Las-Peñas, De-La-Llave-Rincón, et al., 2019; Fernández-de-las-Peñas, Falla, et al., 2019). Postoperative pain control is an important point to discuss, due to the possibility of reducing the costs associated with the use of opioids or narcotics (Gause et al., 2020; Ingram et al., 2018).

Park et al. (2017) in their systematic review found that depression is strongly related to postoperative pain. However, an association between psychological factors and functional outcomes and satisfaction is not so clear. By contrast, Jerosch-Herold et al. (2017), in a large multicenter prospective cohort study with more than 700 CTS patients, reported a significant relationship between anxiety and depression and the quality of life and severity of symptoms reported by the patient. Both pain-related depression and anxiety in CTS patients have been shown to improve significantly in the long term after surgery (Datema et al., 2018; Shin et al., 2018).

In relation to returning to work, Peters, Johnston, Hines, Ross, and Coppieters (2016) in their systematic review studied the prognostic factors for delaying patients in their return to work or for chronic working disability. Prognostic psychosocial factors suggesting a poor work-related outcome included a lower expectation of recovery and a poorer mental health condition. It is thus essential to consider psychosocial variables prior to decision-making in CTS management, since they can predict the results of interventions in highly relevant areas for patient satisfaction, such as pain intensity, functional recovery, and returning to work.

Mini-dictionary of terms

Carpal tunnel syndrome (CTS): Is a compressive peripheral neuropathy characterized by pain, tingling sensation, and paresthesia in the region of the median nerve.

Boston Carpal Tunnel Questionnaire functional scale (CTQ-FS): Self-report measure to objectify functionality.

Boston Carpal Tunnel Questionnaire-symptom severity scale (CTQ-SSS): Self-report measure to objectify the severity of symptoms.

Disabilities of the Arm Shoulder and Hand (DASH): Self-report measure to objectify disability in the upper limb.

Key facts

- Evidence suggests that central and peripheral sensitization mechanisms are involved in the etiology of pain in CTS.
- The most highly associated risk factors are obesity, age, female sex, and tasks involving manual force.
- The severity of symptoms can be assessed using the CTQ-SS.
- Other recommended assessment tools include the use of the Katz hand diagram, the Phalen test, the Tinel sign, the carpal compression test, and Semmes-Weinstein Monofilament test.
- The CTQ-FS or the DASH is recommended for functional assessment.
- Conservative or surgical treatment of CTS may be indicated depending on both the severity and temporality of the condition.
- With regard to the results of surgery, around 70%–88% of patients are satisfied and symptom-free.
- Catastrophic thinking has proven to be one of the most important predictors of upper extremity disability and is associated with wider areas of pain.
- Depression and anxiety are highly associated with the postoperative satisfaction of patients and with the severity of symptoms before and after surgery.

Summary points

- CTS is a compressive peripheral neuropathy characterized by pain, tingling sensation, and paresthesia in the region of the median nerve.
- The patient's medical history, the presence of risk factors, and the characteristics of the symptoms are key aspects for establishing a suitable differential diagnosis.
- The conservative treatment of mild and moderate CTS includes, among the most common treatments in physiotherapy and occupational therapy, the use of orthoses, physical agents, manual therapy, therapeutic exercises, education, ergonomic interventions, and multimodal intervention.
- Surgery is indicated in patients for whom conservative treatment has failed and also when presenting severe conditions, with persistent hypoesthesia of the median nerve region and motor impairment.
- It is essential to consider psychosocial variables prior to decision-making in CTS management.

Acknowledgments

The authors would like to thank Priscila Escobar Gimpel for creating the illustration of the median nerve in the carpal tunnel.

References

Bae, J.-Y., Kim, J. K., Yoon, J. O., Kim, J. H., & Ho, B. C. (2018). Preoperative predictors of patient satisfaction after carpal tunnel release. *Orthopaedics & Traumatology, Surgery & Research, 104*(6), 907–909.

Bickel, K. D. (2010). Carpal tunnel syndrome. *Journal of Hand Surgery, 35*(1), 147–152. https://doi.org/10.1016/j.jhsa.2009.11.003.

Boyd, K. U., Gan, B. S., Ross, D. C., Richards, R. S., Roth, J. H., & MacDermid, J. C. (2005). Outcomes in carpal tunnel syndrome: Symptom severity, conservative management and progression to surgery. *Clinical and Investigative Medicine. Medecine Clinique et Experimentale, 28*(5), 254–260.

Burton, C. L., Chesterton, L. S., Chen, Y., & van der Windt, D. A. (2016). Clinical course and prognostic factors in conservatively managed carpal tunnel syndrome: A systematic review. *Archives of Physical Medicine and Rehabilitation, 97*(5), 836–852. e1.

Cheung, D. K. M., MacDermid, J., Walton, D., & Grewal, R. (2014). The construct validity and responsiveness of sensory tests in patients with carpal tunnel syndrome. *The Open Orthopaedics Journal, 8*(1), 100–107. https://doi.org/10.2174/1874325001408010100.

Crijns, T. J., Bernstein, D. N., Ring, D., Gonzalez, R. M., Wilbur, D. M., & Hammert, W. C. (2019). Depression and pain interference correlate with physical function in patients recovering from hand surgery. *Hand, 14*(6), 830–835. https://doi.org/10.1177/1558944718777814.

Dale, A. M., Harris-Adamson, C., Rempel, D., Gerr, F., Hegmann, K., Silverstein, B., … Evanoff, B. (2013). Prevalence and incidence of carpal tunnel syndrome in US working populations: Pooled analysis of six prospective studies. *Scandinavian Journal of Work, Environment & Health, 39*(5), 495–505.

Datema, M., Tannemaat, M. R., Hoitsma, E., van Zwet, E. W., Smits, F., van Dijk, J. G., & Malessy, M. J. A. (2018). Outcome of carpal tunnel release and the relation with depression. *Journal of Hand Surgery, 43*(1), 16–23.

Davison, P. G., Cobb, T., & Lalonde, D. H. (2013). The patient's perspective on carpal tunnel surgery related to the type of anesthesia: A prospective cohort study. *Hand, 8*(1), 47–53. https://doi.org/10.1007/s11552-012-9474-5.

De Kleermaeker, F. G. C. M., Meulstee, J., Bartels, R. H. M. A., & Verhagen, W. I. M. (2019). Long-term outcome after carpal tunnel release and identification of prognostic factors. *Acta Neurochirurgica, 161*(4), 663–671.

De Kleermaeker, F. G. C. M., Meulstee, J., Claes, F., Bartels, R. H. M. A., & Verhagen, W. I. M. (2019). Outcome after carpal tunnel release: Effects of learning curve. *Neurological Sciences: Official Journal of the Italian Neurological Society and of the Italian Society of Clinical Neurophysiology, 40*(9), 1813–1819.

De, S. D., Vranceanu, A.-M., & Ring, D. C. (2013). Contribution of kinesophobia and catastrophic thinking to upper-extremity-specific disability. *Journal of Bone and Joint Surgery (American Volume), 95*(1), 76–81. https://doi.org/10.2106/jbjs.l.00064.

de-la-Llave-Rincón, A. I., Puentedura, E. J., & Fernández-de-las-Peñas, C. (2012). New advances in the mechanisms and etiology of carpal tunnel syndrome. *Discovery Medicine, 13*(72), 343–348.

Ellis, R., Blyth, R., & Arnold, N. (2016). Is there a relationship between impaired median nerve excursion and carpal tunnel syndrome? A systematic review. *Manual Therapy, 25*, e75. https://doi.org/10.1016/j.math.2016.05.121.

Erel, E., Dilley, A., Greening, J., Morris, V., Cohen, B., & Lynn, B. (2003). Longitudinal sliding of the median nerve in patients with carpal tunnel syndrome. *Journal of Hand Surgery, 28*(5), 439–443.

Erickson, M., Lawrence, M., Jansen, C. W. S., Coker, D., Amadio, P., & Cleary, C. (2019). Hand pain and sensory deficits: Carpal tunnel syndrome. *Journal of Orthopaedic and Sports Physical Therapy, 49*(5), CPG1–CPG85.

Fernández-de-Las-Peñas, C., Arias-Buría, J. L., Cleland, J. A., Pareja, J. A., Plaza-Manzano, G., & Ortega-Santiago, R. (2020). Manual therapy versus surgery for carpal tunnel syndrome: 4-year follow-up from a randomized controlled trial. *Physical Therapy*. https://doi.org/10.1093/ptj/pzaa150.

Fernandez-de-las-Penas, C., de la Llave-Rincon, A. I., Fernandez-Carnero, J., Cuadrado, M. L., Arendt-Nielsen, L., & Pareja, J. A. (2009). Bilateral widespread mechanical pain sensitivity in carpal tunnel syndrome: Evidence of central processing in unilateral neuropathy. *Brain, 132*(6), 1472–1479. https://doi.org/10.1093/brain/awp050.

Fernández-de-Las-Peñas, C., De-La-Llave-Rincón, A. I., Cescon, C., Barbero, M., Arias-Buría, J. L., & Falla, D. (2019). Influence of clinical, psychological, and psychophysical variables on long-term treatment outcomes in carpal tunnel syndrome: Evidence from a randomized clinical trial. *Pain Practice: The Official Journal of World Institute of Pain, 19*(6), 644–655.

Fernández-de-las-Peñas, C., Falla, D., Palacios-Ceña, M., De-la-Llave-Rincón, A. I., Schneebeli, A., & Barbero, M. (2019). Perceived pain extent is not associated with physical, psychological, or psychophysical outcomes in women with carpal tunnel syndrome. *Pain Medicine, 20*(6), 1185–1192. https://doi.org/10.1093/pm/pny248.

Gause, T. M., 2nd, Nunnery, J. J., Chhabra, A. B., & Werner, B. C. (2020). Perioperative narcotic use and carpal tunnel release: Trends, risk factors, and complications. *Hand, 15*(2), 234–242.

Gelberman, R. H., Rydevik, B. L., Pess, G. M., Szabo, R. M., & Lundborg, G. (1988). Carpal tunnel syndrome. A scientific basis for clinical care. *Orthopedic Clinics of North America, 19*(1), 115–124.

Graham, B., Peljovich, A. E., Afra, R., Cho, M. S., Gray, R., Stephenson, J., … Sevarino, K. (2016). The American Academy of Orthopaedic Surgeons evidence-based clinical practice guideline on: Management of carpal tunnel syndrome. *Journal of Bone and Joint Surgery. American Volume, 98*(20), 1750–1754.

Harris-Adamson, C., Eisen, E. A., Kapellusch, J., Garg, A., Hegmann, K. T., Thiese, M. S., … Rempel, D. (2015). Biomechanical risk factors for carpal tunnel syndrome: A pooled study of 2474 workers. *Occupational and Environmental Medicine, 72*(1), 33–41.

Harris-Adamson, C., Eisen, E. A., Neophytou, A., Kapellusch, J., Garg, A., Hegmann, K. T., … Rempel, D. (2016). Biomechanical and psychosocial exposures are independent risk factors for carpal tunnel syndrome: Assessment of confounding using causal diagrams. *Occupational and Environmental Medicine*. https://doi.org/10.1136/oemed-2016-103634. p. oemed – 2016.

Huisstede, B. M., van den Brink, J., Randsdorp, M. S., Geelen, S. J., & Koes, B. W. (2018). Effectiveness of surgical and postsurgical interventions for carpal tunnel syndrome—A systematic review. *Archives of Physical Medicine and Rehabilitation, 99*(8), 1660–1680. e21.

Ingram, J., Mauck, B. M., Thompson, N. B., & Calandruccio, J. H. (2018). Cost, value, and patient satisfaction in carpal tunnel surgery. *Orthopedic Clinics of North America, 49*(4), 503–507.

Jerosch-Herold, C., Houghton, J., Blake, J., Shaikh, A., Wilson, E. C., & Shepstone, L. (2017). Association of psychological distress, quality of life and costs with carpal tunnel syndrome severity: A cross-sectional analysis of the PALMS cohort. *BMJ Open, 7*(11), e017732.

Jerosch-Herold, C., Shepstone, L., Houghton, J., Wilson, E. C. F., & Blake, J. (2019). Prognostic factors for response to treatment by corticosteroid injection or surgery in carpal tunnel syndrome (palms study): A prospective multicenter cohort study. *Muscle & Nerve, 60*(1), 32–40.

Jiménez Del Barrio, S., Bueno Gracia, E., Hidalgo García, C., Estébanez de Miguel, E., Tricás Moreno, J. M., Rodríguez Marco, S., & Ceballos Laita, L. (2018). Conservative treatment in patients with mild to moderate carpal tunnel syndrome: A systematic review. *Neurología, 33*(9), 590–601.

Karl, J. W., Gancarczyk, S. M., & Strauch, R. J. (2016). Complications of carpal tunnel release. *Orthopedic Clinics of North America, 47*(2), 425–433.

Keith, M. W., Masear, V., Chung, K. C., Maupin, K., Andary, M., Amadio, P. C., … McGowan, R. (2009). American Academy of Orthopaedic Surgeons clinical practice guideline on diagnosis of carpal tunnel syndrome. *Journal of Bone and Joint Surgery. American Volume, 91*(10), 2478–2479.

Kozak, A., Schedlbauer, G., Wirth, T., Euler, U., Westermann, C., & Nienhaus, A. (2015). Association between work-related biomechanical risk factors and the occurrence of carpal tunnel syndrome: An overview of systematic reviews and a meta-analysis of current research. *BMC Musculoskeletal Disorders, 16*(1). https://doi.org/10.1186/s12891-015-0685-0.

Kuo, T.-T., Lee, M.-R., Liao, Y.-Y., Lee, W.-N., Hsu, Y.-W., Chen, J.-P., & Yeh, C.-K. (2013). Assessment of median nerve mobility by ultrasound dynamic imaging in carpal tunnel syndrome diagnosis. In *2013 IEEE International Ultrasonics Symposium (IUS)*. https://doi.org/10.1109/ultsym.2013.0225.

Levine, D. W., Simmons, B. P., Koris, M. J., Daltroy, L. H., Hohl, G. G., Fossel, A. H., & Katz, J. N. (1993). A self-administered questionnaire for the assessment of severity of symptoms and functional status in carpal tunnel syndrome. *Journal of Bone and Joint Surgery, 75*(11), 1585–1592. https://doi.org/10.2106/00004623-199311000-00002.

Louie, D. L., Earp, B. E., Collins, J. E., Losina, E., Katz, J. N., Black, E. M., … Blazar, P. E. (2013). Outcomes of open carpal tunnel release at a minimum of ten years. *Journal of Bone and Joint Surgery (American Volume), 95*(12), 1067–1073. https://doi.org/10.2106/jbjs.l.00903.

Luckhaupt, S. E., Dahlhamer, J. M., Ward, B. W., Sweeney, M. H., Sestito, J. P., & Calvert, G. M. (2013). Prevalence and work-relatedness of carpal tunnel syndrome in the working population, United States, 2010 National Health Interview Survey. *American Journal of Industrial Medicine*, *56*(6), 615–624.

Ludlow, K. S., Merla, J. L., Cox, J. A., & Hurst, L. N. (1997). Pillar pain as a postoperative complication of carpal tunnel release: A review of the literature. *Journal of Hand Therapy: Official Journal of the American Society of Hand Therapists*, *10*(4), 277–282.

Moradi, A., Mellema, J. J., Oflazoglu, K., Isakov, A., Ring, D., & Vranceanu, A.-M. (2015). The relationship between catastrophic thinking and hand diagram areas. *Journal of Hand Surgery*, *40*(12), 2440–2446. e5.

Mosegaard, S. B., Stilling, M., & Hansen, T. B. (2020). Higher preoperative pain catastrophizing increases the risk of low patient reported satisfaction after carpal tunnel release: A prospective study. *BMC Musculoskeletal Disorders*, *21*(1), 42.

Nazari, G., Shah, N., MacDermid, J. C., & Woodhouse, L. (2017). The impact of sensory, motor and pain impairments on patient-reported and performance based function in carpal tunnel syndrome. *Open Orthopaedics Journal*, *11*, 1258–1267.

Nora, D. B., Becker, J., Ehlers, J. A., & Gomes, I. (2005). What symptoms are truly caused by median nerve compression in carpal tunnel syndrome? *Clinical Neurophysiology: Official Journal of the International Federation of Clinical Neurophysiology*, *116*(2), 275–283.

Nunez, F., Vranceanu, A.-M., & Ring, D. (2010). Determinants of pain in patients with carpal tunnel syndrome. *Clinical Orthopaedics and Related Research*, *468*(12), 3328–3332.

Núñez-Cortés, R., Cruz-Montecinos, C., Antúnez-Riveros, M. A., & Pérez-Alenda, S. (2020). Does the educational level of women influence hand grip and pinch strength in carpal tunnel syndrome? *Medical Hypotheses*, *135*, 109474. https://doi.org/10.1016/j.mehy.2019.109474.

Núñez-Cortés, R., Espinoza-Ordóñez, C., Pommer, P. P., Horment-Lara, G., Pérez-Alenda, S., & Cruz-Montecinos, C. (2019). A single preoperative pain neuroscience education: Is it an effective strategy for patients with carpal tunnel syndrome? *Medical Hypotheses*, *126*, 46–50.

Page, M. J., O'Connor, D., Pitt, V., & Massy-Westropp, N. (2012). Exercise and mobilisation interventions for carpal tunnel syndrome. *Cochrane Database of Systematic Reviews*, *6*, CD009899.

Parish, R., Morgan, C., Burnett, C. A., Baker, B. C., Manning, C., Sisson, S. K., & Shipp, E. R. (2020). Practice patterns in the conservative treatment of carpal tunnel syndrome: Survey results from members of the American Society of Hand Therapy. *Journal of Hand Therapy: Official Journal of the American Society of Hand Therapists*, *33*(3), 346–353.

Park, J. W., Gong, H. S., Rhee, S. H., Kim, J., Lee, Y. H., & Baek, G. H. (2017). The effect of psychological factors on the outcomes of carpal tunnel release: A systematic review. *Journal of Hand Surgery (Asian-Pacific Volume)*, *22*(2), 131–137.

Peters, S., Johnston, V., Hines, S., Ross, M., & Coppieters, M. (2016). Prognostic factors for return-to-work following surgery for carpal tunnel syndrome: A systematic review. *JBI Database of Systematic Reviews and Implementation Reports*, *14*(9), 135–216.

Pourmemari, M.-H., Heliövaara, M., Viikari-Juntura, E., & Shiri, R. (2018). Carpal tunnel release: Lifetime prevalence, annual incidence, and risk factors. *Muscle & Nerve*, *58*(4), 497–502. https://doi.org/10.1002/mus.26145.

Rigouin, P., Ha, C., Bodin, J., Le Manac'h, A. P., Descatha, A., Goldberg, M., & Roquelaure, Y. (2014). Organizational and psychosocial risk factors for carpal tunnel syndrome: A cross-sectional study of French workers. *International Archives of Occupational and Environmental Health*, *87*(2), 147–154.

Rivlin, M., Kachooei, A. R., Wang, M. L., & Ilyas, A. M. (2018). Electrodiagnostic grade and carpal tunnel release outcomes: A prospective analysis. *Journal of Hand Surgery*, *43*(5), 425–431.

Roquelaure, Y., Mariel, J., Dano, C., Fanello, S., & Penneau-Fontbonne, D. (2001). Prevalence, incidence and risk factors of carpal tunnel syndrome in a large footwear factory. *International Journal of Occupational Medicine and Environmental Health*, *14*(4), 357–367.

Sayegh, E. T., & Strauch, R. J. (2015). Open versus endoscopic carpal tunnel release: A meta-analysis of randomized controlled trials. *Clinical Orthopaedics and Related Research*, *473*(3), 1120–1132.

Schmid, A. B., Bland, J. D. P., Bhat, M. A., & Bennett, D. L. H. (2014). The relationship of nerve fibre pathology to sensory function in entrapment neuropathy. *Brain: A Journal of Neurology*, *137*(Pt 12), 3186–3199.

Schmid, A. B., Fundaun, J., & Tampin, B. (2020). Entrapment neuropathies: A contemporary approach to pathophysiology, clinical assessment, and management. *Pain Reports*, *5*(4), e829.

Shi, Q., Bobos, P., Lalone, E. A., Warren, L., & MacDermid, J. C. (2020). Comparison of the short-term and long-term effects of surgery and nonsurgical intervention in treating carpal tunnel syndrome: A systematic review and meta-analysis. *Hand*, *15*(1), 13–22.

Shin, Y. H., Yoon, J. O., Kim, Y. K., & Kim, J. K. (2018). Psychological status is associated with symptom severity in patients with carpal tunnel syndrome. *Journal of Hand Surgery*, *43*(5), 484.e1–484.e8. https://doi.org/10.1016/j.jhsa.2017.10.031.

Uchiyama, S., Itsubo, T., Nakamura, K., Kato, H., Yasutomi, T., & Momose, T. (2010). Current concepts of carpal tunnel syndrome: Pathophysiology, treatment, and evaluation. *Journal of Orthopaedic Science*, *15*(1), 1–13. https://doi.org/10.1007/s00776-009-1416-x.

Uddin, Z., & MacDermid, J. C. (2016). Quantitative sensory testing in chronic musculoskeletal pain. *Pain Medicine*, *17*(9), 1694–1703. https://doi.org/10.1093/pm/pnv105.

Urits, I., Gress, K., Charipova, K., Orhurhu, V., Kaye, A. D., & Viswanath, O. (2019). Recent advances in the understanding and management of carpal tunnel syndrome: A comprehensive review. *Current Pain and Headache Reports*, *23*(10), 70.

Vranceanu, A.-M., Jupiter, J. B., Mudgal, C. S., & Ring, D. (2010). Predictors of pain intensity and disability after minor hand surgery. *Journal of Hand Surgery*, *35*(6), 956–960.

Westenberg, R. F., Oflazoglu, K., de Planque, C. A., Jupiter, J. B., Eberlin, K. R., & Chen, N. C. (2020). Revision carpal tunnel release: Risk factors and rate of secondary surgery. *Plastic and Reconstructive Surgery*, *145*(5), 1204–1214.

Wilkens, S. C., Lans, J., Bargon, C. A., Ring, D., & Chen, N. C. (2018). Hand posturing is a nonverbal indicator of catastrophic thinking for finger, hand, or wrist injury. *Clinical Orthopaedics and Related Research*, *476*(4), 706–713.

Zanette, G., Cacciatori, C., & Tamburin, S. (2010). Central sensitization in carpal tunnel syndrome with extraterritorial spread of sensory symptoms. *Pain*, *148*(2), 227–236.

Chapter 26

Pain and HIV

Sara Pullen

Department of Rehabilitation Medicine, Emory University School of Medicine, Atlanta, GA, United States

Abbreviations

AIDS	acquired immune deficiency syndrome
ART	antiretroviral therapy
HIV	human immunodeficiency virus
HIV-DSP	HIV-associated distal symmetric polyneuropathy
PT	physical therapy
PWH	people living with HIV

Introduction

The burden of chronic pain—defined as pain lasting over 3 months—is reported by an alarmingly high number of people living with HIV (PWH) in the United States, affecting up to 85% of PWH compared to only 11% of the general population (Bruce et al., 2017; Cunningham, 2018). Chronic pain has emerged as a treatment priority for this patient population, and has been associated with profound psychological, physical, and functional morbidity, decreased adherence to antiretroviral therapy (ART), and reduced retention in HIV primary care (Merlin et al., 2015, 2019). In addition, PWH are more likely to receive opioid analgesic therapy and receive *higher doses* of opioids as well as experience substance use disorders compared with the general population, positioning them at amplified risk for opioid use disorder (Cunningham, 2018). Recent studies reveal that even when prescribed opioids, PWH continue to experience higher levels of pain and associated comorbidities (Cunningham, 2018). This suggests that current strategies to manage chronic pain among PWH are not effective, and that there is a gap in both clinical practice and research exploring nonpharmacological, feasible, effective strategies to mitigate chronic pain in this population.

Despite the extensive increase in life expectancy due to use of highly active antiretroviral therapy (ART), the prevalence of chronic pain in PWH remains high even in those who are virologically stable and whose CD4 counts remain high (Addis, DeBerry, & Aggarwal, 2020). Although pain is widespread and burdensome for many PWH, the specific mechanisms that undergird it remain poorly understood which results in insufficient clinical approaches to pain mitigation and hinders effective treatment and development of effective approaches to decrease suffering and chronic pain sequelae. The past decades have shown increased awareness of pain as a significant problem for PWH, including the creation of the International Task Force on Pain and AIDS (1994) and the HIV Medical Association of Infectious Diseases Society of America's Clinical Practice Guideline for the Management of Chronic Pain in Patients Living With HIV (2017). Despite this, the prevalence of HIV-related chronic pain has only increased since the 1980s, highlighting the fact that life-saving ART and current pain mitigation strategies are insufficient to address the multidimensional burden of pain among PWH (Addis et al., 2020).

This chapter will provide an overview of the etiology of HIV-related pain, current pain trends among PWH, treatment guidelines, and suggestions for how to best fulfill the needs of PWH to effectively diagnose and decrease pain, prevent opioid initiation and subsequent risks, and improve quality of life for this unique patient population.

Etiology of chronic pain in HIV

HIV-related chronic pain can include single or multisite pain which is musculoskeletal, neuropathic, or inflammatory in nature and can affect PWH at any stage of illness (Krashin, Merrill, & Trescot, 2012; Parker, Stein, & Jelsma, 2014).

Features and Assessments of Pain, Anesthesia, and Analgesia. https://doi.org/10.1016/B978-0-12-818988-7.00050-9
Copyright © 2022 Elsevier Inc. All rights reserved.

Pain may be related to the effects of the virus on the body, antiretroviral therapy, opportunistic infections, or a combination of the three (Parker et al., 2014). The complex, multifactorial etiology of the chronic pain that causes significant burden in this patient population highlights both the importance of and the challenges to effective treatment. While the life-saving necessity of ART cannot be understated, some of the associated are widely associated with side effects including painful peripheral neuropathy and other pain syndromes (Winias, Radithia, & Savitri Ernawati, 2020). Consequently, PWH and their providers face the fact that the necessity of ART regimens may directly correlate with significant pain and subsequent distress, impaired function, and decreased quality of life (Pullen, Acker, et al., 2020; Pullen, del Rio, et al., 2020).

Chronic pain among PWH has been attributed to virus and ART drug–induced sensory neuropathy (ART-SN) as well as chronic nonneuropathic inflammatory processes (Merlin et al., 2017). While PWH may certainly present with the same pain syndromes present in the general population stemming from various musculoskeletal, neuropathic, or trauma-related events, we will focus the remainder of this section on HIV-DSP given its prevalence and burden in this patient population.

HIV-associated distal symmetric polyneuropathy

HIV neuropathies can manifest in various ways. Because HIV can affect peripheral sensory nerves, motor nerves, thoracic nerves, cranial nerves, or autonomic nerves, it remains a challenge to treat the resulting pain manifestations effectively (Dudley, Borkum, Basera, Wearne, & Heckmann, 2019). HIV-related nerve disorders include HIV mononeuropathy (affecting only one nerve) or inflammatory neuropathy similar to Guillain-Barre syndrome. The most common nerve-related manifestation of HIV and ART is HIV-related distal symmetric polyneuropathy (HIV-DSP), which affects an estimated 40% of PWH treated with ART (Winias, 2020).

HIV-DSP has become the most common neurologic complication of HIV, which can be largely attributed to the increased lifespan of PWH due to ART. HIV-DSP can be ascribed to the HIV disease process itself *or* to secondary effects of certain ART regimens, and often the two differential etiologies cannot be differentiated (Kaku & Simpson, 2014). Regardless of its origin, HIV-DSP can affect multiple sensory and motor nerves in the distal extremities and is widely associated with chronic, widespread pain syndromes (Winias, 2020). With HIV-DSP, patients may experience painful paresthesias and/or numbness and pain in hands and feet, causing inability to perform essential activities of daily living, work, household, and family duties.

Diagnosis of HIV neuropathies is based on history, clinical examination, and supporting laboratory investigations and can include nerve conduction studies, skin biopsies to evaluate cutaneous nerve involvement, and nerve and muscle biopsies for histopathological evaluation (Kaku & Simpson, 2014).

Treatment of HIV neuropathies depends on the type, individual patient circumstances, and comorbid conditions. Positive patient outcomes of treatment of HIV-DSP require good control of HIV infection, which requires the patient to be therapeutic on their prescribed ART regimen. Pain due to HIV polyneuropathy can be treated with antiseizure medications, antidepressants, or analgesics including opiate drugs. Despite the high prevalence of painful HIV-SN among PWH, evidence for effectively managing it is poor (Centner et al., 2018).

Research and anecdotal patient accounts have shown that the experience of pain is not always exclusively related to tissue damage (Barrett & Chang, 2016; Institute of Medicine of the National Academy of Science, 2011). Social stressors including experiences of discrimination and social instability (such as unemployment, homelessness, and/or food insecurity) can lead to amplification of the perception of tissue-based pain (Buscemi, Chang, Liston, McAuley, & Schabrun, 2019; Maly & Vallerand, 2018). These issues should be carefully considered when diagnosing and managing pain among PWH. On a related note, patients' individual circumstances should be carefully weighed when ascribing a pain management program. Patients' clinic accessibility to the clinic, transportation, child care and housing issues, and other psychosocial factors can drastically impact the patient's ability to adhere to a pain management plan.

Current pain management

While current pain management for PWH varies depending on pain etiology, patient history, and other factors, it is key to consider pharmacological vs nonpharmacological methods used in this patient population.

HIV, pain, and opioid use

More than half of the chronic pain diagnoses have been recorded before SUD diagnoses. Previous studies have reported that most opiate use disorder individuals developed the disorder after chronic pain (Hser et al., 2017) and may exacerbate the desire for opiate-based relief. **People with HIV (PWH) are at especially high risk for developing opiate use disorder**

(OUD) because PWH disproportionately present with the following risk factors identified for OUD (Kaye, Jones, Kaye, et al., 2017): (1) **chronic pain** (up to 85% of PWH report living with unmanaged HIV-related chronic pain) (Bruce et al., 2017); (2) **a history of social stressors** (it is common for PWH, particularly PWH of color, to experience discrimination and social instability leading to anxiety, depression, and/or posttraumatic stress disorder) (Bengtson, Pence, Powers, et al., 2018; Hassan, Le Foll, Imtiaz, & Rehm, 2017; Livingston, Flentje, Heck, Szalda-Petree, & Cochran, 2017; Scherer et al., 2019); and (3) **a current opioid prescription** (stemming from increasingly ineffective efforts to maintain control over HIV-related physical pain, close to one-third of PWH on opioids transition from appropriate opioid use to OUD after as little as 5 days of using the prescription drug) (Baker, 2017; Dowell, Haegerich, & Chou, 2016; Martell et al., 2007; Mosher, Hofmeyer, Hadlandsmyth, Richardson, & Lund, 2018; Vowles et al., 2015) (Fig. 1).

Nonpharmacologic approaches to HIV-related pain

The adverse consequences—including widespread addiction and overdose—of opioid use have resulted in a call for nonpharmacological treatment of chronic pain. Other commonly prescribed, nonopioid forms of pain control also carry risks when considered alongside of some common HIV-comorbidities. **Nonsteroidal antiinflammatory drugs (NSAIDS)** are contraindicated in patients with preexisting renal disease. Up to 30% of people with HIV have abnormal kidney function, and side effects of some HIV medications can cause kidney disease (Risser, Donovan, Heintzman, & Page, 2009). Given the prevalence of abnormal kidney function among PWH along with the recommendation that NSAIDS should be avoided in persons with preexisting renal disease, NSAIDS are not an ideal method of pain control for many people with HIV (Krashin et al., 2012). **Acetaminophen**, another common pain analgesic, is contraindicated for individuals with abnormal liver function. PWH may be at particular risk for acetaminophen-induced hepatotoxicity due both to HIV status alone *and* given the high incidence of coinfection of hepatitis C and HIV. Given the significant risk of liver damage with acetaminophen among PWH, research shows that acetaminophen exposure should be minimized in this patient population (Edelman et al., 2013).

Nonpharmacological approaches to chronic pain management have included aerobic and resistive exercise, acupuncture, and vibratory stimulus (Amaniti et al., 2019), and the HIVMA recommends the following for nonpharmacological treatments: cognitive behavioral therapy, yoga, physical and occupational therapy, and acupuncture. Nonopioid pharmacological recommendations by the IDSA include early initiation of ART to prevent and treat HIV-DSP, gabapentin as a first-line oral pharmacologic, capsaicin as a topical NSAID treatment, alpha lipoic acid, and medical cannabis (Bruce et al., 2017).

A preliminary body of research suggests that when chronic pain among PWH is managed through skilled physical therapy (PT), PWH have reported decreased pain levels and analgesic use as well as increased functional independence and quality of life (Pullen, 2017, 2019). The use of PT in health care settings has an established record of providing and maintaining nonpharmacologic, physiological pain mitigation, and functional recovery (Hylands-White, 2017) and promising research has shown that PT can be effective in significantly reducing or stopping opioid misuse among PWH by eliminating the chronic pain that undergirds a patient's perceived need for escalated or ongoing opioid use (Baker, 2017; Martell et al., 2007; Pullen, 2019; Pullen, Acker, et al., 2020; Pullen, del Rio, et al., 2020).

Pain can affect PWH at any point in their illness. Given the various comorbidities and contraindications to be taken into account when managing pain in PWH and the widespread availability of PT, PT should be considered as a nonpharmacological, safe, and potentially helpful intervention for pain management among PWH.

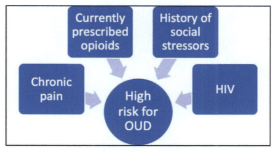

FIG. 1 Opioid risk among people living with HIV. *(From Pullen, S., del Rio, C., Marconi, V. C., Head, C., Nimmo, M., O'Neil, J., et al. (2021). From Silos to Solidarity: Feasibility of patient-centered, integrative approach to opioid tapering and chronic pain mitigation in a multidisciplinary AIDS Clinic. Journal of AIDS and HIV Treatment, 3, 4–11. Unpublished manuscript, Under review, with permission.)*

For PWH who screen positive for chronic pain, a multidisciplinary, biopsychosocial approach should be utilized to cover the complex underlying reasons present. This should—at a minimum—include taking a "pain history": an evaluation of the pain's onset and duration, intensity and character, exacerbating and alleviating factors, past and current treatments, underlying or cooccurring disorders and conditions, and the effect of pain on physical and psychological function. This should be followed by a physical examination, psychosocial evaluation, and diagnostic workup to determine the potential cause of the pain (Bruce et al., 2017).

Future steps

As people live longer with HIV, they will continue to present with both HIV and age-related comorbidities. Chronic pain has emerged as a common clinical presentation and burden for PWH, and it is essential to address this pain from an integrated, multidisciplinary team approach where the complex layers of pain can be addressed. Clinicians should seek a comprehensive understanding of patients' pain experience, including patient medical and social history, current and past substance use, and patient preferences as this may alter management strategies. Current research and clinical guidelines recommend a nonpharmacological approach as the first line of chronic pain management, as this can mitigate the risk of opioid initiation and its subsequent risks for addiction and overdose (IDSA, 2017; Pullen et al., 2021).

It is essential that this team has the patient at the center, so that the patient may feel empowered as an equal decision-maker about their pain management (Pullen et al., 2021). The team should include the primary HIV provider (or primary medical provider if patient's HIV is managed by a generalist), physical therapist or other pain specialist, and mental health provider. Given the psychological burden and, often, psychosocial origin of chronic pain, a mental health component is ideal as part of the pain management team to assist the patient with understanding underlying psychological aspects of chronic pain as well as provide positive pain coping mechanisms. This type of patient-centered collaboration can best address the complexities of chronic pain and provide a clear avenue of recovery for this unique patient population (Fig. 2).

Applications to other areas

In this chapter, we have reviewed the etiology, current management strategies, and suggestions for an integrated approach for PWH with chronic pain syndromes. It is important to highlight how patients presenting with pain must be screened of a wide variety of issues, necessitating the involvement and consultation of a multidisciplinary team of providers. This integrated approach calls for other areas aside from infectious disease and primary care to be involved in pain mitigation, with the patient as an equal member of the care team.

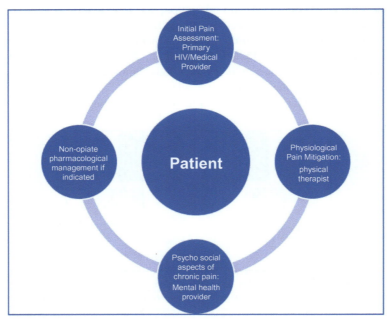

FIG. 2 Integrative model for chronic pain. *(Original work of the author for this publication.)*

Mini-dictionary of terms

Chronic pain: Pain lasting more than 3 months that causes a significant burden and impairs patients' daily function.
HIV-related sensory neuropathy: The most common neurological manifestation of HIV, causing burdensome pain syndromes and paresthesias.
Opioids: Narcotic analgesics historically used for chronic pain, that carry a high risk of addiction and overdose.
Integrative medicine: Clinical approach that incorporates the care of the whole person, including physiological, psychosocial, and individual patient factors. It emphasizes the therapeutic relationship between practitioner and the patient and integrates various disciplines together with the patient as part of the care team.

Key facts

Key facts of HIV and pain

- Chronic pain affects an estimated 85% of people living with HIV (PWH) compared to 11% of the general population.
- PWH frequently present clinically with chronic pain.
- Due to their myriad of risks, opioids should not be used as a first-line treatment against chronic pain.
- Nonpharmacological treatment such as physical therapy has been shown to be effective for chronic pain in PWH.
- Increased life expectancy for PWH necessitates an integrative approach to HIV-related comorbidities.

Summary points

- People living with HIV (PWH) are disproportionately affected by chronic pain.
- Chronic pain remains poorly understood in this patient population.
- Both the HIV disease process and antiretroviral therapy can contribute to the development of HIV distal sensory polyneuropathy.
- To fully address the complexities of chronic pain in this patient population, an integrated, multidisciplinary approach must be utilized.
- As PLH's life expectancies have increased, we can expect to see increased clinical presentations of chronic pain.
- Physical therapy is a safe, effective means of chronic pain mitigation and should be considered as part of the first line of chronic pain treatment.

References

Addis, D. R., DeBerry, J. J., & Aggarwal, S. (2020). Chronic pain in HIV. *Molecular Pain, 16*. https://doi.org/10.1177/1744806920927276, 1744806920927276. 32450765. PMC7252379.

Amaniti, A., Sardeli, C., Fyntanidou, V., Papakonstantinou, P., Dalakakis, I., Mylonas, A., et al. (2019). Pharmacologic and non-pharmacologic interventions for HIV-neuropathy pain. A systematic review and a meta-analysis. *Medicina (Kaunas), 55*(12), 762. https://doi.org/10.3390/medicina55120762.

Baker, D. W. (2017). History of The Joint Commission's pain standards: Lessons for today's prescription opioid epidemic. *Journal of the American Medical Association, 317*(11), 1117–1118.

Barrett, K., & Chang, Y. P. (2016). Behavioral interventions targeting chronic pain, depression, and substance use disorder in primary care. *Journal of Nursing Scholarship, 48*(4), 345–353.

Bengtson, A. M., Pence, B. W., Powers, K. A., et al. (2018). Trajectories of depressive symptoms among a population of HIV-infected men and women in routine HIV care in the United States. *AIDS and Behavior, 22*(10), 3176–3187.

Bruce, D. R., Merlin, J., Lum, P. J., Ahmed, E., Alexander, C., Corbett, A. H., et al. (2017). 2017 HIVMA of IDSA clinical practice guideline for the management of chronic pain in patients living with HIV. *Clinical Infectious Diseases, 65*(10), e1–e37.

Buscemi, V., Chang, W. J., Liston, M. B., McAuley, J. H., & Schabrun, S. M. (2019). The role of perceived stress and life stressors in the development of chronic musculoskeletal pain disorders: A systematic review. *Journal of Pain, 20*(10), 1127–1139.

Centner, C. M., Little, F., Van Der Watt, J. J., Vermaak, J. R., Dave, J. A., Levitt, N. S., et al. (2018). Evolution of sensory neuropathy after initiation of antiretroviral therapy. *Muscle Nerve, 57*(3), 371–379. https://doi.org/10.1002/mus.25710.

Cunningham, C. O. (2018). Opioids and HIV infection: From pain management to addiction treatment. *Topics in Antiviral Medicine, 25*(4), 143–146.

Dowell, D., Haegerich, T. M., & Chou, R. (2016). CDC guideline for prescribing opioids for chronic pain—United States, 2016. *Journal of the American Medical Association, 315*(15), 1624–1645.

Dudley, M. T., Borkum, M., Basera, W., Wearne, N., & Heckmann, J. M. (2019). Peripheral neuropathy in HIV patients on antiretroviral therapy: Does it impact function? *Journal of the Neurological Sciences, 406*, 116451. https://doi.org/10.1016/j.jns.2019.116451.

Edelman, E. J., et al. (2013). Acetaminophen receipt among HIV-infected patients with advanced hepatic fibrosis. *Pharmacoepidemiology and Drug Safety, 22*(12), 1352–1356.

Hassan, A. N., Le Foll, B., Imtiaz, S., & Rehm, J. (2017). The effect of post-traumatic stress disorder on the risk of developing prescription opioid use disorder: Results from the National Epidemiologic Survey on Alcohol and Related Conditions III. *Drug and Alcohol Dependence, 179*, 260–266.

Hser, Y. I., Mooney, L. J., Saxon, A. J., Miotto, K., Bell, D. S., & Huang, D. (2017). Chronic pain among patients with opioid use disorder: Results from electronic health records data. *Journal of Substance Abuse Treatment, 77*, 26–30. https://doi.org/10.1016/j.jsat.2017.03.006.

Hylands-White, N. (2017). An overview of treatment approaches for chronic pain management. *Rheumatology International, 37*(1), 29–42.

Institute of Medicine of the National Academy of Science. (2011). *Relieving pain in America: A blueprint for transforming prevention, care, education, and research* (p. 5). Washington, DC: Institute of Medicine.

Kaku, M., & Simpson, D. M. (2014). HIV neuropathy. *Current Opinion in HIV and AIDS, 9*(6), 521–526. https://doi.org/10.1097/COH.0000000000000103.

Kaye, A. D., Jones, M. R., Kaye, A. M., et al. (2017). Prescription opioid abuse in chronic pain: An updated review of opioid abuse predictors and strategies to curb opioid abuse: Part 1. *Pain Physician, 20*(2S), S93–S109.

Krashin, D. L., Merrill, J. O., & Trescot, A. M. (2012). Opioids in the management of HIV-related pain. *Pain Physician, 15*(3 Suppl.), ES157–ES168.

Livingston, N. A., Flentje, A., Heck, N. C., Szalda-Petree, A., & Cochran, B. N. (2017). Ecological momentary assessment of daily discrimination experiences and nicotine, alcohol, and drug use among sexual and gender minority individuals. *Journal of Consulting and Clinical Psychology, 85*(12), 1131–1143.

Maly, A., & Vallerand, A. H. (2018). Neighborhood, socioeconomic, and racial influence on chronic pain. *Pain Management Nursing, 19*(1), 14–22.

Martell, B. A., O'Connor, P. G., Kerns, R. D., Becker, W. C., Morales, K. H., Kosten, T. R., et al. (2007). Systematic review: Opioid treatment for chronic back pain: Prevalence, efficacy, and association with addition. *Annals of Internal Medicine, 146*(2), 116–127.

Merlin, J. S., Westfall, A. O., Heath, S. L., Goodin, B. R., Stewart, J. C., Sorge, R. E., et al. (2017). Brief report: IL-1β levels are associated with chronic multisite pain in people living with HIV. *Journal of Acquired Immune Deficiency Syndromes, 75*(4), e99–e103.

Merlin, J. S., et al. (2015). Pain self-management in HIV-infected individuals with chronic pain: A qualitative study. *Pain Medicine, 16*(4), 706–714.

Merlin, J. S., et al. (2019). The association of chronic pain and long-term opioid therapy with HIV treatment outcomes. *Journal of Acquired Immune Deficiency Syndromes, 79*(1), 77–82.

Mosher, H. J., Hofmeyer, B. A., Hadlandsmyth, K., Richardson, K. K., & Lund, B. C. (2018). Predictors of long-term opioid use after opioid initiation at discharge from medical and surgical hospitalizations. *Journal of Hospital Medicine, 13*(4), 243–248.

Parker, R., Stein, D. J., & Jelsma, J. (2014). Pain in people living with HIV/AIDS: A systematic review. *Journal of the International AIDS Society, 17*, 18719.

Pullen, S. (2017). Physical therapy as non-pharmacological chronic pain management of adults living with HIV: Self-reported pain scores and analgesic use. *HIV AIDS (Auckl), 9*, 177–182.

Pullen, S. (2019). Chronic pain mitigation and opioid weaning at a multidisciplinary AIDS clinic: A case report. *Rehabilitation Oncology, 37*(1), 37–42.

Pullen, S. D., Acker, C., Kim, H., Mullins, M., Sims, P., Strasbaugh, H., et al. (2020). Physical therapy for chronic pain mitigation and opioid use reduction among people living with human immunodeficiency virus in Atlanta, GA: A descriptive case series. *AIDS Research and Human Retroviruses, 36*(8), 670–675.

Pullen, S., del Rio, C., Brandon, D., Colonna, A., Denton, M., Ina, M., et al. (2020). Associations between chronic pain, analgesic use and physical therapy among adults living with HIV in Atlanta, GA: A retrospective cohort study. *AIDS Care, 32*(1), 65–71.

Pullen, S., del Rio, C., Marconi, V. C., Head, C., Nimmo, M., O'Neil, J., et al. (2021). From Silos to Solidarity: A patient-centered, integrative approach to opioid tapering and chronic pain mitigation in a multidisciplinary AIDS Clinic. *Journal of AIDS and HIV Treatment, 3*(1), 4–11. Unpublished manuscript, Under review.

Risser, A., Donovan, D., Heintzman, J., & Page, T. (2009). NSAID prescribing precautions. *American Family Physician, 80*(12), 1371–1378.

Scherer, M., Weiss, L., Kamler, A., George, M. C., Navis, A., Gebhardt, Y., et al. (2019). Patient recommendations for opioid prescribing in the context of HIV Care: Findings from a set of public deliberations. *AIDS Care, 23*, 1–8.

Vowles, K. E., McEntee, M. L., Julnes, P. S., Frohe, T., Ney, J. P., & van der Goes, D. N. (2015). Rates of opioid misuse, abuse, and addiction in chronic pain: A systematic review and data synthesis. *Pain, 156*(4), 569–576.

Winias, S., Radithia, D., & Savitri Ernawati, D. (2020). Neuropathy complication of antiretroviral therapy in HIV/AIDS patients. *Oral Diseases, 26*(Suppl. 1), 149–152. https://doi.org/10.1111/odi.13398.

Chapter 27

Pain mechanisms in computer and smartphone users

Alberto Marcos Heredia-Rizo[a], Pascal Madeleine[b], and Grace P.Y. Szeto[c]

[a]*Department of Physiotherapy, Faculty of Nursing, Physiotherapy and Podiatry, University of Seville, Seville, Spain,* [b]*Sport Sciences—Performance and Technology, Department of Health Science and Technology, Aalborg University, Aalborg, Denmark,* [c]*School of Medical and Health Sciences, Tung Wah College, Hong Kong, China*

Abbreviations

LBP   low-back pain
MSD   musculoskeletal disorder
NP   neck pain
NSP   neck-shoulder pain
VDU   visual display unit

Introduction

Over the last centuries, the nature of work has influenced the health condition of individuals. Musculoskeletal disorders (MSDs) represent one of the most common occupational problems worldwide (Hoe, Urquhart, Kelsall, Zamri, & Sim, 2018). The number of office workers has increased in the last decades, with approximately 57% of occupations requiring the use of visual display units (VDUs) (Parent-Thirion et al., 2017). Additionally, the trend is for people to spend longer periods of time with computers and smartphones, both at work and during leisure activities. The commonly used electronic devices include laptops and tablet computers in addition to desktop computers.

Compared with other jobs, white-collar workers report the highest incidence (35%–60%) and annual prevalence (40%–70%) of neck-shoulder pain (NSP) (Hoe et al., 2018). Low-back pain (LBP) is also a frequent and disabling condition within this population, with a 1-year prevalence of 35%–40%, and an incidence rate at 14%–23% (Janwantanakul, Sihawong, Sitthipornvorakul, & Paksaichol, 2018). Most evidence indicates a higher prevalence in women, compared with men.

This poses a serious burden to society due to reduced productivity (presenteeism), absenteeism from work, and use of health care resources (Bevan, 2015) (Fig. 1). In 2015, the total costs of work-related MSDs represented up to 2% of the gross domestic product in the European Union (approx. €240 billion annually), and involved 40 million workers (Bevan, 2015). Concomitant NSP and LBP increase the risk for long-term sickness absence by 75% in white-collar workers (Andersen, Mortensen, Hansen, & Burr, 2011), and 10% of them report reduced productivity due to occupational MSDs (Hagberg, Vilhemsson, Tornqvist, & Toomingas, 2007). The financial burden of musculoskeletal pain among computer users is estimated on average $AUD 1520 (€ 895) annually per person (Pereira et al., 2017). Job demands, level of satisfaction, and psychological distress are the causes that mostly influence productivity loss (Hagberg et al., 2007).

Origins and mechanisms of neck-shoulder pain

As pointed out earlier, NSP is common in computer and smartphones users. The cervical spine and the shoulder joint are complex structures having a postural role, i.e., stabilization of the head and the shoulder girdle, and a functional role in the control of head and arm movements (Madeleine, 2010), necessary when using VDUs.

The origin of NSP in MSDs is located in the deep structures around cartilage, tendons, ligaments, and muscles. Musculoskeletal pain is mediated by nociceptive-free nerve endings (Mense, 1993). Muscle nociceptors are located in the connective tissue, linked to primary afferent neurons, and are mostly polymodal, meaning that they can respond to mechanical

---

Features and Assessments of Pain, Anesthesia, and Analgesia. https://doi.org/10.1016/B978-0-12-818988-7.00021-2
Copyright © 2022 Elsevier Inc. All rights reserved.

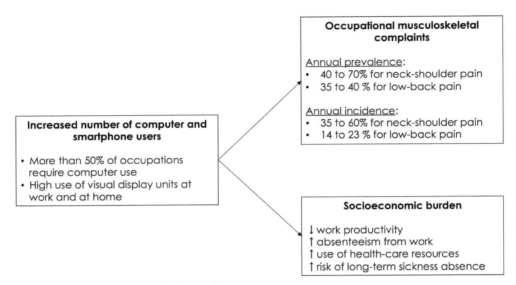

FIG. 1 Burden of occupational neck-shoulder and low-back pain.

and chemical stimuli. Biochemical changes, including altered oxidative metabolism and intramuscular microcirculation (Larsson, Björk, Kadi, Lindman, & Gerdle, 2004), and a higher accumulation of nociceptive substances after sustained and increased muscle load or activity (Rosendal et al., 2004), are reported in the presence of work-related trapezius myalgia. Such changes appear sufficient to activate peripheral nociception. The processing of nociceptive inputs via to the dorsal horn is only partially known. Wide dynamic range neurones located in the spinal cord respond to a large range of noxious and nonnoxious stimuli (Wagman & Price, 1969). A general central inhibition of the γ-motor neurones on agonist and antagonist muscles have been demonstrated in occupational pain syndromes (Johansson & Sojka, 1991).

Peripheral and centrally mediated mechanisms play important roles in muscle pain development (Graven-Nielsen, 2006). Recurrent acute pain episodes may induce sensitization and lead to chronic work-related pain. However, chronic NSP among computer and smartphones users is still a matter of debate due to a lack of clear etiology of MSDs in general (Madeleine, 2008). Further, the relationship between occupational exposure and neck-shoulder disorders remains uncertain (Madeleine, Vangsgaard, Hviid Andersen, Ge, & Arendt-Nielsen, 2013). Overall, the excitability of the pain system is reported to be normal in computer users with low pain intensity. Only those with higher pain intensity and lower pressure pain thresholds have shown decreased efficiency in descending pain modulation (Ge, Vangsgaard, Omland, Madeleine, & Arendt-Nielsen, 2014; Heredia-Rizo, Petersen, Madeleine, & Arendt-Nielsen, 2019). These findings support the idea that chronic NSP can only be partly explained by an increased excitability of the pain system.

Low-back pain mechanisms in computer and smartphone users

Similar to NSP, our understanding of the pain mechanisms related to LBP is limited. Episodes of LBP result from a mix of personal and work-related physical or psychosocial exposure. Apparently, several behavioral patterns may predispose office workers to a higher prevalence of LBP. There is evidence of signs of central sensitization in nonspecific LBP (Den Bandt, Paulis, Beckwée, Ickmans, & Nijs, 2019). Further, altered pronociceptive and antinociceptive mechanisms, i.e., temporal summation of pain and conditioned pain modulation, are present in LBP patients and related to pain severity (McPhee, Vaegter, & Graven-Nielsen, 2020), although scarce research has focused on office workers. All in all, LBP results in adaptative changes impacting the working life of computer and smartphone users.

Office, customer service and call center employees spend an average of 75% of their time seated (Thorp et al., 2012). This has been traditionally linked with increased low-back discomfort, reduced mobility, and with LBP onset. Yet, current evidence on the latter issue is controversial (Bontrup et al., 2019). Although too much sitting is a health concern, the transition to sit/stand workstation resulting in prolonged occupational standing has also been associated with the occurrence of LBP (Coenen et al., 2018). Computer users with nonspecific LBP demonstrate lower flexor and extensor trunk muscles endurance (del Pozo-Cruz et al., 2013), a more asymmetrical seated position, and less frequent postural shifts, compared with healthy workers (Akkarakittichoke & Janwantanakul, 2017). During sustained sitting, people with LBP show reduced temporal variability in the activation of the low-back muscles (Ringheim, Indahl, & Roeleveld, 2019), which could be a risk factor for intervertebral disk changes. Several mechanisms may contribute to the altered sensitivity, modified neuromuscular behavior, and LBP development associated with prolonged sitting. For example, previous LBP episodes, the frequency of breaks at work, an increased low-back muscle activity, and the soft tissue deformation induced by keeping a flexed spine position during long periods of time (Greene et al., 2019). However, no definite conclusions can be drawn due to the high variability of LBP presentations and modifying factors. Future research should investigate pain mechanisms and adaptations in subgroups of workers with LBP of similar clinical features.

Individual risks factors

Among the individual risk factors that affect the development of MSDs in computer and smartphone users, the most commonly examined may be sex/gender and age. A study on over 300 female computer users concluded that the individual factors linked with poorer neck disability were older age and higher body mass index (Johnston, Souvlis, Jimmieson, & Jull, 2008). Later, in a cohort study on office workers in Sudan, age and sex were considered confounding aspects for physical, i.e., workstation setting, and posture; as well as psychological risk factors, i.e., job strain and time pressure (Eltayeb, Staal, Khamis, & De Bie, 2011). In short, age, sex, and history of trauma are significantly associated with NSP onset in users of VDUs (Keown & Tuchin, 2018). A history of LBP is a strong predictor of the onset and chronification of LBP in this population, with inconclusive evidence for the role of age (Ye, Jing, Wei, & Lu, 2017). Individual physical capacity, expressed by muscle strength, endurance, and flexibility, has also been correlated with the occurrence of neck and low-back symptoms in office workers (Cabral, de Moreira, de Barros, & de Sato, 2019).

Sex differences in MSDs have been demonstrated in white-collar employees. For example, Cho, Hwang, and Cherng (2012) observed a higher prevalence of neck symptoms in males, compared with females, whereas females showed higher rate of shoulder compared to neck pain (NP), although these findings are not consistent among studies. Similarly, women report higher NSP intensity, longer pain duration, and more pain locations, compared with men (Madeleine et al., 2013). Laboratory studies suggest that women tend to use higher forces relative to their maximum capacity to type on keyboards, which is related to their anthropometrics (Won, Johnson, Punnett, & Dennerlein, 2009).

In the general population, systematic reviews have concurred that female sex, and older age (for men) are important risk factors for new onset of NP (Mc Lean et al., 2010), with no gender differences for LBP prevalence (Hoy, Brooks, Blyth, & Buchbinder, 2010). Interestingly, women report lower work-ability than men (Madeleine et al., 2013), underlining the complexity of integrating only age or sex/gender as individual factors when studying the impact of NSP and LBP in computer and smartphone users.

Physical risks factors for the development of pain

Workplace physical risk factors concerning computer workstation have been broadly investigated, including computer, keyboard and mouse use duration, and the workstation design, e.g., monitor, keyboard, and mouse location, chair height, sit/stand workstation, and arm support. Prolonged sedentary hours and worker posture are also considered important risk factors.

Jun, Zoe, Johnston, and O'Leary (2017) presented the physical risk factors for computer-related NSP among office workers and concluded that keyboard position and low work task variation were predictors of NSP onset. The risk estimates for MSDs appear to be stronger for mouse use compared to keyboard and total computer use. In fact, increased risk of NSP has been associated with over 3–4 h of self-reported mouse use per day (Ijmker et al., 2011), with inconclusive evidence on this issue (Jun et al., 2017). The role of the speed and force of repetitive movements in performing keyboard typing and mouse work were investigated in a series of laboratory studies by Szeto, Straker, and O'Sullivan (2005a, 2005b, 2005c). These studies found that computer and smartphone users with NSP have increased neck muscle activity during functional tasks (Szeto et al., 2005a; Xie, Szeto, Dai, & Madeleine, 2016) (Fig. 2). There is general agreement that providing arm support during keyboard work is important (Rempel et al., 2006). It is also recommended keeping the keyboard in front and maintaining the forearms and wrist joints in neutral positions to reduce the risk for neck-shoulder-arm symptoms (Jun et al., 2017).

In terms of the display screen position, angled computer location (monitor not in front) is associated with NSP and LBP in office workers (Szeto & Sham, 2008; Ye et al., 2017). Research has generally found that increased neck flexion angles due to a low screen position produces higher muscle activity to maintain the posture (Xie, Madeleine, Tsang, & Szeto, 2017) (Fig. 3). This is suggested to be a risk factor for NP development. For female office workers, the combination of poor working postures and job strain could predispose to LBP (Janwantanakul, Sitthipornvorakul, & Paksaichol, 2012). Prolonged standing computer work due to the use of sit/stand workstation can also result in LBP (Coenen et al., 2018).

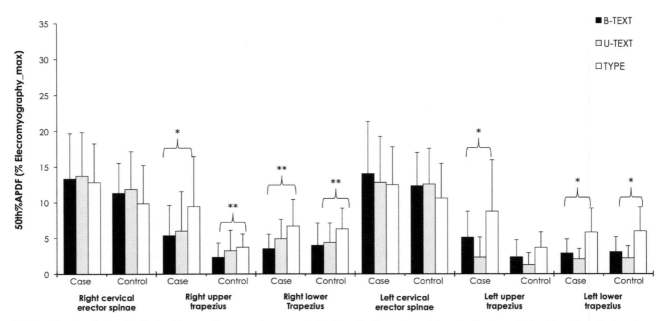

FIG. 2 Muscle activity of neck-shoulder muscles comparing computer keyboard typing and smartphone texting in neck-shoulder pain patients and controls. *APDF*, amplitude probability distribution function; *B-Text*, bilateral smartphone texting; *U-Text*, unilateral texting; *TYPE*, computer keyboard typing. *Significant between-groups differences, *$P < .05$, **$P < .01$.

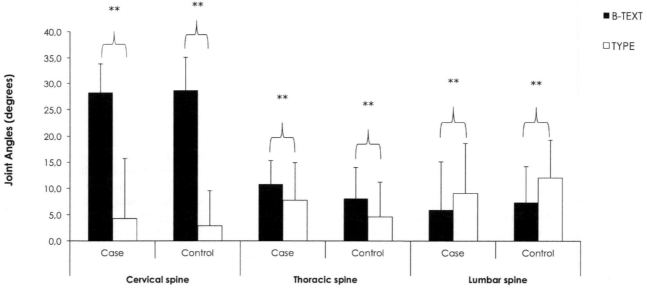

FIG. 3 Median angles of spine flexion/extension comparing computer keyboard typing and smartphone texting in neck-shoulder pain patients and controls. *B-Text*, bilateral smartphone texting; *TYPE*, computer keyboard typing. *Significant between-groups differences, **$P < .01$.

In conclusion, a relatively static flexion posture, sustained muscle activity, repetitive arm movements, and insufficient rest are the most known physical risk factors for occupational MSDs. However, majority of the research on this topic has only produced limited evidence for the associated risk of physical factors involved with NSP and LBP (Jun et al., 2017; Ye et al., 2017).

Psychosocial risks factors

Stress, depressed mood, cognitive functioning, and pain behavior are well-known risk factors in NP and LBP in the general population (Linton, 2000). Work-related psychosocial risk aspects refer to the employee perception about how the work is planned, supervised, and conducted. Various studies have examined the role of these elements, i.e., mental stress, work pace, job autonomy and satisfaction, and support from others.

A systematic review on the psychosocial factors on NSP concluded that job demands, job control and strain, and social support have a strong "incremental" effect on adding to the influence of physical risk factors to the development of NSP among office workers (Kraatz, Lang, Kraus, Münster, & Ochsmann, 2013). Most studies examined physical and psychosocial aspects (Janwantanakul et al., 2012); hence, it is difficult to differentiate the separate causal effects of these two domains. The time course of the interaction between psychological health and job-related behavior also needs to be investigated.

There is conflicting level of evidence for self-perceived medium/high muscular tension as a risk factor for NSP in computer users (Wahlström, Lindegård, Ahlborg Jr., Ekman, & Hagberg, 2003). The phenomenon of perceived muscular tension may reflect an interactive process of physical and psychosocial stress underlining once again the complexity of NSP, although this is still to be explored. The role of other psychosocial aspects, such as job demands, low influence at work, and low satisfaction with the workplace, also remains controversial (Jun et al., 2017; Paksaichol, Janwantanakul, Purepong, Pensri, & Van Der Beek, 2012). This may be due to the large heterogeneity of studies in this field.

Overall, the complex interrelations among internal and external risk factors emphasize the multimodality of work-related MSDs (Fig. 4). The large number of risk factors and their potential relationships explains why when dealing with work-related disorders, one is confronted to heterogeneous patient groups with both diffuse (constant fatigue, stiffness, referred pain) and specific symptoms (increased muscle tone, painful locations, and trigger points).

Physical activity to prevent and manage musculoskeletal pain in computer and smartphone users

Treatment for occupational MSDs may focus on the workers' health status or on the job tasks and environment (Fig. 5). This section summarizes the current evidence on the efficacy of exercise interventions in computer and smartphone users. The idea is to restore balance between the workers ability and the work demand so the relative load is tipping on the safe side (Fig. 6).

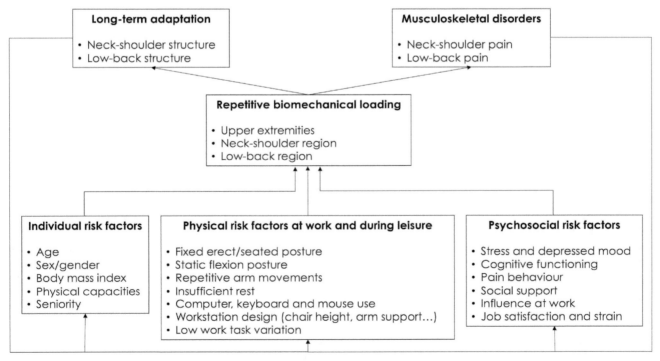

FIG. 4 Main exposures for neck-shoulder and low-back pain among computer and smartphone users.

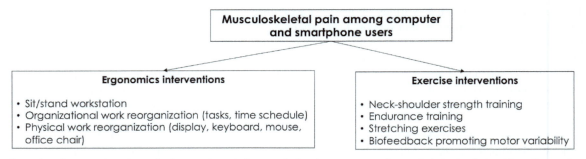

FIG. 5 Examples of ergonomic and exercise interventions aiming at reducing the occurrence of musculoskeletal pain in white-collar employees.

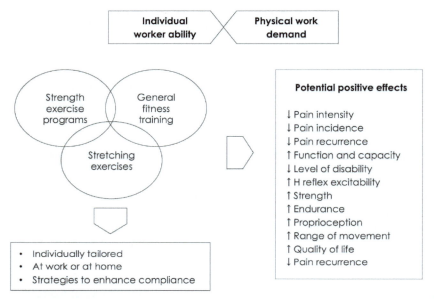

FIG. 6 Effects of exercise therapy for work-related musculoskeletal pain.

Neck-shoulder strengthening exercise programs and general fitness training are effective to manage work-related NSP and disability among white-collar workers, with high-quality evidence for endurance and stretching exercises to help workers "at risk" (Chen et al., 2018). Most protocols include two to three sessions of 20 min weekly up to 52 weeks and show a strong association between treatment adherence and NSP relief (Chen et al., 2018). Five weeks of strength training induce neuromuscular changes facilitating the H-reflex excitability and increase strength and rate of force development of the trapezius muscle (Vangsgaard, Taylor, Hansen, & Madeleine, 2014). Central adaptations of pain responses, as a result of exercise-induced hypoalgesia, are also observed following eccentric training in computer users with NSP (Heredia-Rizo et al., 2019; Heredia-Rizo, Petersen, Arendt-Nielsen, & Madeleine, 2020). However, there are contradictory findings as to which exercises are the most effective and about the importance of supervision when training at the workplace.

Variability is important to prevent MSDs (Madeleine, 2010). Taking active breaks from computer work has been studied to check if active pauses, compared to passive pauses, can change the activation pattern of the shoulder girdle muscles (St-Onge, Samani, & Madeleine, 2017). Active pauses induce a dynamic reorganization of muscle activity that could be beneficial since it changes muscle activation pattern (Samani, Holtermann, Søgaard, & Madeleine, 2009). That approach can be integrated using online biofeedback during computer work as a mean to increase variability in muscle activation (Samani, Holtermann, Søgaard, & Madeleine, 2010).

For LBP, there is high-quality evidence to support exercise therapy (Maher, Underwood, & Buchbinder, 2017), both at home and at the workplace (del Pozo-Cruz et al., 2012). In users of VDUs, a spinal stabilization protocol (Kim, Kim, & Cho, 2015) or the combination of strengthening, mobility, and stretching exercises (del Pozo-Cruz et al., 2012) helps to decrease pain and disability, and improve proprioception, range of movement, and quality of life, with conflicting findings (Suni, Rinne, Tokola, Mänttäri, & Vasankari, 2017). Additionally, exercise training can reduce LBP recurrence (del Pozo-Cruz et al., 2012) and prevent pain in workers with reduced trunk muscle endurance and extension flexibility (Sihawong, Janwantanakul, & Jiamjarasrangsi, 2014). In general, 5–10 min of daily training could be enough to improve LBP-related measures (del Pozo-Cruz et al., 2012), with shorter sessions recommended for those who work long shifts (Bell & Burnett, 2009).

Adherence is one of the main barriers in the real-life scenario. Several strategies are suggested to enhance compliance, such as flexible enrollment and engagement options, constant therapist-worker communication, and a tailored individual approach (Mullane et al., 2019).

Ergonomics to prevent occupational musculoskeletal pain in computer and smartphone users

Ergonomic interventions to adapt the job environment and the working process have been investigated. The issue of prolonged sitting has attracted special attention, and standing at work is promoted as a potential positive approach to improve musculoskeletal health. Along with this, computer and smartphone users should be well educated about proper ergonomics and sit-stand workstations (Agarwal, Steinmaus, & Harris-Adamson, 2018). With no consensus on an "optimal" sitting position, transitioning from a seated to a standing posture appears to be beneficial to prevent and treat MSDs and shows no negative impact on productivity (Waongenngarm, Areerak, & Janwantanakul, 2018). Interestingly, sustained standing may be detrimental for LBP, as previously explained, thus workers need to be individually trained to progressively include workstation adjustments.

A systematic review concluded that there is very low- to moderate-quality evidence for the role of physical, i.e., arm support or mouse based on neutral posture, and organizational ergonomic interventions, i.e., change of work-rest cycle, to manage occupational NSP (Hoe et al., 2018). Still, adapting the working environment is especially effective in symptomatic workers (Chen et al., 2018). All in all, more up-to-date studies are needed, as workstations have evolved from a stationary to a mobile design, including the use of tablets and smartphones for office work, and work in various environment (office and home workplaces). Low levels of compliance with these interventions represent an important concern and may explain the contradictory results among studies. Therefore, strategies to improve compliance are needed.

Applications to other areas

We have reviewed how occupational MSDs impact the health condition of computer and smartphone users. A complex interaction of individual, and personal or job-related physical or psychosocial risk factors explain the high heterogeneity of cases. A society that is becoming more sedentary and a stressful working environment represent a dangerous cocktail in a highly competitive context. This reality is not exclusive of white-collar workers, but also applies to students and blue-collar employees.

Among the modifiable risk factors, poor working postures, job strain, and the individual physical capacity need to be targeted when dealing with occupational musculoskeletal pain. The goal of health promotion at work is to decrease health risks and actively prevent disease incidence.

Lack of physical activity represents a serious health burden beyond the working setting. It is associated with increased morbidity and mortality, and a risk factor for many severe conditions, i.e., cardiovascular diseases, diabetes, and dementia, among others. Regular physical activity helps to modulate the immune and pain inhibitory systems and promotes psychological well-being. Interventions that promote physical activity are safe and cost effective to reduce pain and disability, improve the quality of life, and enhance social participation. Not to forget, exercise training needs to be individually tailored to increase compliance.

Despite all this, exercise programs are rarely a first-line treatment option. This applies to many other chronic pain conditions. A major issue is to close the gap between scientific evidence and clinical practice. Health professionals are to lead a transition from a traditional passive model to a more active and individual-based approach, where patients represent the cornerstone of the intervention.

Other agents of interest

We have described in detail the origin, mechanisms, and adaptations associated with NSP and LBP, as well as the strategies used to prevent and manage MSDs in users of VDUs. Other possible, but less frequent, repetitive strain injuries may include the elbow, wrist, forearm, and hand. Additionally, computer-related visual symptoms are becoming more common in white-collar workers.

298 PART | II The syndromes of pain

Smartphones and tablets represent a major communication tool nowadays, both in adults and among children and adolescents. Despite the clear advantages of the modern technology in many different fields, i.e., health, education, and work, there is increasing interest about the health and safety issues associated with the inappropriate use of these devices. In fact, excessive computer and smartphone use has been linked with increased prevalence of tension type headache and migraine, and is reported to have a negative impact on the psychological, cognitive, and social domains. For instance, problems such as a disorganized life, poor concentration and performance at school or work, reduced sleep quality, and difficulties in written and verbal communication have been observed in those who overuse smartphones. Even the possible harm of the electromagnetic radiations of these devices has been the focus of previous research.

Education for the responsible use of smartphones, tablets, and computers is mandatory and should be stressed in today's society. Keeping an active physical and social life directly benefit mental health and well-being and should be balanced with a rational use of VDUs.

Mini-dictionary of terms

Ergonomics: Applied science promoting safe and efficient interactions between the employee and the physical, organizational and cognitive aspects of the working environment.

Exercise intervention: Planned intervention including structured and repetitive movements to maintain or improve physical fitness.

Occupational risk factor: A physical, psychosocial, or any other agent that may cause harm to an exposed worker in the job setting.

Visual display unit: An electronic device displaying text and images in a screen.

Work-related musculoskeletal disorder: Any condition provoked or worsen by performing a certain job or linked to the working context itself.

Key facts

Key facts of risk factors

- Physical and psychosocial risk factors are of equal importance and need to be considered together.

Key facts of interventions in computer users

- High-quality evidence supports the use of exercise programs in while-collar workers.
- Work variability is important to prevent occupational MSDs.
- Lack of compliance is a critical challenge.
- Ergonomic interventions can be effective.

Summary points

- Occupational NSP and LBP is a health issue in computer and smartphone users.
- Peripheral and centrally mediated mechanisms are involved in work-related musculoskeletal pain.
- Age and sex/gender may play a role in NSP and LBP in office workers.
- Current research has produced limited evidence for the association of workplace physical risk factors with NSP and LBP.
- Psychosocial risk aspects have a strong additive effect combined with physical risk factors.
- Exercise therapy is effective to manage work-related MSDs in computer and smartphone users.
- Ergonomic interventions may help to improve pain, although further evidence is warranted.

References

Agarwal, S., Steinmaus, C., & Harris-Adamson, C. (2018). Sit-stand workstations and impact on low back discomfort: A systematic review and meta-analysis. *Ergonomics, 61*, 538–552.

Akkarakittichoke, N., & Janwantanakul, P. (2017). Seat pressure distribution characteristics during 1 hour sitting in office workers with and without chronic low back pain. *Safety and Health at Work, 8*, 212–219.

Andersen, L. L., Mortensen, O. S., Hansen, J. V., & Burr, H. (2011). A prospective cohort study on severe pain as a risk factor for long-term sickness absence in blue- and white-collar workers. *Occupational and Environmental Medicine, 68*, 590–592.

Bell, J. A., & Burnett, A. (2009). Exercise for the primary, secondary and tertiary prevention of low back pain in the workplace: A systematic review. *Journal of Occupational Rehabilitation, 19*, 8–24.

Bevan, S. (2015). Economic impact of musculoskeletal disorders (MSDs) on work in Europe. *Best Practice and Research: Clinical Rheumatology, 29*, 356–373.

Bontrup, C., Taylor, W. R., Fliesser, M., Visscher, R., Green, T., Wippert, P. M., et al. (2019). Low back pain and its relationship with sitting behaviour among sedentary office workers. *Applied Ergonomics, 81*, 102894.

Cabral, A. M., de Moreira, R. F. C., de Barros, F. C., & de Sato, T. O. (2019). Is physical capacity associated with the occurrence of musculoskeletal symptoms among office workers? A cross-sectional study. *International Archives of Occupational and Environmental Health, 92*, 1159–1172.

Chen, X., Coombes, B. K., Sjøgaard, G., Jun, D., O'Leary, S., & Johnston, V. (2018). Workplace-based interventions for neck pain in office workers: Systematic review and meta-analysis. *Physical Therapy, 98*, 40–62.

Cho, C. Y., Hwang, Y. S., & Cherng, R. J. (2012). Musculoskeletal symptoms and associated risk factors among office workers with high workload computer use. *Journal of Manipulative and Physiological Therapeutics, 35*, 534–540.

Coenen, P., Willenberg, L., Parry, S., Shi, J. W., Romero, L., Blackwood, D. M., et al. (2018). Associations of occupational standing with musculoskeletal symptoms: A systematic review with meta-analysis. *British Journal of Sports Medicine, 52*, 174–181.

del Pozo-Cruz, B., Adsuar, J., Parraca, J., del Pozo-Cruz, J., Moreno, A., & Gusi, N. (2012). A web-based intervention to improve and prevent low back pain among office workers: A randomized controlled trial. *Journal of Orthopaedic and Sports Physical Therapy, 42*, 831–841.

del Pozo-Cruz, B., Gusi, N., Adsuar, J. C., del Pozo-Cruz, J., Parraca, J. A., & Hernandez-Mocholí, M. (2013). Musculoskeletal fitness and health-related quality of life characteristics among sedentary office workers affected by sub-acute, non-specific low back pain: A cross-sectional study. *Physiotherapy (United Kingdom), 99*, 194–200.

Den Bandt, H. L., Paulis, W. D., Beckwée, D., Ickmans, K., & Nijs, J. (2019). Pain mechanisms in low back pain: A systematic review and meta-analysis of mechanical quantitative sensory testing outcomes in people with non-specific low back pain. *Journal of Orthopaedic and Sports Physical Therapy, 49*, 698–715.

Eltayeb, S. M., Staal, J. B., Khamis, A. H., & De Bie, R. A. (2011). Symptoms of neck, shoulder, forearms, and hands: A cohort study among computer office workers in Sudan. *Clinical Journal of Pain, 27*, 275–280.

Ge, H.-Y., Vangsgaard, S., Omland, Ø., Madeleine, P., & Arendt-Nielsen, L. (2014). Mechanistic experimental pain assessment in computer users with and without chronic musculoskeletal pain. *BMC Musculoskeletal Disorders, 15*, 412.

Graven-Nielsen, T. (2006). Fundamentals of muscle pain, referred pain, and deep tissue hyperalgesia. *Scandinavian Journal of Rheumatology, 35*, 1–43.

Greene, R. D., Frey, M., Attarsharghi, S., Snow, J. C., Barrett, M., & De Carvalho, D. (2019). Transient perceived back pain induced by prolonged sitting in a backless office chair: Are biomechanical factors involved? *Ergonomics, 62*, 1415–1425.

Hagberg, M., Vilhemsson, R., Tornqvist, E. W., & Toomingas, A. (2007). Incidence of self-reported reduced productivity owing to musculoskeletal symptoms: Association with workplace and individual factors among computer users. *Ergonomics, 50*, 1820–1834.

Heredia-Rizo, A. M., Petersen, K. K., Arendt-Nielsen, L., & Madeleine, P. (2020). Eccentric training changes the pressure pain and stiffness maps of the upper trapezius in females with chronic neck-shoulder pain: A preliminary study. *Pain Medicine, 21*, 1936–1946.

Heredia-Rizo, A. M., Petersen, K. K., Madeleine, P., & Arendt-Nielsen, L. (2019). Clinical outcomes and central pain mechanisms are improved after upper trapezius eccentric training in female computer users with chronic neck/shoulder pain. *Clinical Journal of Pain, 35*(1), 65–76.

Hoe, V., Urquhart, D., Kelsall, H., Zamri, E., & Sim, M. (2018). Ergonomic interventions for preventing work-related musculoskeletal disorders of the upper limb and neck among office workers. *Cochrane Database of Systematic Reviews, 10*, CD008570.

Hoy, D., Brooks, P., Blyth, F., & Buchbinder, R. (2010). The epidemiology of low back pain. *Best Practice and Research: Clinical Rheumatology, 24*, 769–781.

Ijmker, S., Huysmans, M. A., Van Der Beek, A. J., Knol, D. L., Van Mechelen, W., Bongers, P. M., et al. (2011). Software-recorded and self-reported duration of computer use in relation to the onset of severe arm-wrist-hand pain and neck-shoulder pain. *Occupational and Environmental Medicine, 68*, 502–509.

Janwantanakul, P., Sihawong, R., Sitthipornvorakul, E., & Paksaichol, A. (2018). A path analysis of the effects of biopsychosocial factors on the onset of nonspecific low back pain in office workers. *Journal of Manipulative and Physiological Therapeutics, 41*, 405–412.

Janwantanakul, P., Sitthipornvorakul, E., & Paksaichol, A. (2012). Risk factors for the onset of nonspecific low back pain in office workers: A systematic review of prospective cohort studies. *Journal of Manipulative and Physiological Therapeutics, 35*, 568–577.

Johansson, H., & Sojka, P. (1991). Pathophysiological mechanisms involved in genesis and spread of muscular tension in occupational muscle pain and in chronic musculoskeletal pain syndromes: A hypothesis. *Medical Hypotheses, 35*, 196–203.

Johnston, V., Souvlis, T., Jimmieson, N. L., & Jull, G. (2008). Associations between individual and workplace risk factors for self-reported neck pain and disability among female office workers. *Applied Ergonomics, 39*, 171–182.

Jun, D., Zoe, M., Johnston, V., & O'Leary, S. (2017). Physical risk factors for developing non-specific neck pain in office workers: A systematic review and meta-analysis. *International Archives of Occupational and Environmental Health, 90*, 373–410.

Keown, G. A., & Tuchin, P. A. (2018). Workplace factors associated with neck pain experienced by computer users: A systematic review. *Journal of Manipulative and Physiological Therapeutics, 41*, 508–529.

Kim, T. H., Kim, E. H., & Cho, H. Y. (2015). The effects of the CORE programme on pain at rest, movement-induced and secondary pain, active range of motion, and proprioception in female office workers with chronic low back pain: A randomized controlled trial. *Clinical Rehabilitation, 29*, 653–662.

Kraatz, S., Lang, J., Kraus, T., Münster, E., & Ochsmann, E. (2013). The incremental effect of psychosocial workplace factors on the development of neck and shoulder disorders: A systematic review of longitudinal studies. *International Archives of Occupational and Environmental Health*, *86*, 375–395.

Larsson, B., Björk, J., Kadi, F., Lindman, R., & Gerdle, B. (2004). Blood supply and oxidative metabolism in muscle biopsies of female cleaners with and without myalgia. *Clinical Journal of Pain*, *20*, 440–446.

Linton, S. J. (2000). A review of psychological risk factors in back and neck pain. *Spine*, *25*, 1148–1156.

Madeleine, P. (2008). Functional adaptations in work-related pain conditions. In T. Graven-Nielsen, L. Arendt-Nielsen, & S. Mense (Eds.), *Fundamentals of musculoskeletal pain* (pp. 401–416). Seattle: IASP Press.

Madeleine, P. (2010). On functional motor adaptations: From the quantification of motor strategies to the prevention of musculoskeletal disorders in the neck-shoulder region. *Acta Physiologica*, *199*, 1–46.

Madeleine, P., Vangsgaard, S., Hviid Andersen, J., Ge, H. Y., & Arendt-Nielsen, L. (2013). Computer work and self-reported variables on anthropometrics, computer usage, work ability, productivity, pain, and physical activity. *BMC Musculoskeletal Disorders*, *14*, 226.

Maher, C., Underwood, M., & Buchbinder, R. (2017). Non-specific low back pain. *The Lancet*, *389*, 736–747.

Mc Lean, S. M., May, S., Klaber-Moffett, J., Macfie, D., Gardiner, E., McLean, S. M., et al. (2010). Risk factors for the onset of non-specific neck pain: A systematic review. *Journal of Epidemiology and Community Health*, *64*, 565–572.

McPhee, M. E., Vaegter, H. B., & Graven-Nielsen, T. (2020). Alterations in pronociceptive and antinociceptive mechanisms in patients with low back pain: A systematic review with meta-analysis. *Pain*, *161*, 464–475.

Mense, S. (1993). Nociception from skeletal muscle in relation to clinical muscle pain. *Pain*, *54*, 241–289.

Mullane, S. L., Rydell, S. A., Larouche, M. L., Toledo, M. J. L., Feltes, L. H., Vuong, B., et al. (2019). Enrollment strategies, barriers to participation, and reach of a workplace intervention targeting sedentary behavior. *American Journal of Health Promotion*, *33*, 225–236.

Paksaichol, A., Janwantanakul, P., Purepong, N., Pensri, P., & Van Der Beek, A. J. (2012). Office workers' risk factors for the development of non-specific neck pain: A systematic review of prospective cohort studies. *Occupational and Environmental Medicine*, *69*, 610–618.

Parent-Thirion, A., Biletta, I., Cabrita, J., Vargas-Llave, O., Vermeylen, G., Wilczynska, A., et al. (2017). *Sixth European working conditions survey— Overview report*. Luxembourg: Publications Office of the European Union.

Pereira, M. J., Johnston, V., Straker, L. M., Sjøgaard, G., Melloh, M., O'Leary, S. P., et al. (2017). An investigation of self-reported health-related productivity loss in office workers and associations with individual and work-related factors using an employer's perspective. *Journal of Occupational and Environmental Medicine*, *59*, e138–e144.

Rempel, D. M., Krause, N., Goldberg, R., Benner, D., Hudes, M., & Goldner, G. U. (2006). A randomised controlled trial evaluating the effects of two workstation interventions on upper body pain and incident musculoskeletal disorders among computer operators. *Occupational and Environmental Medicine*, *63*, 300–306.

Ringheim, I., Indahl, A., & Roeleveld, K. (2019). Reduced muscle activity variability in lumbar extensor muscles during sustained sitting in individuals with chronic low back pain. *PLoS One*, *14*, 1–13.

Rosendal, L., Larsson, B., Kristiansen, J., Peolsson, M., Søgaard, K., Kjær, M., et al. (2004). Increase in muscle nociceptive substances and anaerobic metabolism in patients with trapezius myalgia: Microdialysis in rest and during exercise. *Pain*, *112*, 324–334.

Samani, A., Holtermann, A., Søgaard, K., & Madeleine, P. (2009). Active pauses induce more variable electromyographic pattern of the trapezius muscle activity during computer work. *Journal of Electromyography and Kinesiology*, *19*, e430–e437.

Samani, A., Holtermann, A., Søgaard, K., & Madeleine, P. (2010). Active biofeedback changes the spatial distribution of upper trapezius muscle activity during computer work. *European Journal of Applied Physiology*, *110*, 415–423.

Sihawong, R., Janwantanakul, P., & Jiamjarasrangsi, W. (2014). A prospective, cluster-randomized controlled trial of exercise program to prevent low back pain in office workers. *European Spine Journal*, *23*, 786–793.

St-Onge, N., Samani, A., & Madeleine, P. (2017). Integration of active pauses and pattern of muscular activity during computer work. *Ergonomics*, *60*, 1228–1239.

Suni, J. H., Rinne, M., Tokola, K., Mänttäri, A., & Vasankari, T. (2017). Effectiveness of a standardised exercise programme for recurrent neck and low back pain: A multicentre, randomised, two-arm, parallel group trial across 34 fitness clubs in Finland. *BMJ Open Sport and Exercise Medicine*, *3*, 1–12.

Szeto, G. P. Y., & Sham, K. S. W. (2008). The effects of angled positions of computer display screen on muscle activities of the neck–shoulder stabilizers. *International Journal of Industrial Ergonomics*, *38*, 9–17.

Szeto, G. P. Y., Straker, L. M., & O'Sullivan, P. B. (2005a). The effects of speed and force of keyboard operation on neck-shoulder muscle activities in symptomatic and asymptomatic office workers. *International Journal of Industrial Ergonomics*, *35*, 429–444.

Szeto, G. P. Y., Straker, L. M., & O'Sullivan, P. B. (2005b). A comparison of symptomatic and asymptomatic office workers performing monotonous keyboard work—1. Neck and shoulder muscle recruitment patterns. *Manual Therapy*, *10*, 270–280.

Szeto, G. P. Y., Straker, L. M., & O'Sullivan, P. B. (2005c). A comparison of symptomatic and asymptomatic office workers performing monotonous keyboard work—2: Neck and shoulder kinematics. *Manual Therapy*, *10*, 281–291.

Thorp, A. A., Healy, G. N., Winkler, E., Clark, B. K., Gardiner, P. A., Owen, N., et al. (2012). Prolonged sedentary time and physical activity in workplace and non-work contexts: A cross-sectional study of office, customer service and call centre employees. *International Journal of Behavioral Nutrition and Physical Activity*, *9*, 1–9.

Vangsgaard, S., Taylor, J. L., Hansen, E. A., & Madeleine, P. (2014). Changes in H reflex and neuromechanical properties of the trapezius muscle after 5 weeks of eccentric training: A randomized controlled trial. *Journal of Applied Physiology*, *116*, 1623–1631.

Wagman, I. H., & Price, D. D. (1969). Responses of dorsal horn cells of M. mulatta to cutaneous and sural nerve A and C fiber stimuli. *Journal of Neurophysiology*, *32*, 803–817.

Wahlström, J., Lindegård, A., Ahlborg, G., Jr., Ekman, A., & Hagberg, M. (2003). Perceived muscular tension, emotional stress, psychological demands and physical load during VDU work. *International Archives of Occupational and Environmental Health*, *76*, 584–590.

Waongenngarm, P., Areerak, K., & Janwantanakul, P. (2018). The effects of breaks on low back pain, discomfort, and work productivity in office workers: A systematic review of randomized and non-randomized controlled trials. *Applied Ergonomics*, *68*, 230–239.

Won, E. J., Johnson, P. W., Punnett, L., & Dennerlein, J. T. (2009). Upper extremity biomechanics in computer tasks differ by gender. *Journal of Electromyography and Kinesiology*, *19*, 428–436.

Xie, Y., Madeleine, P., Tsang, S. M. H., & Szeto, G. P. Y. (2017). Spinal kinematics during smartphone texting—A comparison between young adults with and without chronic neck-shoulder pain. *Applied Ergonomics*, *68*, 160–168.

Xie, Y., Szeto, G. P. Y., Dai, J., & Madeleine, P. (2016). A comparison of muscle activity in using touchscreen smartphone among young people with and without chronic neck-shoulder pain. *Ergonomics*, *59*, 61–72.

Ye, S., Jing, Q., Wei, C., & Lu, J. (2017). Risk factors of non-specific neck pain and low back pain in computer-using office workers in China: A cross-sectional study. *BMJ Open*, *7*, 9–11.

Part III

Interlinking anesthesia, analgesics and pain

Chapter 28

Health professionals' and lay people's positions regarding the use of analgesics in cancer cases

Etienne Mullet[a] and Paul Clay Sorum[b]

[a]*Department of Ethics, Institute of Advanced Studies (EPHE), Plaisance, France,* [b]*Department of Internal Medicine and Pediatrics, Albany Medical College, Albany, NY, United States*

Abbreviations

TS   terminal sedation

Introduction

Pain is common in adult patients with cancer. Oncology guidelines outline how to treat it, adding opioids if needed (World Health Organization, 2018). Nonetheless, this pain is notoriously poorly controlled, with little progress over the past decade (van den Beuken-van Everdingen, Hochstenbach, Joosten, Tjan-Heijnen, & Janssen, 2016). It is important to know, therefore, why adult patients with cancer continue to suffer so much pain and how to intervene.

The reasons, which have been reviewed extensively (Bouya et al., 2019; Jacobsen, Sjøgren, Møldrup, & Christrup, 2007; Kasasbeh, McCabe, & Payne, 2017; Kwon, 2014; Makhlouf, Pini, Ahmed, & Bennet, 2019; Paice, 2018; Saini & Bhatnagar, 2016; Salim, Al-Attyat, Tuffaha, Higm, & Brant, 2017), are multiple and reinforcing and vary from one location and situation to another. They include clinicians' lack of knowledge of pain treatment and their insufficient awareness of patients' pain, because, in part, of patients' reticence to report their pain and of the insufficient use of reliable pain measurements; the fear of opioids among patients, clinicians, family members, and the public; the intractability of some patients' pain, patients' unrealistic expectations concerning pain relief, their lack of adherence to analgesic regimens, and the inadequacy of current pain treatments; and the lack of specialists in pain management (or in low-income countries, even of physicians), the absence in some areas of accessible pharmacies stocking opioids, the reluctance of insurance companies to pay for opioids, and the legal and regulatory barriers in some countries to prescribing opioids. The epidemic of opioid abuse in recent years has complicated the issue greatly (National Academies of Sciences, Engineering, and Medicine, 2019; Paice, 2018).

Of considerable importance among these impediments to effective pain control are the attitudes about pain treatment and, in particular, about the use of opioids of the clinicians—the physicians who prescribe and the nurses who, in the hospitals and patient homes, assess patients' pain and administer treatment. It is also of importance to understand the seldom-assessed attitudes about pain treatment and opioids of the public that surrounds the clinicians, patients, and family members since these attitudes help to form the context of their decision-making.

Studies of clinicians and the public

The studies of the attitudes of clinicians and the public about cancer pain have used two types of methodologies: first—including all the studies cited above—answering a survey or list of questions about pain treatment or, second, less frequently, responding to sets of scenarios describing patients in pain. The advantages of lists of questions are that, in comparison to sets of scenarios, they are easier to fill out, can include and assess more single items, and can be administered to larger groups of people. The advantages of scenarios are that they can present the issue in a realistic way and that, using factorial designs (Anderson, 2008), they make possible an assessment of the impact of each aspect of the situation and its

Features and Assessments of Pain, Anesthesia, and Analgesia. https://doi.org/10.1016/B978-0-12-818988-7.00026-1
Copyright © 2022 Elsevier Inc. All rights reserved.

interaction with the other aspects and, using cluster analysis (Hofmans & Mullet, 2013), a mapping of the variety of different positions held by clinicians and the public.

Recent survey-based studies of nurses and physicians around the world (Breuer, Fleishman, Cruciani, & Portenoy, 2011; Darawand, Alnajar, Abdalrahim, & El-Aqoul, 2017; Jho et al., 2014; Saifan et al., 2019; Shahriary et al., 2015; Utne, Småstuen, & Nyblin, 2019; Zhang et al., 2015) have shown that most physicians and nurses declare not only that control of pain in cancer patients is very important but that clinicians should believe patients' self-reports about their pain. Nonetheless, these studies have found persisting knowledge deficits. For example, the average correct answers to questions about cancer pain treatment were, in one recent study (Saifan et al., 2019), only 58.37% among physicians and 46.1% among nurses in cancer wards; the main barriers and misconceptions there were identified as "fears related to addiction and side effects of the pain management medications, lack of knowledge of opioid usage and preventive methods for medication side effects, and the influence of nurses' own cultural beliefs about pain."

Lay persons' attitudes about the treatment of cancer patients' pain have, in contrast, seldom been surveyed (Grant, Ugalde, Vafiadis, & Philip, 2015; Larue, Fontaine, & Brasseur, 1999; Levin, Cleeland, & Dar, 1985). In the largest study, Larue et al. (1999) did telephone surveys of the knowledge and attitudes of representative samples of the French public in 1990 (1001 respondents) and 1996 (1006 respondents). During this period, considerable efforts were made to educate French physicians and the public about the importance of pain management, including the use of opioids. The results showed an evolution in public opinion. The percentages in 1996 vs 1990 for selected questions were "believe that pain is frequent at advanced stages of cancer," 84% vs 72%; "associate morphine with pain relief," 80% vs 44%; "associate morphine with drug abuse," 18% vs 26%; "quite or very afraid of becoming addicted if had to take morphine for pain relief, or would refuse morphine," 15% vs 52%; and "agree that morphine can be prescribed to cancer patients with pain," 83% vs 79%. In 1996, 58% declared that their knowledge had improved over the past 5 years. These surveys were performed, of course, before the current opioid epidemic.

Four scenario-based studies in France have been specifically designed to map the views of clinicians and lay persons. Each of these will be discussed in detail.

Mapping general practitioner's positions regarding the use of opioids

Mas, Albaret, Sorum, and Mullet (2010) mapped the differing positions of French general practitioners regarding the use of opioids to manage the pain suffered by terminally ill cancer patients. They used the combination of vignette technique and cluster analysis. The participants—a convenience sample of 115 private practitioners in the southern part of France—were presented with a set of realistic scenarios depicting a person who was dying from cancer. These scenarios varied as a function of five situational factors shown in previous studies to have an impact on the judgments of appropriateness of using opioids in such cases: (a) the patient's age (quite young vs quite old); (b) the patient's gender; (c) the patient's reported pain level (4 or 7, on a 10-point scale); (d) signs of depression (absent vs present); and (e) the patient's request for pain relief (no request, request, or repeated request). A sample scenario is shown in Table 1. The physicians were also presented with an anxiety questionnaire in order to explore whether appropriateness judgments were associated with their personality traits.

Four clusters were identified. The first cluster (11%) was termed *Prescribe in response to request* because, as shown in Fig. 1 (left-hand panel), appropriateness ratings were much higher in cases of repeated requests than in cases of no request. The second cluster (37%) was termed *Prescribe in response to pain level* since, as shown in the second panel, appropriateness ratings were much higher when the pain level was 7 than when it was 4. The third cluster (37%) was termed *Prescribe in response to all signs of distress*; appropriateness ratings were higher (a) when the pain level was 7 than when it was 4; (b) in cases of repeated request than in cases of no request, especially when the pain level was 4; and (c) when signs of depression were present, especially when the pain level was 4. Finally, the fourth cluster (14%) was termed *Always prescribe*. Physicians who had attended courses on pain management as part of continuing medical education were more often in this fourth cluster than physicians who had not.

General practitioners seem to have different styles and approaches to dying patients, and these styles underlie both prescribing opioids, as shown here, and, making home visits, as illustrated in the other part of this study. The different positions were not associated with physicians' degrees of underlying anxiety, nor with their age or gender. Irrespective of their positions, however, the physicians responded to patient's signs of distress without regard for the patients' demographic characteristics. This is reassuring in light of evidence elsewhere of age and gender bias in pain treatment. However, patients' signs of depression were not taken into account to the same extent as their requests or pain levels.

The small cluster of physicians who deemed it necessary to prescribe opioids to all dying patients in pain were in better conformance than the other physicians with the recommendations of the World Health Organization and others to treat patients' pain, especially terminal patients' pain, aggressively. The fact that the physicians in this cluster were more likely

TABLE 1 Examples of scenario used in the four studies specifically designed to map the views of clinicians and lay persons.

Study	Example of scenario	Response scale
Mas et al. (2010)	Mr. Sudres is 73 years old. He is in the terminal phase of cancer. Mr. Sudres suffers from severe pain rated at 7/10. Mr. Sudres appears to be in a state of great sadness, of withdrawal and of deep worry. Mr. Sudres asks systematically for pain relief. To what extent do you deem it appropriate to treat this patient with opioids?	15-Point scale with anchors of "Not appropriate at all" (1) and "Completely appropriate" (15)
Cano Romero, Muñoz Sastre, Quintard, Sorum, and Mullet (2017)	Mrs. Vidotto is a nurse in the department of gastrointestinal surgery of the Pasteur Hospital. Mr. Joeri Mairesse, an 80-year-old patient, has recently been operated on for cancer. The operation was successful. The morphine pump has recently been removed. The patient now experiences a level of pain of 6 on a scale of 10, according to the nurse's own assessment. The surgeon has prescribed opioids. Mr. Mairesse has just requested that another dose of painkiller be given. Mrs. Vidotto does not immediately comply with his request. She considers that since Mr. Mairesse has been administered opioids 2 h ago, it would be too risky to give him another dose of opioids because of possible interactions with other treatments. To what extent do you believe that the nurse's decision is acceptable?	11-Point scale with anchors of "Not acceptable at all" (0) and "Completely acceptable" (10)
Mazoyer, Muñoz Sastre, Sorum, and Mullet (2017)	Mrs. Denis is 82 years old. She suffers from pancreatic cancer; she is terminally ill. She is treated at the hospital and she receives the comfort care that her condition requires. Since yesterday, Mrs. Denis has been declaring that she is suffering more. Her score on the pain scale is 7. Mrs. Denis is currently treated with a mildly potent analgesic: Codoliprane (paracetamol + codeine). She clearly asks the health professionals who take care of her for a stronger pain treatment. The physician in charge of Mrs. Denis decides, in agreement with her team, to prescribe morphine. To what extent do you believe that the physician's decision is acceptable?	11-Point scale with anchors of "Not acceptable at all" (0) and "Completely acceptable" (10)
Mazoyer, Muñoz Sastre, Sorum, and Mullet (2016)	Mrs. Noirot is 85 years old. She has been in the hospital for 10 days. She is in the terminally ill phase of a pancreatic cancer. She suffers horribly, and no pain-relieving treatment, even the strongest ones, can alleviate her pain. Her life expectancy is about 1 week. Mrs. Noirot has requested continuous sedation from her treating physician and her family has supported her claim. After collective discussion with her team, the physician has decided to proceed to terminal sedation in view of allowing Mrs. Noirot to die without pain. To what extent do you believe that the physician's decision is acceptable?	11-Point scale with anchors of "Not acceptable at all" (0) and "Completely acceptable" (10)

than the others to have attended continuing education courses on pain management may merely reflect their greater interest in palliative care. But it may also suggest that educating physicians about palliative care leads to more effective care for dying patients.

Mapping lay people's and health professionals' positions on the use of analgesics in postoperative pain management

Cano Romero et al. (2017) mapped the positions of French lay people and health professionals regarding the use of opioids to manage postoperative pain. The participants—a convenience sample in the south of France of 169 lay people and 226

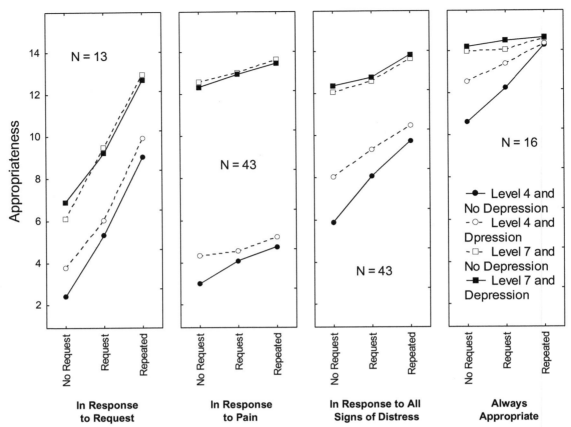

FIG. 1 The four clusters observed. In each panel, the mean appropriateness judgments are on the y-axis, the four curves correspond to a combination of the two levels of patient's reported pain and of the two levels of apparent depression, and the three levels of the patient's request factor are on the x-axis.

health professionals (138 registered nurses, 32 nurse's aides, 33 physicians, and 23 psychologists working in hospitals)—were presented with scenarios (see Table 1) depicting a postoperative patient who requested additional pain treatment. These scenarios varied as a function of four situational factors shown in previous studies to have an impact on the judgments of acceptability of giving another dose of morphine: (a) the patient's age (about 40 years vs about 80 years); (b) the patient's current level of pain, as assessed by the nurse (3, 6, or 9 out of 10); (c) the patient's number of previous requests (four, two, or none); and (d) the level of risk (e.g., interactions with other treatments) associated with the administration of a repeat dose (depending, in particular, on the length of time since the previous dose).

Five clusters were identified. The first cluster (33%) was termed *Not acceptable* since, as shown in Fig. 2 (left-hand panel), the ratings of the nurse's decision not to give pain medication were always quite low. Nurses (57%) were more frequently members of this cluster than nurses' aides (31%) and physicians (30%), who were in turn more frequently members than lay people (17%) and psychologists (13%). The second cluster (26%) was termed *Depends on pain* since, as shown in the second panel, acceptability ratings were considerably higher when the level of pain was low than when it was high. Physicians (52%) were more frequently members of this cluster than other participants (22%).

The third cluster (18%) was termed *Depends on risk* since, as shown in the third panel, ratings were considerably higher when risk was high than when it was low. Psychologists (43%) were more frequently members of this cluster than other participants (19%). The fourth cluster (11%) was termed *Mainly acceptable*. Lay people (19%) and nurse's aides (16%) were more frequently members of this cluster than nurses (3%) and physicians (3%). The fifth cluster (11%, not shown) was termed *Undecided* since the ratings were always close to the center of the response scale. Lay people (18%) and psychologists (13%) were more frequently members of this cluster than other participants (6%).

Most nurses (57%) strictly applied, in their judgments, the autonomy principle of bioethics, i.e., that patients should determine the need for pain relief, and those nurses who did not had reasonable views, considering either that the pain was bearable or that the patient's health was at risk from excessive opioids (34%). A majority of physicians (91%) held similar views. Most potential conflicts between nurses and physicians seem to be analyzable in terms of what differentiates

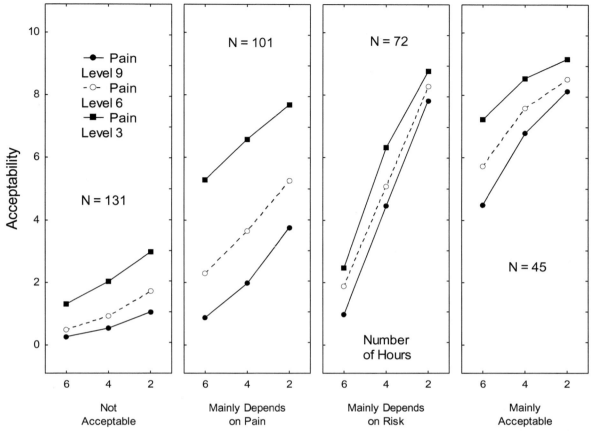

FIG. 2 Four of the five clusters observed. In each panel, the mean acceptability judgments are on the y-axis, the three curves correspond to the three levels of patient's assessed pain, and the three levels of the risk factor are on the x-axis. Mean acceptability ratings have been pooled across number of requests and patient's age. *6*, Last injection 6 h ago; *4*, last injection 4 h ago; *2*, last injection 2 h ago.

the first two positions, *Not acceptable* to withhold another dose and *Depends on pain*. In particular, the potential for conflict is highest when patients ask for a repeat dose of opioids even though they got a dose only 2 h before and their level of pain seems to the nurse to be low. Nurses would be angry at physicians in no few cases for being reluctant to treat pain and in some others for agreeing with patients' requests in spite of an apparent increased risk of adverse effects. It would be useful, therefore, for nurses and physicians in postoperative (and other) settings to meet together for case-based discussions of pain treatment in an effort to reach consensus.

Potential disagreements between nurses and lay people, such as patients' companions, seem to be more diverse in kind. Lay people tended to agree, more often than nurses, with the nontreating behavior depicted in the vignettes. They seemed, therefore, to worry even more about overtreatment with opioids than about undertreatment of pain. Disagreements might also be more complex. On the one hand, nurses could find themselves in the uneasy position of having to explain to patients' companions why physicians are reluctant to treat with opioids despite patients' repeated requests and despite their own conviction that patients' requests should be honored. On the other hand, nurses might have to explain to patients' companions why physicians are taking the risk of ordering repeat doses of opioids in response to patients' requests even though the nurses themselves think the cumulative doses are too high. Accordingly, it is important not only for physicians and nurses to reach consensus, but also for nurses to receive training in how to explain pain treatment in terms both patients and their companions can understand.

Mapping lay people's and health professionals' positions regarding the use of morphine to relieve intense pain

Mazoyer et al. (2017) mapped the positions of French lay people and health professionals regarding the use of morphine to alleviate the pain of terminally ill cancer patients. The participants—a convenience sample in the south of France of 120 lay people and 30 health professionals (23 registered nurses and 7 physicians)—were presented with scenarios (see Table 1) depicting a hospitalized terminally ill woman who was in pain. These scenarios varied as a function of four factors: (a) the level of pain reported by the patient (either 4 or 7), (b) the patient's explicit request for additional administration of analgesics, (c) the final decision (e.g., to continue the current pain treatment or to increase the level of analgesic medication), and (d) the way the decision was taken (collective vs individual). In this study, the patient was treated with conventional analgesics and the issue was the acceptability of adding morphine, whereas in the previous study (Cano Romero et al., 2017), the patient had already been treated with morphine and the issue was the acceptability of giving additional morphine.

Seven clusters were identified (Fig. 3). The first cluster (8% of lay people, 9% of nurses, and 14% of physicians) was called *Tend to disagree with any decision*. As shown in Fig. 3 (first panel), the ratings were always much lower than the center of the acceptability scale. Male participants were more often found in this cluster than female participants. The second cluster (18%, 9%, and 0%) was called *Increase the strength of the painkiller in any case*. As shown in the second panel, acceptability ratings were considerably higher when the physician decided to give a more potent analgesic, either paracetamol/codeine or morphine, than in the other cases. Older participants were more often found in this cluster than younger participants. The third cluster (8%, 9%, and 29%) was a variant of the preceding position. It was named *Give morphine preferentially* because, as shown in the third panel, ratings were very high when morphine was given, low when the treatment was not changed, and intermediate when paracetamol/codeine was given. In addition, ratings were higher when morphine was given and the pain level was 7 rather than 4. Politically right-wing participants were more often found in this cluster than other participants.

The fourth cluster (22%, 30%, and 29%) was named *Partly depends on pain level* because, as shown in the fourth panel, this factor strongly interacted with the decision factor: (a) when the pain level was 7, ratings were high if a stronger pain killer was given and low if the treatment was not changed and (b) when the pain level was 4, ratings were high only if paracetamol/codeine was given. If the treatment was not changed, or if morphine was given, ratings were in the intermediate range. Regular attendees at religious services were less often found in this cluster than other participants. The fifth cluster (6%, 4%, and 14%) was called *Fully depends on pain level*. As shown in the fifth panel, this factor interacted with the decision factor still more strongly than in the preceding cluster. Atheists and participants with tertiary education were less often found in this cluster than other participants. The sixth cluster (21%, 35%, and 0%) was called *Depends on decision process and on pain level* because when the decision did not involve the team, all ratings were low, whereas when the decision was a collective one, the pattern of ratings was similar to the one observed in the partly depends on pain cluster. Younger participants and leftists were more often found in this cluster than other participants. Finally, the seventh cluster (17%, 4%, and 14%) was named *Unsure* because the ratings were generally close to the center of the acceptability scale. Right-wingers were more often found in this cluster than other participants were.

In summary, and taking into consideration all positions, 91% of participants (90% of professionals and 92% of lay people) agreed with the use of morphine in terminally ill cancer patients when the pain level was high and the decision to increase the strength of the painkiller was taken collectively. This percentage dropped to 69% when the team was not involved in the decision and to 40% when the pain level was lower. In other words, if opposition to the use of morphine exists, it is not opposition to morphine itself, but opposition to the circumstances of its use. These results were consistent with previous findings that a majority of lay people in France (Larue et al., 1999) agree with the use of morphine for cancer patient care. Health professionals, unsurprisingly, expressed very diverse positions; this was consistent with findings regarding the way French physicians treat their terminally ill patients' pain discussed above. Most nurses (69%) were, however, of the view that pain level must be taken into consideration before administering morphine.

Whether the patient did or did not request an increase in pain treatment had no impact in any cluster. While this appears to contradict the growing insistence in France, as well as elsewhere, on patient autonomy and shared decision-making, it is understandable in the context of pain. As demonstrated in this study, both lay people and health professionals generally view pain—or at least severe pain—as something that is bad, especially in terminally ill patients, and that should be alleviated whether or not the patient specifically requests it.

Mapping lay people's and health professionals' positions regarding temporary or terminal sedation

Mazoyer et al. (2016) mapped the positions of French lay people and health professional regarding the use of morphine to sedate terminally ill, suffering cancer patients. The participants—a convenience sample in the south of France of 223 lay people and 53 health professionals (32 registered nurses and 21 physicians)—were presented with scenarios (see Table 1) that varied as a function of (a) whether the choice for sedation was terminal, 24-h, or none; (b) the patient's life expectancy;

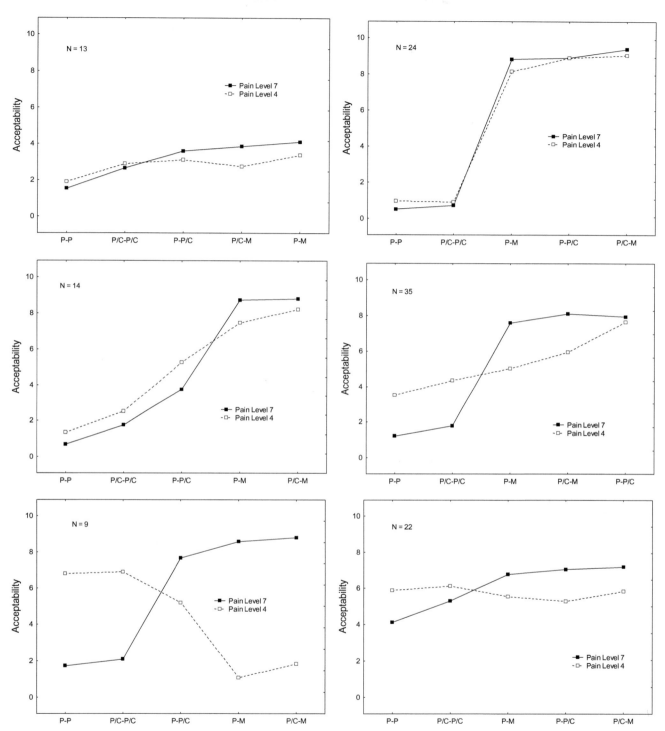

FIG. 3 Six of the seven clusters observed. In each panel, the mean acceptability judgments are on the y-axis, the two curves correspond to the two levels of patient's reported pain, and the five decisions are on the x-axis.

(c) how the decision was made (by the physician alone or by a collective decision of the care team); and (d) what the patient requested and the family wanted.

Through cluster analysis, five clusters were identified. The first cluster (5% of lay people, 6% of nurses, and 9% of physicians) was termed *24-h sedation not acceptable* since, as shown in Fig. 4 (left panel), the ratings for 24-h sedation were systematically low. Terminal sedation (TS) was, however, acceptable when the patient requested it. Absence of

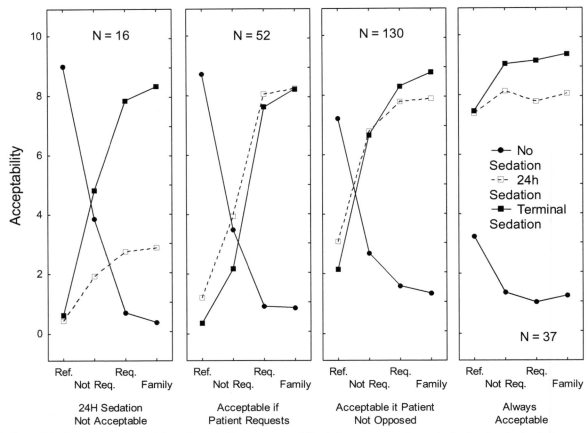

FIG. 4 Four of the five clusters observed. In each panel, the mean acceptability judgments are on the y-axis, the four levels of patient request are on the x-axis, and the three curves correspond to the three levels of sedation.

sedation when the patient requested it was unacceptable. The second cluster (18%, 12%, and 34%) was named *TS acceptable if patient explicitly requested it*. As shown in the second panel, both 24-h and TS were acceptable when the patient requested it, and absence of sedation was unacceptable when the patient requested sedation. The third cluster (50%, 45%, and 24%) was named *TS acceptable if patient did not explicitly oppose it*. As shown in the third panel, when there was no explicit opposition to sedation, ratings were high.

The fourth cluster (14%, 25%, and 9%; not shown) was named *TS acceptable if decision taken collectively* because the decision-making process had by far the strongest effect. When the process was a collective one, ratings were similar to those shown in the second panel. When the process was not a collective one, ratings were always low. Younger participants were more often found in this cluster than older ones. Finally, the fifth cluster (13%, 12%, and 24%) was named *TS always acceptable* because, as shown in the fourth panel, ratings for sedation were high in all cases. Older participants were more often found in this cluster than younger ones.

In summary, (a) lay people and health professionals living in France do not appear to be systematically opposed to TS for terminally ill, suffering cancer patients; (b) the most important factor in increasing acceptability is the patients' request for it; and (c) TS is additionally acceptable when it is the collective decision of the medical team.

General discussion

The surveys of the knowledge and attitudes of physicians, nurses, and the public concerning the treatment of the pain of cancer patients—as well as of their perceptions of the barriers to effective treatment—have highlighted the continued need, around the world, for education in pain management, not only of physicians and nurses (Latchman, 2014) but also of patients, family members, and the general public. Indeed, lay people's education must not be neglected since, after all, the future patients and family caregivers come from them. Moreover, studies have shown that family caregivers often have poor knowledge of and attitudes toward pain treatment and that their scores predict the adequacy, or not, of treatment

(Lou & Shang, 2017; Meeker, Finnell, & Othman, 2011; Saifan, Bashayreh, Batiha, & AbuRuz, 2015; Vallerand, Collins-Bohler, Templin, & Hasenau, 2007). Even though efforts in education have shown limited success (Adam, Bond, & Murchie, 2015; Oldenmenger et al., 2018), they must be continued—along with other interventions to improve pain treatment reviewed by van den Beuken-van Everdingen et al. (2016) and Paice (2018)—especially as the new concern about opioids threatens to limit the use of opioids even when needed.

Yet the scenario-based studies, though done only in France, give reason to be optimistic. They demonstrate that physicians, nurses, and lay people are not homogenous but are composed of groups with different views. Most are, however, sensitive to the suffering of patients—and in particular to the pain of cancer patients—and want to relieve it, especially if patients ask for pain relief and if the pain is severe.

One problem in the era of anxiety about opioids is the heterogeneity of cancer patients in pain, namely in the severity of the cancer and in the cause of the pain. The clinician might be quite willing to treat aggressively, including with high doses of opioids, the pain of a terminal patient no matter whether the pain is caused by the cancer or not, but reticent to treat with opioids the pain of a patient who has a controlled cancer and a long life expectancy whether the pain is caused by the cancer or not. The above studies, whether survey- or scenario-based, did not directly address either of these factors, although the scenario-based study of terminal sedation of cancer patients with intractable pain suggests that physicians and nurses as well as lay people are willing to go to considerable lengths to relieve the suffering of terminal patients whether or not they have cancer.

Applications to other areas

Scenario-based studies using the methods of the psychologist Norman Anderson reveal, much better than survey-based studies, the impact on people's judgments of changes in context. They show how realistic combinations of different levels of multiple important factors affect such judgments as the acceptability of giving opioids for cancer pain. The addition of cluster analysis reveals the multiplicity of distinct positions on pain treatment and other issues within a group of participants. These differences are not appreciated in the typical analyses based on finding the means and standard deviations of the study participants taken as a whole. Cluster analysis allows understanding why issues such as treatment with opioids can be so politically and culturally charged. These methods can be applied to other pain-related questions, such as the differences in attitudes regarding cancer vs noncancer pain, the differences between patients vs their caregivers, and the differences between people in different cultures, as well as to other controversial medical and nonmedical interventions.

Mini-dictionary of terms

Personal position: A cognitive-emotional schema that, through the integration of the multiple and diverse components of a given situation, produces rapid judgments of widely differing kinds (e.g., of acceptability).

Key facts

- The persistence of deficits in knowledge about cancer pain treatment in clinicians around the world is illustrated by the survey of physicians and nurses in cancer wards in Jordan in 2015–16 that found an average of only 58.37% correct answers among physicians and 46.1% among nurses.
- The attitudes of lay people in France evolved considerably between 1990 and 1996, with, for example, a decrease in those who were "quite or very afraid of becoming addicted if had to take morphine for pain relief, or would refuse morphine" from 52% to 15%.
- No survey has investigated lay people's opinions in other countries or the impact of the recent epidemic of opioid addiction, but recent survey-based studies in France have shown considerable support for using opioids for uncontrolled pain in cancer patients.
- The French general practitioners sampled in 2010 were willing to prescribe opioids to patients with terminal cancer in response to higher pain levels and/or patient requests, though only a minority (14%) were willing in all cases.
- In the context of postoperative pain, in 2013–16, most of a sample of French hospital nurses (57%) and physicians (30%)—more than lay people (17%)—supported patients' autonomy, i.e., their right to determine the need for pain relief.
- For terminally ill cancer patients in France, 90% of health professionals and 92% of lay people agreed in 2016 with the use of morphine if the patient's pain level was high and the decision was made collectively by the treatment team.

Summary points

- Pain is common in cancer patients, but continues to be poorly controlled.
- Survey-based studies of the barriers to adequate pain treatment have highlighted, among the multiple barriers at all levels, the inadequate knowledge of many clinicians of how to treat pain effectively and the fears of patients, family members, and clinicians of the adverse effects of opioids.
- Scenario-based studies in France have revealed that clinicians and lay people hold a wide variety of positions about treating pain.
- Most of these clinicians and lay people are, however, in favor of using even high doses of opioids for terminally ill cancer patients in pain, especially if the patients ask for pain relief and if the pain is severe.
- Educational interventions need to ascertain, adapt to, and target the different positions taken by clinicians, patients, family members, and the public.

References

Adam, R., Bond, C., & Murchie, R. (2015). Educational interventions for cancer pain. A systemic review of systematic reviews with nested narrative review of randomized controlled trials. *Patient Education and Counseling, 98*, 269–282.

Anderson, N. H. (2008). *Unified social cognition.* New York, NY: Psychology Press.

Bouya, S., Balouchi, A., Maleknejad, A., Koochakzai, M., AlKhasawneh, E., & Abdollahimohammad, A. (2019). Cancer pain management among oncology nurses: Knowledge, attitude, related factors, and clinical recommendations: A systematic review. *Journal of Cancer Education, 34*, 839–846.

Breuer, B., Fleishman, S. B., Cruciani, R. A., & Portenoy, R. K. (2011). Medical oncologists' attitudes and practice in cancer pain management: A national survey. *Journal of Clinical Oncology, 29*, 4769–4775.

Cano Romero, M. D., Muñoz Sastre, M. T., Quintard, B., Sorum, P. C., & Mullet, E. (2017). The ethics of postoperative pain management: Mapping nurses' views. *International Journal of Nursing Practice, 23*, e12514.

Darawand, M., Alnajar, M. K., Abdalrahim, M. S., & El-Aqoul, A. M. (2017). Cancer pain management at oncology units: Comparing knowledge, attitudes and perceived barriers between physicians and nurses. *Journal of Cancer Education, 34*, 366–374.

Grant, M., Ugalde, A., Vafiadis, P., & Philip, J. (2015). Exploring the myths of morphine in cancer: Views of the general practice population. *Support Care Cancer, 23*, 485–489.

Hofmans, J., & Mullet, E. (2013). Towards unveiling individual differences in different stages of information processing: A clustering-based approach. *Quality & Quantity, 47*, 555–564.

Jacobsen, R., Sjøgren, P., Møldrup, C., & Christrup, L. (2007). Physician-related barriers to cancer pain management with opioid analgesics: A systematic review. *Journal of Opioid Management, 3*, 207–214.

Jho, H. J., Kim, Y., Kong, K. A., Choi, J. Y., Nam, E. J., Choi, J. Y., et al. (2014). Knowledge, practices, and perceived barriers regarding cancer pain management among physicians and nurses in Korea: A nationwide multicenter survey. *PLoS One, 9*(8), e105900.

Kasasbeh, M. A. M., McCabe, C., & Payne, S. (2017). Cancer-related pain management: A review of knowledge and attitudes of healthcare professionals. *European Journal of Cancer Care, 26*, e12625.

Kwon, J. H. (2014). Overcoming barriers in cancer pain management. *Journal of Clinical Oncology, 32*, 1727–1733.

Larue, F., Fontaine, A., & Brasseur, L. (1999). Evolution of the French public's knowledge and attitudes regarding postoperative pain, cancer pain, and their treatments: Two national surveys over a six-year period. *Anesthesia and Analgesia, 89*, 659–664.

Latchman, J. (2014). Improving pain management at the nursing education level: Evaluating knowledge and attitudes. *Journal of the Advanced Practitioner in Oncology, 5*, 10–16.

Levin, D. N., Cleeland, C. S., & Dar, R. (1985). Public attitudes toward cancer pain. *Cancer, 56*, 2337–2339.

Lou, F., & Shang, S. (2017). Attitudes towards pain management in hospitalized cancer patients and their influencing factors. *Chinese Journal of Cancer Research, 29*, 75–85.

Makhlouf, S. M., Pini, S., Ahmed, S., & Bennet, M. I. (2019). Managing pain in people with cancer—A systematic review of the attitudes and knowledge of professionals, patients, caregivers and public. *Journal of Cancer Education.* https://doi.org/10.1007/s13187-019-01548-9. Published online.

Mas, C., Albaret, M.-C., Sorum, P. C., & Mullet, E. (2010). French general practitioners vary in their attitudes toward treating terminally ill patients. *Palliative Medicine, 24*, 60–67.

Mazoyer, J., Muñoz Sastre, M. T., Sorum, P. C., & Mullet, E. (2016). French lay people's and health professionals' views on the acceptability of terminal sedation. *Journal of Medical Ethics, 42*, 627–631.

Mazoyer, J., Muñoz Sastre, M. T., Sorum, P. C., & Mullet, E. (2017). Mapping French people and health professionals' positions regarding the circumstances of morphine use to relieve cancer pain. *Support Care Cancer, 25*, 2723–2731.

Meeker, M. A., Finnell, D., & Othman, A. K. (2011). Family caregivers and cancer pain management: A review. *Journal of Family Nursing, 17*, 29–60.

National Academies of Sciences, Engineering, and Medicine. (2019). *Pain management for people with serious illness in the context of the opioid use disorder epidemic: Proceedings of a workshop*. Washington, DC: The National Academies Press.

Oldenmenger, W. H., Geerling, J. L., Mostovoya, I., Vissers, K. C. P., de Graeff, A., Reyners, A. K. L., et al. (2018). A systematic review of the effectiveness of patient-based educational interventions to improve cancer-related pain. *Cancer Treatment Reviews, 63*, 96–103.

Paice, J. A. (2018). Cancer pain management and the opioid crisis in America: How to preserve hard-earned gains in improving the quality of cancer pain management. *Cancer, 124*, 2491–2497.

Saifan, A. R., Bashayreh, I. H., Al-Ghabeesh, S. H., Batiha, A.-M., Alrimawi, I., Al-Saraireh, M., et al. (2019). Exploring factors among healthcare professionals that inhibit effective pain management in cancer patients. *Central Europe Journal of Nursing and Midwifery, 10*, 967–976.

Saifan, A., Bashayreh, I., Batiha, A.-M., & AbuRuz, M. (2015). Patient- and family caregiver-related barriers to effective cancer pain control. *Pain Management Nursing, 16*, 400–410.

Saini, S., & Bhatnagar, S. (2016). Cancer pain management in developing countries. *Indian Journal of Palliative Care, 22*, 373–377.

Salim, N., Al-Attyat, Tuffaha, M., Higm, H. A., & Brant, J. (2017). Knowledge and attitude of oncology nurses toward cancer pain management: A review. *Archives of Medicine, 9*, 2.

Shahriary, S., Shiryazdi, S. M., Shiryazki, S. A., Arjomandi, A., Haghighi, F., Vakili, F. M., et al. (2015). Oncology nurses knowledge and attitudes regarding cancer pain management. *Asian Pacific Journal of Cancer Prevention, 16*, 7501–7506.

Utne, I., Småstuen, M. C., & Nyblin, U. (2019). Pain knowledge and attitudes among nurses in cancer care in Norway. *Journal of Cancer Education, 34*, 677–684.

Vallerand, A. H., Collins-Bohler, D., Templin, T., & Hasenau, S. M. (2007). Knowledge of and barriers to pain management in caregivers of cancer patients receiving homecare. *Cancer Nursing, 30*, 31–37.

van den Beuken-van Everdingen, M. H. J., Hochstenbach, L. M. J., Joosten, E. A. J., Tjan-Heijnen, V. C. G., & Janssen, D. J. A. (2016). Update on prevalence of pain in patients with cancer: Systemic review and meta-analysis. *Journal of Pain and Symptom Management, 51*, 1070–1090.

World Health Organization. (2018). *WHO guidelines for the pharmacological and radiotherapeutic management of cancer pain in adults and adolescents*. Geneva: World Health Organization.

Zhang, Q., Yu, C., Feng, S., Yao, W., Shi, H., Zhao, Y., et al. (2015). Physicians' practice, attitudes toward, and knowledge of cancer pain management in China. *Pain Medicine, 16*, 2195–2203.

Chapter 29

Manual compression at myofascial trigger points ameliorates musculoskeletal pain

Kouichi Takamoto[a,b], Susumu Urakawa[c], Shigekazu Sakai[a], Taketoshi Ono[a], and Hisao Nishijo[a]

[a]*System Emotional Science, Faculty of Medicine, University of Toyama, Toyama, Japan,* [b]*Department of Sports and Health Sciences, Faculty of human sciences, University of East Asia, Yamaguchi, Japan,* [c]*Department of Musculoskeletal Functional Research and Regeneration, Graduate School of Biomedical and Health Sciences, Hiroshima University, Hiroshima, Japan*

Abbreviations

amPFC	antero-medial PFC
MTrP	myofascial trigger point
PFC	prefrontal cortex
PPT	pressure pain threshold

Introduction

Musculoskeletal pain, which is pain caused by lesions or deficits in the motor system such as bones, muscles, ligaments, and nerves, leads to impaired functional ability, and reduced quality of life and mental well-being (Briggs et al., 2016). About a third of patients with musculoskeletal pain reported chronic complaints about their physical conditions (Wijnhoven, de Vet, & Picavet, 2006). Low back, neck, shoulder, and knee are common sites of musculoskeletal pain (Nakamura, Nishiwaki, Ushida, & Toyama, 2011; Shamsi, Safari, Samadzadeh, & Yoosefpour, 2020). Prevalence of patients with musculoskeletal pain has been increasing over age and time (Ahacic & Kåreholt, 2010; Harkness, Macfarlane, Silman, & McBeth, 2005). These findings indicate that musculoskeletal pain is one of the important healthcare issues to be resolved.

A myofascial trigger point (MTrP) is suggested to cause muscle pain. MTrPs can be identified by the following diagnostic criteria: (i) presence of a localized hypersensitive palpable nodule in a taut band of muscle fibers, (ii) induction of characteristic referred pain when compressing the point (Simons, Travell, & Simons, 1999). Furthermore, MTrPs lead to motor dysfunction (stiffness and limited range of motion) and autonomic phenomena. There are two types of MTrPs: active and latent MTrPs. Active MTrPs are defined as the spots, compression of which reproduces the same muscle pain as subjects usually complain, while latent MTrPs are defined as spots, compression of which induces muscle pain that is not associated with clinical pain (Simons et al., 1999). Epidemiological studies reported that active MTrPs were associated with acute and chronic pain conditions in the neck, shoulder, low back, joints, and head (see the section "Link between MTrPs and musculoskeletal pain"). These findings suggest that treatment of active MTrPs could ameliorate musculoskeletal pain.

Pharmacological therapy, exercise therapy, manual therapy, and surgery have been applied to treat musculoskeletal pain (Babatunde et al., 2017). Physicians (orthopedists, internists, and anesthetists) frequently apply not only injection of analgesics (mainly metamizole/paracetamol, nonsteroidal antiinflammatory drugs (NSAID), or week opioid) but also physical therapy (mainly manual therapy, transcutaneous electrical stimulation (TENS), acupuncture, or dry needling) to treat chronic pain in the muscle (Fleckenstein et al., 2010). Compression at MTrPs is one of the manual therapeutic techniques for musculoskeletal pain (Bialosky, Bishop, Price, Robinson, & George, 2009). Clinical trial studies suggest that compression at MTrP is effective to reduce musculoskeletal pain (see the section on "Effects of compression at MTrPs for musculoskeletal pain").

Features and Assessments of Pain, Anesthesia, and Analgesia. https://doi.org/10.1016/B978-0-12-818988-7.00023-6
Copyright © 2022 Elsevier Inc. All rights reserved.

In this chapter, we discuss (1) prevalence of MTrPs in the musculoskeletal pain, (2) clinical efficacy of MTrP compression for musculoskeletal pain, and (3) peripheral and central mechanisms of MTrP compression effects.

Link between MTrPs and musculoskeletal pain

Epidemiologic studies analyzed prevalence of MTrPs in patients who visited hospitals and pain centers due to pain. Active MTrPs were identified in 30% of patents with pain who visited a primary care unit (Skootsky, Jaeger, & Oye, 1989). Furthermore, 85% of patients who visited a comprehensive pain center due to chronic pain were diagnosed with pain associated with MTrPs (Fishbain, Goldberg, Meagher, Steele, & Rosomoff, 1986). In addition, a nationwide survey study of 332 German physicians in the hospitals and pain centers, who experienced in treating patients with chronic pain muscle (orthopedist, internist, and anesthetist), reported that prevalence of active MTrPs was 46.1% of the overall population (Fleckenstein et al., 2010). These results suggest that MTrPs might account for causes of a large percentage of patients with chronic pain.

Several cross-sectional studies investigated the difference in prevalence of active and latent MTrPs between patients with musculoskeletal pain in the neck, shoulder, low back, and knee pain and healthy subjects (Table 1). In each study, one assessor, who had sufficient experience for diagnosis of MTrPs, diagnosed MTrPs by physical examination combined with palpatory, visual inspection, and patient feedback.

Low-back pain

Iglesias-González et al. (2013) investigated that prevalence of active and latent MTrPs in patients with chronic nonspecific low-back pain and age- and sex-matched healthy subjects. Active MTrPs in the low back and hip muscles were detected only in the patients with low-back pain. Number of latent MTrPs in the patients was significantly higher than the healthy subjects. Active MTrPs were detected mostly in the quadratus lumborum, iliocostalis lumborum, and gluteus medius muscles in the patients. Furthermore, numbers of active MTrPs in the patients were positively correlated with the intensity of subjective feeling of pain and negatively correlated with sleep quality. In addition, MTrPs in the low back and hip muscles (mainly lumbar quadrate, gluteus medius muscles) were identified in 85.7% of the patients with failed back surgery pain syndrome (Teixeira et al., 2011).

Neck pain

Two cross-sectional studies investigated prevalence of MTrPs in patients with mechanical neck pain and healthy subjects (Fernández-de-las-Peñas et al., 2007; Muñoz-Muñoz et al., 2012). Active MTrPs in the neck muscles were identified only in the patients. However, number of latent MTrPs was not different between the patients and healthy subjects. Active MTrPs were detected mostly in the upper trapezius and levator scapulae muscles in patients with neck pain. Furthermore, another study reported that active MTrPs in the trapezius muscle (94%) was most common among the neck-shoulder muscles in patients with chronic nonspecific neck pain (Cerezo-Téllez et al., 2016). In addition, active MTrPs were also frequently identified in the levator scapulae (82%), multifidi (78%), and splenius cervicis (63%) muscles. In acute whiplash-associated disorder (WAD), which is a complex condition with neck pain and disability after car accidents, active MTrPs were found mostly in the upper trapezius and levator scapulae muscles (Fernández-Pérez et al., 2012). The number of active MTrPs was associated with greater pain intensity and longer elapsed days from the accident, and reduced cervical range of motion. Furthermore, the mean number of active MTrPs in patients with chronic whiplash-associated disorder was significantly greater than patients with mechanical neck pain (Castaldo, Ge, Chiarotto, Villafane, & Arendt-Nielsen, 2014). In addition, number of active MTrPs was positively associated with pain intensity and spontaneous pain area.

Shoulder pain

Alburquerque-Sendín et al. (2013) compared MTrPs in the shoulder muscles between patients with unilateral Shoulder impingement syndrome and healthy subjects. Active MTrPs in the shoulder muscles were identified only in patients with this syndrome. Averaged number of active MTrPs was significantly higher in the affected side than the unaffected side in the patients. Active MTrPs were found mostly in the trapezius, infraspinatus, subscapularis, and scalene muscles in the affected side of the patients, and number of active MTrPs was positively correlated with current pain intensity at rest. However, there was no difference in total number of latent MTrPs in the shoulder muscles between the patients and healthy subjects. Another study reported that active MTrPs were identified in the shoulder muscles of the all 72 patients with

TABLE 1 Prevalence of active and latent MTrPs in patients with musculoskeletal pain (neck, shoulder, low back, and knee pain) and healthy subjects.

Author (year)	Patients	HS	Examined muscles (both sides)	Number of active MTrPs		Number of latent MTrPs	
				Patients	HS	Patients	HS
Iglesias-González, Muñoz-García, Rodrigues-de-Souza, Alburquerque-Sendín, and Fernández-de-Las-Peñas (2013)	Chronic nLBP (n = 42; age = 45 ± 8 years; M/F: 21/21)	(n = 42; age = 45 ± 9 years; M/F: 21/21)	Quadratus lumborum, iliocostalis lumborum, psoas, piriformis, gluteus minimus, gluteus medius	3.5 ± 2.3	0	2.0 ± 1.5[#]	1.0 ± 1.5
Fernández-de-las-Peñas, Alonso-Blanco, and Miangolarra (2007)	Chronic neck pain (n = 20; age = 28 ± 7 years; M/F: 7/13)	(n = 20; age = 29 ± 9 years, M/F: 10/10)	Upper trapezius, levator scapulae, sternocleidomastoid, suboccipital	1.8 ± 0.8[*]	0	2.5 ± 1.3	2.0 ± 0.8
Muñoz-Muñoz, Muñoz-García, Alburquerque-Sendín, Arroyo-Morales, and Fernández-de-las-Peñas (2012)	Neck pain (n = 15; age = 39 ± 8 years; M/F: 3/12)	(n = 12; age = 40 ± 7 years; M/F: 3/9)	Upper trapezius, levator scapulae, sternocleidomastoid, scalene, splenius capitis, semispinalis capitis	2.0 ± 2.0[*]	0	1.6 ± 1.4	1.3 ± 1.4
Fernández-Pérez et al. (2012)	Acute WADs (n = 20; age = 28.7 ± 12.4 years; M/F: 10/10)	(n = 20; age = 29.1 ± 12.2 years; M/F: 10/10)	Upper trapezius, levator scapulae, sternocleidomastoid, scalene, temporalis, masseter	3.9 ± 2.5[#]	0	3.4 ± 2.7[#]	1.7 ± 2.2
Alburquerque-Sendín, Camargo, Vieira, and Salvini (2013)	Unilateral SIS (n = 27; age = 35.6 ± 12.1 years; M/F: 13/14)	(n = 20; age = 37.0 ± 11.2 years; M/F: 9/11)	Upper trapezius, levator scapula, scalenes, supraspinatus, infraspinatus, subscapularis, middle deltoid, pectoralis minor	4.3 ± 2.5	0	3.6 ± 3.2	3.8 ± 4.3
Alburquerque-García, Rodrigues-de-Souza, Fernández-de-las-Peñas, and Alburquerque-Sendín (2015)	Knee osteoarthritis (n = 18; age = 85.0 ± 4 years; M/F:0/18)	(n = 18; age = 85.6 ± 4 years; M/F: 0/18)	Tensor fasciae latae, sartorius, rectus femoris, vastus lateralis, vastus medialis, gracilis, biceps femoris, semitendinosus, tibialis anterior, gastrocnemius	1.1 ± 4[#]	0	4.0 ± 2	4.0 ± 3

Data are presented as Mean ± SD.

HS, healthy subjects; nLBP, nonspecific low-back pain; M/F, male/female; WADs, whiplash-associated disorders; SIS, shoulder impingement syndrome.

[*]Significant difference from HS (unpaired t-test, $P < .01$).

[#]Significant difference from HS (Mann-Whitney U test, $P < .01$).

320 PART | III Interlinking anesthesia, analgesics and pain

nontraumatic shoulder pain examined, suggesting that prevalence of MTrP is high (Bron, Dommerholt, Stegenga, Wensing, & Oostendorp, 2011). In this study, number of the muscles with active MTrPs was positively correlated with self-reporting disability scores.

Knee pain

In patients with anterior knee pain, active MTrPs were identified mostly in the vastus medialis and vastus lateralis muscles of the affected side (Rozenfeld, Finestone, Moran, Damri, & Kalichman, 2020). Although there was no significant difference in the number of latent MTrPs between the patients with anterior knee pain and healthy subjects, total number of MTrPs (active plus latent) in the patients with anterior knee pain was positively correlated with self-reporting disability scores in the knee and hip. Another study reported that elderly female patients with painful knee osteoarthritis had a greater number of active MTrPs than healthy female subjects, but there was no difference in number of latent MTrPs between the two groups (Alburquerque-García et al., 2015). The number of the active MTrPs was positively correlated with pain intensity, suggesting that active MTrPs contributes to pain in knee osteoarthritis.

Other body regions

MTrPs were frequently (75%) identified in the trapezius muscle in patients with carpal tunnel syndrome (Azadeh, Dehghani, & Zarezadeh, 2010). In chronic lateral epicondylalgia, active MTrPs were found in the extensor carpi radialis brevis muscle of the affected side in all patients (Fernández-Carnero, Fernández-de-las-Peñas, de la Llave-Rincón, Ge, & Arendt-Nielsen, 2008). In addition, active MTrPs were identified in the iliopsoaos, rectus femoris, and tensor fasciae latae muscle in the patients with hip osteoarthritis (Ceballos-Laita et al., 2019).

These studies indicate that presence of active MTrPs are significantly associated with acute and chronic pain in patients with and without structural changes of the bones, articular ligaments, etc., and suggest that active MTrPs significantly contribute to development of pain.

Diagnosis of MTrPs

Simons et al. (1999) and Gerwin, Shannon, Hong, Hubbard, and Gevirtz (1997) proposed the minimal diagnostic criteria (see section Introduction). The criteria showed the good reliability for MTrP detection. However, a meta-analysis study indicated that reproducibility in MTrP identification was influenced by differences of experience and training periods of examiners, diagnostic criteria used, and measurement muscle sites; consequently, overall reliability of physical examination of MTrPs did not reach acceptable achievement levels (Lucas, Macaskill, Irwig, Moran, & Bogduk, 2009; Myburgh, Larsen, & Hartvigsen, 2008; Rathbone, Grosman-Rimon, & Kumbhare, 2017). On the other hand, a recent study reported good reliability for MTrP detection in the upper quarter muscles by two expert examiners who received the same training course for MTrP detection (Mayoral Del Moral, Torres Lacomba, Russell, Sánchez Méndez, & Sánchez Sánchez, 2018). These findings suggest the importance of standardized diagnostic criteria of MTrPs for reliability and reproducibility of MTrP diagnosis. Furthermore, recent noninvasive imaging techniques using magnetic resonance elastography and sono-elastography allow to objectively detect MTrPs as elastic nodules (Chen et al., 2016; Jafari, Bahrpeyma, Mokhtari-Dizaji, & Nasiri, 2018). Therefore, standardization of MTrP diagnosis combined with reliable imaging of elasticity would contribute to development of MTrP diagnosis.

Effects of compression at MTrPs for musculoskeletal pain

Compression at MTrP is a massage technique for musculoskeletal pain, in which sustained pressure stimuli by the thumb or forefinger is repeatedly applied to MTrP until the MTrP becomes inactive (Simons et al., 1999). Compression at MTrP for 60–90 s with high intensity (between pain threshold and tolerance level) was more effective to reduce pain (Hou, Tsai, Cheng, Chung, & Hong, 2002; Pecos-Martin et al., 2019). Several clinical trial studies reported that effectiveness of compression at MTrP for musculoskeletal pain (Table 2; see below).

Low-back pain

Takamoto et al. (2015) conducted a clinical trial on compression at active MTrP in patients with acute low-back pain. The 63 patients with acute back pain and active MTrPs in the muscles of the lumber region were treated with one of the three

Analgesic effects of MTrP compression Chapter | 29 321

TABLE 2 Effectiveness of MTrP compression for musculoskeletal pain (low-back pain, neck pain, shoulder pain, and knee pain).

Author (year)	Design	Patients	Intervention	Protocol	Results
Takamoto et al. (2015)	RCT	Acute LBP	A: Comp at active MTrPs ($n = 22$) B: Comp at Non-MTrPs ($n = 17$) C: Effleurage massage ($n = 16$)	Three times/week (Max 6)	VAS↓ (A > B, C in changes from baseline at 1 week after and 1-month follow-up); VAS during movement↓ (A > B, C in changes from baseline at 1 week after and 1-month follow-up); PPT↑ (A > B, C in changes from baseline at 1 week after and 1-month follow-up); lumbar ROM↑ (A > B, C in changes from baseline at 1 week after and 1-month follow-up); RMQ (no difference in changes from baseline at 1 week after, and 1-month follow-up among the groups)
Kodama et al. (2019)	RCT	Chronic LBP	A: Comp at active MTrP in lumbar quadrate M ($n = 16$) B: Comp at Non-MTrP in lumbar quadrate M ($n = 16$)	30 s × 5	VAS↓, PPT↑ (A > B in change from baseline at posttreatment)
Cagnie et al. (2013)	No-RCT	Chronic NP	A: Comp at MTrPs in neck and shoulder muscles ($n = 22$) B: No treatment ($n = 25$)	Two times/week (total 8)	NRS↓ (A at posttreatment and at 6-month follow-up > B at posttreatment); PPT↑ (A > B at posttreatment); ROM↑ (A > B at posttreatment); muscle strength↑ (A > B at posttreatment); NDI (A at post-treatment and 6 months follow-up ≮ B at posttreatment)
Morikawa et al. (2017)	RCT	Chronic NP	A: Comp at active MTrP in trapezius M ($n = 11$) B: Comp at Non-MTrP in trapezius M ($n = 10$)	30 s × 4	VAS↓ (A > B at change from baseline to posttreatment)
Hains, Descarreaux, and Hains (2010)	RCT	Chronic SP	A: Comp at MTrPs in shoulder M (supraspinatus, deltoid, infraspinatus, biceps tendon) ($n = 41$) B: Comp at MTrPs in neck and upper back M ($n = 18$)	Three times/week (total 15)	SPADI↓ (A > B at post-treatment) (pre > post-treatment, 1 month, and 6 months follow-up in A); percentages of amelioration↑ (A > B at posttreatment)
Bron, de Gast, et al. (2011)	RCT	Chronic SP	A: Comp at active MTrPs in shoulder M plus SCT ($n = 37$) B: Waiting list (delay of the intervention for 3 months) ($n = 35$)	One times/week (max 12)	VAS↓ (A > B at posttreatment); DASH↓ (A > B at posttreatment); TNAM↓ (A > B at posttreatment); ROM (A B at posttreatment)
Akbaba et al. (2019)	RCT	RCF	A: Comp at active MTrPs plus SCT ($n = 23$) B: Only SCT ($n = 23$)	Two times/week (total 12)	VAS↓ (pre > posttreatment in A); TNAM↓ (pre > posttreatment in A); DASH↓ (pre > posttreatment in A and B; A ≮ B at posttreatment); ROM↑ (pre > posttreatment in A and B; A B at posttreatment); ASES↓ (pre > posttreatment in A and B; A ≮ B at posttreatment)
Hains and Hains (2010)	RCT	PFPS	A: Comp at active MTrPs around patellar regions ($n = 27$) B: Comp at active MTrP located in gluteus maximus, medius, and minimus M ($n = 11$)	Three times/week (total 15)	PGT↓ (A > B at posttreatment); VAS↓ (pre > posttreatment in A at 30 days, and 6 months follow-up) (A ≮ B at posttreatment)

Continued

TABLE 2 Effectiveness of MTrP compression for musculoskeletal pain (low-back pain, neck pain, shoulder pain, and knee pain)—cont'd

Author (year)	Design	Patients	Intervention	Protocol	Results
Behrangrad and Kamali (2017)	RCT	PFPS	A: Comp at MTrP in the vastus medialis obliquus M ($n=15$) B: Lumbopelvic manipulation ($n=15$)	Three times/week (total 3)	VAS↓ (A > B at 1 week after, and 1, 3 months follow up); Kujala questionnaire↑ (A > B at 1 week after, and 1, 3 months follow-up); PPT↑ (A > B at 1 week after, and 1, 3 months follow-up)

RCT, randomized clinical trial; *LBP*, low-back pain; *Comp*, compression; *VAS*, visual analog scale; *RMQ*, Roland Morris disability questionnaire; *PPT*, pain pressure threshold; *ROM*, range of motion; *M*, muscle; *NP*, neck pain; *NRS*, numerical rating scale; *NDI*, neck disability index; *SP*, shoulder pain; *SPADI*, shoulder pain and disability index; *RCF*, rotator cuff tears; *SCT*, standard comprehensive treatment (cold application and home exercise); *DASH*, Disability of the Arm, shoulder, and hand score; *TNAM*, total number of active MTrP; *ASES*, American shoulder and elbow surgeons score; *PGT*, patellar-grinding test scores; *PFPS*, patella-femoral pain syndrome; < or >, significant difference between groups; ≮ or , no significant difference between groups.

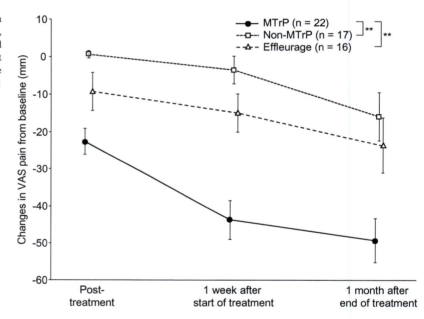

FIG. 1 Effects of compression at active MTrP in patients with acute low-back pain (Takamoto et al., 2015). MTrP compression significantly improved acute low-back pain compared with compression at Non-MTrP and effleurage massage. Error bars indicate SEM. *VAS pain*, visual analog scale of pain; **$P < .01$ (Tukey test).

manual therapies: compression at active MTrP (MTrP group), compression at an area 2 cm away from the MTrP (Non-MTrP group), and superficial massage (effleurage group). The patients received the intervention of one of the three manual therapies three times/week for 2 weeks. The results indicated that compression at active MTrP significantly improved low-back pain compared with the Non-MTrP and effleurage groups (Fig. 1). Furthermore, pressure pain threshold (PPT) and range of motion in the lumbar region were significantly increased in the MTrP group compared with the Non-MTrP and effleurage groups. Another study reported that compression at active MTrP in the quadratus lumborum muscle significantly improved chronic low-back pain compared with Non-MTrP compression (Kodama et al., 2019).

Neck pain

In office workers who had pain or complain in the neck and shoulder regions, compression at MTrP significantly improved neck/shoulder pain just after treatment and at 6 months after the treatment (Cagnie et al., 2013). In patients with chronic neck pain, compression at active MTrPs in the trapezius muscle significantly improved the pain compared with Non-MTrP compression (Morikawa et al., 2017). A meta-analysis study on 15 randomized-controlled trials reported that compression at active MTrPs in the trapezius muscle improved pain, PPT, and range of motion in the cervical region (Cagnie et al., 2015).

Shoulder pain

In patients with chronic shoulder pain, compression at MTrPs in the supraspinatus, infraspinatus, and deltoid muscles, and the biceps tendon significantly improved self-reporting scores for shoulder pain and related disabilities after 15 treatments compared with compression at MTrPs in the cervical and upper dorsal areas (Hains, Descarreaux, et al., 2010). Furthermore, multimodal intervention consisting of compression at MTrP, muscle stretching exercise and intermittent cold application to unilateral nontraumatic chronic shoulder pain improved pain and self-reporting disability scores, and reduced number of muscles with active MTrPs (Bron, de Gast, et al., 2011). In addition, patients with rotator cuff tear, compression at active MTrPs significantly decreased pain and number of MTrPs after 12 treatments (Akbaba et al., 2019).

Knee pain

In patients with anterior knee pain, compression at active MTrPs in peri- and retro-patellar regions improved pain. The reduction of pain was maintained at 6 months after the treatment (Hains & Hains, 2010). Another clinical trial using patients with anterior knee pain reported that compression at MTrPs in the vastus medialis obliquus muscle improved pain and self-reporting functional states of the knee (Behrangrad & Kamali, 2017). The effects of MTrP compression were maintained at 3 months after the treatment.

Pain in other body regions

In chronic nonspecific foot pain, compression at active MTrPs in the foot regions combined with mobilization therapy improved self-reporting pain and disability scores after 15 treatments (Hains, Boucher, & Lamy, 2015). In patients with chronic carpal tunnel syndrome, compression at MTrPs around regions along running of the median nerve (i.e., biceps and pronator teres muscles, etc.) improved self-reporting symptoms and functional states after 15 treatments compared with compression at MTrPs in the shoulder muscles (Hains, Descarreaux, Lamy, & Hains, 2010).

Taken together, MTrPs could cause pain in patients without specific primary disorders in the motor system, and modulate (usually enhance) pain caused by primary disorders. In either case, compression at MTrPs could reduce musculoskeletal pain.

Possible mechanisms of MTrP compression effects

There could be two possible sites of action for MTrP compression effects; peripheral direct effects on muscles and indirect effects mediated through the central nervous system (CNS). In muscles with MTrPs, local contracture (i.e., MTrPs) induces mitochondrial deficits due to local muscle ischemia/hypoxia. Deficits of energy and oxygen supply cause release of algesic substances, which induces hypersensitivity and pain (Gerwin, Dommerholt, & Shah, 2004; Shah et al., 2015). A metabolome study using a rat model of musculoskeletal pain in the gastrocnemius reported that compression on the muscle reduced mechanical hyperalgesia and altered metabolic profiles in the muscle, consistent with ameliorative effects of mitochondrial respiration (Urakawa et al., 2015). Furthermore, compression at MTrP increases regional blood flow and decreases neuromuscular excitability (Moraska, Hickner, Kohrt, & Brewer, 2013). Thus, compression at MTrP may alleviate pain by improving energy balance in the muscle.

Extensive studies reported that various psychological factors modulate musculoskeletal pain such as low-back pain, suggesting an involvement of the central nervous system in pain induction. In patients with chronic musculoskeletal pain with MTrPs, morphological and functional changes in pain-related brain regions such as the prefrontal cortex (PFC), insula, and anterior cingulate cortex have been reported (Coppieters et al., 2016; Niddam, 2009; Niddam, Lee, Su, & Chan, 2017). The PFC is involved in cognition of pain as well as pain chronification (Ong, Stohler, & Herr, 2019). Furthermore, activity in the antero-medial PFC (amPFC) was increased and correlated with spontaneous pain intensity in patients with chronic low-back pain (Baliki et al., 2006), and positively correlated with sympathetic activity in patients with chronic neck pain (Morikawa et al., 2017). Interestingly, increases in sympathetic activity may develop MTrP

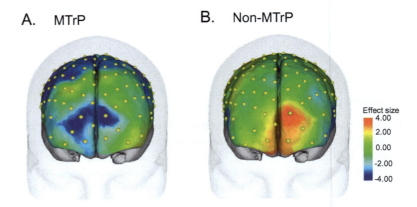

FIG. 2 Effects of compression at active MTrP (A) and Non-MTrP (B) on cerebral hemodynamic activity (Kodama et al., 2019). The figure indicates that cerebral hemodynamic activity decreased in the amPFC during compression at an active MTrP, which was derived from one subject. Cerebral hemodynamic activity was measured using near-infrared spectroscopy (NIRS).

formation (Ge, Fernández-de-las-Peñas, & Arendt-Nielsen, 2006), which might be attributed to facilitation of ACh release in the neuromuscular junction by sympathetic system (Gerwin et al., 2004).

Two imaging studies investigated effects of MTrP compression on CNS activity. Compression at active MTrPs decreased hemodynamic activity in the amPFC in patients with chronic low-back pain (Fig. 2) (Kodama et al., 2019) and in patients with chronic neck pain (Morikawa et al., 2017), and reduced sympathetic activity (Morikawa et al., 2017). Furthermore, amPFC activity was positively correlated with sympathetic nervous activity, which was negatively correlated with pain. In addition, MTrP compression decreased functional connectivity in between the amPFC and insula

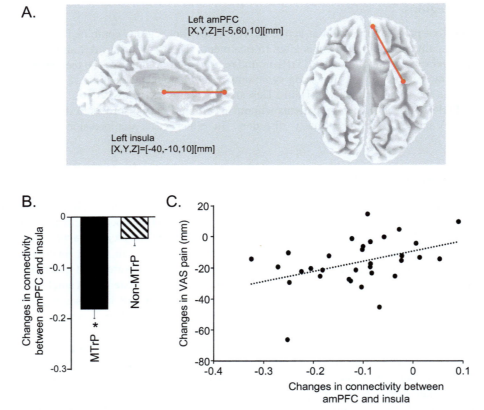

FIG. 3 Effects of MTrP compression on functional connectivity between the left amPFC and left insula cortex (Kodama et al., 2019). (A) Locations of the left amPFC and left insula cortex where functional connectivity between two brain regions was measured based on EEG data. (B) Changes in functional connectivity from the baseline during compression. *$P < .05$. (C) Relationships between changes in functional connectivity and changes in subjective pain from the baseline during compression. *VAS pain*, visual analog scale of pain.

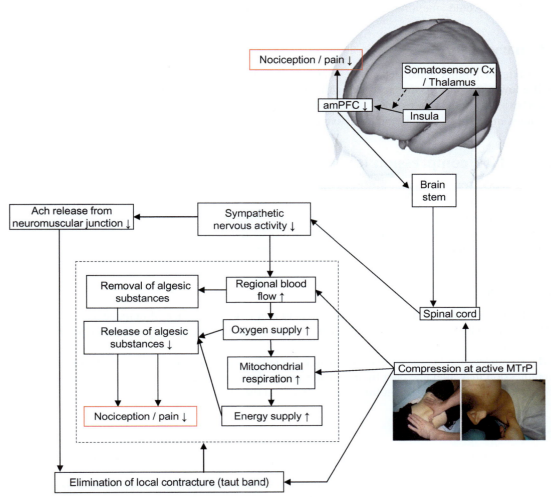

FIG. 4 Diagram showing hypothetical mechanisms of compression at active MTrP. *Dotted line* indicates suppressive effects on functional connectivity (see Kodama et al., 2019 for detailed discussion). MTrP compression may affect the descending pain modulatory system, which is not included in this figure. *Cx*, cortex.

cortex compared with compression at Non-MTrP (Fig. 3A and B) (Kodama et al., 2019). There was significant positive correlation between strength of the functional connectivity and subjective pain (Fig. 3C). These findings suggest that persistent nociceptive inputs from MTrPs may be transmitted to the amPFC through the insula cortex, which may in turn increase sympathetic activity to further develop MTrPs. Compression at active MTrPs may provide pain relief by breaking the vicious circle among the amPFC, sympathetic activity, and MTrPs (Fig. 4).

Applications to other areas

Active MTrPs evoke chronic pain, which could secondarily induce mental disorders. Common neural mechanisms may underline in pain- and depression-induced changes in the PFC and hippocampus (Sheng, Liu, Wang, Cui, & Zhang, 2017). In healthy subjects, presence and number of latent MTrPs in the scapular muscles were associated with depression levels (Celik & Kaya Mutlu, 2012). Major depression was more frequently observed in patients with chronic neck pain and active MTrPs in the trapezius muscle than healthy controls (Altindag, Gur, & Altindag, 2008). Furthermore, injection into active MTrPs in the trapezius muscle improved depression (Ay, Evcik, & Tur, 2010). Thus, MTrP-induced pain may develop depression. On the other hand, sympathetic nervous activity was increased in major depression, and decreased by the administration of antidepressant (Veith et al., 1994), while chronic pain due to active MTrPs was positively correlated with sympathetic activity (Morikawa et al., 2017). These findings suggest that depression is involved in generation and maintenance of MTrPs by increasing sympathetic activity. The close link between depression and MTrPs suggests that compression at MTrPs would ameliorate depression and prevent facilitatory effects of depression on MTrPs.

Mini-dictionary of terms

- **Myofascial trigger points:** Sensitized stiff spots in muscles with chronic local contracture of muscle fibers.
- **Nonspecific low-back pain:** Low-back pain without specific organic pathologies.
- **Insular cortex:** This brain area receives multimodal afferents from sensory (including pain), limbic, and autonomic systems, and is involved in sensory and affective aspects of pain information processing.

Key facts of MTrP compression for musculoskeletal pain

- Prevalence of active MTrPs is high (30%–85% of patients with acute or chronic pain).
- Active MTrPs are associated with clinical pain.
- MTrP compression improves muscle metabolism and blood circulation.
- MTrP compression ameliorates acute and chronic musculoskeletal pain.
- MTrP is associated with other mental disorders such as depression.

Summary points

- Active MTrPs are responsible for acute and chronic pain in a large percentage of outpatients.
- Active MTrPs not only primarily cause musculoskeletal pain but also secondarily enhance pain induced by primary organic disorders.
- Active MTrP compression is a beneficial treatment especially for nonspecific musculoskeletal pain.
- MTrP compression may alleviate pain by improving energy balance in the muscle.
- MTrP compression may alleviate pain through its effects on the prefrontal cortex.

References

Ahacic, K., & Kåreholt, I. (2010). Prevalence of musculoskeletal pain in the general Swedish population from 1968 to 2002: Age, period, and cohort patterns. *Pain*, *151*(1), 206–214.

Akbaba, Y. A., Mutlu, E. K., Altun, S., Turkmen, E., Birinci, T., & Celik, D. (2019). The effectiveness of trigger point treatment in rotator cuff pathology: A randomized controlled double-blind study. *Journal of Back and Musculoskeletal Rehabilitation*, *32*(3), 519–527.

Alburquerque-García, A., Rodrigues-de-Souza, D. P., Fernández-de-las-Peñas, C., & Alburquerque-Sendín, F. (2015). Association between muscle trigger points, ongoing pain, function, and sleep quality in elderly women with bilateral painful knee osteoarthritis. *Journal of Manipulative and Physiological Therapeutics*, *38*(4), 262–268.

Alburquerque-Sendín, F., Camargo, P. R., Vieira, A., & Salvini, T. F. (2013). Bilateral myofascial trigger points and pressure pain thresholds in the shoulder muscles in patients with unilateral shoulder impingement syndrome: A blinded, controlled study. *Clinical Journal of Pain*, *29*(6), 478–486.

Altindag, O., Gur, A., & Altindag, A. (2008). The relationship between clinical parameters and depression level in patients with myofascial pain syndrome. *Pain Medicine (Malden, Mass.)*, *9*(2), 161–165.

Ay, S., Evcik, D., & Tur, B. S. (2010). Comparison of injection methods in myofascial pain syndrome: A randomized controlled trial. *Clinical Rheumatology*, *29*(1), 19–23.

Azadeh, H., Dehghani, M., & Zarezadeh, A. (2010). Incidence of trapezius myofascial trigger points in patients with the possible carpal tunnel syndrome. *Journal of Research in Medical Sciences: The Official Journal of Isfahan University of Medical Sciences*, *15*(5), 250–255.

Babatunde, O. O., Jordan, J. L., Van der Windt, D. A., Hill, J. C., Foster, N. E., & Protheroe, J. (2017). Effective treatment options for musculoskeletal pain in primary care: A systematic overview of current evidence. *PLoS One*, *12*(6), e0178621.

Baliki, M. N., Chialvo, D. R., Geha, P. Y., Levy, R. M., Harden, R. N., Parrish, T. B., et al. (2006). Chronic pain and the emotional brain: Specific brain activity associated with spontaneous fluctuations of intensity of chronic back pain. *Journal of Neuroscience: The Official Journal of the Society for Neuroscience*, *26*(47), 12165–12173.

Behrangrad, S., & Kamali, F. (2017). Comparison of ischemic compression and lumbopelvic manipulation as trigger point therapy for patellofemoral pain syndrome in young adults: A double-blind randomized clinical trial. *Journal of Bodywork and Movement Therapies*, *21*(3), 554–564.

Bialosky, J. E., Bishop, M. D., Price, D. D., Robinson, M. E., & George, S. Z. (2009). The mechanisms of manual therapy in the treatment of musculoskeletal pain: A comprehensive model. *Manual Therapy*, *14*(5), 531–538.

Briggs, A. M., Cross, M. J., Hoy, D. G., Sànchez-Riera, L., Blyth, F. M., Woolf, A. D., et al. (2016). Musculoskeletal health conditions represent a global threat to healthy aging: A report for the 2015 World Health Organization world report on ageing and health. *Gerontologist*, *56*(Suppl. 2), S243–S255.

Bron, C., de Gast, A., Dommerholt, J., Stegenga, B., Wensing, M., & Oostendorp, R. A. (2011). Treatment of myofascial trigger points in patients with chronic shoulder pain: A randomized, controlled trial. *BMC Medicine, 9*, 8.

Bron, C., Dommerholt, J., Stegenga, B., Wensing, M., & Oostendorp, R. A. (2011). High prevalence of shoulder girdle muscles with myofascial trigger points in patients with shoulder pain. *BMC Musculoskeletal Disorders, 12*, 139.

Cagnie, B., Castelein, B., Pollie, F., Steelant, L., Verhoeyen, H., & Cools, A. (2015). Evidence for the use of ischemic compression and dry needling in the management of trigger points of the upper trapezius in patients with neck pain: A systematic review. *American Journal of Physical Medicine & Rehabilitation, 94*(7), 573–583.

Cagnie, B., Dewitte, V., Coppieters, I., Van Oosterwijck, J., Cools, A., & Danneels, L. (2013). Effect of ischemic compression on trigger points in the neck and shoulder muscles in office workers: A cohort study. *Journal of Manipulative and Physiological Therapeutics, 36*(8), 482–489.

Castaldo, M., Ge, H. Y., Chiarotto, A., Villafane, J. H., & Arendt-Nielsen, L. (2014). Myofascial trigger points in patients with whiplash-associated disorders and mechanical neck pain. *Pain Medicine (Malden, Mass.), 15*(5), 842–849.

Ceballos-Laita, L., Jiménez-Del-Barrio, S., Marín-Zurdo, J., Moreno-Calvo, A., Marín-Boné, J., Albarova-Corral, M. I., et al. (2019). Effects of dry needling in HIP muscles in patients with HIP osteoarthritis: A randomized controlled trial. *Musculoskeletal Science & Practice, 43*, 76–82.

Celik, D., & Kaya Mutlu, E. (2012). The relationship between latent trigger points and depression levels in healthy subjects. *Clinical Rheumatology, 31*(6), 907–911.

Cerezo-Téllez, E., Torres-Lacomba, M., Mayoral-Del Moral, O., Sánchez-Sánchez, B., Dommerholt, J., & Gutiérrez-Ortega, C. (2016). Prevalence of myofascial pain syndrome in chronic non-specific neck pain: A population-based cross-sectional descriptive study. *Pain Medicine (Malden, Mass.), 17*(12), 2369–2377.

Chen, Q., Wang, H. J., Gay, R. E., Thompson, J. M., Manduca, A., An, K. N., et al. (2016). Quantification of myofascial taut bands. *Archives of Physical Medicine and Rehabilitation, 97*(1), 67–73.

Coppieters, I., Meeus, M., Kregel, J., Caeyenberghs, K., De Pauw, R., Goubert, D., et al. (2016). Relations between brain alterations and clinical pain measures in chronic musculoskeletal pain: A systematic review. *Journal of Pain: Official Journal of the American Pain Society, 17*(9), 949–962.

Fernández-Carnero, J., Fernández-de-las-Peñas, C., de la Llave-Rincón, A. I., Ge, H. Y., & Arendt-Nielsen, L. (2008). Bilateral myofascial trigger points in the forearm muscles in patients with chronic unilateral lateral epicondylalgia: A blinded, controlled study. *Clinical Journal of Pain, 24*(9), 802–807.

Fernández-de-las-Peñas, C., Alonso-Blanco, C., & Miangolarra, J. C. (2007). Myofascial trigger points in subjects presenting with mechanical neck pain: A blinded, controlled study. *Manual Therapy, 12*(1), 29–33.

Fernández-Pérez, A. M., Villaverde-Gutiérrez, C., Mora-Sánchez, A., Alonso-Blanco, C., Sterling, M., & Fernández-de-Las-Peñas, C. (2012). Muscle trigger points, pressure pain threshold, and cervical range of motion in patients with high level of disability related to acute whiplash injury. *Journal of Orthopaedic and Sports Physical Therapy, 42*(7), 634–641.

Fishbain, D. A., Goldberg, M., Meagher, B. R., Steele, R., & Rosomoff, H. (1986). Male and female chronic pain patients categorized by DSM-III psychiatric diagnostic criteria. *Pain, 26*(2), 181–197.

Fleckenstein, J., Zaps, D., Rüger, L. J., Lehmeyer, L., Freiberg, F., Lang, P. M., et al. (2010). Discrepancy between prevalence and perceived effectiveness of treatment methods in myofascial pain syndrome: Results of a cross-sectional, nationwide survey. *BMC Musculoskeletal Disorders, 11*, 32.

Ge, H. Y., Fernández-de-las-Peñas, C., & Arendt-Nielsen, L. (2006). Sympathetic facilitation of hyperalgesia evoked from myofascial tender and trigger points in patients with unilateral shoulder pain. *Clinical Neurophysiology: Official Journal of the International Federation of Clinical Neurophysiology, 117*(7), 1545–1550.

Gerwin, R. D., Dommerholt, J., & Shah, J. P. (2004). An expansion of Simons' integrated hypothesis of trigger point formation. *Current Pain and Headache Reports, 8*(6), 468–475.

Gerwin, R. D., Shannon, S., Hong, C. Z., Hubbard, D., & Gevirtz, R. (1997). Interrater reliability in myofascial trigger point examination. *Pain, 69*(1–2), 65–73.

Hains, G., Boucher, P. B., & Lamy, A. M. (2015). Ischemic compression and joint mobilisation for the treatment of nonspecific myofascial foot pain: Findings from two quasi-experimental before-and-after studies. *Journal of the Canadian Chiropractic Association, 59*(1), 72–83.

Hains, G., Descarreaux, M., & Hains, F. (2010). Chronic shoulder pain of myofascial origin: A randomized clinical trial using ischemic compression therapy. *Journal of Manipulative and Physiological Therapeutics, 33*(5), 362–369.

Hains, G., Descarreaux, M., Lamy, A. M., & Hains, F. (2010). A randomized controlled (intervention) trial of ischemic compression therapy for chronic carpal tunnel syndrome. *Journal of the Canadian Chiropractic Association, 54*(3), 155–163.

Hains, G., & Hains, F. (2010). Patellofemoral pain syndrome managed by ischemic compression to the trigger points located in the peri-patellar and retro-patellar areas: A randomized clinical trial. *Clinical Chiropractic, 13*, 201–209.

Harkness, E. F., Macfarlane, G. J., Silman, A. J., & McBeth, J. (2005). Is musculoskeletal pain more common now than 40 years ago?: Two population-based cross-sectional studies. *Rheumatology (Oxford, England), 44*(7), 890–895.

Hou, C. R., Tsai, L. C., Cheng, K. F., Chung, K. C., & Hong, C. Z. (2002). Immediate effects of various physical therapeutic modalities on cervical myofascial pain and trigger-point sensitivity. *Archives of Physical Medicine and Rehabilitation, 83*(10), 1406–1414.

Iglesias-González, J. J., Muñoz-García, M. T., Rodrigues-de-Souza, D. P., Alburquerque-Sendín, F., & Fernández-de-Las-Peñas, C. (2013). Myofascial trigger points, pain, disability, and sleep quality in patients with chronic nonspecific low back pain. *Pain Medicine (Malden, Mass.), 14*(12), 1964–1970.

Jafari, M., Bahrpeyma, F., Mokhtari-Dizaji, M., & Nasiri, A. (2018). Novel method to measure active myofascial trigger point stiffness using ultrasound imaging. *Journal of Bodywork and Movement Therapies, 22*(2), 374–378.

Kodama, K., Takamoto, K., Nishimaru, H., Matsumoto, J., Takamura, Y., Sakai, S., et al. (2019). Analgesic effects of compression at trigger points are associated with reduction of frontal polar cortical activity as well as functional connectivity between the frontal polar area and insula in patients with chronic low back pain: A randomized trial. *Frontiers in Systems Neuroscience, 13*, 68.

Lucas, N., Macaskill, P., Irwig, L., Moran, R., & Bogduk, N. (2009). Reliability of physical examination for diagnosis of myofascial trigger points: A systematic review of the literature. *Clinical Journal of Pain, 25*(1), 80–89.

Mayoral Del Moral, O., Torres Lacomba, M., Russell, I. J., Sánchez Méndez, Ó., & Sánchez Sánchez, B. (2018). Validity and reliability of clinical examination in the diagnosis of myofascial pain syndrome and myofascial trigger points in upper quarter muscles. *Pain Medicine (Malden, Mass.), 19*(10), 2039–2050.

Moraska, A. F., Hickner, R. C., Kohrt, W. M., & Brewer, A. (2013). Changes in blood flow and cellular metabolism at a myofascial trigger point with trigger point release (ischemic compression): A proof-of-principle pilot study. *Archives of Physical Medicine and Rehabilitation, 94*(1), 196–200.

Morikawa, Y., Takamoto, K., Nishimaru, H., Taguchi, T., Urakawa, S., Sakai, S., et al. (2017). Compression at myofascial trigger point on chronic neck pain provides pain relief through the prefrontal cortex and autonomic nervous system: A pilot study. *Frontiers in Neuroscience, 11*, 186.

Muñoz-Muñoz, S., Muñoz-García, M. T., Alburquerque-Sendín, F., Arroyo-Morales, M., & Fernández-de-las-Peñas, C. (2012). Myofascial trigger points, pain, disability, and sleep quality in individuals with mechanical neck pain. *Journal of Manipulative and Physiological Therapeutics, 35*(8), 608–613.

Myburgh, C., Larsen, A. H., & Hartvigsen, J. (2008). A systematic, critical review of manual palpation for identifying myofascial trigger points: Evidence and clinical significance. *Archives of Physical Medicine and Rehabilitation, 89*(6), 1169–1176.

Nakamura, M., Nishiwaki, Y., Ushida, T., & Toyama, Y. (2011). Prevalence and characteristics of chronic musculoskeletal pain in Japan. *Journal of Orthopaedic Science: Official Journal of the Japanese Orthopaedic Association, 16*(4), 424–432.

Niddam, D. M. (2009). Brain manifestation and modulation of pain from myofascial trigger points. *Current Pain and Headache Reports, 13*(5), 370–375.

Niddam, D. M., Lee, S. H., Su, Y. T., & Chan, R. C. (2017). Brain structural changes in patients with chronic myofascial pain. *European Journal of Pain (London, England), 21*(1), 148–158.

Ong, W. Y., Stohler, C. S., & Herr, D. R. (2019). Role of the prefrontal cortex in pain processing. *Molecular Neurobiology, 56*(2), 1137–1166.

Pecos-Martin, D., Ponce-Castro, M. J., Jiménez-Rejano, J. J., Nunez-Nagy, S., Calvo-Lobo, C., & Gallego-Izquierdo, T. (2019). Immediate effects of variable durations of pressure release technique on latent myofascial trigger points of the levator scapulae: A double-blinded randomised clinical trial. *Acupuncture in Medicine: Journal of the British Medical Acupuncture Society, 37*(3), 141–150.

Rathbone, A., Grosman-Rimon, L., & Kumbhare, D. A. (2017). Interrater agreement of manual palpation for identification of myofascial trigger points: A systematic review and meta-analysis. *Clinical Journal of Pain, 33*(8), 715–729.

Rozenfeld, E., Finestone, A. S., Moran, U., Damri, E., & Kalichman, L. (2020). The prevalence of myofascial trigger points in hip and thigh areas in anterior knee pain patients. *Journal of Bodywork and Movement Therapies, 24*(1), 31–38.

Shah, J. P., Thaker, N., Heimur, J., Aredo, J. V., Sikdar, S., & Gerber, L. (2015). Myofascial trigger points then and now: A historical and scientific perspective. *PM & R: The Journal of Injury, Function, and Rehabilitation, 7*(7), 746–761.

Shamsi, M., Safari, A., Samadzadeh, S., & Yoosefpour, N. (2020). The prevalence of musculoskeletal pain among above 50-year-old population referred to the Kermanshah-Iran health bus in 2016. *BMC Research Notes, 13*(1), 72.

Sheng, J., Liu, S., Wang, Y., Cui, R., & Zhang, X. (2017). The link between depression and chronic pain: Neural mechanisms in the brain. *Neural Plasticity, 2017*, 9724371.

Simons, D. G., Travell, J. G., & Simons, L. S. (1999). *Myofascial pain and dysfunction, the trigger point manual, the upper extremities. Vol. 1* (2nd ed.). Baltimore, USA: Williams and Wilkins.

Skootsky, S. A., Jaeger, B., & Oye, R. K. (1989). Prevalence of myofascial pain in general internal medicine practice. *Western Journal of Medicine, 151*(2), 157–160.

Takamoto, K., Bito, I., Urakawa, S., Sakai, S., Kigawa, M., Ono, T., et al. (2015). Effects of compression at myofascial trigger points in patients with acute low back pain: A randomized controlled trial. *European Journal of Pain (London, England), 19*(8), 1186–1196.

Teixeira, M. J., Yeng, L. T., Garcia, O. G., Fonoff, E. T., Paiva, W. S., & Araujo, J. O. (2011). Failed back surgery pain syndrome: Therapeutic approach descriptive study in 56 patients. *Revista da Associacao Medica Brasileira (1992), 57*(3), 282–287.

Urakawa, S., Takamoto, K., Nakamura, T., Sakai, S., Matsuda, T., Taguchi, T., et al. (2015). Manual therapy ameliorates delayed-onset muscle soreness and alters muscle metabolites in rats. *Physiological Reports, 3*(2), e12279.

Veith, R. C., Lewis, N., Linares, O. A., Barnes, R. F., Raskind, M. A., Villacres, E. C., et al. (1994). Sympathetic nervous system activity in major depression. Basal and desipramine-induced alterations in plasma norepinephrine kinetics. *Archives of General Psychiatry, 51*(5), 411–422.

Wijnhoven, H. A., de Vet, H. C., & Picavet, H. S. (2006). Prevalence of musculoskeletal disorders is systematically higher in women than in men. *Clinical Journal of Pain, 22*(8), 717–724.

Chapter 30

Multimodal analgesia and postsurgical pain

Martina Rekatsina[a], Antonella Paladini[b], Giorgia Saltelli[c], and Giustino Varrassi[d]

[a]*Chronic Pain Clinical Fellow, Whipps Cross University Hospital, Barts Health NHS Trust, London, United Kingdom,* [b]*Department MESVA, University of L'Aquila, L'Aquila, Italy,* [c]*Sant'Andrea Hospital, "La Sapienza" University of Roma, Roma, Italy,* [d]*Paolo Procacci Foundation, Rome, Italy*

List of abbreviations

PSP   postsurgical pain
CPSP   chronic postsurgical pain
CNS   central nervous system
ERAS   enhanced recovery after surgery programs
MA   multimodal analgesia
NSAIDs nonsteroidal antiinflammatory drugs

Introduction

In the recent years, pain management knowledge, techniques, and medication have significantly evolved. However, a great number of patients still experience a significant amount of postsurgical pain (PSP). More than half of surgical patients are dissatisfied with their pain management and one-third of them are reporting that their pain was inadequately managed (CHANGE-PAIN, 2017).

Now it is widely approved that an effective PSP management plan is not a standardized regime, rather it is tailored to the needs of the individual patient, taking into account their medical, psychological, and physical condition. Age, level of fear and anxiety, the surgical procedure, personal preferences, and response to therapeutic agents used are also of utmost importance (Gupta et al., 2010).

The ultimate goal in the management of PSP is to minimize the dose of each medication in order to lessen side effects while providing adequate analgesia (Gupta et al., 2010).

Definition and classification of PSP

PSP is the result of surgical injury. Inflammation is considered the main mechanism and results from either tissue trauma (i.e., surgical incision, dissection, burns) or direct nerve injury (i.e., nerve transection, stretching, or compression) (Mariano, 2019).

Depending on its origin, PSP could be either somatic (superficial, sharp, pricking, throbbing, or burning) or visceral (deep, dull aching quality, and less well localized) (Rosenquist & Vrooman, 2013). Depending on its duration, it can be categorized as acute (pain present for less than 3 months), subacute (pain present at least 6 weeks but less than 3 months), or chronic (pain present for more than 3 months) (King, 2013).

Katz, Weinrib, and Clarke (2019) updated further the definition of chronic postoperative pain (CPSP), adding specific features such as that pain interferes to a great extend with health-related quality of life, to have a continuation with the acute PSP, to be localized to the surgical field, and not to be caused by other factors.

Incidence of PSP and CPSP

The estimated incidence and severity of PSP and CPSP varies widely in different studies, while most studies include only a limited number of patients and this affects the power of the results. Moreover, the different characteristics of patients, e.g., type of surgery, gender, age, and presence of severe PSP in the first 24 h after surgery interfere with the results (Sansone et al., 2015).

Features and Assessments of Pain, Anesthesia, and Analgesia. https://doi.org/10.1016/B978-0-12-818988-7.00040-6
Copyright © 2022 Elsevier Inc. All rights reserved.

In one study, 12% of patients claimed to experience "severe-to-extreme" pain and 54% "moderate-to-extreme" pain at discharge. Moreover, during the first 2 weeks after discharge, 13% of patients had "severe-to-extreme" pain and 46% had "moderate-to-extreme" pain (Buvanendran et al., 2015). Regarding CPSP, Sansone et al. (2015) claimed that the incidence of CPSP at 6 months was: 45.2% mild, 15.9% moderate, and 2.7% severe, while at 12 months: 35.9% mild, 11.8% moderate, and 2.5% severe. Alerting were their findings regarding neuropathic pain, as 31.9% of patients with moderate CPSP at 6 months and 40.3% of patients with moderate CPSP at 12 months developed this type of pain.

Of note, surgeries with the highest incidence of CPSP have been shown to be amputations (50%–85%), thoracotomies (5%–65%), cardiac surgery (30%–55%), and breast surgery (20%–50%) (Macrae, 2008).

Pathophysiology of postoperative pain

Different surgical procedures involve distinct organs and specific tissues, creating a variety of patterns of nociceptor sensitization and differences in the quality, location, and intensity of PSP. Pain initiated in the periphery or viscera is detected by a specialized subset of sensory neurons called *nociceptors* that convey the information to the central nervous system (CNS), where pain is generated (Raouf, Quick, & Wood, 2010) (Fig. 1).

Pain pathway: The pain is generated as soon as the damaged tissues release chemical mediators, such as prostaglandins, bradykinin, serotonin, substance P, and histamine. These substances activate receptor complexes, which in their turn initiate action potentials through ion channels. The signals are mainly transmitted through small myelinated Aδ, large myelinated Aβ, and unmyelinated C fibers to the dorsal horn of the spinal cord. In the dorsal horn of the spine, these first-order afferent neurons are synapsing with second-order neurons in the gray matter of the ipsilateral dorsal horn. The transmission of the signal is facilitated by neurotransmitters such as substance P, calcitonin, gene-related peptide, and glutamate aspartate adenosine triphosphate (ATP) (Rosenquist & Vrooman, 2013) (Fig. 2), while glycine and gamma-aminobutyric acid (GABA) are important neurotransmitters that are acting at inhibitory interneurons (Reddi, Curran, & Stephens, 2013).

Following, pain is transmitted to the thalamus through ascending tracts within the spinal cord (Fig. 2). Consequently, the secondary afferent neurons ascend in the contralateral spinothalamic tract to nuclei within the thalamus, where third-order neurons ascend to terminate in the somatosensory cortex (Rosenquist & Vrooman, 2013).

Tissue trauma could lead to *hyperalgesia*, which is attributed either to sensitization of the peripheral pain receptors (primary hyperalgesia) or/and increased excitability of CNS neurons (secondary hyperalgesia). Tissue trauma could also result to *allodynia* (Mariano, 2019).

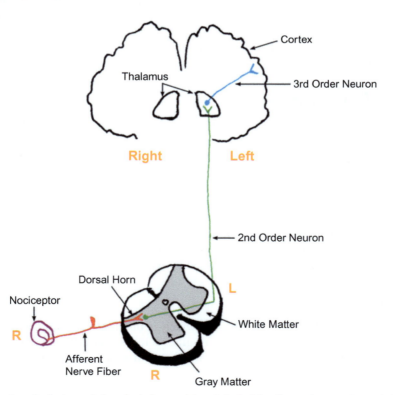

FIG. 1 Basical anatomical structures for the transmission of pain from periphery to brain. https://www.change-pain.com/grt-change-pain-portal/change_pain_home/chronic_pain/physician/physician_tools/picture_library/en_EN/312500026.jsp.

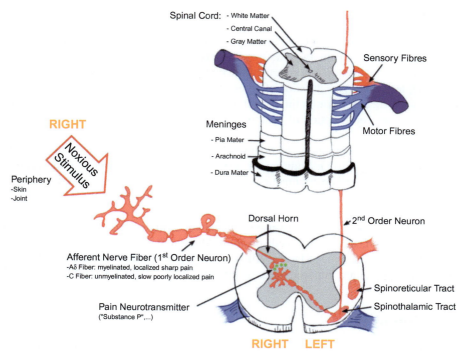

FIG. 2 Complexity of the transmission system, including neurotransmitter of the posterior horn of spinal cord. https://www.change-pain.com/grt-change-pain-portal/change_pain_home/chronic_pain/physician/physician_tools/picture_library/en_EN/312500026.jsp.

The direct trauma of nerves during surgery could lead to spontaneous discharge of them, and this accounts for qualitative features of neuropathic pain that may be present early in the postoperative period and can evolve into chronic neuropathic pain (Martinez et al., 2012).

Peripheral sensitization, central sensitization, and neuroplastic changes in the brain after incision

Important features involved in PSP are the peripheral and central sensitization as well as the neuroplastic changes in the brain. The term peripheral sensitization refers to the increased responsiveness and reduced threshold of nociceptive neurons in the periphery to the stimulation. In contrast, central sensitization is the enhanced responsiveness of nociceptive neurons and the concomitant amplification of pain intensity (Pogatzki-Zahn, Segelcke, & Schug, 2017) and can be possibly prevented by spinal inhibitory mechanisms (Reichl, Augustin, Zahn, & Pogatzki-Zahn, 2012). Regarding the neuroplastic changes that take place in the brain during surgery, a lot of research is still ongoing. Basically, this refers to the ability of the brain to reorganize itself by inducing chemical and structural changes and alter pain perception after surgery. These changes amplify signals arising from tissue damage and can increase the intensity of experienced pain (Pogatzki-Zahn, Segelcke, & Schug, 2017). This phenomenon is not only secondary to surgery, but also to any acute pain even of inflammatory origin (Fusco, Skaper, & Coaccioli, 2016).

Importance of PSP control

PSP relief is of utmost importance in the recovery pathway of patients, not only for its significant physiological benefits (Garimella & Cellini, 2013) but also because it provides early mobilization, reduced length of hospital stay (Mariano, 2019), fewer pulmonary and cardiac complications, and a reduced risk of deep vein thrombosis as well as reduced morbidity and mortality (Garimella & Cellini, 2013; Ramsay, 2000). Moreover, effective PSP control can prevent the activation of the sympathetic branch of the autonomic nervous system and prevent responses such as hypertension, tachycardia, diaphoresis, pallor, shallow respiration, and restlessness (Kagane, Gomes, & Thakur, 2016). In the long term, inadequately controlled acute pain can be one of the factors for the development of chronic pain, extended hospital stays, and readmission (Kagane et al., 2016) as well as development of chronic neuropathic pain (Ramsay, 2000).

Risk factors for CPSP

The most robust risk factor for development of CPSP is the pain itself. The presence and intensity of preoperative chronic pain, the intensity of acute postoperative pain, the time spent with severe pain after surgery, as well as pain intensity in the weeks after surgery and existence of pain in other parts of the body reliably predict CPSP across a range of surgical procedures (Katz et al., 2019). Interestingly, the duration of severe pain in the initial 24 h postoperatively predicted the chance of developing CPSP. For every 10% increase in time spent in severe pain, the risk of developing CPSP went up by 30% (Correll, 2017). Moreover, higher consumption of postoperative analgesics, typically a proxy for intense postoperative pain, is associated with more intense CPSP.

Another risk factor for CPSP development is the preoperative opioid use, in part due to opioid-induced hyperalgesia (Katz et al., 2019; Wardhan & Chelly, 2017). Additionally, psychological patient attitudes (e.g., depression, preoperative distress, anxiety, pain catastrophizing, and posttraumatic stress symptoms, etc.) also serve as a risk factor, while low income and low self-rated health and lack of education have a negative impact in the development of chronic pain after surgery (Shipton & Tait, 2005).

Independent predictors of severe postoperative pain are younger age, female gender, level of preoperative pain, incision size and type of surgery, reoperation, and genetic predisposition (Shipton & Tait, 2005) as well as genome-wide polymorphisms (van Reij et al., 2020) and epigenetics (James, 2013). Finally, postoperative infection, bleeding, organ rupture, or compartment syndrome development are important risk factors for CPSP (Shipton & Tait, 2005) (Table 1).

Evaluation of pain intensity

In order to effectively treat PSP, specific questions have to be answered such as the onset and pattern, location, quality, intensity of pain, as well as aggravating/relieving factors, previous treatments and their effect, or even barriers to pain assessment (Chou et al., 2016).

For assessing the intensity, the tools used to measure pain must be simple, quick to administer, and easily understood by the patients (Younger, McCue, & Mackey, 2009). In general, there are three types of pain scales, the unidimensional, the multidimensional, and the behavioral. The former provides fast measures of pain that can be administered multiple times, with minimal administrative effort and are mostly used in the postoperative period. The second assess several dimensions of pain, such as pain intensity, quality, affect, interference with functioning, and effects on general quality of life. Therefore, multidimensional scales are bridging the commonly observed lack of association between pain intensity and disability and are mostly used for chronic pain. Last, behavioral scales are often used in noncommunicative patients for whom direct assessment of pain self-report is not possible (Younger et al., 2009).

One commonly used unidimensional tool is the numerical rating scale (NRS) with scorings ranging from 0 to 10 or 0 to 100, where 0 represents "no pain" and 10 (or 100) the "worst pain imaginable." NRS's alternative is the visual analogue

TABLE 1 Risk factors for CPSP.

Independent	Preoperative	Postoperative
Younger age	Preoperative chronic pain	Severe pain after surgery
Female gender	Preoperative pain in other body parts	Time spent in severe pain
Level of preoperative pain	Psychological patient attitudes (depression, preoperative distress, anxiety, pain catastrophizing)	Pain in other body parts
Incision size	Preoperative opioid use	Infection
Type of surgery		Bleeding
Reoperation		Organ rupture
Genetic predisposition		Compartment syndrome development
Genome-wide polymorphisms		
Epigenetics		

scale (VAS). Regarding the multidimensional scales, the short-form McGill Pain Questionnaire (SF-MPQ) and the Brief Pain Inventory short form (BPI-SF) are often used (Younger et al., 2009). In addition, behavioral scales often measure facial or bodily movements as proxies for pain. Such scales are existing for both adult and children, e.g., the Faces Rating Scale for children (Beltramini, Milojevic, & Pateron, 2017; Younger et al., 2009), PAINAD—for Pain Assessment In Advanced Dementia (Paulson, Monroe, & Mion, 2014), Behavioral Pain Scale (BPS), or Critical-Care Pain Observation Tool (CPOT) for mechanically ventilated patients after surgery (Rijkenberg, Stilma, Bosman, van der Meer, & van der Voort, 2017).

PSP management: Enhanced recovery after surgery programs (ERAS)-preemptive, preventive, and multimodal analgesia (MA)

Management of postoperative pain is best tailored to the individual, as it is considered a highly personal and subjective experience. The ultimate target of any surgical team should be the fast recovery of their patients and this is best served by adopting ERAS, wherever possible, where multimodal analgesia (MA) is one of their most important components (Gelman et al., 2018).

The main concept of multimodal analgesic techniques was introduced more than 25 years ago by Kehlet and Dahl (1993). The relatively recent developed ERAS programs take place mostly in specialized centers, where early recovery is ensured by effective pain control and by minimizing the impact of surgical stress. A very important feature of ERAS is patient information and education concerning the process and the organization of care (Garin, 2020).

Multimodal analgesia, the fundamental element of ERAS, consists of the administration of two or more medication that act by different analgesic mechanisms (having additive or synergistic effects), as well as the utilization of regional analgesic techniques (peripheral and neuraxial blocks) and a wide variety of multilevel interventions including medical, physical, and psychological therapies (Mugabure et al., 2014). The aim of multimodal analgesia is to improve pain relief while reducing opioid requirements and opioid-related adverse effects (Kehlet & Dahl, 1993; Rosero & Joshi, 2014; Wardhan & Chelly, 2017).

In the past, it was considered important that analgesia is best administered preemptively (before surgical insult). However, this is no longer considered the golden rule, as current general consensus indicates that this does not translate into consistent clinical benefits after surgery. Instead, preventive analgesia, which encompasses a wider concept, where the timing of analgesic administration in relation to the surgical incision is not critical, has taken its place. The aim of preventive analgesia is to minimize both peripheral and central sensitization (Hanna, Ouanes, & Tomas, 2013; Rosero & Joshi, 2014).

The surgery-specific multimodal analgesic protocol should be functioning more like a checklist than a recipe, with options to adapt to the individual patient (Schwenk & Mariano, 2018). Moreover, managing acute PSP with an approach based on the involved pathophysiologic mechanisms could lead to a better pain management plan, since postoperative pain is multidimensional and multifaceted (Rekatsina, Paladini, Cifone, et al., 2020; Rekatsina, Paladini, Piroli, et al., 2020).

Framework of multimodal analgesia (MA)

As it is stated earlier, the selection of the elements of a multimodal therapy is a real challenge. Not only each patient is unique but also each surgical procedure has many potential therapeutic combinations. The MA plan consists of several steps or layers, with elements from each layer chosen based on the specific patient and surgery (Chou et al., 2016; Dunkman & Manning, 2018; Imani & Varrassi, 2019; Schwenk & Mariano, 2018).

The following pyramid can be utilized in order to establish a MA plan (each level is elucidated in detail in text):

FX
Higher Base: patient education +/− nonpharmacological therapy (cognitive–behavioral modalities).

Second Level: regional or local analgesia tailored to the invasiveness of the operation (neuraxial and peripheral nerve blockade, as well as tissue infiltration techniques).

Third Level: nonopioid systemic analgesics. This comprises a strong foundation for analgesia and modulating the stress response, ideally with drugs from several different therapeutic categories, e.g., paracetamol, NSAIDs, ketamine, dexmedetomidine, gabapentin, pregabalin, lidocaine etc.

Lower pick of the pyramid: Systemic opioids should be used sparingly as a rescue medication or to cover areas of inadequate block coverage.

It is important that levels 2 and 3 (systemic nonopioids and regional/local analgesia) are overlapping each other, therefore could also be reversed.

When applying multimodal analgesia, clinicians should be aware of the different side effect profile for each analgesic medication or technique used and provide appropriate monitoring to identify and manage possible adverse events. The basic principle of multimodal anesthesia is that clinicians should routinely try to avoid systemic opioids wherever possible (Chou et al., 2016).

Nonpharmacological interventions

The addition of nonpharmacological interventions might result in additional effects consistent with the biopsychosocial model of pain. Physical modalities such as transcutaneous electrical nerve stimulation (TENS), acupuncture and related interventions (massage, cold therapy, localized heat, etc.) may be used. However, although these therapies are generally considered to be safe, evidence on their effectiveness as adjunctive therapies as part of a MA to perioperative pain management varies substantially. Moreover, a number of cognitive-behavioral modalities (such as guided imagery, relaxation methods, hypnosis, intraoperative music) have been evaluated as adjunctive treatments in patients who undergo surgery. These modalities can be provided to patients by a *multidisciplinary team*. In general, cognitive–behavioral modalities are noninvasive and do not appear to be associated with significant harm (Chou et al., 2016).

Systemic nonopioid analgesics

Paracetamol

Serves as an effective and well-tolerated analgesic for mild-to-moderate pain and is associated with a decrease in pain and opioid usage as part of a multimodal regimen for surgery. It is given orally, intravenously or rectally in doses up to 3–4 g per day. Benefits of paracetamol are reduction of pain at rest or movement, reduced postoperative opioid consumption, as well as PONV (Dunkman & Manning, 2018).

Nonsteroidal antiinflammatory drugs (NSAIDs)

NSAIDs encompass a class of compounds that substantially differ in terms of clinical efficacy and safety (Varrassi et al., 2019; Varrassi, Pergolizzi, Peppin, & Paladini, 2019; Varrassi, Pergolizzi, Dowling, & Paladini, 2020). They are effective analgesic and antiinflammatory medication that inhibit cyclooxygenase (COX), thereby reducing prostaglandin synthesis. NSAIDs are either nonselective, inhibiting both inhibit COX-1 and COX-2 (e.g., ibuprofen or ketorolac), or selective COX-2 inhibitors (e.g., celecoxib, parecoxib). COX-2 is the enzyme induced by pain and inflammation and COX-1 is involved with normal gastrointestinal and platelet function (Dunkman & Manning, 2018).

NSAIDs can cause platelet dysfunction and prolong bleeding times and as such, nonselective NSAIDs in particular are often withheld. However, a recent metaanalysis found that ketorolac did not increase bleeding and COX-2 inhibitors have minimal effect on bleeding even at high doses (Dunkman & Manning, 2018). Other concerns regarding NSAIDs (mostly the nonselective) are the increased risk of anastomotic leak (Dunkman & Manning, 2018), as well as NSAIDs-induced enteropathy which is nowadays being increasingly recognized and is mostly being associated with long history of NSAIDs use (Rekatsina, Paladini, Cifone, et al., 2020; Rekatsina, Paladini, Piroli, et al., 2020).

Selective COX-2 inhibitors have been reported to offer GI safety advantage over nonselective NSAIDs. Regarding patients suffering from cardiovascular diseases, no safe NSAID medication for long-term use exists. Therefore, it is suggested that NSAID therapy should always be taken for the least amount of time and at the lowest effective dosage, with a careful evaluation of any sign and symptom of cardiovascular complication such as hypertension, peripheral edema, or renal insufficiency (Varrassi et al., 2020; Varrassi, Alon, et al., 2019; Varrassi, Pergolizzi, et al., 2019).

Other agents of interest

γ-Aminobutyric acid analogues (Gabapentenoids)

Originally developed as anticonvulsants, gabapentenoids (such as pregabalin and gabapentin) have analgesic properties mediated by the α_2-delta subunits of presynaptic calcium channels and decreased excitatory neurotransmitter release. Both have been shown to reduce pain scores and opioid use and serve as a valuable addition to a multimodal analgesic regimen. Their main side effects are sedation, dizziness, and visual disturbances (Chou et al., 2016; Dunkman & Manning, 2018).

N-methyl-D-aspartate receptor antagonists

NMDA receptor antagonists, such as ketamine and magnesium, are used perioperatively to improve pain and reduce opioid use and related side effects.

Ketamine

Ketamine is a noncompetitive inhibitor of NMDA receptors, having analgesic, antiinflammatory, and antihyperalgesic effects. The prevention of the phenomenon of opioid-induced hyperalgesia thereby preventing hypersensitivity and drug tolerance is an important feature of this medication (Imani & Varrassi, 2019). When used perioperatively, it has been shown to reduce pain scores and opioid use, while it is not a respiratory depressant. It may be of particular use in patients with chronic pain or chronic opioid use, because NMDA receptors are involved in chronic pain development (Dunkman & Manning, 2018; Loftus et al., 2010). The doses that have been proposed vary widely, Loftus et al. (2010) suggested, to be used as a preoperative bolus of 0.5 mg/kg followed by an infusion at 10 mg/kg/min intraoperatively, with or without a postoperative infusion at a lower dosage. Hemodynamic changes, hallucinations, and headaches are the major side effects of this medication (Imani & Varrassi, 2019).

Systemic magnesium

Magnesium also exerts an analgesic effect by antagonism of NMDA receptors. It is typically given intraoperatively as an intravenous bolus and/or infusion. It has been shown to reduce pain both at rest and movement, as well as opioid use. Respiratory muscle weakness and potentiation of nondepolarizing neuromuscular blockade are concerns that require careful monitoring (Dunkman & Manning, 2018).

Systemic lidocaine

Intravenous lidocaine has analgesic, antiinflammatory, and antihyperalgesic effects. A metaanalysis studying the effects of perioperative lidocaine infusion found that it was associated with lower pain scores, reduced opioid use, and faster recovery after abdominal surgery. Lidocaine infusion is a useful adjunct in patients where other local anesthetic approaches, such as regional anesthesia, are not possible (Dunkman & Manning, 2018). It is typically administered as a bolus followed by an infusion through the end of surgery. The recommendation of Chou et al. (2016) is an induction dose of 1.5 mg/kg followed by 2 mg/kg/h intraoperatively; however, an optimal dose is yet not determined.

α_2-Agonists

α_2-Agonists have several properties that are desirable as a perioperative adjunct medication including sedation, hypnosis, anxiolysis, sympatholytic, and analgesia. The commonly used agents are clonidine and dexmedetomidine. Both agents decreased pain intensity, postoperative opioid use, and nausea. Clonidine was associated with an increased risk of hypotension, while dexmedetomidine was associated with an increased risk of bradycardia (Dunkman & Manning, 2018).

TABLE 2 Major pharmacological agents used in postoperative pain management.

Systemic nonopioid analgesics	Often used agents
Paracetamol	
Nonsteroidal antiinflammatory drugs (NSAIDs)	Nonselective (e.g., ibuprofen or ketorolac) or selective COX-2 (e.g., celecoxib, parecoxib)
g-Aminobutyric acid analogues (Gabapentenoids)	Pregabalin, gabapentin
N-methyl-D-aspartate receptor antagonists	Ketamine, Magnesium
Systemic lidocaine	
a_2-Agonists	Clonidine, dexmedetomidine
Glucocorticoids	Dexamethasone

Glucocorticoids

Glucocorticoids are used to reduce inflammation in a variety of conditions and have also well-established antiemetic properties. Dexamethasone is commonly used as prophylaxis for postoperative nausea and vomiting (Dunkman & Manning, 2018). (Table 2)

Systemic opioid analgesics

Oral over intravenous administration of opioids for postoperative analgesia is strongly recommended in patients who can use the oral route, as iv administration does not seem superior (Ruetzler et al., 2014). Opioid minimization is one of the primary tenets of the enhanced recovery pathway. Although opioids do provide excellent analgesia in the short term, they come with a host of undesirable side effects including PONV, urinary retention, constipation, ileus, pruritus, sedation, and respiratory depression. These side effects can substantially delay a patient's recovery from surgery and may be more unpleasant than the pain they are treating, and so other analgesic methods are preferred (Dunkman & Manning, 2018).

Consideration should also be given to the treatment of acute noncancer pain with opioids, as the misuse and abuse of opioids in these patients seem to comprise a huge problem, in some countries, having led to the massive "opioid epidemic" (Varrassi et al., 2018).

Regional anesthesia, local anesthesia, and site infiltration with local anesthetics

Surgical site-specific local anesthetic infiltration or use topical local anesthetics in combination with peripheral nerve blocks in adults and children for specific procedures are strongly recommended (Chou et al., 2016). Moreover, a continuous, local anesthetic-based peripheral block is suggested when the need for analgesia is likely to exceed the duration of effect of a single injection (Richman et al., 2006).

Neuraxial analgesia (epidural and spinal)

Regarding major thoracic and abdominal procedures, neuraxial analgesia is strongly recommended, particularly in patients at risk for cardiac complications, pulmonary complications, or prolonged ileus. Neuraxial analgesia also according to several studies is associated with a decreased risk of postoperative mortality, venous thromboembolism, myocardial infarction, pneumonia and respiratory depression, and decreased duration of ileus versus systemic analgesia (Pöpping et al., 2014; Pöpping, Elia, Marret, Remy, & Tramer, 2008).

Thoracic epidural analgesia (TEA) remains the gold standard for regional anesthesia in major abdominal surgery, having been shown to provide superior analgesia versus parental opioids, while having less side effects and more benefits such as shorter length of stay as well as less morbidity and mortality. Reduction in deep venous thromboembolism/pulmonary embolism, reduction in atelectasis, pneumonia and respiratory depression as well as reductions in myocardial infarction, renal failure, ileus, and postoperative nausea and vomiting are only some of its benefits. However, its benefits do not come without certain important risks such as arterial hypotension, urinary retention, motor blockade, and neurologic injury (including epidural hematoma or abscess) (Dunkman & Manning, 2018).

Spinal analgesia is emerging as an attractive alternative for laparoscopic procedures within an enhanced recovery pathway. Intrathecal opioid alone (without local anesthetic) has been shown to decrease pain following abdominal surgery and is superior to opioid PCA. Spinal analgesia is a safe and well-established technique but it is not without complications or safety concerns of which pruritus and PONV are some of the most common. Delayed respiratory depression is also a risk of opioid used intrathecally (Dunkman & Manning, 2018).

Transversus abdominis plane blocks

Truncal blocks, such as the transversus abdominis plane (TAP) block, offer analgesic options somewhere between neuraxial techniques and local infiltration in terms of invasiveness and efficacy. TAP block has several approaches in order to cover the dermatomes needed for different surgical procedures and is considered generally safe (Dunkman & Manning, 2018).

Paravertebral blocks

Paravertebral nerve blocks (PVB) is a well-established technique for analgesia following thoracotomy, breast surgery, or even abdominal surgery as it can provide analgesia at different levels depending where the block is placed (Dunkman & Manning, 2018).

Local infiltration

Surgical site infiltration (e.g., intraperitoneal or incision field infiltration) with local anesthetic is an appropriate part of multimodal analgesia when other types of regional analgesia are not necessary or possible (Dunkman & Manning, 2018).
Suggestions for studying:
Regional anesthesia/analgesia guidance regarding specific surgical procedures:

1) New York School of Regional Anesthesia (NYSORA) https://www.nysora.com/regional-anesthesia-for-specific-surgical-procedures/.
2) Upper and lower limb blocks in children https://www.wfsahq.org/components/com_virtual_library/media/94d3ffbd1b85aae7e3d3cbfa2a04b884-Upper-and-lower-limb-blocks-in-children.pdf.
3) WFSA's virtual library (individual search for specific blocks) https://www.wfsahq.org/resources/virtual-library.

Populations that need special consideration in treating PSP

Children

The best option for acute pain management in children is considered to be oral analgesia, whenever possible. The mainstay of the management of acute pain remains paracetamol and ibuprofen, while oral morphine can be administered effectively in combination with the two formers. Current evidence is suggesting an oral dose of paracetamol of 15 mg/kg (max 75 mg/kg/24 h or 60 mg/kg/24 h for neonates and 3 g for children over 50 kg). Regarding ibuprofen, for children over 3 months, ibuprofen's dosage is 10 mg/kg and for children over 40 kg, the dosage is 400 mg (max 1.6 g/24 h). Oramorph for mild-to-moderate pain should be given at a dose of 0.1 mg/kg, while for moderate-to-severe pain, it varies with the age. For neonates up to 3 months, the dosage is 0.1 mg/kg, for 3 months to 1 year, the dosage is 0.2 mg/kg, and for children greater than 1 year, the suggested dose is 0.3 mg/kg (in patient dose) (Mason, 2018; Yoshihara, Paulino, & Yoneoka, 2018).

Obese patients and patients with chronic obstructive pulmonary disease (COPD)

Obese patients and patients suffering from COPD and obstructive sleep apnea (OSA) can be studied in the same section as they are sharing a lot of common features (Porhomayon, Leissner, El-Solh, & Nader, 2013). In these patients, regional techniques should be preferably used as they have a greater margin of safety in terms of respiratory side effects; along with NSAIDs (if there are not contraindications) in order to avoid the side effects of opioids. Local anesthetic wound infiltration is also encouraged. Moreover, tramadol has been found advantageous over morphine, while dexmedetomidine showed significant reduction postoperative opioid requirements, providing fewer episodes of desaturation. Ketamine

338 PART | III Interlinking anesthesia, analgesics and pain

has also been reported to be a safe and effective alternative to opioids. If an opioid PCA is used, the dose should be based on lean body mass and continuous infusions should be avoided. Local anesthetic wound infiltration and adjunct (nonnarcotic) analgesic medications are encouraged to lower the opioid requirements (Porhomayon et al., 2013).

Patients receiving long-term opioids

In patients already being treated with opioids, it is important to conduct a preoperative evaluation to determine preoperative opioid use and doses. Important is that postoperative opioid requirements will typically be greater and the pain will be more difficult to control. Local anesthetic peripheral and neuraxial analgesic techniques would be of significant value as will nonpharmacological interventions. Nonopioid systemic medications such as gabapentin, pregabalin, and ketamine should also have a place in the pain management. For difficult to manage pain, a PCA with basal infusion of opioids would be appropriate but always with appropriate monitoring, as well as proper education and instructions on tapering opioids to target dose after discharge (Chou et al., 2016).

Application of MA to other areas

An optimal postoperative pain management plan is important not only to be applied by anesthetists, but also to be in the postoperative plan of all kind of surgeons (general, orthopedic, gynecologists, etc.). As patients are in the center of any MA plan, this will be definitely of their favor, but it will also benefit the society, considering that it helps to save money not only for the health care systems but also for the society, as no money for treating CPSP will be spent.

Conclusions

A well-designed analgesic plan, that takes into account the unique characteristics of pain in the operated patient and also utilizes the huge range of components of a multimodal analgesic plan, is the ultimate target of best practice for postoperative pain management. The specific characteristics of each patient (including their psychologic status) and the operation that they undergone are the basic indicators of the establishment of the plan. Patient education, involvement of a multidisciplinary team, adequate psychological support, nonpharmacological techniques (invasive and noninvasive), as well as in time application of regional techniques and minimizing usage of opioids could work together and allow both fast recovery of patient and nonchronification of the PSP.

Application to other areas

An optimal postoperative pain management plan is important not only to be applied by anesthetists, but also to be in the postoperative plan of all kind of surgeons (general, orthopedic, gynecologists, etc.). As patients are in the center of any MA plan, this will be definitely of their favor, but it will also benefit the society, considering that it helps to save money not only for the health care systems but also for the society, as no money for treating CPSP will be spent.

Other agents of interest

Other agents that have been used for the treatment of pain and could serve in the management of postoperative, mainly neuropathic pain are tricyclic antidepressants, like amitryptiline or serotonin-noradrenaline reuptake inhibitor (SNRIs) like duloxetine. However, their use in acute pain is very limited (Wong et al., 2014).

Key facts

Postoperative pain is difficult to treat and needs a multimodal approach The incidence of PSP and CPSP, although varies among studies is still very high Pathophysiology of PSP is a complex, multilevel process. Different drugs have different targets of action. Populations that need special consideration in treating PSP involve elderly, children, obese patients and patients on long term opioids.

Mini-dictionary of terms

Nociceptor: A high-threshold sensory receptor of the peripheral somatosensory nervous system that is capable of transducing and encoding noxious stimuli.

    Hyperalgesia: Increased pain from a stimulus that normally provokes pain.

    Allodynia: Pain due to a stimulus that does not normally provoke pain.

    Peripheral sensitization: Increased responsiveness and reduced threshold of nociceptive neurons in the periphery to the stimulation of their receptive fields.

    Central sensitization: Increased responsiveness of nociceptive neurons in the CNS to their normal or subthreshold afferent input.

    Multidisciplinary team: A variety of practitioners, including psychologists, psychotherapists, nurses, physicians, social workers, and child life specialists.

Summary points

1. Postoperative pain relief has significant physiological and financial benefits.
2. The goal of postoperative pain management is elimination of pain, minimization of side effects, fast recovery, and fast discharge without chronification of pain.
3. Risk factors chronification of pain should be recognized before establishing a treatment plan.
4. A pathophysiologic approach of postoperative is the best way to establish an effective multimodal analgesic plan.
5. Patient education, a multidisciplinary team, psychological support, nonpharmacological techniques (invasive and non-invasive), as well as in time application of regional techniques and minimization of opioids are important ingredients of an effective multimodal analgesic plan.

For the treatment of PSP, a multimodal analgesic approach, as a part of an enhanced recovery protocol, is the best option for the patient. Beyond nonopioid systematic analgesics, regional analgesia, peripheral blocks, infiltration techniques and opioid analgesics, patient education, and nonpharmacological options have a substantial role. Individuality of pain (different type of surgeries, different population, etc.) is one of the most important characteristic that physicians who treat patients after surgery should take into account.

References

Beltramini, A., Milojevic, K., & Pateron, D. (2017). Pain assessment in newborns, infants, and children. *Pediatric Annals, 46*(10), e387–e395.

Buvanendran, A., Fiala, J., Patel, K. A., Golden, A. D., Moric, M., & Kroin, J. S. (2015). The incidence and severity of postoperative pain following inpatient surgery. *Pain Medicine, 16*(12), 2277–2283.

CHANGE-PAIN. (2017). *Post-operative pain management—Currently unmet needs and deficits Grünenthal Group.* https://www.change-pain.com/grt-change-pain-portal/318500022.jsp. (Accessed 4 May 2020).

Chou, R., Gordon, D. B., de Leon-Casasola, O. A., Rosenberg, J. M., Bickler, S., Brennan, T., et al. (2016). Management of postoperative pain: A clinical practice guideline from the American pain society, the American Society of Regional Anesthesia and Pain Medicine, and the American Society of Anesthesiologists' committee on regional anesthesia, executive committee, and administrative council. *The Journal of Pain, 17*(2), 131–157.

Correll, D. (2017). Chronic postoperative pain: Recent findings in understanding and management. *F1000Research, 6,* 1054. https://doi.org/10.12688/f1000research.11101.1.

Dunkman, J., & Manning, M. (2018). Enhanced recovery after surgery and multimodal strategies for analgesia. *Surgical Clinics of North America, 98*(6), 1171–1184. https://doi.org/10.1016/j.suc.2018.07.005.

Fusco, M., Skaper, S., Coaccioli, S., et al. (2016). Degenerative joint diseases and neuroinflammation. *Pain Practice, 17*(4), 522–532. https://doi.org/10.1111/papr.12551. In this issue.

Garimella, V., & Cellini, C. (2013). Postoperative pain control. *Clinics in Colon and Rectal Surgery, 26*(3), 191–196. https://doi.org/10.1055/s-0033-1351138.

Garin, C. (2020). Enhanced recovery after surgery in pediatric orthopedics (ERAS-PO). *Orthopaedics & Traumatology, Surgery & Research, 106*(1), S101–S107.

Gelman, D., Gelmanas, A., Urbanaitė, D., Tamošiūnas, R., Sadauskas, S., Bilskienė, D., et al. (2018). Role of multimodal analgesia in the evolving enhanced recovery after surgery pathways. *Medicina (Kaunas, Lithuania), 54*(2), 20. https://doi.org/10.3390/medicina54020020.

Gupta, A., Kaur, K., Sharma, S., Goyal, S., Arora, S., & Murthy, R. S. (2010). Clinical aspects of acute post-operative pain management & its assessment. *Journal of Advanced Pharmaceutical Technology & Research, 1*(2), 97–108.

Hanna, M., Ouanes, J. P., & Tomas, V. G. (2013). Postoperative pain and other acute pain syndromes. In *Practical management of pain* (5th ed., pp. 271–297). Elsevier Inc.

Imani, F., & Varrassi, G. (2019). Ketamine as adjuvant for acute pain management. *Anesthesiology and Pain Medicine, 9*(6). https://doi.org/10.5812/aapm.100178, e100178.

James, S. (2013). Human pain and genetics: Some basics. *British Journal of Pain*, 7(4), 171–178. https://doi.org/10.1177/2049463713506408.

Kagane, S., Gomes, M., & Thakur, P. (2016). Management of pain. *Journal of Medical Science and Clinical Research*, 4(05), 10588–10598.

Katz, J., Weinrib, A. Z., & Clarke, H. (2019). Chronic postsurgical pain: From risk factor identification to multidisciplinary management at the Toronto general hospital transitional pain service. *Canadian Journal of Pain*, 3(2), 49–58.

Kehlet, H., & Dahl, J. B. (1993). The value of "multimodal" or "balanced analgesia" in postoperative pain treatment. *Anesthesia & Analgesia*, 77(5), 1048–1056.

King, W. (2013). Acute pain, subacute pain, and chronic pain. In G. F. Gebhart, & R. F. Schmidt (Eds.), *Encyclopedia of pain*. Berlin, Heidelberg: Springer.

Loftus, R. W., Yeager, M. P., Clark, J. A., Brown, J. R., Abdu, W. A., Sengupta, D. K., et al. (2010). Intraoperative ketamine reduces perioperative opiate consumption in opiate-dependent patients with chronic back pain undergoing back surgery. *Anesthesiology: The Journal of the American Society of Anesthesiologists*, 113(3), 639–646.

Macrae, W. A. (2008). Chronic post-surgical pain: 10 years on. *British Journal of Anaesthesia*, 101(1), 77–86.

Mariano, E. R. (2019). Management of acute perioperative pain. In S. Fishman (Ed.), *UpToDate*. Waltham, MA: UpToDate. Available from: https://www.uptodate.com/contents/management-of-acute-perioperative-pain?search=postoperative%20pain%20management&source=search_result&selectedTitle=1~150&usage_type=default&display_rank=1. (Accessed 4 May 2020).

Martinez, V., Ammar, S. B., Judet, T., Bouhassira, D., Chauvin, M., & Fletcher, D. (2012). Risk factors predictive of chronic postsurgical neuropathic pain: The value of the iliac crest bone harvest model. *Pain*, 153(7), 1478–1483.

Mason, D. G. (2018). Fifteen-minute consultation: Pain relief for children made simple—A pragmatic approach to prescribing oral analgesia in the post-codeine era. *Archives of Disease in Childhood. Education and Practice Edition*, 103(1), 2–6. https://doi.org/10.1136/archdischild-2016-311613.

Mugabure, B. B., González, S. S., Uría, A. A., Rubín, N. A., García, S. D., & Azkona, A. M. (2014). *Multimodal analgesia for the management of postoperative pain and treatment*. Available from: https://www.intechopen.com/books/pain-and-treatment/multimodal-analgesia-for-the-management-of-postoperative-pain. (Accessed 1 May 2020).

Paulson, C. M., Monroe, T., & Mion, L. C. (2014). Pain assessment in hospitalized older adults with dementia and delirium. *Journal of Gerontological Nursing*, 40(6), 10–15. https://doi.org/10.3928/00989134-20140428-02.

Pogatzki-Zahn, E. M., Segelcke, D., & Schug, S. A. (2017). Postoperative pain—From mechanisms to treatment. *Pain Reports*, 2(2), e588. https://doi.org/10.1097/PR9.0000000000000588.

Pöpping, D. M., Elia, N., Marret, E., Remy, C., & Tramer, M. R. (2008). Protective effects of epidural analgesia on pulmonary complications after abdominal and thoracic surgery: A meta-analysis. *Archives of Surgery*, 143(10), 990–999.

Pöpping, D. M., Elia, N., Van Aken, H. K., Marret, E., Schug, S. A., Kranke, P., et al. (2014). Impact of epidural analgesia on mortality and morbidity after surgery: Systematic review and meta-analysis of randomized controlled trials. *Annals of Surgery*, 259(6), 1056–1067.

Porhomayon, J., Leissner, K. B., El-Solh, A. A., & Nader, N. D. (2013). Strategies in postoperative analgesia in the obese obstructive sleep apnea patient. *The Clinical Journal of Pain*, 29(11), 998–1005.

Ramsay, M. A. (2000). Acute postoperative pain management. *Proceedings (Baylor University. Medical Center)*, 13(3), 244–247. https://doi.org/10.1080/08998280.2000.11927683.

Raouf, R., Quick, K., & Wood, J. N. (2010). Pain as a channelopathy. *The Journal of Clinical Investigation*, 120(11), 3745–3752.

Reddi, D., Curran, N., & Stephens, R. (2013). An introduction to pain pathways and mechanisms. *British Journal of Hospital Medicine*, 74(Suppl 12), C188–C191.

Reichl, S., Augustin, M., Zahn, P. K., & Pogatzki-Zahn, E. M. (2012). Peripheral and spinal GABAergic regulation of incisional pain in rats. *Pain*, 153(1), 129–141.

Rekatsina, M., Paladini, A., Cifone, M. G., Lombardi, F., Pergolizzi, J. V., & Varrassi, G. (2020). Influence of microbiota on NSAID enteropathy: A systematic review of current knowledge and the role of probiotics. *Advances in Therapy*, 37(5), 1933–1945. https://doi.org/10.1007/s12325-020-01338-6 (Epub 2020 Apr 10).

Rekatsina, M., Paladini, A., Piroli, A., Zis, P., Pergolizzi, J. V., & Varrassi, G. (2020). Pathophysiologic approach to pain therapy for complex pain entities: A narrative review. *Pain and therapy*, 9(1), 7–21. https://doi.org/10.1007/s40122-019-00147-2.

Richman, J. M., Liu, S. S., Courpas, G., Wong, R., Rowlingson, A. J., McGready, J., et al. (2006). Does continuous peripheral nerve block provide superior pain control to opioids? A meta-analysis. *Anesthesia & Analgesia*, 102(1), 248–257.

Rijkenberg, S., Stilma, W., Bosman, R. J., van der Meer, N. J., & van der Voort, P. H. (2017). Pain measurement in mechanically ventilated patients after cardiac surgery: Comparison of the behavioral pain scale (BPS) and the critical-care pain observation tool (CPOT). *Journal of Cardiothoracic and Vascular Anesthesia*, 31(4), 1227–1234.

Rosenquist, R. W., & Vrooman, B. M. (2013). Chronic pain management. In J. Butterworth, D. C. Mackey, & J. D. Wasnick (Eds.), *Morgan & Mikhails clinical anesthesiology* (5th ed., pp. 1023–1085). United States: McGraw-Hill Education.

Rosero, E. B., & Joshi, G. P. (2014). Preemptive, preventive, multimodal analgesia: What do they really mean? *Plastic and Reconstructive Surgery*, 134 (4S-2), 85S–93S.

Ruetzler, K., Blome, C. J., Nabecker, S., Makarova, N., Fischer, H., Rinoesl, H., et al. (2014). A randomised trial of oral versus intravenous opioids for treatment of pain after cardiac surgery. *Journal of Anesthesia*, 28(4), 580–586.

Sansone, P., Pace, M. C., Passavanti, M. B., Pota, V., Colella, U., & Aurilio, C. (2015). Epidemiology and incidence of acute and chronic post-surgical pain. *Annali Italiani di Chirurgia*, 86(4), 285–292.

Schwenk, E. S., & Mariano, E. R. (2018). Designing the ideal perioperative pain management plan starts with multimodal analgesia. *Korean Journal of Anesthesiology*, 71(5), 345.

Shipton, E. A., & Tait, B. (2005). Flagging the pain: Preventing the burden of chronic pain by identifying and treating risk factors in acute pain. *European Journal of Anaesthesiology*, 22(6), 405–412.

van Reij, R., Hoofwijk, D., Rutten, B., Weinhold, L., Leber, M., Joosten, E., et al. (2020). The association between genome-wide polymorphisms and chronic postoperative pain: A prospective observational study. *Anaesthesia, 75*(Suppl 1), e111–e120. https://doi.org/10.1111/anae.14832.

Varrassi, G., Alon, E., Bagnasco, M., Lanata, L., Mayoral-Rojals, V., Paladini, A., et al. (2019). Towards an effective and safe treatment of inflammatory pain: A Delphi-guided expert consensus. *Advances in Therapy, 36*(10), 2618–2637. https://doi.org/10.1007/s12325-019-01053-x.

Varrassi, G., Fusco, M., Skaper, S. D., Battelli, D., Zis, P., Coaccioli, S., et al. (2018). A pharmacological rationale to reduce the incidence of opioid induced tolerance and hyperalgesia: A review. *Pain and therapy, 7*(1), 59–75. https://doi.org/10.1007/s40122-018-0094-9.

Varrassi, G., Pergolizzi, J., Peppin, J. F., & Paladini, A. (2019). Analgesic drugs and cardiac safety. In S. Govoni, P. Politi, & E. Vanoli (Eds.), *Brain and heart dynamics*. Cham: Springer. https://doi.org/10.1007/978-3-319-90305-7_43-1.

Varrassi, G., Pergolizzi, J. V., Dowling, P., & Paladini, A. (2020). Ibuprofen safety at the golden anniversary: Are all NSAIDs the same? A narrative review. *Advances in Therapy, 37*, 61–82. https://doi.org/10.1007/s12325-019-01144-9.

Wardhan, R., & Chelly, J. (2017). Recent advances in acute pain management: understanding the mechanisms of acute pain, the prescription of opioids, and the role of multimodal pain therapy. *F1000Research, 6*, 2065. https://doi.org/10.12688/f1000research.12286.1.

Wong, K., Phelan, R., Kalso, E., Galvin, I., Goldstein, D., Raja, S., et al. (2014). Antidepressant drugs for prevention of acute and chronic postsurgical pain: Early evidence and recommended future directions. *Anesthesiology, 121*(3), 591–608. https://doi.org/10.1097/ALN.0000000000000307. 25222675.

Yoshihara, H., Paulino, C., & Yoneoka, D. (2018). Predictors of increased hospital stay in adolescent idiopathic scoliosis patients undergoing posterior spinal fusion: Analysis of national database. *Spine Deformity, 6*(3), 226–230.

Younger, J., McCue, R., & Mackey, S. (2009). Pain outcomes: A brief review of instruments and techniques. *Current Pain and Headache Reports, 13*(1), 39–43. https://doi.org/10.1007/s11916-009-0009-x.

Chapter 31

Pain, ultrasound-guided Pecs II block, and general anesthesia

A.A. Gde Putra Semara Jaya[a], Marilaeta Cindryani[b], and Tjokorda Gde Agung Senapathi[b]

[a]*Department of Anesthesiology and Intensive Care, Mangusada Hospital, Faculty of Medicine, Udayana University, Badung, Bali, Indonesia,*
[b]*Department of Anesthesiology and Intensive Care, Sanglah Hospital, Faculty of Medicine, Udayana University, Denpasar, Bali, Indonesia*

Abbreviations

BPB	brachial plexus block
ESPB	erector spinae plane block
ICB	intercostal nerve block
mcg	microgram
mL	milliliter
LA	local anesthetic
Pecs block	pectoralis nerve block
PMM	pectoralis major muscle
PmM	pectoralis minor muscle
PONV	postoperative nausea and vomiting
SAM	serratus anterior muscle
SAPB	serratus anterior muscle plane block
TEA	thoracic epidural anesthesia-analgesia
TPVB	thoracic paravertebral block
TTPB	transversus thoracis muscle plane block

Introduction

Surgical procedures might produce mild-to-severe postoperative pain. Breast surgery is a procedure in the thoracic wall area and related to significant acute and chronic postoperative pain. Conventionally, thoracic epidural anesthesia-analgesia (TEA) and thoracic paravertebral block (TVPB) remain the gold standard of regional anesthesia-analgesia (RA) for surgery in the thoracic area. Neurological injury is a relatively rare but devastating complication of TEA and TPVB. Therefore, not every anesthesiologist is confident in performing these procedures. The second version of pectoral nerves block (Pecs II block) is a novel approach of RA to breast surgery. It is simple, reliable, and can be an alternative to TEA and TVPB. Combined ultrasound-guided Pecs II block and general anesthesia have shown an important role regarding perioperative analgesia, opioid consumption, and postoperative nausea and vomiting (PONV).

Acute and chronic postoperative pain in breast surgery

Breast surgery is surgical procedures performed on the breast, ranged from minor to major surgery, and related to significant acute and chronic postoperative pain. The type of breast surgery can be oncological, reconstructive, or aesthetic surgery. Both acute and persistent (chronic) postoperative pain have a detrimental effect on patient. Acute pain could slow the recovery phase, and persistent pain decrease patient' functional status and quality of life. Management of persistent postoperative pain is more challenging than acute postoperative pain.

Nearly 70% of breast cancer surgery patients experience moderate-to-severe acute postoperative pain, with moderate-to-severe pain persisting for 1 month in almost 25% of patients (Fecho et al., 2009; Habib, Kertai, Cooter, Greenup, &

Features and Assessments of Pain, Anesthesia, and Analgesia. https://doi.org/10.1016/B978-0-12-818988-7.00017-0
Copyright © 2022 Elsevier Inc. All rights reserved.

Hwang, 2019) and up to 20% of patients at 6–12 months after surgery (Meretoja, Leidenius, Tasmuth, Sipila, & Kalso, 2014; Spivey et al., 2018; Turan, Karaman, Karaman, Uyar, & Gonullu, 2014). Two to 3 years after surgery for primary breast cancer, 13%–39% patients had moderate-to-severe persistent pain in the surgical area and 58% patients reported sensory disturbances or discomfort (Gartner et al., 2009).

There are several factors that most highly associated with increased severity of acute postoperative pain in breast surgery. These factors are longer duration of surgery, more complicated surgery, higher pain catastrophizing scale, current use of an anxiolytic and/or antidepressant, and surgeon (Armstrong et al., 2016;Fecho et al., 2009; Habib et al., 2019). Meanwhile, preoperative radiotherapy and advanced age were associated with lower acute pain scores (Fecho et al., 2009; Habib et al., 2019). Development of persistent pain after breast cancer surgery is associated with younger age, higher body mass index, higher baseline anxiety and depression, radiotherapy, axillary lymph node dissection, significant acute postoperative pain, and preoperative pain (Habib et al., 2019; Spivey et al., 2018; Wang et al., 2016). Persistent postoperative breast pain is associated with persistent sensory changes, pain interference, and functional impairments (Langford et al., 2014).

Acute to persistent postoperative pain transition is a complex and poorly understood process. Postincisional nociception, peripheral, and central sensitizations have a role in the pathophysiology of transition from acute to persistent postoperative pain. Typical cellular and molecular alterations, distinct from other pain models occur after surgical incision. The postincisional nociception is usually a combination of the inflammatory response and nerve injuries. The peripheral and central sensitizations produce hyperalgesia and allodynia (Richebe, Capdevila, & Rivat, 2018).

Increased severity of acute postoperative pain is one of the risk factors for persisting postoperative pain in breast surgery. Hence, acute postoperative pain management should be optimized for all patients undergoing breast surgery to achieve long-term benefits. Regional anesthesia, general anesthesia, and multimodal analgesia play a role in controlling acute postoperative pain.

General anesthesia: A brief description

General anesthesia is a state of unconsciousness, amnesia, antinociception, and akinesia, with maintenance of physiological (respiratory, cardiovascular, autonomic, and thermoregulatory) stability. It is controlled, reversible, and can be induced by intravenous and/or inhalation anesthetic agents. General anesthesia is a reversible drug-induced coma (Brown, Lydic, & Schiff, 2010; Brown, Pavone, & Naranjo, 2018). General anesthesia is widely used in daily practice to facilitate surgery, as a sole agent or in combination with regional anesthesia.

Intravenous anesthetic, inhalation anesthetic, antinociception agent, and neuromuscular blocking agent or muscle relaxant are commonly used drugs in general anesthesia. Intravenous anesthetic and/or inhalation anesthetic are used to induce general anesthesia and maintain state of unconsciousness and amnesia. Opioid is often used as antinociception and has a sedative effect. Akinesia or immobility is achieved by administration of neuromuscular blocking agent. Intravenous anesthetic and inhalation anesthetic also contribute to muscle relaxation (Brown et al., 2018).

Balanced general anesthesia is a strategy that was developed to avoid the use of ether as a single general anesthetic agent. The principle of this strategy is combining different general anesthetic agent to achieve anesthetic state. This combination uses less dose of each drug to achieve the desired effect, thus is believed to minimize the side effects of the drug rather than used as a sole agent. New balanced general anesthesia is also proposed to reduce or eliminate opioid use by implementation of multimodal pain management intraoperatively, which targets different circuits in the nociceptive system, and this approach should be continued to postoperative period (Brown et al., 2018). Opioid has antinociceptive property with several undesirable adverse effects and might induce tolerance and hyperalgesia. The side effects of opioid administration include muscle rigidity, PONV, postoperative urine retention, pruritus, constipation, sedation, and respiratory depression (Mulier, 2019).

Even though safety assurance is improving, there are still potential complications of general anesthesia. General anesthesia-associated morbidities are rare and ranged from minor to major complications, with or without long-term sequelae, including PONV, dental injury, laryngeal and tracheal injury, bronchospasm, airway management failures, pulmonary aspiration, and cardiac arrest. Even minor complications would make the patient uncomfortable. Major complications could increase hospital cost, length of stay, and unplanned admission, particularly in same-day surgery. The risk of complications can be reduced by performing a conscientious preoperative evaluation. Well-planned anesthesia management can be established if the risks of complication are identified early in the preoperative period (Harris & Chung, 2013).

Technical description of ultrasound-guided Pecs II block

Blanco, Fajardo, and Maldonado (2012) introduced ultrasound-guided Pecs II block as a novel approach to breast surgery. Pecs II block is a modification of Pecs I block with additional a second injection to block the axillary area and lateral mammary area (Fig. 1) (Blanco & Maldonado, 2013). Pectoralis nerve block was inspired by the infraclavicular brachial plexus block (BPB) and transversus abdominis plane block.

Pecs II block is performed under ultrasound guidance using high-frequency linear array probe in awake or anesthetized patients. The following description is an oblique in-plane approach from proximal and medial, to distal and lateral which ensures that the clavicle is away from the area of interest during the block as described by Blanco et al. (2012). Patient in the supine position, the ipsilateral arm abducted 90 degrees, and head rotated to the opposite side. The probe is placed below the lateral third of clavicle in the paramedian sagittal plane. After the PMM, subclavian muscle, axillary artery, and axillary vein are identified (Fig. 2A), the probe is moved distally and rotated slightly toward the axilla until the PmM is identified. Then, the pectoral branch of thoracoacromial artery between the pectoralis muscles is identified using color Doppler. Location of the lateral pectoral nerve is consistently adjacent to the artery. Slight rotation of transducer also allows an in-plane needle trajectory from the proximal and medial toward the lateral side. A first injection, 10 mL of local anesthetic (LA) is placed between the PMM and PmM (Fig. 2B). Distal and lateral movement of the probe is continued until the lateral border of PmM is reached. The first rib is under the axillary artery, the ribs then counted, and there are SAM covering the ribs and the continuation of the suspensory ligament of axilla over the third rib. A second injection, 20 mL of LA is placed between PmM and SAM (Fig. 2C) (Blanco, 2011; Blanco et al., 2012; Senapathi, Widnyana, Aribawa, Jaya, & Junaedi, 2019).

The first injection of Pecs II block targets the lateral and medial pectoral nerves at interfascial plane between PMM and PmM. The second injection targets the intercostobrachial nerve, lateral cutaneous branches of intercostals nerve T3–T6, and long thoracic nerve (Fig. 3). The second injection results in consistent anesthesia of dermatomes from T2 to T4 with variable spread to T6 (Blanco et al., 2012). However, an anatomical study by Versyck, Groen, van Geffen, Van Houwe, and Bleys (2019) found that intercostobrachial nerve is blocked by the first injection of Pecs II block.

Effectiveness of adjuvants to local anesthesia in Pecs II block has been studied. A randomized, double-blind prospective study by Kaur et al. (2017) concluded that addition of 1 µg/kg dexmedetomidine to 0.25% ropivacaine for Pecs II block increases the duration of analgesia and decreases postoperative morphine consumption in oncological breast surgery. Randomized-controlled trial by Ahmed (2018), which investigated the efficacy of Pecs II block using bupivacaine 0.25% with or without 1 mL of magnesium sulfate 50%, found that Pecs II block reduces the intra- and postoperative dose of narcotics with postoperative pain-free, while magnesium sulfate extends the pain-free duration.

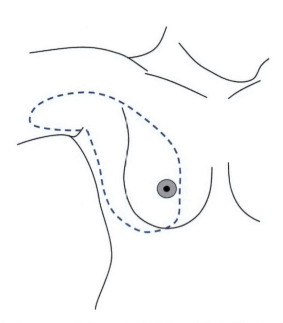

FIG. 1 Illustration of dermatomal distribution pattern of ultrasound-guided Pecs II block. The blue dotted line expresses the expected area of block coverage.

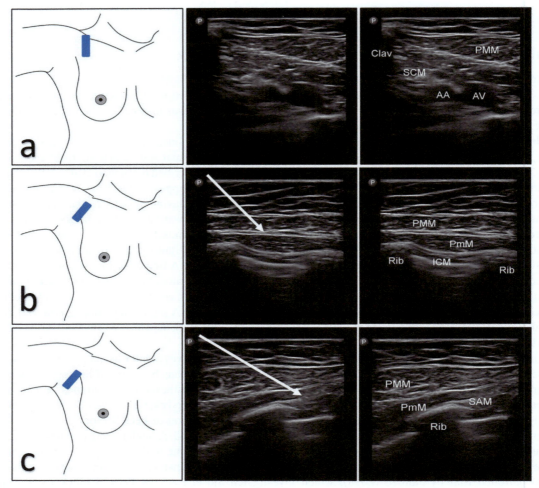

FIG. 2 Performing ultrasound-guided Pecs II block. (A) A high-frequency linear array probe is placed below the lateral third of clavicle, in the paramedian sagittal plane. (B) First injection, 10 mL of LA is placed at interfascial plane between PMM and PmM. (C) Second injection, 20 mL of LA is placed at interfascial plane between PmM and SAM: AA, axillary artery. AV, axillary vein; Clav, clavicle; ICM, intercostal muscles; LA, local anesthetic; PMM, pectoralis major muscle; PmM, pectoralis minor muscle; SAM, serratus anterior muscle; SCM, subclavius muscle.

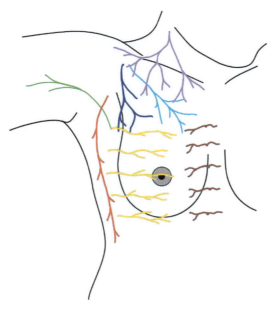

FIG. 3 Illustration of chest wall innervation. Anterior cutaneous branches of intercostal nerves (brown), intercostobrachial nerve (green), lateral cutaneous branches of intercostal nerves (yellow), lateral pectoral nerve (light blue), long thoracic nerve (orange), medial pectoral nerves (blue), and supraclavicular nerve (pink).

Clinical application ultrasound-guided Pecs II block

Ultrasound-guided Pecs II block without (Table 1) or in combination with general anesthesia is indicated for various oncological, reconstructive, and aesthetic surgery, include lumpectomy, wide excisions, axillary clearances, sentinel node dissection, breast reconstruction, several types of mastectomies, breast expanders, and prosthesis insertions. As described by Blanco et al. (2012), the PMM is mainly involved during breast expanders and prosthesis insertions surgery. The intercostal nerves are mostly involved in lumpectomy, mastectomy, and sentinel node dissection.

There are several systematic reviews, and metaanalyses indicate analgesic efficacy of ultrasound-guided Pecs II block for breast surgery. Studies by Jin, Li, Gan, He, and Lin (2020) and Grape, Jaunin, El-Boghdadly, Chan, and Albrecht (2020) concluded that Pecs II block provides significantly better perioperative analgesia, less opioid consumption and reduces the rate of PONV compared to general anesthesia with systemic opioids. When compared to TPVB, in a study by Grape, El-Boghdadly, and Albrecht (2020), there is low-quality evidence that a Pecs II block only provides marginal postoperative analgesic benefit after radical mastectomy at 2 h postoperative and not beyond. A procedure-specific postoperative pain management (PROSPECT) guideline has been released for oncological breast surgery. Pectoralis nerve block is recommended for major breast surgery if no axillary node dissection or paravertebral block is contraindicated as preoperative and intraoperative interventions (Grade A) (Jacobs, Lemoine, Joshi, Van de Velde, & Bonnet, 2020).

The potentiality of Pecs II block to prevent persistent postoperative pain has also been investigated. A prospective observational study by De Cassai et al. (2019) shown that Pecs II block might play an important role in lowering incidence of chronic pain. They found patient in the group received Pecs II block had lower incidence of chronic pain at 3 months compared with the group without Pecs II block. However, the incidence of chronic pain did not differ between groups at 6, 9, and 12 months after surgery (Besch et al. (2018).

The average analgesia onset time of Pecs II block is 3 min, and analgesia lasts for 8 h. A catheter can be placed either at interfascial plane between PMM and PmM or between PmM and SAM to extend the analgesic effect of Pecs II block. Blanco et al. (2012) recommend the catheter is placed in the interfascial plane between both pectoral muscles, leaving 10 cm inside to avoid migration for breast expanders surgery with an infusion of 0.125% levobupivacaine set at 5 mL/h. Tiburzi, Cerotto, Gargaglia, Carli, and Gori (2019) report the Pecs II block with programmed intermittent bolus (PIB) of local anesthetic via catheter placement at the end of surgery into the interfascial plane between PmM and SAM for modified radical mastectomy with sentinel node or axillary dissection and immediate breast reconstruction with prosthesis. They use 0.1% levobupivacaine with PIB of 3 mL every hour for 72 h.

Ultrasound-guided Pecs II block has a low risk of complication. Hematoma, infection, and pneumothorax may occur after Pecs II block (Senapathi et al., 2019). Ueshima and Otake (2017a) report there were eight patients have hematoma around the injection site, out of 498 patients who underwent Pecs II block for breast cancer surgery. Five of these patients had taken oral anticoagulant and antiplatelet drugs. Long thoracic nerve blockage as a result of the second injection of Pecs II block must be our concern, particularly when axillary dissection is planned. Miller, Pawa, & Mariano, 2019 report surgeon's difficulty to stimulate the long thoracic nerve and subsequently activate the serratus anterior muscle. This condition

TABLE 1 Several successful uses of ultrasound-guided Pecs II block without general anesthesia for breast surgery.

Pecs II block for simple mastectomy and lumpectomy in geriatric patients	Murata, Ichinomiya, and Hara (2015)
Pecs II block as primary anesthesia under dexmedetomidine sedation for breast conserving surgery	Moon, Kim, Chung, Song, and Yi (2017)
Pecs II block and internal intercostal plane block for excision of giant fibroadenoma of the breast	Kim, Shim, and Kim (2017)
Combined Pecs II block and TVPB for various type of breast surgery under sedation	Pawa et al. (2018)
Pecs II block for a modified radical mastectomy with axillary lymph node dissection in a comorbid patient	Debbag and Saricaoglu (2019)
Combined Pecs II block, parasternal block, and propofol sedation for unilateral quadrantectomy without axillary dissection	Fusco et al. (2019)
Combined Pecs II block, transversus thoracic muscle plane block, and dexmedetomidine sedation for partial mastectomy and sentinel lymph node biopsy in a patient with achondroplasia	Nakanishi, Yoshimura, and Toriumi (2019)

348 PART | III Interlinking anesthesia, analgesics and pain

is concerning since the correct identification of long thoracic nerve is obscured and potentially expose patients to inadvertent nerve trauma. The injury of the long thoracic nerve results in winged scapula.

Pectoral nerves, intercostal nerve, and long thoracic nerve

The pectoral nerves arise from brachial plexus innervating pectoral muscles. The lateral pectoral nerve arises from the lateral cord of the brachial plexus and contains fibers from cervical spinal nerve C5, C6, and C7. The lateral pectoral nerve runs anteriorly between the PMM and PmM near the pectoral branch of thoracoacromial artery and supply the PMM. The median pectoral nerve arises from the medial cord of the brachial plexus and contains fibers from cervical spinal nerve C8 and thoracic spinal nerve T1. The medial pectoral nerve runs anteriorly under the PmM, crosses the PmM to reach the lower third of the PMM after piercing two layers of the clavipectoral fascia. It innervates the PMM and PmM (Fig. 3) (Blanco et al., 2012; Senapathi et al., 2019).

The intercostal nerves T2–T6 derived from the anterior rami of thoracic spinal nerve T2–T6. They enter the correlated intercostal space at the back between pleura and posterior intercostal membrane and lie in a plane between the intercostal muscles as far as the sternum. The intercostal nerves T2–T6 give off lateral and anterior cutaneous branches (Fig. 3). The lateral cutaneous branches of the intercostal nerves T3–T6 exit at the level of mid-axillary, pierce the external intercostal muscle and the SAM, and give off anterior and posterior terminal branches. The lateral cutaneous branch of the atypical intercostal nerve T2 (intercostobrachial nerve) does not divide into anterior and posterior terminal branches and supplies cutaneous innervation of the axilla and superior region of the medial arm. The anterior cutaneous branches of intercostal nerve T2–T6 cross in front of the internal mammary artery, pierce the internal intercostal muscle, intercostal membranes, and PMM. It divides into medial and lateral terminal branches to supply the medial aspect of breast (internal mammary area) (Blanco et al., 2012).

The long thoracic nerve arises from the anterior rami of cervical spinal nerve C5–C7, enters the axilla behind the rest of the brachial plexus, and lies on the superficial surface of the SAM. It innervates the SAM (Fig. 3) and susceptible to injury during axillary clearances, or radical mastectomy producing a winging scapula (Blanco et al., 2012).

Applications to other areas

Despite ultrasound-guided Pecs II block was originally desired as anesthesia-analgesia for breast surgery, many authors report its use for other procedures as sole technique or in combination with other techniques. The main reason of ultrasound-guided Pecs II block application was avoiding use of general and neuraxial anesthesia in high-risk patients. Furthermore, there were also reports of its use for pain management. There were some reports of successful ultrasound-guided Pecs II block usage for various purposes (Table 2).

The use of continuous Pecs II block was also reported. Shakuo, Kakumoto, Kuribayashi, Oe, and Seo (2017) reported the postoperative analgesia potency of continuous Pecs II block for patients underwent transapical transcatheter aortic valve implantation. At least two randomized-controlled trials were published regarding ultrasound-guided Pecs II block for non-breast surgery. Quek et al. (2018) concluded that the addition of a Pecs II block to a supraclavicular BPB improves regional anesthesia for proximal arm arteriovenous access surgery patients in patients with end-stage renal disease. Reynolds et al. (2019) concluded that the addition of a Pecs II block to an interscalene BPB significantly improved postoperative analgesia and reduced the need for opioids in the postanesthesia care unit after arthroscopic shoulder surgery with an open biceps tenodesis.

Other agents of interest

In this chapter, we have described clinical usefulness of ultrasound-guided Pecs II block for various procedures and pain management, either as sole agent or in combination with other anesthesia-analgesia techniques. Other regional anesthesia–analgesia techniques for chest wall include:

- Thoracic epidural anesthesia-analgesia (TEA)
- Thoracic paravertebral block (TPVB)
- Intercostal nerve block (ICB)
- Serratus anterior muscle plane block (SAPB)
- Transversus thoracis muscle plane block (TTPB)
- Erector spinae plane block (ESPB)

TABLE 2 Reports on the successful use of ultrasound-guided Pecs II block for nonbreast surgery.

Ultrasound-guided Pecs II block for surgery	
Combined Pecs II block and supraclavicular BPB for upper arm fistula creation	Purcell and Wu (2014)
Pecs II block for axillary sentinel lymph node biopsy in malignant tumors on the upper extremities	Yokota, Matsumoto, Murakami, and Akiyama (2017)
Combined Pecs II block and sedation for transthoracic arteriovenous graft repair	Farkas, Weber, Miller, and Xu (2018)
Combined Pecs II block, supraclavicular BPB, continuous spinal anesthesia, and sedation for axillofemoral-femoral bypass	Soberon Jr., Endredi, Doyle, and Berceli (2019)
Pecs II block and sedation for an ICD implantation in the super obese patient	Pai, Shariat, and Bhatt (2019)
Ultrasound-guided Pecs II block for pain management	
Second injection of Pecs II block for the management of herpes zoster-associated pain	Kim, Park, Shim, and Kim (2016)
Pecs II block as postoperative rescue analgesia in a patient undergoing minimally invasive cardiac surgery	Yalamuri et al. (2017)
Pecs II block for intractable postherpetic neuralgia	Oh (2018)

TEA, TPVB, and ICB are classical regional anesthesia-analgesia techniques for chest wall surgery and pain management. TEA is commonly selected for major surgery and bilateral chest wall anesthesia-analgesia. TPVB offers the advantages of unilateral chest wall anesthesia-analgesia. Multiple ICBs are needed for multidermatomal sensory block. Several novel ultrasound-guided interfascial plane block regional anesthesia–analgesia techniques, namely the SAPB, TTPB, and ESPB, have been reported as alternative to TEA, TPVB, and ICB. Those new techniques are considered simple, easy to perform, less invasive, safe, and associated with a low risk of side effects. The injection site and target nerve of these regional techniques are different (Fig. 4).

Blanco, Parras, McDonnell, and Prats-Galino (2013) present SAPB as a novel ultrasound-guided regional anesthesia technique that may achieve complete paresthesia of the hemithorax. The local anesthetic is injected either superficial to the serratus anterior muscle or deep underneath the muscle overlying the fifth rib in midaxillary line, under ultrasound guidance using a high-frequency linear probe. Thirty minutes after LA injection, they demonstrated the dermatomal distribution of sensory loss from T2 to T9.

Ultrasound-guided TTPB was introduced by Ueshima and Kitamura (2015) to block multiple anterior branches of intercostal nerves (T2–T6), which innervate the region of internal mammary area. Using a high-frequency linear probe, LA is placed in the plane between transversus thoracic muscle and internal intercostal muscle, between fourth and fifth ribs next to the sternum (Ueshima & Otake, 2016).

Forero, Adhikary, Lopez, Tsui, and Chin (2016) reported the ESPB as a novel interfascial plane block for analgesia in thoracic neuropathic pain. A high-frequency linear probe is placed in a longitudinal orientation 3 cm lateral to the T5 spinous process and local anesthetic placed in the interfascial plane between rhomboid major and erector spinae muscles. The site of action is likely at the dorsal and ventral rami of the thoracic spinal nerves, and the sensory block was from T2 to T9 dermatomes within 20 min after each injection.

Mini-dictionary of terms

Akinesia. Neuromuscular disturbance in which no movement could be inflicted due to blockade in musculoskeletal system.

Allodynia. An unwanted painful sensation which was caused by nonnoxious or innocuous stimuli. Allodynia should be considered as deranged of physiological process which could lead to several debilitating consequences.

Amnesia. Neurologic dysfunctional state in which a person could not revoke memories ranges from short until long term due to certain intracranial or extracranial cause.

Breast cancer surgery. Keystone of breast cancer management. A surgical technique in oncology specifically as treatment regime for breast cancer ranging from preserves healthy breast tissue, removes cancerous tissue, and sometimes extends to both mammary glands and adjacent lymph nodes removal.

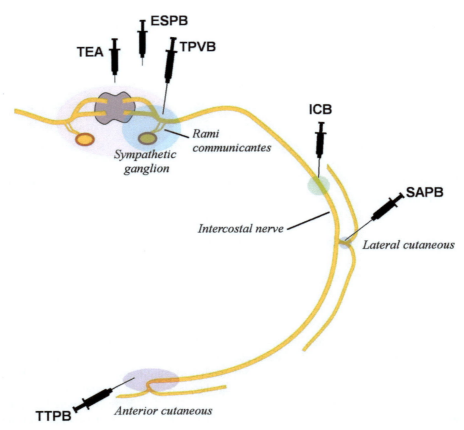

FIG. 4 Schematic diagram of the injection site and target nerve of regional anesthesia-analgesia techniques for chest wall. ESPB and TPVB have different injection site but same nerve targets, and they produce unilateral-multidermatomal sensory block: ESPB, erector spinae plane block; ICB, intercostal nerve block; SAPB, serratus anterior muscle plane block; TEA, thoracic epidural anesthesia-analgesia; TPVB, thoracic paravertebral block; TTPB, transversus thoracis muscle plane block.

Hyperalgesia. Overstimulated or exaggerated painful sensation in or extending over the coverage area. When it comes at the injury site and emerges with mechanical and heat stimulation, it is called primary hyperalgesia, but when it extends over the injury site and usually emerges with mechanical stimulation, it is called secondary hyperalgesia.

Key facts of ultrasound-guided Pecs II block

- Blanco et al. (2012) introduced the ultrasound-guided Pecs II block (modified Pecs I block) as a novel approach to breast surgery.
- Pecs II block involves two injections of LA, the first injection is made at interfascial plane between PMM and PmM; the second injection is made at interfascial plane between PmM and SAM.
- Pecs II block targets the lateral and medial pectoral nerves, the intercostobrachial nerve, the lateral branches of intercostals nerve T3–T6, and the long thoracic nerve.
- Pecs II block can be used as sole agent, in combination with other regional techniques, or with general anesthesia for breast surgery and other procedures involving lateral thoracic wall area.
- Procedure-specific postoperative pain management (PROSPECT) guideline for oncological breast surgery recommends pectoralis nerve block as preoperative and intraoperative interventions for major breast surgery if no axillary node dissection or paravertebral block is contraindicated.
- Pecs II block can be performed as a single shot or continuous techniques, with or without adjuvants to LA. A catheter can be placed either at the interpectoral plane or above the SAM, according to the desired target nerve.

Summary points

- Acute and persistent (chronic) postoperative pain have detrimental effects on patient, acute pain could slow the recovery phase and persistent pain decrease patient' functional status and quality of life.
- Combined general and regional anesthesia are often used to facilitate surgical procedures. The addition of regional anesthesia will reduce the requirement of general anesthetic agents. The regional technique provides anesthesia-analgesia during and after surgery. A regional technique continued into the postoperative period offers postoperative analgesia, decreased opioid consumption, and reduction in PONV.
- Ultrasound-guided Pecs II block is an easy, fast-acting, and reliable superficial block. It is a simple alternative to "gold standard" TEA and TPVB for breast surgery. It can be used as a sole anesthesia agent, an analgesia component of balanced anesthesia or postoperative rescue analgesia.
- Surgical procedures involving lateral thoracic area can be facilitated by ultrasound-guided Pecs II block and extension of block coverage can be achieved by combining Pecs block with other regional techniques.

References

Ahmed, A. A. A. (2018). Efficacy of pectoral nerve block using bupivacaine with or without magnesium sulfate. *Anesthesia, Essays and Researches, 12*, 440–445.

Armstrong, K. A., Davidge, K., Morgan, P., Brown, M., Li, M., Cunningham, L., et al. (2016). Determinants of increased acute postoperative pain after autologous breast reconstruction within an enhanced recovery after surgery protocol: A prospective cohort study. *Journal of Plastic, Reconstructive & Aesthetic Surgery, 69*, 1157–1160.

Besch, G., Lagrave-Safranez, C., Ecarnot, F., De Larminat, V., Gay, C., Berthier, F., et al. (2018). Pectoral nerve block and persistent pain following breast cancer surgery: An observational cohort study. *Minerva Anesthesiologica, 84*, 769–771.

Blanco, R. (2011). The 'Pecs block': A novel technique for providing analgesia after breast surgery. *Anaesthesia, 66*, 847–848.

Blanco, R., Fajardo, M., & Maldonado, T. P. (2012). Ultrasound description of Pecs II (modified Pecs I): A novel approach to breast surgery. *Revista Española de Anestesiología y Reanimación, 59*, 470–475.

Blanco, R., & Maldonado, T. P. (2013). Reply to the article entitled "Ultrasound description of Pecs II (modified Pecs I): A novel approach to breast surgery". Reply of the authors. *Revista Española de Anestesiología y Reanimación, 60*, 296–297.

Blanco, R., Parras, T., McDonnell, J. G., & Prats-Galino, A. (2013). Serratus plane block: A novel ultrasound-guided thoracic wall nerve block. *Anaesthesia, 68*, 1107–1113.

Brown, E. N., Lydic, R., & Schiff, N. D. (2010). General anesthesia, sleep, and coma. *The New England Journal of Medicine, 363*, 2638–2650.

Brown, E. N., Pavone, K. J., & Naranjo, M. (2018). Multimodal general anesthesia: Theory and practice. *Anesthesia & Analgesia, 127*, 1246–1258.

De Cassai, A., Bonanno, C., Sandei, L., Finozzi, F., Carron, M., & Marchet, A. (2019). PECS II block is associated with lower incidence of chronic pain after breast surgery. *The Korean Journal of Pain, 32*, 286–291.

Debbag, S., & Saricaoglu, F. (2019). Pectoral nerve block as the sole anesthetic technique for a modified radical mastectomy in a comorbid patient. *Saudi Medical Journal, 40*, 1285–1289.

Farkas, G., Weber, G., Miller, J., & Xu, J. (2018). Transthoracic arteriovenous graft repair with the pectoralis (PECS) II nerve block for primary intraoperative anesthesia and postoperative analgesia: A case report. *A & A Practice, 11*, 224–226.

Fecho, K., Miller, N. R., Merritt, S. A., Klauber-Demore, N., Hultman, C. S., & Blau, W. S. (2009). Acute and persistent postoperative pain after breast surgery. *Pain Medicine, 10*, 708–715.

Forero, M., Adhikary, S. D., Lopez, H., Tsui, C., & Chin, K. J. (2016). The erector spinae plane block: A novel analgesic technique in thoracic neuropathic pain. *Regional Anesthesia and Pain Medicine, 41*, 621–627.

Fusco, P., Cofini, V., Petrucci, E., Pizzi, B., Necozione, S., & Marinangeli, F. (2019). The anaesthetic and analgesic effects of pectoral nerve and parasternal block combination for patients undergoing breast cancer surgery: A phase II study. *European Journal of Anaesthesiology, 36*, 798–801.

Gartner, R., Jensen, M. B., Nielsen, J., Ewertz, M., Kroman, N., & Kehlet, H. (2009). Prevalence of and factors associated with persistent pain following breast cancer surgery. *JAMA, 302*, 1985–1992.

Grape, S., El-Boghdadly, K., & Albrecht, E. (2020). Analgesic efficacy of PECS vs paravertebral blocks after radical mastectomy: A systematic review, meta-analysis and trial sequential analysis. *Journal of Clinical Anesthesia, 63*, 109745.

Grape, S., Jaunin, E., El-Boghdadly, K., Chan, V., & Albrecht, E. (2020). Analgesic efficacy of PECS and serratus plane blocks after breast surgery: A systematic review, meta-analysis and trial sequential analysis. *Journal of Clinical Anesthesia, 63*, 109744.

Habib, A. S., Kertai, M. D., Cooter, M., Greenup, R. A., & Hwang, S. (2019). Risk factors for severe acute pain and persistent pain after surgery for breast cancer: A prospective observational study. *Regional Anesthesia and Pain Medicine, 44*, 192–199.

Harris, M., & Chung, F. (2013). Complications of general anesthesia. *Clinics in Plastic Surgery, 40*, 503–513.

Jacobs, A., Lemoine, A., Joshi, G. P., Van de Velde, M., & Bonnet, F. (2020). PROSPECT guideline for oncological breast surgery: A systematic review and procedure-specific postoperative pain management recommendations. *Anaesthesia*.

Jin, Z., Li, R., Gan, T. J., He, Y., & Lin, J. (2020). Pectoral nerve (PECs) block for postoperative analgesia-a systematic review and meta-analysis with trial sequential analysis. *International Journal of Physiology, Pathophysiology and Pharmacology, 12*, 40–50.

Kaur, H., Arora, P., Singh, G., Singh, A., Aggarwal, S., & Kumar, M. (2017). Dexmedetomidine as an adjunctive analgesic to ropivacaine in pectoral nerve block in oncological breast surgery: A randomized double-blind prospective study. *Journal of Anaesthesiology Clinical Pharmacology, 33*, 457–461.

Kim, H., Shim, J., & Kim, I. (2017). Surgical excision of the breast giant fibroadenoma under regional anesthesia by Pecs II and internal intercostal plane block: A case report and brief technical description: A case report. *Korean Journal of Anesthesiology, 70*, 77–80.

Kim, Y. D., Park, S. J., Shim, J., & Kim, H. (2016). Clinical usefulness of pectoral nerve block for the management of zoster-associated pain: Case reports and technical description. *Journal of Anesthesia, 30*, 1074–1077.

Langford, D. J., Paul, S. M., West, C., Levine, J. D., Hamolsky, D., Elboim, C., et al. (2014). Persistent breast pain following breast cancer surgery is associated with persistent sensory changes, pain interference, and functional impairments. *The Journal of Pain, 15*, 1227–1237.

Meretoja, T. J., Leidenius, M. H. K., Tasmuth, T., Sipila, R., & Kalso, E. (2014). Pain at 12 months after surgery for breast cancer. *JAMA, 311*, 90–92.

Miller, B., Pawa, A., & Mariano, E. R. (2019). Problem with the Pecs II block: The long thoracic nerve is collateral damage. *Regional Anesthesia and Pain Medicine, 44*, 817–818.

Moon, E. J., Kim, S. B., Chung, J. Y., Song, J. Y., & Yi, J. W. (2017). Pectoral nerve block (pecs block) with sedation for breast conserving surgery without general anesthesia. *Annals of Surgical Treatment and Research, 93*, 166–169.

Mulier, J. P. (2019). Is opioid-free general anesthesia for breast and gynecological surgery a viable option? *Current Opinion in Anaesthesiology, 32*, 257–262.

Murata, H., Ichinomiya, T., & Hara, T. (2015). Pecs block for anesthesia in breast surgery of the elderly. *Journal of Anesthesia, 29*, 644.

Nakanishi, T., Yoshimura, M., & Toriumi, T. (2019). Pectoral nerve II block, transversus thoracic muscle plane block, and dexmedetomidine for breast surgery in a patient with achondroplasia: A case report. *JA Clinical Reports, 5*, 47.

Oh, D. S. (2018). Pecs II block for intractable postherpetic neuralgia. *Journal of Anesthesia, 32*, 460.

Pai, B. H. P., Shariat, A. N., & Bhatt, H. V. (2019). PECS block for an ICD implantation in the super obese patient. *Journal of Clinical Anesthesia, 57*, 110–111.

Pawa, A., Wight, J., Onwochei, D. N., Vargulescu, R., Reed, I., Chrisman, L., et al. (2018). Combined thoracic paravertebral and pectoral nerve blocks for breast surgery under sedation: A prospective observational case series. *Anaesthesia, 73*, 438–443.

Purcell, N., & Wu, D. (2014). Novel use of the PECS II block for upper limb fistula surgery. *Anaesthesia, 69*, 1294.

Quek, K. H., Low, E. Y., Tan, Y. R., Ong, A. S. C., Tang, T. Y., Kam, J. W., et al. (2018). Adding a PECS II block for proximal arm arteriovenous access—A randomised study. *Acta Anaesthesiologica Scandinavica, 62*, 677–686.

Reynolds, J. W., Henshaw, D. S., Jaffe, J. D., Dobson, S. W., Edwards, C. J., Turner, J. D., et al. (2019). Analgesic benefit of pectoral nerve block II blockade for open subpectoral biceps tenodesis: A randomized, prospective, double-blinded, controlled trial. *Anesthesia & Analgesia, 129*, 536–542.

Richebe, P., Capdevila, X., & Rivat, C. (2018). Persistent postsurgical pain: Pathophysiology and preventative pharmacologic considerations. *Anesthesiology, 129*, 590–607.

Senapathi, T. G. A., Widnyana, I. M. G., Aribawa, I. G. N. M., Jaya, A. A. G. P. S., & Junaedi, I. M. D. (2019). Combined ultrasound-guided Pecs II block and general anesthesia are effective for reducing pain from modified radical mastectomy. *Journal of Pain Research, 12*, 1353–1358.

Shakuo, T., Kakumoto, S., Kuribayashi, J., Oe, K., & Seo, K. (2017). Continuous PECS II block for postoperative analgesia in patients undergoing transapical transcatheter aortic valve implantation. *JA Clinical Reports, 3*, 65.

Soberon, J. R., Jr., Endredi, J. J., Doyle, C., & Berceli, S. A. (2019). Novel use of the PECS II block in major vascular surgery: A case report. *A & A Practice, 13*, 145–147.

Spivey, T. L., Gutowski, E. D., Zinboonyahgoon, N., King, T. A., Dominici, L., Edwards, R. R., et al. (2018). Chronic pain after breast surgery: A prospective, observational study. *Annals of Surgical Oncology, 25*, 2917–2924.

Tiburzi, C., Cerotto, V., Gargaglia, E., Carli, L., & Gori, F. (2019). Pectoral nerve block II with programmed intermittent bolus of local anesthetic and postoperative pain relief in breast surgery. *Minerva Anestesiologica, 85*, 201–202.

Turan, M., Karaman, Y., Karaman, S., Uyar, M., & Gonullu, M. (2014). Postoperative chronic pain after breast surgery with or without cancer: Follow up 6 months. *European Journal of Anaesthesiology, 31*, 216.

Ueshima, H., & Kitamura, A. (2015). Blocking of multiple anterior branches of intercostal nerves (Th2-6) using a transversus thoracic muscle plane block. *Regional Anesthesia and Pain Medicine, 40*, 388.

Ueshima, H., & Otake, H. (2016). Where is an appropriate injection point for an ultrasound-guided transversus thoracic muscle plane block? *Journal of Clinical Anesthesia, 33*, 190–191.

Ueshima, H., & Otake, H. (2017a). Ultrasound-guided pectoral nerves (PECS) block: Complications observed in 498 consecutive cases. *Journal of Clinical Anesthesia, 42*, 46.

Versyck, B., Groen, G., van Geffen, G. J., Van Houwe, P., & Bleys, R. L. (2019). The pecs anesthetic blockade: A correlation between magnetic resonance imaging, ultrasound imaging, reconstructed cross-sectional anatomy and cross-sectional histology. *Clinical Anatomy, 32*, 421–429.

Wang, L., Guyatt, G. H., Kennedy, S. A., Romerosa, B., Kwon, H. Y., Kaushal, A., et al. (2016). Predictors of persistent pain after breast cancer surgery: A systematic review and meta-analysis of observational studies. *CMAJ, 188*, E352–E361.

Yalamuri, S., Klinger, R. Y., Bullock, W. M., Glower, D. D., Bottiger, B. A., & Gadsden, J. C. (2017). Pectoral fascial (PECS) I and II blocks as rescue analgesia in a patient undergoing minimally invasive cardiac surgery. *Regional Anesthesia and Pain Medicine, 42*, 764–766.

Yokota, K., Matsumoto, T., Murakami, Y., & Akiyama, M. (2017). Pectoral nerve blocks are useful for axillary sentinel lymph node biopsy in malignant tumors on the upper extremities. *International Journal of Dermatology, 56*, e64–e65.

Chapter 32

Pain control during prostate biopsy and evolution of local anesthesia techniques

Mustafa Suat Bolat[a], Önder Cinar[b], Ali Batur (Furkan)[c], Ramazan Aşcı[d], and Recep Büyükalpelli[d]

[a]*Department of Urology, Gazi State Hospital, Samsun, Turkey,* [b]*Department of Urology, Bülent Ecevit University, Zonguldak, Turkey,* [c]*Department of Urology, Selçuk University, Konya, Turkey,* [d]*Department of Urology, Ondokuzmayıs University, Samsun, Turkey*

List of abbreviations

MetS	metabolic syndrome
HDL	high-density lipoprotein
PSA	prostate-specific antigen
MHz	megahertz
M receptor	muscarinic receptor
TRUS-PB	transrectal ultrasound-guided prostate biopsy
PPNB	periprostatic nerve block
TPLA	transperineal local anesthetic infiltration
IPLA	intraprostatic local anesthetic infiltration
PPB	pelvic plexus blockade
IRLA	intrarectal local anesthesia
TENS	transcutaneous electrical nerve stimulation
EMLA	eutectic mixture of local anesthetic
GTN	glyceryl trinitrate

Introduction

Why is prostate cancer significant?

1. Prostate cancer is the most frequently diagnosed tumor with 14.5% among genitourinary cancers. It is the second most commonly diagnosed cancer among all cancers, with 15.5% after lung cancer (Ferlay et al., 2019).
2. The global increase in obesity contributes to the dramatic rise of MetS. In addition to having a waist circumference > 102 cm, one of the MetS components, including hypertension, increased fasting blood sugar, hypertriglycemia, and low HDL, is a risk factor for prostate cancer (Blanc-Lapierre et al., 2015).
3. Although clinically insignificant prostate cancers can be diagnosed, the overall 5-year survival rate for the early stage of prostate cancer is almost 100%.
4. Irrespective of clinical significance, prostate biopsy remains a unique way to diagnose prostate cancer. It may become a bothering procedure in which patients experience fear of pain and discomfort, and this may affect the patient's decision when future rebiopsy is needed (Alavi, Soloway, Vaidya, Lynne, & Gheiler, 2001).

Features and Assessments of Pain, Anesthesia, and Analgesia. https://doi.org/10.1016/B978-0-12-818988-7.00024-8
Copyright © 2022 Elsevier Inc. All rights reserved.

Neuroanatomy of the prostate

The prostate remains quite small due to low androgen levels during the prepubertal period. However, with the beginning of puberty, dihydrotestosterone, converted from testosterone by 5-alpha reductase, triggers the prostate's rapid growth until it reaches the full mature size by 18–20 years (Berry, Coffey, Walsh, & Ewing, 1984).

The prostate receives rich adrenergic, cholinergic, nonadrenergic–noncholinergic, and peptidergic autonomic innervations (Vaalasti & Hervonen, 1979). Sympathetic fibers arise from the T_{10-12}, and the $L_{1,2}$ spinal segments form the thoracolumbar splanchnic nerves, which subsequently reach the superior hypogastric plexus on the anterior surface of the aortic bifurcation (Kepper & Keast, 1995). The nerves leaving the superior hypogastric plexus, also called the hypogastric nerves, traverse on surfaces of the common iliac arteries, subsequently reach the inferior hypogastric plexuses bilaterally. Preganglionic noradrenergic sympathetic nerve fibers, responsible for smooth muscle contraction, and transport the prostatic fluid into the ejaculate, reach the prostate along with the capsular arteries. Autonomic innervation of the prostate also plays a role in voiding and ejaculation control (Davies, 1997). Smooth muscle contractions occur via the alpha-1 L receptor, a subtype of the alpha-1A receptor (Muramatsu et al., 1994). The human prostate also contains alpha-2-adrenoceptors localized in the vascular and glandular epithelium, which inhibit the release of noradrenaline, causing relaxation.

Parasympathetic fibers arising from the S_{2-4} spinal segments form the left and right pelvic splanchnic nerves reach the inferior hypogastric plexus, also known as pelvic plexus between tips of seminal vesicles and the lateral sides of the rectum (Schlegel & Walsh, 1987). Stromal and glandular epithelial cholinergic fibers, mediated by muscarinic M receptors, are responsible for prostatic secretions (Smith & Lebeaux, 1970). In addition to the autonomic innervation, somatic pudendal nerves originating from the S_{2-4} spinal segments innervate the rectal and urinary sphincter (Fig. 1). In humans, the location of prostatic sensory nerves has not been demonstrated yet. In some mammals, it has been shown that afferent neurons reaching the prostate are transmitted through the pelvic and pudendal nerves from the sacral dorsal root ganglia. A smaller portion of the afferent fibers originates from the inferior hypogastric plexus (also known as pelvic plexus). Differences in pain thresholds and local anesthetic infiltrations to different localizations may affect pain perception of the patients.

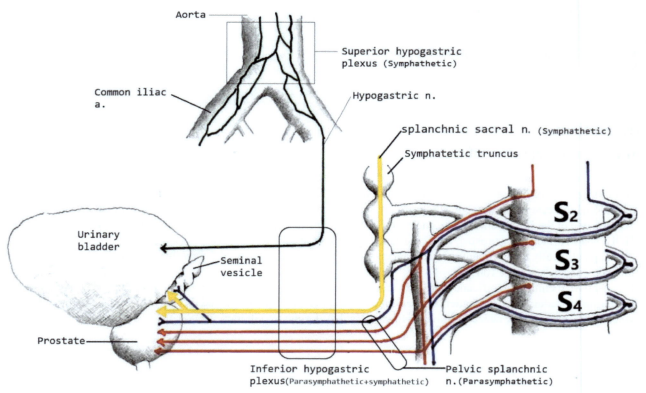

FIG. 1 Autonomous innervation of the prostate and adjacent organs.

TABLE 1 Anesthesia techniques for prostate biopsy.

A. Local anesthesia techniques
  1. Bilateral periprostatic nerve block (PPNB)
  2. Transperineal local anesthetic infiltration (TPLA)
  3. Intraprostatic local anesthetic infiltration (IPLA)
  4. Pelvic plexus blockade (PPB)
  5. Intrarectal local anesthetic use (IRLA)
  6. Caudal block
  7. Combination of any of the above
B. The transcutaneous electrical nerve stimulation with or without IRLA
C. General anesthesia

History of prostate biopsy

Before the PSA and transrectal ultrasound era, Young et al. described his open perineal prostatic biopsy under general anesthesia in 1926, as a frozen section procedure before radical prostatectomy. Although its diagnostic accuracy was higher as 95%, its use remained low due to high comorbidities, such as incontinence and erectile dysfunction (Young & Davis, 1926). Four years later, Ferguson (1930) modified this technique using an aspiration biopsy needle under local anesthesia, but the obtained tissue was not enough to diagnose cancer (Ferguson, 1930). Due to the risk of life-threatening infections, the perineal approach was adopted until 1937, but a finger-guided transrectal biopsy technique was first described and improved by Astraldi (1937). Barringer described the first prostatic biopsy using a screw-tip needle through the perineal route under local anesthesia using novocaine in a patient with hard and irregular prostate (Barringer, 1942). Transrectal ultrasound, showing abnormal areas that cannot be palpated in the prostate, was first introduced as an imaging method in 1963, and the prostate biopsy technique evolved into a diagnostic procedure using the 7 MHz rectal probes after the 1980s (Takahashi & Ouchi, 1963). The standard anesthesia protocol was first described here in 1996 with the application of local anesthetic infiltration to the periprostatic region for prostate biopsy (Nash, Bruce, Indudhara, & Shinohara, 1996).

The primary purpose of the anesthesia during the prostate biopsy is to provide adequate pain control using (a) different number and infiltration sites, (b) infiltration of single or combined anesthetic agents. Also, as a novel technique, transcutaneous electrical nerve stimulation has been introduced.

Today's understanding is to achieve high cancer detection rates with a painless procedure.

In this chapter, local anesthesia techniques used to cope with the pain arising during the prostate biopsy will be discussed in detail (Table 1).

Local anesthesia techniques for prostate biopsy

Bilateral periprostatic nerve block (PPNB)

Transrectal ultrasound-guided prostate biopsy (TRUS-PB) under local anesthesia to the periprostatic nerves is the most widely used method, currently. The blockage of periprostatic nerves (PNB) using local anesthetics allows multiple biopsies in the same session. In this technique, the hypogastric plexus and bilateral periprostatic area are targeted (Fig. 2, marked with "*"). In this technique, the anatomical site has a critical role. Two variations of infiltration sites have been reported: (a) PNB lateral to the base of the prostate, (b) PNB at the apex of the prostate (Rodriguez, Kyriakou, Leray, Lobel, & Guillé, 2003). Meta-analytic studies have shown that bilateral PNB at the base of the prostate provided better analgesia than bilateral apical PNB (Li et al., 2017). However, this technique has limitations, such as an increased risk of septic complications and difficulty accessing the anterior region of the prostate. In general, the rate of infectious complications requiring hospitalization is around 2% (Akdeniz, Bolat, & Akdeniz, 2018).

Transperineal local anesthetic infiltration (TPLA)

The transperineal approach can reduce septic events. In this technique, broader access can be reached to all anatomical parts of the prostate, especially the apical and anterior lobes (Skouteris et al., 2018). The local anesthetic is given into the place between the prostatic apex and perineal skin (Fig. 2, marked with "O") (Meyer et al., 2018). Pain during the prostate biopsy predominately arises from the prostatic capsule or stroma. The somatic sensory fibers of the pudendal nerve and the nerves that arise from the pelvic plexus entering the prostatic capsule are targeted (Shahin & Koch, 2018). Until 2003, the authors believed that the perineal route could provide a comfortable biopsy due to the adequate anesthesia of the prostate's anterior

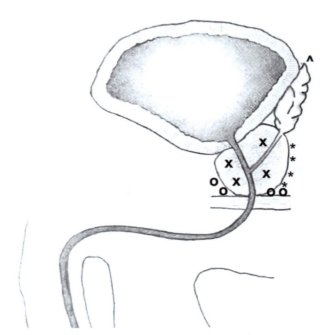

FIG. 2 Different application sites of local anesthetic agents: bilateral periprostatic nerve blockade (*), transperineal infiltration (O), intraprostatic local anesthetic infiltration (X), pelvic plexus blockade ( ).

part. Jones et al. showed that the pain developed secondary to the stimulation of the anal pain fibers, which could be prevented by a rectal sensation test (Jones & Zippe, 2003).

Intraprostatic local anesthetic infiltration (IPLA)

Intraprostatic local anesthesia was first introduced by Mutaguchi et al. (2005). This technique relies on anesthesia of all sensory nerves by injecting a local anesthetic directly into the prostatic parenchyma (Fig. 2, marked with "X"). To date, some researchers have only been used as a part of a local anesthetic combination for prostate biopsy (Bingqian, Peihuan, Yudong, Jinxing, & Zhiyong, 2009; Cam, Sener, Kayikci, Akman, & Erol, 2008; Lee et al., 2007; Singh, Kumar, Griwan, & Sen, 2012). In this technique, pain-sensitive intraprostatic somatic nerves located in both anterior and posterior parts of the prostate are targeted. No sufficient data are present regarding complications associated with this method, and probably, for this reason, this technique has not widely used to date.

Bilateral pelvic plexus blockade (PPB)

In this technique, lateral local anesthetic infiltration is performed at the tip of the seminal vesicles, which allows a larger block area (Fig. 2, marked with " "). The anatomical basis of the PPB leads to a more definite analgesia expectation. Still, it has been shown in several studies that there is no significant difference between PNB and PPB. Also, reaching more proximal is technically more difficult than PNB; the PPB has not widely used in common urology practice (Kim et al., 2019).

Intrarectal local anesthetic use (IRLA)

The close relationship of the anterolateral nerve fibers of the prostate with the anterior rectal wall, the excellent absorption capacity of the rectum, and the friction-reducing effect the local anesthetic gel during probe insertion are the basis of this technique. A recent meta-analysis reported that only 5 of 17 randomized controlled trials showed that lidocaine gel significantly reduced pain compared to traditional ultrasound gel. In contrast, 12 studies reported no difference in pain control (Lee et al., 2020). It is reasonable to assume that the IRLA reduces pain compared to the placebo. Sufficient pain control cannot be provided due to prostatic capsular and parenchymal nerve fibers located far from the rectum mucosa. However, IRLA may support pain control by reducing acute tissue damage and/or cytokine release during probe insertion and dilatation of the anal canal (Yan, Wang, Huang, & Zhang, 2016).

The eutectic mixture of local anesthetic cream (EMLA) is made up of 2.5% lidocaine, 2.5% prilocaine with no additional protective material. Almost 90% of the EMLA remains on the applied surface, and systemic side-effect prevalence is extremely low (Başar et al., 2005). Although EMLA is usually applied to the perianal area and anterior rectal wall corresponding to the prostate, additional oral or parenteral analgesic support is generally needed (Kim et al., 2011). In this technique, only the pain sensation from rectal fibers can be prevented, and these fibers usually do not travel upward above the dentate line (Yang et al., 2017).

Caudal block

The sacral canal contains the terminal part of the medulla spinalis. Kauda equina, formed by the five sacral nerve roots and coccygeal nerve, proceeds in this canal and traverses to the pelvic organs by way of anterior and posterior foramina. As a more invasive procedure, the caudal block is not an appropriate approach for prostate biopsy because the prostate receives multiple innervations from both sympathetic, parasympathetic, and somatic afferents (pudendal nerve). Also, anatomically larger sacral canal volume requiring higher volumes of local anesthetic may lead to an increased incidence of dose-related side effects (Horinaga, Nakashima, & Nakanoma, 2006).

Combination of techniques mentioned above

Several combinations are used in the prostate biopsy, mainly PPNB + IPLA, Caudal block + IRLA, and TENS + IRLA. Meta-analytic studies indicated that PPNB + IRLA provided superior pain control. Although a combination of the PPNB + IPLA provided effective pain control, the direct intraprostatic injection has been reported as a rather complicated technique associated with prolonged procedure time (Yan et al., 2016). For the TENS technique, more comprehensive prospective studies are needed.

Transcutaneous electrical nerve stimulation (TENS)

Pain perception is transferred to the brain through the dorsal horn of the medulla spinalis. Pain sensation is mainly mediated by small nerve fibers (C-fibers), whereas touch and pressure sensations are transferred by large nerve fibers (A-beta fibers). Some conditions may affect this transmission according to the gate-control theory, described by Melzack and Wall in the mid-1960s (Melzack & Wall, 1965). Transmission cells and inhibitory interneurons control this transmission traffic together. Increased stimulation of the large fibers compared to small fibers causes inhibitory neuron stimulation, resulting in less sensation of pain.

Roughly, when a painful stimulus is given, small nerve fibers are activated, transmission cells are stimulated, and the pain sensation is transferred to the upper centers. At that time, stimulation of large nerve fibers, for example, by massaging the trauma site activates the inhibitory interneurons and pain ceases. Gate control theory does not explain the pain control homogeneously in all patients. Psychological factors and the past experiences of pain may affect the sense of pain perception. As a noninvasive technique, TENS is based on the principle of interfering with neural traffic. Stimulation of the thoracal 11–12 skin dermatomes by a TENS device activates A-beta fibers. Thus, C-fibers transmitting the pain sensation to the upper center are inhibited. In a recent study, the INFLATE technique has been described and presented to the literature with satisfactory pain control (Fig. 3A–C). Following the application of IRLA, a two-electrode two-channel TENS device is attached to the anterior suprapubic, and posterior presacral skin surfaces using adhesive electrodes correspond to bilateral T11–12 dermatomes. Following adjustment of the frequency and energy level, 3–6 min of stimulation is provided, and then the biopsy is started. The advantage of this technique is that it does not require penetrating local anesthetic agent applications (Bolat, Cinar, Asci, & Buyukalpelli, 2019).

General anesthesia

There is a lack of sufficient data in the literature regarding general anesthesia during the transrectal biopsy. Physicians usually do not prefer general anesthesia. Because hospitalization may be needed and cardiovascular complications may increase (Tobias-Machado et al., 2006). Particularly, an uncomfortable biopsy experience in the past may require prostate biopsy under intravenous sedation or general anesthesia.

Different numbers and sites of infiltration

The first systematic prostate biopsy was performed using 5 mL 1% lidocaine infiltration into each side and 5 mL 0.9% sodium chloride into the contralateral side, under the guidance of transrectal ultrasound and provided a better pain control

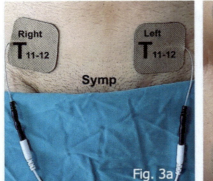  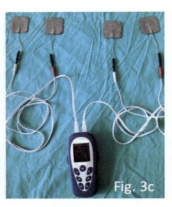

FIG. 3 Transcutaneous electrical nerve stimulation (TENS) according to INFLATE (Infiltration-free local anesthesia technique). Electrode attachment sites are shown on the abdominal wall (A) and posterior surface (B). The TENS device with electrode connections (C). T_{11-12} shows the thoracal 11–12 dermatome lines both on anterior and posterior surfaces. Symp, symphysis pubis.

in favor to lidocaine infiltration (Nash et al., 1996). This can be explained by a small number of biopsies and the fact that almost a quarter of the nerve fibers extending to the contralateral side of the bladder, although not proved for the prostate (Tuttle, Steers, Albo, & Nataluk, 1994). Following this report, several investigators studied the different number of infiltrations for better pain control, and the majority of them showed bilateral basal prostatic-vesical junction infiltration. Although better effective pain control was achieved using six different infiltrations (Soloway & Öbek, 2000), clinicians faced with increasing infectious complications due to increased numbers of infiltration (Bolat et al., 2019).

Bilateral apical infiltration of local anesthetics is another technique to control pain. Ashley et al. compared bilateral apical infiltration, intraprostatic infiltration, and seminal vesicular-prostatic base infiltration during the prostate biopsy. They showed that apical and intraprostatic infiltrations provided better pain control than the seminal vesicular prostatic blockade. They also stated that apical infiltration was the most easily applied technique (Ashley et al., 2007).

Different infiltration techniques can be a more effective choice in pain control than individual applications. These combinations include (a) seminal vesicular-prostatic base + intraprostatic, (b) seminal vesicular-prostatic base +bilateral apical, (c) seminal vesicular-prostatic base + intravenous sedation, and (d) seminal vesicular-prostatic base + intrarectal local anesthetic gel application (Cam et al., 2008).

Studies on bladder innervation showed that the ipsilateral nerve blockage affected 25% of the contralateral side (Tuttle et al., 1994). Although no evidence present for the prostatic innervation, it is reasonable to assume that a similar mechanism may exist for the prostate due to similar autonomous innervation and close relationship with the bladder (Rodrigues, Machado, & Wroclawski, 2002).

Local anesthetics such as lidocaine, articaine, lignocaine, prilocaine, and bupivacaine can be used for prostate biopsy (Table 2).

Lidocaine is the most commonly used local anesthetic agent with a long duration of action. It is generally used in doses of 5 to 10 ml of 1% lidocaine. The maximum dose of this agent should be 20 mL. Adverse effects secondary to lidocaine absorption are infrequent (Raber et al., 2008; Tüfek et al., 2012).

Prilocaine is another local anesthetic agent similar to lidocaine. Five ml of 2% prilocaine is usually sufficient for the PPNB. Methemoglobinemia may occur when a maximum of 500 mg dose is exceeded. Fortunately, very low doses are needed for the local anesthesia during the TRUS-guided prostate biopsy. For each gram of prostate, 0.1 ml 2% prilocaine is required (Gonulalan et al., 2013; Long et al., 2012).

Articaine is a local anesthetic agent generally used for dental anesthesia. Although there is a lack of data on articaine use in the prostate biopsy, 5 ml of 1% articaine was reported for an effective anesthesia (Lee & Woo, 2014).

Local anesthetic gel products containing lidocaine can be used during the biopsy. Instillagel gel, one of the most commonly used drugs, provides both local anesthesia and disinfecting properties with its ingredients of lidocaine hydrochloride and chlorhexidine digluconate. Similarly, Xylocaine 2% jelly, containing lignocaine is applied 10 min before the prostate biopsy. Eutectic mixture of local anesthetics, also known as EMLA cream, contains lidocaine and prilocaine. In general, EMLA cream 15 min before the biopsy is superior to other topical IRLAs (Cormio et al., 2013). Glyceryl trinitrate (GTN) relaxes the anal canal up to 90 min via nitric oxide release and helps to painless insertion of biopsy probe through the anal canal. However, a headache occurs in 10% to 11.8% of patients secondary to vasodilation (McCabe, Hanchanale, Philip, & Javle, 2007; Rochester, Le Monnier, & Brewster, 2005; Skriapas, Konstantinidis, Samarinas, Xanthis, & Gekas, 2011).

TABLE 2 Local anesthetic agents, dosages, and trademarks used in PPNB[a] and IRLA[b]

Route of administration	Local anesthetic agents	Dosage	Trademark
Infiltration	Lidocaine	5, 10, or 15 mL of 1%	Lidopen
	Articaine	5 mL of 1%	Septocaine
	Lignocaine	10 mL of 1%	Xylocaine
	Prilocaine	10 mL of 1%	Citanest
	Bupivacaine	6.6 mL of 0.75%	Marcaine
Intrarectal	Lidocaine	10–15 mL of 2%	Lidocaine cream
	Lignocaine	10 mL of 2%	Xylocaine cream
	Lidocaine-Prilocaine combination	5 mL	EMLA cream
	40% Dimethyl sulfoxide-lidocaine combination	10 mL	Not applicable, magistral drug
	Lidocaine-GTN[c] combination	10 mL of 2% + 1 g of 0.2%	Not applicable, magistral drug

[a]PPNB, periprostatic nerve block.
[b]IRLA, intrarectal local anesthetic gel.
[c]GTN, glyceryl trinitrate.

Compared to their single-use, a combination of PPNB and IRLA or combination of lidocaine and prilocaine may provide better pain control (Wang et al., 2015).

Applications of local anesthetics to other areas

- As an antiarrhythmic agent, lidocaine corrects ventricular tachycardia and preventricular contractions by preventing sodium ion influx into cardiac fibers.
- Incisional infiltration of local anesthetics provides effective pain control and decreases the need for opioid use following general anesthesia.
- In the form of a small catheter and pump combination, local anesthetics can provide adequate pain control for several days. Thus, hospital stay can be decreased.
- Local anesthetics can be combined with general anesthesia. In some orthopedic surgeries and dentistry, this approach provides light general anesthesia with low adverse events.
- Spinal and epidural anesthesia using local anesthetics are frequently applied techniques for urological, obstetric, gynecological, and orthopedical surgery.
- Local anesthetics are also used for specific nerve blockade of upper extremities.

In conclusion, a prostate biopsy is an integral and possibly the most critical part of a prostate cancer diagnosis. A painless and comfortable procedure should be the primary concern that can positively affect patients' future decisions for rebiopsy. Although some new techniques have been developed, local anesthetic agents are successfully applied to different localizations for over a century alone or in combination.

Mini dictionary

- INFLATE technique: Infiltration-free local anesthesia technique based on gate-control theory. This technique has been first described by Bolat MS et al for pain control during the prostate biopsy. In this technique, a two-electrode two-channel TENS device is attached to the anterior suprapubic, and posterior presacral skin surfaces using adhesive electrodes correspond to bilateral T11–12 dermatomes. Stimulation of the thoracal 11–12 skin dermatomes by a TENS device activates A-beta fibers and inhibits C-fibers transmitting the pain sensation to the upper center.

- PSA: Prostate-specific antigen is a protein produced by the prostate gland. A small amount of it passes into the bloodstream. Higher age-adjusted PSA levels may be a predictor of prostate cancer.

Key facts

Prostate cancer is a common disease among men all over the world. Early diagnosis is of great importance in the treatment of the disease. However, a prostate biopsy is the most challenging part of the diagnostic workup. Because an uncomfortable or painful biopsy procedure can increase patients' discomfort, in this chapter, as a new method, the INFLATE technique providing pain control during prostate biopsy will be discussed.

Summary points

- Prostate cancer is the second most commonly diagnosed cancer among all cancers.
- The global increase in obesity contributes to metabolic syndrome which is a risk factor for prostate cancer.
- Early diagnosis of prostate cancer provides almost 100% overall 5-year survival rate.
- Past experiences of biopsy, fear of pain, and discomfort may affect the patient's decision. Therefore, painless prostate biopsy remains a unique way to diagnose prostate cancer.

References

Akdeniz, E., Bolat, M. S., & Akdeniz, S. (2018). A comparison of prilocaine vs prilocaine + bupivacaine in periprostatic block in ambulatory prostate biopsies: A single-blind randomized controlled study. *Pain Medicine (United States)*, *19*, 2069–2076.

Alavi, A. S., Soloway, M. S., Vaidya, A., Lynne, C. M., & Gheiler, E. L. (2001). Local anesthesia for ultrasound guided prostate biopsy: A prospective randomized trial comparing 2 methods. *Journal of Urology*, *166*, 1343–1345.

Ashley, R. A., Inman, B. A., Routh, J. C., Krambeck, A. E., Siddiqui, S. A., Mynderse, L. A., et al. (2007). Preventing pain during office biopsy of the prostate: A single center, prospective, double-blind, 3-arm, parallel group, randomized clinical trial. *Cancer*, *110*, 1708–1714.

Astraldi, A. (1937). Diagnosis of cancer of the prostate:Biopsy by rectal route. *The Urologic and Cutaneous Review*, *41*, 421–427.

Barringer, B. (1942). Prostatic carcinoma. *Journal of Urology*, *47*, 306–310.

Başar, H., Başar, M. M., Özcan, Ş., Akpinar, S., Başar, H., & Batislam, E. (2005). Local anesthesia in transrectal ultrasound-guided prostate biopsy: EMLA cream as a new alternative technique. *Scandinavian Journal of Urology and Nephrology*, *39*, 130–134.

Berry, S. J., Coffey, D. S., Walsh, P. C., & Ewing, L. L. (1984). The development of human benign prostatic hyperplasia with age. *Journal of Urology*, *132*, 474–479.

Bingqian, L., Peihuan, L., Yudong, W., Jinxing, W., & Zhiyong, W. (2009). Intraprostatic local anesthesia with periprostatic nerve block for transrectal ultrasound guided prostate biopsy. *Journal of Urology*, *182*, 479–483.

Blanc-Lapierre, A., Spence, A., Karakiewicz, P. I., Aprikian, A., Saad, F., & Parent, M.É. (2015). Metabolic syndrome and prostate cancer risk in a population-based case-control study in Montreal, Canada chronic disease epidemiology. *BMC Public Health*, *15*, 913.

Bolat, M. S., Cinar, O., Asci, R., & Buyukalpelli, R. (2019). A novel method for pain control: Infiltration free local anesthesia technique (INFLATE) for transrectal prostatic biopsy using transcutaneous electrical nerve stimulation (TENS). *International Urology and Nephrology*, *51*, 2119–2126.

Cam, K., Sener, M., Kayikci, A., Akman, Y., & Erol, A. (2008). Combined periprostatic and intraprostatic local anesthesia for prostate biopsy: A double-blind, placebo controlled, randomized trial. *Journal of Urology*, *180*, 141–144.

Cormio, L., Lorusso, F., Selvaggio, O., Perrone, A., Sanguedolce, F., Pagliarulo, V., et al. (2013). Noninfiltrative anesthesia for transrectal prostate biopsy: A randomized prospective study comparing lidocaine-prilocaine cream and lidocaine-ketorolac gel. *Urologic Oncology: Seminars and Original Investigations*, *31*, 68–73.

Davies, M. R. Q. (1997). Anatomy of the nerve supply of the rectum, bladder, and internal genitalia in anorectal dysgenesis in the male. *Journal of Pediatric Surgery*, *32*, 536–541.

Ferguson, R. S. (1930). Prostatic neoplasms. Their diagnosis by needle puncture and aspiration. *The American Journal of Surgery*, *9*, 507–510.

Ferlay, J., Colombet, M., Soerjomataram, I., Mathers, C., Parkin, D. M., Piñeros, M., et al. (2019). Estimating the global cancer incidence and mortality in 2018: GLOBOCAN sources and methods. *International Journal of Cancer*, *15*, 1941–1953.

Gonulalan, U., Kosan, M., Kervancioglu, E., Cicek, T., Ozturk, B., & Ozkardes, H. (2013). S191 the optimum dose of prilocaine in periprostatic nerve block during transrectal ultrasound guided prostate biopsy. *European Urology Supplements*, *12*, e1299.

Horinaga, M., Nakashima, J., & Nakanoma, T. (2006). Efficacy compared between caudal block and periprostatic local anesthesia for transrectal ultrasound-guided prostate needle biopsy. *Urology*, *68*, 348–351.

Jones, J. S., & Zippe, C. D. (2003). Rectal sensation test helps avoid pain of apical prostate biopsy. *Journal of Urology*, *170*, 2316–2318.

Kepper, M., & Keast, J. (1995). Immunohistochemical properties and spinal connections of pelvic autonomic neurons that innervate the rat prostate gland. *Cell and Tissue Research*, *281*, 533–542.

Kim, H., Kim, K., Lee, H., & Cho. (2019). Is pelvic plexus block superior to periprostatic nerve block for pain control during transrectal ultrasonography-guided prostate biopsy? A double-blind, randomized controlled trial. *Journal of Clinical Medicine*, *557*, 2–9.

Kim, S., Yoon, B. I., Kim, S. J., Cho, H. J., Kim, H. S., Hong, S. H., et al. (2011). Effect of oral administration of acetaminophen and topical application of emla on pain during transrectal ultrasound- guided prostate biopsy. *Korean Journal of Urology, 52*, 452–456.

Lee, H. Y., Lee, H. J., Byun, S. S., Lee, S. E., Hong, S. K., & Kim, S. H. (2007). Effect of intraprostatic local anesthesia during transrectal ultrasound guided prostate biopsy: Comparison of 3 methods in a randomized, double-blind, placebo controlled trial. *Journal of Urology, 178*, 469–472.

Lee, M. S., Moon, M. H., Kim, C. K., Park, S. Y., Choi, M. H., & Jung, S. I. (2020). Guidelines for transrectal ultrasonography-guided prostate biopsy: Korean society of urogenital radiology consensus statement for patient preparation, standard technique, and biopsy-related pain management. *Korean Journal of Radiology, 21*, 422–430.

Lee, C., & Woo, H. H. (2014). Current methods of analgesia for transrectal ultrasonography (TRUS)-guided prostate biopsy-A systematic review. *BJU International, 113*(SUPPL. 2), 48–56.

Li, M., Wang, Z., Li, H., Yang, J., Rao, K., Wang, T., et al. (2017). Local anesthesia for transrectal ultrasound-guided biopsy of the prostate: A meta-analysis. *Scientific Reports, 7*, 404–421.

Long, C. Y., Juan, Y. S., Wu, M. P., Liu, C. M., Chiang, P. H., & Tsai, E. M. (2012). Changes in female sexual function following anterior with and without posterior vaginal mesh surgery for the treatment of pelvic organ prolapse. *The Journal of Sexual Medicine, 9*, 2167–2174.

McCabe, J. E., Hanchanale, V. S., Philip, J., & Javle, P. M. (2007). A randomized controlled trial of topical glyceryl trinitrate before transrectal ultrasonography-guided biopsy of the prostate. *BJU International, 100*, 536–539.

Melzack, R., & Wall, P. D. (1965). Pain mechanisms: A new theory. *Science, 150*, 971–979.

Meyer, A. R., Joice, G. A., Schwen, Z. R., Partin, A. W., Allaf, M. E., & Gorin, M. A. (2018). Initial experience performing in-office ultrasound-guided transperineal prostate biopsy under local anesthesia using the PrecisionPoint transperineal access system. *Urology, 115*, 8–13.

Muramatsu, I., Oshıta, M., Ohmurs, T., Kıgoshı, S., Akıno, H., Gobara, M., et al. (1994). Pharmacological characterization of α1-adrenoceptor subtypes in the human prostate: Functional and binding studies. *British Journal of Urology, 74*, 572–578.

Mutaguchi, K., Shinohara, K., Matsubara, A., Yasumoto, H., Mita, K., & Usui, T. (2005). Local anesthesia during 10 core biopsy of the prostate: Comparison of 2 methods. *Journal of Urology, 173*, 742–745.

Nash, P. A., Bruce, J. E., Indudhara, R., & Shinohara, K. (1996). Transrectal ultrasound guided prostatic nerve blockade eases systematic needle biopsy of the prostate. *Journal of Urology, 155*, 607–609.

Raber, M., Scattoni, V., Roscigno, M., Dehò, F., Briganti, A., Salonia, A., et al. (2008). Topical prilocaine-lidocaine cream combined with peripheral nerve block improves pain control in prostatic biopsy: Results from a prospective randomized trial. *European Urology, 53*, 967–973.

Rochester, M. A., Le Monnier, K., & Brewster, S. F. (2005). A double-blind, randomized, controlled trial of topical glyceryl trinitrate for transrectal ultrasound guided prostate biopsy. *Journal of Urology, 173*, 418–420.

Rodrigues, A. O., Machado, M. T., & Wroclawski, E. R. (2002). Prostate innervation and local anesthesia in prostate procedures. *Revista Do Hospital Das Clínicas, 57*, 287–292.

Rodriguez, A., Kyriakou, G., Leray, E., Lobel, B., & Guillé, F. (2003). Prospective study comparing two methods of anaesthesia for prostate biopsies: Apex periprostatic nerve block versus intrarectal lidocaine gel: Review of the literature. *European Urology, 44*, 195–200.

Schlegel, P. N., & Walsh, P. C. (1987). Neuroanatomical approach to radical cystoprostatectomy with preservation of sexual function. *Journal of Urology, 138*, 1402–1406.

Shahin, O., & Koch, M. (2018). Die transperineale Prostatabiopsie in Lokalanästhesie. *Journal Für Urologie Und Urogynäkologie/Österreich, 25*, 90–94.

Singh, S. K., Kumar, A., Griwan, M. S., & Sen, J. (2012). Comparative evaluation of periprostatic nerve block with and without intraprostatic nerve block in transrectal ultrasound-guided prostatic needle biopsy. *Korean Journal of Urology, 53*, 547–551.

Skouteris, V. M., Crawford, E. D., Mouraviev, V., Arangua, P., Metsinis, M. P., Skouteris, M., et al. (2018). Transrectal ultrasound-guided versus transperineal mapping prostate biopsy: Complication comparison. *Reviews in Urology, 20*, 19–25.

Skriapas, K., Konstantinidis, C., Samarinas, M., Xanthis, S., & Gekas, A. (2011). Comparison between lidocaine and glyceryl trinitrate ointment for perianal-intrarectal local anesthesia before transrectal ultrasonography-guided prostate biopsy: A placebo-controlled trial. *Urology, 77*, 905–908.

Smith, E. R., & Lebeaux, M. I. (1970). The mediation of the canine prostatic secretion provoked by hypogastric nerve stimulation. *Investigative Urology, 7*, 313–318.

Soloway, M. S., & Öbek, C. (2000). Periprostatic local anesthesia before ultrasound guided prostate biopsy. *Journal of Urology, 163*, 172–173.

Takahashi, H., & Ouchi, T. (1963). The ultrasonic diagnosis in the field of urology. *Proceedings of the Japan Society of Ultrasonics in Medicine, 3*, 8.

Tobias-Machado, M., Verotti, M. J., Aragao, A. J., Rodrigues, A. O., Borrelli, M., & Wroclawski, E. R. (2006). Prospective randomized controlled trial comparing three different ways of anesthesia in transrectal ultrasound-guided prostate biopsy. *International Brazilian Journal of Urology, 32*, 172–179.

Tüfek, I., Akpinar, H., Atuğ, F., Öbek, C., Esen, H. E., Keskin, M. S., et al. (2012). The impact of local anesthetic volume and concentration on pain during prostate biopsy: A prospective randomized trial. *Journal of Endourology, 26*, 174–177.

Tuttle, J. B., Steers, W. D., Albo, M., & Nataluk, E. (1994). Neural input regulates tissue NGF and growth of the adult rat urinary bladder. *Journal of the Autonomic Nervous System, 49*, 147–158.

Vaalasti, A., & Hervonen, A. (1979). Innervation of the ventral prostate of the rat. *American Journal of Anatomy, 154*, 231–244.

Wang, J., Wang, L., Du, Y., He, D., Chen, X., Li, L., et al. (2015). Addition of intrarectal local analgesia to periprostatic nerve block improves pain control for transrectal ultrasonography-guided prostate biopsy: A systematic review and meta-analysis. *International Journal of Urology, 22*, 62–68.

Yan, P., Wang, X. Y., Huang, W., & Zhang, Y. (2016). Local anesthesia for pain control during transrectal ultrasound-guided prostate biopsy: A systematic review and meta-analysis. *Journal of Pain Research, 9*, 787–796.

Yang, Y., Liu, Z., Wei, Q., Cao, D., Yang, L., Zhu, Y., et al. (2017). The efficiency and safety of intrarectal topical anesthesia for transrectal ultrasound-guided prostate biopsy: A systematic review and meta-analysis. *Urologia Internationalis, 99*, 373–383.

Young, H., & Davis, D. M. (1926). *Young's practice of urology. Based on a study of 12,500 cases* (1st ed.). Philadelphia/London: W. B. Saunders Co.

Chapter 33

Pain reduction in cosmetic injections: Fillers and beyond

Hamid Reza Fallahi[a,b], Roya Sabzian[c], Seied Omid Keyhan[b], and Dana Zandian[a,d]

[a]*Dental Research Center, Research Institute of Dental Sciences, Shahid Beheshti University of Medical Sciences, Tehran, Iran,* [b]*Founder and Director of Maxillofacial Surgery and Implantology and Biomaterial Research Foundation (www.maxillogram.com), Isfahan, Iran,* [c]*Dental Students Research Center, School of Dentistry, Isfahan University of Medical Sciences, Isfahan, Iran,* [d]*Director of Maxillofacial Surgery and Implantology and Biomaterial Research Foundation(www.maxillogram.com), Isfahan, Iran*

Abbreviations

FDA   The United States Food and Drug Administration
ADR   adverse drug reactions
BTX-A   botulinum toxin A
PABA   para-aminobenzoic acid

Introduction

Cosmetic injections, such as Botox and filler injections, as well as mesotherapy, are receiving increasing attention as people crave a more beautiful and younger appearance. These injections are renowned due to their low cost, results that are both fast and repeatable, and reduced invasiveness and morbidity (Sezgin et al., 2014).

Cosmetic injections are used for face recontouring or rejuvenation. Also, sometimes they are needed to optimize the cosmetic surgeries' result (Feily, Fallahi, Zandian, & Kalantar, 2011). It should be noted that beauty is not the only application of these type of injections, for example, Botox injections are used in the cases of blepharospasm, strabismus, hemifacial spasm, spasticity, and dystonia disorders in the face area (Verheyden & Blitzer, 2002) with numbers of other noncosmetic treatment usage other parts of the human body (Feily et al., 2011; Verheyden & Blitzer, 2002).

Due to the fact that these procedures mostly require a number of injections, a significant issue is patient anxiety (Brandt, Bank, Cross, & Weiss, 2010), whereas all of these injections are not prescribed for elective cosmetic treatment plans.

Needle fear

Generally, the "needle fear/phobia" is attributed to the state of anxiety because of needle-included procedures. The frequency of fearing from needles and suffering from this phobia decrease as age increase, and it is more frequent in women than men (McLenon & Rogers, 2019).

These fear and distress stem from the past needle-included procedures a person or their relevant have experienced during their life (McMurtry, Riddell, et al., 2015). In fact, common causes for needle fear are unpleasant memory of injection side effects like bleeding and discomfort that the patient has experienced/witnessed or were mis-informed by others about the side effects of such procedures (Du, Jaaniste, Champion, & Yap, 2008; Rachman, 1977). The other reasons can be the fear of being contaminated by the needle or fainting due to injection (McMurtry, Noel, et al., 2015).

When pain is mismanaged during needle-involved procedures for pediatric patients, the child may develop exacerbating memories of fear around this procedure leading to considerable anxiety at such procedures in the future (Noel, McMurtry, Chambers, & McGrath, 2009). Unfortunately, this exacerbation of fear memories may turn to a self-perpetuating cycle resulting in greater fear and pain of needles as the time pass (Noel et al., 2012).

In other words, a memory of previous painful procedure can be exaggerated by the patient overtime. The level of pain they are going to perceive during procedure is closer to this memory in comparison with what they have actually sensed previous time (Gedney & Logan, 2006).

Features and Assessments of Pain, Anesthesia, and Analgesia. https://doi.org/10.1016/B978-0-12-818988-7.00009-1
Copyright © 2022 Elsevier Inc. All rights reserved.

364 PART | III Interlinking anesthesia, analgesics and pain

However, it is probable that the healthcare professionals do not pay sufficient attention to patients' stress for needle procedures (Lidén, Olofsson, Landgren, & Johansson, 2012), it is reported that, 74% of surveyed subjects expressed concern about associated pain with cosmetic procedures, and 42% of subjects who had a cosmetic procedure would consider not having other procedures due to concerns about pain (Sarkany, 2012). These results suggest that a sizeable number of subjects are given inadequate measures to control pain during cosmetic procedures (Cohen & Gold, 2014).

Effective pain management seems to be helpful to hinder needle fears development. In fact, successful pain reduction not only can decrease future pain and fear but also may prevent future distresses and impairments, such as noncompliance (McMurtry, Noel, et al., 2015; McMurtry, Riddell, et al., 2015).

In addition, successfully management of pain and discomfort to reach a high-qualified care is of great importance (Hein, Schönwetter, & Iacopino, 2011). Maximizing patient comfort is important to enhance patient satisfaction with all procedures. This is particularly true with elective cosmetic procedures. To provide patient satisfaction with cosmetic injections, several patient values must be considered: the ability to fit treatments into a busy schedule, instant pain relief, and reduced anxiety over injections. The use of an efficacious, cost-effective, fast-acting analgesic technique is thus essential to deliver pain-free facial rejuvenation (Zeiderman et al., 2018). Of characteristics of a desirable topical anesthetic can name effectiveness, rapid duration of action, and minimum side effects (Mally, Czyz, Chan, & Wulc, 2014).

Psychological intervention

Based on the fact that the vast majority of patients experience discomfort and anxiety, psychological approaches may be beneficial for them (Babamiri & Nassab, 2010). Simple psychological interventions can be useful because these kinds of approaches are accessible, cheap, and easy to use (Boerner et al., 2015).

Melzack (2005) introduced a new neuromatrix theory of pain. According to this theory, pain is experienced on a multidimensional level and is created via an extensively distributed neural network. This "body-self neuromatrix" produces neural impulses with special patterns. Such neural impulses, also called neurosignatures, are generally produced through sensory stimuli. They can also be generated independent of them (Melzack, 2005). Therefore, a variety of dimension and the interactions that happen between them can affect pain sensation (Melzack & Katz, 2004).

Based on the neurocognitive model of attention, the perception of pain can be lessened by cognitive load of a distracting task or stimuli which is not related to the pain (Legrain, Crombez, Verhoeven, & Mouraux, 2011). Distraction can be used for cognitive refocusing, which make patient to pay attention to a more pleasant stimuli instead of painful one leading to reduced pain sensation (Ruscheweyh, Kreusch, Albers, Sommer, & Marziniak, 2011).

Music is thought as a useful device to distract patients from the pain during procedure (Johnson, Breakwell, Douglas, & Humphries, 1998). However, review studies do not support its implementation for pain relief (Boerner et al., 2015; Cepeda, Carr, Lau, & Alvarez, 2006) since music application is not adequate to completely eliminate the pain. In addition, the mere approach of explanation and handholding, which suggested to reduce anxiety (Babamiri & Nassab, 2010), is not enough in most of the cases (Chen & Eichenfield, 2001). It seems that these techniques are suitable when they are used beside other approaches.

Vibration

Rubbing and applying pressure to the area close to injection site can reduce pain (Mally et al., 2014). The reason can be explained by the gate control theory. This theory suggests that the sensation of pain is modulated via the nonnoxious stimulus (Mally et al., 2014). The mechanical stimulus activates the mechanoreceptors; this activation results in the stimulation of large-diameter Ab fibers. Consequently, pain signal transmission to the brain will be prevented (Sezgin et al., 2014). As a mechanical stimulus, the vibration anesthesia mechanism is same as applying manual pressure or rubbing (Babamiri & Nassab, 2010). It was Reed (1984) who first introduced this technique. He utilized this method of anesthesia for patients with extensive alopecia who needed to receive frequent corticosteroid injections for hair growth stimulation. At that time, he explained that this mitigating effect may be because of either a depolarization or conduction in peripheral nerve or the so-called gate control theory (Reed, 1984). Nowadays, the vibrator tools are used for anesthetic purposes for cosmetic injection; Fig. 1 presents a vibrator device. This method of anesthesia is compelling because it requires little money and time, and causes minimal side effects (Guney, Sezgin, & Yavuzer, 2017).

In the study conducted by Mally et al. (2014) on patients received dermal filler injections on various facial areas, injection with the application of vibrating stimuli notably led to lesser injection pain than injection without any pain reduction intervention.

FIG. 1 A vibrator device.

FIG. 2 Botulinum toxin injection with vibration anesthesia (with permission).

Sharma, Czyz, and Wulc (2011) performed a split design study on 50 patients who underwent facial injections of botulinum toxin type A (BTX-A) for cosmetic purposes. They reported that when injection was accompanied with on-site vibration not only patients perceived lesser pain in comparison with control areas but also 86% of patients requested vibration anesthesia for their next experience. In yet another research study, patients receiving BTX-A injections for various purposes experienced reduced pain when injections were paired with vibration. Furthermore, 75% of these study patients chose the vibration method of anesthesia for their next injection appointment (Li, Dong, Wang, & Xu, 2017) (Fig. 2).

Vibration safety

Vibration seems a safe method for inducing local anesthesia. There is evidence from occupational studies that indicate chronic exposure to high-intensity whole-body vibration can be correlated with increased risk of spinal degeneration (Bovenzi, 2005; Cardinale & Pope, 2003). Moreover, hand-transmitted vibration over a long period of time may result in neural and vascular changes in the upper limbs (Bovenzi, 2005). It is important to note that short period of exposure to topical vibration has not been linked to neither temporary nor permanent effects. In a study exploring the negative effects of vibration, tingling of the skin as well as teeth was temporary; additionally, the authors believed that the headaches and bruising experienced by participants were independent of the vibration stimulus (Sharma et al., 2011). Guney et al. (2017) conducted a study on the efficacy of vibration on pain reduction during hyaluronic acid filler injection. They noted that in most areas of the face, topical anesthesia and cooling the specific area provide sufficient anesthesia; however, this was insufficient for injections administered in the lip area, which is why the authors suggest vibration anesthesia as a potential parallel approach (Guney et al., 2017).

Disadvantages

Despite the numerous studies supporting vibration anesthesia, this technique does not have the ability to completely erase pain (Mally et al., 2014) and it just modulates the pain. Specifically in areas surrounding the lips, the vibration can be a source of discomfort and anxiety for some. Therefore, prior to administration, it is important to thoroughly explain the procedure to patients (Guney et al., 2017; Mally et al., 2014). Furthermore, vibration may interfere with needle placement and the dermal filler or neurotoxin injection process. Since vibration alone reduces but does not eliminate pain, another form of topical or local anesthetic in conjunction may be necessary to ensure patient comfort (Zeiderman et al., 2018).

Local anesthesia

Although there are a number of substances that are capable of inducing local anesthesia, not all are appropriate for use in a clinical setting (Liu, Yang, Li, & Mo, 2013). The majority of clinically used local anesthetics have a benzene ring connected by amide or ester to an amine group, and the "caine" comes at the end of their names (Scholz, 2002) (Fig. 3).

Amides like lidocaine, bupivacaine, articaine, mepivacaine, prilocaine, and levobupivacaine are metabolized by the hepatic system (Auletta, 1994; Park & Sharon, 2017). Special attention must be paid when amides are used for patients with liver disease (Tucker & Mather, 1998). Esters including procaine, proparacaine, benzocaine, chloroprocaine, tetracaine, and cocaine are metabolized by plasma cholinesterase, which produces para-aminobenzoic acid (PABA), a potentially dangerous metabolite that has been implicated in allergic reactions (Park & Sharon, 2017; Pomerantz, Lee, & Siegel, 2011). Hydrolysis is fast and its by-products are expelled in the urine (Koay & Orengo, 2002).

After topical anesthetics are absorbed through the skin, they serve the same mechanism of action as injectable anesthetics to affect dermal nerve endings (Park & Sharon, 2017). They block ion channels in the neuronal cell membrane leading to impediment of action potential progress, and reduced pain perception (Scholz, 2002). In patients with needle

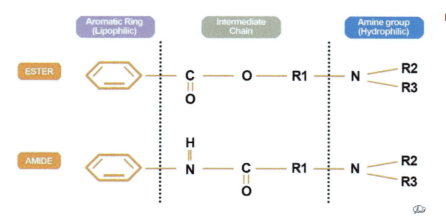

FIG. 3 Basic structure of amids and esters.

phobia, the use of topical anesthesia is helpful. In clinical practice, use of them for filler injection is common (Hashim, Nia, Taliercio, & Goldenberg, 2017).

Various types of topical anesthetics are available to manage pain and provide relief during cosmetic procedures (Sobanko, Miller, & Alster, 2012). Some practitioners apply compounded formulations of anesthetics to provide dermal anesthesia for a painless procedure. However, these products have been found to not be standardized and frequently have higher concentrations of anesthetics than products that are approved by the US Food and Drug Administration (FDA) (Sobanko et al., 2012). Moreover, most of them do not have a standard instruction for use. Some of these compounds may contain ingredients with sympathomimetic actions or anesthetics with unknown strength of action (Kravitz, 2007; Sobanko et al., 2012).

Therefore, it is beneficial that only FDA-approved topical anesthetics be used to increase the safety in clinical practice and reach more predictable outcome (Table 1).

Cohen and Gold (2014) investigated the effectiveness of an FDA-approved cream (Pliaglis Cream) that was formulated with 7% lidocaine and 7% tetracaine. Their study examined its efficacy for relieving pain associated with filler injections. Prior to injection, they used this cream at the injection spots, and noticed a significant reduction in patient discomfort.

In another randomized clinical trial study that compared the efficacy of alkalinized lidocaine solution (10 mL 8.4% sodium bicarbonate and 20 mL 2% lidocaine solution and 22 mL sterile Aquagel) and lidocaine gel (22 mL standard 2% lidocaine gel Instillagel and 30 mL 0.9% normal saline solution) on pain reduction after injections of botulinum toxin, no differences were noted between these anesthetic materials (Nambiar et al., 2016). In theory, the alkanization of an anesthetic solution should serve to increase its ability to diffuse into the cellular membrane; however, in this study, not only it was not better but also it was more expensive than the lidocaine gel (Nambiar et al., 2016).

Disadvantages

Even though local anesthetics are effective in pain reduction, they may have some negative effects (Zeiderman et al., 2018). It is important to be aware of the negative side effects and risks associated with application of local anesthetics, which will aid patient health monitoring and counseling. Additionally, it is also important for practitioners to be well-versed in order to successfully manage adverse situation when they arise (Liu et al., 2013).

TABLE 1 FDA approved topical anesthetics for intact skin (Cada, Arnold, Levien, & Baker, 2006; Eaglstein, 1999; Gammaitoni, Alvarez, & Galer, 2003; McKinlay, Hofmeister, Ross, & MacAllister, 1999; Sobanko et al., 2012; Young, 2015; http://www.synera.com/health-care-professionals/ordering-prescribing/ Prescribing Information; https://www.taro.com/pliaglis; https://www.accessdata.fda.gov/scripts/cder/daf/index.cfm).

Topical anesthetics	Active ingredient	form	Duration	Dosing for adults	Other
Eutectic mixture of local anesthetics (EMLA)	Lidocaine 2.5% + prilocaine 2.5%	Cream	1–2 h	Patients aged 7 and higher, weighing +20 kg, should not have more than 20 g of EMLA applied to skin, it should not cover more than 200 cm^2 of surface area, for less than 4 h	EMLA includes sodium hydroxide, to prevent alkaline chemical injury, must avoid use around eyes
Lidoderm	Lidocaine 5%	Patch		Three patches a day for 12 h>	FDA approved for postherpetic neuralgia
Pliaglis	Lidocaine 7% + tetracaine 7%	Cream	11 h	59 g of cream over 400 cm^2 for up to 120 min produced peak lidocaine concentrations of 220 ng/mL, but undetectable tetracaine	According to manufacturer instructions, lips and eyes must be avoided
Synera	70 mg lidocaine, 70 mg tetracaine	Patch	100 min< Better to start procedure immediately after patch removal	One patch for 30 min before the procedure No more than two patches recommended	A 6.25 *7.5-cm^2 disc It cannot be cut and contain heating component

368 PART | III Interlinking anesthesia, analgesics and pain

An extensive systematic review of local anesthetic adverse effects revealed that lidocaine was concerned with 43.71% and bupivacaine in 16.32% of 723 cases who showed adverse drug reactions (ADR) of local anesthetics (Liu et al., 2013). Moreover, local anesthetics combined with epinephrine were responsible in 45.3% of 723 cases (Liu et al., 2013). According to this study, although the overall incidence of ADRs in single use of local anesthetics was low, practitioners should be careful and be well-versed in the emergency medical procedures because these reaction still can occur (Liu et al., 2013).

It has been reported that allergic reactions are rare among ADRs of local anesthetic. However, it occurs, and immediate-type allergic reactions are more common (Fuzier et al., 2009). Ester anesthetics are metabolized by the plasma enzyme, pseudocholinesterase. Para-aminobenzoic acid (PABA), which is identified as an allergen, is the main metabolite of the hydrolysis reaction. As a result, ester anesthetics are normally correlated with a greater occurrence of allergy in comparison with their amide counterparts (Koay & Orengo, 2002).

Nevertheless, preparation of both ester and amide may include methylparaben and/or sulfonamides acting as preservative agents (Boren, Teuber, Naguwa, & Gershwin, 2007).

In addition, anesthetic ointments might not be suitable for use in combination with BTX-A. It has been reported that topical anesthetics use had an adverse effect on the effectiveness of BTX-A. A potential explanation for this adverse effect may be the interference of the topical anesthetics with the nerve stimulation needed for BTX-A effect. It must be noted that topical anesthetic application was followed by cryoanalgesia, which may have had an additive effect (Sami, Soparkar, Patrinely, Hollier, & Hollier, 2006).

Since topical anesthetics impact terminal neuron fibers, the depth of their penetration through skin is an important factor to reach suitable levels anesthesia (Hashim et al., 2017). However, it would be undesirable for patients to wait some minutes for topical anesthetic-required effect before their injections.

There are a number of cosmetic injectables that do contain local anesthetics (Babamiri & Nassab, 2010). According to a clinical trial study comparing the pain of filler injection in nasolabial ford area, hyaluronic acid–based filler +0.3% lidocaine injections led to lesser pain than hyaluronic acid–based filler (Weinkle et al., 2009).

Monheit et al. (2010) comparing the level of pain associated with the injection of Prevelle SILK (preincorporated 0.3% lidocaine) and Captique (same molecular filler without lidocaine) in the nasolabial fold area through a single-blind control trial study. The result indicated that Prevelle SILK resulted in a 50% pain reduction, but without any significant differences in either the cosmetic outcome or side effects. Although mixing anesthetic substance directly into the filler prevents the pain of tissue expansion and edema for a short time after the injection, it does not help with the pain of needle puncture (Brandt et al., 2010; Zeiderman et al., 2018). Moreover, there is a chance that these anesthetic materials–included fillers result in tissue distortion and a degree of systemic crisis (Brandt et al., 2010; Koay & Orengo, 2002).

Mixing lidocaine with filler before injection by practitioners is not suggested since it increases the possibility of contamination, also by altering the concentration of filler, it could cause failure to attain desirable quality, consistency, and flow characteristics (Busso & Voigts, 2008; Sagrillo, 2008). Therefore, these approaches are inconvenient and time consuming, and may result in changes in the viscosity of the filler preparation (Brandt et al., 2010).

Injecting anesthetics techniques

Considering regional nerve blocks can be helpful when it is vital to prevent tissue distortion for visual monitoring throughout cosmetic injection. Additionally, they provide extensive and deep anesthesia (Hashim et al., 2017).

For nerve block injections, the practitioners must be skillful enough to minimize potential serious risks, such as paralysis (Li et al., 2017) while the infiltration techniques with agents such as lidocaine that are frequently used during cosmetic procedures may be associated with distortion at the area of injection (Koay & Orengo, 2002).

Although local and injected anesthetics provide effective anesthesia, systemic adverse drug reactions and distortion of the tissue are important concerns. Anesthetic materials should be used with caution to decrease the risk of life-threatening side effects, including irregular heartbeat, seizures, and breathing difficulties (Sharma et al., 2011).

Practitioners must monitor the anesthetics dosing carefully to avoid possible anesthetics-induced toxicity symptoms. Fig. 4 demonstrates dose-dependent symptoms according to lidocaine-induced toxicity.

Vapocoolant anesthesia

Travell (1955) first introduced vapocoolant sprays for use as a topical anesthesia. These sprays contain halogenated alkanes. When they are sprayed onto skin, the sprayed material evaporates by absorbing the heat from the skin. The

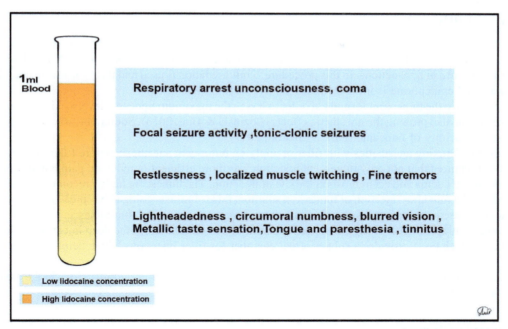

FIG. 4 Toxic lidocaine concentrations in 1 mL of blood and relative signs and symptoms (Torp & Simon, 2019; Walsh & Walsh, 2011).

absorption of heat causes the skin to be cooler than before (Zeiderman et al., 2018). Through lowering the temperature, the conduction velocity in nerve fiber also decreases resulting in pain relief (Waterhouse, Liu, & Wang, 2013).

In a split face design study on 30 patients, 15 patients received neurotoxin injections while the other 15 got filler injections, with or without vapocoolant spray on each half of the face. The results indicated that using vapocoolant spray reduced 59% of the perceived pain scores for neurotoxin injections and 64% of the perceived pain scores for the filler injections (Zeiderman et al., 2018). Another similarly designed study on 52 patients reported that 67% patients who got botulinum toxin injections with vapocoolant spray noted less pain compared to those underwent the procedure without vapocoolant spray. Additionally, 54% of patients noted their preference for this specific type of anesthesia to be used for their future appointment (Weiss & Lavin, 2009). According to Dixit, Lowe, Fischer, and Lim (2013) study, 75% of surveyed Australian dermatologists prefer ice anesthesia pain management methods in cosmetic injections. The second most frequently used technique was topical anesthesia.

Compared to local anesthetics, skin refrigerants are cost effective, but ice remains the cheaper option (Engel, Afifi, & Zins, 2010; Nambiar et al., 2016). Since ice induces anesthesia with the same mechanism of action as skin refrigerants, it could be favored. However, application of ice for anesthetic purpose may cause uneven anesthesia and cold burns (Zeiderman et al., 2018).

Topical coolants do have some drawbacks. In the course clinical application, if skin refrigerant causes blanching, then the likelihood of losing the vessel patterns exists. Therefore, prior to applying vapocoolant, it is vital that the injection sites are marked (Weiss & Lavin, 2009). The application of ice packs is difficult and may not thoroughly eliminate the pain associated with injections. Although vapocoolant sprays or other topical anesthetic skin refrigerants have fast onset of action, they may result in frostbite and tissue necrosis if they are not correctly applied (Weiss & Lavin, 2009). It is especially important when vapocoolants are applied around the periorbit, special precautions like protective eye shields must be used (Weiss & Lavin, 2009). Moreover, the use of vapocoolant sprays has been linked with skin hypopigmentation or hyperpigmentation (Kontochristopoulos, Gregoriou, Zakopoulou, & Rigopoulos, 2006). It must be noted that the anesthesia following ice or cooled air is variable for most of the times because their application cannot be accurate cannot be administered accurately or precisely.

Applications to other areas

These techniques may be used for other types of injections in other parts of human body as well as the oral cavity. The authors of this chapter suggest further investigations on these methods for other types of injections.

Other agents of interest

Tumescent anesthesia has the potential to be used for a number of procedures, including body contouring, liposuction, hair transplantations, endovascular ablation, as well as extensive surgical procedures (Park & Sharon, 2017). Tumescent anesthesia was described for use in liposuctions. In this procedure, a mild sedation is used and the lidocaine in combination with sodium bicarbonate and epinephrine is infiltrated into the hypodermis (Klein, 2000).

Epinephrine is added to the tumescent formula in order to lessen the maximum lidocaine levels and postpone absorption, confer hemostasis. It can also prolong the anesthetic effects (Lozinski & Huq, 2013). Sodium bicarbonate is in the formula since it neutralizes the acidity of lidocaine (Park & Sharon, 2017).

In the comparison of mixing anesthetics with filler or applying tumescent anesthesia before filler only injection on 30 women, Kim (2017) reported that not only tumescent injection was more efficient in reducing pain but also patients experienced lesser amounts of edema and ecchymosis at the site of injection. The tumescent highly recommended by this author and he mentioned that 10 min after injection of tumescent solution, this solution is going to be spread and there may be very less volume change. Consequently, worries about volume changes are not necessary after tumescent solution (Kim, 2017). Moreover, in another study, Kim (2018) suggested the tumescent hydro dissection for the deep injections in forehead area. He explained that this technique provides a subperiosteal chamber for filler injections which helps practitioners to avoid vascular injury.

Since the filler and botulinum toxin injections usually require multiple punctures and injections in the area, tumescent anesthesia use for these procedures seems an appealing topic for further investigations.

Mini-dictionary of terms

Blepharospasm. The uncontrollable blinking and twitching caused by the abnormal tonic or clonic spasm in the orbicularis oculi muscle.
Strabismus. The situation of misalignment of eyes.

Key facts

- Albert Niemann was the first person who extracted the active ingredient from coca leaves and named it cocaine in 1860.
- Later, Koller introduced the application of cocaine to reach local anesthesia for the ophthalmic procedures as an alternative of general anesthesia (Ball & Westhorpe, 2003).

Summary points

- The memory of previous painful procedure can be exaggerated by the patient overtime. The level of pain patients perceive procedure is closer to what they expect according to their memory instead of what they have actually sensed previous time. Therefore, pain management is of great importance.
- Vibrators can reduce pain serving the gate control theory.
- Local anesthetic materials are very effective for pain reduction in needle-involved procedures. However, practitioners have to pay attention to the systemic side effects that these materials may cause.
- Vapocoolant sprays can reduce pain by reducing the conduction velocity. Comparing with ice, they are more useful for clinical practice.

References

Auletta, M. J. (1994). Local anesthesia for dermatologic surgery. *Seminars in Dermatology, 13*(1), 35–42.

Babamiri, K., & Nassab, R. (2010). Back to basics: The evidence for reducing the pain of administration of local anesthesia and cosmetic injectables. *Journal of Cosmetic Dermatology, 9*(3), 242–245.

Ball, C., & Westhorpe, R. (2003). Local anaesthesia—Freud, Koller and cocaine. *Anaesthesia and Intensive Care, 31*(3), 247.

Boerner, K. E., Birnie, K. A., Chambers, C. T., Taddio, A., McMurtry, C. M., Noel, M., … Riddell, R. P. (2015). Simple psychological interventions for reducing pain from common needle procedures in adults: Systematic review of randomized and quasi-randomized controlled trials. *Clinical Journal of Pain, 31*(Suppl. 10), S90.

Boren, E., Teuber, S. S., Naguwa, S. M., & Gershwin, M. E. (2007). A critical review of local anesthetic sensitivity. *Clinical Reviews in Allergy & Immunology, 32*(1), 119–127.

Bovenzi, M. (2005). Health effects of mechanical vibration. *Giornale Italiano di Medicina del Lavoro ed Ergonomia, 27*(1), 58–64.

Brandt, F., Bank, D., Cross, S. L., & Weiss, R. (2010). A lidocaine-containing formulation of large-gel particle hyaluronic acid alleviates pain. *Dermatologic Surgery, 36*, 1876–1885.

Busso, M., & Voigts, R. (2008). An investigation of changes in physical properties of injectable calcium hydroxylapatite in a carrier gel when mixed with lidocaine and with lidocaine/epinephrine. *Dermatologic Surgery, 34*, S16–S24.

Cada, D. J., Arnold, B., Levien, T., & Baker, D. E. (2006). Lidocaine/tetracaine patch. *Hospital Pharmacy, 41*(3), 265–273. https://doi.org/10.1310/hpj4103-265.

Cardinale, M., & Pope, M. H. (2003). The effects of whole body vibration on humans: Dangerous or advantageous? *Acta Physiologica Hungarica, 90*(3), 195–206.

Cepeda, M. S., Carr, D. B., Lau, J., & Alvarez, H. (2006). Music for pain relief. *Cochrane Database of Systematic Reviews*, (2).

Chen, B. K., & Eichenfield, L. F. (2001). Pediatric anesthesia in dermatologic surgery: When hand-holding is not enough. *Dermatologic Surgery, 27*(12), 1010–1018.

Cohen, J. L., & Gold, M. H. (2014). Evaluation of the efficacy and safety of a lidocaine and tetracaine (7%/7%) cream for induction of local dermal anesthesia for facial soft tissue augmentation with hyaluronic acid. *Journal of Clinical and Aesthetic Dermatology, 7*(10), 32.

Dixit, S., Lowe, P., Fischer, G., & Lim, A. (2013). Ice anaesthesia in procedural dermatology. *Australasian Journal of Dermatology, 54*(4), 273–276.

Du, S., Jaaniste, T., Champion, G. D., & Yap, C. S. (2008). Theories of fear acquisition: The development of needle phobia in children. *Pediatric Pain Letter, 10*(2), 13–17.

Eaglstein, F. N. (1999). Chemical injury to the eye from EMLA cream during erbium laser resurfacing. *Dermatologic Surgery, 25*(7), 590–591. https://doi.org/10.1046/j.1524-4725.1999.98289.x.

Engel, S. J., Afifi, A. M., & Zins, J. E. (2010). Botulinum toxin injection pain relief using a topical anesthetic skin refrigerant. *Journal of Plastic, Reconstructive & Aesthetic Surgery, 63*(9), 1443–1446.

Feily, A., Fallahi, H., Zandian, D., & Kalantar, H. (2011). A succinct review of botulinum toxin in dermatology; update of cosmetic and noncosmetic use. *Journal of Cosmetic Dermatology, 10*(1), 58–67.

Fuzier, R., Lapeyre-Mestre, M., Mertes, P. M., Nicolas, J. F., Benoit, Y., Didier, A., & Montastruc, J. L. (2009). Immediate-and delayed-type allergic reactions to amide local anesthetics: Clinical features and skin testing. *Pharmacoepidemiology and Drug Safety, 18*(7), 595–601.

Gammaitoni, A. R., Alvarez, N. A., & Galer, B. S. (2003). Safety and tolerability of the lidocaine patch 5%, a targeted peripheral analgesic: A review of the literature. *Journal of Clinical Pharmacology, 43*(2), 111–117.

Gedney, J. J., & Logan, H. (2006). Pain related recall predicts future pain report. *Pain, 121*(1–2), 69–76.

Guney, K., Sezgin, B., & Yavuzer, R. (2017). The efficacy of vibration anesthesia on reducing pain levels during lip augmentation: Worth the buzz? *Aesthetic Surgery Journal, 37*(9), 1044–1048.

Hashim, P. W., Nia, J. K., Taliercio, M., & Goldenberg, G. (2017). Local anesthetics in cosmetic dermatology. *Cutis, 99*(6), 393–397.

Hein, C., Schönwetter, D. J., & Iacopino, A. M. (2011). Inclusion of oral-systemic health in predoctoral/undergraduate curricula of pharmacy, nursing, and medical schools around the world: A preliminary study. *Journal of Dental Education, 75*(9), 1187–1199.

Johnson, M. H., Breakwell, G., Douglas, W., & Humphries, S. (1998). The effects of imagery and sensory detection distractors on different measures of pain: How does distraction work? *British Journal of Clinical Psychology, 37*(2), 141–154.

Kim, J. (2017). Tumescent anesthesia for reducing pain, swelling, and ecchymosis during polycaprolactone filler injections in the face. *Journal of Cosmetic and Laser Therapy, 19*(7), 434–438.

Kim, J. (2018). Novel forehead augmentation strategy: Forehead depression categorization and calcium-hydroxyapatite filler delivery after tumescent injection. *Plastic and Reconstructive Surgery. Global Open, 6*(9).

Klein, J. A. (2000). *Tumescent technique: Tumescent anesthesia & microcannular liposuction.* Mosby Incorporated.

Koay, J., & Orengo, I. (2002). Application of local anesthetics in dermatologic surgery. *Dermatologic Surgery, 28*(2), 143–148.

Kontochristopoulos, G., Gregoriou, S., Zakopoulou, N., & Rigopoulos, D. (2006). Cryoanalgesia with dichlorotetrafluoroethane spray versus ice packs in patients treated with botulinum toxin-a for palmar hyperhidrosis: Self-controlled study. *Dermatologic Surgery, 32*(6), 873–874.

Kravitz, N. D. (2007). The use of compound topical anesthetics: A review. *Journal of the American Dental Association, 138*(10), 1333–1339.

Legrain, V., Crombez, G., Verhoeven, K., & Mouraux, A. (2011). The role of working memory in the attentional control of pain. *Pain, 152*(2), 453–459.

Li, Y., Dong, W., Wang, M., & Xu, N. (2017). Investigation of the efficacy and safety of topical vibration anesthesia to reduce pain from cosmetic botulinum toxin A injections in Chinese patients: A multicenter, randomized, self-controlled study. *Dermatologic Surgery, 43*, S329–S335.

Lidén, Y., Olofsson, N., Landgren, O., & Johansson, E. (2012). Pain and anxiety during bone marrow aspiration/biopsy: Comparison of ratings among patients versus health-care professionals. *European Journal of Oncology Nursing, 16*(3), 323–329.

Liu, W., Yang, X., Li, C., & Mo, A. (2013). Adverse drug reactions to local anesthetics: A systematic review. *Oral Surgery, Oral Medicine, Oral Pathology, Oral Radiology, 115*(3), 319–327.

Lozinski, A., & Huq, N. S. (2013). Tumescent liposuction. *Clinics in Plastic Surgery, 40*(4), 593–613.

Mally, P., Czyz, C. N., Chan, N. J., & Wulc, A. E. (2014). Vibration anesthesia for the reduction of pain with facial dermal filler injections. *Aesthetic Plastic Surgery, 38*(2), 413–418.

McKinlay, J. R., Hofmeister, E., Ross, E. V., & MacAllister, W. (1999). EMLA cream–induced eye injury. *Archives of Dermatology, 135*(7), 855–856.

McLenon, J., & Rogers, M. A. (2019). The fear of needles: A systematic review and meta-analysis. *Journal of Advanced Nursing, 75*(1), 30–42.

McMurtry, C. M., Noel, M., Taddio, A., Antony, M. M., Asmundson, G. J., Riddell, R. P., ... Shah, V. (2015). Interventions for individuals with high levels of needle fear: Systematic review of randomized controlled trials and quasi-randomized controlled trials. *Clinical Journal of Pain, 31*(Suppl. 10), S109.

McMurtry, C. M., Riddell, R. P., Taddio, A., Racine, N., Asmundson, G. J., Noel, M., ... Shah, V. (2015). Far from "just a poke": Common painful needle procedures and the development of needle fear. *Clinical Journal of Pain, 31*, 3–11.

Melzack, R. (2005). Evolution of the neuromatrix theory of pain. The Prithvi Raj lecture: Presented at the third World Congress of World Institute of Pain, Barcelona 2004. *Pain Practice, 5*(2), 85–94.

Melzack, R., & Katz, J. (2004). The gate control theory: Reaching for the brain. In T. Hadjistavropoulos, & K. D. Craig (Eds.), *Pain: Psychological perspectives* (pp. 13–34). Mahwah, NJ: Lawrence Erlbaum Associates.

Monheit, G. D., Campbell, R. M., Neugent, H., Nelson, C. P., Prather, C. L., Bachtell, N., ... Holmdahl, L. (2010). Reduced pain with use of proprietary hyaluronic acid with lidocaine for correction of nasolabial folds: A patient-blinded, prospective, randomized controlled trial. *Dermatologic Surgery, 36*(1), 94–101.

Nambiar, A. K., Younis, A., Khan, Z. A., Hildrup, I., Emery, S. J., & Lucas, M. G. (2016). Alkalinized lidocaine versus lidocaine gel as local anesthesia prior to intra-vesical botulinum toxin (BoNTA) injections: A prospective, single center, randomized, double-blind, parallel group trial of efficacy and morbidity. *Neurourology and Urodynamics, 35*(4), 522–527.

Noel, M., Chambers, C. T., Petter, M., McGrath, P. J., Klein, R. M., & Stewart, S. H. (2012). Pain is not over when the needle ends: A review and preliminary model of acute pain memory development in childhood. *Pain Management, 2*(5), 487–497.

Noel, M., McMurtry, C. M., Chambers, C. T., & McGrath, P. J. (2009). Children's memory for painful procedures: The relationship of pain intensity, anxiety, and adult behaviors to subsequent recall. *Journal of Pediatric Psychology, 35*(6), 626–636.

Park, K. K., & Sharon, V. R. (2017). A review of local anesthetics: Minimizing risk and side effects in cutaneous surgery. *Dermatologic Surgery, 43*(2), 173–187.

Pomerantz, R. G., Lee, D. A., & Siegel, D. M. (2011). Risk assessment in surgical patients: Balancing iatrogenic risks and benefits. *Clinics in Dermatology, 29*(6), 669–677.

Rachman, S. (1977). The conditioning theory of fearacquisition: A critical examination. *Behaviour Research and Therapy, 15*(5), 375–387.

Reed, M. L. (1984). Mechanoanesthesia for intradermal injections. *Journal of the American Academy of Dermatology, 11*(2), 303.

Ruscheweyh, R., Kreusch, A., Albers, C., Sommer, J., & Marziniak, M. (2011). The effect of distraction strategies on pain perception and the nociceptive flexor reflex (RIII reflex). *Pain, 152*(11), 2662–2671.

Sagrillo, D. P. (2008). Emerging trends with dermal fillers. *Plastic Surgical Nursing, 28*(3), 152–153.

Sami, M. S., Soparkar, C. N., Patrinely, J. R., Hollier, L. M., & Hollier, L. H. (2006). Efficacy of botulinum toxin type A after topical anesthesia. *Ophthalmic Plastic & Reconstructive Surgery, 22*(6), 448–452.

Sarkany, M. (2012). Limitations of currently used topical anaesthetics in daily practice. In *Poster presented at IMCAS* (pp. 26–29).

Scholz, A. (2002). Mechanisms of (local) anaesthetics on voltage-gated sodium and other ion channels. *British Journal of Anaesthesia, 89*(1), 52–61.

Sezgin, B., Ozel, B., Bulam, H., Guney, K., Tuncer, S., & Cenetoglu, S. (2014). The effect of microneedle thickness on pain during minimally invasive facial procedures: A clinical study. *Aesthetic Surgery Journal, 34*(5), 757–765.

Sharma, P., Czyz, C. N., & Wulc, A. E. (2011). Investigating the efficacy of vibration anesthesia to reduce pain from cosmetic botulinum toxin injections. *Aesthetic Surgery Journal, 31*(8), 966–971.

Sobanko, J. F., Miller, C. J., & Alster, T. S. (2012). Topical anesthetics for dermatologic procedures: A review. *Dermatologic Surgery, 38*(5), 709–721.

Torp, K. D., & Simon, L. V. (2019). Lidocaine toxicity. In *StatPearls* StatPearls Publishing (Internet).

Travell, J. (1955). Factors affecting pain of injection. *Journal of the American Medical Association, 158*(5), 368–371.

Tucker, G. T., & Mather, L. E. (1998). Properties, absorption, and disposition of local anesthetic agents. In *Vol. 3. Neural blockade in clinical anesthesia and management of pain* (pp. 55–95). Philadelphia: Lippincott-Raven.

Verheyden, J., & Blitzer, A. (2002). Other noncosmetic uses of BOTOX. *Disease-a-Month, 48*(5), 357–366.

Walsh, A., & Walsh, S. (2011). Local anaesthesia and the dermatologist. *Clinical and Experimental Dermatology: Clinical Dermatology, 36*(4), 337–343.

Waterhouse, M. R., Liu, D. R., & Wang, V. J. (2013). Cryotherapeutic topical analgesics for pediatric intravenous catheter placement: Ice versus vapocoolant spray. *Pediatric Emergency Care, 29*(1), 8.

Weinkle, S. H., Bank, D. E., Boyd, C. M., Gold, M. H., Thomas, J. A., & Murphy, D. K. (2009). A multi-center, double-blind, randomized controlled study of the safety and effectiveness of Juvéderm® injectable gel with and without lidocaine. *Journal of Cosmetic Dermatology, 8*(3), 205–210.

Weiss, R. A., & Lavin, P. T. (2009). Reduction of pain and anxiety prior to botulinum toxin injections with a new topical anesthetic method. *Ophthalmic Plastic & Reconstructive Surgery, 25*(3), 173–177.

Young, K. D. (2015). Topical anaesthetics: What's new? *Archives of Disease in Childhood. Education and Practice Edition, 100*(2), 105–110.

Zeiderman, M. R., Kelishadi, S. S., Tutela, J. P., Rao, A., Chowdhry, S., Brooks, R. M., & Wilhelmi, B. J. (2018). Vapocoolant anesthesia for cosmetic facial rejuvenation injections: A randomized, prospective, split-face trial. *Eplasty, 18*, e6.

Chapter 34

Anesthesia and combat-related extremity injury

Robert (Trey) H. Burch, III
Department of Anesthesiology, Walter Reed National Military Medical Center, Bethesda, MD, United States

Abbreviations

ASRA	American Society of Regional Anesthesia
ATLS	advanced trauma life support
BAS	battalion aide station
CFR	case fatality rate
CSH	combat support hospital
CWMP	combat wound medication pack
DCR	damage control resuscitation
FST	forward surgical team
GSW	gunshot wound
IED	improvised explosive device
LOS	length of stay
MARAA	military advanced regional anesthesia and analgesia initiative
MTF	medical treatment facility
NSAID	nonsteroidal antiinflammatory drug
OEF	Operation Enduring Freedom
OI	osseointegration
OIF	Operation Iraqi Freedom
OTFC	oral transmucosal fentanyl citrate
PCA	patient-controlled analgesia
POI	point of injury
TMR	targeted muscle reinnervation

Introduction

Pain, specifically related to trauma, is inherent to injuries sustained in combat. War is constantly changing and the capabilities of the United States' military must consistently and rapidly adapt novel medical sciences and treatments in order to preserve American (and foreign) life. The origin of "modern" military anesthesia can be traced all the way back to the first United States armed conflict chiefly fought on foreign soil, the Mexican-American War (1846–48). At the time, Dr. Edward Barton utilized diethyl ether to anesthetize a soldier in an effort to facilitate a bilateral lower extremity amputation (Condon-Rall, 1995). An onlooker noted the soldier was "rendered completely insensible to all pain … and the limb was removed without the quiver of a muscle" (Porter, 1984). Although we have come a long way from the utilization of diethyl ether in this way, the continuous evolution of combat medicine to provide care for those wounded has resulted in the lowest case-fatality rates (CFR) in the history of the US military (Howard et al., 2019).

Although lacking in long term, randomized controlled studies, the benefits of adequate analgesia has long been associated with improved clinical outcomes (Wu, Naqibuddin, & Fleisher, 2001). The corollary has also proven true: inadequate analgesia has been linked to adverse outcomes such as thromboembolic incidents, prolonged hospital stays, increased agitation, and unnecessary suffering (Wu et al., 2003). A subsequent increase in the catabolic stress response secondary to pain has been proposed as a mechanism for these resultant adverse outcomes (Helander et al., 2019). This stress response can lead to an increase in the catabolic state, tachycardia, increased inflammatory markers, prothrombic states, and overall

Features and Assessments of Pain, Anesthesia, and Analgesia. https://doi.org/10.1016/B978-0-12-818988-7.00006-6
Copyright © 2022 Elsevier Inc. All rights reserved.

increase in oxygen consumption (De Gaudio, 2018). The prolonged complications of these increased inflammatory markers are not fully understood; however, the cumulative effect of a sustained increase in these markers has the potential to inflict a myriad of consequences leading to neuropathic changes and increased incidence of chronic pain (DeLeo, Tanga, & Tawfik, 2004).

The result of acute pain is mediated by a complex response of signaling pathways resulting from a stimulus exerting its effects on a nociceptor. This results in a cascading effect by releasing multiple local inflammatory and chemical mediators (such as substance P, aspartate, glutamate, prostaglandins, serotonin, and histamine) that are communicated via A delta and C pain fibers (first-order pain fibers) within the dorsal horn of the spinal cord. Additionally, there are inhibitory mechanisms existing at this level of the spinal cord that are modulated by GABA, glycine, enkephalins, norepinephrine, dopamine, and acetylcholine. From here, signals communicated through second-order neurons will transmit input to the thalamus by way of the brainstem within the contralateral spinothalamic tract. Finally, information is transmitted along third-order neurons from the thalamus to the sensory cortex (McDowell, 2019). Although the long-term effects of acute pain on these signaling pathways are not fully understood, evidence suggests poorly managed acute pain leads to the development of chronic pain.

Anesthesia and war

Military conflict within Iraq and Afghanistan brought about a surge in traumatic distal extremity injuries sustained from gunshot wounds (GSW) and improvised explosive devices (IEDs), commonly referred to as "dismounted complex blast injuries." The increase of dismounted complex blast injuries correlated with an increase in penetrating trauma. During Operations Iraqi Freedom (OIF) and Enduring Freedom (OEF), 83% of trauma-related deaths reported among service members can be attributed to penetrating types of trauma, which is drastically different than the 84%–94% of trauma-related deaths reported among the civilian population during the same time period resulting from blunt trauma (Kelly et al., 2008). Blast injuries resulting from penetrating trauma can cause a myriad of complex polytrauma injuries consisting of multiextremity traumatic amputations, major burns, and traumatic brain injuries.

Patient care is broken down into five different levels of care in the military: Role I–Role V. This is not to be confused with civilian trauma level designations. A medical treatment facility (MTF) designated as Role I is generally oriented closest to the point of injury (POI) and ranges from first aid care to life saving measures defined in ATLS. Care of this type may be provided by the injured soldier, a fellow soldier, or a combat medic attached to a unit. Commonly, Role I MTFs are referred to as battalion aid stations (BAS) and provide triage, first aid, and evacuation if needed. Soldiers at Role I MTFs typically have access to basic pain treatment options including nonsteroidal antiinflammatory drugs and acetaminophen. If a combat medic is attached to the unit, capabilities are often expanded to include opioids (typically morphine auto injector), ketamine, and oral transmucosal fentanyl citrate (OTFC). Therapy can be escalated based upon the mechanism of injury, ability to return to duty, and concern for hemodynamic stability.

In contrast, Role II MTFs provide broader care to patients and serve as an intermediary to a Role III MTF. Although each Role II MTF is unique to its respective military branch, these facilities generally offer surgical capabilities, basic radiography, the ability to transfuse of whole blood with walking blood banks, and the potential for inpatient options (Hejl et al., 2015). In the Army, a forward surgical team (FST) can be mobilized when there is a need for immediate, life-saving damage control surgery. FSTs typically consist of 20 personnel, including general and orthopedic surgeons, anesthesia providers, and nursing staff. FSTs have anesthesia personnel capable managing complex injuries and are able to offer further analgesic relief ranging from intravenous sedation to general anesthetics with total intravenous anesthesia or draw over vaporizers.

In the Army, Role III MTFs are known as the combat support hospital (CSH) and serve as a modern Army field hospital. CSH's serve as the eventual destination for in-theater evacuations and have even broader capabilities as needed to escalate casualty care. From here, soldiers can be returned to duty or have their wounds stabilized prior to evacuation to Role IV facilities outside of the combat zone. Role III MTFs have full operative capabilities similar to hospitals for restorative surgery and are further staffed with pharmacist, nutritionist, and respiratory therapist. Role V MTFs are only located within the United States.

Acute pain

Injuries sustained in combat, and subsequent pain management, in acutely wounded distal extremity injuries are often complex; however, recent studies suggest that early intervention can potentially have the greatest impact on patient's recovery (Gallagher et al., 2019). Combat anesthesia often begins on the battlefield at the POI and is usually provided by soldiers actively engaged in combat. It is important to recognize that treating acute pain has been associated with improved outcomes; however, the injured casualty must also be stabilized first. Given that hemorrhagic blood loss continues to be the most common cause of preventable death on the battlefield, soldiers with high mortality injuries must receive life-saving maneuvers immediately, be evacuated to higher care facilities, all while receiving acute pain treatment. Soldiers undergo extensive training in combat casualty care. As a result, medical aid can be provided by the soldier, a fellow soldier, or a soldier formally trained as a combat medic.

Over the course of the US military war efforts, and with the civilian population, there has been a greater focus on providing multimodal analgesic approach to pain management. This approach is used in an effort to enhance postoperative pain and decrease the incidence of deleterious side effects from a single treatment (Prabhakar, Cefalu, Rowe, Kaye, & Urman, 2017). Individual soldiers deployed to war zones often carry a Combat Wound Medication Pack (CWMP), which includes Meloxicam 15 mg (a NSAID, primary Cox-2 inhibitor), Acetaminophen (650 mg extended release, x2 tablets), and Moxifloxacin (Butler, Kotwal, & Buckenmaier, 2014). Acetaminophen exerts its action through the central inhibition of prostaglandin synthesis, while NSAIDS offer combined central and peripheral inhibition of prostaglandin synthesis. In conjunction with NSAIDs, cyclooxygenase 2 selective agents are historically preferred secondary to its more favorable side-effect profile (Malchow & Black, 2008). The purpose of the CWMP is provide pain relief for soldiers with mild-to-moderate pain to allow them to continue their mission, provide self-aid, or assist with their own evacuation by not altering their sensorium. When injuries sustained in combat require medications beyond the CWMP, medics are equipped with opioid medications such as morphine intramuscular auto injectors and oral transmucosal fentanyl citrate (OTFC), as well as ketamine.

Opioid medications are often the first-line therapy in a critically wounded soldier. Most opioids work primarily via the mu receptor that provides analgesic relief; however, also carries the potential for respiratory depression, which could be disastrous in austere battlefield environments. The kappa receptor can also provide analgesic relief with lesser effects on the respiratory drive (Gordon et al., 2016). Opioids are primarily given parenterally in the combat environment. This can vary from different points of care and training of the soldier administering the medication, and can range from intramuscular morphine at the point of injury, oral OTFC, or intravenously either by bolus dosing or patient-controlled analgesic (PCA) devices. PCA delivery has been improved throughout the war efforts and the American Society of Regional Anesthesia and Pain Medicine (ASRA) has endorsed the delivery of opioid medication via PCA to be superior to that of nurse delivered bolus dosing (Malchow & Black, 2008). Methadone, existing as parenteral and enteral dosing, is unique in that while it works as an opioid receptor agonist it also exerts effects as an NMDA antagonist and inhibiting serotonin reuptake in the spinal cord (Malchow & Black, 2008).

Ketamine has emerged as a critical nonopioid analgesic medication utilized in combat exerting its action via NMDA receptor antagonism. It has proven useful in providing rapid analgesic relief in critically wounded while avoiding the disastrous side effects of respiratory depression seen with opioid medications in hemodynamically compromised patients. Ketamine in the combat environment can be administered as an IV bolus (carried by trained combat medic and available at Role II and higher MTFs) or administered by PCA, to include a basal rate. It is important to recognize the potential dissociative properties of ketamine, especially in acutely wounded soldiers, and is often given in conjunction with a benzodiazepine to lessen these effects (Crumb, Bryant, & Atkinson, 2018).

Regional anesthesia in extremity injury

Regional anesthesia has played a role in pain management since the Vietnam War when epidural anesthesia and peripheral nerve analgesia was first utilized in a warfare setting (Condon-Rall, 1995). However, pain control still remained an issue at the beginning of the Global War on Terror. After witnessing first-hand combat related injuries, it was Colonel (Retired) Jack Chiles who initially advocated for the additional benefit of using regional anesthesia in the Iraq War. While deployed to Balad, Iraq, in 2003, COL (Ret) Chester "Trip" Buckenmaier III, MD successfully placed the first in-theater continuous lumbar plexus peripheral nerve catheter in a wounded soldier. The analgesic relief from this catheter allowed that casualty to be air transported back to United States with minimal additional pain requirements. Through the efforts of these pioneers, along with Colonel (Retired) Todd Carter, and Colonel (Retired) Ann Virtis, the Military Advanced Regional Anesthesia and Analgesia Initiative (MARAA) was born in an effort to rapidly disseminate research advances to deployed anesthesia providers (Army Regional Anesthesia and Pain Management Initiative, 2008). The MARAA handbook is still utilized in the regional block bays at multiple department of defense anesthesia training facilities.

Overtime, the US Military has seen a shift from conventional to unconventional warfare within the endeavors of OEF and OIF. An increase in penetrating trauma secondary to blast injuries from improvised explosive devices (IEDs), land mines, and shrapnel has resulted in nearly 65% of injuries sustained in combat (Gallagher et al., 2019). Regional anesthesia provides a plethora of opportunities to treat acute pain extremity injuries. Whether by single injection, neuraxial (or continuous neuraxial), or peripheral nerve catheters, regional anesthesia can be employed quickly and safely by trained

376 **PART | III** Interlinking anesthesia, analgesics and pain

providers. These methods of regional anesthesia allow the provider to further provide a multimodal pain management approach to obtain patient comfort in the critically ill. In their review of critically ill pain management, Malchow and Black (2008), demonstrate that a multitude of improved outcomes with regional anesthesia result in decrease length of stay (LOS), decreased mortality, and improved patient comfort.

Neuraxial anesthetics remain an option in the combat environment; however, the anesthesia provider must remain vigilant of concerns such as hemodynamic stability as well as potential coagulopathy associated inherent with trauma. Bolus injections of local anesthetics into the epidural or intrathecal space can prove problematic while undergoing resuscitation with sustained hemorrhagic blood loss and can have devastating hemodynamic consequences. Often it is the safest practice to proceed with a general anesthetic in a combat trauma patient until completion of the initial surgical course and competition of the initial resuscitation (or Damage Control Resuscitation, DCR). This allows for the best potential chance of initial stabilization and provides the opportunity to evaluate for ongoing injuries, potential nerve injuries, and volume resuscitation. However, after the initial surgical insult, neuraxial techniques can be utilized with lumbar and thoracic epidural with infusions of local anesthetics, and potentially opioids dependent on the patient mental state and ventilator status.

Peripheral nerve analgesia can be achieved by single injection or infusion via in situ catheter. With the advancement of ultrasound imaging, peripheral nerve blocks can be placed more accurately in higher echelons of care. However, in forward operating units, ultrasound capability is not always feasible (although smaller, more compact ultrasounds are actively being produced) and the anesthesia provider must still rely on an expert knowledge of anatomy and potentially a peripheral nerve stimulator. The decision to place a peripheral nerve catheter must be balanced with the duration of potential transport, knowledge of personnel to deal with complications, and risk for sequela from coagulopathies.

The risks in peripheral nerve analgesia for those in combat are the same as for those in the civilian population; however, there is an increased probability of undiagnosed injuries and comorbidities in those wounded by complex blast injuries. Moreover, increased attention to sterility is paramount given the assumed nonsterile environment of combat injuries and the provider must be able to both recognize and treat any signs of local anesthetic systemic toxicity.

Osseointegration and targeted muscle reinnervation

With the advent of unconventional warfare and complex blast injuries, the US Military has seen an increase in amputation rates as a result of complex distal extremity injuries sustained in combat (Stansbury, Lalliss, Branstetter, Bagg, & Holcomb, 2008). This often occurs in previously high functioning individuals who have been left with the inability to achieve their previous baseline level of performance with conventional prosthesis. Amputations have also resulted in an increase in chronic pain from either the residual limb itself, or from perceived pain in the form of phantom limb pain. This pain often results from the amputated nerve endings forming neuromas and is very complex to treat. Patients are often seen in a chronic pain clinic environment and require multiple medications to include NSAIDs, acetaminophen, gabapentinoids, opioids, and potentially multiple invasive neurolytic procedures. Efforts to decrease the amount of medication and help to patients achieve better baseline independence have evolved through surgical procedures, osseointegration, and targeted muscle reinnervation, as surgical capabilities continue to advance.

Osseointegration has evolved to help provide more functional prosthesis to soldiers with amputations. This relies on the integration of a metal implant directly in the distal bone to serve as a bridge between the musculoskeletal system and the implant. The subsequent calcified matrix on the surface of the periimplant bone is similar to early biologic bone fixation (Zaid, O'Donnell, Potter, & Forsberg, 2019). This type of remodeling allows for contact between the bone and the implant to provide mechanical loading comparable to the intrinsic host bone (Mayrogenis, Dimitriou, Parvizi, & Babis, 2009).

Amputated nerves and subsequent development of neuromas continues to be a source of potentially debilitating chronic pain. Commonly, after failed conservative treatment, neuromas are excised with the fresh nerve ending buried into nearby muscle, fat or bone, only to regrow. New advancements in surgical technology have evolved in the form of targeted muscle reinnervation in which neuroma can now be excised with the fresh nerve incorporated to a nearby motor nerve. This evolving surgical proven to be highly effective in the military population in decreasing phantom limb pain in amputees when conservative measures have failed (Dumanian et al., 2019).

Damage control resuscitation

Although a complete discussion of DCR is outside of the purview of this chapter, it bears mentioning briefly as it relates to distal extremity injuries in combat. DCR emphasizes treating and mitigating the "lethal triad" of hypothermia, coagulopathy, and acidosis. Specifically, these factors have been linked to increased mortality in wounded soldiers, as well as civilian trauma. These principles serve to help stabilize wounded while surgical efforts are made to save life or limb.

Hypothermia can predispose to worsening coagulopathy and alterations in drug metabolism and should be rapidly corrected with active warming of the patient either via emergency hypothermia blanket in the field, increasing the temperature of the room, convective warming, or warm irrigation. Intravenous resuscitation should be guided by blood product replacement rather than crystalloid to correct coagulopathy and to prevent potential dilutional anemia. The military has adopted the practice of 1:1:1 ratio of transfusion of PRBCs to plasma to platelets; however, there is still ongoing debate as whether this practice definitively decreases mortality (del Junco et al., 2013). There are also continued efforts to improve transfusion capabilities of whole blood in the deployed setting with walking blood banks and the addition of tranexamic acid to help prevent hyperfibrinolysis. As the US military continues in its war efforts, these resuscitative strategies will continue to evolve in an attempt to decrease mortality and prevent loss of American life.

Conclusion

Distal extremity injuries, and combat trauma in general, provoke unique and complex situations for military anesthesia providers. This is even more evident in the age of unconventional warfare. Throughout the military's time in warfare, this role has continually been redefined and most certainly will continue to evolve in our global involvement. Military anesthesia providers must be able to adapt and improvise in often complex situations in austere environments while providing the highest standard of care for all military personnel and others. The military community deserves this standard and combat anesthesia providers will continue to meet this demand.

Applications to other areas

Anesthesia and pain management for those wounded in combat can be applied to trauma in the civilian population as well. Although the austerity of the environment may be different, that basic principles of basic life support (BLS) and advanced cardiac life support (ACLS) are still applicable. Furthermore, the response to pain is universal regardless of the setting. Untreated pain results in a catabolic stress state that, if left untreated, can predispose to an increased incidence in chronic pain and unnecessary suffering.

Other agents of interest

For peripheral nerve blockade, local anesthetic medications are injected around a target nerve or within a fascial plane to help provide analgesia. Local anesthetics work by binding to the intracellular side of the transmembrane sodium channel. By blocking the sodium channel, subsequent membrane depolarization is inhibited and blocks the propagation of signal transduction.

Mini-dictionary of terms

- Parenteral administration—given by nonoral administration.
- Walking blood bank—a method used in a deployed setting where soldiers are previously tested for their blood type and antibodies that can be called upon to immediately transfuse whole blood to a bleeding casualty.
- Coagulopathy—inability of a person's blood to properly form clots.
- Dissociative—relating to a side effect of ketamine, a transient mental state where a person feels disconnected from their surroundings.
- Oral transmucosal fentanyl citrate (OTFC)—a solid formulation of fentanyl that is manufactured in the form of a lozenge on a stick (lollipop).

Key facts
Key facts of ketamine

- Is a NMDA receptor antagonist
- Preserves respiratory efforts in analgesic doses
- Is a dissociative amnestic

Key facts of point of injury care

- Can be performed by the injured soldier or anyone trained
- Usually occurs while engaged in combat
- Important to perform life-saving measures immediately
- Stabilize until can escalate to higher levels of care, if necessary

Summary points

- This chapter focuses on anesthesia in the combat setting
- Early intervention of pain can have long-term impact on prevention of chronic pain
- Combat trauma is usually associated with penetrating wound injuries
- Osseointegration and targeted muscle reinnervation are emerging surgical techniques to treat chronic pain and help improve limb prosthetic function

References

Army Regional Anesthesia and Pain Management Initiative. (2008). Retrieved from www.arapmi.org/index.html.

Butler, F., Kotwal, R., & Buckenmaier, C. (2014). A triple-option analgesia plan for tactical combat casualty care: TCCC guidlines change. *Journal of Special Operations Medicine, 14,* 13–25.

Condon-Rall, M. E. (1995). *Textbook of military medicine, Part IV, surgical combat casualty care: anesthesia and perioperative care of the combat casualty.* Washington, DC: Office of the Surgeon General.

Crumb, M. W., Bryant, C., & Atkinson, T. J. (2018). Emerging trends in pain medication management: Back to the future: A focus on ketamine. *The American Journal of Medicine, 131*(8), 883–886. https://doi.org/10.1016/j.amjmed.2018.02.037.

De Gaudio, A. R. (2018). The stress response of critical illness: Which is the role of sedation? In A. R. DeGaudio, S. Romagnoli, M. Bonifazi, A. R. DeGaudio, & S. Romagnoli (Eds.), *Critical care sedation* (pp. 9–19). Springer International Publishing.

del Junco, D. J., Holcomb, J. B., Fox, E. E., Brasel, K. J., Phelan, H. A., Bulger, E. M., ... Rahbar, M. H. (2013). Resuscitate early with plasma and platelets or balance blood products gradually: Findings from the PROMMTT study. *The Journal of Trauma and Acute Care Surgery, 75*(1 Suppl 1), S24–S30. https://doi.org/10.1097/TA.0b013e31828fa3b9.

DeLeo, J. A., Tanga, F. Y., & Tawfik, V. L. (2004). Neuroimmune activation and neuroinflammation in chronic pain and opioid tolerance/hyperalgesia. *Neuroscientist, 10*(1), 40–52.

Dumanian, G. A., Potter, B. K., Mioton, L. M., Ko, J. H., Cheesborough, J. E., Souza, J. M., ... Jordan, S. W. (2019). Targeted muscle reinnervation treats neuroma and phantom pain in major limb amputees: A randomized clinical trial. *Annals of Surgery, 270*(2), 238–246. https://doi.org/10.1097/SLA.0000000000003088.

Gallagher, R., Rosemary, P., Giodano, N., Farrar, J., Guo, W., Taylor, L., ... Buckenmaier, C. (2019). Prospective cohort study examining the use of regional anesthesia for early pain management after combat-related extremity injury. *Regional Anesthesia and Pain Medicine.* Epub ahead of print.

Gordon, D., de Leon-Casaola, O., Wu, C., Sluka, K., Brennan, T., & Chou, R. (2016). Research gaps in practice guidelines for acute postoperative pain management in adults: Findings from a review of the evidence for an American Pain Society clinical practice guideline. *Journal of Pain,* 158–166.

Hejl, C. G., Martinaud, C., Macarez, R., Sill, J., Le Golvan, A., Dulou, R., ... De Rudicki, S. (2015). The implementation of a multinational "walking blood bank" in a combat zone: The experience of a health service team deployed to a medical treatment facility in Afghanistan. *Journal of Trauma and Acute Care Surgery, 78*(5), 949–954.

Helander, E. M., Webb, M. P., Menard, B., Prabhakar, A., Helmstetter, J., Cornett, E. M., ... Kaye, A. D. (2019). Metabolic and the surgical stress response considerations to improve postoperative recovery. *Current Pain and Headache Reports, 23*(5), 33.

Howard, J. T., Kotwal, R. S., Stern, C. A., Janak, J. C., Mazuchowski, E. L., Butler, F. K., ... Smith, D. J. (2019). Use of combat casulaty care data to assess the US militar trauma system during the Afghanistan and Iraq conflicts. *Journal of the American Medical Association, 154*(7), 600–608.

Kelly, J. F., Ritenour, A. E., McLaughlin, D. F., Bagg, K. A., Podaca, A. N., Mallak, C. T., ... Holcomb, J. B. (2008). Injury severity and causes of death from Operation Iraqi Freedom and Operation Enduring Freedom: 2003–2004 versus 2006. *Journal of Trauma, 64,* S21–S27.

Malchow, R. J., & Black, I. H. (2008). The evolution of pain management in the critically ill trauma patient: Emerging concepts from the global war on terrorism. *Critical Care Medicine, 36*(7 Suppl), S346–S357.

Mayrogenis, A., Dimitriou, R., Parvizi, J., & Babis, G. (2009). Biology of implant osseointegraion. *Journal of Musculoskeletal and Neuronal Interactions, 9*(2), 67–71.

McDowell, T. (2019). Peripheral mechanisms of pain transmission and modulation. In Abd-Elsayed (Ed.), *Pain* (pp. 37–40). Springer.

Porter, J. B. (1984). Medical and surgical notes of campaigns in the war with Mexico during the years 1845–1848. *Anesthesiology, 61,* 585–588.

Prabhakar, A., Cefalu, J., Rowe, J. S., Kaye, A. D., & Urman, R. (2017). Techniques to optimize multimodal analgesia in ambulatory surgery. *Current Pain and Headache Reports, 21*(5), 24.

Stansbury, L., Lalliss, S., Branstetter, J., Bagg, M., & Holcomb, J. (2008). Amputations in U.S. military personnel in the current conflicts in Afghanistan and Iraq. *Journal of Orthopedic Trauma, 22*(1), 43–46.

Wu, C. L., Naqibuddin, M., & Fleisher, L. A. (2001). Measurement of patient satisfaction as an outcome of regional anesthesia and analgesia: A systematic review. *Regional Anesthesia and Pain Medicine, 26*(3), 196–208. Retrieved from https://doi.org/10.1053/rapm.2001.22257.

Wu, C. L., Naqibuddin, M., Rowlingson, A. J., Lietman, S. A., Jermyn, R. M., & Fleisher, L. A. (2003). The effect of pain on health-related quality of life in the immediate postoperative period. *Anesthesia and Analgesia, 97*(4), 1078–1085.

Zaid, M., O'Donnell, R., Potter, B., & Forsberg, J. (2019). Osseointegration: State of the art. *Journal of the American Academy of Orthopaedic Surgeons, 27*(22), e977–e985.

Chapter 35

Spinal anesthesia: Applications to cesarean section and pain

Reyhan Arslantas

Department of Anesthesiology and Reanimation, Taksim Training and Research Hospital, Istanbul, Turkey

Abbreviations

ASA	American Society of Anesthesiologists
CSF	cerebrospinal fluid
EBP	epidural blood patch
ED95	the 95 percent% effective dose
PDPH	postdural puncture headache

Introduction

Cesarean section is defined as the birth of an infant through laparotomy and hysterotomy. Goals for anesthesia for the cesarean section must include the mother's comfort and safety and the well-being of the fetus and neonate. Options for anesthesia for cesarean section delivery include general and neuraxial anesthesia. Use of neuraxial anesthesia (i.e., spinal anesthesia, combined spinal-epidural anesthesia, and epidural anesthesia) for cesarean section minimizes the neonates' exposure to general anesthetic medications of the mothers, avoids airway manipulation, improves postoperative pain, and allows the mothers to see their children almost immediately after birth.

Spinal anesthesia is a type of neuraxial anesthesia; local anesthetic is injected into cerebrospinal fluid in the intrathecal space (subarachnoid space) to anesthetize nerves that exit the spinal cord. Spinal anesthesia, also called spinal block, subarachnoid block, intradural block, and intrathecal block, is a simple and reliable technique that allows visual confirmation of correct needle placement (by visualization of cerebrospinal fluid) and is technically easier to perform than epidural anesthesia. Spinal anesthesia provides rapid onset of a dense neural blockade that is typically more profound than that provided with epidural anesthesia and resulting in shorter operation time, a reduced need for supplemental intravenous analgesics and anxiolytics (Riley, Cohen, Macario, Desai, & Ratner, 1995). This chapter introduces some of the scientific background and spinal anesthesia management for cesarean section and pain.

History of spinal anesthesia in the field of obstetrics

The use of spinal anesthesia for surgical procedures dates back to 1885, but this analgesia method did not become popular in obstetrics until the 1940s. Over half a million spinal blocks had been performed on parturient in America by the mid-1950s. However, spinal anesthesia is the most frequently used technique for obstetrical anesthesia world in the 1950s, subsequent improvements in epidural anesthesia techniques resulted in a decline in popularity of obstetric spinal anesthesia in the late 1960s (Morgan, 1995).

Spinal technique

Spinal anesthesia is performed by insertion a fine needle between the lumbar vertebrae and through the dura to inject anesthetic medication in the subarachnoid space.

Features and Assessments of Pain, Anesthesia, and Analgesia. https://doi.org/10.1016/B978-0-12-818988-7.00028-5
Copyright © 2022 Elsevier Inc. All rights reserved.

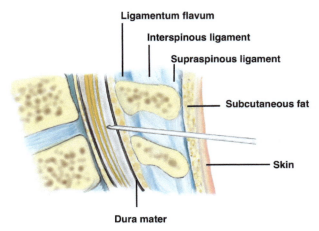

FIG. 1 Cross section of the spinal canal, spinal cord with meningeal layers, and adjacent ligaments.

Anatomy

The medulla spinalis is continuous with the medulla oblongata proximally and terminates distally in the conus medullaris as the cauda equina and the filum terminale. The level of conus medullaris termination varies from L3 in infants to the lower border of L1 in adults because of differential growth rates between the bony vertebral canal and the central nervous system (Macdonald, Chatrath, Spector, & Ellis, 1999). Surrounding the spinal cord in the vertebral column are three covering membranes: the dura, arachnoid, and pia maters (from outermost to innermost). Surrounding the dura mater is the epidural space, which includes nerve roots, blood vessels, lymphatics, fat, and areolar tissue. When performing spinal anesthesia using the midline approach, the spinal needle should traverse (from posterior to anterior) the skin, subcutaneous fat, supraspinous ligament, interspinous ligament, ligamentum flavum, dura mater, subdural space, arachnoid mater, and finally the subarachnoid space (Fig. 1).

Dermatomes are areas of skin that send signals to the brain supplied by the spinal nerves. Surface landmarks that correspond to appropriate dermatome levels include the following:

- Fifth finger—C8
- Nipple—T4
- Tip of the xiphoid process—T7
- Umbilicus—T10
- Inguinal ligament—T12

The sensory level required for Cesarean delivery commonly performed under neuraxial anesthesia is T4.

Preparation

Patient informed consent must be obtained, with adequate documentation of the explanation of complications risk. For the neuraxial anesthesia, the standard and emergency anesthesia equipment, standard American Society of Anesthesiologists (ASA) monitors should be applied (i.e., oxygen saturation, electrocardiography, blood pressure), and medications should be prepared. Anticholinergic drugs (i.e., atropine and glycopyrrolate) and epinephrine should be readily available because a large percentage of patients require the administration of a vasoconstrictor during hypotension and/or bradycardia due to spinal anesthesia.

Preprepared spinal kits are now frequently used and contain fenestrated drapes, syringes, needles, spinal needles, sterilizing solutions, swabs, and towels. Spinal needles have a close-fitting, removable stylet that prevents skin and subcutaneous tissue from plugging the needle during insertion. These needles are classified according to the tip shape and needle diameter. Spinal Needles fall into two main categories by tip shapes: those that Cutting-tip needles have sharp cutting tips, with the hole at the end of the needle (Quincke), and the Pencil-point needles have a closed tip shaped like that of a pencil, with the hole on the side of the needle near the tip (Whitacre and Sprotte) (Fig. 2). Current evidence suggests that the pencil-point spinal needle was significantly superior compared with the cutting-tip spinal needle regarding the frequency of postdural puncture headache (PDPH), PDPH severity, and the use of epidural blood patch (EBP) (Xu et al., 2017).

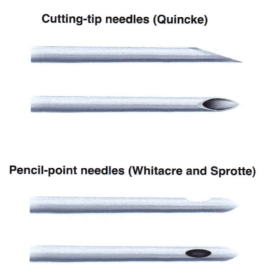

FIG. 2 Different types of spinal needles.

The spinal needle's external diameter is another factor that may be involved in the mechanisms of PDPH. The cross-sectional area of the needle determines the outer diameter. It is hypothesized that larger diameters create larger holes in the dura mater, leading to increased cerebrospinal fluid (CSF) leakage. Smaller numbers represent larger gauge needles (e.g., 16 gauge, 17 gauge), and larger numbers represent smaller gauge needles (e.g., 29 gauge, 32 gauge) (Calthorpe, 2004). Using small needles reduces postdural puncture headache from 40% with a 22 G needle to less than 2% with a 29 G needle. However, the use of larger needles improves the tactile needle insertion sensation, and so although the 29-G needles result in a very low postdural puncture headache rate, the failure rate increases (Flaatten et al., 1989; Morros-Vinoles et al., 2002). Studies indicate the superiority of Pencil-point needles over Cutting-tip needles for women undergoing spinal anesthesia for cesarean section surgery. However, further research may be required to determine the optimal gauge of Pencil-point needles (Lee, Sandhu, Djulbegovic, & Mhaskar, 2018).

Ideal patient positioning is critical to the success of spinal anesthesia for cesarean section. The lateral decubitus and sitting positions are used most commonly. Flexion of the spinal column opens the space between the vertebral spinous processes and is most important when a midline approach technic is used. The sitting position is most common, and the midline may be easier to estimate than the lateral position. However, compared with the sitting position, the induction of spinal anesthesia in the lateral position is associated with reducing vagal reflexes and less hemodynamic change. Therefore, it is concluded that the lateral position provides more stable blood pressure and maybe preferred when hemodynamic stability is desired during the induction of spinal anesthesia for cesarean section with plain bupivacaine (Obasuyi, Fyneface-Ogan, & Mato, 2013).

Choice of spinal medication

For spinal anesthesia, a local anesthetic (LA) maybe applied alone or in combination with an opioid. A combination is usually administered for spinal anesthesia for cesarean delivery.

Local anesthetic agents: The essential factors in determining the distribution of local anesthetics are properties of the local anesthetic solution (e.g., baricity, dose, volume, temperature, concentration) and patient characteristics (e.g., the position of the patient during and just after injection, height, spinal column anatomy). Local anesthetics used in spinal anesthesia include bupivacaine, ropivacaine, levobupivacaine, chloroprocaine, and lidocaine.

Bupivacaine

Bupivacaine is the most commonly used long-acting amide local anesthetic agent for cesarean delivery. Bupivacaine is available as 0.25%, 0.5%, and 0.75% isobaric solutions and also as a hyperbaric 0.5% (in Europe) and 0.75% (in USA) solution containing 80 mg/mL glucose. The intrathecal dose of bupivacaine commonly used for cesarean delivery ranges from 4.5 to 15 mg. The 95% effective dose (ED95) of intrathecal hyperbaric bupivacaine for cesarean delivery is 11.2 mg (Ginosar, Mirikatani, Drover, Cohen, & Riley, 2004). The onset of anesthesia occurs in 5–8 min with bupivacaine and provides an average of 1.5–2.5 h of surgical anesthesia; thus, it is appropriate for cesarean delivery.

Ropivacaine

Ropivacaine is a highly protein-bound amide local anesthetic, and so it is characterized by slow onset and a long duration of action. Compared to spinal bupivacaine, ropivacaine has a shorter surgical anesthesia duration, less cardiotoxicity, and more significant motor-sensory block differentiation, resulting in less motor block and earlier recovery (Gautier et al., 2003). The minimum local anesthetic dose of ropivacaine was 14.22 mg (CI 95%: 13.67–14.77) for Cesarean section (Parpaglioni et al., 2006). Ropivacaine is not approved by the U.S. Food and Drug Administration for intrathecal use even if it did not show neurotoxic effects in clinical studies at dosages from 8 to 22.5 mg (Celleno, Parpaglioni, Frigo, & Barbati, 2005).

Levobupivacaine

Levobupivacaine is the pure S (−) enantiomer of racemic bupivacaine, and it was synthesized to reduce cardiotoxicity. It has a similar onset and duration when used at similar doses as bupivacaine for spinal anesthesia, and there is no significant difference in clinical efficacy (Celleno et al., 2005; Gautier et al., 2003).

Chloroprocaine

Chloroprocaine is an ultra-short duration of action ester local anesthetic. A dose of 3060 mg of preservative-free chloroprocaine preparations can provide reliable, short-duration spinal anesthesia (Goldblum & Atchabahian, 2013; Maes, Laubach, & Poelaert, 2016). Chloroprocaine has limited use for cesarean delivery due to its short duration of action (Goldblum & Atchabahian, 2013).

Lidocaine

Lidocaine is a hydrophilic, relatively weakly protein-bound amide local anesthetic. The use of lidocaine for spinal anesthesia is controversial due to concerns about permanent nerve injury and transient neurological symptoms (Forget, Borovac, Thackeray, & Pace, 2019).

Intrathecal opioids

Opioids, whether administered intrathecally in combination with a local anesthetic or alone, can have a direct analgesic effect on the spinal cord and nerve roots or prolong the sensory and motor blockade duration. The addition of an opioid to the local anesthetic used in spinal anesthesia for cesarean delivery is particularly beneficial to prevent visceral manipulation discomfort (e.g., manipulation of the uterus) and improve perioperative comfort (Dahl, Jeppesen, Jorgensen, Wetterslev, & Moiniche, 1999). High-dose intrathecal opioid analgesia seems to increase the risk of fetal bradycardia, respiratory depression, nausea and vomiting, pruritus, urinary retention, and must, therefore, be avoided (Van de Velde, 2005).

Lipophilic opioids

Lipid-soluble opioids such as fentanyl (10–25 μg) and sufentanil (2.–10 μg) are usually added to local intrathecal anesthetic to improve intraoperative analgesia and for postoperative analgesia (Dahl et al., 1999). Besides analgesic effects, the addition of spinal fentanyl to bupivacaine reduces the incidence of nausea and/or vomiting during cesarean delivery (Dahlgren et al., 1997).

Hydrophilic opioids

Hydrophilic opioid (e.g., preservative-free morphine with 50–150 μg) in spinal anesthesia has a slow onset but provides adequate analgesia for up to 24 h with minimal side effects for cesarean deliveries (Gehling & Tryba, 2009; Girgin, Gurbet, Turker, Aksu, & Gulhan, 2008). Common side effects associated with hydrophilic opioids include nausea, vomiting, and pruritus. Intrathecal hydromorphone 50–100 μg provides comparable analgesia to 100–200 μg of morphine, with a similar duration of action and side effects (Quigley, 2002). There are only limited data related to the use of hydromorphone in spinal analgesia for cesarean delivery. The use of 75 μg of intrathecal hydromorphone for cesarean delivery produces postoperative analgesia of similar effectiveness at 24 h as that produced by 150 µg of intrathecal when combined with a multimodal analgesia regimen (Sharpe et al., 2020). Also, side effects, including nausea, pruritus, and respiratory depression also similar.

Other adjuvants

It has been found that the addition of various intrathecal adjuvants to local anesthetics improves the quality and prolongs the duration of spinal block. Epinephrine (0.1–0.2 mg) and clonidine (75–150 µg) can be added to improve the quality and duration of the block and improve intraoperative analgesia, but concern that it may increase complications such as neurotoxic effects and sedation has hampered widespread use (Crespo, Dangelser, & Haller, 2017).

The addition of neostigmine (50 µg) to bupivacaine as an adjuvant in spinal anesthesia prolongs sensory and motor block duration. However, the high incidence of side effects (nausea, vomiting, pruritus, and sedation) and delayed recovery from anesthesia limit the clinical use of these doses for spinal anesthesia (Liu, Hodgson, Moore, Trautman, & Burkhead, 1999).

Continuous spinal anesthesia

Continuous spinal anesthesia is a neuraxial anesthesia technic that the level of sensory blockage can be titrated with great precision to the desired dermatomal level with intrathecal catheters, allowing control of the hemodynamic consequences of the spinal anesthesia-associated sympathetic blockade (Palmer, 2010). The continuous spinal anesthesia technique can be used to control arterial blood pressure in pregnant women with severe aortic stenosis or complex heart disease, morbid obesity patients, and in situations where previous spinal surgery may prevent local anesthesia from the epidural spread. Spinal catheters can also be used in selected patients for laparotomies, where general anesthesia may be a significant risk (e.g., difficult airway). The drug administration for continuous spinal anesthesia is similar to single-shot spinal anesthesia. Preservative-free 0.5% bupivacaine 5.0 mg (1 mL) fentanyl 15 µg administered for the first dose, then 0.5 mL 0.5% bupivacaine every 5 min until the desired block level is achieved (2.5 mg) bolus are administered. Repeat the dose of 0.5 mL of bupivacaine as needed to maintain the desired block level (Palmer, 2010).

Intraoperative management

During cesarean delivery, the uterus can compress the inferior vena cava and aorta, leading to decreased venous return to the heart, decreased cardiac output, and uteroplacental perfusion. Prolonged aortocaval compression can be harmful to both mother and fetus. Recent studies on healthy term parturient having cesarean section with spinal anesthesia found that they had a similar neonatal acid-base state whether tilted or supine, if the women's blood pressure was maintained at baseline with fluids and vasopressors (Lee & Landau, 2017).

Evidence suggests that prophylactic use of supplemental oxygen during neuraxial anesthesia for cesarean deliveries could prevent maternal desaturation resulting from receiving sedation and intraoperative hypotension (Siriussawakul et al., 2014). But, there is no convincing evidence that supplementation of supplemental oxygen to healthy term pregnant women during elective cesarean section under spinal anesthesia is beneficial or harmful for the short-term clinical outcome of the mother or fetus as assessed by Apgar scores (Chatmongkolchart & Prathep, 2016).

Hypotension is a very common consequence of sympathetic vasomotor block caused by spinal anesthesia for cesarean section, and maternal symptoms such as nausea, vomiting, and dyspnea are often accompanied. It also leads to side effects on the fetus, including depressed Apgar scores and umbilical acidosis, in relation to the severity and duration of the hypotension (Kinsella et al., 2018). The systolic arterial pressure should be maintained at $\geq 90\%$ of the obtained before spinal anesthesia and avoided a decrease to $<80\%$ baseline (Kinsella et al., 2018). Intravenous vasopressors (ephedrine, noradrenaline, or phenylephrine) and volume preloading with crystalloid or colloid should be used routinely and preferably prophylactically maintain systolic arterial pressure (Kinsella et al., 2018). If the hypotension is accompanied by bradycardia, an anticholinergic drug (glycopyrrolate or atropine) may be required. Adrenaline should be used for cardiovascular collapse.

Nausea and vomiting caused by hypotension during spinal anesthesia for cesarean section are significantly more common than during nonobstetric surgery. The etiology of this is that transient brainstem ischemia due to decreased cerebral perfusion activates the vomiting center, and splanchnic hypoperfusion due to spinal anesthesia releases emetogenic factors such as serotonin from the gastrointestinal system. Prophylactic vasopressors' use reduces the incidence of intraoperative nausea and vomiting during the cesarean section (Habib, 2012). Furthermore, the use of supplemental oxygen may relieve this nausea (Hirose et al., 2016).

Perioperative hypothermia or shivering are frequent events during cesarean delivery under spinal anesthesia and increase the risk of complications such as perioperative coagulopathy, infection, and myocardial ischemia. The leading causes of hypothermia are through heat redistribution from the core to the periphery due to vasodilation after the induction of spinal anesthesia, environmental conditions in the operating theatre, and surgical factors (Chen, Liu, Mnisi, Chen, & Kang, 2019). Some active warming strategies, such as preanesthetic forced-air, mattresses, and warmed intravenous fluid infusions, appear to be useful for preventing hypothermia and shivering and can attenuate adverse events (Chen et al., 2019; Jun et al., 2019).

Failed or inadequate spinal anesthesia manifesting as failure to achieve optimum painless operative conditions. The incidence of failed obstetric spinal anesthesia with the newer literature lies between 0.5% and 9.1% (Rukewe, Adebayo, & Fatiregun, 2015). The independent predictors of failure obstetric spinal anesthesia were smaller gauge spinal needles (26G and 27G) than 25G, multiple lumbar puncture attempts, the use of the L4/L5 intervertebral space, and the level of experience of the anesthesia provider (Rukewe et al., 2015). If the painless state is not reached initially, more time may be required for the block to spread. Also, slightly tilting the bed's head down may allow the local anesthetic to spread more cranially. If positioning also does not work or time does not allow, general anesthesia or a second regional technique may be applied.

Total spinal anesthesia may be a complication of spinal or epidural anesthesia for cesarean delivery. It may result from the unintentional injection of medication into the subarachnoid or subdural space (through a misplaced catheter or needle) instead of the epidural space or from an overdose of medication into the subarachnoid space. Signs and symptoms of high or total spinal anesthesia include bradycardia, hypotension, shortness of breath, and difficulty swallowing or phonation associated with rapidly rising sympathetic, sensory, and motor block. Loss of consciousness and respiratory depression may be the first signs of total spinal anesthesia. Loss of consciousness may occur due to brainstem hypoperfusion and/or brainstem anesthesia. Respiratory depression may occur secondary to respiratory muscle paralysis and brainstem hypoperfusion. If there is significant respiratory depression, intubation, and mechanical ventilation may be required. Mechanical ventilation under sedation will need to continue until the block wears off and spontaneous ventilation resumes. If bradycardia and/or hypotension occur, it will need to treat with atropine, vasopressors, and IV fluids.

Recovery from spinal anesthesia

The two-segment regression time of neuraxial blockade for local anesthetics such as bupivacaine, levobupivacaine, and ropivacaine, commonly used in spinal anesthesia, is 60–120 min. Using more massive doses of local anesthetics results in a longer duration of anesthesia. Also, the addition of epinephrine or opioids can prolong the duration of local anesthetics.

The early onset of oral intake after uncomplicated cesarean delivery under subarachnoid block provides an earlier return of bowel movement and greater maternal satisfaction. It is also not associated with an increased incidence of gastrointestinal symptoms or paralytic ileus (Mawson, Bumrungphuet, & Manonai, 2019).

The anesthesiologist should closely assess the recovery of motor and sensory functions after neuraxial anesthesia. Women should be informed that breastfeeding after spinal anesthesia is safe, and those postoperative analgesics have a favorable safety profile.

Postoperative complications

In the postoperative period, the relatively common complications of spinal anesthesia are postdural puncture headache and backache. Cerebrospinal fluid loss is thought to reduce intracranial pressure and lead to headaches. As a mechanism, it is suggested that the traction of pain-sensitive intracranial structures due to pressure drop or development of intracerebral vasodilation to balance the pressure causes headaches (Turnbull & Shepherd, 2003). Factors that can increase headaches after spinal anesthesia include younger age, female gender, pregnancy, larger needle size, and multiple dural punctures. Conservative treatment for headaches includes rest in the supine position, iv hydration, caffeine, and oral analgesics. An epidural blood patch is the most effective treatment for headache after dural puncture. The epidural blood patch is ideally done 24 h after the spinal anesthesia and after developing headache symptoms (Harrington, 2004).

Back injury is the most feared complication of spinal anesthesia among patients. However, evidence shows that back pain incidence after spinal anesthesia is not different from general anesthesia (Benzon, Asher, & Hartrick, 2016). Risk factors for persistent back pain after neuraxial anesthesia include preexisting back pain, surgery time greater than 2.5 h, lithotomy position, BMI greater than 32 kg/m^2, and multiple attempts for the neuraxial block (Benzon et al., 2016).

Conclusion

Spinal anesthesia is the most commonly used neuraxial anesthesia technique for cesarean delivery in the developed world. It provides the onset of intensive, bilateral, reliable anesthesia using a low-dose drug with minimal maternal toxicity or risk of fetal drug transfer.

Application to other areas

Spinal anesthesia provides excellent operation conditions for cesarean operations and other underbelly surgeries. Thus, it is also used in urology, gynecology, lower abdomen, perineal general surgery, lower extremity vascular surgery, and orthopedic surgery.

Other agents of interest

Other neuroaxial anesthesia techniques other than spinal anesthesia during cesarean delivery

- *Epidural anesthesia:* Dural puncture is not required and can be used for labor analgesia management, continuous intraoperative anesthesia, and postoperative analgesia with local anesthetics administered from the inserted catheter.
- *Combined spinal-epidural anesthesia:* It provides the rapid onset of more intense lumbosacral and thoracic anesthesia using low doses of local anesthetics and opioids compared to epidural anesthesia.

Mini-dictionary of terms

Dermatome: An area of skin that send signals to the brain supplied by the spinal nerves.
Neuraxial anesthesia: The administration of medication into the subarachnoid or epidural space around the nerves of the central nervous system to produce anesthesia and analgesia, such as spinal anesthesia, caudal anesthesia, and epidural anesthesia.
Quincke: A cutting-tip spinal anesthesia needles have sharp cutting tips, with the hole at the end of the needle.
Pencil-point needles: A closed tip spinal needles shaped like a pencil with the hole on the side of the needle near the tip.

Key facts

Key facts of spinal anesthesia
- Spinal anesthesia is a type of neuraxial anesthesia; local anesthetic is injected into cerebrospinal fluid in the intrathecal space (subarachnoid space) to anesthetize nerves that exit the spinal cord.
- The sensory level required for Cesarean delivery commonly performed under neuraxial anesthesia is T4.
- The pencil-point spinal needle was significantly superior compared with the cutting-tip spinal needle regarding the frequency of postdural puncture headache.
- For spinal anesthesia, a local anesthetic (LA) may be applied alone or in combination with an opioid.
- Hypotension is a very common consequence of sympathetic vasomotor block caused by spinal anesthesia for cesarean section, and maternal symptoms such as nausea, vomiting, and dyspnea are often accompanied.

Summary points

- This chapter introduces some of the scientific background and spinal anesthesia management for cesarean section and pain.
- Spinal anesthesia for cesarean section minimizes the neonates' exposure to general anesthetic medications of the mothers, avoids airway manipulation, improves postoperative pain, and allows the mothers to see their children almost immediately after birth.
- Current evidence suggests that the pencil-point spinal needle was significantly superior compared with the cutting-tip spinal needle regarding the frequency of postdural puncture headache (PDPH), PDPH severity, and the use of epidural blood patch (EBP).
- The intrathecal dose of bupivacaine commonly used for cesarean delivery ranges from 4.5 to 15 mg.
- Opioids, whether administered intrathecally in combination with a local anesthetic or alone, can have a direct analgesic effect on the spinal cord and nerve roots or prolong the sensory and motor blockade duration.

388   **PART | III** Interlinking anesthesia, analgesics and pain

- Hypotension is a very common consequence of sympathetic vasomotor block caused by spinal anesthesia for cesarean section, and maternal symptoms such as nausea, vomiting, and dyspnea are often accompanied.
- Intravenous vasopressors (ephedrine, noradrenaline, or phenylephrine) and volume preloading with crystalloid or colloid should be used routinely and preferably prophylactically maintain systolic arterial pressure.
- Nausea and vomiting caused by hypotension during spinal anesthesia for cesarean section are significantly more common than during nonobstetric surgery.
- Conservative treatment for PDPH includes rest in the supine position, iv hydration, caffeine, and oral analgesics. An epidural blood patch is the most effective treatment for PDPH.

References

Benzon, H. T., Asher, Y. G., & Hartrick, C. T. (2016, Jun). Back pain and neuraxial anesthesia. *Anesthesia and Analgesia*, *122*(6), 2047–2058. https://doi.org/10.1213/ANE.0000000000001270.

Calthorpe, N. (2004, Dec). The history of spinal needles: Getting to the point. *Anaesthesia*, *59*(12), 1231–1241. https://doi.org/10.1111/j.1365-2044.2004.03976.x.

Celleno, D., Parpaglioni, R., Frigo, M. G., & Barbati, G. (2005, Sep). Intrathecal levobupivacaine and ropivacaine for cesarean section. New perspectives. *Minerva Anestesiologica*, *71*(9), 521–525. https://www.ncbi.nlm.nih.gov/pubmed/16166911.

Chatmongkolchart, S., & Prathep, S. (2016). Supplemental oxygen for caesarean section during regional anaesthesia. *Cochrane Database of Systematic Reviews*, *3*. https://doi.org/10.1002/14651858.CD006161.pub3, CD006161.

Chen, W. A., Liu, C. C., Mnisi, Z., Chen, C. Y., & Kang, Y. N. (2019, Nov). Warming strategies for preventing hypothermia and shivering during cesarean section: A systematic review with network meta-analysis of randomized clinical trials. *International Journal of Surgery*, *71*, 21–28. https://doi.org/10.1016/j.ijsu.2019.09.006.

Crespo, S., Dangelser, G., & Haller, G. (2017). Intrathecal clonidine as an adjuvant for neuraxial anaesthesia during caesarean delivery: A systematic review and meta-analysis of randomised trials. *International Journal of Obstetric Anesthesia*, *32*, 64–76. https://doi.org/10.1016/j.ijoa.2017.06.009.

Dahl, J. B., Jeppesen, I. S., Jorgensen, H., Wetterslev, J., & Moiniche, S. (1999). Intraoperative and postoperative analgesic efficacy and adverse effects of intrathecal opioids in patients undergoing cesarean section with spinal anesthesia: A qualitative and quantitative systematic review of randomized controlled trials. *Anesthesiology*, *91*(6), 1919–1927. https://doi.org/10.1097/00000542-199912000-00045.

Dahlgren, G., Hultstrand, C., Jakobsson, J., Norman, M., Eriksson, E. W., & Martin, H. (1997). Intrathecal sufentanil, fentanyl, or placebo added to bupivacaine for cesarean section. *Anesthesia and Analgesia*, *85*(6), 1288–1293. https://doi.org/10.1097/00000539-199712000-00020.

Flaatten, H., Rodt, S. A., Vamnes, J., Rosland, J., Wisborg, T., & Koller, M. E. (1989). Postdural puncture headache. A comparison between 26- and 29-gauge needles in young patients. *Anaesthesia*, *44*(2), 147–149. https://doi.org/10.1111/j.1365-2044.1989.tb11167.x.

Forget, P., Borovac, J. A., Thackeray, E. M., & Pace, N. L. (2019). Transient neurological symptoms (TNS) following spinal anaesthesia with lidocaine versus other local anaesthetics in adult surgical patients: A network meta-analysis. *Cochrane Database of Systematic Reviews*, *12*. https://doi.org/10.1002/14651858.CD003006.pub4, CD003006.

Gautier, P., De Kock, M., Huberty, L., Demir, T., Izydorczic, M., & Vanderick, B. (2003). Comparison of the effects of intrathecal ropivacaine, levobupivacaine, and bupivacaine for caesarean section. *British Journal of Anaesthesia*, *91*(5), 684–689. https://doi.org/10.1093/bja/aeg251.

Gehling, M., & Tryba, M. (2009). Risks and side-effects of intrathecal morphine combined with spinal anaesthesia: A meta-analysis. *Anaesthesia*, *64*(6), 643–651. https://doi.org/10.1111/j.1365-2044.2008.05817.x.

Ginosar, Y., Mirikatani, E., Drover, D. R., Cohen, S. E., & Riley, E. T. (2004). ED50and ED95of intrathecal hyperbaric bupivacaine coadministered with opioids for cesarean delivery. *Anesthesiology*, *100*(3), 676–682. https://doi.org/10.1097/00000542-200403000-00031.

Girgin, N. K., Gurbet, A., Turker, G., Aksu, H., & Gulhan, N. (2008). Intrathecal morphine in anesthesia for cesarean delivery: Dose-response relationship for combinations of low-dose intrathecal morphine and spinal bupivacaine. *Journal of Clinical Anesthesia*, *20*(3), 180–185. https://doi.org/10.1016/j.jclinane.2007.07.010.

Goldblum, E., & Atchabahian, A. (2013). The use of 2-chloroprocaine for spinal anaesthesia. *Acta Anaesthesiologica Scandinavica*, *57*(5), 545–552. https://doi.org/10.1111/aas.12071.

Habib, A. S. (2012). A review of the impact of phenylephrine administration on maternal hemodynamics and maternal and neonatal outcomes in women undergoing cesarean delivery under spinal anesthesia. *Anesthesia and Analgesia*, *114*(2), 377–390. https://doi.org/10.1213/ANE.0b013e3182373a3e.

Harrington, B. E. (2004). Postdural puncture headache and the development of the epidural blood patch. *Regional Anesthesia and Pain Medicine*, *29*(2), 136–163. discussion 135 https://doi.org/10.1016/j.rapm.2003.12.023.

Hirose, N., Kondo, Y., Maeda, T., Suzuki, T., Yoshino, A., & Katayama, Y. (2016). Oxygen supplementation is effective in attenuating maternal cerebral blood deoxygenation after spinal anesthesia for cesarean section. *Advances in Experimental Medicine and Biology*, *876*, 471–477. https://doi.org/10.1007/978-1-4939-3023-4_59.

Jun, J. H., Chung, M. H., Jun, I. J., Kim, Y., Kim, H., Kim, J. H., et al. (2019). Efficacy of forced-air warming and warmed intravenous fluid for prevention of hypothermia and shivering during caesarean delivery under spinal anaesthesia: A randomised controlled trial. *European Journal of Anaesthesiology*, *36*(6), 442–448. https://doi.org/10.1097/EJA.0000000000000990.

Kinsella, S. M., Carvalho, B., Dyer, R. A., Fernando, R., McDonnell, N., Mercier, F. J., et al. (2018). International consensus statement on the management of hypotension with vasopressors during caesarean section under spinal anaesthesia. *Anaesthesia*, *73*(1), 71–92. https://doi.org/10.1111/anae.14080.

Lee, A. J., & Landau, R. (2017). Aortocaval compression syndrome: Time to revisit certain dogmas. *Anesthesia and Analgesia, 125*(6), 1975–1985. https://doi.org/10.1213/ANE.0000000000002313.

Lee, S. I., Sandhu, S., Djulbegovic, B., & Mhaskar, R. S. (2018). Impact of spinal needle type on postdural puncture headache among women undergoing Cesarean section surgery under spinal anesthesia: A meta-analysis. *Journal of Evidence-Based Medicine, 11*(3), 136–144. https://doi.org/10.1111/jebm.12311.

Liu, S. S., Hodgson, P. S., Moore, J. M., Trautman, W. J., & Burkhead, D. L. (1999). Dose-response effects of spinal neostigmine added to bupivacaine spinal anesthesia in volunteers. *Anesthesiology, 90*(3), 710–717. https://doi.org/10.1097/00000542-199903000-00012.

Macdonald, A., Chatrath, P., Spector, T., & Ellis, H. (1999). Level of termination of the spinal cord and the dural sac: A magnetic resonance study. *Clinical Anatomy, 12*(3), 149–152. https://doi.org/10.1002/(sici)1098-2353(1999)12:3<149::Aid-ca1>3.0.Co;2-x.

Maes, S., Laubach, M., & Poelaert, J. (2016). Randomised controlled trial of spinal anaesthesia with bupivacaine or 2-chloroprocaine during caesarean section. *Acta Anaesthesiologica Scandinavica, 60*(5), 642–649. https://doi.org/10.1111/aas.12665.

Mawson, A. L., Bumrungphuet, S., & Manonai, J. (2019). A randomized controlled trial comparing early versus late oral feeding after cesarean section under regional anesthesia. *International Journal of Women's Health, 11*, 519–525. https://doi.org/10.2147/IJWH.S222922.

Morgan, P. (1995). Spinal anaesthesia in obstetrics. *Canadian Journal of Anaesthesia, 42*(12), 1145–1163. https://doi.org/10.1007/BF03015105. 1995/12/01.

Morros-Vinoles, C., Perez-Cuenca, M. D., Cedo-Lluis, E., Colls, C., Bueno, J., & Cedo-Valloba, F. (2002, Nov). Comparison of efficacy and complications of 27G and 29G Sprottte needles for subarachnoid anesthesia. *Revista espanola de anestesiologia y reanimacion, 49*(9), 448–454. https://www.ncbi.nlm.nih.gov/pubmed/12516488. Comparacion de la eficacia y complicaciones de dos agujas punta Sprotte G27 y G29 para anestesia subaracnoidea.

Obasuyi, B. I., Fyneface-Ogan, S., & Mato, C. N. (2013). A comparison of the haemodynamic effects of lateral and sitting positions during induction of spinal anaesthesia for caesarean section. *International Journal of Obstetric Anesthesia, 22*(2), 124–128. https://doi.org/10.1016/j.ijoa.2012.12.005.

Palmer, C. M. (2010). Continuous spinal anesthesia and analgesia in obstetrics. *Anesthesia and Analgesia, 111*(6), 1476–1479. https://doi.org/10.1213/ANE.0b013e3181f7e3f4.

Parpaglioni, R., Frigo, M. G., Lemma, A., Sebastiani, M., Barbati, G., & Celleno, D. (2006). Minimum local anaesthetic dose (MLAD) of intrathecal levobupivacaine and ropivacaine for caesarean section. *Anaesthesia, 61*(2), 110–115. https://doi.org/10.1111/j.1365-2044.2005.04380.x.

Quigley, C. (2002). Hydromorphone for acute and chronic pain. *Cochrane Database of Systematic Reviews*, (1). https://doi.org/10.1002/14651858. CD003447, CD003447.

Riley, E. T., Cohen, S. E., Macario, A., Desai, J. B., & Ratner, E. F. (1995). Spinal versus epidural anesthesia for cesarean section: A comparison of time efficiency, costs, charges, and complications. *Anesthesia and Analgesia, 80*(4), 709–712. https://doi.org/10.1097/00000539-199504000-00010.

Rukewe, A., Adebayo, O. K., & Fatiregun, A. A. (2015). Failed obstetric spinal anesthesia in a Nigerian teaching hospital: Incidence and risk factors. *Anesthesia and Analgesia, 121*(5), 1301–1305. https://doi.org/10.1213/ANE.0000000000000868.

Sharpe, E. E., Molitor, R. J., Arendt, K. W., Torbenson, V. E., Olsen, D. A., Johnson, R. L., et al. (2020). Intrathecal morphine versus intrathecal hydromorphone for analgesia after cesarean delivery: A randomized clinical trial. *Anesthesiology, 132*(6), 1382–1391. https://doi.org/10.1097/ALN.0000000000003283.

Siriussawakul, A., Triyasunant, N., Nimmannit, A., Ngerncham, S., Hirunkanokpan, P., Luang-Aram, S., et al. (2014). Effects of supplemental oxygen on maternal and neonatal oxygenation in elective cesarean section under spinal anesthesia: A randomized controlled trial. *BioMed Research International, 2014*, 627028. https://doi.org/10.1155/2014/627028.

Turnbull, D. K., & Shepherd, D. B. (2003). Post-dural puncture headache: Pathogenesis, prevention and treatment. *British Journal of Anaesthesia, 91*(5), 718–729. https://doi.org/10.1093/bja/aeg231.

Van de Velde, M. (2005). Neuraxial analgesia and fetal bradycardia. *Current Opinion in Anaesthesiology, 18*(3), 253–256. https://doi.org/10.1097/01.aco.0000169230.53067.49.

Xu, H., Liu, Y., Song, W., Kan, S., Liu, F., Zhang, D., et al. (2017). Comparison of cutting and pencil-point spinal needle in spinal anesthesia regarding postdural puncture headache: A meta-analysis. *Medicine (Baltimore), 96*(14). https://doi.org/10.1097/MD.0000000000006527, e6527.

Chapter 36

Postoperative pain management: Truncal blocks in thoracic surgery

Gulbin Tore Altun

Memorial Health Group, Department of Anesthesiology and Reanimation, Istanbul, Turkey

Abbreviations

ASA	American Society of Anesthesiologists
ERAS	enhanced recovery after surgery
ESPB	erector spinae plane block
LAST	local anesthetic systemic toxicity
MTPB	midpoint transverse process pleural block
NSADIs	nonsteroidal antiinflammatory drugs
PECs	pectoralis nerve blocks
PVB	paravertebral block
RISS	rhomboid intercostal and serratus plane block
RLB	retrolaminar block
SAPB	serratus Anterior Plane Block
VATS	video assisted thoracic surgery

Introduction

Postoperative pain following thoracic surgery begins with surgical trauma; this and many factors make recovery and pain management difficult in these patients. Thoracic surgery patients often have significant comorbidities that limit their pain management options. Effective relief of postthoracotomy pain accelerates postoperative recovery and decreases complication rates, and shortening of hospital stay. There is no single agent or method suitable for every patient, and therefore postoperative pain management after thoracic surgery should be multimodal and for to the patient.

Although opioids are generally sufficient to manage pain after surgical procedures, postoperative thoracic surgery patients have limited therapeutic options due to severe pulmonary pathology. In addition, postoperative cardiovascular complications associated with thoracic operations make pain management difficult. Intercostal incisions, thoracostomy tube placement, and even pleural irritation are also painful procedures. Intraoperative rib fractures and iatrogenic factors such as planned resection, chest tube insertion and injuries, suboptimal intraoperative positioning may also worsen postoperative pain (Elmore, Nguyen, Blank, Yount, & Lau, 2015).

Thoracic surgery patients' pain is triggered by the respiratory cycle and aggravated by movements such as coughing or deep breathing. The effectiveness of analgesic therapy is determined by deep breathing and optimal pain management targets effective coughing. There is a strong relationship between effective postoperative pain management and pulmonary complications such as atelectasis and pneumonia in these patients. However, high doses of opioids can also worsen postoperative respiratory function (Sungur & Şentürk, 2017).

Poorly controlled acute postoperative pain may lead to the development of chronic postoperative pain (Kehlet, Jensen, & Woolf, 2006). "Chronic postoperative pain syndrome" is associated with thoracotomy, but video-assisted thoracic surgery (VATS) incisions can also cause chronic postoperative pain syndrome that can last for years.

Features and Assessments of Pain, Anesthesia, and Analgesia. https://doi.org/10.1016/B978-0-12-818988-7.00016-9
Copyright © 2022 Elsevier Inc. All rights reserved.

Pain management for thoracic surgery patients within ERAS (enhanced recovery after surgery) protocols

Postoperative improved recovery protocols have been shown to enhance patient outcomes for various surgical procedures. Related elements of ERAS care models in thoracic surgery are aggressive pain control, fluid restriction, and early mobilization (Thompson, French, & Costache, 2018).

ERAS guidelines recommend multimodal pain management strategies. Multimodal analgesia aims to use analgesic agents with different mechanisms of action, to avoid opioids as much as possible, to use regional anesthesia, and to switch to oral analgesic agents as soon as possible (Chou et al., 2016).

Multimodal analgesia

Multimodal analgesia is a concept that aims to minimize the side effects caused by analgesics and provides effective analgesia by using analgesic agents with different action mechanisms at optimum doses. This concept should include the use of synergistically effective agents and methods for the prevention and treatment of acute pain in a patient-specific manner.

Optimization of multimodal analgesia is achieved by using combinations of pharmacological and nonpharmacological techniques (Young & Buvanendran, 2012). For multimodal analgesia in thoracic surgery; it is possible to use nonsteroidal anti-inflammatory drugs (NSAIDs), anticonvulsants, acetaminophen, beta-blockers, NMDA receptor antagonists, alpha-2 agonists, glucocorticoids, and opioids, as well as neuraxial techniques, surgical field infiltration, and regional anesthesia.

Truncal blocks

An important component of successful multimodal analgesia is regional anesthesia techniques. Chest wall and abdominal trunk nerve blocks can provide similar analgesic effects with neuraxial techniques, with less risk when used correctly as a component of multimodal analgesia. In addition, truncal nerve blocks are easier to apply and learn than neuraxial techniques (Urits et al., 2019).

As compared to peripheral nerve blocks, there is no need to visualize the nerve in ultrasound-guided truncal block applications. Local anesthetics are injected into the specified muscle plane and they spread in this area and reach the desired nerves. The reason for the ease in implementing and learning field blocks lies in the easy technique.

Numerous thoracic wall blocks have been described, and almost all of them are facial plane blocks. They are based on passive diffusion of local anesthetic in the plane or in adjacent tissue compartments. Information on the most appropriate dosage regimens and block techniques is constantly updated, appropriate options should be determined individually.

Pectoral block type I and II

Introduction

Pectoralis nerve blocks (PECs) are regional anesthesia techniques used for postoperative analgesia in thoracic surgery. The increasing use of ultrasonography to identify tissue layers has led to the development of several new interfascial injection techniques for chest wall analgesia (Blanco, 2011; Blanco, Fajardo, & Parras Maldonado, 2012).

Indications

- Breast surgery (mastectomy, lumpectomy, breast expander/prosthesis, axillary dissection)
- Insertion of port catheters, pacemakers, implantable cardiac defibrillators
- Vascular surgery in the case of an axillofemoral-femoral bypass to prevent general anesthesia and tracheal intubation in patients with comorbidities
- Anterior thoracotomy
- Chest wall surgery
- Anterior shoulder surgery
- Postherpetic neuralgia that is refractory to neuraxial blocks
- Open subpectoral biceps tenodesis

Technique

PECs I and PECs II involve injecting a local anesthetic into the fascial plane between the minor and major pectoralis, and serratus anterior muscles. Thus, both medial and lateral pectoral nerves, T2–T6 lateral cutaneous branches of the intercostal nerves, intercostobrachial, and long thoracic nerves can be blocked (Blanco, 2011; Blanco et al., 2012).

The block is performed with the patient supine, either with the arm next to the chest or abducted 90 degrees.

- *Landmark*
  PECs I: The pectoralis major and pectoralis minor muscles and the pectoral branch of the thoracoacromial artery (Fig. 1).
  PECs II: the anterior axillary line at the level of the fourth rib (Fig. 2).

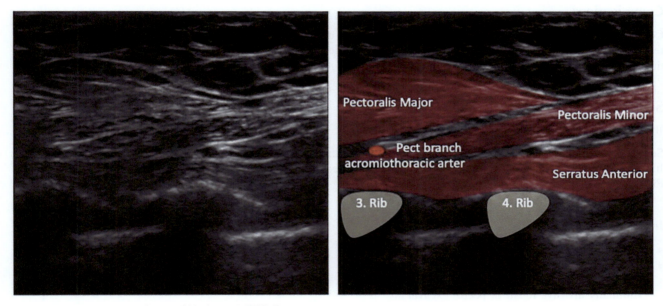

FIG. 1 Ultrasound anatomy of pectoral block type I (PECs I).

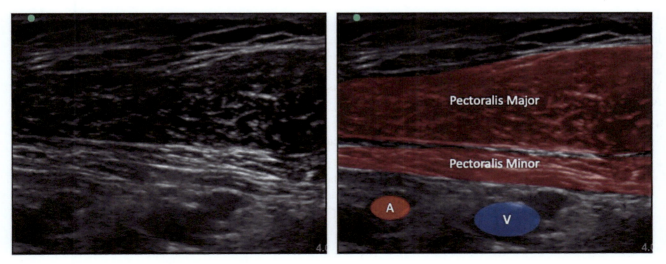

FIG. 2 Ultrasound anatomy of pectoral block type II (PECs II).

- *Ultrasound-assisted block*
  The coracoid process is found using a linear ultrasound probe in the paramedian sagittal plane, and the location of the pectoral branch of the thoracoacromial artery is determined between the pectoralis major and minor muscles with the color Doppler. (Figs. 1 and 2) The opening of the space between the pectoralis major and minor muscles by hydrodissection is the confirmation of the correct interfascial plane (PECs I) (Blanco, Parras, McDonnell, & Prats-Galino, 2013). The probe is placed in the mid-clavicular line and angled inferolateral to visualize the axillary artery, axillary vein, and second rib. The probe is then moved laterally until the minor pectoralis muscle, and serratus anterior muscle, third and fourth ribs are identified. The location is confirmed to be in the correct interfascial plane by hydrodissection (PECs II) (Blanco et al., 2012).

- *Injectate*
  Currently, there is no definitive recommendation based on available evidence, but the recommended volume in studies is 0.2 mL/kg of a long-acting local anesthetic (Battista & Krishnan, 2020).

Complications

Complications are rare as the procedure is performed under ultrasound guidance. Common complications a as in other regional anesthesia techniques; infection, vascular puncture, local anesthetic systemic toxicity (LAST), allergy, pneumothorax, and failed block.

Serratus anterior plane block (SAPB)

Introduction

Serratus plane block is defined as an ultrasound-guided block that can provide complete paresthesia in a single hemithorax. It is safe, effective, and easy to apply, with low risk of side effects. It is applied in a more lateral and posterior position than PECs blocks (Blanco et al., 2013).

Indications

- Rib fractures and contusions
- Posttraumatic intercostal neuralgia
- Thoracoscopic surgery and thoracotomy
- Minimally invasive repair of pectus excavatum
- Breast surgery and postmastectomy pain syndrome
- Laparotomy in the upper abdomen

Technique

In this block, which is usually performed in the lateral, supine, or sitting decubitus, the injectate is applied superficially or deeply to the serratus anterior muscle. The duration of the block differences according the injection site. SAPB provides analgesia to the anterolateral chest wall throughout T2–T9 with a single injection (Chakraborty, Khemka, & Datta, 2016). When analgesia is required in both hemithorax, such as pectus surgery, this block can be applied bilaterally (Tore Altun, Arslantas, Corman Dincer, & Aykac, 2019).

- *Landmark*
  In the midaxillary line, the thoracodorsal artery is between the latissimus dorsi and serratus anterior muscles.

- *Ultrasound-assisted block*
  The transducer is placed in the midaxillary line where the latissimus dorsi muscle appears more prominently. The thoracodorsal artery is visualized between the latissimus dorsi and serratus anterior muscles. Both in-plane and out-of-plane approaches are suitable. The correct inter-surface plane position is verified by hydrodissection (Chakraborty et al., 2016) (Fig. 3).

- *Injectate*

  A long-acting local anesthetic is injectated at a dose of 0.2–0.4 mL/kg (Southgate & Herbst, 2020).

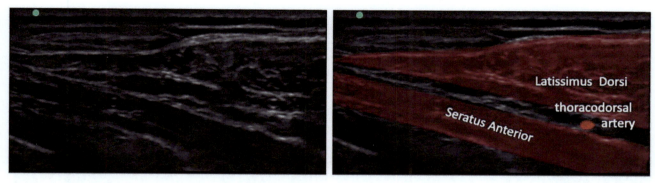

FIG. 3 Ultrasound anatomy of serratus anterior plane block (SAPB).

Complications

Depending on the duration of action of the local anesthetic used, rebound pain may be encountered. The use of dilute anesthetic is recommended to avoid LAST.

Pneumothorax is a potential complication, especially since the pleural line can be clearly visualized on ultrasound in this block, if a pneumothorax is suspected, ultrasound can help confirm the shift of the lung immediately after the procedure.

Nerve injury is unlikely as the needle is applied not directly to the nerves, but to the plane through which the nerves pass (Southgate & Herbst, 2020).

Paravertebral block (PVB)

Introduction

PVB is the application of local anesthetic around the place where the spinal nerves exit the intervertebral foramen. It can be applied at the cervical, thoracic, and lumbar levels, but only thoracic paravertebral block will be discussed in this section. Analgesia, covering 4–6 dermatomes, can be provided by a single-site injection.

Indications

- Breast surgery
- Pectus surgery
- Herniorrhaphy
- Thoracotomy
- VATS
- Thoracoabdominal esophageal surgery
- Fractured ribs
- Lung contusions
- Chest wound exploration
- Minimally invasive cardiac surgery
- Conventional cardiac surgery
- Treatment of hyperhydrosis

Technique

- Landmark
- Block-level is chosen according to the surgery to be applied. Thoracic PVB can be applied in sitting, lateral, or prone position, especially T3–T7. Block should be performed at the T4 level for sternum surgery, T6 for thoracic surgery, and T10 for abdominal procedures. For successful pain relief determination and marking the appropriate level of the spinous process from which the block will be made is essential.

 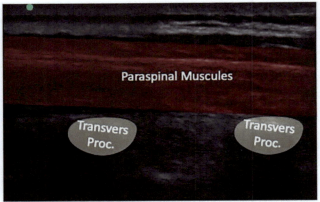

FIG. 4 Ultrasound anatomy of paravertebral block (PVB).

- Ultrasound-assisted block
  Several approaches have been described. The borders of the paravertebral space containing the nerves, pleura, and other tissues are visualized and the vascular structures can be identified with color Doppler. Both para-sagittal and transverse approaches have been described for thoracic PVB application. The needle can be placed in-plane or out-of-plane in one of the two approaches (D'Ercole, Arora, & Kumar, 2018). *Parasagittal approach* with ultrasound guidance; the transducer is placed at a distance of approximately 2.5 cm from the vertical plane of the spinous process. In this approach, since the neural foramina are approached vertically, the spread of local anesthesia can be observed under direct visualization, resulting in less epidural dissemination and systemic complications (Hara, Sakura, Nomura, & Saito, 2009). *Transverse approach* with ultrasound guidance; the ultrasound probe is placed next to the targeted level of spinous process in axial or transverse plane. The transverse out-of-plane approach is similar to the anatomical landmark-based technique. Although it may be easier to visualize the needle tip with ultrasound with this approach, the incidence of epidural dissemination or involuntary intrathecal injection is high (D'Ercole et al., 2018) (Fig. 4).

- Injectate
  Local anesthetic concentration and dosage regimens for PVB vary between institutions. For unilateral block, approximately 2 mL of local anesthetic per dermatome level, i.e., a single injection volume of 10 mL in total in an average of five dermatomes, is required. Since the spread of local anesthetics is unpredictable, it is also recommended to use a 3–5 mL multiinjection technique in each thoracic vertebra or in injections above and below dermatomal levels to block alternative levels. The dose of local anesthetic should be adjusted in the elderly and in cases of cachexia. Since the vascularization of the block area is quite high, the local anesthetic enters the systemic circulation rapidly, peak plasma concentration is reached quickly. Therefore, epinephrine (2.5–5.0 μg/mL) can be used during the first injection as it reduces systemic absorption and hence the potential for toxicity.

Complications

Complications after ultrasound-guided TPVB are rare; vascular puncture, hypotension, and pneumothorax may be encountered. Since sympathetic block is unilateral and segmental, the risk of hypotension is lower than thoracic epidural block. In addition, complications such as respiratory depression, urinary retention, incomplete or unsuccessful block, LAST and rarely permanent neurological damage may be encountered (D'Ercole et al., 2018).

Intercostal block

Introduction

Intercostal nerve blocks for chest and upper abdominal analgesia are useful and are easy to apply, also have a low incidence of complications. Intercostal nerves provide sensory innervation to the trunk and upper abdomen as well as to the intercostal muscles. Intercostal nerves originate from vertebral spinal nerve roots that are at the same level as the adjacent ribs. Unlike other spinal nerves, intercostal nerves follow an autonomous course and do not form a plexus (Lopez-Rincon & Kumar, 2020b).

Indications

- Breast surgery
- Rib and Sternal fractures
- Chest wall surgery and tumors
- Chest tubes, thoracic trauma
- Incisional pain from thoracic surgery

Technique

- Landmark

    Application of local anesthetic on the parietal pleura in the intercostal space can be defined as an intercostal nerve block. An ipsilateral block is expected at the blocked intercostal levels, and it is rare for the block to extend to the upper or lower levels. It is applied in prone, sitting, or lateral position with the arms pulling the scapula sideways. It is usually performed 6–8 cm lateral to spinous procedures where palpation is easy and reduces the possibility of pleural puncture (Baxter, Singh, & Fitzgerald, 2020).

- Ultrasound-assisted block

    In this block, the risk of intravascular injection and pneumothorax can be reduced with ultrasound guidance. In addition, the success of the block increases as the injection is made closer to the midline with the use of ultrasound. The ultrasound probe is placed 4 cm next to the spinous process in the sagittal plane. Ribs, pleura, and lungs are visualized in the intercostal space. Ribs appear hypoechoic and are defined by the intercostal muscles between them. While injecting the local anesthetic the pleura is pushed out. Intercostal nerves can be blocked using an in-plane or out-of-plane approach. It is recommended to insert the needle through upper edge of the rib below during the block, as it reduces the possibility of complications (Bhatia, Gofeld, Ganapathy, Hanlon, & Johnson, 2013; Chakraborty et al., 2016) (Fig. 5).

- Injectate

    Once the needle has been placed under ultrasound guidance and controlled by negative aspiration, a dose of 0.1 to 0.15 mL/kg of the local anesthetic agent can be administered per intercostal space. The maximum dose should not exceed 2–3 mL for each applied area (Lopez-Rincon & Kumar, 2020). In this block, a higher level of blood local anesthesia is encountered than many other regional anesthesia procedures. Therefore, the dose safety margin should be carefully calculated, especially in young children, elderly, and debilitated patients, and patients with organ failure.

Complications

While applying the block to an awake patient, symptoms of pneumothorax or intraneural injection can be inspected. However, the warning symptoms of these complications may not be noticed in patients under sedation or general anesthesia. Complications under ultrasound guidance are rare, but practitioners need to be able to recognize and treat

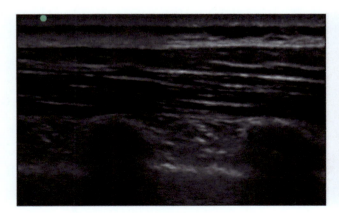 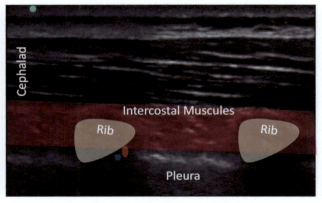

FIG. 5 Ultrasound anatomy of intercostal block.

pneumothorax and LAST. There are case reports of the spinal block after intercostal nerve block. Negative aspiration before the injection is recommended to prevent intravascular and intrapleural injection as well as intrathecal injection, but this application does not prevent complications. In order to eliminate these complications, patients should be followed in the block application area for approximately 30 min after the block is applied (Baxter et al., 2020; Chaudhri, Macfie, & Kirk, 2009).

Retrolaminar block (RLB)

Introduction

Visualization of the ultrasound-guided retrolaminar block (RLB) is relatively easy as compared to paravertebral techniques and is performed in a superficial tissue plane farther from the pleura. It is designed to reduce the risk of injury to the pleura and deep paravertebral structures (Pfeiffer et al., 2006; Voscopoulos et al., 2013).

Indications

As it is an alternative to PVB intervention, the application indications of this block are like PVB.

Technique

- Landmark
  RLB intervention is considered to be anatomically safer than PVB intervention. The risk of vascular or neural damage is much lower. The needle is inserted 1 cm lateral to the spinous process along the parasagittal plane. It can be preferred for pediatric patients and patients who are unconscious or under anesthesia (Murouchi & Yamakage, 2016).

- Ultrasound-assisted block
  The recommended ultrasound transducer placement for this block is the sagittal paramedian. The use of the probe in the transverse plane can also be applied, but the risk of epidural injection is higher with this application. In-plane visualization of the needle is important. Scanning with ultrasonography is carried out in the paramedian sagittal plane, at the level of the ribs, 5–6 cm lateral to the spinous processes of the corresponding level. Spinal laminas are visualized by advancing the probe toward the midline. It spreads between the lamina and deep paraspinous muscles during local anesthetic injection (Voscopoulos et al., 2013) (Fig. 6).

- Injectate
  Studies have shown that the distribution of injected local anesthetic from the retrolaminar plane to the paravertebral area is volume dependent. Although there are no definitive recommendations regarding the dose to be used, case series have shown that high volume local anesthetics (approx. 20 mL solutions) should be used for regional anesthesia and

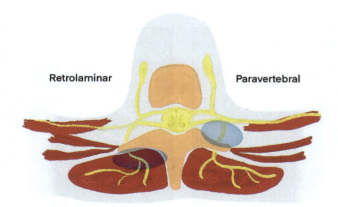

FIG. 6 Illustration of the retrolaminar and paravertebral blocks.

analgesia. Consequently, the amount of local anesthetic applied for RLB should be used in higher volumes for blockage compatible with the PVB area (Damjanovska, Stopar Pintaric, Cvetko, & Vlassakov, 2018; Murouchi & Yamakage, 2016).

Complications

RLB intervention is considered anatomically safe compared to PVB and is a technique that requires less experience and skill. There are no large vessels or nerves in the needle path. Although very low under ultrasound guidance, the only risk may be bleeding. However, this bleeding is intramuscular and does not pose a problem. There is little risk of nerve root damage or intrathecal and epidural injection. When the RLB technique is applied correctly, the risk of pneumothorax is minimized or even eliminated (Pfeiffer et al., 2006; Voscopoulos et al., 2013).

Erector spinae plane block (ESPB)

Introduction

The ESPB is an ultrasound-guided technique in which local anesthesia is injected into the facial plane between the ends of the transverse processes and the erector spine muscle.

With cadaver and imaging studies, in the ESPB, it was observed that the local anesthetic spread to the paravertebral and epidural spaces and to the cutaneous branches of intercostal nerves. With a single injection point, spreading to 3–5 segments can be observed in the epidural space (El-Boghdadly & Pawa, 2017; Forero, Adhikary, Lopez, Tsui, & Chin, 2016).

Indications

After the introduction of ultrasound-guided ESPB, it has been used in the treatment of acute and chronic pain. It has been used for postoperative analgesia in many areas from shoulder surgery to hip surgery, and its effectiveness has been demonstrated by research. There is a lot of ongoing research on the areas where this technique can be used (Krishnan & Cascella, 2020; Tulgar, Selvi, Senturk, Serifsoy, & Thomas, 2019).

- Nuss procedure
- Thoracotomies
- The ESP block can be used for a wide variety of surgical procedures in the anterior, posterior and lateral thoracic and abdominal areas. Since it is a newly applied block, most of the indications are based on case series (Krishnan & Cascella, 2020).

Technique

- Landmark
  The erector spinae muscle is a group of muscles that includes the iliocostalis, longissimus, and spinalis muscles. It extends bilaterally from the skull to the sacral region and one of its main functions is to stabilize the spine. The injection site should be determined by taking transverse processes as the center, paying attention to the craniocaudal spread of the local anesthetic.

- Ultrasound-assisted block
  ESPB is usually administered under in-plane ultrasound guidance. The probe is placed parasagittal and rotated 90 degrees onto the transverse process. Trapezius, rhomboid major, and spinous muscles are visualized at the level of the 5th thoracic vertebra. The needle is applied with an in-plane approach. It can be applied craniocaudal or vice versa depending on the area to be blocked. The lower plane of the erector spina muscle should be displayed with the hydro-dissection technique. A multi dermatomal block is created in the posterior, lateral, and anterior thoracic and abdominal walls (Kot et al., 2019; Krishnan & Cascella, 2020) (Fig. 7).
- Injectate
  Studies have shown different results regarding local anesthetic volume. There is a study found by averaging the different local anesthetic volumes used for this block and saying that 3.4 mL is sufficient for each segment. It has also been shown

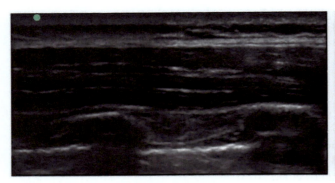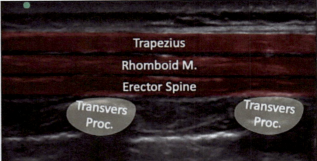

FIG. 7 Ultrasound anatomy of erector spinae plane block (ESPB).

that 20 mL of local anesthetic can spread between 3 and 7 levels. Local anesthesia in a volume of 20–30 mL is considered to be sufficient for the thoracic region. Block success will increase with high volume (De Cassai & Tonetti, 2018; Krishnan & Cascella, 2020).

Complications

Since it is a newly applied block, reported complications are very few. However, it is a block with a low probability of complications because it is performed under ultrasound guidance and the injection site is far from the pleura, main vascular structures, and spinal cord. Infection, LAST and allergy, vascular puncture, pneumothorax, and failed block may be encountered at the intervention site. When the ongoing research on this technique is concluded, data on its complications will also increase (De Cassai et al., 2019).

Truncal blocks contraindications

- Absolute contraindications: infection or malignancy at the injection site, allergy to local anesthetic medications, and patient rejection.
- Relative contraindications: include anatomical variations that make sonographic visualization of landmarks difficult. These may be scarring and fibrosis due to previous surgery.
- Anticoagulation may be a relative contraindication for these blocks.

Truncal blocks postprocedure checks

If you are concerned about pneumothorax or vascular puncture, the postprocedure should be checked by ultrasound (or by chest X-ray).

Block-level can be evaluated by pin-prick test or ice.

Equipment

- USG machine with high-frequency linear probe (5–10 MHz) (6–13 MHz)
- Sterile gloves and gowns (and gels)
- Marking pen
- Antimicrobial skin cleanser (2% chlorhexidine gluconate)
- Peripheral nerve block 20–22 gauge, 4–10 cm needle
- Local anesthetics
- Epidural or peripheric nerve catheter (optional)
- Sterile gauze, sterile syringe, and needle
- Standard American Society of Anesthesiologists (ASA) applied monitors: ECG 3 leads, HR, BP, Pulse oximeter and if possible nasal cannula with $EtCO_2$ monitoring

- Resuscitation equipment and medications (oxygen supply and masks, laryngoscopes, bag-mask, suction, defibrillator, epinephrine, phenylephrine, atropine, etc.)
- Lipid emulsion is available in 20% (intravenous lipid emulsion is an accepted treatment for severe LAST

Conclusion

Truncal nerve blocks for analgesia after surgeries in the thoracic region, such as breast and chest wall surgery, have developed with the introduction of ultrasound technology in recent years. There are currently mostly reports and retrospective clinical studies on the usefulness of these ultrasound-guided interventional analgesia modalities. Truncal blocks effectively provide analgesia in the chest wall and abdomen combined with other analgesic methods, although they may not be sufficient alone. The indications, usage areas, efficacy, applied techniques, and drug doses of these blocks are not yet sufficient. However, the available evidence is promising. Data from upcoming randomized controlled trials will be required to ensure the analgesic benefit, indications, and safety of thoracic plane blocks. Future works will assist anesthesiologists with best practice techniques and strategies by providing powerful analgesia tools in acute and chronic pain settings.

Application to other areas

Ultrasound-guided interfascial plane blocks are a new development in modern regional anesthesia research and practice and represent a new transmission route for local anesthesia to various anatomical sites. Although widely used in thoracic surgery, these blocks can also be used in abdominal surgery. For example, the ESP block can be used as an alternative to epidural anesthesia for operations on the abdominal wall such as laparotomy, appendectomy, laparoscopic surgery, abdominoplasty, cesarean delivery.

Although truncal plane blocks are generally used in treating acute postoperative pain in the chest and abdomen surgery, they can also effectively treat pain due to some other causes (e.g., chronic pains, rib fractures).

Other agents of interest

The midpoint transverse process pleural block (MTPB), described much more recently than the blocks mentioned, is actually a variant of the thoracic paravertebral block. Techniques for thoracic paravertebral block application under ultrasound guidance can also be used for MTPB. With the same technique, MTPB can be injected in front of the costotransverse ligament instead of the paravertebral space. It has been shown with this technique that local anesthesia spreads into the paravertebral space (Costache, 2019; Scimia, Fusco, Droghetti, Harizaj, & Basso Ricci, 2018).

Rhomboid intercostal and serratus plane (RISS) block is a newly described truncal block. RISS is a plane bounded medially by the erector spinae muscle and laterally by the serratus anterior muscle. This field block may be advantageous for thoracic surgery, as it can be applied from an area far from many surgical incisions compared to the SAPB or PECs blocks. It is an easy block to apply under ultrasound guidance (Elsharkawy et al., 2018).

Mini-dictionary of terms

Multimodal analgesia: Multimodal analgesia can be defined as the combined application of drugs or methods acting with different mechanisms to provide effective analgesia.

ERAS: Enhanced Recovery After Surgery; It includes multimodal, perioperative interventions to improve postoperative outcomes.

Video-assisted thoracoscopy: Video-assisted thoracoscopic surgery (VATS) is a surgery performed through a small incision between the ribs with the help of a camera called a thoracoscope. It allows seeing the entire chest cavity without opening the chest or ribs.

Neuraxial block: Neuraxial anesthesia is regional anesthesia methods defined as the application of local anesthetic around the nerves of the central nervous system, including spinal, epidural, combined spinal-epidural, and caudal epidural injections.

Truncal block: Truncal nerve blocks involve local anesthetic infiltration into facial planes around trunk nerves. It can be used alone or as a component of multimodal analgesia.

Dermatome: It is the area of skin innervated by the afferent nerves of the dorsal root of a single spinal nerve.

402 PART | III Interlinking anesthesia, analgesics and pain

Plane block: It is the technique of applying local anesthetic to the muscle planes where the nerve is located, not the sheath and itself, which has emerged with developments in ultrasound-guided block procedures. Nerve injury and other complications are less common and easier to perform. The spread of local anesthesia in the desired area can be monitored during the procedure.

Key facts

Key facts of truncal blocks in thoracic surgery
- Trunk blocks may be preferred instead of epidural analgesia for thoracic surgery, trauma or chronic pain.
- Block preference may vary depending on the patient, the surgery and the incision location, the ultrasound probe-needle placement area, the patient's position, technical possibilities, and the experience of the practitioner.
- Truncal blocks are part of multimodal analgesia. Multimodal analgesia increases patient comfort and satisfaction. It is the cause of early discharge and increase in quality of life.
- Although it varies according to the local anesthetic and volume used with the single injection method, the maximum effect time is approximately 12–18 h. For a longer effect, a catheter can be placed in the area with similar techniques in these blocks.
- One of the most important parameters for field blocks is the volume of local anesthetic used. However, the systemic LAST risk should not be forgotten. Care should be taken in dose calculation and adding epinephrine to the local anesthetic.
- It is also important to prepare areas of block application for resuscitation and general anesthesia, to have equipment and experienced general practitioners to treat complications that may be encountered in these areas.

Summary points

- This chapter gives information about the commonly used truncal blocks in thoracic surgeries.
- Altered lung functions are seen in most of the thoracic surgery patients.
- Multimodal analgesia techniques provide better outcome than systematically applied opioids.
- In truncal blocks less complication and more efficient pain relief is seen with ultrasound guidance.
- The determination of truncal block to be applied depends on the surgery and the provider's skill and education.

References

Battista, C., & Krishnan, S. (2020). *Pectoralis nerve block*. StatPearls. https://www.ncbi.nlm.nih.gov/pubmed/31613471.

Baxter, C. S., Singh, A., & Fitzgerald, B. M. (2020). *Intercostal nerve block*. StatPearls. https://www.ncbi.nlm.nih.gov/pubmed/29489198.

Bhatia, A., Gofeld, M., Ganapathy, S., Hanlon, J., & Johnson, M. (2013). Comparison of anatomic landmarks and ultrasound guidance for intercostal nerve injections in cadavers. *Regional Anesthesia and Pain Medicine, 38*(6), 503–507. https://doi.org/10.1097/AAP.0000000000000006.

Blanco, R. (2011). The 'pecs block': A novel technique for providing analgesia after breast surgery. *Anaesthesia, 66*(9), 847–848. https://doi.org/10.1111/j.1365-2044.2011.06838.x.

Blanco, R., Fajardo, M., & Parras Maldonado, T. (2012). Ultrasound description of pecs II (modified pecs I): A novel approach to breast surgery. *Revista Española de Anestesiología y Reanimación, 59*(9), 470–475. https://doi.org/10.1016/j.redar.2012.07.003.

Blanco, R., Parras, T., McDonnell, J. G., & Prats-Galino, A. (2013). Serratus plane block: A novel ultrasound-guided thoracic wall nerve block. *Anaesthesia, 68*(11), 1107–1113. https://doi.org/10.1111/anae.12344.

Chakraborty, A., Khemka, R., & Datta, T. (2016). Ultrasound-guided truncal blocks: A new frontier in regional anaesthesia. *Indian Journal of Anaesthesia, 60*(10), 703–711. https://doi.org/10.4103/0019-5049.191665.

Chaudhri, B. B., Macfie, A., & Kirk, A. J. (2009). Inadvertent total spinal anesthesia after intercostal nerve block placement during lung resection. *The Annals of Thoracic Surgery, 88*(1), 283–284. https://doi.org/10.1016/j.athoracsur.2008.09.070.

Chou, R., Gordon, D. B., de Leon-Casasola, O. A., Rosenberg, J. M., Bickler, S., Brennan, T., et al. (2016, Feb). Management of Postoperative Pain: A clinical practice guideline from the American pain society, the American Society of Regional Anesthesia and Pain Medicine, and the American Society of Anesthesiologists' Committee on Regional Anesthesia, Executive Committee, and Administrative Council. *The Journal of Pain, 17*(2), 131–157. https://doi.org/10.1016/j.jpain.2015.12.008.

Costache, I. (2019). Mid-point transverse process to pleura block for surgical anaesthesia. *Anaesthesia Reports, 7*(1), 1–3. https://doi.org/10.1002/anr3.12003.

Damjanovska, M., Stopar Pintaric, T., Cvetko, E., & Vlassakov, K. (2018). The ultrasound-guided retrolaminar block: Volume-dependent injectate distribution. *Journal of Pain Research, 11*, 293–299. https://doi.org/10.2147/JPR.S153660.

De Cassai, A., Bonvicini, D., Correale, C., Sandei, L., Tulgar, S., & Tonetti, T. (2019). Erector spinae plane block: A systematic qualitative review. *Minerva Anestesiologica, 85*(3), 308–319. https://doi.org/10.23736/S0375-9393.18.13341-4.

De Cassai, A., & Tonetti, T. (2018, Aug). Local anesthetic spread during erector spinae plane block. *Journal of Clinical Anesthesia, 48*, 60–61. https://doi.org/10.1016/j.jclinane.2018.05.003.

D'Ercole, F., Arora, H., & Kumar, P. A. (2018). Paravertebral block for thoracic surgery. *Journal of Cardiothoracic and Vascular Anesthesia, 32*(2), 915–927. https://doi.org/10.1053/j.jvca.2017.10.003.

El-Boghdadly, K., & Pawa, A. (2017). The erector spinae plane block: Plane and simple. *Anaesthesia, 72*(4), 434–438. https://doi.org/10.1111/anae.13830.

Elmore, B., Nguyen, V., Blank, R., Yount, K., & Lau, C. (2015). Pain management following thoracic surgery. *Thoracic Surgery Clinics, 25*(4), 393–409. https://doi.org/10.1016/j.thorsurg.2015.07.005.

Elsharkawy, H., Maniker, R., Bolash, R., Kalasbail, P., Drake, R. L., & Elkassabany, N. (2018). Rhomboid intercostal and subserratus plane block: A cadaveric and clinical evaluation. *Regional Anesthesia and Pain Medicine, 43*(7), 745–751. https://doi.org/10.1097/AAP.0000000000000824.

Forero, M., Adhikary, S. D., Lopez, H., Tsui, C., & Chin, K. J. (2016). The erector spinae plane block: A novel analgesic technique in thoracic neuropathic pain. *Regional Anesthesia and Pain Medicine, 41*(5), 621–627. https://doi.org/10.1097/AAP.0000000000000451.

Hara, K., Sakura, S., Nomura, T., & Saito, Y. (2009). Ultrasound guided thoracic paravertebral block in breast surgery. *Anaesthesia, 64*(2), 223–225. https://doi.org/10.1111/j.1365-2044.2008.05843.x.

Kehlet, H., Jensen, T. S., & Woolf, C. J. (2006). Persistent postsurgical pain: Risk factors and prevention. *The Lancet, 367*(9522), 1618–1625. https://doi.org/10.1016/s0140-6736(06)68700-x.

Kot, P., Rodriguez, P., Granell, M., Cano, B., Rovira, L., Morales, J., et al. (2019). The erector spinae plane block: A narrative review. *Korean Journal of Anesthesiology, 72*(3), 209–220. https://doi.org/10.4097/kja.d.19.00012.

Krishnan, S., & Cascella, M. (2020). *Erector spinae plane block.* StatPearls. https://www.ncbi.nlm.nih.gov/pubmed/31424889.

Lopez-Rincon, R. M., & Kumar, V. (2020). *Ultrasound-guided intercostal nerve block.* StatPearls. https://www.ncbi.nlm.nih.gov/pubmed/32310360.

Murouchi, T., & Yamakage, M. (2016, Dec). Retrolaminar block: Analgesic efficacy and safety evaluation. *Journal of Anesthesia, 30*(6), 1003–1007. https://doi.org/10.1007/s00540-016-2230-1.

Pfeiffer, G., Oppitz, N., Schone, S., Richter-Heine, I., Hohne, M., & Koltermann, C. (2006, Apr). Analgesia of the axilla using a paravertebral catheter in the lamina technique. *Anaesthesist, 55*(4), 423–427. https://doi.org/10.1007/s00101-005-0969-0. Analgesie der Achselhohle durch Paravertebralkatheter in Laminatechnik.

Scimia, P., Fusco, P., Droghetti, A., Harizaj, F., & Basso Ricci, E. (2018, Jun). The ultrasound-guided mid-point transverse process to pleura block for postoperative analgesia in video-assisted thoracoscopic surgery. *Minerva Anestesiologica, 84*(6), 767–768. https://doi.org/10.23736/S0375-9393.18.12485-0.

Southgate, S. J., & Herbst, M. K. (2020). *Ultrasound guided serratus anterior blocks.* StatPearls. https://www.ncbi.nlm.nih.gov/pubmed/30860711.

Sungur, M. O., & Şentürk, M. (2017). Pain management following thoracic surgery. In *Postoperative care in thoracic surgery* (pp. 243–257). Cham: Springer.

Thompson, C., French, D. G., & Costache, I. (2018). Pain management within an enhanced recovery program after thoracic surgery. *Journal of Thoracic Disease, 10*(Suppl. 32), S3773–S3780. https://doi.org/10.21037/jtd.2018.09.112.

Tore Altun, G., Arslantas, M. K., Corman Dincer, P., & Aykac, Z. Z. (2019). Ultrasound-guided serratus anterior plane block for pain management following minimally invasive repair of pectus excavatum. *Journal of Cardiothoracic and Vascular Anesthesia, 33*(9), 2487–2491. https://doi.org/10.1053/j.jvca.2019.03.063.

Tulgar, S., Selvi, O., Senturk, O., Serifsoy, T. E., & Thomas, D. T. (2019). Ultrasound-guided erector spinae plane block: Indications, complications, and effects on acute and chronic pain based on a single-center experience. *Cureus, 11*(1). https://doi.org/10.7759/cureus.3815, e3815.

Urits, I., Ostling, P. S., Novitch, M. B., Burns, J. C., Charipova, K., Gress, K. L., et al. (2019). Truncal regional nerve blocks in clinical anesthesia practice. *Best Practice & Research. Clinical Anaesthesiology, 33*(4), 559–571. https://doi.org/10.1016/j.bpa.2019.07.013.

Voscopoulos, C., Palaniappan, D., Zeballos, J., Ko, H., Janfaza, D., & Vlassakov, K. (2013). The ultrasound-guided retrolaminar block. *Canadian Journal of Anaesthesia, 60*(9), 888–895. https://doi.org/10.1007/s12630-013-9983-x.

Young, A., & Buvanendran, A. (2012). Recent advances in multimodal analgesia. *Anesthesiology Clinics, 30*(1), 91–100. https://doi.org/10.1016/j.anclin.2011.12.002.

Chapter 37

Postoperative pain management: Truncal blocks in general surgery

Gulbin Tore Altun

Memorial Health Group, Department of Anesthesiology and Reanimation, Istanbul, Turkey

Abbreviations

ASA	American Society of Anesthesiologists
ASIS	anterior superior iliac spine
BP	blood pressure
ECG	electrocardiogram
ERAS	enhanced recovery after surgery
ESPB	erector spinae plane block
EtCO$_2$	end-tidal carbon dioxide
HR	heart rate
IH	iliohypogastric
II	ilioinguinal
LAST	local anesthetic systemic toxicity
PVB	paravertebral block
QL	quadratus lumborum
QLB	quadratus lumborum block
TAPB	transversus abdominis plane block
TFPB	transversalis fascia plane block
TLF	thoracolumbar fascia
US	ultrasound

Introduction

Acute postoperative pain, which is not being managed effectively, causes many complications and related morbidity-mortality. It can cause pain, atelectasis, and pneumonia. It may also cause tachycardia, hypertension, arrhythmia, myocardial ischemia and may even cause thromboembolic complications as it restricts mobilization. It may also cause decreased motility in the bladder and gastrointestinal system and deterioration of blood sugar regulation. All these complications cause prolonged hospital stay, rehospitalization, and delay in recovery. Minimally invasive surgical techniques and multimodal acute postoperative pain management should be used together to avoid these results (Kehlet, 2018). With the development of surgical techniques in recent years, abdominal surgeries have become more frequently performed laparoscopically. Surgery performed with a minimally invasive method has also brought together less invasive analgesic methods. Various abdominal plane blocks and nerve blocks added to the multimodal analgesia plan have been developed. Epidural analgesia, a neuraxial block, is still an essential component of multimodal analgesia after abdominal surgery and thoracic surgery. However, the increase in the patient population using anticoagulants has led to the search for new analgesic methods due to possible serious complications (Hemmerling, 2018).

The basis of analgesia after abdominal surgery is based on multimodal analgesia. The widespread use of ultrasonography in peripheral blocks and avoidance of epidural analgesia by considering the risk-benefit balance has increased abdominal body blocks. When used with nonopioid analgesics, the side effects are avoided by reducing the use of opioids. Since the local anesthetic agent's distribution is limited in regional analgesia techniques applied to the abdominal wall, unwanted hemodynamic effects will be minimal. These blocks, which provide abdominal wall analgesia, are essential components of multimodal analgesia. Enhanced recovery after surgery (ERAS) guidelines recommends multimodal pain

---

Features and Assessments of Pain, Anesthesia, and Analgesia. https://doi.org/10.1016/B978-0-12-818988-7.00042-X
Copyright © 2022 Elsevier Inc. All rights reserved.

management strategies. Multimodal analgesia aims to use analgesic agents with different mechanisms of action, to avoid opioids as much as possible, to use regional anesthesia, and to switch to oral analgesic agents as soon as possible. The main purpose of multimodal analgesia is to avoid side effects by taking advantage of the synergistic effects of analgesic drugs or techniques.

The most common side effects of opioids are nausea, vomiting, decreased bowel motility, and sedation. In the postoperative period, painful warnings may cause nausea-vomiting and ileus, or the main cause of gastrointestinal atony may be the surgery itself. With the choice of postoperative pain management after abdominal surgery, these gastrointestinal symptoms can be reduced or increased depending on the drug or technique used (de Boer, Detriche, & Forget, 2017).

Postoperative pain, which is not treated effectively and timely, negatively affects the recovery, quality of life, and the development of chronic pain after surgery (Chou et al., 2016).

Although the incidence of postoperative chronic pain varies according to many factors, especially the type of surgery, it is almost up to 50% (Bouman et al., 2014; Kehlet, Jensen, & Woolf, 2006). Chronic pain is common after abdominal surgery, especially inguinal hernia and cholecystectomy (Macrae, 2001). Treatment of chronic pain is difficult and costly. Therefore, it is essential to prevent chronic pain with effective acute postoperative pain management (Wylde et al., 2017).

Truncal blocks

Trunk nerve blocks of the abdominal wall are being used more frequently with the widespread use of ultrasound (US). The US is not utilized to visualize the nerves or the neural plexus as in peripheral nerve blocks. It is used to determine the muscle plane where local anesthesia will be injected and to visualize the anesthetic is spread along the fascial plane. When used as part of multimodal analgesia, it has analgesic effects similar to neuraxial techniques (Abrahams, Derby, & Horn, 2016). Also, they have a lower complication rate and are technically easier to apply and learn (Urits et al., 2019).

When the abdominal wall's anatomy is well known, truncal blocks are easier to understand and apply. The abdominal wall consists of three muscle layers and associated fascial sheaths.

US guidance enables determining the area and the correct placement of the needle and monitoring the spread of the local anesthetic. This guidance decreases the complication risk as the anesthetic spread can be controlled after block. In general, these blocks are used as single-shot injections, but if a longer duration of the analgesia is aimed, they can also be applied by inserting a catheter (Yarwood & Berrill, 2010).

Transversus abdominis plane block

Introduction

The transversus abdominis plane block (TAPB) has been one of the most frequently used truncal blocks since it was described in 2001 to obtain a field block through the Petit triangle (Rafi, 2001). With the development of ultrasound technology, TAPBs have become safer and technically easier to apply. Although it is only effective on somatic pain, it provides sufficient analgesia as part of multimodal treatment using an appropriate dose and volume of local anesthetic single shot or via a catheter.

TAPB can be applied unilaterally or bilaterally with different approaches under US guidance. TAPB can be applied to the same fascial plane from different regions, and there are four types of approaches: subcostal, lateral, posterior, and oblique subcostal. This nomenclature is based on the corresponding spinal nerves in the same facial area for clinical benefit (Tsai et al., 2017).

Indications

- Colorectal surgery
- Hernia repairs
- Appendectomy
- Cholecystectomy and colectomy
- Bariatric surgeries
- Nephrectomy
- Renal transplant
- Exploratory laparotomies and laparoscopic surgeries

Technique

The rectus abdominis, external oblique, internal oblique, and transverse abdominis muscles form the anterolateral abdominal wall (Tran, Bravo, Leurcharusmee, & Neal, 2019). The transversus abdominis plane compartment lies between the internal oblique and transversus abdominis muscles (Fig. 1). The T6-L1 thoracolumbar, branches of the lower thoracic intercostal and subcostal nerves, ilioinguinal, and iliohypogastric nerves are located within this plane (Mavarez & Ahmed, 2020). The nerves that travel in this compartment proceed to form plexuses with each other. Thus, anterolateral sensory anesthesia of the anterior abdominal wall can be achieved with the spread of local anesthetic applied to this area (McDonnell et al., 2007). During the posterior approach, except for slight lateralization, the patient is positioned supine for TAPB.

- *Landmark*
  The blind technique was used when TAPB was first defined. Two "pops" are felt as the needle passes through the two muscular fascias (outer and inner oblique) (Rafi, 2001). Later, when it is believed that the block is more effective and safe with US guidance, the blind technique became less used. (O'Donnell, McDonnell, & McShane, 2006).
- *Ultrasound-assisted block*
  It should be kept in mind that the choice of US probe may vary according to the patient (choices will be different for children, elderly, obese, and cachectic patients) during the application. The probe is placed at the xiphoid process level, and the rectus abdominis muscles and linea alba are visualized; the probe is advanced toward the side of the block at the lower costal border. For "Subcostal TAPB" application, a local anesthetic is injected into the area between the posterior sheath of the rectus muscle and the transversus abdominis muscle (Tsai et al., 2017).
  During the "Lateral TAPB" application, the probe is advanced downward through the rectus abdominis muscle and placed in a horizontal position on the midaxillary line, midway between the lower costal border and the iliac crest. Three muscle layers are visualized. Block is applied to the area between the transversus abdominis and the internal oblique muscle (Tsai et al., 2017).
  In the "Posterior approach," the point where the internal oblique and transversus abdominis muscles meet with the quadratus lumborum muscle fascia is displayed. This area is the most posterior point where the block can be applied (Lee, Barrington, Tran, Wong, & Hebbard, 2010).
  For TAPB, in-plane intervention is made 2 to 3 cm lateral to the probe in all approaches. While the needle is being advanced, the feeling of "pop" is taken as in the blind technique, and the place is confirmed with hydrodissection and by visualizing the needle tip. After the location is confirmed by the US, the distribution of the applied local anesthetic can also be visualized (Mavarez & Ahmed, 2020).

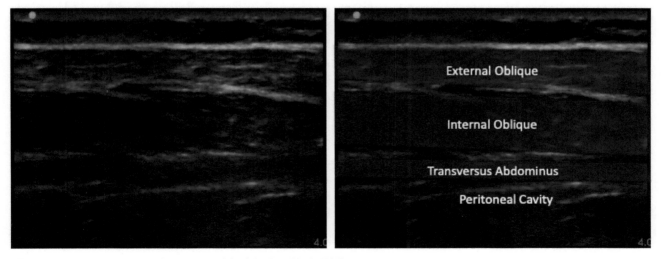

FIG. 1 The ultrasound anatomy of transversus abdominis plane block (TAP).

- *Injectate*
  The volume of a local anesthetic to be applied depends on the patient's weight and the concentration of the local anesthetic. High volume local anesthetic should be used, and the concentration should be calculated according to the patient, paying attention to the toxic dose. Various studies have shown that a local anesthetic volume of 15–20 mL is sufficient for one side. (Bharti, Kumar, Bala, & Gupta, 2011; Tran, Ivanusic, Hebbard, & Barrington, 2009).

Complications

Complications such as intestinal perforation, liver-spleen laceration, vascular injection and injury, intraabdominal-retroperitoneal hematoma, intraperitoneal and intrahepatic injection of local anesthesia may be encountered. In addition, the local anesthetic injected for TAPB can pass through the inguinal ligament, causing femoral nerve palsy. There is a risk of local infection, neurotoxicity, intramuscular injection and related myonecrosis, local anesthetic systemic toxicity (LAST), which can also be encountered as in all regional blocks. When administered under ultrasound guidance, complications of TAPB are rare. There is also a risk of the failed block due to individual anatomical variations (Khan, McAtamney, & Farling, 2010; Salaria, Kannan, Kerner, & Goldman, 2017; Yarwood & Berrill, 2010).

Rectus sheath block

Introduction

The rectus sheath block, when described quite a long time ago, was used to relax abdominal wall muscles during laparotomy. Today, it is used to prevent somatic pain of the structures remaining on the parietal peritoneum as a part of multimodal analgesia. Since it provides midline analgesia between the xiphoid process and the symphysis pubis, it should be preferred for surgeries using a midline median incision (Yarwood & Berrill, 2010).

The rectus abdominis muscle is surrounded by the rectus sheath, which forms approximately 2/3 of the anterior abdominal wall. The obliquus externus, located in the lateral part of the abdominal wall, is formed by the anterior union of the sheaths of the internus and transversus abdominis muscles. The midline of the muscle is marked with linea alba and on the lateral edges by linea semilunaris. The sheath is fixed in the front, free at the back and the lower and upper epigastric vessels pass through it (Ellis, 2009). The block is performed with local anesthetic injection into the space between the rectus muscle and its posterior sheath. (Fig. 2).

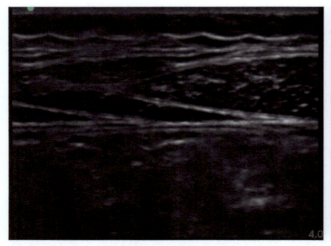
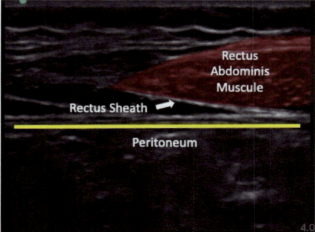

FIG. 2 The ultrasound anatomy of rectus sheath block.

Indications

- Midline incisions of the abdomen
- Umbilical and incisional hernia repair
- Open gastrectomy
- Surgical oncology cases
- Transplantation of the pancreas

Technique

The aim is to block the 9th–11th intercostal nerves that make the skin innervation around the umbilicus, which penetrate the posterior wall of the rectus abdominis. Sometimes the block may fail despite the injection in the right place due to anatomical variations (Willschke & Kettner, 2012).

In the blind, traditional technique, when the patient is in the supine position, a point is chosen 2–3 cm away from the midline based on the umbilicus. A "pop" should be felt as it passes the anterior sheath. When the posterior wall resistance is felt, the local anesthetic injection can be applied. It is repeated for the other side of the midline. Depth cannot be predicted since there is no relationship between the patient's age and weight and rectus sheath depth (Willschke et al., 2006; Yarwood & Berrill, 2010).

- *Landmark*
  The rectus sheath block is applied bilaterally in the medial of the linea semilunaris and just above the umbilicus (Chin et al., 2017).
- *Ultrasound-assisted block*
  In the US-guided block, the transducer is placed transversely just above the umbilicus to visualize the linea alba and rectus abdominis muscle. The hypoechoic rectus muscle and hyperechoic sheath are visualized by sliding the probe laterally, and the needle is inserted between them. There are both "in-plane" and "out-of-plane" approaches. The separation of the muscle and the sheath from each other with the injection is visualized. The pulsation of superior and inferior epigastric arteries is visualized (Chin, McDonnell, et al., 2017; Rucklidge & Beattie, 2018).
- *Injectate*
  A local anesthetic volume of 0.1–0.2 mL/kg is recommended for children for single side and a total volume of 10–20 mL for the adult patient (Chin, McDonnell, et al., 2017; Willschke et al., 2006).

Complications

As with other abdominal wall blocks, intraperitoneal injection, intestinal injury, vascular injury, and intravascular injection can be seen. LAST and failed blocks may also be encountered (Yarwood & Berrill, 2010).

Ilioinguinal nerve and iliohypogastric nerve blocks

Introduction

Blockade of the ilioinguinal (II) nerve and iliohypogastric (IH) nerves is a good analgesic option for operations in the groin area. It has been shown that they reduce the need for opioids and are as effective as the caudal block in pediatric patients (Hannallah, Broadman, Belman, Abramowitz, & Epstein, 1987; van Schoor, Boon, Bosenberg, Abrahams, & Meiring, 2005).

Indications

- Inguinal hernia repair
- Orchiopexy, hydrocele, and varicocele repair surgery
- Renal transplant (combination with T11 and T12 intercostal nerve blocks)

Technique

The II and IH nerves originate from the lumbar plexus and are branches of T12 and L1. They coarse between the internal and external oblique muscles (van Schoor et al., 2005). While applying the block, the patient is in the supine position.

- Landmark
  The injection is made on the line drawn between the anterior superior iliac spine (ASIS) and the umbilicus at a distance of approximately 2.5 cm to the medial of the ASIS. Fascial click or loss of resistance is felt (Willschke et al., 2005).
- Ultrasound-assisted block
  A high-frequency linear probe is needed for a US-guided block. The probe is placed along the line connecting the umbilicus with the ASIS. The peritoneum, transversus abdominis muscle, and internal oblique muscle are visualized. (Fig. 3) II and IH nerves can be visualized as hypoechoic structures between the internal oblique and transversus abdominis muscles (Ford et al., 2009; Willschke et al., 2005).
- Injectate
  Local anesthetics in a volume of 0.075–0.3 mL/kg can be used. The lowest dose was valid for the experienced practitioner and US-guided block, while the maximum dose was given for the blind technique (van Schoor et al., 2005; Willschke et al., 2005).

Complications

Complications such as failed block, bowel puncture, intestinal bleeding-hematoma, pelvic retroperitoneal hematoma, transient femoral nerve palsy, and quadriceps palsy may be encountered. LAST is another complication to be mentioned (Urits et al., 2019).

Paravertebral block

Introduction

In paravertebral block (PVB), the anesthesia of the plexuses or nerve roots is targeted as they exit the spinal canal. It can be applied unilaterally or bilaterally to the cervical, thoracic, lumbar, and sacral regions. In this section, paravertebral blocks that are suitable for general surgery operations will be mentioned (Boezaart, Lucas, & Elliott, 2009).

Indications

- Renal surgery
- Abdominoplasty
- Cholecystectomy
- Appendectomy
- Inguinal hernia repair
- Donor hepatectomy

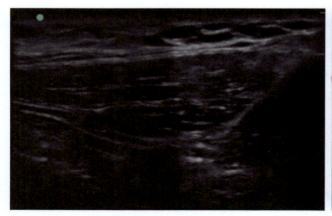
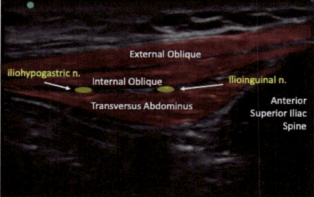

FIG. 3 The ultrasound anatomy of ilioinguinal nerve and iliohypogastric nerve block.

Technique

Although PVB has been studied more and its effect on thoracic surgery has been demonstrated, it is also a useful technique in abdominal surgery. The abdominal wall is innervated with T6–T12, and the block applied at these levels provides abdominal wall analgesia (Boezaart et al., 2009).

It can be applied awake or under sedation/general anesthesia. It can be applied in a sitting position if the patient is awake and in the lateral decubitus position, with the side to be operated on, if under anesthesia (Tighe, Greene, & Rajadurai, 2010).

- Landmark

  If the block effect is four dermatomal levels and less, PVB from the single level is sufficient. If the more dermatomal level spread is aimed, more than one injection area will provide a more reliable effect. When applied in the lumbar region, blockade at all levels with small volumes is recommended because there is no spread like in the thoracic region. At the desired level, approximately 2.5 cm lateral to the spinous processes, just above the transverse processes, is the right intervention site (Tighe et al., 2010).

- Ultrasound-assisted block

  The use of ultrasound offers the advantage of visualizing the boundaries and structures of the paravertebral region. The linear transducer (5–10 MHz) is placed vertically 2.5 cm lateral to the transverse process, and the paravertebral space is visualized from the sagittal paramedian. Transverse processes are observed as hypoechogenic. The pleura, on the other hand, is a bright structure extending deep from neighboring transverse processes (Bondar, Szucs, & Iohom, 2010).

  Ultrasound guidance has advantages over blind techniques: visualization of anatomical structures, ability to visualize needle, catheter, dissemination of local anesthetic, possibly shorter performance-onset time, longer block time, lower local anesthetic volume, lower failure and complication rate, less patient discomfort.

- Injectate

  There is no reliable relationship between the volume of injected local anesthetic agent and the extent of its spread. A local anesthetic is expected to extend in the craniocaudal direction but may also spread to the epidural and intercostal spaces (Tighe et al., 2010).

  Recommendations are made regarding the volume of a local anesthetic to be used based on clinical practice, radiology, and cadaver studies. It has been shown that somatic block in three dermatomes and sympathetic block in eight dermatomes occur with an injection of 15 mL local anesthetic agent from a single area (Cheema, Ilsley, Richardson, & Sabanathan, 1995).

  Another technique is to administer 3–5 mL of multiple injections into each vertebral level to ensure reliable and effective blockage. In children, 0.5 mL/kg volume is useful in an average of four dermatomes (Karmakar, 2001).

  When calculating the volume and concentration, care should be taken that the dose does not exceed the maximum recommended value.

Complications

The reported incidence of complications is extremely low and has decreased even more with US-guided practice. Specifically reported complications are; hypotension, vascular and pleural puncture, pneumothorax, failed block, LAST, dural puncture, and total spinal anesthesia. Ipsilateral Horner syndrome is also a possible complication. The epidural spread is an expected effect and is the cause of bilateral block seen despite unilateral administration (Coveney et al., 1998; Tighe et al., 2010).

Intercostal block

Introduction

Intercostal nerves can be blocked in thoracic surgical procedures as well as in upper abdominal surgeries. Intercostal nerves originate from the vertebral spinal nerve roots at the same level. They do not form plexuses and follow an autonomous course. This block is easy to implement but usually requires multiple injections (Karmakar & Ho, 2003; Lopez-Rincon & Kumar, 2020b).

412 PART | III Interlinking anesthesia, analgesics and pain

Indications

- Upper abdominal surgery

Technique

Thoracoabdominal intercostal nerves T7 through T11 are the nerves that innervate the abdominal wall. Intercostal block can be performed blindly or under ultrasound guidance using anatomical landmarks.

- Landmark
  The ultrasound-guided block can be applied from a point closer to the vertebral column and provides full dermatomal anesthesia as the nerve is blocked before branching. Since there is no sheath around the intercostal nerve branches, the accumulation of local anesthetic solution in the close environment is sufficient to block the nerve (Burns, Ben-David, Chelly, & Greensmith, 2008). This block can be applied when the patient is in the side, prone, or sitting positions.
- Ultrasound-assisted block
  The ultrasound probe is placed parasagittally 4 cm lateral to the spinous process. The rib is visualized hypoechoically, and pleura and lung tissue are observed deeply in the intercostal space. The needle entry should be at the lower edge of the rib. After negative aspiration, 3–5 mL of local anesthetic is injected. Meanwhile, it can be seen that the pleura is pushed.
- Injectate
  After the needle site is confirmed by ultrasound, it is recommended to administer 0.1–0.15 mL/kg of local anesthetic agent per intercostal space (2–3 mL per maximum interval). Care should be taken not to exceed the maximum dose of local anesthetic with multiple injections (Lopez-Rincon & Kumar, 2020).

Complications

Complications that may be encountered in other blocks, such as bleeding and infection can also happen with this block too. When the block is applied to an awake patient, it will be more pronounced in terms of pneumothorax or intraneural injection symptoms than in an anesthetized patient. Pneumothorax is a rare complication, and one must be prepared for the treatment attempt. LAST is an area that needs attention. The concentration of local anesthetic used must be well calculated. It is important for the physician to know LAST and its treatment. Dural puncture and injection cases have also been reported. Aspiration is highly recommended prior to injection to exclude intravascular, intradural, or intrapleural injection. Patients should be followed for 20–30 min after the block (Baxter, Singh, & Fitzgerald, 2020).

Quadratus lumborum block

Introduction

Quadratus lumborum block (QLB) has been described as a variation of the TAP block. It is used as a part of multimodal analgesia after abdominal surgery. It is a block in which new techniques continue to be produced due to the change in the best approach specific to the patient and procedure (Ueshima, Otake, & Lin, 2017).

Indications

- Exploratory laparotomy, laparoscopy
- Bowel resection, gastrostomy
- Cholecystectomy
- Renal surgery, including transplant
- Abdominoplasty

Technique

Thoracolumbar fascia (TLF) is the layer consisting of fascias surrounding the back muscles extending from the thoracic to the lumbar vertebra. TLF is divided into three layers around the posterior muscles: anterior, middle, and posterior. The front layer is located in front of the quadratus lumborum (QL) muscle, the middle layer is between the erector spinae muscle and the QL muscle, and the posterior layer is located around the erector spinae muscle (El-Boghdadly, Elsharkawy, Short, & Chin, 2016; Ueshima et al., 2017) (Fig. 4).

- Landmark

  With QLB, the distribution of local anesthetics is wide, resulting in a large area of sensory inhibition (T7-L1). There are four different techniques (QLB1, QLB2, QLB3, and QLB4) that are named according to the needle's position from different aspects of the QL muscle.

  QLB1 (lateral QLB) is applied posterolaterally to the fascia transversalis and adjacent QL muscle at the level where the aponeurosis of the transversus abdominis muscle is narrowed. QLB2 (posterior QLB) is a posteromedial approach applied between the posterior QL muscle and TLF, which separates the QL muscle from the latissimus dorsi and paraspinal muscles. QLB3 (anterior-transmuscular) involves applying a local anesthetic in front of the QL muscle in the transverse process of the L4 vertebra. Finally, QLB4 (intramuscular) is an injection of the QL muscle itself. These approaches can be best visualized using ultrasound, and a superior approach has not been identified among them (Dhanjal & Tonder, 2020; Hebbard, 2009).

- Ultrasound-assisted block

  This block is an "interfacial plane block" that can only be performed under ultrasound guidance and applied to the posterior abdominal wall (Dhanjal & Tonder, 2020).

  During the lateral QLB application, when the patient is in the supine or lateral position, the ultrasound probe placed between the costal edge and the iliac crest allows the abdominal muscle layers (external oblique, internal oblique, transversus abdominis), TLF, and back muscles (QL, psoas major, erector spinae, and latissimus dorsi) to be visualized. The needle is inserted using an in-plane approach and advances until it reaches the anterolateral edge of the QL. For anterior QLB, the patient must be in the lateral decubitus position. The probe is placed on the iliac crest in the midaxillary line. The "shamrock sign" is observed by moving the probe back. In the posterior QLB application, the patient is in the supine position. Identify the posterior border of the QL and place the needle tip at this point. For intramuscular QLB, the patient can be in a supine or lateral decubitus position. After the QL muscle is visualized, the needle is inserted directly into the muscle from anterolateral to posteromedial direction (Dhanjal & Tonder, 2020; Ueshima et al., 2017).

- Injectate

  In different approaches of the QL block, the spread of local anesthetic is related to both the needle path and the needle tip position, so a standard volume of local anesthetic has not been recommended. There are studies showing that a local anesthetic volume of 15–20 mL is sufficient. Concentration should be carefully calculated in terms of LAST.

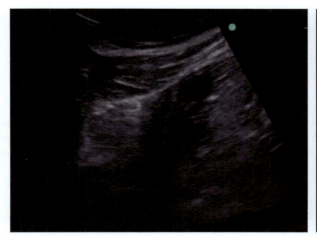 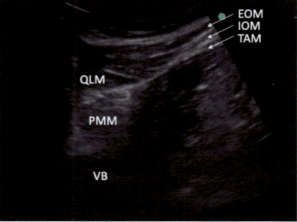

FIG. 4 The ultrasound anatomy of quadratus lumborum block (QLB). *EOM*: external oblique muscle, *IOM*: internal oblique muscle, *PMM*: psoas major muscle, *QLM*: quadratus lumborum muscle, *TAM*: transversus abdominis muscle, *VB*: vertebral body.

414 **PART | III** Interlinking anesthesia, analgesics and pain

Complications

It may cause motor block and delay in mobilization due to the spread of the local anesthetic agent to the lumbar plexus. Hypotension may be encountered, which may be related to local anesthetics spreading into the paravertebral space. LAST is a potential risk due to the vascularity of the area. The proximity of the quadratus lumborum block to the pleura and kidney poses a risk in terms of bleeding and kidney organ injury (Elsharkawy, El-Boghdadly, & Barrington, 2019).

Erector spinae plane block

Introduction

In 2016 this block was used by Forero in thoracic pain treatment (Forero, Adhikary, Lopez, Tsui, & Chin, 2016). Three muscles form erector spinae muscle: spinalis, longissimus, and iliocostalis. The erector spinae plane lies between the erector spinae muscles and the transverse processes of thoracic or lumbar vertebrae. It is relatively safe and simple to apply as it is easy to identify the plane with the US. For the extended duration of analgesia, a catheter can also be used (Forero et al., 2016).

The local anesthetic drug spreads cranio-caudally and diffuses into paravertebral, epidural, and intercostal space (Kot et al., 2019). A median volume of 3.4 mL local anesthetic affects one dermatome (De Cassai & Tonetti, 2018). The widespread area provides somatic and visceral pain relief (Chin, Adhikary, Sarwani, & Forero, 2017). For abdominal surgeries, it is advised to perform ESPB at T7–T8 levels (Tulgar, Selvi, & Kapakli, 2018).

Indications

Acute and chronic pain of thorax, abdomen, and hip surgery may benefit from this block depending on the level the ESPB is applied. As this is a relatively new block, clear indications are still investigated.

- Bariatric surgery
- Major abdominal surgery
- Inguinal hernia repair
- Ventral hernia repair (laparoscopic and open techniques)
- Cholecystectomy (laparoscopic and open techniques)
- Ileostomy closure
- Abdominoplasty
- Duodenal web
- Endoscopic retrograde cholangiopancreatography
- Chronic abdominal pain

Technique

- Landmark
  Sitting, prone, or lateral position of the patient can be used. For abdominal surgeries, the transverse process of the T7 vertebra is palpated or visualized by the US. At this level rhomboid major muscle will not be visualized, and it is used for verification of the point.
- Ultrasound-assisted block
  Depending on the level, a linear or convex probe is positioned on the spinous process then moved 3 cm laterally and rotated 90 degrees to stand in a parasagittal position. With an in-plane technique, the needle is inserted, and the local anesthetic is injected; the linear spread indicates the right position of the needle. The procedure is done to done bilaterally.
- Injectate
  For laparoscopic ventral hernia repair, 20–30 mL ropivacaine 0.5% provided efficient analgesia (Chin, Adhikary, et al., 2017). The total local anesthetic dose must be calculated before applying the drug of choice. Pediatric patient doses range between 0.2 and 0.5 mL/kg (De Cassai et al., 2019).

Complications

It is relatively safe to perform ESPB as there are no main vascular structures and the spinal cord near the fascial plane. Pneumothorax, motor blockage, and block failure were encountered (De Cassai et al., 2019). Infection, allergy, and LAST are also possible complications.

Transversalis fascia plane block

Introduction

This block was first described by Hebbard (2009). The transversalis fascia lies between the transversus abdominis muscle and the extraperitoneal fascia. The anterior and lateral branches of the T12 and L1 nerve, the ilioinguinal and iliohypogastric nerves, are blocked when they coarse between the transversus abdominis fascia and the transversalis fascia.

Posterior TAP block infrequently covers lateral cutaneous branches of the T12 and L1 nerve, and the course of II and IH nerves have anatomical variations, but these are seen less in transversalis fascia plane block(TFPB) (Choudhary, Mishra, & Jadhav, 2018). The difference of TFPB from the QL1 block is that the insertion point is more caudal and anterior (Chin, McDonnell, et al., 2017). For these reasons, II and IH nerves are affected more specifically.

Indications

- Anterior iliac crest bone graft harvesting
- Appendectomy,
- Cecostomy
- Inguinal hernia repair

Technique

- Landmark
  The patient lies in a supine or lateral position. The lateral abdominal wall between the crista iliaca and the subcostal margin is the starting point. External oblique, internal oblique, and transversus abdominis muscles and aponeurosis, quadratus lumborum, the transversalis fascia, retroperitoneal fat, and peritoneum are identified.
- Ultrasound-assisted block
  The linear or curvilinear probe can be used according to the patient's weight. The needle is advanced with an in-plane technique. Following verification of the site by hydrodissection the local anesthetic is injected, and the anterior and posterior spread is visualized.
- Injectate
  The bilateral injection is done in cesarean delivery, but for surgeries located for a single side of the abdomen, 15 mL local anesthetic is used (Choudhary et al., 2018; Tulgar & Serifsoy, 2019).

Complications

Peritoneal penetration, liver trauma, LAST, block failure, quadriceps weakness due to the spread of injectate under transversalis fascia can be seen if (Chin, McDonnell, et al., 2017; Lee, Goetz, & Gharapetian, 2015).

Contraindications in truncal blocks

- Absolute contraindications include infection or malignancy at the injection site, allergy to the drugs (local anesthetics and adjuvant medications) to be used, and patient rejection.
- Relative contraindications include anatomical variations, scarring, and fibrosis due to previous surgeries that make sonographic visualization of landmarks difficult. If the block is applied after the current surgery, the operation can also alter the anatomy of the area. Although these blocks are done via the US guidance, anticoagulation may also be a relative contraindication.

Preparation for truncal blocks

Equipment

- The US machine with linear or curvilinear probes
- Sterile gloves and gowns (and gels)
- Marking pen
- Antimicrobial skin cleanser (2% chlorhexidine gluconate)
- Peripheral nerve block needles (20 to 22 gauge, 4 to 10 cm)
- Local anesthetics
- Serum physiologic for hydrodissection
- Epidural or peripheric nerve catheter (optional)
- Sterile gauze, sterile syringes, and needles
- Standard American Society of Anesthesiologists (ASA) applied monitors: ECG 3 leads, HR, BP, Pulse oximeter, and if possible nasal cannula with $EtCO_2$ monitoring
- Resuscitation equipment and medications (oxygen supply and masks, laryngoscopes, bag-mask, suction, defibrillator, epinephrine, phenylephrine, atropine, etc.)
- Lipid emulsion is available in 20% (intravenous lipid emulsion is an accepted treatment for severe LAST)

Postprocedure

- Minimum 20–30 min after the local anesthetic drug injection block-level can be evaluated by pin-prick test or ice.
- Breaching the peritoneum and pleura, visceral organ, or vasculature injuries are rare with US guidance, but if there is any suspicion about an unwanted event, then radiologic verification is mandatory.

Conclusion

Various truncal blocks are described for pain treatment as an adjunct to other therapies. Truncal blocks target fascial planes to affect the nerves of interest. US guidance offers safer and faster block application. A better understanding of the anatomy and local anesthetic spread is essential.

The aging population with comorbidities and multi-drug usage needs effective analgesia without compromising the current status of the patient. US-guided truncal blocks are easy to learn and apply. Simulation-based education is important with invasive procedures.

Although low concentration high volume local anesthetics are used, the optimal dose and concentrations are still investigated.

Application to other areas

Ultrasound-guided interfascial plane blocks are effectively used in acute pain treatment after thoraco-vascular, gyneco-logic, and orthopedic surgeries, as well. Preoperative application of these blocks decreases the analgesic requirements also during surgery. Chronic pain treatment is an additional application area for truncal blocks. New approaches are still being studied for different surgical indications.

Other agents of interest

When the neuraxial block is contraindicated, truncal blocks offer safe and effective analgesia.

A new approach to an already well-established block may cover different dermatomes like TAPB uses different approaches for various surgeries. To cover more dermatomes, concomitant application of blocks is also possible. The spread of the local anesthetics is studied in cadaver by specific dyes, but some authors suggest that this could be different in living organisms, and new studies are needed.

Mini-dictionary of terms

Multimodal analgesia: Multimodal analgesia can be defined as the combined application of drugs or methods acting with different mechanisms to provide effective analgesia.

Truncal block: Truncal nerve blocks involve local anesthetic infiltration into facial planes around trunk nerves. It can be used alone or as a component of multimodal analgesia.

Dermatome: It is the area of skin innervated by the afferent nerves of the dorsal root of a single spinal nerve.

Plane block: It is the technique of applying local anesthetic to the muscle planes where the nerve is located, not the sheath and itself, which has emerged with developments in ultrasound-guided block procedures. Nerve injury and other complications are less common and easier to perform. The spread of local anesthesia in the desired area can be monitored during the procedure.

Key facts

Key facts of truncal blocks in general surgery
- Multimodal analgesia treatment protocols, including truncal blocks, provide efficient analgesia and better patient outcomes.
- Truncal blocks, when applied before the surgery, decrease the analgesic requirements during the operation.
- US guidance facilitates the block application, and the risk of failure and complication decreases. Good knowledge and skill in the US are essential.
- Type of surgery and patient characteristics play an important role in deciding the block and approach planning.
- Local anesthetic drug concentration and volume are important as they spread in the fascial planes. High volume-low concentration local anesthetics are used.

Summary points

- This chapter gives information about the commonly used truncal blocks in general surgeries.
- Postoperative complications, early mobilization, and discharge from the hospital are positively affected by efficient pain control.
- Multimodal analgesia techniques lead to less systematically applied opioid usage; thus, the side effects are avoided.
- Truncal blocks are becoming more frequently used as ultrasound guidance increases the block's success.
- Surgery, the incision, and the provider's skill and education in truncal blocks determine which block and approach to be applied.

References

Abrahams, M., Derby, R., & Horn, J. L. (2016). Update on ultrasound for truncal blocks: A review of the evidence. *Regional Anesthesia and Pain Medicine*, *41*(2), 275–288. https://doi.org/10.1097/AAP.0000000000000372.

Baxter, C. S., Singh, A., & Fitzgerald, B. M. (2020). *Intercostal nerve block. StatPearls.* https://www.ncbi.nlm.nih.gov/pubmed/29489198.

Bharti, N., Kumar, P., Bala, I., & Gupta, V. (2011). The efficacy of a novel approach to transversus abdominis plane block for postoperative analgesia after colorectal surgery. *Anesthesia and Analgesia*, *112*(6), 1504–1508. https://doi.org/10.1213/ANE.0b013e3182159bf8.

Boezaart, A. P., Lucas, S. D., & Elliott, C. E. (2009). Paravertebral block: Cervical, thoracic, lumbar, and sacral. *Current Opinion in Anaesthesiology, 22* (5), 637–643. https://doi.org/10.1097/ACO.0b013e32832f3277.

Bondar, A., Szucs, S., & Iohom, G. (2010). Thoracic paravertebral blockade. *Medical Ultrasonography*, *12*(3), 223–227. https://www.ncbi.nlm.nih.gov/pubmed/21203600.

Bouman, E. A., Theunissen, M., Bons, S. A., van Mook, W. N., Gramke, H. F., van Kleef, M., et al. (2014). Reduced incidence of chronic postsurgical pain after epidural analgesia for abdominal surgery. *Pain Practice, 14*(2), E76–E84. https://doi.org/10.1111/papr.12091.

Burns, D. A., Ben-David, B., Chelly, J. E., & Greensmith, J. E. (2008). Intercostally placed paravertebral catheterization: An alternative approach to continuous paravertebral blockade. *Anesthesia and Analgesia*, *107*(1), 339–341. https://doi.org/10.1213/ane.0b013e318174df1d.

Cheema, S. P., Ilsley, D., Richardson, J., & Sabanathan, S. (1995). A thermographic study of paravertebral analgesia. *Anaesthesia, 50*(2), 118–121. https://doi.org/10.1111/j.1365-2044.1995.tb15092.x.

Chin, K. J., Adhikary, S., Sarwani, N., & Forero, M. (2017). The analgesic efficacy of pre-operative bilateral erector spinae plane (ESP) blocks in patients having ventral hernia repair. *Anaesthesia, 72*(4), 452–460. https://doi.org/10.1111/anae.13814.

Chin, K. J., McDonnell, J. G., Carvalho, B., Sharkey, A., Pawa, A., & Gadsden, J. (2017). Essentials of our current understanding: Abdominal wall blocks. *Regional Anesthesia and Pain Medicine*, *42*(2), 133–183. https://doi.org/10.1097/AAP.0000000000000545.

Chou, R., Gordon, D. B., de Leon-Casasola, O. A., Rosenberg, J. M., Bickler, S., Brennan, T., et al. (2016). Management of postoperative Pain: A clinical practice guideline from the American Pain Society, the American Society of Regional Anesthesia and Pain Medicine, and the American Society of Anesthesiologists' Committee on Regional Anesthesia, Executive Committee, and Administrative Council. *The Journal of Pain, 17*(2), 131–157. https://doi.org/10.1016/j.jpain.2015.12.008.

Choudhary, J., Mishra, A. K., & Jadhav, R. (2018). Transversalis fascia plane block for the treatment of chronic posthernorrhaphy inguinal pain: A case report. *A&A Practice*, *11*(3), 57–59. https://doi.org/10.1213/XAA.0000000000000730.

Coveney, E., Weltz, C. R., Greengrass, R., Iglehart, J. D., Leight, G. S., Steele, S. M., et al. (1998). Use of paravertebral block anesthesia in the surgical management of breast cancer: Experience in 156 cases. *Annals of Surgery*, *227*(4), 496–501. https://doi.org/10.1097/00000658-199804000-00008.

de Boer, H. D., Detriche, O., & Forget, P. (2017). Opioid-related side effects: Postoperative ileus, urinary retention, nausea and vomiting, and shivering. A review of the literature. *Best Practice & Research. Clinical Anaesthesiology*, *31*(4), 499–504. https://doi.org/10.1016/j.bpa.2017.07.002.

De Cassai, A., Bonvicini, D., Correale, C., Sandei, L., Tulgar, S., & Tonetti, T. (2019). Erector spinae plane block: A systematic qualitative review. *Minerva Anestesiologica*, *85*(3), 308–319. https://doi.org/10.23736/S0375-9393.18.13341-4.

De Cassai, A., & Tonetti, T. (2018). Local anesthetic spread during erector spinae plane block. *Journal of Clinical Anesthesia*, *48*, 60–61. https://doi.org/10.1016/j.jclinane.2018.05.003.

Dhanjal, S., & Tonder, S. (2020). *Quadratus lumborum block. StatPearls*. https://www.ncbi.nlm.nih.gov/pubmed/30725897.

El-Boghdadly, K., Elsharkawy, H., Short, A., & Chin, K. J. (2016). Quadratus lumborum block nomenclature and anatomical considerations. *Regional Anesthesia and Pain Medicine*, *41*(4), 548–549. https://doi.org/10.1097/AAP.0000000000000411.

Ellis, H. (2009). Anatomy of the anterior abdominal wall and inguinal canal. *Anaesthesia & Intensive Care Medicine*, *10*(7), 315–317. https://doi.org/10.1016/j.mpaic.2009.04.009.

Elsharkawy, H., El-Boghdadly, K., & Barrington, M. (2019). Quadratus lumborum block: Anatomical concepts, mechanisms, and techniques. *Anesthesiology*, *130*(2), 322–335. https://doi.org/10.1097/ALN.0000000000002524.

Ford, S., Dosani, M., Robinson, A. J., Campbell, G. C., Ansermino, J. M., Lim, J., et al. (2009). Defining the reliability of sonoanatomy identification by novices in ultrasound-guided pediatric ilioinguinal and iliohypogastric nerve blockade. *Anesthesia and Analgesia*, *109*(6), 1793–1798. https://doi.org/10.1213/ANE.0b013e3181bce5a5.

Forero, M., Adhikary, S. D., Lopez, H., Tsui, C., & Chin, K. J. (2016). The erector spinae plane block: A novel analgesic technique in thoracic neuropathic pain. *Regional Anesthesia and Pain Medicine*, *41*(5), 621–627. https://doi.org/10.1097/AAP.0000000000000451.

Hannallah, R. S., Broadman, L. M., Belman, A. B., Abramowitz, M. D., & Epstein, B. S. (1987). Comparison of caudal and ilioinguinal/iliohypogastric nerve blocks for control of post-orchiopexy pain in pediatric ambulatory surgery. *Anesthesiology*, *66*(6), 832–834. https://doi.org/10.1097/00000542-198706000-00023.

Hebbard, P. D. (2009). Transversalis fascia plane block, a novel ultrasound-guided abdominal wall nerve block. *Canadian Journal of Anesthesia/Journal canadien d'anesthésie*, *56*(8), 618–620. https://doi.org/10.1007/s12630-009-9110-1.

Hemmerling, T. M. (2018). Pain management in abdominal surgery. *Langenbeck's Archives of Surgery*, *403*(7), 791–803. https://doi.org/10.1007/s00423-018-1705-y.

Karmakar, M. K. (2001). Thoracic paravertebral block. *Anesthesiology*, *95*(3), 771–780. https://doi.org/10.1097/00000542-200109000-00033.

Karmakar, M. K., & Ho, A. M. (2003). Acute pain management of patients with multiple fractured ribs. *The Journal of Trauma*, *54*(3), 615–625. https://doi.org/10.1097/01.TA.0000053197.40145.62.

Kehlet, H. (2018). Postoperative pain, analgesia, and recovery-bedfellows that cannot be ignored. *Pain*, *159*(Suppl. 1), S11–S16. https://doi.org/10.1097/j.pain.0000000000001243.

Kehlet, H., Jensen, T. S., & Woolf, C. J. (2006). Persistent postsurgical pain: Risk factors and prevention. *The Lancet*, *367*(9522), 1618–1625. https://doi.org/10.1016/s0140-6736(06)68700-x.

Khan, J. A., McAtamney, D., & Farling, P. A. (2010). Epidural blood patch for hypoactive-hypoalert behaviour secondary to spontaneous intracranial hypotension. *British Journal of Anaesthesia*, *104*(4), 508–509. https://doi.org/10.1093/bja/aeq045.

Kot, P., Rodriguez, P., Granell, M., Cano, B., Rovira, L., Morales, J., et al. (2019). The erector spinae plane block: A narrative review. *Korean Journal of Anesthesiology*, *72*(3), 209–220. https://doi.org/10.4097/kja.d.19.00012.

Lee, T. H., Barrington, M. J., Tran, T. M., Wong, D., & Hebbard, P. D. (2010). Comparison of extent of sensory block following posterior and subcostal approaches to ultrasound-guided transversus abdominis plane block. *Anaesthesia and Intensive Care*, *38*(3), 452–460. https://doi.org/10.1177/0310057X1003800307.

Lee, S., Goetz, T., & Gharapetian, A. (2015). Unanticipated motor weakness with ultrasound-guided transversalis fascia plane block. *A&A Practice*, *5*(7). https://journals.lww.com/aacr/Fulltext/2015/10010/Unanticipated_Motor_Weakness_with.6.aspx.

Lopez-Rincon, R. M., & Kumar, V. (2020). *Ultrasound-guided intercostal nerve block. StatPearls*. https://www.ncbi.nlm.nih.gov/pubmed/32310360.

Macrae, W. A. (2001). Chronic pain after surgery. *British Journal of Anaesthesia*, *87*(1), 88–98. https://doi.org/10.1093/bja/87.1.88.

Mavarez, A. C., & Ahmed, A. A. (2020). *Transabdominal plane block. StatPearls*. https://www.ncbi.nlm.nih.gov/pubmed/32809362.

McDonnell, J. G., O'Donnell, B. D., Farrell, T., Gough, N., Tuite, D., Power, C., et al. (2007). Transversus abdominis plane block: A cadaveric and radiological evaluation. *Regional Anesthesia and Pain Medicine*, *32*(5), 399–404. https://doi.org/10.1016/j.rapm.2007.03.011.

O'Donnell, B. D., McDonnell, J. G., & McShane, A. J. (2006). The transversus abdominis plane (TAP) block in open retropubic prostatectomy. *Regional Anesthesia and Pain Medicine*, *31*(1), 91. https://doi.org/10.1016/j.rapm.2005.10.006.

Rafi, A. N. (2001). Abdominal field block: A new approach via the lumbar triangle. *Anaesthesia*, *56*(10), 1024–1026. https://doi.org/10.1046/j.1365-2044.2001.02279-40.x.

Rucklidge, M., & Beattie, E. (2018). Rectus sheath catheter analgesia for patients undergoing laparotomy. *BJA Education*, *18*(6), 166–172. https://doi.org/10.1016/j.bjae.2018.03.002.

Salaria, O. N., Kannan, M., Kerner, B., & Goldman, H. (2017). A rare complication of a TAP block performed after caesarean delivery. *Case Reports in Anesthesiology*, *2017*, 1072576. https://doi.org/10.1155/2017/1072576.

Tighe, S. Q. M., Greene, M. D., & Rajadurai, N. (2010). Paravertebral block. *Continuing Education in Anesthesia, Critical Care and Pain, 10*(5), 133–137. https://doi.org/10.1093/bjaceaccp/mkq029.

Tran, D. Q., Bravo, D., Leurcharusmee, P., & Neal, J. M. (2019). Transversus abdominis plane block: A narrative review. *Anesthesiology, 131*(5), 1166–1190. https://doi.org/10.1097/ALN.0000000000002842.

Tran, T. M., Ivanusic, J. J., Hebbard, P., & Barrington, M. J. (2009). Determination of spread of injectate after ultrasound-guided transversus abdominis plane block: A cadaveric study. *British Journal of Anaesthesia, 102*(1), 123–127. https://doi.org/10.1093/bja/aen344.

Tsai, H. C., Yoshida, T., Chuang, T. Y., Yang, S. F., Chang, C. C., Yao, H. Y., et al. (2017). Transversus abdominis plane block: An updated review of anatomy and techniques. *BioMed Research International, 2017*, 8284363. https://doi.org/10.1155/2017/8284363.

Tulgar, S., Selvi, O., & Kapakli, M. S. (2018). Erector spinae plane block for different laparoscopic abdominal surgeries: Case series. *Case Reports in Anesthesiology, 2018*, 3947281. https://doi.org/10.1155/2018/3947281.

Tulgar, S., & Serifsoy, T. E. (2019). Transversalis fascia plane block provides effective postoperative analgesia for cesarean section: New indication for known block. *Obstetric Anesthesia Digest, 39*(3). https://journals.lww.com/obstetricanesthesia/Fulltext/2019/09000/Transversalis_Fascia_Plane_Block_Provides.60.aspx.

Ueshima, H., Otake, H., & Lin, J. A. (2017). Ultrasound-guided quadratus lumborum block: An updated review of anatomy and techniques. *BioMed Research International, 2017*, 2752876. https://doi.org/10.1155/2017/2752876.

Urits, I., Ostling, P. S., Novitch, M. B., Burns, J. C., Charipova, K., Gress, K. L., et al. (2019). Truncal regional nerve blocks in clinical anesthesia practice. *Best Practice & Research. Clinical Anaesthesiology, 33*(4), 559–571. https://doi.org/10.1016/j.bpa.2019.07.013.

van Schoor, A. N., Boon, J. M., Bosenberg, A. T., Abrahams, P. H., & Meiring, J. H. (2005). Anatomical considerations of the pediatric ilioinguinal/iliohypogastric nerve block. *Paediatric Anaesthesia, 15*(5), 371–377. https://doi.org/10.1111/j.1460-9592.2005.01464.x.

Willschke, H., Bosenberg, A., Marhofer, P., Johnston, S., Kettner, S. C., Wanzel, O., et al. (2006). Ultrasonography-guided rectus sheath block in paediatric anaesthesia—A new approach to an old technique. *British Journal of Anaesthesia, 97*(2), 244–249. https://doi.org/10.1093/bja/ael143.

Willschke, H., & Kettner, S. (2012). Pediatric regional anesthesia: Abdominal wall blocks. *Paediatric Anaesthesia, 22*(1), 88–92. https://doi.org/10.1111/j.1460-9592.2011.03704.x.

Willschke, H., Marhofer, P., Bosenberg, A., Johnston, S., Wanzel, O., Cox, S. G., et al. (2005). Ultrasonography for ilioinguinal/iliohypogastric nerve blocks in children. *British Journal of Anaesthesia, 95*(2), 226–230. https://doi.org/10.1093/bja/aei157.

Wylde, V., Dennis, J., Beswick, A. D., Bruce, J., Eccleston, C., Howells, N., et al. (2017). Systematic review of management of chronic pain after surgery. *The British Journal of Surgery, 104*(10), 1293–1306. https://doi.org/10.1002/bjs.10601.

Yarwood, J., & Berrill, A. (2010). Nerve blocks of the anterior abdominal wall. *Continuing Education in Anesthesia, Critical Care and Pain, 10*(6), 182–186. https://doi.org/10.1093/bjaceaccp/mkq035.

Chapter 38

Linking analgesia, epidural oxycodone, pain, and laparoscopy

Merja Kokki[a,b] and Hannu Kokki[a,b]

[a]*Department of Anaesthesiology and Intensive Care, Kuopio University Hospital, Kuopio, Finland,* [b]*Faculty of Health Sciences, Clinical Medicine, University of Eastern Finland, Kuopio, Finland*

Abbreviations

AE	adverse event
AUC	area under the concentration-time curve
MOR	mu-opioid receptor
C_{max}	maximum plasma drug concentration
t_{max}	time to reach maximum plasma concentration following drug administration
BBB	blood–brain barrier
CYP	cytochrome P 450 enzyme system
i.v.	intravenous

Introduction

Pain is a common complaint after surgery. Severe and prolonged postoperative pain is recognized risk factor for chronic pain. Adequate pain treatment in early postoperative phase is assumed to prevent risk of persistent postoperative pain (Kehlet, Jensen, & Woolf, 2006; VanDenKerkhof et al., 2012).

Pain after laparoscopic surgery should not be ignored. Laparoscopic surgery is often considered less invasive method of performing surgery than open surgery, but data on that are controversial. Rather extensive surgery can be performed laparoscopically or laparoscopy assisted. Cancer surgery and major gastrointestinal surgery are nowadays common laparoscopic procedures, and postoperative pain is likely outcome. However, even after "minor" laparoscopic surgery early pain is moderate or severe in most patients. Thus, patients should be provided appropriate pain management (Sjövall, Kokki, & Kokki, 2015).

Pain after laparoscopic surgery can be overlooked as the recovery after laparoscopic surgery is usually shorter and patients return to their daily activities sooner than after open surgery (Zhu et al., 2020). However, early postoperative pain is often more intense and unpleasant after laparoscopic surgery than that after open surgery, and patients may have high requirements of analgesia at the early postoperative phase (Ekstein et al., 2006). Mean of the most postoperative pain during the first 24 postoperative hours after routine laparoscopic surgery, e.g., cholecystectomy is moderate to severe. In the Gerbershagen et al. (2013) study, it was 4.8 in numeral rating scale (NRS) 0–10, similar or higher to pain ratings after major surgery. However, two-thirds of patients with laparoscopic surgery did not get any postoperative opioid analgesics, as the procedures were considered a minor surgery.

Carbon dioxide (CO_2) is commonly used in laparoscopic surgery for pneumoperitoneum. It is used to create visual and spatial space within abdominal cavity. Pneumoperitoneum contributes to different factors that can cause pain during and after laparoscopic surgery (Sjövall et al., 2015). These factors are further evaluated.

Oxycodone, a thebaine-derived μ-opioid receptor (MOR) agonist, is highly effective in visceral pain, that is substantial component of laparoscopic postoperative pain. Oxycodone has been evaluated in epidural use and these novel data are critically evaluated in connection with laparoscopic surgery (Kokki, Kokki, & Sjövall, 2012; Kinnunen, Piirainen, Kokki, Lammi, & Kokki, 2019).

Features and Assessments of Pain, Anesthesia, and Analgesia. https://doi.org/10.1016/B978-0-12-818988-7.00030-3
Copyright © 2022 Elsevier Inc. All rights reserved.

Laparoscopy

The gynecologists performed pelviscopy over 100 years ago, and this technique aided development of laparoscopic surgery. The progress of the adaptation of the technique was slow until 1980s when the video technology provided view to other personnel in the operating theatre and enabled the surgical personnel to work as a team. The video laparoscopy technique gained popularity at first as a diagnostic tool in 1980s, when it was increasingly used for therapeutic purposes. Since then, laparoscopic surgery has gained extensive popularity and different operations are performed laparoscopically (Spaner & Warnock, 1997). During the last two decades robotic-assisted laparoscopic surgery has been widely adopted. Other new techniques and devices have been introduced into the clinical practice also. The use of hand-access ports with laparoscopy allows performing major surgeries less invasively. Nowadays, e.g., nephrectomy is often performed laparoscopically assisted with small incision for hand port. Colectomy is another example of major laparoscopic surgery where there is a need to work in multiple areas in the abdominal cavity (Joshi et al., 2013).

This rapid progress to promote laparoscopic approaches is based on minimal solid evidence (Sjövall et al., 2015). One example is cholecystectomy that is nowadays almost entirely performed with laparoscopic technique. The first laparoscopic cholecystectomy was performed in 1985 by Eric Mühe, and by the beginning of the 1990s, laparoscopic cholecystectomy had spread rapidly and had become a golden standard for cholecystectomy technique. However, in randomized, controlled clinical trials it has been proven that minilaparotomy cholecystectomy is similar or superior to laparoscopic cholecystectomy in terms of early and late recovery, and the costs are significantly less with minilaparotomy than those with laparoscopy (Harju, Aspinen, Juvonen, Kokki, & Eskelinen, 2013). This is the case also with robotic surgery. There is some evidence gathering that recovery or outcome may be enhanced with robotic-assisted technique, but the costs are several times higher than that with traditional techniques. Long-term outcomes are awaited (Roh, Nam, & Kim, 2018).

Laparoscopic surgery is considered to be minimally invasive surgery, meaning that endoscopic techniques cause less injury to the bodily homeostasis than open surgery. In general, laparoscopic surgery is deemed safer than open surgery, to allow faster recovery and heal with less pain and better cosmetic outcome. However, in laparoscopic techniques where several ports and hand ports are used and some of them inserted through muscles and ligaments, the tissue trauma can be substantial and early pain after surgery can be more intense and unpleasant than that after meticulously performed open surgery (Ekstein et al., 2006; Harju et al., 2013). There is also a growing body of evidence indicating that the risk for persistent postoperative pain and patients' satisfaction with the cosmetic outcome and health-related quality of life could be similar after open and laparoscopic surgeries (Aspinen et al., 2014).

Pain after laparoscopy

Laparoscopy involves insufflation of the abdominal cavity with gas, usually CO_2, to produce pneumoperitoneum. Gas is insufflated to an intraabdominal pressure of 12–15 mmHg, while the physiological pressure in the abdominal cavity in recumbent position is in average between 4 and 6 mmHg. The pneumoperitoneum can have substantial physiological effects on cardiovascular, respiratory, and renal function especially in elderly patients with comorbidities (Sjövall et al., 2015).

Carbon dioxide is most commonly used gas in creation of pneumoperitoneum. Nitrous oxide, helium, and room air have been used to create visual space during laparoscopy with no benefit considering postoperative recovery (Yu et al., 2017). Gasless laparoscopies using abdominal wall lifting devices have been evaluated, but without improvements in outcome. The visual field is compromised and postoperative pain similar than after conventional laparoscopy (Koivusalo et al., 1998).

Carbon dioxide insufflation into the abdominal cavity is one of the causative agents for postoperative pain after laparoscopy. Pain is often described as a dull, unpleasant pain projected in the shoulder area. Carbon dioxide pneumoperitoneum with intraabdominal pressure less than 15 mmHg may cause irritation of phrenic nerves and pain-related C4 dermatome. Intraabdominal pressure peaks or surgical nerve trauma are involved to neuropathic type of pain after laparoscopy. Carbon dioxide insufflation creates an acidic milieu, which causes irritation in peritoneal cavity. Pneumoperitoneum triggers oxidative stress and inflammatory reaction and contributes to inflammatory pain after laparoscopic procedures (Sjövall et al., 2015).

The tissue trauma from incision and port sites are origins of somatic pain after laparoscopy. Diagnostic laparoscopy and some surgical procedures can be performed with a single port technique, but most surgical procedures are carried out using several ports for instruments. As the port sites penetrate muscles and ligaments this causes nociceptive and inflammatory pain, and if nerves are injured, neuropathic pain can develop (Sjövall et al., 2015).

Taken together, the pain after laparoscopy is multifactorial and originates from multiple mechanisms. Patient may develop nociceptive pain, superficial and deep somatic pain, and visceral pain arising from abdominal organs. Nerve lesions can induce postoperative neuropathic pain and tissue trauma results on inflammatory pain. These different pain categories have distinct intensities and time courses. Thus, the pain management after laparoscopy should reflect the large variation of pain manifestations.

Pain treatment after laparoscopy

Pain after laparoscopic procedures should not be ignored. Pain after laparoscopy is often intense, unpleasant and impairs recovery and quality of life. Patients should be provided appropriate pain management (Sjövall et al., 2015). Enhanced recovery after surgery programs advises use of multimodal pain treatment to ensure calm and quick recovery after surgery. This model applies on laparoscopic procedures also. Pain after laparoscopic surgery has different modalities. Patients have superficial somatic and deep visceral pain, and inflammatory, neuropathic, and referred shoulder pains are also involved. To treat different pain modalities pain management with different mechanisms and sites of action should be employed (Joshi et al., 2013) (Table 1).

Interventions during surgery to prevent postoperative pain

Shoulder and upper abdominal pain are common complaints after laparoscopic procedures, up to 80% of patients have referred pain in shoulder area that persists on average for 2 or 3 days (Lee, Song, Kim, & Lee, 2018). This referred pain can be more intense and unpleasant to the patient than incisional pain (Sjövall et al., 2015).

TABLE 1 Some examples of multimodal analgesia used in treatment of postoperative pain after laparoscopic surgery.

Compound	Type of surgery	Efficacy/outcome	References
Paracetamol 1 g x4 i.v.	Laparoscopic hysterectomy	Oxycodone: Paracetamol, 0.34 mg/kg/24 h; Placebo: 0.43 mg/kg/24 h	Jokela, Ahonen, Seitsonen, Marjakangas, and Korttila (2010)
Paracetamol 1 or 2 g i.v.	Laparoscopic cholecystectomy	Oxycodone: 0.3 vs 0.29 mg/kg to pain relief	Kokki, Broms, Eskelinen, Neuvonen, et al. (2012), Kokki, Kokki, and Sjövall (2012)
Dexketoprofen 10 or 50 mg i.v.	Laparoscopic cholecystectomy	Oxycodone 0.11 vs 0.08 mg/kg to pain relief	Piirainen et al. (2015)
Etoricoxib 120 mg × 1	Laparoscopic cholecystectomy	Similar to placebo in early pain, but less pain at 12 and 24 h	Qiu, Xie, and Qu (2019)
Dexamethasone 5, 10 or15 mg i.v.	Laparoscopic hysterectomy	Oxycodone: Dexamethasone groups: 5 mg, 0.42; 10 mg: 0.44, 15 mg: 0.34 mg/kg/24 h. Placebo: 0.55 mg/kg/24 h	Jokela, Ahonen, Tallgren, Marjakangas, and Korttila (2009)
Pregabalin 150 mg or gabapentin 600 mg	Laparoscopic cholecystectomy	Shoulder pain at 48 h: Pregabalin 13% Gabapentin 12% Placebo 26%	Nakhli, Kahloul, Jebali, Frigui, and Naija (2018)
Quadratus lumborum block with levobupivacaine boluses or infusion	Laparoscopic colorectal surgery	Fentanyl: 12 µg/kg/22 h in both groups	Aoyama et al. (2020)
Dexmedetomidine 1.2 µg/kg + vs morphine 0.08 mg/kg	Laparoscopic bariatric surgery	Morphine: Dexmedetomidine, 40 mg/24 h Morphine, 43 mg/24 h	Zeeni et al. (2019)
Ketamine 0.15 mg/kg, diclofenac 1 mg/kg or combination	Laparoscopic cholecystectomy	Shoulder pain at 24 h: Ketamine 30%; Diclofenac 40%; Keta+Diclo 30%; Placebo 45% Diclofenac: less intense pain first 12 h	Nesek-Adam, Grizelj-Stojčić, Mršić, Rašić, and Schwarz (2012)

Large residual volume of intraabdominal gas and acidic intraperitoneal milieu after CO_2 insufflation are the likely cause of shoulder and upper abdominal pain. Decreasing the residual gas volume by using pulmonary recruitment maneuvers and suction of remaining gas from abdominal cavity at the end of surgery reduced postoperative pain substantially. Pulmonary recruitment maneuvers and gas suction are easy to perform, and not time consuming, and should be done in each laparoscopic procedure (Pergialiotis, Vlachos, Kontzoglou, Perrea, & Vlachos, 2015; Song, Kim, & Lee, 2017) (Table 2).

Selection of insufflation gas does not seem to affect postoperative pain after laparoscopy. Carbon dioxide has been compared with N_2O, He, and room air but with no benefit on postoperative recovery (Yu et al., 2017). Humidification or warming insufflation gas does not seem to enhance the recovery either (Balayssac et al., 2017; Birch et al., 2016).

Intraperitoneal pressure seems to affect postoperative recovery, and thus the lowest pressure that provides sufficient surgical view and surgical condition should be used (Radosa, Radosa, Schweitzer, et al., 2019). Data on efficacy on deep neuromuscular block during laparoscopic surgery is conflicting (Choi et al., 2019; Kim, Lee, Kim, Kim, & Bai, 2019).

TABLE 2 Techniques used during surgery to decrease pain after laparoscopy.

Technique	Study design/patients	Main outcome	N	Adverse events	References
Insufflation gas: CO_2, N_2O, He or room air	Systematic review Three studies N_2O vs CO_2 Five studies He vs CO_2 One study room air vs CO_2/different types of surgery	Similar postoperative pain with all gases	519	Subcutaneous emphysema after He PP	Yu et al. (2017)
Pulmonary recruitment maneuvers or intraperitoneal saline at the end of laparoscopic surgery or standard care	A prospective, randomized, three-arm, controlled trial/gynecological surgery	Intraperitoneal saline less shoulder pain. Both methods less upper abdominal pain	158	Not reported	Tsai et al. (2011)
Pulmonary recruitment maneuver at the end of laparoscopic surgery	Metaanalysis of randomized controlled trials/different types of surgery	Pulmonary recruitment maneuver less shoulder pain at 12, 24 and 48 h	571	Not reported	Pergialiotis et al. (2015)
Pulmonary recruitment maneuver and intraperitoneal saline at the end of laparoscopy	Randomized, blinded, controlled trial, /gynecological surgery	No difference in shoulder or abdominal pain	200	Patients with intraperitoneal saline: vomiting twofold more.	van Dijk et al. (2018)
Humidified and warmed CO_2 vs room temperature and dry insufflation gas	Meta-analysis of 17 clinical trial/gastrointestinal and gynecological surgery	No difference in postoperative shoulder pain, opioid doses, body temperature, length of stay	1139	Not reported	Balayssac et al. (2017)
Humidified and warmed CO_2 vs room temperature and dry insufflation gas	Systematic review of 22 studies/different types of laparoscopic surgery	No difference in postoperative pain, opioid use higher with heated gas at 24 and 48 h	1428	Similar frequency of major adverse events	Birch et al. (2016)
Lowest IP pressure with acceptable surgical view and a standard PP pressure (12 mmHg)	Multicenter randomized controlled trial/colorectal laparoscopic surgery	Low IP higher probability of better physiological recovery	166	Intraoperative adverse events less frequent with low IP	Donatsky, Bjerrum, and Gögenur (2013)

CO_2 = carbon dioxide, He = helium, IP = intraperitoneal, N_2O = nitrous oxide, PP = pneumoperitoneum.

Intraperitoneal saline infusion, 15–20 mL/kg, has been tested in buffering the abdominal cavity acidity, but the results are conflicting. In some studies shoulder and upper abdominal pain has decreased for the first 48 h but some data indicate that the early incisional pain can be more severe in patients with intraperitoneal saline infusion (Phelps, Cakmakkaya, Apfel, & Radke, 2008; Tsai et al., 2011).

Intraperitoneal local anesthetic infusion seems to be superior to no intervention and has similar efficacy to intravenous lidocaine infusion. Port sites local anesthetic infiltrations are well tolerated but the efficacy is short (Dunn & Durieux, 2017).

In conclusion, using low-pressure pneumoperitoneum and decreasing the residual gas volume at the end of procedure are highly effective means in preventing postoperative pain after laparoscopic surgery (Table 2).

Epidural analgesia

Epidural analgesia is a highly effective mode of analgesia in intraabdominal surgery. It provides superior pain relief for dynamic pain after major surgery. In laparoscopic surgery epidural analgesia improves postoperative pulmonary function, and facilitates recovery of bowel function (Joshi et al., 2013).

Epidural analgesia involves insertion of catheter through ligamentum flavum into the epidural space, space outside the dura mater. The epidural space comprises fat, rich network of veins, a few arteries, and lymphatics. The epidural fat serves as a depot for highly lipid-soluble drugs injected into the epidural space and some of the injected drug is absorbed into the veins and lymphatics. Drug transfer from the epidural space to the intrathecal space is greater for less lipophilic compounds and less for more lipophilic compounds. Lipid solubility is low for oxycodone and thus it is likely to provide more segmental analgesic efficacy than fentanyl. Fentanyl is one of the highly lipid-soluble opioids, and may thus have more likely systemic efficacy after absorption of injected drug in systemic circulation (Kinnunen, Piirainen, et al., 2019). When oxycodone is used in epidural analgesia catheter placement is recommended to be inserted at level of desired pain relief (Kinnunen, Kokki, et al., 2019).

Epidural analgesia is invasive method of drug delivery. It is associated with very rare but potentially serious complications. Epidural puncture and catheter insertion and removal may induce epidural bleeding and hematoma formation in the closed space. Catheter and injection solution may induce pyogenic infection and epidural abscess that can mechanically compress the spinal cord also. Thus, the indication to use epidural analgesia should be appropriate, the contraindications should be followed, and the puncture and catheter insertion should be performed with meticulous aseptic technique. However, if the patient has an epidural catheter in place, oxycodone injection would provide highly effective analgesia in laparoscopic procedures (Piirainen, Kokki, Anderson, et al., 2019).

In laparoscopy, pain is most effectively controlled by a multimodal, preventive analgesia approach (Sjövall et al., 2015). In epidural analgesia opioid is commonly coadministered with a low amount of local anesthetic. When α_2-adrenergic agonist, e.g. adrenaline, is combined, this reduces the systemic absorption of opioid and provides enhanced segmental analgesia (Kokki, Ruuskanen, & Karvinen, 2002). Nonopioid analgesics, paracetamol, and nonsteroidal antiinflammatory analgesics are given intravenously or by mouth on regular basis (Joshi et al., 2013).

Oxycodone

Oxycodone was first synthesized in Germany in 1916. It is a semi-synthetic thebaine derivative, MOR-agonist. Oxycodone consists of two aliphatic rings, two planar rings, and four chiral centers and its molecular weight is 351.83 g/mol. Oxycodone is freely water soluble, hygroscopic, and slightly soluble in ethanol. Oxycodone has a long history in clinical use, but more comprehensive data of oxycodone pharmacokinetics originated from the 1990s (Kinnunen, Piirainen, et al., 2019).

Oxycodone is rather selective MOR-agonist, although at higher doses it may bind to other opioid receptors also (Monory et al., 1999; Yoburn, Shah, Chan, Duttaroy, & Davis, 1995). When compared with morphine, it has 5- to 40-fold lower affinity to MOR and four- to eightfold lower MOR mediated G-protein activation. However, oxycodone has similar or superior analgesic efficacy compared to morphine (Kokki, Kokki, & Sjövall, 2012).

Opioid receptors are G-protein-coupled receptors, where the link with effector proteins is mediated by a G-protein. Oxycodone binding of the MOR stimulates the exchange of guanosine-5′-triphosphate (GTP) for guanosine-5′-diphosphate (GDP) on the G-protein complex. This inhibits adenylate cyclase and prevents cyclic adenosine-monophosphate (cAMP) production. As a consequence, the release of several nociceptive neurotransmitters, acetylcholine, dopamine, γ-aminobutyric acid (GABA), noradrenaline, and substance P, is inhibited. Oxycodone inhibits also the release of several hormones like insulin, glucagon, somatostatin, and vasopressin. Oxycodone closes N-type voltage-gated calcium channels and opens G-protein-coupled, inwardly rectifying potassium channels. Thereafter, neuronal excitability is reduced, and hyperpolarization may occur (Kinnunen, Kokki, et al., 2019; Kinnunen, Piirainen, et al., 2019; Kokki, Kokki, & Sjövall, 2012).

Oxycodone taste is neutral, and thus, administration by mouth and administration to mucosal membranes are feasible routes in pain management. It is well absorbed; bioavailability of swallowed oxycodone is from 60% to 70% and transmucosal oxycodone 55%. Different formulations are available, two of the most common ones are immediate release (IR) capsule and controlled release (CR) tablet. Oxycodone exposure is similar after IR and CR formulations but C_{max} values are lower and t_{max} values are reached later with CR tablet. C_{max} values for CR tablets are dose related, after 5 mg CR tablet it is approximately 5 ng/mL and after 10 mg tablet 10–12 ng/mL, and after 40 mg tablet 39–47 ng/mL. After multiple 12-h doses of 10 and 40 mg CR tablets, C_{max} values are little higher, 15 and 57 ng/mL, respectively. After IR capsules C_{max} are twofold higher and t_{max} occurs at 1.5 h compared to 3.8 h after CR tablet (Kinnunen, Kokki, et al., 2019; Kinnunen, Piirainen, et al., 2019; Kokki, Kokki, & Sjövall, 2012) (Table 3).

Oxycodone is metabolized mainly in the liver via cytochrome P450 (CYP) 3A4/5 and 2D6 to noroxycodone (45%), oxymorphone (19%), and noroxymorphone. Oxycodol, noroxycodol, and oxymorphol are reductive metabolites of oxycodone, noroxycodone, and oxymorphone, with each of them having two stereoisomers (α and β) (Fig. 1). Oxymorphone and noroxymorphone have analgesic efficacy. Oxymorphone has a 10- to 45-fold higher affinity for MOR and is 8- to 30-fold more potent compared with oxycodone. Noroxycodone exhibits only weak affinity for MOR and its pain-relieving potency is 5–10-fold less than oxycodone (Kinnunen, Kokki, et al., 2019; Kinnunen, Piirainen, et al., 2019; Kokki, Kokki, & Sjövall, 2012).

Cytochrome P450 2D6 has genetic variability affecting enzyme activity. Four different phenotypes can be categorized according to carried allelic variants; poor (no functional alleles), intermediate (reduced function), extensive (classified also as normal; alleles with normal function), and ultra-rapid (duplication of a functional allele) metabolizers (Stamer et al., 2013). However, in studies concerning pharmacodynamics effects of oxycodone, genetic variability does not seem to affect in pain relieving efficacy.

TABLE 3 Oxycodone pharmacokinetics after intravenous, oral liquid, epidural, and intramuscular administration.

	Intravenous oxycodone	Po oxycodone liquid	Epidural oxycodone CSF	Epidural oxycodone plasma	Intramuscular injection
Dose	0.07 mg/kg	0.28 mg/kg	0.1 mg/kg	0.1 mg/kg	0.14 mg/kg
C_{max} (ng/mL)		38 (14)	10,000 [982–10,000]	28.8 [13.5–77.3]	34
t_{max} (h)		1.0 [0.5–1.0]	0.6 [0.2–4.0]	2.1 [0.6–4.2]	1.0
Clearance (L/min)	0.82				0.78
Volume of distribution at steady state(L/kg)	2.6				
Elimination half-life (h)	3.0	5.1		3.8 [3.1–5.1]	4.9
AUC (ng/mL/h)		245			208
Bioavailability		0.6			
References	Pöyhiä, Olkkola, Seppälä, and Kalso (1991)	Pöyhiä, Seppälä, Olkkola, and Kalso (1992)	Kokki et al. (2014)	Kokki et al. (2014)	Pöyhiä et al. (1992)

Data are expressed as median and [range] or mean. *AUC*: area under concentration time curve, *CSF*: cerebrospinal fluid, C_{max}: maximum concentration, T_{max}: time to maximum concentration.

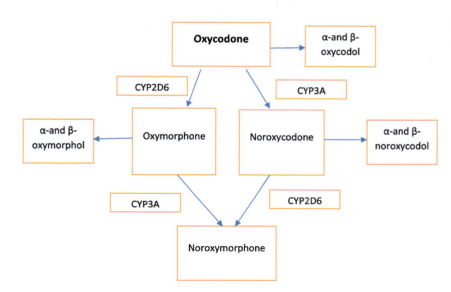

FIG. 1 A schematic figure of oxycodone metabolism in the liver. CYP = Cytochrome P450.

Oxycodone is up to 45% protein-bound, mainly to albumin. Oxycodone and its metabolites are excreted mainly by kidneys as oxycodone from 9% to 11%, noroxycodone 23%, oxymorphone 10%, noroxymorphone 14%, and reduced metabolites $\leq$18% (Lalovic et al., 2006).

The minimum effective concentration (MEC) and minimum effective analgesic concentration (MEAC) of oxycodone in early pain after laparoscopic cholecystectomy are 20–50 ng/mL and 45–90 ng/mL, (Kokki, Broms, Eskelinen, Neuvonen, et al., 2012; Kokki, Broms, Eskelinen, Rasanen, et al., 2012; Piirainen et al., 2015). These values are substantially higher than MEC and MEAC values after breast cancer surgery, 22 and 33 ng/mL (Cajanus et al., 2018) but similar to those obtained after open intra-abdominal surgery, 30–65 and 75 ng/mL, respectively (Choi et al., 2017).

Oxycodone central nervous penetration

The main site of analgesic action of oxycodone is in the central nervous system (CNS), and thus after epidural administration oxycodone should cross the BBB. The ability of compounds to cross the BBB is determined largely by the size of the molecule, lipophilicity, ionization degree at physiological pH, affinity for efflux protein, and uptake mechanisms (Abbott & Friedman, 2012). Blood–brain barrier consists of brain microvascular endothelial cells that are surrounded with pericytes and astrocytes (Fricker & Miller, 2004). The role of BBB is to protect brain homeostasis and neural function in order to regulate substance intake to brain. Glycoprotein P is one of the most important efflux transporters in the BBB. Morphine is a substrate of glycoprotein P, but oxycodone is not, and this might explain oxycodone's high analgesic efficacy despite relatively low affinity and activation on MOR (Kokki & Kokki, 2016).

Oxycodone is a substrate of an energy-dependent, proton-coupled antiporter, H+/OC. This antiporter mediates active blood-brain transport of cationic drugs like oxycodone into the CNS (Kokki & Kokki, 2016). Active transport across the BBB is supported by experimental data showing that the time to reach 50% equilibrium to the deep brain compartment for oxycodone is short, 7 min (Villesen et al., 2006). In human studies, cerebrospinal fluid (CSF) concentrations are used as a surrogate parameter of CNS exposure of compounds. After epidural oxycodone, C_{max} and AUC in the CSF are 320- and 120-fold higher, respectively, compared with intravenous administration (Table 3) (Kokki et al., 2014).

The oxycodone brain:plasma ratio is assumed to be inversely dose-dependent. In an experimental study in rodents, plasma and brain concentrations were measured after 6-day intraperitoneal oxycodone infusions. The brain:plasma ratios were 3.1 in the low-dose group (20 mg/kg/day), 1.5 in the medium-dose group (45 mg/kg/day), and 1.0 in the high-dose group (120 mg/kg/day) (Hill, Dewey, Kelly, & Henderson, 2018).

The metabolites of oxycodone penetrate CNS poorly, oxymorphone may have an active influx into the CNS as the oxymorphone concentrations in the brain are twofold higher than in the plasma (Sadiq, Boström, Keizer, Björkman, & Hammarlund-Udenaes, 2013). After systemic administration noroxymorphone has no analgesic efficacy, because it does not penetrate BBB. However, intrathecally administered noroxymorphone may be more potent than oxycodone (Lemberg, Siiskonen, Kontinen, Yli-Kauhaluoma, & Kalso, 2008).

428   PART | III Interlinking anesthesia, analgesics and pain

To conclude, there is an active transporter system of oxycodone to central nervous system. Other mechanisms may affect oxycodone uptake to the CNS, e.g., glymphatic system enhancing intrathecally administered oxycodone delivery to the brain and spinal cord.

Epidural oxycodone

Opioids are commonly used as an adjuvant to local anesthetic in epidural analgesia. Oxycodone is a hydrophilic opioid, it binds less to epidural fat and absorbs less in systemic circulation, produces longer duration and more precise segmental analgesia than lipophilic opioids like fentanyl or sufentanil. Morphine is another hydrophilic opioid used in epidural analgesia. However, oxycodone is more effective in visceral pain that is a desired characteristic in laparoscopic surgery. Moreover, oxycodone is actively uptaken into the CNS while morphine is a substrate to glycoprotein P, an efflux transporter in the BBB (Kokki & Kokki, 2016).

High antinociceptive efficacy of intrathecal oxycodone was shown for the first time by Pöyhiä and Kalso (1992) in an experimental study in rats. Later, Kinnunen, Kokki, et al. (2019), Kinnunen, Piirainen, et al. (2019) showed in ewes that CSF, spinal cord, and brain tissue concentrations oxycodone are substantially higher after epidural administration than giving the same dose intravenously. Moreover, oxymorphone accumulation in brain tissues was found indicating either CYP2D6 metabolism in the CNS or active uptake through BBB.

Earlier studies on epidural oxycodone were not that encouraging. In Bäcklund, Lindgren, Kajimoto, and Rosenberg (1997) study in patients undergoing major abdominal surgery, the need of epidural oxycodone was 10-fold higher than morphine, and adverse effects were similar with these two compounds. Ten years later Yanagidate and Dohi (2004) showed that epidural oxycodone could be effective in pain management after gynecologic surgery. In that study oxycodone's analgesic dose ratio was 2:1 compared to morphine. In that study adverse effects, nausea and vomiting and pruritus, were less common with oxycodone.

In 2014, Kokki et al. showed that epidural oxycodone has pharmacokinetic characteristics that support its use in postoperative pain management. In that study position of epidural catheter in the epidural space and appropriate segmental level was confirmed before study compound administration. In Kokki's study, oxycodone 0.1 mg/kg was given either epidurally or intravenously to patients having open or laparoscopic gynecologic surgery. The maximum oxycodone concentration in CSF was earlier after epidural administration, at 0.6 h vs 1.1 h, and the exposure to oxycodone in CNS was substantially higher after epidural administration than that after intravenous dosing. C_{max} in CSF was 320-fold and AUC 120-fold higher after epidural administration compared to intravenous dosing. Analgesic efficacy of epidural oxycodone was also superior, and the duration of action was approximately 4 h after a single epidural dosing (Kokki et al., 2014). The CNS PK is shown in Table 3.

After that PK study, Piirainen, Kokki, Anderson, et al. (2019) compared epidural and intravenous oxycodone 0.1 mg/kg in postoperative pain management in women who had undergone gynecological laparoscopic surgery. Piirainen's study supports the feasibility of epidural oxycodone for postoperative pain management. Pain ratings and need of rescue medication were lower in women with epidural oxycodone. The onset of analgesia after a single dose of epidural oxycodone was 30 min and analgesic effect lasted up to 3–4 h. Similar pharmacodynamic characteristics of epidural oxycodone were observed in patients undergoing laparotomy (Piirainen, Kokki, Hautajärvi, Ranta, & Kokki, 2018). In both studies epidural oxycodone was used in multimodal analgesia regimen, for background analgesia patients were given dexketoprofen and paracetamol.

Central nervous system toxicity is an issue for all compounds administered intrathecally. In a human neuroblastoma and mouse motoneuronal cell culture study cells were exposed to oxycodone and morphine in increasing concentrations, cell viability decreased, and cell death was shown only in substantially higher concentrations than used in clinical practice. Moreover, neuronal cell toxicity was similar or less with oxycodone than morphine (Kokki et al., 2016). Safety of clinically used concentrations in epidural pain management was supported in an experimental study in a pregnant sheep-model. No neurodegenerative changes were noted in neuropathological assessments of sheep spinal cord after epidural oxycodone infusion (Piirainen, Kokki, Räsänen, et al., 2019).

Limitation in epidural oxycodone use is that administration method infusion or bolus, dosing intervals, possible adjuvants are to be determined.

Applications to other areas

Oxycodone is commonly used opioid worldwide. However, as oxycodone is an old drug taken to clinical use over 100 years ago, it lacks formal studies compulsory for marketing authorization nowadays. Oxycodone has been studied extensively during last few decades in different patient groups, so application of oxycodone use in other areas is evidence based.

The efficacy of oxycodone to visceral pain provides further uses as abdominal surgery is common procedure. Oxycodone is well absorbed after mucosal administration and bioavailability of mucosal oxycodone is good, up to 55%. Oxycodone is well absorbed also after administration by mouth, bioavailability of swallowed oxycodone is 60%. This enables use of oxycodone in children and elderly. Oxycodone may also be used to treat pain at the latent phase of labor as pain is visceral in this phase of labor.

Oxycodone used in children is evidence based. Pharmacokinetics of different administration formulations have been studied adequately. In preterm babies and infants, oxycodone is feasible opioid. Pharmacokinetics of oxycodone and dosing recommendations based on information have been presented. Oxycodone is preferred in preterm babies, because analgesic effect of oxycodone does not need glucuronidation as morphine.

Epidural oxycodone has been studied in surgical patients. In most recent studies epidural oxycodone has been more effective with less adverse events than intravenous oxycodone. Most of epidural oxycodone data are based on single dose studies, and more data are needed for repeated dosing, infusion and for use with adjuvants.

Other agents of interest

In this chapter, epidural oxycodone use has been evaluated in the treatment of pain after laparoscopic surgery. Oxycodone is not formally approved for intrathecal use and is considered as off license drug use. However, data are accumulating to show high efficacy of epidural oxycodone and that it is well tolerated in intrathecal administration also. Data indicate that neurotoxicity of oxycodone is similar or less than that of morphine.

Morphine, another hydrophilic opioid, is the first opioid approved for intrathecal use. Morphine shares similar pharmacokinetic profile with oxycodone. In a few studies epidural morphine has been used at doses of 2–4 mg with local anesthetics (Hong & Lim, 2008). Main limitation of epidural morphine is its long elimination half-time, 4 h, and that it is associated with potential to delayed respiratory depression up to 24 h after a single dose administration, important safety issues that should be taken into account with this compound.

Sufentanil is another opioid analgesic approved for intrathecal use. Sufentanil has been used in postoperative analgesia combined with local anesthetics (Kokki et al., 2002). For example, epidural infusion of sufentanil 5 µg/h with low dose of local anesthetic provides sufficient pain relief and is well tolerated (Lee et al., 2008).

Despite the fact that fentanyl is one of the most commonly epidurally administered opioid, it does not have approval for intrathecal use. Like sufentanil, fentanyl is highly lipid soluble and the onset of action is relatively quick, but the duration analgesia is short. Thus, in epidural use fentanyl is administered with local anesthetic as continuous infusion (Salomäki, Kokki, Turunen, Havukainen, & Nuutinen, 1996). After laparoscopic surgery epidural fentanyl + local anesthetic infusion provides superior analgesic efficacy and earlier return of gastrointestinal function than intravenous fentanyl with patient-controlled analgesia device. However, epidural fentanyl alone does not provide any benefit compared to intravenous administration if the risks of epidural puncture and catheter-related complications are taken into account (Salomäki, Laitinen, & Nuutinen, 1991).

Hydromorphone belongs to the same phenanthrene group as oxycodone. Hydromorphone is little more lipophilic than oxycodone, but as the lipophilicity of hydromorphone is less than that of sufentanil but higher than that of morphine, its onset and duration of analgesic action is between these two compounds. Hydromorphone has not been approved for intrathecal use, and most of data on epidural use is in obstetrics (Puhto, Kokki, Hakomäki, et al., 2020).

Mini-dictionary of terms

Oxymorphone: A primary metabolite of oxycodone via CYP2D6 metabolism. Approximately 19% of oxycodone is metabolized to oxymorphone. It is also pharmacologically active; oxymorphone has a 10- to 45-fold higher affinity for MOR and is 8- to 30-fold more potent compared with oxycodone.

Noroxymorphone: A metabolite of oxymorphone (via CYP3A) and noroxycodone (via CYP2D6). Noroxymorphone has fourfold higher affinity for MOR and is twofold more potent than oxycodone. However, noroxymorphone does not cross the BBB and is unlikely to contribute to the analgesic action of oxycodone.

Noroxycodone: A metabolite of oxycodone via CYP3A metabolism. Noroxycodone is weakly pharmacologically active, its affinity for MOR is one-third and potency 5- to 10-fold less than oxycodone.

Off-license use: The medicine is being used in a way that is different to that described in the marketing authorization.

Key facts of epidural oxycodone

- Pain after laparoscopy is multifactorial: nociceptive, inflammatory pain caused by surgical trauma, visceral pain arising from abdominal organs and referred pain at shoulder area due to pneumoperitoneum. Neuropathic pain due to surgical nerve trauma or pneumoperitoneum caused nerve distention is also present.
- Oxycodone is a phenanthrene class MOR agonist, that is widely used worldwide. Oxycodone is well absorbed from mucous membranes and after dosing by mouth.
- Epidural oxycodone is off-label use of oxycodone.
- After epidural oxycodone 0.1 mg/kg onset of analgesic action is approximately 30 min and duration of analgesia 3–4 h.
- After epidural oxycodone adverse effects are similar or less common than after intravenous injection.

Summary points

- Pain after laparoscopic procedures should not be ignored.
- Early pain after laparoscopic surgery is moderate or severe in most patients.
- Postlaparoscopic pain is multifactorial; nociceptive and inflammatory pain from tissue trauma, visceral pain from operated abdominal organs, inflammation in abdominal cavity, and referred pain in the shoulder area.
- Pain treatment should be multimodal and include non-pharmacological and pharmacological means.
- Oxycodone is μ-opioid receptor agonist, and highly efficacious in visceral pain.
- Epidural oxycodone is suitable for postoperative pain management in patients with major laparoscopic surgery.

References

Abbott, N. J., & Friedman, A. (2012). Overview and introduction: The blood–brain barrier in health and disease. *Epilepsia, 53*, 1–6.

Aoyama, Y., Sakura, S., Wittayapairoj, A., Abe, S., Tadenuma, S., & Saito, Y. (2020). Continuous basal infusion versus programmed intermittent bolus for quadratus lumborum block after laparoscopic colorectal surgery: A randomized-controlled, double-blind study. *Journal of Anesthesia*. https://doi.org/10.1007/s00540-020-02791-x.

Aspinen, S., Harju, J., Juvonen, P., Karjalainen, K., Kokki, H., Paajanen, H., et al. (2014). A prospective, randomized study comparing minilaparotomy and laparoscopic cholecystectomy as a day-surgery procedure: 5-year outcome. *Surgical Endoscopy, 28*, 827–832.

Bäcklund, M., Lindgren, L., Kajimoto, Y., & Rosenberg, P. H. (1997). Comparison of epidural morphine and oxycodone for pain after abdominal surgery. *Journal of Clinical Anesthesia, 9*, 30–35.

Balayssac, D., Pereira, B., Bazin, J. E., Le Roy, B., Pezet, D., & Gagnière, J. (2017). Warmed and humidified carbon dioxide for abdominal laparoscopic surgery: Meta-analysis of the current literature. *Surgical Endoscopy, 31*, 1–12.

Birch, D. W., Dang, J. T., Switzer, N. J., Manouchehri, N., Shi, X., Hadi, G., et al. (2016). Heated insufflation with or without humidification for laparoscopic abdominal surgery. *Cochrane Database of Systematic Reviews, 19*, CD007821.

Cajanus, K., Neuvonen, M., Koskela, O., Kaunisto, M. A., Neuvonen, P. J., Niemi, M., et al. (2018). Analgesic plasma concentrations of oxycodone after surgery for breast cancer-which factors matter? *Clinical Pharmacology and Therapeutics, 103*, 653–662.

Choi, B. M., Ki, S. H., Lee, Y. H., Gong, C. S., Kim, H. S., Lee, I. S., et al. (2019). Effects of depth of neuromuscular block on postoperative pain during laparoscopic gastrectomy: A randomised controlled trial. *European Journal of Anaesthesiology, 36*, 863–870.

Choi, B. M., Lee, Y. H., An, S. M., Lee, S. H., Lee, E. K., & Noh, G. J. (2017). Population pharmacokinetics and analgesic potency of oxycodone. *British Journal of Clinical Pharmacology, 83*, 314–325.

Donatsky, A. M., Bjerrum, F., & Gögenur, I. (2013). Surgical techniques to minimize shoulder pain after laparoscopic cholecystectomy. A systematic review. *Surgical Endoscopy, 27*, 2275–2282.

Dunn, L. K., & Durieux, M. E. (2017). Perioperative use of intravenous lidocaine. *Anesthesiology, 126*, 729–737. https://doi.org/10.1097/ALN.0000000000001527.

Ekstein, P., Szold, A., Sagie, B., Werbin, N., Klausner, J. M., & Weinbroum, A. A. (2006). Laparoscopic surgery may be associated with severe pain and high analgesia requirements in the immediate postoperative period. *Annals of Surgery, 243*, 41–46.

Fricker, G., & Miller, D. S. (2004). Modulation of drug transporters at the blood–brain barrier. *Pharmacology, 70*, 169–176.

Gerbershagen, H. J., Aduckathil, S., van Wijck, A. J., Peelen, L. M., Kalkman, C. J., & Meissner, W. (2013). Pain intensity on the first day after surgery: A prospective cohort study comparing 179 surgical procedures. *Anesthesiology, 118*, 934–944.

Harju, J., Aspinen, S., Juvonen, P., Kokki, H., & Eskelinen, M. (2013). Ten-year outcome after minilaparotomy versus laparoscopic cholecystectomy: A prospective randomised trial. *Surgical Endoscopy, 27*, 2512–2516.

Hill, R., Dewey, W. L., Kelly, E., & Henderson, G. (2018). Oxycodone induced tolerance to respiratory depression: Reversal by ethanol, pregabalin and protein kinase C inhibition. *British Journal of Pharmacology, 175*, 2492–2503.

Hong, J. Y., & Lim, K. T. (2008). Effect of preemptive epidural analgesia on cytokine response and postoperative pain in laparoscopic radical hysterectomy for cervical cancer. *Regional Anesthesia and Pain Medicine, 33*, 44–51.

Jokela, R., Ahonen, J., Seitsonen, E., Marjakangas, P., & Korttila, K. (2010). The influence of ondansetron on the analgesic effect of acetaminophen after laparoscopic hysterectomy. *Clinical Pharmacology and Therapeutics, 87*, 672–678.

Jokela, R. M., Ahonen, J. V., Tallgren, M. K., Marjakangas, P. C., & Korttila, K. T. (2009). The effective analgesic dose of dexamethasone after laparoscopic hysterectomy. *Anesthesia and Analgesia, 109*, 607–615.

Joshi, G. P., Bonnet, F., & Kehlet, H. PROSPECT collaboration. (2013). Evidence-based postoperative pain management after laparoscopic colorectal surgery. *Colorectal Disease, 15*, 146–155.

Kehlet, H., Jensen, T. S., & Woolf, C. J. (2006). Persistent postsurgical pain: Risk factors and prevention. *Lancet, 367*, 1618–1625.

Kim, H. J., Lee, K. Y., Kim, M. H., Kim, H. I., & Bai, S. J. (2019). Effects of deep vs moderate neuromuscular block on the quality of recovery after robotic gastrectomy. *Acta Anaesthesiologica Scandinavica, 63*, 306–313.

Kinnunen, M., Kokki, H., Hautajärvi, H., et al. (2019). Oxycodone concentrations in the central nervous system and cerebrospinal fluid after epidural administration to the pregnant ewe. *Basic & Clinical Pharmacology & Toxicology, 125*, 430–438.

Kinnunen, M., Piirainen, P., Kokki, H., Lammi, P., & Kokki, M. (2019). Updated clinical pharmacokinetics and pharmacodynamics of oxycodone. *Clinical Pharmacokinetics, 8*, 705–725.

Koivusalo, A. M., Kellokumpu, I., Scheinin, M., Tikkanen, I., Mäkisalo, H., & Lindgren, L. (1998). A comparison of gasless mechanical and conventional carbon dioxide pneumoperitoneum methods for laparoscopic cholecystectomy. *Anesthesia and Analgesia, 86*, 153–158.

Kokki, M., Broms, S., Eskelinen, M., Neuvonen, P. J., Halonen, T., & Kokki, H. (2012). The analgesic concentration of oxycodone with co-administration of paracetamol—A dose- finding study in adult patients undergoing laparoscopic cholecystectomy. *Basic & Clinical Pharmacology & Toxicology, 111*, 391–395.

Kokki, M., Broms, S., Eskelinen, M., Rasanen, I., Ojanperä, I., & Kokki, H. (2012). Analgesic concentrations of oxycodone—A prospective clinical PK/PD study in patients with laparoscopic cholecystectomy. *Basic & Clinical Pharmacology & Toxicology, 110*, 469–475.

Kokki, H., & Kokki, M. (2016). Central nervous system penetration of the opioid oxycodone. V. R. Preedy (Ed.), *Neuropathology of drug addictions and substance misuse.* (1st ed.). Vol. 3. New York: Academic ISBN:9780128006344.

Kokki, H., Kokki, M., & Sjövall, S. (2012). Oxycodone for the treatment of postoperative pain. *Expert Opinion on Pharmacotherapy, 13*, 1045–1058.

Kokki, H., Ruuskanen, A., & Karvinen, M. (2002). A comparison of epidural pain treatment with sufentanil-ropivacaine infusion with or without adrenaline. A double-blind, randomised clinical trial in children. *Acta Anaesthesiologica Scandinavica, 46*, 647–653.

Kokki, M., Pesonen, M., Vehviläinen, P., Litmala, O., Pasanen, M., & Kokki, H. (2016). Cytotoxicity of oxycodone and morphine in human neuroblastoma and mouse motoneuronal cells: A comparative approach. *Drugs in R&D, 16*(2), 155–163. https://doi.org/10.1007/s40268-016-0125-0.

Kokki, M., Välitalo, P., Kuusisto, M., Ranta, V. P., Raatikainen, K., Hautajärvi, H., et al. (2014). Central nervous system penetration of oxycodone after intravenous and epidural administration. *British Journal of Anaesthesia, 112*, 133–140.

Lalovic, B., Kharasch, E., Hoffer, C., Risler, L., Liu-Chen, L. Y., & Shen, D. D. (2006). Pharmacokinetics and pharmacodynamics of oral oxycodone in healthy human subjects: Role of circulating active metabolites. *Clinical Pharmacology and Therapeutics, 79*, 461–479.

Lee, S. J., Hyung, W. J., Koo, B. N., Lee, J. Y., Jun, N. H., Kim, S. C., et al. (2008). Laparoscopy-assisted subtotal gastrectomy under thoracic epidural-general anesthesia leading to the effects on postoperative micturition. *Surgical Endoscopy, 22*, 724–730.

Lee, D. H., Song, T., Kim, K. H., & Lee, K. W. (2018). Incidence, natural course, and characteristics of postlaparoscopic shoulder pain. *Surgical Endoscopy, 32*, 160–165.

Lemberg, K. K., Siiskonen, A. O., Kontinen, V. K., Yli-Kauhaluoma, J. T., & Kalso, E. A. (2008). Pharmacological characterization of noroxymorphone as a new opioid for spinal analgesia. *Anesthesia and Analgesia, 106*, 463–470.

Monory, K., Greiner, E., Sartania, N., Sallai, L., Pouille, Y., Schmidhammer, H., et al. (1999). Opioid binding profiles of new hydrazone, oxime, carbazone and semicarbazone derivatives of 14-alkoxymorphinans. *Life Sciences, 64*, 2011–2020.

Nakhli, M. S., Kahloul, M., Jebali, C., Frigui, W., & Naija, W. (2018). Effects of gabapentinoids premedication on shoulder pain and rehabilitation quality after laparoscopic cholecystectomy: Pregabalin versus gabapentin. *Pain Research & Management*, 9834059.

Nesek-Adam, V., Grizelj-Stojčić, E., Mršić, V., Rašić, Z., & Schwarz, D. (2012). Preemptive use of diclofenac in combination with ketamine in patients undergoing laparoscopic cholecystectomy: A randomized, double-blind, placebo-controlled study. *Surgical Laparoscopy, Endoscopy & Percutaneous Techniques, 22*, 232–238.

Pergialiotis, V., Vlachos, D. E., Kontzoglou, K., Perrea, D., & Vlachos, G. D. (2015). Pulmonary recruitment maneuver to reduce pain after laparoscopy: A meta-analysis of randomized controlled trials. *Surgical Endoscopy, 29*, 2101–2108.

Phelps, P., Cakmakkaya, O. S., Apfel, C. C., & Radke, O. C. (2008). A simple clinical maneuver to reduce laparoscopy-induced shoulder pain: A randomized controlled trial. *Obstetrics and Gynecology, 111*, 1155–1160.

Piirainen, P., Kokki, H., Anderson, B., Hannam, J., Hautajärvi, H., Ranta, V. P., et al. (2019). Analgesic efficacy and pharmacokinetics of epidural oxycodone in pain management after gynaecological laparoscopy—A randomised, double blind, active control, double-dummy clinical comparison with intravenous administration. *British Journal of Clinical Pharmacology, 85*, 1798–1807.

Piirainen, P., Kokki, H., Hautajärvi, H., Ranta, V. P., & Kokki, M. (2018). The analgesic efficacy and pharmacokinetics of epidural oxycodone after gynaecological laparotomy: A randomized, double-blind, double-dummy comparison with intravenous administration. *British Journal of Clinical Pharmacology, 84*, 2088–2096.

Piirainen, A., Kokki, H., Immonen, S., Eskelinen, M., Häkkinen, M. R., Hautajärvi, H., et al. (2015). A dose-finding study of dexketoprofen in patients undergoing laparoscopic cholecystectomy: A randomized clinical trial on effects on the analgesic concentration of oxycodone. *Drugs in R&D, 15*, 319–328.

Piirainen, P., Kokki, M., Räsänen, J., Tuominen, H., Voipio, H. M., Laisalmi, E., et al. (2019). Fetal surgery and asphyxia may disturb spinal cord integrity—An experimental study in a pregnant sheep. *Journal of Veterinary Science & Research, 4*, 044.

Pöyhiä, R., & Kalso, E. A. (1992). Antinociceptive effects and central nervous system depression caused by oxycodone and morphine in rats. *Pharmacology & Toxicology, 70*, 125–130.

Pöyhiä, R., Olkkola, K. T., Seppälä, T., & Kalso, E. (1991). The pharmacokinetics of oxycodone after intravenous injection in adults. *British Journal of Clinical Pharmacology, 32*, 516–518.

Pöyhiä, R., Seppälä, T., Olkkola, K. T., & Kalso, E. (1992). The pharmacokinetics and metabolism of oxycodone after intramuscular and oral administration to healthy subjects. *British Journal of Clinical Pharmacology, 33*, 617–621.

Puhto, T., Kokki, M., Hakomäki, H., et al. (2020). Single dose epidural hydromorphone in labour pain: Maternal pharmacokinetics and neonatal exposure. *European Journal of Clinical Pharmacology, 76*, 969–977.

Qiu, J., Xie, M., & Qu, R. (2019). The influence of etoricoxib on pain control for laparoscopic cholecystectomy: A meta-analysis of randomized controlled trials. *Surgical Laparoscopy, Endoscopy & Percutaneous Techniques, 29*, 150–154.

Radosa, J. C., Radosa, M. P., Schweitzer, P. A., et al. (2019). Impact of different intraoperative CO_2 pressure levels (8 and 15 mmHg) during laparoscopic hysterectomy performed due to benign uterine pathologies on postoperative pain and arterial pCO_2: a prospective randomised controlled clinical trial. *BJOG : An International Journal of Obstetrics and Gynaecology, 126*, 1276–1285.

Roh, H. F., Nam, S. H., & Kim, J. M. (2018). Robot-assisted laparoscopic surgery versus conventional laparoscopic surgery in randomized controlled trials: A systematic review and meta-analysis. *PLoS One, 13*. https://doi.org/10.1371/journal.pone.0191628).

Sadiq, M. W., Boström, E., Keizer, R., Björkman, S., & Hammarlund-Udenaes, M. (2013). Oxymorphone active uptake at the blood-brain barrier and population modeling of its pharmacokinetic-pharmacodynamic relationship. *Journal of Pharmaceutical Sciences, 102*, 3320–3331.

Salomäki, T. E., Kokki, H., Turunen, M., Havukainen, U., & Nuutinen, L. S. (1996). Introducing epidural fentanyl for on-ward pain relief after major surgery. *Acta Anaesthesiologica Scandinavica, 40*, 704–709.

Salomäki, T. E., Laitinen, J. O., & Nuutinen, L. S. (1991). A randomized double-blind comparison of epidural versus intravenous fentanyl infusion for analgesia after thoracotomy. *Anesthesiology, 75*, 790–795.

Sjövall, S., Kokki, M., & Kokki, H. (2015). Laparoscopic surgery: A narrative review of pharmacotherapy in pain management. *Drugs, 75*, 1867–1889.

Song, T., Kim, K. H., & Lee, K. W. (2017). The intensity of postlaparoscopic shoulder pain is positively correlated with the amount of residual pneumoperitoneum. *Journal of Minimally Invasive Gynecology, 24*, 984–989.

Spaner, S. J., & Warnock, G. L. (1997). A brief history of endoscopy, laparoscopy, and laparoscopic surgery. *Journal of Laparoendoscopic & Advanced Surgical Techniques Part A, 7*, 369–373.

Stamer, U. M., Zhang, L., Book, M., Lehmann, L. E., Stuber, F., & Musshoff, F. (2013). CYP2D6 genotype dependent oxycodone metabolism in postoperative patients. *PLoS One, 8*, e60239.

Tsai, H. W., Chen, Y. J., Ho, C. M., Hseu, S. S., Chao, K. C., Tsai, S. K., et al. (2011). Maneuvers to decrease laparoscopy-induced shoulder and upper abdominal pain: A randomized controlled study. *Archives of Surgery, 146*, 1360–1366.

van Dijk, J., Dedden, S. J., Geomini, P., van Kuijk, S., van Hanegem, N., Meijer, P., et al. (2018). Randomised controlled trial to estimate reduction in pain after laparoscopic surgery when using a combination therapy of intraperitoneal normal saline and the pulmonary recruitment manoeuvre. *BJOG, 125*, 1469–1476.

VanDenKerkhof, E. G., Hopman, W. M., Goldstein, D. H., Wilson, R. A., Towheed, T. E., Lam, M., et al. (2012). Impact of perioperative pain intensity, pain qualities, and opioid use on chronic pain after surgery: A prospective cohort study. *Regional Anesthesia and Pain Medicine, 37*, 19–27.

Villesen, H. H., Foster, D. J., Upton, R. N., Somogyi, A. A., Martinez, A., & Grant, C. (2006). Cerebral kinetics of oxycodone in conscious sheep. *Journal of Pharmaceutical Sciences, 95*(8), 1666–1676. https://doi.org/10.1002/jps.20632.

Yanagidate, F., & Dohi, S. (2004). Epidural oxycodone or morphine following gynaecological surgery. *British Journal of Anaesthesia, 93*, 362–367.

Yoburn, B. C., Shah, S., Chan, K., Duttaroy, A., & Davis, T. (1995). Supersensitivity to opioid analgesics following chronic opioid antagonist treatment: Relationship to receptor selectivity. *Pharmacology, Biochemistry, and Behavior, 51*, 535–539.

Yu, T., Cheng, Y., Wang, X., Tu, B., Cheng, N., Gong, J., et al. (2017). Gases for establishing pneumoperitoneum during laparoscopic abdominal surgery. *Cochrane Database of Systematic Reviews, 6*, CD009569.

Zeeni, C., Aouad, M. T., Daou, D., Naji, S., Jabbour-Khoury, S., Alami, R. S., et al. (2019). The effect of intraoperative dexmedetomidine versus morphine on postoperative morphine requirements after laparoscopic bariatric surgery. *Obesity Surgery, 9*, 3800–3808.

Zhu, Z., Li, L., Xu, J., Ye, W., Zeng, J., Chen, B., et al. (2020). Laparoscopic versus open approach in gastrectomy for advanced gastric cancer: A systematic review. *World Journal of Surgical Oncology, 18*, 126–150.

Chapter 39

Levobupivacaine features and linking in infiltrating analgesia

D. Bagatin[a,b], T. Bagatin[a,b], J. Nemrava[a,b], K. Šakić[a,b], L. Šakić[b], J. Deutsch[a], E. Isomura[a], M. Malić[a], M. Šarec Ivelj[a,b], and Z. Kljajić[a]

[a]Polyclinic Bagatin, Department of Surgery and Anesteslology with Ranimatology, Zagreb, Split, Croatia, [b]Faculty of Dental Medicine and Health Osijek, Department of Surgery and Anesteslology With Reanimatology, Depatment of Dental Medicine 1, Univeristy Josip Juraj Strossmayer, Osijek, Croatia

Pain relief (PR) after surgery continues to be a major medical challenge in clinical practice (Rawal, 2016). Acute postoperative pain following abdominal surgeries results in hemodynamic instability, decreased postoperative pulmonary function, delayed recovery, and discharge from hospital (Wu & Fleisher, 2000). Modern anesthesia has advanced, in which all patients can be guaranteed a pain-free intraoperative period. Various modalities of providing postoperative analgesia are being used such as intravenous (IV) nonsteroidal antiinflammatory drugs (NSAIDs) or opioids, epidural analgesia, regional nerve blocks, and also wound infiltration techniques.

Wound infiltration with local anesthetics is a simple concept for providing effective postoperative analgesia for a variety of surgical procedures without any major side effects (Scott, 2010). A significant proportion of surgical pain originates from surgical wound, it is meaningful or effective to use local anesthetics at the site of surgery to manage perioperative pain (McCarthy & Iohom, 2012). Wound infiltration techniques act by blocking the transmission of pain from nociceptive afferents directly from the wound surface and also decreases the local inflammatory response to injury (Liu, Richmann, Thirlby, & Wu, 2006).

A single shot of wound infiltration with a long acting local anesthetics provides analgesia for 4–8 h. Various adjuvants when added with local anesthetics such as opioids, nonopioids, vasoconstrictors, N-methyl D-aspartate antagonists, alpha 2 agonists, and neostigmine can prolong postoperative analgesic effect (Christiansson, 2009).

Pharmacological basis for choice of local anesthetic for central and peripheral blocks

Local anesthetics (LAs) are a different group of agents, which produce reversible inhibition of impulse transmission in peripheral nerves, thus blocking neuronal function in a circumscribed area of the body. LAs may provide anesthesia and analgesia in various parts of the body by different ways of administration. Regional anesthesia may be classified as topical and infiltration anesthesia, peripheral nerve blockade, and central nerve blockade.

LAs are different in their structure and pharmacological properties. All used LAs are either amino esters or amino amides. The important properties of an LA are its potency, onset and duration of action, and safety margin. When used systemically, LAs alter the function of all excitatory tissues and can produce local and systemic toxicity.

Understanding the pharmacology and knowledge of individual agents is essential when choosing a local anesthetic agent. Different techniques have different risks and requirements, and are related to clinical features of the individual agents and to the condition of an individual patient. Except choosing of the agent to be used, decisions have to be made about concentrations, volume, and use of additives.

In this chapter, we will discuss levobupivacaine, its features, and combined use in infiltrating analgesia.

LIA is a new concept of administration of local anesthetic and for alleviating pain. Now creating an adjuvant analgesia therapy with more local pain control is the goal. In this way, a blockade of one pain impulse path is eliminated.

S (−) Bupivacaine is an enantiomer of buvivacaine with a smaller amount of complications, as with racemic type of bupivacaine. It is an amino amide group, anesthetics. Differences between bupivacaine isomers were discovered in 1972 by Aberg and Luduena. The beginnings of its use in long lasting analgesics started around 20 years ago, when technology for production of these specific enantiomers began.

Features and Assessments of Pain, Anesthesia, and Analgesia. https://doi.org/10.1016/B978-0-12-818988-7.00033-9
Copyright © 2022 Elsevier Inc. All rights reserved.

434 **PART | III** Interlinking anesthesia, analgesics and pain

Less toxicity has been seen in the central nervous system and concomitant heart failure. Most frequently cardiac failure has been connected to newly appearing arrhythmias and rhythm failures with possible fatal outcomes which have been connected with ventricular fibrillation. The possibility of this occurring has been very rare with the recommended concentrations used.

In our practice, we use it mostly in breast and body surgery. Working on a large body surface area, topical analgesia is used.

Local anesthetic properties

Local anesthetics have a common molecular structure and similar modes of action. They are different in regards to their potency, onset, duration of effects, and toxicity.

All commonly used LAs have a three-part structure: aromatic ring, intermediate chain, and amino group. The intermediate chain contains either an ester or an amide linkage. Based on this linkage, LAs are classified into amino esters or amino amides.

The aromatic ring of LA's molecular structure is responsible for LA lipophilic properties. Lipophilicity expresses the tendency of a compound to associate with membrane lipids and pass through the membrane, into the axoplasm, on its way to reach the sodium channel. The lipophilic properties of the molecule are directly related to its potency. The lipid solubility can be increased by alkyl substitution for the benzene ring on the remaining structure (Raj, 2000). Adding an aliphatic group to the tertiary amine of mepivacaine creates bupivacaine, which is more lipid-soluble and more potent.

The hydrophilic part of the LA's molecular structure is responsible for agent ionization. Inside the cell the amino group becomes charged, into a cationic form, which is predominantly capable of blocking the sodium channel.

The intermediate linkage of LA molecules determines the resistance, of the molecule to hydrolysis. The amino amides are more resistant than the amino esters.

LAs in solution exist in chemical equilibrium between the unprotonated uncharged basic form and the protonated charged cationic form. At a certain hydrogen ion concentration specific for each agent, the concentration of LA base is equal to the concentration of charged cation. This hydrogen ion concentration is called the pKa. Agents with a higher pKa are more likely to be protonated at a physiologic pH. The onset of LA action is related to pKa. Onset is related to the concentration of the base on the extracellular side of the nerve cell membrane. The higher the pKa, the higher the percentage of the cationic form, and the concentration of the base is lower, which means a slower speed of LA onset.

Both charged and uncharged forms contribute to the diffusion of anesthetic molecules to the sites of action in nerve membranes.

The duration of the LA action is related to protein binding (Table 1). LAs with greater protein binding remain associated with the neuronal membrane for a longer time interval.

The mechanism of LA action

The function of excitable tissue is based on the presence of a cell with a lipid membrane, axoplasm, membrane-integrated, ion-specific channels, and ion gradients maintained by energy-dependent enzymes. Potassium moves freely through the membrane, whereas sodium moves in a semipermeable manner, controlled by gates on the sodium channels. Sodium is restricted to the extracellular space, except when specific ion channels are open. Potassium selectively accumulates inside

TABLE 1 Properties of some local anesthetics.

Agent	Lipid solubility	Relative potency	pKa	Onset of action	Plasma protein binding (%)	Duration (min)
Procaine	1	1	8.9	Slow	6	60–90
Lidocaine	3.6	2	7.7	Fast	65	90–200
Bupivacaine	30	8	8.1	Medium	95	180–600
Ropivacaine	2.8	8	8.1	Medium	94	170–470
Levobupivacaine	30	8	8.1	Medium	95	180–600

the nerve cell to preserve electrical neutrality. In an unexcited state, the electrical potential inside the cell is negative, outside the cell it is positive. This is the resting state potential of the nerve cell membrane. Nerve stimulation opens sodium channels and during the conduction of an impulse (action potential) opens sodium channels allowing sodium to enter, depolarizing the cell. The sodium channels then close and an outward movement of potassium results in a returning to the resting state potential. The sodium–potassium pump restores ion distribution during the resting phase of the action potential.

LAs act primarily by blocking sodium channels and prevent depolarization. The gate, that opens and closes sodium channels, is present on the axoplasmic side of the nerve cell membrane. In the open state, this channel is susceptible to the action of LA molecules, causing it to remain inactive and prevent subsequent depolarization, leading to a conduction blockade. LAs have a higher binding affinity for the open sodium channels than for the closed channels (Tetzlaff, 2000).

To reach its target, the LAs have to be lipid-soluble to diffuse across both the cell membrane and the multiple coverings of a peripheral nerve. Local anesthetics are weak bases with low water solubility. Their formulation of hydrochloride salts increases the ionized fraction, which equilibrates in aqueous solution, with a small amount of free base. After an injection of physiological buffers, this raises the pH and increases the proportion of the unionized lipophilic fraction available, to diffuse into the nerve. Within the intracellular fluid, the slightly more acidic pH, favors the ionized form that blocks the sodium channel.

Pharmacokinetics

Pharmacokinetics describes the movement of an agent through the body. The concentration of an LA in blood is determined by the amount injected, the rate of absorption from the site of injection, the rate of tissue distribution, and the rate of its biotransformation and excretion. Patient-related factors such as age, cardiovascular status, and hepatic function influence the physiologic disposition and the resultant blood concentrations of LAs. An understanding of pharmacokinetics enables us to predict the effect that will be produced with a certain agent.

Absorption

The systemic absorption of an LA is determined by the site of injection, the dosage and volume, addition of a vasoconstrictor agent, and by the pharmacological profile of the LA.

Local vascularity and the proportion of tissue and fat influence the rate of uptake and removal from specific sites. The rate of systemic absorption is an important factor in determining the peak blood level (C_{max}) and the appearance of toxic effects of LAs (Table 2) (Benzon, Raja, Borsook, et al., 1999). The rate of absorption of LAs from various sites decreases in the following order: interpleural > intercostal > caudal > epidural > brachial plexus > sciatic/femoral > subcutaneous > intra-articular > spinal (Avidan, Harvey, Ponte, et al., 2003; Kopatz, Bernards, Allen, et al., 2003). The maximum blood level of an LA is related to the total dose of agent administered for any particular site of administration. Higher blood levels also follow the administration of larger volumes of the same LA dose.

TABLE 2 Peak blood levels (C_{max}) of LAs after various blocks (T_{max} = time until C_{max}).

	Technique	Dose (mg)	C_{max} (µg/mL)	T_{max} (min)	Toxic plasma concentration (µg/mL)
Bupivacaine	Brachial plexus	150.0	1.0	20	3
		100.0	1.50	17	
	Celiac plexus	150.0	1.26	20	
	Epidural	140.0	0.90	30	
	Intercostal	52.5	0.49	24	
	Lumbar sympathetic Sciatic/ femoral	400.0	1.89	15	
Ropivacaine	Brachial plexus	190.0	1.3	53	3
		150.0	1.07	40	
	Epidural Intercostal	140.0	1.10	21	

436 PART | III Interlinking anesthesia, analgesics and pain

A vasoconstrictor agent decreases the rate of absorption of certain LAs from various sites of administration. Physiochemical differences between LAs also significantly affect C_{max}. Bupivacaine and etidocaine appear to produce greater vasodilatation than lidocaine or mepivacaine. Greater vasodilatation should increase the rate of absorption of LAs.

Distribution, biotransformation, and excretion

LAs are distributed throughout all body tissues. The concentration in different tissues varies due to tissue blood flow and the relative blood and tissue solubility of LAs. LAs are rapidly extracted by lung tissue, so that the whole blood concentration of LAs decreases markedly as these agents pass through the pulmonary vasculature. The highest percentage of an injected dose of LA is found in skeletal muscle and it represents the largest reservoir for LAs.

Distribution and elimination are influenced by: protein binding, clearance and metabolism, physiologic effects of absorbed LA, and other physical and pathophysiologic factors.

Pharmacological activity is generally related to unbound or free agent levels. The extent of protein binding varies considerably among the LAs and correlates with the duration of their activity: bupivacaine/etidocaine/levobupivacaine > ropivacaine > mepivacaine > lidocaine > procaine and 2-chlorprocaine. When the concentration of LA increases, the percentage that is bound to proteins decreases, probably as a result of saturation of binding sites.

Amide LAs are primarily bound to α_1-acid glycoprotein (AAG) (high affinity but low capacity) and, to a lesser extent, to albumin (low affinity but high capacity).

Protein binding is also influenced by the pH of the solution so that the percentage of the agent bound decreases as the pH decreases. The influence of pH differs among LAs.

The extent of protein binding varies considerably among patients and is reflected in differences in the effects produced by the same total blood level of LA. AAG plasma concentrations are decreased in pregnant women and in newborns. Protein binding is important in the understanding of placental transfer of agents.

Amino amide LAs are primarily cleared from the bloodstream by metabolism in the liver, with only prilocaine having any significant extrahepatic metabolism. The rate of clearance is largely determined by hepatic blood flow.

Renal clearance of unchanged LAs is a minor route of elimination (3%–10%).

Since the rate of clearance is similar for all amides, the duration of anesthesia varies. Lidocaine and mepivacaine tend to accumulate during continuous techniques, whereas bupivacaine accumulates only minimally.

Amino ester LAs are primarily cleared from the blood by plasma and liver cholinesterase. The rate of hydrolysis of ester-linked LAs depends on the type and location of the substitution in the aromatic ring. Although the amino ester LAs are rapidly metabolized in normal patients, patients with abnormal or deficient plasma cholinesterase can exhibit signs of toxicity from usual dosages of LAs (Raj, 2003). The hydrolysis of all ester-linked LAs leads to the formation of para-aminobenzoic acid (PABA) or a substituted PABA. PABA and its derivatives are associated with a low but real potential for allergic reactions.

Some metabolites of LAs have been shown to be pharmacologically active. Monoethylglycinexylidide (MEGX), the metabolite that arises from N-de-ethylation of lidocaine, is near equipotent with lidocaine. The metabolite from bupivacaine and ropivacaine, PPX, is less active than the parent compound.

The systemic effects of neural blockade, or the absorbed LA itself, can alter the pharmacokinetics of the LA by changes in blood flow in specific regions and organs.

Many diseases influence LAs pharmacokinetics. A patient's age may influence the physiologic disposition of LAs. Elderly patients have demonstrated prolonged lidocaine half-life. Newborns have immature hepatic enzyme systems and prolonged elimination of lidocaine and bupivacaine.

The rate of degradation of amide type LAs is influenced by the hepatic status of the patient. In patients with poor liver function or low hepatic blood flow, significantly higher blood levels of amide LAs occur, due to reduced plasma clearance and prolonged half-life of LAs.

A reduction in cardiac output reduces the volume of distribution and plasma clearance of LAs.

Cholinesterase activity is reduced in newborns, during pregnancy, in patients with renal or liver disease and in patients with genetic changes.

Concomitant administration of some agents can influence protein binding of LAs or their metabolism.

Local anesthetic toxicity

Systemic toxicity of LAs is a result of their action on excitable tissues in the brain and heart, and also depends on systemic concentrations. Plasma concentrations depend on the total dose given, the rate of absorption, the pattern of distribution, and the rate of metabolism. Life-threatening toxicity can result from an inadvertent intravascular injection or from an absolute overdose. There is progressive depression of the central nervous and cardiovascular system as the plasma concentrations increase, but the CNS effects generally appear first. Serious cardiac effects usually become apparent only as a result of hypoxemia, or if a large dose of an agent, such as bupivacaine, enters the circulation rapidly. It has been proven that bupivacaine toxicity is connected with its R-enantiomer. This discovery leads to the formulation of pure S-enantiomer of bupivacaine, creating two of the newer LAs, ropivacaine and levobupivacaine (Dony, Dewinde, Vanderick, et al., 2000; Foster & Markham, 2000; Markham & Faulds, 1996; McClure, 1996; Milligan, 2004; Polley & Columb, 2003; Steward, Kellett, & Castro, 2003).

Lower toxicity is visible in toxicity connected with CNS and cardiotoxicity. There is always most prominent effect on cardiac function and in possibility of disturbance in cardiac rhythm with fatal consequences mostly occurring because of ventricular fibrillation. The possibility of fatal consequences is extremely low with levobupivacaine in doses which are usually used.

Local toxic reactions are rare when using clinically relevant concentrations of the anesthetic solutions, but can lead to severe and permanent complications, such as with an intraneuronal injection. Factors implicated in transient or persistent neuropathy include acidic solutions, additives, the agent itself, needle trauma, compression from hematomas, and inadvertent injection of neurolytic agents (Mcleod, 2001).

Pharmacology and use of levobupivacaine

Based on findings cardiotoxicity is infrequently observed with racemic bupivacaine and shows enantioselectivity, i.e., it is more pronounced with the R(+)-enantiomer, the S(−)-enantiomer (levobupivacaine) has been developed for clinical use, as a long acting local anesthetic. The majority of in vitro, in vivo, and human pharmacodynamic studies of nerve block indicate that levobupivacaine (Table 3) has a similar potency to bupivacaine. However, levobupivacaine had a lower risk of cardiovascular and CNS toxicity than bupivacaine seen in animal studies. In human volunteers, levobupivacaine had less of a negative inotropic effect and, at intravenous doses >75 mg, produced less prolongation of the QTc interval than bupivacaine. Fewer changes indicative of CNS depression on EEG were evident with levobupivacaine.

With an epidural administration, levobupivacaine produced less prolonged motor block than sensory block (Table 4). Conditions satisfactory for surgery and good pain management were achieved by use of local infiltration of levobupivacaine, epidural administration during labor, and was effective for the management of postoperative pain, especially when combined with clonidine, morphine, or fentanyl. The tolerability profiles of levobupivacaine and bupivacaine were very

---

TABLE 3 Levobupivacaine is long acting with a dose-dependent duration of anesthesia.

1. The onset of action is < or = 15 min with various anesthetic techniques.

2. Levobupivacaine provided sensory block for up

    a. 9 h after epidural administration of < or = 202.5 mg,

    b. 6.5 h after intrathecal 15 mg, and

    c. 17 h after brachial plexus block with 2 mg/kg.

TABLE 4 Levobupivacaine hydrochloride.

Proprietary name	Manufacturer	% Local anesthetic	Vaso-constrictor	Duration of analgesia (min)	MRD
Chirocaine	(AbbVie)	0.5	Epinephrine 1:200,000	Soft tissue 240–540 (reports up to 720)	1.3 mg/kg 0.6 mg/lb 90 mg absolute maximum

MRD, maximum recommended dose.

438 **PART | III** Interlinking anesthesia, analgesics and pain

similar and no clinically significant ECG abnormalities or serious CNS events occurred with the doses used. Treatment of hypotension is the most common adverse event associated with levobupivacaine (Foster & Markham, 2000).

Our local infiltration analgesia (LIA) body protocol

Two most frequent procedures in functional and aesthetic surgery of the body are breast augmentation surgery and liposuction which are sometimes combined with abdominoplasty. Breast augmentation surgery is the number one procedure in aesthetic surgery in the world and in 2019, 1,677,320 procedures were performed, while immediately following is liposuction with 1,573,680 procedures.

In the operative procedure of liposuction or the combination of liposuction and abdominoplasty tumescent solution is used which is described above as a 1 L solution (physiological solution or Ringer solution) combined with: 62.5 mg Levobupivacaine, 0.84 g $NaHCO^3$ and 1 mg epinephrine. We usually use 6–8 L of solution for infiltration, waiting for 15 min and then following with VASER ultrasound application and light liposuction. At the end of the operation we fan-shaped infiltration of the abdominal wall muscles in the amount of 40 mL of Levobupivacaine. In this way there is adequate vasoconstriction for blood preservation and adequate liposuction. This infiltration of abdominal wall provides good analgesia of abdominal superficial tissue and wall muscles. Another benefit is early and adequate mobilization of the patient within the early postoperative period and a faster recovery. Mobilization needs to be on the first postoperative day with a full upright position achieved in 2–3 days, the latest after 1 week. In this way, we have almost a full early recovery of patients, with minimal pain within the early postoperative period. We also prevent deep venous thrombosis and pulmonary embolism. It is important for us to reduce the amount of analgesics which our patients use, mainly using them in the first 2 to 3 days up until the first week.

In breast augmentation surgery, we solely use infiltrating analgesia in the area of incision usually the inframammary fold. For each side 8.6 mL of solution with 25 mg Levobupivacaine, 72 mg Lidocaine chloride and 45 µg Epinephrine is used. Postoperativley, we have good analgesia with just the feeling of pressure in the thoracic region, requiring only a little oral analgesia first few days.

The usual duration of levobupivacaine is 3–10 h (Table 5), while in our patients is longer and lasts for 2–3 days which helps our patients recover faster and return to normal activities.

Table of commonly used or experimental analgesics

Discussion

The aim is to prevent origins of pain at surgical sites with the administration of IV analgesics that can produce multiple side effects. Wound infiltration technique is the oldest technique known with better analgesia and the least side effects profile. Postoperative pain relief results in early mobilization of the patient, better hemodynamic stability, oral intake on the 1st postoperative day, and better satisfaction from patient and family.

TABLE 5 Characteristics of local anesthetics-injectable local anesthetics.

Local anesthetic	Onset (min)	Duration (min)	Duration with epinephrine (min)	Maximum dose (mg/kg)	Maximum dose with epinephrine (mg/kg)
Bupivacaine	2–10	120–240	240–480	2.5	3
Chloroprocaine	5–6	30–60	N/A	11	14
Etidocaine	3–5	200	240–360	4.5	6.5
Lidocaine	<1	30–120	60–400	4.5	7
Mepivacaine	3–20	30–120	60–400	6	7
Prilocaine	5–6	30–120	60–400	7	10
Procaine	5	15–90	30–180	10	14
Tetracaine	7	120–240	240–480	2	2

Preincisional infiltration can be planned for shorter duration surgeries it can alter the surgical anatomy and also effect the duration of analgesia dependent upon action of local anesthetic used. Cnar, Kum, Cevizci, Kayaoglu, and Oba (2009) concluded that there was no significant difference in the parameters between pre- and postincisional infiltration. Wound infiltration, administered at the end of surgery during wound closure, results in immediate postoperative pain relief that provides a peak action of infiltrated local anesthetics after extubation. It can be either a single shot wound infiltration with local anesthetics and adjuvants or a continuous wound infiltration technique for postoperative analgesia. Other modes of analgesia such as IV analgesia with NSAIDs and opioids, can result in opioid side effects, while intraperitoneal instillation of local anesthetics was not proved to be superior to wound infiltration (El-labban et al., 2011) in providing analgesia. Epidural or continuous spinal analgesia can cause disturbances in the hemodynamic parameters. However, wound infiltration with local anesthetics not only provides adequate pain control, but reduces the inflammatory responses and catecholamine levels secondary to surgery, which also has an added benefit of enhancing wound healing by increasing wound perfusion and oxygenation (Hopf et al., 1997).

Postoperative analgesia in patients undergoing abdominal surgeries with a single shot wound infiltration technique with Levobupivacaine has similar analgesic efficacy like bupivacaine with reduced cardiovascular and central nervous system toxicity but more effective than ropivacaine. The efficacy of postoperative analgesic duration of levobupivacaine can be doubled (23.4 h) by adding adjuvant clonidine or dexmedetomidine, an additive effect without the rescue analgesic requirement. When levobupivacaine is added to alpha 2 adrenergic agonist this prolongs the duration of analgesia and reduces the total analgesic consumption.

A study conducted by El-labban et al. (2011) showed that postoperative abdominal pain and consumption of rescue analgesia were lower in intraincisional group from 30 min to 24 h postoperatively. Ugley et al. (2015) concluded in their study that in patients undergoing elective total abdominal hysterectomy, the postoperative VAS score was significantly lower in Levobupivacaine/Dexmedetomidine group ($P < 0.003$).

The studies conducted by Mandal et al. (2016) showed nausea and vomiting occurred in few patients with LIA.

Furthermore, in abdominal surgeries involving manipulation of visceral structures, a single shot wound infiltration alone may not be sufficient in providing postoperative analgesia. We suggest further studies on this topic in elderly patients where regional or neuraxial analgesia is difficult or not possible, postoperative analgesia can be provided with long acting local anesthetics by wound infiltration technique. In addition, further studies are required to compare these two alpha 2 adrenergic agonists (clonidine and dexmedetomidine) for wound infiltration technique.

Conclusions

The use of local anesthetics is based on knowledge of basic pharmacology and the patient's condition, as well as, on the skill and experience of the anesthesiologist. The choice of LA is based mainly on time of onset and duration of the block, as well as on the ability to produce a differential sensory motor block and on the safety margin.

With regards to onset of action, there are LAs with slow (procaine, tetracaine, ropivacaine, levobupivacaine, bupivacaine) and fast onset (lidocaine, prilocaine).

With regards to potency, there are LAs of medium (procaine, lidocaine, prilocaine) and high potency (tetracaine, ropivacaine, levobupivacaine, bupivacaine).

With regards to duration of action, there are LAs with short (procaine), intermediate (lidocaine, prilocaine) and long duration (ropivacaine, levobupivacaine, bupivacaine).

Finally, in regards to toxicity, there are LAs with low (prilocaine), medium (procaine, lidocaine, ropivacaine, levobupivacaine) and high toxicity (tetracaine, bupivacaine).

There are many contributing factors that can result, in the overall actions of LAs.

There are always possibilities of combining of levobupivacaine with other medications, in a way to enhance, upgrade, and prolong its efficacy as a long-lasting analgesic, while at the same time lowering its toxicity, by reducing further the dose of levobupivacaine. In this way, good efficacy with a lower dose of analgesic is achieved. Of course, there is always the question: Is this important in levobupivacaine as with bupivacaine? Levobupivacaine toxicity is already low.

Levobupivacaine is a long acting local anesthetic with a clinical profile closely resembling that of bupivacaine. The current preclinical safety and toxicity data show an advantage for levobupivacaine over bupivacaine, and confirms that levobupivacaine is an appropriate choice for use instead of bupivacaine. Appropriate use of wound infiltration technique in patients undergoing abdominal surgeries with levobupivacaine has resulted in the enhancement of postoperative analgesia and excellent to good quality of analgesia with minimal side effect profile. Postoperative pain relief results in early mobilization of the patient, better hemodynamic stability, faster oral intake on the 1st postoperative day, and better patient and family satisfaction.

Applications to other areas

A nose job is one of the most frequent operations at Polyclinic Bagatin. It is the second most common aesthetic operation after breast augmentation. To prevent pain in the perioperative period, infiltration of the septum, dorsum, lateral walls, columella, tip, and alar base is performed. The infiltration is done before the first incision. In total, 6 mL of 0.25% levobupivacaine is used. This amount of anesthetic is sufficient enough to have good pain control.

Other use of anesthetics is with otapostasis surgery, in which the ears placed maximally close to the head. This procedure is done only with local anesthesia and this is the only anesthesia used during this procedure. A combination of 2 mL of 0.5% Levobupivacaine with 3 mL of 2% Xylodonta per ear. The duration is efficient for the entire operation and sometimes even lasts until the next day. The satisfaction rate is very high for both procedures.

There are times when only lidocaine with adrenalin is used for surgical procedures, such as in labiaplasty. In this case, 5 mL of 2% Xylodonta per side is used, and after completing the operation an additional 5 mL of levobupivacaine (25 mg) on both sides is added. This 5 mL is enough for both sides. In this way, good anesthesia and analgesia is achieved after the operation when lidocaine with adrenaline stops working and levobupivacaine starts. This provides patients with no pain for the first 2 days after their operations, which is the most painful period. This helped a lot in the postoperative period since reducing discomfort in the early postoperative period. Patients are extremely grateful for this.

Other agents of interest

Lidocaine is the most widely used and one of the first amino amide type local anesthetic, synthesized originally under the name of xylocaine by Swedish chemist Nils Löfgren in 1943. It is used for superficial infiltration nerve blocks and spinal anesthesia. Intravenously, it is also used for treating arrhythmias, especially during general anesthesia and an induced hypothermic state. Lidocaine is 95% metabolized in the liver. The elimination half-life is biphasic and is around 90 to 120 min in most patients. Prolongation can occur in patients with hepatic impairment or congestive heartfailure. Its metabolites are 90% excreted, while 10% remain unchanged. The metabolites can be either active or inactive, but they have less affinity and potency for blocking sodium channels. The mechanism of action is inactivation of the fast voltage gated Na + channels in the neuronal cell membranes. Blockage of the action potential, blocks the sensory neuron and blocks the sensation of pain. It can be combined with adrenalin to increase the length of its action (up to 180 min). Using adrenalin also increases the onset of analgesia from 4–5 min to 2–3 min.

Amino amide anesthetics are bases, therefore they are inefficient in an acidic (infectious) environment. Exposing them into such an environment neutralizes their effectiveness.

Using combinations with buffers, such as sodium bicarbonate, increases their alkalinity, increases the speed of onset, reduces pain on injection, but decreases the duration of action.

The combining of lidocaine and levobupivacaine is used often. Combining these two agents into one syringe offers the best effects of both drugs: the fast onset of lidocaine and longer duration of action of bupivacaine.

This combination is frequently used for various applications from paravertebral blocks for breast cancer to c-sections.

Ropivacaine was created because of frequent cardiac arrest after using pure bupivacaine, particularly in pregnant women. Ropivacaine has less cardio toxicity than bupivacaine. A reduced cardio toxicity than pure bupivacaine was developed in an S-enantiomer, levobupivacaine. This isomer is more frequently used now, and is known as Chirocaine.

Ropivacaine is preferred for epidural anesthesia due to its longer action, reduced absorption into motor fibers and reduced motor blockade, therefore allowing more involvement of the laboring patient.

Mini-dictionary of terms

Modern anesthesia has advanced, in which all patients can be guaranteed a pain-free intraoperative and postoperative period.

Local anesthetics (LAs) are a different group of agents, which produce reversible inhibition of impulse transmission in peripheral nerves, thus blocking neuronal function in a circumscribed area of the body.

Acute postoperative pain results in hemodynamic instability, decreased postoperative pulmonary function, delayed recovery, and discharge from hospital. This is the reason that a modern, practical, and pain-free anesthesia is needed.

Wound infiltration techniques act by blocking the transmission of pain from nociceptive afferents directly from the wound surface and also decreases the local inflammatory response to injury. This is the most effective local pain control which has been used and investigated in millions of people worldwide.

LIA (local infiltration anesthesia) is anesthesia which provides us good analgesia in the specific parts of the body in which we do surgery. Systemic toxicity of LAs is a result of their action on excitable tissues in the brain and heart, and also depend on systemic concentrations. Plasma concentrations depend on the total dose given, the rate of absorption, the pattern of distribution and the rate of metabolism. Life-threatening toxicity can result from an inadvertent intravascular injection or from an absolute overdose. There is progressive depression of the central nervous and cardiovascular system as the plasma concentrations increase, but the CNS effects generally appear first. Serious cardiac effects usually become apparent only as a result of hypoxemia, or if a larger dose of an agent, such as bupivacaine, enters the circulation rapidly. It has been proven that bupivacaine toxicity is connected with its R-enantiomer. This discovery leads to the formulation of pure S-enantiomer of bupivacaine, creating two of the newer LAs, ropivacaine and levobupivacaine.

Key facts of levobupivacaine

- The S(−)-enantiomer (levobupivacaine) has been developed for clinical use, as a long acting local anesthetic.
- S (−) Bupivacaine is an enantiomer of buvivacaine with a smaller amount of complications, as with racemic type of bupivacaine. It is an amino amide group, local anesthetic. Differences between bupivacaine isomers were discovered in 1972 by Aberg and Luduena. The beginnings of its use in long lasting analgesics started around 20 years ago, when technology for production of these specific enantiomers began.
- Levobupivacaine has a similar potency to bupivacaine.
- Levobupivacaine had a lower risk of cardiovascular and CNS toxicity than bupivacaine seen, in animal studies.
- In human volunteers, levobupivacaine had less of a negative inotropic effect and, at intravenous doses >75 mg, produced less prolongation of the QTc interval than bupivacaine.
- Fewer changes indicative of CNS depression on EEG were evident with levobupivacaine.
- Conditions satisfactory for surgery and good pain management were achieved by use of local infiltration of levobupivacaine, epidural administration during labor, and was effective for the management of postoperative pain, especially when combined with clonidine, morphine, or fentanyl.
- The tolerability profiles of levobupivacaine and bupivacaine were very similar and no clinically significant ECG abnormalities or serious CNS events occurred with the doses used.
- Treatment of hypotension is the most common adverse event associated with levobupivacaine.

Summary points

- Pain relief after surgery continues to be a major medical challenge in clinical practice.
- A single shot of wound infiltration with a long-acting local anesthetics provides analgesia for 4–8 h.
- Local anesthetics (LAs) are a different group of agents, which produce reversible inhibition of impulse transmission in peripheral nerves, thus blocking neuronal function in a circumscribed area of the body. All used LAs are either amino esters or amino amides.
- LIA is a new concept of administration of local anesthetic and for alleviating pain. Now creating an adjuvant analgesia therapy with more local pain control is the goal. In this way, a blockade of one pain impulse path is eliminated.
- S (−) Bupivacaine is an enantiomer of bupivacaine with a smaller amount of complications, as with racemic type of bupivacaine. It is an amino amide group, anesthetics.
- Less toxicity has been seen in the central nervous system and concomitant heart failure. Most frequently cardiac failure has been connected to newly appearing arrhythmias and rhythm failures with possible fatal outcomes that have been connected with ventricular fibrillation. The possibility of this occurring has been very rare with the recommended concentrations used.
- The rate of degradation of amide type LAs is influenced by the hepatic status of the patient. In patients with poor liver function or low hepatic blood flow, significantly higher blood levels of amide LAs occur, due to reduced plasma clearance and prolonged half-life of LAs.
- A reduction in cardiac output reduces the volume of distribution and plasma clearance of LAs.
- The use of local anesthetics is based on knowledge of basic pharmacology and the patient's condition, as well as, on the skill and experience of the anesthesiologist and the surgeon. The choice of LA is based mainly on time of onset and duration of the block, as well as on the ability to produce a differential sensory motor block and on the safety margin.

442 PART | III Interlinking anesthesia, analgesics and pain

- Postoperative pain relief results in early mobilization of the patient, better hemodynamic stability, faster oral intake on the 1st postoperative day, and better patient and family satisfaction and levobupivacaine is a long-acting local anesthetic that provides all that with minimal toxicity and complications and this is the reason that it is our first choice.

References

Avidan, M., Harvey, A. M. R., Ponte, J., et al. (2003). *Perioperative care, anaesthesia, pain management and intensive care* (pp. 88–91). London: Churchill Livingstone.

Benzon, H. T., Raja, S. N., Borsook, D., et al. (1999). *Essentials of pain medicine and regional anesthesia* (pp. 336–349). Philadelphia: Churchill Livingstone.

Christiansson, L. (2009). Update on adjuvants in regional anaesthesia. *Periodicum Biologorum, 111*, 161–170.

Cnar, S. O., Kum, U., Cevizci, N., Kayaoglu, S., & Oba, S. (2009). Effects of levobupivacaine infiltration on postoperative analgesia and stress response in children following inguinal hernia repair. *European Journal of Anaesthesiology, 26*, 430–434.

Dony, P., Dewinde, V., Vanderick, B., et al. (2000). The comparative toxicity of ropivacaine and Bupivacaine at equipotent doses in rats. *Anesthesia and Analgesia, 91*, 1489–1492.

El-labban, G. M., Hokkam, E. N., El-labban, M. A., Morsy, K., Saadl, H., & Heissam, K. S. (2011). Intraincisional vs. intraperitoneal infiltration of local anaesthetic for controlling early post-laparoscopic cholecystectomy pain. *Journal of Minimal Access Surgery, 7*, 173–177.

Foster, R. H., & Markham, A. (2000). Levobupivacaine. *Drugs, 59*, 551–579.

Hopf, H. W., Hunt, T. K., West, J. M., Blomquist, P., Goodson, W. H., 3rd, Jensen, J. A., et al. (1997). Wound tissue oxygen tension predicts the risk of wound infection in surgical patients. *Archives of Surgery, 132*, 997–1004.

Kopatz, D. J., Bernards, C. M., Allen, H. W., et al. (2003). A model to evaluate the pharmacokinetic and pharmacodynamic variables of extended-release products using in vivo tissue microdialysis in humans: Bupivacaine loaded microcapsules. *Anesthesia and Analgesia, 97*, 124–131.

Liu, S. S., Richmann, J. M., Thirlby, R. C., & Wu, C. L. (2006). Efficacy of continuous wound catheters delivering local anaesthetics for postoperative analgesia: A quantitative and qualitative systematic review of randomized controlled trials. *Journal of the American College of Surgeons, 203*, 914–932.

Mandal, D., Das, A., Chhaule, S., Halder, P. S., Paul, J., RoyBasunia, S., et al. (2016). The effect of dexmedetomidine added to preemptive (2% lignocaine with adrenaline) infiltration on intraoperative hemodynamics and postoperative pain after ambulatory maxillofacial surgeries under general anesthesia. *Anesthesia, Essays and Researches, 10*, 324–331.

Markham, A., & Faulds, D. (1996). Ropivacaine. *Drugs, 52*, 429–449.

McCarthy, D., & Iohom, G. (2012). Local infiltration analgesia for postoperative pain control following total hip arthroplasty: A systematic review. *Anesthesiology Research and Practice, 2012*, 709531.

McClure, J. H. (1996). Ropivacaine. *British Journal of Anaesthesia, 76*, 300–307.

Mcleod, D. B. (2001). Levobupivacaine. *Anesthesia, 56*, 331–341.

Milligan, K. R. (2004). Recent advances in local anaesthetics for spinal anaesthesia. *European Journal of Anaesthesiology, 21*, 837–847.

Polley, L. S., & Columb, M. O. (2003). Ropivacaine and bupivacaine: Concentrating on dosing! *Anesthesia and Analgesia, 96*, 1251–1253.

Raj, P. (2000). *Practical management of pain* (pp. 557–573). New York: Mosby.

Raj, P. (2003). *Radiographic imaging for anesthesia and pain management* (pp. 14–22). London: Elsevier.

Rawal, N. (2016). Current issues in postoperative pain management. *European Journal of Anaesthesiology, 33*, 160–171.

Scott, N. B. (2010). Wound infiltration for surgery. *Anesthesia, 65*(Suppl. 1), 67–75.

Steward, J., Kellett, N., & Castro, D. (2003). The central nervous system and cardiovascular effects of levobupivacaine and ropivacaine in healthy volunteers. *Anesthesia and Analgesia, 97*, 412–416.

Tetzlaff, J. E. (2000). The pharmacology of local anesthetics. *Anesthesiology Clinics of North America, 18*, 217–233.

Ugley, A., Gunes, I., Bayram, A., Cihangir, B., Kurt, F. M., Muderis, I., et al. (2015). The analgesic effects of incisional levobupivacaine with dexmedetomidine after total abdominal hysterectomy. *Erciyes Medical Journal, 37*, 64–68.

Wu, C. L., & Fleisher, L. A. (2000). Outcomes research in regional anesthesia and analgesia. *Anesthesia and Analgesia, 91*, 1232–1242.

Part IV

Assessments, screening, and resources

Chapter 40

The pain catastrophizing scale: Features and applications

Turgay Tuna

University Hospital Erasme, Free University of Brussels, Anderlecht, Belgium

Introduction

Pain catastrophizing is a persistent, negative cognitive-affective mental set, that was first coined by Ellis (1962), but earlier records can be found in traditional Chinese medicine or in French literature of the 19th century.

It is characterized by three dimensions, helplessness, magnification, and ruminative thoughts regarding one's pain, be it actual or anticipated (Schütze, Rees, Smith, et al., 2018), pain being already felt by most as a common negative experience that signifies danger, illness, and possible doom.

It is a potent predictor of negative pain-related outcomes in general (Campbell & Edwards, 2009) and in specific pathologies (Tuna, Boz, Van Obbergh, Lubansu, & Engelman, 2018).

Pain catastrophizing has been linked to health outcomes (Tuna, Van Obbergh, Vancutsem, & Engelman, 2018): greater disability (Picavet, Vlaeyen, & Schouten, 2002), work absenteeism (Besen, Gaines, Linton, & Shaw, 2017), opioid misuse (Lazaridou et al., 2017), and healthcare utilization (de Boer, Struys, & Versteegen, 2012).

Depression (Picavet et al., 2002), anxiety (Leung, 2012), sleep disturbance (Buenaver et al., 2012; Campbell et al., 2015), and temporal summation of pain (Edwards, Smith, Stonerock, & Haythornthwaite, 2006) are also influenced by it.

Pain catastrophizing is also distinctly specific to the experience of pain. It powerfully predicts outcomes of pain, including chronic pain intensity (Papaioannou et al., 2009), postsurgical pain intensity (Granot & Ferber, 2005; Pinto et al., 2012), poor response to opioids (Martel et al., 2013), greater use of preoperative opioids (Martel et al., 2013), persistent use of opioids after surgery (Hemelhorst et al., 2014), and persistent postsurgical pain, with a greater likelihood in musculoskeletal surgeries compared to other surgery types (Granot & Ferber, 2005; Pinto et al., 2012; Tuna, Boz, et al., 2018).

Pain catastrophizing is associated with biological processes interacting with the modulation of nociception, including dysregulation of the hypothalamic-pituitary axis (Quartana et al., 2010), reduced descending inhibitory control through endogenic opioid pathways, endogenous alteration of supraspinal pain-inhibitory and pain-facilitating pathways known as the diffuse noxious input controls, and increased activation of brain areas associated with effects of pain (Goodin et al., 2009), and pain facilitating changes in functional connectivity of the brain's default mode network (Kucyi et al., 2014).

Pain catastrophizing is supported by the fear avoidance model, decreasing the treatment efficacy of treatments based on biomedical concepts. Alternatively, treatment efficacy is increased in therapies aiming to reduce fear-avoidance.

In treatment settings, pain catastrophizing is so targeted, particularly in psychological and multidisciplinary interventions for patients with chronic pain (Racine et al., 2016; Smeets, Vlaeyen, Kester, & Knottnerus, 2006; Spinhoven et al., 2004; Turner, Holtzman, & Mancl, 2007).

In 1995, the Pain Catastrophizing Scale (PCS) was developed as an evaluation tool for experimental and clinical research (Sullivan, 2009). It is actually widely used across the world, in different languages: Catalan (Miro, Nieto, & Huguet, 2008), Chinese (Yap et al., 2008), Dutch (Van Damme, Crombez, Bijttebier, Goubert, & Van Houdenhove, 2002), French (Tremblay et al., 2008), German (Meyer, Sprott, & Mannion, 2008), Greek (Papaioannou et al., 2009), Japanese (Matsuoka & Salano, 2006), and modified versions exist for children (Crombez et al., 2003) and adolescents (Tremblay et al., 2008).

Features

Catastrophizing is viewed as a multidimensional construct based on three elements: rumination, magnification, and helplessness. The items on the PCS are drawn from these.

The PCS asks subjects to reflect on past painful experiences through 13 items describing thoughts or feelings when experiencing pain, and to indicate the degree on 5-point scales with the following end points: 0 (not at all) to 4 (all the time).

The PCS yields thus a total score ranging from 0 to 52, and three subscale scores assessing rumination, magnification and helplessness.

The three subscales are composed of the following items:

(1) Rumination, with a score ranging from 0 to 16:
- Item 8: I anxiously want the pain to go away.
- Item 9: I can't seem to keep it out of my mind.
- Item 10: I keep thinking about how much it hurts.
- Item 11: I keep thinking about how badly I want the pain to stop.

As described, subjects having a tendency to ruminate will constantly focalize their attention on it. Pain will become the center of their attention, eliciting layer after layer of other vicious circles, as physical deconditioning, kinesiophobia, anxio-depressive troubles, insomnia, among others. Rumination has been correlated with pain intensity (Wertli et al., 2014), and disability in relation to pain (Craner, Gilliam, & Sherry, 2016).

(2) Magnification, with a score ranging from 0 to 12:
- Item 6: I become afraid that the pain will get worse.
- Item 7: I keep thinking of other painful events.
- Item 13: I wonder whether something serious may happen.

Magnifiers will amplify the perceived menace related to the painful event. Magnification has been linked to physical health-related quality of life, mental health-related quality of life, and depression (Sullivan, Lynch, & Clark, 2005).

(3) Helplessness, with a score ranging from 0 to 24:
- Item 1: I worry all the time about whether the pain will end.
- Item 2: I feel I can't go on.
- Item 3: It's terrible and I think it's never going to get any better.
- Item 4: It's awful and I feel that it overwhelms me.
- Item 5: I feel I can't stand it anymore.
- Item 12: There's nothing I can do to reduce the intensity of pain.

The feeling described in the items is one of being confronted to an impossible to overcome obstacle, with the impression that nothing may reverse anymore the situation.

Helplessness is correlated to pain severity (Kulpa, Kosowicz, Stypula-Ciuba, et al., 2014), daily-life disability due to pain, mental health quality of life, depression, incapacity to cope with life difficulties, anxiety, and passive coping (Sullivan, Thorn, Haythornthwaite, et al., 2001; Sullivan, Tripp, & Santor, 2000).

Rumination and magnification are the results of a first analysis done by subjects when confronted to painful events, with helplessness intervening secondly, growing more and more with pain chronicization (Sullivan et al., 2001). These differences are amplified by the presence of central sensitization (Tuna, Van Obbergh, et al., 2018).

Globally, the higher the score of the PCS and/or its subscales, the higher the probability of developing pain chronicization, with values situated between percentiles 50 and 75 inducing a moderate risk of pain chronicization and values higher than percentile 75 inducing a high risk (Sullivan, 2009).

The PCS is easily applicable in standard clinical practice, its completion and scoring taking less than 5 min (Sullivan, 2009).

Applications

Prevention

The adverse effects of a catastrophizing coping style toward pain have been described and represent a growing burden, in parallel to the mounting evidence related to it, as it increases the probability that the painful condition will persist over an extended period of time (Helmerhorst et al., 2012).

Identifying patients at risk is crucial if caregivers want to positively influence such factors (Vranceanu et al., 2014; Das De, Vranceanu, & Ring, 2013; Hovik, Winther, Foss, et al., 2016; Khan, Ahmed, Blakeway, et al., 2011; Pavlin, Sullivan, Freund, et al., 2005; Roh et al., 2014; Spinhoven et al., 2004; Theunissen et al., 2012; Vissers et al., 2012) as:

- Evolution of postoperative pain;
- Patient satisfaction following surgery;
- Better patient-reported surgical outcomes;
- Reduction in opioid use.

Each of these factors, by itself, is to be taken into account, not only because pain is a vital parameter, but they influence the quality of life of our patients.

Misidentification of levels of catastrophizing represent an undesirable situation, reinforcing the need for accurate assessments (Kwon, Li, & Kim, 2011; Sabo & Roy, 2019). Prevention includes educating the caregivers who, in their daily practice, regularly meet patients prior to surgery, in regard of pain catastrophizing.

Perioperative medicine should include the PCS in their evaluation, among other tools, and define specific care pathways for the patients at moderate and high risk of chronicization. In absence of surgical emergencies, time should be taken to apply corrective measures regarding the maladaptive situation to which they're confronted.

Research

Its use in experimental and clinical research has shown the strong criterion related, concurrent, discriminant validity and reliability. Its internal consistency has been demonstrated and it has a high test-retest correlation (Meints, Stout, Abplanalp, & Hirsch, 2017; Osman et al., 1997, 2000).

Research concerning pain catastrophizing is centered on three categories of subjects: healthy volunteers, patients with acute pain and patients with chronic pain.

Topics covered range from physical disability, relation with gender, age, race, genetic susceptibilities as heritability, neurophysiology through neuroimaging techniques, in particular functional MRI. Association of pain catastrophizing with increased fMRI activities in the medial frontal cortex and cerebellum for anticipation of pain, in the dorsal accumbens and the dorsolateral prefrontal cortex for attention to pain have been described, among others (Leung, 2012).

Different models have been proposed to explain its development, including a preferential and dysfunctional bias of attention toward pain known as the attention-bias model, a specific pain schema with excessively pessimistic beliefs about pain known as the schema-activation model, and the appraisal model, characterized by maladaptive appraisals of the pain subjects experience (Leung, 2012). The fear-avoidance model of pain is actually supported, where negative beliefs about pain or illness leads to a catastrophizing response in which patients imagine the worst possible outcome, followed by fear of activity and avoidance. This causes disuse, distress, reinforcing it in a deleterious cycle (Slepian, Ankaw, & France, 2020; Wertli et al., 2014).

Treatment

Pain catastrophizing is a modifiable variable and many interventions have pointed that its reduction is considered a key factor in determining successful therapeutic interventions in regard to chronic pain.

Cut-offs for the PCS between pretreatment and posttreatment values have been determined, a score reduction of at least 40% (Scott, Wideman, & Sullivan, 2014) being considered significant for a positive evolution of one's painful condition. An association between a decrease in catastrophizing and an increase in daily activities and a decrease in pain has been shown (Jensen, Turner, & Romano, 2001).

Most efficient treatments in regard to managing pain catastrophizing are cognitive-behavioral therapies, mindfulness based stress reduction, self-regulation, relaxation, hypnosis, acceptance and commitment therapies, guided imagery, and multimodal therapies including educational, psycho-social and physical therapies (Gibson & Sabo, 2018; Moseley, 2004; Ostelo, Deyo, Stratford, et al., 2008; Robinson, Brown, Georges, et al., 2005; Smeets et al., 2006; Sullivan, Feuerstein, Gatchel, et al., 2005; Sullivan, Stanish, Sullivan, & Tripp, 2002; Wideman, Adams, & Sullivan, 2009). Resilience and self-efficacy are other resources currently incorporated in the therapeutic management. Nevertheless, the only way to clearly achieve that goal is by associating pain neurophysiology educational programs (Woods & Asmundson, 2008; Lluch, Duenas, Falla, et al., 2008; Moseley, Nicholas, & Hodges, 2004; Thorn, Day, Burns, et al., 2011) as a first step in their multidisciplinary approach. These programs have indeed resulted in better results in regard to the different therapies formerly described (Scott et al., 2014).

Subscale-specific treatments are also beneficial. Intensity of pain and disability due to pain, and depression are reduced thanks to reduction of the levels of rumination and helplessness, while quality of life is favorably influenced by attenuating the feeling of magnification.

Summary-conclusion

With more than 6000 articles published on this topic, researchers' enthusiasm on pain catastrophism hasn't diminished. Key questions remain and will certainly necessitate our collective efforts, as the moderating effects of pain catastrophizing level reduction therapeutic strategies on the different health and pain outcomes.

Prevention is to be privileged. The different cut-off values described help the screening of at-risk patients, integrating them in specific care pathways with the goal of reducing pain chronicization.

These pathways should always include educational programs centered on the neurophysiology of pain, as first-line step, followed by the different therapies previously described, with the optimal goal of developing a virtuous circle through acceptation.

Applications to other areas

Psychology, physiotherapy, neurology, surgery, and peri-operative medicine.

Mini-dictionary of terms

PCS: Pain Catastrophizing Scale.

Key facts

- Pain catastrophizing is also distinctly specific to the experience of pain. It powerfully predicts outcomes of pain.
- Misidentification of levels of catastrophizing represent an undesirable situation, reinforcing the need for accurate assessments.
- Pain catastrophizing is a modifiable variable and many interventions have pointed that its reduction is considered a key factor.
- The only way to clearly achieve that goal is achieved by associating pain neurophysiology educational programs as a first step in their multidisciplinary approach.

Conflict of interest

No conflict of interest.

References

Besen, E., Gaines, B., Linton, S. J., & Shaw, W. S. (2017). The role of pain catastrophizing as a mediator in the work disability process following acute low back pain. *Journal of Applied Biobehavioral Research, 22*, e12085.

Buenaver, L. F., Quartana, P. J., Grace, E. G., Sarlani, E., Simango, M. M., Edwards, R. R., et al. (2012). Evidence for indirect effects of pain catastrophizing on clinical pain among myofascial temporomandibular disorder participants: The mediating role of sleep disturbance. *Pain, 153*(6), 1159–1166.

Campbell, C. M., Buanever, L. F., Finan, P., Bounds, S. C., Redding, M., McCauley, L., et al. (2015). Sleep, pain catastrophizing and central sensitization in knee osteoarthritis patients with and without insomnia. *Arthritis Care and Research, 67*(10), 1387–1396.

Campbell, C. M., & Edwards, R. R. (2009). Mind-body interactions in pain: The neurophysiology of anxious and catastrophic pain-related thoughts. *Transgenic Research, 153*(3), 97–101.

Craner, J. R., Gilliam, W. P., & Sherry, J. A. (2016). Rumination, magnification and helplessness: How do different aspects of pain catastrophizing relate to pain severity and functioning? *The Clinical Journal of Pain, 32*, 1028–1035.

Crombez, G., Bijttebier, P., Eccleston, C., Mascagni, T., Mertens, G., Goubert, I., et al. (2003). The child version of the pain catastrophizing scale (PSC-C): A preliminary validation. *Pain, 104*, 639–646.

Das De, S., Vranceanu, A. M., & Ring, D. C. (2013). Contribution of kinesiophobia and catastrophic thinking to upper-extremity-specific disability. *The Journal of Bone and Joint Surgery. American Volume, 95*, 76–81.

de Boer, M. J., Struys, M. M., & Versteegen, G. J. (2012). Pain-related catastrophizing in pain patients and people with pain in the general population. *European Journal of Pain, 16*, 1044–1052.

Edwards, R. R., Smith, M. T., Stonerock, G., & Haythornthwaite, J. A. (2006). Pain-related catastrophizing in healthy women is associated with greater temporal summation and reduced habituation to thermal pain. *The Clinical Journal of Pain, 22*(8), 730–737.

Ellis, A. (1962). *Reason and emotion in psychotherapy* (p. 442). New York: L. Stuart.

Gibson, E., & Sabo, M. T. (2018). Can pain catastrophizing be changed in surgical patients? A scoping review. *Canadian Journal of Surgery, 61*(5), 311–318.

Goodin, B. R., McGuire, L., Allshouse, M., Stapleton, L., Haythornwaithe, J., Burns, N., et al. (2009). Associations between catastrophizing and endogenous pain-inhibitory processes: Sex differences. *The Journal of Pain, 10*, 180–190.

Granot, M., & Ferber, S. G. (2005). The roles of pain catastrophizing and anxiety in the prediction of postoperative pain intensity: A prospective study. *The Clinical Journal of Pain, 21*(5), 439–445.

Helmerhorst, G. T., et al. (2012). Risk factors for persistent postsurgical pain in women undergoing hysterectomy duet to benign causes: A prospective predictive study. *The Journal of Pain, 13*(11), 1045–1057.

Hemelhorst, G. T., et al. (2014). Risk factors for continued opioid use one to two months after surgery for musculoskeletal trauma. *The Journal of Bone and Joint Surgery. American Volume, 96*(6), 495–499.

Hovik, L. H., Winther, S. B., Foss, O. A., et al. (2016). Preoperative pain catastrophizing and postoperative pain after total knee arthroplasty: A prospective color study with one year follow-up. *BMC Musculoskeletal Disorders, 17*, 214.

Jensen, M. P., Turner, J. A., & Romano, J. M. (2001). Changes in beliefs, catastrophizing, and coping are associated with improvement in multidisciplinary pain treatment. *Journal of Consulting and Clinical Psychology, 69*, 655–662.

Khan, R. S., Ahmed, K., Blakeway, E., et al. (2011). Catastrophizing: A predictive factor for postoperative pain. *American Journal of Surgery, 201*, 122–131.

Kucyi, A., Moayedi, M., Weissman-Fogel, I., Goldberg, M. B., Freeman, B. V., Tenenbaum, H. C., et al. (2014). Enhanced medial prefrontal-default mode network functional connectivity in chronic pain and its association with pain rumination. *The Journal of Neuroscience, 34*, 3969–3975.

Kulpa, M., Kosowicz, M., Stypula-Ciuba, B. J., et al. (2014). Anxiety and depression, cognitive coping strategies, and health locus of control in patients with digestive system cancer. *Gastroenterology Review, 9*, 329–335.

Kwon, B., Li, L., & Kim, D. (2011). Influence of a patient's psychological distress on surgical decision making. *Spine, 1*, 161S-2S.

Lazaridou, A., Franceschelli, O., Buliteanu, A., Cornelius, M., Edwards, R. R., & Jamison, R. N. (2017). Influence of catastrophizing on pain intensity, disability, side effects, and opioid misuse among pain patients in primary care. *Journal of Applied Biobehavioral Research, 22*, e12081.

Leung, L. (2012). Pain catastrophizing: An updated review. *Indian Journal of Psychological Medicine, 34*, 204–217.

Lluch, E., Duenas, L., Falla, D., et al. (2008). Preoperative pain neuroscience education combined with knee joint mobilization for knee osteoarthritis: A randomized controlled trial. *The Clinical Journal of Pain, 34*, 44–52.

Martel, M. O., et al. (2013). Catastrophic thinking and increased risk for prescription opioid misuse in patients with chronic pain. *Drug and Alcohol Dependence, 132*(1–2), 335–341.

Matsuoka, H., & Salano, Y. (2006). Assesment of cognitive aspect of pain: Development, reliability and validation of the Japanese version of pain catastrophizing scale. *Japanese Journal of Psychosomatic Medicine, 47*, 95–102.

Meints, S. M., Stout, M., Abplanalp, S., & Hirsch, A. T. (2017). Pain-related rumination but not magnification or helplessness, mediates race and sex differences in experimental pain. *The Journal of Pain, 18*(3), 332–339.

Meyer, K., Sprott, H., & Mannion, A. F. (2008). Cross-cultural adaptation, reliability, and validity of the German version of the Pain Catastrophizing Scale. *Journal of Psychosomatic Research, 64*, 469–478.

Miro, J., Nieto, R., & Huguet, A. (2008). The Catalan version of the Pain Catastrophizing Scale: A useful instrument to assess catastrophic thinking in whiplash patients. *The Journal of Pain, 9*, 397–406.

Moseley, G. L. (2004). Evidence for a direct relationship between cognitive and physical change during an education intervention in people with chronic low back pain. *European Journal of Pain, 8*, 39–45.

Moseley, G. L., Nicholas, M. K., & Hodges, P. W. (2004). A randomized controlled trial of intensive neurophysiology education in chronic low back pain. *The Clinical Journal of Pain, 20*, 324–330.

Osman, A., Barrios, F. X., Gutierrez, P. M., Kopper, B. A., Merrifield, T., & Grittman, L. (2000). The Pain Catastrophizing Scale: Further psychometric evaluation with adult samples. *Journal of Behavioral Medicine, 23*, 351–356.

Osman, A., Barrios, F. X., Kopper, B. A., Hauptmann, W., Jones, J., & O'Neill, E. (1997). Factor structure, reliability, and validity of the Pain Catastrophizing Scale. *Journal of Behavioral Medicine, 20*, 589–605.

Ostelo, R. W. J. G., Deyo, R. A., Stratford, P., et al. (2008). Interpreting change scores for pain and functional status in low back pain-toward international consensus regarding minimal important change. *Spine, 33*, 90–94.

Papaioannou, M., et al. (2009). The role of catastrophizing in the prediction of postoperative pain. *Pain Medicine, 10*(8), 1452–1459.

Pavlin, D. J., Sullivan, M. J., Freund, P. R., et al. (2005). Catastrophizing: A risk factor for postsurgical pain. *The Clinical Journal of Pain, 21*, 83–90.

Picavet, H. S., Vlaeyen, J. W., & Schouten, J. (2002). Pain catastrophizing and kinesiophobia: Predictors of chronic low back pain. *American Journal of Epidemiology, 156*, 1028–1034.

Pinto, P. R., et al. (2012). The mediating role of pain catastrophizing in the relationship between presurgical anxiety and acute post surgical pain after hysterectomy. *Pain, 153*(1), 218–226.

Quartana, P. J., Buenaver, L. F., Edwards, R. R., Klick, B., Haythornwaithe, J. A., & Smith, M. T. (2010). Pain catastrophizing and salivary cortisol responses to laboratory pain testing in temporomandibular disorder and healthy participants. *The Journal of Pain, 11*, 186–194.

Racine, M., Moulin, D. E., Nielson, W. R., Morley-Forster, P. K., Lynch, M., Clarck, A. J., et al. (2016). The reciprocal associations, between catastrophizing and pain outcomes in patients being treated for neuropathic pain: A cross-lagged panel analysis study. *Pain, 157*, 1946–1953.

Robinson, M. E., Brown, J. L., Georges, S. Z., et al. (2005). Multidimensional success criteria and expectations for treatment of chronic pain: The patient perspective. *Pain Medicine, 6*, 336–345.

Roh, Y. H., et al. (2014). Effect of anxiety and catastrophic pain ideation on early recovery after surgery for distal radius fractures. *The Journal of Hand Surgery, 39*(11), 2258–2264.

Sabo, M. T., & Roy, M. (2019). Surgeon identification of pain catastrophizing versus the Pain Catastrophizing Scale in orthopedic patients after routine surgical consultation. *Canadian Journal of Surgery, 62*(4), 265–270.

Schütze, R., Rees, C., Smith, A., et al. (2018). How can we best reduce pain catastrophizing in adults with chronic noncancer pain? A systematic review and meta-analysis. *The Journal of Pain, 19*(3), 233–256.

Scott, W., Wideman, T. H., & Sullivan, M. J. (2014). Clinically meaningful scores on pain catastrophizing before and after pain multidisciplinary rehabilitation. *The Clinical Journal of Pain, 30*, 183190.

Slepian, P. M., Ankaw, B., & France, C. R. (2020). Longitudinal analysis supports a fear-avoidance model that incorporates pain resilience alongside pain catastrophizing. *Annals of Behavioral Medicine, 54*(5), 335–345.

Smeets, R. J., Vlaeyen, J. W., Kester, A. D., & Knottnerus, J. A. (2006). Reduction of pain catastrophizing mediates the outcome of both physical and cognitive-behavioral treatment in chronic low back pain. *The Journal of Pain, 7*, 261–271.

Spinhoven, P., Ter Kuile, M., Kole-Snijders, A. M., Hutten Mansfeld, M., Den Ouden, D. J., & Vlaeyen, J. W. (2004). Catastrophizing and internal pain control as mediators of outcome in the multidisciplinary treatment of chronic low back pain. *European Journal of Pain, 8*, 211–219.

Sullivan, M. J. L. (2009). *The pain catastrophizing scale user manual.* Montreal, Quebec: McGill University.

Sullivan, M., Feuerstein, M., Gatchel, R. J., et al. (2005). Integrating psychological and behavioral interventions to achieve optimal rehabilitation outcomes. *Journal of Occupational Rehabilitation, 15*, 475–489.

Sullivan, M. J., Lynch, M. E., & Clark, A. J. (2005). Dimensions of catastrophic thinking associated with pain experience and disability in patients with neuropathic pain conditions. *Pain, 113*, 310–315.

Sullivan, M. J. L., Stanish, W., Sullivan, M. E., & Tripp, D. (2002). Differential predictors of pain and disability in patients with whiplash injuries. *Pain Research & Management, 7*, 68–74.

Sullivan, M. J. L., Thorn, B., Haythornthwaite, J. A., et al. (2001). Theoretical perspectives on the relation between catastrophizing and pain. *The Clinical Journal of Pain, 17*, 52–64.

Sullivan, M. J. L., Tripp, D. A., & Santor, D. (2000). Gender differences in pain and pain behavior: The role of catastrophizing. *Cognitive Therapy and Research, 24*, 121–134.

Theunissen, M., et al. (2012). Preoperative anxiety and catastrophizing: A systematic review and meta-analysis of the association with chronic postsurgical pain. *The Clinical Journal of Pain, 28*(9), 819–841.

Thorn, B. E., Day, M. A., Burns, J., et al. (2011). Randomized trial of group cognitive behavioral therapy compared with a pain education control for low-literacy rural people with chronic pain. *Pain, 152*, 2710–2720.

Tremblay, I., Beaulieu, Y., Bernier, A., Crombez, G., Laliberte, S., Thibault, P., et al. (2008). Pain Catastrophizing Scale for Francophone adolescents: A preliminary validation. *Pain Research & Management, 13*, 19–24.

Tuna, T., Boz, S., Van Obbergh, L., Lubansu, A., & Engelman, E. (2018). Comparison of the pain sensitivity questionnaire and the pain catastrophizing scale in predicting postoperative pain and pain chronicisation after spine surgery. *Clinical Spine Surgery, 31*(9), 432–440.

Tuna, T., Van Obbergh, L., Vancutsem, N., & Engelman, E. (2018). Usefulness of the pain sensitivity questionnaire to discriminate the pain behavior of chronic pain patients. *British Journal of Anaesthesia, 121*(3), 616–622.

Turner, J. A., Holtzman, S., & Mancl, L. (2007). Mediators, moderators and predictors of therapeutic change in cognitive-behavioral therapy for chronic pain. *Pain, 127*, 276–286.

Van Damme, S., Crombez, G., Bijttebier, P., Goubert, L., & Van Houdenhove, B. (2002). A confirmatory factor analysis of the pain Catastriophizing scale: Invariant factor structure across clinical and non-clinical populations. *Pain, 96*, 319–324.

Vissers, M. M., et al. (2012). Psychological factors affecting the outcome of total hip and knee arthroplasty: A systematic review. *Seminars in Arthritis and Rheumatism, 41*(4), 576–588.

Vranceanu, A. M., et al. (2014). Psychological factors predict disability and pain intensity after skeletal trauma. *The Journal of Bone and Joint Surgery. American Volume, 96*(3), e20.

Wertli, M. M., Burgstaller, J. M., Weiser, S., Steurer, J., Kofmehl, R., & Held, U. (2014). Influence of catastrophizing on treatment outcome in patients with nonspecific low back pain. *Spine, 39*(3), 263–273.

Wideman, T., Adams, H., & Sullivan, M. J. (2009). A prospective sequential analysis of the fear avoidance model of pain. *Pain, 145*, 45–51.

Woods, M. P., & Asmundson, G. J. G. (2008). Evaluating the efficacy of graded in vivo exposure for the treatment of fear in patients with chronic back pain: A randomized controlled clinical trial. *Pain, 136*, 271–280.

Yap, J. C., Lau, J., Chen, P. P., Gin, T., Wong, T., Chan, I., et al. (2008). Validation of the Chinese pain catastrophizing scale (HK-PCS) in patients with chronic pain. *Pain Medicine, 9*, 186–195.

Chapter 41

Pain-related behavioral scales among a low back pain population: A narrative review

Dalyah Alamam

Physiotherapy, Health Rehabilitation Sciences Department, Collage of Applied Medical Sciences, King Saud University, Riyadh, Saudi Arabia; Arthritis and Musculoskeletal Research Group, Sydney School of Health Sciences, Faculty of Medicine and Health, The University of Sydney, NSW, Australia

Abbreviations

IASP	International Association for the Study of Pain
LBP	low back pain
PaBS	pain behavior scale
PBA	the pain behavior assessment
RCTs	randomized control clinical trials

Introduction

Pain, a globally leading cause of disability that can interfere with daily life, is often associated with psychological distress (Froud et al., 2014; Goldberg & McGee, 2011; Lee et al., 2015; Treede et al., 2019). Pain is recognised by the International Association for the Study of Pain (IASP) to be a complex condition, defining it as "an unpleasant sensory and emotional experience associated with actual or potential damage or described in terms of such damage" (Merskey, 1964; Treede, 2018). This definition indicates the perception of pain results from a complex interaction between physical, psychosocial, and behavioral aspects, which might influence the overall pain experience (Monina, Falzetti, Firetto, Mariani, & Caputi, 2006; O'Sullivan et al., 2018). Major consequences of pain include depression, inability to work, and disrupted social relationships, rendering its treatment and prevention a global health priority (Goldberg & McGee, 2011).

Pain-related behavior refers to the actions or postural displays exhibited by someone when they experience pain (Fordyce, 1977; Hadjistavropoulos & Craig, 1994, 2004). These behaviors are considered important components of coping with or communicating pain, or seeking social support (Keefe, Wilkins, & Cook, 1984; Keefe, Crisson, Maltbie, Bradley, & Gil, 1986; Sullivan et al., 2001). A distinction between communicative pain behaviors such as verbal expressions (e.g., sighs) or facial expressions (e.g., grimaces), and behaviors intended to protect the body from symptom exacerbation or further injury (e.g., guarding and rubbing) exists (Hadjistavropoulos & Craig, 2002; Sullivan et al., 2006; de C Williams & Craig, 2006; Sullivan, 2008). Those pain-related behaviors are common in people with chronic pain (Koho, Aho, Watson, & Hurri, 2001; Meyer et al., 2016; Prkachin, Hughes, Schultz, Joy, & Hunt, 2002; Prkachin, Schultz, & Hughes, 2007) and are associated with important attributes of conditions such as psychological distress (Olugbade, Bianchi-Berthouze, & Williams, 2019), pain severity (Labus, Keefe, & Jensen, 2003), and pain-related disability (Connally & Sanders, 1991; Prkachin et al., 2007; Schultz et al., 2008). Therefore, assessing pain behaviors is important for comprehensive chronic pain management.

Pain behaviors have been assessed in various ways (Main, Keefe, Jensen, Vlaeyen, & Vowles, 2015), such as indirect observation of videotaped interviews of people with chronic lower back pain (LBP) (Keefe & Block, 1982), nonchronic LBP (Jensen, Bradley, & Linton, 1989), and rheumatoid arthritis (McDaniel et al., 1986), observational checklists such as the UAB pain behavior scale (Richards, Nepomuceno, Riles, & Suer, 1982) and pain behavior checklist (Turk, Wack, & Kerns, 1985), or real time assessment procedures of those with LBP (Prkachin, Schultz, Berkowitz, Hughes, & Hunt, 2002) (Appendix 1 adapted from Alamam, 2019). Some procedures are impractical in clinical settings (Feuerstein, Greenwald, Gamache, Papciak, & Cook, 1985) and may delay incorporation of the assessment of pain behaviors into the clinical

---

Features and Assessments of Pain, Anesthesia, and Analgesia. https://doi.org/10.1016/B978-0-12-818988-7.00038-8
Copyright © 2022 Elsevier Inc. All rights reserved.

consultation. Further, evidence is inconclusive for the psychometric properties or appropriateness of various tools to measure pain behaviors.

Pain behavior is commonly recognised as an integral part of models of pain and disability (Fordyce, 1977), therefore, an examination of the properties of different behavioral assessment tools would provide some perspective to account for the complexity of the pain experience. Clinicians require brief and sensible tools to assist with assessment and management of chronic pain conditions such as LBP. A recently developed Pain Behavior Scale (PaBS) was validated on an LBP population (Alamam et al., 2019), and deemed to be a reliable tool for measuring the presence and severity of pain behaviors while performing functional tests. Herein, other behavioral (observational) scales are reviewed to describe the clinical utility of the available behavioral tools in clinical settings. This review examines published behavioral (observational) measures for pain-related behavior in LBP populations, summarizes psychometric qualities of their measures, and compares the clinical characteristics and usefulness of pain behavioral scores identified using each tool.

Methodology

Electronic databases (MEDLINE (OvidSP), EMBASE (OvidSP), CINAHL (EBSCO host), PsycINFO (OvidSP), Cochrane Central Register of Controlled Trials (OvisdSP), Scopus, PEDro, and PROSPERO) were searched using keywords (e.g., pain behavior or behavior, low back pain, assessment, observation, reliability) for studies published from 2000 to March 2020. Google Scholar was also searched for studies yet to be indexed in mainstream databases. Additionally, relevant studies were manually searched for in reference lists of identified studies.

Study selection (eligibility criteria):

1. Full text articles in English published in peer-review journals including observational (cross-sectional or longitudinal) studies or randomized control clinical trials (RCTs).
2. Studies including participants of at least 18 years of age.
3. Studies describing an assessment instrument/scale for LBP population or a subgroup of LBP population with nonspecific LBP which could be acute/subacute (<12 weeks) or chronic (≥ 3 months).
4. Study outcomes reporting measurement of pain-related behavior assessed by direct or indirect observational measures.
5. Study outcomes reporting pain intensity or functional disability during clinical assessment assessed by valid self-report measures.
6. Studies reporting at least one physical activity or physical performance measurement test during clinical assessment of LBP population.

Studies were excluded if they investigated specific LBP pathologies (such as lumbar radiculopathy, spinal stenosis, fracture, osteoporosis, and infection or pregnancy-related LBP), or a population undergoing or having undergone spinal surgery as an intervention.

Keyword/phrase searches identified 213 published English reports between 2000 and 2020. Titles and abstracts were screened for eligibility (conducted in LBP population, with the terms pain-related behaviors and observational assessment measures mentioned as a part of the study design). When not found in the abstract, the methods sections of manuscripts were vetted for these search terms. Articles meeting inclusion criteria were considered for full-text review by one author (DA), and psychometric scales were evaluated. Fig. 1 summarizes the article exclusion process.

Criteria used to evaluate psychometric aspects and clinical utility of scales are based partly on Streiner and Norman's requirements for health measurement scales (Streiner, Norman, & Cairney, 2015), adapted from Zwakhalen, Hamers, Abu-Saad, and Berger (2006). The following (if available) were extracted from each article: type of assessment scale (including items of scale), origin of items, scoring/scaling response, number of participants, and information regarding feasibility, homogeneity, reliability, and validity. In addition to other data such as title, author(s), year of publication, objective, study design, setting and location, sample characteristics, primary outcomes, and main results were extracted.

Results

Four publications met inclusion criteria. The main outcomes of these studies (in chronological order of publication) are presented in Table 1. In these studies, a cross-sectional design was used to examine the reliability and other psychometric properties of tools in addition to an exploratory study conducted in a small population of LBP ($n = 22$) (Alamam et al., 2019). Studies examined pain behavior during a suit of physical performance tests (e.g., timed 5-min walk), standardized physical examination of people with LBP (e.g., range of motion exercise), or during functional capacity evaluation. As suggested by Keefe and Block (1982), pain behaviors included abnormal gait, audible pain behavior (e.g., groaning,

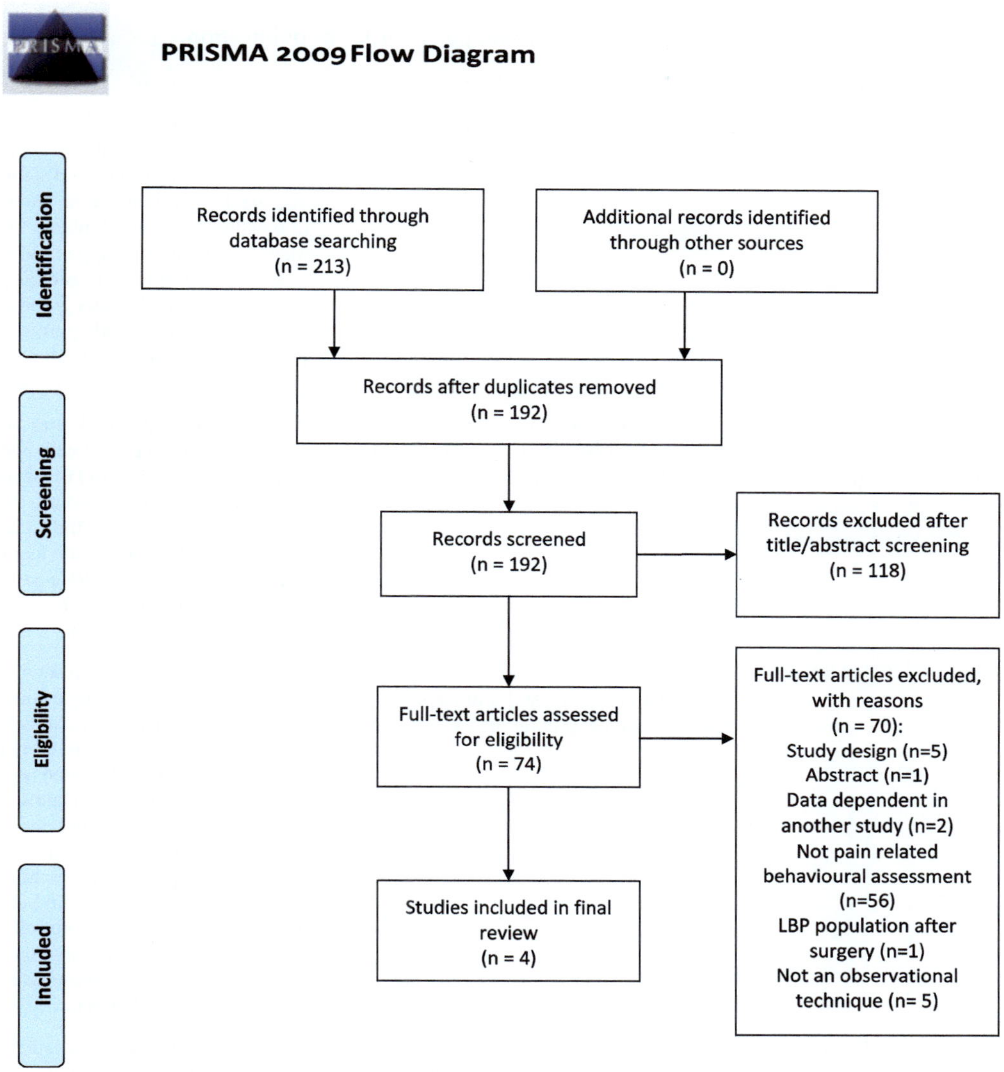

PRISMA 2009 Flow Diagram

FIG. 1 PRISMA flow diagram. A total of 213 references were retrieved by electronic searches, of which 4 references were finally eligible for inclusion in this review.

sighing), facial pain behavior (e.g., grimacing), touching or holding the affected body part, and guarding tense stiff posture. The occurrence of each behavior was rated (Koho et al., 2001; Meyer et al., 2016; Prkachin, Schultz, et al., 2002), in addition to which the severity of observed pain behaviors was assessed on a single scale (Alamam et al., 2019). Two studies directly observed participants (Alamam et al., 2019; Meyer et al., 2016), while two did so indirectly (Koho et al., 2001; Prkachin, Schultz, et al., 2002).

Tables 2 and 3 present the quality criteria and outcomes of tools adapted from Zwakhalen et al. (2006) for these four studies. In terms of tool originality, all were based on or adapted from the Keefe and Block observational approach (Keefe & Block, 1982). Tools reported acceptable reliability and construct validity, and correlations with pain or/and disability that generally exceeded 0.4.

Discussion

We review the (observational) pain assessment tools used to assess pain behaviors in people with LBP and evaluate their psychometric qualities. A substantial gap in literature investigating pain-related behaviors is identified. Of four behavioral pain assessment tools identified, two use indirect observational methods (videotaping and real-time assessment) which may delay integration of the assessment of pain behaviors during the clinical consultation (Alamam, 2019). The pain behavior

454 PART | IV Assessments, screening, and resources

TABLE 1 Main outcomes of included studies ($n = 4$) in chronological order of publication.

Reference	Design	Participant number and setting	Scale description
Koho et al. (2001)	Reliability cross-sectional	51 with chronic pain. Rehabilitation center in Helsinki, Finland	Subjects videotaped and observed during sitting; timed 5-min walk; lie down prone to the floor and roll over 360° and stand up; bending and reaching; filling, lifting, and carrying a box of weights; and stair climbing. Observers identified the following pain behaviors from videotape: distorted gait, audible pain behavior (groaning, sighing), facial grimacing, touching or holding affected body part, stopping or resting, verbal complaints about pain, support and leaning, and guarding tense stiff posture
Prkachin, Hughes, et al. (2002), Prkachin, Schultz, et al. (2002)	Reliability cross sectional	176 with subacute or chronic LBP pain. Workers' Compensation Board of British Columbia claimants, Canada	A system for in vivo, real-time assessment of pain behavior integrated with a standardized physical examination for LBP patients. Five pain behavior categories (guarding, touching/rubbing, words, sounds, and facial expressions) rated by checking a column associated with a behavior. Observers rated facial behaviors meeting criteria for pain expression on a 3-point intensity scale (none, some, much)
Meyer et al. (2016)	Reliability cross sectional	200 with chronic nonspecific LBP. Two rehabilitation clinics and a university hospital outpatient clinic in Switzerland	Pain behaviors rated by observing patients during functional capacity evaluation testing—a standardized battery of clinical tests to measure physical ability to carry out work-related activities safely (often used to evaluate allowances for disability claims or as a measurement prior to functional rehabilitation program). Dichotomous scale ($0 =$ no pain behavior; $1 =$ pain behavior) applied to all items. Criteria for rating pain behavior 'yes/no' were defined by numeric cut-off levels (where applicable), evaluated in previous studies. Total pain behavior scores represent sums of all items rated 1 (with pain behavior)
Alamam et al. (2019)	Reliability exploratory	22 with chronic LBP. A physiotherapy outpatient clinic in Riyadh, Saudi Arabia	Specific pain behaviors included sighing, breath-holding, grimacing, guarding, rubbing, and antalgic gait (assessed during timed up and go and 50-ft walk tests). A 4-point scale ranged 'None' (no observed behavior) to 'Severe' (marked pain behavior), with two measures obtained: the presence/absence of a behavior, and a total score of the severity of all pain behaviors. Total severity (on a scale of 0–15) was determined by summing individual severity ratings for behaviors for each test. The scale was administered during performance of physical tests of 10–15 m, including sit to stand, trunk flexion, timed up and go, loaded reach, and 50-ft walk

assessment (PBA) (Meyer et al., 2016) comprehensively assesses pain behavior, and comprises a range of aspects hitherto uncaptured in any single test (e.g., pain perception), but assessment involves physically demanding tests that are atypical of normal daily function and may be impractical in a clinical environment. The fourth of these tests, PaBS, was adapted from existing measures of physical function in people with LBP following critical reflection of gaps in existing pain scales (Alamam, 2019), and by applying the PaBS in a chronic LBP population during a validated functional test set

TABLE 2 Quality criteria of tools adapted from Zwakhalen et al. (2006).

Items	Description
Origin	2, if items specifically collected for use in people with LBP 1, if items modified for use in people with LBP 0, if items originated from a scale developed for another population
Number of participants with LBP	2, if $n \geq 100$ 1, if $50 \leq n < 100$ 0, if $n < 50$
Content validity	2, if scale covers all important items/dimensions (in the authors' opinion): pain items are collected for a specific population and different sources/methods are used to collect items 1, if scale covers important items/dimensions to a moderate extent: items are adapted to a population and different sources/methods are used to collect items 0, if scale does not cover important items/dimensions
Criterion validity	2, if correlations are acceptable to high ($r \geq 0.60$) according to the 'gold standard,' and sensitivity/specificity is considered acceptable 1, if correlations are moderate–acceptable ($0.40 \leq r < 0.60$) 0, if correlations are low ($r < 0.40$)
Construct validity in relation to other pain tools	2, if correlations with other pain measures are acceptable to high ($r \geq 0.60$) 1, if correlations with other pain measures are moderate ($r \leq 0.40 < 0.60$) 0, if correlations are low ($r < 0.40$) or no information is provided
Homogeneity	2, if $0.70 \leq$ alpha < 0.90 1, if $0.60 \leq$ alpha < 0.70 0, if alpha < 0.60 or no information is provided
Inter-rater reliability	2, if reliability coefficient ≥ 0.80 1, if $0.60 \leq$ reliability coefficient < 0.80 0, if reliability coefficient < 0.60 or no information is provided
Intra-rater and/or test-retest reliability	2, if reliability coefficient ≥ 0.80 1, if $0.60 \leq$ reliability coefficient < 0.80 0, if reliability coefficient < 0.60 or no information is provided
Feasibility	2, if scale is short, manageable with instructions, scoring interpretation 1, if scale is manageable (one format) 0, if scale is more complex

(Simmonds et al., 1998). Because the PaBS is based on direct observation during functional tests that are commonly conducted in the clinic to assess people with LBP, its potential for clinical application to assess and monitor pain behaviors is greater than the PBA test. Further, PaBS measures the severity of observed pain behaviors, potentially supporting a comprehensive exploration of pain, and may address the complexity of human beliefs and behavior.

Observational studies have reported significant variability in the extent to which patients display pain behaviors (Keefe & Block, 1982; Keefe, Crisson, & Trainor, 1987) because of individual (e.g., age) (Keefe et al., 2000), psychological (e.g., depression) (Keefe, Wilkins, Cook Jr, Crisson, & Muhlbaier, 1986; Keefe et al., 2000), or cultural (e.g., social norms) (Hobara, 2005) factors. Therefore, solely measuring the frequency of pain behaviors may miss patients exhibiting different degrees of pain-behavior severity (Von Korff, Ormel, Keefe, & Dworkin, 1992), and therefore not capture the overall pain experience (Vlaeyen, Van Eek, Groenman, & Schuerman, 1987). As such, PaBS may provide a more feasible means of behavioral assessment of patients during those physical performance tests commonly used in assessment of people with LBP. It is, however, important to note that the PaBS study was exploratory and conducted on a limited sample, so other psychometric aspects of this scale require investigation.

Quality judgement criteria relating to homogeneity, validity, reliability, and feasibility reveal all existing tools that are generally reliable with acceptable construct validity. However, two of these tools scored 12 and one 13 points out of a maximum quality score of 20 (Table 3), so their overall psychometric quality is moderate (Zwakhalen et al., 2006). Tools require further research on aspects of their psychometric properties.

TABLE 3 Quality criteria and outcomes of included tools.

Reference	Item origin	N with LBP	Content validity	Criterion validity	Construct validity in relation to other pain tools	H	InterR R (range)	IntraRr R and/or T-r R (range)	Feasibility
Koho et al. (2001)	1, based on Keefe and Block (1982)	1, $n = 51$	2	NA	1, cor with pain $r = 0.40$, disability $r = 0.53$	1	1 (0.58–0.92)	1 (0.6–0.8) observer 1, (0.60–0.85) observer 2	1, functional, video-based assessment
Prkachin, Schultz, et al. (2002)	2, adawpted from Keefe and Block (1982)	2, $n = 176$	2	NA	0, cor with pain mostly r ≤ 0.4	2	2	1 (0.7–0.2)	1, real-time assessment
Meyer et al. (2016)	1, based on Keefe and Block (1982), nonorganic signs of Waddell, and other sources	2, $n = 200$	2, with 4 subscales: pain perception (5 items), overt pain behavior (7 items), effort (9 items), and consistency of behavior (31 items)	NA	2, cor with work ability index $r = 0.59$, fear avoidance $r = 0.39$, and 0.37, and disability $r = 0.70$	2	1 (0.51–0.77) (Trippolini et al., 2014)	1 (0.49–0.68)	1, assessment using battery of clinical tests to measure physical ability to perform work-related activities safely
Alamam et al. (2019)	2, adapted from Keefe and Block (1982)	0, n = 22	2	1, Sensitivity and specificity determined	2, cor with pain intensity and disability $r = 0.6$	0	2 (0.8–1.0)	2 (0.4–0.9)	2, assessment in performance of functional tests

cor, correlations; *NA*, not applicable; *H*, homogeneity; *InterR R*, inter-rater reliability; *IntraR R*, intra-rater reliability; *T-R R*, test-retest reliability.

Current scales are mostly similar in examined behaviors, with few heterogeneous items such as the pain behaviors measured in PBA (e.g., effort and consistency of behaviors). Overlapping items on scales (e.g., sighs, grimace) might be most common and important (Keefe & Block, 1982), while uncommon items might be more characteristic of a target group. Regarding scoring methods and the clinical utility of scales, including the number of items and scoring interpretation (Ramelet, Abu-Saad, Rees, & McDonald, 2004), most studies lack data on sensitivity and specificity, rendering their scales potentially impractical for clinical practice (Zwakhalen et al., 2006). Finally, no behavioral assessment scale has been comprehensively tested in a variety of care settings or among different pain conditions, meaning that all require further investigation (Zwakhalen et al., 2006).

Limitations

Studies are not systematically synthesized, nor are the qualities of studies appraised prior to inclusion. While our eligibility criteria attempt to objectively review these studies, quality judgement scores should be interpreted with caution due to potential review bias. Future research should investigate tools validity, such as evaluation of aspects of PaBS validity in other settings and countries. Longitudinal studies are needed to assess tool predictive validity.

Conclusions

We contribute to an understanding of pain-related behavior observational assessment tools. While we deem existing pain-behavioral tools to be valid and reliable, only two are feasible in clinical outpatient settings, of which PaBS can be most readily applied. Because the PaBS rates two constructs (the presence and severity of common pain-behaviors) it better captures the pain experience. Because this can be achieved while performing physical performance tests that are representative of daily activities, it is more useful as a clinical outcome measure. Further research to examine PaBS is recommended on different patient groups and populations.

Clinical implication/Applications to other areas

One of the main goals of the biopsychosocial approach is to reduce the unhelpful or inaccurate patient beliefs and behaviors that can adversely affect treatment outcomes, e.g., pain, disability, and return to work. Pain behaviors were reported in different health conditions, e.g., rheumatoid arthritis (Waters, Riordan, Keefe, & Lefebvre, 2008), people who are cognitively impaired or people who are unable to self-report (Shega et al., 2008; Horgas, Elliott, & Marsiske, 2009). Healthcare practitioners may benefit from knowledge, awareness, and measurement of the pain-related behaviors in people with different chronic pain conditions. As this may assist into a more holistic biopsychosocial understanding of a health condition (Vlaeyen & Linton, 2000; Waddell, 1992) and may assist to identify key modifiable targets for management (O'Sullivan et al., 2018).

Pain behavior, particularly muscle guarding, might contribute to a feedback cycle that perpetuates pain (Olugbade et al., 2019). It is may possibly targeting these pain behaviors can improve pain levels. For example, a relationship was found between guarding with disability (Prkachin et al., 2007) and with pain and it was mediated by anxiety in people with chronic pain (Olugbade et al., 2019). As such, pain-related guarding is more likely to be effectively addressed by intervention to reduce anxiety (e.g., Yoga) (De Giorgio, Padulo, & Goran, 2018) rather than pain (e.g., analgesics) (Olugbade et al., 2019); interventions that addressing the behavioral aspect of pain as one component, may has the potential to help to regain movement free of pain behaviors.

Mini-dictionary of terms

Pain-related behavior: Pain behaviors as were first described by Fordyce (1977) indicating the verbal and nonverbal behaviors exhibited by pain subjects that serve to communicate the fact that they are experiencing pain.

Sighing: As described by Keefe and Block (1982), a clear exhalation accompanied by a rise and fall of the shoulders.

Grimacing: As described by Keefe and Block (1982), a clear facial expression of pain that may include a furrowed brow, tightened lips, and clenched teeth.

Guarding: As described by Keefe and Block (1982), abnormal and stiff movement during shifting.

Rubbing: As described by Keefe and Block (1982), rubbing the affected body part to relieve pain.

Physical performance tests: As reported by Simmonds et al. (1998), a suite of timed functional tests commonly used in rehabilitation medicine that was able to discriminate people with chronic low back pain (CLBP) from normal healthy people. This suite of tests included repeated trunk flexion, repeated sit to stand, timed up and go, loaded reach, and 50-ft walk tests.

458 PART | IV Assessments, screening, and resources

Key facts

- Pain-related behaviors are considered as an important element of pain experience.
- Clinicians' understanding/analyzing of these behaviors is an important part of a multidimensional patient assessment and treatment.
- Assessment of pain behaviors could help health professionals to predict the severity of a painful condition and may help to understand thoughts and beliefs about pain.

Summary points

- Four behavioral (observational) pain assessment tools were identified in this review, indicating a considerable gap in literature investigating pain-related behaviors.
- Only two observational tools are feasible in clinical outpatient settings, of which PaBS can be most readily applied.
- While we deem existing pain-behavioral tools to be valid and reliable, current scales mostly lack data on sensitivity, specificity or needed to be tested in a larger population.

Appendix

Authors	Procedure	Construct	Sample
Keefe and Block (1982)	Each subject was videotaped for a 10-min interval while performing a standardized and random sequence of sitting, walking, standing, and reclining. Each sequence included l- and 2-min sitting and standing periods, two 1-min reclining periods, and two 1-min walking periods. The total pain behavior score was derived from the summation of eight specific pain behaviors (i.e., guarding, bracing, grimacing, sighing, rigidity, passive rubbing, active rubbing, and self-stimulation)	The presence of pain behavior. A pain behavior was scored only once even though it may have occurred multiple times during the observation interval	LBP patients
Ahles, Coombs, Jensen, et al. (1990)	Patients were videotaped for one 10-min session utilizing the procedure described by Keefe and Block (1982)	The presence of pain behavior	Cancer patients
Richards et al. (1982)	The UAB Pain Behavior Scale consists of 10 target behaviors, each of which contributes equally to the total score (range of possible scores from 0 to 10)	Ratings are based on frequency estimates for observed behaviors: none (0), occasional (l/2) and frequent (1)	Inpatient treatment program for chronic pain patients
Feuerstein et al. (1985)	Adapted from Richards et al. (1982)	The frequency estimates of observed behaviors	Adapted the inpatient pain behavior rating scale modified for outpatient (multiple sites of chronic pain) use
McDaniel et al. (1986)	Adapted from Keefe and Block (1982)	The presence of pain behavior	Rheumatoid arthritis
Kleinke and Spangler (1988)	Administrating an audio-visual taxonomy of pain behavior that was adapted from Keefe and Block (1982)	The presence of pain behavior	LBP
Jensen et al. (1989)	Adapted from Keefe and Block (1982)	The presence of pain behavior	Nonchronic LBP

Continued

Authors	Procedure	Construct	Sample
Watson and Poulter (1997)	Patients were videotaped performing a circuit of functional tasks, walking; sitting into and rising from a low chair; walking stepping over obstacles on the floor; filling, lifting and carrying a box of weights; getting onto and off all fours; walking carrying a weight; getting on and off a bed; bouncing and rolling a ball; climbing stairs and sitting into and rising from a highchair	Frequency of pain behavior	LBP
Koho et al. (2001)	Subjects were observed sitting; timed 5-min walk; lie down prone to the floor and roll over 360° and stand up; bending and reaching; filling, lifting and carrying a box of weights; and stair climbing	Frequency of pain behavior	LBP
Prkachin, Schultz, et al. (2002)	Patients underwent a standardized physical examination that was included; anthropometric measures, physical signs, flexion and extension, medically incongruent signs, reflexes, strength, muscle bulk, sensation, tenderness, and range of motion in a variety of tests were recorded in real time during clinical conditions	Frequency of pain behavior	LBP

References

Ahles, T. A., Coombs, D. W., Jensen, L., et al. (1990). Development of a behavioral observation technique for the assessment of pain behaviors in cancer patients. *Behavior Therapy*, *21*(4), 449–460.

Alamam, D. (2019). *Investigating chronic low back pain-related disability in Saudi Arabia*. Unpublished PhD thesis Australia: University of Sydney.

Alamam, D. M., Leaver, A., Moloney, N., Alsobayel, H. I., Alashaikh, G., & Mackey, M. G. (2019). Pain behaviour scale (PaBS): An exploratory study of reliability and construct validity in a chronic low back pain population. *Pain Research & Management*, *2019*.

Connally, H. G., & Sanders, H. S. (1991). Predicting low back pain patients' response to lumbar sympathetic nerve blocks and interdisciplinary rehabilitation: The role of pretreatment overt pain behavior and cognitive coping strategies. *Pain*, *44*(2), 139–146.

de C Williams, A. C., & Craig, K. D. (2006). A science of pain expression? *Pain*, *125*(3), 202–203.

De Giorgio, A., Padulo, J., & Goran, K. (2018). Effectiveness of yoga combined with back school program on anxiety, kinesiophobia and pain in people with non-specific chronic low back pain: A prospective randomized trial. *Muscle, Ligaments and Tendons Journal*, 104–112.

Feuerstein, M., Greenwald, M., Gamache, M., Papciak, A., & Cook, E. (1985). The pain behavior scale: Modification and validation for outpatient use. *Journal of Psychopathology and Behavioral Assessment*, *7*(4), 301–315.

Fordyce, W. (1977). Behavioral methods for chronic pain and illness. *Pain*, *3*(3), 291–292.

Froud, R., Patterson, S., Eldridge, S., Seale, C., Pincus, T., Rajendran, D., et al. (2014). A systematic review and meta-synthesis of the impact of low back pain on people's lives. *BMC Musculoskeletal Disorders*, *15*(1), 1.

Goldberg, D. S., & McGee, S. J. (2011). Pain as a global public health priority. *BMC Public Health*, *11*, 770.

Hadjistavropoulos, H., & Craig, K. (1994). Acute and chronic low back pain: Cognitive, affective, and behavioural dimensions. *Journal of Consulting and Clinical Psychology*, *62*(2), 341–349.

Hadjistavropoulos, T., & Craig, K. D. (2002). A theoretical framework for understanding self-report and observational measures of pain: A communications model. *Behaviour Research and Therapy*, *40*(5), 551–570.

Hadjistavropoulos, T., & Craig, K. D. (2004). *Pain: Psychological perspectives*. Psychology Press.

Hobara, M. (2005). Beliefs about appropriate pain behavior: Cross-cultural and sex differences between Japanese and euro-Americans. *European Journal of Pain*, *9*(4), 389–393.

Horgas, A. L., Elliott, A. F., & Marsiske, M. (2009). Pain assessment in persons with dementia: Relationship between self-report and behavioral observation. *Journal of the American Geriatrics Society*, *57*(1), 126–132.

460 PART | IV Assessments, screening, and resources

Jensen, I. B., Bradley, L. A., & Linton, S. J. (1989). Validation of an observation method of pain assessment in non-chronic back pain. *Pain, 39*(3), 267–274.

Keefe, F. J., & Block, A. R. (1982). Development of an observation method for assessing pain behavior in chronic low back pain patients. *Behavior Therapy, 13*(4), 363–375.

Keefe, F. J., Crisson, J. E., Maltbie, A., Bradley, L., & Gil, K. M. (1986). Illness behavior as a predictor of pain and overt behavior patterns in chronic low back pain patients. *Journal of Psychosomatic Research, 30*(5), 543–551.

Keefe, F. J., Crisson, J. E., & Trainor, M. (1987). Observational methods for assessing pain: A practical guide. In J. A. Blumenthal, & D. C. McKee (Eds.), *Applications in behavioral medicine and health psychology: A clinician's source book* (pp. 67–94). Professional Resource Exchange, Inc.

Keefe, F. J., Lefebvre, J. C., Egert, J. R., Affleck, G., Sullivan, M. J., & Caldwell, D. S. (2000). The relationship of gender to pain, pain behavior, and disability in osteoarthritis patients: The role of catastrophizing. *Pain, 87*(3), 325–334.

Keefe, F. J., Wilkins, R. H., & Cook, W. A. (1984). Direct observation of pain behavior in low back pain patients during physical examination. *Pain, 20*(1), 59–68.

Keefe, F. J., Wilkins, R. H., Cook, W. A., Jr., Crisson, J. E., & Muhlbaier, L. H. (1986). Depression, pain, and pain behavior. *Journal of Consulting and Clinical Psychology, 54*(5), 665–669.

Kleinke, C., & Spangler, A. (1988). Psychometric analysis of the audiovisual taxonomy for assessing pain behavior in chronic back-pain patients. *Journal of Behavioral Medicine, 11*(1), 83–94.

Koho, P., Aho, S., Watson, P., & Hurri, H. (2001). Assessment of chronic pain behaviour: Reliability of the method and its relationship with perceived disability, physical impairment and function. *Journal of Rehabilitation Medicine, 33*(3), 128–132.

Labus, J. S., Keefe, F. J., & Jensen, M. P. (2003). Self-reports of pain intensity and direct observations of pain behavior: When are they correlated? *Pain, 102*(1), 109–124.

Lee, H., Hübscher, M., Moseley, G. L., Kamper, S. J., Traeger, A. C., Mansell, G., et al. (2015). How does pain lead to disability? A systematic review and meta-analysis of mediation studies in people with back and neck pain. *Pain, 156*(6), 988–997.

Main, C. J., Keefe, F. J., Jensen, M. P., Vlaeyen, J. W., & Vowles, K. E. (2015). *Fordyce's behavioral methods for chronic pain and illness: Republished with invited commentaries.* Lippincott Williams & Wilkins.

McDaniel, L. K., Anderson, K. O., Bradley, L. A., Young, L. D., Turner, R. A., Agudelo, C. A., et al. (1986). Development of an observation method for assessing pain behavior in rheumatoid arthritis patients. *Pain, 24*(2), 165–184.

Merskey, H. (1964). *An investigation of pain in psychological illness.* University of Oxford.

Meyer, K., Klipstein, A., Oesch, P., Jansen, B., Kool, J., & Niedermann, K. (2016). Development and validation of a pain behavior assessment in patients with chronic low back pain. *Journal of Occupational Rehabilitation, 26*(1), 103–113.

Monina, E., Falzetti, G., Firetto, V., Mariani, L., & Caputi, C. A. (2006). Behavioural evaluation in patients affected by chronic pain: A preliminary study. *The Journal of Headache and Pain, 7*(6), 395–402.

O'Sullivan, P. B., Caneiro, J., O'Keeffe, M., Smith, A., Dankaerts, W., Fersum, K., et al. (2018). Cognitive functional therapy: An integrated behavioral approach for the targeted management of disabling low back pain. *Physical Therapy, 98*(5), 408–423.

Olugbade, T., Bianchi-Berthouze, N., & Williams, A. C. C. (2019). The relationship between guarding, pain, and emotion. *Pain Reports, 4*(4), e770.

Prkachin, K. M., Hughes, E., Schultz, I., Joy, P., & Hunt, D. (2002). Real-time assessment of pain behavior during clinical assessment of low back pain patients. *Pain, 95*, 23–30.

Prkachin, K. M., Schultz, I. Z., Berkowitz, J., Hughes, E., & Hunt, D. (2002). Assessing pain behaviour of low-back pain patients in real time: Concurrent validity and examiner sensitivity. *Behaviour Research and Therapy, 40*(5), 595–607.

Prkachin, K. M., Schultz, I. Z., & Hughes, E. (2007). Pain behavior and the development of pain-related disability: The importance of guarding. *The Clinical Journal of Pain, 23*(3), 270–277.

Ramelet, A.-S., Abu-Saad, H. H., Rees, N., & McDonald, S. (2004). The challenges of pain measurement in critically ill young children: A comprehensive review. *Australian Critical Care, 17*(1), 33–45.

Richards, S. J., Nepomuceno, C., Riles, M., & Suer, Z. (1982). Assessing pain behavior: The UAB pain behavior scale. *Pain, 14*(4), 393–398.

Schultz, I. Z., Crook, J. M., Berkowitz, J., Meloche, G. R., Milner, R., & Zuberbier, O. A. (2008). Biopsychosocial multivariate predictive model of occupational low back disability. *Spine (Phila Pa 1976), 27*(23), 2720–2725.

Shega, J. W., Rudy, T., Keefe, F. J., Perri, L. C., Mengin, O. T., & Weiner, D. K. (2008). Validity of pain behaviors in persons with mild to moderate cognitive impairment. *Journal of the American Geriatrics Society, 56*(9), 1631–1637.

Simmonds, M. J., Olson, S. L., Jones, S., Hussein, T., Lee, C. E., Novy, D., et al. (1998). Psychometric characteristics and clinical usefulness of physical performance tests in patients with low back pain. *Spine (Phila Pa 1976), 23*(22), 2412–2421.

Streiner, D. L., Norman, G. R., & Cairney, J. (2015). *Health measurement scales: A practical guide to their development and use.* USA: Oxford University Press.

Sullivan, M. J. (2008). Toward a biopsychomotor conceptualization of pain: Implications for research and intervention. *The Clinical Journal of Pain, 24*(4), 281–290.

Sullivan, M. J., Thibault, P., Savard, A., Catchlove, R., Kozey, J., & Stanish, W. D. (2006). The influence of communication goals and physical demands on different dimensions of pain behavior. *Pain, 125*(3), 270–277.

Sullivan, M. J., Thorn, B., Haythornthwaite, J. A., Keefe, F., Martin, M., Bradley, L. A., et al. (2001). Theoretical perspectives on the relation between catastrophizing and pain. *The Clinical Journal of Pain, 17*(1), 52–64.

Treede, R.-D. (2018). The International Association for the Study of Pain definition of pain: as valid in 2018 as in 1979, but in need of regularly updated footnotes. *Pain Reports, 3*(2).

Treede, R.-D., Rief, W., Barke, A., Aziz, Q., Bennett, M. I., Benoliel, R., et al. (2019). Chronic pain as a symptom or a disease: The IASP classification of chronic pain for the international classification of diseases (ICD-11). *Pain, 160*(1), 19–27.

Trippolini, M. A., Dijkstra, P. U., Jansen, B., Oesch, P., Geertzen, J. H. B., & Reneman, M. F. (2014). Reliability of clinician rated physical effort determination during functional capacity evaluation in patients with chronic musculoskeletal pain. *Journal of Occupational Rehabilitation, 24*, 361–369.

Turk, D. C., Wack, J. T., & Kerns, R. D. (1985). An empirical examination of the "pain-behavior" construct. *Journal of Behavioral Medicine, 8*(2), 119–130.

Vlaeyen, J. W., & Linton, S. J. (2000). Fear-avoidance and its consequences in chronic musculoskeletal pain: A state of the art. *Pain, 85*(3), 317–332.

Vlaeyen, J. W., Van Eek, H., Groenman, N. H., & Schuerman, J. A. (1987). Dimensions and components of observed chronic pain behavior. *Pain, 31*(1), 65–75.

Von Korff, M., Ormel, J., Keefe, F. J., & Dworkin, S. F. (1992). Grading the severity of chronic pain. *Pain, 50*(2), 133–149.

Waddell, G. (1992). Biopsychosocial analysis of low back pain. *Baillière's Clinical Rheumatology, 6*(3), 523–557.

Waters, S. J., Riordan, P. A., Keefe, F. J., & Lefebvre, J. C. (2008). Pain behavior in rheumatoid arthritis patients: Identification of pain behavior subgroups (in Eng.). *Journal of Pain and Symptom Management, 36*(1), 69–78.

Watson, P., & Poulter, M. (1997). The development of a functional task-oriented measure of pain behaviour in chronic low back pain patients. *Journal of Back and Musculoskeletal Rehabilitation, 9*(1), 57–59.

Zwakhalen, S. M., Hamers, J. P., Abu-Saad, H. H., & Berger, M. P. (2006). Pain in elderly people with severe dementia: A systematic review of behavioural pain assessment tools. *BMC Geriatrics, 6*(1), 3.

Chapter 42

The analgesia nociception index: Features and application

Sonia Bansal and Kamath Sriganesh

Department of Neuroanaesthesia and Neurocritical Care, National Institute of Mental Health and Neurosciences, Bengaluru, India

List of abbreviations

ANI	analgesia nociception index
ICU	intensive care unit
HRV	heart rate variability
RSA	respiratory sinus arrhythmia
ECG	electrocardiography
ANS	autonomic nervous system
VLF	very low frequency
LF	low frequency
HF	high frequency
AUC	area under curve
SPI	surgical pleth index
BIS	bi-spectral index
HR	heart rate
BP	blood pressure
SB	scalp block
PSI	pin site infiltration
ISI	incision site infiltration
VAS	visual analogue scale
NRS	numeric rating scale
PACU	post anesthesia care unit
FLACC	facial expression, leg position, degree of activity, quality of cry, consolability
BPS	behavioral pain scale
PD	pupillary diameter
VCPD	variation coefficient of pupillary diameter
PLR	pupillary light reflex
PRD	pupillary reflex dilatation

Introduction

Pain is a subjective experience and hence, self-reported by individuals in various forms. Validated pain scales help in understanding the magnitude of pain and also guide therapeutic interventions. Pain related to injury, surgery, or trauma is mostly acute and nociceptive in nature. In certain populations and scenarios, self-reporting of pain is not possible or reliable such as during general anesthesia or sedation for surgery, in patients in the intensive care unit (ICU) on sedation and/or artificial airway, in patients with diminished consciousness, dementia, or psychiatric disorders, and in young children. In these individuals or situations, surrogate measurements such as movement, tearing, tachycardia, or hypertension are considered as signs of inadequate analgesia. These manifestations are however, not specific to nociception and hence their utility to detect pain is often unreliable.

Features and Assessments of Pain, Anesthesia, and Analgesia. https://doi.org/10.1016/B978-0-12-818988-7.00039-X
Copyright © 2022 Elsevier Inc. All rights reserved.

TABLE 1 Commonly used pain assessment systems in perioperative clinical practice.	
Subjective postoperative pain assessment	Objective intraoperative tools
Visual analog score	Analgesia nociception index
Numerical rating score	Surgical pleth index
No-mild-moderate-severe scoring	Response entropy-state entropy difference
Yes/no scoring	Pupillometry

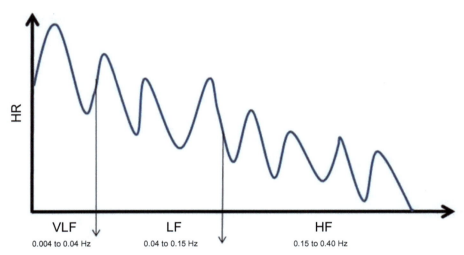

FIG. 1 The spectral content of RR series.

Objective assessment of nociception-antinociception balance has been attempted earlier by using tools such as skin vasomotor reflexes, pulse plethysmographic signal, pupillometry, and heart rate variability (HRV). Both underdosing and overdosing of analgesics are likely during the intraoperative period in the absence of a reliable objective tool to measure pain, and can lead to adverse consequences. Administering optimal dose of analgesia is important during surgery, hence, an objective analgesia/nociception tool is necessary to quantify and manage pain/nociception. The Analgesia Nociception Index (ANI) is one of the recently introduced noninvasive, continuous, objective tools to evaluate analgesia and nociception. In this chapter, we discuss the principles, features, and applications of ANI in clinical practice and assess its comparability with other objective monitors. Table 1 shows other scoring systems employed for perioperative pain assessment.

Basic principle of the ANI

The ANI is computed from the electrocardiography (ECG) data initially sampled at 250 Hz, artefact tested and then resampled at 8 Hz. The data are subsequently structured in 64 s moving windows (512 RR samples) and further analyzed after a wavelet transformation. Analyses of the HRV provide great insights into the activity of the autonomic nervous system (ANS). The very low frequency (VLF) of the heart rate ranges from 0.004 to 0.04 Hz and corresponds to the thermoregulatory system and neuroendocrine activity. The low frequency (LF) ranges from 0.04 to 0.15 Hz and is influenced by sympathetic and parasympathetic activities. The high frequency (HF) ranges from 0.15 to 0.40 Hz and is only affected by the parasympathetic tone (Shaffer & Ginsberg, 2017). Fig. 1 shows the various frequency components of the heart rate obtained from the ECG.

The filtering process allows specifically analyzing of only the HF range of HRV. The respiratory pattern significantly influences this bandwidth of HRV and its effect is estimated using an area under curve (AUC) model. A higher parasympathetic tone is associated with a greater influence of respiration [from respiratory sinus arrhythmia (RSA)] on the HF of HRV. A shift in the autonomic balance toward high sympathetic activity reduces this influence. The level of RSA is expressed numerically as a percentage score from 0 to 100, with 0 implying strong sympathetic tone and minimal influence of RSA on HF, and 100 indicating predominant parasympathetic activity with maximum influence of RSA on HF of HRV (Jeanne, Logier, De Jonckheere, & Tavernier, 2009). Naturally, such a score will be affected by conditions or drugs influencing the RSA or the respiratory pattern. However, initial studies have demonstrated potential for clinical applicability of this HRV-based principle and technology in the assessment of pain, as pain significantly impacts the ANS.

Features of the ANI monitor

The ANI monitor (ANI MetroDoloris, France) has been a result of work from the University Hospital of Lille, France and its Center for Innovative Technologies. The method employed in the ANI monitor provides real-time analysis of the parasympathetic activity of the ANS in a 64 s time window. The ANI computation is based on an algorithm that quantifies the relative amount of parasympathetic activity in the R-R series obtained noninvasively from ECG. The two proprietary single-use sensors (Fig. 2) are applied in V1 and V5 position on the individual's chest (Fig. 3). They record the ECG, detect R waves, and calculate the R-R interval. This R-R series is then filtered using a fast wavelet transform keeping only the HF band, where RSA is expressed. The calculated ANI index varies between 0 and 100, with 0 reflecting a weak, and 100 reflecting a strong parasympathetic tone (Logier et al., 2010).

The ANI monitor displays two values (Fig. 4). The instantaneous ANI, (ANIi) updated every second, is displayed in yellow color and reflects patient's reactions to the nociception during surgery. The ANIi is extremely sensitive as it displays even small changes after mild stimulations such as cold or electric stimulation of a neuromuscular monitor. The mean ANI (ANIm) value results from a 2-min averaging of the ANIi signal, is displayed in orange color and represents effect of analgesia on the patient.

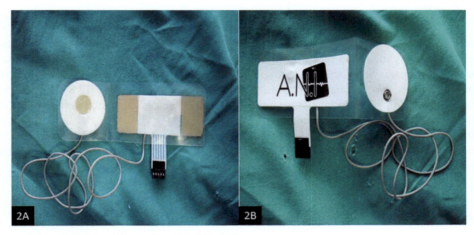

FIG. 2 Two surfaces; patient surface (A) and free surface (B) of the ANI sensor.

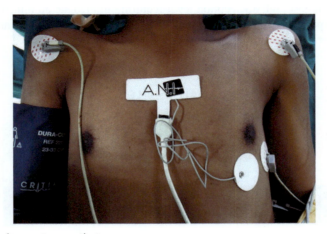

FIG. 3 The position of ANI sensor placement on a patient.

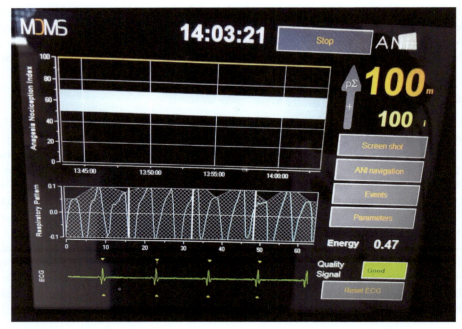

FIG. 4 Display screen of the ANI monitor.

The bluish-green gridded area of the monitor displays the respiratory pattern and represents continuous assessment of RSA (Fig. 4). It is automatically measured by detecting the upper and lower envelopes of the RR series. As the RSA is proportionate to the amount of parasympathetic tone present, a greater surface area indicates a higher parasympathetic tone. Because of the normalization process, the maximum possible area of RSA is 0.2×64 s $= 12.8$ s. This enables to obtain an index between 0 and 100 by dividing the measured area by 12.8. The ANI value is calculated as follows: ANI $= 100 *$ AUC totnu/12.8 where totnu means total normalized units.

The energy of the RR series is displayed continuously as a numerical and normally ranges from 0.05 to 2.5 (Fig. 4). This value represents the spectral power of the entire ANS in the given patient. The ANI values are unreliable when energy values are below or above the normal range (ANI Parameter, 2021; Logier, Jeanne, Tavernier, & De Jonckheere, 2006).

Interpretation of ANI values

The ANI values can range from 0 to 100, where 0 indicates severe nociception and extremely inadequate analgesia, and 100 implies complete absence of nociception and luxurious analgesia. Fig. 5 presents a simple graphical interpretation of ANI values where values below 50 suggest high nociception level requiring immediate administration of an analgesic, values between 50 and 70 imply acceptable nociceptive stimulus and optimal analgesia state, and values above 70 indicate minimal nociception level requiring reduction in analgesia dose (ANI Parameter, 2021).

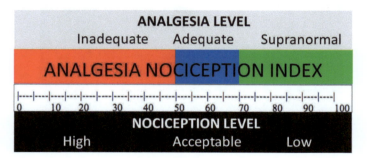

FIG. 5 Interpretation of ANI values.

Clinical applications

Detection of intraoperative nociception

Even to this date, intraoperative administration of analgesics is most often based on the changes in the autonomic reflexes. However, it is a well-known fact that hemodynamic parameters are poor determinants of analgesic requirements and pain management based on hemodynamic perturbations may therefore be unreliable. If the utility of ANI as an objective monitor for the prediction of intraoperative nociception can be established, it would prevent undertreatment of pain, overdosage of opioids and assist in individualized pain management.

Gruenewald et al. studied the ability of objective analgesia monitors ANI and Surgical Pleth Index (SPI) with conventional parameters such as bispectral index (BIS), heart rate (HR), and blood pressure (BP) to detect standardized noxious stimuli during propofol-remifentanil anesthesia in 25 patients. Tetanic stimulation was applied at different effect-site concentrations (Ce_{remi}) of 0, 2 and 4 ng/mL of remifentanil. There was a significant change in the ANI and SPI, but not in the BIS, HR, or BP with tetanic stimulation at different remifentanil levels, indicating their superiority in detecting nociception during anesthesia (Gruenewald et al., 2013).

Jeanne et al. compared ANI with hemodynamics (HR and BP) during various noxious stimuli in 15 patients undergoing laparoscopic surgeries with propofol-remifentanil anesthesia. Induction of anesthesia resulted in lowering of HR and BP, and increase in ANI suggesting predominance of the parasympathetic component of the ANS. Tetanic stimulation of the ulnar nerve resulted in decrease in ANI values without significant changes in the hemodynamics. Creation of pneumoperitoneum for laparoscopy also decreased ANI values, but there were no significant changes in the HR and BP suggesting better sensitivity of ANI compared to hemodynamic parameters in detecting nociceptive stimuli during surgery (Jeanne, Clément, De Jonckheere, Logier, & Tavernier, 2012).

Laryngoscopy and intubation is one of the most noxious stimuli in the intraoperative period. Sixty adult patients undergoing craniotomy were monitored for changes in ANI, HR, and BP during anesthetic induction and tracheal intubation. There was reduction in BP and ANI (but not below the threshold of 50) along with increase in HR with induction of anesthesia. Laryngoscopy and intubation resulted in significant increase in HR and BP and significant decrease in ANI below the critical threshold of 50. A significant negative linear correlation was observed between ANI and hemodynamic parameters during intubation (Sriganesh, Theerth, Reddy, Chakrabarti, & Rao, 2019).

In a study on 21 patients undergoing craniotomy; ANI, hemodynamic variables and spectral entropy values were studied during episodes of noxious stimulation such as intubation, skull pin fixation, skin incision, craniotomy drilling, dural opening, and skin closure. An inverse correlation was observed between ANI and hemodynamic variables during the noxious stimuli. When fentanyl was administered to manage responses to the noxious stimuli, ANI increased without any changes in the state or response entropy values (Kommula, Bansal, & Umamaheswara Rao, 2019).

The association between ANI and hemodynamic changes was studied in 30 patients during sevoflurane anesthesia. ANI decreased with noxious stimuli such as airway management and skin incision and promptly increased after administration of potent opioid analgesic, fentanyl. However, the predictive value to preempt >10% changes in HR and BP, was low (Ledowski, Averhoff, Tiong, & Lee, 2014).

The ability of ANI to preemptively detect nociception and consequently predict hemodynamic reactivity has also been studied. Jeanne et al. evaluated ANI during periods of hemodynamic reactivity (defined as 20% changes in HR or BP) in 27 patients undergoing total knee replacement under propofol anesthesia. They noted that ANI of 63 predicted hemodynamic changes with good sensitivity and specificity (Jeanne et al., 2014). The dynamic variations in ANI ($\triangle$ ANI) might be a better predictor of hemodynamic response than static values. The ANI response to hemodynamic reactivity (20% changes in HR or BP within 5 min) with noxious stimuli was studied in 128 patients undergoing ear-nose-throat or orthopedic surgery with desflurane/remifentanil anesthesia. The authors observed that a decrease in ANI of $\geq$19% in 1 min has a high predictive value for increase in HR and BP by 20% in the next 5 min (Boselli, Logier, Bouvet, & Allaouchiche, 2016).

Spine surgery is considered to be associated with severe pain. While both inhalational and intravenous anesthesia are used for spine surgery, intravenous anesthesia is preferred if intraoperative neurophysiological monitoring is considered. Therefore, it is important to understand if there is a difference in nociception with two different anesthetic techniques. The ANI values did not differ between propofol-remifentanil and sevoflurane-remifentanil technique in patients undergoing spine stabilization procedure indicating that either technique provides satisfactory intraoperative analgesia (Turan et al., 2017).

ANI by virtue of its capability to measure pain, can be utilized to compare the efficacy of different analgesic techniques in day-to-day clinical practice. For example, during craniotomy for neurosurgical procedures, the application of Mayfield skull pins is a very noxious stimulus and elicits severe hemodynamic response as the scalp and periosteum are richly

468 PART | IV Assessments, screening, and resources

innervated with nerve fibers. Hence, several strategies are employed to minimize this response and include loco-regional analgesia with local anesthetic drugs such as scalp block (SB), pin site infiltration (PSI), incision site infiltration (ISI) or systemic administration of drugs such as lignocaine, opioids, alpha 2 agonists, and beta-blockers. In a study on 60 patients undergoing craniotomy, Theerth et al. observed a lower magnitude and shorter duration of decrease in ANI (and increase in HR and BP) with SB than with PSI, indicating higher efficacy of SB in minimizing ANI and hemodynamic response to skull pin insertion (Theerth, Sriganesh, Chakrabarti, Reddy, & Rao, 2019).

ANI also helps to differentiate the cause for hemodynamic changes. During surgery, increases in BP could occur due to prolonged tourniquet application or due to inadequate analgesia. In this setting, ANI might be useful in differentiating hypertension and pain, and in selecting appropriate choice of drugs to manage them (Logier et al., 2011).

ANI monitoring also helps in comparing intraoperative opioid consumption during surgeries with different analgesia techniques. In a study comparing SB with ISI of local anesthetic in 60 patients undergoing craniotomies, a reduced administration of fentanyl was observed with SB (Theerth, Sriganesh, Reddy, Chakrabarti, & Umamaheswara Rao, 2018).

Postoperative pain assessment

Self-assessment of pain using Visual Analogue Scale (VAS) on 0–100 scale or Numeric Rating Scale (NRS) on 0–10 scale is considered the gold standard for monitoring postoperative pain. However, in sedated patients and children with learning difficulties and cognitive deficits, these assessments may not be possible. This calls for objective pain monitoring.

The ANI as a tool to assess postoperative pain was investigated in several studies. In an experimental study on 20 healthy volunteers, four different stimuli (unexpected and expected electrical painful stimuli, and expected nonpainful and sham stimuli) were randomly applied. Blinded volunteers rated stimuli on the NRS, and ANI was also simultaneously recorded. ANI could not differentiate painful, nonpainful or sham stimuli limiting the usefulness of ANI in awake persons (Jess, Pogatzki-Zahn, Zahn, & Meyer-Frießem, 2016). In another volunteer study, Issa et al. administered 2 Hz electrical stimulus to 23 volunteers, increasing current intensity by 5 mA increments up to 30 mA and compared ANI and NRS at each level. A weak negative correlation was seen between ANI and NRS, and also between change in ANI from baseline (ΔANI) and NRS. There was no significant change in HR, BP or respiratory rate (Issa et al., 2017).

Jeanne et al. also observed that ANI poorly predicted pain in conscious awake patients who had undergone total knee replacement under propofol anesthesia (Jeanne et al., 2014). Similar results were reported by Szental et al. where ANI-guided intraoperative analgesia with morphine did not reduce postoperative pain measured using VAS or rescue analgesia in 120 patients undergoing laparoscopic cholecystectomy (Szental et al., 2015). On the contrary, inverse linear relationship was noted between ANI and VAS during labor in 45 awake parturients receiving epidural analgesia. ANI decreased during uterine contractions and an ANI score < 49 indicated a VAS >30 (Le Guen et al., 2012).

The performance of ANI to discriminate mild from severe postoperative pain is important for it to be beneficial in patients where subjective feedback is not possible. This also helps in making choice about the drug, route and dose of an analgesic for treatment of pain. In a study comparing ANI with NRS, 816 pain ratings from 114 patients were studied after surgery under sevoflurane anesthesia. Small but statistically significant negative correlation was observed between ANI and NRS. However, ANI was not able to reliably differentiate severity of acute postoperative pain (NRS 0 from NRS 6–10) (Ledowski et al., 2013). In another study on 200 patients who underwent surgery or endoscopy with general anesthesia, the ability of ANI to detect patients with NRS > 3 and NRS $\geq$ 7 in the postanesthesia care unit (PACU) was studied. A negative linear correlation was noted between ANI and NRS. At a threshold of 57, ANI could discriminate between NRS of $\leq$3 or > 3 with good sensitivity and specificity. The performance of ANI further improved with propofol anesthesia compared to inhalational anesthesia. At a threshold of 48, ANI discriminated between severe (NRS $\geq$7) and nonsevere pain (NRS < 7), irrespective of the anesthetic agent (Boselli et al., 2013). Several factors may affect the ability of ANI to differentiate postoperative pain. It is likely that inhalational anesthesia is associated with higher plasma noradrenaline levels and has different effects on HRV compared to intravenous anesthesia. Similarly, the quality of recovery (anxiety and arousal) from two anesthetic techniques may be different affecting the discriminatory power of ANI in detecting pain severity in the postoperative setting. These aspects may have contributed to contrasting findings in different studies.

Consistent with earlier studies, a negative linear relation was observed between ANI and NRS in the postoperative period after inhalational-opioid anesthesia in 107 patients (Abdullayev, Uludag, & Celik, 2019). Similar negative linear relationship was seen between preextubation ANI and the PACU NRS after ear, nose, and throat or orthopedic surgery with inhalational anesthesia and remifentanil in 200 patients. The authors observed an 86% sensitivity and specificity to differentiate between NRS of $\leq$3 and NRS >3 (mild vs moderate-to-severe postoperative pain) at an ANI threshold of 50 (Boselli et al., 2014). However, in all the above studies, patients received neostigmine and atropine to reverse neuromuscular blockade and the impact of these drugs on ANI is unquantifiable. This problem can probably be overcome by using sugammadex for neuromuscular recovery.

Pain assessment in pediatric patients

Persistent postoperative pain in children may lead to behavioral problems in the future. So adequate pain assessment followed by treatment is crucial. Diagnosing pain is a challenge in pediatric patients. Studies have shown that while hemodynamic parameters demonstrate poor predictive value for nociception, ANI may have a significant role for diagnosing noxious stimulation in children undergoing surgery. The ANI recordings were studied in 49 children aged between 2 and 12 years 5 min before and after the surgical incision under sevoflurane-remifentanil anesthesia. The ANIi and ΔANI were more accurate when compared to ANIm while hemodynamic parameters- HR and BP, were not predictive of noxious stimuli (Julien-Marsollier et al., 2018).

The ANIi has better predictive value than HR for assessing inadequate antinociception in pediatric surgical patients under sevoflurane anesthesia and also reflects the return of sufficient analgesia with administration of opioid. The ANIm did not reflect these changes, probably due to the prolonged averaging period (Weber, Geerts, Roeleveld, Warmenhoven, & Liebrand, 2018).

As far as the effect of age on HRV is concerned, age affects the LF part of the HRV while ANI is entirely based on the HF oscillations (Bobkowski et al., 2017). ANI monitor was observed to be more responsive to nociception [standardized tetanic stimulation] than skin conductance or hemodynamic perturbations in sevoflurane anesthetized children at different concentrations of remifentanil (Sabourdin et al., 2013).

Pain assessment can be particularly demanding in infants or those with cognitive deficits. ANI has been found to be a sensitive indicator of pain in children <7 years and those with cognitive impairment who recovered from general anesthesia. ANI correlated well with the facial expression, leg position, degree of activity, quality of cry, and consolability (FLACC) scale. FLACC score was higher and ANI was lower in patients recovering from anesthesia after surgery as compared to those recovering after anesthesia for imaging procedures. ANI threshold of 56 was most predictive of FLACC score ≥ 4 (Gall et al., 2015).

Pain assessment in ICU setting

Patients in the ICU frequently experience pain especially during procedures like insertion of lines and drains, tracheal suctioning, handling of wounds, positioning, and physiotherapy. Unfortunately, assessment of pain in this population is also difficult as these patients are often sedated, delirious, noncommunicative, and sometimes paralyzed. Even the Behavioral Pain Scale (BPS) is subjective. Detection of pain in ICU patients will help in administering appropriate analgesia rather than excessive sedation for facilitating the conduct of invasive procedures, ambulation, weaning from ventilation, and early extubation, thereby reducing ICU stay and improving patient outcomes.

The effectiveness of ANI was evaluated in 41 deeply sedated and mechanically ventilated patients. ANI values decreased with painful stimulation (turning for wash stand) in all the study patients. ANI was also effective in detecting pain in the presence of norepinephrine but the values of ANI were higher. However, the authors did not find correlation between ANI and BPS (Broucqsault-Dédrie, De Jonckheere, Jeanne, & Nseir, 2016). Similar results were reported in 21 patients with traumatic brain injury during painful stimulation (tracheal suctioning). A negative linear relationship was seen between ANI and BPS; for an ANI cut-off of 50, the sensitivity and specificity to detect BPS ≥ 5 was 73% and 62%, respectively (Jendoubi, Abbes, Ghedira, & Houissa, 2017). In another study in 110 patients, 969 assessments of ANI and BPS were performed during three painful procedures (dressing change, position change, and tracheal suctioning) in the noncomatose noncommunicative ICU patients. An ANI value ≥ 43 had a negative predictive value of 90% to be not associated with BPS of ≥ 5. ANIi was again found to be more reliable than ANIm (Chanques et al., 2017).

Effect on opioid consumption

Opioids are the most commonly used analgesics during surgery. Although they are the most effective analgesics, the associated side-effects limit their use. ANI by quantifying pain may help to optimize their administration and minimize opioid-related adverse events.

ANI-guided intraoperative analgesia (maintaining ANI > 50) resulted in lower NRS pain scores in the postoperative period and lower fentanyl consumption in the recovery room compared to patients receiving intraoperative analgesia by standard practice in 50 patients undergoing lumbar spine surgery with sevoflurane anesthesia. Harmful effects such as shivering and postoperative nausea and vomiting were also less in patients receiving ANI-guided analgesia (Upton, Ludbrook, Wing, & Sleigh, 2017). However, intraoperative predicted effect site fentanyl concentration was higher with ANI-guided intraoperative analgesia during nociceptive time points.

Similar results have been reported by Daccache et al., where ANI guidance led to lower remifentanil consumption, lower postoperative pain scores, reduced number of episodes of intraoperative hemodynamic fluctuations and decreased postoperative rescue opioid use in 180 patients undergoing vascular surgery (Daccache et al., 2017). However, in another study, comparing ANI-guided versus standard opioid administration in obese patients undergoing bariatric surgery, the opioid (sufentanil) consumption was lower in the ANI group but the opioid-related adverse events were similar (Le Gall et al., 2019).

Comparison with other objective analgesia/nociception monitors

There are several objective tools to assess pain and analgesia. It is important to know how they fare against each other. In a study involving 58 children, pupillary diameter (PD) and ANI were compared for early assessment of regional anesthesia success following skin incision under sevoflurane anesthesia. Failure of regional anesthesia, as observed by an increase in $HR \geq 10\%$ with skin incision, was associated with rapid and significant changes in both the monitors (PD and ANI) (Migeon et al., 2013).

Charier et al. compared the ability of ANI, variation coefficient of pupillary diameter (VCPD), pupillary light reflex (PLR) and PD to measure postoperative pain (defined as $VAS \geq 4$). The VCPD >6.4 had the highest sensitivity (0.92) and specificity (0.73), while ANI < 40, had good sensitivity but poor specificity. ANIi had only a weak negative correlation with VAS. There was no correlation between PD and pain (Charier et al., 2019). There is another parameter known as Pupillary Reflex Dilatation (PRD) to measure pain intensity. However, it requires application of a noxious stimulus whereas PLR does not require any stimulation.

A metaanalysis of 10 randomized controlled trials that used ANI, SPI or pupillometry to study the effect of objective nociception monitoring on intraoperative opioid consumption, showed reduced opioid consumption with SPI-guided analgesia compared to conventional analgesia. However, similar difference was not seen with ANI (Jiao et al., 2019).

In 189 conscious postoperative patients, both SPI and ANI were significantly different in patients with (NRS > 0) and without pain (NRS $= 0$). The cut-off values for SPI and ANI for predicting postoperative pain were 44 and 63, respectively (Lee et al., 2019).

Confounders for ANI

Since, the ANI monitor is based on the detection of autonomic tone, confounders to sympathetic/parasympathetic modulation should be taken into account while interpreting ANI values. Drugs such as clonidine and dexmedetomidine with sympatholytic properties may induce changes in ANI independent of nociception, making ANI unreliable in the presence of these drugs. Ketamine is an analgesic with sympathomimetic properties. The changes in ANI after ketamine could be from drug's sympathomimetic effect (decrease in ANI) or analgesic effect (increase in ANI). In a study on 20 women undergoing abdominal hysterectomies, a single bolus of 0.5 mg/kg dose of ketamine during general anesthesia and absent noxious stimulation did not affect the ANI values suggesting ANI monitoring to be feasible during the use of ketamine (Bollag et al., 2015).

ANI monitor may be unreliable in the setting of severe arrhythmias, pacemaker, atrial fibrillation, cardiopulmonary bypass or with antimuscarinic drugs. As ANI is based on detection of RSA, irregular respiration or apnea may also affect the interpretation of ANI values. Other confounders reported by studies are emotion and stress. Therefore, a relaxed state of mind may lead to higher ANI scores (Jess et al., 2016).

Conclusions

Objective assessment of pain is needed to detect and manage nociception during the intraoperative period. The ANI monitoring helps in rapid and reliable detection of nociception and response to administration of analgesia during surgery. It compares reasonably well with other objective nociception monitors during the intraoperative period. Its utility in conscious spontaneously breathing persons is limited when compared to subjective pain scores such as VAS or NRS, though a liner negative correlation is seen. Caution needs to be exercised in interpretation of ANI in presence of confounders. However, the advantages of ANI monitoring are several and include noninvasiveness, absence of need for noxious stimulation and ability for continuous monitoring.

Mini-dictionary of terms

Nociception: Nociception implies the sensory process of providing signals leading to the perception of pain.

Analgesia: Analgesia is a phenomenon of loss of pain sensation mostly from administration of a drug or from a disease process.

Confounders: Confounders are factors or variables other than those studied that affect the studied factor or variable.

Heart rate variability: Heart rate variability is a physiological phenomenon of variation in the time interval between the heart beats or the R-R intervals on the electrocardiogram.

Respiratory sinus arrhythmia: It is the variability in the heart rate occurring in harmony with respiration where the R-R interval of the electrocardiogram is reduced during inspiration and increased during expiration.

Key facts of ANI

- ANI provides real-time analysis of the autonomic nervous system in a 64 s time window.
- ANI computation is based on an algorithm that quantifies the relative amount of parasympathetic activity in the R-R series obtained from an electrocardiogram.
- ANI varies between 0 and 100, with 0 reflecting a weak, and 100 reflecting a strong parasympathetic tone.
- Monitoring of the ANI helps prevent undertreatment of pain, over-dosage of opioids and assist in individualized pain management during surgery.
- ANI monitoring may be unreliable in certain settings where autonomous nervous system is affected from drugs or disease.

Summary points

- Objective monitoring of nociception is important to detect pain and titrate analgesia during the intraoperative period.
- Analgesia Nociception Index is an objective, continuous, noninvasive tool for monitoring nociception and analgesia.
- Analgesia Nociception Index is derived from the heart rate variability of the electrocardiogram.
- Changes in the Analgesia Nociception Index reflect nociceptive stimuli during surgery and response to administration of analgesia such as opioids.
- Analgesia Nociception Index is comparable to other objective pain assessment tools and better than conventional methods for assessment of pain during surgery.

References

Abdullayev, R., Uludag, O., & Celik, B. (2019). Analgesia nociception index: Assessment of acute postoperative pain. *Revista Brasileira de Anestesiologia, 69*(4), 396–402. https://doi.org/10.1016/j.bjan.2019.01.003.

ANI Parameter, 2021, http://www.medconmed.ro/en/wp-content/uploads/sites/3/2015/12/ANI-technology-introduction.pdf. Retrieved 27-06-2020 http://www.medconmed.ro/en/wp-content/uploads/sites/3/2015/12/ANI-technology-introduction.pdf.

Bobkowski, W., Stefaniak, M. E., Krauze, T., Gendera, K., Wykretowicz, A., Piskorski, J., et al. (2017). Measures of heart rate variability in 24-h ECGs depend on age but not gender of healthy children. *Frontiers in Physiology, 8*, 311. https://doi.org/10.3389/fphys.2017.00311.

Bollag, L., Ortner, C. M., Jelacic, S., Rivat, C., Landau, R., & Richebé, P. (2015). The effects of low-dose ketamine on the analgesia nociception index (ANI) measured with the novel PhysioDoloris™ analgesia monitor: A pilot study. *Journal of Clinical Monitoring and Computing, 29*(2), 291–295. https://doi.org/10.1007/s10877-014-9600-8.

Boselli, E., Bouvet, L., Bégou, G., Dabouz, R., Davidson, J., Deloste, J. Y., et al. (2014). Prediction of immediate postoperative pain using the analgesia/nociception index: A prospective observational study. *British Journal of Anaesthesia, 112*(4), 715–721. https://doi.org/10.1093/bja/aet407.

Boselli, E., Daniela-Ionescu, M., Bégou, G., Bouvet, L., Dabouz, R., Magnin, C., et al. (2013). Prospective observational study of the non-invasive assessment of immediate postoperative pain using the analgesia/nociception index (ANI). *British Journal of Anaesthesia, 111*(3), 453–459. https://doi.org/10.1093/bja/aet110.

Boselli, E., Logier, R., Bouvet, L., & Allaouchiche, B. (2016). Prediction of hemodynamic reactivity using dynamic variations of analgesia/nociception index (ΔANI). *Journal of Clinical Monitoring and Computing, 30*(6), 977–984. https://doi.org/10.1007/s10877-015-9802-8.

Broucqsault-Dédrie, C., De Jonckheere, J., Jeanne, M., & Nseir, S. (2016). Measurement of heart rate variability to assess pain in sedated critically ill patients: A prospective observational study. *PLoS One, 11*(1). https://doi.org/10.1371/journal.pone.0147720, e0147720.

Chanques, G., Tarri, T., Ride, A., Prades, A., De Jong, A., Carr, J., et al. (2017). Analgesia nociception index for the assessment of pain in critically ill patients: A diagnostic accuracy study. *British Journal of Anaesthesia, 119*(4), 812–820. https://doi.org/10.1093/bja/aex210.

Charier, D., Vogler, M. C., Zantour, D., Pichot, V., Martins-Baltar, A., Courbon, M., et al. (2019). Assessing pain in the postoperative period: Analgesia nociception index(TM) versus pupillometry. *British Journal of Anaesthesia, 123*(2), e322–e327. https://doi.org/10.1016/j.bja.2018.09.031.

472 PART | IV Assessments, screening, and resources

Daccache, G., Caspersen, E., Pegoix, M., Monthé-Sagan, K., Berger, L., Fletcher, D., et al. (2017). A targeted remifentanil administration protocol based on the analgesia nociception index during vascular surgery. *Anaesthesia, Critical Care & Pain Medicine, 36*(4), 229–232. https://doi.org/10.1016/j.accpm.2016.08.006.

Gall, O., Champigneulle, B., Schweitzer, B., Deram, T., Maupain, O., Montmayeur Verchere, J., et al. (2015). Postoperative pain assessment in children: A pilot study of the usefulness of the analgesia nociception index. *British Journal of Anaesthesia, 115*(6), 890–895. https://doi.org/10.1093/bja/aev361.

Gruenewald, M., Ilies, C., Herz, J., Schoenherr, T., Fudickar, A., Höcker, J., et al. (2013). Influence of nociceptive stimulation on analgesia nociception index (ANI) during propofol-remifentanil anaesthesia. *British Journal of Anaesthesia, 110*(6), 1024–1030. https://doi.org/10.1093/bja/aet019.

Issa, R., Julien, M., Décary, E., Verdonck, O., Fortier, L. P., Drolet, P., et al. (2017). Evaluation of the analgesia nociception index (ANI) in healthy awake volunteers. *Canadian Journal of Anaesthesia, 64*(8), 828–835. https://doi.org/10.1007/s12630-017-0887-z.

Jeanne, M., Clément, C., De Jonckheere, J., Logier, R., & Tavernier, B. (2012). Variations of the analgesia nociception index during general anaesthesia for laparoscopic abdominal surgery. *Journal of Clinical Monitoring and Computing, 26*(4), 289–294. https://doi.org/10.1007/s10877-012-9354-0.

Jeanne, M., Delecroix, M., De Jonckheere, J., Keribedj, A., Logier, R., & Tavernier, B. (2014). Variations of the analgesia nociception index during propofol anesthesia for total knee replacement. *The Clinical Journal of Pain, 30*(12), 1084–1088. https://doi.org/10.1097/ajp.0000000000000083.

Jeanne, M., Logier, R., De Jonckheere, J., & Tavernier, B. (2009). Validation of a graphic measurement of heart rate variability to assess analgesia/nociception balance during general anesthesia. *Conference Proceedings: Annual International Conference of the IEEE Engineering in Medicine and Biology Society, 2009*, 1840–1843. https://doi.org/10.1109/iembs.2009.5332598.

Jendoubi, A., Abbes, A., Ghedira, S., & Houissa, M. (2017). Pain measurement in mechanically ventilated patients with traumatic brain injury: Behavioral pain tools versus analgesia nociception index. *Indian Journal of Critical Care Medicine, 21*(9), 585–588. https://doi.org/10.4103/ijccm.IJCCM_419_16.

Jess, G., Pogatzki-Zahn, E. M., Zahn, P. K., & Meyer-Frießem, C. H. (2016). Monitoring heart rate variability to assess experimentally induced pain using the analgesia nociception index: A randomised volunteer study. *European Journal of Anaesthesiology, 33*(2), 118–125. https://doi.org/10.1097/eja.0000000000000304.

Jiao, Y., He, B., Tong, X., Xia, R., Zhang, C., & Shi, X. (2019). Intraoperative monitoring of nociception for opioid administration: A meta-analysis of randomized controlled trials. *Minerva Anestesiologica, 85*(5), 522–530. https://doi.org/10.23736/s0375-9393.19.13151-3.

Julien-Marsollier, F., Rachdi, K., Caballero, M. J., Ayanmanesh, F., Vacher, T., Horlin, A. L., et al. (2018). Evaluation of the analgesia nociception index for monitoring intraoperative analgesia in children. *British Journal of Anaesthesia, 121*(2), 462–468. https://doi.org/10.1016/j.bja.2018.03.034.

Kommula, L. K., Bansal, S., & Umamaheswara Rao, G. S. (2019). Analgesia nociception index monitoring during supratentorial craniotomy. *Journal of Neurosurgical Anesthesiology, 31*(1), 57–61. https://doi.org/10.1097/ana.0000000000000464.

Le Gall, L., David, A., Carles, P., Leuillet, S., Chastel, B., Fleureau, C., et al. (2019). Benefits of intraoperative analgesia guided by the analgesia nociception index (ANI) in bariatric surgery: An unmatched case-control study. *Anaesthesia, Critical Care & Pain Medicine, 38*(1), 35–39. https://doi.org/10.1016/j.accpm.2017.09.004.

Le Guen, M., Jeanne, M., Sievert, K., Al Moubarik, M., Chazot, T., Laloë, P. A., et al. (2012). The analgesia nociception index: A pilot study to evaluation of a new pain parameter during labor. *International Journal of Obstetric Anesthesia, 21*(2), 146–151. https://doi.org/10.1016/j.ijoa.2012.01.001.

Ledowski, T., Averhoff, L., Tiong, W. S., & Lee, C. (2014). Analgesia nociception index (ANI) to predict intraoperative haemodynamic changes: Results of a pilot investigation. *Acta Anaesthesiologica Scandinavica, 58*(1), 74–79. https://doi.org/10.1111/aas.12216.

Ledowski, T., Tiong, W. S., Lee, C., Wong, B., Fiori, T., & Parker, N. (2013). Analgesia nociception index: Evaluation as a new parameter for acute postoperative pain. *British Journal of Anaesthesia, 111*(4), 627–629. https://doi.org/10.1093/bja/aet111.

Lee, J. H., Choi, B. M., Jung, Y. R., Lee, Y. H., Bang, J. Y., & Noh, G. J. (2019). Evaluation of surgical Pleth index and analgesia nociception index as surrogate pain measures in conscious postoperative patients: An observational study. *Journal of Clinical Monitoring and Computing*. https://doi.org/10.1007/s10877-019-00399-5.

Logier, R., De Jonckheere, J., Delecroix, M., Keribedj, A., Jeanne, M., Jounwaz, R., et al. (2011). Heart rate variability analysis for arterial hypertension etiological diagnosis during surgical procedures under tourniquet. *Conference Proceedings: Annual International Conference of the IEEE Engineering in Medicine and Biology Society, 2011*, 3776–3779. https://doi.org/10.1109/iembs.2011.6090645.

Logier, R., Jeanne, M., De Jonckheere, J., Dassonneville, A., Delecroix, M., & Tavernier, B. (2010). PhysioDoloris: A monitoring device for analgesia/nociception balance evaluation using heart rate variability analysis. *Conference Proceedings: Annual International Conference of the IEEE Engineering in Medicine and Biology Society, 2010*, 1194–1197. https://doi.org/10.1109/iembs.2010.5625971.

Logier, R., Jeanne, M., Tavernier, B., & De Jonckheere, J. (2006). Pain/analgesia evaluation using heart rate variability analysis. *Conference Proceedings: Annual International Conference of the IEEE Engineering in Medicine and Biology Society, 2006*, 4303–4306. https://doi.org/10.1109/iembs.2006.260494.

Migeon, A., Desgranges, F. P., Chassard, D., Blaise, B. J., De Queiroz, M., Stewart, A., et al. (2013). Pupillary reflex dilatation and analgesia nociception index monitoring to assess the effectiveness of regional anesthesia in children anesthetised with sevoflurane. *Paediatric Anaesthesia, 23*(12), 1160–1165. https://doi.org/10.1111/pan.12243.

Sabourdin, N., Arnaout, M., Louvet, N., Guye, M. L., Piana, F., & Constant, I. (2013). Pain monitoring in anesthetized children: First assessment of skin conductance and analgesia-nociception index at different infusion rates of remifentanil. *Paediatric Anaesthesia, 23*(2), 149–155. https://doi.org/10.1111/pan.12071.

Shaffer, F., & Ginsberg, J. P. (2017). An overview of heart rate variability metrics and norms. *Frontiers in Public Health, 5*, 258. https://doi.org/10.3389/fpubh.2017.00258.

Sriganesh, K., Theerth, K. A., Reddy, M., Chakrabarti, D., & Rao, G. S. U. (2019). Analgesia nociception index and systemic haemodynamics during anaesthetic induction and tracheal intubation: A secondary analysis of a randomised controlled trial. *Indian Journal of Anaesthesia, 63*(2), 100–105. https://doi.org/10.4103/ija.IJA_656_18.

Szental, J. A., Webb, A., Weeraratne, C., Campbell, A., Sivakumar, H., & Leong, S. (2015). Postoperative pain after laparoscopic cholecystectomy is not reduced by intraoperative analgesia guided by analgesia nociception index (ANI®) monitoring: A randomized clinical trial. *British Journal of Anaesthesia, 114*(4), 640–645. https://doi.org/10.1093/bja/aeu411.

Theerth, K. A., Sriganesh, K., Chakrabarti, D., Reddy, K. R. M., & Rao, G. S. U. (2019). Analgesia nociception index and hemodynamic changes during skull pin application for supratentorial craniotomies in patients receiving scalp block versus pin-site infiltration: A randomized controlled trial. *Saudi Journal of Anaesthesia, 13*(4), 306–311. https://doi.org/10.4103/sja.SJA_812_18.

Theerth, K. A., Sriganesh, K., Reddy, K. M., Chakrabarti, D., & Umamaheswara Rao, G. S. (2018). Analgesia Nociception Index-guided intraoperative fentanyl consumption and postoperative analgesia in patients receiving scalp block versus incision-site infiltration for craniotomy. *Minerva Anestesiologica, 84*(12), 1361–1368. https://doi.org/10.23736/s0375-9393.18.12837-9.

Turan, G., Ar, A. Y., Kuplay, Y. Y., Demiroluk, O., Gazi, M., Akgun, N., et al. (2017). Analgesia nociception index for perioperative analgesia monitoring in spinal surgery. *Revista Brasileira de Anestesiologia, 67*(4), 370–375. https://doi.org/10.1016/j.bjan.2017.03.004.

Upton, H. D., Ludbrook, G. L., Wing, A., & Sleigh, J. W. (2017). Intraoperative "analgesia nociception index"-guided fentanyl administration during sevoflurane anesthesia in lumbar discectomy and laminectomy: A randomized clinical trial. *Anesthesia and Analgesia, 125*(1), 81–90. https://doi.org/10.1213/ane.0000000000001984.

Weber, F., Geerts, N. J. E., Roeleveld, H. G., Warmenhoven, A. T., & Liebrand, C. A. (2018). The predictive value of the heart rate variability-derived analgesia nociception index in children anaesthetized with sevoflurane: An observational pilot study. *European Journal of Pain, 22*(9), 1597–1605. https://doi.org/10.1002/ejp.1242.

Chapter 43

Pain, anesthetics and analgesics/back pain evaluation questionnaires

Jun Komatsu

Department of Medicine for Motor Organs, Juntendo University Graduate School of Medicine, Tokyo, Japan; Department of Orthopaedic Surgery, Juntendo Tokyo Koto Geriatric Medical Center, Tokyo, Japan

List of abbreviations

LBP	low back pain
MRI	magnetic resonance imaging
VAS	visual analogue scale
RDQ	Roland-Morris Disability Questionnaire
QOL	quality of life
ODI	Oswestry disability index
JOA	Japanese Orthopedic Association
JOABPEQ	Japanese Orthopedic Association Back Pain Evaluation Questionnaire
HRQOL	health-related quality of life
SF-36	the short form 36 questionnaire
BSPOP	the brief scale for psychiatric problems in orthopedic patients
AAOS	The American Academy of Orthopedic Surgeons

Introduction

Low back pain (LBP) is a significant health problem, particularly in industrialized countries, and has led to increases in medical costs and the prevalence of lumbar disability around the world. Moreover, it has been reported that over 80% of people experience LBP at least once in their lifetime (Balagué, Mannion, Pellisé, & Cedraschi, 2012; Fujii & Matsudaira, 2013). Among the various types of LBP, chronic LBP is a musculoskeletal disease that causes problems in daily life if it persists longer than 3 months (Sherafat et al., 2013). In addition, cases of "nonspecific LBP," which is not associated with lower limb symptoms and does not match imaging findings, have been reported. Therefore, diagnostic investigations of LBP (e.g., X-ray, magnetic resonance imaging [MRI]) play no role in the management of 80%–90% of cases of nonspecific LBP. Moreover, findings from a systematic review showed no consistent associations between MRI findings and future episodes of LBP (Steffens et al., 2014).

As LBP has a broad meaning, an exact definition has yet to be adopted. Moreover, LBP may represent a disease condition, rather than a name. The causes of LBP may include psychological and somatoform disorders, degeneration, infection, inflammation, tumor, vascular systems other than the spine, the digestive system, and the gynecological system (Gore, Sadosky, Stacey, Tai, & Leslie, 2012). Therefore, it may be necessary to quantify the clinical outcomes of LBP considering individual differences using an objective disease-specific questionnaire to confirm the presence of any pretreatment effects.

One of the major problems of previous disease-specific questionnaires is that such tools are physician as opposed to patient based. However, the patient's perspective is now widely considered to be essential for medical decision making and evaluating the results of interventions.

Pain evaluation questionnaires for evaluating patients with LBP are used a visual analog scale (VAS; a representative pain scale), the Japanese Orthopaedic Association (JOA) score (JOA score; neurological status and clinical outcomes assessed using the scoring system), the Oswestry Disability Index (ODI), the Roland-Morris Disability Questionnaire (RMQ), and the JOA Back Pain Evaluation Questionnaire (JOABPEQ) (ODI, RMQ, and JOABPEQ; a functional scale that evaluates quality of life [QOL]). This section discusses the importance of objective, disease-specific questionnaires

---

Features and Assessments of Pain, Anesthesia, and Analgesia. https://doi.org/10.1016/B978-0-12-818988-7.00020-0
Copyright © 2022 Elsevier Inc. All rights reserved.

in the assessment of treatment for LBP. The RMQ and the ODI are widely used for the disease-specific measurement of LBP. The JOABPEQ was developed as a new patient-oriented disease-specific outcome measure for patients with LBP to solve the problems associated with the JOA score. It is an objective, patient-oriented test that assesses health-related QOL (HRQOL) among patients with lumbar disorders (Fukui et al., 2009). This section also discusses the importance of using ancillary diagnostic tools, such as the Short Form 36 (SF-36; a 36-item psychosocial scale and comprehensive health assessment) and the Brief Scale for Evaluation of Psychiatric Problems in Orthopaedic Patients (BS-POP) to assess mental health and social life function, which are key points for gaining a better understanding of LBP.

Back pain evaluation questionnaires for disease-specific measures

Oswestry disability index (ODI)

The ODI is a tool that was developed to assess pain-related disability in individuals with acute, subacute, or chronic LBP. Several different versions of the ODI have been developed since its initial publication in 1980 (e.g., ODI version 2.0, ODI AAOS; modified by the American Academy of Orthopaedic Surgeons [AAOS]), the Modified ODI) (Fairbank, Couper, Davies, & O'Brien, 1980; Fairbank & Pynsent, 2000). Among these versions, the Modified ODI is recommended for general use (Fig. 1) (Roland & Fairbank, 2000). The ODI is composed of 10 items, one on pain and nine on activities of daily living (i.e., lifting, walking, social life, personal care, sitting, standing, sleeping, traveling, and sex life), with each item scored on a scale from 0 to 5. Scores on the Modified ODI are calculated the same way as those on the original ODI (Haegg, 2013). The ODI has been recommended as an appropriate back pain-specific measure of disability by researchers in the field (Deyo et al., 1998).

Roland-Morris disability questionnaire (RMQ)

The RMQ is a 24-item self-reported outcome measure initially published in 1983 for use in primary care research to assess physical disabilities resulting from LBP. The RMQ, which enables the level of disability experienced by an individual suffering from LBP to be measured (Fig. 2) (Roland & Morris, 1983), has become one of the most commonly used outcome measures for LBP (Chapman et al., 2011). The original measure has been shortened to 18- and 23-item versions, and has been adapted or translated for use in other countries. Despite the various versions of the RMQ, the original remains the most widely used and validated (Macedo et al., 2011). The 24 items on the RMQ, which include items on physical ability/activity (15), sleep/rest (3), psychosocial (2), household management (2), eating (1), and pain frequency (1), assess individual perceptions of back pain and associated disabilities over the previous 24 h (Izumida & Inoue, 1986). Respondents are shown each statement and asked if he/she feels that it is descriptive of his/her own circumstances on that day.

Japanese Orthopaedic Association (JOA) score for diseases with LBP

Clinical outcomes were assessed using the scoring system proposed by the JOA. In 1986, JOA members developed a short questionnaire, the so-called JOA score, to assess treatments for and evaluate patients with LBP in terms of pain and functionality (Fardon & Milette, 2001; Fujii & Matsudaira, 2013; Fujiwara et al., 2003; Izumida & Inoue, 1986; Jeong & Bendo, 2003; Komatsu et al., 2014, 2018, 2020; Miyakoshi et al., 2000; Takahashi et al., 1990; Tiusanen, Hurri, Seitsalo, Osterman, & Harju, 1996; Yamada et al., 2012; Yorimitsu, Chiba, Toyama, & Hirabayashi, 2001). This criterion is applicable to all backpain-related diseases, including disc herniation, spondylolisthesis, and spinal canal stenosis (Izumida & Inoue, 1986). The JOA score is composed of the following four sections (14 items): subjective symptoms, including items on LBP, leg pain, and tingling gait (score range: 0–9 points); clinical signs, including items on straight leg raising, test sensory disturbance, and motor disturbance (range 0–6 points); restriction of activities, including items on turning over while lying, standing, washing, leaning forward, sitting for about 1 h, lifting or holding a heavy object, and walking (range 0–14 points); and urinary bladder function (range 0–6 points) (Fig. 3).

The JOA score is used to evaluate neurological status and clinical results (Komatsu et al., 2014, 2018, 2020). Therefore, this physician-oriented outcome measure is not sufficiently reliable to describe the objective status of the function and QOL of patients with low back disorders. In addition, to date, no psychometric analyses have been able to confirm sufficiently the validity and reliability of this rating system (Fukui et al., 2007).

Japanese Orthopaedic Association back pain evaluation questionnaire (JOABPEQ)

Members of the Subcommittee on LBP, who also belong to the Clinical Outcomes Committee of the JOA, have developed a new patient-oriented outcome measure for patients with LBP, the so-called JOABPEQ, to solve the problems associated

Section 1 – Pain intensity
I have no pain at the moment
The pain is very mild at the moment
The pain is moderate at the moment
The pain is fairly severe at the moment
The pain is very severe at the moment
The pain is the worst imaginable at the moment

Section 2 – Personal care (washing, dressing etc)
I can look after myself normally without causing extra pain
I can look after myself normally but it causes extra pain
It is painful to look after myself and I am slow and careful
I need some help but manage most of my personal care
I need help every day in most aspects of self-care
I do not get dressed, I wash with difficulty and stay in bed

Section 3 – Lifting
I can lift heavy weights without extra pain
I can lift heavy weights but it gives extra pain
Pain prevents me from lifting heavy weights off the floor, but I can manage if they are conveniently placed (eg. on a table)
Pain prevents me from lifting heavy weights, but I can manage light to medium weights if they are conveniently positioned
I can lift very light weights
I cannot lift or carry anything at all

Section 4 – Walking*
Pain does not prevent me walking any distance
Pain prevents me from walking more than 1mile
Pain prevents me from walking more than 1/2 mile
Pain prevents me from walking more than 1yards
I can only walk using a stick or crutches
I am in bed most of the time

Section 5 – Sitting
I can sit in any chair as long as I like
I can only sit in my favourite chair as long as I like
Pain prevents me sitting more than one hour
Pain prevents me from sitting more than 30 minutes
Pain prevents me from sitting more than 10 minutes
Pain prevents me from sitting at all

Section 6 – Standing
I can stand as long as I want without extra pain
I can stand as long as I want but it gives me extra pain
Pain prevents me from standing for more than 1 hour
Pain prevents me from standing for more than 30 minutes
Pain prevents me from standing for more than 10 minutes
Pain prevents me from standing at all

Section 7 – Sleeping
My sleep is never disturbed by pain
My sleep is occasionally disturbed by pain
Because of pain I have less than 6 hours sleep
Because of pain I have less than 4 hours sleep
Because of pain I have less than 2 hours sleep
Pain prevents me from sleeping at all

Section 8 – Sex life (if applicable)
My sex life is normal and causes no extra pain
My sex life is normal but causes some extra pain
My sex life is nearly normal but is very painful
My sex life is severely restricted by pain
My sex life is nearly absent because of pain
Pain prevents any sex life at all

Section 9 – Social life
My social life is normal and gives me no extra pain
My social life is normal but increases the degree of pain
Pain has no significant effect on my social life apart from limiting my more energetic interests eg, sport
Pain has restricted my social life and I do not go out as often
Pain has restricted my social life to my home
I have no social life because of pain

Section 10 – Travelling
I can travel anywhere without pain
I can travel anywhere but it gives me extra pain
Pain is bad but I manage journeys over two hours
Pain restricts me to journeys of less than one hour
Pain restricts me to short necessary journeys under 30 minutes
Pain prevents me from travelling except to receive treatment

FIG. 1 The Modified Oswestry Disability Index (ODI) (Modified ODI). Instructions: This questionnaire has been designed to give us information as to how your back or leg pain is affecting your ability to manage in everyday life. Please answer by checking ONE box in each section for the statement which best applies to you. We realize you may consider that two or more statements in any one section apply but please just shade out the spot that indicates the statement that most clearly describes your problem (Roland & Fairbank, 2000).

with the JOA score. The JOABPEQ, which was first published in May 2007 (Fukui et al., 2007), is a patient-oriented tool for assessing the HRQOL of patients with low back disorders in the previous week. Five functional scores (i.e., LBP, lumbar function, walking ability, social life function, and mental health) are calculated by the examiner from the answers to 25 questions. The functional score of each domain is calculated according to specific formulas. The score for each domain ranges from 0 to 100, with a higher score indicating a better condition. The five functional scores are designed to be used

478 PART | IV Assessments, screening, and resources

This list contains some sentences that people have used to describe themselves when they have back pain. When you read them, you may find that some stand out because they describe you today. As you read the list, think of yourself today. When you read a sentence that describes you today, mark the box next to it. If the sentence does not describe you, then leave the space blank and go on to the next one. Remember, only mark the sentence if you are sure that it describes you today.

1. I stay at home most of the time because of the pain in my back.
2. I change position frequently to try and make my back comfortable.
3. I walk more slowly than usual because of the pain in my back.
4. Because of the pain in my back, I am not doing any of the jobs that I usually do around the house.
5. Because of the pain in my back, I use a handrail to get upstairs.
6. Because of the pain in my back, I lie down to rest more often.
7. Because of the pain in my back, I have to hold on to something to get out of a reclining chair.
8. Because of the pain in my back, I ask other people to do things for me.
9. I get dressed more slowly than usual because of the pain in my back.
10. I only stand up for short periods of time because of the pain in my back.
11. Because of the pain in my back, I try not to bend or kneel down.
12. I find it difficult to get out of a chair because of the pain in my back.
13. My back hurts most of the time.
14. I find it difficult to turn over in bed because of the pain in my back.
15. My appetite is not very good because of the pain in my back.
16. I have trouble putting on my socks (or stockings) because of the pain in my back.
17. I only walk short distances because of the pain in my back.
18. I sleep less because of the pain in my back.
19. Because of the pain in my back, I get dressed with help from someone else.
20. I sit down for most of the day because of the pain in my back.
21. I avoid heavy jobs around the house because of the pain in my back.
22. Because of the pain in my back, I am more irritable and bad tempered with people.
23. Because of the pain in my back, I go upstairs more slowly than usual.
24. I stay in bed most of the time because of the pain in my back.

FIG. 2 The Roland-Morris Disability Questionnaire (RMQ). The scoring method is based on "Yes/No" response; therefore, the total score ranges from 0 (no disability) to 24 (maximal disability).

independently; thus, no total score is calculated (Fig. 4) (Fukui et al., 2007, 2007). Several studies have reported the use of the JOABPEQ for evaluations of patient in association with several different lumbar diseases (Fukui et al., 2009; Komatsu et al., 2018, 2020; Ohtori et al., 2010).

A verification study (Fukui et al., 2009) reported finding a significant correlation between patient self-ratings and acquired points on the JOABPEQ, in that 20 points could be interpreted as a threshold for substantial clinical benefits.

Conclusion

It is important to evaluate the effects of LBP on the social background and daily life of patients to promote appropriate treatment; thus, the assessment of LBP needs to be multifaceted. Assessing outcomes using a patient-oriented test that evaluates HRQOL, such as the ODI, RMQ, or JOABPEQ, which are commonly used as disease-specific measurements of LBP, is crucial for such patients. Although the JOA score is considered excellent for assessing neurological status and clinical outcomes among patients with LBP, it is a subjective questionnaire based on the medical examiner as opposed to a patient-oriented test (Komatsu et al., 2014, 2018, 2020). Since it is not possible to identify changes in a patient's

Subjective symptom (9 points)

Low back pain	None	3
	Occasionally mild	2
	Always present or sometimes severe	1
	Always severe	0
Leg pain and/or numbness	None	3
	Occasionally mild	2
	Always present or sometimes severe	1
	Always severe	0
Walking ability		
	Normal walking	3
	Able to walk 500 m, pain/numbness/weakness present	2
	Unable to walk 500 m due to pain/numbness/weakness	1
	Unable to walk 100 m due to pain/numbness/weakness	0

Objective finding (6 points)

Straight leg raising	Normal	2
	30°–70°	1
	< 30°	0
Sensory function	Normal	2
	Mild sensory disturbance	1
	< Apparent sensory disturbance	0

Motor function

Normal (MMT normal holds test position against strong pressure or MMT 5)	2
Slightly decreased muscle strength (MMT good holds test position against moderate pressure or MMT 4)	1
Markedly decreased muscle strength (MMT less than fair holds test position (no added pressure) or MMT 3 to 0)	0

Restriction of activities of daily living (14 points)

Activities of daily living include the following: 7 items (turning over while, lying down, standing, washing one's face, leaning forward, ability to sit for approximately 1 h, ability to lift or hold heavy objects, and ambulatory ability.)

None	2
Moderate	1
Severe	0

Bladder function (− 6 points)	Normal	0
	Mild dysuria	− 3
	Severe dysuria	− 6

Total 29

FIG. 3 JOA Criteria of the JOA scoring system. JOA score; neurological status and clinical outcomes assessed using the scoring system. The total score ranges from −6 to +29 points, with a higher score indicating a better condition.

medical condition using only one questionnaire, multiple questionnaires should be used to compare LBP evaluations. Each questionnaire has its own advantages and disadvantages, so it is important to understand the characteristics of each measure. In addition, the ODI has been shown not to correlate with lower limb symptoms such as pain and numbness (Gum, Glassman, & Carreon, 2013).

With regard to your health condition during the last week, please circle the item number of the answer for the following questions that best applies. If your condition varies depending on the day or time, circle the item number when your condition is at its worst.

Q1-1. To alleviate low back pain, you often change your posture. 1) Yes 2) No

Q1-2. Because of low back pain, you do not do any routine housework these days. 1) No 2) Yes

Q1-3. Because of low back pain, you lie down more often than usual. 1) Yes 2) No

Q1-4. Because of low back pain, you sometimes ask someone to help you when you do something. 1) Yes 2) No

Q1-5. Because of low back pain, you refrain from bending forward or kneeling down. 1) Yes 2) No

Q1-6. Because of low back pain, you have difficulty standing up from a chair. 1) Yes 2) No

Q1-7. Your lower back aches most of the time. 1) Yes 2) No

Q1-8. Because of low back pain, turning over in bed is difficult. 1) Yes 2) No

Q1-9. Because of low back pain, you have difficulty putting on socks or stockings. 1) Yes 2) No

Q1-10. Because of low back pain, you walk only short distances. 1) Yes 2) No

Q1-11. Because of low back pain, you cannot sleep well. (If you take sleeping pills because of the pain, select "No.") 1) No 2) Yes

Q1-12. Because of low back pain, you stay seated most of the day. 1) Yes 2) No

Q1-13. Because of low back pain, you become irritated or angry at other persons more often than usual. 1) Yes 2) No

Q1-14. Because of low back pain, you go up stairs more slowly than usual. 1) Yes 2) No

Q2-1. How is your present health condition? 1) Excellent 2) Very good 3) Good 4) Fair 5) Poor

Q2-2. Do you have difficulty in climbing stairs? 1) I have great difficulty. 2) I have some difficulty. 3) I have no difficulty.

Q2-3. Do you have difficulty in any one of the following motions: bending forward, kneeling, or stooping? 1) I have great difficulty. 2) I have some difficulty. 3) I have no difficulty.

Q2-4. Do you have difficulty walking more than 15 minutes? 1) I have great difficulty. 2) I have some difficulty. 3) I have no difficulty.

Q2-5. Have you been unable to do your work or ordinary activities as well as you would like? 1) I have not been able to do them at all. 2) I have been unable to do them most of the time. 3) I have sometimes been unable to do them. 4) I have been able to do them most of the time. 5) I have always been able to do them.

Q2-6. Has your work routine been hindered because of the pain? 1) Greatly 2) Moderately 3) Slightly (somewhat) 4) Little (minimally) 5) Not at all

Q2-7. Have you been discouraged or depressed? 1) Always 2) Frequently 3) Sometimes 4) Rarely 5) Never

Q2-8. Do you feel exhausted? 1) Always 2) Frequently 3) Sometimes 4) Rarely 5) Never

Q2-9. Do you feel happy? 1) Always 2) Almost always 3) Sometimes 4) Rarely 5) Never

Q2-10. Do you think you are in reasonable health? 1) Yes (I am healthy.) 2) Fairly (my health is better than average) 3) Not (very much)/particularly (my health is average) 4) Barely (my health is poor) 5) Not at all (my health is very poor)

Q2-11. Do you feel your health will get worse? 1) Very much so 2) A little at a time 3) Sometimes yes and sometimes no 4) Not very much 5) Not at all

FIG. 4 The JOA Back Pain Evaluation Questionnaire. Twenty-five items selected for the draft of the JOABPEQ evaluated. The functional score of each domain is calculated according to specific formulas. The score for each domain ranges from 0 to 100, with a higher score indicating a better condition.

On the other hand, the SF-36 is able to assess HRQOL along with various diseases and evaluation methods (Mousavi, Parnianpour, Mehdian, Montazeri, & Mobini, 2006). Therefore, it is often used in combination with disease-specific questionnaires. A previous study reported that the ODI and SF-36 were moderately correlated in terms of evaluating LBP (DeVine et al., 2011).

The JOABPEQ is composed of more specific items than the SF-36, and thus, can evaluate lumbar diseases more easily. Therefore, the JOABPEQ can be used as a specific tool for the evaluation of lumbar diseases (Komatsu et al., 2018, 2020). Moreover, as LBP often involves psychological disorders, the use of the BS-POP for assessing psychological problems in orthopedic patients should be considered (Konno & Sekiguchi, 2018; Yoshida et al., 2015). Therefore, gaining a better understanding of the characteristics, validity, reliability, and reproducibility of each back pain questionnaire is needed to apply these tools more effectively.

Suggestions for practical procedures

ODI: The ODI has been translated for use in at least 20 countries, and is freely downloadable and usable without the need to obtain permission from the producer unless the content is changed. The total score is calculated by multiplying the sum of the scores by 2, thereby giving a total score range from 0 to 100. Higher scores reflect greater disability. The Modified ODI is composed of eight items, as the item on sex life was removed.

Interpretation of scores (Roland & Fairbank, 2000).

Zero percent to 20%: minimal disability: The patient can cope with most living activities. Usually no treatment is indicated apart from advice on lifting sitting and exercise.

Twenty-one percent to 40%: moderate disability: The patient experiences more pain and difficulty with sitting, lifting, and standing. Travel and social life are more difficult and they may be disabled from work. Personal care, sexual activity, and sleeping are not grossly affected and the patient can usually be managed by conservative means.

41%–60%: severe disability: Pain remains the main problem in this group but activities of daily living are affected. These patients require a detailed investigation.

61%–80%: crippled: Back pain impinges on all aspects of the patient's life. Positive intervention is required.

81%–100%: These patients are either bed-bound or exaggerating their symptoms.

RMQ: The RMQ is free and does not require registration when used for nonprofit purposes; otherwise, registration is required. The scoring method is based on "Yes/No" response; therefore, the total score ranges from 0 (no disability) to 24 (maximal disability). When this scoring method is used, the result should be converted to a percentage score after excluding unanswered questions from the total when more than one question is left unanswered. No training is required to conduct the RMQ.

JOA: To date, no official site is available for downloading the JOA score. Please refer to the original paper in Japanese (Izumida & Inoue, 1986) or the article in this paper. No registration fee is required. The total score ranges from -6 to $+29$ points, with a higher score indicating a better condition. The recovery rate is calculated based on the JOA score according to the following formula:

$$\text{Recovery rate} = [(\text{postoperative score} - \text{preoperative score})/(29 - \text{preoperative score})] \times 100\%.$$

JOABPEQ: It is prohibited to modify the JOABPEQ or to make additions to the original English translation. The JOABPEQ is published by the JOA, and it is necessary to mention the source when using it. The JOABPEQ is freely available for download from the homepage of the JOA. In the way of scores on the JOABPEQ, treatment is judged as "effective" for a particular patient if: 1) the patient provides answers to all the questions required to calculate a domain score and shows an increase of ≥ 20 points after treatment, or 2) the functional score after treatment is >90 points, even if the answer for an unanswered item is assumed to be the worst possible choice.

Applications to other areas

SF 36 and BS-POP are not back pain evaluation questionnaires. So, I indicate other questionnaires of comprehensive measure of general health status and psychiatric problems measure in orthopedic patients.

Other agents of interest

Comprehensive measure of general health status

SF-36

The SF-36 was derived from the General Health Survey of the Medical Outcomes Study, and was standardized as a self-report measure of functional health and well-being in 1990 (Ware, Snow, Kosinski, & Gandek, 1993). The SF-36, which is widely used as a measure of general health status (Horchner & Tuinebreijer, 1998; Jacoby, Baker, Steen, & Buck, 1998;

Nortvedt, Riise, Myhr, & Nyland, 1998), is a unique questionnaire, as it is not condition specific. It was developed for application in primary care and chronic disease populations, often in combination with disease-specific questionnaires. Within the last few decades, studies in the field of orthopedic surgery have increasingly used the SF-36 to measure outcomes in combination with disease-specific questionnaires. The SF-36 was introduced in spinal surgery in 1997 by Grevitt et al. (Grevitt, Khazim, Webb, Mulholland, & Shepperd, 1997). The SF-36 is composed of eight scales, four each measuring physical health (physical function [PF], physical role [PR], bodily pain [BP], and general health [GH]) and psychological health (vitality [VT], social function [SF], emotional role [RE], and mental health [MH]) (Ware et al., 1993). The results are presented as a profile of the eight scales, with scores ranging from 0 to 100, and a higher score indicating better health. Some advantages of the SF-36 are its extensive validation and the existence of age-matched normative data from large populations. Comparative data for a variety of specific diseases are also available. Moreover, translations of the SF-36 are performed according to standardized principles ensured by the International Quality of Life Assessment Project (Aaronson et al., 1992), making international comparisons possible. The SF-36 total score is not calculated. A high score indicates good overall health.

Psychiatric problems measure in orthopedic patients

BS-POP

The BS-POP was developed as a measure for psychiatric problems in orthopedic patients. Higher scores on the BS-POP suggest a higher probability of lower QOL associated with psychiatric factors. The BS-POP (physician and patient versions) (Figs. 5 and 6) was developed for use by orthopedic surgeons in clinical settings, and is considered more convenient for assessing psychiatric problems than are the abovementioned methodologies (Sato, Kikuchi, Mashiko, Okano, & Niwa, 2000). The BS-POP enables orthopedists to identify psychiatric problems easily in orthopedic patients. However, further study is required regarding its responsiveness (Yoshida et al., 2011). In a study involving patients with spinal disorders on the validity of the Japanese version of the BS-POP, sensitivity and specificity were both high for indicating the presence of psychiatric problems in case of ≥ 11 points on the physician version and ≥ 15 points on the patient version (Watanabe, Kikuchi, Konno, Niwa, & Masiko, 2005).

Mini-dictionary of terms

Health-related quality of life (HRQOL) is a multidimensional concept that includes physical, mental, emotional, and social functioning domains. It focuses on the impact of health status on a patient's quality of life (QOL). It is called health-related QOL (HRQOL) when associated with a disease or how a treatment affects QOL.

A patient-oriented test is a patient-based evaluation method that evaluates the effects of function, health, life, and treatment from the perspective of the patient; it does not include the interpretation of a doctor.

The JOA (Japanese Orthopaedic Association) was established in 1926 with the aim of contributing to the promotion and dissemination of orthopedics and the development of academic culture by disseminating research announcements, communications, alliances, and research related to Japanese orthopedics.

Disease-specific measure is a measure of HRQOL and disorders specific to a particular disease that makes it possible to assess each particular disease or disorder.

Comprehensive measure is a measure of general health status, with or without disease or disability; it enables health status, including functions, to be measured and compared.

Key facts

- The JOA score for diseases with LBP is considered excellent for evaluating neurological status and clinical outcomes; however, it should be noted that the JOA score is a subjective questionnaire for the medical doctor, not a patient-oriented test (Komatsu et al., 2014, 2018, 2020).
- The ODI is not correlated with lower limb symptoms such as pain and numbness (Gum et al., 2013).
- The SF-36 measures HRQOL and can be used to evaluate various diseases and methods (Mousavi et al., 2006); therefore, it is often used in combination with disease-specific questionnaires.

1. The patient's pain appears uninterrupted

    1. That is not the case

    2. The pain is intermittent

    3. The patient appears to be almost always in pain

2. The patient has a specific way of indicating the symptomatic area(s)

    1. That is not the case

    2. They rub the symptomatic area(s)

    3. Without instruction, they begin to remove their clothes and show the symptomatic area(s)

3. The patient appears to have pain over the whole symptomatic area

    1. That is not the case

    2. They sometimes do

    3. Almost all the time

4. When examination or treatment is recommended, the patient becomes badly tempered, easily angered, or argumentative

    1. That is not the case

    2. They show slight resistance

    3. They show significant resistance

5. When having their senses assessed, the patient responds excessively to stimulation

    1. That is not the case

    2. Their response is slightly excessive

    3. Their response is quite excessive

6. The patient repeatedly asks questions regarding their condition or surgery

    1. That is not the case

    2. They sometimes do

    3. Almost all the time

7. The patient changes their attitude depending on the medical staff member

    1. That is not the case

    2. They do somewhat

    3. They do significantly

8. The patient wishes that their symptoms were gone, even with regard to slight symptoms

    1. That is not the case

    2. They do somewhat

    3. They do significantly

Numbers indicate scores

FIG. 5 The Brief Scale for Psychiatric problems in Orthopaedic Patients (BSPOP) (for use by physicians) Questionnaire for medical personnel, English version. Higher scores on the BS-POP suggest a higher probability of lower QOL associated with psychiatric factors. Indicating the presence of psychiatric problems in case of ≥11 points on the physician version.

- The average scores of the five categories of the JOABPEQ were similarly distributed in all patients, and since the JOABPEQ has more specific items compared with the SF-36, it allows lumbar diseases to be evaluated more easily. Thus, the JOABPEQ can be used as a specific tool for the evaluation of lumbar diseases (Komatsu et al., 2018, 2020).
- LBP often involves psychological problems; therefore, it is necessary to consider the use of the BS-POP to assess psychological problems in orthopedic patients (Konno & Sekiguchi, 2018; Yoshida et al., 2015).

1. Do you ever feel like crying, or do you cry? (1. No, 2. Sometimes, 3. Almost all the time)
2. Do you always feel miserable and unhappy? (1. No, 2. Sometimes, 3. Almost all the time)
3. Do you always feel nervous and irritated? (1. No, 2. Sometimes, 3. Almost all the time)
4. Do you feel annoyed and aggravated over small things? (1. No, 2. Sometimes, 3. Almost all the time)
5. Do you have a normal appetite? (1. No, 2. I sometimes lose my appetite, 1. Yes)
6. Are you in your best mood in the morning? (1. No, 2. Sometimes, 3. Almost all the time)
7. Do you get somewhat tired? (1. No, 2. Sometimes, 3. Almost all the time)
8. Are you able to put your usual effort into your work? (3.NO, 1. I sometimes can't, 1. Yes)
9. Do you feel satisfied with the sleep you are getting? (1.No, 2. sometimes don't feel satisfied, 1.Yes)
10. Do you have trouble falling asleep for any reason other than pain? (1. No, 2. Sometimes, 3. Almost all the time)

Numbers next to responses indicate scores

FIG. 6 The Brief Scale for Psychiatric problems in Orthopaedic Patients (BSPOP) (for use by patients) Questionnaire for patients, English version. Higher scores on the BS-POP suggest a higher probability of lower QOL associated with psychiatric factors indicating the presence of psychiatric problems in case of ≥15 points on the patient version.

Summary points

- The assessment of LBP should be multifaceted, as it is important to evaluate the effects of LBP on the patient's social background and daily life to promote appropriate treatment.
- Multiple questionnaires should be used to allow a comparison of evaluations of LBP as it is not possible to capture changes in a patient's medical condition using only a single questionnaire.
- Assessing outcomes with a patient-oriented test that evaluates HRQOL, such as the ODI, RMQ, or JOABPEQ, is thought to be important for patients with LBP; such patient-oriented tests are widely used as disease-specific measurements for LBP.
- Each questionnaire has its own advantages and disadvantages; thus, it is important to understand the characteristics of each method.
- Gaining a better understanding of the characteristics, validity, reliability, and reproducibility of each back pain questionnaire is needed to apply these tools more effectively.

References

Aaronson, N. K., Acquadro, C., Alonso, J., Apolone, G., Bucquet, D., Bullinger, M., et al. (1992). International quality of life assessment (IQOLA) project. *Quality of Life Research, 1*, 349–351.

Balagué, F., Mannion, A. F., Pellisé, F., & Cedraschi, C. (2012). Nonspecific low back pain. *Lancet, 379*, 482–491.

Chapman, J. R., Norvell, D. C., Hermsmeyer, J. T., Bransford, R. J., DeVine, J., McGirt, M. J., et al. (2011). Evaluating common outcomes for measuring treatment success for chronic low back pain. *Spine, 36*, 54–68.

DeVine, J., Norvell, D. C., Ecker, E., Fourney, D. R., Vaccaro, A., Wang, J., et al. (2011). Evaluating the correlation and responsiveness of patient-reported pain with function and quality-of-life outcomes after spine surgery. *Spine, 36*, 69–74.

Deyo, R. A., Battie, M., Beurskens, A. J., Bombardier, C., Croft, P., Koes, B., et al. (1998). Outcome measures for low back pain research: A proposal for standardized use. *Spine, 23*, 2003–2013.

Fairbank, J. C., Couper, J., Davies, J. B., & O'Brien, J. P. (1980). The Oswestry low back pain disability questionnaire. *Physiotherapy, 66*, 271–273.

Fairbank, J. C., & Pynsent, P. B. (2000). The Oswestry disability index. *Spine, 25*, 2940–2952.

Fardon, D. F., & Milette, P. C. (2001). Nomenclature and Classification of lumbar disc pathology recommendations of the combined task forces of the North American Spine Society, American Society of Spine Radiology, and American Society of Neuroradiology. *Spine, 26*, 93–113.

Fujii, T., & Matsudaira, K. (2013). Prevalence of low back pain and factors associated with chronic disabling back pain in Japan. *European Spine Journal, 22*, 432–438.

Fujiwara, A., Kobayashi, N., Saiki, K., Kitagawa, T., Tamai, K., & Saotome, K. (2003). Association of the Japanese Orthopaedic Association score with the Oswestry disability index, Roland-Morris disability questionnaire, and short-form 36. *Spine, 28*, 1601–1607.

Fukui, M., Chiba, K., Kawakami, M., Kikuchi, S., Konno, S., Miyamoto, M., et al. (2007). JOA back pain evaluation questionnaire: Initial report. *Journal of Orthopaedic Science, 12*, 443–450.

Fukui, M., Chiba, K., Kawakami, M., Kikuchi, S., Konno, S., Miyamoto, M., et al. (2007). Japanese Orthopaedic association back pain evaluation questionnaire. Part 2. Verification of its reliability the subcommittee on low back pain and cervical myelopathy evaluation of the clinical outcome committee of the Japanese Orthopaedic association. *Journal of Orthopaedic Science, 12*, 526–532.

Fukui, M., Chiba, K., Kawakami, M., Kikuchi, S., Konno, S., Miyamoto, M., et al. (2009). JOA back pain evaluation questionnaire (JOABPEQ)/JOA cervical myelopathy evaluation questionnaire (JOACMEQ). The report on the development of revised versions April 16, 2007. *Journal of Orthopaedic Science, 14*, 348–356.

Gore, M., Sadosky, A., Stacey, B. R., Tai, K.-S., & Leslie, D. (2012). The burden of chronic low back pain: Clinical comorbidities, treatment patterns, and health care costs in usual care settings. *Spine, 37*, 668–677.

Grevitt, M., Khazim, R., Webb, J., Mulholland, R., & Shepperd, J. (1997). The short form-36 health survey questionnaire in spine surgery. *Journal of Bone and Joint Surgery. British Volume (London), 79*, 48–52.

Gum, J. L., Glassman, S. D., & Carreon, L. Y. (2013). Clinically important deterioration in patients undergoing lumbar spine surgery: A choice of evaluation methods using the Oswestry disability index, 36-item short form health survey, and pain scales: Clinical article. *Journal of Neurosurgery. Spine, 19*, 564–568.

Haegg, O. (2013). *Oswestry disability index. Encyclopedia of pain* (pp. 2559–2562). Berlin Heidelberg: Springer.

Horchner, R., & Tuinebreijer, W. (1998). Improvement of physical functioning of morbidly obese patients who have undergone a lap-band operation: Oneyear study. *Obesity Surgery, 9*, 399–402.

Izumida, S., & Inoue, S. (1986). Assessment of treatment for low back pain. *The Japanese Orthopaedic Association (in Japanese), 60*, 391–394.

Jacoby, A., Baker, G. A., Steen, N., & Buck, D. (1998). The SF-36 as a health status measure for epilepsy: A psychometric assessment. *Quality of Life Research, 8*, 351–364.

Jeong, G. K., & Bendo, J. A. (2003). Lumbar intervertebral disc cyst as a cause of radiculopathy. *The Spine Journal, 3*, 242–246.

Komatsu, J., Iwabuchi, M., Endo, T., Fukuda, H., Kusano, K., Miura, T., et al. (2020). Clinical outcomes of lumbar diseases specific test in patients who undergo endoscopy-assisted tubular surgery with lumbar herniated nucleus pulposus: An analysis using the Japanese Orthopaedic association back pain evaluation questionnaire (JOABPEQ). *European Journal of Orthopaedic Surgery and Traumatology, 30*, 207–213.

Komatsu, J., Muta, T., Nagura, N., Iwabuchi, M., Fukuda, H., Kaneko, K., et al. (2018). Tubular surgery with the assistance of endoscopic surgery via a paramedian or midline approach for lumbar spinal canal stenosis at the L4/5 level. *Journal of Orthopaedic Surgery (Hong Kong), 26*. 2309499018782546.

Komatsu, J., Muta, T., Nagura, N., Obata, H., Sugiyama, Y., Katsube, S., et al. (2014). A study of operation for lumbar spinal canal stenosis through comparative analysis of microendoscopic laminectomy with a unilateral approach and microendoscopic muscle preserving interlaminar decompression. *Orthopedic Surgery (in Japanese), 65*, 1117–1121.

Konno, S. I., & Sekiguchi, M. (2018). Association between brain and low back pain. *Journal of Orthopaedic Science, 23*, 13–17.

Macedo, L. G., Maher, C. G., Latimer, J., Hancock, M. J., Machado, L. A., & McAuley, J. H. (2011). Responsiveness of the 24-, 18- and 11-item versions of the Roland Morris disability questionnaire. *European Spine Journal, 20*, 458–463.

Miyakoshi, N., Abe, E., Shimada, Y., Okuyama, K., Suzuki, T., & Sato, K. (2000). Outcome of one-level posterior lumbar interbody fusion for spondylolisthesis and postoperative intervertebral disc degeneration adjacent to the fusion. *Spine, 25*, 1837–1842.

Mousavi, S. J., Parnianpour, M., Mehdian, H., Montazeri, A., & Mobini, B. (2006). The Oswestry disability index, the Roland-Morris disability questionnaire, and the Quebec back pain disability scale: Translation and validation studies of the Iranian versions. *Spine, 31*, 454–459.

Nortvedt, M. W., Riise, T., Myhr, K. M., & Nyland, H. I. (1998). Quality of life in multiple sclerosis: Measuring the disease effects more broadly. *Neurology, 53*, 1098–1103.

Ohtori, S., Ito, T., Yamashita, M., Murata, Y., Morinaga, T., Hirayama, J., et al. (2010). Evaluation of low back pain using the Japanese Orthopaedic association Back pain evaluation questionnaire for lumbar spinal disease in a multicenter study: Differences in scores based on age, sex, and type of disease. *Journal of Orthopaedic Science, 15*, 86–91.

Roland, M., & Fairbank, J. (2000). The Roland-Morris disability questionnaire the Oswestry disability questionnaire. *Spine, 25*, 3115–3124.

Roland, M., & Morris, R. (1983). A study of the natural history of back pain. Part I: development of a reliable and sensitive measure of disability in low-back pain. *Spine, 8*, 141–144.

Sato, K., Kikuchi, S., Mashiko, H., Okano, T., & Niwa, S. (2000). Rinshou Seikeigeka. *Clinics in Orthopedic Surgery (in Japanese), 35*, 843–852.

Sherafat, S., Salavati, M., Ebrahimi Takamjani, I., Akhbari, B., Mohammadirad, S., Mazaheri, M., et al. (2013). Intrasession and intersession reliability of postural control in participants with and without nonspecific low back pain using the Biodex Balance System. *Journal of Manipulative and Physiological Therapeutics, 36*, 111–118.

Steffens, D., Hancock, M. J., Maher, C. G., Williams, C., Jensen, T. S., & Latimer, J. (2014). Does magnetic resonance imaging predict future low back pain? A systematic review. *European Journal of Pain, 18*, 755–765.

Takahashi, K., Kitahara, H., Yamagata, M., Murakami, M., Takata, K., Miyamoto, K., et al. (1990). Long-term results of anterior interbody fusion for treatment of degenerative spondylolisthesis. *Spine, 15*, 1211–1215.

Tiusanen, H., Hurri, H., Seitsalo, S., Osterman, K., & Harju, R. (1996). Functional and clinical results after anterior interbody lumbar fusion. *European Spine Journal, 5*, 288–292.

Ware, J. E., Snow, K. K., Kosinski, M., & Gandek, B. (1993). *SF-36 health survey: Manual and Interpretation guide.* Boston: Health Institute, New England Medical Center.

Watanabe, K., Kikuchi, S., Konno, S., Niwa, S., & Masiko, H. (2005). Brief scale for psychiatric problems in orthopaedic patients (BS-POP) validation study. Rinshou Seikeigeka. *Clinics in Orthopedic Surgery (in Japanese), 40*, 745–751.

Yamada, H., Yoshida, M., Hashizume, H., Minamide, A., Nakagawa, Y., Kawai, M., et al. (2012). Efficacy of novel minimally invasive surgery using spinal microendoscope for treating extraforaminal stenosis at the lumbosacral junction. *Journal of Spinal Disorders & Techniques, 25*, 268–276.

Yorimitsu, E., Chiba, K., Toyama, Y., & Hirabayashi, K. (2001). Long-term outcomes of standard discectomy for lumbar disc herniation: A follow-up study of more than 10 years. *Spine, 26*, 652–657.

Yoshida, K., Sekiguchi, M., Otani, K., Mashiko, H., Shioda, H., Wakita, T., et al. (2015). Computational psychological study of the brief scale for psychiatric problems in Orthopaedic patients (BS-POP) for patients with chronic low back pain: Verification of responsiveness. *Journal of Orthopaedic Science, 20*, 469–474.

Yoshida, K., Sekiguchi, M., Otani, K., Mashiko, H., Shiota, H., Wakita, T., et al. (2011). A validation study of the brief scale for psychiatric problems in Orthopaedic patients (BS-POP) for patients with chronic low back pain (verification of reliability, validity, and reproducibility). *Journal of Orthopaedic Science, 16*, 7–13.

Chapter 44

The back pain functional scale: Features and applications

Meltem Koç and Kılıçhan Bayar
Department of Physiotherapy and Rehabilitation, Faculty of Health Sciences, Muğla Sıtkı Koçman University, Muğla, Turkey

Abbreviations

LBP	low back pain
BPFS	the back pain functional scale
RMQ	the Roland–Morris Questionnaire
ODI	the Oswestry disability index
WHO	World Health Organization

Introduction

One of the biggest problems for clinicians and researchers in musculo-skeletal disorders is to determine the correct outcome measurement because many musculo-skeletal disorders are not a disease but a syndrome with many symptoms (Chapman et al., 2011; Zanoli, Strömqvist, Jönsson, Padua, & Romanini, 2002). A good example of this situation is chronic low back pain (LBP), which is a complex clinical syndrome. LBP is one of the most prevalent complaints encountered in primary care settings and it is estimated that approximately from 70% to 85% of the western population will develop LBP at least once during their lifetime (Geurts, Willems, Kallewaard, van Kleef, & Dirksen, 2018). Medical costs attributable to LBP are high and rising faster than overall health care spending. It is known that it can affect individuals of all ages in all societies (Fritz, Kim, Magel, & Asche, 2017; Hoy, Brooks, Blyth, & Buchbinder, 2010). However, despite all these, there is no clear consensus on its standard evaluation and optimal treatment (Zanoli et al., 2002).

The common symptom is pain in all patients with LBP. Therefore, pain has been the most common and simplest measurement criterion in the evaluation of therapeutic interventions for years (Koes, van Tulder, & Thomas, 2006; Zanoli et al., 2002). However, in the last 20 years, it has been observed that pain assessment alone does not yield enough results for patients with chronic pain. Along with pain, additional assessment methods such as pain-related quality of life and function are needed, because as all clinicians working with LBP know, there are multiple cases that show very good function or participation despite high pain severity, or on the contrary, show poor functional results despite low pain severity. Therefore, assessment of pain-related functions is the most critical point in determining the current clinical condition of the patient or determining the effectiveness of the treatment. Self-report outcome measures provide the "gold standard" in assessing pain-related outcomes because they reflect the subjective nature of pain, but they should be supplemented by careful assessments (Dworkin et al., 2005).

There are many patient-reported outcome measures for LBP in the literature. These include generic and disease-specific measures. Generic measures are designed to assess functional status regardless of a disease or disorder. Disease-specific measures are designed to be sensitive to the specific disease or disorder of interest (Grotle, Brox, & Vøllestad, 2005; Longo, Loppini, Denaro, Maffulli, & Denaro, 2010). The focus of back-related outcome scores usually has been pain, functional disability, and health-related quality of life. The major domains assessed for chronic pain are pain intensity, pain-related function, and emotional burden (Caumo et al., 2013), while for acute pain it is mainly pain intensity and relief (Gordon, 2015). Two central points should be considered in the selection of these outcome measures by Health Care Professionals. The first is that the outcome measure used must be valid, reliable, and responsive. The ability of the outcome measure to detect clinically significant changes (as distinct from statistically significant changes) when they occur is referred to as

---

Features and Assessments of Pain, Anesthesia, and Analgesia. https://doi.org/10.1016/B978-0-12-818988-7.00002-9
Copyright © 2022 Elsevier Inc. All rights reserved.

488 PART | IV Assessments, screening, and resources

responsiveness (Grotle, Brox, & Vøllestad, 2004). Second, if an evaluation is not made on a specific symptom, multidimensional tools containing all symptoms should be selected. However, symptom-specific unidimensional tools can be selected when symptoms need to be evaluated separately (especially for scientific research purposes) (Dittner, Wessely, & Brown, 2004).

BPFS is a self-report functional status outcome measure appropriate for clinical practice and scientific research and it was developed by Stratford, Binkley, and Riddle (2000). It was conducted according to the World Health Organization's (WHO) model of impairment, disability, and handicap. It consists of 12 items that measure the patient's functional level. These 12 items cover different domains (work, school, home activities, habits, bending, wearing shoes or socks, lifting an object from the ground, sleeping, sitting, standing, walking, climbing stairs, and driving). The last question for patients who do not drive can be answered by thinking of traveling. Each item is scaled on a six-point Likert Scale (range 0–5), with "0" indicating the inability to perform the action due to back pain and "5" indicating no difficulty. The total score ranges from 0 to 60. The lower the score is, the lower the patient's functional status due to pain. It takes less than 5 min to complete and take less than 30 s to score.

BPFS was originally developed in English. The original BPFS with 12 items is shown in Table 1. Up to the present, Turkish (Koç, Bayar, & Bayar, 2018; Meltem & Bayar, 2017) and Persian (Nakhostin Ansari, Naghdi, Habibzadeh,

---

TABLE 1 The back pain functional scale.

On the questions listed below we are interested in knowing whether you are having **ANY DIFFICULTY** at all with the activities **because of your back problem** for which you are currently seeking attention.

Please provide an answer for each activity.

Today, do you or *would you have* any DIFFICULTY at all with the following activities BECAUSE OF YOUR BACK PROBLEM?

	(circle one number on each line)					
	Unable to perform activity	Extreme difficulty	Quite a bit of difficulty	Moderate difficulty	A little bit of difficulty	No difficulty
1. Any of your usual work, housework, or school activities	0	1	2	3	4	5
2. Your usual hobbies, recreational, or sporting activities	0	1	2	3	4	5
3. Performing heavy activities around your home	0	1	2	3	4	5
4. Bending or stooping	0	1	2	3	4	5
5. Putting on your shoes or socks (pantyhose)	0	1	2	3	4	5
6. Lifting a box of groceries from the floor	0	1	2	3	4	5
7. Sleeping	0	1	2	3	4	5
8. Standing for 1 h	0	1	2	3	4	5
9. Walking a mile	0	1	2	3	4	5
10. Going up or down 2 flights of stairs (about 20 stairs)	0	1	2	3	4	5
11. Sitting for 1 h	0	1	2	3	4	5
12. Driving for 1 h	0	1	2	3	4	5
Sub totals =						
	Total Score = /60					

From Stratford, P. W., Binkley, J. M., & Riddle, D. L. (2000). Development and initial validation of the back pain functional scale. *Spine, 25*(16), 2095–2102, with permission.

Salsabili, & Ebadi, 2018) translation and cultural adaptation of BPFS have been made. Internal consistency of the BPFS's original study was 0.93 (95% CI: 0.90–0.96) (Stratford et al., 2000), Turkish version was 0.930 (95% CI: 0.883–0.958) (Meltem & Bayar, 2017), and Persian version was 0.88 (CI 95%: 0.80–0.93) (Nakhostin Ansari et al., 2018). The reliability coefficient of original study was 0.88, Turkish version was 0.899 and Persian version was 0.895. BPFS's ability to detect the change was evaluated by comparing BPFS change scores with Roland–Morris Questionnaire (RMQ) change scores in study of Stratford et al. The RMQ is a frequently cited and studied self-report measure for patients with LBP in the literature (Roland & Fairbank, 2000). The current results demonstrated a correlation of 0.82 between BPFS and RMQ change scores. This correlation result is somewhat greater than those between RMQ change scores and other LBP disability outcome measures (Stratford et al., 2000).

However, study by Stratford et al. have shown that BPFS is superior to RMQ in detecting clinical changes in LBP patients with pain duration less than 2 weeks (Stratford et al., 2000). As is known, in many cases, acute LBP improves very well after the first 2 weeks and remains completely symptom free until the next attack (Pengel, Herbert, Maher, & Refshauge, 2003). Therefore, it is very difficult to evaluate the clinical change or response to treatment with each outcome measures with in the first 2 weeks. The RMQ consists of 24 sentences including functional limitations in LBP. Patients answer each sentence with "yes" if it matches their own situation, and "no" if it does not. Since BPFS scores each item with a six-point Likert Scale (range 0–5), it can detect clinical change better. Another advantage of BPFS is that the items are shorter and more understandable. If the outcome measures contain too many items and have complex sentence structures, the rate of correct answers decreases (Edwards, Roberts, Sandercock, & Frost, 2004). In addition, outcome measures with complex sentence structures reduce the reliability of evaluation in individuals or societies with a low level of education.

Applications to other areas

In this chapter, we have reviewed in detail the BPFS, which is based upon the International Classification of Function model proposed by the WHO, evaluates the functional status in chronic LBP and is a very easy and understandable outcome measure. However, the excess of outcome measurements and the lack of a standard in the literature make it difficult to compare the results of scientific studies with each other. In addition, it makes it difficult for the clinician and the researcher to select the questionnaire. In back pain problems, it is important to standardize and improve these questionnaires because of the increased use of these assessment tools and the difficulty of selection. Therefore, although BPFS is more advantageous in practical use, RMQ is a most commonly used outcome measure in scientific research on LBP (Chapman et al., 2011; Deyo et al., 1998; Garg, Pathak, Churyukanov, Uppin, & Slobodin, 2020). Besides that, The International Consortium for Health Outcomes Measurement standard set for LBP also recommended The Oswestry Disability Index (ODI) to assessment functional level. Because it "is the most heavily studied, providing superior interpretability" and "the most feasible to implement as it has been validated in 14 languages (…) and is relatively short." (Clement et al., 2015). For the reader reading this chapter, it is important that this issue is not overlooked.

Other agents of interest

There is a great deal of interest in outcome measures in scientific research. In a systematic review conducted by Costa, Maher, and Latimer (2007), they found 40 questionnaires developed solely for LBP (Costa et al., 2007). Besides BPFS, some of the other functional scales used to assessment physical function in patients with LBP are listed below (Garg et al., 2020).

The Roland–Morris Questionnaire (RMQ)
The Oswestry Disability Index (ODI)
The Low Back Outcome Score (LBOS)
The Quebec Back Pain Disability Scale (QBPDS)
The Aberdeen Low Back Disability Scale (ALBDS)
The NASS Lumbar Spine Outcome Assessment Instrument (NASS LSO)
The Low Back Pain Rating Scale (LBPRS)
Back Illness Pain and Disability Scale (BIPDS)
Functional Rating Index (FRI)
Bournemouth Questionnaire (BQ)
The Waddell Disability Index (WDI)
The Back-Performance Scale (BPS)

Mini-dictionary of terms

Patient-Reported Outcome Measure: Questionnaires or scales in which the patient's response is reported directly by him/her without being interpreted by a clinician or anyone else.
Validity: It's about to what extent the results really measure what they need to measure.
Reliability: It is about the extent to which results can be reproduced when the research is repeated under the same conditions.
Responsiveness: It concerns the ability of an instrument to perceive change over time.

Key facts of back pain functional scale

- It is clear that LBP is one of the most common problems, which most people experience at some point in their life. In order to understand and recorded the impact of pain and symptoms in the LBP on the patient's life, assessment of functional level has a crucial task.

Summary points

- BPFS is a self-reported outcome measure used in the assessment of pain-related functional loss in chronic LBP.
- BPFS is valid, reliable, and sensitive to clinical change.
- BPFS is highly correlated with the RMQ, the most commonly used for LBP.
- BPFS is more sensitive to clinical changes than other commonly used outcome measures for back pain.
- BPFS is simple and easy to understand and administer by the patients.

References

Caumo, W., Ruehlman, L. S., Karoly, P., Sehn, F., Vidor, L. P., Dall-Ágnol, L., et al. (2013). Cross-cultural adaptation and validation of the profile of chronic pain: Screen for a Brazilian population. *Pain Medicine, 14*(1), 52–61.

Chapman, J. R., Norvell, D. C., Hermsmeyer, J. T., Bransford, R. J., DeVine, J., McGirt, M. J., et al. (2011). Evaluating common outcomes for measuring treatment success for chronic low back pain. *Spine, 36*, S54–S68.

Clement, R. C., Welander, A., Stowell, C., Cha, T. D., Chen, J. L., Davies, M., et al. (2015). A proposed set of metrics for standardized outcome reporting in the management of low back pain. *Acta Orthopaedica, 86*(5), 523–533.

Costa, L. O. P., Maher, C. G., & Latimer, J. (2007). Self-report outcome measures for low back pain: Searching for international cross-cultural adaptations. *Spine, 32*(9), 1028–1037.

Deyo, R. A., Battie, M., Beurskens, A., Bombardier, C., Croft, P., Koes, B., et al. (1998). Outcome measures for low back pain research: A proposal for standardized use. *Spine, 23*(18), 2003–2013.

Dittner, A. J., Wessely, S. C., & Brown, R. G. (2004). The assessment of fatigue: A practical guide for clinicians and researchers. *Journal of Psychosomatic Research, 56*(2), 157–170.

Dworkin, R. H., Turk, D. C., Farrar, J. T., Haythornthwaite, J. A., Jensen, M. P., Katz, N. P., et al. (2005). Core outcome measures for chronic pain clinical trials: IMMPACT recommendations. *Pain, 113*(1), 9–19.

Edwards, P., Roberts, I., Sandercock, P., & Frost, C. (2004). Follow-up by mail in clinical trials: Does questionnaire length matter? *Controlled Clinical Trials, 25*(1), 31–52.

Fritz, J. M., Kim, M., Magel, J. S., & Asche, C. V. (2017). Cost-effectiveness of primary care management with or without early physical therapy for acute low Back pain: Economic evaluation of a randomized clinical trial. *Spine, 42*(5), 285–290.

Garg, A., Pathak, H., Churyukanov, M. V., Uppin, R. B., & Slobodin, T. M. (2020, Mar). Low back pain: Critical assessment of various scales. *European Spine Journal, 29*(3), 503–518.

Geurts, J. W., Willems, P. C., Kallewaard, J. W., van Kleef, M., & Dirksen, C. (2018). The impact of chronic discogenic low back pain: Costs and patients' burden. *Pain Research & Management, 2018*, 4696180.

Gordon, D. B. (2015). Acute pain assessment tools: Let us move beyond simple pain ratings. *Current Opinion in Anesthesiology, 28*(5), 565–569.

Grotle, M., Brox, J. I., & Vøllestad, N. K. (2004). Concurrent comparison of responsiveness in pain and functional status measurements used for patients with low back pain. *Spine, 29*(21), E492–E501.

Grotle, M., Brox, J. I., & Vøllestad, N. K. (2005). Functional status and disability questionnaires: What do they assess?: A systematic review of back-specific outcome questionnaires. *Spine, 30*(1), 130–140.

Hoy, D., Brooks, P., Blyth, F., & Buchbinder, R. (2010). The epidemiology of low back pain. *Best Practice & Research. Clinical Rheumatology, 24*(6), 769–781.

Koç, M., Bayar, B., & Bayar, K. (2018). A comparison of Back pain functional scale with Roland Morris disability questionnaire, Oswestry disability index, and short form 36-health survey. *Spine, 43*(12), 877–882.

Koes, B. W., van Tulder, M. W., & Thomas, S. (2006). Diagnosis and treatment of low back pain. *BMJ, 332*(7555), 1430–1434.

Longo, U. G., Loppini, M., Denaro, L., Maffulli, N., & Denaro, V. (2010). Rating scales for low back pain. *British Medical Bulletin, 94*, 81–144.

Meltem, K., & Bayar, K. (2017). Fonksiyonel Bel Ağrısı Skalası'nın Türkçe uyarlaması: geçerlik ve güvenirlik çalışması. *Journal of Exercise Therapy and Rehabilitation, 4*(2), 82–89.

Nakhostin Ansari, N., Naghdi, S., Habibzadeh, F., Salsabili, N., & Ebadi, S. (2018). Persian translation and validation of the back pain functional scale. *Physiotherapy Theory and Practice, 34*(3), 223–230.

Pengel, L. H., Herbert, R. D., Maher, C. G., & Refshauge, K. M. (2003). Acute low back pain: Systematic review of its prognosis. *BMJ, 327*(7410), 323.

Roland, M., & Fairbank, J. (2000). The Roland–Morris disability questionnaire and the Oswestry disability questionnaire. *Spine, 25*(24), 3115–3124.

Stratford, P. W., Binkley, J. M., & Riddle, D. L. (2000). Development and initial validation of the back pain functional scale. *Spine, 25*(16), 2095–2102.

Zanoli, G., Strömqvist, B., Jönsson, B., Padua, R., & Romanini, E. (2002). Pain in low-back pain: Problems in measuring outcomes in musculoskeletal disorders. *Acta Orthopaedica Scandinavica, 73*(Suppl. 305), 54–57.

Chapter 45

Cognitive impairment, pain, and analgesia

Vanesa Cantón-Habas[a,b], José Manuel Martínez-Martos[c], Manuel Rich-Ruiz[a,b,c], María Jesús Ramirez-Éxposito[c], and María del Pilar Carrera-González[a,b,c,*]

[a]*Department of Nursing, Pharmacology and Physiotherapy, Faculty of Medicine and Nursing, University of Córdoba, Córdoba, Spain,* [b]*Maimónides Institute for Biomedical Research (IMIBIC), University of Córdoba, Reina Sofia University Hospital, Córdoba, Spain,* [c]*Experimental and Clinical Physiopathology Research Group, Department of Health Sciences, Faculty of Experimental and Health Sciences, University of Jaén, Jaén, Spain*
*Corresponding author

Abbreviations

AD	Alzheimer's disease
GDS	global deterioration scale
LBD	Lewy body dementia
sTNFαRII	soluble tumor necrosis factor-α receptor II
MCI	mild cognitive impairment
sIgA	salivary immunoglobulin A

Introduction

Pain is an unsolved problem in older people, especially in those with cognitive impairment who may not be able to verbalize their pain. Additionally, pain can be an important cause of behavioral disorders in elderly subjects with dementia. An adequate evaluation and management of pain as well as the correct use of sedatives, associated with impaired cognition, falls, fractures, and increased risk of death, are essential (Ballard et al., 2009). The solution is to have valid and reliable assessment tools, which consequently allow proper management of analgesia and improve the quality of life of these patients. The determination of biomarkers of pain in saliva would be an enormously useful tool. In fact, some pain biomarkers have been determined in saliva (Singla, Desjardins, & Chang, 2014), such as salivary cortisol, widely studied since 2005, establishing an important relationship between its salivary levels and the level of pain (McLean et al.). Several authors point to sIgA and sTNFαRII as potential salivary markers of pain in healthy people and AD patients (Cantón-Habas, del Carrera-González, Moreno-Casbas, Quesada-Gómez, & Rich-Ruiz, 2019; Sobas et al., 2016).

In this group of patients there are two converging situations that difficult even more the assessment of pain, like age, associated with its own set of pathologies, as well the risk of neurodegenerative pathologies such as AD, which in advanced stages makes the communication impossible. In this context, it is primordial to quickly evaluate the effects of aging over the brain and its relation with the pain process, and to determine some of the factors described to be involved in the perception of pain, like gender, environment or even the aggravation of the cognitive impairment.

Applications to other areas

Pain is a problem in people who cannot communicate, such as those with advanced dementia, aggravated in the case of older adults due to the concomitance with age-associated pathologies that require tools for its detection and consequently assessment of analgesia. These tools would apply to other patient groups whose communication is difficult and complex, such as infants in the neonatal intensive care unit. In this sense, the implementation of a multidimensional pain tool that measures acute and chronic pain is essential for proper pain assessment (Desai, Aucott, Frank, & Silbert-Flagg, 2018), being the biomarkers of pain in saliva a promising tool.

Another of the great applications of pain biomarker determinations in saliva would be in children/adolescents with cerebral palsy, as established by Symons et al. (2015). Assessing and treating pain in children with developmental disabilities who cannot use their verbal skills are a clinical challenge. Current assessment approaches rely on clinical impression and behavioral rating scales completed by proxy report. Furthermore, Sobas et al. (2020) have recently described its possible use in pain perception after advanced corneal surface ablation surgery.

However, pain older adults must also be considered with the modifications typical of aging, both in pain physiology and in pain perception. Therefore, this work requires a brief introduction in this regard. Besides, variables such as gender, rural or urban setting as well as the degree of cognitive deterioration also influence pain and, consequently, analgesia in this group of patients with communication disabilities and will be treated in this study.

Mini-dictionary of terms

- Aging: Progressive loss of physiological integrity, leading to impaired function, and increased vulnerability to death.
- Mild cognitive impairment (MCI): The impairment of cognitive functions within the normal performance of activities of daily living.
- Alzheimer's disease (AD): Progressive, fatal, and currently incurable neurodegenerative disease. It is a slowly progressive brain disease that begins many years before symptoms emerge.
- The Reisberg Global Deterioration Scale (GDS): Scale used to assess the general clinical severity of dementia.
- Biomarkers of pain: Protein determinations associated with a painful process that would allow its detection in people who cannot communicate them in fluids such as saliva.

Key facts

- Aging is associated with age-related diseases that generate pain in the elderly, often not well managed.
- Age is a risk factor for the development of AD.
- Advanced states of AD-type dementia are associated with an inability to communicate and consequently under-treatment of pain.
- Factors such as gender, living in a rural and/or urban medium and the degree of cognitive decline could influence the perception of pain, being able to determine, at least in part, the management of analgesia.
- The determination of pain, and consequently the management of analgesia, are based on observational parameters that require great experience and knowledge on the part of the caregiver.
- Objective and noninvasive tools are required for pain determination in this group of patients, that is, pain biomarkers.
- The development of these tools allows their use in other groups of patients such as adolescents with cerebral palsy and infants in intensive care units, among others.

Physiological and pathological aging: Cognitive impairment

Biology of aging

The reason for aging, as collected by Steven N. Austad (1997), is the decline in the force of natural selection after the peak of reproduction. So, aging is defined and characterized by a progressive loss of physiological integrity, leading to impaired function and increased vulnerability to death. In fact, many studies have confirmed that most physiological processes deteriorate progressively after about 30 years, but some functions are more severely affected than others (Erdő, Denes, & de Lange, 2017). This deterioration is the primary risk factor for major human pathologies including cardiovascular disorders and neurodegenerative diseases (López-Otín, Blasco, Partridge, Serrano, & Kroemer, 2013). Even though, some authors suggest that aging should be defined as a disease (Stambler, 2017). It is true that aging inevitably occurs with time in all organisms and emerges at a molecular, cellular, organ, and organism level with genetic, epigenetic, and environmental modulators. However, it must also be noted that individuals with the same chronological age and their organs exhibit different trajectories of decline related to age (Khan, Singer, & Vaughan, 2017).

Brain aging

Understanding the age-related changes in cognition is important given the growing age of the population and the importance of cognition in maintaining functional independence and effective communication with others. Cognitive abilities often decline with age, but some types of changes in cognition are expected as part of normal aging and some other changes might suggest the onset of a brain disease. Structural and functional changes in the brain correlate with these age-related cognitive changes, including alterations in neuronal structure without neuronal death, loss of synapses, and dysfunction of neuronal networks (Murman, 2015).

In relation to brain structure, the size of the brain decreases with age. Not all brain areas develop atrophy equally with aging, but both gray and white matter regions are affected by it. Gray matter volume loss is most prominent in the prefrontal cortex (Raz, Gunning-Dixon, Head, Dupuis, & Acker, 1998). The temporal lobes, especially the medial temporal lobe, which includes the hippocampus, also show moderate declines in volume with aging. White matter volumes decline with age too (Madden et al., 2009). It has been assumed that gray matter volume loss was due to neuronal loss, but with improvements in neuron-counting techniques, it is now clear that this is not the case. Many studies demonstrate that loss of neurons during normal aging is restricted to specific regions of the nervous system and that this loss is no more than 10% of neurons found in young adults (Pannese, 2011). Cortical neuronal loss is most notable in the dorsal lateral prefrontal cortex and hippocampus, and greater subcortical neuronal loss have been observed in the substantia nigra and cerebellum.

In normal aging, a substantial number of neurons change in structure but do not die. These aging-related structural changes to neurons include a decrease in the number and length of dendrites, loss of dendritic spines, and a decrease in the number of axons, an increase in axons with segmental demyelination, and a significant loss of synapses (Pannese, 2011). In fact, synaptic loss is a key structural marker of aging in the nervous system. In age-related neurodegenerative diseases such as AD, the loss of neurons is much greater, especially in the hippocampus and entorhinal cortex (Morrison, 1997).

In addition to all age-related brain changes too, the progressive brain inflammatory state by microglia alteration and microglial senescence that occur in the aging brain are likely to be two linked events, which could be crucial for the development of aging-related neurodegenerative disease (Mecca, Giambanco, Donato, & Arcuri, 2018).

Biology of pain

Pain is defined as an unpleasant sensory and emotional experience associated with actual or potential tissue damage, whereas nociception is the neural process of encoding noxious stimuli (IASP, 2021). The human body is equipped with different types of sensory neurons and nociceptors, which form the primary unit of pain, which are able to detect stimuli that have the potential to cause damage. When a noxious stimulus activates an ion channel on a nociceptor, it produces a depolarization of the nociceptor, generating an electric potential. If the receptor's potential is of sufficient magnitude to reach the activation threshold for voltage-dependent channels, it will trigger an action potential generation and the transmission of a pain signal to the spinal cord (Dubin & Patapoutian, 2010; Reichling, Green, & Levine, 2013).

Pain perception

Pain perception begins in the periphery, and then ascends in several tracts, relaying at different levels. Pain signals arrive in the thalamus and midbrain structures that form the pain neuromatrix, a constantly shifting set of networks and connections that determine conscious perception. Several cortical regions become active simultaneously during pain perception; activity in the cortical pain matrix evolves over time to produce a complex pain perception network. Dysfunction at any level has the potential to produce an unregulated, persistent pain (Fenton, Shih, & Zolton, 2015). Understanding pain is fundamental to improve the evaluation, treatment, and innovation in the management of acute and persistent pain syndromes.

However, it is well-known that the perception of clinical pain per se seems to vary greatly from person to person in the general population. Nevertheless, intraindividual differences also exist. In fact, several factors, such as spontaneous neuronal fluctuations, attention, expectation of pain, cognitive and emotional states, sleep habits and stress, may influence pain perception (Kröger, Menz, & May, 2016).

Pain in aging

Throughout life, age-related changes occur in relation to pain sensitivity. As a matter of fact, this suggests that pain perception diminishes in old age (El Tumi, Johnson, Dantas, Maynard, & Tashani, 2017). Authors as Dubin and Patapoutian (2010) state that pain perception is correlated with the activation of some but not all subtypes of heat and mechanic nociceptors. Moreover, judgments of pain threshold are also influenced by differences in central nervous system processing of noxious peripheral input from superficial and deep tissue and differences in pain modulator processes. However, pain perception is also influenced by biopsychosocial and age-related environmental factors changes; therefore, pain perception is likely to be a complex phenomenon.

In this context, it has been proposed that sex differences in pain perception are greater during reproductive years in women since pain sensitivity is believed to decline postmenopause (LeResche, Mancl, Drangsholt, Saunders, & Korff, 2005). Authors as Lautenbacher, Peters, Heesen, Scheel, and Kunz (2017) describe that aging decreases sensitivity for pain of low intensity, being the reduced sensitivity especially apparent for heat pain and for pain applied in the head. In contrast, the aging does not seem to have a strong effect on pain tolerance.

Pain in the elderly

Advancing age and multiple comorbidities increase the risk of persistent pain in older adults. Pain is a complex, multifaceted problem for aging adults who are significantly less likely to have any pain assessment or reassessment documented (Herr & Titler, 2009), and this contributes to undertreatment. Hence, pain is common among the elderly due to the increased prevalence of age-related conditions like osteoporosis, arthritis, and cardiovascular disease, and this is also true for people with dementia (Achterberg et al., 2013). Pain that goes untreated or undertreated has a higher chance of causing detrimental consequences, such as functional decline, incapacitation, and frailty (Malec & Shega, 2015). Pain assessment from a prevention perspective should be given as much attention as other primary preventive actions in older adults.

Neurodegenerative disorders: Alzheimer's disease

The world's population is aging considerably due to declining fertility rates and the increase of life expectancy (WHO, 2015). A longer life brings opportunities for individuals and society, but also the progressive deterioration of physical and mental health of older adults, and the consequent need for increased medical and social care (Santoni et al., 2016). The aging population presents the highest risk of the disease, especially in developed countries. Therefore, the number of people affected by age-related diseases is expected to increase dramatically in the coming decades (Wimo et al., 2013). Dementia-related disorders are the major cause of disability and dependency among older adults (Vetrano et al., 2019).

AD is a progressive, fatal, and currently incurable neurodegenerative disease and the most common cause of dementia worldwide, accounting for between 50% and 70% of the cases recorded among people over 65 (Cordero, García-Escudero, Avila, Gargini, & García-Escudero, 2018), being 2/3 of them women (Brookmeyer et al., 2011). It is a slowly progressive brain disease that begins many years before symptoms emerge (Alzheimer's Association, 2020). The gradual loss of cognitive function, including slow deterioration of memory, reasoning, abstraction, language, and emotional stability clinically characterize AD. As a consequence, in the final stages of the disease, the patient is unable to perform any daily task without adequate assistance from family members or social services (Blennow & Hampel, 2003). The hallmark pathologies of AD are the accumulation of the protein fragment beta-amyloid as plaques outside neurons in the brain and twisted strands of the protein tau inside neurons. These changes are accompanied by the death of neurons and damage to brain tissue (Alzheimer's Association, 2020).

TABLE 1 General descriptive table of the study sample.

Sex	
Women	78% ($n = 78$)
Men	22% ($n = 22$)
Marital status	
Single	9% ($n = 9$)
Married	27% ($n = 27$)
Divorced	62% ($n = 62$)
Widowhood	2% ($n = 2$)
Dementia diagnosis	
Alzheimer's disease	79% ($n = 79$)
Vascular dementia	6% ($n = 6$)
Lewy's bodies dementia	–
Mixed dementia	12% ($n = 12$)
Primary degenerative dementia	3% ($n = 3$)
Medium	
Rural	16% ($n = 16$)
Urban	84% ($n = 84$)
Age (in years)	
Mean (SD)	83.80 (SD = 7.82)
Analgesics consumption	
Yes	44% ($n = 44$)
No	56% ($n = 56$)

Elaborated by the author.

In our study, 78% of people with dementia are women, while 22% of people with dementia are men. Regarding the diagnosis of dementia types, the diagnosis of AD was a majority (79%), followed by mixed dementia (12%) and vascular dementia (6%). The age range of the diagnosis of dementia was above 80 years in all cases. In relation to the consumption of analgesia, surprisingly, less than half of the patients (44%) presented some type of analgesic treatment (Table 1).

Pain and Alzheimer's disease

Pain is described as a multidimensional experience consisting of sensory, cognitive, and affective components (Melzack & Casey, 1968). If one or more of these components is altered, the ability to detect and report pain may also be altered. This occurs in people with AD. Therefore, people with AD have a variable pain response that is threshold dependent (Monroe et al., 2014), thus making the treatment of dementia a particular challenge.

So, AD is a risk factor for the undertreatment of pain, due in part to a lack of understanding of the impact of AD on psychophysiological factors that influence the pain experience (Monroe et al., 2016). In fact, and as described by Achterberg et al. (2013) regarding the pain in the person with dementia and/or AD, different aspects must be considered, such as the impact of the neuropathology of dementia on the perception and processing of pain in AD and other dementias, where white matter lesions and cerebral atrophy seem to influence the neurobiology of pain; the evaluation of pain, a complex fact, as it is based on observational data due to the patient's poor capacity for self-reporting; the evidence of efficient treatment; and one of the most complex, pain management, which requires qualified professionals as well as accurate and validated assessment tools that are sensitive to different types of pain and therapeutic effects, backed by better training and support for care staff in all settings. The optimal treatment in these patients is therefore predominantly based on experience (Barry, Parsons, Passmore, & Hughes, 2013). From all this, an important clinical and research dilemma arises: the difficulty of differentiating between typical behavior related to pain and behavior that is caused by the dementia process itself (Husebo & Corbett, 2014). Thus, it is likely that these difficulties end up resulting in both under and overtreatment. In fact, it is common for patients with cognitive impairment to receive less pain relievers than patients with intact cognitive abilities (Bauer et al., 2016). Therefore, efficacy studies of analgesics in patients with dementia are an urgent need.

Pain in the elderly with advanced dementia

Dementia is a progressive disease associated with irreversible impairment and loss of cognitive abilities (Prince et al., 2013). Approximately from 4.5% to 8% of people over 70 and 15% to 64% of people over 80 will experience dementia (Prince et al., 2013). As the population ages, the number of older people who suffer dementia will also increase (Prince et al., 2013). Until recently, little was known about how pain and dementia impacted each other or even if people with dementia continued to experience pain. It is currently known that pain persists in dementia. Authors such as Gagliese, Gauthier, Narain, and Freedman (2018) point out that the prevalence of pain among older people with dementia is comparable to other groups of older people. In fact, these authors consider that pain in older people with dementia can be conceptualized as the final result of the interaction of three heterogeneous phenomena–pain, aging, and dementia–which are created and influenced by the interactions of predisposing, lifelong, and current biopsychosocial factors. Furthermore, these three factors would have reciprocal relationships.

In this context, we describe factors such as gender, living in urban or rural populations, or state of Global deterioration scale (GDS) and their relationship with the management of analgesia in order to determine whether such factors in older people with cognitive impairment could affect pain management and consequently, be considered in the development of future tools.

Gender, pain, and analgesia

It is recognized that psychological and socio-cultural mechanisms can influence pain perception, expression, and tolerance in both sexes, thus confounding gender-related pain analysis. Nevertheless, the overall findings from epidemiological and clinical studies demonstrate that women are at higher risk for many common pain conditions than men, although data on pain intensity are less consistent and could be influenced by several methodological factors and the differences in the effects of pain treatments (Pieretti et al., 2016).

Gender differences in pharmacological therapy and nonpharmacological pain interventions have also been reported, but these effects appear to depend on the treatment type and characteristics. It is becoming very evident that gender differences in pain and its relief arise from an interaction of genetic, anatomical, physiological, neuronal, hormonal, psychological, and social factors that modulate pain differently in the sexes.

Differences between men and women in pain prevalence, the seeking of medical treatment of pain syndromes, pain behavior, and responses to analgesic drugs have long been reported (Bartley & Fillingim, 2013; Pieretti et al., 2016).

Biological factors such as sex hormones are thought to be one of the main mechanisms explaining sex differences in pain perception (Pieretti et al., 2016). In this context, menopause can play an important role in altering pain sensitivity. Interestingly, although the loss of estrogen can lead to a decrease in life-long painful conditions such as headache, menopause can also be accompanied by new painful conditions such as osteoporosis and joint inflammation (Meriggiola, Nanni, Bachiocco, Vodo, & Aloisi, 2012). Regarding mental health, gender differences vary by social context and exposure to risk-factors for mental illnesses, greater in older women, but older men may be more vulnerable to their impacts (Kiely, Brady, & Byles, 2019).

In our study, carried out on a population with dementia, no significant differences could be observed between analgesic consumption in relation to gender (47.44% women; 31.82% men); neither in relation to the type of analgesia, being similar in both groups based mainly on nonopioid analgesics (40% in women, 28.5% in men), followed by adjuvants (34.2% in women, 42.8% in men) (Table 2).

Only 2.86% of the population received analgesia based on powerful opioids and NSAIDs. We also studied the consumption period, being equivalent for both sexes. In this context, we should contemplate that the consumption of analgesia cannot be directly extrapolated to the presence of pain, since in most cases these patients have communication disabilities,

TABLE 2 Analgesic consumption according to sex.

	Women	Men	P value
Analgesics consumption			
Yes	47.44% (n = 37)	31.82% (n = 7)	P = 0.1925
No	52.56% (n = 41)	68.18% (n = 15)	
Analgesics type			
Nonopioids	40% (n = 14)	28.57% (n = 2)	P = 0.56
NSAID and ASA	2.86% (n = 1)	–	–
Minor opioids	–	–	–
Strong opioids	2.86% (n = 1)	14.29% (n = 1)	–
Coadjuvants	34.29% (n = 12)	42.86% (n = 3)	P = 0.66
Nonopioids and coadjuvants	11.43% (n = 4)	14.29% (n = 1)	–
Nonopioids and minor opioids	2.86% (n = 1)	–	–
Nonopioids, minor opioids and coadjuvants	–	–	–
Nonopioids and NSAID	–	–	–
Minor opioids and coadjuvants	2.86% (n = 1)	–	–
Strong opioids and NSAID	2.86% (n = 1)	–	–
Consumption time (in days)			
Mean (SD)	811.06 (SD = 906.48)	842.86 (SD = 764.11)	P = 0.72
Median	500	485	
Max.	4321	2046	
Min.	5	231	

Elaborated by the author.

and therefore such management of analgesia depends on expertise and training from their caregivers, as well as the tools available to assess such pain. The inability of patients with dementia to verbally communicate their pain makes them a vulnerable patient group, dependent on their caregivers (Brorson, Plymoth, Örmon, & Bolmsjö, 2014).

In this sense, it is interesting for us to address the differences of the analgesia according to the rural or urban areas.

Urban and rural area, pain, and analgesia

Given that urban and rural social structures and lifestyles differ significantly, it is possible that the factors affecting health services used by people living in rural areas may diverge from those in urban areas (Goode, Freburger, & Carey, 2013). Rural residency is associated with higher prevalence of chronic pain and other psychiatric and medical comorbidities, especially depression (Rost, Fortney, Fischer, & Smith, 2002). Rural residents with chronic pain report higher pain frequency and intensity, as well as more pain-related disability and depression than people with pain living in urban areas (Goode et al., 2013). These differences could also affect the presence of cognitive impairment and consequently dementia. Recent studies show that the presence of cognitive deterioration in the urban population is associated with age, loss of physical activity, the presence of diabetes and having three or more children. On the contrary, factors associated with cognitive decline in the rural population include being female, age, as well as exposure to pesticides and histories of encephalitis or meningitis and head trauma (Tang et al., 2016).

The sociodemographic characteristics of the patients according to the rural or urban areas included in our study are detailed in Table 3. Both in the urban and rural population, the percentage of women affected by dementia is higher (76.19% in the urban population and 87.5% in rural population) compared to men (23.81% in urban areas and 12.5% in rural areas). This difference could also be related to cultural issues associated with home care, since the study population was recruited in nursing homes.

500   PART | IV Assessments, screening, and resources

TABLE 3 Consumption of analgesics according to the medium.

	Urban	Rural	*P* value
Analgesics consumption			
Yes	46.43% (*n* = 39)	31.25% (*n* = 5)	*P* = 0.262
No	53.57% (*n* = 45)	68.75% (*n* = 11)	
Analgesics type			
Nonopioids	40.54% (*n* = 15)	20% (*n* = 1)	*P* < 0.05
NSAID and ASA	2.70% (*n* = 1)	–	–
Minor opioids	–	–	–
Strong opioids	5.41% (*n* = 2)	–	–
Coadjuvants	32.43% (*n* = 12)	60% (*n* = 3)	*P* < 0.05
Non-opioids and coadjuvants	13.51% (*n* = 5)	–	–
Nonopioids and minor opioids	2.70% (*n* = 1)	–	–
Nonopioids, minor opioids and coadjuvants	–	–	–
Nonopioids and NSAID	–	–	–
Minor opioids and coadjuvants	2.70% (*n* = 1)	–	–
Strong opioids and NSAID	–	20% (*n* = 1)	–
Consumption time (in days)			
Mean (SD)	861.05 (SS = 916.02)	475.60 (SD = 376.26)	*P* = 0.36
Median	507	365	
Max.	4321	1080	
Min.	5	123	

Elaborated by the author.

At this point we should note that the rising prevalence of dementia will also mean that a growing number of families are providing in-home care for older adults with dementia (Dudgeon, and Risk Analytica, and Alzheimer Society of Canada, 2010) and, more than half of family caregivers are females (Xiong, Biscardi, Nalder, & Colantonio, 2018), since gender constitutes the socially constructed roles, behaviors, expressions, and identities of girls, women, boys, and men (Government of Canada, 2015). However, the main civil situation reported was widowhood.

In relation to the diagnosis of the type of dementia, AD is the majority in both media, although we also found diagnoses of mixed, vascular, and primary degenerative dementia in the urban population (Table 4).

Surprisingly, the time of diagnosis of dementia was very similar in both the urban and rural population, around 5 years. Regarding the consumption of analgesia in urban and rural areas in people with dementia (Table 3), although we have not found significant differences among populations, we have observed significant differences between the types of treatment. In particular, about the consumption of nonopioid analgesics (*P* < 0.05) and adjuvants (*P* < 0.05), being higher the consumption of nonopioid analgesics in urban areas (49.5% urban area compared to 20% rural area) and the adjuvants in rural areas (32.4% urban area compared to 60% rural area) (Table 3). These differences could be determined, at least in part, by differences in training as well as experience of health personnel in residences in urban and rural areas.

Global deterioration scale (GDS), pain, and analgesia

AD starts gradually and worsens over several years, therefore creating the notion of progressively passing several stages of severity (Perneczky et al., 2006). In clinical practice, the diagnosis is based on behavioral assessments and cognitive tests that highlight quantitative and qualitative changes in cognitive functions and activities of daily living, which are characteristic of the dementia syndrome and its underlying diseases. Because of this natural course of the disease, dementia has

TABLE 4 Sociodemographic characteristics according to the area.

	Urban (N = 84)	Rural (N = 16)
Sex		
Women	76.19% (n = 64)	87.50% (n = 14)
Men	23.81% (n = 20)	12.50% (n = 2)
Marital status		
Single	7.14% (n = 6)	18.75% (n = 3)
Married	29.76% (n = 25)	12.50% (n = 2)
Divorced	2.38% (n = 2)	–
Widowhood	60.71% (n = 51)	68.75% (n = 11)
Dementia diagnosis		
Alzheimer's disease	75% (n = 63)	100% (n = 16)
Vascular dementia	7.14% (n = 6)	–
Lewy's bodies dementia	–	–
Mixed dementia	14.29% (n = 12)	–
Primary degenerative dementia	3.57% (n = 3)	–
Age (in years)		
Mean (SD)	83.51 (SD = 8.02)	85.31 (SD = 6.72)
Dementia diagnosis time (in years)		
Mean (SD)	5.43 (SD = 3.20)	5.78 (SD = 2.15)

Elaborated by the author.

sometimes been divided into stages (e.g., predementia, mild cognitive impairment, early dementia, moderate dementia, and advanced dementia), but the number and characteristics of these stages vary accordingly based on the assessment scales used (OldeRikkert et al., 2011).

Mild cognitive impairment (MCI) is defined as the impairment of cognitive functions within the normal performance of activities of daily living. It is considered a transitional stage between normal aging and dementia. The Reisberg Global Deterioration Scale (GDS) is used to assess the general clinical severity of dementia (Reisberg, Ferris, de Leon, & Crook, 1982). It consists of the clinical description of seven phases differentiated from normality to the most severe degrees of AD dementia.

In this context, an evaluation of pain parallel to the progression of the disease should be considered necessary in order to establish correct guidelines for correct prescription of analgesia. Results obtained by Monroe et al. (2016) found increased thresholds for detecting moderate pain in people with AD as a group compared to non-AD controls, but it did not appear to influence the worsening global cognitive impairment on detection threshold for moderate pain. Although it does become clear that in the presence of similar painful conditions, when compared to cognitively intact older adults, people with AD have been shown to receive less pain medication (Monroe et al., 2016). Furthermore, in the presence of similar painful conditions, people with AD verbally report pain less frequently but exhibit similar pain-related behaviors when moved (Horgas, Elliott, & Marsiske, 2009).

It is unknown if these findings reflect less perceived pain in AD or an inability to recognize pain or to communicate pain. However, no evidence has been found suggesting that worsening global cognitive function was associated with lower reports of unpleasantness in the presence of mild and moderate pain. In this context, our results show that in GDS stages 5, 6, and 7, the main type of analgesic consumed is nonopioid analgesics, regardless of the severity of the cognitive decline (Table 5).

502 PART | IV Assessments, screening, and resources

TABLE 5 GDS score and analgesic consumption.

	GDS 5	GDS 6	GDS 7	*P* value
Analgesics consumption				
Yes	69.23% (*n* = 9)	42.14% (*n* = 18)	37.78% (*n* = 17)	*P* = 0.13
No	30.77% (*n* = 4)	57.14% (*n* = 24)	62.22% (*n* = 28)	
Analgesics type				
Nonopioids	44.44% (*n* = 4)	35.29% (*n* = 6)	41.18% (*n* = 7)	*P* = 0.94
NSAID and ASA	–	5.88% (*n* = 2)	–	–
Minor opioids	–	–	–	–
Strong opioids	22.22% (*n* = 2)	–	–	–
Coadjuvants	22.22% (*n* = 2)	29.41% (*n* = 5)	47.06% (*n* = 8)	*P* = 0.44
Nonopioids and coadjuvants	–	17.65% (*n* = 3)	11.76% (*n* = 2)	–
Nonopioids and minor opioids	11.11% (*n* = 1)	–	–	–
Nonopioids, minor opioids and coadjuvants	–	–	–	–
Nonopioids and NSAID	–	–	–	–
Minor opioids and coadjuvants	–	5.88% (*n* = 1)	–	–
Strong opioids and NSAID	–	5.88% (*n* = 1)	–	–
Consumption time (in days)				
Mean (SD)	535.89	1059.33	700.44	*P* = 0.28
Median	341	527.50	697.50	
Max.	2046	4321	1840	
Min.	5	87	79	

Elaborated by the author.

We found a decrease in the consumption of analgesia parallel to the worsening of their cognitive situation (Fig. 1), but no significant differences were observed. In all GDS stages, the treatment was complemented with adjuvants in 22.2%, 29.4%, and 47% in GDS 5, 6, and 7, respectively (Table 4). Regarding the sociodemographic variables according to the GDS status, it should be noted, on the one hand, that the proportion of women was higher in the three study groups and, on the other hand, that the predominant marital status was widowhood in those with GDS 5, 6, and 7. In the same way, and contrary to what is expected, the average age was slightly higher in the patients who presented less cognitive decline, particularly this was 84.62 years for people with GDS 5, 83.21 for patients with GDS 6 and 84.11 for those with GDS 7 (Table 6).

We must take into account the aforementioned barriers such as less perceived pain in AD or an inability to recognize pain or to communicate pain, and also that the prevalence of pain among older people with dementia is comparable to other groups of older people as described Gagliese et al. (2018). This data must be taken into consideration by expert caregivers of patients with advanced dementia.

Future

Inadequately managed pain in people with AD is a significant public health concern. AD in general is a risk factor for the undertreatment of pain, due in part to a lack of understanding of the impact of AD on psychophysiological factors that influence the pain experience. As people age, the risk of developing pain increases and as the population of older adults continues to grow, and so will the number of people diagnosed with dementia who suffer from pain (Monroe et al., 2014). Therefore, cognitive disability and dementia will be a pressing problem in a few years, and it is expected to be higher in developed areas. Western health-care systems need to provide solutions for this type of population. In this context, we consider vitally important to develop noninvasive, objective and reliable tools that allow assessing pain, particularly in

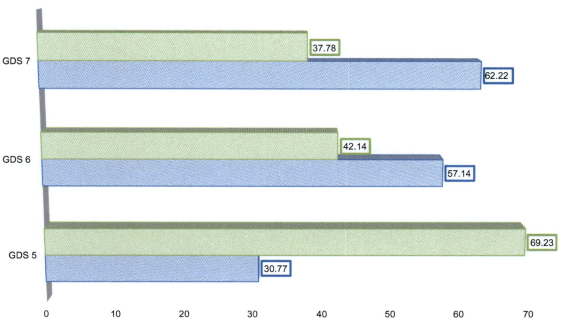

FIG. 1 Consumption of analgesia with respect to the degree of cognitive deterioration. We showed the percentage of patients in each GDS state and the consumption of analgesia (yes/no). We found a decrease in the consumption of analgesia parallel to the worsening of their cognitive situation, but no significant differences were observed. Elaborated by the author.

TABLE 6 GDS score and sociodemographic characteristics.

	GDS 5 ($n = 13$)	GDS 6 ($n = 42$)	GDS 7 ($n = 45$)
Sex			
Women	69.23% ($n=9$)	83.33% ($n=35$)	75.56% ($n=34$)
Men	30.77% ($n=4$)	16.67% ($n=7$)	24.44% ($n=11$)
Marital status			
Single	7.69% ($n=1$)	9.54% ($n=4$)	8.89% ($n=4$)
Married	15.38% ($n=2$)	38.10% ($n=16$)	20% ($n=9$)
Widowhood	76.92% ($n=10$)	50% ($n=21$)	68.89% ($n=31$)
Divorced	–	2.38% ($n=1$)	2.22% ($n=1$)
Medium			
Urban	92.31% ($n=12$)	76.19% ($n=32$)	88.89% ($n=40$)
Rural	7.69% ($n=1$)	23.81% ($n=10$)	11.11% ($n=5$)
Dementia diagnosis			
Alzheimer's disease	69.23% ($n=9$)	71.43% ($n=30$)	82.22% ($n=37$)
Vascular dementia	7.69% ($n=1$)	14.29% ($n=6$)	2.22% ($n=1$)

Continued

504 PART | IV Assessments, screening, and resources

TABLE 6 GDS score and sociodemographic characteristics—cont'd

	GDS 5 (n = 13)	GDS 6 (n = 42)	GDS 7 (n = 45)
Degenerative dementia	–	2.38% (n = 1)	2.22% (n = 1)
Mixed dementia	23.08% (n = 3)	11.9% (n = 5)	13.33% (n = 6)
Lewy's body dementia	–	–	–
Age			
Mean (SD)	84.62 (SD = 6.39)	83.21 (SD = 8.23)	84.11 (SD = 7.83)
Median	83	87	85
Max.	94	94	95
Min.	76	65	65
Dementia diagnosis time			
Mean (SD)	5.24 (SD = 3.09)	4.92 (SD = 2.67)	6.29 (SD = 3.26)
Median	4.31	4.75	5.45
Max.	9.63	8.49	11.48
Min.	1.31	1.43	1.79

Elaborated by the author.

elderly people with advanced dementia who are unable to identify/communicate the pain. These tools would allow all those involved in the care process: family caregivers, involved in the front-line in maintaining the quality of life of these patients; as well as health professionals, whose activity in this regard is based upon their experience, professionalism; and in those involved in decision making, like legislators or managers.

References

Achterberg, W. P., Pieper, M. J., van Dalen-Kok, A. H., de Waal, M. W., Husebo, B. S., Lautenbacher, S., et al. (2013). Pain management in patients with dementia. *Clinical Interventions in Aging, 8*, 1471–1482.

Alzheimer's Association. (2020). 2020 Alzheimer's disease facts and figures. *Alzheimer's & Dementia: The Journal of the Alzheimer's Association, 16*(3), 391–460. https://doi.org/10.1002/alz.12068.

Austad, S. N. (1997). *Why we age: What science is discovering about the body's journey through life.* New York: J. Wiley & Sons.

Ballard, C., Hanney, M. L., Theodoulou, M., Douglas, S., McShane, R., Kossakowski, K., et al. (2009). The dementia antipsychotic withdrawal trial (dart-ad): Long-term follow-up of a randomised placebo-controlled trial. *The Lancet Neurology, 8*(2), 151–157. https://doi.org/10.1016/S1474-4422(08)70295-3.

Barry, H. E., Parsons, C., Passmore, A. P., & Hughes, C. M. (2013). Community pharmacists and people with dementia: A cross-sectional survey exploring experiences, attitudes, and knowledge of pain and its management: Community pharmacists and people with dementia. *International Journal of Geriatric Psychiatry, 28*(10), 1077–1085. https://doi.org/10.1002/gps.3931.

Bartley, E. J., & Fillingim, R. B. (2013). Sex differences in pain: A brief review of clinical and experimental findings. *British Journal of Anaesthesia, 111*(1), 52–58. https://doi.org/10.1093/bja/aet127.

Bauer, U., Pitzer, S., Schreier, M. M., Osterbrink, J., Alzner, R., & Iglseder, B. (2016). Pain treatment for nursing home residents differs according to cognitive state—A cross-sectional study. *BMC Geriatrics, 16*(1), 124. https://doi.org/10.1186/s12877-016-0295-1.

Blennow, K., & Hampel, H. (2003). CSF markers for incipient Alzheimer's disease. *The Lancet Neurology, 2*(10), 605–613. https://doi.org/10.1016/S1474-4422(03)00530-1.

Brookmeyer, R., Evans, D. A., Hebert, L., Langa, K. M., Heeringa, S. G., Plassman, B. L., et al. (2011). National estimates of the prevalence of Alzheimer's disease in the United States. *Alzheimer's & Dementia, 7*(1), 61–73. https://doi.org/10.1016/j.jalz.2010.11.007.

Brorson, H., Plymoth, H., Örmon, K., & Bolmsjö, I. (2014). Pain relief at the end of life: Nurses' experiences regarding end-of-life pain relief in patients with dementia. *Pain Management Nursing, 15*(1), 315–323. https://doi.org/10.1016/j.pmn.2012.10.005.

Cantón-Habas, V., del Carrera-González, M. P., Moreno-Casbas, M. T., Quesada-Gómez, J. M., & Rich-Ruiz, M. (2019). Correlation between biomarkers of pain in saliva and PAINAD scale in elderly people with cognitive impairment and inability to communicate: Descriptive study protocol. *BMJ Open, 9*(11). https://doi.org/10.1136/bmjopen-2019-032927, e032927.

Cordero, J. G., García-Escudero, R., Avila, J., Gargini, R., & García-Escudero, V. (2018). Benefit of oleuropein aglycone for Alzheimer's disease by promoting autophagy. *Oxidative Medicine and Cellular Longevity, 2018*, 5010741. https://doi.org/10.1155/2018/5010741.

Desai, A., Aucott, S., Frank, K., & Silbert-Flagg, J. (2018). Comparing n-pass and nips: Improving pain measurement in the neonate. *Advances in Neonatal Care, 18*(4), 260–266. https://doi.org/10.1097/ANC.0000000000000521.

Dubin, A. E., & Patapoutian, A. (2010). Nociceptors: The sensors of the pain pathway. *Journal of Clinical Investigation, 120*(11), 3760–3772. https://doi.org/10.1172/JCI42843.

Dudgeon, S., & Risk Analytica, & Alzheimer Society of Canada. (2010). *Rising tide: The impact of dementia on Canadian society: A study*. Retrieved from: https://www.deslibris.ca/ID/220673.

El Tumi, H., Johnson, M. I., Dantas, P. B. F., Maynard, M. J., & Tashani, O. A. (2017). Age-related changes in pain sensitivity in healthy humans: A systematic review with meta-analysis. *European Journal of Pain, 21*(6), 955–964. https://doi.org/10.1002/ejp.1011.

Erdő, F., Denes, L., & de Lange, E. (2017). Age-associated physiological and pathological changes at the blood–brain barrier: A review. *Journal of Cerebral Blood Flow & Metabolism, 37*(1), 4–24. https://doi.org/10.1177/0271678X16679420.

Fenton, B. W., Shih, E., & Zolton, J. (2015). The neurobiology of pain perception in normal and persistent pain. *Pain Management, 5*(4), 297–317. https://doi.org/10.2217/pmt.15.27.

Gagliese, L., Gauthier, L. R., Narain, N., & Freedman, T. (2018). Pain, aging and dementia: Towards a biopsychosocial model. *Progress in Neuro-Psychopharmacology and Biological Psychiatry, 87*, 207–215. https://doi.org/10.1016/j.pnpbp.2017.09.022.

Goode, A. P., Freburger, J. K., & Carey, T. S. (2013). The influence of rural versus urban residence on utilization and receipt of care for chronic low back pain. *The Journal of Rural Health, 29*(2), 205–214. Geographic residence and chronic low back pain https://doi.org/10.1111/j.1748-0361.2012.00436.x.

Government of Canada. (2015). *Definitions of sex and gender*. CIHR. Retrieved from: https://cihr-irsc.gc.ca/e/47830.html.

Herr, K., & Titler, M. (2009). Acute pain assessment and pharmacological management practices for the older adult with a hip fracture: Review of ed trends. *Journal of Emergency Nursing, 35*(4), 312–320. https://doi.org/10.1016/j.jen.2008.08.006.

Horgas, A. L., Elliott, A. F., & Marsiske, M. (2009). Pain assessment in persons with dementia: Relationship between self-report and behavioral observation: Pain assessment in persons with dementia. *Journal of the American Geriatrics Society, 57*(1), 126–132. https://doi.org/10.1111/j.1532-5415.2008.02071.x.

Husebo, B. S., & Corbett, A. (2014). Pain management in dementia—The value of proxy measures. *Nature Reviews Neurology, 10*(6), 313–314. https://doi.org/10.1038/nrneurol.2014.66.

IASP. (2021). *International Association for the Study of Pain (IASP)*. https://www.iasp-pain.org/Education.

Khan, S. S., Singer, B. D., & Vaughan, D. E. (2017). Molecular and physiological manifestations and measurement of aging in humans. *Aging Cell, 16*(4), 624–633. https://doi.org/10.1111/acel.12601.

Kiely, K. M., Brady, B., & Byles, J. (2019). Gender, mental health and ageing. *Maturitas, 129*, 76–84. https://doi.org/10.1016/j.maturitas.2019.09.004.

Kröger, I. L., Menz, M. M., & May, A. (2016). Dissociating the neural mechanisms of pain consistency and pain intensity in the trigemino-nociceptive system. *Cephalalgia, 36*(8), 790–799. https://doi.org/10.1177/0333102415612765.

Lautenbacher, S., Peters, J. H., Heesen, M., Scheel, J., & Kunz, M. (2017). Age changes in pain perception: A systematic-review and meta-analysis of age effects on pain and tolerance thresholds. *Neuroscience & Biobehavioral Reviews, 75*, 104–113. https://doi.org/10.1016/j.neubiorev.2017.01.039.

LeResche, L., Mancl, L. A., Drangsholt, M. T., Saunders, K., & Korff, M. V. (2005). Relationship of pain and symptoms to pubertal development in adolescents. *Pain, 118*(1), 201–209. https://doi.org/10.1016/j.pain.2005.08.011.

López-Otín, C., Blasco, M. A., Partridge, L., Serrano, M., & Kroemer, G. (2013). The hallmarks of aging. *Cell, 153*(6), 1194–1217. https://doi.org/10.1016/j.cell.2013.05.039.

Madden, D. J., Spaniol, J., Costello, M. C., Bucur, B., White, L. E., Cabeza, R., et al. (2009). Cerebral white matter integrity mediates adult age differences in cognitive performance. *Journal of Cognitive Neuroscience, 21*(2), 289–302. https://doi.org/10.1162/jocn.2009.21047.

Malec, M., & Shega, J. W. (2015). Pain management in the elderly. *Medical Clinics of North America, 99*(2), 337–350. https://doi.org/10.1016/j.mcna.2014.11.007.

McLean, S. A., Williams, D. A, Harris, R. E., Kop, W. J., Groner, K. H., Ambrose, K., et al. (2005). Momentary relationship between cortisol secretion and symptoms in patients with fibromyalgia. *Arthritis & Rheumatology, 52*(11), 3660–3669. https://doi.org/10.1002/art.21372.

Mecca, C., Giambanco, I., Donato, R., & Arcuri, C. (2018). Microglia and aging: The role of the trem2–dap12 and cx3cl1-cx3cr1 axes. *International Journal of Molecular Sciences, 19*(1), 318. https://doi.org/10.3390/ijms19010318.

Melzack, R., & Casey, K. (1968). *Sensory, motivational, and central control determinants of pain: A new conceptual model in pain*. Springfield (Illinois): Charles C. Thomas.

Meriggiola, M. C., Nanni, M., Bachiocco, V., Vodo, S., & Aloisi, A. M. (2012). Menopause affects pain depending on pain type and characteristics. *The Journal of The North American Menopause Society, 19*(5), 517–523. https://doi.org/10.1097/gme.0b013e318240fe3d.

Monroe, T. B., Gibson, S. J., Bruehl, S. P., Gore, J. C., Dietrich, M. S., Newhouse, P., et al. (2016). Contact heat sensitivity and reports of unpleasantness in communicative people with mild to moderate cognitive impairment in Alzheimer's disease: A cross-sectional study. *BMC Medicine, 14*(1), 74. https://doi.org/10.1186/s12916-016-0619-1.

Monroe, T. B., Misra, S. K., Habermann, R. C., Dietrich, M. S., Cowan, R. L., & Simmons, S. F. (2014). Pain reports and pain medication treatment in nursing home residents with and without dementia: Pain reports and treatment in dementia. *Geriatrics & Gerontology International, 14*(3), 541–548. https://doi.org/10.1111/ggi.12130.

Morrison, J. H. (1997). Life and death of neurons in the aging brain. *Science, 278*(5337), 412–419. https://doi.org/10.1126/science.278.5337.412.

Murman, D. (2015). The impact of age on cognition. *Seminars in Hearing, 36*(03), 111–121. https://doi.org/10.1055/s-0035-1555115.

OldeRikkert, M. G. M., Tona, K. D., Janssen, L., Burns, A., Lobo, A., Robert, P., et al. (2011). Validity, reliability, and feasibility of clinical staging scales in dementia: A systematic review. *American Journal of Alzheimer's Disease and Other Dementias, 26*(5), 357–365. https://doi.org/10.1177/1533317511418954.

Pannese, E. (2011). Morphological changes in nerve cells during normal aging. *Brain Structure and Function*, *216*(2), 85–89. https://doi.org/10.1007/s00429-011-0308-y.

Perneczky, R., Wagenpfeil, S., Komossa, K., Grimmer, T., Diehl, J., & Kurz, A. (2006). Mapping scores onto stages: Mini-mental state examination and clinical dementia rating. *The American Journal of Geriatric Psychiatry*, *14*(2), 139–144. https://doi.org/10.1097/01.JGP.0000192478.82189.a8.

Pieretti, S., Di Giannuario, A., Di Giovannandrea, R., Marzoli, F., Piccaro, G., Minosi, P., et al. (2016). Gender differences in pain and its relief. *Annali Dell'Istituto Superiore Di Sanita*, *52*(2), 184–189. https://doi.org/10.4415/ANN_16_02_09.

Prince, M., Bryce, R., Albanese, E., Wimo, A., Ribeiro, W., & Ferri, C. P. (2013). The global prevalence of dementia: A systematic review and metaanalysis. *Alzheimer's & Dementia*, *9*(1), 63–75.e2. https://doi.org/10.1016/j.jalz.2012.11.007.

Raz, N., Gunning-Dixon, F. M., Head, D., Dupuis, J. H., & Acker, J. D. (1998). Neuroanatomical correlates of cognitive aging: Evidence from structural magnetic resonance imaging. *Neuropsychology*, *12*(1), 95–114. https://doi.org/10.1037/0894-4105.12.1.95.

Reichling, D. B., Green, P. G., & Levine, J. D. (2013). The fundamental unit of pain is the cell. *Pain*, *154*, S2–S9. https://doi.org/10.1016/j.pain.2013.05.037.

Reisberg, B., Ferris, S., de Leon, M., & Crook, T. (1982). The global deterioration scale for assessment of primary degenerative dementia. *American Journal of Psychiatry*, *139*(9), 1136–1139. https://doi.org/10.1176/ajp.139.9.1136.

Rost, K., Fortney, J., Fischer, E., & Smith, J. (2002). Use, quality, and outcomes of care for mental health: The rural perspective. *Medical Care Research and Review*, *59*(3), 231–265. https://doi.org/10.1177/1077558702059003001.

Santoni, G., Marengoni, A., Calderón-Larrañaga, A., Angleman, S., Rizzuto, D., Welmer, A.-K., et al. (2016). Defining health trajectories in older adults with five clinical indicators. *The Journals of Gerontology Series A: Biological Sciences and Medical Sciences*. https://doi.org/10.1093/gerona/glw204, glw204.

Singla, N. K., Desjardins, P. J., & Chang, P. D. (2014). A comparison of the clinical and experimental characteristics of four acute surgical pain models: Dental extraction, bunionectomy, joint replacement, and soft tissue surgery. *Pain*, *155*(3), 441–456. https://doi.org/10.1016/j.pain.2013.09.002.

Sobas, E. M., Reinoso, R., Cuadrado-Asensio, R., Fernández, I., Maldonado, M. J., & Pastor, J. C. (2016). Reliability of potential pain biomarkers in the saliva of healthy subjects: Inter-individual differences and intersession variability. *PLoS One*, *11*(12). https://doi.org/10.1371/journal.pone.0166976, e0166976.

Sobas, E. M., Vázquez, A., Videla, S., Reinoso, R., Fernández, I., Garcia-Vazquez, C., et al. (2020). Evaluation of potential pain biomarkers in saliva and pain perception after corneal advanced surface ablation surgery. *Clinical Ophthalmology (Auckland, N.Z.)*, *14*, 613–623. https://doi.org/10.2147/OPTH.S225603.

Stambler, I. (2017). Recognizing degenerative aging as a treatable medical condition: Methodology and policy. *Aging and Disease*, *8*(5), 583. https://doi.org/10.14336/AD.2017.0130.

Symons, F. J., ElGhazi, I., Reilly, B. G., Barney, C. C., Hanson, L., Panoskaltsis-Mortari, A., et al. (2015). Can biomarkers differentiate pain and no pain subgroups of nonverbal children with cerebral palsy? A preliminary investigation based on noninvasive saliva sampling. *Pain Medicine*, *16*(2), 249–256. https://doi.org/10.1111/pme.12545.

Tang, H.-D., Zhou, Y., Gao, X., Liang, L., Hou, M.-M., Qiao, Y., et al. (2016). Prevalence and risk factor of cognitive impairment were different between urban and rural population: A community-based study. *Journal of Alzheimer's Disease*, *49*(4), 917–925. https://doi.org/10.3233/JAD-150748.

Vetrano, D. L., Rizzuto, D., Calderón-Larrañaga, A., Onder, G., Welmer, A.-K., Qiu, C., et al. (2019). Walking speed drives the prognosis of older adults with cardiovascular and neuropsychiatric multimorbidity. *The American Journal of Medicine*, *132*(10), 1207–1215.e6. https://doi.org/10.1016/j.amjmed.2019.05.005.

Wimo, A., Jönsson, L., Bond, J., Prince, M., Winblad, B., & Alzheimer Disease International. (2013). The worldwide economic impact of dementia 2010. *Alzheimer's & Dementia*, *9*(1), 1–11.e3. https://doi.org/10.1016/j.jalz.2012.11.006.

World Health Organization. (2015). *World report on ageing and health 2015*. (s. f.). Retrieved from http://www.who.int/ageing/events/world-report-2015-launch/en/.

Xiong, C., Biscardi, M., Nalder, E., & Colantonio, A. (2018). Sex and gender differences in caregiving burden experienced by family caregivers of persons with dementia: A systematic review protocol. *BMJ Open*, *8*(8). https://doi.org/10.1136/bmjopen-2018-022779, e022779.

Chapter 46

Biomarkers in endometriosis-associated pain

Deborah Margatho and Luis Bahamondes

Department of Obstetrics and Gynaecology, University of Campinas Medical School, Campinas, SP, Brazil

Abbreviations

ASRM	American Society for Reproductive Medicine
CA	cancer antigen
CD	cluster of differentiation
CRP	C-reactive protein
CYP19	cytochrome P450 family 19
DNA	deoxyribonucleic acid
ERα	estrogen receptor alpha
ERβ	estrogen receptor beta
GnRH	gonadotropin releasing hormone
hTERT	human telomerase reverse transcriptase
IL	interleukin
LN-1	laminin-1
MALDI-TOF-MS	matrix-assisted laser desorption/ionization time-of-flight mass
miRNAs	microRNAs
MRI	magnetic resonance imaging
NK	natural killer
PGP	protein gene product
PROK1	prokineticin 1
RNA	ribonucleic acid
sICAM-1	soluble form of the intercellular adhesion molecule-1
TNFα	tumor necrosis factor alpha
VDBP	vitamin D-binding protein
VEGF	vascular endothelial growth factor

Background

Endometriosis is defined as the presence of endometrial glandular or stromal cells outside the endometrial cavity, provoking a pathologic inflammatory response (Dunselman et al., 2014). The first description of the disease was published in 1860 (Nezhat, Nezhat, & Nezhat, 2012) and many theories have since been proposed to explain its pathogenesis. The most commonly accepted theories are that of retrograde menstruation proposed by Sampson in 1921 (Nisolle & Donnez, 1997) and the coelomic metaplasia theory in which the peritoneal mesothelium is transformed into endometrial glandular cells (Giudice & Kao, 2004; Zondervan Krina, Becker Christian, & Missmer, 2020). Genetic, environmental, immunological, proinflammatory, and hormonal factors have been linked to the disease and its progression (Ahn et al., 2015; Borghese, Zondervan, Abrao, Chapron, & Vaiman, 2017; Subramanian & Agarwal, 2010; Zondervan Krina et al., 2020). This benign estrogen-dependent disease may undergo malignant transformation, albeit in less than 1% of cases, with transformation occurring particularly when the pathology involves the ovary (Dunselman et al., 2014). The rare cases in which the disease has been described in men, in prepubertal girls and in postmenopausal women remain as yet unexplained. Early menarche, short menstrual cycles, and heavy menstrual bleeding may constitute risk factors for endometriosis (Zondervan Krina et al., 2020) (Fig. 1).

Features and Assessments of Pain, Anesthesia, and Analgesia. https://doi.org/10.1016/B978-0-12-818988-7.00036-4
Copyright © 2022 Elsevier Inc. All rights reserved.

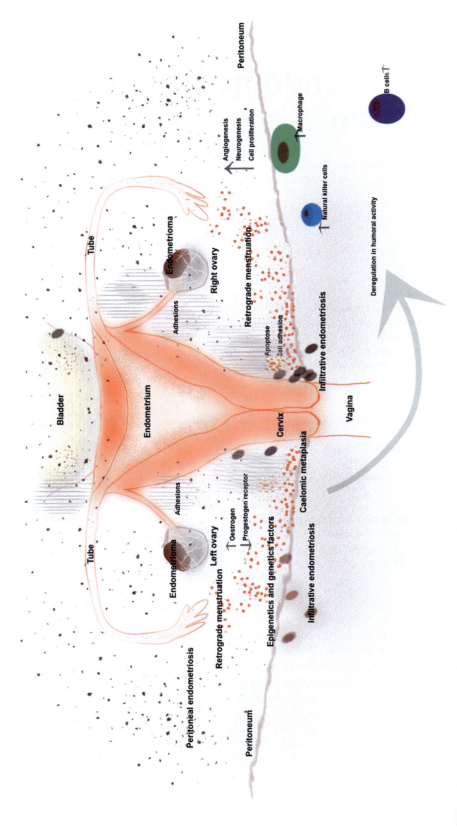

FIG. 1 The most widely accepted theories on the pathophysiology of endometriosis.

Prevalence

The incidence of endometriosis is difficult to estimate due to the different phenotypic presentations associated with the disease, which include not only cases of chronic pelvic pain that may or may not be associated with infertility, but also asymptomatic cases. There are no specific or pathognomonic symptoms of endometriosis. The disease occurs in around 10% of women in the general population; however, prevalence may reach as high as 50% among infertile women with chronic pelvic pain (Buck Louis et al., 2011; Dunselman et al., 2014; Giudice & Kao, 2004; Subramanian & Agarwal, 2010).

Diagnosis

Painstaking analysis of the patient's medical history together with meticulous clinical examination may not raise suspicion of endometriosis, since the association between the severity of the disease and the clinical symptoms can be weak. The main symptoms suggestive of endometriosis include dysmenorrhea; progressively frequent dysmenorrhea culminating in daily pelvic pain, chronic pelvic pain, dyspareunia, dysuria, fatigue, premenstrual abdominal distension, and infertility. These symptoms have a negative impact on both the quality of life and the emotional load of the women affected (Sundell, Milsom, & Andersch, 1990; Zondervan Krina et al., 2020). Symptoms may occur 6 to 12 years prior to surgical diagnosis of the disease (Matsuzaki et al., 2006; Petta et al., 2005) because many healthcare professionals often fail to associate these symptoms with the possibility of the disease.

Notwithstanding, early diagnosis with proper treatment represents the best strategy for the control of symptoms and to slow down progression of the disease, and surgery, including biopsy of the lesions, is the gold standard for reaching a definitive diagnosis (Dunselman et al., 2014). The benefit of surgery lies in the "see and treat" management strategy to reduce the symptoms of pain. Conversely, its drawback is that it may reduce ovarian reserve in cases in which coagulation is performed and/or ovarian endometriomas are removed (Singh & Suen, 2017). Other disadvantages of recommending surgery for patients with pain symptoms include the high associated costs, the inherent risks associated with anesthesia and surgery, and the risk of pelvic organ injuries (2%) and large vessel injuries (0.01%), in addition to the risk of the lesions recurring in around 40% to 50% of cases over a 5-year period (Kodaman, 2015). Therefore, the disproportionate delay in reaching diagnosis lies in the fact that surgery is an invasive diagnostic test, while physical examination alone does not permit a diagnosis of endometriosis to be reached (Eskenazi et al., 2001).

Imaging methods have been proposed as complementary tests to the initial investigation in order to assess the site and extent of endometriotic lesions and as a means of counteracting the high costs and risks associated with surgery. Transvaginal ultrasound and transabdominal ultrasound, with or without prior bowel preparation, can be used to investigate endometriomas and sites of deep endometriosis in the anterior and posterior compartments of the pelvis. In addition, these ultrasound techniques are useful when screening cases for further assessment with magnetic resonance imaging (MRI) (Haas, Shebl, Shamiyeh, & Oppelt, 2013; Johnson et al., 2017).

Endometriomas are more easily identified by imaging methods, although it can be difficult to distinguish between endometriomas and ovarian tumors. When located in the posterior pelvic compartment, deep endometriosis is generally multifocal (Kinkel, 1999). The principal regions affected by deep endometriosis are the uterosacral ligaments, rectovaginal septum, vaginal wall, pouch of Douglas, and the rectosigmoid, all of which form part of the posterior compartment of the pelvis, the most common site of the deep form of the disease. Deep endometriosis involving the bladder, ureters, and/or the anterior compartment of the pelvis is much less common (Nisenblat et al., 2016) and it is even more unusual for the disease to affect the diaphragm or lungs (Gordts, Koninckx, & Brosens, 2017).

Classification

The American Society for Reproductive Medicine (ASRM) classifies endometriosis into four stages based on the appearance and depth of the lesions, whether they are present bilaterally, and whether adhesions are identified during surgery. These stages are: I/minimal disease; II/mild disease; III/moderate disease and IV/severe disease (American Society for Reproductive Medicine, 1996). From a histological viewpoint, ectopic endometrial lesions can be purely glandular, purely stromal, or mixed. With respect to the site and depth of the lesions, endometriosis can present in at least three different forms: peritoneal, ovarian, and deep infiltrating endometriosis in which tissue infiltration exceeds 5 mm in depth (Koninckx, Ussia, Adamyan, Wattiez, & Donnez, 2012; Nisolle & Donnez, 1997).

Treatment

Treatment should be individualized, taking into consideration the patient's age, the presence of pelvic pain, their quality of life, reproductive wishes, and the site of the disease (Dunselman et al., 2014; Singh & Suen, 2017). Treatment can be surgical and/or clinical, with both aimed at reducing the size of the lesions and the inflammatory process characteristic of the disease. The objective of clinical treatment is to reduce the proliferative action of estrogen on endometriotic lesions and suppress or reduce menstruation. Combined hormonal contraceptives, progestin-only methods, and GnRH agonists can be used for this purpose. With surgical treatment, the direct goal is cytoreduction (Dunselman et al., 2014).

Biomarkers

Since endometriosis is a complex disease with heterogeneous phenotypic and clinical presentations, since the etiology of the disease remains unclear, and since it can affect women's quality of life and potentially impair their fertility, a great number of studies have focused on identifying a noninvasive biomarker or a panel of biomarkers that would be useful for early diagnosis. However, the main issue is to determine the relationship between the biomarker/panel and the clinical parameters of endometriosis, since it is impossible to determine the incidence of the disease. To identify a biomarker with high sensitivity, its positive predictive value has to be high. The Biomarkers Definitions Working Group of the United States National Institutes of Health defines a biomarker as "*a characteristic that is objectively measured and evaluated as an indicator of normal biological processes, pathogenic processes, or pharmacologic responses to a therapeutic intervention*" (Atkinson et al., 2001).

Noninvasive biomarkers have been evaluated in systematic reviews since 2010, involving studies on potential candidates in serum, urine, and in tissue samples from the eutopic and ectopic endometrium (Fassbender, Burney, Dorien, D'Hooghe, & Giudice, 2015; Gupta et al., 2016; Liu et al., 2015; May et al., 2010; Nisenblat et al., 2016). However, up to this moment, there is insufficient evidence that the sensitivity and specificity of any noninvasive biomarker or of any panel of noninvasive biomarkers are sufficiently high to enable them to be used as a diagnostic tool or for monitoring patients after treatment.

Cochrane reviews (Fassbender et al., 2015; Gupta et al., 2016; Liu et al., 2015; May et al., 2010; Nisenblat et al., 2016) have analyzed several studies on noninvasive biomarkers. These biomarkers are described here and listed in the attached tables, which include the number of the relevant studies and year of publication, as well as the sensitivity and specificity of the biomarker.

The following paragraphs provide a brief explanation of each biomarker investigated and its function, according to its category. The tables below list the studies included in the present review, with sample sizes, number of studies, year of publication, and whether the biomarker is clinically applicable, according to previous Cochrane reviews (Gupta et al., 2016; Liu et al., 2015; Nisenblat et al., 2016). In those systematic reviews, however, there were significant methodological differences in the studies evaluating the same biomarkers.

Blood biomarkers

Because blood samples are simple to obtain and minimally invasive, requiring only peripheral intravenous access, attempts to identify biomarkers have been mostly conducted using serum and plasma samples. The most recent metaanalysis (Nisenblat et al., 2016) lists the blood biomarkers already evaluated for which differential expressions in women with endometriosis were identified (Fassbender et al., 2015; May et al., 2010; Nisenblat et al., 2016).

Biomarkers of angiogenesis, growth factors, and growth factor receptors

Glycodelin-A is found in the endometrial glands in the secretory phase and may be associated with neovascularization and cell proliferation. It has an immunosuppressive effect and may inhibit natural killer (NK) cell activity (Kocbek, Vouk, Mueller, Rižner, & Bersinger, 2013; Nisenblat et al., 2016). Growth hormone is the main regulator of insulin-like growth factor-binding protein-3 (Nisenblat et al., 2016). The hormone leptin, which affects metabolism and plays a role in metabolic regulation and obesity, may also regulate angiogenesis and immune response (Nisenblat et al., 2016; Viganò et al., 2002). Vascular endothelial growth factor (VEGF) may be the major stimulus for angiogenesis and for the increased vascular permeability observed in endometrial tissue invasion (Mohamed, El Behery, & Mansour, 2013). The peptide urocortin can increase uterine contractility induced by endometrial prostaglandins (Wu, Yuan, Larauche, Wang, & Million, 2013).

Apoptosis markers

Annexin-V is a cellular protein used to detect apoptotic cells. It binds to the cell plasma membrane by interacting with phosphatidylserine (Schutte, Nuydens, Geerts, & Ramaekers, 1998). Survivin is an essential protein for cell division can inhibit cell death and can be expressed only in cells undergoing active proliferation (Wheatley & Altieri, 2019) (Table 1).

Cell adhesion molecules and other matrix-related proteins

The soluble form of the intercellular adhesion molecule-1 (sICAM-1) is expressed on the surface of various types of cells including leukocytes and endothelial cells. This glycoprotein belongs to a superfamily involved in the adhesion of neutrophils to endothelial cells and in the extravascular migration of neutrophils (Lawson, Ainsworth, Yacoub, & Rose, 1999); however, the data available are as yet insufficient to enable it to be included as a biomarker. Laminin-1 (LN-1) is a protein that appears to be important in epithelial development and is limited to endometrial basal membranes, possibly playing a role in endometriosis (Ekblom, Lonai, & Talts, 2003); however, the data areas yet insufficient to confirm it as a biomarker.

Cytoskeleton molecules

No differential expression was found forcytokeratin-19 in endometriosis.

Molecules involved in DNA repair/telomere maintenance

Telomere length in peripheral blood cells has been reported to remain unchanged in endometriosis (Hapangama et al., 2008; Kuessel et al., 2014) (Table 1).

High-throughput molecular markers

The metabolome is the quantitative measure of the metabolic response of a biological system after pathophysiological stimuli or genetic modifications (Nicholson, Lindon, & Holmes, 1999). Some studies on the proteome and metabolome serum profile of women with endometriosis have reported a high expression of various proteins with immunological characteristics or that are activated in inflammatory responses (Seeber et al., 2010).

Hormonal markers

In relation to hormones, only prolactin has been reported as a potential inducer of angiogenesis, and it has been studied following the finding that prolactin levels are higher in women with endometriosis. However, the results are insufficient to allow any conclusions to be reached regarding the benefit of analyzing prolactin levels in the diagnosis of endometriosis or as a means of monitoring cases of the disease (Bilibio et al., 2014; Laganà et al., 2020) (Table 1).

Immune system and inflammatory biomarkers

Some studies have detected an increase in autoantibodies, chemokine recruitment, monocytes, neutrophils, and lymphocytes in patients with endometriosis, through a proinflammatory response (Ozhan, Kokcu, Yanik, & Gunaydin, 2014). With respect to the immune cells, neutrophils, the neutrophil-lymphocyte ratio, and white blood cells, which are used to predict prognosis in some types of cancer, have also been evaluated in relation to endometriosis; however, there is no evidence of their clinical applicability in endometriosis (Yavuzcan et al., 2013). Interleukins (IL-1β, IL-4, IL-6, and IL-8) and other immune and inflammatory markers such as soluble CD23 (Margatho, Carvalho, & Bahamondes, 2020; Margatho, Mota Carvalho, Eloy, & Bahamondes, 2018) and copeptin, a surrogate marker of vasopressin, all of which participate in the regulation of the immune system, have also been found to be of limited value in endometriosis. C-reactive protein (CRP) increases in response to situations of acute inflammation, inflammatory diseases, and trauma; however, although studies have been conducted to investigate CRP in endometriosis using high-sensitivity CRP tests, no differences have been found in CRP expression levels in endometriosis (Akdis et al., 2016; Bedaiwy, 2002; Ozhan et al., 2014; Podgaec et al., 2007; Tuten et al., 2014).

TABLE 1 Some blood biomarkers studied for diagnosis of endometriosis.

Test	N of cases (Studies)	Year	Sensibility (CI 95%)	Specificity (CI 95%)	Comments
1. Angiogenesis and growth factors and their receptors					
Glycodelin-A cut-off threshold >2.07 ng/mL	99 (1)	2013	0.82	0.79	Without evidence for its potential clinical use
Glycodelin cut-off threshold >9.0 ng/mL	45 (1)	2012	0.71	0.35	
Glycodelin cut-off threshold >18 ng/mL	99 (1)	2012	0.62	0.44	
IGFBP-3 (insulin-like growth factor-binding protein-3) cut-off threshold >200 ng/mL	45 (1)	2012	0.71	0.29	
IGFBP-3 cut-off threshold >210 ng/mL	99 (1)	2012	0.55	0.44	
VEGF (vascular endothelial growth factor) cut-off threshold >1.5 pg/mL	99 (1)	2012	0.50	0.61	
VEGF cut-off threshold >236 pg/mL	95 (1)	2012	0.92	0.77	
VEGF cut-off threshold >680 pg/mL	60 (1)	2013	0.93	0.97	
Urocortin cut-off threshold >29 pg/mL	80 (1)	2007	0.97	0.85	
Urocortin cut-off threshold >33 pg/mL	80 (1)	2007	0.88	0.90	
Urocortin cut-off threshold >41.6 pg/mL	88 (1)	2011	0.88	0.90	
2. Apoptosis markers					
Survivin cut-off threshold not reported	60 (1)	2012	0.07	0.90	Without evidence for its potential clinical use
3. Cell adhesion molecules and other matrix- related proteins					
sICAM-1 (soluble form of intercellular adhesion molecule-1) cut-off threshold <243 ng/mL	99 (1)	2012	0.55	0.50	Without evidence for its potential clinical use
sICAM-1 cut-off threshold <254.6 ng/mL	28 (1)	2012	0.73	0.29	
sICAM-1 cut-off threshold >241.46 µg/mL	60 (1)	2006	0.60	0.87	
LN-1 (laminin-1) cut-off threshold >1110.0 pg/mL	73 (1)	2014	0.72	0.70	

4. High- throughput molecular markers

Metabolome by ESIMS/MS (SMOH C16:1 + PCaa C36:2/ PCae C34:2)	92 (1)	2012	0.90	0.85	Without evidence for its potential clinical use
Proteome by SELDITOF-MS (3 peaks with the MW 3956.00, 11,710.00 and 6986.00 Da)	31 (1)	2009	0.88	0.80	
Proteome by SELDITOF MS (5 peaks with MW 4159.00,5264.00, 5603.00,9861.00 and 10,533.00 Da)	90 (1)	2009	0.78	0.59	
Proteome by SELDITOF MS (5 peaks with MW 9926.31, 10,072.2, 6753.04, 4302.67, 9328.49 Da)	67 (1)	2012	0.40	0.82	
Proteome by SELDITOF MS (5 peaks with MW 2831. 02, 7554.66, 4241.29, 2953.25, 9927.73 Da)	98 (1)	2012	0.38	0.85	
Proteome by SELDITOF MS (5 peaks with MW 11,366.3, 5712.69, 10,070. 7, 3017.68, 3824.44 Da)	88 (1)	2012	0.53	0.82	
Proteome by SELDITOF-MS (6 peaks with MW1629, 3047, 3526,3774, 5046 and 5068 Da)	139 (1)	2010	0.66	0.99	

5. Hormonal markers

Prolactin cut-off threshold >14.8 ng/mL	97 (1)	2014	0.44	0.94	Without evidence for its potential clinical use
Prolactin cut-off threshold >20 ng/mL	97 (1)	2014	0.21	1.00	

6. Immune system and inflammatory markers

Anti-endometrial Abs	759 (4)	1991 1994 1996 2007	0.81	0.75	Without evidence for its potential clinical use
Anti-endometrial Abs (MW of 26/ 34/ 42 kd)	36 (1)	1993	1.00	0.39	
Anti-laminin auto Abs	68 (1)	2003	0.40	0.88	

Continued

TABLE 1 Some blood biomarkers studied for diagnosis of endometriosis—cont'd

Test	N of cases (Studies)	Year	Sensibility (CI 95%)	Specificity (CI 95%)	Comments
sCD23 (soluble CD23)	200 (3)	1996 2018 2020	0.25	0.93	
MCP-1 (monocyte chemotactic protein-1) cut-off threshold >100 pg/mL	101 (1)	1996	0.65	0.61	
Copeptin cut-off threshold >251.2 pg/mL	87 (1)	2014	0.65	0.58	
hs-CRP (high sensitive C-reactive protein) cut-off threshold >0.62 mg/L	295 (1)	2011	0.62	0.56	
hs-CRP cut-off threshold >0.73 mg/L	60 (1)	2011	0.68	0.47	
hs-CRP cut-off threshold >0.61 mg/L	119 (1)	2011	0.54	0.50	
hs-CRP cut-off threshold >438 µg/mL	95 (1)	2012	0.83	0.87	
hs-CRP cut-off threshold >0.70 mg/L	116 (1)	2011	0.59	0.64	
hs-CRP	116 (1)	2010	0.41	0.71	
IFN-γ (interferon gamma) cut-off threshold <76 pg/mL	45 (1)	2012	0.68	0.65	
MIF (macrophage migration inhibitory factor) cut-off threshold >0.57 ng/mL	93 (1)	2005	0.65	0.66	
TNF-α (tumor necrosis factor alpha) cut-off threshold >12.45 pg/mL	95 (1)	2012	0.89	0.87	
TNF-α cut-off threshold <45.6 pg/mL	45 (1)	2012	0.68	0.35	
TNF-α cut-off threshold not reported	116 (1)	2010	0.79	0.74	
Neutrophils cut-off threshold >4058 cells/mL	100 (1)	2013	0.68	0.60	
NLR (neutrophil-to lymphocyte ratio) cut-off threshold >2.19	100 (1)	2013	0.76	0.82	
WBC (white blood cells) cut-off threshold >6400 cells/mL	100 (1)	2013	0.64	0.54	
IL-1β (interleukin-1beta) cut-off threshold <0.9 pg/mL	45 (1)	2012	0.82	0.35	
IL-4 cut-off threshold ≥3 pg/mL	50 (1)	2012	0.64	0.65	
IL-6 cut-off threshold >1.03 pg/mL	138 (1)	2008	0.81	0.51	
IL-6 cut-off threshold >15.4 pg/mL	78 (1)	2013	0.89	0.82	
IL-6 cut-off threshold >25.75 pg/mL	84 (1)	2007	0.73	0.83	
IL-6 cut-off threshold not specified	116 (1)	2010	0.59	0.76	

IL-8 cut-off threshold >24 pg/mL	101 (1)	2013	0.76	0.73	
IL-8 cut-off threshold >25 pg/mL	91 (1)	2008	0.71	0.81	
IL-8 cut-off threshold not specified	116 (1)	2010	0.49	0.71	
IL-6 cut-off threshold >15.4 pg/mL	78 (1)	2013	0.89	0.82	
Anti-endometrial Abs	759 (4)	1991 1994 1996 2007	0.81	0.75	
Anti-endometrial Abs (MW of 26/ 34/ 42 kd)	36 (1)	1993	1.00	0.39	
Anti-laminin auto Abs	68 (1)	2003	0.40	0.88	
sCD23 (soluble CD23)	200 (3)	1996 2018 2020	0.25	0.93	
MCP-1 (monocyte chemotactic protein-1) cut-off threshold >100 pg/mL	101 (1)	1996	0.65	0.61	
7. Other peptides and proteins shown to influence key events implicated in endometriosis					
Follistatin cut-off threshold >1433 pg/mL	104 (1)	2009	0.92	0.92	Without evidence for its potential clinical use
STX-5 (syntaxin-5) cut-off threshold >55 ng/mL	80 (1)	2014	0.78	0.70	
PON-1(paraoxonase-1) cut-off threshold <141.5 U/L					
Thiols cut off threshold <396 U/L	108 (1)	2014	0.73	0.80	Without evidence for its potential clinical use
Follistatin cut-off threshold >1433 pg/mL	104 (1)	2009	0.92	0.92	
STX-5 (syntaxin-5) cut-off threshold >55 ng/mL	80 (1)	2014	0.78	0.70	
8. Posttranscriptional regulators of gene expression (microRNAs)					
miR-9[a]cut-off threshold not specified	85 (1)	2013	0.68	0.96	Without evidence for its potential clinical use
miR-17-5 cut-off threshold <0.9057	40 (1)	2013	0.70	0.70	
miR-20[a] cut-off threshold <0.6879	40 (1)	2013	0.60	0.90	
miR-22 cut-off threshold <0.5647	40 (1)	2013	0.90	0.80	

Continued

TABLE 1 Some blood biomarkers studied for diagnosis of endometriosis—cont'd

Test	N of cases (Studies)	Year	Sensibility (CI 95%)	Specificity (CI 95%)	Comments
miR-122mcut-off threshold not specified	85 (1)	2013	0.80	0.76	
miR-141[a] cut-off threshold not specified	85 (1)	2013	0.72	0.96	
miR-145[a] cut-off threshold not specified	85 (1)	2013	0.70	0.96	
miR-199[a] cut-off threshold not specified	85 (1)	2013	0.78	0.76	
miR-532-3p cut-off threshold not specified	85 (1)	2013	0.80	0.92	
10. Tumor markers					
CA-15.3 (cancer antigen-15.3) cut-off threshold >15.04 U/mL	88 (1)	2014	0.65	0.62	Without evidence for its potential clinical use
CA-15.3 cut-off threshold >30 U/mL	119 (1)	1992	0.04	0.92	
CA-19.9 (cancer antigen-19.9) cut-off threshold >7.5 IU/mL	76 (1)	2012	0.73	0.56	
CA-19.9 cut-off threshold >9.5 IU/mL	198 (1)	2012	0.55	0.58	
CA-19.9 cut-off threshold >10.67 IU/mL	88 (1)	2014	0.65	0.62	
CA-19.9 cut-off threshold ≥12 IU/mL	119 (1)	1996	0.62	0.70	
CA-19.9 cut-off threshold >37 IU/mL	330 (3)	2002 2004 2009	0.36	0.87	
CA-19.9 cut-off threshold not specified	60 (1) 116 (1)	2010 2012	0.53 0.36	0.90 0.71	
CA-72 (cancer antigen-72) cut-off threshold >4 U/mL	35 (1)	1994	0.05	0.75	
CA-72 cut-off threshold >6 U/mL	119 (1)	1992	0.09	0.89	
CA-125 (cancer antigen-125) cut-off threshold >10–14.7 U/mL	733 (5)	2003 2007 2012	0.70	0.64	
CA-125 cut-off threshold >11.5 U/mL	45 (1)	2012	0.86	0.65	
CA-125 (cancer antigen-125) cut-off threshold >13.5 U/mL	35 (1)	2012	0.79	0.31	
CA-125 cut-off value >16–17.6 U/mL	430 (5)	1989 1990 1991 1994 2012	0.56	0.91	

CA-125 cut-off value >20 IU/mL	1304 (6)	1994, 1999, 2005, 2007, 2011, 2014	0.67	0.69
CA-125 cut-off value >25–26 U/mL	963 (3)	1996, 2005, 2014	0.73	0.70
CA-125 cut-off value >30–33 U/mL (1 study >33 U/mL)	1206 (6)	1989, 1996, 2005, 2007, 2013, 2014	0.62	0.76
CA-125 cut-off value >35–36 U/mL (1 study >36 U/mL)	3550 (29)	1986, 1989, 1991, 1993, 1994, 1995, 1996, 1998, 2002, 2003, 2005, 2007, 2009, 2013, 2018, 2020	0.40	0.91
CA-125 cut-off value >42 U/mL	104 (1)	2009	0.44	0.90
CA-125 cut-off value >43 U/mL	63 (1)	2014	1.00	0.80
CA-125 cut-off value not specified	59 (1)	2010	0.72	0.79
	119 (1)	2010	0.65	0.72
	60 (1)	2012	0.82	0.90
	116 (1)	2010	0.68	0.71

[a]Data obtained from "Blood biomarkers for the noninvasive diagnosis of endometriosis. Cochrane Database of Systematic Reviews (Nisenblat et al., 2016).

Oxidative stress markers

Carbonyls, paraoxonase-1, and thiols were evaluated in no more than one study each. There ported sensitivity and specificity for carbonyls were 94% and 51%, respectively. Further studies are required to determine the role of oxidative stress markers in endometriosis (Rosa e Silva et al., 2014) (Table 1).

Posttranscriptional regulators of gene expression

MicroRNAs or miRNAs are the most widely studied class of small noncoding RNA. They are able to modulate up to 60% of the protein-encoding genes in the human genome at translational level. With a role in gene regulation, miRNAs are involved in physiological processes; however, their dysregulation has been associated with several pathological disorders. The evaluation of miRNAs may lead to the identification of a potentially highly accurate marker for the diagnosis of endometriosis; nevertheless, these markers need additional validation in a large and well-defined population of women, including women with all the different stages of endometriosis (Wang, Ya-Nan, Bo-Wei, & Hong Shun-Jia, 2013).

Tumor markers

Tumor markers are glycoproteins usually produced by glandular cells. These markers have been widely studied in endometriosis, with different cut-off point shaving been established to evaluate the accuracy of diagnostic tests. However, no set of tumor markers has yet been identified with sensitivity and specificity that are high enough to constitute a sufficiently accurate test for the diagnosis of endometriosis. CA125 (cancer antigen 125) is the most widely studied biomarker in endometriosis. It is detected in the epithelium of normal and neoplastic coelomic origin such as the endometrial epithelium, the endocervical epithelium, the Fallopian tube epithelium, and the epithelial cells of ovarian cancer (Niloff, Knapp, Schaetzi, Reynolds, & Bast, 1984). Many studies have been carried out to assess the usefulness of CA125 for the diagnosis of endometriosis and to monitor patients during treatment, particularly in women with endometriomas (Abrão, Podgaec, Pinotti, & De Oliveira, 1999; May et al., 2010; Margatho et al., 2018, 2020). Other tumor markers evaluated include CA-15.3 (cancer antigen 15.3), which can be found in the breast epithelium, particularly in breast cancer, CA-19.9 (cancer antigen 19.9), the levels of which may be higher in pancreatic and rectal cancer, and CA-72 (cancer antigen-72), which may be present in cases of endometrial adenocarcinoma. As with CA125, these glycoproteins have also been studied in women with endometriosis; however, they cannot be considered promising biomarkers for diagnosis of the disease (Table 1).

Urine biomarkers

Since urine is noninvasively accessible and less complex to analyze than blood, there has been an active search for biomarkers in this organic human waste material. For many years, the analysis of urinary biomarkers was limited to renal pathologies, and to bladder and prostate cancer. However, 70% of urinary peptides come from the kidney and 30% from circulation. In addition, plasma filtration may reflect the most abundant proteins in urine. Recent studies have shown that it may be possible to identify urine biomarkers for systemic diseases (Yun et al., 2014).

The following is a description of biomarkers that have already been evaluated in urine for endometriosis, according to the 2015 Cochrane review (Liu et al., 2015). However, there were no statistically significant differences between women with and without endometriosis. The urinary biomarkers evaluated in the studies included enolase 1, which is an enzyme associated with inflammation, vitamin D-binding protein (VDBP) (Cho et al., 2012) and proteomics analysis using matrix-assisted laser desorption/ionization time-of-flight mass spectrometry (MALDI-TOF-MS). Only a few studies have evaluated the performance of proteomics for the detection of endometriosis in its different clinical presentations; however, one study failed to find any conclusive evidence regarding the role of these techniques in endometriosis (El-Kasti et al., 2011). The evaluation of cytokeratin-19 fragments, VEGF, VEGF-A, and tumor necrosis factor alpha (TNFα), even when studied individually, suggested that these markers are not reliable for use in a diagnostic test for endometriosis (Table 2).

Endometrial biomarkers

The identification of endometrial biomarkers in biopsy specimens represents another alternative, although the process involved is more invasive compared to the use of peripheral blood or urine samples. On the other hand, the endometrium has a unique peculiarity in that it responds differently to sex hormones at the different stages of the menstrual cycle. In addition, biomarkers already studied in the endometrium have been found to exhibit differential expressions in women with endometriosis (Gupta et al., 2016).

Biomarkers in endometriosis Chapter | 46 519

TABLE 2 Some urinary biomarkers studied for diagnosis of endometriosis.a

Test	N of cases (Studies)	Year	Sensibility (CI 95%)	Specificity (CI 95%)	Comments
Enolase 1 (NNE) cut-off >0.96 ng/mgCr	1(59)	2014	0.56	0.70	Without evidence for its potential clinical use
Vitamin D binding protein (VDBP) cut-off >87.83 ng/mgCr	1(95)	2012	0.58	0.55	Without evidence for its potential clinical use
MALDI-TOF-MS proteomics Proteome: peptide m/z 1824.3 Da cut-off $\geq$29.34 au	1(28)	2011	0.77	0.73	Without evidence for its potential clinical use
Proteome: peptide m/z 1767.1 Da cut-off $\geq$35.22 au	1(27)	2011	0.75	0.87	
Proteome: peptide m/z 2052.3 Da cut-off not reported	1(122)	2014	0.83	0.69	
Proteome: peptide m/z 3393.9 Da cut-off not reported	1(122)	2014	0.85	0.71	
Proteome: peptide m/z 1579.2 Da [collagen alpha 6(IV) chain precursor] cut-off not reported	1(122)	2014	0.83	0.69	
Proteome: peptide m/z 891.6 Da [collagen alpha1 chain precursor] cut-off not reported	1(122)	2014	0.82	0.65	
Proteome: 5 peptides m/z 1433.9 + 1599.4 + 2085.6 + 6798.0 + 3217.2 Da cut-off not reported	1(25)	2014	0.91	0.93	
Cytokeratin-19 fragments cut-off >5.3 ng/mL	1(98)	2014	0.11	0.94	Without evidence for its potential clinical use
Vascular endothelial growth factor (VEGF) or vascular endothelial growth factor-A (VEGF-A)	1(62) 1(70)	2004 2007			Without evidence for its potential clinical use
Tumor necrosis factor-alpha (TNFα)	1(70)	2007			Without evidence for its potential clinical use

^{a}Data obtained from "Urinary biomarkers for the noninvasive diagnosis of endometriosis. Cochrane Database of Systematic Reviews (Yun et al., 2014).

Angiogenesis and growth factors/cell adhesion molecules/DNA repair molecules/endometrial and mitochondrial proteome/posttranscriptional regulators of gene expression

The endometrial biomarkers studied have involved angiogenesis and growth factors such as prokineticin 1 (PROK1), cell adhesion molecules such as α6 integrin, α3β1 integrin, α4β1 integrin and β1 integrin, DNA repair and telomere maintenance molecules such as human telomerase reverse transcriptase (hTERT) and high-throughput molecular markers (endometrial proteome). The expression of endometrial microRNAs may be different in women with endometriosis as well as in endometriomas when compared to the eutopic endometrium. However, further studies and validations need to be carried out in the different phases of the menstrual cycle before endometrial microRNAs can be considered useful for diagnosis (Kuokkanen et al., 2010).

Hormonal markers

Hormonal markers such as estrogen receptor alpha (ERα) and estrogen receptor beta (ERβ) have also been evaluated in the endometrium of women with endometriosis. The study that evaluated these biomarkers failed, however, to yield any satisfactory results. The studies conducted to evaluate the performance of aromatase cytochrome P450 (CYP19) gene expression were heterogeneous, and, individually, failed to achieve sufficient sensitivity and specificity to enable this

520 PART | IV Assessments, screening, and resources

marker to be used as a screening test for endometriosis (Gupta et al., 2016). Another study evaluated the role of 17 beta-hydroxysteroid dehydrogenase type 2 at different stages of the cycle; however, those authors failed to find any statistically significant differences.

Inflammatory and myogenic markers

Interleukin-1 receptor type II and myogenic markers (caldesmon) have been analyzed as immune system and inflammatory markers in the endometrium. Although these markers are promising, evidence for diagnostic accuracy in a clinical setting remains insufficient.

Neural markers

After the role of neuroangiogenesis in endometriosis was established, initial studies were conducted to evaluate neural biomarkers and some studies were performed based on the evaluation of neural biomarkers and dysmenorrhea in women with endometriosis (Tamburro et al., 2003). Protein gene product (PGP) 9.5 is a highly specific pan-neuronal marker, which was initially described in peritoneal endometriotic lesions. The density of this marker was analyzed in the eutopic endometrium, with results showing greater density in women with endometriosis compared to women without the disease (Tokushige, Markham, Russell, & Fraser, 2006). However, the increase in nerve fiber density may also occur in other uterine pathologies such as fibroids and adenomyosis (Ellett et al., 2015; Newman et al., 2013).

Tumor markers

Only one study compared CA125 levels in menstrual fluid with serum CA125 levels in women with different stages of endometriosis, with reported sensitivity and specificity of 66% and 89%, respectively (Table 3).

TABLE 3 Some endometrial biomarkers studied for diagnosis of endometriosis.a

Test	N of cases (Studies)	Year	Sensibility (CI 95%)	Specificity (CI 95%)	Comments
1. Angiogenesis and growth factors					
Prot-1 mRNA (prokineticin 1)	1 (24)	2010	0.67	0.83	Without evidence for its potential clinical use
2. Cell adhesion molecules					
depolarized α-6 integrin (glandular)	1(49)	2006	0.67	0.84	Without evidence for its potential clinical use
α3β1 integrin (glandular)	1(32)	2003	1.00	0.27	
α3β1 integrin (stroma)	1(32)	2003	0.53	0.27	
α4β1 integrin (glandular)	1(32)	2003	0.65	0.40	
α4β1 integrin (stroma)	1(32)	2003	0.59	0.20	
β1 integrin (glandular)	1(32)	2003	0.18	0.87	
β1 integrin (stroma)	1(32)	2003	0.76	0.00	
3. DNA repair and telomere maintenance molecules					
hTERT mRNA	1(69)	2014	0.28	0.80	Without evidence for its potential clinical use
4. High-through put molecular markers					
Endometrial proteome	1(27)	2012	0.88	0.80	Without evidence for its potential clinical use
	1(26)	2010	0.92	0.92	
Mitochondrial proteome	1(53)	2010	0.88	0.86	

Biomarkers in endometriosis Chapter | 46 521

TABLE 3 Some endometrial biomarkers studied for diagnosis of endometriosis.a</ce:cross-ref—cont'd

Test	N of cases (Studies)	Year	Sensibility (CI 95%)	Specificity (CI 95%)	Comments
5. Hormonal markers					
CYP19 (aromatase P450)	8(444)	2002 2004 2005 2006 2007 2008 2011	0.77	0.74	Without evidence for its potential clinical use
17βHSD2 mRNA	1(53)	2008	0.53	0.91	
Estrogen Receptor-α (glandular)	1(90)		0.73	0.43	
Estrogen Receptor-α (stroma)	1(90)		0.77	0.50	
Estrogen Receptor-β (glandular)	1(90)		0.67	0.63	
Estrogen Receptor-β (stroma)	1(90)		0.63	0.73	
6. Immune System and inflammatory markers					
IL-1R2 mRNA (glandular)	1(31)	2008	1.00	0.53	Without evidence for its potential clinical use
IL-1R2 mRNA (stroma)	1(32)	2008	0.93	0.76	
IL-1R2 mRNA (glandular secretory)	1(19)	2008	1.00	0.67	
IL-1R2 mRNA (stroma secretory)	1(20)	2008	0.90	0.90	
7. Myogenic markers					
Caldesmon (proliferative)	1(35)	2013	0.95	1.00	Without evidence for its potential clinical use
Caldesmon (secretory)	1(35)	2013	0.90	0.93	
CALD1 mRNA (proliferative)	1(35)	2013	0.60	0.87	
CALD1 mRNA (secretory)	1(35)	2013	0.75	0.67	
8. Nerve sheath and nerve growth markers					
PGP 9.5	8(429) 7(361)	2007 2009 2011 2013	0.96	0.86	Without evidence for its potential clinical use
9. Tumor markers					
CA 125 (menstrual fluid)	1(104)	1990	0.66	0.89	Without evidence for its potential clinical use

[a]*Data obtained from Endometrial biomarkers for the noninvasive diagnosis of endometriosis.* Cochrane Database of Systematic Reviews *(Gupta et al., 2016).*

Discussion

The expression of the noninvasive biomarkers evaluated may differ as a function of the different phenotypic presentations of endometriosis (peritoneal, ovarian, and deep infiltrating endometriosis) and, consequently, findings may vary. The gold-standard techniques for the diagnosis of endometriosis are laparoscopy and biopsy of the lesions. A systematic review reported sensitivity and specificity of 94% and 79%, respectively, for the accuracy of laparoscopy as confirmed by histology of biopsy specimens (Wykes, Clark, & Khan, 2004). However, the potential of finding a noninvasive biomarker that would eliminate the need to perform surgery and that would have high sensitivity (a low rate of false positives) and high specificity (a low rate of false negatives) would increase accessibility to treatment and assist healthcare professionals in their decision regarding whether or not to indicate surgery.

A systematic review was conducted with a set of studies involving more than 100 biomarkers; however, none was found to have any clinical significance for potential use (May et al., 2010). There is immunological and inflammatory dysfunction at the site of endometriotic lesions and it is uncertain whether endometriosis develops in an environment in which the inflammatory response is exacerbated, promoted by the continuous presence of retrograde menstruation, or if the onset of the disease itself promotes an increase in local inflammatory response. Many studies have reported high levels of inflammatory cytokines, inflammatory response factors, both in peritoneal fluid and in serum, as well as in endometrial tissue in ectopic foci outside the uterus and in the eutopic endometrium. Endometriotic implants provoke a local inflammatory response and immune dysfunction, leading to an increase in the substances involved in this process (Giudice & Kao, 2004), including growth factors, hormones, proteolytic enzymes, autoantibodies, glycoproteins, and adhesion molecules. Moreover, some studies have shown that in the endometrium of women with endometriosis there is a pattern of molecular expression that can facilitate invasion and progression to its ectopic manifestation (Ahn, Singh, & Tayade, 2017; Burney et al., 2007). Many biological markers have been investigated in an attempt to identify correlations, improve the diagnosis of endometriosis and to monitor patients in treatment (Fassbender et al., 2015; May et al., 2010). Several of these biomarkers (inflammatory cytokines, angiogenesis factors) are the final or intermediate products of inflammatory response. Menstruation itself is an inflammatory process, as is endometriosis, and, based on Sampson's theory, the eutopic endometrium would be the source of ectopic implantation.

In 1984, Niloff et al. found that the levels of the glycoprotein CA125 may be high in 80% of women with ovarian tumors and in 16% of first-trimester pregnant women (Niloff et al., 1984). In 1988, CA125 levels were found to vary as a function of the phase of the menstrual cycle, with a possible increase in women with stage IV endometriosis, but not in women with mild or minimal endometriosis (Masahashi et al., 1988). Since then, CA125 has been the most widely studied serum biomarker (Mol et al., 1998; Margatho et al., 2018, 2020; Nisenblat et al., 2016; Rosa e Silva, Rosa e Silva, & Ferriani, 2007). Mainly cells originating from the normal epithelial cells of the coelomic epithelium during the period of embryonic development produce this high-molecular-weight glycoprotein. Increased CA125 levels have been associated with several gynecological pathologies.

Data from studies included in recent metaanalyses on biomarkers identified in blood, urine, and endometrial tissue samples showed poor specificity and sensitivity; therefore, they should not be used in clinical practice as biomarkers for endometriosis but must be restricted to a research setting (Fassbender et al., 2015; Gupta et al., 2016; Liu et al., 2015; May et al., 2010; Nisenblat et al., 2016).

Some studies have identified an aberrant pattern of molecular expression in the eutopic endometrium of women with endometriosis that may facilitate invasion and implantation. In addition, aberrant regulatory mechanisms may be present in the endometrium of women with endometriosis due to epigenetic changes in response to steroid hormones (Guo, 2009).

The fact that it may take as long as ten years after the onset of symptoms for a diagnosis of endometriosis to be reached (Hadfield, Mardon, Barlow, & Kennedy, 1996) has been widely reported and discussed. Considering that the disease occurs in approximately 10% of women of reproductive age and since endometriosis may affect mental health, with an impact on quality of life, identifying a noninvasive means of diagnosing the disease at an early stage is crucial. Identification of a biomarker or a panel of biomarkers could help when screening women with symptoms suggestive of the presence of endometriosis, thus providing support for the healthcare professional in his/her decision regarding whether or not to recommend surgery. Moreover, the availability of a biomarker could serve to guide clinical treatment, monitoring the success or failure of treatment interventions and of recurrence following surgery. Imaging methods such as MRI and ultrasound with prior bowel preparation (Dunselman et al., 2014; Zondervan Krina et al., 2020) can be useful in evaluating the site of lesions in the anterior and posterior compartments of the pelvis and in identifying endometriomas. However, these techniques are of no use in the diagnosis of peritoneal endometriosis.

Mini-dictionary of terms

- **Biomarker** is a molecule that is measured and can be used to diagnose or monitor the disease.
- **Chronic pelvic pain** is a pain reported by the woman that lasts for more than six months, with involvement of the pelvic site.
- **Coelomic metaplasia** is one of the theories of the pathogenesis of endometriosis. Cells from peritoneum and endometrium derive from the coelomic cells.
- **Dysmenorrhea** is the term used to describe pain and cramps during menstrual bleeding days or some days before. Usually, menstrual cramps stop at the end of the menstrual flow.
- **Endometriosis** is benign estrogen-dependent pathology defined as the presence of endometrial glandular or stromal cells outside the endometrial cavity, provoking a pathologic inflammatory response.

Key facts of endometriosis

- The ancient Egyptians (1855 BCE) represented, through symbols, the suffocation of the womb (allusion to the symptoms of dysmenorrhea and chronic pelvic pain)
- Karl von Rokitansky, a pathologist, was the first to describe microscopically the endometriosis in 1860.
- Jonh Sampson, in his article from 1927, introduced the most accepted theory of the pathogenesis of endometriosis "the retrograde menstruation theory".
- Women with endometriosis were among the first women to use birth control pills (for treatment off label) when they were launched in 1957.
- Endometriosis can occur in around 10% of women in the general population; however, prevalence may reach as high as 50% among infertile women with chronic pelvic pain.

Summary points

- This chapter focus on noninvasive biomarker in women with endometriosis- associated pelvic pain.
- The main symptoms that suggest endometriosis include dysmenorrhea; progressively frequent dysmenorrhea culminating in daily pelvic pain.
- The American Society for Reproductive Medicine (ASRM) classifies endometriosis into four stages based on surgery findings.
- Treatment should be individualized. Hormonal contraceptives, GnRH-analog, and surgery are the most used.
- Noninvasive biomarkers have been evaluated in studies on potential candidates in serum, urine, and in tissue samples from the eutopic and ectopic endometrium.
- The availability of a biomarker could serve to guide clinical treatment, monitoring the success or failure of treatment interventions and of recurrence following surgery.

Applications to other areas

In this chapter, we have reviewed the useful of different biomarkers in the diagnosis and follow-up of treatment of women with endometriosis-associate pelvic pain. We presented in detail many urinary, serum, and endometrial tissue biomarkers described to help clinicians in the identification of women with endometriosis. Unfortunately, up today all the biomarkers described in our chapter cannot achieve high sensibility and specificity to be incorporated in the diagnosis and follow-up of treatment of endometriosis.

References

Abrão, M. S., Podgaec, S., Pinotti, J. A., & De Oliveira, R. M. (1999). Tumor markers in endometriosis. *International Journal of Gynecology & Obstetrics*, 66(1), 19–22.

Ahn, S. H., Monsanto, S. P., Miller, C., Singh, S. S., Thomas, R., & Tayade, C. (2015). Pathophysiology and immune dysfunction in endometriosis. *BioMed Research International*, 1–12.

Ahn, S. H., Singh, V., & Tayade, C. (2017). Biomarkers in endometriosis: Challenges and opportunities. *Fertility and Sterility*, 107(3), 523–532.

524 PART | IV Assessments, screening, and resources

Akdis, M., Aab, A., Altunbulakli, C., Azkur, K., Costa, R. A., Crameri, R., et al. (2016). Interleukins (from IL-1 to IL-38), interferons, transforming growth factor β, and TNF-α: Receptors, functions, and roles in diseases. *Journal of Allergy and Clinical Immunology, 138*(4), 984–1010.

American Society for Reproductive Medicine. (1996). Revised American Society for Reproductive Medicine classification of endometriosis. *Fertility and Sterility, 67*(5), 817–821.

Atkinson, A. J., Colburn, W. A., DeGruttola, V. G., DeMets, D. L., Downing, G. J., Hoth, D. F., et al. (2001). Biomarkers and surrogate endpoints: Preferred definitions and conceptual framework. *Clinical Pharmacology and Therapeutics, 69*(3), 89–95.

Bedaiwy, M. A. (2002). Prediction of endometriosis with serum and peritoneal fluid markers: A prospective controlled trial. *Human Reproduction, 17*(2), 426–431.

Bilibio, J. P., Souza, C. A. B., Rodini, G. P., Andreoli, C. G., Genro, V. K., De Conto, E., et al. (2014). Serum prolactin and CA-125 levels as biomarkers of peritoneal endometriosis. *Gynecologic and Obstetric Investigation, 78*(1), 45–52.

Borghese, B., Zondervan, K. T., Abrao, M. S., Chapron, C., & Vaiman, D. (2017). Recent insights on the genetics and epigenetics of endometriosis. *Clinical Genetics, 91*(2), 254–264.

Buck Louis, G. M., Hediger, M. L., Peterson, C. M., Croughan, M., Sundaram, R., Stanford, J., et al. (2011). Incidence of endometriosis by study population and diagnostic method: The ENDO study. *Fertility and Sterility, 96*(2), 360–365.

Burney, R. O., Talbi, S., Hamilton, A. E., Kim, C. V., Nyegaard, M., Nezhat, C. R., et al. (2007). Gene expression analysis of endometrium reveals progesterone resistance and candidate susceptibility genes in women with endometriosis. *Endocrinology, 148*(8), 3814–3826.

Cho, S. H., Choi, Y. S., Yim, S. Y., Yang, H. I., Jeon, Y. E., Lee, K. E., et al. (2012). Urinary vitamin D-binding protein is elevated in patients with endometriosis. *Human Reproduction, 27*(2), 515–522.

Dunselman, G. A. J., Vermeulen, N., Becker, C., Calhaz-Jorge, C., D'Hooghe, T., De Bie, B., et al. (2014). ESHRE guideline: Management of women with endometriosis. *Human Reproduction, 29*(3), 400–412.

Ekblom, P., Lonai, P., & Talts, J. F. (2003). Expression and biological role of laminin-1. *Matrix Biology, 22*, 35–47.

El-Kasti, M. M., Wright, C., Fye, H. K. S., Roseman, F., Kessler, B. M., & Becker, C. M. (2011). Urinary peptide profiling identifies a panel of putative biomarkers for diagnosing and staging endometriosis. *Fertility and Sterility, 95*(4), 1261–1266.

Ellett, L., Readman, E., Newman, M., McIlwaine, K., Villegas, R., Jagasia, N., et al. (2015). Are endometrial nerve fibres unique to endometriosis? A prospective case-control study of endometrial biopsy as a diagnostic test for endometriosis in women with pelvic pain. *Human Reproduction, 30*(12), 2808–2815.

Eskenazi, B., Warner, M., Bonsignore, L., Olive, D., Samuels, S., & Vercellini, P. (2001). Validation study of nonsurgical diagnosis of endometriosis. *Fertility and Sterility, 76*(5), 929–935.

Fassbender, A., Burney, R. O., Dorien, F. O., D'Hooghe, T., & Giudice, L. (2015). Update on biomarkers for the detection of endometriosis. *BioMed Research International, 2015*, 1–14.

Giudice, L. C., & Kao, L. C. (2004). Endometriosis. *The Lancet, 364*, 1789–1799.

Gordts, S., Koninckx, P., & Brosens, I. (2017). Pathogenesis of deep endometriosis. *Fertility and Sterility, 108*(6), 872–885.

Guo, S. (2009). Epigenetics of endometriosis. *Molecular Human Reproduction, 15*(10), 587–607.

Gupta, D., Hull, M. L., Fraser, I., Miller, L., Bossuyt, P. M. M., Johnson, N., et al. (2016). Endometrial biomarkers for the non-invasive diagnosis of endometriosis. *Cochrane Database of Systematic Reviews, 4*, 1–231.

Haas, D., Shebl, O., Shamiyeh, A., & Oppelt, P. (2013). The rASRM score and the Enzian classification for endometriosis: Their strengths and weaknesses. *Acta Obstetricia et Gynecologica Scandinavica, 92*, 3–7.

Hadfield, R., Mardon, H., Barlow, D., & Kennedy, S. (1996). Delay in the diagnosis of endometriosis: A survey of women from the USA and the UK. *Human Reproduction, 11*(4), 878–880.

Hapangama, D. K., Turner, M. A., Drury, J. A., Quenby, S., Saretzki, G., Martin-Ruiz, C., et al. (2008). Endometriosis is associated with aberrant endometrial expression of telomerase and increased telomere length. *Obstetrical and Gynecological Survey, 63*(11), 711–713.

Johnson, N. P., Hummelshoj, L., Adamson, G. D., Keckstein, J., Taylor, H. S., Abrao, M. S., et al. (2017). World endometriosis society consensus on the classification of endometriosis. *Human Reproduction*, 1–10.

Kinkel, K. (1999). Magnetic resonance imaging characteristics of deep endometriosis. *Human Reproduction, 14*(4), 1080–1086.

Kocbek, V., Vouk, K., Mueller, M. D., Rižner, T. L., & Bersinger, N. A. (2013). Elevated glycodelin-A concentrations in serum and peritoneal fluid of women with ovarian endometriosis. *Gynecological Endocrinology, 29*(5), 455–459.

Kodaman, P. H. (2015). Current strategies for endometriosis management. *Obstetrics and Gynecology Clinics of North America, 42*, 87–101.

Koninckx, P. R., Ussia, A., Adamyan, L., Wattiez, A., & Donnez, J. (2012). Deep endometriosis: Definition, diagnosis, and treatment. *Fertility and Sterility, 98*, 564–571.

Kuessel, L., Jaeger-Lansky, A., Pateisky, P., Rossberg, N., Schulz, A., Schmitz, A. A. P., et al. (2014). Cytokeratin-19 as a biomarker in urine and in serum for the diagnosis of endometriosis—A prospective study. *Gynecological Endocrinology, 30*(1), 38–41.

Kuokkanen, S., Chen, B., Ojalvo, L., Benard, L., Santoro, N., & Pollard, J. W. (2010). Genomic profiling of MicroRNAs and messenger RNAs reveals hormonal regulation in MicroRNA expression in human Endometrium1. *Biology of Reproduction, 82*(4), 791–801.

Laganà, A. S., Garzon, S., Casarin, J., Raffaelli, R., Cromi, A., Franchi, M., et al. (2020). Know your enemy: Potential role of cabergoline to target neoangiogenesis in endometriosis. *Journal of Investigative Surgery*, 1–2. https://doi.org/10.1080/08941939.2020.1725191.

Lawson, C., Ainsworth, M., Yacoub, M., & Rose, M. (1999). Ligation of ICAM-1 on endothelial cells leads to expression of VCAM-1 via a nuclear factor-κB-independent mechanism. *Journal of Immunology, 162*(5), 2990–2996.

Liu, E., Nisenblat, V., Farquhar, C., Fraser, I., Bossuyt, P. M. M., Johnson, N., et al. (2015). Urinary biomarkers for the non-invasive diagnosis of endometriosis. *Cochrane Database of Systematic Reviews, 12*.

Margatho, D., Carvalho, N. M., & Bahamondes, L. (2020). Endometriosis-associated pain scores and biomarkers in users of the etonogestrel-releasing subdermal implant or the 52-mg levonorgestrel-releasing intrauterine system for up to 24 months. *The European Journal of Contraception & Reproductive Health Care, 25*(2), 133–140.

Margatho, D., Mota Carvalho, N., Eloy, L., & Bahamondes, L. (2018). Assessment of biomarkers in women with endometriosis-associated pain using the ENG contraceptive implant or the 52 mg LNG-IUS: A non-inferiority randomised clinical trial. *The European Journal of Contraception & Reproductive Health Care, 23*(5), 344–350.

Masahashi, T., Matsuzawa, K., Ohsawa, M., Narita, O., Asai, T., & Ishihara, M. (1988). Serum CA 125 levels in patients with endometriosis: Changes in CA 125 levels during menstruation. *Obstetrics and Gynecology, 72*(3), 328–331.

Matsuzaki, S., Canis, M., Pouly, J. L., Rabischong, B., Botchorishvili, R., & Mage, G. (2006). Relationship between delay of surgical diagnosis and severity of disease in patients with symptomatic deep infiltrating endometriosis. *Fertility and Sterility, 86*(5), 1314–1316.

May, K. E., Conduit-Hulbert, S. A., Villar, J., Kirtley, S., Kennedy, S. H., & Becker, C. M. (2010). Peripheral biomarkers of endometriosis: A systematic review. *Human Reproduction Update, 16*(6), 651–671.

Mohamed, M. L., El Behery, M. M., & Mansour, S. A. E. A. (2013). Comparative study between VEGF-A and CA-125 in diagnosis and follow-up of advanced endometriosis after conservative laparoscopic surgery. *Archives of Gynecology and Obstetrics, 287*(1), 77–82.

Mol, B. W. J., Bayram, N., Lijmer, J. G., Wiegerinck, M. A. H. M., Bongers, M. Y., Van Der Veen, F., et al. (1998). The performance of CA-125 measurement in the detection of endometriosis: A meta-analysis. *Fertility and Sterility, 70*(6), 1101–1108.

Newman, T. A., Bailey, J. L., Stocker, L. J., Woo, Y. L., MacKlon, N. S., & Cheong, Y. C. (2013). Expression of neuronal markers in the endometrium of women with and those without endometriosis. *Human Reproduction, 28*(9), 2502–2510.

Nezhat, C., Nezhat, F., & Nezhat, C. (2012). Endometriosis: Ancient disease, ancient treatments. *Fertility and Sterility, 98*(6), S1–S62.

Nicholson, J. K., Lindon, J. C., & Holmes, E. (1999). "Metabonomics": Understanding the metabolic responses of living systems to pathophysiological stimuli via multivariate statistical analysis of biological NMR spectroscopic data. *Xenobiotica, 29*(11), 1181–1189.

Niloff, J. M., Knapp, R. C., Schaetzi, E., Reynolds, C., & Bast, R. C. J. (1984). Ca 125 antigen levels in obstetric and gynecologic patients. *Obstetrics & Gynecology, 64*(5), 703–707.

Nisenblat, V., Bossuyt, P. M. M., Shaikh, R., Farquhar, C., Jordan, V., Scheffers, C. S., et al. (2016). Blood biomarkers for the non-invasive diagnosis of endometriosis. *Cochrane Database of Systematic Reviews, 5.*

Nisolle, M., & Donnez, J. (1997). Peritoneal endometriosis, ovarian endometriosis, and adenomyotic nodules of the rectovaginal septum are three different entities. *Fertility and Sterility, 68*(4), 585–596.

Ozhan, E., Kokcu, A., Yanik, K., & Gunaydin, M. (2014). Investigation of diagnostic potentials of nine different biomarkers in endometriosis. *European Journal of Obstetrics and Gynecology and Reproductive Biology, 178*, 128–133.

Petta, C. A., Ferriani, R. A., Abrao, M. S., Hassan, D., Rosa e Silva, J. C., Podgaec, S., et al. (2005). Randomized clinical trial of a levonorgestrel-releasing intrauterine system and a depot GnRH analogue for the treatment of chronic pelvic pain in women with endometriosis. *Human Reproduction, 20*(7), 1993–1998.

Podgaec, S., Abrao, M. S., Dias, J. A., Rizzo, L. V., de Oliveira, R. M., & Baracat, E. C. (2007). Endometriosis: An inflammatory disease with a Th2 immune response component. *Human Reproduction, 22*(5), 1373–1379.

Rosa e Silva, A. C. J. S., Rosa e Silva, J. C., & Ferriani, R. A. (2007). Serum CA-125 in the diagnosis of endometriosis. *International Journal of Gynecology & Obstetrics, 96*(3), 206–207.

Rosa e Silva, J. C., do Amaral, V. F., Mendonça, J. L., Rosa e Silva, A. C., Nakao, L. S., Poli Neto, O. B., et al. (2014). Serum markers of oxidative stress and endometriosis. *Clinical and Experimental Obstetrics and Gynecology, 4*(41), 371–374.

Schutte, B., Nuydens, R., Geerts, H., & Ramaekers, F. (1998). Annexin V binding assay as a tool to measure apoptosis in differentiated neuronal cells. *Journal of Neuroscience Methods, 86*(1), 63–69.

Seeber, B., Sammel, M. D., Fan, X., Gerton, G. L., Shaunik, A., Chittams, J., et al. (2010). Proteomic analysis of serum yields six candidate proteins that are differentially regulated in a subset of women with endometriosis. *Fertility and Sterility, 93*(7), 2137–2144.

Singh, S. S., & Suen, M. W. H. (2017). Surgery for endometriosis: Beyond medical therapies. *Fertility and Sterility, 107*(3), 549–554.

Subramanian, A., & Agarwal, N. (2010). Endometriosis—Morphology, clinical presentations and molecular pathology. *Journal of Laboratory Physicians, 2*(1), 1–9.

Sundell, G., Milsom, I., & Andersch, B. (1990). Factors influencing the prevalence and severity of dysmenorrhoea in young women. *BJOG: An International Journal of Obstetrics and Gynaecology, 97*(7), 588–594.

Tamburro, S., Canis, M., Albuisson, E., Dechelotte, P., Darcha, C., & Mage, G. (2003). Expression of transforming growth factor β1 in nerve fibers is related to dysmenorrhea and laparoscopic appearance of endometriotic implants. *Fertility and Sterility, 80*(5), 1131–1136.

Tokushige, N., Markham, R., Russell, P., & Fraser, I. S. (2006). High density of small nerve fibres in the functional layer of the endometrium in women with endometriosis. *Human Reproduction, 21*(3), 782–787.

Tuten, A., Kucur, M., Imamoglu, M., Kaya, B., Acikgoz, A. S., Yilmaz, N., et al. (2014). Copeptin is associated with the severity of endometriosis. *Archives of Gynecology and Obstetrics, 290*(1), 75–82.

Viganò, P., Somigliana, E., Matrone, R., Dubini, A., Barron, C., Vignali, M., et al. (2002). Serum leptin concentrations in endometriosis. *Journal of Clinical Endocrinology and Metabolism, 87*(3), 1085–1087.

Wang, W.-T., Ya-Nan, Z., Bo-Wei, H., & Hong Shun-Jia, C. Y.-Q. (2013). Circulating MicroRNAs identified in a genome-wide serum MicroRNA expression analysis as noninvasive biomarkers for endometriosis. *The Journal of Clinical Endocrinology and Metabolism, 1*(98), 281–289.

Wheatley, S. P., & Altieri, D. C. (2019). Survivin at a glance. *Journal of Cell Science, 132*(7), 1–8.

Wu, V., Yuan, P. Q., Larauche, M., Wang, L., & Million, M. (2013). Chapter 183 – Urocortins. In A. J. Kastin (Ed.), *Handbook of biologically active peptides* (2th ed., pp. 1346–1353). Boston: Academic Press.

Wykes, C. B., Clark, T. J., & Khan, K. S. (2004). Accuracy of laparoscopy in the diagnosis of endometriosis: A systematic quantitative review. *BJOG: An International Journal of Obstetrics and Gynaecology, 111*(11), 1204–1212.

Yavuzcan, A., Çağlar, M., Üstün, Y., Dilbaz, S., Özdemir, I., Yildiz, E., et al. (2013). Evaluation of mean platelet volume, neutophil/lymphocyte ratio and platelet/lymphocyte ratio in advanced stage endometriosis with endometrioma. *Journal of the Turkish German Gynecology Association, 14*(4), 210–215.

Yun, B. H., Lee, Y. S., Chon, S. J., Jung, Y. S., Yim, S. Y., Kim, H. Y., et al. (2014). Evaluation of elevated urinary enolase i levels in patients with endometriosis. *Biomarkers, 19*(1), 16–21.

Zondervan Krina, T., Becker Christian, M., & Missmer, S. A. (2020). Endometriosis. *The New England Jounal of Medicine, 382*, 1244–1256.

Chapter 47

Biomarkers in bladder pain syndrome: A new narrative

Thais F. de Magalhaes and Jorge Haddad

Urogynecology Division, Discipline of Gynecology, Clinics Hospital at University of Sao Paulo, Sao Paulo, SP, Brazil

Abbreviations

APF	antiproliferative factor
AUA	American Urology Association
BPS	bladder pain syndrome
CRP	c-reactive protein
EGF	epidermal growth factor
ESSIC	European Society for the Study of IC/BPS
HB-EGF	heparin-binding growth factor-like growth factor
GAG	glycosaminoglycan
GP51	glycoprotein 51
HIF	H
HIP/PAP	hepatocarcinoma-intestine-pancreas/pancreatitis-associated protein
IC	interstitial cystitis
ICI	International Consultation on Incontinence-Research Society
ICS	International Continence Society
IL	interleukin
MAPK	mitogen-activated protein kinase
MIF	macrophage migration inhibitory factor
MCP	monocyte chemotactic protein-1
NGF	nerve growth factor
NIDDK	National Institute of Diabetes and Digestive and Kidney Diseases
NT	neurotropins
PBS	painful bladder syndrome
QoL	quality of life
TNF	tumor necrosis factor
ZO-1	zonula occludens-1

Introduction

Bladder pain syndrome (BPS) is chronic pain disorder of unknown etiology that negatively affects quality of life (QoL).

In medical literature, several other terms, such as painful bladder syndrome or interstitial cystitis (IC), have been used interchangeably to describe this condition, but it is not clearly defined neither that bladder inflammation is the primary mechanism of the disease, nor that the disease originates in the bladder interstitium. As such, the term IC seems inappropriate; and, since pain is the hallmark of this condition, the new nomenclature "BPS" was proposed (van de Merwe et al., 2008). However, because of its historic importance in the diseases' recognition as a legitimate medical condition, the term "IC" been kept in use and, throughout this chapter, this condition will be referred to as "IC/BPS."

In the last 40 years, disease definition has changed considerably (Pape, Falconi, De Mattos Lourenco, Doumouchtsis, & Betschart, 2019). In 1987–1988, the National Institute of Diabetes and Digestive and Kidney Diseases (NIDDK) established the first diagnostic criteria for IC/BPS (Fig. 1). Such criteria was initially intended as a research tool, but ended

---

Features and Assessments of Pain, Anesthesia, and Analgesia. https://doi.org/10.1016/B978-0-12-818988-7.00014-5
Copyright © 2022 Elsevier Inc. All rights reserved.

FIG. 1 A summary of the first established criteria for the diagnosis of IC/BPS, developed by the NIDDK. Specific setups for the cystocopic examination and characteristics of glomerulations are also required (Hanno, 2002).

up being far too restrictive to be clinically useful (Hanno, Erickson, Moldwin, & Faraday, 2015). After, in 2002, the International Continence Society (ICS) published a revised definition of IC/BPS (Abrams et al., 2002); but it was also shown to be limiting. More recently, emphasizing the superiority of clinical over cystoscopic findings, since the latter is variable and inconsistent with symptom severity, the American Urological Association (AUA) adopted a nomenclature and clinical criteria that would serve as a working basis for the present knowledge on the disease, defining IC/BPS as "an unpleasant sensation (pain, pressure, or discomfort), related to the bladder, in the presence of urinary tract symptoms of over 6 weeks' duration, in the absence of other identifiable causes" (Hanno et al., 2015).

Challenges in IC/BPS knowledge

Despite their divergences, what societies agree on is that IC/BPS can have a profound effect on QoL, is associated with other pain syndromes and difficult to treat (Pape et al., 2019). As disease pathophysiology is not well understood, there are no targeted treatment options, and millions of patients worldwide suffer from this disease (Magalhaes, Baracat, Doumouchtsis, & Haddad, 2019). The absence of standardized, well-accepted diagnostic criteria imposes challenges on estimating the true prevalence of the disease, describing disease mechanisms, discovering useful biomarkers, and developing effective treatments.

Currently, excluding confounding conditions is one of the main goals of the diagnostic evaluation in a patient with suspected IC/BPS (Pape et al., 2019). A question that remains unresolved is whether IC/BPS is a disease in itself, or whether it encompasses multiple not-yet-identified, different pathological conditions that have a nonspecific clinical presentation that can fulfill the current diagnostic criteria.

In order to answer this question, we must extensively study the pathophysiologic mechanisms involved in IC/BPS, compare its biomarkers with clinical presentation and how they vary in different responses to treatment. Studying IC/BPS biomarkers has a huge role in achieving this knowledge; and this type of research should be encouraged. Conjointly, we should aim to identify variables that make it possible to diagnose IC/BPS according to its own pathomechanisms so that, in the future, it is no longer a diagnosis of exclusion. But, in order for reports to be comparable and for us to progress in determining which biomarkers are associated with each possible facet of the disease, the definition of IC/BPS used in research should be standardized. We urge that societies come to an evidence-based agreement on what the standard definition of IC/BPS should be.

This chapter will discuss the main biomarkers associated with IC/BPS, but we reinforce the importance that researchers agree on a definition for IC/BPS that should be used throughout studies, in order for us to have comparable data and evolve on the knowledge of the pathophysiology of this burdensome dysfunction. This, we believe, is one of the primordial steps in understanding the disease and ultimately improving patients' lives.

Biomarkers

The ideal biomarker should be capable of detecting pathological onset, differentiate IC/BPS from other similar conditions, determine disease prognosis, and correlate with disease intensity and symptom assessment (Hanno, 2002; Patnaik et al.,

2017), all with high accuracy. If, at least, there were a biomarker that could help predict response to a specific therapy, it would aid in the choice of treatment. Although, to date, there is no such marker, there is ongoing investigation on the subject. Recent findings on biomarkers for IC/BPS are explored below, and have been subdivided according to sample source (whether urine, bladder biopsy specimens, blood, stool, or other).

In the following sections, we provide a review of promising biomarkers associated with IC/BPS. Be aware that the use of different criteria for IC/BPS affects the interpretation of findings—the different definitions used by the studies reporting each biomarker are listed in Tables 1–3. Systematic reviews of biomarkers in IC/BPS according to specific criteria have been previously published elsewhere (Magalhaes et al., 2019).

TABLE 1 Potential biomarkers in bladder wall specimens described for IC/BPS.

Biomarker	Status in IC/BPS (compared to controls)	IC/BPS definition used by reference
Apoptotic signaling molecules (Bad, Bax, cleaved caspase-3)	Increased	Characteristic symptoms of IC/PBS and glomerulation after cystoscopic hydrodistension (Lee, Jiang, & Kuo, 2013; Shie & Kuo, 2011)
ARID1A	Increased	AUA (Shahid et al., 2018)
Chondroitin sulfate	Increased	NIDDK (Slobodov et al., 2004)
CTLA-4, CD20	Increased	ESSIC (Gamper et al., 2013)
CXCR3-binding chemokines and TNFSF14	Increased	A revised form of the Japanese guidelines by East Asian urologists (Ogawa et al., 2010)
E-cadherin	Increased	NIDDK (Slobodov et al., 2004)
HIF-1α	Increased	NIDDK (Lee & Lee, 2011)
HIP/PAP	Increased	NIDDK (Makino, Kawashima, Konishi, Nakatani, & Kiyama, 2010)
IL-16, VCAM-1, ICAM-1	Increased	ICS (Corcoran et al., 2013)
Mast cells	Increased	NIH-NIDDK (Peeker, Enerbäck, Fall, & Aldenborg, 2000), ESSIC (Grover, Srivastava, Lee, Tewari, & Te, 2011)
Nerve density	Increased	AUA 2011 (Tonyali et al., 2018)
NGF	Increased	Characteristic symptoms and cystoscopic hydrodistention (Kuo, Liu, Tyagi, & Chancellor, 2010); not reported (Lowe et al., 1997)
PDECGF/TP	Increased	Frequency, urgency, and bladder pain for more than 6 months and resistant to oral anticholinergic agents; or urinary frequency more than 8 times daily or an average voided volume of less than 100 mL, other pathological conditions ruled out (Tamaki, Saito, Ogawa, Yoshimura, & Ueda, 2004)
Pronociceptive inflammatory proteins (TRPV1, 2, 4, ASIC1, NGF, CVCL9, TRPM2)	Increased	A revised form of the Japanese guidelines by East Asian urologists (Homma et al., 2013)
Tryptase signaling	Increased	Characteristic symptoms of IC/PBIS and diffused glomerulation in cystoscopic hydrodistention (Shie & Kuo, 2011)
Uroplakin III	Increased	NIDDK (Zeng et al., 2007)
VEGF	Increased	NIDDK (J. D. Lee & Lee, 2011; Peng, Jhang, Shie, & Kuo, 2013); ESSIC (Kiuchi et al., 2009)
ZO-1	Increased	NIDDK (Slobodov et al., 2004)
Biglycan, perlecan	Decreased	NIDDK (Hauser et al., 2008)

Continued

530 **PART | IV** Assessments, screening, and resources

TABLE 1 Potential biomarkers in bladder wall specimens described for IC/BPS—cont'd

Biomarker	Status in IC/BPS (compared to controls)	IC/BPS definition used by reference
Chondroitin sulfate proteoglycans	Decreased	Long-standing history of pain and urinary urgency and frequency without other known causes, and petechial bleeding upon bladder dilation (Hurst et al., 1996)
E-cadherin	Decreased	Characteristic symptoms of IC/PBIS and diffused glomerulation in cystoscopic hydrodistention (Shie & Kuo, 2011); symptoms of frequency, urgency, and bladder pain and characteristic glomerulation hemorrhages on cystoscopic hydrodistension under general anesthesia (Lee et al., 2013; Liu, Shie, Chen, Wang, & Kuo, 2012)
IL-8	Decreased	NIDDK (Tseng-Rogenski & Liebert, 2009)
WNT-11	Decreased	AUA (Choi et al., 2018)
ZO-1	Decreased	Symptoms of frequency, urgency, and bladder pain and the characteristic glomerulations on cystoscopic hydrodistension under general anesthesia (Liu et al., 2012)

TABLE 2 Potential urinary biomarkers described for IC/BPS.

Biomarker	Status in IC/BPS (compared to controls)	IC/BPS diagnostic criteria or definition used by reference
A1BG, ORM1	Increased	Not reported (Goo, Tsai, Liu, Goodlett, & Yang, 2010)
APF	Increased	NIDDK (Erickson et al., 2002; Keay et al., 2001; Zhang et al., 2003; Zhang, Li, & Kong, 2005); not reported (Chai et al., 2000)
BK polyoma virus	Increased	Not reported (Van der Aa, Beckley, & de Ridder, 2014)
CXCL-1 and -10, IL-6, NGF	Increased	NIDDK (Tyagi et al., 2012)
EGF	Increased	NIDDK (Erickson et al., 2002; Keay et al., 1997, 2001; Zhang et al., 2003, 2005)
Etio-S	Increased	Chronic unpleasant sensation in the urinary bladder, associated with lower urinary tract symptoms, in the absence of infection or other identifiable causes (Parker et al., 2016)
Glucosamine, galactosamine	Increased	ESSIC (Buzzega, Maccari, Galeotti, & Volpi, 2012)
HIP, PAP	Increased	NIDDK (Makino et al., 2010)
Histamine, methylhistamine	Increased	Clinical presentation and presence of mucosal glomerulations and inflammation on bladder distention (El-Mansoury, Boucher, Sant, & Theoharides, 1994); ICDB criteria (Lamale, Lutgendorf, Zimmerman, & Kreder, 2006)
IGF-1	Increased	NIDDK (S. Keay et al., 1997)
IGF-binding protein-3	Increased	NIDDK (Erickson et al., 2002; S. Keay et al., 1997)
IL-2, IL-8	Increased	NIDDK (Peters, Diokno, & Steinert, 1999)
IL-6	Increased	ICDB (Lamale et al., 2006); NIDDK (Erickson et al., 2002; Felsen et al., 1994; Peters et al., 1999); NIH (Lotz, Villiger, Hugli, Koziol, & Zuraw, 1994)

TABLE 2 Potential urinary biomarkers described for IC/BPS—cont'd

Biomarker	Status in IC/BPS (compared to controls)	IC/BPS diagnostic criteria or definition used by reference
Macrophage-derived chemokine, IL-4	Increased	Urinary frequency and suprapubic or bladder pain for at least 6 months, other diagnoses excluded (Abernethy et al., 2017)
MIF	Increased	AUA (Vera et al., 2018)
NGF	Increased	Characteristic symptoms and cystoscopic findings of glomerulation, petechia, or mucosal fissure after hydrodistension (Liu, Tyagi, Chancellor, & Kuo, 2009); characteristic symptoms and cystoscopic hydrodistension (Kuo et al., 2010; Liu, Tyagi, Chancellor, & Kuo, 2010)
PGE-2	Increased	A revised form of the Japanese guidelines by East Asian urologists (Wada et al., 2015)
Tyramine, 2-oxoglutarate	Increased	Not reported (Wen et al., 2015)
Uronate, sulfated GAG	Increased	NIDDK, except glomerulations (Lokeshwar et al., 2005)
VCAM-1, ICAM-1	Increased	ICS (Corcoran et al., 2013)
GP51	Decreased	Symptom complex, glomerulations, and/or Hunner's ulcers and negative urine culture (Byrne et al., 1999); NIDDK (Rofeim, Shupp-Byrne, Mulholland, & Moldwin, 2001)
HB-EGF	Decreased	NIDDK (S. Keay et al., 1997; Keay et al., 2001; Zhang et al., 2005; Zhang et al., 2003); not reported (Chai et al., 2000)
MAPK pathway	Decreased	AUA (Bradley et al., 2018)

TABLE 3 Potential serum biomarkers described for IC/BPS.

Biomarker	Status in IC/BPS (compared to controls)	IC/BPS definition used by reference
CRP	Increased	Characteristic symptoms and cystoscopic findings of glomerulation, petechiae, or mucosal fissure after hydrodistension under general anesthesia (Chung, Liu, Lin, & Kuo, 2011; Jiang, Peng, Liu, & Kuo, 2013)
IL-1β, IL-6, IL-8, TNF-α	Increased	Characteristic cystoscopic findings after hydrodistention (Jiang et al., 2013)
NGF	Increased	Symptoms of frequency, urgency, bladder pain as well as presence of glomerulations during cystoscopic hydrodistention performed under general anesthesia (Liu & Kuo, 2012)
HB-EGF	Decreased	NIDDK (Keay, Kleinberg, Zhang, Hise, & Warren, 2000)

Bladder wall biopsy specimens

Major roles of the urothelium are said to be to prevent penetration of urinary solutes into the bladder wall and to act as a barrier in preventing bacterial and crystal adherence (Hurst et al., 1996). Changes in this barrier, whether at the structural or molecular level, may play a role in IC/BPS and, as such, the bladder wall has been extensively studied. Potential biomarkers described in bladder wall specimens, acquired through biopsies, are listed in Table 1.

Urine

The urine itself of IC/BPS patients might carry a substance that accounts for or reflects pathological changes occurring in the disease. Some theories of pathogenesis suggest access of a component of urine to the bladder wall interstitium, ultimately resulting in an inflammatory response, whether by toxic, allergic, or immunologic means (Hanno, 2002). Potential biomarkers found in urine are described in Table 2.

Serum

Serum is a noninvasive, easy-to-obtain source for potential biomarkers. However, since it reflects systemic changes, finding accurate serological biomarkers for urological conditions in serum is challenging. Nonetheless, a few substances have been studied and are further described in Table 3.

Stool

Patients with IC/BPS have shown significantly reduced levels of some intestinal bacteria (namely, *C. aerofaciens*, *E. sinensis*, *F. prausnitzii*, and *O. splanchnicus*); while showing increased levels of glyceraldehyde in stool (Braundmeier-Fleming et al., 2016).

IC/BPS biomarkers and their potential relationship to disease pathophysiology

As previously mentioned, pathophysiology for IC/BPS is not established; and several hypotheses for its mechanism have taken place based on previous studies (Pape et al., 2019). Central neurologic mechanisms likely play a role, and there is general consensus that IC/BPS apparently results from inflammation caused by a defective or dysfunctional bladder epithelium; in the majority of cases, no injury to justify inflammation or neuro-endocrine activity is identified. Mechanisms that have been suggested to contribute to symptoms include defects in the urothelial glycosaminoglycan (GAG) layer, infection, autoimmune dysregulation, microcirculation changes, neurogenic inflammation, and hypoxia (Pape et al., 2019). The most studied or promising biomarkers and their relationship with possible disease pathomechanisms are explored below.

NGF

NGF is a member of the neurotropins (NT), a family of trophic factors. The NTs are produced in peripheral tissues and in the central nervous system and are required for the establishment of neuronal connections, regulation of established synaptic structures, and neuronal survival. They are well-established mediators of pain, particularly in chronic conditions. Tissue concentration of NGF has been shown to be increased in visceral dysfunctions associated with chronic pain. The cellular sources of NGF in the bladder include smooth muscle cells and the urothelial layer; in cultures, these cells were able to release NGF following cycles of stretch and relaxation; secretion was very low in normal conditions. Urothelial NGF secretion appears to be regulated by estrogens; if NGF is indeed involved in pathophysiology of IC/BPS, this could be one of the reasons why this disease is more prevalent in women. NGF is also increased following bladder ischemia, inflammation, and stressful conditions; bladder surgery and stress are known triggers for IC/BPS symptom. In addition to being elevated in patients IC/BPS, a significant decline in urinary NGF was reported after treatment with cystoscopic hydrodistension, oral pentosane polysulphate, hyaluronic acid, or botulinum toxin, indicating that there might be a correlation between symptom severity and levels of urinary NGF and that NGF levels can reflect treatment response (Coelho, Oliveira, Antunes-Lopes, & Cruz, 2019).

On the other hand, NGF is also increased in urine samples of patients with overactive bladder and bladder cancer, suggesting it may not be specific enough to serve as a diagnostic biomarker for IC/BPS (Evans et al., 2011). Its use as a target therapy, with Tanezumab—a humanized monoclonal antibody that binds to and inhibits NGF activity—was disappointing: when tested in IC/BPS patients, it showed only modest results at the cost of adverse effects such as hyperestesia, paresthesia, and migraines (Evans et al., 2011). Moreover, there are also issues with the use of urinary NTs as biomarkers: there is no uniform protocol for urine collection and sampling or standardized analysis kits for measuring urinary NTs, and commercial kits need improvements in variability (Coelho et al., 2019).

In conclusion, NGF may contribute as a biomarker for disease prognosis as well as prognosis, but it is far from being the ideal biomarker.

Inflammatory and angiogenic markers

Cytokine signaling is highly complex in nature, as cytokines' function may overlap and the same cytokine can generate multiple downstream responses, making it difficult to understand their role in IC/BPS. The cytokines TNFSF-14, TNF-α, and IFN-γ are associated with Th1 activation and effector function; CTLA-4, CD20, and CD79A are B- and T-cell markers. These findings suggest that IC might have an adaptive immune-mediated component (Duh et al., 2018).

IL-33, in turn, acts as an activator of mast cells and a mediator of innate immune response (Duh et al., 2018). Mast cells have been found especially in areas of weakened urothelium, of greater vascularity, of inflammatory granulation tissue or in the presence of Hunner lesions (Marcu, Campian, & Tu, 2018). They are involved in immune response and can be activated by neuropeptides, proinflammatory cytokines, acetylcholine, and bacterial or viral pathogens. After activation, they produce and release proinflammatory mediators such as histamine, VEGF, prostaglandins, proteases, IL-6, and IL-8 (Grover et al., 2011).

The finding of increased IL-6 in some studies in spontaneously voided urine but not in ureteral urine (Lotz et al., 1994) suggests that it is a product secreted in the bladder and that inflammation may indeed play a role in IC/BPS pathogenesis. These relationships suggest a link between urothelial damage and autoimmunity (Duh et al., 2018). IL-8, in addition to being a proinflammatory cytokine, appears to be essential for normal urothelial cell survival (Tseng-Rogenski & Liebert, 2009); its activity in IC/BPS and role in its pathogenesis needs to be further studied to be better determined.

PDECGF/TP appears to plays a role in long-term bladder inflammation; its coexpressed protein CD44 enhances the action of heparin-binding growth factors (Marcu et al., 2018).

In turn, overexpression of VEGF was highly associated with the finding of glomerulations during bladder hydrodistension; the neovascularization it promotes might play a role in the pathogenesis of IC/BPS (Tamaki et al., 2004).

In sum, although triggering mechanisms are not clear, inflammation and angiogenesis appear to play a role in IC/BPS. Related biomarkers have been studied, but their usefulness in clinical practice is yet to be demonstrated.

APF and HB-EGF

APF is one of the most studied biomarkers in IC/BPS. This peptide regulates important signaling pathways, leading to increases in transurothelial cellular permeability and suppression of proliferation of the bladder epithelium by decreasing incorporation of thymidine through normal bladder epithelial cells. It also inhibits the production of HB-EGF, an epithelial growth factor (Patnaik et al., 2017). These findings support the hypothesis that APF can disrupt the normal process of urothelial proliferation and repair, and may contribute to the pathogenesis of IC/BPS. Additionally, increased APF and decreased HB-EGF have been described in several studies (Tables 1 and 2), strengthening their possibility as potential biomarkers.

Etio-S

Etiocholan-3α-ol-17-one sulfate (Etio-S), a sulfoconjugated 5-β reduced isomer of testosterone. This substance can act as a positive allosteric modulator of the receptor GABA-A, and high local concentrations of this substance can stimulate acute phase reactions; in addition, Etio-S levels have shown a great association with IC/BPS symptoms (Magalhaes et al., 2019). These findings suggest that Etio-S may have a role in disease pathogenesis and/or pain processing, but further research is required, as its potential role in IC/BPS has only been described in one report (Parker et al., 2016).

Urothelial dysfunction

The urothelium is considered a barrier between the underlying bladder wall and urine. Dysfunction in this layer may lead to an abnormal migration of urinary solutes, which can cause tissue injury and bladder pain in IC/BPS. Bladder dysfunction can be a consequence of inflammation, and a cause of it, such as in the case of trauma. Several studies have reported decreased expression of GAG components in the bladder wall of patients with IC/BPS (Table 1); although such changes appear to play a role in disease pathogenesis, its use as clinically useful biomarkers remains to be determined (Slobodov et al., 2004).

GP51

GP51 is a urinary antibacterial glycoprotein, and its decreased presence in patients with IC/BPS supports the hypothesis that infection may play a role in this disease's pathophysiology, such as being a trigger for chronic inflammation (Marcu et al., 2018).

Microbiome

Establishing a direct relation between microbiome (whether in urine or stool) and disease is difficult, because it naturally changes seasonally and throughout the life cycle, as well as in response to environmental changes. The finding of increased BK polyoma virus described in the urine of patients with IC/BPS (Van der Aa et al., 2014) suggests, but not determines, that this virus can have a role in triggering disease symptoms; it is indeed an interesting finding, but needs to be further investigated. In turn, stool microbiome changes described by Braundmeier-Fleming et al. (2016) are even less promising as biomarkers, as spontaneous dietary changes made by patients can affect the intestinal flora (Magalhaes et al., 2019).

Applications to other areas

In this chapter, we have reviewed the current knowledge on IC/BPS definition and biomarkers, and have described how such biomarkers can relate to possible mechanisms in disease pathogenesis. IC/BPS is one cause of chronic pelvic pain in both men and women, and other causes of chronic pelvic pain should be considered as confounding or concomitant diagnoses for patients with suspected IC/BPS. They include:

- Adenomyosis
- Adhesions
- Chronic prostatitis
- Endometriosis
- Fibromyalgia
- Inflammatory bowel disease
- Irritable bowel syndrome
- Malignancies
- Myofascial pain syndrome
- Neuropathic pain
- Pelvic congestion syndrome
- Pelvic inflammatory disease
- Recurrent urolithiasis
- Recurrent urinary tract infection

Evaluation of chronic pelvic pain is complex and must include a thorough clinical history, a complete physical examination, especially concerning the abdomen and pelvis; assessment of comorbid conditions; and assessment of psychosocial health, so that all factors contributing to CPP can be identified. Laboratory or imaging exams should be ordered based on findings in the clinical history/physical examination.

Likewise, treatment for CPP is also complex, as multiple conditions may occur concomitantly and contribute to worsening the vicious cycle of pain perception and central sensitization. Treatment should ideally include a multidisciplinary team including physician, physical therapist, and psychologist, as well as other specialties, as needed. Initial therapy aims at addressing the presumed cause, but central pain amplification may also need treatment, whether pharmacological or non-pharmacological. In such cases, after potentially life-threatening, reversible causes have been discarded, treatment should aim on improving patients' QoL and can be maintained for years if necessary.

Other agents of interest

This section broadens the chapter and authors have a great deal of flexibility. So in a chapter on the adverse effects of a particular anesthetic, the author may wish to list those agents that are less prone to be the subject of adverse reporting. A chapter on a particular general anesthetic used in a specific procedure may list other general anesthetics. A chapter on a particular class of agonists may describe other classes and agents.

Not applicable—this chapter does not involve anesthetic agents.

Mini-dictionary of terms

- **Biomarker:** A substance found in a body tissue or fluid that marks the presence of a disease, that can be objectively measured, and is ideally very specific.
- **Bladder pain syndrome**: A chronic syndrome of pain associated with the urinary bladder, with several definitions published by different societies.
- **Confounding conditions**: In this case, other pathological conditions that may have a similar clinical presentation as IC/BPS, causing mistakes in diagnosis.
- **Diagnostic criteria**: A listing of symptoms, findings, and conditions that defines diagnosis for a specific disease, ideally with high accuracy.
- **Interstitium**: A complex structure within organs that is fluid-filled and supported by collagen, which has structural, transport, and immune functions, among others.

Key facts

Key facts of biomarkers in bladder pain syndrome

- A condition similar to IC/BPS was first documented in the early 19th century, but its first consensus criteria were only published in 1988.
- Prevalence of IC/BPS in the United States can reach 6.5% of the population, but that depends on the source, the type of research, and the criteria used to define disease.
- IC/BPS can be accompanied by other pain syndromes, such as endometriosis or fibromyalgia, and by psychological conditions, such as anxiety or depression.
- Treatment options for IC/BPS include dietary changes, simple analgesics, nonsteroidal antiinflammatory drugs, amitriptyline, antihistamines, pentosane polysulfate sodium, and bladder hydrodistension, among others.
- Novel technologies, such as the use of mass spectrometry for proteomics or metabolomics approaches, show potential for the search of IC/BPS biomarkers.

Summary points

- IC/BPS is chronic pain disorder of unknown etiology that negatively affects QoL; its definition has changed several times in the past 40 years and is not homogeneous throughout guidelines.
- Proposed pathophysiologic mechanisms for IC/BPS include central neurologic changes, inflammation, urothelial dysfunction, autoimmunity, infection, microcirculation changes, and hypoxia, among others.
- The ideal biomarker should be capable of detecting pathological onset, differentiate IC/BPS from other similar conditions, determine disease prognosis and response to treatment, and correlate with disease intensity and symptom assessment.
- Potential biomarkers for IC/BPS found in the bladder wall, urine, serum, and stool have been reported, but none has proven to be clinically useful so far.
- Some biomarkers are elevated, while others are decreased, in comparison to controls; their possible clinical applications and role in disease pathomechanisms are explored.

References

Abernethy, M. G., Rosenfeld, A., White, J. R., Mueller, M. G., Lewicky-Gaupp, C., & Kenton, K. (2017). Urinary microbiome and cytokine levels in women with interstitial cystitis. *Obstetrics and Gynecology*, *129*(3), 500–506.

Abrams, P., Cardozo, L., Fall, M., Griffiths, D., Rosier, P., Ulmsten, U., ... Wein, A. (2002). The standardisation of terminology of lower urinary tract function: Report from the standardisation sub-committee of the international continence society. *American Journal of Obstetrics and Gynecology*, *187*(1), 116–126.

Bradley, M. S., Burke, E. E., Grenier, C., Amundsen, C. L., Murphy, S. K., & Siddiqui, N. Y. (2018). A genome-scale DNA methylation study in women with interstitial cystitis/bladder pain syndrome. *Neurourology and Urodynamics*, *37*(4), 1485–1493.

Braundmeier-Fleming, A., Russell, N. T., Yang, W., Nas, M. Y., Yaggie, R. E., Berry, M., ... Klumpp, D. J. (2016). Stool-based biomarkers of interstitial cystitis/bladder pain syndrome. *Scientific Reports*, 6, 26083.

Buzzega, D., Maccari, F., Galeotti, F., & Volpi, N. (2012). Determination of urinary hexosamines for diagnosis of bladder pain syndrome. *International Urogynecology Journal*, 23(10), 1367–1372.

Byrne, D. S., Sedor, J. F., Estojak, J., Fitzpatrick, K. J., Chiura, A. N., & Mulholland, S. G. (1999). The urinary glycoprotein GP51 as a clinical marker for interstitial cystitis. *The Journal of Urology*, 161(6), 1786–1790.

Chai, T. C., Zhang, C. O., Shoenfelt, J. L., Johnson, H. W., Jr., Warren, J. W., & Keay, S. (2000). Bladder stretch alters urinary heparin-binding epidermal growth factor and antiproliferative factor in patients with interstitial cystitis. *The Journal of Urology*, 163(5), 1440–1444.

Choi, D., Han, J. Y., Shin, J. H., Ryu, C. M., Yu, H. Y., Kim, A., ... Choo, M. S. (2018). Downregulation of WNT11 is associated with bladder tissue fibrosis in patients with interstitial cystitis/bladder pain syndrome without Hunner lesion. *Scientific Reports*, 8(1).

Chung, S. D., Liu, H. T., Lin, H., & Kuo, H. C. (2011). Elevation of serum c-reactive protein in patients with OAB and IC/BPS implies chronic inflammation in the urinary bladder. *Neurourology and Urodynamics*, 30(3), 417–420.

Coelho, A., Oliveira, R., Antunes-Lopes, T., & Cruz, C. D. (2019). Partners in crime: NGF and BDNF in visceral dysfunction. *Current Neuropharmacology*, 17(11), 1021–1038.

Corcoran, A. T., Yoshimura, N., Tyagi, V., Jacobs, B., Leng, W., & Tyagi, P. (2013). Mapping the cytokine profile of painful bladder syndrome/interstitial cystitis in human bladder and urine specimens. *World Journal of Urology*, 31(1), 241–246.

Duh, K., Funaro, M. G., DeGouveia, W., Bahlani, S., Pappas, D., Najjar, S., ... Stern, J. N. H. (2018). Crosstalk between the immune system and neural pathways in interstitial cystitis/bladder pain syndrome. *Discovery Medicine*, 25(139), 243–250.

El-Mansoury, M., Boucher, W., Sant, G. R., & Theoharides, T. C. (1994). Increased urine histamine and methylhistamine in interstitial cystitis. *The Journal of Urology*, 152(2 Pt 1), 350–353.

Erickson, D. R., Xie, S. X., Bhavanandan, V. P., Wheeler, M. A., Hurst, R. E., Demers, L. M., ... Keay, S. K. (2002). A comparison of multiple urine markers for interstitial cystitis. *The Journal of Urology*, 167(6), 2461–2469.

Evans, R. J., Moldwin, R. M., Cossons, N., Darekar, A., Mills, I. W., & Scholfield, D. (2011). Proof of concept trial of tanezumab for the treatment of symptoms associated with interstitial cystitis. *The Journal of Urology*, 185(5), 1716–1721.

Felsen, D., Frye, S., Trimble, L. A., Bavendam, T. G., Parsons, C. L., Sim, Y., & Vaughan, E. D., Jr. (1994). Inflammatory mediator profile in urine and bladder wash fluid of patients with interstitial cystitis. *The Journal of Urology*, 152(2 Pt 1), 355–361.

Gamper, M., Viereck, V., Eberhard, J., Binder, J., Moll, C., Welter, J., & Moser, R. (2013). Local immune response in bladder pain syndrome/interstitial cystitis ESSIC type 3C. *International Urogynecology Journal*, 24(12), 2049–2057.

Goo, Y. A., Tsai, Y. S., Liu, A. Y., Goodlett, D. R., & Yang, C. C. (2010). Urinary proteomics evaluation in interstitial cystitis/painful bladder syndrome: A pilot study. *International Brazilian Journal of Urology*, 36(4), 464–478. discussion 478–469, 479.

Grover, S., Srivastava, A., Lee, R., Tewari, A. K., & Te, A. E. (2011). Role of inflammation in bladder function and interstitial cystitis. *Therapeutic Advances in Urology*, 3(1), 19–33.

Hanno, P. M. (2002). Interstitial cystitis-epidemiology, diagnostic criteria, clinical markers. *Revista de Urología*, 1(Suppl 1), S3–S8. 4 Suppl.

Hanno, P. M., Erickson, D., Moldwin, R., & Faraday, M. M. (2015). Diagnosis and treatment of interstitial cystitis/bladder pain syndrome: AUA guideline amendment. *The Journal of Urology*, 193(5), 1545–1553.

Hauser, P. J., Dozmorov, M. G., Bane, B. L., Slobodov, G., Culkin, D. J., & Hurst, R. E. (2008). Abnormal expression of differentiation related proteins and proteoglycan core proteins in the urothelium of patients with interstitial cystitis. *The Journal of Urology*, 179(2), 764–769.

Homma, Y., Nomiya, A., Tagaya, M., Oyama, T., Takagaki, K., Nishimatsu, H., & Igawa, Y. (2013). Increased mRNA expression of genes involved in pronociceptive inflammatory reactions in bladder tissue of interstitial cystitis. *The Journal of Urology*, 190(5), 1925–1931.

Hurst, R. E., Roy, J. B., Min, K. W., Veltri, R. W., Marley, G., Patton, K., ... Parsons, C. L. (1996). A deficit of chondroitin sulfate proteoglycans on the bladder uroepithelium in interstitial cystitis. *Urology*, 48(5), 817–821.

Jiang, Y. H., Peng, C. H., Liu, H. T., & Kuo, H. C. (2013). Increased pro-inflammatory cytokines, C-reactive protein and nerve growth factor expressions in serum of patients with interstitial cystitis/bladder pain syndrome. *PLoS One*, 8(10), e76779.

Keay, S., Kleinberg, M., Zhang, C. O., Hise, M. K., & Warren, J. W. (2000). Bladder epithelial cells from patients with interstitial cystitis produce an inhibitor of heparin-binding epidermal growth factor-like growth factor production. *The Journal of Urology*, 164(6), 2112–2118.

Keay, S., Zhang, C. O., Kagen, D. I., Hise, M. K., Jacobs, S. C., Hebel, J. R., ... Warren, J. W. (1997). Concentrations of specific epithelial growth factors in the urine of interstitial cystitis patients and controls. *The Journal of Urology*, 158(5), 1983–1988.

Keay, S. K., Zhang, C. O., Shoenfelt, J., Erickson, D. R., Whitmore, K., Warren, J. W., ... Chai, T. (2001). Sensitivity and specificity of antiproliferative factor, heparin-binding epidermal growth factor-like growth factor, and epidermal growth factor as urine markers for interstitial cystitis. *Urology*, 57(6 Suppl 1), 9–14.

Kiuchi, H., Tsujimura, A., Takao, T., Yamamoto, K., Nakayama, J., Miyagawa, Y., ... Okuyama, A. (2009). Increased vascular endothelial growth factor expression in patients with bladder pain syndrome/interstitial cystitis: Its association with pain severity and glomerulations. *BJU International*, 104(6), 826–831 (discussion 831).

Kuo, H. C., Liu, H. T., Tyagi, P., & Chancellor, M. B. (2010). Urinary nerve growth factor levels in urinary tract diseases with or without frequency urgency symptoms. *Low Urinary Tract Symptoms*, 2(2), 88–94.

Lamale, L. M., Lutgendorf, S. K., Zimmerman, M. B., & Kreder, K. J. (2006). Interleukin-6, histamine, and methylhistamine as diagnostic markers for interstitial cystitis. *Urology*, 68(4), 702–706.

Lee, C. L., Jiang, Y. H., & Kuo, H. C. (2013). Increased apoptosis and suburothelial inflammation in patients with ketamine-related cystitis: A comparison with non-ulcerative interstitial cystitis and controls. *BJU International*, 112(8), 1156–1162.

Lee, J. D., & Lee, M. H. (2011). Increased expression of hypoxia-inducible factor-1α and vascular endothelial growth factor associated with glomerulation formation in patients with interstitial cystitis. *Urology, 78*(4), 971.e11-5.

Liu, H. T., & Kuo, H. C. (2012). Increased urine and serum nerve growth factor levels in interstitial cystitis suggest chronic inflammation is involved in the pathogenesis of disease. *PLoS One, 7*(9), e44687.

Liu, H. T., Shie, J. H., Chen, S. H., Wang, Y. S., & Kuo, H. C. (2012). Differences in mast cell infiltration, E-cadherin, and zonula occludens-1 expression between patients with overactive bladder and interstitial cystitis/bladder pain syndrome. *Urology, 80*(1), 225.e213-8.

Liu, H. T., Tyagi, P., Chancellor, M. B., & Kuo, H. C. (2009). Urinary nerve growth factor level is increased in patients with interstitial cystitis/bladder pain syndrome and decreased in responders to treatment. *BJU International, 104*(10), 1476–1481.

Liu, H. T., Tyagi, P., Chancellor, M. B., & Kuo, H. C. (2010). Urinary nerve growth factor but not prostaglandin E2 increases in patients with interstitial cystitis/bladder pain syndrome and detrusor overactivity. *BJU International, 106*(11), 1681–1685.

Lokeshwar, V. B., Selzer, M. G., Cerwinka, W. H., Gomez, M. F., Kester, R. R., Bejany, D. E., & Gousse, A. E. (2005). Urinary uronate and sulfated glycosaminoglycan levels: Markers for interstitial cystitis severity. *The Journal of Urology, 174*(1), 344–349.

Lotz, M., Villiger, P., Hugli, T., Koziol, J., & Zuraw, B. L. (1994). Interleukin-6 and interstitial cystitis. *The Journal of Urology, 152*(3), 869–873.

Lowe, E. M., Anand, P., Terenghi, G., Williams-Chestnut, R. E., Sinicropi, D. V., & Osborne, J. L. (1997). Increased nerve growth factor levels in the urinary bladder of women with idiopathic sensory urgency and interstitial cystitis. *British Journal of Urology, 79*(4), 572–577.

Magalhaes, T. F., Baracat, E. C., Doumouchtsis, S. K., & Haddad, J. M. (2019). Biomarkers in the diagnosis and symptom assessment of patients with bladder pain syndrome: A systematic review. *International Urogynecology Journal, 30*(11), 1785–1794.

Makino, T., Kawashima, H., Konishi, H., Nakatani, T., & Kiyama, H. (2010). Elevated urinary levels and urothelial expression of hepatocarcinoma-intestine-pancreas/pancreatitis-associated protein in patients with interstitial cystitis. *Urology, 75*(4), 933–937.

Marcu, I., Campian, E. C., & Tu, F. F. (2018). Interstitial cystitis/bladder pain syndrome. *Seminars in Reproductive Medicine, 36*(2), 123–135.

Ogawa, T., Homma, T., Igawa, Y., Seki, S., Ishizuka, O., Imamura, T., ... Nishizawa, O. (2010). CXCR3 binding chemokine and TNFSF14 over expression in bladder urothelium of patients with ulcerative interstitial cystitis. *The Journal of Urology, 183*(3), 1206–1212.

Pape, J., Falconi, G., De Mattos Lourenco, T. R., Doumouchtsis, S. K., & Betschart, C. (2019). Variations in bladder pain syndrome/interstitial cystitis (IC) definitions, pathogenesis, diagnostics and treatment: A systematic review and evaluation of national and international guidelines. *International Urogynecology Journal, 30*(11), 1795–1805.

Parker, K. S., Crowley, J. R., Stephens-Shields, A. J., van Bokhoven, A., Lucia, M. S., Lai, H. H., ... Henderson, J. P. (2016). Urinary metabolomics identifies a molecular correlate of interstitial cystitis/bladder pain syndrome in a multidisciplinary approach to the study of chronic pelvic pain (MAPP) research network cohort. *eBioMedicine, 7*, 167–174.

Patnaik, S. S., Lagana, A. S., Vitale, S. G., Buttice, S., Noventa, M., Gizzo, S., ... Dandolu, V. (2017). Etiology, pathophysiology and biomarkers of interstitial cystitis/painful bladder syndrome. *Archives of Gynecology and Obstetrics, 295*(6), 1341–1359.

Peeker, R., Enerbäck, L., Fall, M., & Aldenborg, F. (2000). Recruitment, distribution and phenotypes of mast cells in interstitial cystitis. *The Journal of Urology, 163*(3), 1009–1015.

Peng, C. H., Jhang, J. F., Shie, J. H., & Kuo, H. C. (2013). Down regulation of vascular endothelial growth factor is associated with decreased inflammation after intravesical OnabotulinumtoxinA injections combined with hydrodistention for patients with interstitial cystitis- -clinical results and immuno-histochemistry analysis. *Urology, 82*(6), 1452.e1451-6.

Peters, K. M., Diokno, A. C., & Steinert, B. W. (1999). Preliminary study on urinary cytokine levels in interstitial cystitis: Does intravesical bacille Calmette-Guérin treat interstitial cystitis by altering the immune profile in the bladder? *Urology, 54*(3), 450–453.

Rofeim, O., Shupp-Byrne, D., Mulholland, G. S., & Moldwin, R. M. (2001). The effects of hydrodistention on bladder surface mucin. *Urology, 57*(6 Suppl 1), 130.

Shahid, M., Gull, N., Yeon, A., Cho, E., Bae, J., Yoon, H. S., ... Kim, J. (2018). Alpha-oxoglutarate inhibits the proliferation of immortalized normal bladder epithelial cells via an epigenetic switch involving ARID1A. *Scientific Reports, 8*(1), 4505.

Shie, J. H., & Kuo, H. C. (2011). Higher levels of cell apoptosis and abnormal E-cadherin expression in the urothelium are associated with inflammation in patients with interstitial cystitis/painful bladder syndrome. *BJU International, 108*(2 Pt 2), E136–E141.

Slobodov, G., Feloney, M., Gran, C., Kyker, K. D., Hurst, R. E., & Culkin, D. J. (2004). Abnormal expression of molecular markers for bladder impermeability and differentiation in the urothelium of patients with interstitial cystitis. *The Journal of Urology, 171*(4), 1554–1558.

Tamaki, M., Saito, R., Ogawa, O., Yoshimura, N., & Ueda, T. (2004). Possible mechanisms inducing glomerulations in interstitial cystitis: Relationship between endoscopic findings and expression of angiogenic growth factors. *The Journal of Urology, 172*(3), 945–948.

Tonyali, S., Ates, D., Akbiyik, F., Kankaya, D., Baydar, D., & Ergen, A. (2018). Urine nerve growth factor (NGF) level, bladder nerve staining and symptom/problem scores in patients with interstitial cystitis. *Advances in Clinical and Experimental Medicine, 27*(2), 159–163.

Tseng-Rogenski, S., & Liebert, M. (2009). Interleukin-8 is essential for normal urothelial cell survival. *American Journal of Physiology. Renal Physiology, 297*(3), F816–F821.

Tyagi, P., Killinger, K., Tyagi, V., Nirmal, J., Chancellor, M., & Peters, K. M. (2012). Urinary chemokines as noninvasive predictors of ulcerative interstitial cystitis. *The Journal of Urology, 187*(6), 2243–2248.

van de Merwe, J. P., Nordling, J., Bouchelouche, P., Bouchelouche, K., Cervigni, M., Daha, L. K., ... Wyndaele, J. J. (2008). Diagnostic criteria, classification, and nomenclature for painful bladder syndrome/interstitial cystitis: An ESSIC proposal. *European Urology, 53*(1), 60–67.

Van der Aa, F., Beckley, I., & de Ridder, D. (2014). Polyomavirus BK—a potential new therapeutic target for painful bladder syndrome/interstitial cystitis? *Medical Hypotheses, 83*(3), 317–320.

Vera, P. L., Preston, D. M., Moldwin, R. M., Erickson, D. R., Mowlazadeh, B., Ma, F., ... Fall, M. (2018). Elevated urine levels of macrophage migration inhibitory factor in inflammatory bladder conditions: A potential biomarker for a subgroup of interstitial cystitis/bladder pain syndrome patients. *Urology*, *116*, 55–62.

Wada, N., Ameda, K., Furuno, T., Okada, H., Date, I., & Kakizaki, H. (2015). Evaluation of prostaglandin E2 and E-series prostaglandin receptor in patients with interstitial cystitis. *The Journal of Urology*, *193*(6), 1987–1993.

Wen, H., Lee, T., You, S., Park, S. H., Song, H., Eilber, K. S., ... Kim, J. (2015). Urinary metabolite profiling combined with computational analysis predicts interstitial cystitis-associated candidate biomarkers. *Journal of Proteome Research*, *14*(1), 541–548.

Zeng, Y., Wu, X. X., Homma, Y., Yoshimura, N., Iwaki, H., Kageyama, S., ... Kakehi, Y. (2007). Uroplakin III-delta4 messenger RNA as a promising marker to identify nonulcerative interstitial cystitis. *The Journal of Urology*, *178*(4 Pt 1), 1322–1327 (discussion 1327).

Zhang, C. O., Li, Z. L., & Kong, C. Z. (2005). APF, HB-EGF, and EGF biomarkers in patients with ulcerative vs. non-ulcerative interstitial cystitis. *BMC Urology*, *5*, 7.

Zhang, C. O., Li, Z. L., Shoenfelt, J. L., Kong, C. Z., Chai, T. C., Erickson, D. R., ... Keay, S. (2003). Comparison of APF activity and epithelial growth factor levels in urine from Chinese, African-American, and white American patients with interstitial cystitis. *Urology*, *61*(5), 897–901.

Chapter 48

Biomarkers of statin-induced musculoskeletal pain: Vitamin D and beyond

Michele Malaguarnera

Research Centre "The Great Senescence", University of Catania, Catania, Italy

Abbreviation

LDL	low-density lipoprotein
SAMS	statin associated muscle symptoms
HMG-CoAR	hydroxy-3-methyl-glutaryl-coenzyme A reductase
CK	creatine phosphokinase
VDR	vitamin D receptor
UNL	upper normal limit
7DHC	7-dehydrocholesterol
2HG	2-hydroxyglutarate
miRNAs	micro ribonucleic acids
myomiRs	miRNAs expressed in skeletal muscle

Introduction

Statins or hydroxy-methylglutaryl-CoA reductase inhibitors are the most prescribed drugs in the USA and the world. These drugs are most effective for managing elevated concentrations of low-density lipoprotein (LDL) cholesterol and reducing both cardiovascular diseases and cardiac events in coronary artery disease patients. The number of US individuals eligible for statin therapy has been estimated to increase from 37.5% to 48.6% in the next years (Pencina et al., 2014). Statins differ in their absorption, bioavailability, plasma protein binding, excretion, and solubility. They are obtained after fungal fermentation: lovastatin, simvastatin (prodrugs), and pravastatin; others by synthesis: fluvastatin, atorvastatin, and cerivastatin. Statins are metabolized through cytochrome P450 system and the majority of them by CYP3A4 isoenzyme.

Statin are well tolerated, although common symptoms like headache and gastrointestinal manifestations have been reported. Thus, even a rare side effect may occur relatively commonly. The statins are a common reason for hospital admission with myopathies or rhabdomyolysis. In a meta-analysis of trials involving 100,000 subjects, the incidence of muscle symptoms was 12.7% in the statin-treated group which shows an insignificant difference (12.4%) compared to placebo group. Despite these numbers, statin-induced myopathy is one of the primary reasons for statin discontinuation that contributes to adverse cardiovascular outcomes (Law & Rudnicka, 2006). Statin-associated muscle symptoms (SAMS) include fatigue, muscle tenderness, muscle weakness, muscle pain, and nocturnal cramping (Baker, 2005). The muscle symptoms tend to be generalized and worse with exercise. The mechanism of statin involved in muscle disease is unknown and remains elusive.

Statin associated muscle symptoms

SAMS include discomfort (myalgia), muscle weakness (myopathy), tenderness to palpation, with or without muscle inflammation (myositis), and/or myonecrosis. The latter may occur in a subtle form, just with increasing serum levels of creatine phosphokinase (CK) or in a more severe presentation called rhabdomyolysis with CK elevations >10 times upper normal limit (UNL). This acute and wide lysis of skeletal muscle cells is also characterized by a relevant release

Features and Assessments of Pain, Anesthesia, and Analgesia. https://doi.org/10.1016/B978-0-12-818988-7.00015-7
Copyright © 2022 Elsevier Inc. All rights reserved.

TABLE 1 SAMS classification.

SAMS classification	Phenotype	Incidence	Definition
SAMS 0	CK elevation $<4\times$ ULN	1.5%–26%	No muscle symptoms
SAMS 1	Myalgia, tolerable	190/100,000 Patient-years; 0.3%–33%	Muscle symptoms without CK elevation
SAMS 2	Myalgia, intolerable	0.2–2/1000	Muscle symptoms, CK $<4\times$ ULN, complete resolution on dechallenge
SAMS 3	Myopathy	5/100,000 Patient-years	CK elevation $>4\times$ ULN $<10\times$ ULN $\pm$ muscle symptoms, complete resolution on dechallenge
SAMS 4	Severe myopathy	0.11%	CK elevation $>10\times$ ULN $<50\times$ ULN, muscle symptoms, complete resolution on dechallenge
SAMS 5	Rhabdomyolysis	0.1–8.4/ 100,000 Patient-years	CK elevation $>10\times$ ULN with evidence of renal impairment + muscle symptoms or CK $>50\times$ ULN
SAMS 6	Autoimmune-mediated necrotizing myositis	$\sim$2/million per year	HMGCR antibodies, HMGCR expression in muscle biopsy, incomplete resolution on dechallenge

This table was modified with permission from Alfirevic, A., Neely, D., Armitage, J., Chinoy, H., Cooper, R. G., Laaksonen, R., Carr, D. F., Bloch, K. M., Fahy, J., Hanson, A., Yue, Q. Y., Wadelius, M., Maitland-Van Der Zee, A. H., Voora, D., Psaty, B. M., Palmer, C. N. A., & Pirmohamed, M. (2014). Phenotype standardization for statin-induced myotoxicity. *Clinical Pharmacology and Therapeutics* 96(4), pp. 470–476. doi:10.1038/clpt.2014.121. The numeric classification was developed by an expert group; descriptive nomenclature was adapted from the recommendation of the American College of Cardiology/American Heart Association/National Heart, Lung, and Blood Institute Clinical Advisory Board.

of electrolytes in the extracellular fluid, a large quantity of CK and myoglobin in blood that can produce an acute renal failure. In order to assess and diagnose SAMS, The European Society of Atherosclerosis suggested to take into account "the nature of muscle symptoms, increased CK levels and their temporal association with initiation of therapy with statin, and statin therapy suspension and rechallenge" (Toth et al., 2018) (Table 1).

The blinded, controlled effect of statins on Skeletal Muscle Function and Performance (STOMP) study administered atorvastatin 80 mg or placebo for 6 months on 420 healthy, statin-naive subjects. The results of this study showed that atorvastatin increased average CK, suggesting that statins produce mild muscle injury even among asymptomatic subjects (Parker et al., 2013).

SAMS are believed to affect around 1% to 25% of the statin-treated people and it is estimated that more than 1.5 million people per year experience these symptoms (Sathasivam & Lecky, 2008). These symptoms can be classified in different categories with an increase in the degree of impairment of patient wellness (Rosenson, Baker, Jacobson, Kopecky, & Parker, 2014) (see Table 1).

However, there is no indication that SAMS proceed gradually from myalgia to more severe manifestations of myopathy. Every kind of statin can cause muscle toxicity if high enough dose is used, although the degree of risk due to the therapeutic dose may change depending on the statin, hence their physicochemical properties and pharmacokinetic profile. Genetic factors and drug interactions could also influence these differences. For what concerns the genetic predisposition, common polymorphic variants within SLCO1B1 on chromosome 12 are strongly associated with an increased risk of statin-induced myopathy, perhaps by increasing statin blood levels. Myotoxicity does not seem too much different from the statin approved. It is an exception simvastatin (at high daily doses established as 80 mg daily) that showed a higher incidence of myotoxicity compared with maximum approved doses of other statins (Backes, Howard, Ruisinger, & Moriarty, 2009). In addition, the higher incidence of fatal rhabdomyolysis was the cause of the withdrawal of cerivastatin, another approved statin drug.

SAMS are often a specific and referred to calves, buttocks, and thighs. Female gender, Asian ethnicity, high levels of physical activity, hypothyroidism, diabetes, renal and liver function deficiency have been confirmed as possible risk factors. The co-administration of drugs that share the same metabolic pathway via the cytochrome P450, overall through CYP3A4 isoenzyme, may account for up to 60% of SAMS. In fact, drugs like glucocorticoids, protease inhibitors, anti-psychotics, immunosuppressives, erythromycin, azole antifungals, and the fibric acid derivative gemfibrozil have been associated to SAMS during statin treatment. Other substances involved were orange or cranberry juice and alcohol consumption. Other statins such as fluvastatin and rosuvastatin are metabolized by CYP2C9 isoenzyme, subjected to less interaction with other drugs.

The most frequent associations are with fibrates or colchicine, metabolized through CYP3A4. Colchicine could induce myopathy per se (Kuncl et al., 1987). In any case, to avoid or solve muscle toxicity, they could be associated with statins metabolized by CYP2C9 (Goh, How, & Tavintharan, 2013).

More often, skeletal muscles become vulnerable to drug-related injury, due to their high mitochondrial energy metabolism, their mass and the high degree of blood flow, and mitochondrial energy metabolism. Other suggested etiological factors related to SAMS are the genetic predisposition, low body mass index (BMI), parathyroid dysfunction, underlying fibromyalgia or polymyalgia rheumatica, autoimmune disorders, disturbances in muscle metabolism, and low plasma vitamin D level (Michalska-Kasiczak et al., 2015). On top of that, statin acts through the reversible block of the HMG-CoA reductase that produces a downsizing of the cholesterol production. The decreased cholesterol may lead to changes in the myocyte membranes, changing their behavior and chemistry, making skeletal muscle cells more vulnerable to damage. Even though, the extraction of membrane cholesterol in muscle cells confirms this hypothesis (Draeger et al., 2006), the inhibition of squalene synthase does not cause myotoxicity in vitro (Johnson et al., 2004). Hence, it has been suggested that the etiological mechanism leading to SAMS is due to the suppression of a downstream factor of HMG-CoA pathway. A third mechanism is the depletion of isoprenoids lipids product of HMG-CoA pathway involved in a series of reaction called geranylgeranylation and/or farnesylation. The shortage of these mechanism can contribute to mitochondrial-mediated apoptotic signaling cascade observed during statin myotoxicity (Dirks & Jones, 2006).

Another proposed mechanism is related to the deficiency of vitamin D that has been linked to advanced age, female sex, muscle-skeletal pain, and genetic predisposition. The possible mitochondrial dysfunction and other mechanisms during SAMS could be reversible with vitamin supplementation and subsequent normalization of serum vitamin D levels (Glueck et al., 2011).

Cholesterol and vitamin D

Cholesterol and vitamin D are interconnected due to the existence of 7-dehydrocholesterol (7DHC), the precursor in the synthesis of both molecules. Statins act by blocking HMG-CoAR thereby reducing cholesterol synthesis and vitamin D production. The 7DHC, present in the epidermidis, can be converted to previtamin D3 via exposure to UVB light and then to vitamin D (cholecalciferol), that is then it is converted in the liver and kidneys where it is converted into active forms of vitamin D, respectively 25-hydroxycholecalciferol and then 1,25-dihydroxycholecalciferol.

Serum vitamin D levels influence muscle strength, growth, development, contractility, and postural stability. Its low serum concentrations contribute to muscular weakness and the increase in falls and bone fractures of the elderly. Additionally, a positive correlation between higher vitamin D levels and muscle strength has been described in older subjects (Zhu, Austin, Devine, Bruce, & Prince, 2010). Its receptor (VDR) is expressed in myosatellite cells, involved in muscle regeneration after muscle mass injury. An in vitro study showed that vitamin D exerts a promyogenic effect on satellite cells after injury (Braga, Simmons, Norris, Ferrini, & Artaza, 2017). These results are in accordance with Barker, Schneider, Dixon, Henriksen, and Weaver (2013) with a trial in male adults. The administration of vitamin D speeds short-term recovery in peak isometric force after an intense exercise protocol and lowers the muscle-damage circulating biomarkers. Skeletal muscle specific VDR KO mice showed reductions in type II muscle fibber diameter (Chen, Villalta, & Agrawal, 2016). Moreover, vitamin D supplementation resulted in an improvement in muscle mitochondrial function, oxidative phosphorylation and lowered the levels of CK following exercise. For these reasons, reduced serum vitamin D levels have been associated with a possible cause of SAMS.

Half of older adults use statins and about half are vitamin D deficient, hence the interaction between these two factors is of great clinical importance (Bischoff-Ferrari et al., 2017). Holding in account that about 30%–50% of patients with self-reported SAMS, it is even possible that other factors like vitamin D deficiency may contribute to muscle pain rather than the use of statins (Taylor, Lorson, White, & Thompson, 2017).

Other study aligns itself supporting the hypothesis that vitamin D and SAMS are interconnected and indicates vitamin D as a possible solution for treating patients with muscle symptoms that present already a vitamin D deficiency or insufficiency (Pennisi et al., 2019). Furthermore, vitamin D supplementation may enhance muscle strength, it could be already an easy and inexpensive way to introduce a treatment of a nonspecific musculoskeletal pain.

SAMS biomarkers

The difficulties to diagnose correctly SAMS have led different working group to find a definition for these symptoms. For example, the American College of Cardiology (ACC), the American Heart Association (AHA), and the Canadian Working

Group (CWG) stated that myopathy is a general term referring to any disease of muscle, while myalgia is the muscle ache or weakness without the elevation of a specific marker. Instead myositis requires biopsy to be confirmed and is characterized by muscle ache or weakness with serum creatine phosphokinase (CK) levels higher than the normal limit and rhabdomyolysis has similar muscle symptoms with CK ten times upper the limit (10,000 U/L) and creatinine elevation (Mancini et al., 2013). Despite CK is particularly useful to describe and define these symptoms and diseases, there are new and old markers that are convenient to use and describe into the diagnosis, prognosis and follow-up of SAMS. Those biomarkers, studied and evaluated in muscle diseases and could have a clinical validity in the evaluation of SAMS, are listed here.

Creatine

Creatine is a nonprotein amino acid produced in the human body from the amino acid glycine and arginine, with requirement for methionine to catalyze the transformation of guanidinoacetate.

Creatine is transported through the blood and taken up by tissues with high energy demands, such as the brain and skeletal muscle. This molecule improves the cellular energy state and muscle performance, facilitating intracellular energy transport, reducing oxidative damage to DNA, lowering homocysteine levels, and lipid peroxidation (Greenhaff et al., 1993).

Creatine deficiency has been associated to SAMS. It has been demonstrated in vitro that Atorvastatin inhibits guanidinoacetate methyltransferase (Phulukdaree, Moodley, Khan, & Chuturgoon, 2015), the last enzymatic step to the synthesis of creatine.

Lactic acid

Lactic acid is an organic acid and its conjugate base is called lactate. Lactic acid is chiral consisting of two enantiomers. During the normal metabolism and overall exercise, l-(+)-lactic acid is produced via lactate dehydrogenase (LDH) from pyruvate. Despite its energetic function, lactate tissue and blood concentration have been associated with pain in both physiological and pathological conditions, such as muscle fatigue induced by exercise, discogenic back pain, chronic Achilles tendinopathy, incisional pain, complex regional pain syndrome, and chronic inflammatory pain (Huang et al., 2019). The molecular mechanism of pain caused by increased lactate levels is related to internal and peripheral effect on nervous system. Lactate increases reactive oxygen species and the activity of acid-sensing ion channel-3 during lactic acidosis (Immke & McCleskey, 2001). These receptors are highly expressed on sensory neurons, important transducers for nociception, that innervate the muscles. Several studies showed that statins could cause lactic acidosis or increase the lactate/pyruvate blood ratio (Goli, Goli, Byrd, & Roy, 2002; Seachrist, Loi, Evans, Criswell, & Rothwell, 2005). Other articles did not find any association between lactate and statins (Delliaux et al., 2006; Galtier et al., 2012). Further investigations are needed to disclose the potential of lactate as a marker in SAMS subjects.

Creatinine

Creatinine is produced through the reaction with CK by the disruption of creatine, phosphocreatine, and adenosine triphosphate from muscle and protein metabolism. Creatinine is then removed from the blood by the kidneys via both glomerular filtration and proximal tubular secretion.

Intense exercise and muscle damage are usually accompanied by the increase of CK and creatinine. Likewise, increase in dietary intake of protein rich aliments or the use of creatine as a supplement augments the production and the daily excretion of creatinine.

Troponin

Troponin is a complex of three regulatory proteins, that modulate the muscle contraction, making actin-myosin interactions sensitive to cytosolic calcium levels. These three proteins are characterized by different molecular weight and functions. Troponin C binds calcium ion producing a conformational change in troponin I. Troponin I instead binds actin and forms the actin-tropomyosin complex and Troponin T, which in turn binds tropomyosin forming the troponin-tropomyosin complex. Troponin complex exists in different isoforms (Giannoni, Giovannini, & Clerico, 2009).

Although evidence reports that statin therapy could decrease cardiac troponin T, the isoforms of Troponin I have been used as markers of muscular damage (Sorichter et al., 1997). In particular, the fast isoforms of skeletal troponin I (fsTnI) increase in patients with SAMS (Trentini et al., 2019).

Myoglobin

Myoglobin is found at high concentration in muscle tissue. Its main function is to store and release oxygen, but also regulate nitric oxide at the microvascular and tissue level (Kamga, Krishnamurthy, & Shiva, 2012).

Myoglobin has extensively been used as a marker of muscle damage. The release during muscle damage could also lead to the liberation of the heme group and its disruption promotes the peroxidation of mitochondrial membranes (Jürgens, Papadopoulos, Peters, & Gros, 2000).

Fatty acid-binding proteins

Another promising plasma marker is Fatty Acid-Binding Proteins 3 (FABP3). Fatty acid-binding proteins are a family composed by 12 types identified so far. Their main function is the transportation as intracellular carrier of different fatty acid, eicosanoids, and retinoids. Therefore, they could serve as metabolic energy sources, substrates for membranes, and signaling molecules for metabolic regulation. Among them FABP3 has been inserted in different batteries of marker evaluating the muscular damage. Despite it has been shown to increase in patients with muscular dystrophy (Burch et al., 2015), its possible use as a biomarker of SAMS should be proven.

Coenzyme Q_{10} (CoQ_{10})

CoQ_{10}, a lipid-soluble benzoquinone, is an essential cofactor of the mitochondrial electron transport chain, carrying electrons from the first to the second and then to the third complex. It is mainly concentrated in the mitochondria and is ubiquitous. CoQ_{10} also serves as an antioxidant.

The reduced level of CoQ_{10} has been shown to be associated with the SAMS. Since, cholesterol is a precursor of CoQ_{10}, blocking HMG-CoAR should determine a reduction of CoQ_{10} synthesis. Animal and clinical study shows that statin therapy reduces these cofactors in serum, although different report does not confirm the reduction in skeletal muscle biopsies. CoQ_{10} supplementation may reduce the incidence of the SAMS (Caso, Kelly, McNurlan, & Lawson, 2007), although no reduction in skeletal muscle biopsies of simvastatin users has been found (Laaksonen, Jokelainen, Sahi, Tikkanen, & Himberg, 1995) and its administration did not reduce muscle pain in patients with statin myalgia (Taylor, Lorson, White, & Thompson, 2015).

Enzymes

The amount of enzyme efflux from muscle tissue to serum can be influenced by physical exercise in healthy subjects or in muscle diseases. Among them serum CK, alanine aminotransferase (ALT), aspartate aminotransferase (AST), and lactic dehydrogenase (LDH) are involved in muscle metabolism and have been extensively used as biomarker of muscle injury.

Creatine kinase

CK is a ubiquitous enzyme present in tissue and energy-demanding cells and it catalyzes the reaction that leads creatine to its conversion in phosphocreatine and vice versa. Among its isoforms the cytosolic have been more studied. They are formed by two subunits constituting three isoforms: CK-MM, CK-MB, and CK-BB. CK-MM and CK-MB are predominantly present in skeletal muscle. CK-MB content is also higher in cardiac muscle, where it represents approximatively 20% of total CK, and in the skeletal muscle representing up to 3% of total CK (Christenson & Azzazy, 1998).

CK has been used as a reference for the definition of myalgia, myositis, and rhabdomyolysis (see the classification in Table 1). Following this classification many studies set serum CK cut-off on >10 times or >5 times ULN to indicate SAMS. Although there are subjects that could suffer a mild grade of myositis with lower serum CK levels, as well as there are also asymptomatic subjects that show higher CK levels (>10 times ULN). Clinical monitoring of SAMS includes baseline CK levels of patients, especially when there are risk factors, such as impaired renal function, genetic myopathy, or significant alcohol abuse (Malaguarnera et al., 2011).

Recently Nogueira, Strunz, Takada, and Mansur (2019) did not find correlation between SAMS and CK, but with CK-MB on a sample of 6698 patients treated with statins. This isoenzyme has been found elevated also in other muscular diseases (Erlacher et al., 2001).

Few studies took in consideration the relationship between vitamin D and CK. Recently, a double-blind placebo-controlled study using vitamin D at 2000 UI on runners showed a significative decrease in CK in the vitamin D group (Zebrowska et al., 2020). Another similar study on rats showed a decrease in plasma CK levels in the group receiving supplemental vitamin D (Choi, Park, Cho, & Lee, 2013). CK was recovered by the treatment with vitamin D in mice model of simvastatin-induced myopathy (Ren et al., 2020). A retrospective analysis of vitamin D levels in 120 patients with SAMS, found associations between baseline and posttreatment vitamin D levels and in CK with statin therapy, as well as between vitamin D levels and pain with placebo (Taylor et al., 2017). In case of persistent CK levels after cessation of statin medication other causes of elevated CK levels should be considered and investigated including anti-HMGCoAr antibodies, that can be detected in serum and muscle of affected patients.

Transaminase AST and ALT

AST and ALT are intracellular enzymes responsible for the catalysis of transamination between amino acids and ketonic acid. Increased AST and ALT usually indicates liver injury. Nevertheless, hepatotoxicity has been reported mainly at high doses of Atorvastatin (Clarke, Johnson, Hall, Ford, & Mills, 2016), transaminases have been measured during statins clinical trials. AST and ALT are both increased in chronic muscle diseases (Nathwani, Pais, Reynolds, & Kaplowitz, 2005). AST is mainly distributed in the myocytes followed by hepatocytes, making it a potential biomarker of muscle damage.

Nogueira et al. (2019) found higher AST and ALT in patients with a history of statin intolerance. The authors of the study excluded any hepatotoxic effect, since the ALT levels were slightly above the limit, while AST levels were three times higher with respect to the limit levels since it is more present in muscle. Vitamin D supplementation attenuated the increase of circulating ALT and AST, normally experienced after exercise in healthy humans (Barker et al., 2013).

Lactate dehydrogenase (LDH)

Lactate dehydrogenase is an enzyme that interconverts pyruvate and lactate, with concomitant interconversion of NADH and NAD. It is enconded by LDHA and LDHB genes. If LDHA is mutated, high intensity physical activity leads to an insufficient amount of energy during anaerobic phase, the weakening of the muscle and eventually break down, conducing to myopathy or rhabdomyolysis (Kanno et al., 1988). CK and LDH serum levels are specifically increased in inflammatory myopathies. The symptoms of SAMS are often proportional with the CK and LDH serum levels. Lactate/pyruvate ratio has also been used as a possible marker of myotoxicity, because it could reflect a dysfunction in the mitochondrial respiratory chain. De Pinieux et al. showed higher lactate/pyruvate ratios in statin-treated patients compared with fenofibrate-treated patients, untreated patients, or healthy controls (De Pinieux et al., 1996).

Despite LDH has been recommended as a biomarker of muscle impairment and its efficacy has been proven in rats (Tonomura, Mori, Torii, & Uehara, 2009), its clinical utility has not been validated in cases of SAMS (Toth et al., 2018) or vitamin D-related studies. In any case, it appears that vitamin D does not change LDH levels after intense physical activity (Zebrowska et al., 2020).

Carbonic anhydrases III

Carbonic anhydrases III (CAIII) are metalloproteinases present in all living beings. CAIII is almost located in skeletal muscle, where it is the major soluble protein and is possibly involved in the diffusion of CO_2 into the tissue. This enzyme is released into the circulation after muscle injury. Statins seem to inhibit the enzymes of carbonic anhydrase family at micro molar concentration (Parkkila et al., 2012). CAIII has been found to not relate with SAMS, but recently patients with CK-MB displayed higher serum values of CAIII (Nogueira et al., 2019).

Aldolase

Aldolase is a glycolytic enzyme which catalyzes the transformation of fructose 1–6 biphosphate in glyceraldehyde and dihydroxyacetone phosphate. Among aldolase isozymes, aldolase A is the prevalent form in human muscle (Brancaccio, Lippi, & Maffulli, 2010). In the older adults, its activity is a predictor of lower physical function (Kusakabe, Motoki, & Hori, 1997). Although, aldolase serum levels seem not to be altered in a pilot study on statin users (Trentini et al., 2019), no study has studied the alteration of this enzyme in patients with SAMS.

2-Hydroxyglutarate (2HG)

2HG is a specific biomarker for skeletal muscle injury, produced mainly in the mitochondria by malate dehydrogenase and lactate dehydrogenase A from α-ketoglutarate. Its plasma levels may increase for a mitochondrial dysfunction. A study in rats affected by muscle injury caused by cerivastatin revealed the increase in 2HG, suggesting its possible use as a clinical marker in SAMS patients (Obayashi et al., 2017).

Urinary biomarkers

24-h creatinine excretion has been considered a marker of muscle mass, but it has been shown high variability and high sensibility to diet, particularly to meat intake. Furthermore, the sampling result is quite challenging.

3-methyl histidine has also been used as a marker of muscle protein degradation. This molecule is produced when myosin and actin are methylated in the muscle, then it is not recycled for protein synthesis, which makes it an ideal urine biomarker. However, its validity has been questioned in recent years as it has been displayed to not respond to any intervention known to increase protein degradation (Nedergaard, Karsdal, Sun, & Henriksen, 2013).

Myoglobin in urine has also been associated with rhabdomyolysis or muscle destruction. Myoglobinuria is characterized by several symptoms such as dark urine, calcium ion loss, fever, and nausea. This condition could lead to kidney complications and it has been found in severe cases of statin intolerance.

Micro RNA

Micro RNAs (miRNAs) are small, noncoding RNA molecules that regulate cellular function at the posttranscriptional level, influencing normal biological processes, as well as various pathological conditions (Bartel, 2004). miRNA may be released into bloodstream in response to tissue stress. For this reason, different studies have proposed the use miRNA as emerging biomarkers (Yan, Qian, Chen, Chen, & Shen, 2016).

The miRNAs, expressed in skeletal muscle, called "myomiRs", participate in the regulation of myoblast proliferation, differentiation contractility, and stress responsiveness (van Rooij, Liu, & Olson, 2008). These myomiRs have been examined as markers of acute muscle damage and chronic muscle disease. To date, the identified myomiRs are: miR-1, miR-133a, miR-133b, miR-206, miR-208a, miR-208b, miR-486, and miR-499 (Sempere et al., 2004). MyomiRs are expressed in both cardiac and skeletal muscle with the exception of miR-206, which is skeletal muscle specific, miR-208a, cardiac muscle specific, and miR-486, that is muscle-enriched rather than muscle-specific (McCarthy, 2008). Recently, myomiRs have been studied in relation with statins users. Min et al. (2016), measured the circulating levels of miR-133a, miR-206, and miR-499-5p in runners treated or not with statins. Muscle injury developed during exercise was accompanied with augmented extracellular release of different myomiRs, among them miR-499-5p was more increased in the treated runners.

Conclusion

SAMS are the common reason for hospital admission related to statin toxicity and are one of the primary reasons for statin discontinuation.

The most used serum marker is the serum CK level, but its exclusive use as a diagnostic marker is inadequate and not specific. Moreover, hyperCKemia is not always associated with myopathy (Di Stasi, MacLeod, Winters, & Binder-Macleod, 2010). Routine liver function analysis is no longer recommended and liver abnormalities are rare and dose related during this type of therapy.

Few investigations have been concluded about the beneficial effect of vitamin D on these biomarkers. Vitamin D supplementation lowered creatine kinase, troponin I, and lactic acid dehydrogenase activity, the main used biomarkers for muscle injury, in healthy adults (Al-Eisa, Alghadir, & Gabr, 2016) and runners (Zebrowska et al., 2020).

Further clinical examination, investigations, and biomarkers analysis should be performed to identify the possible causes of muscle pathology as well as the implication of vitamin D in SAMS.

New biomarkers for SAMS are emerging. Unfortunately, most of them cannot be commonly used because of the high cost and the complexities in methodology and their sensitivity and specificity still need to be defined.

Other agents of interest

Energy metabolism has been indicated as one of the key mechanism underlying SAMS. Indeed, statins reduced respiratory capacity in skeletal muscle cells and mitochondrial dysfunction has been linked to SAMS. Statins appeared to be strong inhibitors of mitochondrial complex and induced mitochondrial dysfunction that lead to cytoplasm Ca^{2+} overload and mitochondrial DNA depletion. Hence, procedures or treatments that ameliorate energy metabolism-related myotoxicity by statins could improve the physical state of SAMS subject. Cholesterol reduction by statins results in changes in membrane fluidity in muscle, alterations in ionic channels and a subsequent damaging myopathy.

546 **PART | IV** Assessments, screening, and resources

There are several drugs that can increase the risk of mitochondrial dysfunction such as: antiviral nucleoside reverse transcriptase inhibitors (zalcitabine, ganciclovir, acyclovir, and zidovudine), antibiotics (tetracycline and ciprofloxacine), chemotherapy agents (cisplatin and cyclophosphamide), antipsychotic agents (clozapine), antidepressant selective serotonin reuptake inhibitors (fluoxetine and sertraline), or sodium valproate are drugs that could cause mitochondrial toxicity. Although the association with statins and the majority of these agents have been demonstrated to be safe until today, it could limit their potential therapeutic efficacy with high adverse effects.

As described in the "Statin associated muscle symptoms" section statins are metabolized mainly by CYP3A4 and CYP2C9 isoenzymes. The co-administration with glucocorticoids, erythromycin, fibrates and colchicine, but also substance like alcohol or orange and cranberry juice could determines the decrease of the elimination of statins, leading to SAMS.

In addition, low levels of vitamin D have been related to SAMS. Drugs like anticonvulsant and steroids, used also during myopathies, can reduce vitamin D levels.

Applications to other areas

Pain represents the body alarm system to self-preservation and serves as an alert of danger, becomes instead incapacing to the individual and is then seen as nature's course. Abnormal conditions of chronic pain caused by lesion or disease of the somatosensory nervous system are generally intractable to conventional analgesia and become debilitating.

Among the various types of chronic pain, chronic musculoskeletal pain is one of the most experienced. SAMS is a toxic condition that could be also used to study this kind of pain. Moreover, SAMS may be considered an experimental target from many point of view such as: etiologic, pathogenetic, diagnostic, and therapeutic.

The pain caused by SAMS depend from the dosage and the physicochemical characteristics of the drug, but could also is influenced by lifestyle, genetic and immune hypersensitivity, the type of diet, the co-administration of other drugs, age, kidney, muscular, and hepatic.

In order to avoid SAMS, it is important to consider that the use of statins represent a prevention tool of cardiovascular diseases. Hence, this treatment must reduce the risk without determine other possible pathologies. The suspension of the statin therapy is the first measure to stop the muscular pain. Other solution may be the reduction of the dose or the substitutions with other similar therapy.

The causes of SAMS are not completely known and the low serum levels of vitamin D have been linked to SAMS. This relationship has also been extended to muscular chronic musculoskeletal pain. The possible explanations of this association have been extensively discussed along the chapter, although other concomitant diseases affecting liver or kidney may decrease the synthesis of this vitamin and affect its metabolism. However, the causes could not completely due to other organs and it is necessary to find biomarkers of muscular damage and pain in this condition. Although the main biomarkers such as CK, ALT, and AST, myoglobin and lactic acid have been extensively used for muscular disease or pain condition, the others reported may be of great interest in the next years.

Mini-dictionary of terms

- **Myalgia**: Muscle discomfort.
- **Myopathy**: muscle weakness with variable histopathological findings as atrophy, inflammation, and mitochondrial changes.
- **Myositis**: muscle inflammation with macrophages infiltration.
- **Myonecrosis**: muscle pain with a marked CK elevation and necrosis.
- **Rhabdomyolysis**: clinical syndrome characterized by muscle injury associated with myoglobinuria, electrolyte abnormalities, elevated CK, and often acute kidney injury.

Key facts

- Statins are a low-density lipoprotein (LDL) cholesterol lowering class of drugs. Despite this class is well tolerated, 12.4% of treated subjects show muscular symptoms known as Statin Associated Muscle Symptoms.
- Statin Associated Muscle Symptoms (SAMS) comprehend discomfort (myalgia), muscle weakness (myopathy), tenderness to palpation, with or without muscle inflammation (myositis), and/or myonecrosis.
- Older subjects, female, high physical activity users, but also patients affected by neuromuscular diseases, endocrinopathies, renal and liver diseases, diabetes or cutaneous diseases have higher risk to develop these symptoms. Likewise,

subjects that are using treatments metabolized by the isoenzyme CYP3A4, could prolong statins activity and could be detrimental for muscle health.

- Since cholesterol is a molecule involved in several biochemical process, its decrease could lead to the diminution of vitamin D or Coenzyme Q.
- The biomarker analysis to correctly diagnose SAMS takes into account Creatine Kinase levels, but other factors have to be taken in consideration.

Summary points

- This chapter focuses on old and new biomarkers of Statin myopathy with a particular focus on vitamin D.
- SAMS include fatigue, muscle weakness, muscle cramps, and pain.
- Statins act by blocking HMG-CoAR thereby reducing cholesterol synthesis and vitamin D production. Statins patients with a lower concentration of vitamin D are much more vulnerable to SAMS.
- Although CK is particularly useful to describe and define SAMS, its exclusive use as a diagnostic marker is inadequate and not specific. Therefore, a battery of specific tests is recommended.

References

Al-Eisa, E. S., Alghadir, A. H., & Gabr, S. A. (2016). Correlation between vitamin d levels and muscle fatigue risk factors based on physical activity in healthy older adults. *Clinical Interventions in Aging*, *11*, 513–522. https://doi.org/10.2147/CIA.S102892.

Backes, J. M., Howard, P. A., Ruisinger, J. F., & Moriarty, P. M. (2009). Does simvastatin cause more myotoxicity compared with other statins? *Annals of Pharmacotherapy*, *43*(12), 2012–2020. https://doi.org/10.1345/aph.1M410.

Baker, S. K. (2005). Molecular clues into the pathogenesis of statin-mediated muscle toxicity. *Muscle and Nerve*, *31*(5), 572–580. https://doi.org/10.1002/mus.20291.

Barker, T., Schneider, E. D., Dixon, B. M., Henriksen, V. T., & Weaver, L. K. (2013). Supplemental vitamin D enhances the recovery in peak isometric force shortly after intense exercise. *Nutrition and Metabolism*, *10*(1), 69. https://doi.org/10.1186/1743-7075-10-69.

Bartel, D. P. (2004). MicroRNAs: Genomics, biogenesis, mechanism, and function. *Cell*, *116*(2), 281–297. https://doi.org/10.1016/S0092-8674(04)00045-5.

Bischoff-Ferrari, H. A., Fischer, K., Orav, E. J., Dawson-Hughes, B., Meyer, U., Chocano-Bedoya, P. O., et al. (2017). Statin use and 25-hydroxyvitamin D blood level response to vitamin D treatment of older adults. *Journal of the American Geriatrics Society*, *65*, 1267–1273. https://doi.org/10.1111/jgs.14784.

Braga, M., Simmons, Z., Norris, K. C., Ferrini, M. G., & Artaza, J. N. (2017). Vitamin D induces myogenic differentiation in skeletal muscle derived stem cells. *Endocrine Connections*, *6*(3), 139–150. https://doi.org/10.1530/ec-17-0008.

Brancaccio, P., Lippi, G., & Maffulli, N. (2010). Biochemical markers of muscular damage. *Clinical Chemistry and Laboratory Medicine*, *48*, 757–767.

Burch, P. M., Pogoryelova, O., Goldstein, R., Bennett, D., Guglieri, M., Straub, V., et al. (2015). Muscle-derived proteins as serum biomarkers for monitoring disease progression in three forms of muscular dystrophy. *Journal of Neuromuscular Diseases*, *2*(3), 241–255. https://doi.org/10.3233/JND-140066.

Caso, G., Kelly, P., McNurlan, M. A., & Lawson, W. E. (2007). Effect of coenzyme Q10 on myopathic symptoms in patients treated with statins. *American Journal of Cardiology*, *99*(10), 1409–1412. https://doi.org/10.1016/j.amjcard.2006.12.063.

Chen, S., Villalta, S. A., & Agrawal, D. K. (2016). FOXO1 mediates vitamin D deficiency-induced insulin resistance in skeletal muscle. *Journal of Bone and Mineral Research*, *31*(3), 585–595. https://doi.org/10.1002/jbmr.2729.

Choi, M., Park, H., Cho, S., & Lee, M. (2013). Vitamin D3 supplementation modulates inflammatory responses from the muscle damage induced by high-intensity exercise in SD rats. *Cytokine*, *63*(1), 27–35. https://doi.org/10.1016/j.cyto.2013.03.018.

Christenson, R. H., & Azzazy, H. M. E. (1998). Biochemical markers of the acute coronary syndromes. *Clinical Chemistry*, *44*, 1855–1864. https://doi.org/10.1093/clinchem/44.8.1855.

Clarke, A. T., Johnson, P. C. D., Hall, G. C., Ford, I., & Mills, P. R. (2016). High dose atorvastatin associated with increased risk of significant hepatotoxicity in comparison to simvastatin in UK GPRD cohort. *PLoS One*, *11*(3), e0151587. https://doi.org/10.1371/journal.pone.0151587.

De Pinieux, G., Chariot, P., Ammi-SaïD, M., Louarn, F., Lejonc, J. L., Astier, A., et al. (1996). Lipid-lowering drugs and mitochondrial function: Effects of HMG-CoA reductase inhibitors on serum ubiquinone and blood lactate/pyruvate ratio. *British Journal of Clinical Pharmacology*, *42*(3), 333–337. https://doi.org/10.1046/j.1365-2125.1996.04178.x.

Delliaux, S., Steinberg, J. G., Lesavre, N., Paganelli, F., Oliver, C., & Jammes, Y. (2006). Effect of long-term atorvastatin treatment on the electrophysiological and mechanical functions of muscle. *International Journal of Clinical Pharmacology and Therapeutics*, *44*(6), 251–261. https://doi.org/10.5414/CPP44251.

Di Stasi, S. L., MacLeod, T. D., Winters, J. D., & Binder-Macleod, S. A. (2010). Effects of statins on skeletal muscle: A perspective for physical therapists. *Physical Therapy*, *90*(10), 1530–1542. https://doi.org/10.2522/ptj.20090251.

Dirks, A. J., & Jones, K. M. (2006). Statin-induced apoptosis and skeletal myopathy. *American Journal of Physiology. Cell Physiology*, *291*, C1208–C1212. https://doi.org/10.1152/ajpcell.00226.2006.

Draeger, A., Monastyrskaya, K., Mohaupt, M., Hoppeler, H., Savolainen, H., Allemann, C., et al. (2006). Statin therapy induces ultrastructural damage in skeletal muscle in patients without myalgia. *Journal of Pathology, 210*(1), 94–102. https://doi.org/10.1002/path.2018.

Erlacher, P., Lercher, A., Falkensammer, J., Nassonov, E. L., Samsonov, M. I., Shtutman, V. Z., et al. (2001). Cardiac troponin and β-type myosin heavy chain concentrations in patients with polymyositis or dermatomyositis. *Clinica Chimica Acta, 306*(1–2), 27–33. https://doi.org/10.1016/S0009-8981(01)00392-8.

Galtier, F., Mura, T., Raynaud de Mauverger, E., Chevassus, H., Farret, A., Gagnol, J. P., et al. (2012). Effect of a high dose of simvastatin on muscle mitochondrial metabolism and calcium signaling in healthy volunteers. *Toxicology and Applied Pharmacology, 263*(3), 281–286. https://doi.org/10.1016/j.taap.2012.06.020.

Giannoni, A., Giovannini, S., & Clerico, A. (2009). Measurement of circulating concentrations of cardiac troponin i and T in healthy subjects: A tool for monitoring myocardial tissue renewal? *Clinical Chemistry and Laboratory Medicine, 47*, 1167–1177. https://doi.org/10.1515/CCLM.2009.320.

Glueck, C. J., Budhani, S. B., Masineni, S. S., Abuchaibe, C., Khan, N., Wang, P., et al. (2011). Vitamin D deficiency, myositismyalgia, and reversible statin intolerance. *Current Medical Research and Opinion, 27*, 1683–1690. https://doi.org/10.1185/03007995.2011.598144.

Goh, I. X. W., How, C. H., & Tavintharan, S. (2013). Cytochrome P450 drug interactions with statin therapy. *Singapore Medical Journal, 54*, 131–135. https://doi.org/10.11622/smedj.2013044.

Goli, A. K., Goli, S. A., Byrd, R. P., & Roy, T. M. (2002). Simvastatin-induced lactic acidosis: A rare adverse reaction? *Clinical Pharmacology and Therapeutics, 72*(4), 461–464. https://doi.org/10.1067/mcp.2002.127943.

Greenhaff, P. L., Casey, A., Short, A. H., Harris, R., Soderlund, K., & Hultman, E. (1993). Influence of oral creatine supplementation of muscle torque during repeated bouts of maximal voluntary exercise in man. *Clinical Science, 84*, 565–571. https://doi.org/10.1042/cs0840565.

Huang, J., Du, J., Lin, W., Long, Z., Zhang, N., Huang, X., et al. (2019). Regulation of lactate production through p53/β-enolase axis contributes to statin-associated muscle symptoms. *eBioMedicine, 45*, 251–260. https://doi.org/10.1016/j.ebiom.2019.06.003.

Immke, D. C., & McCleskey, E. W. (2001). Lactate enhances the acid-sensing NA+ channel on ischemia-sensing neurons. *Nature Neuroscience, 4*(9), 869–870. https://doi.org/10.1038/nn0901-869.

Johnson, T. E., Zhang, X., Bleicher, K. B., Dysart, G., Loughlin, A. F., Schaefer, W. H., et al. (2004). Statins induce apoptosis in rat and human myotube cultures by inhibiting protein geranylgeranylation but not ubiquinone. *Toxicology and Applied Pharmacology, 200*(3), 237–250. https://doi.org/10.1016/j.taap.2004.04.010.

Jürgens, K. D., Papadopoulos, S., Peters, T., & Gros, G. (2000). Myoglobin: Just an oxygen store or also an oxygen transporter? *News in Physiological Sciences, 15*(5), 269–274. https://doi.org/10.1152/physiologyonline.2000.15.5.269.

Kamga, C., Krishnamurthy, S., & Shiva, S. (2012). Myoglobin and mitochondria: A relationship bound by oxygen and nitric oxide. *Nitric Oxide: Biology and Chemistry, 26*(4), 251–258. https://doi.org/10.1016/j.niox.2012.03.005.

Kanno, T., Sudo, K., Maekawa, M., Nishimura, Y., Ukita, M., & Fukutake, K. (1988). Lactate dehydrogenase M-subunit deficiency: A new type of hereditary exertional myopathy. *Clinica Chimica Acta, 173*(1), 89–98. https://doi.org/10.1016/0009-8981(88)90359-2.

Kuncl, R. W., Duncan, G., Watson, D., Alderson, K., Rogawski, M. A., & Peper, M. (1987). Colchicine myopathy and neuropathy. *New England Journal of Medicine, 316*(25), 1562–1568. https://doi.org/10.1056/nejm198706183162502.

Kusakabe, T., Motoki, K., & Hori, K. (1997). Mode of interactions of human aldolase isozymes with cytoskeletons. *Archives of Biochemistry and Biophysics.* https://doi.org/10.1006/abbi.1997.0204.

Laaksonen, R., Jokelainen, K., Sahi, T., Tikkanen, M. J., & Himberg, J. J. (1995). Decreases in serum ubiquinone concentrations do not result in reduced levels in muscle tissue during short-term simvastatin treatment in humans. *Clinical Pharmacology and Therapeutics, 57*(1), 62–66. https://doi.org/10.1016/0009-9236(95)90266-X.

Law, M., & Rudnicka, A. R. (2006). Statin safety: A systematic review. *American Journal of Cardiology, 97*(8 Suppl. 1). https://doi.org/10.1016/j.amjcard.2005.12.010.

Malaguarnera, M., Vacante, M., Russo, C., Gargante, M. P., Giordano, M., Bertino, G., et al. (2011). Rosuvastatin reduces nonalcoholic fatty liver disease in patients with chronic hepatitis C treated with α-interferon and ribavirin: Rosuvastatin reduces NAFLD in HCV patients. *Hepatitis Monthly, 11*(2), 92–98. http://www.pubmedcentral.nih.gov/articlerender.fcgi?artid=3206670&tool=pmcentrez&rendertype=abstract.

Mancini, G. B. J., Tashakkor, A. Y., Baker, S., Bergeron, J., Fitchett, D., Frohlich, J., et al. (2013). Diagnosis, prevention, and management of statin adverse effects and intolerance: Canadian working group consensus update. *Canadian Journal of Cardiology, 29*(12), 1553–1568. https://doi.org/10.1016/j.cjca.2013.09.023.

McCarthy, J. J. (2008). MicroRNA-206: The skeletal muscle-specific myomiR. *Biochimica et Biophysica Acta, Gene Regulatory Mechanisms, 1779*(11), 682–691. https://doi.org/10.1016/j.bbagrm.2008.03.001.

Michalska-Kasiczak, M., Sahebkar, A., Mikhailidis, D. P., Rysz, J., Muntner, P., Toth, P. P., et al. (2015). Analysis of vitamin D levels in patients with and without statin-associated myalgia—A systematic review and meta-analysis of 7 studies with 2420 patients. *International Journal of Cardiology, 178*, 111–116. https://doi.org/10.1016/j.ijcard.2014.10.118.

Min, P. K., Park, J., Isaacs, S., Taylor, B. A., Thompson, P. D., Troyanos, C., et al. (2016). Influence of statins on distinct circulating microRNAs during prolonged aerobic exercise. *Journal of Applied Physiology, 120*(6), 711–720. https://doi.org/10.1152/japplphysiol.00654.2015.

Nathwani, R. A., Pais, S., Reynolds, T. B., & Kaplowitz, N. (2005). Serum alanine aminotransferase in skeletal muscle diseases. *Hepatology, 41*(2), 380–382. https://doi.org/10.1002/hep.20548.

Nedergaard, A., Karsdal, M. A., Sun, S., & Henriksen, K. (2013). Serological muscle loss biomarkers: An overview of current concepts and future possibilities. *Journal of Cachexia, Sarcopenia and Muscle, 4*(1), 1–17. https://doi.org/10.1007/s13539-012-0086-2.

Nogueira, A. A. R., Strunz, C. M. C., Takada, J. Y., & Mansur, A. P. (2019). Biochemical markers of muscle damage and high serum concentration of creatine kinase in patients on statin therapy. *Biomarkers in Medicine*, *13*(8), 619–626. https://doi.org/10.2217/bmm-2018-0379.

Obayashi, H., Kobayashi, N., Nezu, Y., Yamoto, T., Shirai, M., & Asai, F. (2017). Plasma 2-hydroxyglutarate and hexanoylcarnitine levels are potential biomarkers for skeletal muscle toxicity in male fischer 344 rats. *Journal of Toxicological Sciences*, *42*(4), 385–396. https://doi.org/10.2131/jts.42.385.

Parker, B. A., Capizzi, J. A., Grimaldi, A. S., Clarkson, P. M., Cole, S. M., Keadle, J., et al. (2013). Effect of statins on skeletal muscle function. *Circulation*, *127*, 96–103. https://doi.org/10.1161/CIRCULATIONAHA.112.136101.

Parkkila, S., Vullo, D., Maresca, A., Carta, F., Scozzafava, A., & Supuran, C. T. (2012). Serendipitous fragment-based drug discovery: Ketogenic diet metabolites and statins effectively inhibit several carbonic anhydrases. *Chemical Communications*, *48*(29), 3551–3553. https://doi.org/10.1039/c2cc30359k.

Pencina, M. J., Navar-Boggan, A. M., D'Agostino, R. B., Williams, K., Neely, B., Sniderman, A. D., et al. (2014). Application of new cholesterol guidelines to a population-based sample. *The New England Journal of Medicine*, *370*, 1422–1431. https://doi.org/10.1056/nejmoa1315665.

Pennisi, M., Di Bartolo, G., Malaguarnera, G., Bella, R., Lanza, G., & Malaguarnera, M. (2019). Vitamin D serum levels in patients with statin-induced musculoskeletal pain. *Disease Markers*, *2019*. https://doi.org/10.1155/2019/3549402, 3549402.

Phulukdaree, A., Moodley, D., Khan, S., & Chuturgoon, A. A. (2015). Atorvastatin increases miR-124a expression: A mechanism of Gamt modulation in liver cells. *Journal of Cellular Biochemistry*, *116*(11), 2620–2627. https://doi.org/10.1002/jcb.25209.

Ren, L., Xuan, L., Han, F., Zhang, J., Gong, L., Lv, Y., et al. (2020). Vitamin D supplementation rescues simvastatin induced myopathy in mice via improving mitochondrial cristae shape. *Toxicology and Applied Pharmacology*, *401*. https://doi.org/10.1016/j.taap.2020.115076, 115076.

Rosenson, R. S., Baker, S. K., Jacobson, T. A., Kopecky, S. L., & Parker, B. A. (2014). An assessment by the statin muscle safety task force: 2014 update. *Journal of Clinical Lipidology*, *8*, S58–S71. https://doi.org/10.1016/j.jacl.2014.03.004.

Sathasivam, S., & Lecky, B. (2008). Statin induced myopathy. *BMJ*, *337*(7679), 1159–1162. https://doi.org/10.1136/bmj.a2286.

Seachrist, J. L., Loi, C. M., Evans, M. G., Criswell, K. A., & Rothwell, C. E. (2005). Roles of exercise and pharmacokinetics in cerivastatin-induced skeletal muscle toxicity. *Toxicological Sciences*, *88*, 551–561. https://doi.org/10.1093/toxsci/kfi305.

Sempere, L. F., Freemantle, S., Pitha-Rowe, I., Moss, E., Dmitrovsky, E., & Ambros, V. (2004). Expression profiling of mammalian microRNAs uncovers a subset of brain-expressed microRNAs with possible roles in murine and human neuronal differentiation. *Genome Biology*, *5*, R13. https://doi.org/10.1186/gb-2004-5-3-r13.

Sorichter, S., Mair, J., Koller, A., Gebert, W., Rama, D., Calzolari, C., et al. (1997). Skeletal troponin I as a marker of exercise-induced muscle damage. *Journal of Applied Physiology*, *83*(4), 1076–1082. https://doi.org/10.1152/jappl.1997.83.4.1076.

Taylor, B. A., Lorson, L., White, C. M., & Thompson, P. D. (2015). A randomized trial of coenzyme Q10 in patients with confirmed statin myopathy. *Atherosclerosis*, *238*(2), 329–335. https://doi.org/10.1016/j.atherosclerosis.2014.12.016.

Taylor, B. A., Lorson, L., White, C. M., & Thompson, P. D. (2017). Low vitamin D does not predict statin associated muscle symptoms but is associated with transient increases in muscle damage and pain. *Atherosclerosis*, *256*, 100–104. https://doi.org/10.1016/j.atherosclerosis.2016.11.011.

Tonomura, Y., Mori, Y., Torii, M., & Uehara, T. (2009). Evaluation of the usefulness of biomarkers for cardiac and skeletal myotoxicity in rats. *Toxicology*, *266*(1–3), 48–54. https://doi.org/10.1016/j.tox.2009.10.014.

Toth, P. P., Patti, A. M., Giglio, R. V., Nikolic, D., Castellino, G., Rizzo, M., et al. (2018). Management of Statin Intolerance in 2018: Still more questions than answers. *American Journal of Cardiovascular Drugs*. https://doi.org/10.1007/s40256-017-0259-7.

Trentini, A., Spadaro, S., Rosta, V., Manfrinato, M. C., Cervellati, C., Corte, F. D., et al. (2019). Fast skeletal troponin i, but not the slow isoform, is increased in patients under statin therapy: A pilot study. *Biochemia Medica*, *29*(1). https://doi.org/10.11613/BM.2019.010703, 010703.

van Rooij, E., Liu, N., & Olson, E. N. (2008). MicroRNAs flex their muscles. *Trends in Genetics*, *24*(4), 159–166. https://doi.org/10.1016/j.tig.2008.01.007.

Yan, W., Qian, L., Chen, J., Chen, W., & Shen, B. (2016). Comparison of prognostic microRNA biomarkers in blood and tissues for gastric cancer. *Journal of Cancer*, *7*(1), 95–106. https://doi.org/10.7150/jca.13340.

Zebrowska, A., Sadowska-Krępa, E., Stanula, A., Waśkiewicz, Z., Łakomy, O., Bezuglov, E., et al. (2020). The effect of vitamin D supplementation on serum total 25(OH) levels and biochemical markers of skeletal muscles in runners. *Journal of the International Society of Sports Nutrition*, *17*(1). https://doi.org/10.1186/s12970-020-00347-8.

Zhu, K., Austin, N., Devine, A., Bruce, D., & Prince, R. L. (2010). A randomized controlled trial of the effects of vitamin D on muscle strength and mobility in older women with vitamin d insufficiency. *Journal of the American Geriatrics Society*, *58*(11), 2063–2068. https://doi.org/10.1111/j.1532-5415.2010.03142.x.

Chapter 49

Performance-based and self-reported physical fitness in musculoskeletal pain

Cristina Maestre-Cascales[a,*], Javier Courel-Ibáñez[b,*], and Fernando Estévez-López[c]

[a]*LFE Research Group, Department of Health and Human Performance, Polytechnic University of Madrid, Madrid, Spain,* [b]*Department of Physical Activity and Sport, University of Murcia, San Javier, Spain,* [c]*Department of Pediatrics, Wilhelmina Children's Hospital, University Medical Center Utrecht, Utrecht, The Netherlands*

Abbreviations

1RM   one repetition maximum
IFIS   International Fitness Scale
US   The United States of America

Physical fitness

Physical fitness is defined as a number of qualities that people have or achieve in relation to their ability to be physically active (U.S. Department of Health and Human Services, 1996). Broadly, this term is attributed to a state of well-being with low risk of premature health problems and high energy to participate in a variety of activities (Howley & Franks, 1997), including morphological fitness, bone strength, muscle fitness, flexibility and motor fitness, cardiovascular fitness, and metabolic fitness (Bouchard, Shephard, & Stephens, 1994).

From a practical point of view, physical fitness is an integrated measure for most of body functions—usually all—that are involved in the practice of daily physical activity, including cardiorespiratory, hematocrine, psychoneurological, and enkocrine/metabolic responses. Thus, when physical fitness is analyzed, the functional capacity of all these systems is indirectly assessed. This is one of the main reasons why physical fitness is considered one of the most important health markers (Blair et al., 1989; Metter, Talbot, Schrager, & Conwit, 2002; Mora et al., 2003). Fitness level can be influenced by both environmental conditions, genetics but, above all, physical exercise (Ortega, Ruiz, Castillo, & Sjöström, 2008). However, a safe and effective physical exercise practice requires an individual assessment of fitness and health status, particularly in people with physical diseases, pain, or musculoskeletal disorders.

Musculoskeletal disorders

The World Health Organization defines musculoskeletal disorders as "health problems of the locomotor apparatus, i.e. muscles, tendons, bone skeleton, cartilage, ligaments, and nerves. This includes any type of complaint, from slight transitory discomforts to irreversible and incapacitating injuries" (Miranda, Kaila-Kangas, & Ahola, 2011). Chronic musculoskeletal pain is a predominant complaint of people living with these conditions. A number of biopsychosocial factors influence the impact of pain, including but not limited to the presence of comorbid health problems, social support, gender/sex, health education and knowledge, socioeconomic status, and personality (Hawker, 2017).

Musculoskeletal conditions and pain together are the main pathologies leading to disability worldwide and a cause of a number of physical impairments, such as low levels of physical fitness, poor agility, slow gait speed, weak grip strength, poor balance (ability to maintain tandem position), limited mobility, frailty, and risk of falling. In addition, musculoskeletal conditions are associated with cognitive impairments, depression and poor sleep quality (Chen, Hayman, Shmerling, Bean,

---

* The first and second authors have equally contributed as first authors of the chapter.

Features and Assessments of Pain, Anesthesia, and Analgesia. https://doi.org/10.1016/B978-0-12-818988-7.00012-1
Copyright © 2022 Elsevier Inc. All rights reserved.

552 PART | IV Assessments, screening, and resources

& Leveille, 2011; Stubbs et al., 2013; Van Der Leeuw et al., 2016). These problems contribute to reducing physical activity and functional capacity while increasing the risk of disability of those who suffer from musculoskeletal conditions (Blyth & Noguchi, 2017).

Widespread musculoskeletal pain

Widespread musculoskeletal pain is defined as "pain on the left side of the body, pain on the right side of the body, pain above the waist, pain below the waist and pain on axial skeletal (cervical spine, anterior chest, thoracic spine or low back)" (Wolfe et al., 1990). Widespread musculoskeletal pain has physical and psychological consequences. Sleep disorders are among the most frequent, disturbing 78% of people with fibromyalgia, 60% of people with rheumatoid arthritis, and 50% of those with chronic low-back pain. In addition, a high percentage of people suffering from chronic musculoskeletal pain present a high degree of physical fatigue that makes it difficult to perform their daily activities (physical function): this occurs in 76% of people with fibromyalgia and 41% of people with osteoarthritis or arthritis (Riedemann, 2008).

Previous evidence has shown that there is a strong association between pain and decreased physical activity (Hawley & Wolfe, 1991), and therefore the severity, duration, or location of pain plays an important role in a person's physical functioning (Ang, Kroenke, & McHorney, 2006; Kovacs, Abraira, Zamora, Fernández, & Network, 2005). As a consequence, this decrease in physical activity contributes to the deterioration of physical fitness levels, in particular, the progressive loss of strength and flexibility (Tüzün, 2007). Hence, levels of physical fitness act as a marker of health in musculoskeletal pain.

Physical fitness can be assessed subjectively using self-reports or objectively using performance-based tests. Both types of measures (subjective and objective) provide distinct and complementary information (Estévez-López et al., 2018, 2017; Munguía-Izquierdo et al., 2019; Pulido-Martos et al., 2020).

Assessment of physical fitness in widespread musculoskeletal pain

One of the most commonly used batteries for the evaluation of the physical fitness in widespread musculoskeletal pain is the **Senior Fitness Test**, which was originally designed for the elderly (Rikli & Jones, 1999a, 1999b). This battery has shown to be feasible and reliable in people with widespread musculoskeletal pain (Carbonell-Baeza et al., 2015) and has been widely used for because of three main reasons:

(i) The battery has shown no ceiling and floor effects, which is a relevant aspect considering the high heterogeneity of people with widespread musculoskeletal pain (Estévez-López et al., 2020, 2017; Pérez-Aranda et al., 2019; Wilson, Robinson, & Turk, 2009).

(ii) People with widespread musculoskeletal pain have levels of physical fitness similar to older adults from the general population.

(iii) The use of this battery allows comparisons with population-based samples (Alvarez-Gallardo et al., 2017; Carbonell-Baeza et al., 2011, 2015; Jones, Rakovski, Rutledge, & Gutierrez, 2015; Ofei-Dodoo et al., 2018).

In addition to the **Senior Fitness Test**, there are a variety of tests available to measure physical fitness in people with widespread musculoskeletal pain. Generally, the most widely used tests for their reliability and accuracy are the objective tests as shown below. A summary of the guidelines, variations, and instruments is described in Table 1 (upper- and lower-limb strength), Table 2 (upper- and lower-limb range of motion/flexibility), Table 3 (static and dynamic balance), Table 4 (gait speed), and Table 5 (aerobic endurance and fitness level). All these tests are safe and easy to perform and score, and presents adequate psychometric properties. It is advisable to previously screen patients and undergo a medical examination to identify physical, cardiovascular, or metabolic conditions to confirm they can complete the test in a secure manner.

At the same time, it is very important to know some of the subjective tests used occasionally in this population (self-perception of physical fitness). Among the most widely used is the International Scale of Physical Fitness (IFIS), validated and specific for people with widespread musculoskeletal pain. This instrument consists of five questions from the Likert scale on how participants perceive their general physical condition, cardiorespiratory status, muscle strength, speed-agility, and flexibility with the answer options being: 1 (very bad), 2 (bad), 3 (average), 4 (good), and 5 (very good).

Applications to other areas

It would be of special interest to apply this type of measure to favor family and/or collective support, increase adherence to the program, and encourage the use of new technologies through monitoring and control of the effects of training. Furthermore, the use of these technologies could also be extrapolated to the design of home-based programs with remote control, or the use of alternative therapies such as Tai-Chi.

TABLE 1 Upper- and lower-limb strength tests commonly used in musculoskeletal pain.

Test	Aim and description	Data collection	Score	Material
Upper-limb strength				
Handgrip strength[a,b,c]	Aim: To assess general strength levels The participant presses gradually, keeping at least 2″, performing the test with the right and left hands in turn. Two attempts with each hand, with the arm fully extended, forming a 30 degree angle with respect to the trunk. 1′ rest between measurements.	Mean score of the left and right hand, or, the dominant and nondominant side (as independent variables) *Variation:* only with their dominant hand	Higher scores indicate better performance	Dynamometer (e.g., TKK 5101 Grip-D; Takey, Tokyo)
Arm curl test[a,c]	Aim: To assess the strength of the upper limb The test measures the number of times the subject performs full elbow bends in 30″, with the person sitting upright, (2.3 kg for women and 3.6 kg for men). Patients perform a test with both hands.	They perform a trial after familiarization. The number of repetitions carried out within the time interval in dominant and nondominant hand is recorded. If the arm is more than halfway down at the end of 30″ it is counted as a complete bend. *Variation:* 2 kg for women	Higher scores indicate better performance	Dumbbell or free weights
Lower-limb strength				
30-s- chair stand test[c]	Aim: To assess the strength of the lower limb The test involves counting the number of times within 30s that an individual can rise to a full stand from a seated position with back straight and feet flat on the floor, without pushing off with the arms. Arms crossed at chest level.	Patients perform a trial after familiarization and note the number of repetitions performed. If the participant has done more than half the movement, it is counted as complete. It should be counted aloud. *Variation:* count the number of times within 30 s	Higher scores indicate better performance	Chair and stopwatch
10-step-stair climbing[d]	Aim: Assessing lower limb strength and balance, agility, and fall risk Participants have to climb stairs (step height 20 cm) "as fast and safe as possible". The use of handrails when necessary is allowed for safety reasons.	The time it takes to go up (or up and down) 10 steps is measured. *Variation:* Change the task requirement, ascent only or combined ascent/descent	Lower scores indicate better performance	Step height 20 cm and stopwatch
One repetition maximum (1RM) test- leg extension machine[e]	Aim: To assess general strength levels. Assessing Trunk Extender and Bending Strength. 15′ pre-warm-up on the bike. The test is administered within a range of motion from 0 to 90 degree, considering the full extension of the knee joint as the anatomical zero point. The trunk, hip and knee are stabilized. Similar auditory and visual stimuli should be given.	Subjects are familiarized before 1RM measurement. The 1RM is considered the highest weight lifted using the appropriate form. After a minimum of 72 h, the 1RM is checked again. The highest measurement of the test is considered the 1RM.	Higher RM scores indicate better performance	Dynamometer Cybex 6000; Lumex, Albertson, NY Progressive loading test on leg-extension bench
Subjective test				
Muscular strength dimension[c]	Aim: To assess muscular strength (self-perception) Dimension number 3 of the	*Variation:* Possibility of analyzing it including all dimensions.	Higher scores indicate a	

Continued

TABLE 1 Upper- and lower-limb strength tests commonly used in musculoskeletal pain—cont'd

Test	Aim and description	Data collection	Score	Material
	International Fitness Scale (IFIS) entitled *"My muscular strength is..."* assesses the participant's perception of overall strength levels (reported self-perception). It is scored on a 5-point scale (1: very poor, 2; poor, 3; average, 4; good, 5; very good)		better performance.	

[a]Maestre-Cascales, C., Peinado Lozano, A.B., & Rojo González, J.J. (2019). Effects of a strength training program on daily living in women with fibromyalgia. Journal of Human Sport and Exercise, *https://doi.org/10.14198/jhse.2019.144.03.*
[b]Maestre-Cascales, C., Girela-Rejón, M., Sánchez-Gallo, D., et al. (2019). Association of handgrip strength and well-being in women with fibromyalgia. Revista Internacional de Ciencias del Deporte, 15(58), 307–322. https://doi.org/10.5232/ricyde.
[c]Estévez-López, F., Segura-Jiménez, V., Álvarez-Gallardo, I.C., et al. (2017). Adaptation profiles comprising objective and subjective measures in fibromyalgia: The al-Ándalus project. Rheumatol (United Kingdom), 56(11), 2015–2024. https://doi.org/10.1093/rheumatology/kex302.
[d]Collado-Mateo, D., Adsuar, J.C., Olivares, P.R., Dominguez-Muñoz, F.J., Maestre-Cascales, C., & Gusi, N. (2016). Performance of women with fibromyalgia in walking up stairs while carrying a load. PeerJ, 2016(2). https://doi.org/10.7717/peerj.1656.
[e]Wong, A., Figueroa, A., Sanchez-Gonzalez, M.A., Son, W.M., Chernykh, O., & Park, S.Y. (2018). Effectiveness of Tai Chi on cardiac autonomic function and symptomatology in women with fibromyalgia: A randomized controlled trial. Journal of Aging and Physical Activity, 26(2), 214–221. https://doi.org/10.1123/japa.2017-0038.

TABLE 2 Upper- and lower-limb range of motion tests commonly used in musculoskeletal pain.

Test	Aim and description	Data collection	Score	Material
Upper-limb range of motion				
Back-scratch[a,b]	Aim: Measure the overall range of motion of the shoulders and assess the flexibility of the upper body In a standing position, the participant places one hand behind the shoulder on the same side, palm down and fingers extended, reaching halfway down the back (elbow up). The participant places the other hand behind the back, palm out, reaching as far as possible in an attempt to touch or overlap the extended middle (or longer) fingers of both hands. After the demonstration, the assessor gives the participant a test time and then performs the test with two attempts (one with each hand).	The distance of overlap or between the tips of both middle fingers (or the longer ones) is measured to the nearest centimeter. A negative (−) score is given for the remaining distance and a positive (+) score is given for the overlap distance. Record both values and circle the best value and use it to evaluate test performance. Be sure to record "minus" (−) or "plus" (+) on the record sheet.	Higher scores indicate a better performance	Scaled measuring stick or measuring tape
Lower-limb range of motion				
Chair sit and reach[c,d]	Aim: To assess the extensibility of the ischial musculature and evaluate the flexibility of the lower body The participant sits on the edge of the chair keeping one leg in flexion (foot resting on the floor) and the other in	It records either the number of centimeters the subject is missing to touch the toes (negative score) or the score is exceeded (positive score). Record both tests to the nearest cm and circle the best score. The best score is used to	Higher scores indicate a better performance	Approved box (35 cm × 45 cm × 32 cm)

TABLE 2 Upper- and lower-limb range of motion tests commonly used in musculoskeletal pain—cont'd

Test	Aim and description	Data collection	Score	Material
	maximum extension (heel resting on the floor and foot flexed 90 degree). The hands are placed one on top of the other, so that the longest fingers coincide with each other. They should move down over the extended leg trying to touch the toes. A ruler is recommended as a guide to slide the hands and at the same time measure the distance reached. It should be held at least 2″.	evaluate the test. Be sure to indicate "minus" (−) or "plus" (+) on the score card. *Variation:* A test variant: "box sit-and-reach"		
Subjective test				
Flexibility dimension[e]	Aim: To assess flexibility (self-perception) Dimension number 3 of the International Fitness Scale (IFIS) entitled "*My flexibility is…*" assesses the participant's perception of flexibility levels (reported self-perception). It is scored on a 5-point scale (1: very poor, 2; poor, 3; average, 4; good, 5; very good)		Higher scores indicate a better performance.	

[a]Soriano-Maldonado, A., Estévez-López, F., Segura-Jiménez, V., et al. (2016). Association of Physical Fitness with Depression in Women with Fibromyalgia. Pain Medicine, 17(8), 1542–1552. https://doi.org/10.1093/pm/pnv036
[b]Alvarez-Gallardo, I.C., Carbonell-Baeza, A., Segura-Jimenez, V., et al. (2017). Physical fitness reference standards in fibromyalgia: The al-Andalus project. The Scandinavian Journal of Medicine & Science in Sports, 27(11), 1477–1488. https://doi.org/10.1111/sms.12741.
[c]Segura-Jiménez, V., Soriano-Maldonado, A., Estévez-López, F., et al. (2017). Independent and joint associations of physical activity and fitness with fibromyalgia symptoms and severity: The al-Ándalus project. Journal of Sports Sciences, 35(15), 1565–1574.
[d]Soriano-Maldonado, A., Ruiz, J.R., Aparicio, V.A., et al. (2015). Association of physical fitness with pain in women with fibromyalgia: The al-andalus project. Arthritis Care & Research, 67(11), 1561–1570. https://doi.org/10.1002/acr.22610.
[e]Munguía-Izquierdo, D., Pulido-Martos, M., Acosta, F.M., et al. (2019). Objective and subjective measures of physical functioning in women with fibromyalgia: what type of measure is associated most clearly with subjective well-being? Disability and Rehabilitation, https://doi.org/10.1080/09638288.2019.1671503.

TABLE 3 Static and dynamic balance tests commonly used in musculoskeletal pain.

Test	Aim and description	Data collection	Score	Material
Static balance				
One-leg stance test[a]	Aim: To assess functional status and fall risk. The participant remains in a balanced position on one leg with eyes open and arms on hips. When the participant loses the balance position, they are allowed to touch the ground with their foot.	It is timed in seconds from the time a foot is bent on the ground to the time it touches the ground, or the standing leg or arm leaves the hips. In addition, the number of test attempts is recorded. The test is performed with right and left extremities leaving 30″ of rest between them. A single attempt. *Variations;* eyes open or closed, arms crossed or attached to the body, bare feet or shoes, etc.	Higher values indicate better results. < 5″ indicates a higher risk of falling	Stopwatch

Continued

PART | IV Assessments, screening, and resources

TABLE 3 Static and dynamic balance tests commonly used in musculoskeletal pain—cont'd

Test	Aim and description	Data collection	Score	Material
Blind flamingo test[b]	Aim: Evaluate the strength of the lower limb, pelvis and trunk, as well as the body's dynamic balance. Go up to the surface without shoes. Keep your balance with the help of the instructor. While balancing on the supporting leg, the free leg flexes (90 degree) bringing the heel closer to the buttock. Start the watch when the instructor releases it. Stop the watch each time the person loses balance. Start again, timing it again until the person loses his or her balance. Count the number of falls in 60 s of balance.	The number of attempts required by the performer (not falls) to maintain balance for one minute is counted. If there are more than 15 falls in the first 30 s, the test is terminated, and the score is given. Example: if the person has needed 5 attempts, 5 points are assigned	Higher scores indicate poorer performance	A metal beam 50 cm long, 4 cm high and 3 cm wide. Stopwatch
The stork balance stand test[c]	Aim: To assess static balance. On a hard, flat surface, the participant stands on one foot without shoes and hands on hips with the opposite foot placed against the inside of the knee pad. At the signal, the subject lifts the heel of the foot off the ground and tries to maintain balance as long as possible for 1'. The test ends if the participant moves his or her hands from the hips, the supporting foot, or if contact with the knee is lost. The test is timed in seconds using a stopwatch. The stopwatch starts when the heel is lifted off the ground.	Two attempts are made with each leg and the best results are recorded. Data are collected for the right and left foot, and the average was used.	Higher scores indicate a better performance	A stopwatch and a hard, flat surface.
Motor agility/dynamic balance				
8-ft up-and-go[d,e]	Aim: To assess speed-agility. The test begins with the participant sitting in the chair, hands on quads and feet supported. At the signal, the participant stands up, walks as fast as possible but without running around to the cone, turns around, and returns to the chair. The evaluator sets the time at the signal and stops it when the participant sits in the chair. After a demonstration, the participant performs the test twice.	The score is the time completed. Record the score of both tests to within a tenth of a second and circle the best score (the shortest time).	Lower scores indicate better performance.	A cone, a stopwatch and a chair.

[a]Tomas-Carus, P., Gusi, N., Hakkinen, A., Hakkinen, K., Raimundo, A., & Ortega-Alonso, A. (2009). Improvements of muscle strength predicted benefits in HRQOL and postural balance in women with fibromyalgia: an 8-month randomized controlled trial. Rheumatology, 48(9), 1147–1151. https://doi.org/10.1093/rheumatology/kep208.
[b]Aparicio, V.A., Carbonell-Baeza, A., Ruiz, J.R., et al. (2013). Fitness testing as a discriminative tool for the diagnosis and monitoring of fibromyalgia. The Scandinavian Journal of Medicine & Science in Sports, 23(4), 415–423. https://doi.org/10.1111/j.1600-0838.2011.01401.x.
[c]Santos, E., Campos, M.A., Párraga-Montilla, J.A., Aragón-Vela, J., Latorre-Román, P.A. (2020). Effects of a functional training program in patients with fibromyalgia: A 9-year prospective longitudinal cohort study. The Scandinavian Journal of Medicine & Science in Sports, 30(5), 904–913. https://doi.org/10.1111/sms.13640.
[d]Estevez-Lopez, F., Segura-Jimenez, V., Alvarez-Gallardo, I.C., et al. (2017). Adaptation profiles comprising objective and subjective measures in fibromyalgia: the al-Andalus project. Rheumatology, 56(11), 2015–2024. https://doi.org/10.1093/rheumatology/kex302.
[e]Álvarez-Gallardo, I.C.C., Carbonell-Baeza, A., Segura-Jiménez, V., et al. (2017). Physical fitness reference standards in fibromyalgia: The al-Ándalus project. The Scandinavian Journal of Medicine & Science in Sports, 27(11), 1477–1488. https://doi.org/10.1111/sms.12741.

TABLE 4 Gait speed tests commonly used in musculoskeletal pain.

Test	Aim and description	Data collection	Score	Material
Gait speed				
Walking speed test [a,b]	Aim: To assess the gait speed walking. The race starts with the participant standing with both feet at the same height in front of the starting line. At the signal, the participant must walk to the finish line to cover the distance of 30 m as quickly as possible without running. There will be two attempts with a 1' rest. It is advisable to put an additional mark 2 or 3 m beyond the 30 m to prevent participants from stopping before crossing the finish line.	The score is the time from the signal until the participant passes the finish line at 30 m. The best of the two attempts will be taken into account.	Lower scores indicate better performance.	A cone and a stopwatch
Subjective test				
Speed/ agility dimension[c]	Aim: To assess speed/agility (self-perception). Dimension number 3 of the International Fitness Scale (IFIS) entitled "My speed/agility is..." assesses the participant's perception of speed/agility levels (reported self-perception). It is scored on a 5-point scale (1: very poor, 2; poor, 3; average, 4; good, 5; very good)			Higher scores indicate a better performance.	

[a]Pierrynowski, M.R., Tiidus, P.M., & Galea, V. (2005). Women with fibromyalgia walk with an altered muscle synergy. Gait Posture, 22(3), 210–218. https://doi.org/10.1016/j.gaitpost.2004.09.007.
[b]Costa I da, S., Gamundí, A., Miranda, J.G.V., França, L.G.S., De Santana, C.N., & Montoya, P. (2017). Altered functional performance in patients with fibromyalgia. Frontiers in Human Neuroscience, 11, https://doi.org/10.3389/fnhum.2017.00014.
[c]Álvarez-Gallardo, I.C., Soriano-Maldonado, A., Segura-Jiménez, V, et al. (2016). International fitness scale (IFIS): Construct validity and reliability in women with fibromyalgia: The al-ándalus project. The Archives of Physical Medicine and Rehabilitation, 97(3), 395–404. https://doi.org/10.1016/j.apmr.2015.08.416.

Other agents of interest

Specific instruments to measure physical fitness may be applicable to other conditions associated with pain and physical limitations. Moreover, researchers should reach a consensus for harmonising the measurement of physical fitness across studies. Besides, physical fitness is not only considered a relevant health marker in people with widespread musculoskeletal pain, but it is recognized at a general level.

Mini-dictionary of terms

- **Fibromyalgia**: Chronic disease characterized by widespread musculoskeletal pain as the main symptom, associated with the presence of multiple locations of tender points.
- **Frailty**: Weakness or tendency to deteriorate.
- **Disability**: Lack or limitation of any physical or mental faculty that limits or hinders the normal development of a person's activity.
- **Physical activity:** Any body movement produced by skeletal muscles that requires energy expenditure.
- **Objective measures**: Information based on performance-based tests.
- **Subjective measures**: Information self-reported by people by means of questionnaires; also referred to as patient-reported outcomes.

558 PART | IV Assessments, screening, and resources

TABLE 5 Aerobic endurance and fitness level tests commonly used in musculoskeletal pain.

Test	Aim and description	Data collection	Score	Material
Aerobic endurance				
6-min walk test[a,b]	Aim: To assess aerobic fitness. The participant should walk as fast as possible without running the distance for 6 min, avoiding the group and conversation during the test. If necessary, the participants can stop and rest (two chairs should always be placed in the opposite corners of the rectangle so that the older ones can rest), and then continue walking. The participants should be warned when there are 2 and 1 min to go, respectively, and they should be motivated during the test. At the end of the 6 min, participants are asked to stop and move to the right, where an evaluator will record the score.	The score is the total number of meters walked in the 6 min. The assistant records the mark of the nearest cone. *Variation*: Changing the walking speed	Higher scores indicate a better performance.	Cones and stopwatch
Incremental cycling test[c]	Aim: To assess aerobic fitness. Participants perform the maximum incremental test on an ergometer following the protocol: Warm-up period of 6 min of steady state low intensity cycle followed by 15 W of initial power for 2 min increasing 15 W every 2 min. Perceived effort (RPE) is evaluated every minute throughout the test with a standard 100 mm VAS.	Record the perceived effort, the number of intervals, the power and the duration achieved *Variation*: The test starts with a 3′ warm-up at 50 W. The intensity is then increased by 20 W at 2′ intervals until exhaustion. It can be used, lactate samples, electromyograph, among others.	Higher scores indicate a better performance.	Electromyograph and effort perception scale.
2-min step test [d]	Aim: To assess aerobic fitness. This test requires that the participant to walk in place, lifting her knees to an intermediate point between her kneecap and the iliac crest. This test consists of completing cycles (1 cycle is 2 knee lifts, one with each leg) for 2 min. Participants are asked to do so as quickly as possible, keeping in mind their functional limitations	The number of cycles performed in 2 min is recorded. *Variation*: Participants could use a wall or chair to maintain balance, if necessary.	A higher step count indicates better cardiovascular fitness. Thus, higher scores indicate a better performance.	

TABLE 5 Aerobic endurance and fitness level tests commonly used in musculoskeletal pain—cont'd

Test	Aim and description	Data collection	Score	Material
Subjective test				
Cardiorespiratory dimension[e]	Aim: To assess cardiorespiratory fitness (self-perception). Dimension number 3 of the International Fitness Scale (IFIS) entitled "*My cardiorespiratory fitness (capacity to do exercise, for instance long running) is...* "assesses the participant's perception of overall cardiorespiratory levels (reported self-perception). It is scored on a 5-point scale (1: very poor, 2; poor, 3; average, 4; good, 5; very good)	*Variation*: Possibility of analyzing it including all dimensions.	Higher scores indicate a better performance.	

[a]Ericsson, A., Palstam, A., Larsson, A., et al. (2016). Resistance exercise improves physical fatigue in women with fibromyalgia: a randomized controlled trial. Arthritis Research & Therapy, 18(1), 176. https://doi.org/10.1186/s13075-016-1073-3.

[b]Carbonell-Baeza, A., Ruiz, J.R., Aparicio, V.A., Ortega, F.B., & Delgado-Fernandez, M. (2013). The 6-minute walk test in female fibromyalgia patients: Relationship with tenderness, symptomatology, quality of life, and coping strategies. Pain Management Nursing, 14(4), 193–199. https://doi.org/10.1016/j.pmn.2011.01.002.

[c]Bardal, E.M., Roeleveld, K., & Mork, P.J. (2015). Aerobic and cardiovascular autonomic adaptations to moderate intensity endurance exercise in patients with fibromyalgia. Journal of Rehabilitation Medicine, 47(7), 639–646. https://doi.org/10.2340/16501977-1966.

[d]Bohannon, R.W., Bubela, D.J., Wang, Y.C., Magasi, S.S., & Gershon, R.C. Six-minute walk test Vs. three-minute step test for measuring functional endurance. The Journal of Strength and Conditioning Research, 29(11), 3240–3244. https://doi.org/10.1519/JSC.0000000000000253.

[e]Munguía-Izquierdo, D., Pulido-Martos, M., Acosta, F.M., et al. (2019). Objective and subjective measures of physical functioning in women with fibromyalgia: What type of measure is associated most clearly with subjective well-being? Disability and Rehabilitation, https://doi.org/10.1080/09638288.2019.1671503.

Key facts

- Chronic musculoskeletal pain and all the conditions it encompasses have a strong impact on the quality of life of patients suffering from it.
- Objective and subjective assessments of physical fitness provide distinct and complementary information.
- In people with chronic musculoskeletal pain, physical fitness is an important marker of health.

Summary points

- Physical fitness is an integrated practical measure for assessing most body functions involved in physical activity. Therefore, it assesses the functional capacity of all systems. For this reason, it is one of the most important health markers and a predictor of morbidity and mortality.
- Musculoskeletal disorders such as widespread chronic pain are leading cause of disability worldwide.
- The Senior Fitness Test battery is one of the most used batteries in people with widespread musculoskeletal pain.

References

Alvarez-Gallardo, I. C., Carbonell-Baeza, A., Segura-Jimenez, V., Soriano-Maldonado, A., Intemann, T., Aparicio, V. A., ... Ortega, F. B. (2017). Physical fitness reference standards in fibromyalgia: The al-Andalus project. *Scandinavian Journal of Medicine & Science in Sports*, 27(11), 1477–1488. https://doi.org/10.1111/sms.12741.

Ang, D. C., Kroenke, K., & McHorney, C. A. (2006). Impact of pain severity and location on health-related quality of life. *Rheumatology International*, 26 (6), 567–572.

Blair, S. N., Kohl, H. W., Paffenbarger, R. S., Clark, D. G., Cooper, K. H., & Gibbons, L. W. (1989). Physical fitness and all-cause mortality: A prospective study of healthy men and women. *JAMA*, *262*(17), 2395–2401.

Blyth, F. M., & Noguchi, N. (2017). Chronic musculoskeletal pain and its impact on older people. *Best Practice & Research. Clinical Rheumatology*, *31*(2), 160–168.

Bouchard, C. E., Shephard, R. J., & Stephens, T. E. (1994). Physical activity, fitness, and health: International proceedings and consensus statement. In *International consensus symposium on physical activity, fitness, and health, 2nd, May, 1992, Toronto, ON, Canada*Human Kinetics Publishers.

Carbonell-Baeza, A., Alvarez-Gallardo, I. C., Segura-Jimenez, V., Castro-Pinero, J., Ruiz, J. R., Delgado-Fernandez, M., & Aparicio, V. A. (2015). Reliability and feasibility of physical fitness tests in female fibromyalgia patients. *International Journal of Sports Medicine*, *36*(2), 157–162. https://doi.org/10.1055/s-0034-1390497.

Carbonell-Baeza, A., Aparicio, V. A., Ortega, F. B., Cuevas, A. M., Alvarez, I. C., Ruiz, J. R., & Delgado-Fernandez, M. (2011). Does a 3-month multidisciplinary intervention improve pain, body composition and physical fitness in women with fibromyalgia? *British Journal of Sports Medicine*, *45*(15), 1189–1195. https://doi.org/10.1136/bjsm.2009.070896.

Chen, Q., Hayman, L. L., Shmerling, R. H., Bean, J. F., & Leveille, S. G. (2011). Characteristics of chronic pain associated with sleep difficulty in older adults: The maintenance of balance, independent living, intellect, and zest in the elderly (MOBILIZE) Boston study. *Journal of the American Geriatrics Society*, *59*(8), 1385–1392.

Estévez-López, F., Álvarez-Gallardo, I. C., Segura-Jiménez, V., Soriano-Maldonado, A., Borges-Cosic, M., Pulido-Martos, M., ... Geenen, R. (2018). The discordance between subjectively and objectively measured physical function in women with fibromyalgia: Association with catastrophizing and self-efficacy cognitions. The al-Ándalus project. *Disability and Rehabilitation*, *40*(3), 329–337. https://doi.org/10.1080/09638288.2016.1258737.

Estévez-López, F., Maestre-Cascales, C., Russell, D., Álvarez-Gallardo, I. C., Rodriguez-Ayllon, M., Hughes, C. M., ... McVeigh, J. G. (2020). Effectiveness of exercise on fatigue and sleep quality in fibromyalgia: A systematic review and meta-analysis of randomised trials. *Archives of Physical Medicine and Rehabilitation*. https://doi.org/10.1016/j.apmr.2020.06.019.

Estévez-López, F., Segura-Jiménez, V., Álvarez-Gallardo, I. C., Borges-Cosic, M., Pulido-Martos, M., Carbonell-Baeza, A., ... Delgado-Fernández, M. (2017). Adaptation profiles comprising objective and subjective measures in fibromyalgia: The al-Ándalus project. *Rheumatology (Oxford, England)*, *56*(11), 2015–2024. https://doi.org/10.1093/rheumatology/kex302.

Hawker, G. A. (2017). The assessment of musculoskeletal pain. *Clinical and Experimental Rheumatology*, *35*(5), S8–S12.

Hawley, D. J., & Wolfe, F. (1991). Pain, disability, and pain/disability relationships in seven rheumatic disorders: A study of 1,522 patients. *The Journal of Rheumatology*, *18*(10), 1552–1557.

Howley, E. T., & Franks, B. D. (1997). *Health fitness instructors handbook*. Champaign, IL: Human Kinetics.

Jones, C. J., Rakovski, C., Rutledge, D., & Gutierrez, A. (2015). A comparison of women with fibromyalgia syndrome to criterion fitness standards: A pilot study. *Journal of Aging and Physical Activity*, *23*(1), 103–111. https://doi.org/10.1123/japa.2013-0159.

Kovacs, F. M., Abraira, V., Zamora, J., Fernández, C., & Network, S. B. P. R. (2005). The transition from acute to subacute and chronic low back pain: A study based on determinants of quality of life and prediction of chronic disability. *Spine*, *30*(15), 1786–1792.

Metter, E. J., Talbot, L. A., Schrager, M., & Conwit, R. (2002). Skeletal muscle strength as a predictor of all-cause mortality in healthy men. *The Journals of Gerontology. Series A, Biological Sciences and Medical Sciences*, *57*(10), B359–B365.

Miranda, H., Kaila-Kangas, L., & Ahola, K. (2011). *Ache and melancholy: Co-occurence of musculoskeletal pain and depressive symptoms in Finland*. Finnish Institute of Occupational Health.

Mora, S., Redberg, R. F., Cui, Y., Whiteman, M. K., Flaws, J. A., Sharrett, A. R., & Blumenthal, R. S. (2003). Ability of exercise testing to predict cardiovascular and all-cause death in asymptomatic women: A 20-year follow-up of the lipid research clinics prevalence study. *JAMA*, *290*(12), 1600–1607.

Munguía-Izquierdo, D., Pulido-Martos, M., Acosta, F. M., Acosta-Manzano, P., Gavilán-Carrera, B., Rodriguez-Ayllon, M., ... Estévez-López, F. (2019). Objective and subjective measures of physical functioning in women with fibromyalgia: What type of measure is associated most clearly with subjective well-being? *Disability and Rehabilitation*. https://doi.org/10.1080/09638288.2019.1671503.

Ofei-Dodoo, S., Rogers, N. L., Morgan, A. L., Amini, S. B., Takeshima, N., & Rogers, M. E. (2018). The impact of an active lifestyle on the functional fitness level of older women. *Journal of Applied Gerontology*, *37*(6), 687–705. https://doi.org/10.1177/0733464816641390.

Ortega, F. B., Ruiz, J. R., Castillo, M. J., & Sjöström, M. (2008). Physical fitness in childhood and adolescence: A powerful marker of health. *International Journal of Obesity*, *32*(1), 1–11. https://doi.org/10.1038/sj.ijo.0803774.

Pérez-Aranda, A., Andrés-Rodríguez, L., Feliu-Soler, A., Núñez, C., Stephan-Otto, C., Pastor-Mira, M. A., ... Luciano, J. V. (2019). Clustering a large Spanish sample of patients with fibromyalgia using the fibromyalgia impact questionnaire-revised: Differences in clinical outcomes, economic costs, inflammatory markers, and gray matter volumes. *Pain*, *160*(4), 908–921. https://doi.org/10.1097/j.pain.0000000000001468.

Pulido-Martos, M., Luque-Reca, O., Segura-Jiménez, V., Álvarez-Gallardo, I. C., Soriano-Maldonado, A., Acosta-Manzano, P., ... Estévez-López, F. (2020). Physical and psychological paths toward less severe fibromyalgia: A structural equation model. *Annals of Physical and Rehabilitation Medicine*, *63*(1), 46–52. https://doi.org/10.1016/j.rehab.2019.06.017.

Riedemann, P. (2008). Impacto del dolor musculoesquelético. *Medwave*, *8*(5). https://doi.org/10.5867/medwave.2008.05.1754.

Rikli, R. E., & Jones, C. J. (1999b). Functional fitness normative scores for community-residing older adults, ages 60–94. *Journal of Aging and Physical Activity*, *7*(2), 162–181.

Rikli, R. E., & Jones, C. J. (1999a). Development and validation of a functional fitness test for community-residing older adults. *Journal of Aging and Physical Activity*, *7*(2), 129–161.

Stubbs, B., Binnekade, T. T., Soundy, A., Schofield, P., Huijnen, I. P. J., & Eggermont, L. H. P. (2013, September 1). Are older adults with chronic musculoskeletal pain less active than older adults without pain? A systematic review and meta-analysis. *Pain Medicine (United States)*, *14*, 1316–1331. https://doi.org/10.1111/pme.12154.

Tüzün, E. H. (2007). Quality of life in chronic musculoskeletal pain. *Best Practice & Research. Clinical Rheumatology, 21*, 567–579. https://doi.org/10.1016/j.berh.2007.03.001.

U.S. Department of Health and Human Services. (1996). *Physical activity and health: A report of the surgeon general*. http://www.Cdc.Gov/Nccdphp/Sgr/Pdf/Execsumm.Pdf.

Van Der Leeuw, G., Eggermont, L. H. P., Shi, L., Milberg, W. P., Gross, A. L., Hausdorff, J. M., … Leveille, S. G. (2016). Pain and cognitive function among older adults living in the community. *The Journals of Gerontology. Series A, Biological Sciences and Medical Sciences, 71*(3), 398–405.

Wilson, H. D., Robinson, J. P., & Turk, D. C. (2009). Toward the identification of symptom patterns in people with fibromyalgia. *Arthritis and Rheumatism, 61*(4), 527–534. https://doi.org/10.1002/art.24163.

Wolfe, F., Smythe, H. A., Yunus, M. B., Bennett, R. M., Bombardier, C., Goldenberg, D. L., … Clark, P. (1990). The American College of Rheumatology 1990 criteria for the classification of fibromyalgia. *Arthritis & Rheumatism: Official Journal of the American College of Rheumatology, 33*(2), 160–172.

Index

Note: Page numbers followed by *f* indicate figures and *t* indicate tables.

A

Abdominal pain
  appendicitis (*see* Appendicitis)
  in gastroparesis (*see* Gastroparesis)
Acute Pain Service (APS)
  agents of interest, 76
  application to other areas, 76
  characteristics
    criteria to activate APS, 71, 71*t*
    multidisciplinary teams, 71
    postoperative evaluation flow chart, 71, 72*f*
    preoperative pain optimization, 71–72
  clinical conditions fit for discharge, 70*t*
  clinical experiences
    adverse event rate, 75
    commonly used/experimental agents for pain control, 75*t*
    elective unilateral total hip arthroplasty, 76
    epidural analgesia, 75–76
    pain intensity and side effects, 74–75
    published experiences, 75*t*
  clinical outcome
    adverse events prevention, 73
    awareness in patient care, 73
    clinical advantages, 72
    clinical monitoring, 73
    cost reduction, 74
    timing of postoperative pain evaluation, 73, 73*f*
  diffusion, 74
  Enhanced Recovery After Surgery (ERAS) approach, 69, 70*t*
  history, 70–71
  key facts, 76–77
  postoperative pain control, 69
Alzheimer's disease (AD)
  age and analgesic treatment, 497
  characterization, 496
  dementia-related disorders, 496
  future research in pain management, 502–504
  gender-related pain analysis, 498–499, 499*t*
  Global deterioration scale (GDS)
    and analgesic consumption, 501–502, 502*t*, 503*f*
    barriers, 502
    behavioral assessments and cognitive tests, 500–501
    degree of cognitive deterioration, 502, 503*f*

mild cognitive impairment (MCI), 501
    pain-related behaviors, 501
    Reisberg Global Deterioration Scale (GDS), 501
    score, 501, 502*t*
    and sociodemographic characteristics, 502, 503–504*t*
  pain response, 497–498
  pathology, 496
  psychophysiological factors, 497–498
  sociodemographic characteristics
    analgesics consumption, 499–500, 500*t*
    sex, 500, 501*t*
  study sample, 497*t*
Analgesia nociception index (ANI)
  confounders, 470
  features, 465–466, 465–466*f*
  ICU setting, 469
  intraoperative nociception detection
    craniotomy, 467–468
    hemodynamics, 467–468
    laryngoscopy and intubation, 467
    sevoflurane anesthesia, 467
    spine surgery, 467
    and Surgical Pleth Index (SPI), 467
    total knee replacement, 467
  key facts, 471
  monitor/display screen, 465–466, 466*f*
  nociception-antinociception balance, 464
  vs. objective analgesia/nociception monitors, 470
  opioid consumption, 469–470
  pediatric pain assessment, 469
  perioperative clinical practice, 464*t*
  postoperative pain assessment, 468–469
  principle, 464–465, 464*f*
  sensor placement, 465, 465*f*
  terminology, 471
  validated pain scales, 463
  values interpretation, 466, 466*f*
Analgosedation
  commonly used opioids, 62–63, 62*t*
  definition, 62
  Opioid-Related Adverse Drug Events (ORADES), 63
  results, 63
  surgical patients, 63
Appendicitis
  agents of interest, 194–197

applications to other areas, 194
  atypical presentation, 192
  causes, 189
  clinical prediction rules, 193
  common and rare causes of abdominal pain, 196*t*
  complications, 194
  etiology, 189–190
  imaging tests, 193
  incidence, 189
  indications, 189
  inflammatory markers, 192–193
  interval appendectomy, 194
  key facts, 197
  lower abdominal pain
    in adults, 195*t*
    in children, 196*t*
  management, 193–194
  natural history models, 190–191, 191*f*
  open and laparoscopic appendectomies, 194
  pathophysiology, 190, 190*f*
  pelvic abdominal pain, 195*t*
  surgical approach, 194
  terminology, 197
  typical presentation, 192
Attachment
  attachment theory, 15
  concept, 15
  internal working models (IWM), 15
  key facts, 22
  meanings and measurement
    attachment style, 15
    dismissing attachment style, 16, 17*f*
    fearful attachment style, 16, 17*f*
    insecure attachment patterns, 16, 17*f*
    preoccupied attachment style, 16, 17*f*
    secure attachment pattern, 16, 16*f*
    self-protective strategies, 16
  and pain
    in adult, 20
    agents of interest, 21–22
    applications to other areas, 21
    in child and adolescence, 18–20
    models, 18, 19*f*
    psychological predisposing factors, 18
    typical individual features, 18, 19*t*
  terminology, 22
Auditory evoked potential (AEP), 83–84

563

564 Index

B

Back pain functional scale (BPFS)
  agents of interest, 489
  applications to other areas, 489
  key facts, 490
  original BPFS, 488–489, 488*t*
  outcome measures, 487–488
  vs. Roland–Morris Questionnaire (RMQ), 489
  terminology, 490
  12 items, 488
  World Health Organization's (WHO) model, 488
Bilateral pelvic plexus blockade (PPB), 356
Bilateral periprostatic nerve block (PPNB), 355, 356*f*
Biomarkers
  bladder pain syndrome (BPS)
    antiproliferative factor (APF) biomarker, 533
    bladder wall specimens, 529–530*t*, 531
    Etio-S, 533
    GP51, 534
    inflammatory and angiogenic markers, 533
    microbiome, 534
    nerve growth factor (NGF), 532
    serum biomarkers, 531*t*, 532
    stool, 532
    urinary biomarkers, 530–531*t*, 532
    urothelial dysfunction, 533
  blood biomarkers
    angiogenesis, growth factors, and growth factor receptors, 510
    apoptosis markers, 511
    cell adhesion molecules and other matrix-related proteins, 511
    cytoskeleton molecules, 511
    diagnosis, 512–517*t*
    high-throughput molecular markers, 511
    hormonal markers, 511
    immune system and inflammatory biomarkers, 511
    molecules in DNA repair/telomere maintenance, 511
    oxidative stress markers, 518
    posttranscriptional regulators of gene expression, 518
    tumor markers, 518
  definition, 510
  endometrial biomarkers
    angiogenesis and growth factors, 519
    diagnosis, 520–521*t*
    hormonal markers, 519–520
    inflammatory and myogenic markers, 520
    neural markers, 520
    tumor markers, 520
  noninvasive biomarkers, 510
  statin associated muscle symptoms (SAMS)
    aldolase, 544
    carbonic anhydrases III (CAIII), 544
    coenzyme Q_{10} (CoQ_{10}), 543
    creatine, 542
    creatine kinase (CK), 543–544
    creatinine, 542
    diagnosis, 541–542

enzyme, 543–544
fatty acid-binding proteins, 543
2-hydroxyglutarate (2HG), 544
lactate dehydrogenase (LDH), 544
lactic acid, 542
micro RNAs (miRNAs), 545
myoglobin, 543
transaminase AST and ALT, 544
troponin, 542
urinary biomarkers, 545
  urinary biomarkers, 518, 519*t*
Biopsychosocial (BPS) model, 226, 226*f*
Bispectral index (BIS)
  limitations and influence on, 82–83
  monitoring, 82, 82*f*
Bladder pain syndrome (BPS)/interstitial cystitis (IC)
  agents of interest, 534
  biomarkers
    bladder wall specimens, 529–530*t*, 531
    serum biomarkers, 531*t*, 532
    stool, 532
    urinary biomarkers, 530–531*t*, 532
  challenges, 528
  characterization, 527
  diagnostic criteria, 527–528, 528*f*
  key facts, 535
  pathophysiology
    antiproliferative factor (APF) biomarker, 533
    Etio-S, 533
    GP51, 534
    inflammatory and angiogenic markers, 533
    microbiome, 534
    nerve growth factor (NGF), 532
    urothelial dysfunction, 533
  terminology, 535
Blast injuries, 374
Body mass
  airway management, 38
  anesthesia, anesthetic agents, and muscle relaxants, 39
  anesthesia cessation, 42
  body mass index (BMI), 37, 38*t*
  cardiovascular system assessment, 38
  endocrine and musculoskeletal system assessment, 38–39
  fentanyl, 40
  heart hypertrophy, 38
  intraoperative management, 39
  key facts, 42
  neostigmine, 40
  neuromuscular-blocking agents and antagonists, 40
  obesity hypoventilation syndrome (OHS), 37
  obstructive sleep apnea (OSA), 37
  opioids, 40
  peripheral venous access, 38
  postoperative care, 42
  preoperative assessment, 37
  prevalence, 37
  propofol, 40
  regional anesthesia
    advantage, 40–41
    antithrombotic prophylaxis, 41–42

hypotension, 41
parasagittal oblique view, 41*f*
peripheral nerve blocks, 42
respiratory system assessment, 37–38
sugammadex, 40
terminology, 42
thiopental sodium, 39
type II diabetes, 38–39
venous thromboembolism, 38
volatile anesthetics, 40
Boston Carpal Tunnel Questionnaire-symptom severity scale (CTQ-SSS), 277
Botox injections, 363
Breast cancer
  agents of interest, 252–253
  applications to other areas, 252
  arthrotomy models, 250, 250*t*
  bone inoculation models
    distinct bone inoculation models, 251
    history, 251
  bone pain
    analgesic treatment, 249
    bisphosphonates, 249
    cancer-induced bone pain (CIBP), 249
    denosumab, 249
    external beam radiotherapy (EBR), 249
    factors, 249
    intensity *vs.* time, 249
    neurodestruction/neuromodulation, 249
    osteogenic pain, 249
    osteonecrosis, 249
    radiopharmaceuticals, 249
    side effects, 249–250
  breast inoculation models, 250–251, 250*t*
  cancer pain treatment
    acetaminophen, 248
    adjuvant therapies, 248
    antidepressants/anticonvulsants, 248
    drowsiness and dizziness, 248
    gastrointestinal toxicity, 248
    limitations, 248
    nonopioid analgesics, 248
    nonpharmacological and interventional practices, 248
    opioid-induced constipation (OIC), 248
    serotonin syndrome, 248
    skin inflammation, 248
    tramadol, 248
    WHO analgesic scale, 247–248
  chronic pain management, 247
  key facts, 253
  nonpharmacological treatment, 252
  pharmacological targets
    cannabinoid receptor 2 (CB2), 252
    neurotrophins, 251–252
    transient receptor potential vanilloid 1 (TRPV1), 252
    tumor necrosis factor-α (TNFα), 252
  prevalence, 247
  terminology, 253
Breast surgery. *See also* Ultrasound-guided Pecs II block
  acute and chronic postoperative pain, 343–344
  general anesthesia, 344

thoracic epidural anesthesia-analgesia (TEA) and thoracic paravertebral block (TVPB), 343
Brief Scale for Psychiatric problems in Orthopaedic Patients (BSPOP), 481–482, 483–484*f*

C

Cancer-induced bone pain (CIBP), 249. *See also* Breast cancer
Cancer pain
  agents of interest, 6
  assessment and reassessment, 4
  etiologies, 4–5, 5*t*
  global assessment, 5–6
  management, 3
  multimorphic cancer pain
    analgesics, 7
    concept, 6–7
    factors influencing, 7
    treatment strategies, 7–8
  physiopathological mechanism, 4, 5*t*
  supportive care, 5–6
  WHO cancer pain management guiding principles
    adverse events with opioids, 9
    analgesic drugs, 8
    bone pain, 10
    cancer-related neuropathic pain, 10
    cessation of opioid use, 10
    drug-based treatment, 8–10
    goal, 8
    initiation of cancer-related pain relief, 8
    interventional and surgical analgesic techniques, 11
    maintenance of cancer-related pain relief, 8
    nociceptive cancer-related pain, 8–10
    non-drug-based approaches and noninvasive techniques, 11
    opioid-induced constipation (OIC), 9
    opioid switching, 9–10
    palliative and end-of-life situations, 11
    strong opioids, 9
Cannabinoid receptor 2 (CB2), 252
Carpal tunnel syndrome (CTS)
  agents of interest, 278
  applications to other areas, 276
  carpal tunnel structure, 275, 276*f*
  central and peripheral sensitization mechanisms, 276
  characterization, 275
  conservative treatment, 278
  considerations for evaluation
    acute condition, 277
    Boston Carpal Tunnel Questionnaire-symptom severity scale (CTQ-SSS), 277
    diagram, 278*f*
    grip strength, 277
    risk factors, 276–277
    symptoms, 277
  key facts, 281
  median nerve compression, 275–276
  multidimensional etiology, 276, 277*f*
  orthoses, 278
  prevalence, 275

  psychosocial aspects
    catastrophic thinking, 280
    depression and anxiety, 280
    kinesiophobia, 280
    occupational biomechanical risk factors, 280
    prognostic factors, 280
    psychological variables, 280
  quantitative sensory tests, 276
  splints, 278
  surgical treatment, 278–279, 279*f*
Catastrophizing, pain
  applications
    prevention, 447
    research, 447
    treatment, 447
  biological processes, 445
  characterization, 445
  experience of pain, 445
  fear avoidance model, 445
  health outcomes, 445
  helplessness, 446
  key facts, 448
  magnification, 446
  Pain Catastrophizing Scale (PCS), 445
  rumination, 446
  score and subscales, 446
  terminology, 448
  treatment efficacy, 445
Caudal block, 357
Cerebral state index (CSI), 86
Cesarean delivery, 381. *See also* Spinal anesthesia
Cholecystectomy, 422
Chronic pain in military veterans
  applications to other areas, 231
  biopsychosocial (BPS) model, 226, 226*f*
  catastrophic injuries, 227
  combat-related injuries, 225, 227
  comorbid conditions, 227
  key facts, 232
  military basic training and rates of injury, 226–227
  musculoskeletal injuries, 227
  noncombat-related injuries, 225
  nonpharmacologic pain modalities, 229
  pain assessment and treatment
    clinical pain guidelines and policies, 229
    Defense and Veterans Pain Rating Scale (DVPRS), 229, 230*f*
    multimodal treatment, 229
    Pain, Enjoyment, and General Activity (PEG) scale, 229, 231*f*
  polytrauma, 227
  prevalence, 225
  psychological comorbidities and social factors
    anxiety, 228
    depression, 228
    PTSD, 228
    socioeconomic challenges, 228–229
    substance use disorder (SUD), 228
  terminology, 231
  theoretical models, 226, 226*f*
  types, 227

Cinchona officinalis (quinine), 49, 49*f*
Cloves, 48–49, 49*f*
Cluster headache
  agents of interest, 101
  applications to other areas, 100
  chronic cluster headache, 93
  diagnostic criteria, 93, 94*t*
  epidemiology, 93–94
  episodic cluster headache, 93
  key facts, 101
  pathophysiology
    autonomic system, 95
    genetics, 95–96
    hypothalamus, 95
    trigeminal nerve, 95
  rhythmicity, 94
  terminology, 101
  treatment
    deep brain stimulation (DBS), 100
    Greater Occipital Nerve (GON-injection), 100
    high-flow oxygen therapy, 98
    neuromodulatory treatments, 100
    noninvasive vagus nerve stimulation (NVNS), 100
    Occipital Nerve Stimulation (ONS), 100
    pharmacological treatments, 98–99
    principles, 98
    prophylactic treatment, 98–99
    SPG stimulation/blockade, 100
    sumatriptan, 98
    transitional treatment, 99
  trigeminal autonomic cephalalgias (TACs)
    attack frequency and duration, 97*f*
    diagnostic criteria, 96, 96–97*t*
    differentiation, 97, 98*t*
    paroxysmal hemicrania, 97
    treatment recommendations, 99*t*
Cocaine, 48, 48*f*
Cognitive impairment. *See also* Alzheimer's disease (AD)
  applications to other areas, 493–494
  behavioral disorders and pain, 493
  key facts, 494
  in older people, 493
  pain in the elderly, 496
  physiological and pathological aging
    biology of aging, 494–495
    biology of pain, 495–496
    brain aging, 495
    pain in aging, 496
    pain perception, 495–496
  salivary pain biomarkers, 493
  terminology, 494
Combat-related extremity injury, 225, 227
  acute pain, 374
    early intervention, 374–375
    ketamine, 375
    multimodal analgesic approach, 375
    opioid medications, 375
  agents of interest, 377
  anesthesia and war
    battalion aid stations (BAS), 374
    combat support hospital (CSH), 374
    dismounted complex blast injuries, 374

566 Index

Combat-related extremity injury *(Continued)*
    forward surgical team (FST), 374
      patient care, 374
  applications to other areas, 377
  case-fatality rates (CFR), 373
  clinical outcomes, 373–374
  damage control resuscitation (DCR), 376–377
  diethyl ether, 373
  inflammatory markers, 373–374
  key facts, 377–378
  military anesthesia, 373
  osseointegration, 376
  regional anesthesia
    critically ill pain management, 375–376
    Military Advanced Regional Anesthesia and Analgesia Initiative (MARAA), 375
    neuraxial anesthetics, 376
    peripheral nerve analgesia, 376
  targeted muscle reinnervation, 376
  terminology, 377
Complex regional pain syndrome (CRPS)
  agents of interest, 122
  characterization, 117
  diagnosis, 117–118, 118*t*
  interventional treatment
    intravenous regional anesthesia, 122
    spinal cord stimulation (SCS), 122
    sympathetic nerve blocks (SNB), 121–122
  key facts, 123
  management, 118–119
  pathophysiology, 118
  pharmacologic treatment
    anticonvulsants, 119
    antidepressants, 119
    antihypertensives and α-adrenergic antagonists, 119
    antiinflammatory drugs, 120
    bisphosphonates, 120
    calcitonin, 121
    cannabis, 121
    gabapentin, 119
    intravenous immunoglobulin (IVIG), 121
    ketamine, 121
    medications, 120*t*
    naltrexone, 121
    nifedipine, 119
    NMDA receptor antagonists, 121
    opioids, 121
    phenoxybenzamine, 119
    pregabalin, 119
    tricyclic antidepressants (TCAs), 119
  terminology, 123
  variable symptoms, 122
Computer and smartphone users. *See* Occupational musculoskeletal disorders (MSDs)
Cosmetic injections
  agents of interest, 370
  application, 363
  applications to other areas, 369
  key facts, 370
  local anesthesia
    adverse drug reactions (ADR), 368

    allergy, 368
    amides and esters, 366, 366*f*
    contamination, 368
    disadvantages, 367–368
    efficacy, 367
    FDA approved topical anesthetics, 367, 367*t*
    injecting anesthetics techniques, 368, 369*f*
    mechanism of action, 366–367
  needle fear, 363–364
  psychological intervention, 364
  tumescent anesthesia, 370
  vapocoolant anesthesia
    ice anesthesia, 369
    mechanism of action, 368–369
    topical coolants, 369
  vibration anesthesia
    botulinum toxin injection with, 365, 365*f*
    disadvantages, 366
    safety, 366
    vibrator device, 364, 365*f*

D

Defense and Veterans Pain Rating Scale (DVPRS), 229, 230*f*
Dementia, 497–498
Diabetic neuropathy, 139–140. *See also* Painful diabetic neuropathy (PDN)
Dismounted complex blast injuries, 374
Distal symmetric polyneuropathy (DSP), 286

E

Early-life stress (ELS), 149, 159
Electroencephalography (EEG)
  applications to other areas, 86
  bispectral index (BIS)
    limitations and influence on, 82–83
    monitoring, 82, 82*f*
  cerebral state index (CSI), 86
  entropy, 84
  evoked potential
    auditory evoked potential (AEP), 83–84
    somatosensory evoked potential (SSEP), 83
    visual evoked potential (VEP), 83
  key facts, 87
  Narcotrend, 84–86, 85*t*
  raw EEG activity
    classification, 80, 80*f*
    electrodes positioning, 80*f*
    during general anesthesia, 80*f*, 81
    influence on, 81–82
    during routine EEG monitoring, 80*f*
    waveforms, 80, 81*f*
  terminology, 87
Endometriosis-associated pain
  applications to other areas, 523
  biomarker, 510
  blood biomarkers
    angiogenesis, growth factors, and growth factor receptors, 510
    apoptosis markers, 511
    cell adhesion molecules and other matrix-related proteins, 511

    cytoskeleton molecules, 511
    diagnosis, 512–517*t*
    high-throughput molecular markers, 511
    hormonal markers, 511
    immune system and inflammatory biomarkers, 511
    molecules in DNA repair/telomere maintenance, 511
    oxidative stress markers, 518
    posttranscriptional regulators of gene expression, 518
    tumor markers, 518
  classification, 509
  definition, 507
  diagnosis, 509
  endometrial biomarkers
    angiogenesis and growth factors, 519
    diagnosis, 520–521*t*
    hormonal markers, 519–520
    inflammatory and myogenic markers, 520
    neural markers, 520
    tumor markers, 520
  key facts, 523
  pathophysiology, 507, 508*f*
  prevalence, 509
  terminology, 523
  theory, 507
  treatment, 510
  urinary biomarkers, 518, 519*t*
Enhanced recovery after surgery (ERAS), 69, 70*t*, 392
Entropy, 84
Erector spinae plane block (ESPB)
  complications, 400, 415
  indications, 399, 414
  technique, 399–400, 400*f*, 414
Erythroxylum coca (cocaine), 48, 48*f*
Eutectic mixture of local anesthetic cream (EMLA), 357
Evoked potential
  auditory evoked potential (AEP), 83–84
  somatosensory evoked potential (SSEP), 83
  visual evoked potential (VEP), 83
Exercise, 110–113

F

Filler injections, 363

G

Gastroparesis
  abdominal pain
    antiepileptics, 182
    antipsychotics, 182
    antispasmodics, 182
    assessment, 181*t*
    buspirone, 182
    cannabinoid, 182–183
    causes, 174–175
    celiac plexus block, 184
    chronic abdominal wall pain, 184
    etiology, 174, 176*f*
    haloperidol, 182
    mirtazapine, 180
    neuromodulators, 180, 181*t*, 182

neuropathic and nociceptive pain, 174
nonsteroidal antiinflammatory drugs, 183
opiate analgesics, 183–184
quality of life, 175
tricyclic antidepressants (TCAs), 180
delayed gastric emptying, 173
diabetes, 174
diagnosis, 177
idiopathic gastroparesis, 174
key facts, 184
occurrence, 173
physical examination, 177
symptom assessment, 176–177
symptoms, 174, 176t
terminology, 184
treatment
botulinum toxin, 179
dietary modification, 177–178
domperidone, 178–179
endoscopic and surgical treatments, 179–180
erythromycin, 179
gastrectomy, 180
gastric electrical stimulation, 179–180
jejunostomy and venting gastrostomy tubes, 180
management, 178t
metoclopramide, 178
pharmacologic treatments, 178–179
pyloroplasty and pyloromyotomy, 179
types, 174, 175t
Ginseng, 51–52, 52f
Global deterioration scale (GDS)
and analgesic consumption, 501–502, 502t, 503f
barriers, 502
behavioral assessments and cognitive tests, 500–501
degree of cognitive deterioration, 502, 503f
mild cognitive impairment (MCI), 501
pain-related behaviors, 501
score, 501, 502t
and sociodemographic characteristics, 502, 503–504t

H

Health professionals and lay people
analgesics in postoperative pain management
bioethics, 308–309
participants, 307–308
patients' companions, 309
scenarios/clusters, 308, 309f
applications to other areas, 313
cancer pain, 305–306
key facts, 313–314
knowledge of pain treatment, 305
morphine for intense pain relief
clusters, 310, 311f
pain level, 310
participants, 309–310
opioids for pain management
clusters, 306, 308f
palliative care, 306–307
participants, 306
sample scenario, 306, 307t

styles and approaches, 306
vignette technique and cluster analysis, 306
scenario-based studies
analgesics in postoperative pain management, 307–309, 309f
morphine for intense pain relief, 309–310, 311f
opioids for pain management, 306–307, 307t
temporary/terminal sedation, 310–312, 312f
survey-based studies, 305–306
temporary/terminal sedation
acceptability, 312
clusters, 311–312, 312f
participants, 310–311
terminology, 313
Heart and pain
antihypertensive therapies, 217–218, 218f
applications to other areas, 217–219, 218f
areas of interest, 219
arousal-dampening mechanism, 219
bradycardiac baro-responses, 217
central detaching mechanism, 219
chronic pain severity regulation, 215–217, 217f
effective pain control, 213
energy metabolism regulation, 219
health-centric approach, 218
history, 212
hypertension-related hypoalgesia, 212–213
key facts, 220
lifestyle behaviors, 218–219
nociception, 211
pain adaptation mechanism, 211
pain-killing etiology, 213
pain management effectiveness, 213
pain-o-meter technology, 214, 215f
pain-related effects on baroreflex symmetry mechanism, 214–215, 216f
terminology, 219–220
Herbal medicine
application, 53
general anesthetics herbs
Hypericum perforatum (St. John's wort), 52–53, 52f
mechanism of action, 50
Panax (ginseng), 51–52, 52f
Passiflora incarnata, 50–51
Valerian officinalis, 51, 51f
hazards, 53
key facts, 54
local anesthetics herbs
Cinchona officinalis (quinine), 49, 49f
Erythroxylum coca (cocaine), 48, 48f
mechanisms of action, 47–48
Spilanthes acmella, 50, 50f
Syzygium aromaticum (clove), 48–49, 49f
terminology, 54
HIV-related pain
acetaminophen effects, 287
applications to other areas, 288
chronic pain etiology, 285–286
current pain management, 286
distal symmetric polyneuropathy (DSP), 286

future research, 288, 288f
HIVMA recommendation, 287
integrative model, 288, 288f
key facts, 289
life expectancy, 285
nonopioid pharmacological treatment, 287
nonpharmacologic approaches, 287–288, 287f
NSAIDS effects, 287
opioid risk, 285–287, 287f
physical therapy (PT), 287
terminology, 289
Hypericum perforatum (St. John's wort), 52–53, 52f
Hypertension
applications to other areas, 207
blood pressure-nociception interaction
baroreflex, 203
descending inhibitory pathways, 203
hyponociception, 204
odontogenic pain, 204
opioidergic pathway, 204
orofacial nociception, 204
pain correlation, 204
temporomandibular disorders, 204
key facts, 208
nociception mechanisms and modulation
hyponociception, 202
orofacial nociception, 202
variables, 202–203, 203f
noxious stimulus, 201, 202f
older adults, 201
pain *vs.* nociception, 201, 202f
prevalence, 201
sex differences and ovarian hormones
clinical and preclinical studies, 205–206, 206t
normotensive patients and rodent models, 204–205
spontaneously hypertensive rats (SHR) rodent model, 207
sign and symptoms, 201
terminology, 207
Hyponociception, 202, 204

I

Ilioinguinal nerve and iliohypogastric nerve blocks
complications, 410
indications, 409–410
technique, 410, 410f
Intercostal block
complications, 397–398, 412
indications, 397, 412
technique, 397, 397f, 412
Intraprostatic local anesthetic infiltration (IPLA), 356
Intrarectal local anesthetic use (IRLA), 356

J

Japanese Orthopaedic Association back pain evaluation questionnaire (JOABPEQ), 476–478, 480f
Japanese Orthopaedic Association (JOA) score, 476, 479f

568 Index

K

Kinesiophobia, 280

L

Labor pain
  assisted/instrumental vaginal delivery, 163
  cesarean section, 164
  induced vaginal delivery, 163
  key facts, 170
  modes of delivery, 163–164
  neuraxial analgesia and anesthesia
    combined spinal epidural block (CSE), 168
    continuous spinal analgesia, 167
    dermatomes, 167
    dural puncture epidural block, 168
    epidural block, 167–168
    general anesthesia (GA), 168
    neuraxial ultrasound (US), 167
    spinal anesthesia, 167, 168t
  nonpharmacological treatment, 169
  pharmacologic treatment, 165t
    application to other areas, 170
    inhalation anesthetics, 166
    locoregional techniques, 166–167
    mixed opioid agonist/antagonists, 166
    nonopioid drugs, 166
    opioids, 165–166
    paracervical block, 167
    phencyclidine derivative, 166
    pudendal nerve block, 166–167
    systemic drug administration, 165–166,
      166t
  postpartum pain
    acute pain management, 168–169
    chronic pain after delivery, 169
    postpartum depression, 169
  spontaneous vaginal delivery, 163
  stages of delivery, 164, 164f
  terminology, 170
  vaginal birth after C-section (VBAC), 164
  vaginal delivery, 163
Laparoscopy
  agents of interest, 429
  applications to other areas, 428–429
  cholecystectomy, 422
  colectomy, 422
  epidural analgesia, 425
  fentanyl, 429
  hydromorphone, 429
  interventions, 423–425, 424t
  key facts, 430
  minimally invasive surgery, 422
  morphine, 429
  nephrectomy, 422
  oxycodone
    central nervous penetration, 427
    epidural oxycodone, 428
    formulation, 426
    metabolism, 426–427, 427f
    minimum effective concentration (MEC)
      and minimum effective analgesic
      concentration (MEAC), 427
    opioid receptors, 425–426
    pharmacokinetics, 426t
    selective MOR-agonist, 425

  structure, 425
  visceral pain, 421
  pain
    carbon dioxide insufflation, 422
    early postoperative pain, 421
    nerve lesions, 423
    pneumoperitoneum, 421–422
    tissue trauma, 422
    treatment, 423, 423t
  quality of life, 422
  robotic-assisted laparoscopic surgery, 422
  sufentanil, 429
  terminology, 429
  video laparoscopy, 422
Levobupivacaine
  abdominal surgeries, 439
  agents of interest, 440
  amino amide anesthetics, 440
  applications to other areas, 440
  key facts, 441
  levobupivacaine hydrochloride, 437t
  local anesthetics (LAs)
    absorption, 435–436, 435t
    biotransformation, 436
    for central and peripheral blocks, 433–434
    distribution and elimination, 436
    mechanism of action, 434–435
    pharmacokinetics, 435–436, 435t
    properties, 434, 434t
    toxicity, 436–437
  local infiltration analgesia (LIA) body
    protocol, 438, 438t
  pain relief, 433
  pharmacology and use, 437–438, 437t
  postoperative analgesia, 439
  preincisional infiltration, 439
  ropivacaine, 440
  terminology, 440–441
  wound infiltration, 433
Low back pain (LBP). *See also* Back pain
    functional scale; Occupational
    musculoskeletal disorders (MSDs)
  agents of interest
    comprehensive measure of general health
      status, 481–482
    psychiatric problems measure in
      orthopedic patients, 482, 483–484f
  BS-POP, 481–482, 483–484f
  causes, 475
  clinical outcomes, 478–479
  disease-specific questionnaires
    Japanese Orthopaedic Association back
      pain evaluation questionnaire
      (JOABPEQ), 476–478, 480f
    Japanese Orthopaedic Association (JOA)
      score, 476, 479f
    Oswestry disability index (ODI), 476, 477f
    Roland-Morris disability questionnaire
      (RMQ), 476, 478f
    suggestions for practical procedures, 481
  generic and disease-specific measures,
    487–488
  incidence, 291
  key facts, 482–483
  mechanisms, 292–293

  medical costs, 487
  pain evaluation questionnaires, 475–476
  prevalence, 487
  SF-36, 480–482
  symptom, 487
  terminology, 482
  types, 475

M

Maternal deprivation (MD)
  agents of interest, 158
  animal models, 151
  applications to other areas, 158
  early life stress in humans, 156, 157f, 159
  environmental enrichment (EE), 156, 157f,
    158
  hypothalamic–pituitary–adrenal (HPA) axis,
    151, 151f
  key facts, 159
  and nociception
    cortical NGF levels, 154
    in early adulthood, 151–153
    Ephrin-B2/EphB receptors, 156
    hormonal responses, 156
    nefastin-1/NUCB2 expression, 156
    neonatal period, 154
    nociceptive outcome, 152t, 153, 153f
    preclinical perspectives, 151–153, 152t,
      153f
    sex-dependent, 153
    signaling pathways, 154–156, 154–155t
    visceral outcome, 154–155t
  nociceptive pathways, 150–151, 150f
  stress-hyporesponsive period (SHRP), 151
  terminology, 159
Median nerve compression, 275–276.
    See also Carpal tunnel syndrome (CTS)
Mesotherapy, 363
Midpoint transverse process pleural block
    (MTPB), 401
Migraine
  acute migraine treatment
    adjuvant medications, 107
    attack time, 106
    CGRP receptor antagonists (GEPANTS),
      108
    5-HT1F receptor agonists (DITANS), 108
    nonspecific acute treatment, 107, 107t
    specific acute treatment, 107, 108t
    stratified treatment, 106
    tailored approach, 106
    triptans, 107, 108t
  aura, 105
  chronic migraine, 106, 106f
  cognitive behavioral therapy, 110–113
  exercise, 110–113
  headache, 106
  incidence, 105
  migraine hangover, 106
  nonpharmacological approach, 110–113
  postdrome, 106
  premonitory phase, 105
  prophylactic migraine treatment
    angiotensin converting enzyme inhibitors,
      and angiotensin II receptor blockers, 110

antidepressants, 109
antiepileptics/neuromodulators, 109
β-blockers, 109
calcium channel blockers, 110
CGRP monoclonal antibodies, 110
OnabotulinumtoxinA (Botox), 110, 111*f*,
112*t*
preventive treatment, 108–109
types, 105
Mild cognitive impairment (MCI), 501
Military veteran injuries. *See* Chronic pain in
military veterans; Combat-related
extremity injury
Minimally invasive surgery, 422.
See also Laparoscopy
Multimodal analgesia in intensive care unit
analgosedation
commonly used opioids, 62–63, 62*t*
definition, 62
Opioid-Related Adverse Drug Events
(ORADES), 63
results, 63
surgical patients, 63
applications to other areas, 65
assessment-driven pain protocols, 58–59, 59*f*
critically ill adults assessment, 58
nonopioid analgesics
efficacy and safety, 65
opioid effects, 63
recommended for use, 64*t*, 65
regional analgesia, 65
opioid choice, route of administration, and
dosing
factors to be consider, 59, 60–61*t*
opioid use disorder, 62
rotating opioids, 59–62
opioids and nonopioids, 57–58
PADIS guidelines, 58–59
pain patterns, 57
terminology, 65–66
Multimorphic cancer pain. *See also* Cancer pain
analgesics, 7
concept, 6–7
factors influencing, 7
treatment strategies, 7–8
Musculoskeletal disorders (MSDs).
See Occupational musculoskeletal
disorders (MSDs)
Musculoskeletal injuries, 227
Musculoskeletal pain
aerobic endurance and fitness level tests,
558–559*t*
definition, 552
gait speed tests, 557*t*
International Scale of Physical Fitness (IFIS),
552
vs. physical activity, 552
Senior Fitness Test, 552
static and dynamic balance tests, 555–556*t*
upper- and lower-limb range of motion tests,
554–555*t*
upper- and lower-limb strength tests,
553–554*t*
Myofascial trigger points (MTrPs)
applications to other areas, 325–326

compression mechanisms
at active MTrP, 324–325, 324*f*
diagram, 325*f*
at non-MTrP, 324–325, 324*f*
psychological factors, 323–324
sites of action, 323
sympathetic activity, 323–324
diagnosis, 320
diagnostic criteria, 317
key facts, 326
musculoskeletal pain, 317
effectiveness, 320, 321–322*t*
foot regions, 323
knee pain, 323
low-back pain, 320–322, 322*f*
neck pain, 322–323
shoulder pain, 323
prevalence
active and latent, 318, 319*t*
knee pain, 320
low-back pain, 318
neck pain, 318
shoulder pain, 318–320
trapezius muscle, 320
terminology, 326
treatment, 317
types, 317

N

Narcotrend, 84–86, 85*t*
Neck-shoulder pain (NSP).
See also Occupational musculoskeletal
disorders (MSDs)
incidence, 291
mechanisms, 292
muscle activity, 294*f*
origins, 291–292
Neonatal chronic stress, 149
Neonates and infants
agents of interest, 272
application to other areas, 272
CNS effects, 265–266, 269*f*
Cries scale, 268*f*
face leg activity cry consolability scale
(FLACC), 267*f*
facial expression PIPP scale, 267*f*
key facts, 272
neonatal pain assessment
problems, 268
tools, 266
neonatal pain communication, 265
nonverbal pain expression, 265
pain pathophysiology, 265–266, 269*f*
pain relief measures
behavior state score, 270*t*
brow bulge score, 271*t*
eye squeeze score, 271*t*
heart rate and saturation score, 270*t*
lidocaine, 269–270
nasolabial furrow score, 271*t*
nonpharmacological agents, 269
opioids and NSAIDs, 269–270
remifentanil, 268
topical anesthetic agents, 269–270
total pain score, 271*t*

venipuncture, 269–270, 270*f*
premature infant pain profile (PIPP) scale,
266, 266*f*
spinothalamic tract, 265, 269*f*
terminology, 272
Neuropathic pain, 139–140
Neurotrophins, 251–252
Nociception
agents of interest, 242
analgesia nociception index (ANI), 237
in awake state, 240
definition, 235
under general anesthesia, 237–238, 238*f*
intraoperative nociception
C-reactive protein (CRP), 240
and postoperative complications, 239–240,
239*f*
on postoperative pain, 240
key facts, 242
monitors, 235, 236*t*
nociceptive pathway during surgery,
236–237, 237*f*
noxious stimuli, 235
somatosensory processing, 241–242, 241*f*
terminology, 242

O

Obesity. *See* Body mass
Occupational musculoskeletal disorders
(MSDs)
agents of interest, 297–298
applications to other areas, 297
burden to society, 291, 292*f*
ergonomic interventions, 296*f*, 297
individual risks factors, 293
key facts, 298
low-back pain (LBP)
incidence, 291
mechanisms, 292–293
muscle activity, 294*f*
neck-shoulder pain (NSP)
incidence, 291
mechanisms, 292
origins, 291–292
physical activity
adherence, 297
exercise interventions, 295, 296*f*
exercise therapy, 295–297, 296*f*
neck-shoulder strengthening exercise
programs, 296
spinal stabilization protocol, 296–297
variability, 296
physical risks factors, 293–294, 294*f*
psychosocial risks factors, 294–295, 295*f*
sex differences, 293
spine flexion/extension posture, 294*f*
terminology, 298
OnabotulinumtoxinA (Botox), 110, 111*f*, 112*t*
Orofacial nociception, 202, 204
Osteogenic pain, 249. *See also* Breast cancer
Oswestry disability index (ODI), 476, 477*f*
Oxycodone
central nervous penetration, 427
epidural oxycodone, 428
formulation, 426

570 Index

Oxycodone *(Continued)*
  metabolism, 426–427, 427f
  minimum effective concentration (MEC) and minimum effective analgesic concentration (MEAC), 427
  opioid receptors, 425–426
  pharmacokinetics, 426t
  selective MOR-agonist, 425
  structure, 425
  visceral pain, 421

P

PADIS guidelines, 58–59, 65
Pain, Enjoyment, and General Activity (PEG) scale, 229, 231f
Painful diabetic neuropathy (PDN)
  agents of interest, 145–146
  applications to other areas, 145
  causes, 140
  central glia-neuronal interaction
    JAK/STAT3 pathway, 143–144, 145f
    neuroimmune communication, 143
    in spinal dorsal horn, 143, 144f
    synaptic plasticity, 143
  diabetes mellitus complication, 139–140, 140t
  glial cell activation, 140
  hyperglycemia, 140, 141f
  key facts
    duloxetine, 146
    pregabalin, 146
  medication effects, 141
  microglia
    function, 141–142
    resting to activated microglia, 142, 142f
    substances, 142
  microglial activation
    KCC2 expression, 143
    mechanisms of action, 143
    p38, 143
    spinal cord pain–related neurons, 141
  minocycline, 145–146
  neuropathic pain, 139
  pathological mechanisms, 140, 141f
  terminology, 146
Pain in older people
  agents of interest, 33
  application to other areas, 33
  assessment
    cognitive impairment, 29
    numerical rating scale (NRS), 28–29, 29f
    simple descriptive pain intensity scale, 28–29, 28f
    system and patient-specific factors, 28
    Wong-Baker FACES Pain Rating Scale, 28–29, 29f
  cannabinoids, 33
  complication in management, 27
  management
    adjuvant analgesics, 32
    duloxetine, 32
    gabapentinoids, 32
    nonsteroidal antiinflammatory drugs (NSAIDs), 31
    opioid analgesics, 32

order of treatment, 30–31, 30f
    paracetamol, 31
    recommendation, 30
    topical lidocaine and capsaicin, 32
    tricyclic antidepressants (TCAs), 32
  pain perception, 28
  pharmacokinetic and pharmacodynamic changes, 29–30, 30t
  pharmacological and nonpharmacological strategies, 27, 30
  prevalence, 27
  terminology, 33
Pain-related behaviors
  assessment, 451–452
  description, 451
  direct observation, 453–455
  indirect observational methods, 453–455
  key facts, 458
  limitations, 457
  methodology
    article exclusion process, 452, 453f
    criteria, 452
    databases, 452
    study selection (eligibility criteria), 452
  outcomes
    publications, 452–453, 454t
    quality criteria, 453, 455–456t
  pain and disability model, 452
  quality judgement criteria, 455
  scoring methods and scales, 457
  terminology, 457
Panax (ginseng), 51–52, 52f
Paravertebral block (PVB)
  complications, 396, 411
  indications, 395, 410–411
  technique, 395–396, 396f, 411
Passiflora incarnata, 50–51
Pectoralis nerve blocks (PECs)
  complications, 394
  indications, 392–393
  technique, 393–394, 393f
Phantom limb pain
  diagnosis, 127–128, 128f
  incidence, 127
  key facts, 136
  limb amputation, 127
  management, 132
  pathophysiology
    afferent C-fibers, 128–129
    altered interneuron control, 130
    bulbospinal pathways, 130
    dorsal root ganglion (DRG), 130
    homunculus reorganization, 131, 131f
    management approaches, 129t
    mechanism, 128
    neuromas, 129
    NMDA receptor systems, 130
    peripheral pain, 128–130
    postganglionic/sympathetic sprouting, 130
    spinal pain, 130
    subcortical reorganization, 132
    supraspinal (brain) pain, 130–132, 131f
  post-amputation symptom clarification algorithm, 127–128, 128f
  prevention

epidural analgesia, 132
    peripheral regional anesthesia, 133
    systemic pharmacotherapy, 132
    targeted muscle reinnervation (TMR), 133
  terminology, 136
  treatment
    acupuncture, 134
    application to other areas, 135
    cryoneurolysis, 136
    dorsal root ganglion (DRG) stimulation, 135
    eye movement desensitization and reprocessing therapy (EMDR), 134
    gabapentin, 133
    graded motor imagery (GMI), 134
    invasive interventions, 134–135
    ketamine, 133
    microcurrent electrotherapy stimulation, 135
    mirror therapy, 133–134
    noninvasive interventions, 135
    opioids, 133
    peripheral nerve cuff integration into prosthetics, 135–136
    peripheral nerve stimulation (PNS), 134–135
    pharmacotherapy, 133
    spinal cord stimulation (SCS), 135
    therapeutic modalities/psychological/other treatments, 133–134
    transcranial direct current stimulation (tDCS), 135
    transcranial magnetic stimulation (TMS), 135
    transcutaneous electrical nerve stimulation, 134
Physical fitness
  agents of interest, 557
  applications to other areas, 552
  body functions, 551
  definition, 551
  key facts, 559
  musculoskeletal disorders
    definition, 551
    pathologies, 551–552
  terminology, 557
  widespread musculoskeletal pain
    aerobic endurance and fitness level tests, 558–559t
    definition, 552
    gait speed tests, 557t
    International Scale of Physical Fitness (IFIS), 552
    vs. physical activity, 552
    Senior Fitness Test, 552
    static and dynamic balance tests, 555–556t
    upper- and lower-limb range of motion tests, 554–555t
    upper- and lower-limb strength tests, 553–554t
Polytrauma, 227
Postoperative analgesia. *See* Acute Pain Service (APS)

Postsurgical pain (PSP)
agents of interest, 338
α_2-agonists, 335
γ-aminobutyric acid analogues
(gabapentenoids), 335
glucocorticoids, 336
ketamine, 335
N-methyl-D-aspartate receptor antagonists,
335
systemic lidocaine, 335
systemic magnesium, 335
allodynia, 330
application to other areas, 338
chronic postoperative pain (CPSP), 329, 332,
332t
classification, 329
definition, 329
hyperalgesia, 330
importance, 331
incidence, 329–330
intensity evaluation, 332–333
key facts, 338
management
enhanced recovery after surgery programs
(ERAS), 333
multimodal analgesia, 333
multimodal analgesia framework, 333–334
neuroplastic changes, 331
nociceptors, 330
nonpharmacological interventions, 334
pain pathway, 330, 331f
pathophysiology, 330–331, 330–331f
peripheral and central sensitization, 331
pharmacological agents, 336t
regional anesthesia, local anesthesia, and site
infiltration with local anesthetics
local infiltration, 337
neuraxial analgesia (epidural and spinal),
336–337
paravertebral nerve blocks (PVB), 337
transversus abdominis plane (TAP) block,
337
risk factors, 332, 332t
special consideration for treatment
children, 337
chronic obstructive pulmonary disease
(COPD), 337–338
obese patients, 337–338
patient receiving long-term opioids, 338
systemic nonopioid analgesics
nonsteroidal antiinflammatory drugs
(NSAIDs), 334–335
paracetamol, 334
systemic opioid analgesics, 336
terminology, 339
Prostate biopsy
anesthesia techniques, 355, 355t
anesthetic agents, 358–359, 359t
applications of to other areas, 359
general anesthesia, 357
history, 355
local anesthesia techniques
bilateral pelvic plexus blockade (PPB), 356
bilateral periprostatic nerve block (PPNB),
355, 356f

caudal block, 357
combined techniques, 357
eutectic mixture of local anesthetic cream
(EMLA), 357
intraprostatic local anesthetic infiltration
(IPLA), 356
intrarectal local anesthetic use
(IRLA), 356
transperineal local anesthetic infiltration
(TPLA), 355–356
prostate cancer, 353
prostate neuroanatomy, 354, 354f
sites of infiltration, 357–358
terminology, 359–360
transcutaneous electrical nerve stimulation
(TENS), 357, 358f

Q

Quadratus lumborum block (QLB)
complications, 414
indications, 412–413
technique, 413–414, 413f
Quinine, 49, 49f

R

Rectus sheath block
complications, 409
indications, 409
technique, 409
ultrasound anatomy, 408, 408f
Reisberg Global Deterioration Scale (GDS), 501
Retrolaminar block (RLB)
complications, 399
indications, 398
technique, 398–399, 398f
Rhinoplasty and rhinologic surgery
agents of interest, 262
applications to other areas, 262
characteristics, 257
endoscopic sinus surgery (ESS), 258
functional disorders, 258
intraoperative analgesia and anesthesia
cottonoids, 258–259
gabapentinoids, 258
general anesthesia (GA), 258
intraoperative endoscopic sphenopalatine
ganglion block, 258
local anesthesia, 259
remifentanil, 259
vasoconstriction and decongestion, 259
key facts, 262–263
manipulation, 257
plastic surgical procedures, 257
postoperative pain management
acetaminophen, 260–261
hydrocodone-acetaminophen, 261
nonopioid analgesics, 260–261
nonsteroidal antiinflammatory drugs,
260–261
opioid analgesics, 261
oxycodone-acetaminophen, 261
pain assessment after surgery, 260
sources of postoperative pain, 259–260
septoplasty, 257–258
terminology, 262

Rhomboid intercostal and serratus plane (RISS)
block, 401
Roland-Morris disability questionnaire (RMQ),
476, 478f

S

Senior Fitness Test, 552
Septoplasty, 257–259. *See also* Rhinoplasty and
rhinologic surgery
Serratus anterior plane block (SAPB)
complications, 395
indications, 394
technique, 394, 395f
SF-36, 480–482
Somatosensory evoked potential (SSEP), 83
Spilanthes acmella, 50, 50f
Spinal anesthesia
agents of interest, 387
anatomical site, 382, 382f
application to other areas, 387
back pain, 386
continuous spinal anesthesia, 385
history, 381
intraoperative management
aortocaval compression, 385
failed/inadequate spinal anesthesia, 386
hypotension, 385
nausea and vomiting, 385–386
perioperative hypothermia/shivering, 386
total spinal anesthesia, 386
key facts, 387
local anesthetic agents
adjuvants, 385
bupivacaine, 383
chloroprocaine, 384
hydrophilic opioids, 384–385
intrathecal opioids, 384
levobupivacaine, 384
lidocaine, 384
lipophilic opioids, 384
ropivacaine, 384
neuraxial anesthesia, 381
obstetrics, 381
postdural puncture headache, 386
postoperative complications, 386
preparation
informed consent, 382
patient positioning, 383
spinal needles, 382–383, 383f
recovery, 386
spinal technique, 382, 382f
terminology, 387
Spinal cord stimulation (SCS), 122
Statin associated muscle symptoms (SAMS)
biomarkers
aldolase, 544
carbonic anhydrases III (CAIII), 544
coenzyme Q_{10} (CoQ_{10}), 543
creatine deficiency, 542
creatine kinase (CK), 543–544
creatinine, 542
diagnosis, 541–542
enzyme, 543–544
fatty acid-binding proteins, 543
2-hydroxyglutarate (2HG), 544

572 Index

Statin associated muscle symptoms (SAMS) *(Continued)*
  lactate dehydrogenase (LDH), 544
  lactic acid, 542
  micro RNAs (miRNAs), 545
  myoglobin, 543
  transaminase AST and ALT, 544
  troponin, 542
  urinary biomarkers, 545
 cholesterol, 541
 classification, 539–540, 540*t*
 etiological factors, 541
 key facts, 546–547
 myotoxicity, 540
 risk factors, 540–541
 Skeletal Muscle Function and Performance (STOMP) study, 540
 statin metabolism, 539
 terminology, 546
 vitamin D deficiency, 541
St. John's wort, 52–53, 52*f*
Supportive care, 5–6
Sympathetic nerve blocks (SNB), 121–122
Syzygium aromaticum (clove), 48–49, 49*f*

T
Therapeutic exercise (TE), 111–113
Transcutaneous electrical nerve stimulation (TENS), 357, 358*f*
Transient receptor potential vanilloid 1 (TRPV1), 252
Transperineal local anesthetic infiltration (TPLA), 355–356
Transversalis fascia plane block
  complications, 415
  indications, 415
  technique, 415
Transversus abdominis plane block (TAPB)
  approaches, 406
  complications, 408
  indications, 406–407
  technique, 407–408, 407*f*
Triptans, 107, 108*t*
Truncal blocks in general surgery
  acute postoperative pain, 405
  agents of interest, 416
  application to other areas, 416
  contraindications, 415–416
  epidural analgesia, 405
  erector spinae plane block
   complications, 415
   indications, 414
   technique, 414
  ilioinguinal nerve and iliohypogastric nerve blocks
   complications, 410
   indications, 409–410
   technique, 410, 410*f*
  intercostal block

complications, 412
  indications, 412
  technique, 412
 key facts, 417
 nonopioid analgesics, 405–406
 paravertebral block (PVB)
  complications, 411
  indications, 410–411
  technique, 411
 postoperative chronic pain, 406
 preparation
  equipment, 416
  postprocedure, 416
 quadratus lumborum block (QLB)
  complications, 414
  indications, 412–413
  technique, 413–414, 413*f*
 rectus sheath block
  complications, 409
  indications, 409
  technique, 409
  ultrasound anatomy, 408, 408*f*
 terminology, 416–417
 transversalis fascia plane block
  complications, 415
  indications, 415
  technique, 415
 transversus abdominis plane block (TAPB)
  approaches, 406
  complications, 408
  indications, 406–407
  technique, 407–408, 407*f*
 ultrasonography, 405–406
Truncal blocks in thoracic surgery
 agents of interest, 401
 application to other areas, 401
 chronic postoperative pain, 391
 contraindications, 400
 coughing and deep breathing, 391
 enhanced recovery after surgery (ERAS), 392
 equipment, 400–401
 erector spinae plane block (ESPB)
  complications, 400
  indications, 399
  technique, 399–400, 400*f*
 intercostal block
  complications, 397–398
  indications, 397
  technique, 397, 397*f*
 key facts, 402
 midpoint transverse process pleural block (MTPB), 401
 multimodal analgesia, 392
 opioids, 391
 paravertebral block (PVB)
  complications, 396
  indications, 395
  technique, 395–396, 396*f*
 pectoralis nerve blocks (PECs)

complications, 394
  indications, 392–393
  technique, 393–394, 393*f*
 postprocedure checks, 400
 regional anesthesia technique, 392
 retrolaminar block (RLB)
  complications, 399
  indications, 398
  technique, 398–399, 398*f*
 rhomboid intercostal and serratus plane (RISS) block, 401
 serratus anterior plane block (SAPB)
  complications, 395
  indications, 394
  technique, 394, 395*f*
 terminology, 401–402
Tumescent anesthesia, 370
Tumor necrosis factor-α (TNFα), 252

U
Ultrasound-guided Pecs II block
 agents of interest, 348–349, 350*f*
 applications to other areas, 348
 clinical application
  analgesic efficacy, 347
  average analgesia onset time, 347
  complication, 347–348
  potentiality, 347
  without general anesthesia, 347, 347*t*
 continuous Pecs II block, 348
 general anesthesia, 344
 intercostal nerves, 348
 key facts, 350
 long thoracic nerve, 348
 for nonbreast surgery, 348, 349*t*
 pectoral nerves, 348
 regional anesthesia-analgesia techniques, 349, 350*f*
 technical description
  adjuvants, 345
  chest wall innervation, 345, 346*f*
  dermatomal distribution pattern, 345, 345*f*
  performance, 345, 346*f*
 terminology, 349–350

V
Valerian officinalis, 51, 51*f*
Vapocoolant anesthesia
 ice anesthesia, 369
 mechanism of action, 368–369
 topical coolants, 369
Vibration anesthesia
 botulinum toxin injection with, 365, 365*f*
 disadvantages, 366
 safety, 366
 vibrator device, 364, 365*f*
Visual evoked potential (VEP), 83
Vitamin D, 541